Proceedings of the 45th
INDUSTRIAL WASTE CONFERENCE
May 8, 9, 10, 1990

Proceedings of the 45th INDUSTRIAL WASTE CONFERENCE May 8, 9, 10, 1990

Purdue University
West Lafayette, Indiana

The Purdue Industrial Waste Conference is under the direction of:

The School of Civil Engineering
Continuing Education, Purdue University

in cooperation with the

Indiana Department of Environmental Management
Indiana Department of Natural Resources
Indiana Environmental Management Board
Indiana Section of American Society of Civil Engineers
Indiana Section of the American Water Works Association
Indiana State Board of Health
Indiana Stream Pollution Control Board
Indiana Water Pollution Control Association

LEWIS PUBLISHERS

ISBN 0-87371-412-1

ISSN 0073-7682

LEWIS PUBLISHERS, INC.
121 South Main Street, Chelsea, Michigan 48118

PRINTED IN THE UNITED STATES OF AMERICA

PREFACE

The 45th Industrial Waste Conference was sponsored by the School of Civil Engineering of Purdue University.

Ninety-seven papers were presented during the three days of the conference. Ninety-one of these papers, including the Keynote Address, were submitted for publication and are included herein.

The papers are divided into the following 10 major sections: 1. Site Remediation; 2. Laws, Regulations, and Training; 3. Hazardous Waste Minimization and Treatment; 4. Biological Systems; 5. Respirometry; 6. New Waste Processes; 7. New Equipment Application; 8. Landfills; 9. Industrial Wastes; and 10. Keynote Address.

The conference keynote speaker was Dr. James E. Etzel, Professor of Environmental Engineering, School of Civil Engineering, Purdue University, W. Lafayette, Indiana. Dr. Etzel talked about various incidents he recalled from previous Purdue Waste Conferences. He then looked to the future and prognosticated what he thought environmental engineers would be dealing with as they serve society.

At the luncheon on Tuesday, Arthur Busch, Environmental Engineering Consultant, Dallas, Texas, spoke on the topic entitled "Common Sense—An Endangered Species."

A quartet, called "Tony Zamora's Jazz Quartet," from Lafayette, Indiana, entertained the audience with various musical selections at the conference banquet on Wednesday evening. In addition, Dr. J. Tim Stivers, Podiatrist, Anchorage, Kentucky, gave an interesting talk entitled "Effects of Orthopedic Hats on People With Warped Minds."

At the end of this proceedings is a comprehensive 10-year index (current and previous volumes). This index is cross-referenced by author and various subjects for the utility of the reader.

The efforts of all who gave freely of their time toward the success of the 45th Purdue Industrial Waste Conference are gratefully acknowledged.

JOHN M. BELL, Editor
Associate Professor of Environmental Engineering
School of Civil Engineering
Purdue University

CONTENTS

Section One. Site Remediation

Section Two. Laws, Regulations, and Training

Section Three. Hazardous Waste Minimization and Treatment

Section Four. Biological Systems—C. Anaerobic

Section Four. Biological Systems—D. Sludge Treatment

Section Five. Respirometry

Section Six. New Waste Processes

Section Seven. New Equipment Application

Section Eight. Landfills – A. General

Section Eight. Landfills – B. Leachates

Section Nine. Industrial Wastes – A. Petroleum Wastes

Section Nine. Industrial Wastes – B. Brewery and Dairy Wastes

Section Nine. Industrial Wastes – C. Metal Processing Wastes

Section One
SITE REMEDIATION

1 GROUNDWATER AND SOIL REMEDIATION BY TREATING AND RECYCLING GROUNDWATER

Kostas Dovantzis, Environmental Engineer
Kurt O. Thomsen, Project Manager
Majid A. Chaudhry, Project Manager
Ronald R. Riesing, Group Leader
PRC Environmental Management, Inc.
Chicago, Illinois 60601

BACKGROUND

A refrigerator manufacturing facility is located on the lower end of two peninsulas formed by a reverse "S" meander of a river (Figure 1). In the late 1960s, local regulatory agencies granted the manufacturer permission to use the upper peninsula as a waste disposal area. Site contamination resulted from the waste disposal activities.

Subsequently, a remedial investigation (RI) was conducted at the site to determine the nature and extent of contamination and to assess potential adverse effects to human health and the environment that could be associated with the site contamination.

The RI was conducted in three phases. The first phase was the exploratory phase, during which six well nests were installed. The results of this phase defined the general site stratigraphy. During the second phase of the RI, sampling was conducted to further define the site stratigraphy, hydrogeologic units, and extent of contamination. Trichloroethylene (TCE), tetrachloroethylene (PCE), and 1,1,1-trichloroethane (TCA) were the major contaminants found in the groundwater. In addition, trace levels (<15 μg/L) of benzene, chloroform, 1,1-dichloroethane, 1,1-dichloroethylene, cis-1,2-dichloroethylene, trans-1,2-dichloroethylene, methylene chloride, toluene, and 1,1,2-trichloroethane were found in the groundwater. Trichloroethylene (TCE) was the contaminant found in highest concentrations (as high as 35 mg/L) at the site.

Based on the results of the first and second phases, a consent judgment was negotiated between the state and manufacturer to remedy the contamination. The consent judgment required that groundwater be treated until it consistently yielded a TCE concentration of no more than 15 μg/L. TCE was selected as the indicator parameter because, in most cases, its concentration in the groundwater was an order of magnitude higher than either TCA or PCE. The 15 μg/L TCE cleanup criterion was based on the background concentration of TCE in the river adjacent to the site. The consent judgment also provided that the system was to be operated annually from April through October. As a result of the consent judgment, a third phase of the RI was conducted to better define the stratigraphy of the central portion of the site where most of the disposal activities took place (Figure 1). These data were needed to initiate a feasibility study of possible remedial alternatives.

The results of Phase 3 indicated that the complexity of site stratigraphy is responsible for variable groundwater movement at the site. Groundwater contamination is limited to an unconfined aquifer in communication with the river adjacent to the site. Discontinuous lenses of aquitard/aquiclude materials are generally present at two levels within the aquifer (Figure 2). The groundwater contaminants are mostly confined to the saturated zone above the upper aquitard with significantly lower amounts of contaminants in the groundwater between the two aquitard/aquiclude layers. No contamination was found between the lower aquitard/aquiclude layer and the confining layer underlying the aquifers. Groundwater flows to the river (Figure 1) and vertically around and through the aquitard/aquiclude layers present at the site. Vertical groundwater movement is upward to the river along the site boundary. However, at other locations both upward and downward groundwater movement has been

45th Purdue Industrial Waste Conference Proceedings, © 1991 Lewis Publishers, Inc., Chelsea, Michigan 48118. Printed in U.S.A.

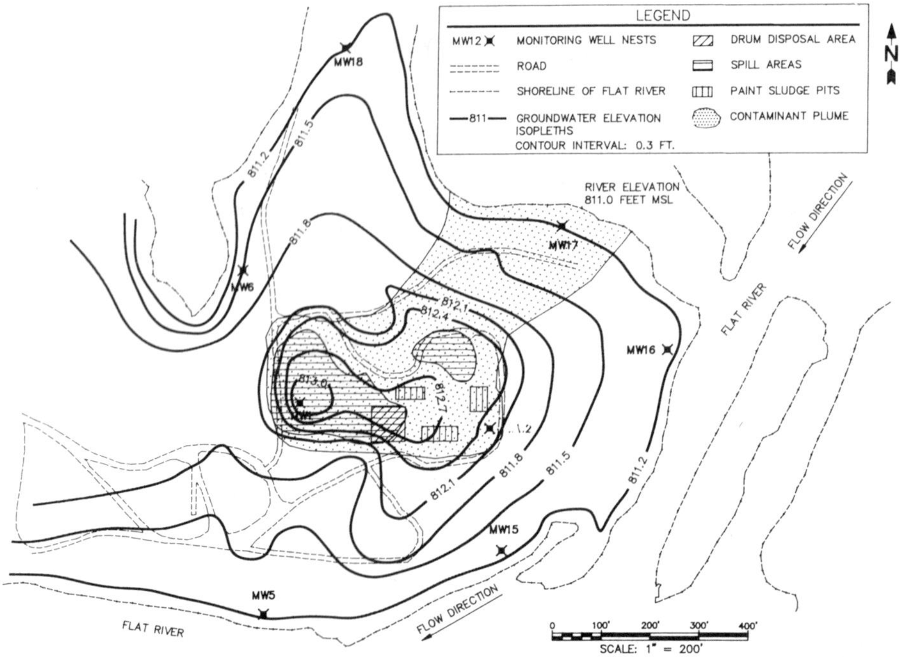

Figure 1. Site plot plan showing location of source areas, potentiometric surface, and contaminant plume boundary.

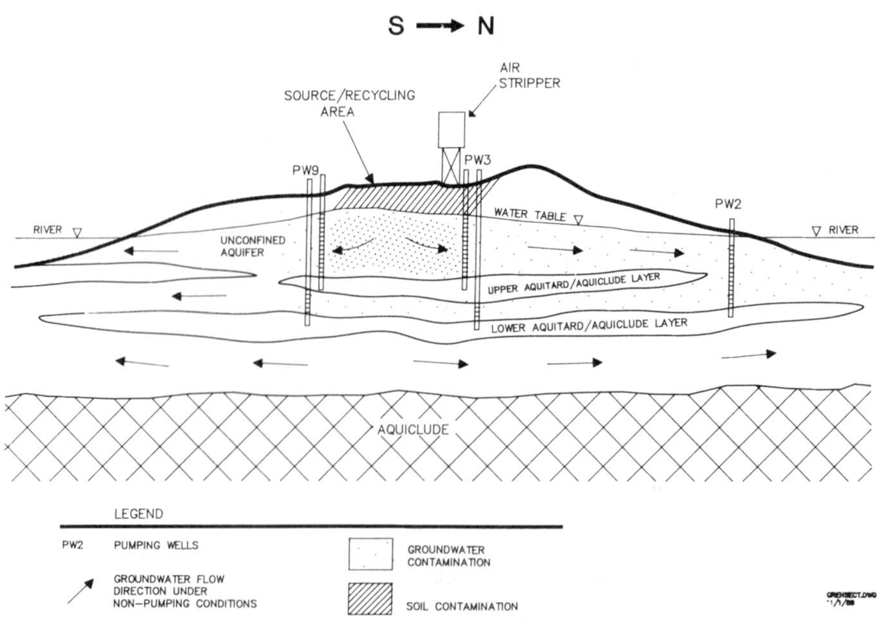

Figure 2. Schematic of a north-south cross-section.

identified. Lateral groundwater movement at the site is directed toward the river (Figure 1). Figure 1 also shows the boundary of the contaminant plume.

A feasibility study (FS) was conducted to identify, evaluate, and select a feasible alternative to remedy site contamination. The remedial alternatives considered in the evaluation included: 1) no action (natural attenuation and/or dilution); 2) pump and treat using granular activated carbon (GAC); and 3) pump and treat using air stripping. The selected remedial action for the site consisted of groundwater collection, treatment by air stripping, and discharge to a recycling basin. Pump tests were conducted to obtain the information necessary to determine the number, location, and depth of groundwater collection wells and the pumping rate needed to capture the groundwater contaminant plume. Instead of discharging the treated groundwater to the adjacent river, it was decided to discharge the treated groundwater to a recycling basin in the area where most of the disposal took place (Figure 1), and thus flush the residual contaminants from the soils. After selecting the remedial action for the site, pilot testing was conducted. Pilot testing included groundwater collection tests, groundwater treatability tests, and treated groundwater recycling tests. The results of the treatability tests showed an overall TCE removal efficiency of 93.9%[1].

The following sections present: 1) the pilot tests conducted at the site; 2) the design and operating characteristics of the full-scale collection, treatment, and recycling system; 3) a performance evaluation of the system based on initial operating data (July to October 1987); and 4) total capital costs and O & M costs. A site plot plan showing the location of the various components is presented in Figure 3.

PILOT TESTS

Pilot tests were conducted concurrently with the third phase of the RI. The data collected during Phase 3 indicated that an effective remediation program would have to be significantly larger than originally envisioned. Therefore, the capacity of the remediation system had to be increased to approximately 600 gpm from the 30-to 60-gpm system proposed in the consent judgment.

Groundwater Collection Tests

Pumping tests were conducted in five pumping wells at three locations and at two depths. The pumping well locations are designated PW-1 to PW-3 on Figure 2. The two depths coincided with the upper and middle portion of the upper outwash aquifer. Contamination within these two portions of the aquifer was documented during the third phase of the remedial investigation. Location PW-1 was chosen because it was representative of the central spill area (Figure 1). PW-2 was chosen to be within the identified contamination plume extending from the waste disposal area to the river, and PW-3 was chosen to represent conditions in the paint sludge pit area.

All pump tests were 48 hours long; a minimum of 4 and a maximum of 6 observation wells were monitored during each test. Observation wells screened at depths other than the pumping depth were included in the group of wells monitored for each test. These wells were included so that the effect of pumping on other portions of the aquifer could be evaluated. A summary of the pump testing results is presented in Table 1.

Groundwater Treatability Tests

The major contaminant identified in the groundwater through the above mentioned remedial investigation is trichloroethylene (TCE). Based on the physical and chemical properties of this volatile organic compound, the two treatment processes suitable for its removal from water are air stripping and carbon adsorption. Although these two treatment processes are both feasible and dependable, air stripping was selected due to its recognized lower capital, operation, and maintenance costs.

Having selected the type of treatment process applicable, the next phase was to determine the optimum engineering parameters for design. Accordingly, groundwater treatability tests were planned and executed. An induced draft air stripper equipped with nozzles having an air-to-water ratio of 575 was used for these tests. While the air-to-water ratio remained constant during each experimental run, the environmental air temperature did vary. Temperature variation is expected during the actual treatment operation.

The objective of the treatability tests was to obtain design data by simulating the actual treatment process. Accordingly, all experiments were conducted with full-scale equipment using the contaminated groundwater at the site. For the induced draft stripper, the objectives were to determine the removal efficiency and the number of strippers needed to attain a treatment level less than or equal to 15 μg/L of TCE. Another objective was to determine the number of passes required to achieve the treatment level using only a single induced draft stripper.

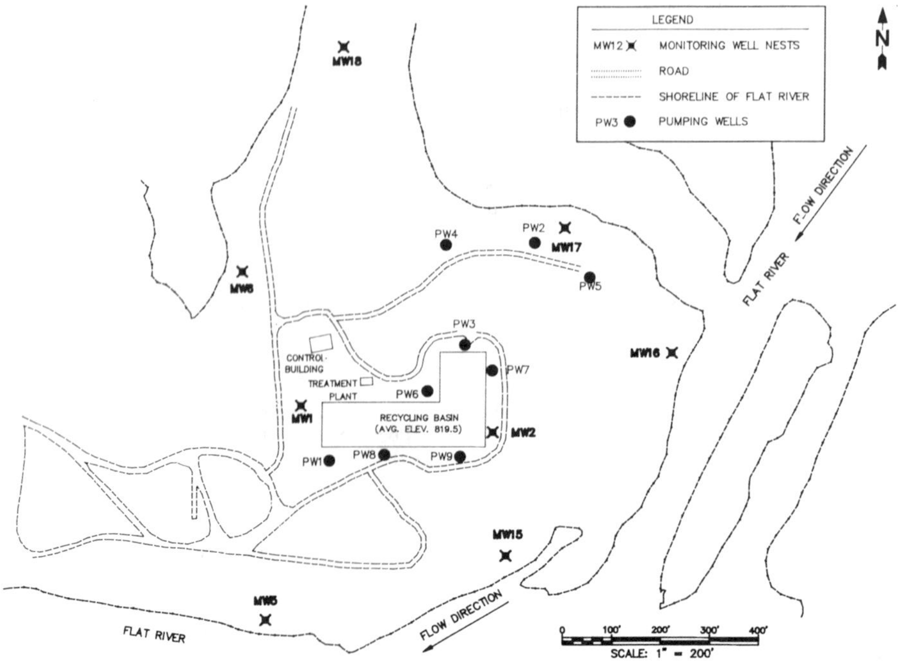

Figure 3. Site plot plan showing location of remediation system.

Figure 4 depicts the field experimental setup. First, groundwater at a temperature of 55° F was pumped to fill the lower reservoir (Tank #1, approximately 500 gallons). The groundwater was then pumped up to the stripper, circumventing the nozzles, and allowed to gravity flow to the upper reservoir (Tank #2, approximately 500 gallons) and then to the lower reservoir where it was recirculated back to the stripper. This was done for about 10 minutes to equilibrate the system. After equilibration when the lower reservoir was filled, two samples were taken and measured for TCE concentration. The average TCE level in the two samples was designated as the initial concentration or C_o. The water in the lower reservoir was then pumped (40 to 50 gpm) through the stripper nozzles and the effluent was collected in the upper reservoir. After all the water had passed through the stripper and into the upper reservoir, two samples were taken for analyses from a sampling bibb. The average level in the two samples was designated as effluent concentration or C_e. The removal efficiency was calculated as follows:

$$\% \text{ removal efficiency} = \frac{C_o - C_e}{C_o} \times 100 \qquad (1)$$

Table I. Results of Groundwater Pumping Tests

Well	Transmissivity (gal/day/ft)	Storativity (unitless)	Local Boundary Condition	Local Aquifer Type
PW-1A[a]	22900	0.0045	Impermeable	Unconfined
PW-1B[a]	12400	0.000015	Recharge	Leaky Confined
PW-2B	54500	0.00005	Impermeable	Unconfined
PW-3A	15900	0.032	Recharge	Unconfined
PW-3B	10100	0.00047	Recharge	Leaky Confined

[a] The suffix "A" denotes a pumping well screened in the upper portion of the aquifer and "B" denotes a pumping well screened in the middle portion of the aquifer.

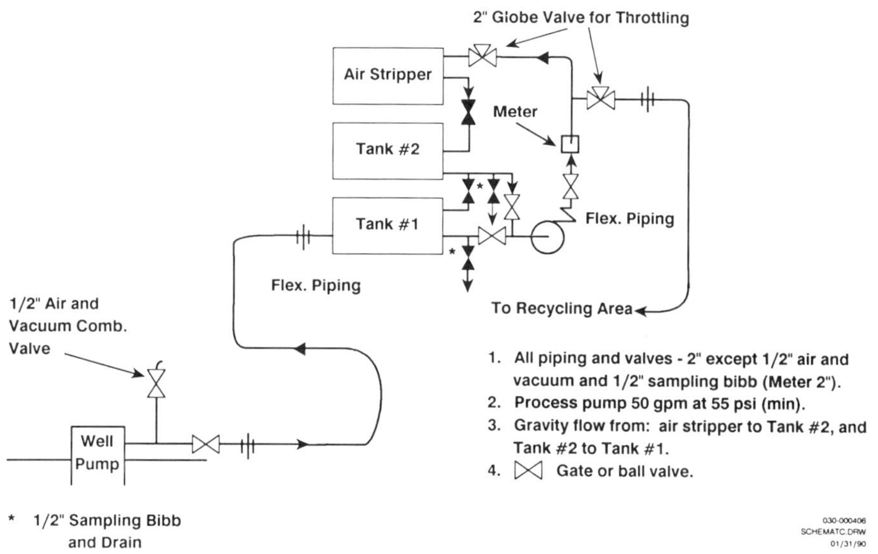

Figure 4. Field experimental setup of pilot treatability tests.

The above procedure was repeated for the number of times required to achieve the treatment level.

Figure 5 depicts typical results of the induced draft air stripper. Plotted are the concentrations as a function of the number of passes through the stripper. These results indicate that almost all the trichloroethylene in the water can be removed. However, to achieve the near zero effluent concentration, five passes through one stripper or one pass through five strippers in series are required. Figure 6 shows the cumulative removal efficiency versus number of passes. As previously indicated, this type of plot can be used to determine number of passes required to achieve a desired removal. As was verified in the field, higher TCE levels (greater than 400 μg/L) would require greater number of passes to achieve a near zero concentration level.

Treated Groundwater Recycling Tests

A 20-by 40-foot pilot recycling area was constructed on-site. The area was excavated, leveled, filled with pea gravel to a depth of approximately 1.0 to 1.5 feet, and bermed. The gravel was used to provide greater distribution of the treated water and to provide additional air/water contact before the water infiltrated into the ground.

Several methods were used to distribute the treated groundwater on the gravel bed of the recycling area. These methods ranged from using perforated flexible field tile to using a rigid system constructed of PVC pipe and equipped with spray nozzles. The latter method proved to be the most effective, both in terms of efficiency and cost. Figure 7 shows a cross-section of the pilot recycling area using the spray distribution system. In constructing the spray distribution system, the gravel bed provided direct structural support. The spray nozzles were directed upward to provide optimal distribution of the treated water over the recycling area and to optimize residence time for additional air stripping of the TCE. This additional air stripping removal efficiency was documented to be 94% at an ambient temperature of 77° F. Removal of 79% of the influent TCE was achieved between the time the water left the nozzles and the time it came in contact with the gravel bed. Another 15% removal was achieved as the water traveled through the gravel bed.

Soil Flushing Tests

Due to the complex nature of residual soil contamination, no pilot tests were conducted addressing contaminant flushing from the soils. The flushing of contaminants from the soils will be a direct result of a combination of treated groundwater recycling and groundwater collection. The groundwater collection system will facilitate the movement of the treated water through the contaminated

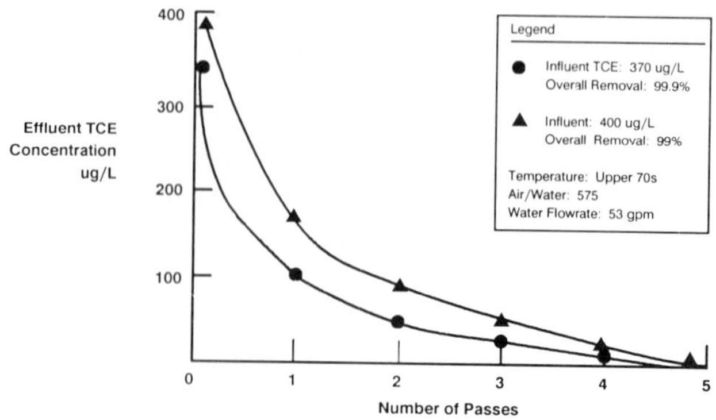

Figure 5. Effect of number of passes on effluent TCE concentration.

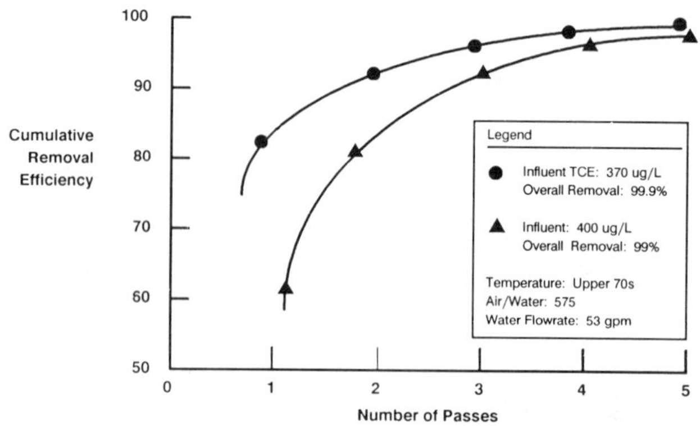

Figure 6. Cumulative TCE removal efficiency as function of number of passes.

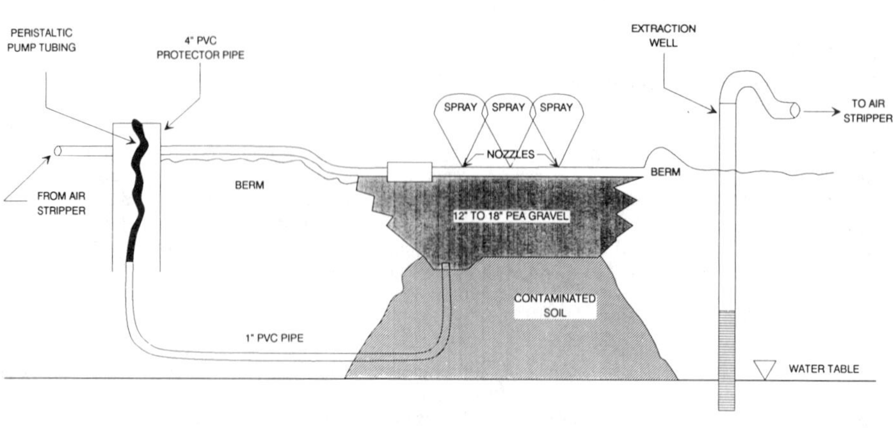

Figure 7. Cross-section of treated groundwater recycling area.

soils. The effectiveness of this remediation system will be evaluated by monitoring the TCE concentrations of influents from wells in key areas over time.

Conclusions of Pilot Tests

The results of the pilot tests indicated that the removal efficiency of the combined groundwater treatment and recycling system ranged from 97.3 to 99.1%.

The results of the pumping tests during the RI also indicated that because of the complex nature of the disposal area stratigraphy, a larger number of collection wells would be required than originally estimated. These wells would be placed at two levels. The collection system would be designed to collect contaminated groundwater and prevent its migration to the adjacent river. The system would also provide a flushing mechanism for the soil contaminants by recycling treated groundwater. It was estimated that a combined pumping rate of 600 gpm would be optimal for the system.

The groundwater treatability tests also indicated that the desired level of TCE removal could be achieved using a series of induced draft air strippers or a combination of the tested induced draft air stripper with the distribution and recycling system. Since the expected flow to the treatment plant was much higher than the previously anticipated flow (600 gpm compared to 30 to 60 gpm), 12 air strippers of the type used in the pilot tests would be required to treat the expected flow. The proposed use of the induced draft air stripper was re-evaluated because of the number of strippers needed.

Finally, the pilot recycling tests showed that the treated groundwater distribution and recycling system was very efficient and cost effective and, under certain circumstances, could stand alone as a groundwater treatment system.

FULL-SCALE SYSTEM DESIGN

Groundwater Collection System

The groundwater extraction system consists of 14 purge wells at nine locations around the perimeter of the site. At five of these locations, a total of 10 wells pump from two levels within the upper glacial outwash aquifer (Figure 2); wells at the remaining four locations (four wells) pump from only one level. The maximum groundwater withdrawal rate from the 14 wells is 600 gpm.

Shallow purge wells were drilled to the aquifer-aquitard layer interface, with the screen placed directly above the layer. A sump is located beneath the well screen to facilitate groundwater flow toward the well. A submersible pump located in the sump lifts groundwater to the front end of the air stripping unit. Groundwater from all purge wells is transported to the treatment system via 2-or 3-inch PVC pipes connected to an 8-inch header pipe. As a result, groundwater from all 14 wells is combined before entering the air stripper. Each influent connection contains a strainer basket to filter out particulates, a flow meter, a sampling port, a check valve, and a globe valve. Design and operating parameters of the collection system are presented in Table II. A process flow diagram of the collection, treatment, and recharge system is presented in Figure 8.

Groundwater Treatment System

The treatment system consists of a cooling tower used to air strip TCE from the contaminated groundwater and a fiberglass storage tank used to store treated water prior to distribution. A cooling tower was used for several reasons: 1) a removal efficiency slightly better than the induced draft air stripping unit; 2) a greater capacity; and 3) approximately half the cost of a packed tower air stripper having a comparable capacity. In the cooling tower, TCE is removed from the groundwater with air that flows counter-current to the water flow. A fan located near the top of the tower draws air into the tower from two sides. Water is distributed over the tower and flows downward into the unit at an angle. Startup testing data collected between September and November 1986 showed an average TCE removal efficiency of 78%; this removal efficiency was obtained at an average groundwater flow rate of 210 gpm and influent TCE concentration of 4,000 μg/L.

The treatment system was designed to accommodate a maximum flow of 600 gpm. Once operation began, however, it was discovered that the recharge capacity of the recycling area was approximately 210 gpm, one-third of the initially estimated capacity. Therefore, recharge capacity became the limiting factor in determining the actual treatment capacity.

A fiberglass storage tank, with an approximate holding capacity of 3,300 gallons, is located underneath the stripping tower (Figure 8). Treated water is stored in this tank prior to distribution over the recycling area. To maintain the water at a preset level, the tank is equipped with a primary water level control that transmits a signal to a modulating influent control valve located on the 8-inch header

Table II. Design and Operating Parameters of Collection, Treatment, and Recharge System

Influent Flow Rate and Concentration:	
Maximum flow rate	600 gpm
Average flow rate	210 gpm
Average TCE concentration	4,000 μg/L
Effluent TCE Goal:	
Pump, treat, and recharge until the TCE concentration in the groundwater is 15 μg/L	
Collection System:	
No. of purge wells	14
Well diameter	4 to 6 inches
Well depth	20 to 30 feet
Submersible pumps	
Maximum flow rate	90 gpm
Total dynamic head	80 feet
Motor power	2 horsepower
PVC piping diameter	2 to 3 inches
Treatment System:	
Air Stripping Tower	
Length	17 feet
Width	6.5 feet
Height	9.5 feet
Maximum hydraulic loading rate	5.5 gpm/sq. ft.
Fan capacity	43,000 cfm
Fan motor power	10 horsepower
Fiberglass storage tank capacity	3,300 gallons
Effluent discharge pump	
Maximum flow rate	330 gpm
Total dynamic head	65 feet
Motor power	10 horsepower
Recycling System:	
Recycling area	40,000 sq. ft.
Berm slope, horizontal:vertical	2:1 to 3:1
Gravel bed depth	1 foot
PVC pipes with spray nozzles	
Diameter	2 inches
Pipe spacing	10 feet
Number of pipes	40
Nozzles	
Nozzle spacing	5 feet
Nozzles per pipe	13
Total number of nozzles	520
Nozzle capacity	0.60 gpm @ 40 psi
Orifice diameter	0.062 inches

pipe. The tank also has a backup unit that automatically interrupts power supply to the entire system, including purge well pumps, when the water in the storage tank reaches a predetermined high or low level. The treatment system is restarted manually after the cause of the shutdown is determined.

Flow from the storage tank can be periodically recirculated to the top of the treatment system for maintenance purposes, such as removing biological growth from the pipe lines. This is achieved via a 4-inch manual control valve and a 6-inch butterfly valve; opening the 4-inch valve allows flow to recirculate, while closing the 6-inch valve prevents flow from reaching the recycling area. To date, biological fouling has not been a problem, but deposits of precipitates caused by oxidation of minerals

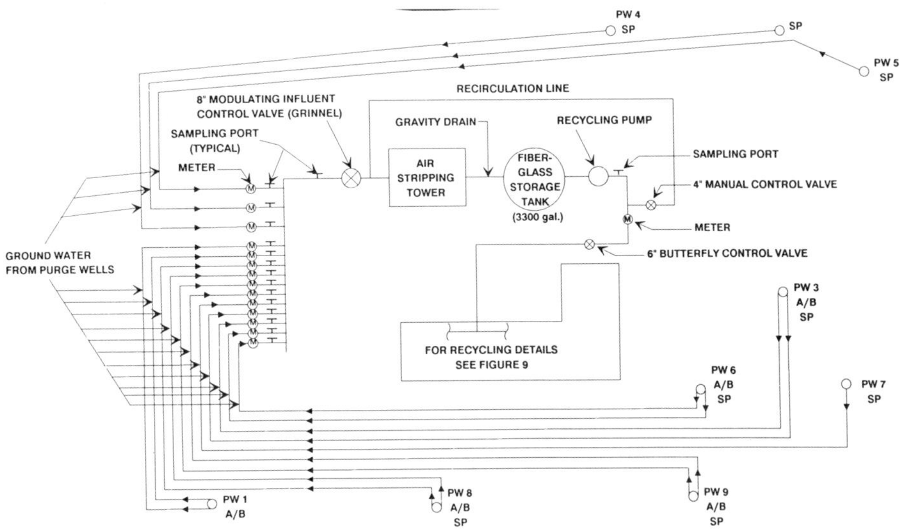

Figure 8. Process flow diagram.

in the influent waste stream may be a future problem. Design and operating parameters of the treatment system are presented in Table II.

Flow rate is measured before the water enters the header pipe and stripping tower (influent) and as the water is pumped from the storage tank to the recycling area. Water quality samples are collected at sampling ports located near the point where the influent and the recycling area flows are monitored and analyzed for the indicator parameter TCE.

Treated Groundwater Recycling System

The treated groundwater is distributed over a 40,000-square-foot recycling area located where the major disposal activities took place (Figure 1). The L-shaped area (Figure 3) is level and surrounded by a 3-to 4-foot berm providing a slope ranging from 3:1 to 2:1 around the perimeter to prevent runoff from the recharge area. The recycling area is subdivided into quadrants, each of which has similar piping and sample collection layouts (Figure 9). The purpose of this configuration is to provide the flexibility to direct and regulate the flow to areas based on the variation of recharge rates within the recycling area. A layer of pea gravel, 1 foot in thickness, covers the recycling area. This layer supports the piping network and acts as a drainage medium for water to percolate into the ground. Design parameters of the recycling system are presented in Table II.

Groundwater pumped from the storage tank to the recycling area is sprayed over the gravel layer through a PVC piping system. The piping system in each quadrant of the recycling area consists of two 4-inch header pipes located at each end of the quadrant and 10 lateral stringers at 10-foot intervals fitted with nozzles to provide upward (spray) water distribution. Aeration that takes place during spraying accounts for additional TCE removal. Water drains through the pea gravel layer and percolates into the ground. Once the water percolates into the ground, the collection, treatment, and recycling cycle is complete.

Each quadrant has a flow interceptor basin, located beneath the gravel layer, that collects some of the treated groundwater before it percolates into the ground. Samples of treated water taken from these basins are used to monitor the overall TCE removal efficiency of the remediation system. For sampling purposes, each basin is connected to the surface with a flexible hose. Each time treated groundwater samples are collected from these flexible hoses, water in the interceptor basins is pumped until the basins are empty. Emptying the interceptor basins each time samples are collected ensures that the sampled water is representative of the treated water entering the soil at the time the samples are taken. A cross-section of the recycling area is presented in Figure 7.

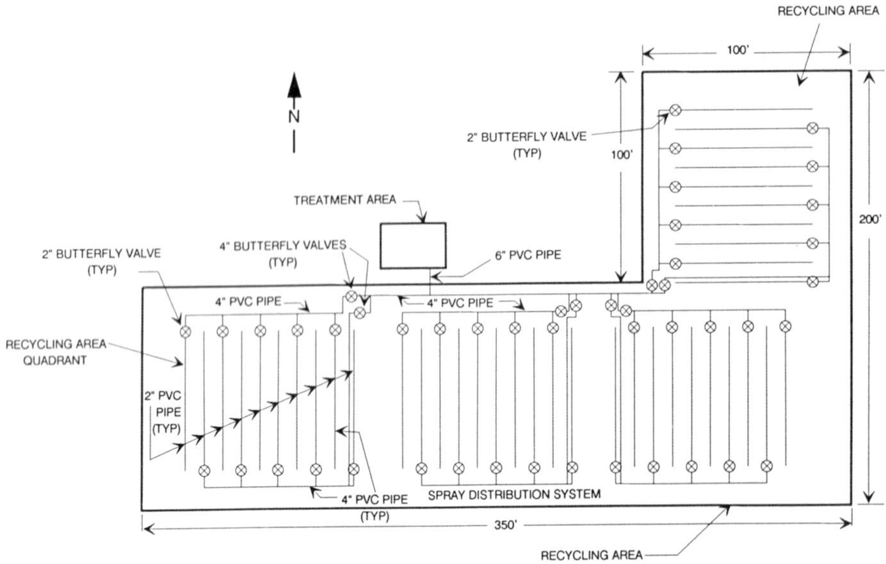

Figure 9. Treated groundwater recycling area layout.

EVALUATION OF SYSTEM PERFORMANCE

The remedial action system was designed and implemented in the summer of 1986, and startup testing of the system began in September 1986. After the winter shutdown and operating permit acquisition during spring, the system began to operate in the summer of 1987. The monitoring data obtained from July to October 1987 were used to evaluate system performance. These data were collected to meet regulatory monitoring requirements.

Monitoring requirements consisted of determining influent and effluent TCE concentrations and flow rates. The influent TCE concentration and flow rate were measured at the header pipe influent to the stripping tower. The header pipe carries combined flow from the 14 groundwater purge wells. The effluent flow rate was measured at the pipe influent to the distribution system, and the TCE

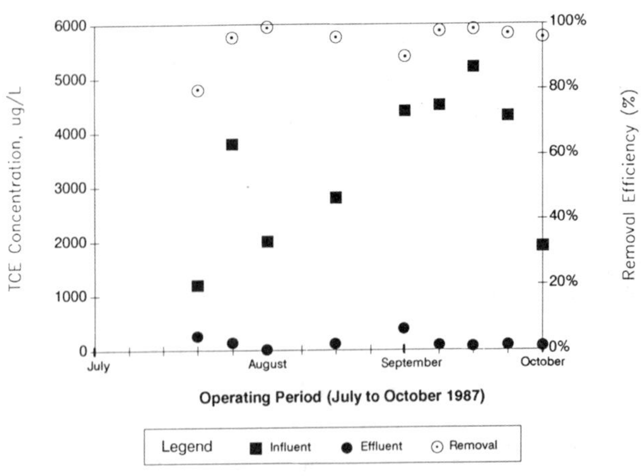

Figure 10. Influent and effluent TCE concentration data and overall system removal efficiency.

Table III. Breakdown of Total Capital and Estimated Annual O & M Costs of Remediation System

		Total Cost
Capital Cost		
Site Work Site cleanup, removal of debris, miscellaneous material handling, fencing, surveying, and access road construction		$ 300,000
Collection System 14 purge wells drilling and installation, pumps, instrumentation, piping, and hydrogeologic studies		650,000
Treatment System Stripping tower, equipment delivery and installation, foundation work, discharge pumps, electrical controls and instrumentation, monitoring equipment, backwash tank, and accessories		500,000
Recycling System		90,000
Building		80,000
Sampling Collection, Monitoring, and Laboratory Equipment		80,000
Miscellaneous Costs		60,000
	Subtotal	$1,760,000
Engineering (20%)		350,000
Construction Management, including health and safety during construction (15%)		270,000
Permitting and Legal		220,000
Contingencies (20%)		400,000
	Total Capital Cost	$3,000,000
Annual O & M Costs		
Pumps and Blowers (electrical)		$ 10,000
System Maintenance and Material		25,000
Material		35,000
Monitoring		20,000
Permit Renewal		15,000
Administration		20,000
	Subtotal	$ 120,000
	Contingencies	25,000
	Total Annual O & M Costs	$ 145,000

concentration was measured at the gravel bed-soil interface. The influent and effluent TCE concentration and the overall removal efficiency of the system are shown in Figure 10.

From the initial operating data, shown in Figure 10, the overall TCE removal efficiency of the system varied from 80 to 99%. The average TCE removal was 94.5%. This is slightly lower but in general agreement with the 97.3 to 99.1% removal efficiency observed during the pilot tests. The data showed that approximately 75% TCE removal may be attributed to air stripping and an additional 20% to the spray recycling system.

From July to October 1987, approximately 24.5 million gallons of groundwater were treated. Approximately 775 pounds of TCE were removed based on an average TCE concentration of 4,000 μg/L in the groundwater entering the treatment system. If the system had been in operation for a full annual operating period (April through October), about 35 million gallons of groundwater would have been treated, removing an estimated 1,100 pounds of TCE.

It is obvious from the design of the system that flushing of the contaminated soils and containment of the groundwater contaminant plume occur. Unfortunately, the efficiency of the soil flushing and plume containment were not established because this was not required by the consent judgment.

REMEDIAL ACTION COSTS

The total capital cost of the remedial action was approximately \$3,000,000. This amount includes costs for site work, purge wells, stripping tower, piping, electrical controls and instrumentation, pumps, treated groundwater recycling, and gravel bed. The capital cost also includes fees for legal, permitting, engineering, and construction oversight services. The estimated annual O & M cost is \$145,000. This includes costs for labor, electrical power, maintenance, replacement parts, influent and effluent monitoring, permitting, and administration. Breakdown of the total capital cost and the estimated annual O & M costs is presented in Table III.

The groundwater will continue to be treated until the cleanup goal is met. The total period of operation could not be estimated since the efficiency of contaminated soil flushing and the effect on groundwater conditions were not determined. Since the time required to remedy site contamination may be several years, the total annualized capital and O & M costs were estimated for three operating periods (15, 20, and 30 years) at a 10% discount rate. The annualized capital and O & M costs of the remedial action for treating 1,000 gallons of groundwater are \$1.70 for 15 years, \$1.60 for 20 years, and \$1.50 for 30 years.

SUMMARY

As a remedy for contamination at a spent solvent disposal site, a groundwater collection, treatment, and recycling system was designed and implemented. The objectives of the remedial action were to: 1) capture the groundwater contaminant plume to minimize off-site contaminant migration; 2) treat groundwater until the remedial action goal (15 μg/L of TCE in groundwater) is met; and 3) flush contaminated soils with treated groundwater to control the source of contamination and enhance site cleanup.

The initial operating data demonstrated that the system can effectively remove TCE from groundwater (approximately 95 percent overall removal efficiency). The estimated annualized capital and O & M costs for treating 1,000 gallons range from \$1.70 for 15 years to \$1.50 for 30 years.

REFERENCE

1. Thomsen, K.O., Khara, B.H., and Aguwa, A.A., "A Collection/Treatment/ Recharge/Flushing Groundwater Remediation Program," *Proceedings of the 7th National Conference on Management of Uncontrolled Hazardous Waste Sites, Superfund '86*, 220–223 (1986).

2 REMEDIAL ACTION CASE HISTORY: FOREST WASTE SITE LAGOONS

James B. Russell, Geotechnical Engineer
CH2M HILL
Milwaukee, Wisconsin 53201

INTRODUCTION

The process of cleaning up abandoned hazardous waste sites under CERCLA is often a long and complicated one. Completing the steps necessary for a site to go from being ranked on the National Priorities list to being remediated can often take many years, as illustrated by the flow diagram in Figure 1.

This paper presents the case history of remedial design and remedial action at the Forest Waste site and discusses the implementation and constructibility issues encountered. The paper concludes with a summary of lessons learned during remedial design and remediation.

The opinions expressed by this case history do not necessarily reflect the opinions of the United States Environmental Protection Agency (U.S. EPA) or the Michigan Department of Natural Resources (MDNR). The EPA will not be bound by any assertions or conclusions in this document.

BACKGROUND

Site Description

The Forest Waste site is in Genesee County, Michigan, about 12 miles north of the City of Flint. The 112-acre property includes a landfill covering about 11 acres and nine surface impoundments or lagoons that cover a collective surface area of about 1 acre. The site was first operated in 1972 as an industrial landfill and starting in 1973 the lagoons were used for industrial liquids disposal. Landfill operation and liquids disposal continued until 1978. The site was concurrently operated as an airport for light aircraft throughout this period.

Remedial Action

A three phase Remedial Investigation (RI) was performed at the site from June 1984 through April 1987. A phased feasibility study (PFS) addressing the lagoons as a separate operable unit was completed in April 1986. The EPA selected a "clean closure" remedy for the lagoons that required removing contaminated liquids, sludges, and soil for offsite disposal at a RCRA facility.

CH2M HILL received a Remedial Design/Remedial Action (RD/RA) work assignment in mid-1987 for the lagoons operable unit. A RD field investigation was performed by CH2M HILL in March and April 1988 to better described the limits of contamination and to test solidifying reagents and mixing ratios for contaminated sludges and soils. The final draft design documents, for removing an estimated 80,000 gallons of contaminated liquids and 5,000 cubic yards of contaminated sludge and soil, were completed in June 1988.

The EPA attached the prepared draft design documents, which included contract plans and specifications, to a unilateral order to a group of Potentially Responsible Parties (PRPs) to complete the RA work. The PRPs decided to comply with the order and started work through their engineer and construction manager, Geraghty & Miller, in October 1988. The PRPs substantially completed the RA in July 1989.

SITE FEATURES, LIMITATIONS, AND CONSTRAINTS

The Forest Waste site is located in a rural area of east central Michigan. The site is generally flat, except in the lagoon areas where elevated berms were constructed to contain the liquids. Vegetation

45th Purdue Industrial Waste Conference Proceedings, © 1991 Lewis Publishers, Inc., Chelsea, Michigan 48118. Printed in U.S.A.

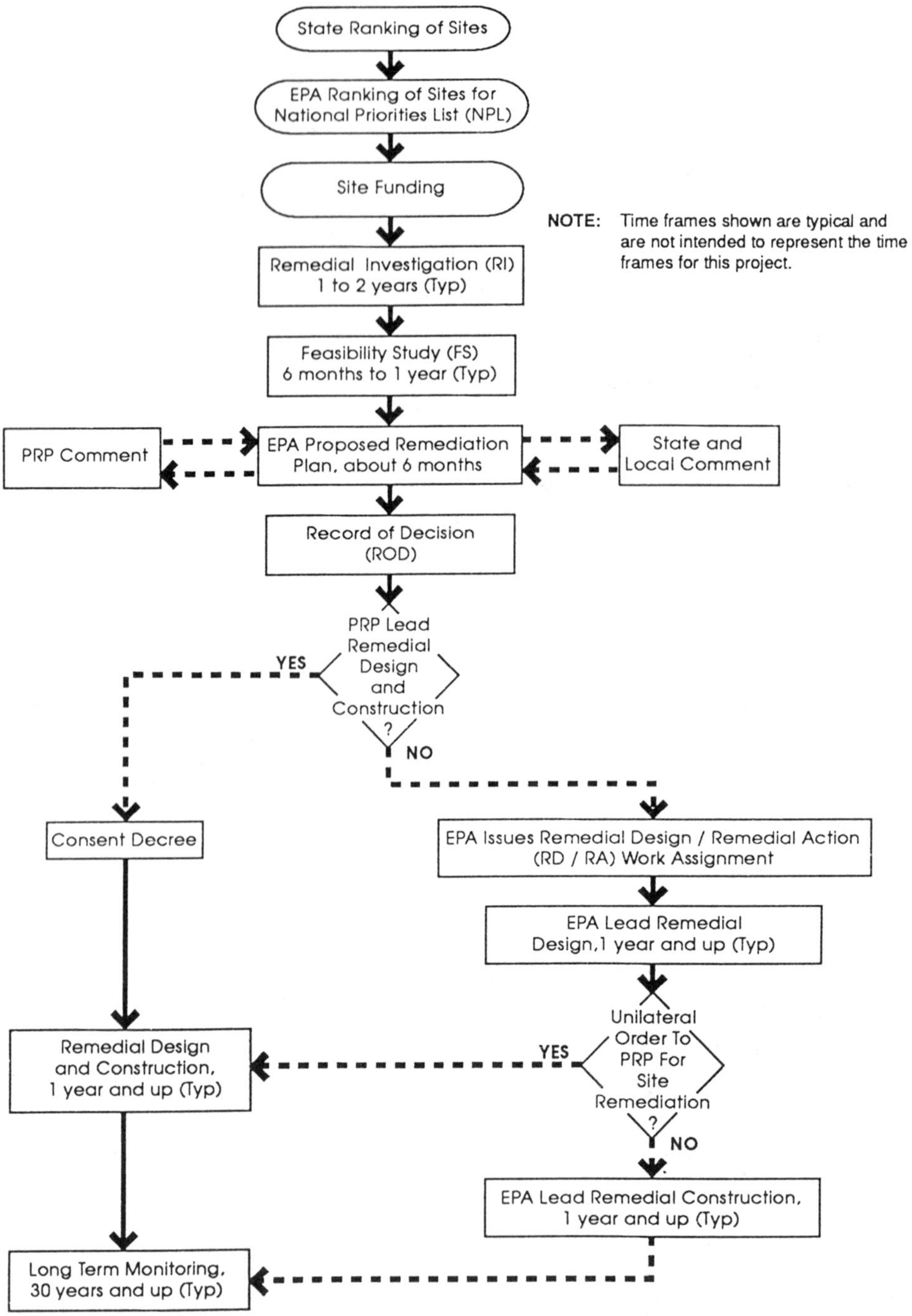

Figure 1. General route to CERCLA site remediation.

consists of grass and weeds, low shrubs, and a few scattered trees. Onsite structures included an office building, airplane hangar, and asphalt runway.

The lagoon area was located approximately 3,000 feet from the nearest county road, a lightly-traveled gravel road. Access to the lagoon area was originally provided by a dirt road, which was impassable during wet weather in the spring and fall. This lagoon access road was overgrown and in poor condition. To provide access to the site lagoons during the RA, a new access road was needed. Arrangements were made with the County Highway Department for the regular grading of the nearby County road to provide a suitable haul road for the trucks which would haul waste from the site.

Residents lived about 3,000 feet from the lagoon area and were aware of the site's history. To protect them and comply with State of Michigan air regulations, EPA required implementation of a perimeter air monitoring plan during all intrusive site activities. This plan included regular perimeter monitoring with direct time instruments such as photoionization detectors and organic vapor analyzers (i.e., HNu and OVA) as well as real time monitoring using portable sampling pumps.

Site hazards included sludges and soils contaminated with volatile organics, polyaromatic hydrocarbons, pesticides, PCBs, and inorganics. The potential of volatilization of hazardous chemicals and generation of contaminated dust during intrusive activities was addressed in a site specific health and safety plan that was followed for all site activities.

REMEDIAL DESIGN FIELD INVESTIGATION

The purpose of the remedial design field investigation was to better estimate the lateral and vertical extent of contamination within and around the lagoon area. The investigation included a lagoons liquid survey of the four onsite lagoons containing standing liquid as well as extensive soil sampling and analyses.

Soil sampling within and around the lagoons was performed using conventional geotechnical exploration methods. An all-terrain drill rig was used due to poor accessibility. Samples were collected at 1-foot and 2-foot intervals with 3-inch O.D. split spoon samplers driven with a 140-pound hammer. About 40 borings were advanced in and around the lagoons from about 1 to 15 feet below ground surface. In addition, surface sampling was performed in and around the lagoons.

To obtain samples from those lagoons containing standing water it was necessary to use a surface casing to prevent lagoon liquids from migrating down the bore hole during drilling and sampling activities. This surface casing generally consisted of a 3-foot diameter steel casing about 4 feet high which was driven into the lagoon bottom and sealed with bentonite pellets around its perimeter. The interior of this casing was bailed dry prior to advancing the borehole.

Samples were analyzed for the primary inorganic constituents identified during the PFS (barium, chromium (Cr 3+ and Cr 6+), lead, and nickel) at an onsite close support laboratory (CSL). Analyses for target compound list volatile organic compounds (VOCs), acid base/neutral extractable compounds, pesticides, and PCBs were performed through the contract laboratory program (CLP). Samples for total organic carbon (TOC), cationic exchange capacity (CEC), alkalinity, and oil and grease were also collected and analyzed through the CLP.

The EPA unilateral order stated that CERCLA material must be disposed of in an offsite policy CERCLA compliant RCRA permitted facility. For contaminated sludges and soils to be disposed of at a RCRA facility they must, among other things, be free of standing liquids. Typically, this is accomplished by stabilizing the waste material using physical or chemical means. The absence or presence of free liquids can be demonstrated using the Paint Filter Liquids Test (PFLT). An onsite treatability study was conducted on lagoon sludges and soils to identify those wastes, which when excavated, could fail the PFLT and require stabilization. Lagoon sludges and soils were mixed with different reagents to determine optimum reagent mixing ratios, estimate volume increases after reagent mixing, observe any exothermic reactions during mixing, and determine which reagents improved the handling characteristics of the waste material. A total of five reagents were examined in the study: fly ash, cement, bentonite, and hydrated and unhydrated lime.

The extensive sampling effort during the RD was performed to better predict the quantity of contaminated material which would require offsite disposal. The "upfront" sampling and treatability study was intended to result in more competitive contractor bids and would add support to the RA cost estimate to be prepared by the Engineer.

REMEDIAL ACTION DESIGN

After the RD field investigation, the sample data was evaluated by CH2M HILL. The "clean closure" solution chosen by EPA required excavation of lagoon sludges and soil with concentrations that exceeded the action levels determined in the PFS for the selected inorganics (barium, total chromium, hexavalent chromium, lead, and nickel) and selected organic compounds. The inorganic action levels were based on Acceptable Intake Chronic (AIC) criteria, media intake of 0.1 g/day, and body weight of 15 kg (child). The PFS risk assessment also presented risk scenarios for health based (ingestion) action levels for four organic compounds (methylene chloride, tetrachloroethene, trichloroethene, and bis(2-ethyl-hexyl)phthalate). However, a total of 50 organic compounds were detected in the lagoon soil and sludge during the RD field investigation which were not addressed in the PFS. Furthermore, the ROD stated that an evaluation of groundwater quality would be made in terms of leachate discharge to the aquifer. Therefore, action levels for the 50 organic compounds detected during the RD field investigation were also calculated based on existing drinking water standards, criteria, guidelines, target concentrations, and cancer potency factors.

To evaluate water quality impacts below the lagoons from the infiltration of surface water through lagoon sludges and soils, leachate concentrations had to be estimated from the soil and sludge sampling data. For the purposes of this estimation, the adsorption coefficient, Koc, of each of the 50 identified chemical compounds was determined from published literature or estimated from published regression equations. Chemical specific partitioning coefficients, Kds, were then calculated using site specific total organic carbon (TOC) data (Kd = Koc x TOC with corrections made for proper units). Soil action levels were then calculated using existing water quality standards or criteria and the final estimated relationship:

$$Ct = Cw(Kd + 1)$$

where Ct = the soil concentration action level for the organic compound
Cw = existing water quality standard or criteria

and,

Kd = the partitioning coefficient

The soil concentration action levels calculated using this method were then compared directly to the sample analyses data collected during the RD field investigation to determine lagoon excavation limits.

DRAFT CONTRACT DOCUMENTS

Draft Contract Documents, including plans and specifications, were prepared in June 1988 by CH2M HILL for the EPA to complete the RA under federal procurement and acquisition regulations. The EPA design included the following technical components:

- Construction of an onsite gravel access road and decontamination facility
- Removal of RI and RD derived wastes which had accumulated onsite (e.g., drill cuttings, monitoring well purge water, personnel protective equipment)
- Removal of an estimated 80,000 gallons of lagoon liquids to an offsite RCRA facility
- Removal of an estimated 5,000 cubic yards (truck volume) of contaminated lagoon sludges and soils to an offsite RCRA facility
- Backfilling, final grading, and seeding of the lagoon areas

PROJECT EXECUTION AND ORGANIZATION

In June 1988, before starting the federal procurement process, the EPA issued a unilateral order to the PRP group to perform the RA based on the draft EPA design or be liable for triple damages (three times the cost for the EPA to complete the RA under federal procurement and acquisition regulations) according to the provisions of CERCLA. The PRP group complied with this order and started work in October 1988 by incorporating the draft EPA design into bidding contract documents.

The PRP group hired Geraghty & Miller of Milwaukee, Wisconsin to serve as the engineer and construction manager. Geraghty & Miller had served as the PRP group's consultant during the RI/FS stages performed by EPA. The contractor selected by the PRP group was Chemical Waste Management's remediation group, ENRAC, of Alsip, Illinois. With its role changed after the unilateral order, CH2M HILL was assigned by U.S. EPA, Region 5, to perform oversight during the remediation.

REMEDIAL ACTION CONSTRUCTION

Chemical Waste Management's ENRAC group and Geraghty & Miller mobilized equipment and personnel to the site starting in mid-October 1988. Approximately 2 weeks were required to set up the field office, and construct the access road and decontamination pad. During this time RI, and RD investigation generated waste was collected and staged for disposal. Excavation and solidification of lagoon sludges and soils began in late October. The contractor, based on experience, proposed to use cement kiln dust as the mixing reagent for lagoon sludge and soil stabilization. Small scale mixing tests (about 5 cubic yards) were performed in each lagoon using the proposed reagent. Based on Michigan Department of Natural Resources (MDNR) concerns, the small scale mixing tests were monitored to ensure that excess heat was not generated during waste stabilization. Excess heat could increase chemical volatility and pose a potential fire hazard. To monitor the exothermic reaction created by mixing lagoon sludges and soils with cement kiln dust, a household meat thermometer was placed in the mixed waste. Monitoring with an HNu was also performed over the mixed material to determine if the heat of reaction caused unacceptable volatile emissions. Both of these field tests indicated that the cement kiln dust was an acceptable reagent for stabilization and the MDNR was satisfied with the results.

As the materials in the lagoons were excavated to the preliminary depths specified as pay limits in the design drawings, confirmatory samples were collected and submitted to a laboratory per an approved Quality Assurance Project Plan (QAPP). If the resulting contaminant concentrations in the lagoon materials at the base of the excavation exceeded the action levels calculated during design then an additional 6-inch layer of material was excavated and tested. Excavation and confirmatory sampling continued until soil action levels at the base of the excavation were not exceeded.

Excavation was performed using two excavators each having a 1 cubic yard bucket. Cement kiln dust was trucked to the site in bulk and end dumped into lagoons requiring waste stabilization. Cement kiln dust was first trucked to the site in pneumatic tankers and pumped into the lagoons. However, the dust generated by pumping was excessive and bulk dumping was found to produce less dust. The cement kiln dust was mixed directly into the lagoons using a kneading action via the excavator bucket. Stabilized waste was stockpiled for eventual loading into licensed commercial hazardous waste hauling trucks. Liquids were pumped from the lagoons and transferred to stationary 20,000 gallon holding tanks prior to being loaded to licensed commercial hazardous waste hauling trucks.

Prior to leaving the site, the exteriors of all trucks carrying contaminated liquids or solids were decontaminated using a portable steam cleaner at the decontamination station. All personnel performing decontamination and excavation activities wore full-face air purifying respirators.

During mid-November, rough grading was started in those lagoons which had undergone confirmatory sampling and had been certified clean. In late November work was halted pending the results of additional confirmatory samples. Excavation activities were resumed and halted three additional times between December 1988 and August 1989 pending results of confirmatory sampling. Final grading and seeding was completed in September 1989. A summary of approximate project construction costs are shown in Table I.

LESSONS LEARNED

Several valuable lessons were learned which may be applied to larger and more complex remediation projects. These lessons include the following:

- The scope and objectives of a remedial design field investigation must be clearly defined and tied to the contract management approach to be used during construction. In this case an extensive "upfront" sampling effort was performed to better define the contaminated liquids and solids removal volumes. A refined estimate of the removal volumes would result in more competitive contractor bids and would reduce the number of confirmatory samples required during construction. The "upfront" sampling resulted in draft design documents which had estimated disposal volumes of about 80,000 gallons of liquids and 5,000 cubic yards of solids (truck measure). Actual disposal quantities at completion of the remediation were about 50,000 gallons and 9,100 tons (waste materials were bid in tons instead of cubic yards by the PRPs).

  An alternative to this "upfront" sampling could have been to use the "observational method." This method recognizes that uncertainty is present and uses a structured approach to determine the appropriateness of the design as it is being implemented. In this case visibly contaminated materials would be removed prior to performing any sampling. Upon reaching "clean material," samples could have been taken and scanned in the field using a gas chromatograph or other method before submitting to a laboratory. The observational method generally requires a relatively loose contract with payment

Table I. Approximate Project Construction Costs

Activity	Unit of Payment	Quantity	Approximate Unit Cost*	Extended Amount	Comments
GENERAL REQUIREMENTS	Lump Sum	1	$36,000	$36,000	Included contractor health and safety requirements
MOBILIZATION/DEMOBILIZATION	Lump sum	1	$15,000	$15,000	Approximately 1 week was required to mobilize to the site.
ACCESS ROAD AND DECONTAMINATION PAD	Lump sum	1	$29,000	$29,000	Approximately 1 week was required to construct the access road and decontamination pad.
REMEDIAL INVESTIGATION AND REMEDIAL DESIGN DERIVED WASTE DISPOSAL					
Hazardous Waste/Contaminated Waste	Per 55-gallon drum	57	$185	$10,500	Drums emptied into rolloff boxes and trucked to Chemical Waste Management's CID Landfill in Calumet City, IL.
Noncontaminated Waste	Per 55-gallon drum	69	$44	$3,000	Drums emptied into rolloff box and trucked to municipal landfill.
Personnel Protective Gear and Clothing	Per 55-gallon drum	17	$100	$1,700	Drums emptied into rolloff box and trucked to municipal landfill.
LAGOON LIQUIDS REMOVAL, TREATMENT, AND DISPOSAL	Per gallon	57,000	$0.50	$28,500	Hauled in tanker trucks to Cyanokem facility in Detroit, MI.
LAGOON SOLIDS REMOVAL AND DISPOSAL	Per ton	9,100	$118	$1,073,800	Hauled in commercial hazardous waste hauling trucks to Chemical Waste Management's CID Landfill in Calumet City, IL. A total of about 400 trucks left the site over the removal period hauling an average of about 20 tons each. Average time for a truck to be loaded, decontaminated, and processed offsite: about 20-25 minutes.
REAGENTS					
Cement Kiln Dust	Per ton	250	$33	$8,300	Kiln dust hauled in bulk from JTM industries in River Rouge, MI.
EARTHWORK	Lump sum	1	$13,000	$13,000	
FINISH GRADING	Lump sum	1	$4,000	$4,000	
SEEDING	Lump sum	1	$5,000	$5,000	
			TOTAL	$1,228,000	

* Unit costs derived from ENRAC's original proposal to Geraghty & Miller, Inc. (dated September 21, 1988). Actual and final project costs were not made available for this paper and, therefore, may vary from the costs shown. Project costs shown do not include air monitoring and associated analytical costs or confirmatory sampling analytical costs. Construction management and oversight labor costs are also not included.

typically on a unit cost basis. Provisions and contingencies must be included when using this method to account for possible quantity overruns.

- A treatability study must be based on the nature of the waste and the requirements of the specific disposal facility to be used. In this case, the reagent was cement kiln dust based on the contractor's preference and previous experience.
- Confirmatory sampling for determining if calculated action levels are met requires that careful attention be given to properly sequencing and staging the removal of waste material to avoid having personnel and equipment on standby while samples are being analyzed by a laboratory. Laboratory turnaround time must be planned and accounted for in any project involving the removal of waste material.
- Splitting off a Remedial Design to one contractor (in this case CH2M HILL) and a Remedial Action to another (in this case the PRPs) is in general less efficient than having one contractor or party perform the entire RD/RA. In this case, the PRPs complied with a unilateral order at the end of the Remedial Design stage. The draft contract documents received by the PRP's engineer, Geraghty & Miller, Inc., were prepared by the EPA's contractor, CH2M HILL. Geraghty & Miller, Inc., was, therefore, required to adopt a set of design documents which were unfamiliar to them. In addition, CH2M HILL performed the RD field investigation under the assumption that they were going to be the construction manager during the RA. The federal procurement rules CH2M HILL followed during the process are very different from those rules which the PRPs may operate under.

In summary, the scope and objectives of a remedial design field investigation must be clearly defined and tied to the contract management approach to be used during construction. In addition, splitting off a remedial design to one contractor and a remedial action to another is in general less efficient than having one contractor or party perform the entire RD/RA. Finally, the roles and responsibilities of EPA, oversight personnel, the PRP's and their agents, and State and local agencies are different, yet all must converge if a remediation project is to be successful. The responsibilities of each party and the lines of communication between parties must be clearly defined prior to beginning and during construction.

3 CLEAN-UP OF TOLUENE-CONTAMINATED SOIL AND GROUNDWATER USING SOIL VAPOR EXTRACTION

David R. Hale, Senior Project Engineer
Qsource Environmental Services

Philip L. Hayden, Chairman of the Board
Qsource Environmental Services
Miamisburg, Ohio 45342

INTRODUCTION AND BACKGROUND

In 1986, a large commercial printing facility experienced a large leak of toluene solvent from a faulty pressure gauge on an underground storage tank (UST) system. Although exact amounts could not be determined, estimates of the amount spilled ranged up to 3,000 gallons. An emergency response action was undertaken to remove a quantity of heavily contaminated soil, and a site investigation to determine subsurface impacts was begun.

SITE INVESTIGATION

The site investigation consisted of the installation and sampling of 14 shallow groundwater monitoring wells. During the installation of these wells, it was observed that the shallow geology of the site consisted of sand and gravel outwash deposits overlying a glacial till layer. This aquitard till layer separates a thin shallow layer of groundwater from the extensive sand and gravel deep aquifer beneath the site. This deep layer is used as the sole source drinking water aquifer for a large metropolitan area.

Shallow groundwater was detected at depths from 15 to 22 feet below the ground surface. The thickness of this groundwater layer ranged from approximately one to three feet. Samples of this shallow groundwater revealed toluene contamination ranging from non-detectable away from the UST spill area to above solubility limits (>450 mg/L) beneath the spill area. No observations or measurements of floating free product were made during the site investigation, although the substantial concentrations observed in several of the well samples (up to 220,000 mg/L) were undoubtedly the result of recovered free product.

INITIAL GROUNDWATER REMEDIATION

Based on the results of the groundwater investigation, a groundwater remediation system consisting of four recovery wells was installed. These wells were located in the area of highest groundwater contamination. Each well had a 4″ diameter PVC casing, and was installed with a air-operated bladder pump to recover groundwater and free product. Recovered groundwater was then pumped to an existing solvent recovery system for treatment prior to discharge.

Operation of the groundwater recovery system indicated that a total of approximately 1 gallon per minute (GPM) of groundwater was being recovered by the four wells. Based on a this recovery rate, and an average concentration of 300 mg/L toluene, this resulted in less than 0.5 gallons per day of toluene being recovered. Because of the low pumping rates, cleanup of the site and control of contamination by groundwater pump and treat was deemed not suitable.

REMEDIAL ALTERNATIVE SELECTION

At this stage, a decision was made to independently review the known conditions at the site and assess alternative remedial strategies for cleaning up the site. Another environmental engineering firm was retained, and after all available information on the site had been reviewed, a short-list of applicable alternatives was developed. These included:

45th Purdue Industrial Waste Conference Proceedings, © 1990 Lewis Publishers, Inc., Chelsea, Michigan 48118. Printed in U.S.A.

Table I. Costs of Groundwater Treatment Alternatives

Alternative	Capital Cost	Operating Cost
UV-Ozone	$125,000	$45,000
In-Situ Biological	$75,000	$35,000
Above-Ground Biological	$25,000	$20,000
Carbon Adsorbtion	$8,000	$7,000
Air Stripping	$5,000	$2,000

Groundwater recovery using trench interceptors
Groundwater treatment using pure-compound recovery and biological treatment
In-situ biological treatment of soil and groundwater
Soil vapor extraction

GROUNDWATER RECOVERY

Based on observations from the existing groundwater recovery efforts, it was initially thought that horizontal trench interceptors might be more efficient than vertical wells in recovery of groundwater from the thin layer of shallow groundwater. However, an assessment of the long-term steady-state flow conditions of this groundwater layer indicated that it would not be able to yield substantially greater volumes of water than were being presently recovered. This calculated aquifer yield was less than 2 GPM for a 100' wide width of aquifer. Thus, a more efficient recovery system would be largely moot and not cost-effective, since available water supply was the rate-limiting factor in groundwater recovery.

Subsequent sampling of groundwater monitoring wells over a period of 18 months indicated that migration of the contaminant plume was very slow, and that installation of additional downgradient interceptor wells was not necessary. Consequently, no changes were undertaken with regards to the existing groundwater recovery system.

GROUNDWATER TREATMENT

Because of the presence of apparent floating layers of toluene, it was thought that some form of free-phase separation step would be necessary prior to treatment of groundwater for dissolved phase contamination. Based on the anticipated flow rates, a small oil/water separator unit would be sufficient, since actual free-phase recovery volumes would be small, estimated to be no more than 1–2 gallons per day.

For treatment of dissolved phase contamination, several options were available and considered. Toluene contamination is readily treated by several methods, including air stripping, activated carbon adsorbtion, biological treatment, and catalytic oxidation by UV/ozone or hydrogen peroxide/ozone. Based on the groundwater flow rates and concentrations at this site, treatment by biological or catalytic oxidation techniques was cost prohibitive from a capital and operating standpoint. Because of the low flow volumes, air stripping using a classical packed tower aerator was not feasible, although the use of small diffused basin aerator was appropriate. Finally, treatment by small, refillable activated carbon adsorbtion units was feasible. Capital and operating costs for each of these groundwater treatment alternatives is presented in Table I.

Based on operational and hazardous waste-generation considerations, groundwater treatment using a small diffused basin air stripper was selected. Because of the low flow rates and subsequent emission rates, no emission controls were required by local air pollution control authorities. The air stripper was designed using a 1500 gallon, two-chamber concrete septic tank located above-grade as the containment vessel. A 100 SCFM regenerative blower was utilized as the aeration source, discharging through several fine-bubble stick diffusers located in the bottom of the aeration basin chambers. A schematic of the air stripper is shown in Figure 1. Since the air stripper would also remove low volumes of any recovered free product, no provision for free product recovery was required with this system.

Total capital cost for this air stripper unit, not including electrical installation, was less than $3,500. Operating costs for the air stripper are approximately $3.50 per day for electricity. Total capital and operating costs for the air stripper system are approximately $10 per day or $0.007 per gallon treated at a 1 GPM flow rate.

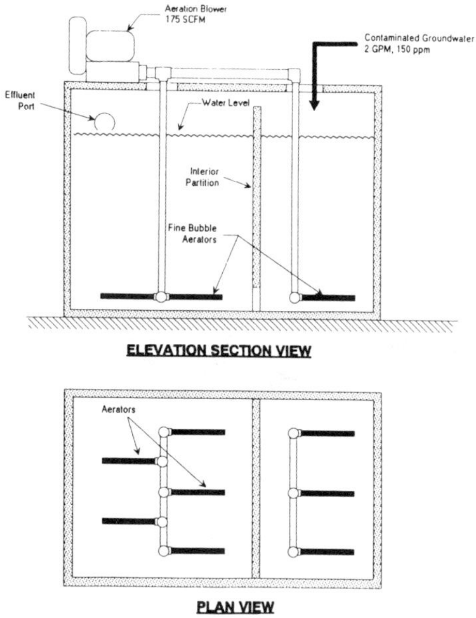

Figure 1. Air stripper schematic

In-Situ Biological Treatment of Soil and Groundwater

Since toluene is a readily biodegradable compound, consideration was given to in-situ bioremediation as a means of treating contaminated soil and groundwater. However, because of anticipated limitations on the amount of natural groundwater recovery, some form of closed-loop groundwater recharge system would have to be installed. There were limitations related to local regulatory approval of a system that would introduce water of less than drinking water quality into the contaminated aquifer, as well as likely regulatory concerns over the potential for spreading the contamination as a result of inadequate groundwater capture. Finally, in a comparative cost analysis, in-situ bioremediation did not appear to be as cost-effective as other alternatives.

Soil Vapor Extraction

Since toluene is a volatile compound, it is amenable to in-situ soil remediation through the use of soil vapor extraction (SVE), or in-situ air stripping. This process involves the placement of one or more vacuum extraction sources, usually subsurface wells, into the area of vadose zone soil contamination, and extracting contaminated soil gas. Clean air subsequently infiltrates into the soil through the surrounding ground surface or through several passive or active injection wells located around the extraction well. It is this author's opinion that passive injection wells are largely ineffective and should not be used. Active injection wells, where clean air is pumped into the subsurface at discrete locations, should be used very carefully, since there is a potential to create channeling of subsurface air flow. In certain situations, this effect can be used to advantage to overcome geological feature constraints.

A schematic representation of a multi-source soil vapor extraction process with surface infiltration is shown in Figures 2 and 3. Note the behavior of subsurface air flow between the uncovered outside the building compared to that beneath the building. For this particular site, there existed the potential that some contamination may have extended beneath the facility. A SVE system located outside and adjacent to the facility would not influence air flow beneath the building, and thus would not remove any soil contamination present there.

Soil vapor extraction was anticipated to be effective in removing floating toluene from the groundwater surface as well as residual soil contamination. Other users of the technology have reported that dissolved phase contaminant concentrations in groundwater have decreased under the influence of soil vapor extraction alone. This effect is theoretically achievable in thin layers of groundwater since VOC's would volatilize through the groundwater surface layer into the air stream, establishing a resultant concentration gradient through the thickness of the groundwater layer. Diffusion of contam-

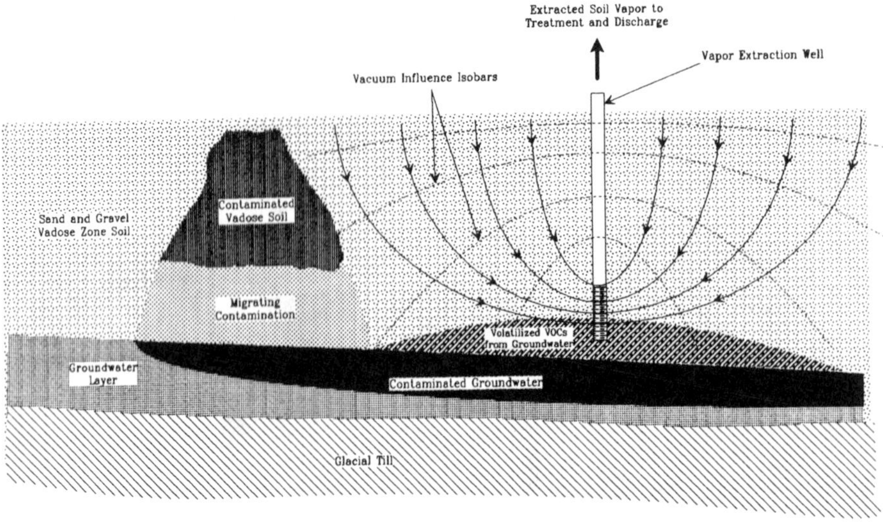

Figure 2. Soil vapor extraction influence on VOC removal from groundwater.

inants from the more contaminated lower water to the upper layers would thus occur, gradually reducing overall groundwater concentrations as cleanup progressed. Thus, this method of soil vapor extraction was selected as a means of achieving integrated remediation of soil, groundwater, and free-phase contamination.

SOIL VAPOR EXTRACTION SYSTEM DESIGN

Because of the requirement for integrated remediation of soil and groundwater, the soil vapor extraction system design was substantially upgraded to allow it to substantially influence the shallow groundwater. This was accomplished by increasing the normal extraction rates from each vacuum well

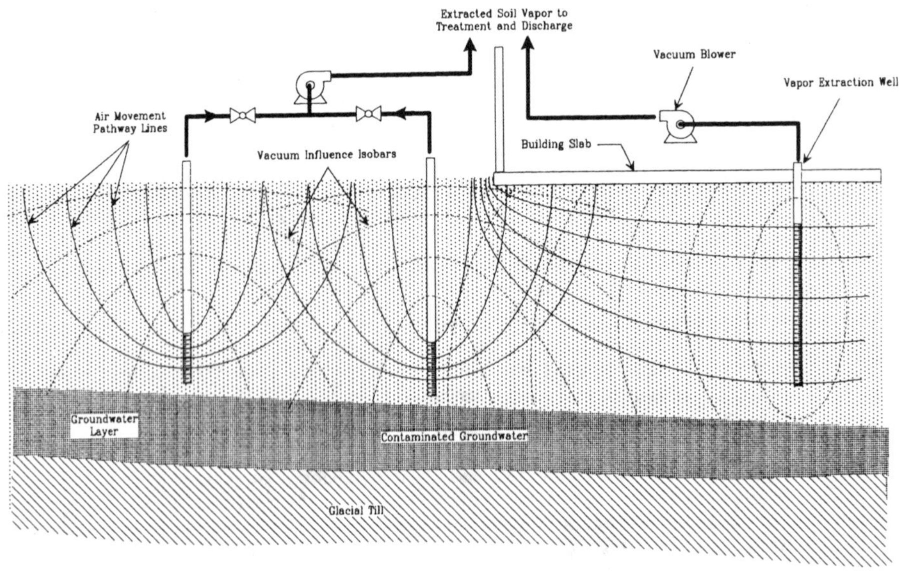

Figure 3. Influence behavior of multiple soil vapor extraction systems.

by 200–300% compared to normal soil-only remediations, and by placing the vacuum wells closer than normal for these extraction rates.

Six 4″ diameter PVC-cased extraction wells located in two rows of three each were placed at nominal spacings of 50 feet apart. The vacuum wells were screened for a total height of 5', and the bottom of the screen section was placed at three feet above the groundwater surface. The screen section was fabricated by drilling 1/2″ diameter holes four places around at a vertical spacing of 3″.

A high pressure centrifugal blower capable of a nominal extraction rate of 2,000 SCFM, or 333 SCFM per extraction well, was used as the vacuum extraction source. A maximum vacuum of approximately 40″-50″ of H_2O was anticipated for these extraction rates in this soil matrix. Because of the high extraction rates and resultant subsurface soil vapor velocities, this system would potentially remove substantial quantities of entrained water from the vadose zone soils. To protect the extraction equipment and emission control system, an air/water separator was designed and installed immediately prior to the vacuum blower. This separator was designed to reduce the nominal air velocity to approximately 400 ft/min, thereby causing any entrained water particles to drop out into the separator. The separator was constructed as a 3' diameter by 4' tall nominal vacuum vessel with internal baffles and drain ports.

Ducting from the vacuum wells consisted of various size PVC piping laid along the ground surface. As additional wells were plumbed into the piping, pie sizes increased to accommodate the larger air flows so as not to excessively increase pressure losses in the piping system. Duct sizes ranged from 4″ at the wells increasing to 6″, 8″, and finally 10″ immediately prior to the blower. Butterfly control valves and vacuum gauges were located at each wellhead, allowing individual monitoring and subsequent balancing of the extractions from each well.

It was anticipated that initial extractions from the soil would result in a very heavily contaminated and potentially explosive air stream. This condition was expected to be short-term and only in the initial stages of the system operation. To provide protection against any catastrophic explosions or fires, a valved fresh air bleed was provided on the vacuum leg prior to the blower. This allowed for selective bleed-in of dilution air to the system prior to its exposure to the blower and downstream emission control system. In addition, the blower was provided as an explosion-proof unit with suitable electrical motor and non-sparking fan section materials.

EMISSION CONTROL SYSTEM

Because the site facility is a commercial printing operation, it already had a large activated carbon emission control and solvent recovery system operational at the plant. Currently, the system had 1/3 available capacity because of the decommissioning of a printing line. This available capacity totaled over 10,000 SCFM, more than adequate for any anticipated needs from the vapor extraction system. It also allowed for future growth of the system should initial operations warrant. As a consequence, a total system consisting of four "segments" of vapor extraction units was designed, with only the first segment located at the UST area to be installed and operated initially. The other three segments were planned for outside downgradient and inside locations. Ducting to the emission control system was designed and constructed to accommodate future tie-ins should future SVE segments be installed. This ducting downstream of the SVE blower to the emission control system was constructed of round galvanized ducting of various diameters to accommodate design air flows.

Because of the daily variable nature of the facility's printing operations, measurement of incremental additions to normal solvent recovery volumes was not possible. Thus monitoring of the SVE system's performance was accomplished by monitoring SVE discharge concentrations and calculating the resultant contaminant extraction rates.

SYSTEM STARTUP, OPERATION, AND MONITORING

After installation was completed and all necessary air permits had been received from the regulatory agency, the system was started up. Since concentrations in the extracted soil gas stream were expected to be initially very high, monitoring of the discharge gases was performed using a combination LEL/OVA (lower explosive limit/organic vapor analyzer) meter. This meter measured the concentration of contaminant in the extraction stream as a percentage of the LEL for toluene (1.3%) or as parts per million (ppm) calibrated as toluene.

Initial startup indicated that the extraction concentration was in excess of 100% of the LEL. Thus, bleed-in air was introduced to the system, estimated to be approximately 65–70% by volume. This reduced the initial LEL readings to approximately 35%, slightly above the 25% operating target. This concentration dropped below 25% within two days, allowing for the quantity of bleed air to be

Table II. SVE Startup Concentrations

Day	Soil Vacuum (In H_2O)	LEL (%)	Conc. (ppm)
1	10″	37	4,800
2	10″	35	4,600
3	10″	32	4,200
4	10″	25	3,200
5	10″	15	1,950
6	22″	26	3,400
7	22″	21	2,700
8	22″	16	2,100
9	30″	17	2,200
10	30″	17	2,200
11	30″	13	1,700
12	30″	12	1,600
15	30″	7	910
20	30″	5	650
25	30″	3	390
30	30″	2	250
40	30″	2	175

reduced and more soil vapor extraction. As concentration dropped within the first weeks, bleed air was steadily reduced and extraction increased. Full extraction was begun after 12 days of operation. Table II lists the extraction concentrations and vacuums associated with system startup.

Calculated extraction rates were approximately 400 gallons per day at system startup. As expected, these extraction rates decreased exponentially with increasing system operation, reducing to approximately 50 gallons per day after 35 days of operation. Figures 4 and 5 are graphical representations of start up concentrations, weekly extration rates, and cumulative contaminant removal.

Once extraction concentrations had decreased below 4% LEL, the monitoring was switched to PPM readings. The meter used was calibrated to register ppm concentrations of toluene above 5 ppm. However, since it was an LEL meter, it also registers any other flammable organic constituents in the

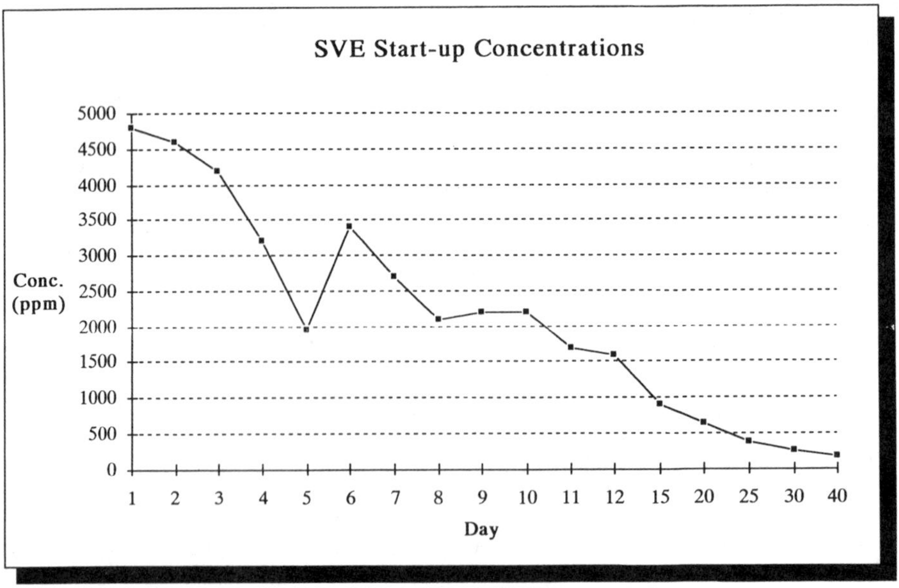

Figure 4. SVE start-up concentrations.

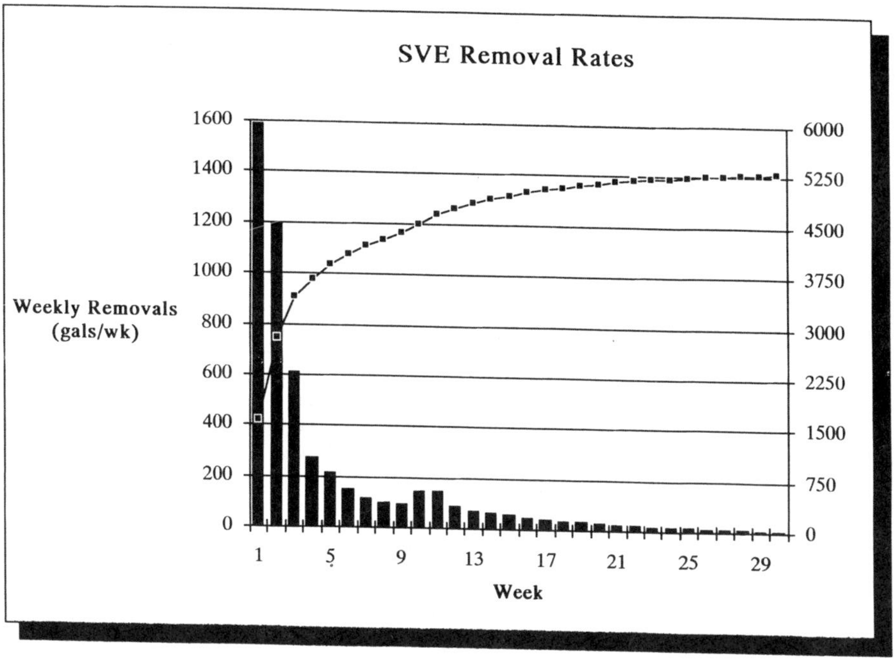

Figure 5. SVE removal rates.

air stream. This is not a significant potential error source when the extraction concentrations where above 100 ppm. However, longer-term monitoring of the extraction stream with the LEL/PPM meter indicated that the concentrations seemed to level off around 100 ppm. This did not seem totally unrealistic, since there was a substantial floating layer of toluene on the groundwater surface that would tend to volatilize at a fairly constant rate, as would any undetected leaks from underground storage tanks or piping. However, it was expected that this leveling of concentrations would not persist for an extended period of time, which it apparently did.

After several months of this level extraction concentration, the air stream was monitored using two other organic vapor analyzers, a photo-ionization detector (PID) and an OVA/FID (flame ionization detector). The PID would measure toluene contaminants while not registering any methane-related constituents that were suspected of incorrectly elevating concentration readings with the LEL/PPM meter. The FID would provide an independent confirmation of the LEL results. These instruments revealed that the actual toluene-based contaminant concentrations were approximately than 15% of the LEL readings at lower concentrations. Thus, the calculated removal efficiencies of the system were revised to reflect this apparent error source in monitoring.

GROUNDWATER REMEDIATION AND HYDRAULIC INFLUENCES

As with any SVE system whose vacuum sources are located close to the groundwater surface, the system altered the normal groundwater surface elevations by causing a mounding effect centered on the extraction well. It is important to anticipate this mounding influence around the well screen, since seasonally-rising water table levels combined with this influence can cause the extraction well to become choked off. In certain cases, this effect can become self-perpetuating, since increased resistance from water in the screen section causes additional vacuum, which raises the water level further, which causes even more vacuum, and so forth until the well is completely restricted by raised groundwater. To counterbalance this effect, it is recommended that the bottoms of extraction wells be placed at least the maximum vacuum above the highest anticipated groundwater elevation, and in no cases less than 36 inches above groundwater.

The SVE system appears to have removed all noticeable floating toluene from the surface of the groundwater. Since these levels were not measured prior to operation of the system, actual removals of this material cannot be estimated. In any case, observations from field well samplers indicated that

Table III. Groundwater Concentrations (mg/L)

Well	12/89	2/89	4/89
MW-6	0	0	0
MW-7	487	352	274
MW-8	138	75	67
MW-12	0	0	0
MW-14	480	164	128

as much as 1″ inch of floating product may have been removed within the first 3–4 months of operation.

Once this floating material was removed, gradual removal of dissolved phase contamination from the groundwater was effected. Subsequent sampling of this effect is noted by comparing before-operation monitoring well sample results in the UST vicinity with subsequent sample results. These indicated that dissolved concentrations have dropped as much as 300 ppm during the operation of the SVE system, with an approximate average reduction of 120 ppm. This removal is expected to continue with increasing operation of the system, although at increasingly slower rates. Groundwater sampling results from selected monitoring wells are summarized in Table III.

CONCLUSIONS

Based on the observations and results of the SVE remediation, it appears that remediation of toluene-contaminated soil has been largely accomplished after 6 months of operation. Current removals are calculated to be approximately 1 gallon of toluene per day, and are assumed to be significantly comprised of dissolved phase contaminants. Remediation of contaminated groundwater and removal of over 1″ of floating free-phase contamination has been effected as anticipated, demonstrating that the selective usage of SVE to remediate contaminated groundwater is feasible. Dissolved phase contamination in groundwater has been reduced by over 50% since the removal of floating free phase product, with additional reductions expected to continue, although at a slower rate.

A calculated total of approximately 5,200 gallons of toluene has been recovered from the site, for a total cost of approximately $110,000. This results in a remediation cost of $2.90 per pound of contaminant recovered. Contaminated soil at the site has been cleaned to a point where removal and replacement of the existing UST's can now be effected without undue concerns over the handling and disposal of potential hazardous waste.

4 USE OF INNOVATIVE TECHNOLOGIES DURING THE CLEANUP OF THE DENVER RADIUM SUPERFUND SITE

Marie Bushway Zanowick, Project Manager
J. Brink, Project Manager
H. Fliniau, Project Manager
T. Rehder, Project Manager
P. Nyberg, Radiation Specialist
J. Goldfarb, Attorney
S. Pennock, Community Relations Specialist
U.S. Environmental Protection Agency Region VIII
Superfund Remedial Branch
Denver, Colorado 80202

INTRODUCTION

The Denver Radium Site consists of 31 properties contaminated by low level radioactive waste that could endanger public health, welfare, and/or the environment. The contamination is believed to have come from residues of the radium processing in Denver from the beginning of World War I to the early 1920's.

The Denver Radium Site was placed on the National Priorities List in September, 1983. It was divided into 11 Operable Units (OU) which contain both open space land areas and structures. The contaminated soils from all but one of the OU are being excavated and hauled by rail to a permanent disposal facility.

During the course of the Denver Radium Site cleanup, the Denver Radium team has investigated and employed a variety of new or infrequently used technologies. The application of these technologies is described in the following paper.

BACKGROUND

Radiation emitting compounds were recognized in the late 1800's. This led to the discovery in 1898 of a new radioactive element named radium. Radium was first found in pitchblende, a rare mineral which is found in Bohemia, Saxony, Cornwall, and Colorado. Early research in the 1900's identified the ability of radiation to destroy or inhibit the growth of selected cells, even though the danger of radiation to living cells was not understood. This discovery led to the widespread use of radium as a cancer treatment and created a strong demand for radium in the United States and Europe, resulting in an increased demand for pitchblende. New sources of radium were sought, and another radium-bering mineral, carnotite, was identified in Colorado. This material was mined and some of the highest ores were sent to Europe for processing.

The need to establish a domestic source of processed radium intensified after the outbreak of World War I in Europe threatened domestic supplies. This concern led to the establishment of the National Radium Institute (NRI) in Denver in 1913. The NRI successfully demonstrated that radium could be economically processed in the United States. After producing 8.5 grams of radium from approximately 16,000 tons of ore, the NRI closed in 1916. Other radium processing plants were also active in Denver during this time. The collapse of the radium processing industry is attributed to the discovery of an extremely rich deposits of radium bearing ore in Africa. This put the U.S. radium producers at a financial disadvantage and the Denver radium industry closed almost overnight as the refineries went bankrupt or were converted to other uses.

45th Purdue Industrial Waste Conference Proceedings, © 1991 Lewis Publishers, Inc., Chelsea, Michigan 48118. Printed in U.S.A.

At the NRI, carnotite ore was treated with a nitric acid leaching process to produce radium chloride, iron vanadate, and sodium uranate. The by-product sodium nitrate was recovered for the manufacture of nitric acid; barium chloride was reused in the process; and an iron-calcium precipitate and leaching/precipitation residues which were discarded. This process was designed to recover more than 90% of the radium, 85% of the uranium, and about 30% of the vanadium.

Although much of the radium and uranium was recovered, process residues, sandy tailings containing uranium, radium, and other radioactive materials were discarded. This material contributed to the radiation currently associated with the Denver Radium Site. Due in part to dilution from surrounding clean soil, radioactivity values for this material is about half that found in typical uranium mill tailings. No efforts were made to recover radium during uranium processing.

SETTING CLEANUP STANDARDS

The Records of Decision (ROD) for all but one of the Denver Radium Site OUs were signed in 1986 and 1987. EPA's preferred remedial action alternative is offsite permanent disposal. The remedy involves excavation of the contaminated soil from open areas and from beneath buildings and transportation to an offsite repository.

The Superfund Amendment and Reauthorization Act of 1986 (SARA) requires that the selected remedial action shall attain a degree of cleanup which, at a minimum, assures protection of human health and the environment. Section 121 (d) also requires that the remedial action attain, at its completion, a level or standard of control that at least attains legally applicable or relevant and appropriate standards, requirements, criteria, or limitations (ARAR's) under Federal or State law.

The degree of cleanup necessary to protect human health and the environment at the Denver Radium Site properties has been determined by evaluating Federal and State ARAR's. Contaminant-specific ARAR's for the Site includes 40 CFR 192 for radium contaminated soils, indoor gamma radiation, and radon decay products. The National Council on Radiation Protection and Measurements (NCRP) and International Commission on Radiological Protection (ICRP) recommended gamma radiation dose was considered for protecting members of the general public during remedial action and in selection of the degree of cleanup.

40 CFR 192, Subpart B—Standards for Cleanup of Land and Buildings Contaminated with Residual Radioactive Materials from Inactive Uranium Processing Sites, provides the standards for cleanup of open lands or buildings was determined to be relative and appropriate in the case of the Denver Radium Site. Radium concentrations are 5 pCi/g above background within 15 cm (6 inches) of the surface, and 15 pCi/g above background in any subsequent 15 cm layer below the surface averaged over a 100 square meter area. Indoor gamma radiation standards are set at 20 μR/hr above background. For radon decay products, a standard of 0.02 working levels (WL) (annual average) has been set, with a maximum level of 0.03 WL.

In addition, the U.S. Department of Energy's Uranium Mill Tailings Remedial Action Project (UMTRAP) implementing these standards has developed guidance for the cleanup of properties that have become contaminated. These guidelines are described in the Vicinity Properties Management and Implementation Manual (DOE, 1984) and have been applied to the Denver Radium Site. For contaminated open land properties, these guidelines direct that, wherever possible, all contamination exceeding EPA standards be removed, regardless of areal extent or degree once clean up of the property was considered to be necessary.

The Nuclear Regulatory Commission regulates gamma radiation doses and airborne concentrations of uranium, thorium-230, and radium-226 for areas in NRC licensed facilities that have either restricted or unrestricted access (NRC, 10 CFR 20). Although the Denver Radium Site properties for which RODs have been signed are not licensed NRC facilities, the NRC regulations are used by the Occupational Safety and Health administration for worker safety in unrestricted areas. As such, the NRC regulations on gamma radiation doses are relevant and appropriate requirements to these properties. The maximum occupational gamma radiation dose for an unrestricted area is 500 millirems per year (mrem/yr). For a restricted area, this dose is 5 mrem/yr.

LOCATION OF BURIED RADIUM CONTAMINATION USING THE RADON SURFACE FLUX AND RADON-IN-SOIL-GAS MEASUREMENTS

The preliminary assessment of the depth and extent of radium contaminated soils at the Denver Radium Site was done using surface gamma, bore-hole gamma, and delta-gamma surveys and laboratory analysis of soil samples. To compensate for the horizontal and vertical shielding by the soil and the confidence level for the instruments, EPAs contractors use a value of 10 pCi/g for gamma surveys, instead of the EPA standard of 15 pCi/g, as the cutoff value for any layer below the first 6 inches. Shielding effects also cause gamma surveys to be ineffective in finding deposits with low activities that are more than a couple of feet below or to the side of the gamma detector. Radon-in-soil-gas and radon flux detector methods have been demonstrated to be more sensitive in locating these deposits.

Radon-in-soil-gas and radon-surface-flux measurements provide different but complementary information. The radon-flux measurements are sensitive to the contamination found in the upper two to three feet while the four foot-deep soil gas measurements are sensitive to deeper deposits. These methods can be successfully combined to extract rough depth information about the radium concentration in soil.

Radon-surface-flux and radon-in-soil-gas measurements were performed by EPA's contractors, UNC/Geotech, at OU X from August 24, 1988 to August 26, 1988. A 50-foot grid laid-out by UNC/Geotech's Land/Survey Department was used for sampling locations. The radon-flux measurements were made using a modified U.S. Army M-11 Gas Mask charcoal canister sealed to the ground with soil. The canisters were left in the field for 48 hours to minimize diurnal effects. At the end of the 48 hour exposure period, metal lids were taped over the open ends of the canisters. This procedure allowed the short-lived radon daughters to reach equilibrium with the radon adsorbed on the charcoal.

The canisters were analyzed using the following techniques: The gamma emissions produced from the radioactive decay of radon daughters Pb-214 and Bi-214 were counted using a three inch NaI detector and a Nucleus PCA-1000 multichannel analyzer board in a Zenith 159 microcomputer.

The radon-in-soil-gas measurements were made using a soil-gas probe and a portable EDA Instruments Model RD-200 alpha scintillation counting system. The soil-gas probe was driven into the ground to a depth of four feet. Shallower samples had to be accepted at some locations due to buried debris on site. Although the measurement of soil-gas concentration from these samples tend to be less responsible to deep radium contamination and more responsive to shallow contamination, the results of the testing can be used to fill in data gaps in the radon survey.

Aliquot quantities of soil gas were pumped into the portable scintillation counter through the hollow probe and counted for one minute. Samples were taken until three consecutive sample results were within counting statistics; usually this took only four to five samples. These three values were averaged and recorded as the location value. Calibration of the portable scintillation unit was conducted prior to the field work at the UNC/Geotech Grand Junction Project Office Radon/Radon Daughter Chamber.

Results of the radon-surface-flux and radon-in-soil gas measurements were compared to the Extent of the Contamination Map for OU X. There was generally good agreement between the two. All deposits found during the original assessment were located by the radon survey. However, both the radon-flux and the soil-gas measurements indicated larger areas of contamination. During the Fall of 1988, OU X was excavated as part of the Remedial Action. During the remediation contaminated soil was removed past the boundaries identified by the original assessment and near the boundaries suggested by the radon-flux and soil-gas survey. No additional contamination missed by either the gamma or radon surveys was found during remediation.

Comparison of soil gas, radon flux, gamma surveys, and delta-gamma measurements shows that radon-in-soil-gas method has the highest signal-to-background ratio for detecting buried radium contamination. Radon surface flux has the second greatest sensitivity, while delta-gamma has the least.

MEASURING RADIUM CONCENTRATION IN SOIL DURING CLEANUP USING THE OPPOSED CRYSTAL SYSTEM (OSC)

In order to verify that the EPA standards for radium-226 are being met during excavation of contaminated soils, excavation control is performed using total count gamma field instrumentation. Assumptions are made as to what fraction of the total gamma emissions originate from radium-226 in the soil and what fraction is due to the presence of other radionuclides. While the measurement capability of the total count instruments is sufficient for detecting the excavation work, EPA Region VIII believes that more accurate measurement of radium concentration is necessary to verify that the cleanup criteria have been met. However, laboratory analysis for radium-226 takes a minimum of seven to twenty-one days, depending upon the analytical method employed. This could result in significant delays in the remedial action.

In order to promptly determine radium-226 concentrations in the soil samples, EPA Region VIII has equipped a number of vans with the Opposed Crystal System (OSC). The OSC is an apparatus that indirectly measures radium-226 activity by determining the activity of Bismuth-214. The analysis is performed by placing a sample containing approximately 600 grams of soil between two scintillation-type detectors (sodium iodide) which are connected to a multi-channel analyzer. The system counts the Bismuth-214 activity 609 keV gamma event for a period of 500 seconds. Radium-226 activity (in pCi/g) is then calculated by multiplying the net count rate by a calibration factor and dividing that number by the sample weight. The entire process, including sample collection, sample preparation and analysis, can be performed in less than 30 minutes.

Accurate measurement of bismuth-214 by the OSC system, and accuracy of the resultant radium-226 concentration values, is affected by the following: 1) moisture content of the soil; 2) sample inhomogeneity; and 3) disequilibrium between bismuth-214 in the soil and the radium-226 due to radon loss. Therefore, the OCS results are calibrated to laboratory values through the application of a regression equation which is based on over 2,000 OCS/laboratory data pairs that have been gathered during the course of the Denver radium project and the Uranium Mill Tailings Remedial Action Program cleanup work underway in Grand Junction, Colorado.

It should be noted that the presence of cesium-137 in the sample will distort the 609 keV photo peak and, consequently, result in erroneous values for radium-226. Care must be taken to identify distortions caused by cesium-137 interference and to make the necessary corrections.

OPERABLE UNIT VIII – BACKGROUND

Operable Unit VIII of the Denver Radium site is currently in the Remedial Investigation/Feasibility Study phase. It was the site of metal processing activities in the early 1920's. Throughout the 1930's vanadium ore, uranium ore, and radium salts were being received at the site. From 1939 to 1941 this property was one of two radium producers in the United States. In 1942 radium production stopped and emphasis shifted to the production of molybdenum; however, the plant continued to receive shipments of vanadium ore until 1943. The plant was actively producing uranium compounds in the 1960's and ceased operations in 1984 due to a drop in the market price of molybdenum.

Processing involved dissolution of the ores with acids and caustics, and the extraction using various processes, some involving organic compounds now considered hazardous or potentially hazardous. Because of the nature of the production process, wherein reagents, solvents, and processing solutions were stored in vats and pumped to different reaction vessels, the chance of spills and leaks was high and resulted in soil and possibly groundwater contamination.

SITE CONDITIONS

Studies of the radiological conditions on the site have identified radiation fields and radionuclide concentrations in various media due to the presence of uranium, radium, and associated decay products.

Metals used on the property include, but are not limited to:

Arsenic	Lead	Selenium
Barium	Magnesium	Silver
Chromium	Manganese	Uranium
Cobalt	Mercury	Vanadium
Copper	Molybdenum	Zinc
Iron	Nickel	

Chemicals or other materials that were either used in the ore processing, used in the laboratories on the property, or stored on the property include:

Inorganic Reagents	**Organic Reagents**
Ammonia	Acetic acid
Ammonium hydroxide	Chloroform
Ammonium nitrate	Methyl isobutyl ketone
Barium carbonate	Tributyl phosphate
Calcium hydroxide	1, 1, 1-trichloroethane
Hydrochloric acid	Tertiary amines
Hydrogen peroxide	
Hydrogen sulfide	**Others**
Nitric acid	Asbestos
Sodium hydroxide	Propane gas
Sodium sulfate	Kerosene
Sulfuric acid	

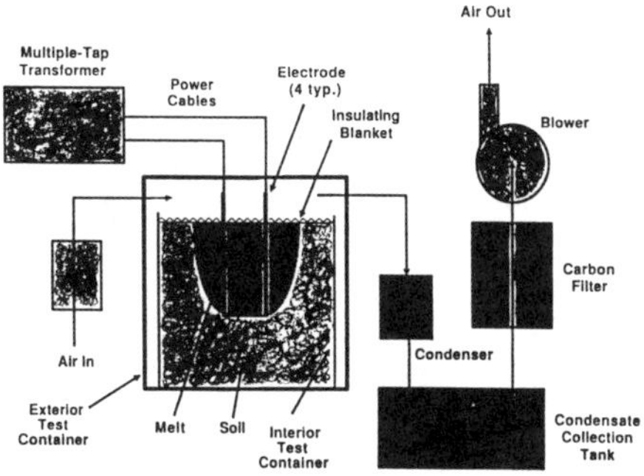

Figure 1. Simplified schematic of the In Situ Vitrification equipment system.

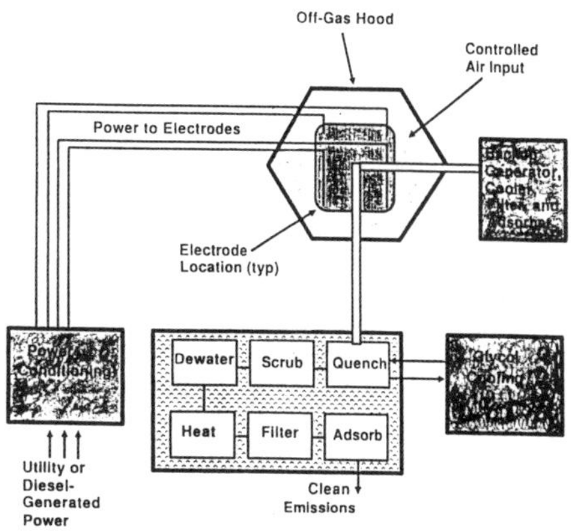

Figure 2. Simplified schematic of the In Situ Vitrification bench scale test equipment.

IN-SITU VITRIFICATION

Due to the combination of mixed organic, inorganic, metal, and radioactive waste, contaminated soils and debris from OU VIII may not meet the acceptance criteria of existing commercial disposal facilities. During the remedial investigation in-situ vitrification (ISV) was identified as a technology which could provide a permanent solution to the mixed waste on site.

ISV was developed in 1980 by Battelle Memorial Institute and Geosafe Corporation. ISV is a thermal treatment process involving electrical melting of soil. The process utilizes an equipment system, as illustrated in simplified form in Figure 1. A square array of four electrodes is inserted into the ground to the desired treatment depth. In order to establish a path of conductance a mixture of flaked graphite and glass frit is placed around the electrodes. An electrical potential of over 12,000 volts is applied to the electrodes which establishes an electrical current in the graphite and glass frit. The resultant power heats the path and surrounding soil to above fusion temperature. The graphite is consumed by oxidation and the current is transferred to the molten soil. Convection currents within the melt serve to transfer heat to the molten mass. Heat is transferred from the molten mass to the adjacent soils by conduction. The melt advances at a rate of one to two inches per hour. A steep thermal gradient exists in adjacent soils, with the 100 C isotherm (the boiling point of water at sea level) usually within nine inches of the melt. The zone between the 100 C isotherm and the melt is devoid of moisture and therefore, has a relatively high vapor permeability.

Most contaminants in the soil typically exist in a solid or liquid phase. During processing, most undergo a change of state to a liquid or gaseous phase. The physical state of contaminants is directly related to their movement during ISV processing.

Many organic components vaporize during processing. Organic vapors increase in temperature to their pyrolysis temperature where they break down into successively smaller chains of molecules and eventually reach the state of elemental or diatomic gases. Upon pyrolysis, the concentration of the original compound vapor is diminished, resulting in a concentration gradient of the original vapor toward the melt. Some of these gases may dissolve into molten mass; the remainder may be expected to escape the treatment volume. Such vapors are collected and treated to below air emission requirements in an off-gas treatment system.

The behavior of inorganic material upon exposure to the advancing thermal gradient is similar to that of the organics. Inorganic compounds may thermally decompose or otherwise enter into reactions with the melt. Nitrates and sulfates yield decomposition products which may dissolve into the melt or may evolve through it and be collected at the surface. Some amount of volatile or semi-volatile inorganics (Pb, Hg) are removed from the treatment volume and recovered in the off-gas system. Pyrolytic destruction of hazardous inorganic elements (CN, for example) is possible; however, destruction of heavy metals is not possible. Heavy metals are ultimately dissolved and incorporated or encapsulated in the residual product. Radioactive elements cannot be destroyed, but are incorporated (immobilized) into the final product.

The processing area is covered by an octagonal shaped off-gas collection hood which is much larger than the treatment area. Flow of air through the hood is controlled to maintain a negative pressure. Air provides oxygen for combustion of pyrolysis products and vapors. Off-gases, combustion products and air are drawn from the hood by induced draft blower into a treatment system. Unit processes within the treatment system include quenching, pH controlled scrubbing, dewatering, heating, particulate filtration, and activated carbon adsorption.

A self controlled cooling system is utilized to cool the quenching/scrubbing solution. A substantial quantity of the off-gas contaminants are removed from the off-gas stream at the quenching and scrubbing stages. The filters and carbon adsorption columns are utilized as secondary or backup stages to ensure safe air emissions. After processing for a time, the scrubber solution, filters and activated carbon likely contain sufficient contaminants to warrant treatment or disposal themselves. Typical treatment may include passing the scrubber water through a diatomaceous earth filter and activated carbon. The water may be recycled or discharged to sanitary sewer, and activated carbon filters placed in a subsequent ISV setting for reprocessing. This way, the only secondary waste resulting from the ISV processing is that contained in the off gas treatment system after the last setting at a site. A backup off gas treatment system consisting of a diesel powered generator, blower, mist cooler, filter and activated carbon column is available and ready to employ in the event of power failure.

Upon completion of the ISV processing, the cooling time of the molten mass affects the relative extent of glassy phase and crystalline phase present in the residual product. Typically, the mass solidifies into a silicate glass and microcrystalline product which is physically and chemically equivalent to natural obsidian. The ISV silicate glass is durable relative to environmental exposure and will hold a wide variety of materials in non-leachable form. Heavy metals and radio nuclides are immobilized due to their high stability in glass. Tests have shown it is unaffected by wet/dry and freeze/thaw cycles. The ISV residual product has been subjected to strength tests in tension (4,000 to 8,000 pounds per square inch) and compression (30,000 to 45,000 psi). It has also been demonstrated to be capable of passing Extraction Procedure Toxicity Test (EP Tox), Solid Waste Leaching Procedure (SWLP), and Toxicity Characteristic Leaching Procedure (TCLP) testing protocols.

BENCH SCALE TESTING OF ISV

The performance objective of the bench-scale ISV tests is to identify the effectiveness of this treatment alternative for the soil contaminants at this site. As previously discussed, the contaminants in the surface and subsurface soils at this site include radionuclides, non-radioactive heavy metals, and organic compounds. The bench-scale test was conducted on April 12, 1990. Results of the test are not available at the time of publication.

A bench-scale vitrification unit, shown in simplified form in Figure 2, was used to conduct the bench or engineering-scale treatability test. Due to the high levels of radioactivity in the test soils, the contaminated soils could not be accepted by the Geosafe laboratory located at the University of Washington and the test equipment was transported to the site.

The test was conducted in an 85 gallon drum which was placed inside a steel box for double containment safety purposes. The four 0.5 inch diameter by 24 inch long molybdenum electrodes were sheathed by 1.5 inch diameter by 20 inch long graphite coolers to prevent oxidation of the molybdenum and to induce subsidence to the melt zone. The molybdenum and graphite electrodes were placed in the soil to a depth of 16 inches. The electrode assemblies were designed to have a center to center spacing of eight inches in a square array. Due to an unexpected high iron content of the contaminated soils, this spacing array was moved out to a 10 inch center to center square array. This revision served to include some of the background soils into the melt, thus diluting the iron, along with other contaminants, to an acceptable level. This revision was necessary because if contaminated soil contains a volume above 15 percent of iron, there is a risk that molten iron would form at the bottom of the melt during the test. This molten pool could serve as a path for the electric current, thus short circuiting the vitrification process.

The contaminated soil was stratified in a two layer zone and placed in an eight inch square area in the center of the drum. The first contaminated zone consisted of the surface soils (SS), which were primarily contaminated with heavy metals, and had a height of five inches. The second contaminated zone consisted of borehole (BH) soils, which were primarily contaminated with radionuclides, also to a height of five inches. The contaminant zone had a height of ten inches and was covered with three inches of clean soil. The clean soil was added to ensure that a full vitreous zone is established prior to encountering any contaminants. The ISV boundaries (adjacent to and beneath the contaminated soils) was surrounded by background soils obtained from the site vicinity. These soils are being analyzed to determine the disposition of the original chemical constituents.

In order to monitor processing depth, type K thermocouples were installed in the center of the test drum extending from the surface down to the bottom of the drum (32 inches) in two inch intervals. In addition, three type K thermocouples were installed on a horizontal axis, 10 inches below the surface grade at two inch intervals to measure horizontal melt dimensions and the temperature isotherms in the surrounding uncontaminated soil. Geosafe continued to vitrify the soil until the thermocouple located at the 16 inch depth reached the 1200 C (2192 F) temperature.

A flaked graphite/glass frit mixture was placed in a one inch by one inch path laid in an "X" and square pattern on the surface of the soil to provide direct electrical conductive path between the electrodes. This is consistent with full-scale remediation operations. Approximately two hours after the start-up of the test, it was evident that a melt was not being established. The vitrification process was stopped and an additional layer of graphite and glass frit was added to promote electrical conductance. A melt was established and the test was completed in the estimated time of five hours.

During the ISV testing, the vitrification zone was covered with a two inch kawool insulation blanket, leaving a small gap around the electrode for venting. The insulation helps promote subsidence of the molten surface and improves the melting efficiency of the operation. This technique is consistent with full-scale operations.

The power system consists of a 30 kW capacity Scott-Tee transformer. The transformer is equipped with 12 voltage taps and two saturable core reactors for controlling power input to the melt. The test was run at a target control power of 15kW to closely simulate full-scale operations (power density). The transformer was equipped with metering on the secondary side (output to the electrodes) for power, voltage, and amperage. Electrical and thermocouple data was recorded at 0.5 hour intervals.

The off-gas stream from the test unit is equipped with a condensate collection tank for the collection of water and particulates released from the melt. This tank simulates the performance of the quencher and scrub tank in the Geosafe large scale system. In addition, the system contains a disposable activated vapor phase carbon sorption column to remove any particulate or gaseous emissions from the melt.

Off-gases generated from the melt were sampled continuously during the test. A total of three side stream samples were drawn through two inch ports located in the off-gas line. EPA Methods 1 and 2 were used to determine stack gas velocity, temperature, and volumetric flow rate. EPA Methods 3 and 3A were used to determine the molecular weight of the stack gas using the following instrumental analyzers. An Infrared Industries Model 2200 oxygen analyzer was used to measure the percent oxygen. An Infrared Industries Model 702D non-dispersive infrared analyzer (NDIR) was used to measure the percent carbon dioxide. An Automated Custom Systems (ACS) Model 3300 NDIR analzyer was used to measure the parts per million (ppm) carbon monoxide. EPA Method 4 will be performed to determine the moisture content of the stack gas. EPA Method 5 techniques were employed to collect samples of particulates for the determination of arsenic, selenium, and gross alpha and beta radioactivity concentrations.

The bench scale tests are designed to challenge the vitrification process with the highest contaminant concentration on site. In that the highest radioactive and non-radioactive soil concentrations are not coincident with the highest organic concentrations, the contaminated soil was placed in two distinct layers in the test drum.

The soil layer for inorganic contaminants (SS) was representative of the highest concentrations of radium, thorium-230, arsenic, and selenium that were observed on the site. The SS also exhibited a substantial concentration of molybdenum. Various inorganic soil contaminants along with parameters which affect the test operation were measured. These parameters were trifurcated into the following pre-test, test, and post-test measurements:

The pre-test measurements which were conducted include various physical, non-radioactive nuclide, and radionuclide measurements of the background soil and of the contaminated soil. The physical measurements include moisture, dry density, fusion temperature, 100 poise temperature, electrical conductivity at 100 poise temperature, melting temperature, and test sample mass. Chemical measurements include HSL metals, Al, Cu, Fe, K, Li, Na, and Si by modified EPA method 6010. The radiochemical measurements include isotopic U, Ra-226, Th-230, Pb-210, gross alpha and gross beta/gamma radiation.

The test measurements which were conducted with the bench test include gas pressure at three points in the test apparatus, air flow at one point in the test apparatus, temperature at 18 points in the test matrix and one point in the test apparatus, electrical power, voltage, and amperage supplied to the electrodes, and total organic vapors in the off-gas effluent control discharge.

The post test measurements which were be conducted include bulk density as the only physical measurement, Chemical measurements including HSL metals, As, and Se, Radiochemical measurements including isotopic U, Ra-226, Th-230, Pb-210, radon flux on the full block, and gross alpha, beta, and gamma.

The top layer of the bench scale test soils consisted of soils contaminated with organic contaminants (BH). The indicator chemicals in this soil were selected based on contaminant abundance, toxicity, mobility, and volatility. The BH soils exhibited the highest concentrations of these indicator chemicals that were observed on the site. Various waste/soil contaminant and operating specific parameters were measured for the organic soil layer. These parameters were trifurcated into the following pre-test, test, and post-test measurements:

The pre-test measurements which were conducted include tributylphosphoric acid ester, toluene, alkanes, and C-10 to C16 hydrocarbons.

The measurements conducted concurrent with the test are basically the same as those performed during the inorganic test.

The post-test measurements which were conducted included density testing of the vitrified product and organic contaminant analysis of the background soil surrounding the product for the same organics listed in the pre-test sampling.

Numerous leaching tests have been developed to test solid wastes, including those developed specifically for stabilized/solidified nuclear and hazardous wastes. Leaching tests and associated analysis will be performed on the vitrified product using the Toxicity Characteristic Leaching Procedure (TCLP) and the Extraction Procedure Toxicity Test (EP Tox). The TCLP is being used by the EPA as the basis for the promulgation of best demonstrated available technologies (BDAT) treatment standards under the land disposal program. The EP Tox procedure is used by EPA for characterizing solid municipal wastes. These leach tests will be performed pursuant to the standard procedures formalized by EPA. The laboratory tests and analysis will be performed to meet the Level III analytical category requirements.

Pre-test, test and post test data were not available at the time of publication. All test data and associated data analysis and interpretations will be presented in the treatability study report. EPA will use this information to determine if a full-scale pilot study on-site is warranted.

REFERENCES

1. CH2M Hill, *Remedial Investigation, Denver Radium Site*, 51-8L01.0, April 1986.
2. CH2M Hill, *Feasibility Study, Denver Radium Site Operable Unit II*, WA # 112-8Lo1.0, August 1987.
3. Geosafe Corporation, *Test Specific Engineering-Scale Test of In Situ Vitrification for Shattuck Chemical Superfund Site*, 883-1551.006, April 1990.
4. Spangler, R. R., Langner, Jr., Rangel, M. J., *Capability of Radon Flux and Radon-in-Soil Gas Measurements to Locate Buried Radium Contamination*, DE-AC07-86ID12584, U.S. Department of Energy, Grand Junction Project Office, Grand Junction Colorado, May 1989.
5. United States Environmental Protection Agency, *Record of Decision, Denver Radium Site Operable Unit II*, September 1987.
6. Versar, Inc., *Work Plan for Bench-Scale Vitrification Treatability Studies at Denver Radium Site—Operable Unit VIII*, Contract # C377624, January 1990.
7. Versar, Inc., *Draft Remedial Investigation Report for Denver Radium Site Operable Unit VIII*, January 1990.

5 REMEDIATION OF GMC-CENTRAL FOUNDRY DIVISION – DEFIANCE FOUNDRY HAZARDOUS WASTE SETTLING BASIN

Ajit K. Chowdhury, Senior Chemical Engineer
Richard C. Krueger, Project Engineer
Michael A. Warner, Project Engineer
RMT, Inc.
Madison, Wisconsin 53708

Thomas R. Wirth, Senior Environmental Engineer
General Motors Corporation
Central Foundry Division
Defiance, Ohio 43512

INTRODUCTION

General Motors Corporation – Central Foundry Division – Defiance (CFD – Defiance) operates a ferrous foundry in Defiance, Ohio. Particulate emissions from the cupola operations are collected by wet cap and venturi scrubber emission control systems. The wastewater from the emission control systems is conveyed to a settling basin, and the effluent is reused in the plant. The accumulated sludge in the settling basin has been periodically removed by hydraulic dredging, and disposed in an on-site, state-approved, solid waste landfill.

In the process of complying with NPDES regulations and Federal Metal Molding and Casting Effluent Guidelines, several sludge samples from different parts of the basin were collected and analyzed. Slightly more than half of these samples released lead in the EP Toxicity test above the hazardous waste criterion of 5.0 mg/L.

CFD – Defiance notified the United States Environmental Protection Agency (USEPA) and the Ohio Environmental Protection Agency (OEPA) that the basin contained EP-Toxic sludge. As a result, CFD – Defiance and the OEPA negotiated a consent order for management of the sludge.

To prevent generation of new hazardous sludge, an in-line wastewater treatment system was installed upstream from the basin to treat the incoming solids. For remediation of the hazardous solids in the basin, a continuous treatment process using a hydraulic dredging system with in-line addition of phosphoric acid was developed by RMT, Inc and implemented by CFD – Defiance. The EP-Toxic sludge in the basin (approximately 400,000 cubic yards) was remediated with this system (slurried and treated) to render the sludge nonhazardous, according to an OEPA-approved "Closure Plan" for achieving RCRA clean closure. The basin remained in use during the remediation process.

BENCH-SCALE TREATABILITY TESTING

Several composite samples of the basin sludge, with initial EP-Toxicity lead concentration varying from 8 to 38.5 mg/L, were treated with different chemical additives to evaluate various treatment possibilities for rendering the sludge non-EP-Toxic. The bench-scale treatability tests were designed to test the following variables:

- Concentration of solids in test slurry.
- Type and dosage of chemical additive.
- Final treated slurry pH.
- Additives for control of final slurry pH.

Treatment additives evaluated during initial screening included:

- Phosphates (three forms)
- Magnesium hydroxide (two grades)

45th Purdue Industrial Waste Conference Proceedings,

- Lime
- Other on-site waste materials (e.g., quenched slag)

Significant chemical treatment effectiveness was observed with the following additives:

- Magnesium hydroxide
- Triple superphosphate (TSP)
- Phosphoric acid

The additives containing phosphate (TSP and phosphoric acid) were effective in rendering the sludge non-EP-Toxic at dosages significantly lower than the magnesium hydroxide dosage. Therefore, bench-scale treatability continued with phosphate as the treatment additive of choice.

The effect of sludge solids concentration, final slurry pH after treatment, pH control additives, reagent dosage, etc., were then evaluated to optimize conditions for the most effective treatment. Figure 1 presents the results of the treatability tests with TSP and phosphoric acid.

The test results with phosphates are summarized as follows:

- The lead concentration of treated solids in the EP Toxicity test decreased sharply to approximately 2 mg/L with increasing phosphate dosage. EP Toxicity lead concentrations were reduced to less than 2 mg/L with TSP and phosphoric acid dosages of 2.5 and 1.5% by weight (dry basis), respectively. EP Toxicity reductions were incremental at higher phosphate dosages. Sludge treatability was not influenced by initial EP Toxicity lead concentration.
- Treatment with phosphoric acid resulted in a slightly lower treated slurry pH and a higher residual dissolved phosphate concentration in comparison to the TSP treated sludge. However, pH control had no measurable effect on the treatability of the sludge at the dosages tested.
- The treatability of the sludge was not significantly influenced by the initial concentration of sludge solids in the test slurry (5 to 20% by weight).
- The residual dissolved phosphate in the treated slurry correlated well with reduction in lead EP Toxicity of the treated sludge.

ASSESSMENT OF TREATMENT PROCESS ALTERNATIVES

The bench-scale treatability results showed that phosphates were the most effective treatment additives. Therefore, several alternatives for treatment of the sludge with phosphates were evaluated for economic, technical, and regulatory considerations. The treatment alternatives included continuous in-line treatment of the hydraulically dredged slurry using the following phosphate additives:

- Phosphoric acid (Alternative 1)
- Slurried triple superphosphate (Alternative 2)
- Dry granular triple superphosphate (Alternative 3)

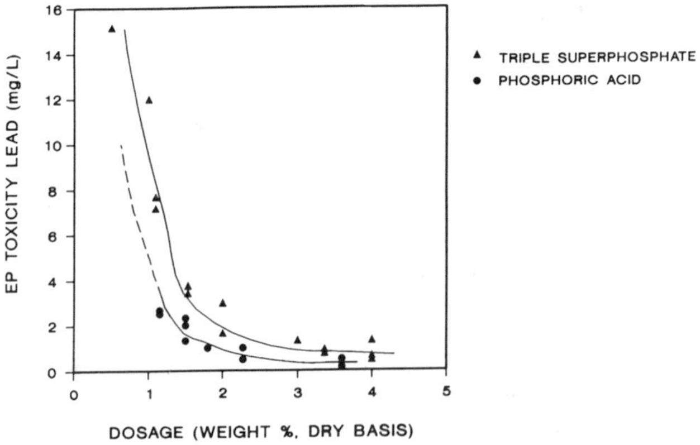

Figure 1. Reagent dosage vs. EP toxicity.

Hydraulic dredging rate, treatment chemical requirements, availability, delivery, ease of material handling, unit costs, etc., were considered in the assessment of the alternatives.

Figure 2 shows the conceptual phosphate treatment process. The treatment reagent was to be delivered to the site in tank trucks and unloaded into storage tanks. The sludge would be dredged and conveyed via a pipeline to a reaction tank. The treatment reagent was to be added to the dredged slurry upstream of the reaction tank. Phosphates were to react with the solids in the pipeline and in the reaction tank downstream of the acid injection point. During full-scale operation, the treated slurry was to flow to a state-approved, on-site, solid waste landfill; the treated solids would settle out, and the supernatant water would be recycled back to the settling basin. During field trials, the treated slurry flowed from the mixing tanks back to the basin.

Preliminary cost estimates were developed with the different alternatives. This assessment concluded that agricultural grade phosphoric acid would be the most cost-effective option. The projected costs with technical grade phosphoric acid and TSP were higher by margins of 27% and 47%, respectively. Phosphoric acid appeared more cost-effective because of low dosage requirements, anticipated ease of material handling, and a higher rate of sludge processing. These factors pointed to a short duration for remediation and a projected lower cost.

None of the alternatives considered exhibits a significant advantage over the others with respect to regulatory/environmental considerations.

THE CLOSURE PLAN

A Closure Plan was developed to address the applicable portions of 40 CFR Part 265 and the Ohio Administrative Code (OAC), Chapter 3745. The plan identified how CFD-Defiance intended to remove and treat the sludge and dispose it, as non-hazardous solid waste, in the on-site, state-approved solid waste landfill. The Closure Plan summarized the following:

- Previous waste sampling and analysis.
- Bench-scale treatability results.
- Field trials and start-up operations.
- Treatment effectiveness monitoring and evaluation during field trials.
- Quality control monitoring during full-scale operations.
- Health and safety issues related to remedial activities.
- Methods for performing clean closure sampling and for documenting clean closure.

The plan was approved by the OEPA and the USEPA after several meetings at which the various aspects of the plan were reviewed and discussed in detail.

FIELD TRIALS

Field trials were conducted for approximately four weeks during March–April 1989. The major objectives of the field trials were as follows:

- Evaluate performance of the full-scale equipment treatment system.
- Demonstrate treatment effectiveness in continuous, full-scale operation.
- Determine operating conditions for full-scale remediation.

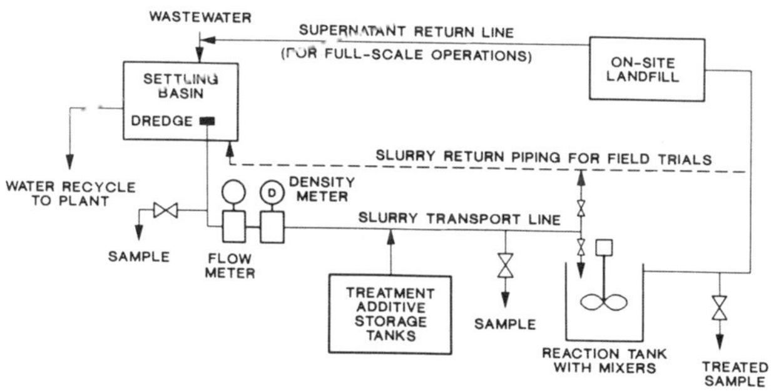

Figure 2. Conceptual treatment process flow schematic settling basin sludge treatment.

The Treatment System

A schematic of the treatment system is shown on Figure 3. The components included the following:

1. Hydraulic floating dredge
2. Slurry pipeline—approximately 1.2 miles long
3. Mixing tanks
4. Phosphoric acid storage and feed pumps
5. Instrumentation and process controls for acid injection into slurry pipeline

In operation, the sludge was dredged continuously and pumped through the pipeline. Concentrated phosphoric acid was injected into the pipeline to render the solids nonhazardous. Reaction occurred in the pipeline and in the mixing tank. During the field trials, the treated slurry was returned to the settling basin. This was done to eliminate any possible discharge of untreated or under-treated solids into the landfill area during this trial period.

Agricultural grade phosphoric acid, with a 70% concentration by weight, was used for sludge treatment. The flow rate of phosphoric acid was controlled by a Programmable Logic Control (PLC) System.

To evaluate the relationship between acid dosage and EP Toxicity using full-scale treatment equipment, the phosphoric acid dosage rate was varied in the range of from 1.54 to 4.10 gallons of 70% acid per ton of solids (gpt) in dredged sludge during field trials.

Field Trial Treatment System Operation

In general, the various mechanical components, instruments, and controls of the treatment system performed satisfactorily during the field trials. However, breakdowns and required maintenance of the hydraulic floating dredge resulted in interruptions as did the calibration and re-calibration of the slurry density and flow meters.

The mixing tanks had considerable mechanical difficulties (e.g., solids buildup, shearing of an agitator shaft) when rocks and large debris were encountered. The concentration of solids in the dredged slurry varied considerably as the cutter head of the hydraulic dredge made a swath through the settled sludge. The approximate range of operating conditions for the treatment system during the field trials was as follows:

- Dredged slurry flow rate = 4,000 gpm (typical)
- Concentration of solids in slurry = 10 to 20% by weight (average)
- Solids mass flow = 3 tons/minute (typical)
- Acid dosage rate = 1.54 to 4.1 gallons/ton

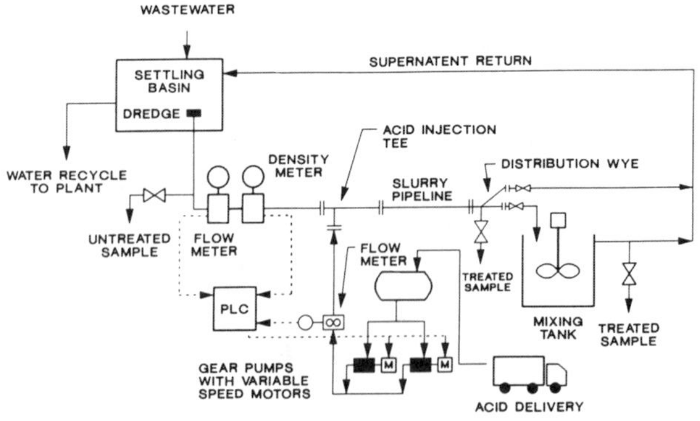

Figure 3. Treatment process flow schematic field trials.

Field Trial Treatability Results

37 untreated samples and 50 treated sludge samples were analyzed for EP Toxicity lead. The field trial results indicate the following:

- The average EP Toxicity lead concentration for untreated solids was 7.2 mg/L, with the maximum being 36.0 mg/L.
- Lead EP Toxicity decreased sharply at acid dosages up to approximately 2 gpt. At dosages above 2 gpt, little additional EP Toxicity reduction was observed. These results confirmed the observations on EP Toxicity reduction during bench-scale treatability testing discussed earlier.
- There was no significant difference in EP Toxicity lead concentration of samples collected before the mixing tank and the samples collected after the mixing tank.
- None of the treated solids samples (50 samples were analyzed) showed EP Toxicity lead concentrations higher than the hazardous waste limit of 5.0 mg/L. Only one treated solids sample had an EP Toxicity lead concentration above 2 mg/L (at an acid dosage of 1.54 gpt), and no results were above 2 mg/L at higher dosages.

The EP Toxicity lead concentration and acid dosage data were statistically analyzed to develop confidence intervals for treatability versus acid dosage. A log transformation of the parameter values was performed and fitted by least-squares linear regression. The difference between observed and predicted values (residuals) showed a normal distribution. From this, various confidence intervals were computed for acid dosages ranging from 0 to 4.5 gpt. This data analysis showed that an acid dosage of approximately 2.1 gpt would provide a 99.9% confidence that the average lead EP Toxicity results would be below the action limit of 2 mg/L.

The pH of the untreated and treated slurries ranged from approximately 8 to 9 and 6 to 7, respectively. Generally, samples with lower pH values after treatment were associated with relatively high residual dissolved phosphate concentrations in the treated slurry and with low EP Toxicity lead concentrations in the treated solids. However, residual dissolved phosphate data varied widely and did not correlate closely with the EP Toxicity lead concentration, unlike the observations during bench-scale treatability testing. The in-field residual phosphate measurements may have been subject to greater analytical difficulties, due to field conditions that were experienced in the laboratory. Nevertheless, both pH and residual dissolved phosphate measurements provided a positive indication that acid was being injected into the slurry pipeline and that the sludge was undergoing treatment.

The field trials using full-scale equipment demonstrated that the settling basin sludge could be effectively treated in continuous operation by injecting phosphoric acid into a dredged slurry pipeline, and determined the operating conditions for full-scale remediation.

FULL-SCALE REMEDIATION

Treatment System Operation

Based on the field trial results, the treatment system was modified for full-scale operation. These modifications included the following:

- Elimination of the mixing tanks and extension of slurry pipeline to provide additional treatment contact time (Figure 4).
- Sampling and analysis consisting of the following:
  - Hourly collection of untreated slurry for pH analysis.
  - Hourly collection of treated slurry for pH and residual dissolved phosphate analysis.
  - Collection and mixing of eight consecutive treated a slurry samples to prepare a composite sample for EP Toxicity lead analysis.
  - Standard quality control procedures, including collection of initial and duplicate composite samples for EP Toxicity lead analysis.
  - Modification of field techniques for residual dissolved phosphate analysis.
- Operation in a control mode where acid injection rate was proportional to the mass flow rate of the solids in the slurry; the acid dosage rate to be based on the correlation between EP Toxicity lead and the dosage rate tested during field trials.
- Installation of an emergency acid feed system.

Full-scale remediation began on May 1, 1989, with operation of 24 hours per day and 6 days per week. Approximately 1,000 to 4,000 cubic yards of material were being treated per day during a good day. However, frequent breakdowns of the dredge resulted in significantly lower average throughput.

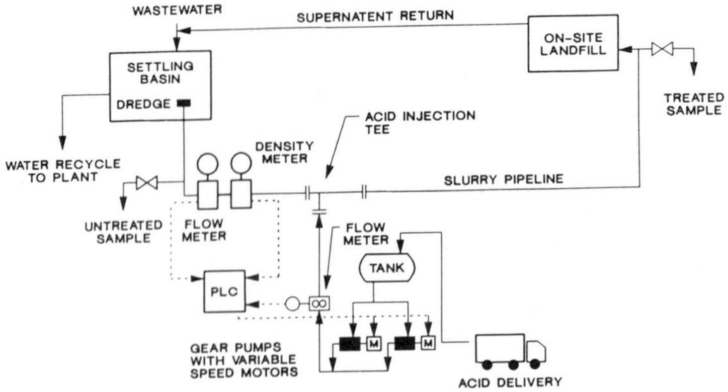

Figure 4. Treatment process flow schematic settling basin remediation.

The RCRA Closure Plan included the establishment of two action limits for EP Toxicity lead concentration in the treated solids. The Lower Action Limit (LAL) was set at 40% (2.0 mg/L) of the Hazardous Waste Limit, and the Upper Action Limit (UAL) was set as 80% (4.0 mg/L) of the Hazardous Waste Limit. An EP Toxicity lead value result above the LAL resulted in additional analyses of the field duplicate sample and a thorough check of the operational data. A result above the UAL required that the treatment system be shut down until the cause for the relatively high value could be detected and corrected. Since the EP Toxicity lead analysis required approximately one week, a significant quantity of material could be introduced into the landfill before an exceedance of the LAL or UAL was detected. Therefore, "real time" process control during the full-scale remediation was extremely important.

Process Control

As the database of treated slurry pH, residual dissolved phosphate, and EP Toxicity lead results expanded, statistical correlations between these parameters were updated. There was little apparent correlation between the raw data, but linear correlations were observed following log transformations of the residual dissolved phosphate and EP Toxicity lead data. Using the relationship between EP Toxicity lead and residual dissolved phosphate, and assuming a conditional distribution approximated by the t-distribution, phosphate values corresponding to EP Toxicity lead values at various confidence levels could be calculated. The computed residual dissolved phosphate concentration corresponding to the 99% confidence of the LAL of EP Toxicity lead was selected as a target value for initial acid dosage control.

Using this residual dissolved phosphate target concentration and standard process control procedures, a CUSUM V-mask control system was developed, based on phosphate results in the treated slurry. The control system tracked deviations of phosphate concentrations from the target value, in order to establish long-term tendencies. When the tendencies became statistically significant, other technical qualitative factors were considered, and the dosage rate was adjusted.

The system was operated successfully for approximately one month with the target residual dissolved phosphate concentration determined by the 99% confidence of the LAL. The target phosphate concentration was then revised to equal the 99% confidence of the UAL, and the treatment system was operated for approximately five months to complete the initial remediation of the settling basin.

Full-Scale Remediation Results

Over 200 composite treated solid samples were analyzed by the EP Toxicity test for lead. The average EP Toxicity lead concentration for these samples was 0.49 mg/L, with approximately 98% of the values below the Hazardous Waste Limit (HWL) of 5.0 mg/L.

Duplicate samples were analyzed for the remaining 2% of the composite samples with EP Toxicity lead exceeding the HWL. All of the duplicate samples had lower EP Toxicity lead and residual dissolved phosphate concentrations than the initial samples, indicating that the reaction of lead and phosphate continued without the agitation present in the pipeline. In fact, with the duplicate sample results, none of the reanalyzed (releached) treated slurry samples exceeded the HWL for lead.

CLEAN CLOSURE STATUS

Approximately 300 samples from the bottom of the settling basin were analyzed to document that the basin met the RCRA clean closure performance standards. The results indicated that the basin sludge had not significantly impacted the underlying soil, based on a statistical comparison of metals concentrations between background soil and underlying soil. The results have also indicated that the Clean Closure Performance Standards for removing EP-Toxic sludge from the basin have been satisfied. The settling basin Clean Closure Documentation Report is expected to be transmitted to the Ohio EPA during May 1990.

CONCLUSIONS

A continuous process, developed by RMT and implemented by CFD-Defiance, where phosphoric acid was injected into a dredged slurry pipeline for treatment of hazardous cupola emission control sludge, was successfully implemented to remediate approximately 400,000 cubic yards of EP-Toxic sludge at the GMC—Central Foundry Division—Defiance foundry.

A feedback control system developed by correlating residual dissolved phosphate and EP Toxicity of the treated sludge and using a CUSUM V-mask control program was used successfully to monitor performance of the treatment process.

Approximately 98% of the treated sludge samples analyzed initially tested non-EP-Toxic for lead. Repeat analysis of the remaining 2% of the treated solids samples also tested non-EP-Toxic, indicating that the treated sludge continued to react with the residual phosphate as time progressed. Sampling within the settling basin indicated that the Clean Closure Performance Standards have been met. CFD-Defiance expects to transmit the Clean Closure Documentation Report to the Ohio EPA during May 1990.

6 A RISK APPRAISAL PROTOCOL FOR DEVELOPING CLEANUP CRITERIA IN THE RESTORATION OF CONTAMINATED SITES

Daniel Kofi Asante-Duah, Senior Project Engineer,
International Technology (IT) Corporation,
Irvine, California 92714.

INTRODUCTION AND BACKGROUND

Hazardous waste sites may pose significant risks to the public because of the potential health and environmental effects, and to the Potentially Responsible Parties (PRPs) because of the financial liability stemming from their effects. To ensure public health and environmental sustainability, decisions relating to hazardous waste management should be based on a systematic and scientifically valid process, such as is offered by risk assessment.

The Comprehensive Environmental Response, Compensation, and Liability Act of 1980, as ammended (CERCLA, or "Superfund"), establishes a national program for responding to releases of hazardous substances into the environment. The U.S. EPA recognizes the use of risk assessment to facilitate decisions on whether or not remedial actions are needed to abate site-related risks, and also in the enforcement of regulatory standards. Risk assessment techniques have been used in various regulatory programs employed by Federal, State and local agencies.

The main objectives of this presentation are to:

- Discuss an approach for strategizing health and environmental risk assessment and site restoration plans for potentially contaminated sites.
- Evaluate a case-specific risk assessment and develop target cleanup criteria for this subject site.

The emphasis will be on the use of health risk assessment techniques to support management decisions in the development of Remedial Action Plans (RAPs) in hazardous waste site restoration programs.

THE RISK APPRAISAL PROTOCOL

Although specific forms of risk assessment may differ considerably in their detail, they share the same general logic. The overall risk assessment hierarchy is comprised of the elements shown in Figure 1. Further discussion of the basic elements are presented in the risk analysis literature.[1–4]

Health Risk Assessment

Important factors to consider in assessing risks and in setting up site cleanup criteria for a potentially contaminated site are discussed by several authors.[2–6] The health risk assessment consists of the process flow shown in Figure 2.[4] The toxicity assessment is conducted as part of a risk assessment to quantitatively and qualitatively assess the potential for adverse human health effects from exposure to the chemicals of potential concern. The quantitative portion of the toxicity assessment entails identifying the relevant toxicity indices such as the Acceptable Daily Intakes (ADIs), the Reference Doses (RfDs), and the Slope Factors (SF), or Cancer Potency Factors (CPFs). The qualitative aspect of the

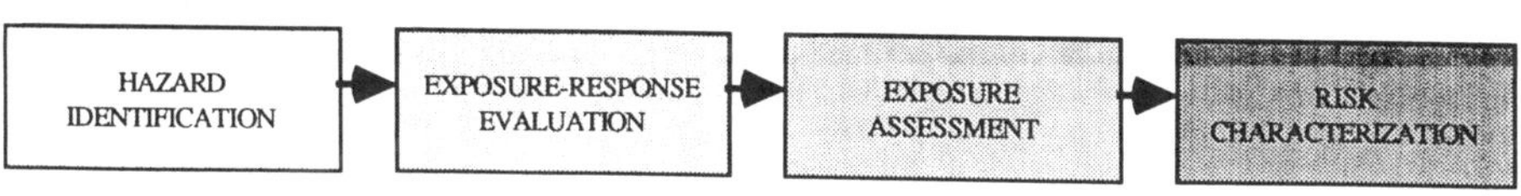

Figure 1. Basic elements of risk assessment.

45th Purdue Industrial Waste Conference Proceedings, © 1991 Lewis Publishers, Inc., Chelsea, Michigan 48118. Printed in U.S.A.

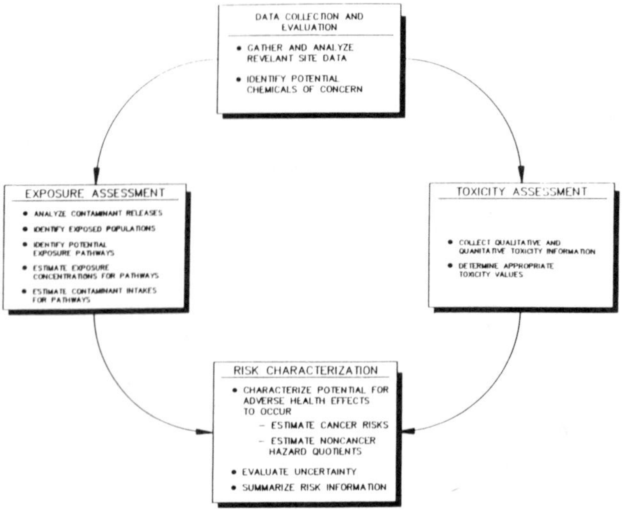

Figure 2. Health risk assessment process flow.

assessment includes summaries of the profiles of the chemical of potential concern. The exposure and toxicity assessments are integrated to characterize risk levels, both quantitatively and qualitatively.

An adequate characterization of risks and hazards at a potentially contaminated site allows a site remediation process to be better focused. Cleanup criteria can then be developed based on the "acceptable" level of risks to potential receptors.

Exposure Scenarios for Contaminated Site Problems

An exposure scenario is the qualitative connection between a contaminant source through one or more environmental media to some receptor population(s) (Figure 3). The event tree concept offers an efficient way to develop exposure scenarios.[3]

The following potential exposure scenarios may be considered representative of the complete exposure pattern anticipated from a typical contaminated site problem:

- Direct human exposure on-site via ingestion of dirt, inhalation of air-borne contaminants, and/or absorption through the skin after dermal contact with contaminated soil
- Direct human exposure off-site via inhalation of fugitive dust and/or dermal contact
- Direct human exposure resulting from on-site use of ground water; exposure may be via ingestion of ground-water, inhalation and/or dermal contact (from use of the ground water for washing)
- Direct human exposure resulting from off-site use of ground water; exposure may be via ingestion of ground water, inhalation and/or dermal contact from use of the ground-water for washing
- Direct human exposure resulting from off-site use of surface water (that have been contaminated from surface runoff and/or ground water discharge); exposure may be via ingestion of surface water and/or dermal contact (from use of the surface water for washing)
- Direct human exposure resulting from off-site recreational use of surface water (that has been contaminated from surface runoff and/or ground-water discharges); exposure may be via ingestion of surface water and/or dermal contact
- Indirect human exposure resulting from bioaccumulation in river fish that is consumed by humans; aquatic life may be exposed to contaminants as a result of run-off and/or ground-water discharges into river(s)
- Indirect human exposure resulting from ingestion of game or livestock (as a result of bioaccumulation through the food chain).

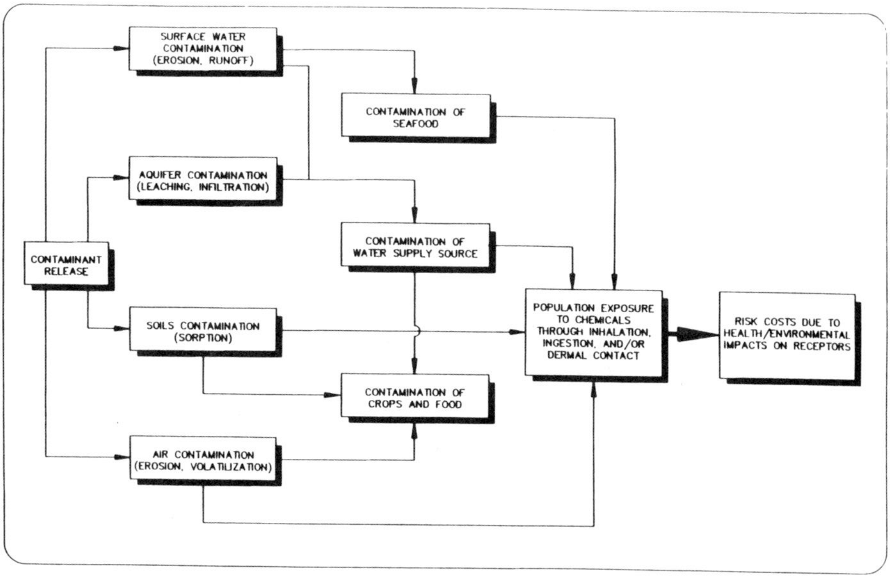

Figure 3. Conceptualization of exposure scenarios.

For exposure to occur a complete pathway is necessary; a risk assessment must address all such exposure pathways. *The accuracy with which exposure is characterized could be a major determinant of the ultimate validity and usefulness of a risk assessment.*

Environmental Modeling in Risk Assessment

Transport and fate of contaminants can be predicted through various methods, ranging from simple mass-balance and analytical procedures to multidimensional numerical solution of coupled differential equations. Mathematical algorithms are used to predict the potential for contaminant migration from a contaminated site to potential receptors, using pathways analyses concepts. Typically, the following scenarios will be simulated and evaluated using the appropriate models for a contaminated site problem (e.g., Figure 4):

- Land Infiltration of Rain Water
- Erosion/Surface Runoff Release of Chemicals
- Emission Rates of Particulates and Vapors
- Chemical Fate and Transport through the Unsaturated Zone
- Mixing of Unsaturated-Saturated Zones
- Chemical Transport through the Aquifer System
- Mixing of Ground Water with Surface Water.

A thorough review of site background information will confirm the need for any modeling effort, in order to more clearly define contaminant migration and exposure pathways.

The risks associated with the site can be calculated for the critical pathways. Once the degree of risk or hazard resulting from existing levels of contamination has been determined, the extent to which contaminant levels at the source must be reduced to not exceed some acceptable risk level can be obtained by "back-modeling" from an acceptable exposure limit of concentration at the most critical potential receptor. This type of calculations are most appropriate for determining acceptable site cleanup levels that will adequately protect the soil, ground water, surface water and atmosphere pathways.

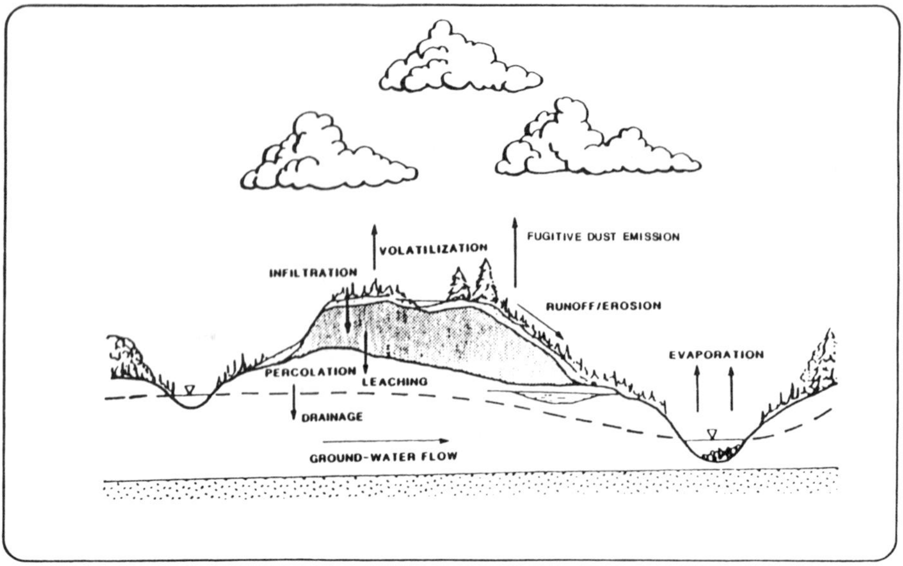

Figure 4. Selected processes affecting contaminant migration.

Exposure Quantification: Estimation of Intakes and Doses

Intake is defined as the amount of chemical contacting body boundaries (such as the skin or gastrointestinal tract), and dose is the amount of chemical absorbed by the body into the bloodstream. The methods by which each type of exposure is estimated is well documented in standard materials prepared under the auspices of the U.S. EPA.[4,5] The generic equation for calculating chemical intakes and/or doses is given by[4]:

$$I = C \times CR \times CF \times FI \times ABS_s \times EF \times ED \times 1/BW \times 1/AT$$

where
- I = Intake, adjusted for absorption (mg/kg-day)
- C = Media-specific chemical concentration (e.g., mg/kg)
- CR = Contact rate (e.g., mg soil/day; liters water/day)
- CF = Conversion factor
- FI = Fraction of intake from contaminated source
- ABSs = Bioavailability/Absorption factor (%)
- EF = Exposure frequency (days/years)
- ED = Exposure duration (years)
- BW = Body weight (kg)
- AT = Averaging time (period over which exposure is averaged—days)

This equation may be replaced by equivalent analytical relationships in their applications to the health risk assessment of a given hazardous waste site.[7]

Estimation of Risk and Hazard Levels

The risks to potentially exposed populations from exposure and subsequent intake of the chemicals of potential concern are characterized by the calculation of noncarcinogenic hazard quotients and indices and/or carcinogenic risks. These parameters form a basis for risk decisions in hazardous waste management.

Carcinogenic Risks, CR, may be estimated by multiplying the cancer slope factor (SF), which is the upper 95% confidence limit of the probability of a carcinogenic response per unit intake over a lifetime of exposure, by the estimated intakes—yielding incremental risk values. The carcinogenic effects of the chemicals of concern are calculated according to the following relationship[4]:

$$\text{Risk, CR} = CDI \times SF$$

where Risk = Probability of an individual developing cancer (unitless)
CDI = Chronic Daily Intake averaged over 70 years (mg/kg-day)
SF = Slope Factor (1/[mg/kg-day])

This represents the linear low-dose cancer risk model, and is valid only at low risk levels (i.e., below estimated risks of 0.01).

For sites where chemical intakes are high (i.e., potential risks above 0.01), the one-hit model is used. The one-hit equation for high carcinogenic risk levels is given by the following model[4]:

$$\text{Risk} = 1 - \exp(-\text{CDI} \times \text{SF})$$

For multiple compounds and multiple pathways, the aggregate total cancer risk for all exposure pathways and all contaminants considered in the risk evaluation will be:

$$\text{Overall Total Risk} = \Sigma\Sigma\,(\text{CDI}_{ij} \times \text{SF}_{ij})$$

for the linear low-dose model for low risk levels, or

$$\text{Overall Total Risk} = \Sigma\Sigma\,(1 - \exp(-\text{CDI}_{ij} \times \text{SF}_{ij}))$$

for the one-hit model used at high carcinogenic risk levels,

where CDI_{ij} = CDI for the ith contaminant and j^{th} pathway
SF_{ij} = SF for the ith contaminant and j^{th} pathway

The CDIs are estimated from the equations given previously for chemical intakes, whereas the SF values are obtained from various sources/databases, including the Integrated Risk Information System (IRIS) and the Health Effects Assessment Summary Tables (HEAST) available through the USEPA. As a rule-of-thumb, incremental risks of between 10^{-4} and 10^{-7} are generally perceived as acceptable levels for the protection of human health and the environment.

Noncarcinogenic Hazards are parameters by which the overall potential noncarcinogenic effects posed by the chemicals of concern are usually expressed in—called the Hazard Index (HI). The noncarcinogenic effects of the chemicals of concern are calculated according to the following relationship[4]:

$$\text{Hazard Quotient, HQ} = \text{E} / \text{RfD}$$

where E = Chemical Exposure Level or Intake (mg/kg-day)
RfD = Reference Dose (mg/kg-day)

The sum total of the hazard quotients for all the chemicals of concern gives the hazard index for a given exposure pathway. The applicable relationship is:

$$\text{Total Hazard Index, HI} = \Sigma\, \text{E}_i / \text{RfD}_i,$$

where E_i = Exposure level (or intake) for the ith contaminant
RfD_i = Acceptable intake level (or Reference Dose) for i^{th} contaminant

For multiple compounds and multiple pathways, the overall total noncancer risk for all exposure pathways and all contaminants considered in the risk evaluation will be:

$$\text{Total Hazard Index, HI} = \Sigma\Sigma\, \text{E}_i / \text{RfD}_i,$$

where E_{ij} = Exposure level (or intake) for the ith contaminant and j^{th} pathway
RfD_{ij} = Acceptable intake level (or Reference Dose) for i^{th} contaminant and j^{th} pathway

The E values are estimated from the equations given previously for chemical intakes, whereas the RfD values are obtained from databases such as IRIS and HEAST available through the USEPA. In interpreting the hazard index (HI) values, a reference value of HI less than or equal to 1 should be taken as the acceptable reference or standard. For HI values greater than unity (i.e., HI > 1), the higher the value, the greater is the likelihood of adverse health impacts.

DEVELOPMENT OF MEDIA CLEANUP CRITERIA

An important consideration in developing a RAP for a contaminated site is the level of cleanup to be achieved—i.e., "how clean is clean?" This could be the driving force behind remediation costs. The cleanup level is a site-specific criterion that a remedial action would have to satisfy in order to keep exposure to potential receptors at or below an "Action Level" (AL). The AL is the concentration of a chemical in a particular medium that, when exceeded, presents significant risk of adverse impact to potential receptors. The ALs tend to drive the cleanup process for a contaminated site. But then, the ALs will not always result in acceptable risk levels. It is therefore prudent to recalculate the residual risks that will remain if remediation is carried out to the ALs, and if necessary, develop more stringent and health-protective levels that will meet the acceptable risk level/criteria.

A site-specific "Recommended Site Cleanup Limit" (RSCL) can be applied to contaminated sites to determine the level of cleanup warranted. The type of exposure scenarios envisioned may drive the RSCL. The RSCLs derived for the various pathways from defined exposure scenarios aid in developing cleanup options, such that public health and/or the environment are not jeopardized by any residual contamination.

Site Cleanup Limits Based on Attenuation-Dilution Factors

Several authors[6–9] have given various relations for computing cleanup limits that either account for attenuation alone or dilution only. It is suggested here that the cleanup level, RSCL, be derived from the following relationship:

$$RSCL = Std \times AF \times \Pi_{(media)} \{DF\}$$

where RSCL = Cleanup level in soil (mg/kg)
Std = Receiving media criteria to be met
AF = Attenuation factor, defining the loss of contaminant during transport—i.e., the distribution constant
$\Pi_{(media)} \{DF\}$ = Cumulative dilution factor during transport, i.e.,product of ratios of (concentration at source) to (concentration at receiving media).

Site Cleanup Levels By Mass Balance Analyses

Simple mass balance analytical relationships can be applied between various environmental compartments, by coupling current contaminant loadings with required or anticipated loadings together with available media standards, in order to derive cleanup limits. The back-calculation is carried out according to the following simple relationship:

$$C_{max} = (C_{std}/C_r) \times C_s$$

where C_{max} = Maximum acceptable source concentration
C_{std} = Receiving media criteria for receptors
C_r = Receiving media concentration
C_s = Source concentration prior to cleanup.

Media Action Levels

The media ALs are calculated for for both systemic toxicants and carcinogens; the more stringent of the two, where both exist, is selected as the site cleanup limit.

The governing equation for calculating action levels for systemic toxicants present at waste sites is given by[6]:

$$C_m = (RfD * BW) / (I * A)$$

where C_m = Action level in medium of concern, Soil (mg/kg)
RfD = Reference Dose (mg/kg/day)
BW = Body Weight (kg)
I = Intake Assumption, i.e., Soil Ingestion Rate (g/day)
A = Absorption Factor (dimensionless).

The governing equation for calculating action levels for carcinogenic constituents present at a hazardous waste site is given by[6]:

$$C_m = (R * BW * LT) / (SF * I * A * ED)$$

where
C_m = Action Level in Medium of Concern, Soil (mg/kg)
R = Assumed/Acceptable Risk Level (dimensionless)
BW = Body Weight (kg)
LT = Assumed Lifetime (years)
CPF = Cancer Slope Factor (1/[mg/kg/day])
I = Intake Assumption, i.e., Soil Ingestion Rate (g/day)
A = Absorption Factor (dimensionless)
ED = Exposure Duration (years).

Health-Protective Cleanup Limits

It is proposed here that, where the computed "residual" risks following remediation will exceed allowable limits, then the RSCL be estimated in the same way as the soil action level, but with a consideration for the aggregation of the total number of chemicals present. Thus, assuming each compound contributes proportionately to the hazard index or cancer risk, the RSCL is estimated according to the following relationship:

$$RSCL = C_{max}/N$$

where
C_{max} = Action level in medium of concern, Soil (mg/kg)
N = Number of chemical contributors to overall hazard index or cancer risk, as appropiate.

Alternatively, the acceptable risk level may be apportioned between the chemical contributors to the overall target risk or hazard, assuming each constituent contributes equally to the total acceptable risk or hazard. The risk fraction obtained for each compound can then be used to obtain the acceptable limit, RSCL, again working from the relationships established previously for the computation of the media action levels.

RISK ASSESSMENT APPLICATION TO A SITE RESTORATION PROGRAM

The health risk assessment methodology discussed earlier is followed to develop cleanup criteria for a hypothetical contaminated site located in a suburban community in Southern California.

Overview of the Hypothetical Case

It has been determined that, there is inorganic metals contamination at a property—the XYZ site—located in Southern California. The XYZ is fenced with barbed wire to prevent unauthorized entry to the property. An environmental site assessment conducted for the XYZ site indicated the presence of Antimony (Sb), Cadmium (Cd), Copper (Cu), Lead (Pb), Nickel (Ni) and Zinc (Zn) at this property. It is suspected that several off-site population groups may be impacted by these chemicals under both current and future land-use conditions, and that the XYZ site may have to be remediated. A risk assessment process is to be used to evaluate the risk information pertinent to the selection of appropriate site remediation techniques for the XYZ site.

Study Design for the XYZ Site

The following potential exposure scenarios are considered to be representative of the XYZ site under current land-use conditions, given the present level of knowledge about this site and vicinity:

- Human residential exposure off-site via inhalation of fugitive dust with site-related contaminants
- Human residential exposure off-site via ingestion of soils with site-related contaminants
- Human residential exposure off-site through dermal contact with soils containing site-related contaminants.

Under future land-use conditions, the critical receptors (i.e., the population groups likely to be maximally exposed to residual contamination that remains after site remediation) are expected to be within the boundary of the XYZ site.

The difference in sensitivities between adults and children demands that they be treated separately in evaluating their exposure intakes and doses of chemicals. Also, due to the variance in activity and behavior of children at the different ages, child exposure of soils is broken down into two categories for this risk assessment, including:

- Children aged up to 6 years (to include infants and pre-school children)
- Children aged between 6 and 12 years (to include young children of school-going age).

For the purpose of this risk assessment and consistent with EPA guidance[4], all population groups aged more than 12 years are included in the adult category.

Table I presents a summary statistics for the chemicals of potential concern in the various environmental compartments for the XYZ site. Site-specific exposure parameters used in the computation of receptor intakes and doses for the XYZ site are given in Table II. The calculation of potential carcinogenic risks under the existing conditions at the XYZ site were performed for the different population groups. The pathway-specific and total cancer risks for the receptor groups are summarized in Table III.

It was determined that cadmium is the only site-related chemical likely to pose any cancer risks to the potential receptors. However, cancer risks of 7E-06, 8E-06 and 6E-05 estimated for children under 6, children aged 6–12, and adult populations, respectively, are all within the acceptable range of 1E-07 to 1E-04. Carcinogenic risks to human health are not indicated, even under the present conditions at the XYZ site. However, the additional carcinogenic effects of lead have not been evaluated in this assessment due to the lack of information on slope factors or carcinogenic potency for this chemical.

The calculation of potential noncarcinogenic hazards under the existing conditions at the XYZ site were performed for the different population groups. The pathway-specific and total hazard indices for the various receptor groups are also summarized in Table III. For children aged up to 6 years, the total exposure hazard index of 20.4 is well above the acceptable level of 1. Exposure through dermal contact is not a significant contributor to the hazard index for this population group. However, ingestion and inhalation exposures are significant to the potential health threats posed by the contaminants of concern. The single greatest contributor to the high hazard level for this receptor group is from lead (Pb).

For children aged between 6 and 12 years, it is apparent that the total exposure hazard index of 15.4 is well above the acceptable level of 1. Exposures through soil ingestion and dermal contact do not appear to be significant contributors to the hazard index for this population group. However, inhalation exposure is very critical to the potential health threats posed by the contaminants of concern.

For adult populations, the total exposure hazard index of 10.9 is well above the acceptable level of 1. Exposures through soil ingestion and dermal contact do not appear to be a problem for this population group. However, inhalation exposure is critical to the potential health threats posed by the contaminants of concern.

Risk Characterization at Soil Action Levels for the RAP

The soil action levels developed for the XYZ site according to the relations discussed earlier on for systemic toxicants and carcinogenic constituents. For the XYZ site, noncarcinogenic effects are the significant contributors to health risks indicated by this site. Furthermore, cancer slope factor exist only for inhalation of cadmium, which does not indicate any carcinogenic risks to potential receptors. Consequently, in determining the health-based cleanup criteria for this site, the relationship for developing soil action levels for systemic toxicants is used.

Residual risks that remain after the implementation of a remedial action to bring contaminant concentrations at the XYZ site to the action levels was assessed. Based on the estimates obtained for soil action levels, Table IV gives a summary of the risk characterization associated with on-site exposure to contaminants at the computed action levels for the XYZ site. This indicates reduced but significant health threats if the site is cleaned up to the estimated soil action levels. It may therefore be necessary to remediate the XYZ site to levels lower than the action levels, especially if direct on-site exposure to children is anticipated under the future land-use conditions.

Site Remediation to the Recommended Soil Cleanup Limit

The RSCL was estimated in the same way as the soil action level, but with a consideration for the aggregation of the six chemicals present. Assuming each compound contributes proportionately to the hazard index, the RSCL was estimated as

Table I. Summary Statistics for the Chemicals of Potential Concern Present at the XYZ Site

Chemical	Carcinogen Classification*		On-Site Soil Concentration Range (mg/kg)	Off-Site Soil Concentration Range (mg/kg)	Mean Concentrations		
	ORAL ROUTE	INHALATION ROUTE			On-Site Soils (mg/kg)	Off-Site Soils (mg/kg)	Air** (u/m3)
Antimony (Sb)	NC	NC	3.0–1,500	0.3–27	2.03E+02	4.70E+00	1.10E+00
Cadmium (Cd)	NC	B1	0.4–210	0.3–3.9	1.51E+01	1.36E+00	8.16E-02
Copper (Cu)	NC	NC	19.0–39,600	6.0–310	3.40E+03	7.55E+01	1.84E+01
Lead (Pb)	B2	B2	32.0–16,500	20.0–1,110	3.96E+03	3.43E+02	2.14E+01
Nickel (Ni)	NC	NC	8.0–12,600	5.0–150	1.13E+03	2.96E+01	6.11E+00
Zinc (Zn)	NC	NC	300.0–299,000	57.0–8,660	4.36E+04	1.15E+03	2.36E+02

* Chemical carcinogen classification by EPA groups.
** Refers to the respirable concentration obtained from air modeling.

Table II. Site-Specific Parameters for Doses and Intakes Calculation for the XYZ Site

Parameter	Children Aged Up to 6	Children Aged 6–12	Adult	Reference Sources
PHYSICAL CHARACTERISTICS				
Average Body Weight	16 kg	29 kg	70 kg	(a,b,c)
Average Total Skin Surface Area	6980 cm2	10,470 cm2	18,150 cm2	(a,b,e,h)
Average Lifetime			70 yrs	(a,b,c,e)
Average Lifetime Exposure Period	5 yrs	6 yrs	58 yrs	(b,e)
ACTIVITY CHARACTERISTICS				
Inhalation Rate	0.25 m3/hr	0.46 m3/hr	0.83 m3/hr	(b,e)
Retention Rate of Inhaled Air	100%	100%	100%	(e)
Frequency of Fugitive Dust Inhalation				
–Off-site residents, schools and passers-by	365 days/yr	365 days/hr	365 days/yr	(b,e)
–Off-site workers	–	–	260 days/yr	(b,e)
Duration of Fugitive Dust Inhalation (Outside)				
–Off-site residents, schools and passers-by	12 hrs/day	12 hrs/day	12 hrs/day	(b,e)
–Off-site workers	–	–	8 hrs/day	(b,e)
Amount of Soil Ingested Incidentally	200 mg	100 mg	50 mg	(a,b,c,e,h,i)
Frequency of Soil Contact				
–Off-site residents, schools and passers-by	330 days/yr	330 days/yr	330 days/yr	(b,e)
–Off-site workers	–	–	260 days/yr	(b,e)
Duration of Soil Contact				
–Off-site residents, schools and passers-by	12 hrs/day	8 hrs/day	8 hrs/day	(b,e)
–Off-site workers	–	–	8 hrs/day	(b,e)
Percentage of Skin Area Contacted by Soil	20%	20%	10%	(b,e,h)
MATERIAL CHARACTERISTICS				
Soil to Skin Adherence Factor	0.75 mg/cm2	0.75 mg/cm2	0.75 mg/cm2	(a,b,e,f,g)
Soil Matrix Attenuation Factor	15%	15%	15%	(d)

(a)–USEPA, 1989: "Risk Assessment Guidance for Superfund, Vol. I—Human Health Evaluation Manual (Part A)," EPA/540/1-89/002.

(b)–USEPA, 1989: "Exposure Factors Handbook," EPA/600/8-89/043.

(c)–USEPA, 1988: "Superfund Exposure Assessment Manual," EPA/540/1-88/001.

(d)–Hawley, J. K., 1985: "Assessment of health Risk from Exposure to Contaminated Soil," Risk Analysis, Vol. 5, No. 4: 289–302.

(e)–Estimate based on Site-Specific Conditions.

(f)–Lepow, M. L., L. Bruckman, M. Gillette, S. Markowitz, R. Robino, J. Kapish, 1975: "Investigations into Sources of lead in the Environment of Urban Children," Environmental Research 10:415–426.

(g)–Lepow, M. L., M. Bruckman, L. Robino, S. Markowitz, M. Gillette, J. Kapish, 1974: "Role of Airborne Lead in increased Body Burden of Lead in Hartford Children," Environmental Health Perspectives 6:99–101.

(h)–Sedman, R., 1989: "The Development of Applied Action Levels for Soil Contact: A Scenario for the Exposure of Humans to Soil in a Residential Setting," Environmental Health Perspectives 79:291–313.

(i)–Calabrese, E. J. et al, 1989: "How Much Soil Do Young Children Ingest: An Epidemiologic Study," Regulatory Toxicology and Pharmacology 10, 123–137.

Table III. Risk Characterization Summary Current Land Use Conditions (Off-Site Scenario)

Population	Exposure Type/Pathway	Pathway Carcinogenic Risk*	Pathway Hazard Index**
Child, Up to 6 Years	INHALATION (of particulates)	7E–06	14.1
	INGESTION (of soil – incidental and pica)	–	5.5
	DERMAL (contact with soil)	–	0.8
Total Exposure Carcinogenic Risk to Child Aged 1-6 Years:		7E–06	
Total Exposure Chronic Hazard Index to Child Aged 1-6 Years:			20.4
Child, 6–12 Years	INHALATION (of particulates)	8E–06	14.3
	INGESTION (of soil – incidental and pica)	–	0.5
	DERMAL (contact with soil)	–	0.6
Total Exposure Carcinogenic Risk to Child Aged 6–12 Years:		8E–06	
Total Exposure Chronic Hazard Index to Child Aged 6–12 Years:			15.4
Adult Populations	INHALATION (of particulates)	6E–05	10.7
	INGESTION (of soil – incidental and pica)	–	0.1
	DERMAL (contact with soil)	–	0.1
Total Exposure Carcinogenic Risk to Adults:		6E–05	
Total Exposure Chronic Hazard Index to Adults:			10.9

*Standard and Acceptable Level = 1E–07 to 1E–04
**Standard and Acceptable Level = 1

$$RSCL = C_{max}/N$$

where N = Number of chemical contributors to overall hazard index, 6.

Based on the values obtained for RSCLs, Table V gives a summary of the risk characterization associated with on-site exposure to contaminants at the RSCLs for the XYZ site. This indicates significantly reduced hazard indices. No significant health threats are indicated if the site is cleaned up to the RSCLs.

Table IV. Risk Characterization Summary Based on Soil Action Levels Future Land Use Conditions (On-Site Scenario)

Population	Exposure Type/Pathway	Pathway Carcinogenic Risk*	Pathway Hazard Index**
Child, Up to 6 Years	INHALATION (of particulates)	–	2.3
	INGESTION (of soil – incidental and pica)	–	5
	DERMAL (contact with soil)	–	1.8
Total Exposure Carcinogenic Risk to Child Aged 1–6 Years:		–	
Total Exposure Chronic Hazard Index to Child Aged 1–6 Years:			9.1
Child, 6–12 Years	INHALATION (of particulates)	–	2.3
	INGESTION (of soil – incidental and pica)	–	1.3
	DERMAL (contact with soil)	–	1.5
Total Exposure Carcinogenic Risk to Child Aged 6–12 Years:		–	
Total Exposure Chronic Hazard Index to Child Aged 6–12 Years:			5.1
Adult Populations	INHALATION (of particulates)	–	1.7
	INGESTION (of soil – incidental and pica)	–	0.3
	DERMAL (contact with soil)	–	0.3
Total Exposure Carcinogenic Risk to Adults:		–	
Total Exposure Chronic Hazard Index to Adults:			2.3

*Standard and Acceptable Level = 1E–07 to 1E–04
**Standard and Acceptable Level = 1

Table V. Risk Characterization Summary Based on Recommended Soil Cleanup Limits Future Land Use Conditions (On-Site Scenario)

Population	Exposure Type/Pathway	Pathway Carcinogenic Risk*	Pathway Hazard Index**
Child, Up to 6 Years	INHALATION (of particulates)	–	0.4
	INGESTION (of soil–incidental and pica)	–	0.8
	DERMAL (contact with soil)	–	0.3
Total Exposure Carcinogenic Risk to Child Aged 1–6 Years:		–	
Total Exposure Chronic Hazard Index to Child Aged 1–6 Years:			1.5
Child, 6–12 Years	INHALATION (of particulates)	–	0.4
	INGESTION (of soil–incidental and pica)	–	0.2
	DERMAL (contact with soil)	–	0.2
Total Exposure Carcinogenic Risk to Child Aged 6–12 Years:		–	
Total Exposure Chronic Hazard Index to Child Aged 6–12 Years:			0.8
Adult Populations	INHALATION (of particulates)	–	0.29
	INGESTION (of soil–incidental)	–	0.04
	DERMAL (contact with soil)	–	0.04
Total Exposure Carcinogenic Risk to Adults:		–	
Total Exposure Chronic Hazard Index to Adults:			0.4

*Standard and Acceptable Level = 1E–07 to 1E–04
**Standard and Acceptable Level = 1

Summary of Recommended Soil Cleanup Limits

Table VI presents the estimated soil action levels and the recommended soil cleanup limits for the chemical constituents present at the XYZ site. No significant potential health impacts is indicated if remediation is carried out to the recommended levels estimated for this site. In the implementation of corrective actions, it may become apparent that some chemicals can be more easily removed than others. In that case, it still may be possible to remediate to lower levels than the RSCL for some constituents and to higher levels than the RSCL for others, and still attain an acceptable hazard index for residual contamination remaining at the XYZ site, following site remediation.

Conclusions

The risk assessment for the XYZ site did not indicate off-site impacts from the chemicals present at the XYZ property located in suburban Southern California. However, site remediation would be necessary before the site is used for future developments. Remediation to residual contamination levels corresponding to the recommended soil cleanup limits will be protective of human health, due to future on-site land-uses.

The levels of cancer risks estimated for all potential receptor groups were within the acceptable range of 1E-07 to 1E-04. No cancer health risks to the population in the vicinity of the XYZ site are therefore indicated. Long-term noncancer risks, however, are indicated in the assessment for the XYZ site. Corrective actions to reduce risks for the site would therefore be necessary. This will be achieved by remediating the site to the proposed RSCLs.

SUMMARY

Contaminants released to the environment are controlled by a complex set of processes including various forms of transport, transformation and biological uptake. To reduce costs in planning for environmental cleanup of contaminated sites, it is important that the decision-making process involved in hazardous waste management be well defined. An approach is presented that focuses on determining "acceptable" residual contaminant levels as goals for remedial action. The concepts presented will facilitate decisions about the effective use of limited funds to clean up a site to a level appropriate/safe for its intended use.

The overall goal of the human health risk evaluation in site restoration is to provide a framework for developing the risk information necessary to assist the decision-making relating to site remediation of potentially contaminated sites. The assessment of health and environmental risks play an important

Table VI. Recommended Soil Cleanup Limits Future Land-Use Conditions (On-Site Scenario)

Chemnical of Concern	Soil Maximum Chemical levels (mg/kg)	Estimated Soil Action level (mg/kg)	Trace Chemical Content of Natural Soils* (mg/kg)	Recommended Soil Cleanup Limit (mg/kg)
Antimony (Sb)	1.50E+03	3.20E+01	2–10 (–)**	5.33E+00
Cadmium (Cd)	2.10E+02	4.64E+02	0.01–0.7 (0.06)	7.73E+01
Copper (Cu)	3.96E+04	2.96E+03	2–100 (30)	4.93E+02
Lead (Pb)	1.65E+04	1.12E+02	2–200 (10)	1.87E+01
Nickel (Ni)	1.26E+04	8.00E+02	5–500 (40)	1.33E+02
Zinc (Zn)	2.99E+05	1.68E+04	2–10 (50)	2.80E+03

*Source: Pollution Control Engineer's Handbook, 1985, Shields, E. J.
**2–10 is Common Range; (–) is Average.

role in the remedial investigation/feasibility study (RI/FS), the RAP, and also the risk mitigation and risk management strategies for potentially contaminated sites.

REFERENCES

1. USEPA: Risk Assessment and Management: Framework for Decision Making. *EPA 600/ 9-85-002*, Washington, DC, 1984.
2. Paustenbach, D.J. (ed.): *The Risk Assessment of Environmental Hazards: A Textbook of Case Studies*. John Wiley & Sons, N.Y., 1988.
3. Asante-Duah, D.K.: "Quantitative Risk Assessment as a Decision Tool for Hazardous Waste Management," In *Proceedings of 44th Annual Purdue Industrial Waste Conference*, Chelsea, MI: Lewis Publishers (1989).
4. USEPA: Risk Assessment Guiadance for Superfund, Volume I—Human Health Evaluation Manual (Part A), *EPA/540/1-89/002*, 1989.
5. USEPA: Superfund Exposure Assessment Manual. Report No. *EPA/540/1-88/001, OSWER Directive 9285.5-1*, USEPA, Office of Remedial Response, Washington, DC, 1988.
6. USEPA: RCRA Facility Investigation (RFI) Guidance, *EPA/530/SW-87-001*, 1987.
7. DOE: The Remedial Action Priority System (RAPS): Mathematical Formulations. U.S.Dept. of Energy, Office of Environ., Safety, & Health, 1987.
8. Brown, H.S.: A Critical Review of Current Approaches to Determining "How Clean is Clean" at Hazardous Waste Sites. *Hazardous Waste & Hazardous Materials*, Vol.3, No.3, pp233–260, 1986.
9. Dawson, G.W. and D. Sanning: "Exposure-Response Analysis for Setting Site Restoration Criteria," In *Proc. National Conf. on Mgmnt of Uncontrolled Hazardous Waste Sites*, Washington, DC, 1982.

7 INVESTIGATIONS INVOLVING OXIDATION-REDUCTION (REDOX) PRETREATMENT IN CONJUNCTION WITH BIOLOGICAL REMEDIATION OF CONTAMINATED SOILS

Carlo D. Montemagno, Program Manager
Robert W. Peters*, Environmental Systems Engineer
Allen Tyree, Coop Student
Energy Systems Division
Argonne National Laboratory
Argonne, Illinois 60439

INTRODUCTION

Oxidation-reduction (REDOX) reactions are among the most important reactions involved in the environmental engineering field. Oxidation is a reaction in which the oxidation state of the treated compound is increased, i.e., the material loses electrons. Reduction involves the addition of a chemical (reducing) agent which lowers the oxidation state of a substance, i.e., the material gains electrons. Both processes of oxidation and reduction occur together. All REDOX reactions are thermodynamically based.

There are a number of oxidizing agents which have been reported in the technical literature[1-4] for treatment of refractory organic compounds. Common oxidizing agents include: hydrogen peroxide, ozone, ultraviolet (UV) irradiation, and combinations thereof, such as UV/ozone and UV/peroxide. A gradient of REDOX reactions is possible, depending on such factors as the oxidation-reduction reaction conditions, the availability of electron donors and acceptors, and the nature of the organic compounds involved.[5]

BACKGROUND

A review of the technical literature revealed that the majority of the oxidation-reduction applications have been in the areas of wastewater treatment and groundwater remediation, with very little attention devoted to the potential of using REDOX technologies for remediation of hydrocarbon contaminated soils. In this particular study, feasibility studies were performed on gasoline-contaminated soil. These studies focused on three major phases: 1) containment of the contamination by addition of tailoring agents to the soil, 2) biological remediation either performed in situ or on-site (using a slurry reactor system), and 3) pretreatment of the contaminated soils using REDOX systems, prior to biological remediation. This particular paper focuses on the third phase of the project, aimed at "softening" the refractory organics resulting in the formation of organic compounds which are more amenable to biological degradation. This paper focuses its attention on the use of hydrogen peroxide treatment and ultraviolet irradiation as a means of pretreating the contaminated soil prior to biological remediation.

The next section provides a description of the site and the contamination problem, and summarizes the results from the soil characterization portion of our study. This section is followed by the section which describes applications of REDOX systems for treatment of hydrocarbons.

Site Description

The soil samples for this study came from a U.S. military installation located on the west coast. Two 20,000-gallon underground storage tanks had been used since the early 20th century to store a number of organic materials. This site is located in a major metropolitan district overlooking a bay area. These tanks were known to leak. Gasoline fuel contaminated soil was found in the vicinity

45th Purdue Industrial Waste Conference Proceedings,

**Correspondence should be directed to this author. The viewpoints expressed here are not necessarily those of the Laboratory or its research sponsors.*

beneath these tanks; contamination was observed at depths of 20 to 50 feet below the surface. The installation contacted Argonne National Laboratory in a effort to perform a series of treatability studies aimed at cleaning up the site. The overall project investigated the following technologies in a series of treatability studies: tailoring the soil with quaternary ammonium salts to contain the organic contamination in-place, biotreatment (in situ and on-site using a slurry reactor system), and pretreatment with oxidation-reduction (REDOX) techniques as a means to "soften" the difficult-to-degrade compounds prior to biotreatment. This paper addresses the REDOX treatability studies performed on natural soil from the site that had been spiked with selected organic contaminants.

The biotreatment studies were performed using the contaminated soil collected from the site. The soil pH ranged from 7.3 to 7.8; the percent organic matter ranged from 1 to 2%. The cation exchange capacity was 13 to 14 meq/100 gm of soil. Total available nitrogen and phosphorus concentrations were 470–529 mg/kg and 9–10 mg/kg, respectively. The soil moisture content was about 35%. The samples were taken with a 6-inch O.D. by 5-ft long splitspoon sampler. The core sleeves were 5 1/2-inch I.D. by 2 1/2-ft long clear plexiglas. Two core sleeves were placed in the splitspoon sampler and the sampler was driven into the undisturbed soil below the bottom of a 10-inch diameter hollow stem auger. The primary purpose for this continuous core was to evaluate remediation options for the contaminated soil. Hydraulic conductivities ranged from 0.7 to 17.7 ft/day for various soil samples collected in the area. The soil was from the Monterey Formation, which is comprised of uncemented clayey sediments with thin lenses of cemented siltstones. The soil texture was a clay loam to a clay. Typical size fractions, determined from sieve analyses indicated the following soil composition: Sand—22 to 29%; Silt—31 to 35%; and Clay—36 to 47%. Total petroleum hydrocarbon (TPH) analyses performed on various soil borings ranged from nondetectable to 1100 ppm. Two samples collected at a depth of 25-ft indicated total petroleum hydrocarbon concentrations of 891 and 294 mg/kg.

Oxidation-Reduction Systems

Conventional biological, physical, and chemical treatment methods are often ineffective in removing hazardous organics from contaminated water and soil. Biological processes are commonly used to remove organic compounds, but a large number of toxic, refractory compounds are only partially degraded by biological treatment. Tertiary treatment processes (such as carbon adsorption) have limitations in removing many toxic and refractory compounds. Activated carbon adsorption often has very low adsorption capacities for polar aliphatic compounds. Further, the activated carbon must be replaced or regenerated when the adsorption capacity is exhausted.

One promising technology involves use of oxidation-reduction techniques, employing hydrogen peroxide, ozone, ultraviolet irradiation, or combinations thereof to treat the toxic, refractory organic compounds. These various REDOX systems are described briefly below.

Hydrogen Peroxide Studies. Hydrogen peroxide can be used either as a bio-enhancer or as a chemical oxidant in site clean-ups.[6] As a bio-enhancer, H_2O_2 does not act directly on contaminants, but rather provides dissolved oxygen for the microorganisms to degrade the organic contaminants.[2, 6]

Hydrogen peroxide can also be used with or without catalysts as a chemical oxidant to degrade organics to carbon dioxide and water or as a pretreatment step. The hydroxyl radicals act as electrophilic reagents in their attack of benzene rings. Electron-donating groups such as OH on an aromatic ring activate the ring for electrophilic substitution reactions.[7] Therefore, phenol is more susceptible to electrophilic attack by hydroxyl radicals than benzene; the presence of hydrogen peroxide accelerates the phenol reaction more than the benzene reaction.

In activated systems, a catalyst (such as ferrous ions or ultraviolet light) breaks down hydrogen peroxide to form highly reactive hydroxyl radicals which attack the hydrocarbons yielding carbon dioxide and water. Hydrogen peroxide also breaks down contaminants into less-toxic intermediates which are easier to degrade than their parent compounds .[6, 8] Oxidation reactions involving H_2O_2 are influenced by pH, the presence of catalysts, temperature, peroxide dosage, and reaction time .[9] One of the most attractive features of H_2O_2 oxidation is its ability to provide available oxygen for biological treatment .[9] Hydrogen peroxide can detoxify a broad range of organic and inorganic wastes including sulfides, thiosulfates, sulfites, mercaptans, disulfides, phenols and substituted phenols, chlorine, cyanide, organic amines, aldehydes, and hydroquinones .[9] By properly adjusting the reaction conditions, the selectivity of H_2O_2 for the contaminant of concern can be increased. For example, H_2O_2 normally does not readily react with phenols. However, by adjusting the pH in the range of 4 to 5 with a ferrous ion catalysts, phenol destruction occurs within a few minutes. A relatively small amount of peroxide is normally sufficient to cleave organic molecules, which are then easier to degrade biologically than their parent compounds. Hydrogen peroxide can also be used as a post-treatment (polishing) technique following biological treatment to eliminate refractory compounds that would normally slip through biological systems unchanged. Since most of the detoxification has already been done biologically, only small amounts of peroxide are required to treat the residual refractory compounds.

Additional benefits cited for this technology as compared with conventional treatment processes include: production of little, if any, sludge; ease of system control; and destruction of organic matter rather than simply removing the material from the liquid phase (which would necessitate subsequent disposal).[8]

Ozone Studies. Ozone has been used to remove phenols, cyanides, alcohols, pesticides, aldehydes, and sulfides from waters.[10] The mechanisms of ozone-pollutant reactions depend on the solution pH. The parameters having the greatest influence on the kinetics of the ozonation reaction system are pH, initial contaminant concentration, and ozone application rate.[10]

UV/Peroxide Studies. For many toxic and refractory organic compounds, the reaction rates are slow for UV and H_2O_2 treatment alone. The combination of H_2O_2 and UV irradiation provides free radicals and excited state species that enhance the rate of oxidation.[11] This technology utilizes the oxidation potential of hydroxyl radicals formed when UV light at less than 400 nm reacts with H_2O_2 to effectively destroy hazardous organic wastes.[12]

Yost [12] classified the variables affecting UV/peroxidation treatment into two types, as listed below:

Type I – related to chemical and physical characteristics of the target waste stream:

- molecular structure of contaminants and relative concentrations
- transmittance/absorbance properties of the waste
- final treatment effluent discharge limit criteria

Type II – related to process equipment and process design:

- reactor sizing and reactor configuration
- waste delivery
- UV irradiation and H_2O_2 dosages
- temperature
- pH conditions
- catalyst utilization

Weir et al.[13] studied the ultraviolet light-catalyzed hydrogen peroxide treatment of benzene-contaminated water. Benzene was more rapidly destroyed by the combination of hydrogen peroxide and UV light than by UV light alone. The rate of oxidation could be increased by either increasing the concentration of H_2O_2 initially present in the reactor or by increasing the UV light intensity.[11, 13] Over 98% of the benzene initially present was removed after 90 minutes of treatment. The benzene was probably attacked by both UV photons and by hydroxyl radicals generated from the hydrogen peroxide. By increasing the hydrogen peroxide concentration, more hydroxyl radicals are available to attack the benzene ring. A higher lamp intensity increases the decomposition of H_2O_2 to hydroxyl radicals. The increase in reaction rate with temperature is the result of higher fluid temperature and an enhanced supply of reactive species.[11] Alkaline pH slowed the reaction rate, which the authors speculated was the result of base-catalyzed decomposition of H_2O_2. Examination of a plot of the fraction of benzene remaining versus reaction time (in a semi-logarithmic fashion) shows that this reaction is linear for the first 10 to 20 minutes, indicating that benzene oxidation is first-order with respect to benzene. Deviations from linearity at later times may have been due to the decrease of hydrogen peroxide concentrations with time or due to a buildup of reaction intermediates which absorb UV light and/or react with hydroxyl radicals.[13] At least four intermediates were identified from both HPLC and GC/MS analyses; these intermediates included phenol, catechol, hydroquinone, and resorcinol.

Sundstorm et al.[11] investigated the destruction of various hazardous organics (including trichloroethylene, tetrachloroethylene, tetrachloroethane, and ethylene dibromide) using UV-catalyzed oxidation with hydrogen peroxide. With H_2O_2 and no UV light, the rate of reaction for destruction of tetrachloroethylene was negligible. When exposed to UV light with no H_2O_2 present, a moderate rate of reaction was observed, indicating that UV light can generate free radicals from the organic species. With both H_2O_2 and UV light, the rate of reaction increased substantially as evidenced by the decreasing concentration of trichloroethylene from an initial value of 58 mg/L to less than 1 mg/L in 50 minutes at 20°C. Higher temperature also increased the reaction rate substantially. At 41°C, the trichloroethylene concentration decreased from 58 to 1 mg/L in about 10 minutes. For chloroform and tetrachloroethane, UV irradiation alone was nearly as effective as the combination of H_2O_2 and UV light. Sundstrom et al.[11] observed that chlorinated compounds with carbon-carbon double bonds (such as trichloroethylene and tetrachloroethylene) degraded much faster than other compounds which contain carbon-carbon single bonds (such as ethylene dibromide and 1,1,2,2-tetrachloroethane). The carbon-carbon double bond is readily attacked by the UV light and reactive species. The rates of decomposition increase with increasing H_2O_2 concentration and temperature and were highly dependent on the chemical structure.

Sundstrom et al.[7] used ultraviolet catalyzed oxidation by hydrogen peroxide of benzene, toluene, chlorobenzene, phenol, 2-chlorophenol, 2,4,6-trichlorophenol, dimethylphthalate, and diethylphthalate. They selected three monosubstituted benzenes to study the effect of a single substituent group on the rate of reaction of benzene. The chlorophenol series was examined to study the effect of adding chlorine atoms to a ring. The phthalate esters were studied due to their common occurrence in industrial discharges. With UV light alone, all of the reaction rates were much lower than for the combined UV light/H_2O_2 system, especially for phenol. The phthalates were the slowest reacting compounds (of those studied) and were especially unreactive with UV light alone. With UV light only, addition of substituent chlorine atoms to phenol enhanced the susceptibility of the aromatic ring to attack by the photons. With both H_2O_2 and UV light, the effect of H_2O_2 on reaction rates became less significant as chlorine atoms were added to the phenol. Reaction intermediates were determined by HPLC and GC/MS. Of particular significance was the fact that the peaks of the intermediates disappeared as the reaction time was extended.

The UV/H_2O_2 system successfully treated dimethylaminomethyl-substituted phenols (DMP) and isophorone-contaminated water.[8] The DMP concentration was reduced from 50 mg/L to less than 0.05 mg/L in 60 minutes treatment time when dosed with 125 mg/L of H_2O_2 and irradiated with UV irradiation (254 nm) with an average intensity of 1200 microwatts/cm^2. The isophorone concentration was reduced from 50 mg/L to less than 0.05 mg/L in 60 minutes when dosed with 250 mg/L of H_2O_2 under the same conditions. The UV/H_2O_2 system has also been successfully used to degrade tannery waste effluents and keratin solutions[14]; 90% of the total organic carbon reduction of tannery waste was achieved in a pilot system with 2 minutes of UV irradiation.

UV/Ozone Studies. Zeff et al.[15] used an UV/ozone process for destruction of various pesticides. At the highest level of ozone dosage (120 mg O_3/min), all the ethylene dibromide disappeared in 2 hours. After 5 hours, only 1% or the original 1,2-dibromo-3-chloropropane (DBCP) and 0.17% of the original diazinon remained. Reducing the ozone concentration to 60 mg/min increased the DBCP and diazinon concentrations after 5 hours by 3-fold to 0.14 and 0.03 mg/L, respectively. Reducing the ozone concentration still further (to 30 mg/min) resulted in DBCP and diazinon concentrations of 0.23 and 0.06 mg/L, respectively.

Combined use of ozone and UV-irradiation induced rapid photochemical oxidation of halogenated organic compounds.[16] The system, developed by Ultrox International, has been used to treat wastes such as pentachlorophenol (PCP), hexachlorobenzene (HCB), 1,2-dibromo,3-chloropropane (DBCP) and gamma 1,2,3,4,5,6-hexachlorocyclohexane (lindane). Ozone alone was effective only for the oxidation of PCP, reducing the concentration from 50 to 6.5 μg/L in 15 minutes. The Ultrox process readily oxidized each of the chlorinated hydrocarbons. PCP was reduced from 50 μg/L to less than 1 μg/L in 5 minutes. Lindane and DBCP were oxidized to less than 1 μg/L in 15 and 30 minutes, respectively, while HCB was reduced to less than 1.5 μg/L in 60 minutes.

Boeve[17] described the OUUU process which utilizes ozone, ultraviolet irradiation, ultrasonics, and ultrapure water to remove and destroy organic contaminants. The process is a two-stage system. In the first stage, the contaminants are removed from the soil slurry (30–40% solids) using ultrapure water as the solvent. In the second stage, the resulting solution is treated to convert the contaminants into carbon dioxide, water, and harmless salts. Advantages cited for the process include: no incineration, landfilling, or chemical requirements, is portable, can be operated on-site, and converts the contaminants into carbon dioxide, water, and harmless salts. Ozone is a very strong oxidizer, and decomposes in water to form oxygen, free radicals, and hydrogen peroxide. The ozone and free radicals are particularly reactive with aliphatic carbon-carbon double bonds and can cause aromatic ring cleavage. Ozone alone seldom results in complete oxidation, but leads to partial oxidation products. The use of ultraviolet irradiation and ultrasonics as catalysts with ozone can reduce petroleum products three times faster than the use of ozone alone. The use of ultrasonics reduces the thickness of the liquid film, enhances gas transfer, and reduces bubble coalescence which increases the interfacial area available for mass transfer. This process has been used to remove polychlorinated biphenyls, pentachlorophenol, 2,4,5-trichlorophenol, benzene, and cyanides from water and soils. The size of the systems ranged from one cubic foot per hour of inert material with a water flow rate of one gallon/minute to 27 cubic yards per hour with a water flow rate of 50 gallons/minute.

OBJECTIVES

The overall goals and objectives of this project are summarized below:

1) Investigate the feasibility of in situ biological remediation for the treatment of contaminated soils resulting from leaking underground storage tanks.
2) Determine optimal biological degradation conditions and time required resulting in degradation to acceptable levels.
3) Investigate an innovative, emerging technology employing a slurry reactor system for on-site treatment of contaminated soils.
4) Investigate the effects of type and concentration/intensity of oxidizing agent, contact time, pH, temperature, presence of visible light spectrum, etc., on the "softening" and resulting degradation of organics in the contaminated soil, through a series of semi-batch reactor studies.
5) Investigate and compare the application mode for introduction of the oxidizing agent to the contaminated soil in the slurry reactor system.
6) Investigate the combined treatment employing oxidation and biological degradation for its effect on the "softening" and removal or organics from contaminated soil.
7) Develop design criteria for field application.

The particular paper focuses on Objectives 4 and 6, described above, to investigate the use of oxidation-reduction (REDOX) systems as a means to pretreat gasoline contaminated soil prior to biological remediation.

EXPERIMENTAL PROCEDURE

Hydrogen Peroxide Pretreatment Studies

These studies involved pretreatment of the contaminated soil using hydrogen peroxide as a means to "soften" the difficult-to-degrade organics contained in the soil.

The experimental procedure for these studies are summarized below. Approximately 10 grams each of uncontaminated soil were weighed out in replicate 250-mL Erlenmeyer flasks. The contaminants were then added to the soil to adsorb the organic contaminants. The samples were covered and in the hood during this adsorption phase. The contaminants employed in these studies involved benzene, toluene, o-xylene, and n-heptane. The organic concentrations employed resulted in 100 mg/g of soil for these experiments.

Sample controls were also prepared in which 100-mL of deionized water was added to the flasks which contained the contaminants.

The contaminated soils were then subjected to hydrogen peroxide pretreatment, employing a hydrogen peroxide concentration of 1.0 and 3.24%, respectively. A 1.0% treatment consisted of 30.8-mL of the stock hydrogen peroxide solution (3.24%) and diluting with deionized water to form a 100-mL solution. The 3.24% solution employed the stock reagent grade hydrogen peroxide solution. Controls were also performed using deionized water (0.00% H_2O_2 solution) on the contaminated soil.

The hydrogen peroxide solutions were added to the contaminated soils to form 10% soil slurries. After sealing each flask with parafilm, the flasks were then placed on a thermostatically-controlled incubator/shaker table, which rotated at 125 revolutions per minute. The flasks were agitated for periods of 1, 3, 8, or 24 hours. Afterwards, the flasks were removed from the shaker table. The contents from each flask were poured into two 50-mL test tubes and were centrifuged for 3 minutes at 2500 rpm. The percentage of hydrogen peroxide was determined on the resulting supernatant solution using the Idiometric method.[18] In this technique, 2.0-mL of the supernatant was transferred from the centrifuge tube to a 250-mL flask, to which 50-mL of deionized water, 10-mL of sulfuric acid (25%, 4.5 N), 10-mL of potassium iodide (3 wt%), and two drops of ammonium molybdate were added. This resulting solution was titrated with 0.1 N sodium thiosulfate solution to a faint yellow color. Then 2.0-mL of starch indicator was added and the titration was continued until the blue color just disappeared. The controls were tested in the same manner except that the hydrogen peroxide was omitted from the sample treatment.

The concentration of the various organic species was determined using gas chromatography techniques. The operating conditions of the GC are listed below:

Injector Port Temperature, C,	200
Ion Detector Temperature, C	200
Column Temperature, C	80
Air Flow, mL/min	240
Hydrogen Flow, mL/min	30
Helium Flow, mL/min	30

Calibration curves were determined for each organic contaminant (benzene, toluene, o-xylene, and n-heptane), as well as for total petroleum hydrocarbon analyses. The Quality Assurance/Quality Control QA/QC) procedures for this study employed replicate sample treatments performed on all different treatment conditions. From each replicate sample prepared, two samples were prepared for the analyses; each sample was subjected to four separate gas chromatograph analyses. The QA/QC protocol is shown schematically in Figure 1. The results presented in the next section summarize the average results obtained from this pretreatment study.

Ultraviolet (UV) Irradiation Studies

These studies involved pretreatment of the contaminated soil using UV irradiation. The experimental procedure for these studies is described below.

Approximately 10 grams each of uncontaminated soil were weighed out into two 250-mL Erlenmeyer flasks. The liquid contaminants (benzene, toluene, o-xylene, and n-heptane) were added to the soil providing the equivalent of 100 mg/g of each contaminant in the soil. The flasks were covered with parafilm and allowed to set for 30 minutes to saturate the soil with the contaminants. Then, 10-mL of deionized water was added to each flask; these flasks were shaken momentarily to suspend the soil. Afterwards, the flasks containing the contaminated soil were placed in the UV irradiation light source apparatus for the chosen length of time for each test. The total UV light source wattage was 16 and 32 watts. The U.V. irradiation source apparatus contained four 8-watt bulbs with a wavelength of 253.7 nm. For the 32-watt source, two lights were located above the soil samples and two lights were located beneath the soil samples. The pretreatment times used in these experiments were 1, 3, 8, and 24 hours.

The organic concentrations in the supernatant and the contaminated soil were determined using gas chromatography techniques, similar to those described in the Hydrogen Peroxide Pretreatment Studies.

RESULTS AND DISCUSSION

Results from the pretreatment of the contaminated soils (initially containing 100 mg/g each of benzene, toluene, o-xylene, and n-heptane) are presented below for hydrogen peroxide treatment and ultraviolet irradiation.

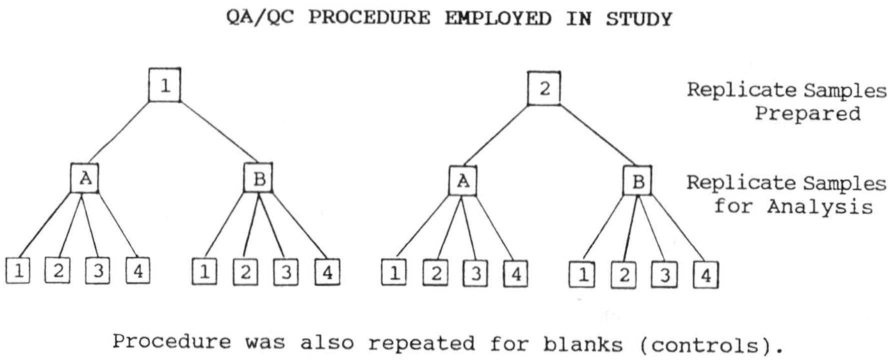

Figure 1. Quality assurance/quality control protocol employed study.

Table I. Experimental Results for Hydrogen Peroxide Pretreatment of the Contaminated Soil.

Organic Contaminant	Peroxide Concentration (%)	Residual Organic Contration, (mg/L) Treatment Time, (hr) 1	3	8	24
Benzene	0.00	523.6	670.8	694.4	637.6
	1.00	537.7	644.7	731.2	–
	3.24	293.6	586.3	657.3	642.4
Toluene	0.00	26.4	41.2	27.6	39.6
	1.00	28.8	40.7	36.3	–
	3.24	31.5	39.2	40.9	42.6
O-Xylene	0.00	9.9	20.4	21.4	20.7
	1.00	16.7	22.5	16.2	–
	3.24	20.2	30.7	33.3	30.5
n-Heptane	0.00	0.23	1.04	0.46	0.59
	1.00	0.70	1.06	0.48	–
	3.24	1.06	0.85	0.74	0.92
Peroxide	0.00	0.00	0.00	0.00	0.00
	1.00	0.49	0.29	0.19	–
	3.24	1.94	1.62	1.04	0.49

Hydrogen Peroxide Pretreatment

The concentrations of the organic contaminants were determined from GC analyses for the three peroxide concentrations (0.0, 1.0, and 3.24%) for treatment times of 1, 3, 8, and 24 hours. The results of benzene, toluene, o-xylene, n-heptane, and residual peroxide concentrations are presented in Table 1.

Figure 2 shows the residual peroxide concentration plotted as a function of the treatment time for the two initial hydrogen peroxide concentrations of 1.0 and 3.24%. Figure 3 presented the percentage removal of peroxide as a function of the treatment time. The decay in peroxide concentration exhibits an exponential type decay and shows a rapid drop in peroxide concentration during the first four hours.

Figures 4 through 7 show the concentration of benzene, toluene, o-xylene, and n-heptane, respectively, in solution for all three initial hydrogen peroxide concentrations used (0.0, 1.0, and 3.24%) as a function of the treatment time. For benzene, the concentration in solution rises quickly and remains relatively constant (about 650 mg/L) after 3 hours treatment time. The concentration of toluene exhibits a similar behavior, with a concentration of about 40 mg/L after 3 hours treatment time. For o-xylene, the concentration in solution remains relatively constant after 8 hours treatment time. However, the higher initial peroxide concentration (3.24%) results in an o-xylene concentration in solution which is about 50% higher than for the other two peroxide concentrations used (0.0 and 1.0%). For n-heptane, the concentration slowly decreases with increasing treatment time, which might indicate that volatilization plays a significant role in the concentration of n-heptane in solution. The concentration of n-heptane in solution was about 1 mg/L.

Figures 8 through 11 show the concentration for each contaminant plotted as a function of the residual peroxide concentration for benzene, toluene, o-xylene, and n-heptane, respectively. The initial peroxide concentration for these figures was 3.24%. The treatment time increases from right to left in these figures. These figures show that the concentration of each contaminant in solution remains relatively constant after about 3 hours treatment time (corresponding to a residual peroxide concentration of about 1.6%).

Extractions were performed on the contaminated soils before and after the REDOX pretreatments. As an example of the capability of this technique, for the 10% soil-slurry system employing a 1.0% H_2O_2 solution for a treatment time of 1.0 hour, the removal of benzene, toluene, o-xylene, and n-heptane averaged 88.2, 83.9, 77.7, and 85.5%, respectively, resulting in an overall removal efficiency of 83.8%. The results of the other cases exhibited similar trends.

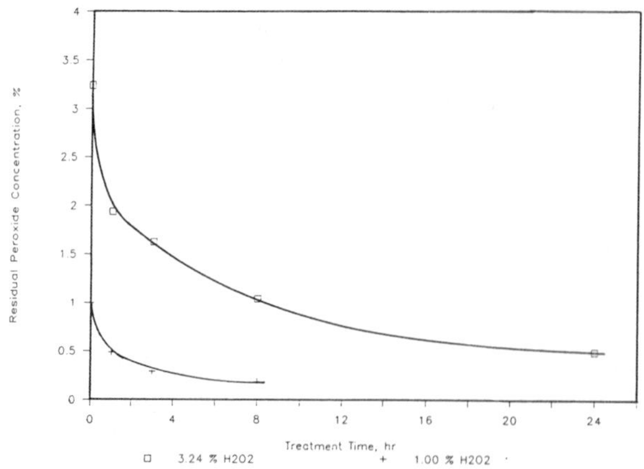

Figure 2. Residual peroxide concentrations as a function of treatment time. Initial peroxide concentrations = 1.0 and 3.24%.

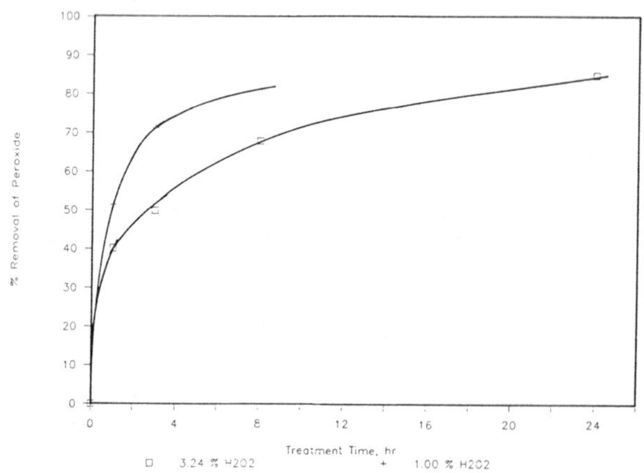

Figure 3. Percent removal of peroxide as a function of the treatment time.

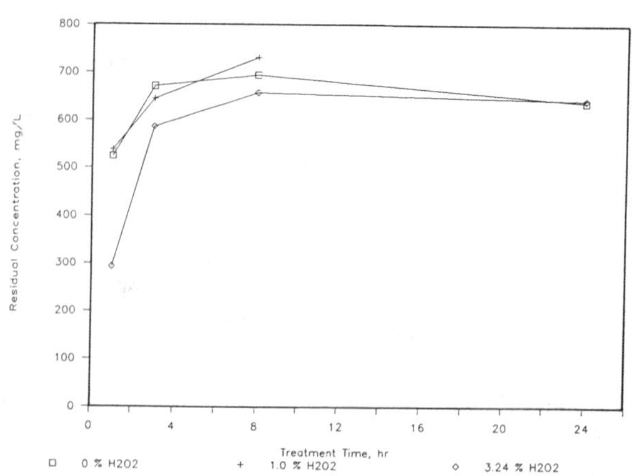

Figure 4. Effect of treatment time and peroxide concentration on residual benzene concentration in solution.

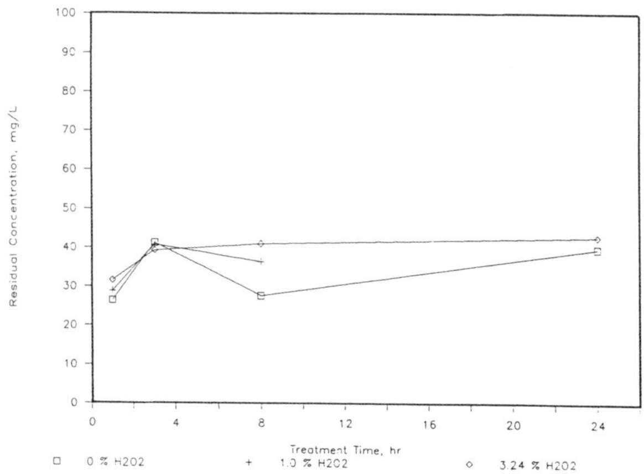

Figure 5. Effect of treatment time and peroxide concentration on residual toluene concentration in solution.

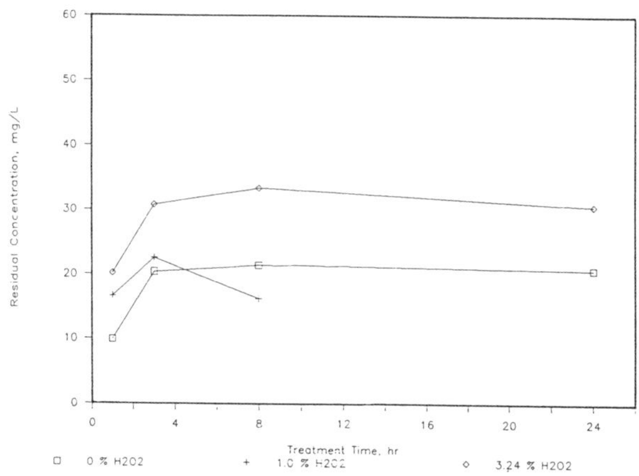

Figure 6. Effect of treatment time and peroxide concentration on residual o-xylene concentration in solution.

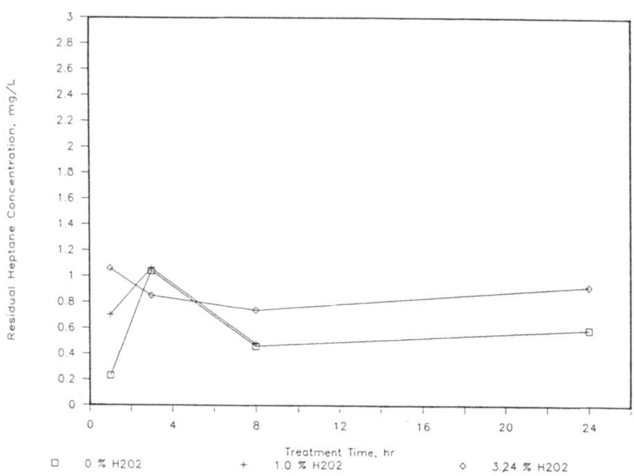

Figure 7. Effect of treatment time and peroxide concentration on residual n-heptane concentration in solution.

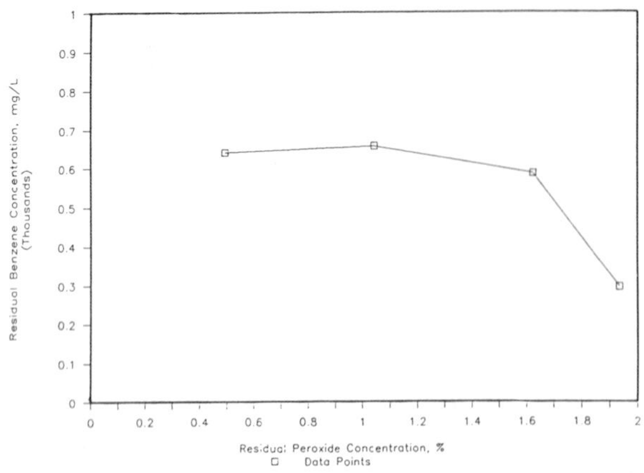

Figure 8. Effect of residual peroxide concentration on the residual benzene concentration in solution.

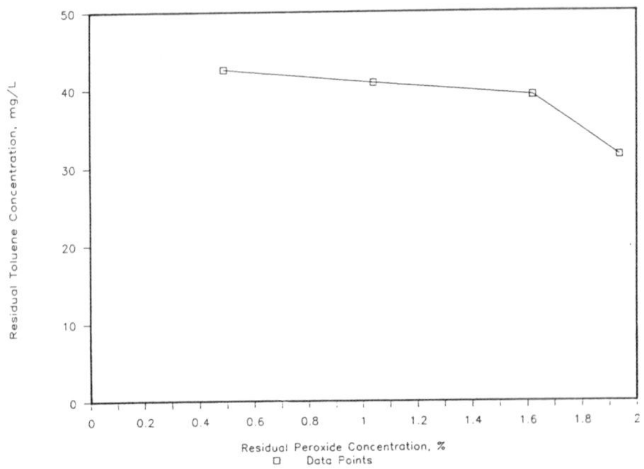

Figure 9. Effect of residual peroxide concentration on the residual toluene concentration in solution.

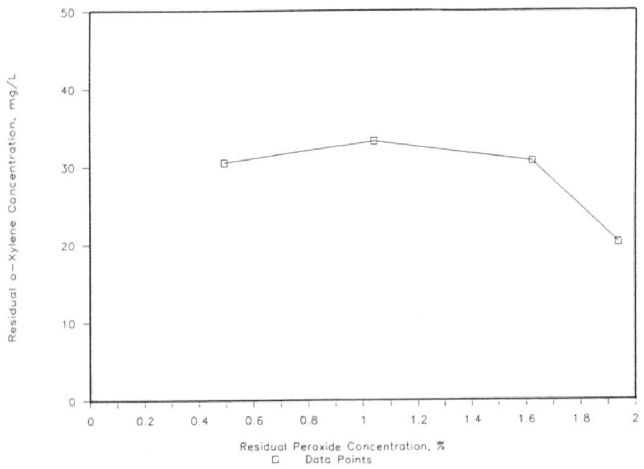

Figure 10. Effect of residual peroxide concentration on the residual o-xylene concentration in solution.

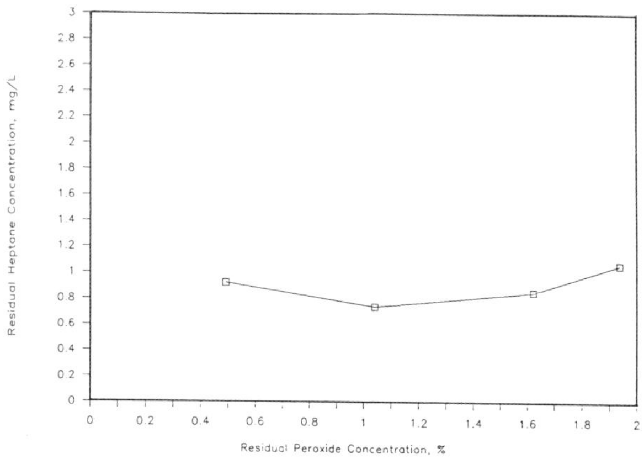

Figure 11. Effect of residual peroxide concentration on the residual n-heptane concentration in solution.

Table II. Results for the Ultraviolet Irradiation Portion of the Study.

Organic Contaminant	UV Wattage	Concentration Difference (Contam-Blank) (mg/L) Treatment Time, (hr) 1	3	8	24
Benzene	16	34.1	–	271.2	27.0
	32	157.1	164.8	48.9	9.4
Toluene	16	28.8	–	24.6	14.8
	32	8.9	20.8	5.6	1.8
o-Xylene	16	9.8	–	10.3	0.9
	32	2.7	1.9	2.5	3.3
n-Heptane	16	0.3	–	0.3	0.5
	32	0.1	0.4	0.4	0.05

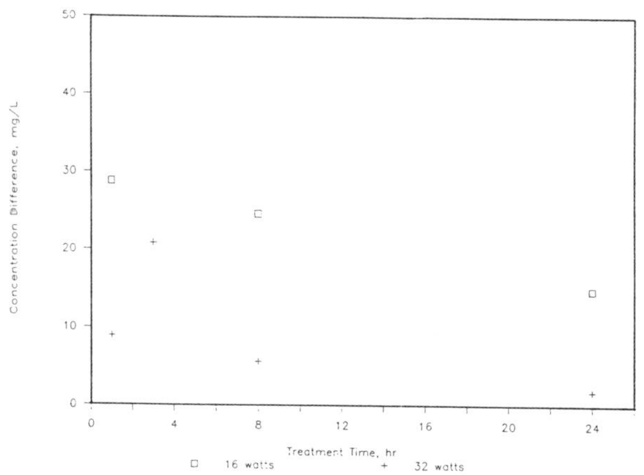

Figure 12. Effect of treatment time and ultraviolet wattage on the residual toluene concentration in solution.

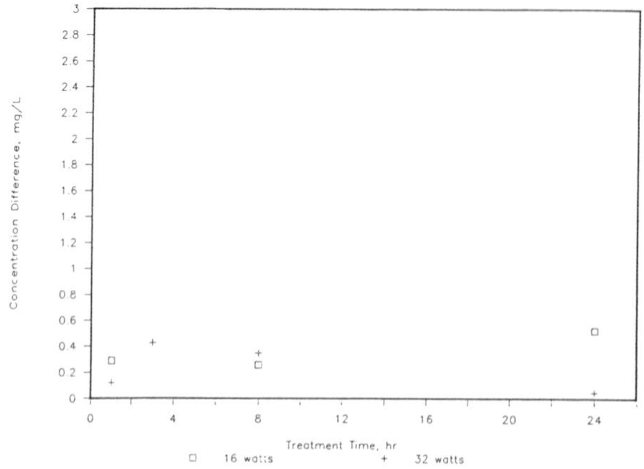

Figure 13. Effect of treatment time and ultraviolet wattage on the residual n-heptane concentration in solution.

Ultraviolet Irradiation Pretreatment

Table II presents the results from the UV irradiation pretreatment. Blanks (controls) were run in which contaminants were added to the solution phase (in the absence of soil). The results presented in Table II represent the concentration difference for UV irradiation of the contaminated soil minus the blank, for 16 and 32 watts. The variability is larger for this pretreatment system than for the hydrogen peroxide pretreatment system. The scatter of the data in this system is larger than for the H_2O_2 system.

Figures 12 and 13 show the concentration difference results as a function of treatment time for toluene and n-heptane, respectively. The concentration of toluene in solution slowly decreases with increasing treatment time, whereas the concentration of n-heptane in solution remains relatively constant over the entire treatment period.

Pretreatment Evaluation

Both hydrogen peroxide and UV irradiation were investigated as a possible oxidation-reduction pretreatment system prior to bioremediation of the contaminated soils. UV irradiation alone did not show significant contaminant removal capability, whereas hydrogen peroxide pretreatment results in excess of 80% of the contaminants being removed from the soil. The use of a combined UV/H_2O_2 pretreatment system may result in even better contaminant removals. This approach will be explored in the future.

As stated earlier in this paper, the use of the REDOX pretreatment technique was planned to be an integral part of the remediation strategy, performed in conjunction with bioremediation. As an integral part of the project team, researchers from the University of Notre Dame performed the relevant biotreatment studies. Due to the fractural nature of the contaminated soil, in situ bioremediation did not seem to be a viable option. However, biological degradation of the contaminated soil

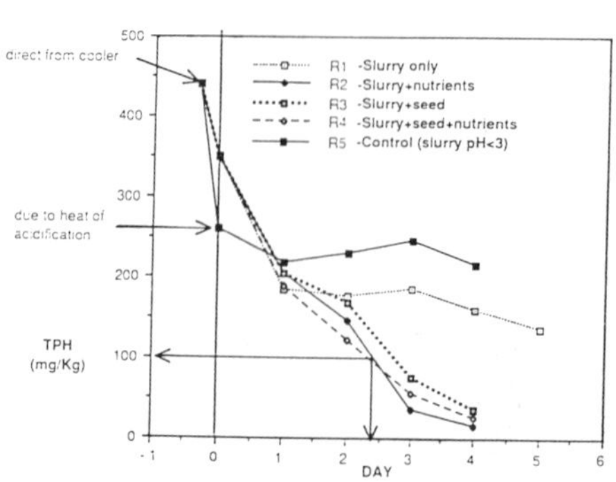

Figure 14. Biological degradation results for treatment of the contaminated soil.

using a slurry reactor system showed very favorable results, as shown in Figure 14. This figure presents the total petroleum hydrocarbon (TPH) concentration of the soil as a function of the biotreatment time. The goal of 100 mg/kg TPH could be met within about 2 1/2 days treatment time using the slurry reactor system with acclimated microorganisms and nutrients. Using the slurry reactor system in conjunction with either nutrients or acclimated micro-organisms results in the 100 mg/kg limit being met within about 3 days. Because of these very favorable treatment times, Phase II of this study investigates the pilot-scale demonstration using only the on-site slurry reactor system; the REDOX pretreatment system was not required for Phase II of this project.

SUMMARY AND CONCLUSIONS

Two oxidation-reduction pretreatment systems (hydrogen peroxide and ultraviolet irradiation) were investigated for their potential as a means of "softening" refractory organic compounds. Residual peroxide concentrations decreased with increasing treatment time in an exponential fashion and is a function of the initial peroxide concentration.

Pretreatment with peroxide for 1.0 hour using a 10% soil slurry treated with a 1.0% hydrogen peroxide solution resulted in the following removal efficiencies for the various organic contaminants studied:

Contaminant	Removal Efficiency, %
Benzene	88.2
Toluene	83.9
o-Xylene	77.7
n-Heptane	85.5
Overall	83.8

Peroxide pretreatment enhanced desorption of the organics from the contaminated soil resulting in nearly 80% of the contaminants being transformed. Pretreatment with hydrogen peroxide slightly enhanced the desorption of organics from the contaminated soil while pretreatment with ultraviolet irradiation resulted in minimal enhancement for removal of the organic contaminants.

Due to the very favorable results obtained from the bioremediation studies employing a slurry reactor system, the REDOX pretreatment system will not be employed in Phase II of this study. Laboratory studies indicated that the 100 mg/kg TPH criteria could be met within about 2.5 to 3 days with the slurry reactor system with acclimated microorganisms and small amounts of nutrient addition.

ACKNOWLEDGMENT

The authors want to acknowledge the research support of the United States Air Force in performing this research.

REFERENCES

1. Kokoszka, L.C., and J.W. Flood, *Environmental Management Handbook: Toxic Chemical Materials and Wastes*, Marcel Dekker, Inc., New York, NY (1989).
2. Calabrese, E.J., and P.T. Kostecki, *Petroleum Contaminated Soil, Volume 2: Remediation Technologies, Environmental Fate, Risk Assessment, Analytical Methodologies*, Lewis Publishers, Inc., Chelsea, MI (1989).
3. Tedder, D.W., and F.G. Pohland, *Emerging Technologies in Hazardous Waste Management*, ACS Symposium Series No. 422, American Chemical Society, Washington, D.C. (1990).
4. Wentz, C.A., *Hazardous Waste Management*, McGraw-Hill, Inc., New York, NY (1989).
5. Ram, N.M., R.F. Christman, and K.P. Cantor, *Significance and Treatment of Volatile Organic Compounds in Water Supplies*, Lewis Publishers, Inc., Chelsea, MI (1990).
6. Roy, K.A., "Interox America: Hydrogen Peroxide Producer Ready for Expanding Environmental Applications", *HAZMAT World*, 26–28, 30–31 (March 1990).
7. Sundstrom, D.W., B.A. Weir, and H.E. Klei, "Destruction of Aromatic Pollutants by UV Light Catalyzed Oxidation with Hydrogen Peroxide," *Environ. Prog.*, *8*(1): 6–11 (1989).
8. Borup, M.B., and E.J. Middlebrooks, "Photocatalyzed Oxidation of Organic Wastestreams," *Proc. 18th Mid-Atlantic Indus. Waste Conf.*, *18*: 554–568 (1986).
9. Konk, R., "Hydrogen Peroxide Applications," *HAZMAT World*, 28–29 (March 1990).
10. Fletcher, D.B., "UV/Ozone Process Treats Toxics," *Waterworld News*, *3*(3) (1987).
11. Sundstrom, D.W., H.E. Klei, T.A. Nalette, D.J. Reidy, and B.A. Weir, "Destruction of Halo-

genated Aliphatics by Ultraviolet Catalyzed Oxidation with Hydrogen Peroxide," *Haz. Waste & Haz. Mater.*, *3*(1): 101–110 (1986).

12. Yost, K.W., "Ultraviolet Peroxidation: An Alternative Treatment for Organic Contamination Destruction in Aqueous Waste Streams," *Proc. 43rd Purdue Indus. Waste Conf.*, *43*: 441–447 (1988).
13. Weir, B.A., D.W. Sundstrom, and H.E. Klei, "Destruction of Benzene by Ultraviolet Light-Catalyzed Oxidation with Hydrogen Peroxide," *Haz. Waste & Haz. Mater.*, *4*(2): 165–176 (1987).
14. Chang, I.L., and N.S. Zaleiko, "Ultraviolet Oxidative Degradation of Protein-Containing Wastewater: Tannery Unhairing Waste," *Proc. 36th Purdue Indus. Waste Conf.*, *36*: 814–820 (1981).
15. Zeff, J.D., E. Leitis, and J.A. Harris, "Chemistry and Application of Ozone and Ultraviolet Light for Water Reuse -Pilot Plant Demonstration," *Proc. 38th Purdue Indus. Waste Conf.*, *38*: 105–116 (1983).
16. Whitlow, J.E., and J.A. Roth, "Heterogeneous Ozonation Kinetics of Pollutants in Wastewater," *Environ. Prog.*, *7*(1): 52–57 (1988).
17. Boeve, L., "Removing Petroleum Products from Soils with Ozone, Ultraviolet, Ultrasonics, and Ultrapure Water," pp. 279–282 in *Petroleum Contaminated Soils, Volume 2: Remediation Techniques, Environmental Fate, Risk Assessment, Analytical Methodologies*, E.J. Calabrese and P.T. Kostecki, Ed., Lewis Publishers, Inc., Chelsea, MI (1989).
18. FMC Corporation, *The Analysis of Hydrogen Peroxide Solutions*, Technical Bulletin No. 59 Philadelphia, PA.

8 PERMEABLE BARRIERS: A NEW ALTERNATIVE FOR TREATMENT OF CONTAMINATED GROUNDWATER

Bruce M. Thomson
Stephen P. Shelton
Department of Civil Engineering
University of New Mexico
Albuquerque, NM 87131

Edward Smith
U.S. Army Construction Engineering Research Laboratory
Urbana, IL 61820

INTRODUCTION

Restoration of aquifier systems which have become contaminated with toxic or hazardous compounds is one of the most challenging problems currently facing environmental engineers and scientists. Conventional aquifer restoration alternatives can be broken into four categories:

1. Contaminant of the aquifier contaminants.
2. Removal of mobile pollutants, followed by surface treatment of contaminated water or recovery of free product, and subsequent disposal or reuse of treated water.
3. Removal of contaminated soil, followed by treatment and/or disposal.
4. In situ stabilization of aquifer contaminants.

Frequently a combination of these methods is used to maximize the performance of the treatment process. These alternatives are briefly described below.

Contaminant of groundwater pollutants is the most immediate concern following determination of a contamination problem. It can be achieved either through the use of a physical barrier such as a grout curtain, slurry cutoff wall or sheet piles, or by creation of a hydraulic barrier resulting from a combination of pumping and injection wells. Containment technology is reviewed by Spooner, et al.[1], and Kelly.[2]

Removal of mobile contaminants from the aquifer system is achieved by directing pollutant migration toward wells or trenches from which it can be recovered or removed. Variations of this process include free product recovery of petroleum products floating on a water table, vacuum extraction of volatile organic compounds, and pump-and-treat processes for soluble constituents. These processes are limited to volatile, liquid, or soluble pollutants; insoluble compounds remain attached to soil particles in the aquifer or vadose zone. Bouwer, et al [3], Wagner, et al.[4], and Guswa et al.[5] have prepared reviews of processes for management and treatment of groundwater contamination problems involving mobile contaminants.

In situ stabilization of hazardous wastes is a relatively new treatment process which shows considerable promise as an alternative to conventional pump and treat methods. Its principal application to date has been in groundwater systems which are contaminated with biodegradable organic materials. In situ treatment is accomplished by stimulating the growth of naturally occurring soil microorganisms by introducing essential nutrients (e.g. nitrogen and phosphorous) and appropriate electron acceptors (e.g. oxygen or hydrogen perioxide) required for the organism's growth. Lee, et al.[6], Wilson, et al.[7], Borden, et al.[8], and Amdurer, et al.[9] have prepared reports and reviews of in situ biological waste stabilization technology, while Sims, et al.[10] discussed in place remediation of contaminated soils. In situ technology is limited to applications in which the contaminants are degradable.

Although conventional groundwater restoration problems, when properly designed and implemented, are generally effective, they have numerous problems associated which include:

45th Purdue Industrial Waste Conference Proceedings, © 1991 Lewis Publishers, Inc., Chelsea, Michigan 48118. Printed in U.S.A.

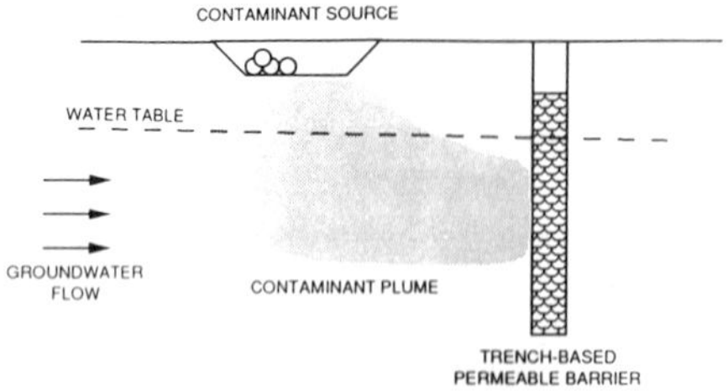

Figure 1. Diagram of trench-based permeable barrier concept.

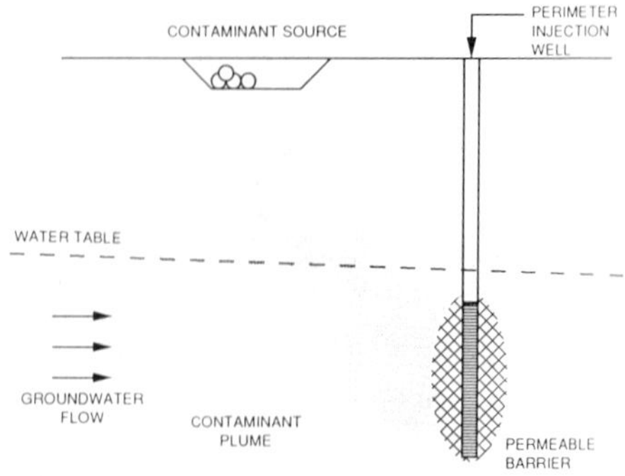

Figure 2. Diagram of well-based permeable barrier concept.

1. They require management of large volumes of water which generally are contaminated at very low levels.
2. There frequently produce undesirable by-products (i.e. sludges) associated with the treatment process which are difficult to manage.
3. Conventional pump and treat alternatives require large surface disruptions for long time periods.
4. Conventional alternatives may affect hydraulic characteristics in uncontaminated parts of the aquifer.
5. Methods which remove mobile contaminants (i.e. vapors, free liquids, or those which are soluble) may not work in aquifers with low hydraulic conductivity.
6. Surface disposal of large volumes of treated groundwater may pose institutional obstacles such as requirements for ground or surface water discharge permits and possible purchase of groundwater rights.

An alternative to conventional groundwater treatment processes is the use of barriers which are permeable to water, but prevent the migration of pollutants. They are referred to here as permeable barriers. This paper examines the hypothesis that it is possible to design permeable barriers which will stabilize, remove or degrade groundwater contaminants. Much of the effort in development of this concept is focused on adapting proven surface treatment technologies for application in an aquifer environment. These technologies include air stripping, adsorption and ion exchange, chemical precipitation, and biological stabilization. The following sections describe the permeable barrier concept more completely, and summarize recent experimental and theoretical work leading towards application of the technology.

PERMEABLE BARRIER CONCEPTS

As in hydraulic containment barriers, it is suggested that there are two general classes of permeable barriers based on their application to: 1) shallow aquifer systems (< 20 m) which are accessible by trenching equipment, and 2) deep aquifer systems which are only accessible by wells. Schematic diagrams of trench-based and well-based barriers are presented in Figures 1 and 2. This distinction is important for two reasons. First, permeable barriers constructed by trenching can make use of particulate materials which cannot be installed by injection wells. In addition, greater technical flexibility is available for solute delivery or solvent recovery systems in a trench-based barrier in that chemical delivery systems and barrier monitoring devices can be easily installed at the time of construction.

A second advantage of a trench-based system is that recovery of the barrier material is readily accomplished by re-excavation of the trench as the media becomes saturated with hazardous constituents. This is especially important when dealing with groundwaters polluted with non-degradable contaminants. A list of possible barrier materials for trench-based barriers is contained in Table 1.

Well-based barriers rely upon the ability to alter the interactions between the aquifer media and soluble contaminants in the groundwater. These changes are induced through injection and subsequent dispersion of soluble constituents into the aquifer strata. Thomson and Shelton (1987) provide a description of this concept. It is not believed to be practical with present technologies due to the lack

Table I. Possible Barrier Materials for Trench-Based Barriers

Degradable Contaminants	Non-Degradable Contaminants
1. Nutrient addition systems	1. Sorption/exchange media
2. Oxidizing agents	2. Precipitation systems
3. Aeration/deareation systems	3. Air stripping systems
	4. Chemical addition systems

of understanding and ability to control dispersion in geologic media resulting in uncertainties in subsurface chemical delivery systems needed to produce such a barrier.

A permeable barrier may be classified as active, or passive. A passive system would be one in which there is little or no operation or maintenance required after the barrier is installed, except for monitoring and possible replacement if the media looses its treatment effectiveness. Examples of a passive system might include neutralization of acidic groundwater by carbonate materials or adsorption of organic pollutants onto sorbing media such as peat. In contrast, an active system will have continuous operation and maintenance requirements, such as aeration or chemical feed.

There is a large body of literature which has been developed regarding both barrier technologies and in situ pollutant stabilization methods (see previously cited references). However, most of this material deals with organic compounds which are generally either biodegradable or volatile and therefore amenable to air stripping. In contrast there is very little information published on permeable barrier concepts, and those studies which have been done primarily address removal of inorganic contaminants from solution through chemical neutralization, precipitation and adsorption.

Two of the most directly related studies are by Fuller[11] and Park[12]. Fuller investigated the use of limestone, clays, iron oxides, and pecan hulls to reduce the migration of Be, Cd, Cr, Fe, Ni, Zn and total organic carbon (TOC) in leachate from municipal solid waste landfills. Crushed limestone was effective at increasing metal rention on soils, but had little effect on TOC. Park examined the use of coal, soil, and fly ash as possible sorptive media. Fly ash and limestone were not very effective at removing contaminants, but were useful in preventing channeling through the media and in neutralizing acidic solutions.

EXAMPLES

This paper will present and discuss two examples of applications of the permeable barrier concept based on previous fundamental work by the authors. The discussion is limited to the trench-based barrier alternative. Each of these examples is based on a hypothetical situation in which a barrier is placed down-gradient from a contaminated zone. The two barriers considered in this analysis are: 1) an active trench based barrier using bubble aeration to strip volatile organic contaminants from groundwater, and 2) a passive barrier in which geochemical modifiers are utilized to achieve immobilization of metals associated with leachates from uranium mill tailings piles.

Trench Based Air Stripping

Ring compounds (benzene, toluene and xylene isomers) are frequent contaminants from spills of petroleum products (Kreamer and Stetzenbach[13]). Conventional remediation procedures are either air stripping or adsorption of these pollutants onto granular activated carbon, or possibly in situ biodegradation as these compounds are readily degradable in aerobic systems (Bouwer, et al.[3]). An alternative approach would be to construct a permeable barrier consisting of a perforated pipe placed in the bottom of the trench which is then backfilled with an inert, porous media such as coarse gravel (Figure 3). Air would be bubbled up through the gravel so that the trench would function as an in situ air stripper.

Conventional stripping tower analysis is not applicable to this configuration as the mas transfer characteristics are significantly different. Instead an estimation of the stripping performance can be developed based on the mass transfer relationships for gas bubbles rising through a liquid solution. This analysis begins with the work by Matter-Müller, et al.[14] for the mass transfer rate of a volatile compound stripped from solution in an aeration system:

$$F = Q_G \, H \, C_L \, [1 - \exp(- \frac{K_L a}{V_L})] \quad (1)$$

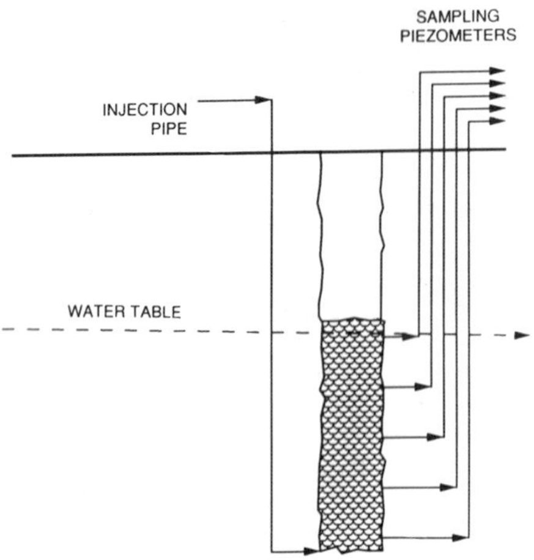

Figure 3. Schematic diagram of an active trench-based barrier.

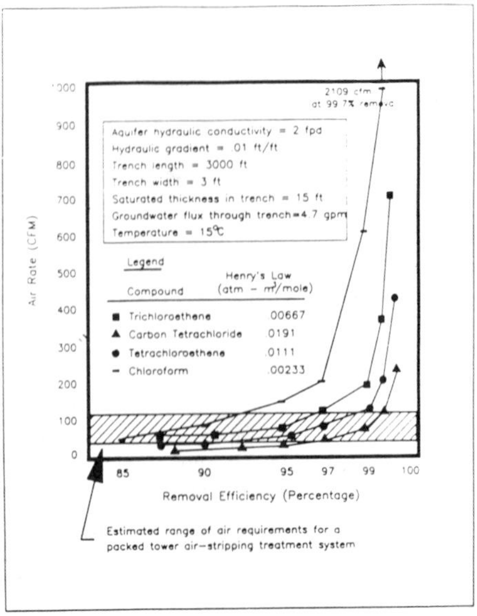

Figure 4. Removal efficiency as a function of air flow rate to an aeration trench for four groundwater contaminants at 15°C.

where
F = mass transfer rate (g/sec)
C_L = Bulk liquid solute concentration
K_La = Mass transfer coefficient (cm^2/sec)
Q_G, Q_L = Volumetric gas and liquid flow rates respectively (m^3/sec)
V_L = Volume of liquid in control volume (m^3)

Plugging this expression into a steady state mass balance over a stripping system gives an equation for the ratio of influent and effluent solute concentrations (C_{Li} and C_L):

$$\frac{C_L}{C_{Li}} = [1 + \frac{H\,Q_G}{Q_L}[1 - \exp(-\frac{K_La\,V_L}{H\,Q_G})]^{-1} \qquad (2)$$

The mass transfer coefficient (K_L) can be estimated using the following equation developed by Akita and Yoshida[15]:

$$K_L = 0.5\, g^{5/8}\, D_a^{1/2}\, \rho^{3/8}\, \sigma^{-3/8}\, d_b^{1/2} \qquad (3)$$

To calculate the bubble surface area, the bubble holdup (ε) is needed:

$$\frac{\epsilon}{(1-\epsilon)^4} = 0.0268\, v^{1/6}\, U_G \left(\frac{\sigma}{\rho L}\right)^{-1/8} \qquad (4)$$

The interfacial bubble surface area (a) is then determined with the following equation:

$$a = \frac{6\,\epsilon}{d_b} \qquad (5)$$

where
d_b = Bubble diameter (cm)
D_a = Bulk diffusivity of the gas in water (cm^2/sec)
V_L = Kinematic viscosity of the liquid (cm^2/sec)
ρ_L = Fluid density (g/cm^3)
U_G = Superficial gas velocity (cm/sec)
σ = Surface tension (gm/sec)

Table II. Aquifer Conditions for Comparison of Trench Based Air Stripping and Packed Tower Air Stripping.

Trench Length = 1,000 m
Trench Width = 1 m
Hydraulic Conductivity (K) = .6 m/d
Hydraulic Gradient ($^{dh}/_{dx}$) = .01
Contaminants & Respective Henry's Law Constants at 15 °C (atm-m^3/mole):
Trichloroethene, H = .00667
Carbon Tetrachloride, H = .0191
Tetrachloroethene, H = .0111
Chloroform, H = .00233

Lowy and Thomson[16] describe a solution of these equations which allows rapid sensitivity analysis of process variables. One of the most obvious design parameters is the bubble diameter. In a trench based system this is effectively determined by the size of the interstitial areas in the backfilled media which is determined by the diameter of the media, as well as its angularity (or sphericity) and packing. Maximum bubble diameter will equal the maximum diameter of the internal pores; actual bubble diameters will be considerably smaller as large bubbles break into smaller ones as they rise through narrow necks within the media. For this analysis, a bubble diameter of 0.2 cm is assumed in water at 20°C. The combined mass transfer coefficient $K_La = 1.22 \times 10^{-3}$ cm^2/sec. In this size range K_La can be shown to be relatively insensitive to bubble diameter.

The major factor which affects the performance of a barrier stripping system is the Henry's law constant (H) which is a measure of the solubility of the constituent and ranges over many orders of magnitude for different pollutants. The variation is best illustrated by calculating the gas-to-liquid volumetric flow ratio (Q_G/Q_L), which is one of the most important design parameters in conventional stripping tower design. Equation 2 can be used to determine Q_G/Q_L for different pollutants.

A comparison of trench based air stripping with conventional packed tower air stripping performance is possible using this stripping theory. A hypothetical system is analyzed for the conditions shown in Table II. The results are presented in Figure 4 as a plot of air flow rate versus percent removal (Lowy and Thomson[16]).

Other factors of relevance in comparing the technical merits of tower and trench-based air stripping are the equipment requirements. Although a trench-based stripping scheme would require comparatively high pressure air compressors (corresponding to a pump head of approximately 8 m in the present example compared to a few cm in a tower stripper), there are no water pumping requirements. Furthermore, equipment needed at the surface with this alternative is limited to an air supply and distribution system. Even at a Q_G/Q_L of 650 the total mass of material (air and water) pumped in a trench-based system is less than in a tower-based system; therefore, operating costs should be less.

Another factor to be considered in trench-based air stripping is biological degradation or organic contaminants. Literature reviews in in situ treatment of contaminated aquifers find that natural soil microorganisms can destroy many organic compounds through biological degradation (Lee, et al.[6]; Thomas, et al.[17]). Biostimulation is achieved by adding limiting nutrients (oxygen, nitrogen and phosphorous) to the contaminated zone, and that oxygen is the most difficult to supply due to its low solubility. In a trench-based air stripping application oxygen is continually supplied to the system, thus it is probable that an air stripping barrier will result in combined air stripping and biodegradation of organic pollutants. Biodegradation may be further enhanced by addition of nutrients through the air distribution system.

Geochemical Barriers for Immobilizing Metals from Uranium Mill Tailings

In contrast to most organic pollutants, inorganic contaminants cannot be destroyed; for example there is no process, short of nuclear transmutation, which can eliminate the toxicity of lead or mercury. Conventional remediation alternatives for aquifers contaminated with inorganic pollutants are therefore limited to those which either remove them from the groundwater system (e.g., soil washing), or permanently immobilize them in place (e.g. in situ vitrification). The trench-based permeable barrier concept suggests a third alternative consisting of concentration and temporary immobilization of the contaminants, followed by excavation and proper disposal at a future time. A strong candidate for application of this technology in the remediation of tailings piles resulting from processing of uranium ores.

Table III. New Mexico Mill Tailings Pond Raffinate Quality (concentrations in mg/L unless noted - Thomson, et al.[20])

	4 Acid Leach Mills			1 Alkaline Leach Mill		
Constituent	Min.	Median	Max.	Min.	Median	Max.
As	.18	1.3	5.6	2.1	5.0	7.2
Mo	0.2	0.9	29.5	72.	98.	105.
Se	.006	.21	6.97	22.1	29.5	51.2
SO4	300	29,700	56,000	5,500	8,400	16,700
U	1.1	15.	69.	4.2	53.	70.
V	39.	74.	107.	1.2	14.	16.
NH3 (as N)	3.3	400.	3,960.	1.1	16.	335
TDS	17,900	39,800	72,801	17,000	25,400	39,700
pH	0.3	1.05	2.15	9.9	10.1	10.3
Gross α (pCi/L)	3,200	38,000	73,000	3,400	6,700	10,000
Ra-226 (pCi/L)	15.	70.	1,800.	56.	58.	90.

Uranium milling is accomplished by grinding the host rock, dissolving uraniferous minerals in either acidic or alkaline solutions, and finally recovering U from solution by ion exchange or solvent extraction processes. The leaching process also result in dissolution of numerous other metals thus producing a raffinate characterized by high concentrations of total dissolved solids (TDS), metals, and metalloids, and high or low pH (Table III). After U is recovered from solution, the spent tailings and raffinate solutions are disposed of in large tailings piles which do not have physical barriers to limit leachate migration, thus groundwater contamination is common.

Most U milling in the U.S. has been done by the acid leach process. Evaluation of the aqueous geochemistry of the acidic leachate solutions suggests that simple neutralization would result in precipitation of many of the contaminants present (Longmire, et al.[18]). Furthermore, the presence of anionic metalloids (As and Se) suggests that adsorption onto an organic sorbent might achieve their removal. Accordingly a batch neutralization and adsorption study was conducted of geochemical modifiers to determine their effectiveness for application in a barrier system (Thomson[19]). The modifiers selected were hydrated lime ($Ca(OH)_2$), limestone ($CaCO_3$) and sphagnum moss peat.

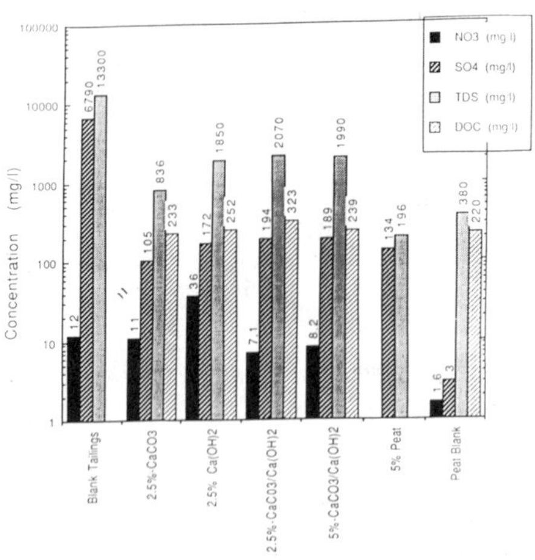

Figure 5. Major ions, total dissolved solids (TDS), and dissolved organic carbon (DOC) concentrations of leachate solutions from uranium mill tailings with 5% peat and various geochemical modifiers (Thomson, 1989).

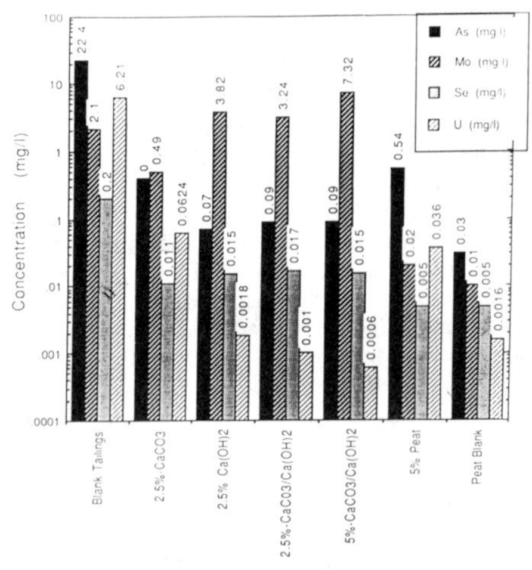

Figure 6. Arsenic (As), molybdenum (Mo), selenium (se), and uranium (U) concentrations of leachate solutions from uranium mill tailings with 5% peat and various geochemical modifiers (Thomson, 1989).

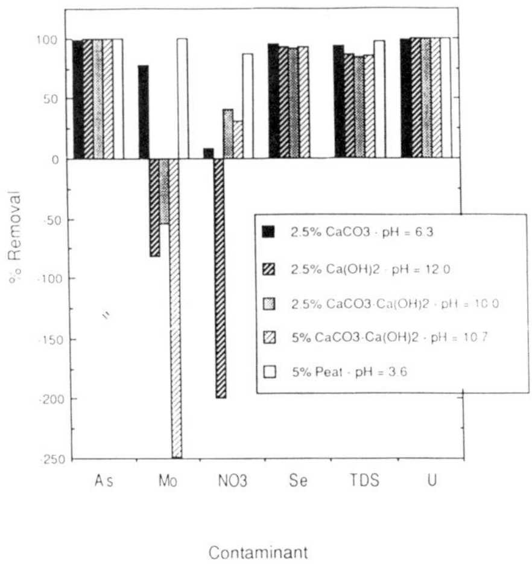

Figure 7. Removal of soluble contaminants from uranium mill tailings leachates mixed with alternative geochemical modifiers (Thomson, 1989).

Briefly, the study consisted of mixing various concentrations of the modifiers with tailings from an acid leach mill, allowing the solutions to equilibrate, and comparing the solution characteristics with those of a blank prepared from tailings without any modifiers present. The results are summarized in Figures 5–7 for one set of analyses in which modifiers were added at 2.5% of the mass of the dry tailings. Note the excellent removal for all pollutants including TDS, except Mo and NO_3^- in Figure 7.

Nitrate does not have an insoluble solid phase and can not be removed from solution by precipitation (Longmire, et al.[18]). Similarly, Mo is predicted to be present in solution as molybdate ($HMoO_4^-$) at high pH values. The negative removal (i.e. increase in concentration) following addition of lime for both NO_3^- and Mo is believed to be due to dissolution of salts formed through evaporation of tailings deposits. Of particular interest is removal of TDS, Mo, NO_3^-, and SO_4^{2-} by peat alone (Figures 5 and 7). This removal occurs at pH of less than four. Peat's effectiveness is believed to be due to adsorption and/or ion exchange of the oxyanions $HMoO_4^-$, NO_3^-, and SO_4^{2-} by protonated functional groups on the surface of the peat. Reduction of TDS is achieved principally by removal of SO_4^{2-} which constitutes a major function of the TDS in tailings raffinates.

Application of these modifiers as a trench-based permeable barrier at a mill tailings pile appears to be relatively straightforward. A trench, located down gradient from the tailings pile and backfilled with low grade limestone and peat, would provide immobilization of nearly all soluble pollutants associated with leachates. Note that these constituents are not destroyed, thus it is likely that a low level of effort monitoring program would be required to assure that they do not become re-mobilized once the acid neutralizing capacity of the limestone had been exhausted. It is envisioned that the barrier would be temporary to contain and concentrate the contaminants. Upon exhaustion it would be re-excavated and the inorganic pollutants, now in a less mobile and more concentrated form, would be disposed of as a hazardous waste.

CONCLUSIONS

This paper describes the permeable barrier concept for remediation of contaminated groundwaters. Potential advantages of this concept include reduced capital, and operations and maintenance costs, improved reliability, and a reduced volume of treatment by-products. The systems may be classified as either trench-based or well-based depending upon how the barrier is to be constructed. They may be further categorized as either active or passive, depending on whether continuing operation and maintenance is required. The paper presents examples of two trench-based systems including both an active and a passive system.

The permeable barrier concept has been tested in field scale applications, however, it appears to hold significant promise for groundwater restoration. The focus of future work should be on identifying conventional water and wastewater treatment technologies which can be adapted to applications in trench-based groundwater environments. An advantage that a barrier treatment process will have over a conventional surface treatment process is that it will operate at a much lower process loading rate due to very small groundwater velocities and low contaminant concentrations. Therefore, it is expected that candidate barrier alternatives can utilize low cost natural materials to achieve treatment.

Examples presented here include use of river run gravel in a trench-based air stripper instead of plastic media, and use of limestone and peat in place of hydrated lime and activated carbon to achieve precipitation and adsorption of inorganic pollutants.

Acknowledgments

This research has been supported in part by the U.S. Army Construction Engineering Research Laboratory, Champaign, IL, and the U.S. Department of Energy through the Waste Management Education and Research Consortium, Las Cruces, NM.

REFERENCES

1. Spooner, P. A., Wetzel, R. S., Spooner, C. E., Furman, C. A., Tokarski, E. F. 1985. *Slurry Trench Construction for Pollution Migration Control*, Noyes Pubs., Park Ridge, NJ, 237 p.
2. Kelly, J. F. 1984. "Optimizing pumping strategies for contaminant studies and remedial actions," *Ground Water Mon. Rev.*, vol. 4, no. 3, pp. 63–74.
3. Bouwer, E., Mercer, J., Kavanaugh, M., DiGiano, F., 1988. "Coping with groundwater contamination," *J. Water Poll. Control Fed.*, vol. 60, no. 8, pp. 1415–1427.
4. Wagner, K., Boyer, K., Claff, R., Evans, M., Henry, S., Hodge, V., Mahmud, S., Sarno, D., Scopino, E., Spooner, P., 1987. *Remedial Action Technology for Waste Disposal Sites*, 2nd ed., Noyes Pubs., Park Ridge, NJ, 642 p.
5. Guswa, J. H., Lyman, W. J., Donigian, A. S. Jr., Lo, T. Y. R., Shanahand, E. W., 1984. *Groundwater Contamination and Emergency Response Guide*, Noyes Pubs., Park Ridge, NJ, 490 p.
6. Lee, M. D., Thomas, J. M., Bedient, P. B. 1988. "Biorestoration of aquifiers contaminated with organic compounds," *CRC Crit. Revs. in Envi. Control*, vol. 18, issue 1, pp. 29–89.
7. Wilson, J. T., et al., 1986. "In situ biorestoration as a groundwater remediation technique," *Ground Water Monit. Rev.*, vol. 6, no. 1, pp. 56–63.
8. Borden, R. C., Lee, M. D., Thomas, J. M., Bedient, P. B., Ward, C. H. 1989. "In situ measurement and numerical simulation of oxygen limited biotransformation," *Ground Water Mon. Rev.*, vol. 9, no. 1, pp. 83–91.
9. Amdurer, M., Fellman, R. T., Roetzer, J., Russ, C., 1986. Systems to accelerate in situ stabilization of waste deposits, EPA/540/2-86/002, Haz. Waste Engr. Res. Lab, Cincinnati, OH, 264 p.
10. Sims, R., Sorensen, D., Sims, J., McLean, J., Mahmood, R., Dupont, R., Jurinak, J., 1984. Review of in-place treatment techniques for contaminated surface soils, Vol. 1: Technical evaluation, EPA-540/2-84-003a, Muni. Envi. Res. Lab., Cincinnati, OH, 163 p.
11. Fuller, W. H., 1981. Liners of natural porous materials to minimize pollutant migration, EPA-600/S2-81-122, Muni. Envi. Res. Lab., Cincinnati, OH, 73 p.
12. Park, J. E., 1984. Testing and evaluation of permeable materials for removing pollutants from leachates at remedial action sites, final report no. 68-03-3183 to USEPA Haz. Waste Engr. Res. Lab., Cincinnati, OH, 45 p.
13. Kreamer, D. K., Stetzenbach, K. J., 1990. "Development of a Standard, Pure-Compound Base Gasoline Mixture for Use as a Reference in Field and Laboratory Experiments," *Ground Water Mon. Rev.*, vol. 10, no. 2, pp. 135–145.
14. Matter-Müller, C., Gujer, W., Giger, W., 1981. "Transfer of volatile substances from water to the atmosphere," *Water Research*, vol. 15, pp. 1271–1279.
15. Akita, K., Yoshida, F., 1974. "Bubble size interfacial area and liquid phase mass transfer coefficient in bubble columns," *Ind. Eng. Chem. Process Des. Devel.*, vol. 13, no. 1.
16. Lowy, R. M., Thomson, B. M., 1988. "In-Situ aeration for stripping volatile organic contaminants from shallow ground water systems," proc. of Joint ASCE-CSCE conf. on Envi. Engr., Am. Soc. of Civil Engrs., Washington, DC, pp. 335–345.
17. Thomas, J. M., Lee, M. D., Bedient, P. B., Borden, R. C., Canter, L. W., Ward, C. H., 1987. Leaking underground storage tanks: Remediation with emphasis on *in situ* biorestoration, EPA/600/S2-87/008, Robt. S. Kerr Envi. Res. Lab., Ada, OK.
18. Longmire, P. A., Brookins, D. G., Thomson, B. M. 1990. "Hydrogeochemical interactions and evolution of acidic solutions in soil," in *Chemical Modeling of Aqueous Systems II*, D.C. Melchior and R. L. Bassett, eds., ACS Symp. Series No. 416, Am. Chem. Soc., Washington, D.C., pp. 154–168.
19. Thomson, B. M. 1988. Neutralization and chemical reduction for immobilizing inorganic contaminants in uranium mill tailings, Gunnison, Colorado, final report submitted to Jacobs Engineering Group, Inc., Albuquerque, NM, 128 p.
20. Thomsom, B. M., Longmire, P. A., Gallaher, B. M., 1982. "The hydrologic environment and uranium production in New Mexico," in *Water and Energy: Technical and Policy Issues*, F. Kilpatrick and D. Matchett, eds., Am. Soc. of Civil Engrs., New York, NY, pp. 525–531.

Section Two
LAWS, REGULATIONS, AND TRAINING

9 SO YOU ARE A PRP!

James D. Edwards, Member
Burgess & Niple, Limited
Columbus, Ohio 43220

Michael S. McMahon, Partner
Benesch, Friedlander, Coplan & Aronoff
Cleveland, Ohio 44114

INTRODUCTION

Being named as a Superfund site Potentially Responsible Party (PRP) is the beginning of a difficult experience. Although participating in a Superfund cleanup is an extended, trying, and costly undertaking, there are ways to minimize the costs involved. You can best deal with the situation by understanding what it means to be a PRP, by being aware of your options, and by taking appropriate action. This paper briefly discusses the PRP experience and offers suggestions on ways to effectively deal with the situation.

COMPREHENSIVE ENVIRONMENTAL RESPONSE, COMPENSATION, AND LIABILITY ACT

Liability Scheme

Liability provisions under the Comprehensive Environmental Response, Compensation and Liability Act (CERCLA or "Superfund") are extremely broad. The following persons are liable for cleanup of any contamination:[1]

1. Current owners of contaminated property or current operators of facilities located on the property;
2. Past owners of the contaminated property or operators of a facility located on it, who were owners or operators at the time hazardous substances were disposed of on the site;
3. Persons who arranged for disposal or treatment or arranged with a transporter for transport for disposal or treatment of hazardous substances owned or possessed by them or anyone else to another location (according to the government, this means anyone who generated hazardous waste which found its way on the contaminated property); and
4. Transporters of hazardous substances who selected the site of disposal or treatment of such hazardous substances.

The type of liability which CERCLA imposes on these persons is joint and several,[2] which means that one liable party can be held responsible for the entire cleanup cost even though that party may have only contributed a small fraction of the contamination. The federal government has used this type of liability to target PRPs, leaving them with the responsibility of suing other PRPs to force them to contribute to the cleanup costs.

Liability under CERCLA is also viewed as strict liability,[3] meaning that there are few legal defenses. The amount of proof needed by the federal government to demonstrate that a party is subject to this liability is very slim. In many cases, it may simply be a pickup ticket from a hauler who frequently used the Superfund site. The government may not be required to prove that the particular waste actually ended up at the Superfund site. Once there is some connection between a party and a site, the burden falls on the party to prove where the waste actually went.

The costs for which PRPs may be liable include the costs of removal of the contamination or remedial action, whether it is conducted by the PRP, the government or others, the cost of health

45th Purdue Industrial Waste Conference Proceedings, © 1991 Lewis Publishers, Inc., Chelsea, Michigan 48118. Printed in U.S.A.

assessments and the costs for damages to natural resources.[4] It is not an exaggeration to state that, at the outset of a matter, CERCLA liability costs can be viewed as potentially unlimited.

Life of a Superfund Site from the United States Environmental Protection Agency's (U.S. EPA) Perspective

1. Site is discovered. Emergency work is conducted (often by the U.S. EPA) to stabilize the site.
2. U.S. EPA conducts preliminary identification of major PRPs. Use of Section 104(e) letters.
3. U.S. EPA completes technical evaluation in order to rank site.
4. U.S. EPA or PRPs perform Remedial Investigation/Feasibility Study (RI/FS). Use of notice letters.
5. U.S. EPA evaluates remediation alternatives and cleanup levels.
6. Record of Decision (ROD) is prepared.
7. Settlement process (Notice Letters).
8. Remediation implemented.
9. Five-year review of remediation.

WHAT TO DO IF YOU THINK YOU ARE A PRP

How to Find Out

There are several ways to find out whether you are considered a PRP for a particular site. The earlier that you learn about your involvement increases the options for minimizing the total cost. Strictly speaking, identification as PRP is made by a letter from the U.S. EPA, and which usually invites the party to conduct a cleanup or contribute to the government's cleanup. Identification as a PRP is not a binding determination of liability. It may, however, trigger certain disclosure obligations under securities laws or generally accepted accounting principles.

There are several ways to find out earlier if a party is under consideration to be a PRP. First, a party may receive a Request for Information (Section 104(e) letter) from the U.S. EPA. This is a letter similar to a set of interrogatories, which the U.S. EPA uses to locate PRPs. There are penalties for failure to respond to several questions which will be used to establish your liability. This is a critical phase in the PRP determination process, and each party should consult an environmental attorney in preparing their response.

Second, a party may receive a letter from another party, or their law firm, already involved in the site requesting voluntary cooperation, generally under threat of a lawsuit. These letters are based upon some evidence uncovered by the sender that the new party is involved in the site.

If you have not been named as a PRP at a Superfund site but suspect you might become one, you can prepare yourself. The U.S. EPA has established a list of Superfund inventory sites (CERCLIS list) which can alert you that a site where you may have contributed wastes may be considered as a potential problem. You may also want to conduct an audit of your records to verify where and how your wastes have been disposed in the past.[5] This audit may reveal sites for which you could be liable. If these searches reveal a site where your involvement is clear, it may make sense to begin immediate remedial action in order to prevent the contamination from spreading. If you never sent wastes to a site for which you are later accused, such an audit can provide the factual basis to verify your noninvolvement.

Initial Response

Once you have learned that you are or may be a PRP at a site, your first response is to make a determination of the extent of potential liability. This is only a rough estimate, but it is necessary to determine the overall strategy for approaching the matter. If the initial investigation reveals that your involvement is minimal, then your best strategy may be to monitor the matter and try to keep legal and engineering costs as small as possible. In some cases, the cost of monitoring the matter, particularly if the site involves litigation, may exceed your ultimate settlement costs.

If the initial investigation reveals that you may have a significant involvement with the site, then the strategy should be to explore methods of minimizing the total cost. These efforts involve legal or engineering expenditures up front, but they are designed to lower the total cost.

To make this preliminary assessment of liability, you should:

1. Discover from the U.S. EPA or other PRPs what is known about the site. Where is the U.S. EPA in their procedural process? Has the RI/FS been completed? Has the ROD been issued? What are the known or suspected contaminants of concern? What is the projected cleanup cost?
2. Discover who else is a PRP. Are there many or just a few? How much investigation has been done to locate others? Who are the owners, operators, and transporters who used the site, including all of their business names?
3. What is your involvement with the site, as known by others? Other PRPs are generally willing to share this information. The U.S. EPA generally requires the filing of a Freedom of Information Request and then may take months to respond. When the U.S. EPA notifies you of your potential liability for future response costs, they should include an information release detailing the names of all PRPs found during their search, the amount of wastes each contributed, and a ranking by volume of substances found at the site.[6]
4. Thoroughly review your own files and records to verify or dispute the facts you have learned. Often this means reconstructing events that occurred 20 or 30 years earlier. Be sure to confirm that the types and amounts of waste attributed to you are also correct.

From this list and your records, determine what you believe is your fair percentage of liability at the site, considering volume, toxicity, mobility, and any other pertinent hazardous characteristics. Do not neglect to interview employees, even retirees, who may have known about the site or the waste involved. A Section 104(e) letter may require such interviews. Time and effort spent at the stage is critical in determining your strategy. It is important that you know as much about the site and your involvement as early as possible so that there are no unpleasant surprises later.

Unfortunately, once you have assessed your potential liability and determined whether you are a "major" or "minor" PRP, the decision to litigate or settle is not yours alone, unless you are the sole PRP. The U.S. EPA may not be in a position to settle until the ROD is completed. Other parties may have determined that they wish to litigate. A minor PRP, willing to settle, may be dragged through litigation because there is no one with whom they can settle and cut off liability. Unless the settlement is crafted correctly, it may be worthless because the settling parties, based upon the principle of joint and several liability, may be open to a lawsuit by a nonsettling third party. Therefore, it is important at this stage to evaluate the types of settlements for which you may ultimately be eligible.

SETTLEMENT MECHANISMS

De Minimis Settlement[7]

A PRP may qualify as a *de minimis* party if it contributes relatively minimal portions of waste to a site based on the amount and toxic or other hazardous effects. *De minimis* settlements can be an effective means for PRPs to achieve resolution of their liability. By "cashing out" of their liability for past and projected response costs, *de minimis* PRPs can relieve themselves of legal fees and other transaction costs. The U.S. EPA has discretion to establish the qualifications for PRPs to be classified as *de minimis* contributors and does so on a site-specific basis. The U.S. EPA also has discretion when to entertain an offer of a *de minimis* settlement although, most often, this decision is deferred until the ROD is issued.

De minimis parties should form a coalition and present one multiparty settlement offer to the U.S. EPA. The U.S. EPA itself may also propose an offer. Any offer must at least equal the volumetric share of the total past and projected response costs. The U.S. EPA may consider granting releases from future liability if it is consistent with the public interest. However, such releases probably require the *de minimis* PRPs to pay a premium and will be subject to reopeners.

Convenants Not to Sue[8]

A covenant not to sue may be granted by the U.S. EPA as an incentive for PRPs to settle. The issuance of such a covenant protects the PRP from liability for additional cleanup costs other than that already agreed upon. However, the covenant not to sue will not extend to potential future liability. At a minimum, it is subject to reopeners which require PRPs to take future action under the following conditions: (1) when there is subsequent discovery of unknown conditions at the site; and (2) additional information indicates that the cleanup remedy no longer protects human health or the environment.

Tips For PRPs

General Tips

1. Participate on committees of PRPs with similar interests in order to share costs and maintain a flow of information.
2. Consider jointly retaining technical or legal experts.
3. Consider estabilishing a joint document repository.

Technical Tips

1. Consider conducting the RI/FS yourself.
2. Scrutinize and comment on the U.S. EPA's RI/FS.
3. Comment on the element in the ROD.
   a. Selection of cleanup level.
   b. Selection of remedial technologies.

Legal Tips

1. Evaluate the "innocent landowner" and other legal defenses.
2. Determine that all PRPs are identified and brought to the negotiating table.
3. Consider alternative dispute resolution techniques rather than litigation.

Mixed Funding[9]

The U.S. EPA may be willing to conduct and/or pay for a portion of the response actions at a Superfund site if it would promote a settlement with PRPs. Although this mixed funding arrangement conflicts with the U.S. EPA's goal of collecting the total response costs from PRPs, the U.S. EPA may do so at its discretion. The amount of fund financing is not dependent on any fair-share allocation among PRPs, but it may more likely be used to pay for orphan shares (e.g., the liability of a bankrupt company).

Carve-Outs

Often it is possible for the settlement with the U.S. EPA to "carve out" certain portions of remedial work or potential liability which will not be required of the settling parties. The U.S. EPA either assumes that work or liability, or more likely, assumes responsibility to impose that liability on nonsettling parties. The ultimate liability under such a scheme for the nonsettlors may be larger than that for which they could have settled. This is a good weapon to encourage settlement by all PRPs.

CONCLUSION – AVOIDING AN UNCONDITIONAL SURRENDER

If a party has not availed itself of opportunities to participate in the investigation or decision making process, then it may be faced with a demand by the U.S. EPA to implement or pay for a certain remedial action. If the party has effectively waived its rights throughout the process, there may be little it can do but comply. The key to avoiding this situation is to become involved (or at least knowledgeable) as early as possible in the process.

ACKNOWLEDGMENT

Mr. Edwards would like to acknowledge Mr. Terry Nishimoto's efforts in preparing this paper. Mr. Nishimoto is a graduate of Oregon State University and is working in the Burgess & Niple, Limited's Columbus, Ohio, office.

REFERENCES

1. CERCLA Section 107(a)(1)–(4).
2. 50 FR 5034 (1985).
3. Ibid
4. CERCLA Section 107(a)(4)(A)–(D).
5. Atlas, J.D., "A Neutral Party Looks at Options of Potentially Responsible Parties at a Superfund Site." Part 1, *American Paint and Coatings Journal*, March 5, 1990, page 42.
6. CERCLA Section 122(e)(1).
7. CERCLA Section 122(g), 52 FR 24,333 (1987).
8. CERCLA Section 122(f), 52 FR 28,038 (1987).
9. CERCLA Section 122(b)(1), 53 FR 8,279 (1988).

10 RCRA CORRECTIVE ACTION: STRATEGIES FOR CONTROLLING COSTS

John W. Donley, President
Donley Environmental Management, Inc.
Stafford, Virginia 22554

INTRODUCTION

The 1984 Hazardous and Solid Waste Amendments (HSWA) to the Resource Conservation and Recovery Act (RCRA) give the U.S. Environmental Protection Agency (EPA) new authority to require cleanup of environmental contamination at operating industrial facilities. The amendments include three provisions which address corrective action for releases of hazardous waste and hazardous constituents. Under section 3004(u), a RCRA permit must include requirements for corrective action for releases from solid waste management units (SWMUs). Section 3008(h) gives the EPA authority to issue corrective action enforcement orders at RCRA interim status facilities: facilities which have managed hazardous waste under RCRA but have not obtained an operating permit. Section 3004(v) extends the EPA's authority beyond the facility boundary.

The EPA is preparing to propose regulations which will implement section 3004(u).[1] In reality, these regulations will only legitimize a process which is already well defined, especially in the early phases, through various guidance documents and the EPA's internal memoranda.[2,3,4,5,6,7,8] In fact, the EPA has already conducted investigations to identify potential corrective action problems at approximately 25% of the 5700 facilities subject to the statute. However, only about 30% of these have been issued corrective action permits or orders. Because of the nature of the process, many of the remaining facilities may not be aware of their involvement in the process or what corrective action requirements will ultimately cost.

By the EPA's own estimates,[1] the average cost of corrective action is likely to be between $6.3 M and $26.9 M at each affected facility. The range in costs is less a function of the proposed cleanup standards than how successful the facility is in gaining control over the process.

THE PROCESS

The EPA has developed a four-step process to accomplish its corrective action mandate (Figure 1). It begins with a facility audit, called the RCRA Facility Assessment (RFA), designed to identify units or activities with a potential to have released hazardous wastes or hazardous constituents into the environment.[2] The RFA consists of three tasks: a Preliminary Review (PR) of file material in the possession of government agencies and the facility; a Visual Site Inspection (VSI) of the facility to verify information obtained during the file review and to observe, first-hand, waste management practices; and a Sampling Visit (SV), when the result of the PR and VSI indicate there is a need for additional data. The EPA conducts the RFA using government resources.

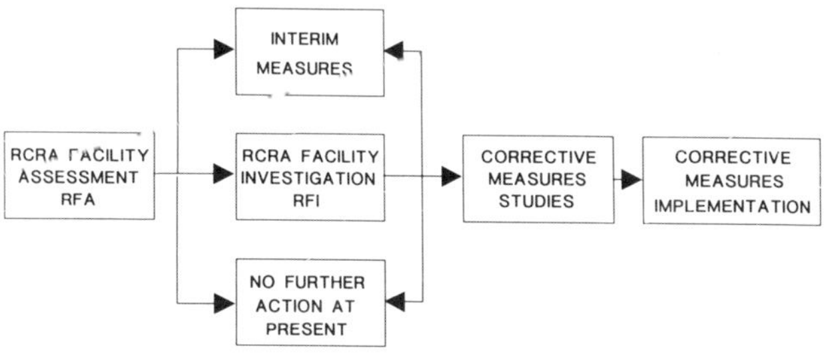

Figure 1. The corrective action process.

45th Purdue Industrial Waste Conference Proceedings, © 1991 Lewis Publishers, Inc., Chelsea, Michigan 48118. Printed in U.S.A.

If the EPA finds that a release of hazardous waste or hazardous constituents may have occurred, it will issue a permit (3004(u)) or order (3008(h)) for the facility to conduct a RCRA Facility Investigation (RFI). The RFI is similar to a Remedial Investigation (RI) under the Comprehensive Environmental Response, Compensation, and Liability Act (Superfund). It typically consists of environmental sampling and analysis to determine the nature and extent of the release.[3] All activities are conducted by the facility, subject to approval and review by the EPA.

Based on the results of the RFI, the facility will then conduct a Corrective Measures Study (CMS)[5] to evaluate and compare remedies which may be suitable to clean up the contamination. This process is similar to, but typically less formal than, a Feasibility Study (FS) under the Superfund program. At the end of the CMS, the facility will propose a remedy, or combination of remedies, for review and approval by the EPA.

Once the final remedy has been selected, the EPA will modify the facility's permit or issue a new corrective action order for the facility to implement the remedy. In theory, corrective action is complete when the facility has met certain pre-defined cleanup levels which will be prescribed in the final regulations.[1]

From the facility's perspective, the process begins with what might seem to be a routine inspection, the VSI. Once the inspection is over, the facility may not hear any more about corrective action until it receives a draft permit or order. Many facilities negotiate the permit or order as they would any other routine environmental requirement, agreeing to conduct a RFI without fully understanding the long-term implications. In so doing, they lose the best opportunity they will have for gaining control over the process and influencing the cost of corrective action. From that point, the cost and manpower requirements begin to increase almost exponentially. And the process moves so rapidly that the facility has little time to develop a long-term strategy that takes into consideration the business implications as well as the environmental requirements of corrective action.

CONTROLLING THE PROCESS

To gain control over the process and, thus, minimize the costs, the facility must first understand the program. This is no small task, as the EPA has developed several volumes of guidance on the subject, most of which remain in draft form.[2,3,4,5,6,7,8] As a result, exceptions have become the rule and the requirements are commonly subject to interpretation by the EPA, regions, states, and even individual permit writers and enforcement personnel. Obviously, a thorough treatment of this subject is beyond the scope of this paper. I will focus, instead, on the most clearly defined components of the process, as well as those which are most flexible. I encourage you to develop in-house technical and legal expertise and to become familiar with the reference sources at the end of this paper.

Before you can develop a strategy for controlling corrective action, you must know where you are in the process. If you have received a permit or order containing corrective action requirements, you probably already know. If not, you may want to take a moment to review the flow chart of the early stages of the process (Figure 2). Pay particular attention to those steps which are likely to result in an interim or final report. These documents are seldom given to the facility unless the facility asks for them.

Facilities Not Subject to Corrective Action

If you believe your facility is not yet in the process, you should determine whether it is subject to corrective action. In general, corrective action applies to any facility which has, or has had, RCRA interim status, and any facility that has applied for or received a RCRA permit to treat, store, or

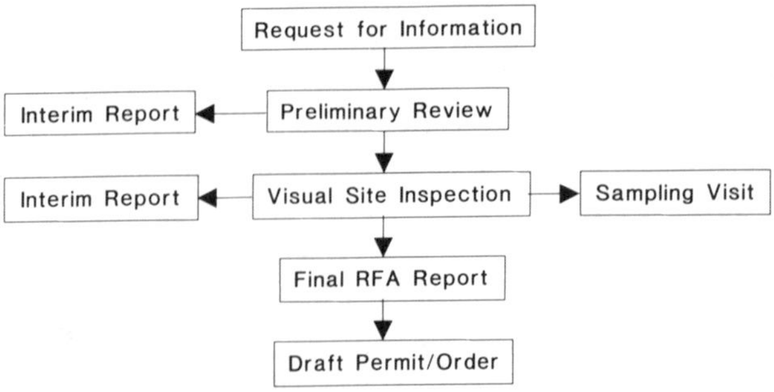

Figure 2. The early stages of the process.

dispose of hazardous waste or to close a unit used for this purpose. In other words, any facility that has, or should have, filed a RCRA "Part A" or "Part B" application is subject to the program.

Some facilities filed a "Part A" application in the early 1980s as a precautionary measure, but never operated a RCRA treatment, storage, or disposal (TSD) facility. These are known as "protective filers" and they are *not* subject to corrective action. If your facility is not presently subject to corrective action, you should try to avoid any future activities which would change this status. Even if you are subject to corrective action, you should consider any changes you can make to your current waste management program that may prevent or reduce corrective action requirements in the future.

Permits Versus Orders

If you are subject to corrective action, but the EPA has not yet conducted a RFA at the facility, you may want to consider whether to seek a RCRA operating permit. Of course, some facilities, due to the nature of their operations, have no choice but to obtain a permit. On the other hand, many facilities have gone to great expense to avoid permitting, only to find that there are important advantages to conducting corrective action under a permit, as opposed to a 3008(h) enforcement order.[6,7,8]

Perhaps the greatest advantage to permitting for most facilities is that enforcement orders, both compliance and unilateral orders, are viewed as a negative thing by the public. The public often sees permitting as a routine practice, an indication that the facility is complying with the law and subject to tight controls by the regulators. But the public views enforcement orders as evidence of the facility's failure to comply, and may be concerned that the facility's actions might have created a health hazard for the community. Failure to comply with an order may also carry a monetary penalty which can also create negative publicity.

Similarly, you should consider how easy it will be to negotiate with the EPA personnel who will oversee the program. Your permit writer may be a patient negotiator, while the regional enforcement personnel may see themselves as policemen whose job it is to catch violators. This can influence not only the reasonableness of the corrective action requirements, but how much flexibility the facility has in selecting between several options with the potential to achieve the same cleanup levels.

The most significant statutory difference between the permit authority (3004(u)) and the enforcement authority (3008(h)) is that the EPA's permit authority is limited to release from solid waste management units. I will discuss the definition of a SWMU later in the paper but, for now, it is sufficient to say that 3008(h) is a broader authority.

Of course, the EPA has the ability to issue both an order and a permit at the same facility. Most facilities seeking a permit also have, or had, RCRA interim status and are, therefore, subject to both statutes. In practice, however, the EPA usually categorizes facilities into one program or the other early in the process and this has not often been an issue. Whether you may be able to change your status for the purpose of corrective action is a question that can only be answered by your EPA regional office. In general, the deeper you are into the process, the more difficult it will be to change your status without receiving both a permit and an order.

Act Early

Once in the process, your best opportunity to gain control is *before* the EPA begins drafting a permit or order: the earlier the better. If the EPA has already prepared a draft, they will be less willing to negotiate. Once the permit or order is formally issued, your only recourse may be a formal appeal.

If you have determined that you are somewhere in the RFA process, now is the time to ask for copies of draft and final reports. These can often be obtained through an informal request of the permit writer or enforcement personnel who conducted the VSI; by asking permission to conduct a general search of the facility's state or EPA files, which typically contain copies of the reports; or through the Freedom of Information Act. The last option should be used only as a last resort, as it is seldom successful early enough in the process to be of value and may create antagonistic feelings between the EPA and the facility.

Once you receive the reports, review them carefully. Send the EPA written comments on the accuracy and completeness of the information, as well as any issues which you feel are inconsistent with the draft proposed regulations or the agency's internal guidance and policy statements. If you are unable to obtain copies of the reports, comment on your observations of the EPA's actions during the VSI and SV.

Areas of Flexibility

In your review of the agency's reports, take advantage of any flexibility in the draft proposed rule or in regional interpretations of EPA Headquarters' policies. In particular, consider the definition of a SWMU if your facility is seeking a RCRA permit and is likely to be subject to corrective action under section 3004(u). The official EPA definition, contained in the draft proposed rule, includes: "any discernible unit at which solid wastes have been placed at any time, irrespective of whether the unit was intended for the management of solid or hazardous waste."[1] In practice, regions disagree over whether certain units meet the definition. For example, some regions include product tanks in the definition, where they can be shown to have leaked. Others adopt a more conservative definition which includes only those tanks known or suspected to have contained a RCRA solid waste. Make it your business to know what types of units have been exempted in other regions or at other facilities in the same region. And remember that, to be a SWMU, a unit must have been used to manage a RCRA solid waste.

Before challenging, consider the cost of the effort required to convince the EPA of your position, against the likely cost of corrective action for that unit. If the unit is not likely to require corrective action, it is probably not worth challenging. On the other hand, if the questionable unit is the only one likely to bring the facility into the corrective action process, it will probably be worth the effort.

Consider that, according to the statute, the EPA's authority is limited to releases "at [the] facility." If the unit or release in question did not occur on contiguous property under the control of the owner or operator of the RCRA TSD facility, it may not be subject to corrective action.

The agency also has a stated policy not to use its corrective action authorities to address releases permitted by others. For example, your NPDES discharge would not ordinarily be subject to corrective action, except for constituents in the waste stream which are not regulated by the permit. In some cases, the EPA may use corrective action to address releases which resulted from non-compliance with permit conditions where the appropriate authority is unwilling or unable to take the necessary actions to correct the problem.

Similarly, by statute[9] and policy, the EPA is supposed to cooperate with other agencies, such as state environmental agencies in an attempt to avoid duplication of effort. In practice, this seldom occurs and facilities may find that they must conduct the same investigations in two different ways to please both authorities. In most cases, the facility must take the lead to get the various agencies to meet and discuss a practical approach to prevent duplication of effort.

Volunteer New Information

If you have not yet received a permit or order, but have a copy of the RFA report, consider whether you may be able to obtain additional data which will prevent needless formal investigations. Perhaps you already have information in your possession which would demonstrate to the agency that a past spill was cleaned up, or that an abandoned tank has been tested and found not to leak. Similarly, consider whether you may be able to discredit the agency's sampling data, either as a result of your observations during sampling activities or by citing poor quality control practices in the sampling plan or reports. Also, remember that a release is only subject to corrective action if it involves a RCRA hazardous waste or a hazardous constituent [see 40 CFR 261 for a list of substances that qualify].

Some facilities attempt to minimize formal corrective action activities by conducting a voluntary cleanup. That can be effective, provided the EPA has not yet conducted a RFA. My experience has been that the agency nearly always requires additional study for all problems identified in the RFA report, regardless of any actions that may have been taken by the facility subsequent to the RFA. If you still desire to undertake voluntary corrective actions, I urge you to obtain the EPA's concurrence in advance.

Prepare Work Plans

If you are certain that you will eventually receive a corrective action permit or order, now is the time to begin preparing the work plan and associated documents. If you complete them early enough, you may be able to convince the agency to simply attach your work plan to the permit or order, in lieu of developing detailed requirements. Even if this can't be done, you will find it much easier to plan the investigation at your leisure, rather than rushing to meet the tight deadlines the EPA usually sets.[6,7,8]

Table I. Case Study

	Draft Permit	Final Permit
TOTAL SWMUs	57	33
No Further Action	11	11
Unit/Waste Characterization	40	22
Ground-Water Investigations	facility-wide	none
Surface Water Investigations	6	1
Air Monitoring	16	none
Soil Sampling	1	none

Read Before Signing

Before accepting a draft permit or order, consider carefully what it is likely to cost. Ask yourself if the requirements are technically feasible and if small investigative costs might be masking substantial long-term liabilities. Consider the possible consequences of failing to accomplish the requirements of the permit or order: stipulated penalties; loss of your operating permit; or additional investigation or cleanup activities.

Seek Other Protection

If you haven't already done so, you should check to see whether you have insurance to cover corrective action costs. Some facilities may also have obtained contractual protection against pre-existing conditions when they acquired the property. In either case, you probably have important reporting obligations to qualify for protection.

After the Permit or Order

Most of my suggestions so far have dealt with steps you can take early in the process. There are two reasons for this. First, the corrective action program is still young; the regulations are still in draft form and few facilities have progressed beyond the RFI, so we have a limited experience base in the latter stages of the process. Second, the facility has less control over the latter stages, as they are more clearly defined by the permit or order.[6,7,8] The agency has been very reluctant to grant appeals, except for administrative issues.

Nevertheless, the draft proposed regulations give us some potential tools which may eventually prove useful for controlling cleanup costs. The agency has indicated it may entertain such options as:

- allowing partial cleanup, or cleanup to less stringent levels, until the owner/operator closes the facility;
- clustering SWMUs into a single "corrective action management unit" (CAMU) to avoid the complications of meeting the other regulatory requirements of RCRA for wastes generated in the cleanup process;
- exempting temporary units used in the cleanup process from the other regulatory requirements of RCRA; and
- extending the time required to complete the cleanup to allow the use of existing on-site capacity.

Other potential cost-savings measures include expediting cleanup activities to beat the regulatory deadlines in the "land ban" and performing formal "closure" of RCRA-regulated units to avoid future corrective action involvement.

CASE STUDY

Because the EPA continues to use its corrective action authorities, despite having no formal regulations upon which to base its decisions, we are seeing a great deal of variability in how this authority is interpreted, especially at the regional level. For example, I was recently involved in negotiating a corrective action permit for a facility in the midwest. In the RFA, EPA identified 57 SWMUs. The draft permit required additional study for 46 of those units (Table I), ranging from waste analysis to a comprehensive ground-water investigation which would essentially duplicate work already underway as part of a Superfund RI/FS.

The permit was appealed and we are still awaiting the final outcome. However, the EPA has tentatively agreed to remove 24 of the units from the permit on the grounds that they do not meet the definition of a SWMU. Seven others are not located within the facility boundaries. All of the units for which air monitoring was required fit into one of these categories, so we do not expect to have to conduct any air monitoring at the facility. The region has also tentatively agreed to remove the requirement for a ground-water investigation unless, or until, such time as we find evidence that a SWMU released to the ground water.

SUMMARY

The corrective action provisions of the 1984 amendments to RCRA give the EPA authority to compel cleanup of releases of hazardous wastes and hazardous constituents at RCRA facilities. These facilities must become more proactive if they are to ensure that the cost of corrective action is justified by the relative threat these releases pose to human health and the environment. This involves developing a greater understanding of the EPA guidance and policies which govern the program; taking steps to keep the facility out of corrective action where possible; monitoring the status of the corrective action process at a particular facility; becoming involved early in the process; challenging or appealing agency decisions which are not justified by the facts or consistent with the agency's own guidance; and taking advantage of any flexibility the program may allow.

REFERENCES

1. "Corrective Action for Solid Waste Management Units (SWMUs) at Hazardous Waste Management Facilities," Draft Proposed Rule, 40 CFR, Part 264, Subpart S, U.S. EPA, September 1988.
2. "Draft RCRA Facility Assessment Guidance," U.S. EPA, October 1986.
3. "Interim Final RCRA Facility Investigation Guidance," U.S. EPA, July 1989.
4. "RCRA Corrective Action Interim Measure Guidance," EPA/530-SW-88-029, U.S. EPA, June 1988.
5. "RCRA Corrective Action Plan," EPA/530-SW-88-028, U.S. EPA, June 1988.
6. "Model Permit for Hazardous Waste Management Facilities, Module XII(B), Corrective Action for Solid Waste Management Units," U.S. EPA.
7. "Model Administrative Order on Consent," 9902.5, U.S. EPA.
8. "Model Administrative Order," 9902.5, U.S. EPA.
9. Resource Conservation and Recovery Act, section 1006(b), *Federal Register* 28714, July 15, 1988.

11 STORMWATER IN THE 1990's: PLANNING FOR THE EPA'S NEW PERMIT REQUIREMENTS

John A. Lytle, Section Director, Environmental Services
Terry M. Nishimoto, Chemical Engineer
Daniel R. Johnson, Civil Engineer
Burgess & Niple, Limited
Columbus, Ohio 43220

INTRODUCTION

"The Water Quality Act of 1987" (WQA) was enacted by Congress in 1987 amending the previously established Clean Water Act (CWA). Among other things, these amendments required municipalities and industries to begin a permitting process for their stormwater discharges. The WQA of 1987 was passed on February 4, 1987, and regulations implementing the stormwater portions of the Act were proposed December 7, 1988. These proposed regulations would add a new Section 122.26 to Title 40 of the Code of Federal Regulations (40 CFR 122.26).

The amendments to the CWA and the implementing regulations will have wide-ranging and dramatic effects on the permitting burden of industries and municipalities. It has been estimated that these regulations may affect 170 cities, 125,000 businesses, and 10,000 construction sites. This paper presents the requirements of the Acts and the regulations, and some actions which industries can take to prepare for these permitting requirements and to comply with the proposed regulations and the law.

HISTORY OF THE STORMWATER PERMITTING REQUIREMENTS

In 1972, Congress amended the Federal Water Pollution Control Act with what has come to be known as the CWA. CWA empowers and requires the Environmental Protectional Agency (EPA) to regulate the discharge of any pollutants to navigable waters from a point source. Some point sources such as industrial process wastewater and municipal sewage were easily identified as requiring immediate attention. Point sources consisting of stormwater runoff from urban and industrial areas were also soon identified as significant sources of pollutants. In fact, by 1986, the "National Water Quality Inventory, 1980 Report to Congress" concluded that pollution in runoff from agricultural and urban areas is the leading cause of water quality impairment.

The EPA began specifically addressing stormwater runoff in 1973; however, stormwater regulations have yet to be finalized. Regulations for stormwater runoff have evolved as a result of a considerable number of court cases involving industry, environmental groups, and the EPA. The rocky road leading to today's version of stormwater regulations is described below and is extracted from EPA's December 7, 1988 proposed stormwater regulations (FR 3376-8).

1972 – CWA Enacted

Congress enacted the CWA. Under the CWA, EPA was to regulate all point sources of pollutants to navigable waters.

1973 – Exemption

EPA exempted from permit requirements those conveying carrying stormwater runoff uncontaminated by industrial or commercial activity unless the particular stormwater discharge had been identified by the National Pollutant Discharge Elimination System (NPDES) Director as a significant contributor of pollution (38 FR 13530 [May 22, 1973]).

45th Purdue Industrial Waste Conference Proceedings, © 1991 Lewis Publishers, Inc., Chelsea, Michigan 48118. Printed in U.S.A.

1975 – Exemption Disallowed by Lawsuit

As a result of a suit brought by the National Resources Defense Council (NRDC), the U.S. District Court of Columbia held that EPA could not exempt discharges identified as point sources from regulations under the NPDES permit program. NRDC v. Train, 396 F. Supp. 1393 (D.D.C. 1975), aff'd, NRDC v. Costle, 568 F.2d 1369 (D.C. Cir. 1977). However, in acknowledging the administrative burden placed on the Agency by requiring individual permits, the court recognized EPA's discretion to use certain administrative devices, such as area permits, to help manage its workload. In addition, the court recognized some discretion on EPA's part to define what constitutes a point source.

1976 – Comprehensive Permitting Program

In response to the District Court's decision in NRDC v. Train, EPA issued a rule on March 18, 1976 (41 FR 11307) establishing a comprehensive permitting program for all stormwater discharges except for rural runoff uncontaminated by industrial or commercial activity. Individual permit applications were not required for municipal storm sewers at that point in time. EPA planned to study such discharges and issue "general" or area permits to such sources as these discharges were expected to be less significant than stormwater contaminated by industrial wastes.

1979-1980 – Permitting Program Revised

On June 7, 1979 and May 19, 1980, EPA published comprehensive revisions to the NPDES regulations (44 FR 32854 [June 7, 1979]; 45 FR 33290 [May 19, 1980]). Under these regulations, the same application information required of all industrial and commercial process wastewater discharges would be required of all stormwater point sources. These new requirements included testing under certain circumstances for a substantially greater number of pollutants identified in the 1977 amendments to the CWA which stressed the control of toxic pollutants.

1982 – Lawsuit and Proposed Revisions to Permitting Program

Several suits were brought to court as a result of the 1979 and 1980 revisions to the NPDES regulations. After 2 years of intensive settlement negotiations with many major trade associations, several of their member companies, NRDC, and Citizens for a Better Environment, the Agency and industry petitioners signed a settlement agreement on July 7, 1982. Under the terms of the agreement, EPA agreed to propose changes to the stormwater regulations (47 FR 52073 [November 18, 1982]). The proposal defined stormwater point sources as consisting only of conveyances of stormwater contaminated by process wastes, raw materials, toxics, hazardous pollutants, or oil and grease. Stormwater discharges were proposed to be divided into two groups based on their potential for significant pollution problems.

1982-1984 – Storm of Comments to Proposed Revisions

EPA's 1982 proposal again generated considerable reaction and comment from industrial groups and trade associations. They asserted that the proposal did not go far enough in restricting the definition of stormwater point sources. States and environmental groups contended that the proposal went too far in narrowing the scope of coverage and questioned whether EPA had a legally sufficient or technically supportable basis for the Group I/Group II designations in the proposal.

1984 – Final Stormwater Regulations

EPA considered public comments and published final stormwater regulations on September 26, 1984 (49 FR 37996). The September 26, 1984 rule defined a stormwater point source as a channelized conveyance of stormwater runoff that: (1) is located in an urbanized area as defined by the Bureau of Census; or (2) discharges from lands of facilities used for industrial or commercial activities; or (3) is designated by the Director.

The final rule set forth two categories of stormwater point sources with different application requirements for each. Group I stormwater point sources were defined as those subject to effluent limitations guidelines, located at an industrial plant or plant associated area, or designated by the Director. All other stormwater point sources were classified as Group II.

1984–1985 – Comments and Lawsuits

The 1984 final stormwater regulations generated considerable controversy and, once again, suits were filed. Many comments expressed the view that requiring full sampling from every single Group I discharger was excessive in terms of providing sufficient data for general permits. They argued that the Agency would be overwhelmed with an unmanageable amount of data that would only be outdated by the time EPA and the States were able to issue permits.

1985 (a) – Proposed Revisions

EPA published proposed changes to the stormwater regulations on March 7, 1985 (50 FR 9362). For industrial facilities, the proposed system would rely primarily on voluntary, written commitments from trade associations to submit quantitative data from selected representative Group I sources. The Agency envisioned that these data would supplement existing data available to it and could provide a basis for establishing permitting priorities and setting permit terms and conditions.

1985 (b) – Comment Period Reopened

On August 12, 1985 (50 FR 32548), EPA reopened the comment period on the March 7 proposal. Although EPA had received 29 commitment letters from trade associations (and a few individual companies) that indicated a willingness to voluntarily submit representative stormwater data from their memberships, the Agency was concerned that such letters might not provide a sufficient basis for suspending the Application Form 2C requirement for all Group I sources.

The August 12 proposal stated the Agency's view that municipal stormwater sewers designed only to convey stormwater runoff (publicly-owned separate storm sewers) are Group I stormwater point sources.

In the August 12 notice, EPA solicited public comment on the appropriateness of relying on the issuance of permits to the municipality or public entity responsible for the separate storm sewer system, thereby relieving all dischargers of stormwater into the system of the need to apply for and obtain individual NPDES permits.

1987 – WQA Enacted

On February 4, 1987, Congress passed the WQA of 1987.

THE PROPOSED PERMITTING PROGRAM

Requirement of the Law

"The Water Quality Act of 1987" was enacted on February 4, 1987, amending the CWA already in place. Among other things, the amendments added Section 402(p) to the CWA. This section specifically addressed stormwater discharges and consisted of six parts:

1. The General Rule
2. Exceptions to the Rule
3. Permit Requirements
4. Permit Application Requirements
5. Studies of other Discharges
6. Regulations of other Discharges.

The General Rule stated that no permit for stormwater discharges would be required before October 1, 1992. The exceptions to that, however, were very broad. Those exceptions were:

1. If an industry or municipality had a discharge permit for their stormwater discharges prior to February 4, 1987, that permit would remain in effect.
2. If stormwater discharges were associated with industrial activity, then a permit would be required before October 1, 1992.
3. Large municipal separate sewer systems, defined in the regulations as populations greater than 250,000, would be required to obtain stormwater discharge permits before October 1, 1992.
4. Medium municipal separate sewer systems, defined as greater than 100,000 but less than 250,000 persons, would be required to obtain permits before October 1, 1992.
5. If a stormwater discharge contributed to a violation or was a significant contributor of

Table I. Industrial Facilities Subject to Stormwater Permitting

1. SIC Codes 10 through 14 and 20 through 39
2. Facilities Subject to
   Effluent Limitation Guidelines
   New Source Performance Standards
   Toxic Pollutant Effluent Standards
3. Contractors (SIC Codes 15 and 16)
4. Hazardous Waste Treatment, Storage, or Disposal Facilities
5. Receivers of Industrial Waste on Land
6. Recyclers
7. Steam Electric Power Generators
8. Some Transportation Facilities
9. POTW Land Where Sludge or Chemicals are Applied or Stored

pollutants, it could be required to obtain a stormwater discharge permit prior to October 1, 1992.

Permit program and application requirements will be discussed in detail below. "Studies" referred to evaluations to be conducted by EPA into stormwater sources not yet required to obtain discharge permits, such as small municipalities and agricultural areas.

Permit Deadlines

The deadline date stated in the WQA for industries to submit their discharge permit applications in February 4, 1990. The proposed regulations provide several deadlines depending on category. For group industrial applications, Part 1 must be submitted 120 days after the final rule is published. Approval must occur within 60 days after submission and Part 2 must be submitted 1 year after Part 1 approval.

Individual industrial permits must be submitted 1 year after the final rule is published. Indirect discharges must submit the required data to municipalities 180 days after the final rule is published.

Large municipalities were required in the WQA to submit permit applications by February 4, 1990. However, EPA has proposed that Part 1 of the permit application be submitted 1 year after the final rule is published and Part 2 containing the analytical data be submitted 2 years after the final rule is published.

Medium municipalities are subject to a WQA deadline of February 4, 1992. EPA has proposed that Part 1 of the permit application be submitted by November 4, 1990 and approval must occur within 90 days. Part 2 must be submitted 1 year after the Part 1 application has been approved.

Municipalities smaller than 100,000 persons are not required to obtain a permit prior to October 1, 1992. EPA will perform studies to determine if regulations are necessary for stormwater discharges from small municipalities.

Industrial Permitting Requirements

The second exception to the "general rule" in the WQA requires permits for sources associated with industrial activity. Stormwater "associated with industrial activity" is defined in the regulations as point source discharge of rainwater runoff from areas that are related to manufacturing, processing, or raw material storage. Manufacturing building roofs, areas outside of those manufacturing buildings including raw material storage areas, are all areas associated with industrial activity. Not included in the requirements for permitting are parking lots, administrative buildings, employee buildings, and other locations that are not directly part of industrial manufacturing, processing, or storage. Table 1 lists industrial facilities included in the requirements for stormwater permitting under the proposed rule.

SIC Codes 10 through 14 are the mineral industry, and Codes 20 through 39 are manufacturing. Reclamation areas are not subject to permitting requirements. Contractors working on land development must comply with permitting requirements unless the area being developed is less than 1 acre for residential or 5 acres for commercial development. Receivers of industrial waste on land means landfills, land application sites, and open dumps that have received any industrial wastes. Transportation facilities with SIC Codes 40 through 45, and 47 which have vehicle maintenance shops, material

Table II. Industrial Categories with Stormwater Discharge Limitations

1. Cement Manufacturing
2. Feed Lots
3. Fertilizer Manufacturing
4. Petroleum Refining
5. Phosphate Manufacturing
6. Steam Electric Power Plants
7. Coal Mining
8. Ore Mining and Dressing

handling facilities, equipment cleaning operations, or airport deicing operations are subject to stormwater permitting requirements.

Some industrial categories have existing limitations located in the pretreatment standards for categorical industries. These are listed in Table 2.

Two types of direct discharge industrial permits are possible. The first type is the group permit. The proposed regulations provide for groups of industries with common discharges to submit a group application, reducing the overall burden of the regulations. A group application would typically be filed by a trade association or other industrial organization already formed, but could be a group of industries who form a new organization for that purpose. Generally, the industries must be in the same pretreatment category or similar in some other way. The other type of permit for an industry is an individual permit. These are submitted by industries which do not fit into an identified group or which do not wish to include themselves in an identified group.

Group Application Requirements. The group application requirements are divided into two parts. Part 1A of the NPDES permit states the participants in the group, summarizing the industrial activities of those participants, explains their similarities, describes their materials management practices, and lists the participants submitting data under Part 2.

Part 1B must be submitted by each individual participant in the group. The requirements are such things as a site map, site information including stormwater control measures, a certification that all outfalls contain only stormwater, a history of significant leaks or spills, a list of all existing NPDES permit constituents, and a statement of activities at this facility different from the group activity description.

Parts 1A and 1B will be submitted by the group and then within 18 months the group must submit Part 2 which contains the analytical data. Under the group application requirements, just 10 percent of the facilities with a minimum of 10 and a maximum of 100 facilities must submit actual sampling data. These data are submitted in Part 2. Part 2 must also contain the name of the facility sampled, storm event data, and the flow rates at the outfalls during the storm event. Once these data are received and compiled by the group, they will be submitted to the U.S. EPA.

Individual Application Requirements. The requirements for individual applications are very similar to the group applications, except each individual facility must submit its own permit application. The requirements include a site map of the facility, information including existing stormwater control measures, certification that all outfalls contain only stormwater, a history of significant leaks and spills, outfall analytical data, storm event data, and flow rates.

Industrial facilities located within a municipality and discharging stormwater to the municipal storm sewer system may be required to submit a stormwater discharge permit. If the municipality does not require an individual discharge permit, then the industrial discharger must submit to the municipality the name of the facility, location of the discharge, description of the products or services produced at the facility, any existing discharge data, and stormwater analytical results. In any case, the municipality is responsible for compliance of the discharge from its storm sewer system with its limitations.

Construction sites must submit a permit application which includes the nature of the construction activity, the area of the site, proposed stormwater control measures, and estimate of the increase of the impervious area and the runoff coefficient, the nature of the fill material, and the name of the receiving water.

Industrial Application Requirements. The regulations propose using new Form 2F, of the NPDES application. In addition to the standard administrative identification information, the most important data to be included are analytical results of stormwater sampling. Table 3 indicates the parameters which must be analyzed.

Table III. Form 2F Analytical Parameters

Effluent Guidelines Subcategory Parameters
Process NPDES Permit Parameters
Oil and Grease
Total Organic Carbon
Total Suspended Solids
Chemical Oxygen Demand
Biochemical Oxygen Demand
pH
Total Phosphorus
Total Nitrogen
Chemicals Known or Believed to be Present

As Table 3 indicates, all industries in a pretreatment subcategory must include all categorical limitation parameters in their sampling. In addition, they must include all parameters for which they test their process discharge. They must also include the list of tests given in Table 3. The sampling and analysis for these parameters will impose the greatest burden of the proposed regulations.

Sampling for the stormwater must be performed during a storm event. For most parameters the samples must be flow-weighted composite samples taken up to the first 3 hours of discharge. Temperature, pH, cyanide, total phenols, residual chlorine, oil and grease, and fecal coliform must be grab samples taken during the first 20 minutes of the discharge. All composite samples must begin with a grab sample collected in the first 20 minutes of discharge and a flow-weighted composite sample.

PLANNING FOR COMPLIANCE

Keeping in mind that the regulations are currently only proposed and will likely change as a result of the 450 comments received by the U.S. EPA, detailed compliance with the regulations would not be prudent nor possible at this time. However, several measures can be taken to ensure that a facility is in compliance as far as possible.

The objective of the regulation is to reduce pollution from stormwater runoff. Therefore, stormwater control measures may be put into place which will reduce or eliminate stormwater runoff and its associated pollution. Stormwater management measures such as maintaining stormwater control structures, controlling erosion, developing a spill response program, constructing a spill containment structure, and sweeping up contaminated dust and debris from the plant site can greatly reduce the contamination from stormwater runoff.

Specifically to comply with the proposed regulations and the deadline date which was February 4, 1990, all industrial facilities should contact their state environmental agency to determine what steps they should take to comply with the regulations. Although it would not be prudent to perform the sampling and analysis as specified in the proposed rule, some data can be collected and submitted to the agency on Form 1 of the NPDES permit.

Things that a facility may wish to perform now include:

1. Locate and identify all outfalls from the facility
2. Perform screening tests to verify that these outfalls contain only stormwater.
3. Eliminate all non-stormwater discharges from these outfalls.

These measures can provide valuable information to a facility not wishing to wait until the final rule is published before determining its situation.

If a group application might apply, a facility should contact its trade association or other organization to determine how to approach these requirements.

In conclusion, these proposed regulations will impose a significant burden upon industries and municipalities. It is anticipated that the final rule will be published around August 8, 1990. Facilities should not wait for the final rule to initiate a program of locating their outfalls and determining for themselves the extent of their stormwater problem.

12 TRANSPORT MODELING OF CONTAMINATED INDUSTRIAL ABRASIVES FROM A SMALL URBANIZED WATERSHED

Paul D. Kuhlmeier, Senior Staff Consulting
Timothy J. Champagne, Civil Engineer
International Technology Corporation
Knoxville, Tennessee 37923

INTRODUCTION

An unprecedented public awareness of pollution problems in the United States is evident and growing. Until the advent of the U.S. Environmental Protection Agency (EPA) in 1970, only one major law addressing contamination had been passed, the Solid Waste Act in 1965. In 1986, then-EPA-Administrator Lee Thomas commissioned a study by senior staff members to address the question of where America's most pressing pollution concerns stand. One recurring problem that ranked near the top of the study's list was nonpoint source pollution. A separate EPA study[1] estimates that 65 percent of stream pollution is produced from nonpoint sources. A large body of information has been gathered pertaining to farmland, feedlot, and industrial runoff pollution. However, one component of nonpoint source pollution has only recently been identified for study; that is pollution caused by abrasives blasting of materials such as bridges, ships, and buildings. This study addresses this issue as it relates to sandblasting ship components and the operation impact on the environment.

At NSY Mare Island (Site 900), sandblasting practices are carried out in open areas. The 900 Area is in a relatively small watershed measuring approximately 990 feet long and 500 feet wide at its widest point. It is completely paved with asphalt. The integrity of the asphalt is poor in several places. Sandblasting of large and small ship parts is practiced throughout the entire watershed. An abrasive, consisting of a nickel-copper slag, called greensand is used to strip paint from metal surfaces. Uncontrolled runoff from the site deposits onto an estuary which comprises the rim of the Mare Island peninsula. Mare Island Strait borders the strip estuary (it occupies less than 2,000 linear feet of coastline).

A wide range of design storm frequencies (1-year to 100-year) were selected for the purpose of estimating the amount of flow which could be expected from the site. Synthetic hydrographs for the drainage basin were developed using the U.S. Army Corps of Engineers HEC-1 model[2], one of the most widely employed hydrologic simulation codes. Sediment transport was simulated through the use of the U.S. Army Corps of Engineers HEC-6 model.[3] The model, which is well suited for classical river sediment transport, is postulated here to hold promise for estimating sheet flow sediment transport. The HEC-6 model was also selected because of its broad acceptability.

RUNOFF MODELING

To determine the potential runoff under different conditions, average meteorological conditions were compiled using storm frequency information from 1947 through 1987 and isohyetal precipitation charts. The storm distribution type for the study area has also been defined by the SCS as type IA.

Three primary subbasin areas were defined that required independent modeling:

- Subbasin 1 (shown in Figure 1) defines the asphalt area which is further subdivided into four blocks for runoff analysis. The Kinematic Wave Model was used for flow modeling.
- Upgradient Subbasin 2 defines the watershed behind the 900 Area site. This is an area of steep grass-covered slopes.
- Crossgradient and upgradient, Subbasin 3 is located south of the containment area.

45th Purdue Industrial Waste Conference Proceedings,

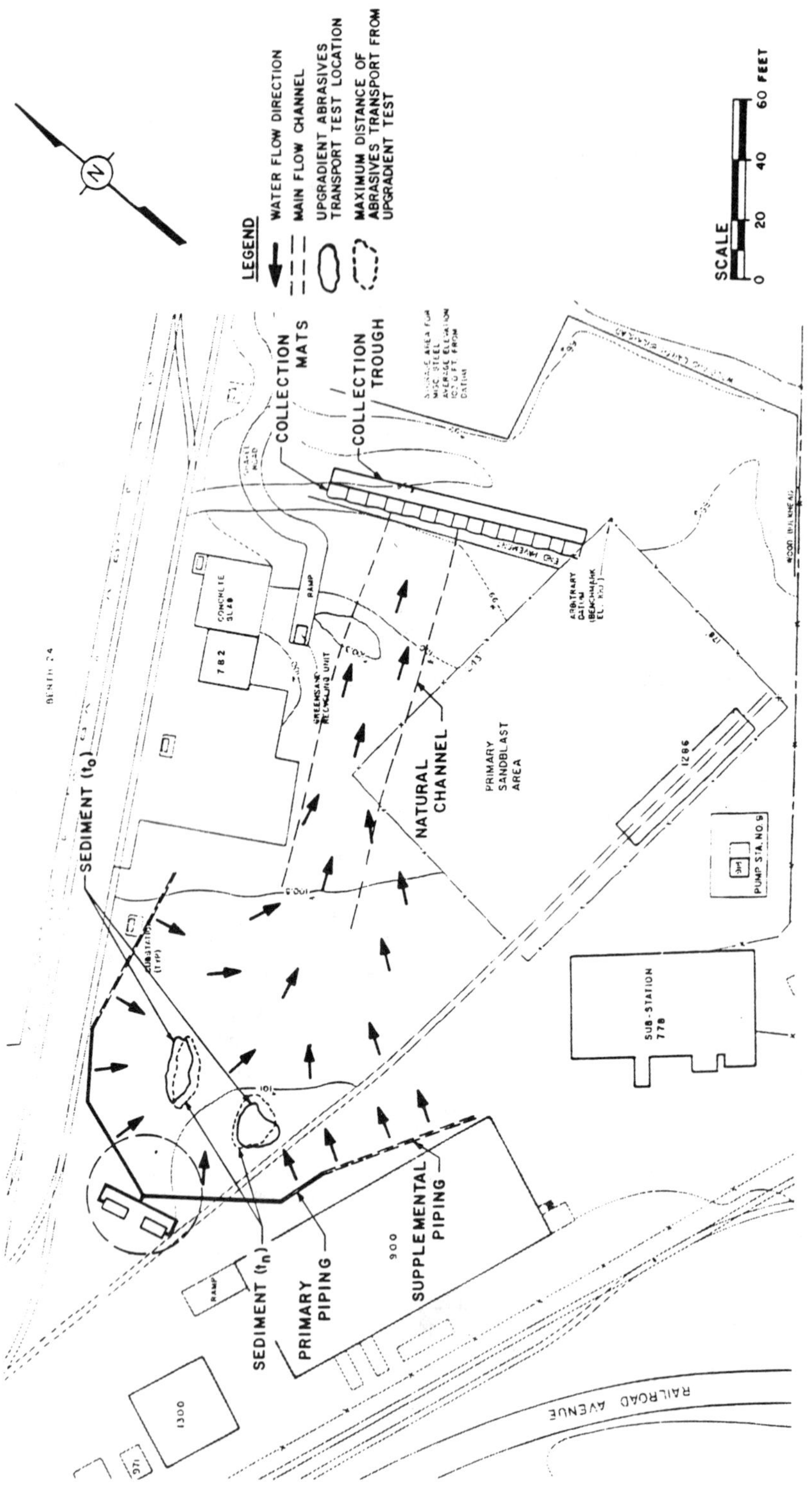

Figure 1. Study watershed (subbasin 1).

Subbasins 2 and 3 were modeled collectively using the SCS curve number technique.[4] Subbasin 1, which is the subbasin of greatest interest, was subdivided into five separate blocks to facilitate more accurate sediment transport calculations. The kinematic wave equation was employed on this subbasin. A schematic of the overland flow patterns as modeled by HEC-1 and HEC-6 are shown in Figure 1.

Physically, rain falling on the watershed is carried to a collecting stream channel, or in the case of a Subbasin, to a simulated central gutter. Flow from the uppermost reach is then routed through the next subbasin and combined with the downgradient subbasin's overland flow. The recombining and flow routing continues until the stream intersects a river or, in this case, ends as the flow enters Mare Island Strait. A total of six separate simulations were developed and executed. In Subbasins 2 and 3, 1-year, 5-year, 25-year, and 100-year return period storms were considered. Two return period storms, 25-year and 100-year, were simulated for Subbasin 1.

A sand area near the bay water was modeled as a channel area for collection of runoff. The total area was estimated at 0.0104 square miles. The overland flow element of this model assumes a rectangular channel of unit width to simplify modeling of runoff. This implies a unit form of Manning's equation:

$$Q = 1.49/n \; S^{1/2} \; R^{2/3} \times A \tag{1}$$

where
- Q = the flow in cubic feet per second
- S = the channel bed slope in foot/foot
- R = the hydraulic radius (area/wetted perimeter)
- A = the cross-sectional area in square feet
- n = Manning's resistance factor.

This equation can be simplified to:

$$Q = 1.49 \; S^{1/2} \; N \tag{2}$$

where
- S = the slope of the channel
- N = the Manning's coefficient of roughness (for overland flow).

This replacement N for n roughness factor for channel flow was determined by Crawford and Linsley[5] to be between 0.05 and 0.10 inclusive. When applying equations to an overland flow element, the lateral inflow is rainfall excess and the outflow is a flow per unit width. Outflow hydrographs for the watershed Subbasins 2 and 3 were developed using the SCS Curve Number Procedure and the HEC-1 program. The SCS procedure was used to create outflow hydrographs for 25-year and 100-year storms with a 24-hour duration. A peak flow of 8 cfs for a 100-year storm was estimated by HEC-1 across Subbasin 1.

SEDIMENT TRANSPORT MODELING

There are three major factors affecting the transport rate of sediment under shallow flow conditions: (1) fluid forces, (2) rainfall impact forces, and (3) particle-to-particle interactions. The first force is relatively well understood and is the basis for most transport equations. The second force has been investigated empirically by Young and Wiersma[6], and the third force has been discussed by Bagnold.[7] Together, these forces represent the physical transport system. The U.S. Corps of Engineers HEC-6 Model[3] employs the modified Einstein equation option as described by Toffaleti[8] and is considered appropriate for use in this study because of its ability to address the discrete transport components of bed load and suspended load independently. Other researchers have offered discrete equations; however, the basic method advanced by Einstein[9] appears to have the most broad-based acceptance by practitioners in the field.

Einstein's Theory

Einstein[9] postulated that a particle moves if the instantaneous hydrodynamic lift force overcomes the submerged particle weight. Therefore, both the beginning and ceasing of sediment motion are probabilistic events and do not depend on whether or not the average fluid force exceeds some threshold. In laminar flow viscous effects dominate. Coleman[10] reported that the lift force coefficient decreases as the Reynolds number decreases. In fact, at very low Reynolds numbers, negative lift coefficients were observed. The ratio of the lift coefficient to drag coefficient was 0.04 to 0.12 when the Reynolds number was within the range of 100 to 500. Therefore, in laminar flow such as demonstrated in the hydrologic modeling for the 900 Area, the lift force acting on a particle is mainly composed of the component of drag force, and the first type of lift force used by Einstein[9] is negligible. Further, the computed total sediment discharge is calculated as the sum of the bed load discharge and the suspended sediment discharge. The bed load discharge is considered to be through a sediment layer that is two-grain diameters thick, and the suspended sediment is considered to be discharge from the water column to the bed layer over distance.

The quantity of sediment transport in a stream by rolling, sliding, or skipping within the bed layer is called the bed load. By assuming that the bed is neither eroding nor scouring (similar to an impervious surface such as asphalt), Einstein[9] developed a bed load equation that can be represented as:

$$i_B q_B = \phi_* i_b p_s g^{3/2} D^{3/2} (S_s - 1)^{1/2} \quad (3)$$

in which: $i^B q_B$ = the rate at which a particle of diameter D in a mixture moves through a unit width of the bed layer per unit of time; i_B is the fraction of bed load in a given size range; q_B is the rate at which the bedlayer sediment moves through the unit width of cross section

ϕ_* = intensity of transport for an individual grain size

i_b = fraction of bed material in a given grain size

p_S = density of the solids

S_s = specific gravity of the sediment.

For a given particle of diameter D, Einstein assumed that all parameters on the right side of Equation 3 are constant except for ϕ_*. The intensity of transport, ϕ_*, is a function of the intensity of shear, ψ_*, for the individual grain size. A curve relating ψ_* and ϕ_* is given in Einstein.[9]

Einstein found that ϕ_* was a unique function of the hydraulic shear intensity parameter ψ_*, which may be defined by:

$$\psi_* = \xi\, Y \left[\frac{\log 10.6}{\log\left(\frac{10.6X}{d_{65}}\right)}\right]^2 \frac{p_s - p}{p} \, \frac{D}{R'_b S_e} \quad (4)$$

where ξ = a correction of effective flow for various grains

Y = a correction of lift force in transition between hydraulically rough and smooth beds

X = a reference grain size for the particular bed

R'_b = hydraulic radius of the bed for grain roughness or surface drag

S_e = slope of the total energy line for the given flow conditions.

HEC-6 MODELING

The model considers an open channel of finite width. Sediment transport is calculated for a reach of river which is described by a series of representative cross sections. Geometry of the model is represented by cross sections, station locations along the river reach, elevation drop, and distance between cross sections. Water discharge is handled as a series of steady state flows over a given period of time. Friction losses are a function of Manning's "n" that describes the hydraulic roughness. Both granular and cohesive materials may be input, but the present version of the model does not allow for reentrainment of silts and clays once they settle to the stream bed. Main water flow through the channel may be augmented by discrete tributary input locations along the modeled reach.

Sediment available for movement is determined in the main channel to be the volume of material above an impermeable boundary elevation and is limited laterally by the width of the channel. Transport capacity is calculated at each cross section based on hydraulic data and grain size information at a given location. The left and right banks of the main channel are considered no-flow boundaries for water and sediment input. As such, the model can define both movable and nonmovable beds. Overland water and sediment runoff normal to the main channel may be modeled by performing a separate submodel that simulates sediment transport across the width of a subreach or subsection with the main channel representing the final location of deposition. These discrete submodels may then be input in the main channel models as tributaries. Information is input in three type sets: geometric data, sediment data, and hydrologic data. The present version of HEC-6 output is generated at each cross section for elevation change, water surface level, flow rate, and sediment discharge by clay, silt and sand fractions.

Eight representative samples of both virgin and spent greensand were collected and a grain size analysis performed. Both classes of abrasives were very uniformly graded and the unused abrasive was found to be larger than the spent abrasive. Characteristic grain size distribution curves for spent greensand are given in Figure 2. Grain size of new greensand ranged from 3 mm to 0.5 mm with the median being approximately 1.7 mm. Spent greensand ranged in size from 8 mm to 0.075 mm with a median of about 0.45 mm.

The model developed for the 900 Area was subdivided into five cross sections coincident with the downstream face of the subwatersheds used in the hydrologic model previously discussed. Data is input from the downstream end of the channel. The cross sections used are at 170, 330, 490, 650, and 810 feet from the bulkhead line on the beach. An arbitrary, triangular-shaped, 40-foot-wide channel was selected for the HEC-1 model and subsequently for the HEC-6 model since this width provides adequate channel to accommodate the peak runoff from a 100-year storm. Utilizing a wide triangular channel is representative of actual site conditions and produces shallow depths for runoff which are coincident with promoting sheet flow. A nonscouring depth was set at 0.1 foot, which represents the average depth of greensand on the asphalt. To accommodate sediment load flowing from remote parts of the watershed to the channel, separate submodels were developed that simulated overland flow across a subwatershed to the central main channel.

It was discovered during this modeling effort that sheet flow from the peak 25- and 100-year storms dispersed over the entire watershed did not provide sufficient energy for significant transport of sediment using the HEC-6 model. This prevented generating sediment load rating curves for tributary inflow to the main channel and providing subsequent steady-state conditions. To compensate for this deficiency, steady-state conditions were abandoned and the runoff area of the channel was evaluated for its potential to transport greensand at various discharge flows. Somewhat conservative values for discharge of sediment at the channel outlet will be predicted. This observation suggests that greensand runoff is minimal in all areas of the watershed where sheet flow is dominant. The effective contribution of the operation area is confined to the portions of the watershed where runoff combines to form rivlets or an equivalent and sufficient energy is developed to move sediment and/or paint materials.

Modeling Rationale and Results

An evaluation of the two worst-case discharge hydrographs, 25- and 100-year storms, generated using the HEC-1 Kinematic Wave model indicates that flow divisions fall into the following three categories:

- Approximately 4 to 6 hours is required for discharge before a rapid slope change occurs in the rising limb of each hydrograph.
- The time span over which peak discharge is achieved (length of rising limb on hydrograph) ranges from 2 to 4 hours.
- Runoff on the receding limb of each hydrograph encompasses a time frame of approximately 12 to 14 hours.

For each of the above time frames, a time-discharge relationship was established.

Using the HEC-6 model, values of discharge were varied for each section of the hydrograph. Each station along the channel is geometrically similar and differs only in the amount of runoff expected at each location. This relationship was used to generate an average sediment discharge relationship for the entire Subbasin 1. Results from the HEC-6 program for each station were averaged for a given discharge. A graph showing the relationship of average sediment transport and discrete station transport and cumulative sediment load at the base of the watershed is presented in Figure 3. To support these findings, a linear regression analysis was calculated and equations generated for each station along the channel length. Values obtained through regression closely compare to actual values used to calculate average discharge-runoff relations. A comparison of the regression and average results is shown in Figure 4. HEC-6 discharge-sediment load output along with regression values for each cross-section are tabulated in Table 1. A listing of the linear equations that were used to derive the regression values are also shown in Table 1. The correlation coefficient ranged from 0.95 to 0.99 for all linear equations derived. The expression given for Station 1 + 70 represents the cumulative sediment loading from the entire basin.

VERIFICATION

It was postulated noted that three conditions appear to limit the general applicability of the sediment yield estimates to small urban watersheds:

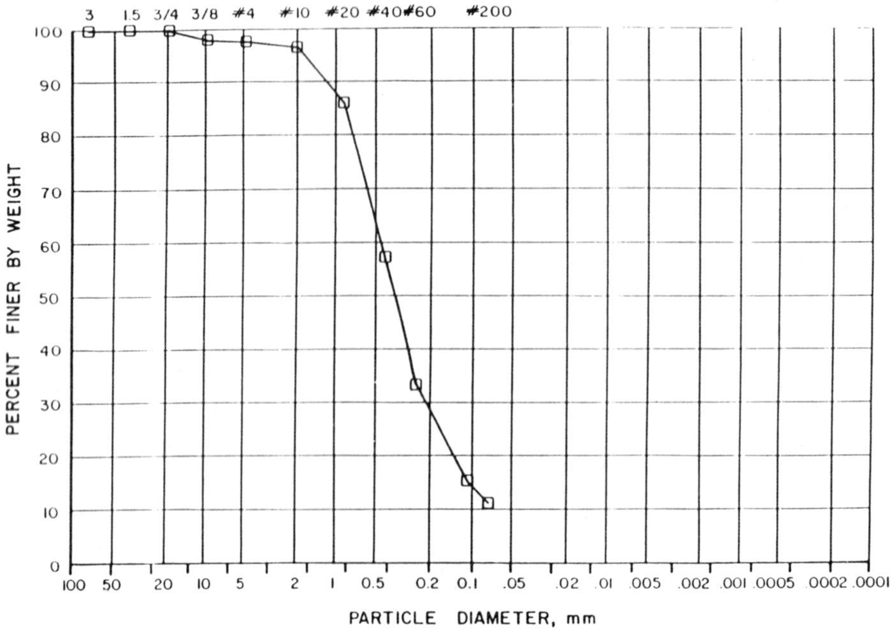

Figure 2. Grain size distribution: spent abrasive.

- Sediment along the outermost portions of the watershed are immobile because of the insufficient energy to move sediment and intermixed paint flakes.
- Runoff water funnels into a definable central channel determined from topographic surveying.
- Sediment transport is limited to the central channel which creates the largest energy potential.

The focus of the field verification thus considered two fundamental objectives: (1) test the validity of the three limiting model assumptions and (2) measure the total sediment (abrasives) yield and compare them to those predicted by the HEC-6 model. Continuous flow runoff tests were performed on the watershed in order to calibrate the numerical model.

Two steady-state flow rates were chosen, 0.5 and 1.0 cfs. The 0.01-square mile watershed requires an output of almost 450 gpm to simulate a rainfall that produces 1.0 cfs maximum runoff. This flow volume was determined to be the largest that available equipment could deliver on a sustainable basis.

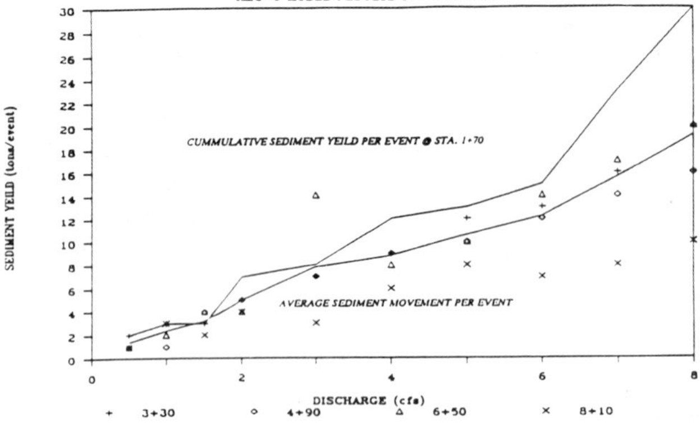

Figure 3. Sediment load rating curve.

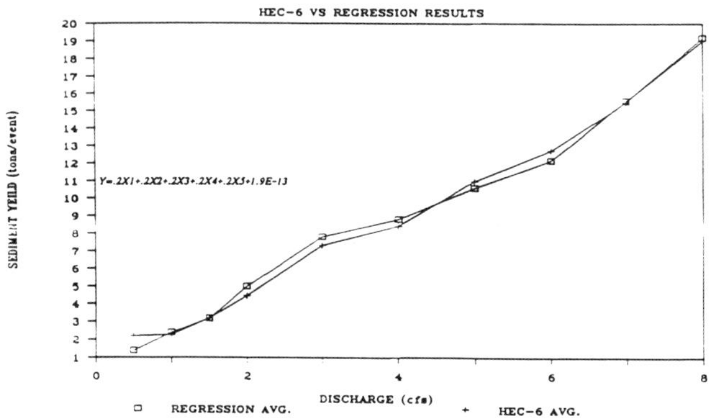

Figure 4. Sediment load comparison curve.

The second flow volume (0.5 cfs) was chosen to further validate the measurements under different conditions. A sustained 450 and 225 gpm of water was added to subbasin 1 at its most upgradient edge through 250 linear feet of perforated 6-inch diameter polyvinyl chloride (PVC) pipe.

Table I. HEC-6 Results

DISCHARGE (cfs)	STATION 170	330	490	650	810	AVG (tons/day)
0.5	2	2	1	1	1	1.4
1	3	3	1	2	3	2.4
1.5	3	3	4	4	2	3.2
2	7	5	5	4	4	5
3	8	7	7	14	3	7.8
4	12	9	9	8	6	8.8
5	13	12	10	10	8	10.6
6	15	13	12	14	7	12.2
7	23	16	14	17	8	15.6
8	30	20	16	20	10	19.2

LINEAR REGRESSION RESULTS

DISCHARGE (cfs)	STATION 170	330	490	650	810	AVG (tons/day)
0.5	3.1	1.2	1.3	3.4	2.0	2.2
1	2.5	3.1	2.8	0.5	2.4	2.3
1.5	2.9	3.1	2.9	4.9	2.3	3.2
2	5.2	5.6	4.4	3.6	3.5	4.5
3	8.0	7.3	7.5	11.1	2.7	7.3
4	10.9	9.5	7.9	8.1	5.6	8.4
5	15.0	11.7	10.5	10.1	7.7	11.0
6	17.4	12.5	11.5	15.1	7.5	12.8
7	22.4	16.0	13.6	17.4	8.2	15.5
8	28.7	20.0	16.7	19.8	10.0	19.0

STATION	REGRESSION EQUATION
1 + 70 = X1	Y = 2.56X2 − .086X3 − .353X4 − 1.39X5 − .224
3 + 30 = X2	Y = .256X1 + .098X3 + .720X4 + .194X5 − .303
4 + 90 = X3	Y = .416X1 − .037X2 + .434X4 + .267X5 − .189
6 + 50 = X4	Y = 2.10X1 − .381X2 + .679X3 − 2.31X5 + 1.63
8 + 10 = X5	Y = −.172X1 + .885X2 + .125X3 − .263X4 + .697
GENERAL	Y = .2X1 + .2X2 + .2X3 + .2X4 + .2X5 + 1.9E-13

A water storage reservoir was created by two 10,000-gallon capacity Baker tanks. Water levels were maintained in the tanks from fire hydrants in the area. A 10-horsepower supply pump produced effluent velocities between 3 and 8 feet per second at 450 gpm. The plumbing also included two upstream gate valves for flow control, a downstream check valve, needle valve, and inline flow meter, accurate to ±2.5 gpm.

A temporary light gauge metal dike was constructed along the bottom periphery of the watershed, approximately 170 feet in length. A synthetic liner, 60 mil-thick and cut in 10-foot-wide sections, was bolted to the metal curb for the purpose of collecting runoff solids (collection mats). An aluminum runner was nailed over the upstream end of the fabric. This runner was removed and replaced after each test. The width of the sections was chosen because it was considered most economical for lifting and weighing purposes. High-density polystyrene energy breaks were placed at each 10-foot interval in an effort to reduce the amount of sediment moving laterally along the collection trough.

Testing Model Assumptions

Two experiments were performed to address the three model assumptions. The first test conducted considered the significance of sediment (abrasives) in the outer regions of the watershed. Three tons of spent abrasives were spray-painted bright orange and spread throughout the outreaches of the watershed downstream of the distribution piping. A minimum distance of 75 feet was maintained from the water pipe to allow for a velocity reduction that would be most consistent with those found at the top of a watershed. The value of 75 feet was established by first adding a small amount of dye to the water system at an inline tap and estimating where the velocity of sheet flow slowed to less than 1.5 feet per second. Spent abrasives were distributed over an area shown in Figure 1 and ranged in depth from less than 0.2 inch to 2.0 inches.

Because the most critical condition would be realized at the higher flow rate, only the 1.0 cfs case was considered. The entire watershed was lightly sprinkled with garden hoses to prewet the test area for purposes of minimizing initial friction influences of dry pavement. The needle valve, having been previously set to 450 gpm during the velocity measurements, was unlocked and the pump started. The test was run until it appeared from visual observations that no significant sediment transport was occurring. The first run took 27 minutes. Total distance of the sediment front was mapped and after the pavement had dried, the abrasives were swept up by a street sweeper and then redeposited in approximately the same original position.

The test was run a second time at 450 gpm to check the test method. This run was stopped at 27 minutes, the sediment front mapped, and the test restarted and continued for 33 minutes to ensure significant sediment movement was not overlooked. The extended second text produced no additional discernable sediment movement at the front of the advance wave.

Results of mapping from both test runs are also depicted in Figure 1. It can be noted that the maximum distance traveled by sediment over the duration of either test was only 3.6 feet. Visual observation noted that more than one-half of this movement occurred in the first 5 minutes of each run. As noted, no significant movement occurred throughout the additional 33 minutes of the second test. Sediment advanced slightly farther in the second run versus the first run, 3.6 feet and 3.1 feet, respectively. It appears quite evident that even though relatively high energy was applied to the particles, there is not enough to create a condition whereby outer reach sediments contribute to the primary sediment yield of the basin. The model assumption is valid and is considered transferrable to other similar watersheds.

The second test was designed to validate the runoff patterns and qualify the 40-foot width used in the HEC-6 model. A more encompassing aspect of this test is its contention that small watersheds may be modeled accurately by knowing the surface contours of a semi-impervious basin, large or small. An additional 200 feet of pipe was added to the system to encompass as much of the watershed periphery as possible. Effluent along the final 50 feet of either side was noticeably minor. The concept of the test was to observe the water flow patterns through the watershed. This was accomplished by adding dye to the water supply and observing the flow patterns. Bright red and orange nonstainable dyes were chosen. Red was injected on one side of the piping system and orange on the other, and the dye advance was monitored.

Again, two runs were made switching the dye colors to opposite sides after approximately one-half of each test had been completed. Here, however, one run was made at 450 gpm and one at 225 gpm to determine if there was any marked difference between relatively low and relatively high overland flow rates. Each run lasted 20 minutes with four injections of dye, two of both colors on each side.

Results support the topographic mapping almost exactly. Two notable observations were made. First, distinct eddys formed around the cracks in the pavement and may produce local accumulations

Table II. Comparison of Field and Model Results for Sediment Yield From Water Shed

MAT NO.	WEIGHT (lbs) Q = 1.0 cfs RUN 1	RUN 2	Q = 0.5 cfs RUN 1	RUN 2
1	41	30	27	19
2	203	245	123	94
3	271	303	231	250
4	403	409	297	317
5	298	289	215	189
6	145	164	111	125
7	29	36	31	20
TOTAL	1390	1476	1035	1014
MODEL	1620		1080	

of sediments. In general, sheet flow moved in a linear fashion toward the central portion of the watershed where the water depth reached a maximum of 1.2 inches at the 450-gpm flow rate and 0.7 inches at a lower flow rate. Significant mixing became noticeable across a central channel width that varied from 17 feet at the top of the watershed to 47 feet at the collection trough. Overall, the average mixing zone (channel width) was estimated by virtue of five cross-section measurements to be 38 feet. It was evident that sheet flow from small urban watershed can be modeled by a buildup from a topographic survey.

Sediment Transport Measurements

From the experiments described above, it was evident that the initial choice of a 40-foot width was adequate within the confidence margin of modeling and measurement. Once the watershed had been allowed to dry, spent abrasives were reapplied to original positions to the extent feasible. A series of ten random locations was chosen to measure sediment depth. Depth, measured with a caliper, varied from 0.05 to 0.19 feet with an overall average of 0.12 feet. Computer simulations performed before these field tests used 0.10 feet. The difference between these two values, was considered to be insignificant.

Two runs at 450 gpm (1.0 cfs) and two runs at 225 gpm (0.5 cfs) were completed. Each run lasted 6.5 hours. At the conclusion of each run, the collection mats were detached and weighed. After each run, the mats were also tiered again. Although there were a total of 17 mats along the downstream face of the watershed, significant amounts of sediment were collected in only the centrally located seven mats. Because of the relatively heavy weights involved with three of the mats, a backhoe was used to lift the mats and place them on the truck scale used for weighing. The scale used was accurate to ±1.0 pound.

Results of the sediment transport tests are presented in Table 2. At 1.0 cfs, the average yield was 1,433 pounds and was 1,024.5 pounds at 0.5 cfs. By comparison, the HEC-6 model predicted approximately 3 tons per day yield rate for 1.0 cfs and 2 tons per day yield rate for 0.5 cfs. These latter values translate to approximately 1,620 pounds of sediment in 6.5 hours at 1.0 cfs and 1,080 pounds of sediment in 6.5 hours at 0.5 cfs. The model based on these field tests estimated sediment yields within +11.5 percent of actual at the higher flow rate and +5.1 percent of actual at the lower flow rate.

CONCLUSIONS

This study sought to accomplish one fundamental objective: investigate the suitability of a large-scale predictive sediment transport model for estimating the yield of spent blasting abrasives off a small urban watershed.

Sediment yield was measured at two flow rates, 0.5 cfs and 1.0 cfs. Each test was performed twice for the purpose of checking the accuracy of the first runs. Results of the field investigation closely fit the predicted sediment yield from computer simulations. Although the computer model overpredicted sediment/abrasives transport for all flow runs by 5.1 to 11.5 percent, this difference is well within acceptable engineering limits. At 0.5 cfs HEC-6 predicted a 1,080-pound yield and the average measured value was 1,024 pounds. At 1.0 cfs HEC-6 predicted 1,620 pounds of sediment would be yielded over a 6.5-hour period. Field measured amounts averaged 1,433 pounds of sediment-abrasive material. It is readily apparent given that most runoff estimates try to achieve an accuracy of an order of magnitude or less, that this investigation has shown acceptable correlation between predicted and measured values. Although additional testing is always warranted, the HEC-6 model may not only be used successfully for large natural watersheds, but apparently also for small-scale conditions and situations such as where hazardous waste transport may be involved.

REFERENCES

1. U.S. EPA (Environmental Protection Agency), *National Water Quality Inventory: 1982 Report to Congress*, Washington, D.C. (1984).
2. U.S. Army Corps of Engineers, *HEC-1 Flood Hydrograph Package Users Manual*, Hydrologic Engineering Center, Davis, California, 293 pp. (1981).
3. U.S. Army Corps of Engineers, *HEC-6 Scour and Deposition in Rivers and Reservoirs Users Manual*, Hydrologic Engineering Center, Davis, California, 274 pp. (1977).
4. U.S. Soil Conservation Service, *National Engineering Handbook, Section 4*, U.S. Department of Agriculture, Washington, D.C. 278 pp. (1972).
5. Crawford, N. H. and R. K. Linsley, "Digital Simulation in Hydrology, Stanford Watershed Model IV," *Civil Engineering Technical Report No. 39*, Stanford University, Palo Alto, California (1966).
6. Young, R. A. and J. L. Wiersma, "The Role of Rainfall Impact in Soil Detachment and Transport," *Water Resour. Res.*, 9(6), pp. 1629–1636 (1973).
7. Bagnold, R. A., "The Nature of Saltation and of Bed-Load Transport in Water," *Proc. Royal Soc. of London*, London, England, A(332), pp. 473–504 (1972).
8. Toffaleti, F. B., "A Procedure for Computation of Total River Sand Discharge and Detailed Distribution, Bed to Surface," Committee on Channel Stabilization, U.S. Army Corps of Engineers (1966).
9. Einstein, H. A., "The Bedload Function for Sediment Transport in Open Channel Flows," *Tech. Bull. 1026*, Soil Conserv. Serv., U.S. Department of Agriculture, Washington, D.C. (Sept. 1950).
10. Coleman, D. R., "A Theoretical and Experimental Study of Drag and Lift Forces Acting on a Sphere Resting on a Hypothetical Streambed," *Proc. 12th Congress IAHR*, Colorado State University, Fort Collins, Colorado, H3, pp. 185–192 (1967).

13 REGULATORY OPTIONS UNDER WATER QUALITY BASED NATIONAL POLLUTANT DISCHARGE ELIMINATION SYSTEM PERMITTING

Philip M. Simmons, Environmental Engineer
Tennessee Department of Health and Environment
150 Ninth Avenue North, Fourth Floor
Nashville, Tennessee 37247-3420

Robert G. O'Dette, Environmental Engineer
Tennessee Department of Health and Environment
150 Ninth Avenue North, Fourth Floor
Nashville, Tennessee 37247-3420

INTRODUCTION

Since the passage of the Federal Water Pollution Control Act (Clean Water Act, as amended) in 1972, it has been the goal of the federal government as interpreted by the Environmental Protection Agency (EPA) to "restore and maintain the chemical, physical, and biological integrity of the Nation's waters"[1]. In implementing this goal and the goal of toxic pollutant control as mandated by the Clean Water Act of 1987, EPA and state water pollution control regulatory agencies have developed options to be used by the permit issuing authority for the protection of water quality through the issuance of National Pollutant Discharge Elimination System (NPDES) permits to dischargers to the nation's waters. The purpose of this paper is to discuss these approaches: the chemical-specific approach to water quality permitting and the whole-effluent approach to water quality permitting. Industrial and municipal facilities that discharge to zero flow receiving streams will be particularly influenced by these options.

THE CHEMICAL-SPECIFIC APPROACH

The chemical-specific approach to toxics control involves the use of laboratory-generated water quality criteria or state standards to limit specific toxicants directly. The toxicity analysis of those chemicals is done in a comprehensive testing program that, unlike whole-effluent testing, attempts to consider a range of toxic endpoints including human health impact and bioaccumulation. Once a criteria is developed, the number is applied, through an exposure analysis, as a permit limit to ensure that the level of that toxicant is not exceeded after discharge. Fate modeling of the toxicant to estimate its behavior after discharge can be an important step in establishing water quality-based permit limits.

There are advantages and disadvantages to this approach. Advantages to this approach are: (1) Treatment systems are more easily designed to meet chemical requirements because more treatability data are available and treatment engineers and permit writers are more familiar with the procedures; (2) The fate of a pollutant can be measured through modeling; (3) Chemical analyses, in simple cases, can be less expensive than toxicity testing; and (4) Specific problem chemicals such as carcinogens or bioaccumulative chemicals can be directly limited. Disadvantages to this approach are: (1) All toxicants in complex wastewaters may not be known and, therefore, control requirements for all potential toxicants cannot be set; (2) It is not always clear which is the toxic compound(s) in the mixture; (3) It can become expensive to measure individual toxicants, particularly where many are present in the mixture. Organic chemicals can, in particular, be costly to measure; and (4) The bioavailability of the toxicants at the discharge site are not assessed, and the interactions between toxicants (e.g., additivity, antagonism) are not measured or accounted for.[2]

All waters in the nation have been classified for different uses that are either based on historical record or anticipated uses. These uses are usually domestic water supply, industrial water supply, fish and aquatic life, recreation, irrigation, livestock watering and wildlife, and navigation. Most waters are classified for more than one use. Each classification has an ambient water quality criteria estab-

45th Purdue Industrial Waste Conference Proceedings, © 1991 Lewis Publishers, Inc., Chelsea, Michigan 48118. Printed in U.S.A.

lished for a particular toxicant for a particular use. This criteria may be numerical or narrative in nature. However, the Clean Water Act of 1987, specifically Section 303 (c)(2)(B) requires states to adopt criteria for priority pollutants (307a list), where there is a reasonable expectation of an adverse impact on water quality. EPA interprets that Section 303 (c)(2)(B) can be met under either of the following options or a combination of those options:

"1. Adopt Statewide numeric water quality standards for all EPA criteria for section 307(a) toxic pollutants regardless of whether the pollutants are known to be present;
2. Adopt specific numeric water quality standards for section 307(a) toxic pollutants as necessary to support designated uses where such pollutants are discharged or are present in the affected waters and could reasonably be expected to interfere with classified uses;
3. Adopt a procedure to be applied to a narrative water quality criterion. This procedure shall be used by the state regulatory agency in calculating derived numeric criteria, which criteria shall be used for all purposes under section 303 (c) of the Clean Water Act. Such criteria need to be developed for section 307(a) toxic pollutants, as necessary to support classified uses, where these pollutants are discharged or present in the affected waters and could reasonably be expected to interfere with classified uses."

EPA published a notice in the Federal Register[3] that listed state compliance with the Clean Water Act of 1987 requirements for adoption of water quality criteria for toxic pollutants. Twenty-two states have adopted the Option 1 approach, twenty-three states have adopted the Option 2 approach, and one state has adopted the Option 3 approach. Eleven states are doing combinations of the options. One item of concern is that most of the water quality criteria are expressed in terms of dissolved fraction while permit limitations are expressed in terms of the total pollutant.

Once the criteria has been identified for the most restrictive classified use, and background concentrations for a particular parameter have been established, permit derivation may begin. Most effluent limitations are derived using the following equation:

$$C_m = (((C_W)(Q_W)) + ((C_S)(Q_S)))/(Q_W + Q_S) \tag{1}$$

where:
C_m = resultant instream concentration after mixing
Cw = effluent concentration of a particular pollutant
Qw = effluent flow
C_s = background concentration of a particular pollutant
Qw = stream flow

This approach may be overly conservative. This equation assumes the total amount of the pollutant being discharged is free and available to an aquatic organism. Various pollutants may be bound or chelated by suspended material in the effluent and receiving stream.

There are methodologies to account for the binding of a pollutant and to bridge between the criteria in the dissolved phase and the permit limits in the total form. Two of these methods will be discussed below.

1. MINTEQA2 [4]

MINTEQA2 is a thermodynamic equilibrium model that computes aqueous speciation, equilibrium adsorption/desorption, and the mass of metal transferred into or out of solution as a result of the dissolution or precipitation of solid phases. It was developed by contractors for EPA to assess the fate of selected priority pollutant metals in aqueous systems. MINTEQA2 alone, however, does not have the capability of computing kinetic, transfer or transport processes.

The program requires two types of data: (1) thermodynamic data and (2) water quality data. The user of the program is only required to provide the water quality data, thermodynamic data is contained in a MINTEQA2 data base. The thermodynamic data is equilibrium constants, enthaplies of reaction, and other basic information required to predict the formation of each species or solid phase. The supplemental data include charge, gram formula weight, carbonate alkalinity factor, extended Debye-Huckel parameters, and name and ID number of each species. Although the MINTEQA2 data base is probably the most thoroughly documented and evaluated thermodynamic data base used in any currently available geochemical model, it is suggested that it should be updated when new and more reliable information is published, or when data for reactions not presently included in the data base become available.

Table I. Linear Partition Coefficients for Priority Metals in Streams and Lakes[6]

Metal	Stream Coefficients K_{PO}	Stream Coefficients a	Lake Coefficients K_{PO}	Lake Coefficients a
Arsenic	0.48×10^6	-0.73	(Same as stream)	
Cadmium	4.00×10^6	-1.13	3.52×10^6	-0.92
Chromium	3.36×10^6	-0.93	2.17×10^6	-0.27
Copper	1.04×10^6	-0.74	2.85×10^6	-0.90
Lead	0.31×10^6	-0.19	2.04×10^6	-0.53
Mercury	2.90×10^6	-1.14	1.97×10^6	-1.17
Nickel	0.49×10^6	-0.57	2.21×10^6	-0.76
Zinc	1.25×10^6	-0.70	3.34×10^6	-0.68

There are several limitations for efficient use of the model. First, the data base contains equilibrium constants for a limited number of heavy metals and organic ligands. Equilibrium constants of heavy metals included in the data base are those of arsenic, cadmium, chromium, copper, lead, mercury, nickel, selenium, thallium, and zinc. The other metals' constants can be inserted into the data base by the user as they become available. A number of organics can form complexes with heavy metals in natural waters, but equilibrium constants for these complexes are widely varying in the literature. Second, the program treats every reaction as if it were at chemical equilibrium. In fact, chemical reactions of precipitation/dissolution and oxidation/reduction are often not at chemical equilibrium. The kinetics of these reactions are slow. Third, the program has not been verified for reactions that would occur in natural waters. Equilibrium constants are based on thermodynamic relationships, assuming a certain set of environmental conditions. If conditions vary, as they do from site to site, conditional stability constants are needed to account for the special chemistry of the site, which is not considered explicitly. There are analytical chemistry problems with verifying the speciation model also. Current analytical techniques do not provide high enough precision to measure separately the activity of individual metal species and complexes.

2. State of Texas Approach to Setting Permit Limitations

The Texas Water Commission in its document "Implementation of Texas Water Commission Standards Via Permitting"[5] outlines a procedure whereby dissolved criteria for metals may be converted to a criteria in the total form. The following formulas are employed:

$$C/CT = \text{Fraction of Metal Dissolved} \tag{2}$$

where: C = concentration of dissolved metal
C_T = concentration of total metal

$$C/C_T = 1/(1 + ((K_P)(SS)(10^{-6})) \tag{3}$$

where: K_P = Linear Partition Coefficient, 1/kg
SS = Total Suspended Solids Concentration, mg/L

K_P can be derived from the following equation:

$$K_P = (K_{PO})(SS)^a \tag{4}$$

where: K_{PO} = Linear Partition Coefficient, 1/kg (from Table 1)
a = constant (from Table 1)

Example

Find the water quality criteria for zinc in a stream with a total suspended solids concentration of 75 mg/L and a hardness of 100 mg/L as CaCO3.

Water Quality Criteria (C) = 0.11 mg/L
K_{PO} = 1.25 X 10^6
a = -0.70

$$K_P = (1.25 \times 10^6)(75)^{-0.70}$$
$$= 0.06 \times 10^6$$
$$C/C_T = 1/((0.06 \times 10^6)(75)(10^{-6}))$$
$$= 0.18$$
$$C_T = C/0.18 = (0.11 \text{ mg/L})/0.18 = 0.61 \text{ mg/L}$$

Permit limits can now be derived using Equation 1.

This methodology provides a mechanism to raise the water quality criteria for low or zero flow streams. These also show that the higher the total suspended solids in a receiving stream, the higher the water quality criteria in the total form of the metal.

THE WHOLE-EFFLUENT APPROACH

The whole-effluent, or toxicity-based, approach to toxics control involves the use of toxicity tests to measure the toxicity of industrial and municipal wastewater discharges. A toxicity test is a relatively simple laboratory procedure. An effluent sample is collected and diluted in test chambers; usually the dilutions used are 100%, 30%, 10%, 3%, 1%, and a control. Often the receiving water is used to dilute the effluent because it more closely simulates effluent/receiving water interactions. Test organisms are placed in these test chambers for specified periods of time. At various points during the testing period, the number of organisms affected in each chamber is counted (the endpoint can be mortality, lower reproduction, reduced growth rates, etc.) and the lowest effluent concentration that causes that endpoint is calculated. This endpoint concentration becomes a quantified measure of the concentration that would cause instream impact if exceeded for a particular period of time. It is usually stated either as an LC50 (the effluent concentration at which 50% of the test organisms are killed) which is an acute exposure or a No Observed Effect Level or NOEL (the highest effluent concentration at which no unacceptable effect will occur even at continuous exposure) which is a chronic exposure.

The measurement of whole-effluent toxicity can then be used to limit the discharge of toxicants in an effluent. Toxicity itself is used as the effluent parameter. The toxicants creating that toxicity need not be specifically identified or controlled where the effluent toxicity is limited. An analogy between effluent toxicity and biochemical oxygen demand (BOD) can be drawn. Both are measurements of a biological effect. Both can be quantified. In neither case are the causative agents of the biological effect specifically identified. Thus, whole-effluent toxicity is like BOD in that it is a useful parameter for reducing an undesirable effect caused by the discharge of a complex mixture of waste materials.

There are also advantages and disadvantages to this approach. Advantages to this approach are: (1) The aggregate toxicity of all constituents in a complex effluent is measured, and toxic effect can be limited by limiting one parameter—effluent toxicity; and (2) the bioavailability of the toxic constituents is assessed, and the effects of interactions of constituents are measured. The disadvantages to this approach are: (1) Effluent toxicity treatability data are lacking, and treatment engineers and permit writers are unfamiliar with the procedures; (2) Where there are chemical/physical conditions (pH changes, photolysis, etc.) present that act on toxicants in such a way as to "release" toxicity downstream, such toxicity may not be measured in the effluent; and (3) Properties of specific chemicals in complex effluents (such as bioaccumulation, carcinogenicity) are not assessed.[2]

1. Acute Toxicity Testing In Tennessee

   Acute toxicity testing shall be performed in accordance with procedures established in *Methods for Measuring the Acute Toxicity of Effluents to Freshwater and Marine Organisms*, EPA/600/4-85/013. These static tests shall be of 48 hours in duration and be conducted on a manually composited sample consisting of four separate grab samples of final effluent which are equal in volume, are collected at evenly spaced (6 hour) intervals over a single 24 hour period. Testing should be initiated within 24 hours of collection of the final grab sample. In order for a test to be considered valid, control organisms must survive at a level greater than or equal to 95%.

   Test organisms shall be fathead minnows (Pimephales promelas) no greater than ninety (90) days in age and or one of three daphnids (*Daphnia magna, Daphnia pulex*, or *Ceriodaphnia dubia*) no greater than 24 hours in age.

   Testing shall be initiated within 90 days of the effective date of permit issuance and be conducted once per quarter for the first year of the permit and once every six months thereafter for the duration of the permit. Testing during the first year of the permit shall be conducted on both the fathead minnow and a daphnid species. After the first year, the permittee may request the Division of Water Pollution Control to review the data to determine the most sensitive species for use in future tests. Other measurement frequencies may be established based on the permit writer's best professional judgement.

Permit Limits shall be established using the dilution ratio of the discharge to the low flow of the receiving stream. This dilution ratio shall be expressed as the ratio between the low flow of the receiving stream and the average daily flow. The permit limit will be based upon an LC50 value computed as follows:

$$\text{Dilution Factor} = \text{Stream Low Flow/Effluent Flow} \tag{5}$$

$$LC_{50}\ \text{Permit Limit} = ((100\%)/((\text{Dilution Factor})(0.3))) \tag{6}$$

Example

Find the LC_{50} permit limit for a facility with a receiving stream low flow of 30 million gallons per day (MGD) and an effluent flow of 1.0 MGD.

$$\text{Dilution Factor} = 30/1.0 = 30$$

$$LC_{50} = ((100\%)/((30)(0.3))) = 11.1\%$$

A violation of the permit for whole effluent acute toxicity testing occurs whenever the effluent at the end-of-pipe exceeds this permit limit.

2. Chronic Toxicity Testing in Tennessee

All test organisms, procedures and quality assurance criteria used shall be in accordance with *Short-term Methods For Estimating the Chronic Toxicity Of Effluents and Receiving Waters to Freshwater Organisms, Section 12: Ceriodaphnia Survival and Reproduction Test Method 1002.0 and Section 10: Fathead Minnow (Pimephales promelas) Larval Survival and Growth Test Method 1000.0*, EPA/600/4-89/001.

All tests will be conducted on 24-hour composite samples of final effluent. All test solutions shall be renewed daily. If, in any control, more than 10% of the test organisms die in 96 hours or more than 20% of the test organisms die in 7 days, that test (control and effluent) shall be repeated. Such testing will determine whether the effluent affects the survival, reproduction, and/or growth of the test organisms.

Test organisms shall be the fathead minnow (*Pimephales promelas*) and the *Ceriodaphnia dubia*.

Testing frequencies shall be established using the permit writer's best professional judgement. Generally that means the dischargers will be required to test monthly for the first year of the permit and then every six months for the life of the permit. After the first year, the permittee may request the Division of Water Pollution Control to review the data to determine the most sensitive species for use in future tests. Testing shall be initiated within 90 days of the effective date of the permit.

Permit limits shall be established using the Instream Waste Concentration factor which is as follows:

$$\text{NOEL} = \text{IWC} = (\text{Qw}/(\text{Qw} + \text{Qs})) \tag{7}$$

Acute toxicity permit limits shall be set in addition to the chronic limit as described in the acute toxicity testing section of this paper.

SUMMARY

The chemical-specific approach to water quality based permitting is used by the permitting agency for certain situations for the protection of water quality. The whole-effluent approach to water based permitting is used by the permitting agency for other situations for the protection of water quality. In most cases, since the advantages of one approach conteracts the disadvantages of the other approach, the permit agency will use both appoaches when a facility is discharging to a low or zero flow stream to insure water quality is being protected in the receiving stream.

REFERENCES

1. Water Pollution Control Act of 1972.
2. *Technical Support Document for Water Quality-based Toxics Control*, September 1985, pages 1-2.
3. *Federal Register*, Vol 55, No. 74, pages 14350–14355.
4. "Processes, Coefficients, and Models for Simulating Toxic Organics and Heavy Metals in Surface Waters," *EPA/600/3-87/015*, (June 1987).
5. Implementation of Texas Water Commission Standards Via Permitting, (May 1989).
6. Delos, C.G., W.L. Richardson, J.V. DePinto, R.B. Ambrose, P.W. Rogers, K. Rygwelski, J.P. St. John, W.J. Shaughnessy, T.A. Faha, W.N. Christie. "Technical Guidance Manual for Performing Waste Load Allocations. Book II: Streams and Rivers. Chapter 3: Toxic Substances," *EPA/440/4-84/022* (1984).

14 HOW WILL INDUSTRIAL PRETREATMENT BE IMPACTED BY EPA'S TECHNICAL SLUDGE REGULATIONS?

Robert G. O'Dette, Environmental Engineer
Tennessee Department of Health and Environment
Nashville, Tennessee 37219

INTRODUCTION

The Federal Water Quality Control Act of 1987 (P.L. 100–4) made several significant amendments to the Clean Water Act (The Act). One of these amendments pertained to municipal (sewage) sludge in Section 405(d). The overall intent of this particular amendment was to insure that municipal wastewater treatment plant sludges would be controlled by regulations promulgated under the authority of the Act.

On February 6, 1989, EPA's technical sludge regulations, 40 CFR Part 503 were proposed. These regulations will affect every municipality in these United States as well as the majority of industrial indirect discharges to public owned treatment works (POTW's). The 503 regulations are perhaps the agency's most controversial set of regulations to date. However, the ink had hardly dried on the paper before the scientific sludge experts from around the country were saying the proposed rule was technically flaw. EPA's lack of peer review is the main reason for this, and although the agency says it wants to promote the beneficial reuse of sludge, the proposed regulations are counter to this policy.

Mr. Bob Perry representing the Water Pollution Control Federation (WPCF) made this point in his public hearing statement on July 6, 1989, in Washington, D.C. Bob said EPA's proposed regulations, if implemented, would result in the elimination of beneficial reuse as a viable sludge disposal alternative. This of course would be in direct conflict with the goals of the Act and EPA's own 1984 Policy on Municipal Sludge Management.

Bob further pointed out that in light of this policy, it was ironic that EPA would propose regulations which could result in the curtailment or elimination of several projects recognized by the agency as model reuse programs. These include composting facilities operated by Los Angeles County (California), Washington Suburban Sanitary Commission in Montgomery (Maryland), and Sussex County (Delaware). All three of these facilities received awards in EPA's 1988 Beneficial Reuse of Sludge Award Program. If the proposed regulations are put into effect, these facilities and many others like them could be out of business. (A fact which is not included in EPA's analysis of the cost impact of the regulations.)

Why is this issue important to industries discharging to POTWs? For the answer all one has to do is look at Section 307(b) of the Act and the General Pretreatment Regulations (40 CFR Part 403). The stated purposes of pretreatment programs are to fulfill three primary objectives. First, to prevent the introduction of pollutants into POTWs that will interfere with the operation of the POTW, including interference with the use or disposal of municipal sludge. Second, to prevent the introduction of pollutants into POTWs that will pass through the treatment works or be otherwise incompatible with such works. Third, to improve opportunities to recycle and reclaim municipal and industrial wastewater and sludges.[1]

In addition to the general prohibition of interference or pass through, EPA has established several specific prohibitions on the introduction of pollutants that create a fire or explosion hazard, are excessively corrosive, may obstruct flow, are slug loads, or are excessively hot. Although categorical restrictions are established for certain types of indirect industrial discharges, such as electroplaters, the basic framework of the pretreatment program relies heavily on each municipality identifying its own problems and proposing solutions. EPA promotes the development of local limits to account for site-specific factors that may concern the performance of a POTW or the disposal of its sludge.

The rationale behind this three-part strategy is, first, that categorical standards provide nationally uniform effluent limits affording a technology-based degree of environmental protection for discharges from particular categories of industry. Second, the prohibited discharge standards recognize

45th Purdue Industrial Waste Conference Proceedings,

the site-specific nature of the problems they are intended to address at sewage treatment works and provide a broader baseline level of control that applies to all industrial users (IUs) discharging to any POTW, regardless of whether or not the IUs fall within particular industrial categories. Third, local limits are specific requirements developed and enforced by individual POTWs implementing the general and specific prohibitions, and also going beyond them as necessary to meet State and local regulations.

This approach ensures that site-specific protections necessary to meet pretreatment objectives are developed by those agencies best placed to understand local concerns—namely POTWs. In this scheme, POTW development and implementation of local limits is the critical link in ensuring that pretreatment standards protecting both the local treatment works and local receiving environment are applied.[2]

The Pretreatment Implementation Review Task Force (PIRT, a work group made up of representatives from municipalities, industries, states, environmental groups and EPA Regions to provide the Agency with recommendations on day-to-day problems faced by POTWs, States, and industries in implementing the Pretreatment Program) suggested in its findings that many POTWs misunderstood the relationship between local limits and categorical standards, thereby hindering effective implementation of pretreatment standards. Categorical standards and local limits are distinct and complementary types of pretreatment standards. Promulgation of a categorical standard by EPA in no way relieves a municipality from its obligations to evaluate the need for, and to develop, local limits to meet the general and specific prohibitions in the General Pretreatment Regulation. As suggested earlier, categorical standards are developed to achieve a nationally-uniform degree of water pollution control for selected industries and pollutants. Local limits are intended to prevent site-specific plant and environmental problems resulting from any non-domestic user.

In many cases POTWs may impose local limits which regulate categorical industries more stringently and/or for more pollutants than are regulated in the applicable categorical standard to afford additional plant or environmental protection. In this case, the local limit supersedes the categorical standard as the applicable pretreatment standard. As a corollary, however, a less stringent local limit does not relieve a categorical industry from its obligation to meet the Federal standard. The central point to be remembered is that the existence of a Federal categorical standard should not deter a city from its obligation to evaluate discharges from all nondomestic users, to identify problem pollutants and to adopt more stringent technicall-based local limits, where necessary.[2] This is especially true regarding the sludge questions.

POTWs are required to prohibit IU discharges in amounts that cause violation of applicable sludge disposal or use regulations, or restrict the POTW from using its chosen sludge disposal or use option. The importance of this requirement is underscored by the recent Clean Water Act amendments which require the incorporation of sludge criteria and requirements into all NPDES permits when they are issued or reissued. EPA has prepared interim guidance on what presently must be incorporated into permits to comply with these amendments. However, the big ticket item is the technical sludge regulations and what the final restraints imposed by these 503 regulations will do as POTW's re-evaluate local limits. The 503 regulations will set forth pollutant-specific criteria relevant to disposal and use practitioners. Thus, POTWs applying sludges to cropland or composting for example, must develop local limits to avoid violations of applicable State and Federal sludge disposal limitations. When an IU's discharge renders sludge unsuitable for land application and necessitate landfilling, incineration, or additional treatment of sludges, the POTW not only must pay the costs of additional treatment, but may lose the revenue obtained from selling sludge. This is considered interference.[2]

Receipt of some industrial wastes may interfere with POTW operations, resulting in a violation of NPDES permit conditions calling for specific removal efficiencies to be achieved and for the plant to be well-operated and maintained. Moreover, some discharges of pollutants, while not causing POTW NPDES permit violations or violations of sludge disposal regulations, can nevertheless disrupt POTW operations, increase POTW operation and maintenance costs, and may cause violations of specific prohibitions. For example, IU discharges that inhibit the POTW's biological treatment systems result in reduced POTW efficiency and, as a result, increased operating costs. At worst, process inhibition may necessitate reseeding and stabilization of the treatment unit. In addition, process inhibition or upset may result in the production of sludges that require either special treatment before disposal, or disposal in a manner not generally practiced by the POTW. This would also be considered interference.[2]

The question of interference relative to contaminated sludge and the lack of sludge criteria and/or regulations is a major point of concern with the Natural Resources Defense Council (NRDC). EPA's removal credits provision was invalidated in NRDC, Inc. V. EPA, 790 F.2d 289 (3rd Cir, 1986), due to

Table I. D/M Criteria v. Selected Tennessee POTWs[5]

PARAMETER	503[4] Criteria MG/KG	Range[5] MG/KG	Percent (%) Failed
Cd	18	2 to 89	26
Cr	530	21 to 7,820	19
Cu	46	53 to 2,010	**100**
Pb	130	21 to 930	63
Ni	76	11 to 1,610	48
Zn	170	277 to 6,414	**100**

this concern. Additionally, it appears as though this same concern is the driving force behind NRDC's actions and comments relative to the 503 regulations.

In NRDC's comments on the 503 regulations, they stated that removal credits should be allowed only to the extent that the increase quantity of toxic contaminants allowed in the affected sludge do not interfere with sludge's beneficial reuse. Otherwise, Congress' goal of encouraging *beneficial reuse* will be undercut. To ensure that removal credits do not undercut either the national pretreatment program or National Sludge Policy, NRDC said that credits should be available only up to the numerical limits set for pollutants in beneficial reuse scenarios (i.e. land application and distribution and marketing). Otherwise, NRDC believes that EPA creates a perverse incentive for industrial users to pressure POTWs to steer clear of composting and land application (for which more stringent numerical limits generally would apply), in favor of non-productive disposal methods such as incineration, monofilling and co-disposal.[3]

In summary, NRDC stated that removal credits should be available on a pollutant-by-pollutant basis only up to the levels of sludge contamination that would not interfere with beneficial reuse of sludge. NRDC says that removal credits for pollutants not otherwise regulated are illegal, for the same reason that the Court of Appeals forbade them in NRDC v. EPA, namely, that if EPA has not defined a safe level of a particular pollutant in a particular use/disposal setting, it cannot authorize removal credits for that pollutant. NRDC believes that EPA's approach turns the law on its head: while declining to regulate these 17 pollutants as part of the required "comprehensive framework" for protection against sludge hazards, the agency in their opinion proposes to give industry permission to increase its discharge of these toxics. NRDC says that EPA cannot have it both ways. If it wishes to define an adverse effects level of a toxic contaminant for purposes of issuance of a removal credit, it must define a safe level (i.e., a numeric concentration limit) for regulatory purposes, as well.[3]

The key point with the discussion about interference is that with the technical sludge criteria, the missing link is essentially put in place for total evaluation of local limits. It will be shown that sludge criteria as proposed in the 503 regulations on February 6, 1989, if finalized will be devastating to IUs and most POTW Pretreatment Programs. The dilemma lies in the fact that the sludge criteria in almost all cases will be the controlling factor relative to pretreatment limitations. To demonstrate this point, let's evaluate the proposed 503 regulations and their requirements for distributing and marketing (D/M) of sludge in relationship to existing sludge characteristics. Table I shows a comparison of the sludge regulations requirements for D/M (40 CFR Part 503.23) for six metals versus actual sludge characteristics from 27 selected POTWs in Tennessee with pretreatment programs.[4,5]

Note that 100% of POTWs failed to meet the requirements for D/M for copper and zinc. This means that not one of those POTWs could D/M their sludge, which would in turn mean a violation of the concept of "interference" and additional pretreatment would be mandated.

Table II. D/M Criteria v. 1980 POTW Study[6]

PARAMETER	503[4] Criteria MG/KG	Range[6] MG/KG	Median MG/KG
Cd	18	1.1 to 59	12
Cr	530	63 to 1,762	278
Cu	46	100 to 1,427	539
Pb	130	39 to 1,169	300
Ni	76	12 to 803	86
Zn	170	420 to 8,468	1,148

Table III. D/M Criteria v. EPA's 44 POTW Study and EPA's Survey of 282 POTWs[7]

PARAMETER	503[4] Criteria MG/KG	44 POTWs[7] Average MG/KG	282 POTW Survey[7] Average MG/KG
Cd	18	32	44
Cr	530	428	785
Cu	46	562	910
Pb	130	378	520
Ni	76	134	217
Zn	170	1,409	2,194

Table II does a similar comparison using date from an October 1980 EPA Study.[6]

Table III makes this same comparison using EPA's 44 POTW study and their survey of 282 POTWs.[7]

Obviously, there is a potentially serious problem facing IUs and POTW authorities. EPA's Guidance Manual recommends that the POTW perform at least a limited amount of effluent and sludge monitoring as part of its preliminary sampling program, in order to screen for pollutants which have concentrated to detectable levels in effluent or sludge even though not detectable in the influent. EPA also recommends that sludge disposal criteria-based local limits should insure that the maximum concentration of the pollutant in the POTW sludge not exceed one-half of the sludge disposal criterion. Following this recommendation cuts the values in half and would really tighten up industrial pretreatment requirements, especially for copper and zinc.

Some often asked questions are: How did the criteria get so restrictive?; and was this the original intent of Congress? It is the author's opinion the EPA violated the intent of Congress in a number of ways. Under Section 405 (d) of the Clean Water Act, as amended, EPA must, on the basis of available information, establish numerical limits and specify management practices for toxic pollutants in sewage sludge that may adversely affect public health or the environment. The key statement is "on the basis of available information", because EPA has violated not only the intent, but also the letter of the law by going beyond available information and synthetically developing their own "information" through complex computer models, based upon assumptions. Assumptions are not existing, documented "available data". Furthermore, the Act says nothing about these criteria having to be risk-based. In fact, it is clear from Section 405(d)(2)(A), that Congress not only wanted these proposed regulations on the "basis of available data". . . ..they wanted them quickly. To prove this, all one has to do is look at the Congressional Record and note President Reagan pocket-vetoed the bill November 6, 1986. This was a bill in which Congress had mandated proposed sludge regulations by November 30, 1986. Therefore, it seems Congress believed EPA had enough data to issue these proposed rules immediately; otherwise, they would have provided more time for regulation development.

Why did Congress mandate such action on such a tight time schedule? Because EPA said they had the data and could do it. The Conference Report accompanying the 1987 Clean Water Act amendments states:

"Agency officials informed the Congress (as long ago as 1983) that there were at that time sufficient data to regulate comprehensively approximately 20 of the most common toxic sludge pollutants. The conferees intend these 20 to be regulated in the 1987 regulations."

Because EPA said it had sufficient information to move quickly in regulating the first group of pollutants, the law instructed that agency to propose and promulgate the first round of sludge rules expeditiously, "on the basis of available information." The second round of regulations (for which all statutory deadlines have also passed) was to cover other pollutants for which then-existing data were insufficient. The law's provision calling for biennial review and revision, "for the purpose of identifying additional toxic pollutants and promulgating regulations", is intended to allow for future changes in the regulations as technical knowledge increases." With this statutory scheme, Congress took great pains to ensure the first round of regulations could go forward without delay. "The deadlines imposed, when viewed against the various statements of purpose (in the law), show simply that Congress concluded that prompt, though imprecise, regulations were preferable to no regulations during the time period, while further studies were being conducted to provide the Administrator with more complete information." State of New York v. Gorsuch, 554 F.Supp. 1060 (S.D.N.Y. 1983).[8]

Furthermore, it is this author's opinion that pursuant to the statutory directive in Section 405 of the Act, EPA went beyond its regulatory authority by proposing new risk-based technical sludge regulations. These regulations will eventually establish minimum Federal standards for use and disposal of sewage sludge when:

1. Applied to Land (Agricultural use)
2. Distributed and Marketed (D/M)
3. Monofilled (sludge-only landfill)
4. Disposed of on the land surface (dedicated sites)
5. Incinerated

The co-landfilling of sludge with municipal solid waste and ocean dumping (banned by Congress after 1991) are covered under other provisions and rulemaking in 40 CFR Part 258, as proposed on August 30, 1988. Future rulemaking under Section 405 authority will address additional sludge pollutants and sewage use and disposal practices.

The proposed rule contains standards for each end use and disposal method consisting of sludge concentration or pollutant loading limits, management practices, and other requirements such as monitoring, record keeping, reporting, and management controls over users and contractors by treatment works. EPA is proposing to establish numerical limits for 25 pollutants in sludge to be land applied, 22 for distribution and marketing, 18 pollutants for monofilling, 18 for surface disposal, and 8 for incineration. The Natural Resources Defense Council (NRDC) says that two important questions are raised here regarding the scope of pollutants proposed for regulation in Round I: (1) has EPA incorporated an appropriate safety factor to account for the limited scope of pollutants to be addressed in Round I, and (2) are there additional pollutants that should be addressed in Round I, but are not?

As proposed, the requirements would apply to both publicly owned treatment works (POTWs) and privately-owned treatment works which treat domestic wastewater as well as septage. However, sludges produced by privately owned industrial facilities, which trap domestic sewage along with industrial process waste, will not be covered in the first round of regulation.

On page 5857, column 3, of the February 6, 1989, Federal Register (proposed 40 CFT Part 503), EPA estimates the benefits of this proposed rule will be an annual reduction of 9.5 cancer cases and 5,266 cases of other health effects. A summary of these data is given in Table XI-1 on page 5858. In order to make such a statement, there would have to be actual, documented cancer and/or other health effects to which a scenario using the term "cases avoided" could be validly used. But "on the basis of available data" there are no documented cancer and/or other health effects related to current sewage sludge use and disposal practices. The only numbers which EPA have are those developed from their computer models. The agency should produce the names and dates of death of any individuals who have died as a direct results of any sewage sludge use or disposal practice. The facts of the matter are that all the available data and evidence show there are no problems. EPA has enough available data to show we are not at risk, and current sludge practices, coupled with continued emphasis on pretreatment programs, protect human health and the environment.

One should compare the proposed sludge restrictions with the restrictions for the agricultural community. Feed lots, dairy farms and manure spreading operations are virtually unregulated. Fertilizer needs are based upon the crop and prevailing soil and climate conditions. The crop must meet certain standards to be sold or used as feed and/or food. The same concept should be used when municipal sludge is involved. As far as pathogens are concerned, from a safety standpoint, compare manure dropped in a feed lot, dairy farm or hay field versus stabilized or composted municipal sludge. The key to safety is in proper treatment/stabilization coupled with good site management and controls on the end products. Basing sludge criteria (ultimate disposal restrictions) on this type of risk assessment methodology is unsound and unfair. . . . .**IF IT AIN'T BROKE. . . .DON'T FIX IT!**

By their own admission on page 5763, column 1, EPA has major problems with some of the data base, namely, the "40 City Study". The problems noted were: (a) the study did not involve processed/stabilized sludge; (b) the "40 City study" was NOT designed to assess the quality of the sludge leaving a POTW; (c) the data collected in 1979–1980 were analyzed using antiquated methods; and (d) the data were collected before most pretreatment programs being effective in reducing toxic pollutants to POTWs. Therefore, can EPA use these data to portray a current sludge profile? Is it appropriate and reasonable, or does it represent a much worse than "reasonable worse case scenario"? Additionally, total metals were used rather than the soluble fraction which makes the numbers even more restrictive.

Again, on the "basis of available data", where are the documented human health and environmental problems that require criteria as restrictive as those proposed? If sewage sludge is so dangerous, surely one documented case can be presented to substantiate EPA's allegations.

We must not forget many industrial pretreatment facilities have been built and put into operation since the "40 City Study" of 1979–80. The pretreatment job is not yet complete, but overall the sludge is "cleaner", i.e., less potential for "interference", human health and environmental problems than before. Thus, based upon available data (municipal sludge inventories, pretreatment mass-balance data, and sludge analyses) it seems where effective pretreatment programs exist or where there are no indirect dischargers. . .there are no sludge problems. In the State of Tennessee, we have used the EPA required municipal sludge inventory to discover industrial pretreatment offenders. In essence "the sludge doesn't lie"! If someone is dumping a toxic substance into the sewerage system, it will show up in the sludge analyses. The best rule to follow is if nitrogen or phosphorus does not control land application rates, then look for a pretreatment problem. And this is as far as the federal regulations should go or need to go. Because Congress stated in Section 405(e) of the Act: "The determination of the manner of disposal for use of sludge is a *local* determination". Let it be this way, provided nitrogen or phosphorus controls and good site management practices or D/M techniques are followed. Otherwise, what this proposed rule does is give no one any prerogative other than dedicated site, Municipal Sanitary Waste Landfill (MSWLF), or incineration. All of these options are of course counter to EPA's beneficial reuse policy.[8]

What about the question of the assumptions used by EPA to develop the technical sludge criteria. As far as this author is concerned EPA's rationale and assumptions contain more sawdust blarney than a carnival midway!

It goes beyond the scope of this paper to look at all assumptions or attempt to evaluate all pathways. However, a quick look at a couple of items should be enough to cast a long shadow of doubt over the foundation on which the 503 regulations are built. This author believes the foundation is not good old solid bedrock. Consider Pathway 2F for example, where the assumption is made that a child from the age of 1 through 5 would ingest some sludge/soil mixture every day of his/her life for these 5 years. This would happen after his house was constructed on land that had been sludged. Only the first 15 cms of earth are considered to be effected. . . .But would this 15 cms survive during construction with site preparation, grading and drainage and building construction?

Another unrealistic limiting pathway was toxicity to ducks eating worms which was referenced on page 5778, column 3. This was the limiting pathway for lead in both land application (agriculture) and D/M. Are these assumptions based upon "available data"?[6]

Another major problem is that EPA has based its numeric standards on modeling the amount of a pollutant that reaches a target organism. The selection of a target organism is therefore a major determinant of the numeric limits. EPA set out to select as its target organism the "most exposed individual (MEI), plant, or animal; one that remains for an extended period of time at or adjacent to the site where the maximum exposure occurs.

The questions remains: how does one assess whether a given MEI adequately represents the most exposed target organism? The enabling Clean Water Act Legislation for the proposed rule requires:

> "the management practices and numerical criteria established (under Section 405) shall be adequate to protect public health and the environment from any reasonably anticipated adverse effects of each pollutant."

Senator Stafford, author of the provision, explained during floor debate on the 1987 amendments that "EPA's rules must protect public health and the environment with an ample margin of safety, and must take care to protect the health of individuals or populations which are at a higher risk than the population as a whole". The language and legislative history of section 405 and the enabling legislation provide EPA with a clear mandate to protect that individual or population reasonably anticipated to have the highest exposure.

Based upon previous discussions, it should be apparent that EPA's MEI (for sludge) could not "reasonably" exist. While one could support EPA's range of acceptable risk levels, it is believed the human exposure routes in the model are unrealistic and therefore do not provide valid risk reduction. Human exposure routes to any significant amounts of sludge are less direct than to other substances regulated by other EPA programs. It is important to establish acceptable risk levels on the basis of "good science" and the most recent data.

For land application, the assumed level of risk is 1 in 10,000. For the small number of individuals in the U.S. population who will be affected by land where sludge is applied, this level of risk translates to orders of magnitude lower risk than for the U.S. population as a whole. The risk numbers should be calculated for a similar population base when comparing them with other EPA standards (e.g. drinking water standard of 1 in 1,000,000 for the whole U.S. population).

A reasonable level of risk for the "MEI" should be developed to correspond to other EPA risk-based regulations. The proposed MEI should reflect the concept of the "Most *Probably* Exposed Individual" in order to provide a realistic basis for establishing pollutant and management practices.

Before leaving the MEI or "Sludged-Man", let's also look at the pathogen reduction requirements in the proposed regulations. Unless "Class A" is achieved, severe restrictions are placed upon food crops with harvested parts that touch the sludge/soil mixture (18 months for above ground crops and 5 years for crops below ground). Such a conservative requirement will again destroy beneficial reuse options currently available. This becomes very apparent when one evaluates the requirements for Class A where essentially a product safer than pasteurized milk has to be produced. Note the following comparison:

PASTEURIZATION

Milk	Sludge (Class A)
145 degrees F for 30 minutes	158 degrees F for 30 minutes

Another interesting comparison is sludge v. top soil commonly found around the United States:

PARAMETER	503 SLUDGE MG/KG	COMMON TOP SOIL MG/KG
As	0.20	1–50
Cd	0.040	0.05–0.70
Cu	8.4	2–100
Pb	0.35	2–200
Ni	7.0	5–500

Looking at these two comparisons and realizing that sludge is going to have to be "clean" for children 1–6 years of age to eat every day of their life, ask yourself this questions: Will sludge have to better than:

(a) Top Soil
(b) Pasteurized Milk
(c) Cookies (what kids eat)
(d) All of the above.

The answer is (d)-all of the above! Therefore, are we at risk?

Will we be at risk unless these proposed regulations become effective and are enforced? This author believes that common sense will tell one that these proposed regulations are overly restrictive. This fortifies the need to have these regulations peer-reviewed by scientific experts such as the W-170 committee. Most of us are not at this level of scientific expertise. However, just because we can't lay an egg, doesn't mean we can't tell a good one from a bad one!

As Bob Perry pointed out, one of the biggest challenges for municipalities operating or contemplating beneficial reuse projects is winning public acceptance. This author, along with the WPCF believes a strong federal regulatory framework can be an important tool in building such acceptance. However, these regulations must be based on valid and correctly applied scientific evidence.[8]

Ralph Waldo Emerson once said: "Society is always taken by surprise with any new example of common sense." The environmental society involved with sludge is due a surprise!

Given the current set of circumstances, all POTWs and IUs are in limbo, as are the regulators who set this whole thing in motion on February 6, 1989. We are telling municipalities in Tennessee to:

1. Land apply sludge
2. Own your own land
3. Apply a dry, manure-type of product

Yes, we are telling them to land apply, mainly because it is the cheapest, most beneficial and productive option. Besides, no one knows for certain what the final 503 requirements will dictate. By owning their own land, they control their destiny and can economically plan for the future, knowing that if 503 comes back to reasonable requirements, they will be okay. The third point is to apply a dry manure-type product which helps their public relations department and reduces the number of hauls which have to be made to their sludge sites.

From our experience in Tennessee, the best interim to long term solution would be to employ the N-VIRO (cement kiln dust) system or lime stabilization/pasteurization. These systems are relatively cheap to install, can be put in place rather quickly, and produce a product that has the best chance of being land applied even under the proposed 503 regulations. Currently, these operations meet the PFRP requirements under 257, and would also meet the Class A requirements of 503. This is a tremendous advantage because you will be free of any site or crop restrictions relative to pathogens.[8]

Industries and POTW authorities need to take a close look at the proposed 503 requirements and follow the progress of the final rule. Currently, EPA is scheduled to have the final rule by October 1991 and so far NRDC seems to be going along with time schedule, since they have not pressed the issue any further in the courts.

REFERENCES

1. Zorc, Joseph M.; Hall, John C.; Rissetto, Christoper L., "Minimizing Liabilities Facing POTWs", *Journal Water Pollution Control Federation*, Volume 60, Number 1 (January 1988).
2. "Guidance Manual on the Development and Implementation of Local Discharge Limitations Under The Pretreatment Program", U.S. Environmental Protection Agency, (December 1987).
3. Comments of The Natural Resources Defense Council, Inc. and the Environmental Defense Fund on EPA's Proposed Rule, "Standards for the Disposal of Sewage Sludge", 54 Fed. Reg. 5746 *et seq*. (February 6, 1989).
4. "Standards for the Disposal of Sewage Sludge" -40 CFR Parts 257 and 503, 54 Fed. Reg. 5746 et seq. (February 6, 1989).
5. State of Tennessee Municipal Sludge Inventory 1989.
6. Naylor, L.M. and Loehr, R.C., "Priority Pollutants in Municipal Sewage Sludge", *BIOCYCLE* (July/August 1982).
7. *BIOCYCLE* (January/February 1985).
8. O'Dette, R.G., "EPA's Technical Sludge Regulations: A Dliemma for Municipalities", Presented at the 62nd Annual Water Pollution Control Federation Conference, San Francisco, California (October 17, 1989).

15 REGULATION OF LIQUID HAULED WASTES IN LOS ANGELES COUNTY

Suzanne S. Wienke, Project Engineer
Sanitation Districts of Los Angeles County
Whittier, CA 90607

INTRODUCTION

The County Sanitation Districts of Los Angeles County (Districts) are a group of special districts serving the wastewater and solid waste management needs of over 4 million people and 8,500 industrial users in Los Angeles County. Most of the southern and eastern portions of Los Angeles County as well as the Santa Clarita Valley and Lancaster/Palmdale areas to the north are included in the Districts' service area. The City of Los Angeles separately operates their own wastewater and solid waste management facilities. Unincorporated portions of the county which are not in the Districts' or the City's service area are under the jurisdiction of the County Department of Public Works (DPW).

Most of the material being discharged into the Districts treatment system comes from sewer-connected homes and businesses. However, there are areas in the Los Angeles basin which are not served by sewers but instead use septic tanks and cesspools. Riverside and San Bernardino Counties to the east of the Districts service area and the rapidly developing areas of northern Los Angeles County have many homes and businesses which are not sewer-connected but instead use septic tanks or cesspools. Periodically, these systems must be pumped out and the septage hauled away. Disposal of septage and other types of nonhazardous liquid wastes is generally accomplished by discharge to an accessible publicly owned treatment works (POTW).

In 1985, the Districts implemented a program to control the discharge of liquid hauled wastes to the sewerage system at a series of designated manhole stations. Prior to 1985, trucks used the designated dumping manholes in an uncontrolled manner which generated concerns about the potential to discharge toxic and/or incompatible materials to the sewerage system. Uncontrolled use of the system also lead to concerns about Districts' users subsidizing the treatment and capital costs for dischargers outside the Districts' service area. Moreover, with the reauthorization of the Resource Conservation and Recovery Act (RCRA), POTWs were faced with the mandate that acceptance of hazardous waste via truck, even unknowingly, would require the POTW to comply with regulations for a Treatment, Storage and Disposal Facility (TSDF). The significance of the RCRA regulations is discussed in more detail below.

RCRA REGULATIONS

Currently, mixtures of domestic sewage and other wastes which commingle in the POTW's collection system prior to reaching the treatment plant are excluded from RCRA regulation under the Domestic Sewage Exclusion. However, hazardous materials discharged to a POTW by rail, truck, vessel or dedicated pipeline are not part of the domestic sewage exclusion and their authorized discharge can only occur if a POTW complies with RCRA requirements for a TSDF.

The Environmental Protection Agency (EPA) recognized that POTWs are required to comply with extensive regulations under the Clean Water Act. Consequently EPA adopted a special TSDF provision known as "permit-by-rule," □40 CFR 270.60(c), for POTWs accepting hazardous wastes by truck, rail or dedicated pipeline. California operates the permit-by-rule similarly. These rules are less comprehensive than those rules which apply to non-POTW TSDFs.

Under the permit-by-rule regulation, a POTW must 1) have an NPDES permit, 2) comply with that permit, 3) obtain a RCRA ID number, 4) comply with certain manifest and reporting requirements, 5) satisfy corrective action requirements and 6) only accept hauled wastes which meet all federal, state and local pretreatment requirements. Many of these steps are easily accomplished, however the manifesting and reporting requirements involve significant amounts of time and manpower on the part of the POTW.

45th Purdue Industrial Waste Conference Proceedings, © 1991 Lewis Publishers, Inc., Chelsea, Michigan 48118. Printed in U.S.A.

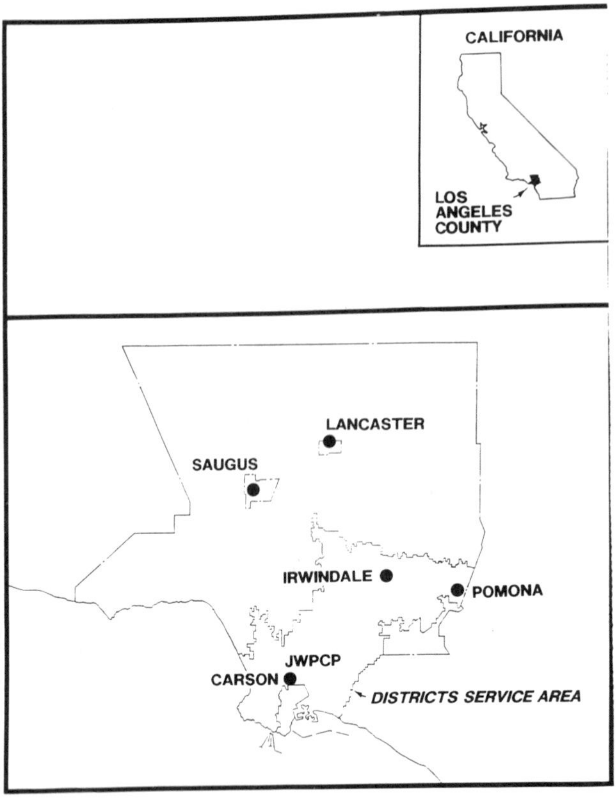

Figure 1. Location of liquid waste disposal stations managed by the sanitation districts of Los Angeles County.

The lists of materials which are considered hazardous are extensive. Basically, if a waste is a listed waste in 40 CFR 261; exhibits characteristics of listed wastes such as ignitability, corrosivity, reactivity or EP toxicity; or is a mixture of a listed and nonhazardous waste, then it is considered hazardous. Some materials such as wastes generated in a household are not hazardous waste by definition.

In June 1987, EPA published *Guidance Manual for the Identification of Hazardous Wastes Delivered to Publicly Owned Treatment Works by Truck, Rail, or Dedicated Pipe* which describes various methods for complying with RCRA requirements for POTWs. Of these methods, the most restrictive is to prohibit the discharge of any waste to the POTW via rail, truck or dedicated pipeline. However, for communities with nonsewered areas such as portions of southern California, this policy is overly restrictive and could endanger public health and environment by creating incentives for improper disposal of septage.

A second method is to prohibit the discharge of all wastes hauled from an industrial facility. This approach allows for continued service to the residential community while protecting against the discharge of hazardous industrial wastes. However, there is some potential for industrial wastes to be illegally mixed with residential wastes for disposal. Implementation of a good monitoring and screening program would ensure that hazardous wastes are not received by the POTW.

A third method is to prohibit the discharge of industrial process water only. This allows residential and domestic-type sewage from industrial facilities to be discharged at the POTW. This method also poses the problem of possibly having industrial process wastes mixed with domestic sewage, however a good monitoring program would enable such practices to be detected and proper enforcement action to be taken.

THE DISTRICTS' PROGRAM

The Districts have basically adopted the third method for control of liquid hauled wastes by designating appropriate disposal locations, limiting the types of wastes which are acceptable, charging a fee for the opportunity to use the manhole stations, and instituting a screening program.

All of the transporters using any of the Districts' five locations are required to obtain permits in which they acknowledge that wastewater from septic tanks, cesspools, and chemical toilets can be discharged. Transporters purchase coupons to pay for discharging the loads and agree to complete the generator information on the coupon. Currently, the Districts do not charge a fee to obtain a permit.

Currently, the permit period for transporters is five years. About 400 trucks have been issued permits by the Districts, though only about 200 are current. As part of the permit approval process, a hauler is required to obtain a County of Los Angeles Health Services registration number. To obtain a number, the hauler's vehicle must be inspected by Health Services and appropriate fees paid. This requirement provides assurance that the vehicle is capable of hauling sewage in a safe manner.

DISTRICTS' FACILITIES

The Districts operate five disposal stations located throughout its service area. All of the facilities are manned by attendants. The facility at the Joint Water Pollution Control Plant (JWPCP) in Carson is open 24 hours per day, seven days per week. In 1989, 429 loads were accepted, with an average volume of 4,221 gallons per load. This usage represents 3.9 percent of the overall system loads accepted and 9.2 percent of the total volume. This facility is located in an industrialized area which has been sewered for quite some time. Most of the material discharged here is from sewage holding tanks at large industrial projects and thus required more vigorous screening.

The Pomona disposal station is located adjacent to the Pomona wastewater treatment plant. Although it is open only for limited hours, the Pomona manhole receives 23 percent of the total system loads and 23.2 percent of the volume. Most material discharged here is domestic sewage. Because it is located on the eastern border of Los Angeles County, it receives many out-ofcounty loads from adjacent San Bernardino and Riverside counties farther east.

The Los Angeles County DPW operates a disposal station in a sewer maintenance yard located in Irwindale. Because this facility discharges to the Districts' treatment facilities, it is officially considered to be part of the Districts' program. This facility is also open limited hours and receives a variety of wastes; it is the recipient of 24.9 percent of the loads and 32.4 percent of the volume of the program. Because this facility is not under the Districts' direct control, special protocols must be developed identifying each agency's obligations for proper operation of the facility.

In the outlying areas of the County, the Districts operate two disposal manholes, one at the Saugus and one at the Lancaster wastewater treatment plant. Both the facilities receive small volume loads of primarily septage and chemical toilet waste.

Overall, the Districts received 11,156 loads from the five facilities in 1989, which represents a 7 percent increase over loads received in 1988. The total volume discharged in 1989 was 19,621,675 gallons which is an average of 1,759 gallons per load.

TYPES OF ACCEPTABLE WASTES

The Districts currently accept five types of wastes at the disposal manholes: 1) septage, 2) cesspool wastes, 3) chemical toilet wastes, 4) grease traps, and 5) special nonhazardous industrial wastes. Septic tanks and cesspool wastes are generated by underground treatment/holding basins for domestic wastes. In order for these systems to operate properly, the contents of these holding basins are periodically removed, usually by a pump truck.

Chemical toilet wastes are usually much higher in suspended solids and chemical oxygen demand (COD) than septage. The waste is often tinted blue or green from the disinfectant chemicals. Although of higher strength, the relative impact of these wastes is diminished because the volume of each load of chemical toilet waste is generally small (usually less than 850 gallons).

Although grease trap wastes are accepted, it is likely that the practice will be discontinued in the near future because it has caused some operational problems within the treatment system such as excessive odors at the disposal sites and abnormal amounts of floating oil on primary clarifiers. Also, most of the waste loads rejected at the disposal stations are grease trap wastes with low pH. Because the disposal manholes discharge into large, fast flowing sewer lines or headworks to the treatment plants, clogging problems are unlikely. However, the premise of installing a trap to keep grease out of the sewer at one location only to introduce it back into the system at another location seems inconsistent, especially since some rendering and other processing facilities are willing to accept or purchase these materials.

Special nonhazardous industrial wastes are accepted at the disposal stations on a case-by-case basis. Prior to disposal the generators of these wastes must obtain a special permit from the Districts after having the waste certified as nonhazardous. This procedure is specified by the California Department of Health Services and includes chemical characterization as well as toxicity testing. Currently there are only two companies with special permits. One manufactures shampoo and the other makes toilet bowl cleaner. In both cases, some of the wastes generated at these companies presented operational problems at the treatment plants which the companies normally discharged to. The special permit provides a mechanism which can remove those wastes from the influent to the affected plants and discharge it to a facility which can adequately handle the waste.

There are no other types of acceptable wastes. Clean water, drilling mud, spoiled food products or anything outside the five specified categories is prohibited. Nevertheless, generators continue to try to bring unacceptable material to the disposal facilities, necessitating careful front-end screening of waste loads.

DISPOSAL PROCEDURES

For each waste load, the attendant carefully reviews the information contained on the coupon provided by the hauler. This must include the name, address and telephone number of the generator and the type of waste. Consequently, the coupon acts both as a method of payment and as a mini-manifest.

Each hauler is required to remove a sample of their waste for the attendant to examine. Examination usually consists of a visual check, odor check and pH determination. If the pH is between 6 and 8 and otherwise appears to be acceptable, the load is discharged at the disposal station. If the pH is below 6, the load is rejected, the sample is saved for further testing and the determination is made whether further investigation and/or enforcement actions are warranted. If the pH is between 8 and 10, the attendant will run field tests for metals and cyanide and either accept or reject the load based on results obtained. For pH greater than 10, the load is rejected, the sample retained for further testing and the Industrial Waste Supervising Inspector is contacted.

The manhole attendants are trained in the use of various field testing methods including heavy metals, cyanide and selected organic solvents. However, common sense and knowledge about the types of wastes which are acceptable are the best defense. Random testing of samples for a broader range of constituents will be implemented in the near future to effectuate a more rigorous screening process. In addition, generators of nonhazardous industrial wastes will be required to recertify their wastes as nonhazardous on an annual or semiannual basis.

DISPOSAL CHARGES

In order to recover treatment and administrative costs, the Districts charge the hauler a fee for each load based on the truck volume. Truck volumes are divided into four classes: Class I—up to 850 gallons, Class II—up to 2,000 gallons, Class III—up to 4,000 gallons and Class IV—over 4,000 gallons. The cost for Classes I and II is $20 per load and $35 per load for Classes III and IV. No cash is accepted at the facilities, only prepurchased coupons. Charges are modified as needed to accommodate treatment and administrative costs.

FUTURE CHALLENGES

Because some wastes come from outside the District's boundaries, there are a variety of policy issues associated with the continuance of this practice. First, there is the question of who has jurisdiction for enforcement if an unacceptable waste load is discharged. According to the Districts' *Wastewater Ordinance* Industrial Waste Inspectors have the right to inspect facilities which discharge to the Districts' sewerage system. However, if a suspicious load is received which was generated in another jurisdiction, it may be difficult to conduct an investigation and determine the origin of the material. Moreover, without a contractual agreement as to which agency has primary enforcement authority, effective enforcement of such incidents is jeopardized.

The second issue concerns the transfer of treatment and disposal responsibilities from one jurisdiction to another by restricting geographical service areas or terminating disposal services. For example, the Chino Basin Municipal Water District (CBMWD), located in San Bernardino County, recently closed its disposal facilities to waste generated outside its service area. This was done after giving a one-year notice to local haulers. For several years preceding this notification, other agencies within

San Bernardino County had been closing their facilities to wastes generated outside their service area. As a result, CBMWD became overburdened with wastes that previously had been handled by these other agencies. After CBMWD imposed its restriction, the Districts' facilities began receiving more out-of-county wastes. For the time being, wastes from outside the Districts' service area are still being accepted; however, a regional disposal plan should be adopted which addresses jurisdictional responsibilities for waste disposal.

Another potential source for noncontrolled discharges is the recreational vehicle (RV) dump station. Within the next year, an RV dump station control program will be implemented. It will probably require that all such facilities obtain a permit, provide a locking cover for the discharge point and maintain a log book with information on the RV operator.

SUMMARY

The Districts' program regulating liquid hauled waste disposal has been ongoing for five years. The program has been effective in ensuring that hazardous materials are not discharged into the Districts' sewerage system at any of the five designated liquid waste disposal stations. In the near future, a program for control of discharges at recreational vehicle dump stations will be implemented and the current practice of accepting restaurant grease trap wastes will be discontinued. The ultimate goal of the control program is to continue service to the local residential community while ensuring that all regulations are met for facilities accepting nonhazardous liquid wastes via truck. This is a small but important part of the Districts' continuing efforts to protect the local environment.

16 DEVELOPMENT OF A STANDARD SAMPLING SYSTEM FOR THE DETECTION OF VOLATILE COMBUSTIBLE MATERIALS IN INDUSTRIAL DISCHARGES TO PUBLIC SEWERAGE SYSTEMS

Linda M. Shadler, Project Engineer
Thomas E. Knapp, Project Engineer
Ross C. Caballero, Head, JWPCP Research Section
Paul C. Martyn, Head, Monitoring Section
County Sanitation Districts of Los Angeles
Whittier, California 90607

INTRODUCTION

The County Sanitation Districts of Los Angeles County (Districts) are a group of special districts serving the wastewater and solid waste management needs of over four million people and 8,500 industrial users in Los Angeles County. The Districts own and operate eleven wastewater treatment plants which treat over 1.9 million m^3/d (500 mgd) of wastewater. Ninety-five percent of this flow is transported and treated in the Joint Outfall System (JOS), which consists of six treatment plants, 1,900 kilometers (1,200 miles) of trunk line sewers and fifty pumping plants. The treated effluents from the JOS are discharged to the ocean or to inland surface waters, and are used for irrigation, groundwater recharge or industrial process water.

In 1972, Districts adopted a Wastewater Ordinance to protect and finance the operation of collection and treatment facilities. The Ordinance established the Industrial Waste Permit Program. The Permit Program requires industrial dischargers to obtain a discharge permit for each industrial wastewater connection to the sewerage system. Over 8,500 permits have been issued to date, for industrial discharges totaling 269,000 m^3/d (71 mgd). As part of the permit approval, dischargers must provide plans that detail all monitoring and pretreatment facilities required by Districts. To maintain a valid permit, dischargers must comply with all federal, state and local effluent limitations, and all policies and guidelines for the measurement, monitoring or restriction of wastewater flows and prohibited materials.

In 1971, Chicago, Illinois, instituted a plan to eliminate explosive conditions in their sewerage system, after an explosion caused by a spill of coking wastes. Following the February 1981 explosion of spilled hexane in the Louisville, Kentucky, sewerage system, in which four people were injured and damages amounted to $51 million, there were attempts nationwide and at the Districts to implement some form of industrial combustibles monitoring program. The Districts' response was the issuance of the Combustible Gas Monitoring System (CGMS) Guidelines which require all significant potential dischargers of flammable substances to install and maintain an accurate CGMS to protect downstream Districts' facilities and the local environment.

In August 1988, an explosion resulted from gasoline dumping in the Gloucester, Massachussetts, sewerage system. The downtown area was evacuated for a time and several homes received minor damage but no one was injured.

Although the federal government took no action in 1981, in 1988, the Environmental Protection Agency proposed several amendments to the general pretreatment regulations (40 CFR 403) for industries which discharge to publicly owned treatment works (POTW). The proposed regulations included provisions for the prohibition of any discharge with a closed cup flash point of less than 60°C (140°F) or which causes concentrations greater than 10 percent of the Lower Explosive Limit (LEL) as methane anywhere in the POTW. The proposed regulation may also require POTWs to develop numerical limits or other controls to eliminate discharges which could produce "toxic gases, vapors, or fumes within a POTW in a quantity that may cause acute worker health and safety problems."

45th Purdue Industrial Waste Conference Proceedings, © 1991 Lewis Publishers, Inc., Chelsea, Michigan 48118. Printed in U.S.A.

The EPA November 1988 proposal is still under review and has not yet been finalized. The Districts are again considering revision to eliminate the flow restriction criteria and to include tank truck washing facilities for tank trucks that haul flammable substances. The EPA has indicated that the 10 percent LEL limit may be deleted from final regulation due to enforceability problems. Flammable discharges are of concern and the closed cap flash point limit will probably still remain.

Prevention of explosive or toxic conditions is best obtained by control of prohibited materials at their point of origin, prior to their discharge to the sewers. For this reason, the Districts' CGMS Guidelines have, since 1981, been aimed at discharge prevention, by monitoring and correction at the industrial site.

COMBUSTIBLE GAS MONITORING SYSTEM PROGRAM

The Districts' CGMS program was established in 1981, following several petrochemical spill episodes in 1979-80, and the Louisville incident in February 1981. The program requires all significant potential dischargers of flammable compounds to install and maintain an accurate CGMS.

The definition of significant potential discharger originally included all petroleum refineries, as well as fuel storage/transfer facilities and chemical plants which discharged more than 94.6 m^3/day (25,000 gpd). The definition was expanded in 1986 to include oil and gas extraction facilities in the latter group. The Districts are again considering revision to eliminate the flow restriction criteria and to include tank truck washing facilities for tank trucks that haul flammable substances.

The CGMS is required to utilize a permanently installed, catalytic combustible gas detector, calibrated for methane detection, and equipped with a two-stage alarm system, chart recorder, real-time readout and automatic flow diversion for alarm episodes. The Districts' CGMS program currently requires the installation of CGMS at 37 companies with 40 outfalls discharging a total of 91,000 m^3/d (24 mgd) to the Districts collection system.

As defined by the Districts Ordinance, a flammable substance is ". . .any liquid, solid, or gas that would cause or tend to cause flammable or explosive conditions to result in the sewer." The most reliable continuous, on-line method for detection of such substances is the catalytic combustible gas detector. Using such a detector, discharge limits can be expressed in terms of percentage of LEL. The Districts have used an action level of 20 percent LEL since the inception of the program, based on OSHA standards for confined space entry. Currently, any discharge stream with 20 percent or greater fraction LEL is deemed unacceptable, and must be diverted from the collection system.

The two-stage alarm requirement was established to ensure the prevention of explosive conditions in the sewerage system. A lower alarm level, based on background levels but not exceeding 15 percent LEL, is intended to give notice to the discharger, allowing correction of pretreatment system faults, or manual diversion prior to reaching the upper alarm level of 20 percent LEL. Automatic diversion is required at the upper alarm level, to prevent the formation of hazardous collection system conditions.

The CGMS Guidelines require regular maintenance and calibration of the CGMS. The normal calibration procedure involves the application of a zero percent LEL gas, and a span gas, usually 25 or 50 percent LEL methane. Calibration is required at least twice per month, and records of the maintenance and calibration must be submitted semiannually. In addition, these records must be maintained by the company for a period of three years.

The required chart recorder, which indicates percentage of LEL, must be located in an immediately accessible location for inspection by Districts' personnel. In an emergency situation, this requirement allows rapid and accurate determination of the source of a potentially flammable discharge.

The CGMS Guidelines require that the detector/sensor head assembly (SHA) be located in an area of significant wastewater turbulence, near the point of final discharge, with sufficient oxygen concentration for the proper operation of the SHA. This vague definition led to six distinctly different design concepts. SHAs were installed in manholes, industrial monitoring boxes, flow metering vaults, interceptor and holding tanks, standpipes and coarse bubble sparging units. In addition, several companies installed air sampling pumps and desiccant systems, after reported high sensor failure rates due to moisture. Because the accurate response of the CGMS is dependent on the location of the SHA, or the location of the sample draw for the remote sampling configurations, the Districts' Industrial Waste Section requested that research be conducted to evaluate the existing sampler configurations, and to recommend a standard CGMS.

A pilot testing area, utilizing full-scale sampling devices, was constructed at the Districts' Joint Water Pollution Control Plant (JWPCP) in Carson, California. Tests conducted at the pilot site by the Districts' Research Section since 1985 demonstrated that most of the original samplers were of questionable value for the detection of volatile compounds in potentially hazardous liquid concentrations. These findings led to the design of a new sampler with greater sensitivity than previously tested devices. This new device holds great promise as the sampler of choice for future combustibles and volatiles monitoring.

INITIAL PILOT TESTING

The initial pilot testing was conducted using four active sampling devices intended to model the six different industrial design concepts (see Figure 1). These four samplers were:

- 1.2 m (48 in) diameter manhole base, with a 1.8 m (72 in) riser structure (Figure 2.1).
- 460 mm (18 in) square Districts' standard industrial sampling box (Figure 2.2).
- 305 mm (12 in) diameter enclosed fountain device, intended to simulate the response at flow meters, interceptor tanks, and standpipes, all of which employ only water turbulence (Figure 2.3).
- A 305 mm (12 in) diameter coarse bubble sparged tank (CBST), with multiple overflow levels. The multiple overflows allowed the evaluation of sparging depth effects on sampler response (Figure 2.4).

The pilot site received JWPCP primary effluent through a 152 millimeter (6 inch) line from a pumping station some 183 meters (200 yards) distant from the site. Unleaded gasoline was injected into the water stream at a venturi reduction. A static mixer was installed in the reduced line when testing showed that the venturi injection was not creating a uniform dispersion of the injected contaminants.

A sidestream, located shortly after the static mixer, was directed to a 305 millimeter (12 inch) counter-current packed column stripper. With a spiral heat exchanger and boiler in line, this column was intended as a reference for the performance of the other pilot site devices.

The main stream of the liquid flowed through a 152 millimeter (6 inch) line alongside the sampler's bulkhead. A sidestream at that point delivered water to the three test devices on the bulkhead. The liquid stream ended at the manhole base sampler.

A "constant temperature"-type catalytic combustible gas sensor, manufactured by Delphian Corporation, was mounted in each device. These sensors were calibrated using a two-point zero-span procedure. Zero was established by applying a hydrocarbon-free synthetic air mixture, while the span was set with a 50 percent LEL methane in synthetic air mixture. This calibration was performed at least bi-weekly while tests were in progress, and at the start and end of any period when tests were not scheduled. The signals from these sensors were recorded on a six-channel strip chart recorder, along with continuous flow data from a magnetic flow meter in the main water delivery line.

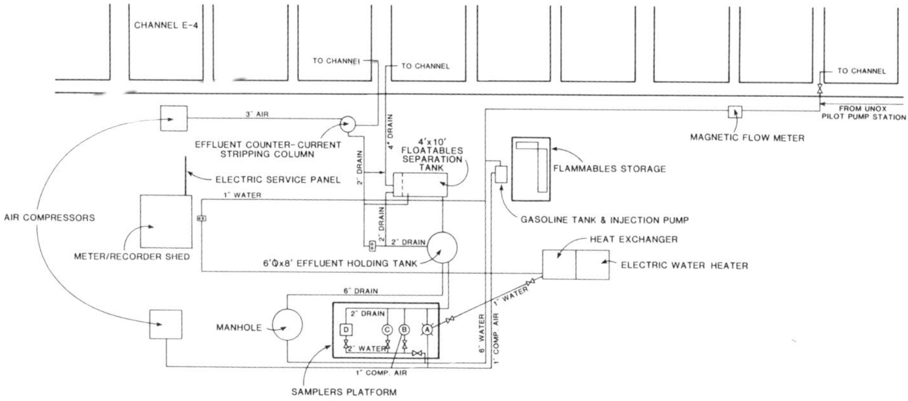

Figure 1. Original pilot plan.

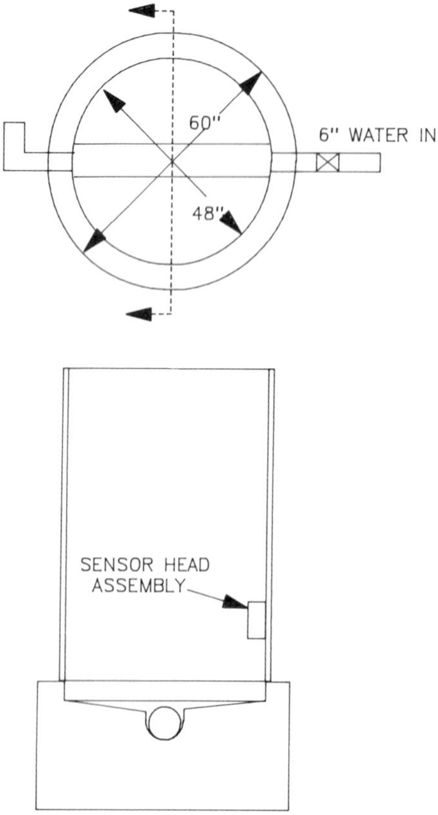

Figure 2.1. Pilot manhole.

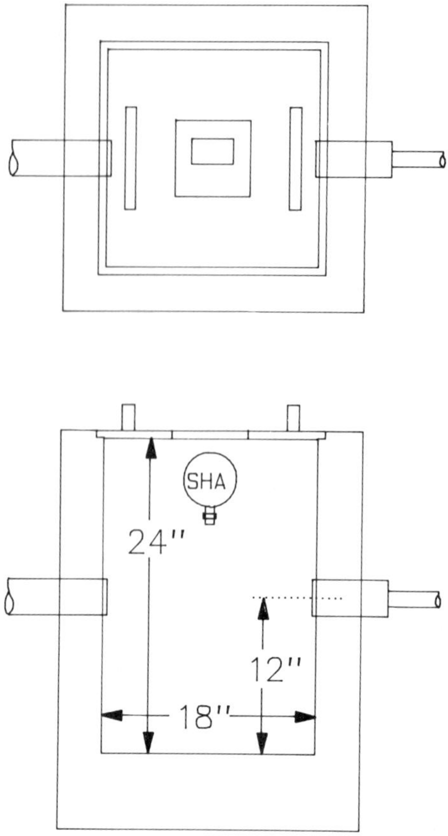

Figure 2.2. Pilot monitoring box.

Test Method

Throughout the pilot testing, the test method utilized remained essentially unchanged. The test compound was metered into the full plant flow on a volumetric basis, using a magnetically coupled air drive motor and gear pump assembly manufactured by Tuthill Pump Company, and a variety of glass tube low flow meters manufactured by King Instrument Company. Mixing was accomplished in the flow line by means of a static mixer. After the mixing section, flow was split as described previously and delivered to the active samplers.

The sensor responses were originally recorded on a six-channel discrete point strip recorder. Data analysis was accomplished by manually entering data from the strip charts into a spreadsheet/ graphics system.

Early Test Results

Tests in 1985, conducted with unleaded gasoline, showed that samplers with static headspaces were generally slower to respond to a combustibles episode than the actively aerated sampler. Testing in 1986, again with gasoline, showed that the manhole and the enclosed fountain device were unreliable even for high concentration (1000 ppmv), long-term combustibles episodes, and generally failed to detect slug doses (3000 ppmv, five-minute pulse). The failure of these two samplers to respond in preliminary tests led to their exclusion from further testing.

After cessation of flow through the enclosed fountain sampler, the industrial monitoring box became unreliable. Investigation showed that the previous marginally acceptable performance was due to the formation of a standing surface layer of contaminant due to downstream hydraulic choking of the gravity drain from the bulkhead. When the enclosed fountain device was abandoned, and the

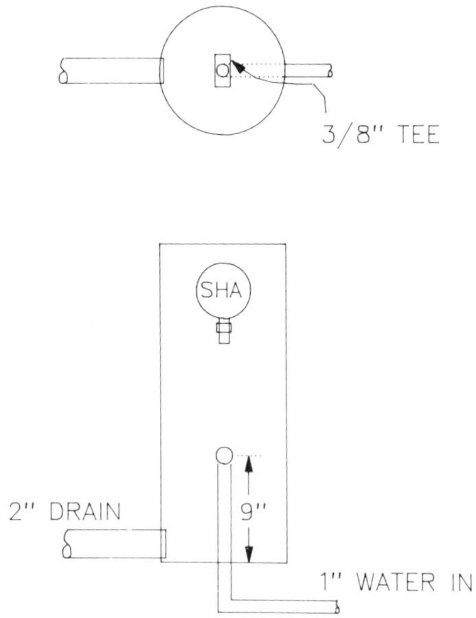

Figure 2.3. Enclosed fountain.

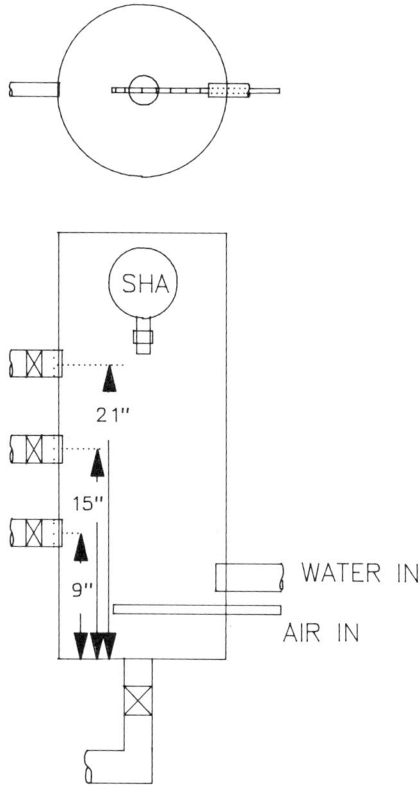

Figure 2.4. Coarse bubble sparged tank

gravity drain was relieved from the choking condition, no standing layer was formed, and the response of the industrial monitoring box CGMS became erratic.

By early 1987, sufficient testing results were available to eliminate the three static headspace pilot devices and the five industrial samplers they represented from consideration. Work concentrated on characterizing the response of the remaining original unit, the coarse bubble sparged tank.

COARSE BUBBLE SPARGED TANK CHARACTERIZATION

The characterization of the CBST was complicated by a number of factors. Early characterization tests using gasoline showed that the CBST had its highest response when operated with a water depth of 380 millimeters (15 inches). Further tests were conducted over a range of liquid flow rates, to determine the effect of air-to-liquid ratio and liquid retention time on CBST response. Figure 3 presents the result of four of these characterization tests. The maximum response of the tank occurred at a liquid flow of 15.1 L/m (4 gpm) and a sparging air flow of 11.3 L/m (24 scfh). This yielded an air-to-liquid ratio of 0.75, with a liquid retention time of 1.7 minutes.

As shown in Figure 3, the response levels were very different for the February test, although the gasoline concentration was the same for all tests. The lower response of the February test was due to dilution of the headspace by ambient winds. This observation made it clear that the eventual sampler of choice would have to be enclosed, to eliminate this dilution problem.

REMOTE SAMPLING/DESICCANT TESTS

Concurrently with the CBST characterization, a control panel was installed to evaluate the effects of remote sampling and desiccant conditioning on CGMS response. The control panel is shown schematically in Figure 4.

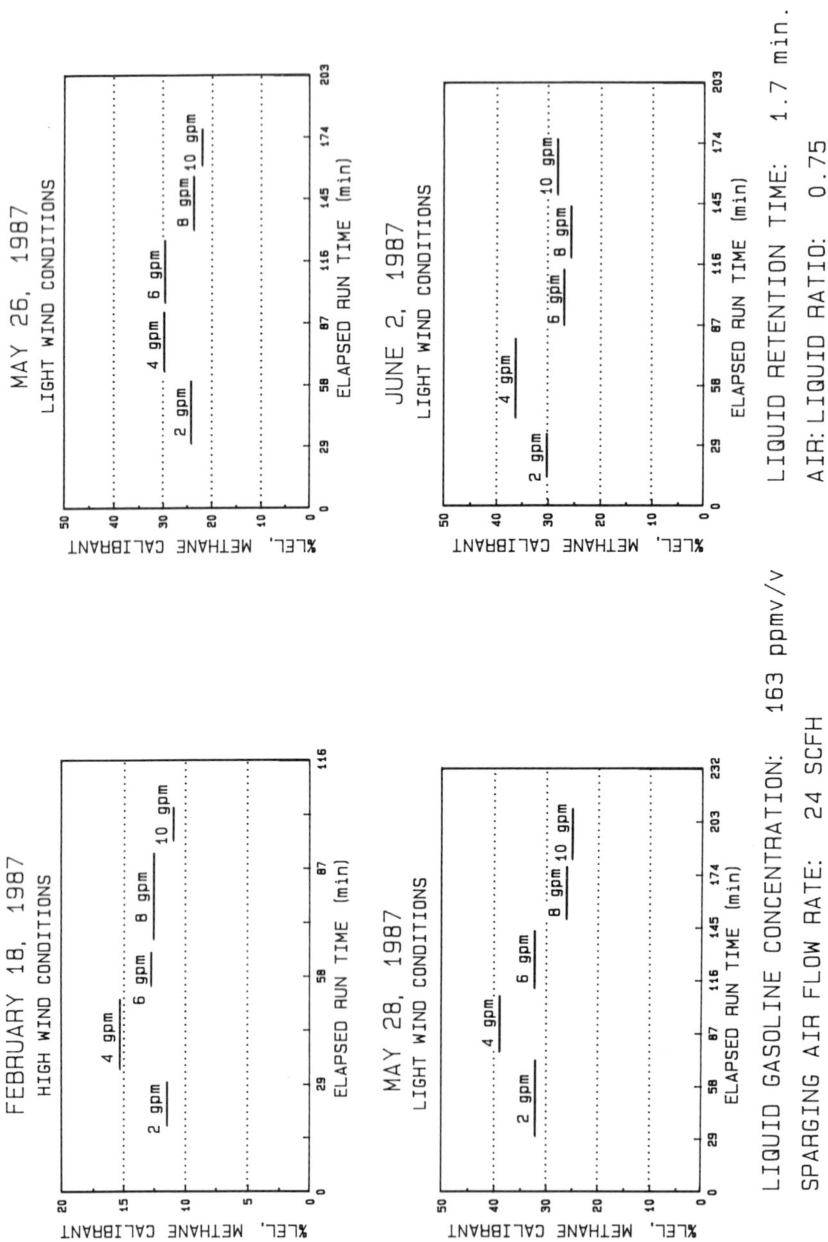

Figure 3. CBST response to flow variation.

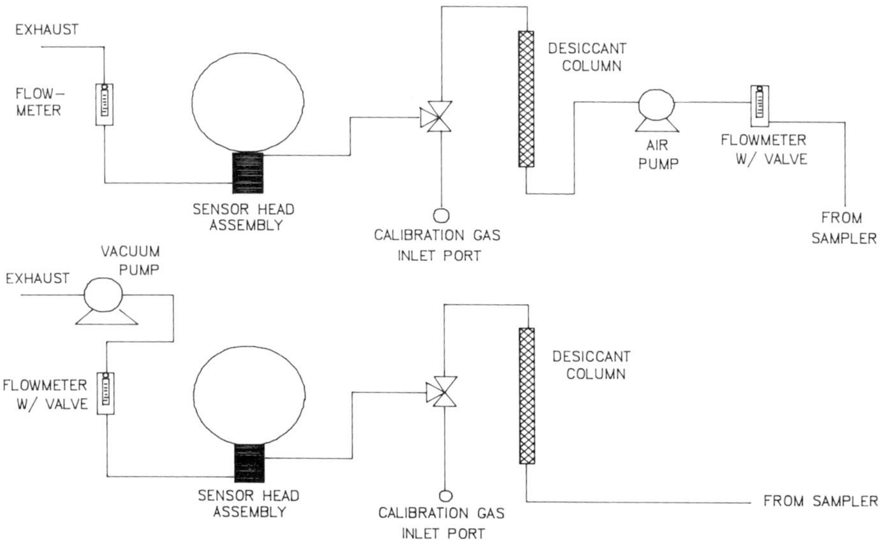

Figure 4. Flow schematic – remote sampling/desiccant study.

Air was drawn from the CBST headspace and was passed through parallel columns of fresh and expended $CaSO_4$ desiccant and through two remote SHAs designed for flow through applications. Test results showed no difference in response for the desiccated and control streams. Neither stream, however, showed a response as great as the CBST response. In addition, both columns exhibited a distinct odor of gasoline when they were opened to recharge the desiccant, indicating that the lack of response may have been due, at least in part, to the sorption of gasoline vapors onto the desiccant.

Additional tests, utilizing blank columns in the control panel, evaluated the difference between sample streams which passed through the remote SHAs under pressure and under vacuum. These tests also showed a reduced response when sample air flows were maintained at the manufacturer's recommended sampling rate. After changing the sample flow rate to match the recommended calibration gas flow rate, both remote SHAs showed higher response than the CBST SHA. This dependence on air flow rate was somewhat unexpected, and may partially explain the earlier low response of the remote SHAs during the desiccant testing.

Shortly after the characterization tests, the pilot system was out of service for maintenance, and the opportunity was taken to perform some modifications by constructing enclosed samplers which would be free from wind effects.

PILOT SITE RENOVATION

The revised pilot system (see Figure 5) used a 76 millimeter (3 inch) flexible line to deliver reclaimed JWPCP secondary effluent as the feed liquid supply. The use of reclaimed secondary effluent eliminated previous problems with solids in the primary effluent, which caused recurring rotameter fouling and interruption of tests.

Another significant aspect of the modified system was the use of an onsite pump for water distribution. This eliminated the less frequent but no less annoying problem of remote pump shutdown. Total plant flow was maintained by use of a manifold flow bypass structure, equipped with three variable area flow meters.

Several of the original samplers were abandoned in place. The only original device which remained active was the CBST.

The two new test devices added at the site were (1) an enclosed fine bubble sparged tank/stripping column (STSC), and (2) an enclosed fine bubble sparged tank (FBST). In addition, the multi-channel strip chart recorder failed during the system renovation. As a result, all sensor outputs were recorded on a four-channel, adaptive memory datalogger. Data analysis was accomplished as before, after downloading and databasing the datalogger memory.

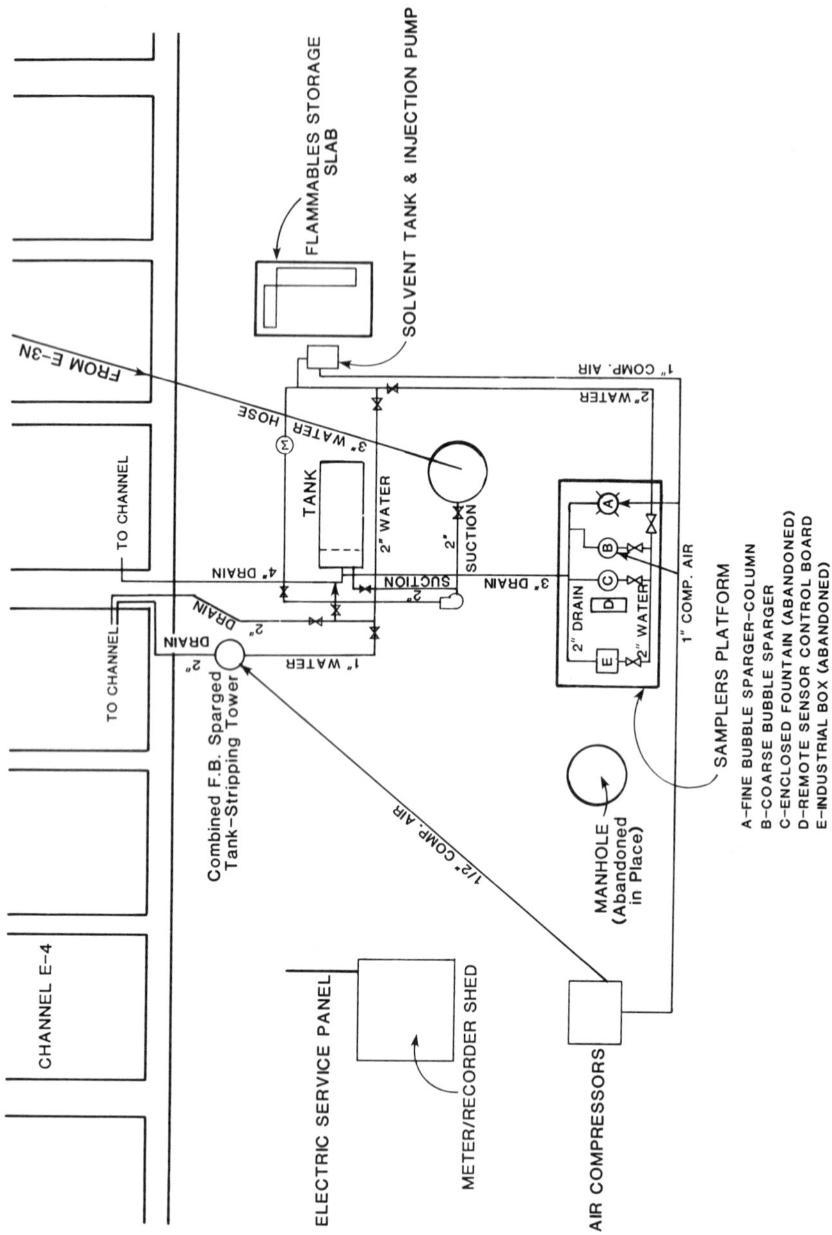

Figure 5. Revised pilot plan.

The STSC was constructed from a section of 457 millimeter (18 inch) steel pipe. Its design was based on the assumption that combining the sparging and stripping actions would aid removal of both liquid-phase resistant and gas-phase resistant compounds from the water stream, resulting in higher concentrations in the exhaust air stream and greater SHA response. In addition, the completely enclosed nature of the device eliminated wind dilution effects in the column response, and made it possible to direct the air stream to a volatiles removal system. The pilot system employed an activated carbon drum filter, due to the need to create long period responses for testing. In an industrial monitoring situation, the requirement for automatic flow diversion would eliminate all significant volatile emissions.

The FBST was a refit of the original "reference" stripping column, and was included to determine if fine bubble sparging would increase the sensitivity of the coarse bubble sparging device.

FINE BUBBLE SPARGED TANK/STRIPPING COLUMN STUDY

The STSC was a 457 millimeter (18 inch) diameter column, shown schematically in Figure 6. The depth of sparging could be adjusted to either of two levels by means of the external drain system. The SHA was mounted externally in the exhaust air line.

Initial characterization tests showed the STSC to achieve maximum response at a liquid flow rate of 1.26 L/s (20 gpm), with air flow of 0.39 L/s (50 scfh). Although these flows were considered somewhat high for a monitoring device, it was decided to determine the effect of sparging on the STSC, to aid in downsizing efforts.

The results of testing are presented in Figure 7. As can be seen, the use of sparging actually had a slight detrimental effect on the STSC response. Testing continued without use of the sparging capability of the STSC, while tests were conducted to evaluate the FBST.

Fine Bubble Sparged Tank Evaluation

The FBST evaluation was conducted with the dual purpose of evaluating the effectiveness of fine bubble sparging, as opposed to coarse bubble sparging, and to develop data to aid in downsizing the STSC. The FBST is presented in Figure 8.

Testing showed no significant difference in the CBST and FBST responses to gasoline in the waste stream. Many of the tests showed no alarm level responses from either device, as gasoline concentrations were being decreased to accomodate the STSC. When the test solvent was changed from gasoline to ethanol, however, both of the sparged tanks' SHA responses dropped to background levels. Figure

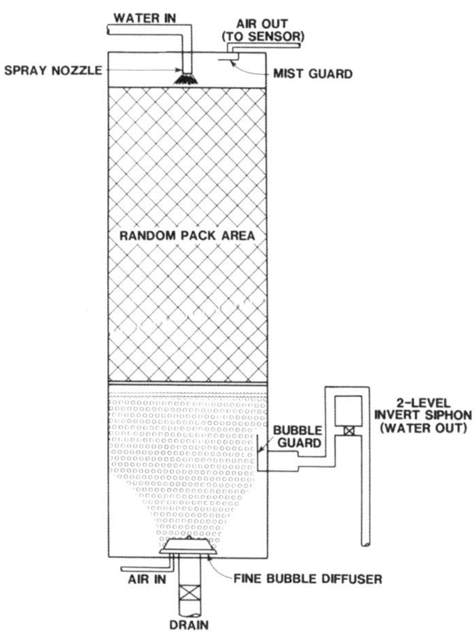

Figure 6. Combined sparged tank/stripping column.

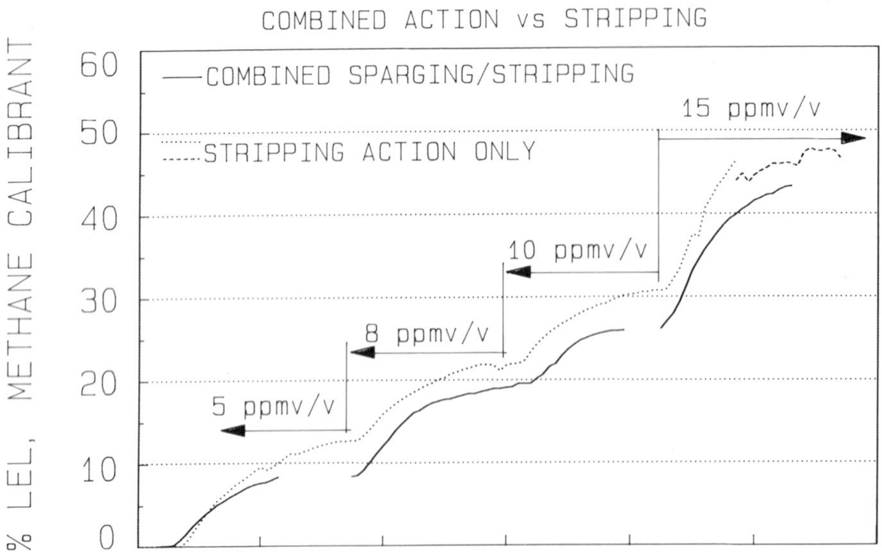

Figure 7. STSC comparison—combined action vs stripping.

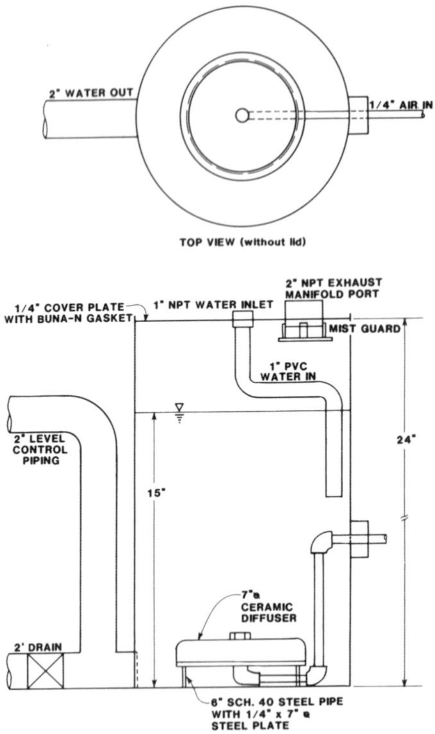

Figure 8. Fine bubble sparged tank.

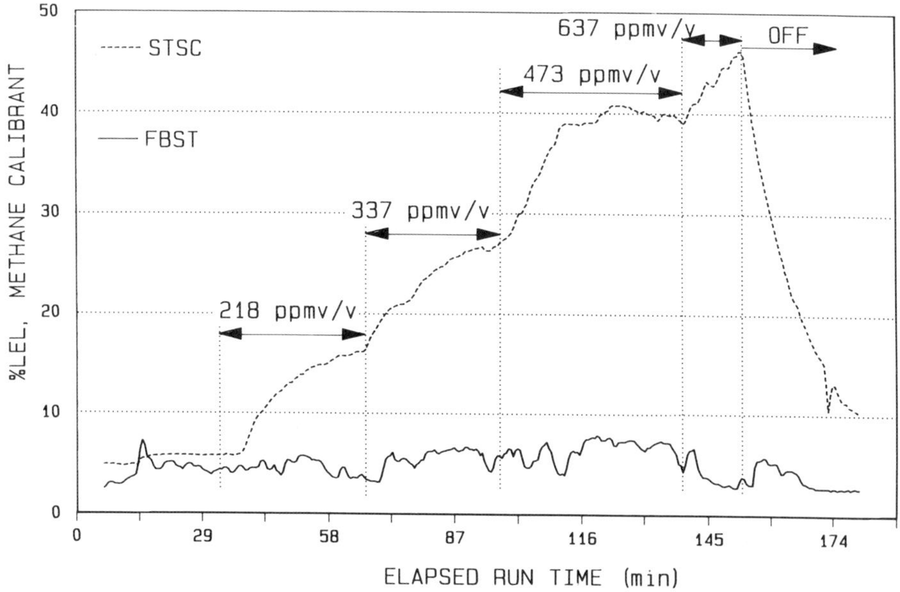

Figure 9. Ethanol response comparison STSC vs FBST.

9 shows a graph of the relative responses of the STSC and the FBST to a series of ethanol concentrations in water.

With the emphasis placed on oxygenated fuels in the South Coast Air Quality Management District's (SCAQMD) 1989 Air Quality Management Plan (AQMP), it can be expected that refineries will have to add methanol capacity to their production facilities, leading to a greater chance of methanol-containing wastes. A sampler which could not detect water soluble volatiles would be of little use in implementing the flammability and acute worker health provisions of the November 1988 proposed 40 CFR 403 amendments. Due to the lack of response to water soluble volatiles, the sparged tank

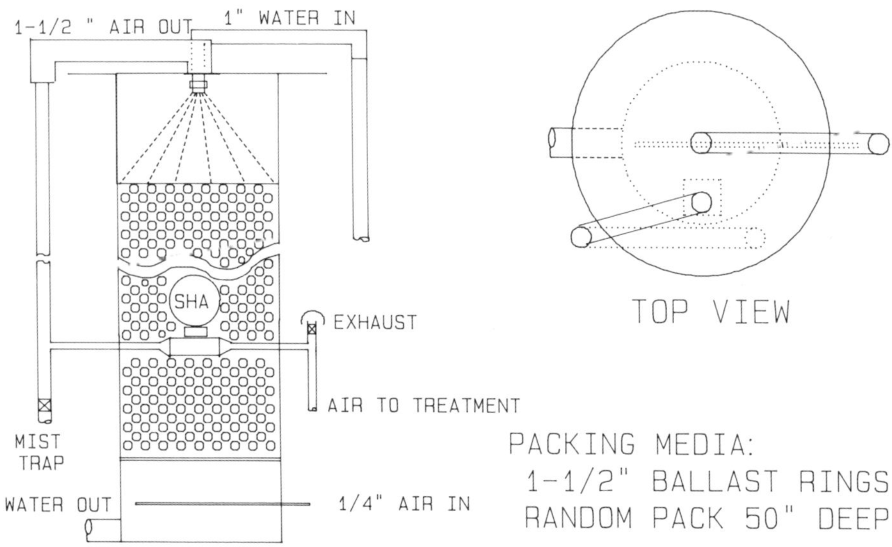

Figure 10. Enclosed column sampler.

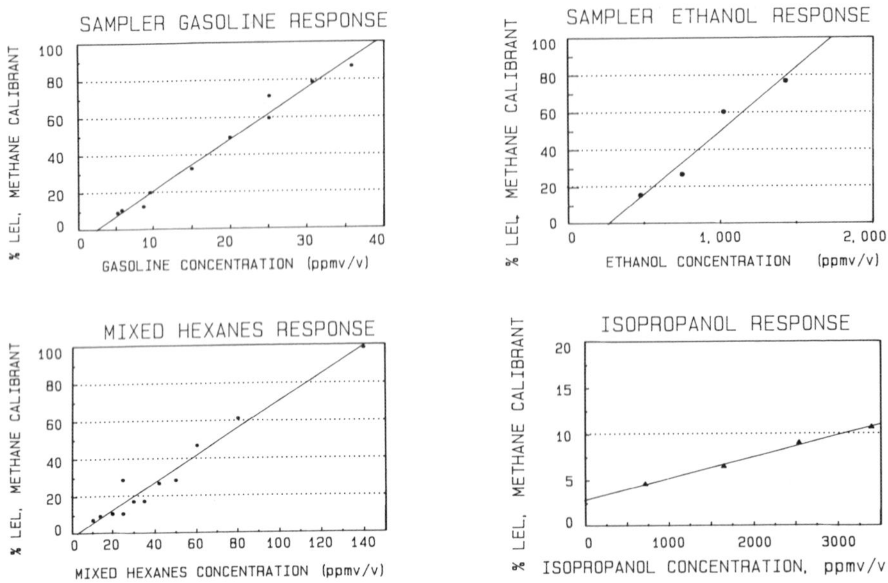

Figure 11. Enclosed column responses.

devices were dropped from consideration, and the FBST was refitted to serve as an enclosed counter-current stripping column.

ENCLOSED STRIPPING COLUMN SAMPLER

The enclosed stripping column sampler is a 305 millimeter (12 inch) diameter version of the STSC, without sparging capabilities, as seen in Figure 10. In this first stage of the downsizing process, liquid and air flows were reduced proportionately with cross-section area, to evaluate possible increased wall effects.

The results to date are presented in Figure 11. The 305 millimeter column has shown response nearly equal to (occasionally exceeding) the response of the 457 millimeter column. As can be seen, all of the compounds show the linear response that is desirable from a monitoring device. Based on the excellent performance of the 305 millimeter column, a 152 millimeter (6-inch) column was constructed to minimize the airstream volume. Since the goal of the sampler testing and design has been to create the maximum practical concentration in the airstream, conventional stripping column models have been ineffective in predicting performance. This lack of effectiveness is due to the model's assumptions of maximum solute removal, leading to very low air phase concentrations. Due to the model's lack of effectiveness, pilot testing has been required for each step in the column downsizing effort. Testing, to be conducted in the spring and summer of 1990, will compare the response of the 152 millimeter column to an enclosed sparged tank, to establish equivalent alarm levels. Tests conducted to date with the 152 millimeter column indicate that it will have a packing depth much less than the 127 centimeter (50 inch) depth of the earlier pilot units.

SUMMARY OF RESULTS

Tests completed at this time indicate that samplers with static headspaces are unacceptable for flammable compound monitoring in industrial discharges. Their response is universally delayed, due to the time necessary for volatiles to diffuse to the SHA location. Whether they respond at all is often a factor of liquid flow stagnation, and they often fail to note slugs of pollutant such as would be experienced in a spill situation.

The coarse bubble sparged tank and the enclosed fine bubble sparged tank showed essentially equal performance. The slightly higher response of the fine bubble tank may be due either to the use of fine bubble diffusion or to the complete enclosure of the air stream, eliminating wind effects. Either of these samplers, if designed for complete enclosure, would seem to be acceptable for monitoring discharges for immiscible petrochemical compounds. It should be remembered, however, that they showed virtually no response to water soluble organics, such as ethanol or methanol, which will be more likely to be discharged as the movement to oxygenated fuels progresses.

The sampler of choice appears, at this time, to be an enclosed, counter-current packed bed stripping column device. Tests indicate that this device is capable of detecting both immiscible and soluble compounds at concentrations much lower than flammability protection alone would require. The added sensitivity of this sampler may make it useful for implementation of the proposed acute worker health and safety monitoring.

CONTINUING STUDIES

As mentioned previously, efforts are underway to establish reliable alarm levels for the enclosed column. This work should be completed by the end of the summer of 1990 and will be the final research step prior to use of the enclosed column in the CGMS program.

If the 40 CFR 403 amendments are finalized without the originally proposed percentage LEL language, it will be necessary to conduct tests to relate the sampler response to closed cup flash point. Such testing would be undertaken if the November 1988 proposal is finalized.

As a result of more stringent regulation by the South Coast Air Quality Management District, the Districts' treatment plants will almost undoubtedly be required to reduce of VOC emissions. In the near future it may be necessary to expand the role of this sampling device to the detection of industrial volatile compounds at concentrations below their POTW flammability hazard levels.

ACKNOWLEDGMENTS

Douglas Johnson and Vladimir Novy designed and built the JWPCP pilot site in 1983-1984. Malcolm Noriega performed pilot site shakedown testing and the early response testing in 1985.

Significant portions of this paper were to have been presented at the 62nd Annual Water Pollution Control Federation Conference at San Francisco. Copies of an October 1989 paper on this subject are available through the WPCF.

DISCLAIMER

Mention of brand names in this report does not constitute endorsement by the County Sanitation Districts of Los Angeles County as to the use and/or performance of the models mentioned.

17 ENVIRONMENTAL ASSESSMENTS AS COMPONENTS OF PRE-ACQUISITION DUE DILIGENCE

Douglas F. Rohrman, Senior Partner
Keck, Martin and Cate
Chicago, Illinois 60606

Michael J. Hoffman, Manager of Industrial Services
Environmental Science and Engineering, Inc.
Peoria, Illinois 61615

INTRODUCTION

Alteration in state and federal environmental regulations and statutes in the last ten years has had unexpected impacts on the acquisition, sale, and financing of businesses and commercial real property. Corporate management and real estate owners are expending significant temporal and financial resources to assess environmental consequences of transactions which had been previously ignored. Businesses are finding that poorly timed or inadequate assessments of environmental liability can have unforeseen and costly effects. Indifference to contaminated ground and surface water, soil and buildings can lead to exposure far exceeding the purchase price of the property or the assets or stock of a business. The casual, unthinking and negligent disposal of hazardous wastes and substances was an ingrained behavior so common in this country that many corporate managers, even to this day, are genuinely surprised at the impact of new environmental laws. Businessmen and their advisors are having to recognize and negotiate transactions of particular environmental vulnerability, and the legal and factual issues pertaining to them. They are having to devise ways in which pre-purchase assessments by legal and engineering consultants should be organized so that environmental contamination and risks can be evaluated prior to closing. The form and substance of transactions are being altered to address environmental issues which impact on corporate and property values.

What transactions may be susceptible to environmental liability? In what ways has the law been so altered to depart from traditional approaches? How should businessmen efficiently and effectively assess the environmental condition of a business or a parcel of real property? This article examines briefly the motivation for the environmental component of the pre-acquisition due diligence process, why it has become a necessity, and how prudent buyers and sellers should implement the process.

Property Transfers

In years past informed sellers of real estate generally sought to transfer environmental liability to purchasers. The principle of *caveat emptor*, except in cases of outright fraud, was the acceptable means of transferring all manner of responsibility to a buyer who was unwary or unwilling to exercise a diligent examination of the property.[1] However, new statutes and case law have created expansive responsibility for the past and present ownership of contaminated land. Sellers cannot, by passage of title, simply erase the consequences of problems they have caused or tolerated on their land. Even "as-is" sales are no longer an absolute assurance that the new owner will not have some recourse against the seller.[2] Purchasers are unwise to assume that they are responsible only for their own, post-closing environmental maintenance. Some of the significant environmental laws impose potential liability on virtually all past and present owners and operators of real property for contamination of soil, buildings, surface water and ground water, and the actual and potential dispersion of pollutants elsewhere. This potential for liability applies not only to the costs of property remediation, but also to reparation and compensation costs for injury to humans, animals and vegetation.

The authors gratefully acknowledge the assistance of Michael J. Hughes in preparing this article.

45th Purdue Industrial Waste Conference Proceedings,

Business Acquisitions

Of the many deal-breaking points in corporate and partnership transactions, environmental liability (even in the recent past) was seldom considered a matter of consequence. Until 1970, there were few environmental laws which impacted transactions. Enforcement was sporadic until the 1980's. However, the explosion of new laws in the last decade and increased agency vigilance has changed the nature of the business transaction. Environmental concerns have either transformed or made not feasible many business purchases which would have been otherwise consummated routinely years ago. Acquisitions which were closed in the past are now encumbered by unexpected liability.

Transaction Structure

The fact that the method of acquiring a business may be critical to liability is not innovative. What is new is that environmental problems relating to mergers and acquisitions are a crucial consideration in structuring transactions. The acquisition of stock rather than assets and the use of statutory merger techniques invariably raise issues of successor liability. Even the careful asset transaction with retention of liabilities by the seller may not preclude environmental liability.[3] No longer can a business be acquired without diligent consideration of the long term liabilities pertaining to the seller's environmental exposure. Purchasers will continue to attempt to structure asset transactions to avoid assumption of such liabilities; but they must nevertheless be fully prepared to confront unavoidable issues such as hazardous substance disposal and releases.[4]

Exposure of Lenders

Banking and other institutions financing acquisitions of businesses and real estate have also become involved in environmentally liability. Despite statutory exemptions from liability for security interest holders[5], case law has focused on lender liability in instances of foreclosure and/or active day-to-day management control of a debtor's business. Thus, lenders have possible exposure to the same environmental liability normally associated with the owner or operator of a business or parcel of property.[6] Not unlike buyers, lending institutions now frequently require careful legal and scientific analyses of those businesses, buildings and land over which they may exert control or obtain ownership. The advent of laws which create environmental liens superior to that of the mortgagee also impact on the necessity for financial institutions to make detailed examinations of the mortgagor's property.[7]

The Environmental Assessment

Heightened environmental awareness on the part of transaction parties has given rise to a more formalized and structured procedure, the environmental "assessment" or "audit". The environmental assessment did not appear overnight; corporate audits have been commonly conducted as part of internal appraisals of compliance with federal and state permits and in response to due diligence requests in the most obvious and egregious circumstances. Environmental assessments also have been routinely performed, in the case of a public company, to comply with securities disclosure obligations, and generally undertaken as part of internal financial planning and economic forecasting goals. U.S. EPA has had a policy statement encouraging routine, systematic and objective environmental auditing.[8] The Agency encourages voluntary audits and, in some cases, mandates individual inquiry. However, it has not been EPA's policy generally to request audit reports; EPA acknowledges the usefulness of audits as incentives to compliance and has stated that routine disclosure of reports would inhibit voluntary assessments.

Only recently have assessments taken on new characteristics, (a) as an essential and routine part of a due diligence review of transactions and (b) as a part of a "baseline" to determine liability status at a particular time period, especially by a landowner attempting to fix the status of the property in contemplation of a buyer who will use property for an unknown, incompletely understood, or possibly hazardous purpose. Both of these characteristics are recognition of the fact that low awareness levels respecting environmental issues carry high costs. Parties to transactions must consider a new set of costs of doing business: avoidance costs for proper management, handling and disposal of hazardous substances; remediation, reparation and compensation costs for improper acts and omissions; and transaction costs for the control, allocation and inquiry into such issues. These costs are applicable potentially to all transaction parties in the form of contamination response and remediation costs and possible civil penalties, personal injury and environmental damage awards and possible criminal sanctions in the case of willful acts or omissions or repeated violations. Avoidance and transaction costs, which are reflected in part in the environmental assessment process, are relatively minor when compared to those for remediation, reparation and compensation.

What laws provide the impetus for the environmental assessment of costs of avoidance and remediation? Why have transaction parties incurred extra expenses in determining the past practices and ofttimes vague histories of a business or a parcel of land? The answers are found in environmental laws and judicial decisions which create expansive, unique liabilities.

Statutory Disclosures

One new type of state law (such as the new Illinois environmental disclosure statute, the Responsible Property Transfer Act, and the Indiana Responsible Property Transfer Law[9]) requires close attention by sellers of certain types of real property. These statutes establish affirmative obligations on the part of sellers and mortgagors to reveal to transferees and mortgagees environmental defects in property prior to closing. The authors suggest that obligations probably will be forced more and more often by careful and cost-conscious management of real estate and corporate acquisitions than by compliance with the disclosure laws. However, disclosure statutes are evidence of important trends: prior to closing, buyers and lenders will compel sellers to divulge information about and give them full access to businesses and property.

The Scope of Superfund and RCRA

The impact of the "Superfund" statute, properly known as the Comprehensive Environmental Response, Compensation and Liability Act ("CERCLA") and the Solid Waste Disposal Act of 1965, as amended by the Resource Conservation and Recovery Act ("RCRA")[10], on transactions has been considerable. While neither law compels disclosure property conditions or business practices to transaction parties, both laws act effectively as indirect inducement for disclosure of environmental conditions and practices of businesses and landowners.

Generally, CERCLA was enacted to address in a meaningful way, for the first time in history, the unregulated area of past releases and disposal of hazardous substances. The law has numerous state counterparts which, in somewhat differing approaches, attempt to accomplish the same end. RCRA, Superfund's predecessor statute, deals primarily with EPA's contemporary and prospective enforcement powers regarding the handling of hazardous waste and the closure of solid waste management units.[11] RCRA is enforced by the states in varying degrees. CERCLA permits to numerous potentially liable parties both past and present; RCRA, on the other hand, generally impacts on current owners and operators, but it is also applied to past improper management and disposal of hazardous waste.[12] Whereas Superfund enforcement has received the most notariety, RCRA is an important law for diligent inquiry by transaction parties.[13] Superfund regulates the expansive term "hazardous substance" (which includes, among other things, "hazardous waste", the regulatory focus of RCRA), and the two statutes are used coextensively to address those issues that are often interwoven by the facts.[14] Superfund's scope, then, is to impose liablity on those responsible for releases or threatened releases of hazardous substances and to establish a public fund (from which it draws its popular name), supplied by a tax on the chemical industry, to finance hazardous substance cleanups when a financially responsible party cannot be found.[15] RCRA creates a comprehensive system which regulates the generation, handling, storage, treatment, disposal and actual releases of hazardous waste. The definition of hazardous waste is exceedingly complex; a waste may be hazardous if it is listed as such in the regulations or if it exhibits specific hazardous characteristics.[16] An owner or operator of a facility which generates, stores, treats, transports or disposes of hazardous waste is subject to RCRA standards, permitting authority and corrective action enforcement.[17]

Transaction parties must focus their due diligence investigative efforts on the many federal and state environmental laws, as they all may potentially impact on the value and costs associated with a business or property. No properly conducted assessment will concentrate just on CERCLA and RCRA, other issues such as water effluents, air emissions, toxic substances, etc. may in many instances have a greater effect on the costs associated with an acquisition. Nevertheless, CERCLA and RCRA are the driving forces behind environmental assessments. Economic deterrents of protracted and complex litigation and substantive penalties provide the impetus for careful examination and assessment of possible exposure by transaction parties. This article focuses on those statutes and their effects on transactions by way of actual and threatened enforcement of their strictures.

Superfund Enforcement

The Superfund law creates several ways in which hazardous substance releases and disposal sites can be addressed. The government has authority, first, to respond to releases by way of short-term removal action or by way of long-term remediation.[18] Next, parties can be ordered to abate impending and substantial dangers caused by releases or threatened releases.[19] This section of the law has a significant effect on the typical landowner. The government must prove an imminent and substantial endangerment posed by an actual or threatened release of a hazardous substance from a facility, but EPA has not been held to particularly rigorous standards of proof in such cases. Only a "substantial likelihood" of human exposure and environmental contamination need be shown; neither crisis or disaster need be evident.[20] Third, liability may be imposed on waste generators, site owners and operators, and transporters for governmental and private party response costs and damages and damage to the environment.[21] This section of the law also carries implications to the owner or operator of real property, but only in the context of a governmental response and reimbursement action. Fourth, government response costs may simply be paid out of the fund when no reimbursement is available. Fifth, there exists a private right of action; e.g., for current owners against person who owned or operated a facility at the time hazardous substances were disposed of.[22] The law thus is intended to create an incentive to the use of care in the handling and disposal of hazardous substances. It also provides a means for allocating costs and financing remediation of and response to releases of hazardous substances which constitute dangers to human health and the environment.

Superfund-Potentially Responsible Parties

While RCRA enforcement focuses on past and present generators, transporters and owners and operators who have created an imminent hazard from hazardous waste activities, CERCLA enforcement centers on placing financial responsibility on a defined group of parties from which the government can seek reimbursement costs for cleanups. Any assessment of potential liability under Superfund requires recognition of the parties which may be liable. Generally, corporate and individual owners and operators, lenders and others may be responsible for remediation costs.[23] These persons are commonly known as "potentially responsible parties" or "PRFs", which, according to the statute, can be more specifically divided into four basic groups: 1. "Owners" and "operators" of facilities – the present or current owner is liable regardless of when the hazardous substance was deposited at the facility or whether or not the present owner did anything to contribute to the pollution.[24] 2. Those who, *at the time hazardous substances were disposed of*, owned or operated the facility at which hazardous substances were deposited or released (despite single or multiple transfers of title).[25] 3. Persons ("generators") who have created a hazardous substance and arranged to have it transported to, disposed of, or treated at a facility owned or operated, by another party.[26] 4. Transporters that chose the site for disposal of hazardous substances.[27]

Superfund liability pertains not only of the classic waste dump, landfill or impoundment, but also any contaminated property whether or not the hazardous substances on the site came to be there purposely, by accident or by unknown means. Superfund sites should not necessarily be characterized as comprising only large landfills where large quantities of hazardous substances and wastes were deposited; any real property on which hazardous substances were released or spilled has the potential for becoming the basis of a statutory action.[28] The same pertains generally to RCRA enforcement, except that the elements of an imminent or substantial endangerment must be found.[29] Superfund authority is often used to take charge of cleanups on otherwise RCRA regulated facilities. Also, RCRA corrective action authority has been used at sites where no hazardous waste has been disposed of as long as there has been a release of hazardous constituents.

When those responsible under Superfund are ordered to respond, they have a choice: comply or not. The refusal to comply carries with it the risk of treble damages and fines of up to $25,000 per day.[30] Compliance for the innocent party carries exceptional problems; in order to be reimbursed, he must prove by a preponderance of the evidence that he is not liable for response costs.[31]

Strict Liability

Of significance to these potentially responsible parties is the peculiar nature of the liability which the statute imposes. Liability under Superfund is "strict", that is, without regard to fault or the negligence of a particular PRP, subject to certain statutory defenses; in other words, liability addresses the failure to reduce risks that could not have been discovered even through the exercise of reasonable care.[32] All the government need show is the necessity for a cleanup, and that the person meeting one of the PRP definitions was somehow casually connected with the wastes at the site.[33] It is not necessary, however, for the government to show that a PRP's particular wastes caused the contamination.[34] Costs of cleanup include all costs of removal and remediation incurred by the federal or state government and any other necessary response costs.[35]

Joint and Several Liability

Not only are PRPs subject to strict liability principles, but also, to joint and several liability. Case law has generally held that joint and several liability pertains to those sites where multiple causes are indivisible and cannot be apportioned among defendants.[36] A federal common law has evolved in the case law to the effect that joint and several liability with rights of contribution among PRPs exists, despite legislative history to the contrary, which the courts have pragmatically viewed as consistent with the aims of CERCLA.[37] Joint and several liability creates a special uncertainty; liability is affected by the actions of those whom a defendant may not be able to identify, much less control. It is typical to find PRPs at a particular site that are varied in their size, sophistication, and ability to pay. Even though the manner of disposal by one PRP may have been entirely proper and others may be objectively far more culpable, that PRP or any subsequent purchaser may be shackled with the sins of the negligent predecessor landowners. This is particularly troublesome for the small PRP implicated with other PRPs that are bankrupt or otherwise unable to contribute. Responding to this problem, one court devised a *de minimis* exception to the joint and several imposition of liability.[38] However, this exception may not afford any comfort to the landowner having no recourse against the seller PRP or prior owner PRPs. In 1986, CERCLA was amended by the Superfund Amendments and Reauthorization Act ("SARA") which, among other things, established an express right of contribution among PRPs.[39] Further, joint and several liability was softened somewhat by allowing settlements with certain PRPs for response costs or performance of response actions.[40] One settlement category involves a PRP whose contribution to the waste at the site is *de minimis*.[41]

U.S. EPA, despite these amendments, bolstered by case law and the statute, has dealt with hazardous substance disposal and releases from a position of almost absolute power. Few litigants win such cases and when they choose to litigate, the path to avoiding liability is arduous and costly. Though there is some theoretical assistance from the SARA amendments, a close and practical reading reveals that there is not a great incentive to settle without a reliable formula under which liability can be allocated. Even then, *de minimis* treatment may not be available when other PRPs are insolvent and/or uninsured.

RCRA Enforcement

The RCRA law establishes a "cradle to grave" system for careful regulation of hazardous wastes. Facilities which treat, store or dispose of hazardous waste in violation of the law are subject, under certain circumstances to "corrective action".[42] Of significant importance are those sections of the law which allow citizens' suits to force EPA action[43] and government suits against "any person (including any past or present generator, past or present transporter, or past or present owner or operator of a treatment, storage, or disposal facility) who has contributed or who is contributing to . . . handling, storage, treatment, transportation or disposal" of a hazardous waste in such manner that an "imminent or substantial endangerment to health or the environment" is found.[44] Case law under RCRA makes it clear that the statute may be used as an alternative or correlative enforcement method to address improper disposal practices. The government may use the "imminent and substantial endangerment" provision to supplement CERCLA enforcement or cost recovery actions.[45] To establish a case under this section of RCRA three elements must be proven: 1. endangerment to human health or welfare or to the environment, 2. which results from the treatment, storage, or disposal of hazardous waste, 3. to which the defendant has "contributed to".[46] Thus, RCRA liability is similar to, but conceptually not as broad as, CERCLA's notion of "potential responsibility". RCRA actions seek to establish causation by placing responsibility on those "contributing to" hazardous waste contamination, regardless of a party's negligence or fault.[47] RCRA enforcement is somewhat less publicized (and is thus perceived as less threatening), nevertheless, transaction parties should consider the impact of present and past treatment, storage and disposal practices of the current and previous owners/operators at the site in question. Sellers must also consider that they will continue to be liable for their past harzardous waste practices well after they transfer title. Buyers and their lenders are advised to focus on these issues as well, especially when a buyer plans to continue the business and hazardous waste treatment, storage and disposal practices of the seller. Not unlike CERCLA, RCRA has been held to impose both strict and joint and several liability is not otherwise divisible between parties that may have contributed to contamination.[48]

Retroactive Liability for Arts of Prior Landowners

One characteristic of Superfund and RCRA liability, that of "retroactive" application, is key to purchasers and lenders. Liability under these laws attaches to owners and operators of sites regardless of the time when the hazardous substance or waste was deposited or discarded. CERCLA imposes liability both on the present owner of property and the past owner of the site at the time of the disposal of hazardous waste. This is the case *even though the present owner did not cause or contribute to the release of such hazardous material at the site.*[49] RCRA imposes what is in essence retroactive liability in that it implies liability for the *present and future* conditions resulting from past acts.[50]

A purchaser of a business or real property, thus, would be liable for the improper release or disposal of hazardous substances of all those who preceded him in the chain of title.[51] Similarly, if a tenant causes a release of hazardous substance, the landlord may be liable for CERCLA costs and perhaps RCRA corrective action.[52] Superfund assures that owners and operators of and generators or depositors to sites are held liable if a later cleanup becomes necessary, even if they used state-of-the-art technology and all other actions were reasonable. Environmental assessments inevitably focus, therefore, on the past acts and omissions of owners and operators of property. Liability may be imposed many years later for environmental contamination that was undiscovered and perhaps undiscoverable at the time of its creation.

Superfund Defenses

CERCLA does establish certain defenses available to sellers, buyers and lenders. Except for one, the so-called "innocent purchaser" defense, most of them are not particularly helpful. 42 U.S.C. §9607(b) provides a defense to a PRP who can establish by a preponderance of the evidence that a release of a hazardous substance was caused solely by an act of God, an act of war, or, in some situations, an act or omission of a third party. The third party defense on its surface is of immediate appeal to purchases of real property; however, it is available only if the third party is not an employee or agent of the defendant, the defendant took precautions with regard to the hazardous substance and foreseeable acts or omissions of the third party, *and* the third party's act or omission did occur "in connection with a contractual relationship, existing directly or indirectly, with the defendant."[53] Another defense is available; the law recognizes that the parties may contractually commit to indemnity provisions between themselves, however, that technique cannot be used to *transfer* CERCLA liability.[54]

Innocent Purchaser "Defense" Background

Buyers have been especially attracted to the "innocent purchaser" or "landowner" defense. SARA added a definition of "contractual relationship" as used in the exception to the third-party defense to include, among other things, "land contracts, deeds or other instruments transferring title or possession."[55] Congress decided to clarify that "under limited circumstances landowners who acquire property without knowing of any contamination at the site and without reason to know of any contamination . . . may have a defense to liability . . . ."[56] Absent such limited circumstances, the law will still find an exception to the third party defense and impose liability.

Innocent Purchaser "Defense" Practical Application

In order to establish a viable third party defense, therefore, the prospective purchaser must endeavor to come under the "limited circumstances" by which no contractual relationship can be found to void the defense. This is the heart of the due diligence inquiry and formal audit the authors suggest is essential to all transactions involving industrial or commercial real estate. The definition of "contractual relationship" will not apply, and a defendant/subsequent purchaser of property will have a defense to Superfund liability, if the following are proven: 1. the defendant acquired the property *after* hazardous substances were disposed of on the property; 2. the defendant, at the time of purchase, did not know and had no reason to know of the presence of contamination by hazardous substances on the property (or the defendant is a governmental entity that acquired the property by escheat or eminent domain, or the defendant acquired the property by inheritance or bequest); *and*, 3. the defendant can satisfy the due care and reasonable precaution requirements of the third party defense.[57] The defense has not provided the safe-harbor that Congress may have intended. Further, it does not apply to significant areas of enforcement liability, such as RCRA and some state Superfund counterparts.[58] On June 6, 1989, the U.S. EPA announced a new policy regarding de minimis settlements with landowners who show satisfactorily their "innocence" in the acquisition of contaminated property. The EPA Guidance does not address the problems inherent in establishing the defense through pre-purchase environmental assessments.[59] The Guidance does, however, present some intriguing possibilities for purchasers of property with contamination already known to the government.

"All Appropriate Inquiry"

Establishing the fact that the landowner "did not know" and "had no reason to know" of the presence of hazardous substances prior to acquisition is key to the viability of the third party defense and is a significant force behind many environmental assessments. The defendant that gains actual knowledge of or contributed to or caused the contamination cannot take advantage of the third party defense.[60] The law sets forth the test the landowner must meet before closing in order that he might establish the defense.[61] Central to this test is the due diligence requirement. It is imperative that purchasers undertake "all appropriate inquiry into the previous ownership and uses of the property consistent with good commercial practice."[62] The statute requires courts to consider several factors in applying the "all appropriate inquiry" standard: 1. any specialized knowledge or experience on the part of the defendant (any purchaser who has the ability to spot CERCLA-associated issues, by way of staff expertise or by experience gained through numerous other similar landholdings, will be held to a higher standard than a purchaser not having that experience and expertise), 2. the price of the property as compared to its value if it were uncontaminated (the purchaser is required to inquire further when the asking price is inordinately or suspiciously low, as compared to comparable properties, or when the property has inexplicitly been on the market for an unusually long period), 3. commonly known or reasonably ascertainable information regarding the property (as discussed below, environmental consultants have a number of routine sources and techniques for obtaining detailed information about past uses of a particular piece of property), and 4. the obviousness of the presence of likely presence of contamination, and the ability to detect the contamination by inspection.[63] A prospective purchaser must make reasonably thorough efforts to obtain reliable information regarding the disposal or release of hazardous substances on the property while considering the factors a court might review.

Organizing the Appropriate Inquiry

The nature of the transaction and role of the client in the transaction will dictate the depth and nature of the assessment process. In a corporate acquisition, the past practices of the seller, both on-site and off-site, are liability issues. For the buyer in a real estate acquisition, on-site activities are the key element. For a lender, financial institution, or an investor, the practices and operations of both the buyer and seller are critical because the value of the investment can be affected by past as well as future actions. This discussion will be all encompassing with the recognition that some elements may not apply in all transactions.

The completeness and accuracy of an environmental assessment is heavily dependent on a knowledge and understanding of past activities on the site. Engineering and legal consultants cannot be expected to work in a vacuum. Access to responsible individuals with factual knowledge of past property use is an important element of a successful audit. Equally important is an understanding by the individuals performing the audit of the processes and operations used in the manufacturing, processing, or storage activities performed at the facility. Engineers with experience in environmental, chemical, or manufacturing operations will often recognize related environmental issues which could be missed by less experienced persons. Finally, the time periods during which industrial and storage activities or waste management practices were utilized can provide insights into potential environmental damage, materials and techniques employed at a given time, and possible insurance protection for past damage.[64]

The scope of an environmental assessment is ultimately influenced by the client's role in the transaction and his objective. The bottom line expectation will direct the role of counsel and the consultant. An attitude of buy or sell at any cost can lessen the client's objective concerns with environmental exposure. When recognized by the other parties in the transaction, this attitude can be used to undermine the client's bargaining power and confidence of the client in the opinions and findings of consultants. Thus, mutual trust and a confidential bond between client and counsel is essential to effective representation and protection of the client's best interests.[65]

Technical Issues

The following paragraphs describe the technical issues often encountered in environmental audits:

On-Site Contamination. This element is the principal focus of the majority of real estate transactions as a result of potential liability under CERCLA, RCRA and similar state laws. As described previously, the owner of real property which is contaminated by hazardous substances or wastes is responsible to prevent the releases or migration to the environment. Most notable in the authors' experience are the continual problems created by leaking underground fuel storage tanks[66] and improper treatment, storage and disposal of hazardous waste. On-site assessments should also consider the nature of other material use, storage, or processing which could have resulted in sudden (spills) or non-sudden (long-term) releases. For example, product storage areas, manufacturing locations, outdoor operations, and building sumps can be sources of damage. Also, off-site contamination should be examined as possible sources of soil, surface water and groundwater contamination.

Off-Site Contamination. This article has centered the reader's attention on on-site issues; but consultants should examine the impact of off-site issues, especially in acquisition transactions in which waste materials were generated. Liability for past management practices, particularly wastewater and solid waste management which can be sources of off-site contamination, are key elements in corporate acquisitions. Discharges from industrial activities or storm water runoff result in damage to adjacent property and to waters of the state. Similarly, groundwater contamination from leaking tanks or other sources of on-site contamination can impact and damage other parties. Finally, waste disposal practices present liability exposure in the event of releases from inadequate facilities.

On-Going Compliance. Environmental regulations are in a constant state of change. Failure to comply, regardless of the excuses, results in potential liability for fines and third-party action. Key issues for examination in all environmental audits are:

1. Air pollution control permits and controls.[67]
2. Water pollution control permits and treatment systems for discharges to public treatment works or waters of the state or United States.[68]
3. Hazardous waste management in compliance with RCRA and CERCLA: permits, records, training, emergency planning, etc.[69]
4. Non-hazardous waste management: compliance with state programs and exposure under proposed regulations.
5. Underground and above-ground storage tanks: compliance with current state and federal regulations, leak testing, corrosive protection, spill preparedness, spill history, incidents requiring reports to EPA, etc.[70]
6. Asbestos in buildings, insulation and heating systems, etc.; potential liability for employee exposure (past and present) and conformance with regulations.[71] Formaldehyde foam insulation has been a residential dwelling issue which may impact on commercial buildings as well.
7. Polychlorinated biphenyls ("PCBs") that were used in electrical equipment are regulated by USEPA under the Toxic Substances Control ACT ("TSCA"). Utilization of PCB-containing equipment on site and past PCB spills present regulatory and financial liability.[72]
8. Hazard Communication Standard compliance under the Occupational Safety and Health Act ("OSHA") and SARA Title III (Emergency Planning and Community Right-to-Know) notification for employees and state agencies.[73]
9. OSHA compliance and employee chemical exposure liabilities.[74] These issues can be a financial disaster decades after actual exposure.
10. In areas known to have radon contamination, routine monitoring may be appropriate.
11. In areas in which wetlands are situated, special problems regarding development should be addressed.

Assessment Protocols. Decent case law suggests that merely hiring an engineering consultant to conduct a rudimentary review is not enough to constitute an appropriate inquiry.[75] An environmental assessment must be planned carefully. Ideally, the environmental audit should be conducted as early in the transaction as possible. Often, this is impossible and consultants must be able to respond very quickly. If possible, a "scope of work" or assessment "protocol" should be developed by the client and counsel in broad terms before an environmental consultant is hired to determine objectives. The protocol should then be refined by the consultant in some detail in conjunction with the client. Most consultants and clients will prefer some sort of contract and cost estimate at this stage. On very large projects that may involve several sites, the client, counsel and consultant should work out the most practical approach, e.g., dividing the work between several consultants. When time is less a factor, counsel and client may consider soliciting bids from consultants and contractors. The value and complexity of the transaction and the importance of time will invariably influence the method of arranging the depth of the protocol and scope of the audit. Ideally, a site and locale reconnaissance and inspection, interviews (of current and past employees, neighbors and regulatory officials), and records review (including public files, maps, aerial photographs, permits, data bases, etc.) will constitute the first phase, or preliminary audit. This is often termed the "walk-through" or "Phase I" assessment. This phase may or may not generate a report depending on the client's needs and strategy.

The next, "Phase II" is sometimes dependent on the timing of the transaction and the results of the walk-through review. This second phase is not reached unless the consultant encounters or reasonably suspects environmental problems. This portion of the audit process involves field and analytical work and more detailed inspection of the properties, records, manufacturing processes, waste and raw material control. For example, sampling of on-site wastes, suspected asbestos-containing materials, suspected spill locations, underground tank locations, soil, groundwater, air emissions, or wastewaters is performed as a second phase in the assessment process.

In this manner, a well-directed investigation can be performed, and costs can be controlled by abandoning an ill-advised transaction without incurring unnecessary sampling and testing fees. Ultimately, cost estimates for response to and correction of environmental problems will be required. These costs should be segregated for capital improvements (such as treatment facilities, and compliance programs) and remedial activities (such as removal of contaminated soil and groundwater treatment).

The response and remediation phase of an environmental project (sometimes referred to as "Phase III") is beyond the scope of this article. However, the final phase in an environmental review and action is subject to difficult issues and decisions; e.g., whether remediation plans should be reviewed by an agency, the point at which an agency should be notified, clean-up criteria to be employed, methodologies for clean-up, timetables, permits, etc.

Negative Impacts. The environmental assessment has evolved as an attempt to create a cost-reducing procedure and to place the parties on an even basis for negotiating the transaction based on realistic values of the assets involved. The process is not without its flaws. First, transactions are vulnerable to over-zealous and unnecessarily costly assessments, field testing and analyses. A common observation about environmental assessments is that if skilled consultants look hard enough, they will find *something* which may or may not have a significant bearing on the value of the transaction. Second, a momentous environmental clean-up problem may necessitate the costly and time-consuming consideration of several competing bids from reliable engineering consultants. Third, a buyer may conduct an assessment, find significant environmental issues and walk away from the negotiations. The seller is then left with knowledge of problems which may make subsequent negotiations difficult or impossible. Fourth, assessment results are on occasion incorrect or incomplete because of time or protocol restrictions or lack of competence. Fifth, results are often mischaracterized or misunderstood by the parties and their representatives. Sellers' reports are especially vulnerable to careful scrutiny by buyers' consultants and lawyers. All of these flaws can be militated by careful planning and prudent use of qualified environmental engineers and lawyers on both sides. Nevertheless, abuses and mistakes occur when parties thoughtlessly employ the environmental assessment.

Selection of Engineering Consultants. The selection of the consultant to perform the assessment and his role in the transaction are difficult issues. Regardless of any cost-sharing arrangements agreed upon between the buyer, seller, and lender, the consultant's role, responsibilities and budget should be clearly defined. In general, the buyer or lending institution should have primary responsibility for performing the audit. Most transactions will require the seller to retain a consultant to protect his interests in negotiation. An astute buyer, his counsel and consultant, can effectively use environmental liability, or its potential, as a negotiating too in the business elements of the transaction. Since few environmental issues are cut-and-dry, a seller should be ready to hire his own consultants to avoid receiving only the buyer's engineering viewpoint.

Consultant qualifications are often difficult to evaluate. Most engineering consultants claim an expertise in this field but few can deliver what most sellers and purchasers need. Generally those hiring such consultants contact a firm by word of mouth or general reputation, less often by direct experience. Individuals within a firm should be carefully interviewed and references checked. Elements which should be considered in the selection process are:

1. Familiarity with industry methods, practices, problems and processes.
2. Knowledge of state and federal regulations and policies, generally and specifically with regard to the particular site classification, release or disposal type or practice, hazardous substances involved, enforcement agency personnel, if any, to be involved and response and remediation methodologies and strategies available.
3. Expertise in environmental engineering and sub-specialties such as PCB and asbestos abatement, soils analysis and hydrogeology.

4. In-house analytical laboratory capability and credentials; i.e., participation in the EPA "Contract Lab Program", certifications, etc.
5. Credentials and experience of the actual individuals performing the work. Corporate credentials can be insignificant if the assigned personnel are inadequate.
6. Experience in dealing with similar problems and with the size or complexity of the particular transaction (response under pressure can be important when the audit must be conducted within a tight time frame of the transaction).

One of the most important issues in the selection of a consultant and the working relationship with counsel and the client is the confidence in his abilities—both technical expertise and sound engineering judgment. Those attributes must be the focus of interviews and reference inquiries. The consultant must be able to communicate technical issues to non-technical individuals and translate them into business elements. A client should ask probing questions of the consultant. If answers cannot be communicated to the client's satisfaction, clearly and succinctly, then the consultant should not be hired. A consultant who does not understand the goals of the client, will invariably frustrate the purpose of the audit. A balance must be maintained to focus properly and direct the audit process on both the technical and business issues. It is the role of the consultant to tailor the audit to the client's needs and desires, tempered by the consultant's expertise and judgment.

NOTES

1. The principle of *caveat emptor* has little or no bearing on the relationship of transaction parties today. *Sunnen Products Co. v. Chemtec Industries*, 658 F.Supp. 276 (E.D. Mo. 1987).
2. Any person has the right to seek contribution against any one who is potentially liable under the Superfund statute. 42 U.S.C. §9613(f) and *see Southland Corp. v. Ashland Oil Co., Inc.*, 696 F.Supp. 994 (D.N.J. 1988). *But see, FMC Corp. v. Northern Pump*, 668 F.Supp. 1292 (D.Minn. 1987). "As-is" sales may only negate warranties of title, not, without more, equitable responsibilities relative to the on-site contamination known to the seller.
3. The majority rule is that assets transactions do not transfer environmental liability unless liabilities are expressly or impliedly assumed by the successor, where the transaction is actually a de facto merger, where the successor is a "mere continuation" of the predecessor, where the transaction is a fraudulent effort to avoid liabilities. *See*, e.g. *In re Acushnet River & New Bedford Harbor Proceedings*, 712 F.Supp. 1010 (D.Mass.1989). New case law indicates a split in judicial authority regarding successor corporate environmental liability. See, *Anspec Co. v. Johnson Controls Inc.*, 30 ERC 1672 (E.D. Mich. 1989), *appeal docketed*, (6th Cir. #89-2393, 12/18/89). *But see, Smith Land & Improvement Corp. v. Celotex Corp.*, 851 F.2d 86 (3d Cir. 1988) and *United States v. Crown Roll Leaf Inc.*, 29 ERC 2018 (D.N.J.1988). Corporate managers must confront issues of parent-subsidiary environmental liability as well. *See Juryln Corp. v. T. L. James & Co.*, 696 F.Supp. 222 (W.D. La, 1988) *aff'd*, No. 88-4901 (5th Cir. 1990).
4. *See*, e.g., *N.J. Dept. of Trans. v. PSC Resources*, 175 N.J. Super. 447, 419 A.2d 1151 (1980) and *National Wood Preservers, Inc. v. Commonwealth*, 489 Pa. 221, 414 A.2d 37 (1980); and *see Green v. Firestone Tire and Rubber Co.*, 122 Ill.App.3d 204, 460 N.E.2d 895 (2d Dist. 1984). (Illinois rule on asset purchases.) Even dissolved corporations may be revived under state corporate law to defend environmental claims. *See In re Alleged PCB Pollution* 26 ERC 2088 (D. Mass. 1987).
5. The Superfund statute specifically excludes from the definition of "owner or operator" those "who, without participating in the management of a . . . facility, [hold] indicia of ownership primarily to protect his security interest in the . . . facility." 42 U.S.C. §9601(20)(A). *See also, United States v. Mirabile*, 15 Envtl.L.Rep. 20994 (E.D. Pa. 1985) and *In re T. P. Long Chemical, Inc.*, 45 Bankr. 278 (N.D. Ohio 1985).
6. *See Guidice v. BFG Electroplating and Mfg.*, 30 ERC 1665 (W.D. Pa. 1989) and *United States v. Maryland Bank & Trust Co.*, 632 F.Supp. 573 (D. Md. 1986). Congress has considered some measures which would provide relief to foreclosing lenders. *See* Representative LaFalce's bill H.R. 2085 introduced in 1989.
7. The Superfund Amendments and Reauthorization Act ("SARA") created a federal lien under CERCLA which is subordinated to rights of purchasers and other perfected lien holders against a responsible party for cleanup costs under 42 U.S.C. §9607(a). *See* 42 U.S.C. §9607(l)(1). Numerous states have enacted cleanup lien laws or are considering them. Several states, notably New Jersey, Massachusetts, Arkansas, New Hampshire, Connecticut and Tennessee have enacted "superlien" laws where cleanup obligations take priority over other claims on the property. Other states, such as Iowa, Kentucky, Michigan, Ohio, Oregon and Texas have enacted lien laws which are non-priority in nature. *See also Kessler v. Tarrats*, 194 N.J. Super. 136, 476 A.2d 326 (App. Div. 1984) and *Chicago Title Ins. Co. v. Kumar*, 506 N.E.2d 154 (Mass. App. 154 (1987)).

Bankruptcy proceedings are another important consideration for the owner/operator of problem site. Several courts have ruled that a CERCLA suit is not stayed by a later bankruptcy filing by the owner of a Superfund site. *See, e.g., United States v. Nicolet Inc.*, 27 ERC 1078 (R.D. Pa. 1988) and *see* 11 U.S.C. § 362 (b)(4). A bankruptcy trustee's abandonment power is limited for contaminated property. *See Midatlantic National Bank v. New Jersey DEP*, 474 U.S. 494 (1986) and *In re Smith-Douglass*, 26 ERC 1893 (E.D.N.C. 1987). A state's response costs are entitled to priority in payment from a bankrupt's estate. *See In re Wall Tube & Metal Products Co.*, 26 ERC 1673 (6th Cir. 1987).

8. *See* 51 Fed. Reg. 25004 (July 9, 1986).
9. Illinois: *Ill.Rev.Stat.* Ch. 30, ¶900 *et seq.*; Indiana *Ind. Code* 13-7-22.5, *et seq.* Illinois and Indiana laws are a version of New Jersey's Environmental Cleanup Responsibility Act ("ECRA"), N.J. Stat. §13:1K-6 and the Connecticut Transfer Act, Pub. Act 85-586 (1985). Illinois for some time has required notice to IEPA and the transferee of the transfer of hazardous waste disposal sites in the state subject to state disposal permit conditions required under Section 39(g) of the Illinois Environmental Protection Act. *Ill.Rev.Stat.* Ch. 111 1/2, §1021(n). Similarly, a closure in place, e.g., of a landfill under RCRA requires notification to the local recorder or registry of deeds.
10. 42 U.S.C. §§9601-9675 and §§6901-6991. Practically every state has similar legislation and elaborate rules which adopt the federal regulatory scheme or augment it. *See*, e.g., *Ill.Rev.Stat.* Ch.111 1/2, ¶1001 *et seq.* and regulations at 35 Ill. Admn. Code Subt. G.
11. 42 U.S.C. §§6901-6991(i). RCRA site owners are not exempt from CERCLA compliance. *See Mardan Corp. v. C. G. C. Music, Ltd.*, 600 F.Supp. 1049 (D. Ariz. 1984), *aff'd.* 840 F.2d 1454 (9th Cir. 1986).
12. 42 U.S.C. §6973.
13. *See United States v. Northeastern Pharmaceutical and Chemical Co.*, 810 F.2d 726 (8th Cir. 1987) and *United States v. Conservation Chemical Co.*, 619 F.Supp. 162 (N.D. Mo. 1985).
14. 42 U.S.C. §9601(14). The definitions of hazardous waste and hazardous substance require thorough analysis. *See* Stensvaag, *Hazardous Waste Law and Practice*, Vol. 1 (1986) and Vol. 2 (1989).
15. 42 U.S.C. §4661.
16. *See* n. 14, *supra* and 40 C.F.R. §261.
17. 42 U.S.C. §§6924 and 6925.
18. 42 U.S.C. §9604 and the Illinois counterpart, *Ill.Rev.Stat.* Ch. 111 1/2, §1022.2(k); and *see T&E Indust. Inc. v. Safety Light Corp.*, 680 F.Supp. 696, 706 (D.N.J. 1988).
19. 42 U.S.C. §9606.
20. *See United States v. Northeastern Pharmaceutical and Chemical Co.*, 579 F.Supp. 823 (W.D. Mo. 1984). Failure of a PRP to carry out a government mandated removal or remediation can result in punitive damages at three times the costs of such response action. *See* 42 U.S.C. §9607(c)(3) and *Solid State Circuits v. EPA*, 812 F.2d 383 (8th Cir. 1987).
21. *See* 42 U.S.C. §§9604, 9607 and 9659.
22. Private suits for damages were specifically authorized by the SARA amendments in 1986, 42 U.S.C. §9607(a). *Cadillac Fairview/California v. Dow Chemical Co.*, 840 F.2d 691 (9th Cir. 1988). Prior to 1986, only necessary costs of response were allowed in private actions; however, private parties have been allowed to present owners for response costs of medical testing and monitoring. *Brewer v. EPA*, 27 ERC 1352 (M.D. Tenn. 1988). Citizens suits are also authorized under 42 U.S.C. §9659.
23. *See* 42 U.S.C. §9607(a). Illinois law fairly well mirrors the federal statute. *See Ill.Rev. Stat.* Ch. 111 1/2, §1022.2(f).
24. *New York v. Shore Realty Corp.*, 759 F.2d 1032 (2d Cir. 1985).
25. 42 U.S.C. Section 9607(a). *See FMC Corp. v. Northern Pump Co.*, 26 ERC 1566 (D. Minn. 1987). See *also Cadillac Fairview/Cal Inc. v. Dow Chemical Co.*, 21 ERC 1108 (1985), *rev'd on other grounds*, 840 F.2d 691 (9th Cir. 1988) and *Ecodyne Corp. v. Shah*, Civ. Action No. C-88-4813-JPV (N.D. Cal. August 28, 1989). Contrary to the *Cadillac* and *Ecodyne* decisions, EPA takes the view that prior owners intervening between prior owners responsible for disposal and the current owner are liable as well on the theory that contaminants continued to leach or spread during the intervening owner's possession.
26. 42 U.S.C. §9607(a)(3). Title to hazardous waste passes but generators are still liable. *U.S. v. Wade*, 577 F.Supp. 1326 (Ed. Pa. 1983).
27. 42 U.S.C. §9607(a)(4).
28. 42 U.S.C. §9606.
29. 42 U.S.C. §6973.
30. *See* 42 U.S.C. §§9607(c)(3) and 9606 (b).

31. *Ibid.* Private party cost reimbursement requires a showing of compliance with the National Contingency Plan. 40 C.F.R. §300.1 *et seq.*
32. *See United States v. Stringfellow*, 661 F.Supp. 1053 (C.D. Cal. 1987).
33. *See United States v. Wade*, 577 F.Supp. 1326, 1331–34 (E.D. Pa. 1983).
34. *See New York v. Shore Realty Corp.*, 759 F.2d 1032 (2d Cir. 1985).
35. 42 U.S.C. §§9607(a)(4)(A) and (D).
36. *See United States v. Chem-Dyne Corp.*, 572 F.Supp. 802 (S.D. Ohio 1983); *United States v. Conservation Chemical Co.*, 589 F.Supp. 59 (W.D. Mo. 1984) and *United States v. Outboard Marine Corp.*, 556 F.Supp. 54 (N.D. Ill. 1982).
37. *United States v. Conservation Chemical Co., supra.*
38. *See United States v. A&F Materials Co.*, 578 F.Supp. 1249 (S.D. Ill. 1984).
39. 42 U.S.C. §9613(f)(1).
40. 42 U.S.C. §9622.
41. *Id.* at 9622(b).
42. 42 U.S.C. §§6924(u) and (v) and 6928 (h). EPA has yet to promulgate corrective action regulations.
43. 42 U.S.C. §6972.
44. 42. U.S.C. §6973.
45. *See United States v. Aceto Agricultural Chemicals Corp.*, 872 F.2d 1373 (8th Cir. 1989).
46. *See United States v. Bliss*, 667 F.Supp. 1298 (E.D. Mo. 1987).
47. *See United States v. Northeastern Pharmaecutical and Chemical Co.*, 810 F.2d 726, 740 (8th Cir. 1986).
48. *See United States v. Aceto, supra* n. 45 at 1377 and *United States v. Conservation Chemical Co.*, 589 F.Supp. 59 (W.D. Mo. 1984).
49. *See* the important ruling in *New York v. Shore Realty Corp.*, 759 F.2d 1032, 1044 (2d Cir. 1985) in which the court stated that Superfund "unequivocally imposes strict liability on the current owner of a facility from which there is a release or threat of release without regard to causation." *See also Oner II. Inc. v. EPA*, 597 F.2d 184 (9th Cir. 1979).
50. *United States v. Northeastern Pharmaceutical and Chemical Co.*, 810 F.2d 726 at 741 (8th Cir. 1986).
51. The extreme case is *United States v. Carolawn Co.*, 21 Env. Rept. (BNA) 2124 (D.S.C. 1984) in which a company acting as a conduit was held responsible even though it held title for only one hour.
52. *See United States v. Argent Corp.*, 21 E.R.C. 1354 (D.N.M. 1984) and *United States v. South Carolina Recycling and Disposal, Inc.*, 21 E.R.C. 1577 (D.S.C. 1984).
53. 42 U.S.C. §9607(b)(3). The defendant's obligation to have exercised due care with regard to the property is important to the defense. *United States v. Sharon Steel,* (No. 86-C-0924J) (D.Utah 1988) and *United States v. Monsanto Inc.*, 858 F.2d 160 (4th Cir. 1988).
54. 42 U.S.C. §9607(e) and *Ill.Rev.Stat.* Ch. 111 1/2, §1022.2(g)(1). *See Mardan Corporation v. C. G. C. Music*, 804 F.2d 1454 (9th Cir. 1986). The use of indemnity provisions in transactions is routine, as are purchase price escrows, holdbacks and put/call options. Only recently have these techniques been used widely to address environmental concerns. *See*, e.g., *Nunn v. Chemical Waste Management, Inc.*, 27 ERC 1655 (D. Kan. 1988).
55. 42 U.S.C. §9601(35)(A).
56. H.R. Rep. No. 962, 99th Cong., 2d Sess. at 186 (1986).
57. 42 U.S.C. §9601(35)(A).
58. Many states have not adopted innocent landowner or purchaser language. Illinois, however, has, for instance, enacted such a defense. *See Ill. Rev. Stat.* Ch. 111 1/2 §1022.2(j)(1)(c).
59. *See U.S. EPA Guidance on Landowner Liability Under Section 107(a)(1) of CERCLA. De Minimis Settlements Under Section 122(g)(1)(B) of CERCLA, and Settlements With Prospective Purchasers of Contaminated Property* (U.S. EPA 6/6/89).
60. 42 U.S.C. §§9601(35)(C) and (D).
61. 42 U.S.C. §9601(35)(B).
62. *Ibid.* One court has stated that the appropriateness of the inquiry depends in each case on the facts. *United States v. Pacific Hide & Fur Depot, Inc.*, 30 ERC 1082 (D. Idaho 1989). *See also, Washington v. Time Oil Co.*, 17 C.W.L.R. 1261 (W.D. Wash. 1988) and *Wickland Oil Terminals v. ASARCO*, 15 C.W.L.R. 1225 (N.D. Calif. 1988). *See also* an extreme case, *United States v. Serafini*, 706 F.Supp. 346 (M.D. Pa. 1988) in which the court seemed reluctant to impose purchaser liability even though the buyers made little inquiry in the face of strong evidence of contaminated property.
63. There has been some attempt in Congress to provide a more specific definition of the requirement to make "all appropriate inquiry". *See* H.R.2787 (9/25/89), the "Innocent Landowner Defense Amendment of 1989". This amendment would add a rebuttable presumption that all appropriate

inquiry was made if the buyer obtained a Phase I environmental audit (defined as a professional survey to determine or discover environmental problems on the property). *See* also *Smith Land & Improvement Co. v. Rapid-American Corp.*, 26 ERC 2023 (M.D. Pa. 1987)(when pollution was obvious at the time of purchase, the land owner may not have a private right of action for reimbursement against prior owners who caused the pollution; however, EPA could have proceeded under CERCLA against both parties. *See* n.22, *supra* and *Philadelphia Electric Co. v. Hercules, Inc.*, 762 F.2d 303 (3d Cir 1985), *cert. denied*, 106 S.Ct. 184 (1985).

64. *See*, for example, *International Minerals and Chem. Corp. v. Liberty Mutual Ins. Co.*, 522 N.E.2d 758 (1st Dist. Ill.App. 1988) and *U.S. Fidelity & Guaranty Co. v. Specialty Coatings Co.*, 535 N.E.2d 1071 (1st Dist. Ill.App. 1989). CGL insurance policies may or may not cover environmental liability, depending on the jurisdiction and the type and form of the policy. *See* Abraham, "*Environmental Liability and the Limits of Insurance*", 88 *Col. L. Rev.* 942 (1988) and *see Continental Ins. Cos. v. Northeastern Pharmaceutical & Chemical Co.*, 842 F.2d 977 (8th Cir. 1988) (cleanup costs are not "damages" under a CGL policy) and compare *Pepper's Steel & Alloys, Inc. v. United States Fidelity and Guaranty Co.*, 668 F.Supp. 1541 (S.D. Fla. 1987); *Buckeye Union Ins. Co. v. Liberty Solvents and Chemicals Co. Inc.*, 17 Ohio App. 127, 477 N.E.2d 1227 (1984) and *Jackson Twp. Mun. Utilities Auth. v. Hartford Accident and Indemnity Co.*, 168 N.J. Super. 156, 451 A.2d 990 (Law Div. 1982) (in which the CGL exclusion for "sudden and accidental" pollution was handled in somewhat surprising fashion).
65. In anticipation of litigation, counsel should hire environmental consultants to preserve the expert discovery immunity as long as possible, however, the parties should not rely on the consultant's opinions and conclusions being undiscoverable. *See* F.R.Civ.P. 26(b)(4) and Ill. Sup. Ct. Rule 220. Consultants should write as little as possible and make periodic oral reports to counsel.
66. Underground tanks are regulated pursuant to 42 U.S.C. §6991 and 40 C.F.R. §280, et. seq. and in Illinois pursuant to *Ill.Rev.Stat.* Ch. 111 1/2 ¶1022.12 *et seq.* Fed.Reg. 12662 (April 17, 1987). The regulations deal with design and construction criteria, installation, filling and repair standards, strict closure and release rules, release detection and reporting, corrective actions, and upgrade and replacement rules.
67. *See* Clean Air Act, 42 U.S.C. §7401 *et seq.*
68. *See* Clean Water Act, 33 U.S.C. §1251 *et seq.*
69. *See* 40 C.F.R. Parts 262–265.
70. *See supra*, n.45 and *see also*, the Federal Spill Prevention, Control and Countermeasure (SPCC) Plan regulations at 40 C.F.R. Part 112.
71. Asbestos is not federally regulated in private buildings. Schools [*see* 15 U.S.C. §2643 and 52 Fed.Reg. 41826 (10-30-87)] and public buildings are subject to some federal controls. The National Emission Standard for Asbestos is found in Subpart M of the air pollution regulations. *See* 42 U.S.C. §7412(b); 40 C.F.R. §61.140–.156 and 40 CFR Part 763. The OSHA asbestos standard is 0.2 fibers/cc for an 8-hour time-weighted average. 29 C.F.R. §1910.1001. *See also Retirement Community Dev., Inc. v. Merine*, No. N-87-2464 (D. Md. 1987) a private damage suit under 42 U.S.C. §9659 for asbestos removal.
72. *See* 15 U.S.C. §2601 et seq.; the PCB regulations are found at 40 CFR Part 761. *See also In re Penn Central Trans. Co.*, 27 ERC 1397 (E.D. Pa. 1988).
73. *See* 29 C.F.R. §1910.1200 and 42 U.S.C. §11001 *et seq.*
74. *See* generally 29 C.F.R. §1910.
75. *See BCW Associates Ltd. v. Occidental Chemical Co.* Civ. #86-5947 (E.D. Pa. 9/30/88).

18 H.R. 2878 – DEFINING THE SCOPE OF ALL APPROPRIATE INQUIRY TO ESTABLISH AN INNOCENT PURCHASER DEFENSE UNDER SUPERFUND

Katherine E. Imbrock, Regulatory Specialist
Burgess & Niple, Limited
Columbus, Ohio 43220

INTRODUCION

Federal bill H.R. 2787, now in the hands of the House Energy and Commerce Committee, is a proposal to amend the so-called "innocent purchaser" provisions of the Comprehensive Environmental Response, Compensation, and Liability Act (CERCLA or Superfund). If enacted, the bill would establish a rebuttable presumption that by having a "Phase I Environmental Audit" performed by "environmental professionals" prior to or at the time of acquisition, a property owner will have satisfied his or her obligation to conduct "all appropriate inquiry" into the past ownership and uses of the property before claiming an "innocent purchaser" defense against CERCLA liability. H.R. 2787 would also establish minimum federal standards for conducting Phase I Environmental Audits and define the term "environmental professional." This paper will consider how well H.R. 2787's Phase I Environmental Audit is likely to function as a national baseline for preacquisition investigations, and considers its implications for consultants and prospective property purchasers entering into agreements to conduct such investigations.

H.R. 2787 was introduced to the House of Representatives June 28, 1989 by its chief sponsor, Representative Curt Weldon (R-PA), and assigned to the House Energy and Commerce Committee's Subcommittee on Transportation and Hazardous Materials, September 25, 1989. H.R. 2787 is now awaiting a hearing date before the Subcommittee, hopefully to be scheduled for late spring or early summer of 1990.

EXPLANATION OF KEY TERMS

The Innocent Purchaser Defense

Generally speaking, the liability provisions of CERCLA Section 107 impose strict, joint and several liability upon all parties bound together in the complex web of "contractual relationships" constituting responsibility for the generation, storage, transport, treatment, and disposal of hazardous substances. [Strict liability means fault need not be established. Joint and several liability means more than one party may be held liable, and that liability may be imposed collectively upon all parties or selectively upon only one or a few, at the enforcer's discretion.] Liability extends to the costs incurred by government agencies or private parties in responding to a release or cleaning up contaminated sites, and to associated natural resource damages. The National Contingency Plan (NCP) is the "blueprint" for conducting such operations and must be adhered to by those wishing to recover costs.

There are three explicit defenses against CERCLA liability, set out in Section 107(b) of the Act. The burden of establishing any one of these defenses is upon the defending party or parties. Contamination or a release resulting solely from an Act of God or Act of War are not subject to CERCLA liability (107[b][1] and [2], respectively). Neither are releases or contamination resulting solely from the unforeseeable acts or omissions of a third party not bound as an agent, employee, or in another "contractual relationship" with the party claiming a third party defense (107[b][3]). Property transfers normally constitute a "contractual relationship" for purposes of establishing such a defense, but there are a number of qualifying exceptions to this presumption set out in CERCLA Section 101(35), one of which is the innocent purchaser defense.

Section 101(35)(A)(i) provides that a party who ". . . did not know and had no reason to know . . ." that hazardous substances were released or disposed of on the property at the time he or she acquired it need not consider himself or herself contractually bound to previous owners for purposes of

45th Purdue Industrial Waste Conference Proceedings, © 1991 Lewis Publishers, Inc., Chelsea, Michigan 48118. Printed in U.S.A.

establishing a third party defense. To qualify for this "innocent purchaser" defense, however, the acquiring party must have conducted ". . . all appropriate inquiry into the previous ownership and uses of the property consistent with good commercial or customary practice . . ." at the time of purchase or other conveyance of the property (101[35][B]). Section 101(35)(B) lists five factors for a court to consider in deciding whether a given investigation satisfies the "all appropriate inquiry" standard; but beyond this, the language of the Act provide no guidance as to the proper scope of such investigations. These five factors are:

1. Any specialized knowledge or experience possessed by the acquiring party.
2. How the purchase price of the property compares to its value assuming it was not contaminated.
3. Common or reasonably ascertainable information about the property.
4. The obviousness of any indications the property was contaminated.
5. The likelihood that an "appropriate inspection" of the property would have detected the contamination.

Innocent purchasers must also fulfill the other conditions of a third party defense, including the exercise of "due care" upon discovery of contamination or a release and precautions against foreseeable third party acts.

Phase I Environmental Audit.

Inherent in the language of CERCLA Section 101(35)(B) is an assumption that an established standard of "customary practice" for conducting preacquisition investigations was already in place in 1986 when the innocent purchaser provisions were added to CERCLA by the Superfund Amendments and Reauthorization Act (SARA). This was, and still is, *not* the case. Attempts to respond to the largely unqualified language of 101(35)(B) have given rise in recent years to the many variations on a theme known as "preacquisition site assessments," 'property transfer assessments," "due diligence investigations," "environmental audits," and by other names. H.R. 2787 sponsors have settled on the term "Phase I Environmental Audit" to describe their version of this type of assessment, now commonly preformed prior to the acquisition of industrial or commercial properties, typically as a condition to securing purchase financing from the buyer's lending institution. By outlining seven minimum elements to be included in every "Phase Environmental Audit", H.R. 2787 proposes to both narrow the scope of "all appropriate inquiry" and establish a much needed "customary practice" baseline against which to compare individual assessments. These seven minimum elements are:

1. Examination of chain-of-title records going back at least 50 years.
2. Examination of "reasonably obtainable" aerial photographs of the property and environs.
3. An investigation designed to unearth existing cleanup liens against the property, either under Superfund or state "mini-Superfund" laws.
4. Examination of "reasonably obtainable" government records related to hazardous substance activity at the site, such as prior assessment reports, spill reporting records, underground storage tank records, hazardous waste generator records, and hazardous waste disposal manifests.
5. A visual inspection of the property and immediately adjacent properties.
6. A report or other compilation of all the information reviewed during the course of the audit.
7. Any other "reasonable steps" as may be necessary to confirm or rule out indications that hazardous substances have been released or disposed of on the property.

Environmental Professional. H.R. 2787 stipulates that the Phase I Environmental Audit described above must be performed by an "environmental professional." The term is defined in the proposed amendment as ". . . an individual, or an entity, managed or controlled by such individual who, through academic training, occupational experience, and reputation (such as engineers, environmental consultants, and attorneys), can objectively conduct one or more aspects of a Phase I Environmental Audit."

Rebuttable Presumption. This is a legal term meaning "until proven otherwise." Under H.R. 2787, a Phase I Environmental Audit meeting the minimum standards set out in the proposed amendment would be presumed to constitute "all appropriate inquiry" for purposes of establishing an innocent purchaser defense. The burden of proving otherwise would be upon the accusing party, not the party claiming the defense.

COMPARISON WITH EXISTING STANDARDS

How well is H.R. 2787's Phase I Environmental Audit likely to function as a national baseline for conducting preacquisition investigations? In this section, four examples of existing standards familiar to the author are summarized for comparison with H.R. 2787's Phase I Environmental Audit. The examples are designed to reflect the range of institutions which have become involved in the preacquisition site assessment process.

The Federal National Mortgage Association

The Federal National Mortgage Association, dubbed "Fannie Mae" for short, is the nation's largest secondary mortgage market. Now operated entirely as a private corporation, the organization began with a federal charter in 1938. By purchasing mortgages and mortgage-backed securities at a discount from member institutions such as the Farmers' Home Administration (FHA) and the Veterans' Home Administration (VHA), Fannie Mae frees capital which would ordinarily remain inaccessible for the term of the loan to be lent out again. Member institutions typically continue to administer the mortgage for Fannie Mae after purchase.

Recognizing its vulnerability to CERCLA liability through foreclosures on contaminated properties and losses to first-priority Superfund or State Superfund liens, Fannie Mae established its "Multi-Family Environmental Hazards Management Procedures" for assessing the environmental risks associated with a given mortgage purchase in August 1988.

Before Fannie Mae will issue a purchase commitment, the member lending institution must submit the results of an environmental assessment which, at minimum, addresses the general and hazard-specific concerns identified in a Phase I information checklist and questionnaire series (see Table 1). The five hazard categories are: Asbestos, PCBs, Radon, Underground Storage Tanks (UST), and Waste Sites. If Phase I results are incriminating or inconclusive for any category, a Phase II assessment must be performed for that category. There is no checklist or set format for Phase II assessments, but Fannie Mae provides examples of the kind of work it expects to see performed in a Phase II assessment, including: bulk asbestos sampling and analysis, soil and groundwater sampling and analysis, UST tightness testing, and further investigation of Superfund or Resource Conservation and Recovery Act (RCRA) enforcement actions on neighboring properties. All environmental assessments must be performed by an independent consultant "not unacceptable" to Fannie Mae. Consultants must certify that work was performed diligently, is complete and accurate, and was performed in accordance with all applicable regulatory and good management standards.

The Procedures also list ten "terminal" environmental conditions which, if present at a site, virtually preclude purchase approval. Unusually among site assessment guidelines, the ten conditions include numerical contamination thresholds for specified soil and groundwater parameters. Other "terminal" conditions include, but are not limited to:

1. Structures built over waste sites
2. The presence of friable asbestos-containing materials in large amounts
3. High-risk neighbors (evidence of spills, soil or groundwater contamination)
4. A documented UST leak greater than 0.05 gallons/hour
5. Outstanding enforcement actions against owners or operators of the site
6. High radon levels which cannot reasonably be corrected
7. Violations of environmental or health regulations or laws
8. Physical constraints which would make remediation impossible or prohibitively expensive.

Despite their severe tone, the Procedures do provide for remediation of some "terminal" conditions, at Fannie Mae's discretion and subject to certain conditions.

The Ohio Department of Transportation (ODOT)

State, federal, and local government entities are not immune to CERCLA liability. Even government entities which acquire land involuntarily or through eminent domain must exercise "due care" upon discovery of any contamination, release, or threat of release, and take precautions against the foreseeable acts of third parties in order to obtain relief from CERCLA liability.

State highway departments regularly acquire large tracts of land in assembling the necessary right-of-ways (ROW) for construction of new interchanges, road widenings, and other highway projects. In response to guidelines issues by the Federal Highway Administration, ODOT prepared its "Interim

Table I. Multi-Family Environmental Hazards Management Procedures Digest of Phase I Information Checklist

Overall Property Description
Building Specifications
Historical Aerial Photographs
Current Aerial Photographs
Title History
Site Survey
Interviews with Local Fire, Health, Land Use, or Environmental Enforcement Officials
Neighborhood Zoning Maps
Neighborhood Land Use Maps
List of Commercial Tenants On-Site
Verification of Public Water and Sewer
Interviews with Builder and/or Property Manager
Other
Asbestos
Dated Building Construction or Rehabilitation Specifications
Engineer's/Consultant's Asbestos Report
Other
Polychlorinated Biphenyls
Utility Transformer Records
Site Survey of Transformers
Site Soil and Groundwater PCB Test Results
Other
Radon
Water Utility Records
Gas Utility Records
On-Site Radon Test Results
Other
Underground Storage Tanks
Oil, Motor Fuel and Waste Oil Systems Reports
CERCLIS/HWDMS Results on Neighborhood (within radius of one mile)
Site Soil and Groundwater Tests
Site Tank Survey
Other

Guidelines: Dealing With Hazardous Waste Sites During Project Development" in August 1989. The document was last revised December 1989.

The ODOT guidelines stress avoidance of contaminated sites early in the project development process. Following project classification based upon the need for additional ROW and other factors, a preliminary screening checklist is applied by ODOT personnel to projects which cannot be eliminated at the classification stage. Depending upon the results of the preliminary screening, a formal two-phase Environmental Site Assessment may be undertaken. Environmental site assessments must be undertaken by a qualified consultant. ODOT's Phase I Environmental Site Assessment is a more detailed records screening which also includes interviews with property owners and tenants and a site visit (see Table 2). The Phase II Assessment is a sampling and testing program undertaken only when Phase I work indicates a need for further investigation and there are no readily available alternate sites. ODOT's Remedial Investigation/Feasibility Study (RI/FS) procedures, originally developed for use by the Agency's Corrective Actions Office, are the model for conducting Phase II Environmental Site Assessments.

The Ohio State Bar Association

The Ohio State Bar Association (OSBA) published its "Practical Guidance for Determining the Impact of Environmental Requirements on Real Estate and Business Transactions" in mid-1988. OSBA's perspective is one of protecting the interests of buyers in commercial and industrial property

Table II. Ohio Department of Transportation Phase I Environmental Site Assessment Digest of Minimum Elements

Geographical/Geological Setting

Property Documentation
- Title Search
- Building Permits
- Zoning Commission
- Fire Department/Emergency Management Agency
- State Fire Marshal's Office (Bureau of UST) Sanborn Maps
- SARA 313 Form R's (National Toxic Chemical Release Inventory)

Interviews

Ohio EPA/U.S. EPA Records (RCRA and CERCLA lists)

Aerial Photographs

Site Visit

Report (to include recommendations for further investigations and attach raw data as appendices)

transfers. OSBA's two-phase investigation is explicitly intended to satisfy the demanding "all appropriate inquiry . . . consistent with good commercial or customary practice" standard of CERCLA's innocent purchaser defense. Like Fannie Mae and ODOT, OSBA makes use of a detailed checklist and adopts a two-phase approach, but unlike them, execution of OSBA's Phase II investigation does not appear to be contingent upon the results of Phase I, except as the latter helps to refine the Phase II scope. OSBA appears to be indicating that some degree of sampling and testing is necessary to satisfy CERCLA's "all appropriate inquiry" standard. The Phase II checklist also provides for a detailed building inspection which includes, among other items, tracing floor drains, sampling insulation material, and sampling transformer oil (see Table 3). The OSBA process culminates in either a negative declaration or a cleanup proposal for the site. It is not clear who is expected to make the declaration, the environmental consultant performing the work, or the client's attorney.

Table III. Digest of OSBA Two-Phase "Customary Practice" Investigation Checklist

Phase I - Preliminary Site Investigation
- Description of Current Operations
- Current and Expired Permits
- Administrative, Judicial or Citizen Actions
- Review of Title Abstract
- Historical Society Records
- Interviews with Current and Former Employees
- Regulatory/Emergency Response Agency Records
- Aerial Photographs
- Neighbor Complaints
- High-Risk Neighbors
- Walkabout
- Report of Findings

Phase II - Detailed Audit
- Product, Equipment and Waste Sample Analyses
- Building Inspection
- Waste/Product Storage Area Sampling and Testing
- Additional Investigations (as needed; e.g., methane sampling at landfill sites, stormwater discharge investigation, well analysis)
- Certificaiton or Clean-Up Proposal

CEC/MW and ASFE

The Consulting Engineers Council of Metropolitan Washington (CEC/MW) and the Association of Engineering Firms Practicing in the Geosciences (ASFE) have been among the first to publish preacquisition site assessment guidelines for consulting engineers actually performing the work. CEC/MW issued its "Guidelines: Environmental Site Assessments" in mid-1989. ASFE's "Preacquisition Site Assessments: Recommended Management Procedures for Consulting Engineering Firms" followed later in the same year.

Both organizations are adamant that no engineer is in a position to judge whether a particular work scope will satisfy the requirements of CERCLA's innocent purchaser defense. Nor, they assert, may any engineer reasonably be expected to certify or otherwise warrant a site "hazard free", even following the most rigorous investigation. Both support the tenet that some degree of soil and groundwater sampling and testing is probably necessary to successfully establish an innocent purchaser defense.

CEC/MW implicitly adopts a three-phase approach (Historical Review and Site Reconnaissance, Subsurface Exploration and Testing, and Remedial Investigation) and suggests that each phase be negotiated as a separate undertaking. CEC/MW's "Phase I" investigation consists of a historical records review, "site reconnaissance" and preparation of a written report. As the risk belongs ultimately to the client, CEC/MW places the burden of decision regarding further investigation squarely with the client, although the engineer will, of course, offer his or her technical opinion in the Phase I report.

The ASFE guidelines echo many of these same themes. Not surprisingly, ASFE places a greater reliance on subsurface exploration and testing. ASFE provides a detailed work checklist and an accompanying decision tree, both of which imply a two-phase approach. The guidelines include a cautionary tale regarding the use of "PSA specialist" firms which routinely perform only a limited records review and site "walkabout" at the rate of two or three per day. While acknowledging that a blending of disciplines is required to do the work well, ASFE takes the position that preacquisition investigations are first and foremost engineering studies.

H.R. 2787's Phase I Environmental Audit

H.R. 2787's Phase I Environmental Audit shares several of the elements common to the four illustrative examples: the two-phased approach, a search of historical and regulatory records, a site visit, and a written report. To this extent, it is a representative reflection of "good commercial or customary practices" which have evolved in response to the language of the innocent purchaser defense. But it also raises new questions.

Beyond a vague reference to taking "reasonable steps" should the Phase I Audit disclose the likelihood of a release, H.R. 2787 does not elaborate on the "customary" components of a Phase II investigation or address the important question of whether or not some degree of subsurface exploration, and possibly other sampling and testing, is necessary to satisfy the "all appropriate inquiry" standard. Yet, if these "reasonable steps" are not taken, the rebuttable presumption set up by the proposed amendment is lost. It appears that one ambiguity has simply been exchanged for another; that "all appropriate inquiry . . . consistent with good commercial or customary practice" has been traded in for "reasonable steps in accordance with current technology available, existing regulations, and generally acceptable engineering practices." In all fairness, this has more to do with difficulties inherent to the innocent purchaser defense itself than with the language of the proposed amendment. Just how *do* you establish an unequivocal standard against which to judge the completeness of a process which is by nature contingent, and whose outcome is ultimately uncertain?

Choice of the term "Environmental Audit" to describe a preacquisition site assessment seems inappropriate. The term "Environmental Audit" is also used to refer to an internal audit performed to assess the state of compliance with environmental permitting requirements, management standards, and other regulations. Although a review of regulatory files and a tour of the facility are likely to be a part of both, the motivation for each is so different that use of a different term to characterize the preacquisition investigation seems in order. The term "Preacquisition Site Assessment" is the most precise of several commonly employed alternatives. Although less precise, the term "Environmental Site Assessment" has gained wide acceptance.

The scope of H.R. 2787's review of regulatory agency files is limited to records which are "reasonably obtainable". "Reasonably obtainable" aerial photographs must also be reviewed. "Reasonably obtainable" is defined in the proposed amendment as ". . . obtainable from the government agency upon request." No time frame is specified. Because more regulatory files are a matter of public record, they are generally available, in theory anyway, upon request. In practice, however, the time it takes for agencies to respond to those requests often exceeds the time frame allotted to the entire preacquisition investigation. As agencies establish new routines for coping with the flood of requests now overwhelming them, and as prospective buyers become more educated about the time it takes to complete an adequate site assessment, this uncomfortable situation should ease. But in the short term, a requirement that all pertinent government records available upon request be reviewed, without regard to time frame, is likely to create an uncomfortable dilemma for consulting engineers and others performing preacquisition site assessments.

IMPLICATIONS OF THE REQUIREMENT THAT "ENVIRONMENTAL PROFESSIONALS" PERFORM THE WORK

In practice, this requirement simply codifies the status quo. Because of their complexity, most preacquisition site assessments are already being performed by "environmental professionals", typically by consulting engineering firms. It is gratifying to see the increasingly multidisciplinary face of this emerging professional class acknowledged, but a few words of caution are in order. Only engineers, geologists, and hydrogeologists have the necessary background to enable them to assess the likelihood and extent of soil or groundwater contamination, and to recommend remedial measures. Only attorneys, on the other hand, are qualified to judge whether a particular scope of work constitutes "all appropriate inquiry" for purposes of establishing an innocent purchaser defense against Superfund liability. This will remain true with or without passage of H.R. 2787.

The myriad of skills required to perform an adequate preacquisition site assessment are not likely to be found together in a single individual. Leonardo Da Vincis are a rare occurrence. Even if a principal individual leads the investigation, he or she should have supporting expertise to draw upon, should the investigation proceed into Phase II, for example, or if remediation costs need to be developed. For this reason, preacquisition site assessments are likely to remain the province of consulting engineering firms. These firms typically already have, or can most readily acquire, the necessary "multidisciplinary face."

At the present time, there is no established licensing or certification program for individuals performing preacquisition site assessments, although a handful of professional societies and the State of California have experimental registration programs. Passage of H.R. 2787, with its requirement that the work be performed by "environmental professionals," will likely lead to state level licensing or certification programs on a wider scale.

CONCLUDING REMARKS

In conclusion, it appears that while H.R. 2787 makes an admirable attempt to put to rest the troublesome issue of what constitutes "all appropriate inquiry" for purposes of establishing an innocent purchaser defense against CERCLA liability, the proposed amendment's Phase I Environmental Audit provides a largely academic answer to a largely academic question. Even with a minimum scope of inquiry outlined, the conditions of the innocent purchaser defense may well be too absolute for the "real world." U.S. EPA appears to echo this perspective in its issued guidance for settling with "de minimus" landowners under CERCLA Section 12(g)(1)(B) (54 FR 34235). Although it acknowledges the language the CERCLA 122(g)(1)(B) differs from that of the Act's innocent purchaser provisions, the Agency equates "de minimus" and "innocent" landowners in the guidance, and explicitly offers the settlement option to innocent landowners "prior to or in the early stages of litigation." The conflicting opinions rendered thus far in the handful of innocent purchaser cases to come before the courts suggest even they are confused.

Preacquisition site assessments are not likely to "go away"; however, even if the real world chooses to move on and leave CERCLA's innocent purchaser defense to sink into obscurity. The environmental integrity of real property is an issue here to stay. Lending institutions have led the way toward integrating the financial risks arising out of environmental concerns into existing frameworks for evaluating the overall financial risk of a given transaction. As a tool for evaluating these risks, the preacquisition site assessment is indispensable, and this is perhaps its true calling. Preacquisition site assessments are essentially risk assessments, not warranties or certifications that a site is "clean", or that all which could have been discovered about a site, was. Even if lending institutions succeed in limiting their own liability under CERCLA, as seems likely through one or more amendments now before the House, they will remain reluctant to become involved with environmentally troubled firms or properties. As long as industrial and commercial properties continue to be acquired with borrowed money, it appears likely that preacquisition site assessments will continue to serve leaders and prospective buyers as the primary vehicle by which the environmental integrity of real property is assessed.

REFERENCES

1. Association of Engineering Firms Practicing in the Geosciences (ASFE), *Preacquisition Site Assessments: Recommended Management Procedures for Consulting Engineering Firms*, 1989.
2. "Bill to Clarify 'Appropriate Inquiry' for Innocent Landowner Defense Introduced," *The Environment Reporter*, July 7, 1989, p. 510.
3. Consulting Engineers Council of Metropolitan Washington (CEC/MW), *Guidelines: Environmental Site Assessments*, 1989.
4. Federal National Mortgage Association, *Conventional Selling*, August 1988, Chapter 5, "Environmental Hazards Management Procedures," pp. 61–84.
5. Environmental Publications, Inc., *1990 Site Assessment Management Sourcebook*, 1990.
6. Hardy, Michael J., J. Jeffrey McNealey, and Douglas A. Haynam, "Practical Guidance for Determining the Impact of Environmental Requirements on Real Estate and Business Transactions," *Proceedings of the 108th Annual Meeting of the Ohio State Bar Association*, May 1988.
7. H.R. 2787.
8. Leifer, Steven L., "EPA's Innocent Landowner Policy: A Practical Approach to Liability Under Superfund," *The Environment Reporter*, August 4, 1989, pp. 646–649, Bureau of National Affairs, Washington, D.C.
9. Leifer, Steven L., *Lender Liability Under CERCLA*, 1988.
10. Mauch, James C., "Site Assessment Standards Sorely Needed," *Hazmat World*, March 1990, pp. 36–39.
11. Mays, Richard H., "The Blessed State of Innocence: The Innocent Landowner Defense Under Superfund," *The Environment Reporter*, September 8, 1989, pp. 809–814, Bureau of National Affairs, Washington, D.C.
12. Ohio Department of Transportation, Bureau of Environmental Services, *Interim Guidelines: Dealing with Hazardous Waste Sites During Project Development*, December 21, 1989.
13. Priznar, Frank J., "Trends in Environmental Auditing," *Environmental Law Reporter*, May 1990, pp. 10179–10183.
14. U.S. Environmental Protection Agency, "Reporting Hazardous Substance Activity When Selling or Transferring Federal Real Property; Final Rule," *Federal Register*, Vol. 55, No. 73, April 16, 1990, pp. 14208–14212.
15. U.S. Environmental Protection Agency, Office of Policy, Planning, and Evaluation, *Selected Current Practices in Property Transfer Environmental Assessment*, September 1989.
16. U.S. Environmental Protection Agency, "Superfund Program; De Minimus Landowner Settlements, Prospective Purchaser Settlements," *Federal Register*, Vol. 54, No. 159, August 18, 1989, pp. 34235–34247.

Section Three
HAZARDOUS WASTE MINIMIZATION AND TREATMENT

19 REMOVAL OF TOXIC ORGANICS FROM WASTEWATER GENERATED IN TANK TRUCK WASHING FACILITIES

Choong Hee Rhee, Senior Engineer
Paul C. Martyn, Head, Monitoring Section
Margaret H. Nellor, Head, Industrial Waste Section
County Sanitation Dist. of Los Angeles County
Whittier, California 90607–4998

INTRODUCTION

The Sanitation Districts of Los Angeles County (LACSD) provide sewerage service to most of Los Angeles County outside the City of Los Angeles. Within this service area are over 9,000 industrial companies and in excess of four million people. Included within the group of industrial companies are nineteen tank truck washing facilities.

Tank trucks are known to transport non-sewerable hazardous materials often containing toxic organics and organic solvents. Between loads, the operators of these vehicles generally have both the interior and exterior of the tanks cleaned at tank truck washing facilities. Wastewaters discharged from these facilities typically contain high concentrations of purgeable toxic organics and oil and grease, which (unless adequately treated) may be discharged to the sewer.

In California, liquid and semi-solid wastes have been effectively banned from solid waste landfills by revised regulations for the discharge of waste to land [California Administrative Code, Title 23, Chapter 3, Subchapter 15, Article 2, Section 2520(d)]. This restriction in the disposal of liquid and semi-solid wastes affects all hazardous and nonhazardous landfills (Class I, II and III sites) and ultimately now wastes are transported and treated. Because of this restriction, there are increased concerns that incompatible or partially treated wastes could be discharged to the public sewer system in part by discharges and from tank truck washing facilities.

There are neither federal nor state pretreatment standards or effluent limitations specifically developed for tank truck washing facilities. Consequently, to insure proper regulation, LACSD initiated studies to control toxic organics found in the wastewaters from these facilities. The study results indicated that rotary vacuum drums, activated carbon filters, dissolved air flotation units and ultrafiltration processes are capable of satisfactorily removing both toxic organics and oil and grease from tank truck washing wastewaters.

LACSD is now imposing effluent limitations on tank truck washing facilities which rely on the above-mentioned treatment processes. Two of the nineteen facilities have completed such treatment systems. The other seventeen facilities are currently being reviewed and, as needed, will be required to meet similar effluent limits.

CHARACTERISTICS OF WASTEWATER

The interiors of tank trucks are cleaned between loads when different types of materials are transported. For example, a tank truck which has hauled crude oil needs to be cleaned thoroughly before diesel oil or gasoline is loaded. The materials typically transported by tank trucks include crude oil, diesel oil, gasoline, latex, resin, ethylene glycol, deforming agents, various organic and inorganic chemicals, and organic solvents such as benzene, toluene and xylene.

In addition to the various types of materials which may have been hauled and subsequently sewered, LACSD's investigation indicates that most tank truck washing facilities are utilizing cleaning

45th Purdue Industrial Waste Conference Proceedings, © 1991 Lewis Publishers, Inc., Chelsea, Michigan 48118. Printed in U.S.A.

Table I. Cleaning Agents Utilized in Tank Truck Washing Facilities

Name of Materials Transported	Cleaning Agent Utilized
•Crude oil, diesel fuel, gasoline	Caustic soda, detergent
•Liquid latex	Cold water rinse, caustic
•Rosin, resin	Methylene chloride, caustic, detergent
•Benzene, toluene, xylene	Steam, hot water
•Tar, tallow, asphalt	TCE, diesel, caustic, hot water
•Acid (HCl, H_2SO_4)	Caustic soda, water
•Pitch, pipe coating material	Methylene chloride, xylene, caustic
•Dry cleaning agent	Steam, caustic, detergent
•Sulfur	Hot caustic

chemicals which may be incompatible with sewerage treatment plant operations; these are shown in Table I. Due to the nature of this industry, the pollutants contained in tank truck wash wastewater generally include high concentrations of purgeable hydrocarbons (EPA Test Method 601), purgeable aromatics (EPA Test Method 602), oil and grease, COD and high or low pH wastes. Some of the observed pollutant concentrations contained in discharges from tank truck washing facilities are shown in Table II.

WASTEWATER TREATMENT SYSTEM CONFIGURATION

Before 1987, wastewater treatment systems at tank truck washing facilities were generally limited to sand and grease interceptors. The purpose of these simple clarifiers was to remove the majority of sand, grit and floating oil and grease from the wastewater before discharge to a public sewerage system. Traditionally, vacuum trucks have pumped the accumulated materials from the sand and grease interceptors for landfill disposal.

The nineteen tank truck washing facilities in LACSD's service area are classified as either hazardous or nonhazardous facilities. Nonhazardous facilities may be able to adequately treat their wastewater discharges with simple treatment systems and proper influent controls. However, for facilities handling hazardous wastes, a front-end control program was devised to provide for acceptance or rejection of waste at tank truck washing facilities.

LACSD's industrial waste pretreatment limits were developed for all tank truck washing facilities as shown in Table III. These limits were established to ensure that the wastewater is treated adequately as well as meeting effluent qualities that could be reasonably achieved with available pretreatment systems.

Toxic wastewaters generated by tank truck washing facilities are typically treated by sedimentation, oil and grease separators, rotary vacuum drums (coated by volcanic ash or diatomaceous earth), activated carbon filters and ultrafiltration units. A simplified flow diagram for a segregated treatment system configuration is shown in Figure 1. It can be seen from Figure 1 that incoming tank trucks are segregated to two parts: toxic and nontoxic. Toxic wastes, which contain volatile organic compounds (VOCs) are treated by rotary vacuum drum and activated carbon columns. The filtrate is recycled for

Table II. Range of Pollutants Before Treatment

Parameter	Concentration (μg/L)
pH	1.9–12.2 units
Oil and grease	500–360,000 mg/L
Methylene chloride	300–540,000
1,1,1-Trichloroethane	95,000–470,000
Carbon tetrachloride	500–1,000
Tetrachloroethylene	100–27,000
Benzene	500–57,000
Toluene	2,000–290,000
Ethylbenzene	700–220,000
O-xylene	100–9,900
m & p-xylene	4,700–1.7×10^6
Phenol	600

Table III. Effluent Concentration Limits[a]

Parameter	Discharge Limit
1. pH	6.0–9.0 units
2. Oil and grease	75 mg/L
3. Toxic Volatile toxic organics per EPA Test Methods 601 and 602	5 mg/L
4. Soluble sulfide	0.1 mg/L
5. Districts' Phase 1 Limits:	
Cyanide (total)	10 mg/L
Arsenic	3 mg/L
Cadmium	15 mg/L
Chromium (total)	10 mg/L
Copper	15 mg/L
Lead	40 mg/L
Mercury	2 mg/L
Nickel	12 mg/L
Silver	5 mg/L
Zinc	25 mg/L
6. Excessive odor	None

[a]All limits shown are instantaneous maximums.

tank truck washing. Nontoxic wastes (which do not contain VOCs) are generally treated with belt skimmers, oil and grease separators and ultrafiltration units. Figure 2 shows a mixed treatment system in which various types of wastewaters, generated by tank truck washing operations, are commingled

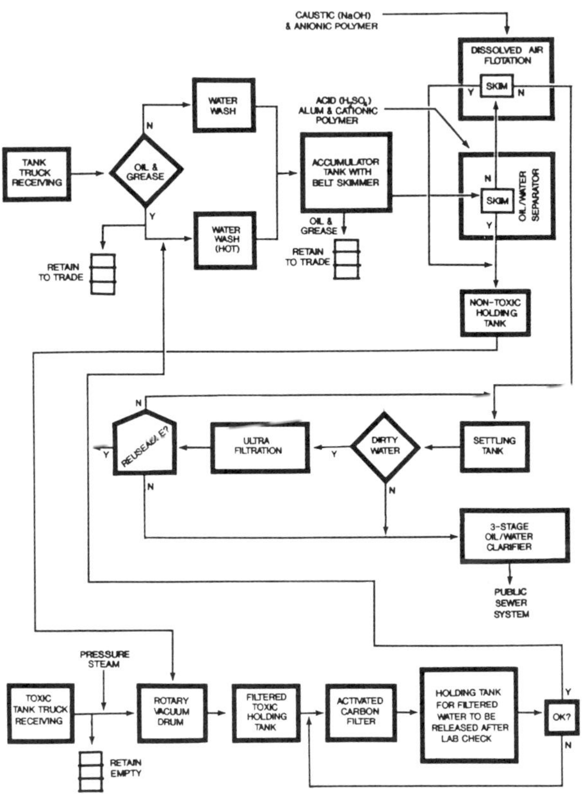

Figure 1. Segregated tank truck washing wastewater treatment system.

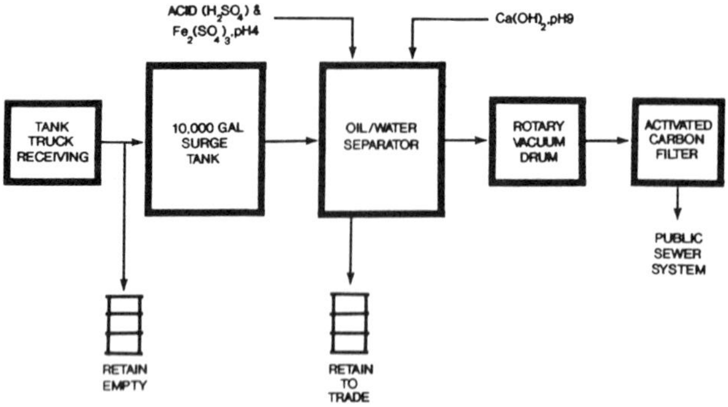

Figure 2. Mixed treatment system for tank truck washing system.

Table IV(a). Segregated Treatment System Performance Toxic Organic Stream

	Before Rotary Vacuum	After Rotary Vacuum	After Carbon Filter
Methylene chloride	310,000	80,000	1,800
1,1,1-trichloroethane	120,000	<25,000	<100
Tetrachloroethylene	550,000	<25,000	<100
Benzene	<25,000	<25,000	<100
Toluene	<25,000	<25,000	<100
Ethylbenzene	<25,000	<25,000	<100
Xylenes	707,000	<25,000	<100

(All units = micrograms/liter)

Table IV(b). Segregated Treatment System Performance Non-toxic Waste Stream

	Column Sump	After Oil & Grease Separator	After DAF	After UF
Methylene chloride	560	<100	110	<100
1,1,1-trichloroethane	–	100	65	61
Tetrachloroethylene	1,200	850	710	<50
Benzene	<100	<100	<50	<50
Toluene	<100	<100	62	<50
Ethylbenzene	<100	<100	<50	<50
Xylenes	240	<240	<180	<50

(All units = micrograms/liter)

Table V. Improved Effluent Quality - System (a) + (b)

	1985–86	1989	1990
Methylene chloride (1)	2,000–29,000	<50	<100
1,1,1-trichloroethane (2)	400–1,000	75	<100
Benzene (3)	530	400	<100
Toluene (4)	1,200	5,000	<100
Chromium (5)	.1–.7	<.05	<.05
Copper (6)	.1–1	<.05	<.05
Lead (7)	.4–.5	<.48	<.48
Oil and grease (8)	100–800	21	54

Units (1)–(4): Micrograms/liter
(5)–(8): Milligrams/liter

Table VI. Mixed Treatment System Performance

	10^3 Gal Surge	Oil & Grease Separator	Rotary Vacuum	Carbon Filter
Methylene chloride	530	350	<100	<100 (ND)
1,1,1-trichloroethane	<100	<100	<100	<100 (ND)
Tetrachloroethylene	100	<100	<100	<100 (ND)
Benzene	<100	<100	<100	<100 (ND)
Toluene	880	670	<100	<100 (ND)
Ethylbenzene	5,400	4,000	630	<100 (ND)
Xylene	61,000	21,500	8,600	<100 (ND)

(All units = micrograms/liter)

in a 10,000 gallon surge tank and then treated by an oil and grease separator, rotary vacuum drum and activated carbon columns.

EFFECTS OF THE TREATMENT SYSTEM

As shown in Tables IV(a) and (b), the wastewater treatment system illustrated in Figure 1 can produce a rather high quality discharge. Table V shows the combined wastewater quality which was obtained from the segregated toxic and nontoxic wastewater streams. As seen in Table V, the 1990 data show a drastic improvement when compared to 1985–86 discharge data. Table VI shows the step-by-step changes in wastewater quality that were obtained through the mixed treatment system, shown in Figure 2. As seen in Tables IV (a) and (b), V and VI, the toxic-rich influent is adequately treated within these facilities. A pretreatment system that had previously been a waste disposal problem now yields a high quality discharge and is able to recycle many of the cleaning compounds previously sewered.

CONCLUSIONS AND SUGGESTIONS

Based on the success of the treatment systems discussed, LACSD will continue to require similar controls for all tank truck washing facilities. At the same time, other sources of toxic organic discharges to the sewer will face more stringent regulations.

The newly installed treatment process at the truck washing facilities studied removes most of the volatile toxic organics and oil and grease from the discharge, and produces a clarified wastewater suitable for reuse. These treatment systems need to be properly designed and operated to ensure that:

1. Regulatory requirements and effluent discharge limits are met to prevent problems in the sewerage system, and
2. The wastes received are compatible with the treatment system in use.

20 EVALUATION OF DISCHARGES FROM A HAZARDOUS WASTE FACILITY TO A POTW

Joseph M. Wong, Project Manager
Brown and Caldwell Consultants
Walnut Creek, California 94596

Antoinette M. Bertolero, Deputy Director of Public Works Department
City of Benicia, California 94510

INTRODUCTION

Since 1980, hazardous waste disposal has become an increasingly critical problem in the United States. More and more publicly owned treatment works (POTWs) are being asked to help solve industrial and hazardous waste disposal problems. These problems may range from disposal of contaminated groundwater from a leaking gas station storage tank to acceptance of effluent from a centralized waste treatment facility[1].

One of the better known examples is the discharge of treated groundwater from the Stringfellow hazardous waste disposal facility, a Superfund site, to two wastewater treatment plants operated by the County Sanitation Districts of Orange County in southern California. From December 1985 through the end of 1988, nearly 26 million gallons of water from the Stringfellow site was treated and discharged to these POTWs[2].

A POTW may accept wastes from these facilities for many good reasons. For example, disposal to a POTW may be the best technical solution to an environmental problem, because a POTW generally can provide a large dilution ratio for the waste. A POTW also has the resources to tackle treatment reliability problems. In some cases, the POTW board and management may feel a public responsibility to help to solve hazardous waste treatment problems. They are willing to accept the risk to fulfill a public obligation. In other cases, a POTW may accept these wastes for a fee that will defray some of the rising costs of operation and maintenance (O&M).

Before accepting these wastes, however, a POTW should conduct a careful evaluation of the impact the discharge may have on the facilities both technically and legally. This paper presents a summary of such an evaluation. The proposed discharge was to a POTW in California. It consisted of treated groundwater, landfill leachate and impounded wastewater from a hazardous waste disposal facility. This evaluation focused primarily on the technical aspects.

BACKGROUND

International Technology (IT) Corporation's Panoche facility, located approximately two miles northeast of the City of Benicia in an unincorporated area of Solano County, California, is a Class 1 hazardous waste disposal facility with a land area of approximately 240 acres. The facility began operation in 1968 to receive, treat, and dispose of hazardous industrial solid and liquid wastes in evaporation ponds and landfills. IT purchased the facility in 1975 and continued to receive hazardous wastes for disposal until December 1986. IT voluntarily stopped receiving solids at the facility in July 1986 and ceased receiving all wastes in December 1986 to comply with an order from the California Department of Health Services (CDHS) related to drum burial areas at the facility.

Using the authority established by the California Toxic Pits Cleanup Act (TPCA), the San Francisco Bay Regional Water Quality Control Board (RWQCB) required IT to close its evaporation ponds. These ponds contained several million gallons of water contaminated with various pollutants. IT used the ponds to evaporate landfill leachate and groundwater recovered from the site. Pond closure will involve dewatering the ponds and finding other means to dispose of the leachate and groundwater that will continue to be generated.

45th Purdue Industrial Waste Conference Proceedings,

After evaluating various disposal options, the RWQCB suggested that IT investigate the option of discharging the water to the City of Benicia's POTW. This option was considered the best technical solution to the environmental problem.

IT proposed to discharge the streams to the City's sewer system after appropriate pretreatment. Because of the potential risks involved, the City engaged Brown and Caldwell to conduct a comprehensive evaluation of the effects of the proposed discharge on its wastewater treatment and sewerage system. The evaluation included a regulatory review, waste characterization, a review of the proposed pretreatment system design, an analysis of effects of the discharges would have on the wastewater treatment plant and sewerage system, treatability and toxicity reduction testing, and recommendations for monitoring and discharge schedules.

IT Corporation funded the evaluation and the City's Public Works Department administered it. A summary of the evaluation follows.

DISCHARGE STREAMS AND THEIR CHARACTERISTICS

Figure 1 shows a layout of the hazardous waste facility. The RWQCB order required Ponds 1, 1A, 2, and 8 to be dewatered. Ponds 1, 1A, and 2 contained surface runoff that contacted hazardous materials (contact water), landfill leachate and groundwater recovered from the site. These ponds had been used for storage and solar evaporation of acidic wastes. Pond 8 also contained contact water. It had been used for storage and evaporation of caustic wastes. The total volume of wastewater contained in these ponds was estimated to be 4.6 million gallons as of July 1988, although natural evaporation was gradually decreasing the volume.

IT had installed groundwater and leachate collection systems under the landfill and in the vicinity of the ponds to intercept flows that might reach the underlying weathered bedrock. Table I shows the estimated groundwater and leachate volumes from each source.

IT had been monitoring and reporting the characteristics of the pond water, groundwater and leachate to the RWQCB periodically. Brown and Caldwell conducted a sampling and analysis program to verify the historical data. The parameters analyzed included BOD_5, COD, general minerals, O&G, TSS, and priority pollutants. Brown and Caldwell also conducted 96-hour static toxicity bioassays with stickleback.

In general, the analytical and historical results were similar in orders of magnitude. The pond liquids and leachate had similar characteristics: low pH (2.4 to 3.0), high total dissolved solids content (50,000 mg/L to 93,000 mg/L), moderate levels of heavy metals (for example, 1.1 mg/L to 82 mg/L Cu), moderate levels of COD (1,900 mg/L to 4,300 mg/L), but relatively low levels of organic priority

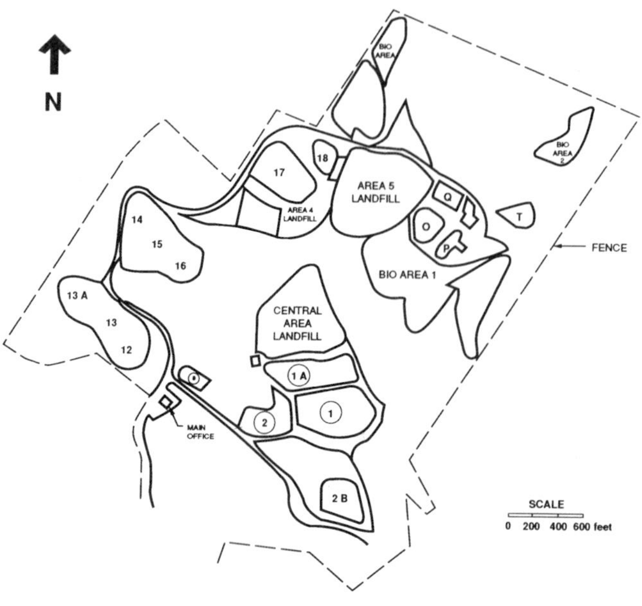

Figure 1. Facility layout.

Table I. Summary of Estimated Groundwater and Leachate Discharge Volume

Source	Gallons per month		
	High	Average	Low
Area 5 wells	25,000	20,000	15,000
Area 5 leachate	5,000	3,000	1,000
17P slope drain (leachate)	6,000	2,000	0
12 and 13 recovery system	8,000	5,000	2,000
Central area recovery system	600,000	450,000	330,000
1, 1A/2B recovery system	190,000	80,000	24,000

pollutants (< 100 ug/l). These samples were quite toxic to stickleback. Static bioassays showed 96-hour median tolerance limits (LC50) of < 10 to 13 percent.

The groundwater was much less contaminated than the leachate and pond liquids, but some of the volatile organic priority pollutant concentrations were relatively higher (1,2-dichloroethane, chlorobenzene, trichloroethylene, and trans-1,2-dichloroethylene from 50 to 500 ppb). The pH was 6.3 and the TDS was 38,000 mg/L. The groundwater was relatively nontoxic. Bioassays showed an LC50 of 75 percent.

Note that these were raw wastewater characteristics. IT proposed to pretreat the streams to meet the City's industrial local discharge limits before discharge. The proposed treatment processes would include pH adjustment, hydroxide precipitation, clarification, filtration, and granular activated carbon (GAC) adsorption. The final pH would be adjusted if necessary. Figure 2 is a schematic flow diagram of the proposed pretreatment system. IT proposed to discharge a maximum rate of 30 to 80 gpm during the pond dewatering period and less afterwards.

EFFECTS ON THE WASTEWATER TREATMENT PLANT

Even if the proposed discharge were treated to meet the local discharge limits, it could have effects on the wastewater treatment plant that might not be covered by the pretreatment ordinance. These include effects on capacity, operation, and several other factors.

Treatment Plant Capacity

The potential effect on plant capacity was investigated first. The Benicia wastewater treatment plant (WWTP) provides secondary treatment for municipal and industrial wastewaters. Treatment processes include comminution, grit removal, primary sedimentation, rotating biological contactors (RBC), secondary sedimentation, and disinfection. Sludge handling processes include primary and secondary anaerobic digestion and dewatering of the digested sludge before disposal to sanitary landfills. The design capacity of the WWTP is 3.0 mgd average dry-weather flow and 12.0 mgd wet-weather peak flow.

The average dry-weather flow to the WWTP is currently about 2.6 mgd. However, the typical diurnal flow pattern, as shown on Figure 3, varies from a low of 1.1 mgd to a high of 4.5 mgd. During periods of wet weather, the flow often reaches 8 mgd and occasionally exceeds 10 mgd, because of

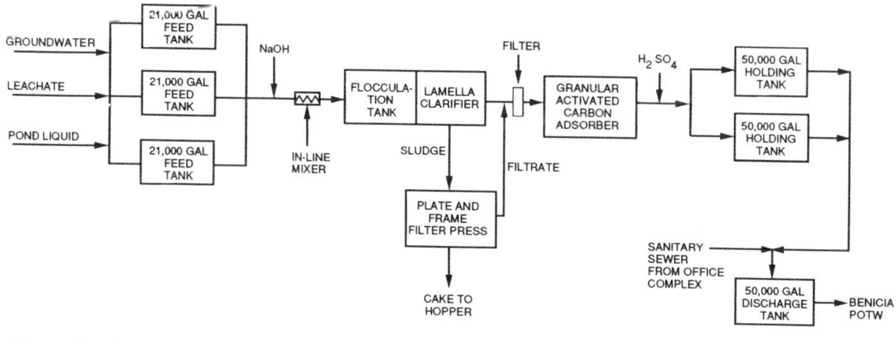

Figure 2. Schematic flow diagram of proposed pretreatment system.

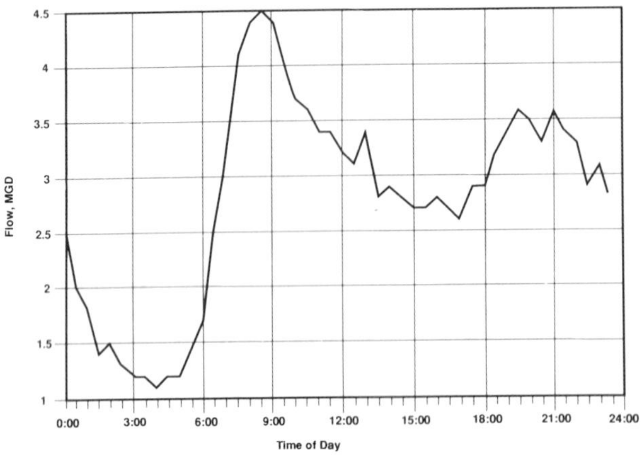

Figure 3. Typical WWTP flow pattern.

infiltration and inflow (I/I) into the sewerage system. The City has identified the I/I problems and is in the process of correcting it. In addition, the City is expanding the WWTP to handle a dry-weather average flow of 4.5 mgd.

Brown and Caldwell reviewed the operations of the existing WWTP and determined that the plant would be able to accommodate additional flows of 5 to 10 percent without much difficulty, even before the plant expansion. The expansion would provide some excess capacity to accommodate future growths, so capacity would not be a limiting factor in this case.

Treatment Plant Operation

Normally, heavy metals and toxic organics have a high potential for upsetting biological treatment processes because of their effects on microorganisms. The proposed pretreatment system would be designed to remove these pollutants to acceptable levels. Total dissolved solids (TDS) removal, however, would be very expensive, so IT's proposed system did not include this capability.

TDS are generally not considered toxic, but at high concentrations, TDS can affect treatment plant operations in many ways. For example they can disrupt clarifier performance because of density differences, decrease biological treatment efficiency, inhibit nitrification, increase biological sludge yield, decrease gas production in anaerobic digestion, and corrode metallic surfaces.

Brown and Caldwell conducted a literature review regarding the effects of TDS concentration on biological treatment plant operations. Many of the references reviewed indicated that with adequate acclimation and lower organic loading rates, biological processes (including activated sludge, trickling filter, and RBC) could treat saline wastewater satisfactorily at TDS concentrations up to 20,000 mg/L to 30,000 mg/L[3,4,5,6,7,8,9]. The shock of changeover from fresh water to saline water can lead to a period of decreased efficiency while the microbial population becomes reacclimated. One study indicated that the changeback from saline water to fresh water led to a longer decrease in efficiency because of extensive cell disintegration[10]. Ludzack and Noran[11] found that anaerobic digesters were much more sensitive to chlorides (or TDS) than activated sludge treatment units.

Because the City's WWTP is an existing system that cannot afford major interruptions to its operations, we made an attempt to determine the maximum acceptable TDS level that would cause no significant adverse effect on the treatment operations. Literature review evidence on this issue includes the following:

1. In one study, no detectable changes in sustained performance of activated sludge units were observed at chloride (TDS expressed as chloride) concentrations of 5,000 to 8,000 mg/L, but temporary clarification problems were observed[11]. The same study reported that chloride additions of 2,000 or 5,000 mg/L to a bench-scale anaerobic digester resulted in a serious gas production loss with some recovery when the load ratio was reduced. In the digester, biota adaptation to chloride changes required more time and were more limited in scope than similar changes in the activated sludge process.
2. A salt level of 8,000 mg/L caused a sustained increase in sludge yield (75 percent) that would increase the volume of sludge requiring handling[10].
3. One industrial activated sludge plant could not consistently meet its discharge limits for

suspended solids, BOD, and ammonia because of high salt content in the wastewater. Plant effluent quality improved significantly after plant operations limited the amount of salt entering the plant and maintained a sodium ion concentration below 1,500 mg/L.

4. A percolating filter (trickling filter) study indicated that continuous addition of 6,600 mg/L of sodium chloride caused little, if any, upset in the performance of the filters as judged by BOD removal. However, intermittent application of salt appeared to increase the effluent BOD_5 from an average concentration of 8 mg/L to 25 mg/L for a testing period of 2 weeks[4].
5. Two studies on anaerobic digestion indicated that cations such as sodium, potassium, ammonium, cadmium, and magnesium at higher concentrations were toxic to the methane-producing organisms[12,13]. The sodium ion at 4,000 mg/L was found to be toxic but the toxicity could be eliminated by adding an equivalent amount of calcium ion. The toxicity effect was quite complex because of synergistic and antagonistic relationships in a multicomponent system.

On the basis of these studies, Brown and Caldwell recommended a conservative maximum TDS concentration of 2,500 mg/L in the WWTP influent to avoid the potential for significant upsets in treatment and sludge handling processes. This TDS concentration will limit the flow rate of the proposed discharge to the POTW. The present plant influent has a TDS level of approximately 600 mg/L.

High sulfate concentrations may also be detrimental to the anaerobic digesters because they may cause increased generation of sulfides during digestion. However, the sulfate concentration in the discharge was only 1.3 percent of the TDS, so limiting the TDS increase in plant influent will also limit its sulfate increase.

Other Effects

The proposed discharge would slightly increase the organic load to the RBCs because of the moderately high levels of BOD and COD in the streams. The sludge production may be increased.

However, the WWTP should be able to handle the extra loads without violating its NPDES permit conditions. Release of volatile organics to the atmosphere would be insignificant because the small amount of VOCs in the waste streams would be removed by pretreatment GAC adsorbers before discharge. A temporary increase of suspended solids concentration may occur in the clarifier effluent during the start and stop of discharge because of water density difference caused by TDS variations.

EFFECTS ON THE SEWERAGE SYSTEM

The high TDS concentration of the proposed discharge could cause corrosion of part of the sewerage system (including pumping stations) where a large dilution is not available. The main concerns were the concentrations of two anions, chloride and sulfate, which could be as high as 30,000 mg/L and 600 mg/L, respectively, during the pond dewatering period. Chloride could cause pitting of metallic surfaces. Sulfate could be reduced to hydrogen sulfide (H_2S) gas under anaerobic conditions in the sewers. The H_2S gas could then be oxidized to sulfuric acid in the vapor space above the wastewater, which would attack the pipe surface above water.

With the City's approval, IT retained an independent corrosion consultant to conduct a corrosion effect study. The consultant[14] concluded that because of the low oxygen concentration in the wastewater, the chloride should have no significant effect on the corrosion of metallic materials in the system as long as the pH of the water was maintained in the range of 4 to 10. Sulfate would also not be a problem because this part of the sewerage system either has steep slopes or is under pressure. The consultant recommended that a corrosometer be installed in the system to monitor the corrosion rates both before and after the discharge if the City wants further assurance on the corrosion effect of the discharge. He also recommended that the pH of the water entering the affected pumping station be monitored and that the bronze wear ring of the pump be inspected periodically.

TREATABILITY AND TOXICITY REDUCTION TESTING

IT proposed to pretreat the discharge for metal removal by hydroxide precipitation and organics removal by GAC adsorption. The treatment processes would include equalization, pH adjustment, flocculation, sedimentation, filtration, pH readjustment (if necessary), and GAC adsorption. The City wanted assurance that the pretreated discharge would meet the City's pretreatment limits (see

Table II. Local Pretreatment Limits for Benicia POTW

Pollutant	Local limit, mg/L
Arsenic	0.5
Cadmium	0.1
Chromium (total)	2.0
Copper	3.0
Lead	1.0
Mercury	0.02
Nickel	1.0
Silver	0.3
Zinc	1.0
Phenolic compounds	10
Polynuclear aromatic hydrocarbons	3.0
Cyanide	0.3
Oil and grease	100
Sulfide (dissolved)	0.1
pH	6 to 9

Table II) and would have reduced toxicity. Brown and Caldwell conducted bench-scale tests to evaluate the effectiveness of the proposed treatment system.

Researchers tested two composite samples—a pond composite that included 25 percent each from Ponds 1, 1A, 2, and 8 and a site composite that included 70 percent groundwater, 10 percent leachate, and 20 percent composite pond liquid. The site composite sample characterized the worst case mixture of the untreated streams during operation. The composite pond sample was tested in case IT had to pretreat and discharge the pond liquid alone because of unforeseeable circumstances.

The testing procedure had two parts. The first part consisted of jar testing to determine metals precipitation and develop activated carbon adsorption isotherms. Researchers used varying pH levels and carbon dosages to determine the degree of metal and organic removal under varying conditions. In the second part of the test, researchers simulated the proposed pretreatment system at bench-scale and tested untreated and treated waste samples at each stage in 96-hour static bioassays with three-spine stickleback.

Results of the treatability and toxicity tests indicated that the proposed pretreatment system could enable the composite discharge stream to meet the City's local limits and reduce bioassay toxicity to an acceptable level. The system removed the heavy metals to below the local limits as long as pH was adjusted to 8.0 and above. It also reduced the toxicity of the site composite sample from an LC50 of 23 percent to 75 percent. The toxicity of the pond sample was not reduced, probably because of the high TDS concentration. Figures 4 through 10 show some of the test results.

REGULATORY REVIEW

The Panoche site was generally permitted for treatment, storage, and disposal of hazardous wastes under the RCRA Part A interim status with California Department of Health Services (CDHS) and had a Land Use Permit from Solano County, although it had already ceased receiving wastes. The following permits were needed for the construction and operation of the pretreatment and conveyance facilities:

1. Review and revision of the Land Use Permit (Solano County Planning Department).
2. Revision of RCRA Part A Permit and Operations Plan (CDHS).
3. Authority to Construct and Authority to Operate (Bay Area Air Quality Management District).
4. Land Use and Marsh Development Permit (Solano County Planning Department).
5. County Grading Permit (Solano County Public Works Department).
6. Environmental Impact Assessment (Negative Declaration).
7. Encroachment Permit (City of Benicia).
8. Industrial Discharge Permit (City of Benicia).

IT proceeded with the applications for the required permits. No major problem was expected in these permitting activities. The Operations Plan submitted to CDHS for revision of the Part A Permit included a description of the proposed facilities, operation procedures, emergency procedures, per-

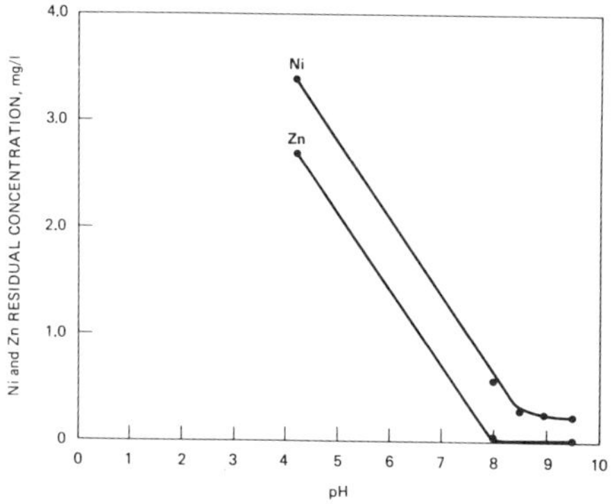

Figure 4. Relationship of residual metals concentrations with pH (site sample).

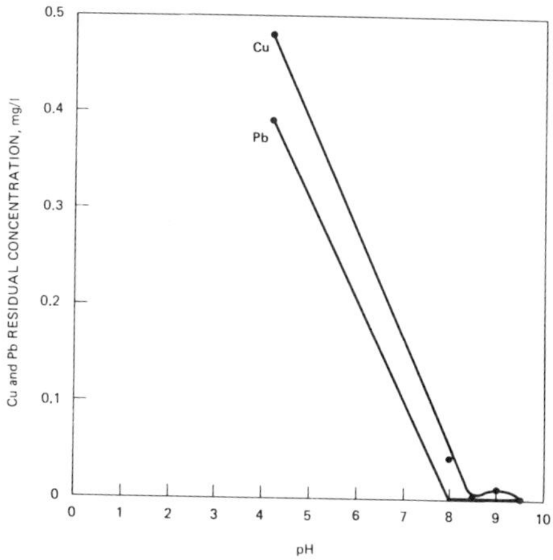

Figure 5. Relationship of residual metals concentrations with pH (site sample).

sonnel training requirements, and recording and reporting procedures, which were generally required for hazardous waste handling facilities.

Note that the receipt of the waste by the POTW should not give rise to federal RCRA regulations or enforcement liability because of the domestic sewage exclusion provision in these regulations.[1]

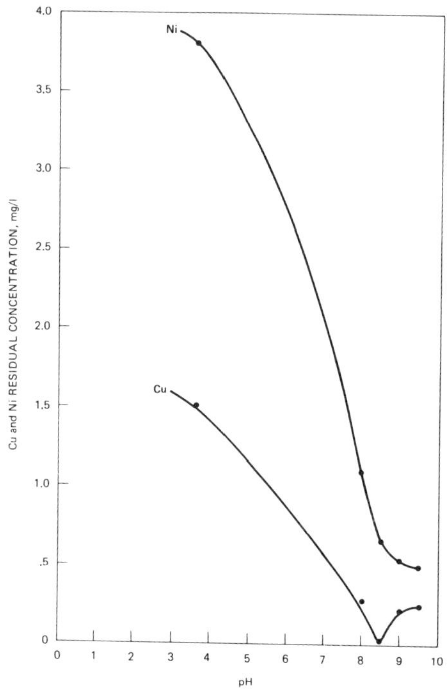

Figure 6. Relationship of residual metals concentrations with pH (pond sample).

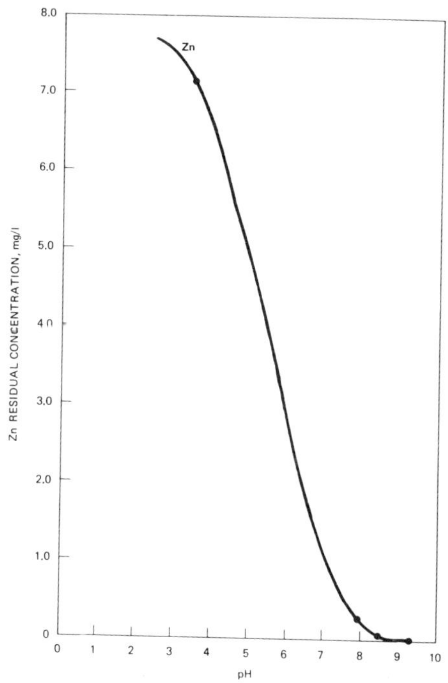

Figure 7. Relationship of residual metals concentrations with pH (pond sample).

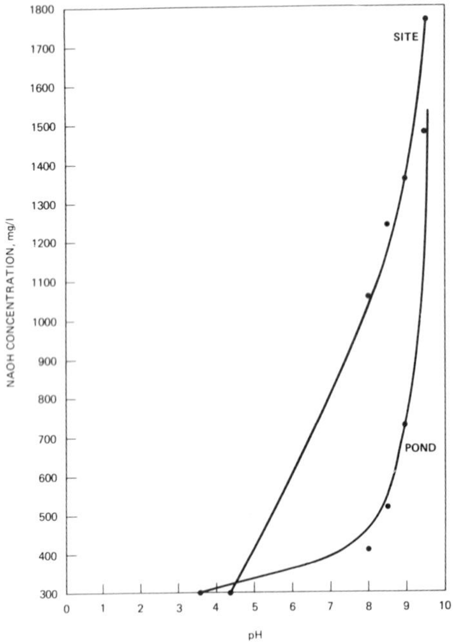

Figure 8. NaOH dosage vs. pH.

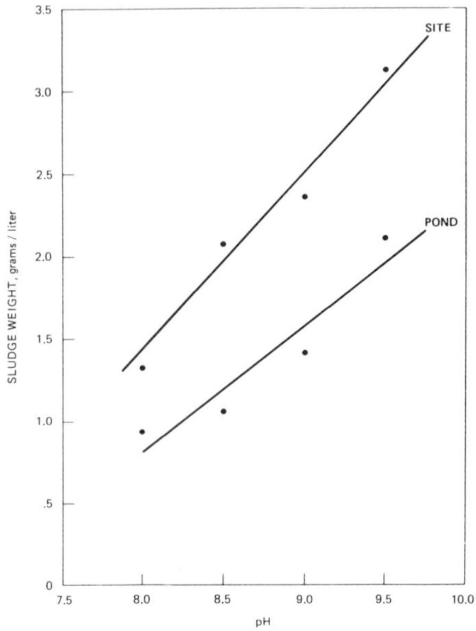

Figure 9. Sludge production vs. pH.

RECOMMENDED MONITORING AND DISCHARGE SCHEDULES

On the basis of the evaluation described above, Brown and Caldwell recommended that the City accept the proposed discharge of pretreated composite waste streams. Discharge of pond liquid alone was not recommended. IT would be responsible for monitoring and reporting to the City. The discharge schedule (flow rate) would depend mainly on the TDS level in the combined influent to the WWTP.

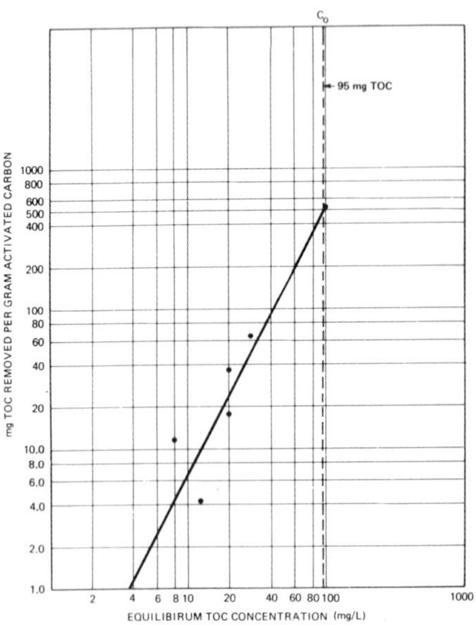

Figure 10. Freundlich adsorption isotherm plot (site sample).

Table III. Recommended Discharge Monitoring Schedule

Monthly	Daily (batch)	Quarterly
Arsenic	pH	EPA organic priority pollutants
Cadmium	TDS	
Chromium (total)	TOC (GAC influent)	
Copper		
Lead	TOC (GAC effluent)	
Mercury		
Nickel	Toxicity	
Silver	Chloride	
Zinc		
Phenolic compounds		
Polynuclear aromatic hydrocarbons		
Cyanide		
Oil and grease		
Sulfide (dissolved)		
Sulfate		
BOD_5		

Monitoring Schedule

From the City's viewpoint, the main water quality requirements for the discharge were the City's local pretreatment limits. However, other concerns such as the effects of high TDS on the WWTP, corrosion, toxicity, and GAC bed performance also had been considered. The recommended monitoring schedule is shown in Table III. The parameters to be monitored included all the local pretreatment limits (see Table II) as well as TDS, chloride, sulfate, BOD_5, TOC, priority organic pollutants, and toxicity as measured by toxicity bioassays. Dioxin analysis was waived because historical and new analytical results did not indicate the presence of dioxin.

Before discharge, every batch in the discharge tank would be tested for pH, TDS, TOC, chloride, and toxicity (bioassays). A 24-hour LC50 (three-spined stickleback) of at least 50 percent would be required. Other alternative tests such as the Microtox test might be substituted if a satisfactory correlation could be developed.

As mentioned earlier the treatability tests established that the concentrations of the regulated heavy metals would be below the local limits as long as the pH in the clarifier was maintained at 8.0 or above. As a result frequent pH monitoring would be adequate to ensure heavy metal removal after the initial comprehensive analyses. The heavy metals can be monitored monthly.

Brown and Caldwell recommended quarterly monitoring of the organic priority pollutants. A more frequent schedule was not recommended because the concentrations of these pollutants were relatively low in the waste streams. The analyses for these pollutants require long turnaround time and are very expensive. Daily monitoring of GAC bed performance by measuring influent and effluent TOC concentration would be adequate.

During initial startup all of the parameters should be monitored at least every other day for 10 days to establish baseline data for treatment effectiveness. The City could modify the toxicity monitoring frequency after the initial monitoring period because of the logistics of bioassay testing.

Discharge Schedule

As mentioned earlier, Brown and Caldwell recommended that the TDS concentration in the combined plant influent not exceed 2,500 mg/L. During pond dewatering the maximum TDS would be approximately 50,000 mg/L, and hence the discharge rate should not exceed 3.8 percent of the plant flow. This percentage would be equivalent to 40 gpm during the early morning (from midnight to 7 a.m.) period and about 80 gpm during the rest of the day. After the pond dewatering period, the flow rate could theoretically be increased because the TDS level in the discharge would decrease. However, a higher flow would not be needed at that time.

CONCLUSIONS

POTWs can play an important role in disposing hazardous wastes when POTW discharge is the best technical solution to the environmental problem. However, risk is always involved, especially if a

POTW cannot provide a very large dilution ratio for the discharges. With careful evaluation, planning, and monitoring, this risk can be minimized. The POTW should have a financial incentive to compensate for taking on the risk of treating the waste. The POTW should also obtain assurance that regulatory agencies recognize the risk so that they would not take enforcement actions against the POTW if unforeseeable interference occurred.

REFERENCES

1. "White Paper, California Association of Sanitation Agencies Industrial Waste Discharge," *The Bulletin*, California Water Pollution Control Association, July 1987.
2. Ullensvang, B. J., and U. P. Singh, "Treatment and Discharge to a POTW: The Stringfellow Experience," *Water Environment & Technology*, Vol. 2, No. 1, January 1990.
3. Burnett, W. E., "The Effect of Salinity Variations on the Activated Sludge Process," *Water and Sewage Works*, Vol. 121 (March 1974).
4. Mills, E. V., and A. B. Wheatland, "Effect of Saline Sewage on the Performance of Percolating Filters," *Water and Waste Treatment*, November/December 1962.
5. Poon, C.P.C., and W. J. Mikucki, "Rotating Biological Contactors Treat Island's Saline Sewage," *Water and Sewage Works* (February 1978).
6. Poon, C.P.C., Y. L. Chao, and W. J. Mikucki, "Factors Controlling Rotating Biological Contactor Performance," *Jour. WPCF*, Vol. 51 (1979).
7. Bishop, P. L., and N. E. Kinner, "Rotating Biological Contactor Treatment of Hypersaline Wastes," 36th Purdue Industrial Waste Conference, 644 (1981).
8. Davis, E. M., J. R. Bishop, and R. K. Guthrie, "Microbial Behavior in Hypersaline Wastes: Effects on BOD and Degradability," 33rd Purdue Industrial Waste Conference, 457 (1978).
9. Tokuz, R. Y., and W. W. Eckenfelder, Jr., "The Effects of F/M on an Activated Sludge System Treating High Salinity Wastewater," 33rd Purdue Industrial Waste Conference, 200 (1978).
10. Kincannon, D. F., and A. F. Gaudy, Jr., "Response of Biological Waste Treatment Systems to Changes in Salt Concentrations," *Biotechnology and Bioengineering*, 10:483–496 (1968).
11. Ludzack, F. J., and D. K. Noran, "Tolerance of High Salinities by Conventional Wastewater Treatment Processes," *Jour. WPCF*, 37:10:1404 (October 1965).
12. McCarty, P. L., and R. E. McKenney, "Salt Toxicity in Anaerobic Digestion," *Jour. WPCF*, 33:4:399 (April 1961).
13. Kugelman, I. J., and P. L. McCarty, "Cation Toxicity and Stimulation in Anaerobic Waste Treatment," *Jour. WPCF*, 37:1:97 (January 1965).
14. Bechtel National, Inc., *An Analysis of the Effect of Discharging Water from the Panoche Facility into the City of Benicia Waste Water System*, Report for City of Benicia, June 1989.

21 BIODEGRADATION/PARTITIONING OF BENZO(A)PYRENE IN LIQUID/SOLIDS SUSPENDED GROWTH SYSTEMS

Robert E. Marks, Graduate Research Assistant
Stephen D. Field, Associate Professor
Department of Civil Engineering
Andrew K. Wojtanowicz, Associate Professor
Department of Petroleum Engineering
Louisiana State University
Baton Rouge, Louisiana 70803.

INTRODUCTION

The scope of this research is to study the biodegradation and partitioning of benzo(a)pyrene (B[a]P) contained in sludges in continuous growth systems with the longer term goal of developing engineering parameters for design of pilot plant systems which will be more effective than existing alternative technologies in reducing accumulations of persistent polynuclear aromatics (PNA's) at Superfund sites. Design parameters will also be developed for the addition of an online waste reduction process as a cleanup stage for detoxification of hazardous organic compounds contained in the waste sludge generated from petroleum refineries and petrochemical plants.

Deposition rates of PNA's from all sources are exceeding the natural destruction rates by microbial decomposition and photo-oxidation, and accumulations in the land/water ecosystems are occurring.[1] The majority of the accumulations are from anthropogenic sources and contain the semi-volatile condensed 4–5 aromatic ring compounds. Five of these PNA's are considered by the Environmental Protection Agency (EPA) to be carcinogens and include: benzo(a)anthracene (B[a]A), benzo(b)fluoranthene (B[b]F), B[a]P, chrysene, and dibenzo(a,h)fluoranthene (DB[a,h]F). B[a]P is considered strongly carcinogenic based on cancers induced in small animal laboratory experimentation.[2] These compounds are typically solids at ambient temperatures, and are characterized by their high organic partition coefficients, very low aqueous solubility, low vapor pressures and tendency to bioaccumulate in the natural environment.

EPA has developed a listing of national sites containing hazardous materials and has established an order of priority for cleanup of these sites—National Priorities List (NPL). Many of these sites are contaminated with PNA's. The EPA Office of Emergency and Remedial Response (OERR) has the responsibility for developing treatment strategies which can be applied to these sites. Three basic treatment mechanisms are favored by EPA.[3] The treatment options include: a) destruction of the contaminant through chemical alteration to less toxic compounds (thermal destruction, dechlorination, and bioremediation); b) transfer of the contaminants to other waste streams for subsequent treatment (low temperature thermal desorption, chemical extraction, and soil washing); and c) contaminant bonding to waste media (immobilization). The current preferred methodology is thermal combustion (incineration), but costs are high and available capacity falls far short of the accumulated and projected industrial production of hazardous wastes. The second choice is bioremediation where costs factors are low but biodegradation times are comparatively long. A bioremediation solution must insure that the contaminated soils or sludges are detoxified to permit safe release to the environment, and all other products associated with the process such as recovered oils, surface scums, aqueous discharges, and decontaminated solids also meet current disposal regulations or are directed to further treatment.

The two hazardous wastes selected for this study are sludges from API separators. The first sludge was obtained from one of 16 API separators from a fully integrated crude oil refinery while the second sludge was obtained from an API separator in a petrochemical complex.

45th Purdue Industrial Waste Conference Proceedings,

Table I. Properties of Refining and Petrochemical Sludges

	Density	Oil Content	Oil Density	Solids
Refinery sludge	1.87 g/mL	3.7% v/v	0.84 g/ml	30.1% w/w
Petrochemical sludge	1.073 g/mL	32.6% v/v	0.76 g/ml	10.2% w/w

BACKGROUND

Constraints on Biodegradation and Partitioning of B[a]P

Microbial populations which contain the genetic material to produce the necessary extracellular enzymes, will proceed to biodegrade some PNA's to phenols.[4] The biodegradation of the simpler aromatic compounds usually proceeds by: removal and hydolysis of side chains, cleavage of the aromatic ring, and finally degradation of the aliphatic side products. The strong bioresistance of B[a]P is attributed to the alignment of the condensed ring structure.[5] It is reported that no microorganism has been isolated that will grow exclusively on B[a]P, but both procaryotic and eucaryotic organisms will biodegrade low concentrations in the presence of a common growth substrate.[6] Unpublished data from EPA (OERR) showed that all reported successful biodegradations used initial B[a]P substrate concentrations of <20 mg/L.

The highest concentrations of B[a]P in waste accumulations, typically are found dissolved in organic phases present in these wastes and partitioning occurs between the organic phase, soil, or other solid particles present in sludges. The strong hydrophobicity of B[a]P prevents partitioning into an available aqueous phase and solubility of B[a]P in water is reported at only 0.0012 mg/L. The octanol-water partition coefficient (K_{ow}) is reported as 6.06 while the organic carbon partition coefficient (K_{oc}) is reported at 5,500,000 ml/g.[7]

This research project will examine a range of design and operating parameters aimed at accelerating the biodegradation of B[a]P into benign or lower toxicity constituents and to determine the quantitative partitioning of B[a]P between the phases that exist in suspended growth systems.

Previous Research in Batch and Suspended Growth Systems

The use of bioengineering for the processing and destruction of petrochemical wastes containing high oil and solids, has been under investigation in the environmental laboratories in the Department of Civil Engineering at Louisiana State University (LSU) for the past 5 years. Microbial cultures have been developed in batch reactor testwork which have demonstrated the capability to biodegrade recalcitrant and bioresistant organic compounds at accelerated removal rates. The cultures have biodegraded both long chains and aromatic ring compounds found in natural crude oil residues and drilling oil muds.[8,9,10] The complex microbial cultures contain several species of bacteria, actinomycetes, fungi, molds, and protozoa and have biodegraded >80% of the oil present in slurries containing up to 24% oil v/v and 30% solids w/w. Successful biodegradations of hazardous wastes have also been demonstrated from API separators and silt pond sources of oil refineries.[11] The maximum growth rate for the suspended growth systems in batch reactor tests was reported as high as 0.41 hr^{-1} which compares favorably to municipal waste water maximum growth rates of 0.38–0.41 hr^{-1}.[12]

Longevity culture studies were run in open continuous reactors on hazardous wastes from both oil refinery and petrochemical plant sources. This program was carried out for a four month period and confirmed the continued viability of the microbial culture to efficiently destroy these two hazardous wastes.[13]

MATERIALS AND METHODS

Characterization of Oil Refinery and Petrochemical Plant Sludges

The properties of the two sludges used in the program are shown in Table I. The high solids (30% w/w) of the API separator refinery sludge consists of granular sands and the oil concentration (3.7% v/v) is relatively low and the oil can be readily separated from the solids fraction. The petrochemical sludge, however, is characterized by a relatively high oil content (32.6% v/v). The solids content (10% w/w) is much lower in the petrochemical sludge than in the refinery sludge and exhibits a clayey consistency. Both sludges are from currently operating API separators and significant weathering has not occurred thus the oil densities are in the 0.76–0.84 g/mL range.

Experimental Design for P[a]B Testwork in Continuous Reactors

The schematic design for the sealed continuous stirred tank reactor is shown in Figure 1. The reactor consisted of a 2 liter capacity sealed reaction kettle where all joints are either teflon or ground glass. Feed sludge, inorganic nutrients, and makeup water were added on a uniform chronological basis. Laboratory air was scrubbed to remove carbon dioxide (CO_2) and was pumped into the vapor space above the stirred liquid. A mixed liquor suspended solids (MLSS) recycle system permitted biomass recycle. An outlet foam trap ensured capture of any escaping organic compounds of interest. Uniform stirring was maintained throughout the testing period. The design shape of the reactor and designated mixing intensity, resulted in a relatively uniform distribution of the solids, and guaranteed that aerobic conditions prevailed throughout the liquid phase. The culture, however, has demonstrated a facultative capability, but biodegradation rates as expected are much slower under anaerobic conditions. One of the four entry ports in the reactor cover is used for sampling of the reactor head space and the MLSS. When a continuous flow reactor system was determined to be in a state of equilibrium during a specific experiment, the reactor was completely disassembled and all vapor, liquid and solid phases in the reactor or downstream systems were tested quantitatively for B[a]P and other PNA's of interest.

The refinery sludge was used as feedstock in the first phase (phase 1) of the testwork and the petrochemical sludge was used in the second phase (phase 2). The key reactor operating conditions are shown on Table II for both phases. The mean cell residence time (MCRT) for both test phases was set at 5 days. Biomass recycle was not used in phase 1 but 9–10% cell recycle was used in the second phase to increase MLSS from an estimated 7,500 mg/L to 10,000–15,000 mg/L.

The composition of the inorganic nutrient broth is shown on Table III and is an adaption from similar media described in Manual of Methods for General Bacteriology.[14] The reactor pH was

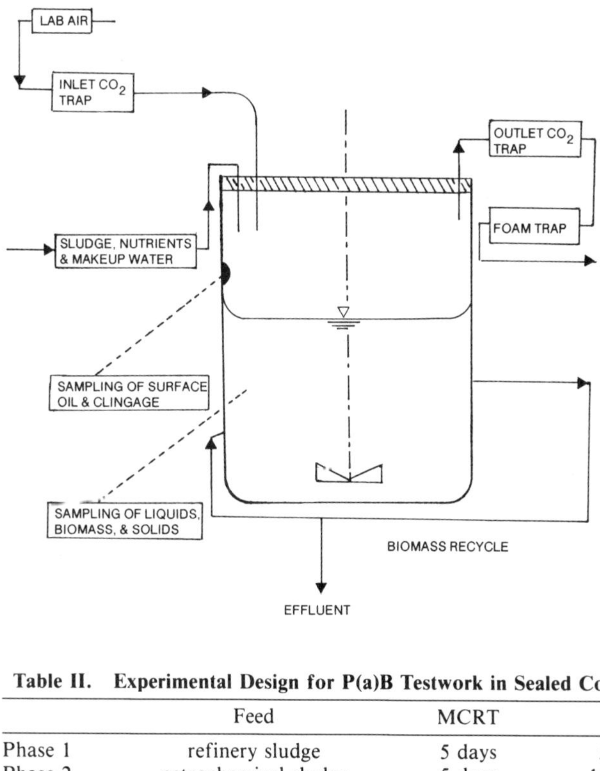

Figure 1. Schematic drawing of continuous suspended growth system.

Table II. Experimental Design for P(a)B Testwork in Sealed Continuous Stirred Tank Reactors

	Feed	MCRT	MLSS	Cell Recycle
Phase 1	refinery sludge	5 days	5000–10000	0%
Phase 2	petrochemical sludge	5 days	10000–15000	10%

MCRT = mean cell residence time
MLSS = mixed liquor suspended solids (mg/L)

Table III. Sterilized Basal Inorganic Salt Media

Macronutrients
- sodium chloride 0.15g, ammonium sulfate 1.8g,
- calcium chloride 0.13g, magnesium sulfate 0.12g,
- potassium monohydrogenphosphate 2.5g,
- potassium dihydrogen phosphate 0.15g,

Micronutrients
- ferrous sulfate 5 drops (1g/100 ml water)
- zinc sulfate 7 drops (5g/100 ml water)
- cobalt chloride 7 drops (5g/100 ml water)
- cupric chloride 3 drops (1g/100 ml water)
- sodium molybdate 12 drops (1g/100 ml water)
- boric acid 25 drops (4g/100 ml water)
- manganese chloride 12 drops (1g/100 ml water)
- makeup total volume to 1800 ml with water

buffered by adding additional monohydrogen phosphate to the media over and above theoretical nutrient requirements. Previous experience with biodegradation of sludges showed that typically sludges contain the required growth factors for biodegradation and shortages of the macronutrients nitrogen and/or phosphorus occurred in only two instances which resulted in slowdown of the degradation rate.[15]

Duplicate reactors were used for each test and the criteria for equilibrium conditions for the continuous suspended growth reactor systems were defined by monitoring of the following variables:

Variable	*Acceptable Repeatability over 3 days*
pH	+/- 0.5 units from mean
CO_2 respiration	+/- 10% from mean
culture plate counts	+/- one magnitude from mean
MLSS	+/- 2000 mg/L from mean
effluent hydrocarbons	+/- 15% v/v from mean

The reactors were disassembled at the test steady state conditions and the following samples were analyzed to provide PNA and solids mass balances.

1. foam trap and overhead reactor vapor system
2. surface oils, scums, wall clingage
3. suspended solids (biomass)
4. aqueous phase
5. solids (sand, silt & clays)
6. reactor feed sludge

The analytical procedures used to obtain PNA balances by high resolution gas chromatography by analysis of all reactor feed streams, product streams, and accumulations are shown on Table IV. All sampling and analytical procedures were performed in accordance with Test Methods for Evaluating Solid Wastes – EPA SW-846 (third edition).[16] All aqueous samples were immediately acidified to 3 pH to inhibit further biodegradation. Separatory funnels were used to separate the biomass, other solids, and aqueous phases. Approved EPA surrogate and matrix spike standards were added to all raw samples. The organic components were extracted from the samples and cleanup preceded separation of the organic fraction into polar and nonpolar fractions via silica gel liquid column chromatography. The fraction containing the PNA's was volume chromatography. The fraction containing the PNA's was volume adjusted and then analyzed by capillary column chromatography using either mass spectrometry or flame ionization detectors.

RESULTS AND DISCUSSION

Refinery Sludge – Phase 1

The PNA removal efficiencies for 7 major PNA's and the constituent concentration levels (CCL) for safe release to the environment for allowable land disposal are shown in Table V. The data indicates that the average reduction was 87.7% for the 7 PNA's shown on Table V. The total oil

Table IV. Analytical Procedures for P(a)B Mass Balances From Reactor Feed, Product Streams and Accumulations

EPA Test Methods for Evaluating Solid Wastes - 1986
surrogate and matrix spike standards method 3500
organics from aqueous solutions method 3510
extract organics from solids/sludges method 3550
silica gel cleanup for PNA's method 3630
capillary column chromatography method 8100
Referee methods
soxhlet extraction of solids and sludges method 3540
gas chromatography/mass spectrometry method 8270

Table V. Phase 1 Refinery Sludge PNA Removals MCRT 5 Days, Oil Removal 73%

	Feed mg/kg	Effluent mg/kg	CCL mg/kg
Naphthalene	24.0	0.24	(reserved)
Fluorene	24.2	0.24	–
Phanthracene	8.6	1.2	7.7
Pyrene	9.9	2.48	2.0
Chrysene	13.9	2.52	2.2
Benzo(a)anth	6.9	0.96	1.4
Benzo(a)pyrene	17.5	2.28	0.84

PNA removal percent: pyrene 75, chrysene 82, B(a)A 86, B(a)P 87
CCL = constituent concentration level

reduction was 73% and the MLSS was measured at 4500 mg/L. The reduction in B[a]P from 17.5 mg/kg in the feed to 2.28 mg/kg in the total effluent for a reduction of 87% was considered significant. The allowable CCL is 0.84 mg/kg and dilution would still be required to meet land disposal requirements but each kg of sludge feed would require mixing with 19.8 kg of inert material to meet the 0.84 mg/kg CCL while the biodegraded material would only require mixing with 1.7 kg of inert material to meet the 0.84 mg/kg standard.

Petrochemical Sludge – Phase 2

The average of 10 PNA removal rates for the petrochemical sludge was 74.1% and the individual removal rates are shown in Table VI. The operating conditions were: MCRT 4.3 days, overall oil removal 75%, biomass recycle 9%, and MLSS 13,000 mg/L. All PNA's with designated EPA CCL limits met those limits and the total effluent was environmentally safe for land disposal from a PNA standpoint. The removal rate for B[a]P was 77%.

Table VI. Phase 2 Petrochemical Sludge PNA Removals MCRT 4.3 Days, Oil Removal 75%, Cell Recycle 9%

	Feed mg/kg	Effluent mg/kg	CCL mg/kg
Naphthalene	41.3	7.3	(reserved)
Fluorene	8.3	2.9	–
Phanthracene	13.1	3.5	7.7
Anthracene	11.8	1.7	6.2
Fluoranthene	2.7	0.1	–
Pyrene	4.4	1.4	2.0
Chrysene	5.2	0.1	2.2
Benzo(a)anth	2.5	0.1	1.4
Benzo(a)pyrene	3.0	0.7	0.84

B(a)P removal 77%
CCL = constituent concentration level

Table VII. Operation/Performance of Suspended Growth Systems

	Sludge	MCRT	MLSS	Removal %
CSTR phase 1	refining	5	4500	73
CSTR phase 2	petrochem	4.3	13000	75
Conventional	sewage	5–15	2500	85–95
CSTR	sewage	5–15	4500	85–95
High rate	sewage	5–10	7000	75–90

MCRT = mean cell residence time - days
MLSS = mixed liquor suspended solids - mg/L
CSTR = continuous stirred tank reactor

Comparisons with Conventional Wastewater Suspended Growth Systems

Table VII shows a comparison between operation and performance of the suspended growth reactors used in this research and conventional wastewater treatment plants (WWTP's). Conventional sewage typically contains low solids (< 10% w/w) and organic compound removal by activated sludge is 85–95%. The API separator sludges contain higher solids and higher organic contents and the type of organics are more recalcitrant and bioresistant to degradation. Mean cell residence times for the two sludges was at the lower limit used by WWTP's and removals were 10% lower (73–75) than the typical conventional sewage plant. Microbial populations, however, were higher in the test reactors than in conventional WWTP's. MCRT in the test reactors was held at the lower limit used in sewage treatment plants. A typical biomass recycle in a WWTP may be 25–40% and tests will be run in the test reactors at these elevated recycle rates to see if degradation of PNA's is enhanced.

Initial Partitioning of B[a]P in Phase 2 Test Reactor

Partitioning data from one set of biodegradation tests using the petrochemical sludge as feed, indicated that the only significant B[a]P was detected in the extracted organic fraction from the microbial solids phase. The detection limit of B[a]P for the 30 meter (0.25 mm I.D.) fused silica capillary column is 50–100 ng using a flame ionization detector.[17] The microbial culture has demonstrated the ability to rapidly separate oil strongly bonded to drill cuttings and then emulsify the oil fractions to enhance metabolic activity. Additional partitioning tests will be run on both the refining and petrochemical sludges. Radio labeled B[a]P feedstock will be used to provide confirmation of the gas chromatography testwork. If further research confirms that the bulk of the residual PNA's are partitioning into the microbial sludge, this would provide a large reduction in the original volume of hazardous waste and accordingly, significantly reduces further disposal costs.

CONCLUSIONS

From the results and discussions of this research, it is concluded that sludges (and sludge dilutions) containing PNA's such as B[a]P can readily be destroyed under optimized conditions in suspended growth systems. If further testwork confirms that bioresistant PNA residuals are concentrating in the microbial sludge, then a significant reduction in volume of hazardous waste has occurred, thus mitigating the land disposal problem. The operating/design characteristics of the benchscale suspended growth systems appear to be similar to conventional WWTP's and this should accelerate development of operating design parameters for construction of online sludge reduction units in the operating trains of refineries and petrochemical plants.

Table VIII. Initial Partitioning of B(a)P in Phase 2 Reactors
B(a)P in Petrochemical Sludge Feed 35.4 mg/L

Product Streams	B(a)P mg/L
Reactor overhead system	not detected
Total surface clingage	not detected
Clarified effluent	not detected
Microbial solids	8.1 mg/L
Other solids	not detected

REFERENCES

1. Jones, K. C., Stratford, J. A., Tidridge, P., and Waterhouse, K. S., "Polynuclear Aromatic Hydrocarbons in an Agricultural Soil: Long-term Changes in Profile Distribution," *Environmental Pollution*, 56, 337–551, (1989).
2. Hartwell, J. L., "Survey of Compounds Which Have Been Tested for Carcinogenic Activity," U.S. Public Health Service, Washington, D.C., pub. 149, (1951).
3. U.S. EPA, "Seminar on Treating Contaminated Soil, Conference on Hazardous Waste and Hazardous Materials, New Orleans, LA, April, (1989).
4. Lee, M. L., Novotny, M. L., and Bartle, K. D., "Toxicology and Metabolism," In *Analytical Chemistry of Polycyclic Aromatic Compounds*, Academic Press, p.62–70 (1981).
5. Dipple, A., "Polycyclic Aromatic Hydrocarbon Carcinogenesis," In *Polycyclic Hydrocarbons and Carcinogenesis*, Ed. R. G. Harvey, American Chemical Society, Washington, D.C., p.1–17, (1985).
6. Coover, M. P., and Sims, R. C., "The Rate of Benzo[a]pyrene Apparent Loss in a Natural and Manure Amended Clay Loam Soil," In *Hazardous Waste & Hazardous Materials*, Pub. Ann Liebert Inc., 152–157, (1987).
7. Sherman, D., Smith, J., and Lynch, J., "Physical/Chemical Properties of Selected PAH Compounds," in *Effectiveness and Cost of Various Bioremediation Technologies*, Remediation Technologies Inc., Austin, TX., (1989).
8. Marks, R. E., Field, S. D., and Wojtanowicz, A. K., "Biodegradation of Oil-based Drill Cuttings and Production Pit Sludges," Proceedings of 10th Annual Energy Source Technology Conference, ASME, Dallas, TX (1987).
9. Marks, R. E., Field, S. D. and Wojtanowicz, A. K., "Biodegradation of Oilfield Production Pit Sludges," Proceedings of 42nd Annual Industrial Waste Conference, Purdue University, West Lafayette, IN (1987).
10. Marks, R. E., Field, S. D., and Wojtanowicz, A. K., "Oil Reduction in Spent Drilling Muds by Biotreatment," Proceedings of a National Conference on Drilling Muds, Environmental and Groundwater Institute, University of Oklahoma, Norman, OK (1987).
11. Field, S. D., and Wojtanowicz, A. K., "Biological Treatment of Petroleum and Petrochemical Hazardous Wastes," Proceedings of the Second Annual Symposium on Hazardous Waste Research, Hazardous Waste Research Center, Louisiana State University, Baton Rouge, LA, p. 62–76 (1988).
12. Gaudy, A. F., and Gaudy, E. T., "Quantitative Description of Growth," in *Microbiology for Environmental Scientists and Engineers*, Ed. McGraw-Hill, p. 245–250 (1980).
13. Field, S. D., Marks, R. E., Ours, D., and Wojtanowicz, A. K., "Biological Treatment of Petrochemical Sludges," Proceedings of Third Annual Symposium on Hazardous Waste Research, Hazardous Waste Research Center, Louisiana State University, Baton Rouge, LA (1989).
14. American Society for Microbiology, "Manual of Methods for General Bacteriology," Ed. P. Gerhardt, Washington, D.C., p. 130–131 (1981).
15. Marks, R. E., "Biotreatment of Oilmuds and Production Sludges," Thesis for degree of Master of Science in Civil Engineering, Louisiana State University, Baton Rouge, LA, p. 47–48 (1987).
16. EPA, "Test Methods for Evaluating Solid Waste, Office of Solid Waste and Emergency Response, Washington, D.C., Vol.1B, Laboratory Manual Physical/Chemical Methods, Methods 3500, 3510, 3520, 3540, 3550, 3630, 8100, and 8270 (1986).
17. J & W Scientific, "High Resolution Chromatography Products," pages 5–10 (1989).

22 DETAILED ENGINEERING DESIGN AND FULL-SCALE TESTING FOR TREATMENT OF AN EP-TOXIC SLUDGE

Robert Chiesa, Group Leader
Ajit K. Chowdhury, Senior Chemical Engineer
James E. Morse, Senior Engrg. Technical
RMT, Inc.
Madison, Wisconsin 53708–8923

Thomas R. Wirth, Senior Environ. Engineer
General Motors Corporation - Central Foundry Division
Defiance, Ohio 43512–0070

INTRODUCTION

General Motors Corporation-Central Foundry Division (GMC-CFD) operates a ferrous foundry in Defiance, Ohio. Particulate emissions from the cupola operations are collected by wet cap and venturi scrubber emission control systems. The wastewater from the emission control systems is conveyed to a settling basin, and the effluent is reused in the plant. The accumulated sludge in the settling basin has been periodically removed by hydraulic dredging, and disposed in an on-site, state-approved, solid waste landfill.

Previous testing of the wet cap and scrubber emissions (at point of origin) verified that they were nonhazardous. However, in the process of complying with NPDES permit requirements, several sludge samples from different parts of the basin were collected and analyzed. Slightly more than half of these samples released lead in the EP Toxicity test above the hazardous waste criterion of 5.0 mg/L.

Bench-scale treatability tests were conducted with different chemical additives to evaluate various treatment possibilities for rendering the sludge non-EP Toxic. Because hydraulic dredging would likely be used to remove the sludge from the basin, these bench-scale tests were conducted over a range of solids contents. Results indicated that the EP Toxicity lead concentration could be reduced to less than 2 mg/L using a phosphoric acid (H_3PO_4) dosage of 2.06 gallons H_3PO_4 per ton of dry sludge solids. The bench-scale tests also showed that both the pH and the residual dissolved phosphate concentration of the wet sludge after treatment were good indicators of treatment effectiveness.

A process assessment was performed to evaluate several treatment options and develop a conceptual design for the selected treatment system, based on technical, economic, and regulatory considerations. Based on the bench-scale treatability testing and process assessment, the selected treatment process included hydraulically dredging the basin sludge (approximately 400,000 cubic yards) to an existing on-site landfill designed to handle dredged slurries, with in-line chemical treatment in the slurry pipeline to render the solids non-EP Toxic. The treated solids would settle in the landfill, and the supernatant would be returned back to the settling basin. Because the settling basin was a critical part of the foundry's wastewater treatment system, the dredging and treatment were required to occur while continuously operating the settling basin. Detailed engineering was performed, the system was constructed, and field trials using full-scale equipment were run to verify the ability of the treatment system to render the solids non-EP Toxic. Based on the field trial results, modifications were made to the treatment system, and the basin was remediated by dredging and treating the sludge.

This paper focuses on the detailed engineering design and full-scale field trial phases of the project.

45th Purdue Industrial Waste Conference Proceedings,

DETAILED DESIGN

Treatment Process Description

Detailed design was based on the conceptual design developed during the process assessment phase of the project. The following is a brief description of the treatment system as initially designed and constructed.

The H_3PO_4 solution would be delivered to the site in bulk trucks and unloaded into bulk storage tanks. The solids in the basin would be removed by hydraulic dredging and conveyed by a pipeline to two mixing tanks in parallel. H_3PO_4 would be injected into the dredged slurry at an acid injection tee upstream of the mixing tanks. During the actual remediation, the treated slurry would flow by gravity from the mixing tanks to a licensed on-site landfill. This landfill was designed to function as a drying bed for disposal of dredged solids, and has a liner, a leachate collection system, and a supernatant removal system. The treated solids would settle in the landfill, and the supernatant would be recycled back to the basin. During field trials, the treated slurry would bypass the landfill and be routed by a pipeline back to the basin. A slurry flow meter and a density meter would be used to continuously measure the mass flow rate of solids in the dredged slurry, with automatic feed-forward control of the H_3PO_4 feed rate proportional to the mass flow rate of the dry solids in the slurry. Samples of the untreated slurry would be used for periodic calibration of the density meter. Hourly measurements of the pH and residual dissolved phosphate in the treated slurry would be used to obtain "real time" treatment effectiveness information. Samples of both the untreated slurry and the treated slurry would also be collected hourly throughout the period of dredging and treatment, and analyzed daily for their EP Toxicity lead concentration to confirm the success of the treatment. Figure 1 is a schematic diagram of the system as initially designed and constructed.

Design Criteria

Detailed engineering was based on the following design criteria:

• Dredged slurry flow rate:	4,000–5,000 gal/min
• Slurry solids concentration:	5–25% by weight
• Acid dosage range:	1.54–4.10 gal/ton of dry solids

These criteria resulted in a design solids mass flow rate range of 0.9 to 6.4 tons/min (as dry solids), an acid feed rate of 1.3 to 26.3 gal/min, and a maximum daily acid consumption rate of over 37,000 gallons per day. Due to the massive quantity of acid to be used, a key concern was the design of a system that would allow acid to be stored and continuously fed into the system safely and reliably.

Because of uncertainty concerning the accuracy, precision, and reliability of the slurry flow meter and slurry density meter under the conditions expected during remediation, the acid feed system was designed to operate in any of the four modes described in Table I. By allowing for operating flexibility in control system design, it was expected that the reliability of the control system would be maintained even during periods when one or both of these instruments were inoperable. The system was designed to be in the automatic mode under normal operating conditions.

The field trials and some of the actual remediation occurred during the winter, so the detailed design also addressed system operation and maintenance during sub-freezing weather conditions.

Detailed Engineering

To resolve these and other concerns, detailed engineering was performed in several steps. First, detailed specifications and initial design drawings of the dredged slurry pipeline and the treatment system were prepared. Bids were then solicited from contractors for performing the remediation, including the following:

- Supplying the system components (including the hydraulic dredge) and constructing the treatment system.
- Dredging and operating the treatment system during the full-scale field trials.
- Dredging and operating the treatment system during the actual remediation.

In addition to bid prices, the bidders were required to provide specific technical information concerning the hydraulic dredge and the treatment system design and equipment to be supplied by the contractor.

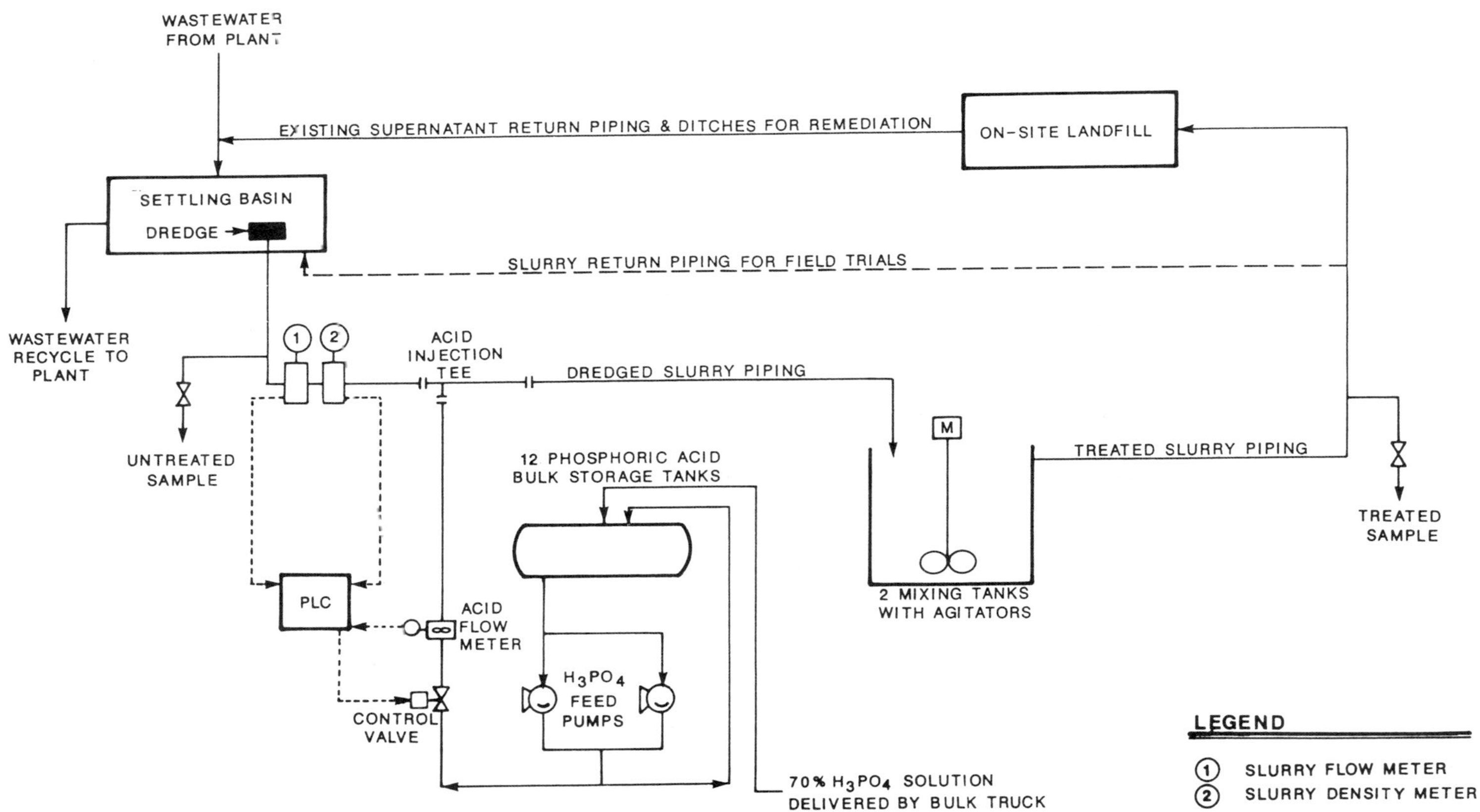

Figure 1. Treatment process flow schematic-design and initial construction for field trials.

Table I. Treatment System Control Modes

Control Mode	Slurry Flow Meter	Slurry Density Meter	Comments
Automatic	In operation	In operation	Normal system mode of operation
Flow	In operation	Out of service	Set slurry solids concentration = 30%
Suspended solids	Out of service	In operation	Set slurry flow rate = 5,000 gal/min
Manual	Out of service	Out of service	Set slurry flow rate = 5,000 gal/min and slurry solids concentration = 30%

Following award of the dredging and treatment contract, engineering details of the design were finalized based on the project specifications, initial design drawings, and the contractor's own technical knowledge. These engineering details included the following:

- Selecting treatment system equipment, controls, and instrumentation, including the acid feed pumps, slurry flow meter and density meter, acid flow meter, acid control valve, mixing tanks, and agitators.
- Preparing detailed drawings of the outdoor bulk acid truck unloading area, storage tanks, acid piping, and acid spill containment.
- Preparing detailed drawings showing the arrangement of the indoor acid injection equipment and piping, and the instrumentation and controls.
- Preparing detailed drawings showing the design and installation of the mixing tanks.

To minimize the problems due to cold weather, the final design located the acid feed pumps, all instrumentation and controls, the acid injection tee, and critical acid feed system pumps and piping indoors. All other piping, the acid storage tanks, and the acid mixing tanks were located outdoors. The acid concentration used (70 percent by weight) was selected to minimize the potential for acid freezing. The layout of the dredge slurry pipeline was designed so that, during field trials, it would be primarily self-draining after a dredge shutdown. The mixing tanks were also designed to be drained.

The remediation contractor provided a control system based on a programmable logic controller (PLC). This allowed the system controls to be developed and changes to be made to the controls by programming of the PLC software via a microcomputer rather than by physical wiring of electrical components. It also allowed the use of a graphics monitor for operator interface with the treatment system controls. The PLC and graphics monitor allowed viewing of any of the several monitor screens to rapidly check the status of various parts of the treatment system, respond to system alarms, or adjust treatment system operation (including changing the system control mode, adjusting the system setpoint dosage, and starting or stopping the acid feed pumps) as desired.

To prevent corrosion due to the acid, type 316 stainless steel was specified for all pumps, piping, valves, and instrumentation in contact with H_3PO_4, and polyethylene was selected for the acid storage tanks. Welding of all acid piping joints (except for flange joints at equipment, control valve, and instrument connections) was specified to minimize the potential for acid leaks. High-density polyethylene (HDPE) was selected for the 14-inch-diameter, 1.2-mile slurry pipeline. The slurry was expected to be abrasive, so the design included a large wall thickness (0.824 inch) and butt-fusing of the 38-foot HDPE pipe sections to minimize the potential for spills from leaks or breaks in the slurry pipeline.

The specifications required that the remediation contractor be responsible for treatment system operation and maintenance (O&M), as well as for construction, during both the field trials and the actual remediation. Therefore, the specifications also contained detailed requirements for the contractor concerning the following O&M issues:

- Instrument calibration during initial start-up and periodic recalibration.
- Process monitoring, including process sampling and on-site analysis.
- Procedures to ensure proper treatment during each system start-up and shut-down.
- Data reporting and notifications to GMC and RMT.
- Responses to emergency conditions (e.g., equipment and instrumentation failures, other treatment process upsets, and spills of acid, slurry, or fuel oil).

The specifications also required the remediation contractor to prepare an O&M manual and a spill plan to further define the contractor's O&M procedures. The contractor's O&M manual included detailed operating directions under different treatment system control modes, specific responses to system alarms, and operator log sheets. The contractor's spill plan included a description of detailed equipment and procedures to be used in response to a spill. Both the O&M manual and the spill plan included procedures (including names and telephone numbers of key contacts) in case of an emergency.

CONSTRUCTION

To expedite construction, the following treatment system items were shop fabricated off-site and then shipped to the remediation site and placed in position:

- A trailer-mounted unit that included the following:
  -Acid feed pumps, piping, and instrumentation.
  -Treatment system electrical components, including the PLC and graphics monitor.
- The two mixing tanks and agitators.

While these components were being shop fabricated and delivered to the site, site construction activities included the following:

- Site grading to accommodate the treatment system equipment.
- Laying of the 1.2-mile slurry pipeline.
- Erecting 12 acid storage tanks (with a combined capacity of 78,000 gallons) and associated acid piping.
- Constructing a bulk acid truck unloading area.
- Installing the concrete spill containment and collection around the truck unloading area, the acid storage tanks, and outdoor acid piping.

Before field trials dredging and treatment were initiated, the 1.2-mile pipeline was hydrostatically tested to demonstrate the absence of leaks. Upon completion of construction, the system was ready for initial start-up and field trials testing.

FIELD TRIALS

Field Trials Plan

Concurrent with development of the specifications and initial drawings for the treatment system, a Field Trials Plan was developed to systematically evaluate the following variables which were expected to significantly affect the treatment system performance:

- Slurry flow rate and slurry solids concentration.
- Acid dosage (in gal H_3PO_4 solution/ton dry solids).
- Ability of the instrumentation to accurately and precisely monitor process variables (i.e., slurry flow rate, slurry solids concentration, and acid flow rate).
- Ability of the controls to maintain the acid dosage within an established range of the setpoint dosage.
- Reliability of the mechanical components (e.g., pumps, valves, mixing tank mixer).

Table II is a summary of the Field Trials Plan. This plan was used to guide the field trial test runs, make decisions concerning additional test runs needed, and provide a basis for evaluating the success of the field trials. The Field Trials Plan and the remediation contractor's drawings, data, and O&M manual were then used as a basis for developing detailed checklists and procedures to be used in conducting the field trials.

Initial Start-up and Debugging

The first task during the field trials was system start-up and debugging. A considerable amount of time was spent during initial start-up in debugging the slurry flow meter and in calibrating the density meter to the actual slurry solids concentration. In addition, the mechanical seals on the acid feed pumps repeatedly failed after a short period of operation, resulting in significant downtime of the acid feed system. These pumps were subsequently replaced by vane pumps, and the acid feed rate was controlled by variable speed motors on the pumps instead of by the acid feed control valve.

During initial start-up, the dredge often pumped large rocks and other debris through the pipeline into the mixing tanks, which resulted in reduced mixing due to the build-up of material within the tanks and damage to an agitator in one of the tanks. A bypass was installed around the mixing tanks

Table II. Field Trials Plan

Field Trial Number	Objectives	System Control Mode	Acid Dosage (gal/ton)
1	• Start up and debug system equipment • Calibrate system instrumentation and controls	Manual	–
2	• Demonstrate system operation in the automatic control mode • Evaluate treatment results at an acid dosage of 2.06 gal/ton	Automatic	2.06
3	• Evaluate treatment results at other acid dosages	Automatic	1.54, 2.54, 3.08, 3.60, 4.10
4	• Demonstrate system operation in the flow proportional control mode	Flow	2.06
5	• Demonstrate system operation in the suspended solids control mode	Suspended Solids	2.06

to eliminate the problems with the mixing tanks, and the field trials were run both with and without the mixing tanks to determine the effect of the mixing tanks on treatment system performance.

The acid feed rate controller was also adjusted during the initial start-up period to optimize the control system's ability to maintain the actual H_3PO_4 dosage at the setpoint dosage. Controller adjustments were made during initial start-up to maintain the actual dosages within the ±25% of the setpoint dosage specified during detailed design.

Figure 2 shows the treatment system as modified after the initial startup and debugging were completed.

Field Trials Description

Field trials were conducted on a 24-hour-per-day, 7-day-per-week basis to minimize the length of time spent on field trials. The field trials were conducted over a 6-week period, including the time for initial start-up and debugging of the treatment system. The field trials generally followed the plan initially developed, except that additional testing was performed to determine the effect on treatment both with and without the mixing tanks. Samples of treated slurry were collected hourly during each of the field trial runs, and composite samples collected for EP Toxicity testing were analyzed on an expedited schedule so that the EP Toxicity test results of the initial field trial runs could be used to guide the planning of the later field trial runs.

Throughout the field trials, a series of operational problems were simulated to evaluate the response of the treatment system and the contractor's work force.

Field Trial Results

Dredging rates during the field trials ranged from less than 2,700 gpm to over 4,200 gpm, with slurry solids concentrations varying from less than 1% to over 30% by weight. Typical solids production rates were from 1.5 to 6 tons/min of dry solids, but ranged from less than 0.5 ton/min to over 10 tons/min. Despite this wide range of operating conditions, the equipment, instrumentation, and controls worked reliably after the initial start-up and debugging. This increased the confidence that the system would provide the reliability needed during the actual remediation. A key consideration in evaluating the field trial results was the ability of the control system to track variations in slurry mass flow rate and to maintain a sufficient acid feed rate to achieve the desired dosage (i.e., the setpoint dosage) without overfeeding acid. Based on 473 readings comparing actual versus setpoint dosage, the average actual dosage was 100.06% of the setpoint dosage, with a standard deviation of 8.70% when the acid feed rate was above 3 gpm. At acid feed rates below 3 gpm (corresponding to low slurry mass flow rates, typically less than 1 ton/min), the acid flow meter became increasingly inaccurate, resulting in variable dosage rates. This was resolved prior to the actual remediation by reprogramming the PLC to maintain a minimum acid feed rate of 3 gpm, which provided sufficient acid for treatment at low slurry mass flow rates without greatly overfeeding acid under these conditions. Figure 3 shows a

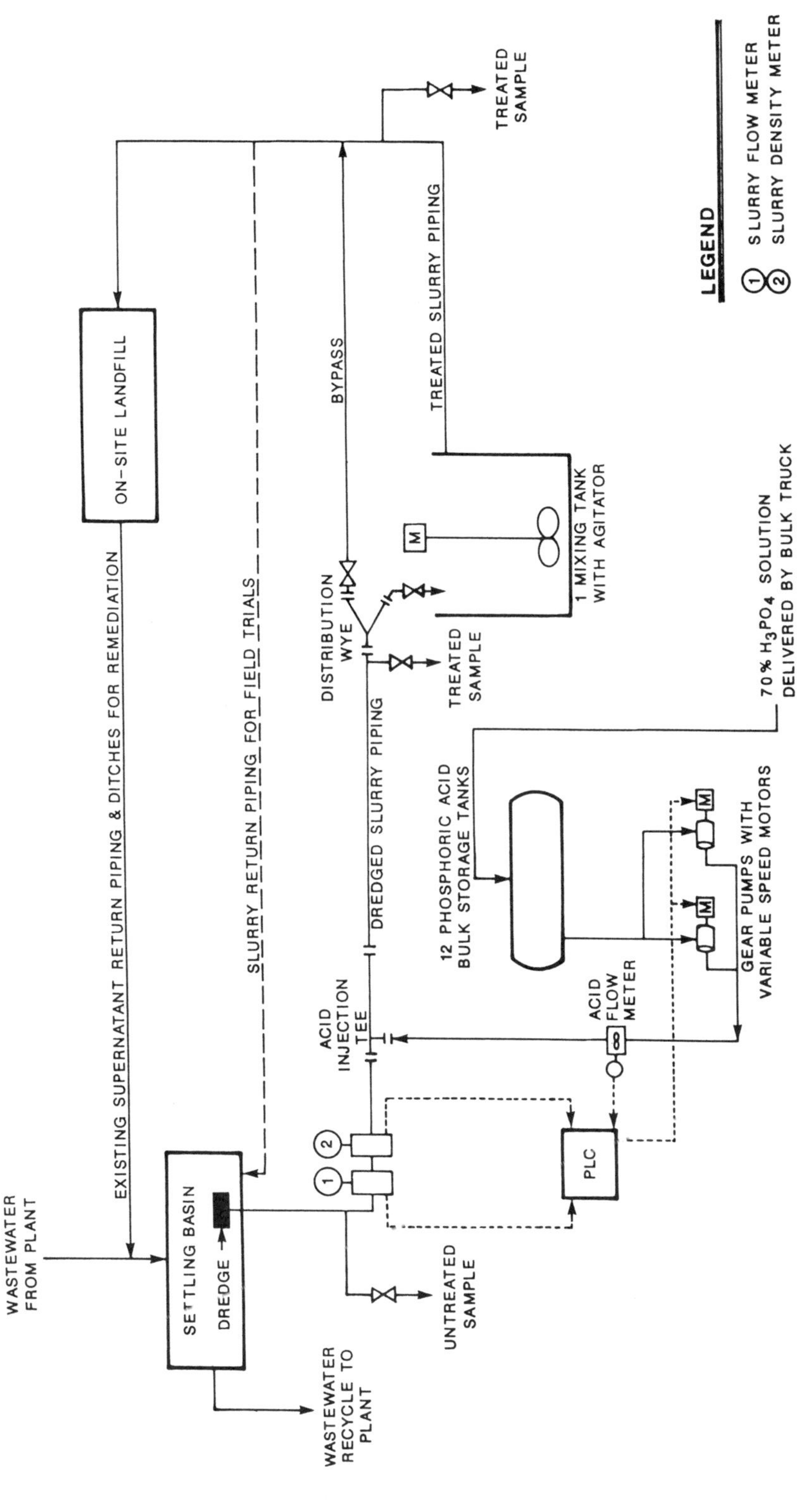

Figure 2. Treatment process flow schematic-modified system after field trials start-up.

graph of the actual versus setpoint dosage during one of the field trial runs. This is typical of the setpoint control achieved during the other field trial runs after the controller had been adjusted to optimize setpoint control.

The treatability testing results showed that the lead EP Toxicity could be reduced to less than 2 mg/L at a H_3PO_4 dosage of 2.06 gal/ton. The results at various dosages are shown in Table III. The results also indicated that there were no significant differences in lead EP Toxicity between the treated slurry samples collected prior to the mixing tanks and those collected after the mixing tanks.

At the request of the regulatory agencies, approximately 30% of the treated slurry samples were also tested for barium, cadmium, and chromium EP Toxicity. All samples tested for these parameters were well below the hazardous waste limits for these parameters.

Due to the variability in the alkalinity and the PO_4 uptake of the solids, pH and residual dissolved phosphate were not precise indicators of treatment effectiveness. However, the results indicated that these parameters could be used as "real time" monitors of general trends in treatment performance.

Recommendations for Remediation

Based on the field trial results, it was decided to go forward with remediation of the basin. To improve system performance and reliability, the following system modifications were recommended:

- Eliminating the mixing tanks and increase the length of piping from the acid injection tee to the discharge into the landfill to obtain equivalent mixing and reaction time as in the system used during field trials.
- Using a H_3PO_4 dosage of 2.1 gal/ton initially for the actual remediation.
- Reprogramming the PLC to maintain a minimum acid dosage of 3 gpm.
- Installing an emergency acid feed system to prevent untreated solids from entering the landfill in the event of an acid feed system failure.
- Implementing statistical process control procedures to monitor treatment performance and control acid dosage.

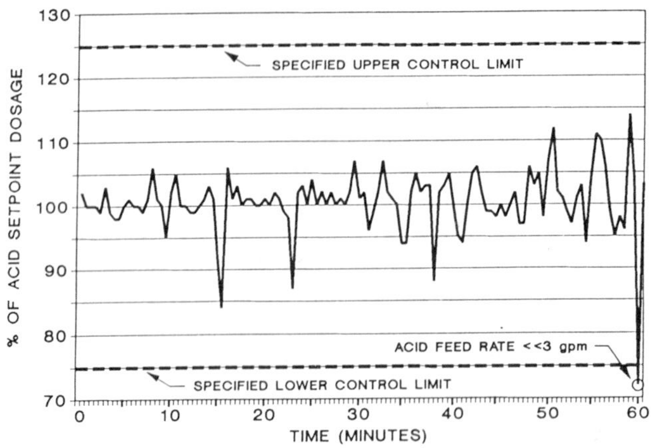

Figure 3. Actual vs. setpoint dosage for a one-hour period during field trial number 2.

Table III. Treatability Testing Results From Field Trials

Acid Dosage (gal/ton)	Lead EP Toxicity Test Result		
	Number of Samples	Average (mg/L)	Maximum (mg/L)
0	37	7.2	36
1.54	9	1.0	4.1
2.06	10	0.6	1.3
2.54	9	0.1	0.2
3.08	3	0.3	0.5
3.60	8	0.4	1.4
4.10	4	0.1	0.1

FULL-SCALE REMEDIATION

The recommended modifications were then made to the system prior to the actual remediation. Remediation dredging and treatment lasted approximately 6 months and resulted in successful remediation of approximately 400,000 yd^3 of sludge in the basin. Statistical process control procedures were used to monitor treatment performance and periodically adjust the setpoint acid dosage during remediation. Based on over 200 samples of treated sludge collected during remediation, the average lead EP Toxicity concentration of the treated sludge was 0.49 mg/L, compared to the regulatory limit of 5.0 mg/L.

Approximately 300 samples from the bottom of the settling basin were analyzed to document that the basin met the RCRA clean closure performance standards. The results indicated that the basin sludge had not significantly impacted the underlying soil, based on a statistical comparison of metals concentrations between background soil and underlying soil.

23 HAZARDOUS WASTE DISPOSAL FIRM INCREASES TREATMENT CAPACITY BY INSTALLING NEW BIOLOGICAL TREATMENT PROCESS

Karl P. Kriger, Production Manager
Waste Conversion, Division of Stout Environmental
Hatfield, Pennsylvania 19440

James E. Carroll, Marketing Manager
Envirex Ltd.
Waukesha, Wisconsin 53186

INTRODUCTION

Stout Environmental, Inc. (S.E.I.) is a full service environmental corporation operating in the Eastern half of the United States. The corporation is comprised of field service and remedial action groups, water and well drilling operations, and hazardous waste treatment plants. The business activities of S.E.I. are centered around two regional hubs, one in the Midwest, Dayton, Ohio, and one in the Mid-Atlantic region, Philadelphia, Pennsylvania.

S.E.I.'s treatment plants consist of four entities specializing in fuel blending, oil recovery, and wastewater treatment. The parent plant under the S.E.I. umbrella is Waste Conversion, Inc. (W.C.I.) of Hatfield, Pennsylvania. W.C.I. is a fully permitted treatment, storage, and disposal facility specializing in the wastewater treatment of hazardous and toxic wastes.

Waste Conversion, Inc. accepts and treats a large variety of hazardous and non-hazardous wastes ranging from landfill leachates, washwaters, oil and water emulsions, acids, caustics, heavy metal wastes, cyanide bearing wastes and electro-plating wastes. W.C.I.'s treatment procedures are designed to minimize the quantity and toxicity of residues placed in landfills.

W.C.I.'s process creates two by-products for disposal; a solid residue from dewatering for landfilling and a liquid effluent which is discharged to the local P.O.T.W. W.C.I. is regulated by permit on its discharge, and is limited on several parameters, most importantly BOD_5, COD, and nitrogen (nitrate and ammonia). Since W.C.I. accepts many waste streams high in nitrogen, BOD_5, and organic compounds, bio-treatment technologies have been researched for viability within W.C.I.'s process. W.C.I. is the first member of S.E.I. to employ bio-treatment technology.

W.C.I. has chosen to operate two fluidized bed bio-reactors to remove organic carbon and nitrogen from its process effluent. W.C.I. plans to use biological treatment to denitrify its effluent, looking primarily to remove nitrate ions, and organic compounds such as alcohols, acetate and petroleum hydrocarbons.

The utilization of fluidized bed reactors has created several advantages for W.C.I. First, the treatment plant is centered in an industrial park complex and does not have expansion capabilities. The bio-treatment system at W.C.I. occupies only approximately 1400 $ft.^2$ of production space, while providing 50,000 gpd treatment capacity. Second, the limits set in W.C.I.'s discharge permit have controlled the acceptance of high BOD_5 wastes. Prior to the installation of bio-treatment, W.C.I. shipped excessive BOD_5 and nitrogen bearing effluent off site to an alternate treatment facility, raising W.C.I.'s treatment costs and reducing competitiveness on high strength BOD_5 and nitrogen wastes.

Since the advent of bio-treatment at W.C.I., treatment costs have been reduced and savings of $2,000 to $3,000 per week have been realized. Also, the capacity for the treatment of high strength BOD_5 and nitrogen wastes has been substantially increased, and at the same time, effluent discharge limits have been maintained.

45th Purdue Industrial Waste Conference Proceedings, © 1991 Lewis Publishers, Inc., Chelsea, Michigan 48118. Printed in U.S.A.

Table I. Loading Rate = 1400 lbs. COD/1000 ft.^{3} bed volume

	Influent (mg/L)	Effluent (mg/L)	% Reduction
Soluble BOD_5	13,147	1,633	88
Soluble COD	19,312	6,487	66
NO_3	5,602	1,385	75
NH_3	206	78	62

PILOT STUDY

The use of fluidized bed reactors was first researched via a pilot plant prototype which was put into operation January, 1987. The Fluid Bed pilot was a self-contained, frame mounted unit requiring a floor space of approximately 3 foot by 4 foot, and an overall height of about 14 feet. The system included a 6 inch diameter by 10 foot tall reactor column, a fluidization pump, raw feed pump, media abrasion pump, sand media, temperature control system, and a control panel. The pilot unit was shipped to the site complete and required only minimal set-up time. Connections were made to provide influent, effluent, waste sludge, and biogas measurement.

The objectives of the pilot plant study were first to evaluate its performance on the treatment of W.C.I. process effluent. Waste was chosen from three sources:

1. The neutralization of nitric acid.
2. The neutralization of acetic acid.
3. The treatment of waste containing petroleum hydrocarbons.

The effluent was combined and stored as an influent stream for the pilot plant. As the pilot study continued other specific waste streams were researched to prove the system's viability for W.C.I.'s operation.

The pilot study included extensive testing of many parameters on both influent and effluent, including BOD_5, COD, nitrates, total nitrogen, ammonia, TSS, VSS, phosphates, and pH. Nitrogen, BOD_5, and COD testing was performed daily, while solids, phosphate and ammonia were tested three times per week.

As the pilot study proceeded, some problems were soon brought to light. The first of these problems was a salt buildup on the sand particles and an eventual clogging of flow distribution nozzles. At the time of the study, calcium hydroxide was utilized as a neutralizing agent for the acidic materials. The W.C.I. effluent contained dissolved Ca^{+2} ions in concentrations of 6000 ppm to 10000 ppm over the course of the study. Since the biological reduction creates CO_2 as a by-product, and the sand media has a large surface area in the reactor, formation of calcium carbonate on the sand particles took place. Over time, the sand particles contained enough $CaCO_3$ to cement together and eventually clog the flow distribution system.

This problem was avoided by mixing the feed solution with excess sodium carbonate causing the precipitation of Ca^{+2} ions through the exchange. The mixed feed solution was dewatered a second time preparing an influent stream that did not contain significant concentrations of calcium. The pretreatment was very effective in eliminating the salting problem, however, created a second problem. The pretreatment with excess $(Na)_2CO_3$ left enough $(Na)_2CO_3$ ions present to raise the alkalinity of the reactor contents. Due to this rise in alkalinity, combined with the production of hydroxide ions through denitrification, reactor pH fluctuated and tended to be raised beyond the range of optimum efficiency. The optimum pH range determined during this study was a pH of 7.0 to 7.5. (Figure 2) this problem was overcome by installing an automatic pH control system.

As these problems were solved, the performance of the pilot plant prototype continued to improve. During the first portion of the pilot study, prior to $(Na)_2CO_3$ pretreatment, loading rates of up to 1400 lbs. COD/1000 ft.^{3} of bed volume were attained. Table I lists measured parameters in the influent and effluent. It can be noted from these results that BOD_5 removal was the most impressive, while the COD reductions remained lower. It has been determined through our research that approximately 18% of the soluble COD was found not amenable to biological degradation, at least within the study's retention times. When adjusting the COD data to account for this, an 81% reduction of biodegradable COD was achieved.

Table II lists parameters from the second phase of the study which was averaged over a six month time span. During this portion of the study, operational and chemical problems were solved and the results overall were improved. However, loading rates were also decreased to approximately 400 lbs.

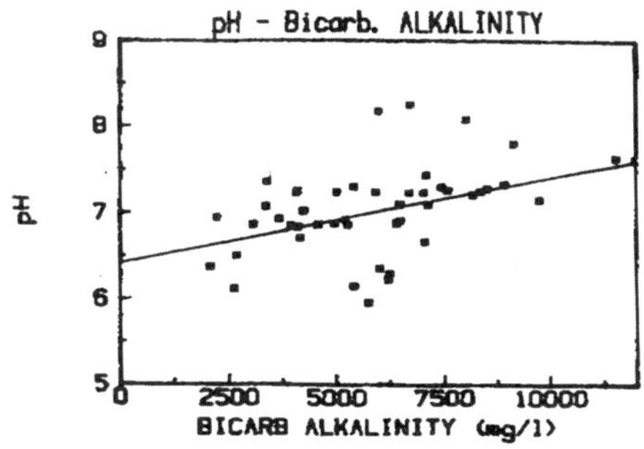

Figure 1. pH versus bicarbonate alkalinity.

COD/1000 ft.^{3} of bed volume. Once again, COD results can be adjusted for the portion of the influent COD not degradable and showed a 99% reduction of biodegradable COD. Figures 3 and 4 show some day-to-day results during this time period. It can be noted that effluent COD tended to fluctuate while effluent NO_3 concentration remained low and consistent. Both influent COD and NO_3 concentrations fluctuated more widely due to variations in feed solutions.

Based on the results of the two phases of the pilot study, excellent results were achieved on loading rates ranging from 400 lbs. COD/1000 ft.^{3} to 1400 lbs. COD/1000 ft.^{3} bed volume. W.C.I. at this time was pleased with the performance of the system and decided to purchase two (2) 25000 gallon fluidized beds from Envirex Inc. for use as a full scale biological treatment system.

FULL SCALE OPERATION

The full scale equipment provided for this project included two (2) 12 foot diameter by 25 foot tall Fluid Bed reactors, three (3) fluidization pumps, three (3) abrasion pumps, three (3) chemical feed pumps, chemical day tank, heat exchanger, and control panel. Also installed was a 50,000 gallon equalization/feed/mix tank.

The two (2) reactors were shop fabricated and shipped to the site complete with internals installed. The pumps, heat exchanger, and controls were all mounted on two (2) shop fabricated skids. All skid piping was completed and ready for connection to the interconnecting piping.

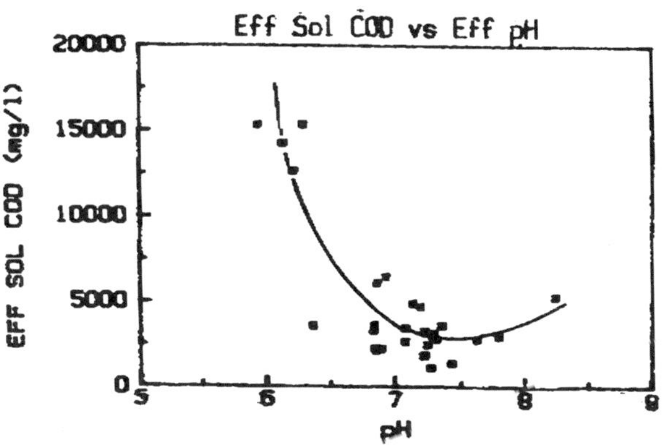

Figure 2. Effluent COD versus effluent pH.

Table II. Loading Rate = 400 lbs. COD/1000 ft.^{3} bed volume

	Influent (mg/L)	Effluent (mg/L)	% Reduction
Soluble BOD_5	13,617	1,309	90
Soluble COD	17,253	3,132	82
NO_3	5,866	274	95
NH_3	205	12.5	94

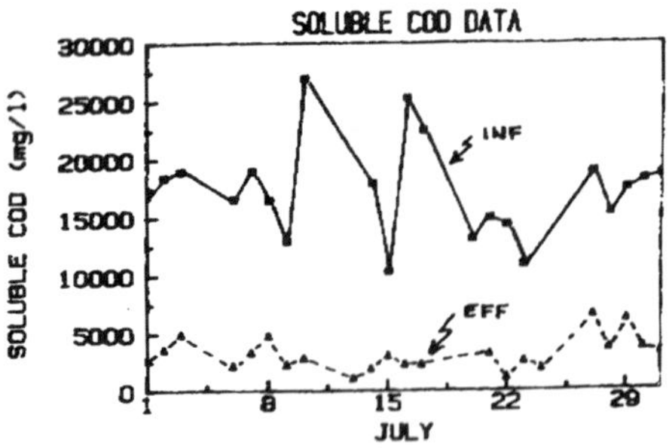

Figure 3. Soluble COD data (July).

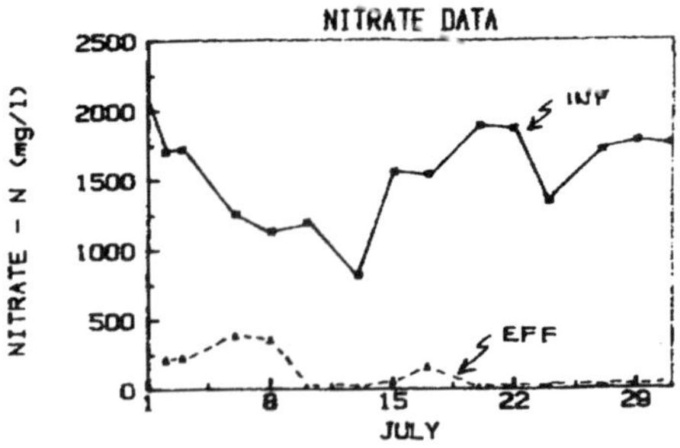

Figure 4. Nitrate data (July).

The installation of the full scale equipment began April 1989, starting with the placement of the reactors. A 10 ft. deep pit was constructed due to height constraints and to provide containment for the system. A corner of the existing building was removed and the tanks were placed via crane. The equalization tank, 50,000 gallons, was installed next. Pumping equipment for recycle, biomass shearing, and the chemical reagent feed system were provided by Envirex Inc. on skids and installed as such. The installation was completed and the system was ready for start-up in early September 1989.

The reactor columns were filled with approximately 30 cubic yards of sand, waste effluent, and 5000 gallons of mixed liquor from the local P.O.T.W. Initial start-up began September 19, 1989. The unit was recirculated for one week for initial seeding purposes and influent feed began on September 26, 1989 at 1000 GPD. The columns were run in series with raw influent going to reactor #1, and the discharge of reactor #1 feeding reactor #2.

Operation of the unit continued in series with flow rates increasing to 5000 gallons/day fed to reactor #1 by December 25, 1989. Results during this initial start-up were inconclusive with reductions of 46% on COD and 45% on NO_3 being achieved. By December, reactor temperatures reached 41°F on the average, with the lowest temperature of 35°F on December 24, 1989.

During this start-up, W.C.I. decided to operate the system exposed to the environment and without the capacity to heat the reactor contents. This decision was made to test the full scale unit and to work out operation problems that might have occurred. Also, this start-up period provided a convenient time to train W.C.I. personnel on the efficient use of the equipment and problem solving techniques. The start-up was approached gradually and full scale operation was not a requirement during this time period. Operation at design capacity was not achieved due to a shutdown of the unit over the Christmas holiday.

This shutdown was caused by an electric power outage on December 25, 1989, and occurred while the W.C.I. plant was shutdown for the holiday. At the plant re-start on December 26, 1989, the system shutdown was noted and pipe rupture due to freezing was discovered.

December's temperatures were well below average, and an extremely cold winter was expected. W.C.I. Management then decided to leave the system down for the remainder of winter. This would allow time to reconstruct the building housing the unit, thus protecting it from the weather.

FUTURE PLAN

W.C.I. is planning the second start-up for early May 1990, and will proceed to full capacity as quickly as possible. The unit will once again be run as a denite reactor, however, the unit is planned to switch to an anaerobic reactor as the volumes of organic carbon bearing wastes are expected to increase. Design capacity is set at 50,000 GPD and represents one-half of W.C.I.'s daily discharge.

W.C.I. expects treatment capacity of high strength BOD_5 and nitrogen waste streams to increase significantly. Also, the use of alternate facilities for treatment of effluent will be eliminated. The initial start-up in the Fall has already proven the reduction of operational overhead and shows great promise for the future. W.C.I. is pleased to have purchased the Envirex Inc. unit and will continue to be on the cutting edge of hazardous waste treatment technology for many years to come.

24 AN AT-SOURCE TREATMENT FOR ORGANOMERCURY-CONTAINING HAZARDOUS LIQUID WASTE

Ralph J. Magliette, Chemical Engineer
D. McKinney, Chemical Engineering Technician
E. S. Venkataramani, Research Fellow
S. Bacher, Director, Developmental Technology
B. Brian, Senior Project Process Engineer
Merck & Co., Inc.
Rahway, NJ 07065-0900

INTRODUCTION

Waste reduction is an economically sensible and sound response to the perceived hazardous waste crisis. Several thousand pounds of hazardous waste are generated annually for every person in the U.S.[1,2] Also, the total costs of waste management is escalating at a rate of 20–30%/year.[3] Many pollution control methods do little more than transfer waste from one phase to another. Thus, establishing a comprehensive multimedia approach to reduce the level of wastes that enter the air, land and water is essential. Waste reduction is cost effective, especially when compared to the time, effort and resources required to comply with regulations. Though enforcement sometimes provides the impetus for waste reduction, cost savings are almost always a consequence.

Reducing the generation of waste and improving the overall efficiency and economics of the manufacturing process are fundamental to successful chemical enterprises. The production of waste from a chemical and biochemical process is a function of the process design and the manner in which it is operated.[3] Thus, reducing waste in an industrial process requires intimate knowledge of the process technology, in contrast to waste treatment which essentially is an add-on at the end of the process. Ultimately, waste minimization has to be carried out at the process level.[4] Hence, waste minimization is a 'process-related opportunity' and failure to reduce the generation of waste at the source only increases the embedded problem.[5]

In general, waste reduction has historically yielded to waste management under the press of more immediate concerns. The predominance of treatment over waste reduction is not new and is difficult to reverse. The most certain means of minimizing the environmental risk is through waste reduction. Waste reduction is also preferable to most waste management practices because it can lead to lower direct costs and higher indirect benefits. Manufacturers like DuPont, Dow Chemical, 3M, Hoechst Celanese, IBM and Merck have waste reduction programs that have achieved dramatic results with regard to both waste minimization and savings from avoidance of treatment and disposal costs and liabilities.[2,3,6,12] Many smaller companies have also adopted pollution source reduction with similar success. "Avoiding the creation of pollution seems to be the best way to fight against it, technically and economically."[13]

BACKGROUND

Merck & Co., Inc. is the world's largest pharmaceutical company. Merck's environmental policy acknowledges the same respect for the quality of human life in minimizing environmental risk through reduced generation of waste as in discovering new life-saving drugs. The research effort that leads to the development of new life-saving pharmaceuticals is also utilized to develop environmentally sound processes that result in reduced waste generation.[14,15] At Merck, it was realized several years ago that it made good technical and business sense to continually maintain pressure on efforts to eliminate or at least reduce waste generation whenever possible. Several examples of our process-related waste minimization practices can be found elsewhere.[11,12,16]

45th Purdue Industrial Waste Conference Proceedings, © 1991 Lewis Publishers, Inc., Chelsea, Michigan 48118. Printed in U.S.A.

The specific example discussed in this paper deals with the development of an at-source treatment technology for the removal and recovery of mercury from a spent fermentation waste. An organomercury compound (Thimerosal, Figure 1) is used as a slow kill biocide in one fermentation process. The Company's waste minimization policy mandated a thorough evaluation of various at-source treatment technologies to render the waste non-hazardous. It has been shown that it is possible to reduce the mercury concentration in the waste to below 10 ppb from 56 ppm by sodium borohydride reduction followed by ultrafiltration of the colloidal mercury metal followed by Granular Activated Carbon (GAC) polishing. Reclamation of the metallic mercury for recovery and non-process reuse was also found to be visible. The Environmental Engineering Laboratory's proposed process for the at-source reduction and removal of mercury, an environmentally sound waste minimization approach, is described in detail.

REVIEW OF PROMISING TECHNOLOGIES

Three promising technologies were evaluated prior to the selection of metal hydride reduction as the method of choice. These were metal hydride reduction, free radical oxidation, and organo-sulfur precipitation.[17-19] Each will be briefly discussed.

Metal Hydride Reduction

Metals can easily be reduced and precipitated from both organic and inorganic media by reducing agents. Commonly used reducing agents include aluminum, zinc, hydrazine, and a group of reducing agents known as metal hydrides. The most common reducing agent utilized is sodium borohydride. In the application to organomercury compounds, sodium borohydride stabilizes the mercury and removes it from the organic portion of the molecule. Sodium borohydride will reduce the mercury present in thimerosal because the mercury atom is in the +2 oxidation state. The elemental mercury formed as a result of sodium borohydride reduction is in a colloidal state and must be separated from solution by filtration.[20]

Free Radical Oxidation

Free radical oxidation fragments large organic compounds into smaller ones. It is the commonly used method to treat compounds such as phenols and organo metallic compounds that contain metals in a +2 oxidation state. The reagents for this procedure can be ozone or hydrogen peroxide. The use of ultraviolet light or ultrasonics can speed up the rate of reaction. If free radical oxidation of thimerosal goes to completion, the products of the reaction would be water, carbon dioxide, and sulfur compounds containing mercury.[21] If elemental mercury is produced by the reaction, it can be removed by conventional post treatment methods such as sulfide precipitation or ultrafiltration. The high capital costs involved and unproven technology made this a less desirable choice for this application.

Precipitation by Organo-Sulfur Reagents

This technology is used to remove strongly chelated and complexed metal compounds. Precipitation by complex organo-sulfur reagents is similar to conventional sulfide precipitation. Both rely upon the insolubility of mercuric sulfide or mercuri-mercapto compounds to achieve removal of mercury from aqueous waste streams. The organo-sulfur precipitation technology does successfully remove thimerosal from solution. Removal efficiencies were found to be approximately 92%. The

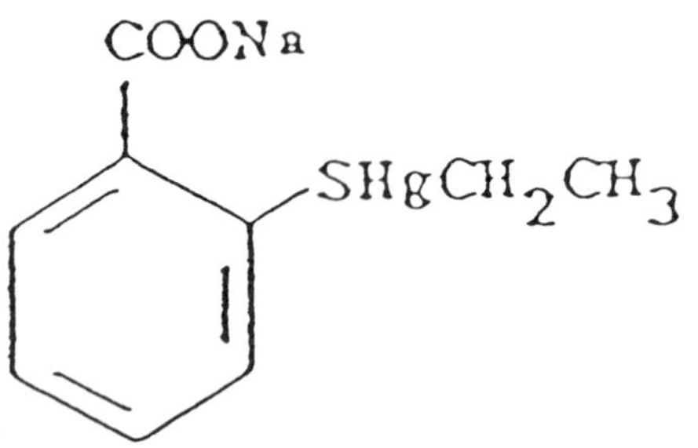

Figure 1. Thimerosal.

filter cake from the filtration contains mercury salts up to 10% by weight. This cake, being hazardous and unacceptable for reclamation, would require additional treatment or disposal. Land disposal of mercury is restricted and is not a desired disposal option. The only alternative for the disposal of mercury salts formed from the precipitation reaction would be incineration. Incineration of mercury compounds is normally not an environmentally sound disposal option as most of it exits in the stack gas as mercury oxides with a small fraction exiting with the incinerator ash.[22-23]

MATERIALS AND METHODS

Reagents and Equipment Requirements

Chemical Requirements

1. 12% solution of Sodium Borohydride in a 40% caustic solution. VenMet solution (Ventron Corporation).
2. Aluminum Metal Powder—Fine (Fisher Scientific, Inc.).
3. Hydrochloric Acid, Concentrated.
4. 50% Sodium Hydroxide Solution.
5. Antifoam B Emulsion (Sigma Chemical Co.)

Equipment Requirements—Pilot Scale

1. Ultrafiltration System with 30,000 Molecular Weight Cut Off (30K MWCO) hydrophilic modified regenerated cellulose membrane #MCR-07 (Bio-Recovery, Inc., Northvale, New Jersey)—A similar membrane may be obtained from other suppliers, e.g. Amicon.
2. F-300 granular activated charcoal unit (Calgon Carbon, Pittsburgh, Pennsylvania).

Procedure

Pretreatment of Cell Paste

1. Add 3N NaOH solution to the cell paste (100 ml/gm), heat to 75°C and stir for 24 hours maintaining a temperature of 75°C to achieve completion of hydrolysis.

Treatment Procedure

1. a. Cell paste: add an amount of aluminum metal powder equivalent to a one-to-one ratio of aluminum to mercury present in the cell paste.
   b. Spent fermentation wastewater: adjust pH to 11.5 with caustic and add an amount of aluminum metal powder at least equivalent to a one-to-one ratio (weight basis) aluminum to mercury present in the wastewater.
2. Mix for 24 hours.
3. Adjust the pH to 7.0 with concentrated hydrochloric acid.
4. Add VenMet solution until a pH of 10.0 is obtained and continue intermittent addition of VenMet solution until the pH is stable at 10.0 for at least five minutes. Note: if solution starts to foam, antifoam/defoamer—e.g., Antifoam B Emulsion, may be added.
5. Mix well for 1 hour.
6. Filter the sample using crossflow ultrafiltration. Filtration parameters—flux 191 liter/meter square/hour at 50 psi. Recycle retentate to achieve a concentration factor of at least 50X.
7. Collect retentate for mercury reclamation.
8. Permeate is passed through a bed of activated carbon (F-300) and the effluent discharged to a holding tank.
9. Upon validation that the mercury content of the effluent in the holding tank is less than 10 ppb, the effluent is sent to neutralization and equalization before discharge to a wastewater treatment facility.

DISCUSSION

The methodology for the reduction and removal of thimerosal (organo-mercury compound) and plausible reclamation and reuse of mercury, an environmentally sound option, is detailed in the following sections. The process is based on the use of aluminum metal reduction at high pH to break the S-H bond and hydrolyze the organomercury fragments. Sodium borohydride is then used to reduce the mercury cations and organomercury fragments to elemental mercury. The process is described in Section A. Ultrafiltration is then employed to capture the mercury in the retenate which can be collected and sent offsite for reclamation and non-process reuse; the specifics are outlined in Section B. The transparent, clear and sparkling permeate was passed through a fixed-bed of granular activated carbon. The GAC provides tertiary treatment to remove residual Hg from the UF permeate and produces an effluent with a mercury concentration of less than 10 ppb. The details are furnished in Section C. A detailed kinetic evaluation was not performed. However, the laboratory experiments furnish qualitative/semi-quantitative kinetic information as outlined in Section D. Comments as to heat effects and hydrogen gas generation are given in Section E and F, respectively. Most of the biocide has been found to be associated with the dead cell mass and, hence, necessitated the release of bound thimerosal from cell paste before sodium borohydride reduction. The details of caustic hydrolysis of cell paste to release the thimerosal are discussed in Section G. The flowsheet and material balance for a proposed large-scale application of the technology is shown in Figures 2 and 3.

SECTION A. PROCESS CHEMISTRY: ALUMINUM POWDER AND SODIUM BOROHYDRIDE REDUCTION

The chemistry of the method is shown in Table I. Aluminum powder is used to desulfurize the organosulfur bond. A catalytic amount of aluminum is required. The minimum charge was found to be one gram of aluminum powder for each gram of mercury present. Use of excess aluminum does not increase the reaction rate. The base catalyzed hydrolysis, using caustic, is carried out at a pH of 11.5.

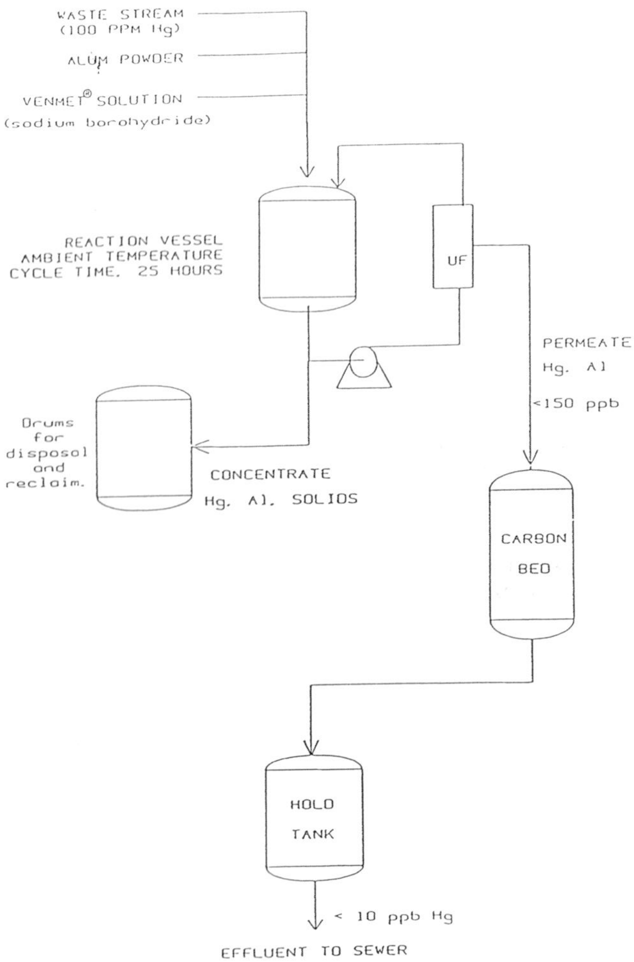

Figure 2. Flow sheet for Thimerosal removal process.

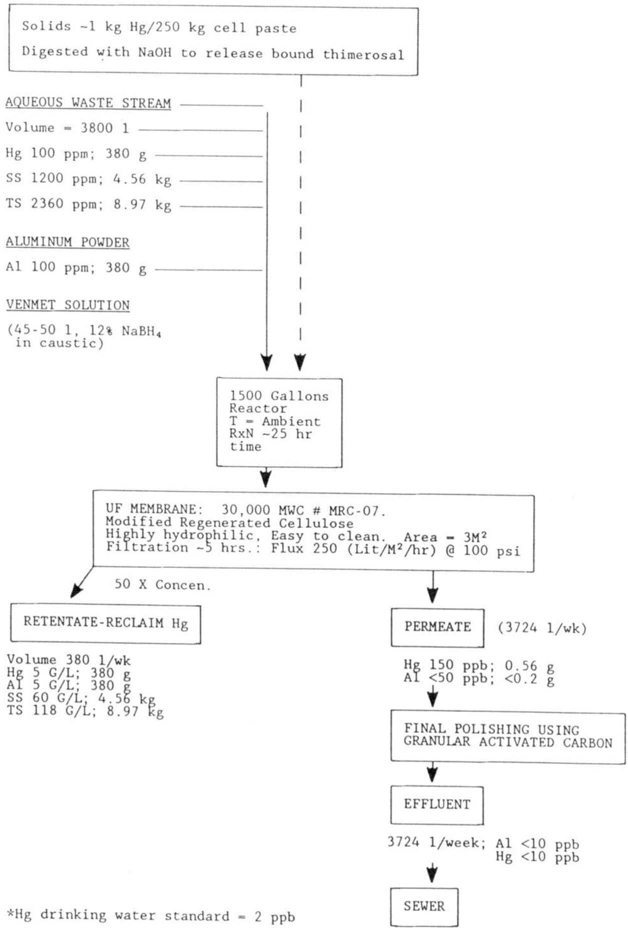

Figure 3. Material balance for large-scale application of the developed technology.

Table I. Chemistry of the Method

Step I. Treatment with Al

Al : Hg = 1:1, pH = 11.5, t = 24 hrs.

THE ALUMINUM ACTS AS A REDUCING AGENT

$Al \text{------>} Al^{+3} + 3e^-$

THE REDUCTION INVOLVES ONE OR MORE OF THE FOLLOWING REACTIONS:

1. $2e^- + RSHgEt + 2H_2O \text{----->} RH + HSHgET + 2OH^-$
2. $2e^- + RSHgEt + H_2O \text{----->} RH + HOHHgEt + S^{-2}$
3. $4e^- + RSHgEt + 2H_2O \text{----->} RH + [HHgEt] + S^{-2} + 2OH^-$

Step II. Treatment with Sodium Borohydride

@ pH = 10

$4\ Hg^{+2} + NaBH_4 + 8OH^- \text{---->} 4Hg + NaBO_2 + 6H_2O$

Note: Organomercury fragments undergo base catalyzed hydrolysis in Step I. The resultant smaller organomercury fragments are reduced to form Hg ions. The final reduction of ionic Hg to elemental mercury is accomplished in Step II.

Table II. Performance of Cross-Flow Ultrafiltration Membrane and Granular Activated Carbon Bed

Initial Hg Concentration in the Wastewater = 56 ppm

MWCO (K)	Type*	Flux-LMH @50 PSI	Concent (X)	Ease of Cleaning	Hg in Permeate	Hg After Post Treat with GAC	% REM Eff
30	1	119	100	No	110 PPB	2 PPB	>99.99
30	2	191	100	Yes	110 PPB	2 PPB	>99.99
80	3	200	100	No	150 PPB	5 PPB	>99.99
100	4	119	100	No	180 PPB	5 PPB	>99.99
100	5	119	100	No	210 PPB	3 PPB	>99.99

*Membrane type

1. Slightly hydrophilic, polysulfone
2. Most hydrophilic, modified regenerated cellulose
3. Most hydrophobic, polysulfone
4. Slightly hydrophobic, polysulfone
5. Slightly hydrophilic, polysulfone

The reaction is heterogeneous and requires 24 hours to reach completion. During the reaction, a substantial amount of the organic material present in the waste undergoes hydrolysis.

Sodium borohydride in a stabilized alkaline solution containing 12% $NaBH_4$, 42% NaOH, and the remainder water is used in this process. In organomercury applications, once the S-Hg bond is broken, the sodium borohydride will stabilize the mercury and cleave it from any remaining organic portions of the molecule. The mercury ions thus formed can then be reduced to the elemental state directly by sodium borohydride. In theory, the amount of sodium borohydride added is based on the ability of sodium borohydride to donate 8 electrons. A mercuric cation requires two electrons to be completely reduced. Therefore, one molecule of sodium borohydride can theoretically reduce four mercuric ions to the elemental state.

$$4\ Hg^{+2} + NaBH_4 + 8\ OH^- \rightarrow 4\ Hg^0 + NaBO_2 + 6\ H_2O$$

In practice, the actual amount of sodium borohydride added is variable due to the various oxidation states of the mercury cations present in the waste stream. It is also dependent on the amount of organomercury fragments and other organic/inorganic compounds present in the wastewater which can undergo reduction. All of these contribute to the sodium borohydride demand. The amount of sodium borohydride required for the complete reduction of mercury can be monitored by pH measurement. The reduction reaction reaches completion when the pH of the wastewater is brought to 10 and remains stable at pH 10.0 for at least ten minutes without need for further addition of sodium borohydride solution. As sodium borohydride is consumed, the pH of the solution will decrease. Therefore, a stable pH reading of 10 indicates that the reaction has gone to completion. The treated wastewater is stirred for one hour and the colloidal mercury is separated by ultrafiltration.

SECTION B. ULTRAFILTRATION

The ultrafiltration system used was of plate and frame design and is manufactured by Bio-Recovery, Inc., Northvale, New Jersey. The lab unit can be directly scaled up to a larger pilot plant unit. A 30K MWCO—modified regenerated cellulose membrane, #MCR07 (the most hydrophilic membrane tested) was found to give the best overall performance. The membrane yielded a flux of 191 liters/meter squared/hour at 50 psi. The membrane was easily cleaned by running a 0.5N NaOH solution through the system for 15 minutes. There was no evidence of decreased flux in filtration before and after multiple uses following cleaning. Visual inspection indicated negligible change in morphology of the membrane. The membrane was used for three consecutive runs with no loss of membrane performance. After the final run, the membrane was cleaned and again visually inspected; it looked morphologically very similar to a new membrane. There was no discoloration or evidence of damage due to abrasion, etc.

The permeate was assayed for mercury content and found to contain 110 ppb of mercury. The initial mercury concentration of the waste was 56 ppm. Therefore, the removal efficiency of the ultrafiltration membrane (#MCR07) is 99.7%. Other membranes were also evaluated and their performance is summarized in Table II.

Table III. GAC Treatment to 10 Liters of Permeate Containing 56 ppm Hg (100K MWCO/hydrophilic)

Total Volume	Color of Permeate from GAC	Mercury (mg/L)
1 liter	clear, no amber color detected	<0.01
2 liters	clear, no amber color detected	<0.01
3 liters	slight amber color detected	<0.01
4 liters	slight amber color detected	<0.01
5 liters	slight amber color detected	<0.01
6 liters	slight amber color detected	<0.01
7 liters	amber color	<0.01
8 liters	amber color	<0.01
9 liters	amber color	>10.0*
10 liters	amber color	>10.0*

*Indicates saturation of the bed

SECTION C. POLISHING OF ULTRAFILTRATION PERMEATE USING GRANULAR ACTIVATED CARBON

The permeate from the ultrafiltration was passed through a fixed bed granular activated carbon filter. F-300 GAC, available from Calgon Carbon, is recommended. A 4.5% maximum loading of mercury has been determined for the F-300 GAC. A pH of 10 is required for maximum mercury adsorption. Less than 10 ppb of mercury was detected in the effluent after the GAC filtration/polishing step.

The adsorption of thimerosal onto carbon depends on three factors: 1. the solubility of thimerosal, 2. the pH of the wastewater stream, and 3. the presence of other compounds in the wastewater stream which may interfere with the adsorption/solubility of thimerosal. Usually, the more soluble a compound is, the less likely it will be adsorbed onto carbon. Thimerosal is a very soluble compound and, therefore, is only slightly adsorbable on carbon. Mercury metal is highly adsorbed on the carbon and experimental results at different carbon loadings are shown in Table III.

The following experiment was utilized to determine the mercury loading on the carbon before breakthrough. The permeate contained soluble organics. Literature indicates that soluble organics are not retained well on GAC at a pH of 10.0, which is the optimal pH for mercury adsorption. A 9 cm funnel was utilized as a support for 10 grams of Calgon F-300 GAC. Hg was added to the treated UF (#PSH07, 100K MWCO) permeate to result in a 56 ppm solution which was utilized for breakthrough experiments. A total of 10 liters was passed through the carbon bed. The data indicates mercury breakthrough at 4.5% loading. This is similar to the results obtained in previous experiments. The organics, thus, do not interfere in the adsorption of the mercury on the carbon. The data is furnished in Table IV.

SECTION D. KINETICS

Kinetic parameters were not specifically evaluated in this study. However, some information of a qualitative nature is available from the laboratory experiments. The concentration of aluminum powder required for the metal hydride reduction of the S-Hg bond was investigated. The data indicate that an aluminum to mercury ratio of 1 to 1 (weight basis) is required. This catalytic amount of aluminum is required for the reaction to go to completion within 24 hours. The initial ratio tested, based on the literature reported procedure, was 250 to 1. Excess aluminum has been found to not interfere in the reaction. Ratios less than 1 to 1 produced incomplete reduction reactions.

Table IV. Total Organic Carbon Removal by the GAC Bed from the UF Permeate

Sample	Membrane	TOC of UF Permeate "as is"	TOC after "GAC"	% TOC Removal by GAC
1	30K/MWCO mod. regen. cel./#MRC07	8900 mg/L	4175 mg/L	53
2	80K/MWCO polysulfone/#PS012	11150 mg/L	4100 mg/L	63.2
3	100K/MWCO polysulfone/#PSO07	15570 mg/L	4825 mg/L	69.0
4	100K/MWCO polysulfone/#PSH07	15975 mg/L	11175 mg/L	30.0

Table V. Caustic Hydrolysis (3N NaOH) of Cell Paste Effect of Temperature and Time on Bound Thimerosal (Hg)

		Hg in solution		
Expt No	Time (Hours)	25°C Hg(ppm)	50°C Hg(ppm)	75°C Hg(ppm)
1	0	0.00	0.0	0.0
2	1	5.33	23.5	30.3
3	2	16.00	25.3	30.8
4	4	22.10	25.5	31.1
5	6	35.30	31.5	35.5
6	8	30.80	35.3	35.9
7	12	33.50	38.8	41.1
8	24	40.30	40.5	42.0
Hg in Initial Solids (g)		43.4	43.4	43.4
Hg in Final Solids (g)		3.57	1.12	0.55
Hg in Liquid Phase (g)		39.83	42.28	42.85
% Hg in Liquid Phase (g)		91.8	97.4	98.7
% Hg Recovery		101.1	95.9	98.0

The time required for the reaction to reach completion at a mercury ration of 1 to 1 was also investigated. The data indicate that a minimum time of 18 hours is required under these conditions. This was experimentally determined by exploiting the property that thimerosal does not readily adsorb onto carbon. The volume of solution to mass of carbon was kept constant throughout the experiments. Mercury loading was also kept below the breakthrough concentration except when breakthrough was evaluated.

SECTION E. HEAT EFFECTS

The overall heat of reaction was monitored in the pilot scale runs (15 gallons each). Batch-to-batch variability in the buffering capacity of the spent fermentation broth resulted in marked differences in the amount of caustic required for pH adjustment of the wastewater. The overall heat of reaction was much greater for the first run than the second. The data given below are for the first pilot scale batch (worst case). The initial pH adjustment with 50% caustic resulted in a 4.2°C increase in temperature. No heat of reaction was observed after the addition of aluminum powder or during the 24-hour period the solution was mixed with the aluminum powder. The second adjustment of pH to 7.0 with concentrated HCl resulted in a 6.0°C increase in temperature. The addition of sodium borohydride solution was accompanied by a small heat of reaction (0.5°C). The total increase in temperature resulting from the overall heats of reactions is 10.7°C. It should be noted that the reported temperature increases were obtained under non-adiabatic conditions.

SECTION F. HYDROGEN GAS GENERATION

Sodium borohydride is unstable at low pH (below pH 7.0). At pH less than 7.0, the compound undergoes hydrolysis and hydrogen gas is evolved. In the proposed process, the pH of the solution to which sodium borohydride is added is adjusted with concentrated hydrochloric acid to a pH of 7.0. Thus, the amount of hydrogen gas evolved is minimized; however, venting of the vessel is recommended so as to allow any hydrogen gas evolved to be removed safely. At pH less than 7.0, the following hydrolysis reaction is favored:

$$BH_4^- + 4H_2O \rightarrow B(OH)_3 + OH^- + 4H_2$$

SECTION G. CELL PASTE

The material balance data indicate that greater than 50% of the thimerosal is associated with the dead cell mass. The cell paste requires 3N NaOH and heating at 75°C for 24 hours to solubilize most of the thimerosal for further treatment (Table V). The release of mercury from cell paste into liquid phase as a function of time at different temperatures is shown in Figures 4 through 6. The hydrolyzed cell debris contain little if any mercury and, therefore, the cell paste is amenable to treatment by this process.

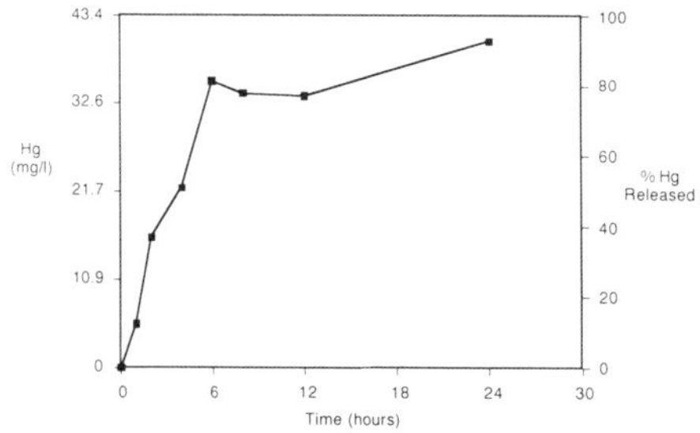

Figure 4. Release of mercury from cell paste into liquid phase @ 25 C, 3 N NaOH.

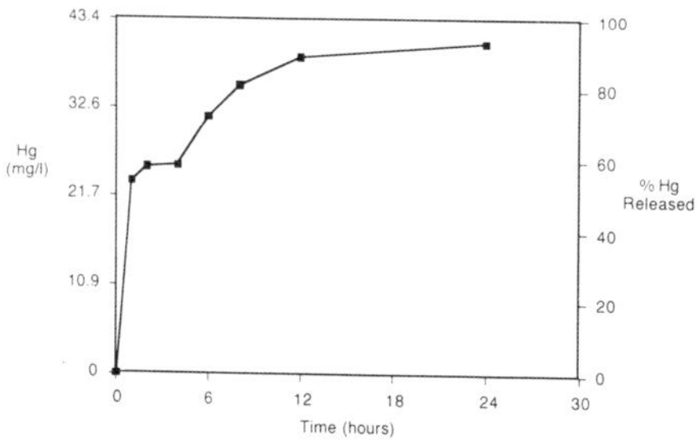

Figure 5. Release of mercury from cell paste into liquid phae @ 50 C, 3 N NaOH.

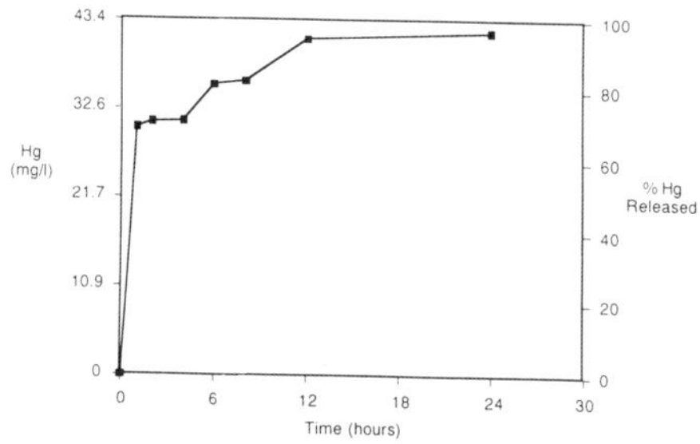

Figure 6. Release of mercury from cell paste into liquid phase # 75 C, 3 N NaOH.

CONCLUSION

An environmentally sound waste minimization approach which utilizes an at-source treatment technology for the removal and recovery of mercury from a spent fermentation waste to render it non-hazardous was developed. The organomercury containing fermentation wastewater is initially treated with aluminum metal powder at pH 11.5 and ambient temperature to reduce the sulfur-hydrogen bond to form mercury cations. Since most of the biocide has been found associated with the cell mass, caustic hydrolysis is used to release the organomercury compound from cell paste before treatment. Subsequent treatment with sodium borohydride reduces the mercury cations to the colloidal metal. Ultrafiltration effectively removes the colloidal metallic mercury.

Final polishing with granular activated carbon yields a transparent, clear and sparkling effluent containing less than 10 ppb mercury. The colloidal Hg present in the UF retentate as well as bound to granular activated carbon is amenable to reclamation and subsequent reuse. The process has been designed so as to mitigate heat effects and hydrogen gas generation.

The at-source treatment technology developed is capable of achieving a consistently high removal efficiency of 99.99%, thus, providing an environmentally sound option for mercury reclamation and reuse. As has been the experience with other processes, this provides yet another classic example of a process related waste minimization opportunity.

ACKNOWLEDGMENTS

Assistance extended by the staff from the Merck Pharmaceutical Manufacturing Division is acknowledged. Ms. Jayne Hojecki is thanked for the skillful typing of the manuscript.

REFERENCES

1. "Serious Reduction of Hazardous Waste," Congress of the United States, Office of Technology Assessment, Washington, D.C., OTA-ITE-318, September 1986.
2. Nichols, A. B., "Hazardous Waste—Industry Initiates Source Prevention," *Jl. Water Pollution Control Federation*, 60(1), 36–44, 1988.
3. Hollod, G. J. and, McCartney, R. F., "Waste Reduction in the Chemical Industry—DuPont's Approach," *Jl. APCA*, 38(2), 174–9, 1988.

4. Hanson, D., "Industry Urged to Accelerate Waste Reduction Programs," *Chemical and Engineering News*, 65, 21-22, November 30, 1987.
5. Valentino, F. W. and Walmet, G. E., "Industrial Waste Reduction: The Process Problem," *Environment*, 28(7), 16-34, September 1986.
6. Tilton, H., "Waste Reduction is the Real Priority," *Chemical Marketing Reporter*, 232(21), 32, 33, 46, November 23, 1987.
7. Wilkes, D. R. and Young, M. K., "Study and Implementation of Waste Minimization at IBM Austin," Proceedings of the 42nd Purdue Industrial Waste Conference, (Ed., J. M. Bell, Lewis Publishers, Inc., 1988) pp. 331-341.
8. Enander, R. T. and Nester, D. J., "A Structured Approach to Solid Waste Management," *Chemical Engineering Progress*, 84(4), 27-31, 1988.
9. Sarokin, D. J., Muir, W. R., Miller, C. G., and Sperber, S. R., "Cutting Chemical Wastes," Inform Inc., NY, (Eds. P. Stryker and P. Lone) pp. 370-398, 1985.
10. Naldi, L., "Waste Minimization Via Process Development," 28th Annual Symposium—North Jersey AIChE, East Brunswick, NJ, May 9, 1988.
11. Venkataramani, E. S., "Process Modification Options and Early Environmental Assessment of Processes for Pollution Source Reduction," Pharmaceutical Manufacturers Association Workshop on "Waste Minimization Practices in the Pharmaceutical Industry," Parsipanny, NJ, November 17-18, 1987.
12. Venkataramani, E. S. and Bacher, S., "Waste Minimization in the Pharmaceutical Industry," AIChE Pollution Prevention Topical Meeting, Washington, D.C., December 4-5, 1989.
13. Overcash, M. R., "Techniques for Industrial Pollution Prevention," (Chelsea, MI, Lewis Publishers Inc., 1986) p. 29.
14. King, M. L., Forman, A. L., Orella, C. and Pines, S. H., "Extractive Hydrolysis for Pharmaceuticals," *Chemical Engineering Progress*, 81(5), 36-89, 1985.
15. "Merck's Environmental Priorities—Protecting People and Reducing Waste," *Merck World*, 10(3), 15, July/August 1989.
16. Venkataramani, E. S., Reynolds, S. D., Kaelin, L. P., Clougherty, B. P., Bacher, S. and Andrews, A. T., "Process Modification Options to Reduce Environmental Impact of High Strength Industrial Wastes," AIChE Summer National Meeting, Boston, MA, August 24-27, 1986.
17. Faust, S. D. and Aly, O. M., "Chemistry of Water Treatment," pp. 26-27 and 517-520, Butterworth Publishers, Woburn Massachusetts, 1983.
18. Eilbeck, W. J. and Mattock, G., "Chemical Process in Wastewater Treatment," pp. 197-282, J. Wiley and Sons, New York, New York, 1987.
19. Martinetz, D., "Chemical Waste Handling and Treatment—Detoxification and Decomposition," pp. 114-256, (Ed. K. R. Muller), Springer-Verlag Pub. Inc., New York, New York, 1985.
20. Cook, M. M., "Correspondence on Morton-Thiokol, Inc. Process for Organomercury Reduction to Elemental Mercury," Ventron, Inc., Woburn Massachusetts, October 21, 1988.
21. Morrison, R. T. and Boyd, R. N., "Organic Chemistry 3rd Edition," pp. 503-504, Allyn and Bacon, Inc., Boston, Massachusetts, 1973.
22. Ammen, C. W., "Recovery and Refining of Precious Metals," pp. 99-100 and 262-267, Nostrand-Reinhold, Co., Inc., New York, New York, 1984.
23. Bonner, T., et al., "Hazardous Waste Incineration Engineering: Pollution Technology Review #88," Noyes Data Corp., Park Ridge, New Jersey, 1983.

25 OVERVIEW OF THE AIR FORCE ENGINEERING AND SERVICES CENTER'S HAZARDOUS WASTE MINIMIZATION PROGRAM

Charles J. Carpenter, Environmental Research Engineer
Headquarters, Air Force Engineering and Services Center
HQ AFESC/RDVS
Tyndall AFB, FL 32403-6001

INTRODUCTION

The United States Air Force operates many facilities which generate hazardous waste. Five of these, called Air Logistics Centers (ALCs), are major depot level repair facilities. They contain all of the necessary industrial equipment and processes required to overhaul and repair the Air Forces aircraft and ground support equipment. Electroplating, painting, depainting, machine shops, and cleaning/degreasing operations are just some of the processes that contribute to the generation of hazardous wastes at these facilities. The electroplating facility at Tinker Air Force Base (AFB) alone covers 40,000 square feet. That shop along with the other equivalently sized industrial shops generate 1.4 million gallons of wastewater per day. Used electroplating baths, spent solvents and paints, and rinse waters from electroplating and other industrial operations contain heavy metals and organics which make those substances hazardous wastes. One of the principal contaminants of the wastewaters from these facilities is chromium, primarily in the hexavalent form. Other metal contaminants include nickel, copper, cadmium, lead, and zinc. Currently these contaminants pose environmental hazards and must be treated as hazardous wastes and the sludge from treatment of these wastes must be disposed of at hazardous waste facilities at significant cost ($300/ton at Tinker AFB).

The Air Force Engineering and Services Center (AFESC) has developed an overall plan for providing eventual elimination of all hazardous waste generation from Air Force electroplating facilities. The plan divides total hazardous waste generation into five main technical areas; cleaning and degreasing wastes; metallic wastes; protective coatings wastes; paint stripping wastes; and air toxic contaminants. Under the program, projects are being developed and completed in each of these technical areas which will reduce or eliminate the generation of hazardous wastes from Air Force facilities. Some of these technologies are advanced end of pipe treatment processes such as ferrous sulfate/sodium sulfide metals treatment and metals recovery from industrial sludges, other projects are technologies to treat the problem at the source by chemical or process substitution or modification. All of these technologies are designed to provide cost-effective methods of meeting compliance requirements and eliminating future liabilities. The following are summaries representative of the work that is accomplished at the Air Force Engineering and Services Laboratory, Tyndall AFB, FL.

BIODEGRADABLE SOLVENT SUBSTITUTION

Objective

The purpose of this program is to: (a) identify solvents for removing wax, grease, and oil that can be replaced by biodegradable solvents, (b) identify the biodegradable solvents that can be used, and (c) develop procedures for, and implement, their use.

Solvents and cleaners are used at the Air Force Air Logistics Centers (ALCs) to remove wax, grease, oil, carbon, and paint from aircraft parts before repairing or electroplating. Most of these solvents are classified as toxic, and many cannot be treated in industrial waste treatment plants (IWTPs) that remove organic chemicals by biological processing. Process wastes must be shipped to approved landfills for disposal. Many solvents will soon be placed under strict EPA control because of their possible ozone-depleting. Expensive control technologies will be necessary as the new Amendment to the Clean Air Act becomes law and new substances are made subject to monitoring and control. To stay in business, a facility must implement new chemicals and processes which result in a significant

45th Purdue Industrial Waste Conference Proceedings, © 1991 Lewis Publishers, Inc., Chelsea, Michigan 48118. Printed in U.S.A.

decrease in environmental consequence. Towards those goals, AFESC has initiated the "Biodegrada ble Solvent" project.

Approach

This project is being conducted in three phases. Figure 1 is a flow chart of the screening process for the Phase I work. Phase I evaluated and screened solvents using four criteria: biodegradability, solubility, cleaning, efficiency, and corrosivity. Solvent were required to pass the first three evaluations before being tested for corrosiveness. Solvent Selection and Performance Evaluation, has been completed, and Phase II, extended performance tests, are nearly complete at the field test facility at Tinker AFB, Oklahoma. Tests include enhancement methods (effects of temperature, mixer agitation, and ultrasonic agitation); cleaning capacity, using aircraft parts, for wax, oil, and grease as a function of solvent loading; rinsing and drying requirements; and impact on the biological treatment plant at Tinker AFB's IWTP. Phase III, full-scale implementation, is implementation of the biodegradable solvents in actual process lines at Tinker Air Force Base in Oklahoma. Phase III has begun and a report will be available April 1990. This research effort is being completed by Ms. Penny Wikoff of EG&G Idaho through a DOE contract.

Status

Five solvents are selected out of over 250 screened. Preliminary tests showed that process enhancement is needed if aircraft parts are to pass the cleaning test. Hence, tests were conducted using ultrasonic and mixer agitation at various temperatures, with and without rinsing. Ultrasonic agitation requires solvent cooling to maintain solvent temperature 50° below the solvent's flash point.

Solvent number one removed 100% of an oil-xylene mixture from aircraft parts with 20 minutes immersion of 120°F and mixer agitation, and can remove wax with ultrasonic agitation at 600 watts, a temperature of 140°F, and immersion for 20 minutes. Solvent number two can remove oil/xylene or wax with ultrasonic agitation at 600 watts, heating to 140°F for 20 minutes, and rinsing. Solvent number three can remove hydraulic fluid with mixer agitation at 100°F or ultrasonic agitation at 300 watts and 77°F; in both cases, a 10-minute immersion and rinsing is needed.

The third solvent can also remove molybdenum disulfide grease with 20 minutes immersion and mixer agitation at 140°F or 300 watts ultrasonic agitation at 77°F.

To test solvent performance, the selected solvents were loaded with various amounts of masking wax, carbonized oil/xylene, or hydraulic fluid and their cleaning efficiency was measured as a function of solvent loading. At 80% of the estimated capacity of solvent number one for oil/xylene, cleaning efficiency decreased to less than 95%. At 50% of the estimated capacity of solvent number two for oil/xylene, cleaning efficiency decreased to less than 80%.

Bioacclimation tests were started on solvent number one loaded with oil/xylene. In the solids contact clarifier at Tinker AFB, the metal sludge floated to the surface. Subsequent jar tests showed that all of the selected solvents either float or disperse the sludge. Methods are being explored to solve this problem.

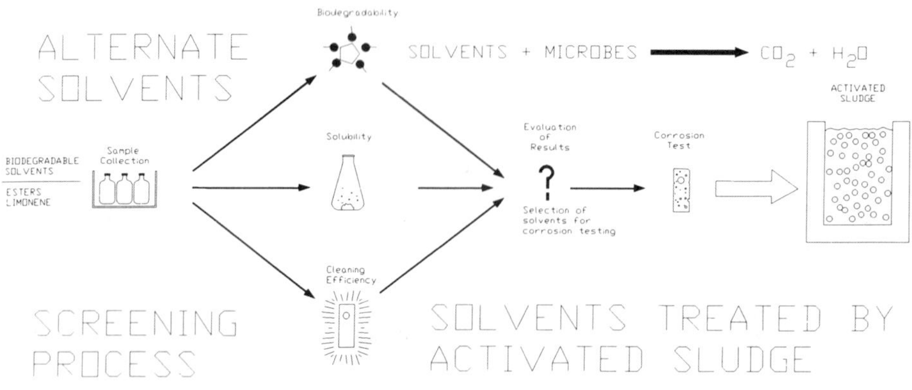

Figure 1. Concept drawing.

Table I. Tinker AFB NPDES Permit Requirements

Constituent	Concentration (mg/L) 1988	1989
Cadmium, total	0.03	0.02
Chromium, total	1.0	0.1
Chromium, hexavalent	0.1	0.01
Copper, total	0.1	0.05
Lead, total	0.1	0.1
Nickel, total	1.0	0.5
Zinc, total	1.0	0.5

An ASTM standard for biodegradability is being developed. Protocol testing began by examining the selection of phenol as a test control compound. Also, tests were conducted to define the percentage of error associated with chemical oxygen demand (COD) measurements. The relative error increases as the lower limit of detection is approached, and decreases at higher COD analyses. The error appears to be linear, and additional characterization of the error is being conducted. Repeatability tests were also conducted, and COD and adenosine triphosphate (ATP) averages were plotted. The data from the TIC/TOC (total inorganic/total organic carbon) analyses had less variability than the data from COD and TOC analyses. The test protocol was developed and submitted to the committee for review.

Although it is too early in Phase III to recommend solvents and conditions, solvents currently being tested are expected to provide the Air Force with biodegradable substitutes for currently used halogenated type solvents and cleaners. The names of the solvents tested are available in ESL-TR-89-04, Substitution of Wax and Grease Cleaners with Biodegradable Solvents, P.M. Wikoff, available through the National Technical Information Service, 5285 Port Royal Road, Springfield Virginia 22161.

FULL-SCALE FERROUS SULFATE/SODIUM SULFIDE REDUCTION OF CHROME (VI) AND METALS PRECIPITATION

Objective

The objective of this project is to show the feasibility of using sodium sulfide and ferrous sulfate to reduce the hexavalent chromium, precipitate the metals, and decrease sludge production and chemical usage over conventional treatment methods of industrial wastewater treatment. The industrial waste treatment plant at Tinker AFB treats 1.4 million gallons of wastewater per day. Principal contaminants are chromium, nickel, copper, cadmium, lead, and zinc and complexing and chelating agents including tartrates, ethylenediaminetetraacetic acid (EDTA), phosphates, and ammonia. Effluent must meet stringent National Pollutant Discharge Elimination Standards (NPDES) shown in Table I.

Approach

The initial Fe/S program was developed in three phases. Phase I of the program determined the feasibility of the process in a bench-scale field verification study using the influent to the tinker AFB Industrial Waste Treatment Plant (IWTP). Phase II of the program was a 5 gpm pilot-scale field study on actual industrial wastewater at Tinker AFB. A pilot-scale field verification plant for the Fe/S process was constructed at the Tinker AFB IWTP. Phase III of the program was the full-scale, 1.4 million gallon per day field verification study and implementation at Tinker AFB. The principal investigator for this project is Ms. Penny Wikoff of EG&G Idaho.

Status

Phase I was completed and a report published March 1988. The report recommended continued investigations. Phase II report was published September 1988. Tests conducted at the pilot plant included determination of the optimum sulfide, ferrous, and polymer concentrations; determination of the effect of phosphate, cyanide, EDTA, and added metals on the reduction of hexavalent chromium and total metal removal; and determination of the optimum parameters for operation of the solids contact clarifier. It was necessary to optimize these requirements, as well as the pH, for this process. In addition, the SCC operation was optimized. The effects of chemical and physical parameters on the process operation was also investigated in the pilot plant. The optimum operating condi-

Table II. Chemical Requirements

Sodium sulfide	2 mg/L S^{-2} per 1 mg/L Cr^{+6}
Ferrous sulfate	1.5 mg/L Fe^{+2} per 1 ml/L Cr^{+6}
Betz 1195 cationic polymer	20 mg/L
Betz 1120 anionic polymer	0.5 mg/L
pH mixer-2	7.2–7.5

tions are summarized in Table II. Phase II recommended full-scale implementation. The Tinker facility was completely converted over in June 1989.

The cost comparison between the acid/sulfur dioxide/lime process and the sodium sulfide/ferrous sulfate process are based on operating logs at Tinker AFB for January through June 1988. During this time, 4,454 lb of hexavalent chromium and 141,913,000 gallons of industrial wastewater was treated. The sludge produced by the sodium sulfide/ferrous sulfate process is 180 tons per year compared to 1880 tons per year produced by the acid/sulfur dioxide/lime process.

Tinker AFB is considering recycling the treated industrial waste back to their cleaning and electroplating processes. With the lime process, the wastewater requires softening to remove calcium to 150 mg/L calcium carbonate. Jar tests were run to determine the volume of soda ash (Na_2CO_3) required for softening. Approximately 828 tons/yr of soda ash would be required and would produce approximately 828 tons of sludge. These costs are included in the cost comparison. The cost for the sodium sulfide/ferrous sulfate process is based on the use of dry chemicals and the treatment of the same column of water and hexavalent chromium as in the acid/lime process. Potential savings for the process are $370,000/yr if the treated water is discharged, as it is presently. If the treated water was to be recycle back to the industrial processes, the potential savings due to the softening requirement would be $655,000/yr. A report was available December 1989.

Further studies are currently ongoing which will optimize the process to a point where the effluent quality may match that of ion exchange, but at a much lower purchase and operation and maintenance cost. An additional study is underway which is determining the feasibility and economics of recovering the metals from the sludges generated by the Fe/S process. That study is discussed later in this chapter. Also, a study has been funded to develop a skid mounted unit with in line mixers which can service facilities that generate low volumes of wastewater.

METALS RECOVERY TECHNOLOGY FROM INDUSTRIAL SLUDGES

Objective

The project will develop technology to remove heavy metals from the sulfide sludges produced by the ferrous sulfate/sodium sulfide metals treatment process developed by the Air Force. The goal is to minimize industrial sludges that must be disposed of as hazardous waste by the Air Logistics Centers (ALCs). Conventional metals treatment and clarification processes result in large quantities of hazardous sludges that must be disposed of as hazardous waste. Usually it is shipped to a landfill at an exorbitant price ($300 per ton). Robins AFB, on the other hand, has stockpiled their sludge for the past 10 years in a specially prepared building. Their idea that by the time the building is filled to capacity, a method will have been developed to treat this waste. Both methods are costly and result in the possibility of future liabilities.

The United States is one of the largest consumers of chromium, primarily as chromite and ferrochromium alloy. Presently, the strategic material ferochromium is not produced domestically, but is imported and stockpiled.

Approach

Figure 2 is the concept drawing for the process. This research will evaluate the feasibility of using technologies such as solvent extraction chemical precipitation and smelting to recover metals from sludges and allow the biodegradation and/or delisting of those sludges. AFESC recently developed a ferrous sulfate/sodium sulfide method of treating industrial waste water for heavy metals, discussed previously, that produces a sulfide sludge for which these technologies will be extremely applicable. Recent literature on techniques used in the mining industry show successful extraction of metals from the sulfide ores. This fact suggests that those techniques will work on chemically similar sludges generated at Tinker AFB. Private industry would stand to save millions of dollars each year. If successful, this project would encourage the change to the ferrous sulfate/sodium sulfide process instead of the more expensive options such as ion exchange, and reverse osmosis. The process would

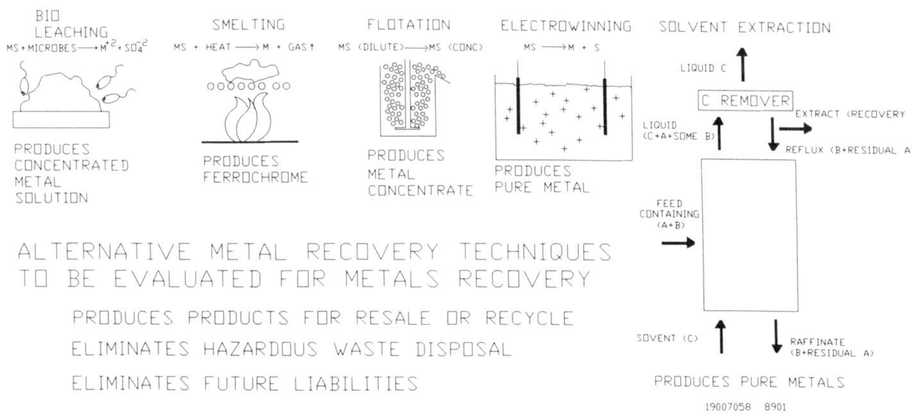

Figure 2. Metal recovery concept drawing.

have tremendous application to DOD, DOE, and private industry. This research effort is being conducted at Tinker AFB by John M. Espinosa, EG&G Idaho.

Status

A market survey was conducted to establish the types of metal compounds that are saleable, the types of complex ores the market would accept, and the penalties associated with undesirable metals. Several possible products were evaluated for transportation and recovery costs to establish the economic recovery that can be realized by processing sludge. The federal and state regulations that may affect transport of the sludge/ore were reviewed. In addition, the conditions that must be met to transport the sludge have been identified and reviewed. The material will be a hazardous material rather than a hazardous waste. The data is being used to define the economic requirements for a given process versus existing disposal costs. This task will also serve as a guide in determining which processing techniques will have to be evaluated to produce a marketable product.

Samples of the Thickener sludge from Tinker AFB were submitted for Scanning Electron Microscopy (SEM). The photomicrographs showed particle sizes ranging from 2 to 15 microns in diameter. Various particle shapes were also noted. The majority of the sludge samples appeared amorphous in nature although there was evidence of some dendritic material. The dendrites ranged from 1 to 10 microns in length. In addition, moisture studies were performed on samples of the sludge from the Solids Contact Clarifier (SCC). The results are shown in Table III.

Energy Dispersive Spectroscopy (EDS) was performed on replicate sludge samples. The results indicated that significant amounts of aluminum (6.77%), silicon (8.16%), phosphorus (5.78%), sulfur (12.95%), calcium (6.94%), chromium (7.18%), iron (11.89%) were present in the sludge.

Thickened SCC metal and activated sludge was obtained, then centrifuged to eliminate as much liquid as possible. Due to unknown metal content of the sludge, samples were digested in nitric acid and analyzed for Cu, Fe, Cr, Ni, Mn, Pb, Cd, and Zn by flame AA spectroscopy.

The first biological test was conducted by using *Thiobacillus ferrooxidans*, an acidophilic iron-oxidizing bacteria. The study was a 7 day batch shake-flask test using wet sludge pulp densities of 1, 10 and 100 g/L in 100 mL media. A pH of 2.0 was used in the presence and absence of 1500 mg/L Fe. Inoculated (using a mixture of *Thiobacillus ferrooxidans* strains) and uninoculated (control) flasks were used in duplicate for each pulp density. Samples were drawn at 0, 2, 4 and 7 days. They were

Table III. Sludge Weights (grams)

	Wet	After 1 hr at 100°F	After 20 hr at 2000°F
Ave Wt	24.791	0.610	0.316
Std Dev	0.077	0.012	0.016

25-ml samples from Sludge Recirculation Line

analyzed for pH, ferrous iron and the metals. (The effects of the addition of sludge on the pH on the leaching medium had been tested to develop a means to accommodate any noted pH changes pertinent to the varying pulp densities).

Results indicated that leaching of the metals under these conditions was minimal. This test was conducted with the two other sludge types. Because of unsatisfactory preliminary results with the thickened sludge, the next approach was to use the microorganisms already present in the sludge. A carbon source was added to some of the test flasks, which were placed under aerobic and anaerobic conditions. Samples were taken and analyzed at 0 and 7 days. Results of these tests indicated that metals cannot be biologically leached from sludge.

A chemical metal separation scheme was devised and carried out for a synthetic sulfide sludge, the sludge from the pilot SCC and the sludge from the Tinker AFB IWTP sludge thickener. Sludge samples from the Solids Contact Clarifier (SCC) and the Tinker AFB Thickener were extracted in accordance with the metal separation scheme. Results indicate that the metals were solubilized during the process but separation and selectivity were somewhat limited. The metal separation scheme was modified so that each step of the process was carried out with a new sludge sample. There was a significant improvement in solubilization with the concentrated acids.

Two types of tests were conducted with different ion exchange resins from Rohm and Haas: IR-120, IRC-718 and DP-1. The first tests are categorized as stirred beaker tests. A sludge sample was weighed and the metals extracted with a nitric acid solution. The resins were placed in a beaker in contact with extract solution. Samples were collected at geometric intervals. The tests were conducted in two sets to observe the effect of pH on the resin effectiveness and/or selectivity.

The second type of test used a nitric acid extract prepared as before except that it was passed through a resin bed of approximately 10 mL. Samples were collected every minute. The column tests with the other resins are performed in a similar method. The conditions were varied to explore their effect on selectivity and metal ion retention.

The results of the analyses to date have shown that under the test conditions the IRC-718 has a slight performance edge. An additional column test with IRC-718 will be performed to conclude the work.

There has also been some research on a bipolar electrode, a Nafion membrane and a dialysis membrane. The sample analyses are currently underway, hence, there are no reportable results at this time. A membrane was prepared with a sputtered gold coating to enhance its conductivity. However, the testing vessel developed a crack during testing and testing has been halted until repairs can be made.

Smelting tests have shown that it is probable that ferrochrome or various grades of stainless steel may be feasible to generate using the sulfide sludge as a metals source. Impurities such as phosphates and sulfides will be removed using fluxes designed for that purpose. A report is expected to be published in July 1990 which will recommend pilot-scale testing for metals recovery.

SPRAY CASTING AS AN ALTERNATIVE FOR ELECTROPLATING

Objective

The project will develop technology to replace conventional electroplating processes with a spray-casting metallizing process, which has the potential to reduce or eliminate hazardous waste generation from many electroplating operations. The spray forming process requires no chemical solutions in order to apply the coating material to the base material. The goal is to minimize or eliminate hazardous wastes generated by the electroplating shop; particularly chromium containing wastes. This process is expected to be much cheaper to operate than laser-enhanced jet electroplating, and will provide a replacement coating process for parts which laser-enhanced technologies may not be practical.

Approach

Figure 3 is the concept drawing for the process. Spray-casting is a process in which molten metal is aspirated into the gas stream by the Bernouli principle, where it is broken up into fine droplets and sprayed onto the part to be plated. The size, temperature, velocity and concentration of these droplets can be controlled by varying the operating parameters (i.e., gas temperature, supply pressure and melt temperature). The droplets are partially cooled in-flight, and compacted against a base metal to form a thin coating. Recent studies have shown the spray-casting process to be technically feasible. The principal investigator for this project is Lloyd Watson of EG&G Idaho.

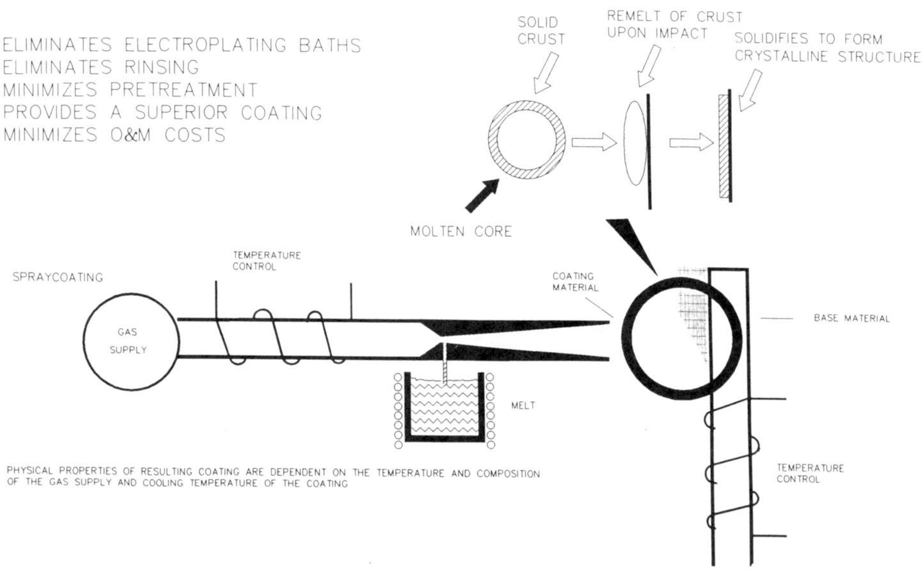

Figure 3. Concept drawing.

Status

A gas heater has been developed for supplying the nozzle and preheating base metal. It controls gas temperatures with pressures up to 50 pounds per square inch and at flow rates up to 500 liters per minute. Remotely operated fixturing has been installed for pouring molten tin from outside the chamber.

A furnace assembly was constructed from nickle-chromium wire to operate up to 1000 degrees Celsius. Pressure profiles were obtained for three nozzle designs at several exit-cone lengths, and an exit-cone angle of 10 degrees was selected. A nozzle/tundish assembly was designed, constructed, and characterized. Background atmosphere measurements were made in the spray chamber. Oxygen levels can be varied from 21% to 1.2% by purging with inert nitrogen gas.

Computer interfaces and software were implemented to monitor these signals and operate a spray-coating system. All instruments are interfaced to a dedicated computer (IBM PC/AT), and considerable software was written and debugged for on-line data recording and later reduction.

Tests were performed with tin to confirm intended operation of the furnace and its remote tilting apparatus. Particle concentrations were measured by an aerosol spectrometer at the chamber exit sampling point. Circumstances included without a purging gas flow, with a purging flow with no chamber component activity, with purging while pouring molten metal, and with purging while operating the rotating drum (base metal fixturing). These measurements constituted a baseline for airborne chamber particulates before metal is sprayed.

Testing encompassed complete exploration of the process range over which the bench-scale spray-coating system can deposit tin onto low carbon steel base metal. Coated coupons have been sectioned from the grit-blasted carbon steel base metal to investigate effects of grit size and base metal temperature, along with coating thickness on coating quality.

Work that has already been finalized is being compiled and arranged for presentation in a final report. The report is expected to be published in July 1990.

LASER-ENHANCED ELECTROPLATING

Objective

The project will develop technology to replace conventional electroplating processes (emphasis on chromium) with a laser-enhanced jet electroplating process which will reduce or eliminate liquid electroplating wastes. Conventional electroplating processes result in large volumes of used electroplating baths and rinse water which require treatment. The treatment process results in the generation of large quantities of hazardous waste, which must be disposed of at an approved hazardous waste disposal facility. This project will help eliminate hazardous waste generation from electroplating processes. Phase I of the project provided proof-of-concept for the laser enhancement of electroplating. The effort showed that the use of a laser beam during an electroplating process can increase the rate of efficiency of the operation. Phase II of the program is continued testing and evaluation of the system. The major thrust of Phase II is to refine system parameters and obtain the basic scientific and engineering data required to scale up the system for pilot work.

Approach

Figure 4 is the concept drawing for the process. In our system, a freestanding jet of plating solution emanates from a pressurized chamber through a nozzle to impinge on the cathode. A laser beam, from a Nd/YAG laser, is directed to the point where the solution strikes the cathode. Electrical contact is from the anode, housed in the pressure chamber, through the jet to the cathode. Anode-cathode current flow is thus restricted to the stagnation region of the jet stream, an arrangement that provides an *in situ* mask. Without the laser, metal is deposited mainly in the stagnation region of the jet, the region directly in front of the nozzle. While liquid does spread away from this region in a thin film, little plating is seen in this area. The current density in this thin film of liquid (called the wall jet) is much smaller than in the stagnation region so plating rates here are low. Since the laser beam is directed to the stagnation region, the beam enhances plating rates at this point only. Any peripheral plating (in the wall jet) is not enhanced. Early data indicates that not only is plating not enhanced, but may well be reduced to zero. This observation furthers the possibility of developing a completely maskless plating system.

Electroplating from a jet of electrolyte increases plating rates by increasing mass transfer at the cathode. The laser further enhances plating rates by shifting the local rest potential and increasing local temperatures which in turn increase the mobility of the electroactive species. Combining of laser, nozzle, and electroplating technologies offers remarkable possibilities in the area of electrochemical metallization.

The process results in minimization or elimination of plating waste through more efficient and controlled use of plating solutions, and elimination of toxic solution additives, such as cyanide, required in plating baths but not necessary for jet plating, which cause considerable trouble for the industrial waste treatment plant. This research effort is being conducted at EG&G Idaho, under the direction of David F. Glenn.

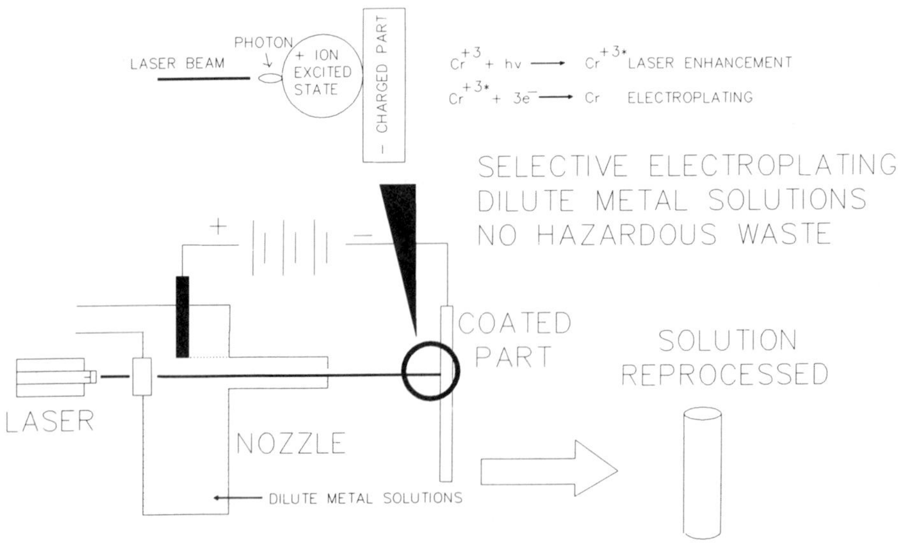

Figure 4. Concept drawing.

Table IV. Operating Parameters

Solution	1.0M CrO_3
CrO_3/SO_4 ratio	100 to 1
Temperature	Room temp. 24°C
Jet orifice	0.05cm dia., 0.5cm long
Linear flow velocity	1.0×10^3 cm/sec
Reynolds number	5.5×10^3
Nozzle-cathode spacing	1cm
Plating-current density	25 A/cm^2
Cw laser power	50W

Status

A summary of the solution, hydrodynamic, and laser parameters used in the preliminary tests of the system are shown in Table IV. Using the parameters shown in this table, chromium can be electroplated at a rate of 50 mils per hour. This is 50 times faster than the 1 mil per hour for the average chromium plating bath. Current efficiency for these parameters was found to be 45%. This is a threefold increase in efficiency over the 15% reported for a bath system. It should be noted that the solution used in these experiments is only one third the concentration used in conventional bath systems, and plating was accomplished at room temperature.

Sensors and diagnostics equipment were identified and installed to provide data for modeling, thermodynamic and kenetic studies, information on nucleation, electrocrystallization, and transport phenomenon. nLaser parameters such as wavelength, power density, whether pulsed or continuous wave, and optical and thermal properties of the material irradiated has been investigated. Typical electrolyte solutions have been used with various substrates to determine the effect of these laser parameters on the local rest potential, Soret potential, surface temperatures of substrate, oscillations in current or voltage, and transport phenomena. This phase of the project has evaluated plating from standard plating solutions and determined structural characteristics of the deposits. These results clearly show that laser radiation during an electroplating process can be effective in enhancing both plating rates and current efficiencies. A report detailing the results of the Phase II work is expected to be published in January 1991. A full technical presentation of the results is presented as another paper at this conference.

NONCHROMATE CORROSION INHIBITOR

Objective

The objective is to provide a demonstration of the use of nonchromate corrosion inhibitors to reduce, or eliminate hazardous waste. This will be done by eliminating chromate corrosion inhibitors in paints and anticorrosion coatings, and thereby eliminating the major component that results in the generation of hazardous waste from paint stripping operations.

The use of chromate corrosion inhibitors in paints and anticorrosion coatings by DOD has been restricted to those systems for which substitute, nonchromate inhibitors are either unavailable or ineffective. The Air Force, as with it's civilian counterparts, is confronted with the challenge of protecting products from corrosion and its deleterious effects. The stringent demands of new technologies associated with modern aircraft and support equipment for improved product life and performance have resulted in continuous investigation of new corrosion control systems. The development of corrosion inhibitors for DOD paints has followed the application of inorganic solids, including strontium and barium chromates, principally because the solids are insoluble in, and nonreactive with, the components of the coatings. These solids have a controlled aqueous solubility when the coatings are exposed to salt water. By contrast, if organic inhibitors are added in greater than minute quantities, the properties of the cured coatings will change, sometimes significantly, and as a result, the application of organic anticorrosion agents has been limited to direct contact with the metal parts to be protected. Currently, there are no nonchromate inhibitors for coating aluminum with primers or top coats. No alternative for paints and other coatings for aluminum presently exists that does not contain a chromate. The result is the generation of large quantities of hazardous wastes for painting and paint stripping operations.

Approach

The formulations of organic inhibitors for paints will be developed for use with microencapsulation technology. The release of an inhibitor core will be controlled by shell thickness and diffusibility. As long as the shell remains impermeable during the drying and curing stages of the coating, the organic inhibitors will remain stable. It will not significantly affect the working properties of the paints when added in quantities similar to those of inorganic inhibitors. It is also probable that the currently restrictive epoxy coating thickness levels can be relaxed, since the 0.6 to 0.9 mil thickness requirement of MIL-P-23377, Epoxy-Polyamide Primer Coatings, is based on the use of the strontium chromate inhibitor, which produces brittle, failed coatings when applied at thickness levels greater than specified. The proposed microcapsules that will be investigated in this project are liquid cores surrounded by a thin shell, and are significantly less brittle or friable.

Chromates are coming more and more under the scrutiny of the EPA, with more stringent controls becoming increasingly costly and difficult to accomplish. Wastes generated by the painting and stripping processes have to be treated as hazardous waste, and consequently require costly disposal. This project will provide a production system that doesn't generate hazardous waste, increases worker safety by eliminating a major source of carcinogenic Cr^{+6}, and provide a production system which would serve as a significant advance in reducing corrosion in high strength alloys, such as 7075 Al, which currently suffers from stress corrosion cracking, corrosion fatigue, and/or pitting. The principle investigator is Dr. Bailen, Lockheed Missiles & Space Company, Research and Development Division, Palo Alto, CA.

Status

The contract for the project has just been awarded (later than expected) and results are not available at this time.

CONCLUSION

These projects are currently being conducted in an effort to minimize or eliminate the generation of hazardous wastes for electroplating operations. Each of the projects is scheduled for demonstration under production conditions. The Air Force intends to reduce the generation of hazardous waste by 50% by 1992. It is expected that a 0-discharge goal will soon be established. The projects discussed as well as others being completed by other Air Force organizations will enable the Air Force to reach these goals.

26 ALTERING TEXTILE MANUFACTURE TO MINIMIZE TREATMENT NEEDED FOR WASTEWATER RECLAMATION

Maria Eduarda de C. Beja Neves, Assistant Professor
Technical University of Lisbon
Lisbon, Portugal

James E. Etzel, Professor
School of Civil Engineering
Purdue University
West Lafayette, Indiana 47907

INTRODUCTION

Purpose

The research involved two major areas of water management in the textile industry: wastewater reclamation and chemical substitutions.

The approach used relates to important management decisions faced by the industry today: minimization of the use of external sources of water, and substitution of process chemicals in order to make their removal from wastewaters easier and, ultimately, to protect the environment.

A linear programming model was developed, capable of minimizing the cost of treatment needed for wastewater reclamation. This was accomplished through the selection of the optimum set of chemicals in respect to their removal from a wastewater by biodegradation, and by so doing, preventing the accumulation of inorganic dissolved solids in the reused wastewater, even with prolonged reuse.

The need for minimizing the accumulation of inorganic dissolved solids is that they cause excessive wear and corrosion on machinery. In addition, the cost of removing dissolved solids, and the problems associated with their final disposal, are difficult and costly, especially if the solids are inorganic in nature. Hence, the general basic approach was to establish the interrelationship between the characteristics of the chemicals used in the production processes with their removal from the plant's wastewaters, along with the suitability of the treated wastewaters for reuse within production processes.

The in-plant approach was the selection of the chemicals used in the various industrial processes based on consumption and their organic and inorganic composition. In this approach, which has been used full scale in other industries, the means of controlling dissolved solids was through the use of biological wastewater treatment.

Biological wastewater treatment not only removes the organic dissolved solids, but converts them into a biological sludge which is disposable in the environment much more readily than would be a concentrated brine coming from processes such as deionization or reverse osmosis. The cost of biological treatment is also lower than the cost of other dissolved solids removal processes and much less complicated to operate.

In other full scale uses of this approach to water reuse, reuse rates as high as 96 to 98% have been achieved on a sustained basis.

Different values were tested for the wastewater reclamation rate, and for the percentage of biodegradable materials in the chemicals used by the industry.

The industrial processes themselves were considered as blocks, with different manufacturing objectives and using different kinds and quantities of chemicals.

The cost of wastewater treatment was influenced by the organic content of the chemicals and by the wastewater reclamation rate.

45th Purdue Industrial Waste Conference Proceedings,

Summary of Methodology

The methodology used to develop the research work can be summarized as follows:

- collection of information and data from the literature and from industries themselves, which were related to the nature of the research work
- evaluation of the information and descriptive overview of the textile industry
- delineation of a process flow diagram, to use as a basis for the research
- establishment of the amount of processing chemicals used, and of the costs of removing biodegradable organics, refractory organics, and inorganic materials from textile wastewaters.
- approximation of the costs associated with the final disposal of organic and inorganic sludges
- development of a linear programming model, including the definition of fixed and decision variables, the definition of the constraints to the model, and the estimation of all the model parameters
- application of the linear programming model, followed by evaluation of the results obtained and the drawing of conclusions

Data and Information

Given the extreme variability shown by the textile industry, from the raw materials used to its final products and the many manufacturing ways to obtain them, the research was aimed to prove the viability and general applicability of a specific approach to water management in the textile industry, rather than to solve a particular problem in the industry. Hence, almost all the data and information used in this research work was obtained from the literature, so as not to incur in the risk of restricting the area of the research to a particular plant in the textile industry.

The linear programming model was designed to be simple, flexible and, at the same time, complete enough to be adapted to different practical situations as a management tool.

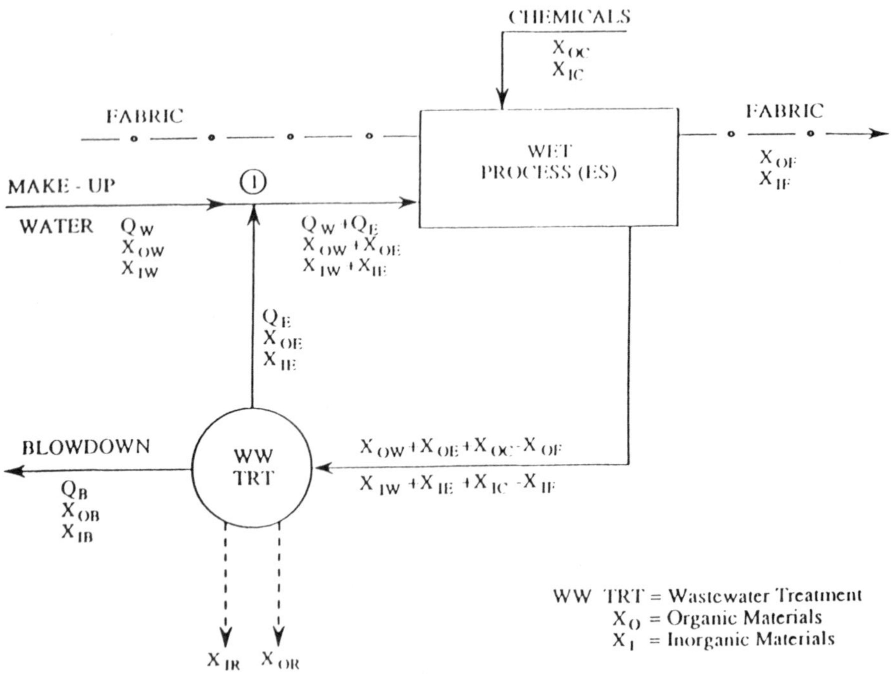

Figure 1. Linear programming model flow diagram.

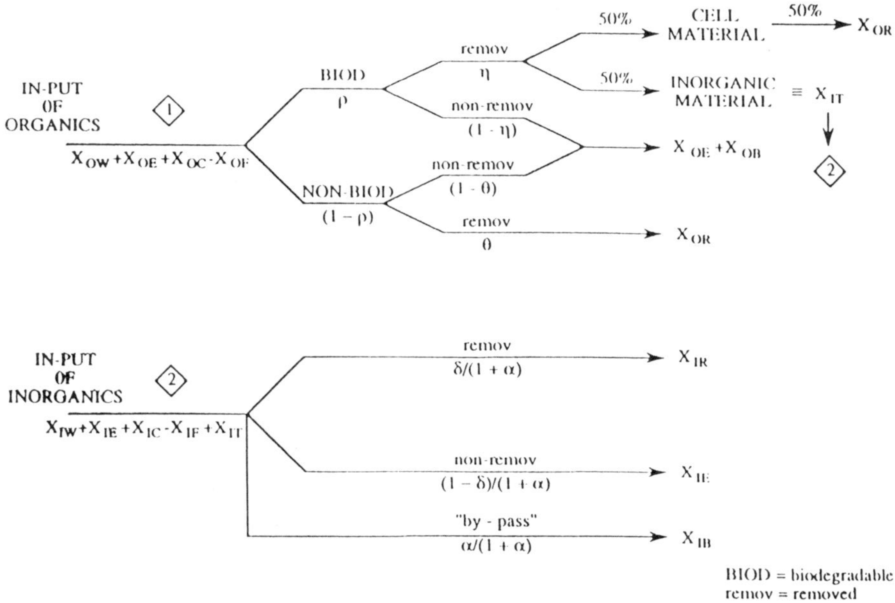

Figure 2. Wastewater treatment input and output of materials.

FLOW DIAGRAM

The flow diagram presented in Figure 1 was the basis for the formulation of the linear programming model.

The wet manufacturing processes, typical for the woven fabric finishing subcategory of the textile industry, were assimilated into a "black box", which could represent from one to several different processing steps. This "black box" could also be subdivided into as many others as necessary, depending on the detail used in the analysis of a specific problem.

Although in this research the model was applied only to the woven fabric finishing subcategory, the flow diagram was generic enough to be used for any of the other subcategories in the industry.

The wet manufacturing process(es) box, as represented in Figure 1, shows three inputs and two outputs, defined as follows:

- Inputs: fabric (or raw material), process water resulting from the mixing of water from a natural source with reclaimed wastewater, and processing chemicals
- Outputs: processed fabric and the combined wastewater resulting from the uses of water in manufacturing.

The combined wastewater was conveyed to a wastewater treatment plant from which three outputs were considered: the liquid treated effluent which constitutes the reclaimed wastewater, a fraction of the treated effluent which was discarded as blowdown, the organic residue and the inorganic residue from treatment.

Organic materials and inorganic materials contents were attributed to each of the inputs and outputs to the manufacturing block.

Also shown are the flows of make-up water and blowdown, and the total flow of water needed for manufacturing.

LINEAR PROGRAMMING MODEL

Assumptions

The following general assumptions were made in order to develop the linear programming model:

- the total amount of organic and inorganic materials entering the manufacturing system as chemicals included the amount of those materials entering the system with the fabric (natural impurities and/or add-ons)
- there were no inhibitory conditions to the performance of the biological treatment system
- the combined wastewater originated in any set of wet manufacturing processes was equalized prior to treatment
- the only chemicals leaving the system with the processed fabric were a fraction of the dyestuffs used for dyeing; all the other process chemicals left the manufacturing system in the wastewater, without any chemical transformation during processing
- the processing steps of printing and finishing were considered to be dry processes.

Model Definition
Decision Variables and Fixed Variables

Nine decision variables were defined, based on the flow diagram in Figure 1, as follows:

- X_{OW} & X_{IW}: organic and inorganic materials contents of the makeup water (lbs/1,000 lbs of fabric)
- X_{OE} & X_{IE}: organic and inorganic materials contents of the effluent from wastewater treatment (lbs/1,000 lbs of fabric)
- X_{OC} & X_{IC}: organic and inorganic materials contents of the process chemicals (lbs/1,000 lbs of fabric)
- X_{IT}: inorganic materials produced during biological treatment (lbs/1,000 lbs of fabric)
- X_{OR}: organic residue from the removal of organic materials from the wastewater (lbs/1,000 lbs of fabric)
- X_{IR}: inorganic residue from wastewater treatment (lbs/1,000 lbs of fabric).

Sixteen fixed variables were defined as follows:

- ρ: percentage of biodegradable materials in the total organics entering wastewater treatment (decimals)
- β: percentage of the organic content of the process chemicals (X_{OC}) leaving with the processed fabric (decimals)
- γ: percentage of the inorganic content of the process chemicals (X_{IC}) out with the processed fabric (decimals)
- α: wastewater reclamation rate (ratio of make-up water flow and wastewater reclamation flow)
- η: percent removal of biodegradable organic materials in the wastewater treatment system (decimals)
- θ: percent removal of non-biodegradable organic materials in the wastewater treatment system (decimals)
- δ: percent removal of inorganic materials in the wastewater treatment system (decimals)
- l_1: upper limit on the concentration of organic materials in the process water (lbs/gallon)
- l_2: upper limit on the concentration of inorganic materials in the process water (lbs/gallon)
- l_2: upper limit on the concentration of inorganic materials in the process water (lbs/gallon)
- X: total amount of organic and inorganic process chemicals desirable for manufacturing (lbs/1,000 lbs of fabric)
- Q_T: total flow of process water desirable for manufacturing (gallon/1,000 lbs of fabric)
- c_1: unit cost of biological treatment followed by multimedia filtration ($/lb of biodegradable organics into treatment)
- c_2: unit cost of removal of refractory organics ($/lb of refractory organics into treatment)
- c_4: unit cost of removal of inorganic materials ($/lb of inorganic materials into treatment)
- c_4: unit cost of organic sludge (organic residue) hauling and disposal ($/lb of biological sludge)

- c_5: unit cost of inorganic sludge (inorganic residue) hauling and disposal ($/lb of non-biological sludge).

These variables were called fixed variables because their values did not result from the application of the linear programming model (as it is the case with the decision variables), but they were pre-defined and incorporated into the attribution of values to the model parameters.

Objective Function

The objective function "C" of the linear programming model represented the total cost of wastewater treatment and sludges hauling and disposal, expressed in U.S. dollars per 1,000 pounds of fabric. It was of the following form:

$$\min C = a_1*X_{OW} + a_2*X_{IW} + a_3*X_{OE} + a_4*X_{IE} + a_5*X_{OC} + a_6*X_{IC} + + a_7*X_{IT} + a_8*X_{OR} + a_9*X_{IR}$$

where:

$$a_1 = a_3 = [c_1*\rho + c_2*(1-\rho)]$$
$$a_2 = a_4 = a_7 = c_3/(a+\alpha)$$
$$a_5 = [(1-\beta)[c_1*\rho + c_2(1-\rho)]]$$
$$a_6 = [c_3*(1-\gamma)/(1+\alpha)]$$
$$a_8 = c_4*0.50$$
$$a_9 = c_5$$

Model Constraints

A total of 18 constraints were introduced into the linear programming model. These constraints were based both on the flow diagram presented in Figure 1 and on the materials transformations occurring during wastewater treatment as illustrated in Figure 2.

The numbers shown in Figure 2 were based on the assumption of a wastewater treatment system including the following processes:

- activated sludge system with anaerobic digestion of the sludges, followed by multimedia filtration of the settled wastewater
- treatment with granular activated carbon
- reverse osmosis treatment of all or a part of the carbon adsorption effluent.

The calculations involved in the definition of the constraints to the model can be summarized as follows:

- mass balance on X_O around the total system
- mass balance on X_I around the total system
- mass balance on X_O around "point 1 – Figure 1", defining the quality of the water into manufacturing
- mass balance on X_I around "point 1-Figure 1", defining the quality of the water into manufacturing
- organic residue production during wastewater treatment
- organic materials left in the treated effluent
- inorganic materials resulting from biological treatment
- inorganic residue production during wastewater treatment
- inorganic materials left in the treated effluent
- total amount of organic and inorganic process chemicals used in manufacturing
- nine non-negativity constraints for each of the decision variables.

MODEL APPLICATION AND SENSITIVITY ANALYSIS

General

This section refers to application of the linear programming model of the general form previously presented.

Before the model was actually used, all the parameters were calculated by manual means, for each different application, based on values attributed to the fixed variables. Hence, the first task was to estimate the fixed variables values, including those of wastewater treatment and sludge hauling and disposal costs. Then it was possible to make actual applications of the model as it was defined for each set of conditions.

The sensitivity of the model was also analyzed, making some of its parameters values change.

A structured library of high quality and reliable fortran subroutines for mathematical programming, called "XMP", was used to solve the model. "XMP" was run on the Engineering Computer Network (ECN) at Purdue University.

Treatment Cost Estimates

Except for the estimations of the cost of removal of inorganic materials by reverse osmosis, which was obtained by personal communication with Romicon, Inc., all other cost estimates were based on the Development Document published by EPA for the textile industry. The main reason for choosing this document as a source of information for cost estimates was that the basic assumptions and the rationale supporting the estimates were based primarily on data collected from an actual industry survey. It presented the calculations of capital and annual costs for the treatment technologies which were considered for the development of effluent limitations and standards for the control of toxic, conventional, and non-conventional pollutants. It also presented cost estimates for individual treatment processes, such as adsorption on activated carbon, as well as the estimated costs for sludge hauling and disposal. Eight different flow levels, ranging from 0.05 to 0.5 MGD, were selected by EPA for the detailed calculation of investment and annual costs. These eight flow levels provided a sufficient range to represent three different sizes in most subcategories, and thus properly represent the actual range of plant sizes.

As the plants were flow sized, they could be related to production levels by the respective values for flow per unit of production for the respective subcategory.

EPA characterized the woven fabric finishing subcategory of the industry with three different sizes of, respectively, 0.11, 0.60, and 1.5 MGD.

Several distinct technologies comprise the treatment and control options considered for the textile industry. In this research, and for the purpose of cost estimation alone, the technologies selected for detailed calculation of costs were as follows:

> screening, followed by extended aeration activated sludge, with thickening and vacuum filtration of the sludges; the effluent would be filtered in a multimedia filter (overall system cost), followed by adsorption on granular activated carbon (without regeneration), and then reverse osmosis; hauling and final disposal of sludges was the last item considered.

The investment costs included the cost of purchasing and installing the components of each treatment scheme, and allowance for contingencies and engineering.

The annual costs included the cost of capital depreciation, operation and maintenance labor, maintenance materials, sludge disposal, energy, chemicals, and monitoring.

The costs obtained from the Development Document represented fourth quarter 1979 dollars. To update these costs to December of 1988, annual construction cost indexes were used with the capital costs, and a combination of annual cost indexes for skilled labor (80%) and materials (20%) was used for the annual costs.

Only the total annual costs were used in the linear programming model. The costs were grouped in two different size ranges: 0.11 to 0.60 MGD, and 0.60 to 1.5 MGD. In each of these ranges, the relationship between costs and size was assumed linear, complying with the proportionality assumption of linear programming.

The same assumption could be made for the variation with size of the costs estimated for individual technologies.

Model Parameters

In order to define the model parameters, it was necessary to attribute values to all the fixed variables.

The values attributed to the treatment efficiencies were based on the removal efficiencies found in the literature for the different treatment methods applied to textile wastewaters, and were as follows:

- η (biological) = 95%
- θ (carbon adsorption) = 80%
- δ (reverse osmosis) = 98%

These values were kept constant in all the applications of the model, since the wastewater treatment system components were unchanged.

The values attributed to "β" and "γ" (percent of organic and inorganic materials that left with the processed fabric) did change for different applications of the model; however, the values of "β" and "γ" for each particular application were considered equal. The actual values were $\beta = \gamma = 0.22$, $\beta = \gamma = 0.40$, $\beta = \gamma = 0.67$.

For the attribution of values to "ρ" (percentage of biodegradable organics into treatment) and "α" (wastewater reclamation rate), two different situations were considered: use of the same values ($\alpha = 1.0$ and $\rho = 0.5$) in all the different applications of the model (basic analysis); use of the following combinations of values in the sensitivity analysis of the model, for only one type of application:

- $\rho = 0.30$ and $\alpha = 0.5$
- $\rho = 0.30$ and $\alpha = 2.0$
- $\rho = 0.80$ and $\alpha = 0.5$
- $\rho = 0.80$ and $\alpha = 2.0$

The variables l_1 and l_2, representing the upper limits of, respectively, the organic and inorganic contents of the process water, were assigned the following constant values:

- $l_1 = 0.00017$ lbs/gallon (20 mg/L)
- $l_2 = 0.0084$ lbs/gallon (1,000 mg/L)

The variables X and Q_T representing, respectively, the total amount of process chemicals and of total water flow into the manufacturing system, were given different values in each application of the model.

The costs referring to the mid-size flow level were used in each application of the model (basic analysis). For the sensitivity analysis of the linear programming model to changes in cost, the values of $\alpha = 1$ and $\rho = 0.5$ were used together with the costs estimated for the lower level, for a single type of application.

This way, the research was restricted to the range of plant sizes between 0.11 and 0.60 MGD, where the costs were considered to vary linearly with size.

Applications Definition

Two basically different applications were considered in using the linear programming model. The first application—"Application 1"—referred to the vat-dyeing of 100-percent cotton woven fabric. The second application—"Application 2"—referred to the wet processing of cotton textiles for the woven fabric finishing subcategory.

In this second application, the total amount of process chemicals used was calculated based on the characteristics of the combined wastewater produced, assuming that only a fraction of the dyestuffs left the system with the processed fabric; all the other chemicals used were assumed to have left the system in the wastewater, without undergoing any chemical transformations. The amount of total solids (TS) in the combined wastewater was assumed to be representative of the total amount of materials leaving the manufacturing system. This application was subdivided into two others—"Application 2.1" and "Application 2.2"—corresponding to a different strength for their wastewaters.

Only "Application 2" was used in the sensitivity analysis of the model

RESULTS AND DISCUSSION

A feasible solution was found for each application of the model.

All the results were obtained using a wastewater reclamation rate (α) of one, and 0.5 expressed as a decimal of biodegradable organic materials (ρ).

The total cost of treatment for applications 1, 2.1 and 2.2 was, respectively, 7.52, 39.16, and 85.98 dollars per 1,000 pounds of fabric. These results showed an increase in costs, as expected, since the total amount of chemicals and water used increased for the same values of "α" and "ρ". However, these results favored the use of inorganic instead of organic chemicals. This was due to the high cost attributed to the removal of nonbiodegradable organics with granular activated carbon that was assumed to be used on a one time basis.

Also, the cost of biological treatment included the cost of the use of multimedia filtration to prepare the water for reverse osmosis, resulting in a higher cost for biological treatment than for the removal of inorganic materials. When forcing the amount of organic chemicals to be, at least, 20% of the total amount of process chemicals used, keeping the values of "α" and "ρ", respectively, 1 and 0.5, the impact of the total cost of treatment assignable to the removal of non-biodegradable organics was again evident.

In order to evaluate the impact of the different component costs on the total cost of treatment, "application 2.1" was tested for the use of organic materials alone, of which 50% were biodegradable and for "α" equal to one, and considering two different wastewater treatment schemes: biological treatment followed by multimedia filtration, and these treatment units followed by adsorption on granular activated carbon. The lower cost was obtained for biological treatment followed by multimedia filtration, without the use of activated carbon.

In further applications of the model, the cost of removal of biodegradable organics from the wastewater (biological treatment) may be totally separated, not only from the cost of removal of inorganic materials, but also from the cost of any other type of treatment used (such as filtration and adsorption on activated carbon). In this way, the amount of biodegradable organic materials calculated by the model will not be affected by the cost of removal of any other type of materials from the wastewater.

When using the model for "application 2.1", changing the values of the wastewater reclamation rate (α) and the percentage of biodegradable organics (ρ), a definite decrease in the total cost of wastewater treatment was observed when the value of "ρ" increased from 0.3 to 0.8, while keeping the value for the wastewater reclamation rate constant.

For the same value of "ρ", the total cost of treatment decreased when the wastewater reclamation rate increased from 0.5 to 2 (which represented a decrease in the reclaimed wastewater flow). The largest decrease in the total cost of treatment was observed when increasing both the wastewater reclamation rate (from 0.5 to 2) and the percentage of biodegradable organics (from 0.3 to 0.8).

Thus, these results seemed to favor the use of increased amounts of biodegradable organics. However, once again the use of inorganic materials instead of organic materials was favored, since the same structure of costs was used.

As expected, when using the treatment cost parameters for the lower level of the textile plant sizes (0.11 MGD), the total cost of wastewater treatment per 1,000 lbs of fabric ($46.21) increased when compared to the use of the "mid-size" (0.6 MGD) costs ($39.16).

CONCLUSIONS

The linear programming model should have wide application in the textile industry.

It was developed to provide the industry with a useful management tool to support important decisions: how much wastewater to recycle and what type of chemicals to use in the manufacturing processes, in order to minimize the cost of wastewater treatment and reclamation.

The model was flexible enough to be used in all different categories of the industry, with more or less detail depending on each particular situation. In fact, the flow diagram presented in Figure 1 may include only one "black box" representing one or more unit manufacturing processes, or may detail different processes in different "boxes".

The large number of fixed variables added flexibility to the overall approach, enabling the user to define different practical scenarios.

It is expected that the use of different unit costs of wastewater treatment, and of hauling and disposal of biological sludges, will enable the model to produce different results in respect to the relative amounts of organic and inorganic materials to be used by the industry.

In this way, local situations involving costs and disposal options can be readily factored into a given textile plant.

It is also expected that the future addition of non-linear programming into the formulation of the model will expand its applicability to different practical situations, namely as applications expand to other parts of the textile manufacturing process.

In any case, the model, as formulated, is widely applicable and is able to provide useful information concerning changes in wastewater treatment costs, when changes in the nature of the chemicals used occur, and/or a new wastewater reclamation rate is desired.

REFERENCES

1. WPCF Statement, "Waste Minimization and Waste Reduction", *Journal of WPCF*, Vol. 61, No. 2, pp. 184, Feb. 1989.
2. "Waste Minimization Opportunity Assessment Manual", EPA/625/7-88/003, Hazardous Waste Engineering Research Laboratory, Office of Research and Development, U.S. Environmental Protection Agency, Ohio, July 1988.
3. Smith, B., and J. Rucker, "Water and Textile Wet Processing—Part 1", *American Dyestuff Reporter*, Vol. 76, No. 7, pp. 15-24, July 1987.
4. Smith, B., and J. Rucker, "Water and Textile Wet Processing—Part 2", *American Dyestuff Reporter*, Vol. 76, No. 8, pp. 68-73, 77-78, August 1987.
5. Monma, Clyde L., "Recent Breakthroughs in Linear Programming Methods", New Jersey, 1987.
6. Engineering News-Record, The McGraw-Hill Construction Weekly, Oct.-Dec. 1980 to Oct.-Dec. 1988.
7. Ashkenazi, B. S., "The Use of Personal Computers in the Textile Industry", *Colourage*, Vol. 31, No. 1, pp. 11-13, Jan. 12, 1984.
8. Chvatal, V., "Linear Programming", W. H. Freeman and Company, New York, 1983.
9. Beckmann, W., and J. Pflug, "Reuse of Weakly Loaded Liquors from Textile Processing Operations", *Textil Praxis International*, 38 No. 3, pgs. 2-7, 1983.
10. Kremer, F., B. Broomfield, and L. Fradkin, "Energy and Materials Recovery Options for the Textile Industry", Industrial Waste Conference, *Proceedings*, Vol. 37, pp. 157-161, Purdue University, 1982.
11. "Development Document for Effluent Limitations, Guidelines, and Standards for the Textile Mills Point Source Category", EPA 440/1-82/022, Sept. 1982.
12. Etzel, J. E., "Inorganic Salt Cement of Materials Used in Production and Biodegradability of Materials Used in Production. An Evaluation of the Feasibility of Biological Treatment of Actual Clarithickner Effluent by Activated Sludge and Fixed Film Contact", for Bedford Plant, Central Foundry Division of GMC, Bedford, Indiana, Jan. 1981.
13. "Industrial Reuse and Recycle of Wastewaters—Literature Review", EPA 600/2-80-183, U.S. Environmental Protection Agency, Dec. 1980.
14. Etzel, J. E., "Chemical and Biological Treatment of Die Water System to Produce Recyclable Water and Identification of Source of Contamination of Mill Water System", for Massena Plant, Central Foundry Division of GMC, Massena, New York, 1979.
15. Nemerow, N. L., "Industrial Water Pollution: Origins, Characteristics, and Treatment", Addison-Wesley Pub. Co., 1978.
16. Montgomery, V., "Reclaiming Water", *Modern Textiles*, Vol. 57, pp. 36-38, August 1976.
17. Abrams, E. F., D. K. Guinan, and D. Derkics, "Assessment of Industrial Hazardous Waste Practices, Textile Industry", Versar Inc., EPA Hazardous Waste Management Division, Report PB-258953, June 1976.
18. Porter, J. J., and C. Brandon's "Zero Discharge as Exemplified by Textile Dyeing and Finishing", *Chemtech*, pp. 402-407, June 1976.
19. "Wastewater Treatment Systems. Upgrading Textile Operations to Reduce Pollution", EPA Technology Transfer Seminar Publication, EPA-625/3-74-004, Oct. 1974.
20. Cooper, L., and D. Steinberg, "Methods and Applications of Linear Programming", W. B. Saunders Company, Philadelphia, London, Toronto, 1974.
21. "In-Plant Control of Pollution. Upgrading Textile Operations to Reduce Pollution", EPA Technology Transfer Seminar Publication, EPA-6233/3-74-004, Oct. 1974.
22. "State of the Art of Textile Waste Treatment", Department of Textiles, Clemson University for Water Quality Office, EPA Water Pollution Control Research Series 12090 ECS 02/71, Feb. 1971.
23. Lever, N. A., A. T. Alexander, R. J. Buttery, B. Moffat, and M. R. Henzen, "The Application of a Linear Programme Model to Water Conservation and Effluent Management at Modderfontein Factory", 5th International Pollution Research Conference, *Proceedings*, Pergamon Press, Ltd., Spring 1971.
24. Committee RA58, Stream Sanitation Technology, "The BOD of Textile Chemicals, Updated List-1966", *American Dyestuff Reporter*, Vol. 55, pp. 39-42, Aug. 29, 1966.
25. "An Industrial Waste Guide to the Cotton Textile Industry", U.S. Department of Health, Education, and Welfare, Public Health Service Publication No. 677, 1959.
26. Masselli, J. W., and G. Burford, "Pollution Reduction Program for the Textile Industry", *Sewage and Industrial Wastes*, Vol. 28, No. 10, pgs. 1273-1290, Oct. 1956.

27 MINIMIZATION OF HAZARDOUS WASTE GENERATION: PRELIMINARY INVESTIGATION OF WASTE MANAGEMENT PROPERTIES OF A SOLVENT SUBSTITUTE FOR TCE

Ricardo B. Jacquez, Professor
Meei-Huey Li, Graduate Research Assistant
Department of Civil, Agricultural, and Geological Engineering
Walter H. Zachritz II, Program Manager
Southwest Technology Development Institute
New Mexico State University
Las Cruces, New Mexico 88003-0083
Martha I. Beach, President
N-CON Systems Company, Inc.
Larchmont, New York 10538

INTRODUCTION

Halogenated solvents such as trichlorethylene (TCE), freon, and methylene chloride, are routinely used in a wide variety of degreasing and cleaning operations.[1] These highly volatile solvents remove unwanted oils, greases, metal residues, and other organic films leaving a clean dry surface. Federal regulations promulgated over the past ten years have classified many of these substances as hazardous (Resource Conservation and Recovery Act – RCRA) or toxic (Toxic Substance Control Act – TSCA). Halogenated solvents release vapors[2] that accumulate in the work place and produce acute and sometimes chronic health effects resulting from continuous exposure through inhalation.[3] Spent halogenated solvents released from process washwater or in concentrate can adversely impact industrial wastewater treatment plants, industrial pretreatment programs and may pose a threat to ground water.[4] Various strategies for minimizing the use and disposal costs of halogenated solvents have been suggested, including: use of solvent substitutes; recycling through filtration and distillation; or sophisticated vapor containment and recovery systems.[1]

Non-halogenated solvent substitutes with positive attributes that pose no exposure related health threats to workers, reduce disposal and treatment costs, eliminate the need for costly equipment to capture volatile organic compounds (VOC's), and result in the hazardous declassification of some mixed wastewaters and sludges are being strongly considered for many industrial operations. However, solvent substitutes may suffer from the following disadvantages: increased washwater use due to decreased rinseability; increased solvent use due to the lower solvent capacity; contaminant carryover; unknown impact on the industrial wastewater treatment plant such as resistance to biodegradation; and physical/chemical interaction with waste components. Viable substitutes must also meet the rigorous manufacturing standards for cleaning effectiveness, material compatibility, component aging and stress, and cost per solvent capacity, while minimizing environmental impacts.

Several types of solvent substitutes which are now available on the market include: terpene-based materials containing d-limonene; aqueous-based materials containing monoethanol amines; alkaline aqueous-based material; and hydrochlorofluorocarbon formulation. Preliminary investigations have indicated that terpene-based formulations appear to be effective for many industrial cleaning operations. Thus, a typical terpene-based formulation was selected for this study. Terpenes are widely distributed in nature and occur in nearly all living plants. The two most abundant natural sources of terpenes are turpentine and other essential oils.[5] Terpenes are generally regarded as a derivative of isoprene (2-methyl-1, 3-butadiene)[6]. Wood and citrus products are the principal commercial sources of terpenes. Solvent substitute applications for terpenes have grown substantially in recent years[5].

Information obtained from the manufacturer's material safety data sheets indicated that the major component of the terpene-based solvent selected for the study is d-limonene (Pentone Corporation,

45th Purdue Industrial Waste Conference Proceedings, © 1991 Lewis Publishers, Inc., Chelsea, Michigan 48118. Printed in U.S.A.

Table I. Composition of the Peptone Synthetic Feed

Compound	Concentration mg/L
Peptone	855
$MnSO_4H_2O$	15
$MgSO_47H_2O$	150
$CaCl_2$	11.25
$FeCl_36H_2O$	0.75
KH_2PO_4 & K_2HPO_4	0.5 mole of each, dissolve in 1l of water, 15 mL/L
$(NH_4)_2SO_4$	750
Tap water	to volume

Tenafly, NJ). Additional components include: ethanolamine, glycol ether, EDTA, surfactants, and water. The pH of the concentrated form was reported to be alkaline (10.4) and the Biochemical Oxygen Demand (BOD) and Chemical Oxygen Demand (COD) concentrations were reported as 250,000–400,000 mg/L and 1,500,000–3,700,000 mg/L, respectively[7]. The flash point and freezing point were reported to be 165°F (concentrate), 175°F (1:3 dilution), and 32°F, respectively. The solvent substitute forms a milky emulsion in water and the rinseability is considered excellent with water. The biodegradability of this terpene-based solvent is reported as complete by the manufacturer. The manufacturer recommends disposal of the spent solvent by simple flushing to the sewer after soil release.

The objective of this study was to determine preliminary waste characteristics for the selected terpene-based solvent substitute. These characteristics include: volatilization properties; biodegradability properties; and toxic inhibition properties.

MATERIALS AND METHODS

Characterization of Factors Affecting Volatilization

Natural terpenes are highly volatile and are found in nearly all living plants. Plants release hundreds of million tons of terpenes into the air each year[8]. The potential volatile nature of the solvent can impact waste characteristics and biodegradability determinations. The following experiments were designed to characterize the basic volatilization properties of the terpene-based solvent as well as evaluate the influence of pH and the presence of suspended and dissolved organic material on volatilization.

Effect of pH

The objective of this experiment was to determine the volatilization rate of the terpene-based solvent under natural and adjusted pH conditions. The COD of the undiluted solvent was determined to be 1,500,000 mg/L. Based on the manufacturers' recommended use rates and expected process dilution, samples were made by diluting the solvent 1:2000 with distilled water. A 500 mL portion of this solution was placed in a 600 ml beaker and was stirred continuously for 10 hours with a 6.0 cm magnetic bar at 300 rpm. The rate of volatilization was determined by monitoring total COD over the test period. The COD was measured in accordance with *Standard Methods* procedure 508C (APHA, 1985). The pH was adjusted to range from 2 to 12 using either 2N H_2SO_4 or NaOH. Baseline volatilization data was determined using a control sample at the natural pH (9.2). Three additional samples were prepared and pH adjusted to compare to this control. The samples included: diluted solvent alone, mixture of peptone combined with diluted solvent, and primary influent combined with diluted solvent. The peptone solution (Table I) and diluted solvent were mixed by proportioning COD 1:1 on a concentration basis. The primary influent and diluted solvent were mixed by proportioning COD 1:2,5, respectively, on a concentration basis. The COD of the undiluted peptone and primary influent were 850 mg/L and 450 mg/L, respectively. Sewage samples for all tests were obtained from the local wastewater treatment plant.

Effect of Sewage Solids

The objective of this experiment was to determine the effect of dissolved and suspended organic solids on the volatilization rate of the solvent substitute.

Table II. Composition of Synthetic Feed for Terpene-Based Solvent Acclimated Seed

Compound	Concentration mg/L
Terpene-Based Solvent	855 (as COD) 570 ppm (as solvent)
$MnSO_4H_2O$	15
$MgSO_47H_2O$	150
$CaCl_2$	11.25
$FeCl_36H_2O$	0.75
KH_2PO_4 & K_2HPO_4	0.5 mole of each, dissolve in 1 l of water, 15 mL/L
$(NH_4)_2SO_4$	750
Tap Water	to volume

Effect of Dissolved Organic Solids. In this experiment, three sample mixtures were used to determine the effect of the interaction between the diluted solvent and dissolved organic solids on the rate of volatilization. The three test samples included: diluted solvent alone; diluted solvent mixed with peptone (Table 1); and diluted solvent mixed with filtered primary with peptone (Table 1); and diluted solvent mixed with filtered primary influent. The tests were monitored at the natural pH of the mixtures. The primary influent was prepared by filtering the sample in accordance with *Standard Methods* procedure 209C (APHA 1985) to reduce suspended solids to less than 5 mg/L. The initial COD for the diluted solvent was 650 mg/L. The COD of the solvent-peptone mixture and the solvent-primary influent mixture were 927 and 828 mg/L, respectively. The experimental procedures previously described were used to determine volatilization rates.

Effect of Suspended Solids. In this experiment, the effect of high and low levels of suspended solids on the volatilization rate of the solvent was evaluated. Diluted solvent was used as a control. To measure the effect of high suspended solids, diluted solvent was mixed with primary influent and primary effluent. The suspended solids concentrations for the primary influent and primary effluent mixtures were 182 mg/L and 47 mg/L, respectively. To measure the effect of a low suspended solids concentration, the diluted solvent was mixed with filtered primary influent and secondary effluent. The final suspended solids concentrations for both mixtures were 3 mg/L. The influence of suspended solids on the rate of volatilization was determined by monitoring soluble COD. Samples for soluble COD analysis were prepared by filtering in accordance with *Standard Methods* procedure 209C.[9] The COD of the solvent-primary influent mixture and the solvent-primary effluent mixture were 846 and 866 mg/L, respectively. The COD of the solvent-filtered primary influent mixture and the solvent-secondary effluent mixture were 828 and 778 mg/L, respectively.

Characterization of Biodegradability

The biodegradability of the terpene-based solvent was evaluated by making respirometric measurements with microbial seeds acclimated to a acclimated heterogenous microbial seeds were prepared for the biodegradability tests. The first seed was obtained from the local wastewater treatment plant and was acclimated to peptone. This seed was maintained at a sludge age of 10 days, in a 10 l reactor mixed with air and received a synthetic feed (Table I) which contained peptone as the sole carbon source. The culture acclimated to the solvent was developed from the peptone culture and received a daily feed (Table II) in which the terpene-based solvent was the sole carbon source. The two cultures received the same amount of carbon source (725 mg as COD) each day.

Biodegradability by Respirometric Measurements

The objective of this experiment was to evaluate the biodegradability of the terpene-based solvent by using an automated respirometric technique to determine BOD and Oxygen Uptake Rates (OUR). The instrument used was an eight station COMPUT-OX respirometer (N-CON Systems Company, Inc., Larchmont, NY). The operating principles for this instrument have been previously described.[10] The cultures used for these tests included: solvent acclimated seed; peptone acclimated seed; and activated sludge obtained from the local wastewater treatment plant. For each test series, the synthetic feed (Table II) was diluted to a concentration range 1000 to 0 ppm in proportion to the carbon source. In order to minimize the loss of solvent in the head space during mixing, 500 mL of each solution was placed in the 690 mL respirometer bottle. Each reactor was seeded to a concentration of 10 mg/L with the test culture and the contents were stirred with a 6.0 cm magnetic stir bar to provide adequate mixing.

Evaluation of Monod Kinetic Coefficients by Batch Growth Methods

The objective of this experiment was to determine the specific growth rate (μ) and yield coefficient (Y) for the terpene-based solvent by using conventional batch growth methods. The synthetic feed (Table 2) was diluted between 400 to 25 ppm in proportion to the carbon source. A 500 mL portion of each substrate concentration was placed in a 600 mL beaker and was mixed with a 6.0 cm magnetic stirring bar at 300 rpm. Each dilution was seeded with biomass acclimated to the solvent to a concentration of 10 mg/L. Degradation activity was evaluated by monitoring soluble COD and turbidity (Ratio Turbidimeter Model 18900, Hach Company, Inc.). Turbidity readings were converted to biomass concentration using a calibration curve. For the 400 ppm substrate concentration, biomass growth was monitored by direct measurement of suspended solids in accordance with *Standard Methods* procedure 209C.[9]

Evaluation of Monod Kinetic Coefficients by Respirometric Measurements

The objective of this experiment was to determine the Monod kinetic coefficients for the culture acclimated to the solvent substitute using respirometric techniques as described by Gaudy et al.[11] The synthetic feed (Table II) was diluted between 400 to 25 ppm in proportion to the carbon source. The results were used to determine the maximum specific growth rate (μ_{max}) and the substrate saturation constant (K_S). The respirometer and experimental protocol used in this experiment are described in the respirometer section presented above.

Characterization of Toxicity

The objective of this experiment was to evaluate the toxicity of the terpene-based solvent using the commercially available test culture POLYTOX, (Polybac Corporation, Bethlehem, PA). POLYTOX is reported by the manufacturer to be a freeze dried culture composed of 12 strains of common sewage bacteria and is recommended as an inexpensive tool to test for effluent toxicity. The seed was prepared by dispersing one vial of POLYTOX in 250 ml of distilled water, mixing for 30 minutes, and settling for 5 minutes. The synthetic feed (Table II) was diluted between 400 to 25 ppm in proportion to the carbon source. Each reactor was seeded with 50 mL of supernatant from the POLYTOX mixture. Toxicity was evaluated by measuring OUR through respirometric technique and determining the Effective Concentration (EC_{50}). EC_{50} was defined as the concentration required to induce a 50% reduction in OUR. The respirometer and the experimental protocol used in this test are described in the respirometer section presented above.

RESULTS AND DISCUSSION

Characterization of Factors Affecting Volatilization

Effect of pH

The results of evaluating the effect of pH on volatilization of the terpene-based, solvent mixtures are shown in Figures 1–3. In each case, the decrease in COD appeared to follow first-order, exponential decay. The residual COD level for the mixtures was reached at a time ranging from 6.5 to 9.0 hours. The COD reductions at these times were 46, 20, and 41% for the solvent, solvent/peptone, and solvent/primary influent mixtures, respectively. Over a pH range 2–12, the loss of COD in each mixture followed a similar, consistent pattern, but pH did not appear to affect the volatilization rate of the solvent. These results indicate that the terpene-based solvent formulation is stable over a wide pH range. Thus, the effect of pH on volatilization would be predicted to be minimal for many industrial applications. However, as the COD data indicates, the loss of unknown product component(s) was significant. Where capture of volatile organic compounds (VOC's) is a requirement, the need for appropriate control technology must be considered. In addition, accumulation of substances such as d-limonene, may present a workplace health hazard.

Effect of Sewage Solids

The sewage solids tested in this experiment were classified into two groups: dissolved and suspended organic solids. The results showing the effect of dissolved organics on solvent volatilization are shown in Figure 4. The corresponding volatilization rates are summarized in Table III. The normalized COD

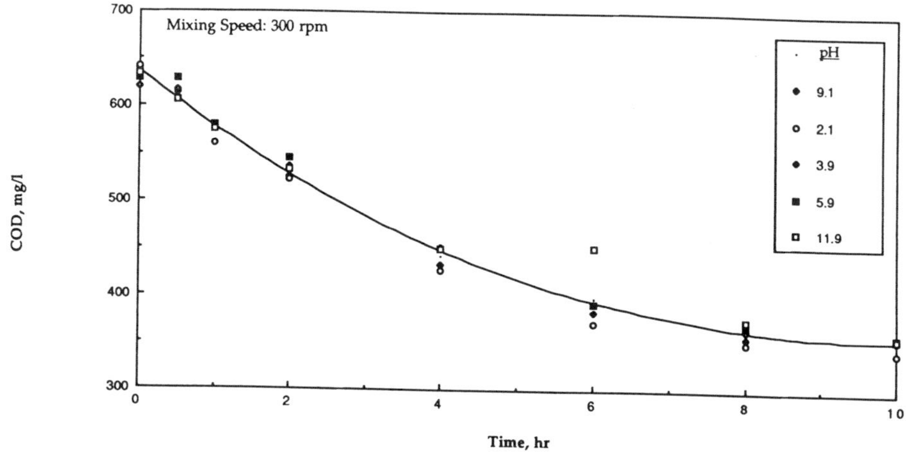

Figure 1. Effect of pH on volatilization of the terpene-based solvent mixture.

data and the rates indicate that the organics in filtered primary influent did not increase the degree of volatilization. Therefore, it is reasonable to speculate that dissolved organics derived from domestic sewage, will not interact with the volatile components of a terpene-based solvent. However, the organics in peptone reduced COD volatilization by approximately 50%. This difference is further supported by comparing the rates of COD reduction (Table 3). The volatilization rate for the solvent/peptone mixture was 27% less than the rate determined for the solvent mixture. Peptone, a complex mixture of simple organic compounds derived from beef extract, is readily biodegradable, but does not volatilize to any degree. It is reasonable to assume that an active ingredient in the solvent formulation interacted with a component(s) in the peptone and reduced volatilization. Based on these results it is possible to speculate that in more complex industrial wastewaters, these interactions may be beneficial or detrimental. Reducing loss of product would extent product life, while stripping of active ingredients may reduce the solvent's capacity to function as an effective cleaning agent. In addition component interaction may result in the formation of more complex organics.

The results measuring the effect of domestic sewage suspended solids on volatilization rates are summarized in Table III. At the suspended solids concentrations tested, no difference in volatilization rates were measured. Therefore, it appears that suspended solids did not influence the volatilization rate of the terpene-based solvent. However, a possible correlation between suspended solids concen-

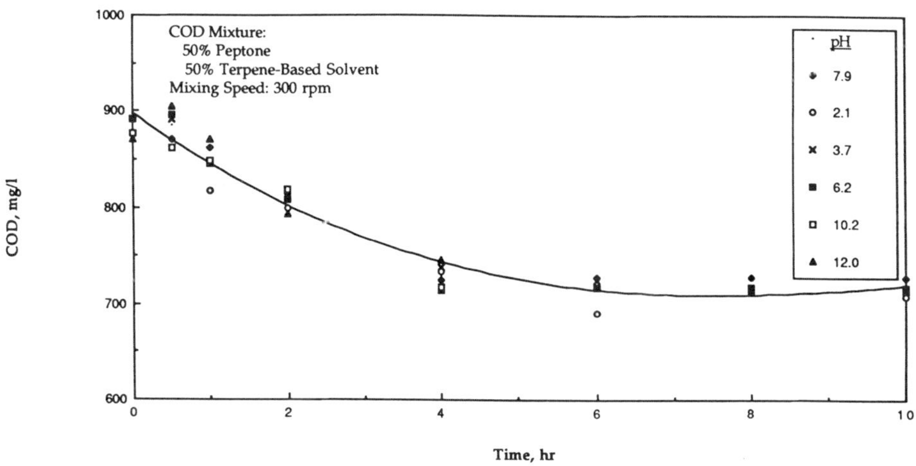

Figure 2. Effect of pH volatilization of the terpene-based solvent and peptone mixture.

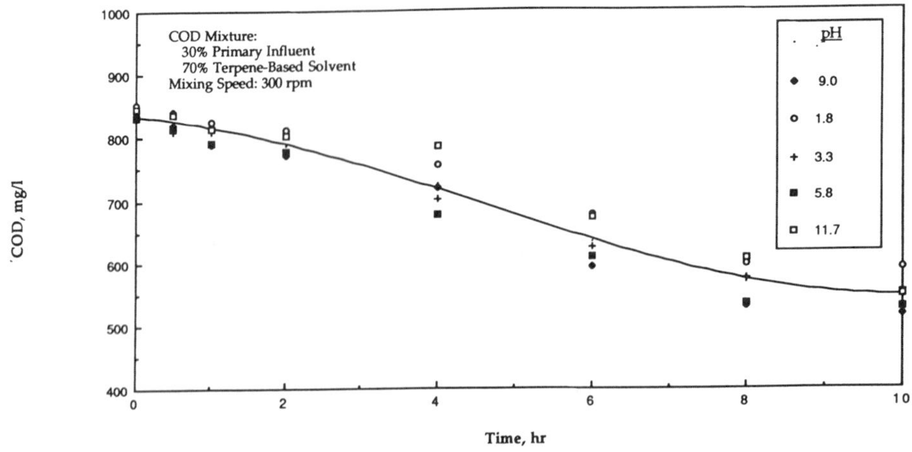

Figure 3. Effect of pH on volatilization of the terpene-based solvent and primary influent mixture.

trations and the rate of volatilization is suggested. The conditions tested suggest that increasing suspended solids decreased the rate of volatilization. Possible interactions which may explain this observation include two mechanisms; adsorption and chemical reaction between the suspended solids and solvent components. Further testing at higher suspended solids concentrations is needed to verify this possible correlation.

The volatile nature of this material suggests several practical implications. If solvent removal from the waste stream is required, physical stripping by conventional methods may provide an effective waste management strategy. However, additional treatment, such as adsorption, would be required to prevent the release of volatile components to the atmosphere. When mixed with a wastewater containing high suspended solids, the oily nature of the terpene-based solvent may prove to be beneficial. The natural oils may increase solids removal through physical adsorption and flotation.

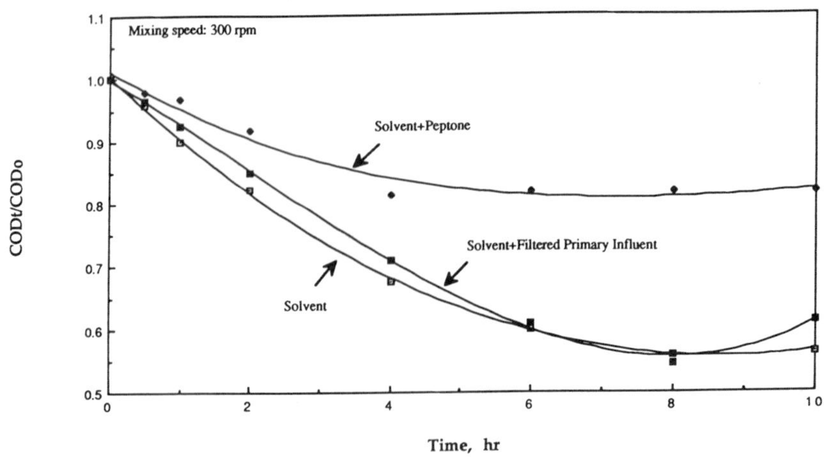

Figure 4. Effect of dissolved organic solids on volatilization of the terpene-based solvent.

Table III. Volatilization Rates for Various Test Mixtures of the Terpene-Based Solvent Substitute

Mixture	Volatilization Rate, hr^{-1}	Coefficient of Determination, r^2
Terpene-Based Solvent	0.062	0.92
Terpene-Based Solvent + Peptone	0.046	0.96
Terpene-Based Solvent + Primary Influent (SS 182 mg/l)	0.055	0.96
Terpene-Based Solvent + Primary Effluent (SS 47 mg/l)	0.060	0.94
Terpene-Based Solvent + Filtered Primary Influent (SS 3 mg/l)	0.064	0.88
Terpene-Based Solvent + Secondary Effluent (SS 3 mg/l)	0.064	0.94

Characterization of Biodegradability

Evaluation of Monod Kinetic Coefficients by Batch Growth Methods

The objective of this experiment was to determine the maximum specific growth rate (μ_{max}) and yield coefficient (Y) for an acclimated culture using the terpene-based solvent as the sole carbon source. The results of these tests were difficult to replicate and interpret. Soluble COD and turbidity measurements did not follow a pattern typical of a highly biodegradable substrate. At all substrate concentrations, the milky emulsion formed by mixing the solvent with water interfered with turbidity measurements and the specific growth rate (μ) and Y were impossible to calculate. At a substrate concentration of 400 ppm, suspended solids were used to monitor the growth of the culture. From these measurements, μ and Y were estimated to be 0.030 and 0.4, respectively. It is commonly reported that for a predominantly carbohydrate carbon source in a natural wastewater, Y ranges from 0.3 to 0.7 with 0.5 as a typical average.[12] Since Y is not substrate dependent and the measured value falls within the range reported above, the experimental value determined by this test is considered reasonable.

Biodegradability by Respirometric Measurements

The results of the first series of experiments clearly demonstrated that the terpene-based solvent was highly volatile. The problems associated with the batch growth study could be attributed in part to the volatile nature of the substrate. Significant loss of substrate through volatilization would minimize the degree of biodegradation measured and therefore the validity of the derived kinetic coefficients. The advantage of using a pressurized respirometer when assessing biodegradabilty of VOC's is that the closed headspace prevents loss of potentially degradable substrate.[13]

The results of the biodegradation studies for the three cultures are shown in Figures 5–7. For concentrations exceeding 400 ppm, the respiration curves show that oxygen uptake was greater than the maximum oxygen transfer capabilities of the instrument. This characteristic has been previously reported for this instrument.[14] Comparison of the three figures indicates that initiation of respiration occurred at 3, 9, and 18 hours for the cultures acclimated to the solvent, peptone, and domestic wastewater, respectively. This increased lag period is a reflection of the specificity of the culture population developed through the acclimation process. For the culture acclimated to peptone and domestic wastewater, the 1000 ppm substrate concentration reflected a shock load response. The acclimated culture was not affected by this same high substrate concentration.

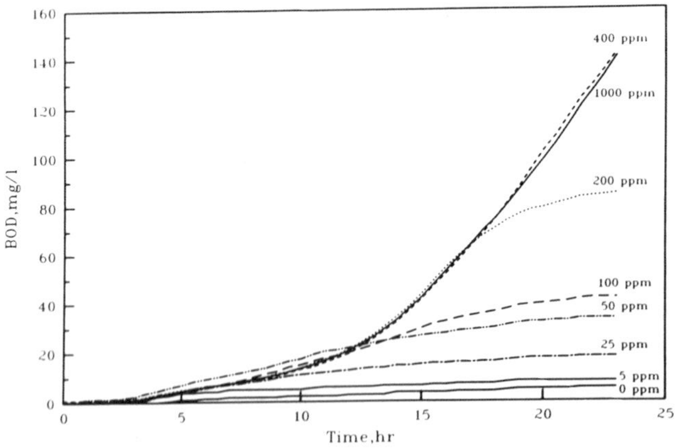

Figure 5. Respirometric measurements for the terpene-based solvent using an acclimated culture.

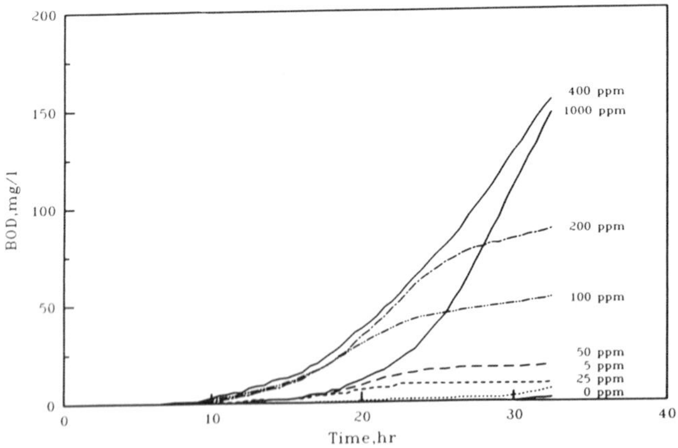

Figure 6. Respirometric measurements for a terpene-based solvent using a culture acclimated to peptone.

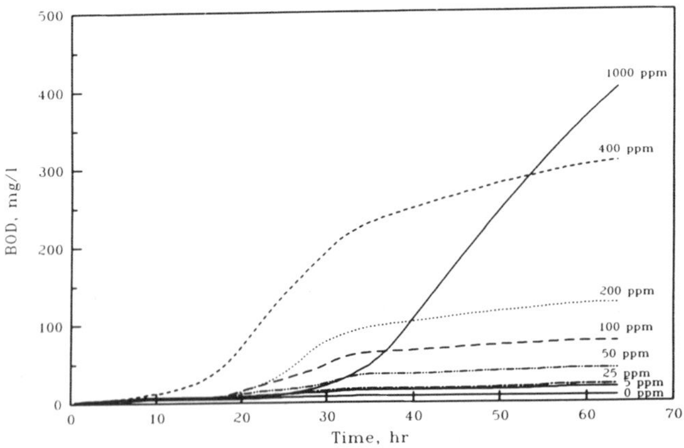

Figure 7. Respirometric measurements for a terpene-based solvent using a culture acclimated to domestic wastewater.

Table IV. Specific Growth Rates for the Culture Acclimated to the Terpene-Based Solvent Determined Through Respirometric Measurements

Concentration (ppm)	Specific Growth Rate (μ)	Coefficient of Determination (r^2)
400	0.048	0.98
350	0.042	0.98
300	0.041	0.98
250	0.040	0.98
200	0.038	0.97
100	0.030	0.92
50	0.031	0.93
25	0.021	0.92

In each case, biodegradation was apparent because the oxygen uptake was basically proportional to the solvent concentration. The following analysis is based on the BOD values recorded approximately 24 hours after initiation of respiration for the solvent concentrations less than 200 ppm. In general, the BOD measured was approximately 50% of the substrate concentration. Based on this measurement and the COD concentration of the solvent, it is estimated that the BOD was approximately 35% of the COD concentration. This ratio compares favorably with the well accepted ratio known for a highly degradable waste such as domestic sewage. The BOD_5/COD ratio for domestic sewage is generally reported as approximately 0.45. Combined with the respirometric data (Figures 5–7), this analysis indicates that the terpene-based solvent substitute is relatively biodegradable.

Evaluation of Monod Kinetic Coefficients by Respirometric Measurements

Evaluation of Monod kinetic coefficients using respirometric measurements is a recent but well established procedure.[11,13] By this technique, oxygen uptake data is translated directly to biomass growth. Biomass growth is then used to determine the Monod kinetic coefficients (μ_{max} and K_S). In order to evaluate these coefficients, three parameters must be measured. These parameters include: initial biomass concentration (X_o), Y, and the biomass COD equivalent (O_x, mg COD/mg biomass). From the batch growth study, Y was estimated to be 0.4. O_x was determined to be 1.45 for the solvent acclimated culture. The specific growth rates determined for this experiment are summarized in Table IV. From this data, μ_{max} was determined to range between 0.045 to 0.050 hr^{-1} and K_s ranged between 45 and 70 mg/L as COD. The magnitude for these parameters is comparable to estimates determined by Dang, et al.[13] using respirometric technique and chlorobenzene as the substrate. The mean values for μ_{max}, K_s, and Y were reported to be 0.060, 3.52, and 0.42, respectively.

Compared to the results obtained from the batch growth study, it is clear that measuring kinetic parameters for a volatile and colored substrate through respirometric technique is more reliable. The batch growth method was not suitable because of the volatile nature of the solvent. During mixing, solvent was lost to the atmosphere thereby reducing available substrate. Under these conditions, measurement of biomass growth was scattered and substrate concentration could not be correlated to specific growth rate. In addition, the color (milky white) associated with the solvent-water mixture interfered with turbidity measurements which were used to monitor the growth of the cultures. Under these conditions, measurements of biomass growth was scattered and substrate concentration could not be correlated to specific growth rate. In addition, the color (milky white) associated with the solvent-water mixture interfered with turbidity measurements which were used to monitor the growth of the cultures. The color further interfered with growth measurements because the milky appearance decreased in proportion to volatilization of the solvent. This interference resulted in measurement of high initial turbidity readings which then decreased due to solvent volatilization. Although monitoring of biomass growth was improved by direct measurement of suspended solids this modification did not offset the lack of correlation between concentration of the volatile substrate and specific growth rate. The respirometer has the capability to offset these factors. In a respirometer flask a volatile substrate will partition between the gas and liquid phases. Then, as biodegradation occurs and the compound in the liquid is consumed, the portion in the gas phase will be transferred back into the liquid phase for decomposition by the culture. The oxygen uptake measurements are then translated directly into biomass growth as described by Gaudy, et al[11] and Dang, et al.[13] This feature makes the respirometer a useful instrument for assessing the characteristics of a volatile and/or turbid/colored waste component such as a terpene-based solvent substitute.

Characterization of Toxicity

The last experiment in this study evaluated the potential toxicity of the terpene-based solvent by using POLYTOX as the seed for respirometric measurements. The results presented in Figure 8 show that the terpene-based solvent did exert temporary inhibition on POLYTOX. First, the data indicates that BOD was exerted in inverse proportion to the solvent concentration. Secondly, the BOD curves can be grouped into two categories according to OUR and solvent concentration. While a solvent concentration range of 5 to 50 ppm produced OUR's similar to the control (0 ppm), the concentration range 100 to 1000 ppm temporarily inhibited the respiration of POLYTOX. Thirdly, it is evident that the bacterial species present in POLYTOX were able to quickly acclimate to the higher solvent concentrations.

Since the OUR's for the second group (100–1000 ppm) were lower than the first group (0–50 ppm) it is possible to use the concept of EC_{50} to evaluate the toxicity of the solvent. Using data for the second group EC_{50} values were determined for the time periods 4–8 hours, 9–12.5 hours, and beyond 15 hours. This grouping of data was based on similarity of the OUR during the specified time periods. For the time periods 4–8 hours and 9–12.5 hours the EC_{50} values for the solvent substitute were 350 and 100 ppm, respectively. For the time period beyond 15 hours the EC_{50} evaluation showed that POLYTOX was not inhibited by the solvent. The results of the EC_{50} evaluation are explained by the following analysis. During the first eight hours of respiration the readily degradable substrate, made available by the bulk, carrier agent (bran), was preferentially consumed by POLYTOX. During the 9–12.5 hour period the POLYTOX culture was forced to acclimate to the solvent substrate, therefore the respiration rate decreased. For these reasons the EC_{50} was higher during the first period of the text. Beyond 15 hours, the cultures were acclimated to the solvent components and respiration reached the maximum oxygen transfer capabilities of the instrument. This analysis clearly indicates that the inhibition exerted by the terpene-based solvent to POLYTOX was only temporary.

The three respiration studies have clearly demonstrated that the solvent is readily degraded by a variety of bacterial cultures. Therefore, at the concentrations tested, the terpene-based solvent is not expected to pose a threat to either an industrial treatment plant or a Publicly Owned Treatment Works (POTW) which utilizes biological processes in the treatment scheme. At a maximum, the treatment plant might experience a short lived acclimation period which may extend 12–24 hours. The minimal impact predicted by this study should be implemented with caution because carryover of known hazardous materials (heavy metals, toxic organics, etc.) resulting from the use of the solvent substitute have not been evaluated.

CONCLUSIONS

1. The terpene-based solvent is highly volatile in a dilute state. Volatilization, measured by COD reduction, appeared to follow first-order, exponential decay. COD reduction by volatilization could exceed 50% depending on the method of agitation.
2. Variation of pH did not affect volatilization of the terpene-based solvent.
3. The degree of physical/chemical interaction by the components of the terpene-based solvent is dependent on the characteristics of the waste source. Dissolved organics derived from domestic sewage did not interact with the volatile components of the terpene-based solvent. However, the

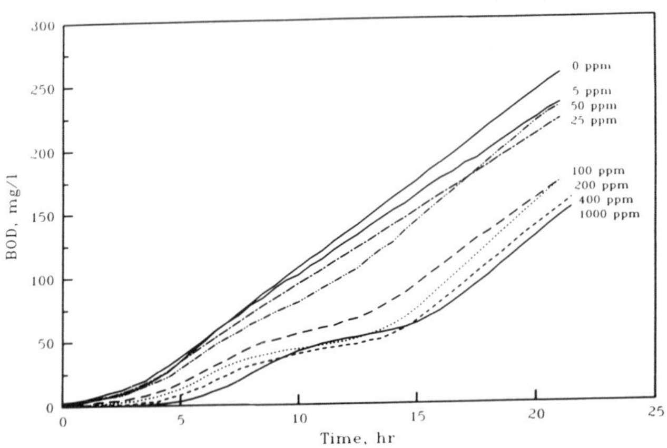

Figure 8. Respirometric measurements for a terpene-based solvent using POLYTOX.

soluble organics in peptone reduced COD volatilization by about 50%. Up to a concentration of 180 mg/L, suspended solids from domestic sewage did not appear to influence the volatilization rate of the terpene-based solvent.

4. The respiration studies demonstrated that the terpene-based solvent is biodegradable by a variety of bacterial cultures.
5. The toxicity impacts of the terpene-based solvent are temporary. Satisfactory acclimation can be achieved in a time period less than 24 hours.

REFERENCES

1. Higgins, J. (1989) *Handbook of Hazardous Waste Minimization*, CRC Press Inc., Boca Raton, Florida.
2. Weiss, G. (1980) *Hazardous Chemicals Data Book*, Noyes Data Corporation, Park Ridge, New Jersey.
3. Astrand, I., and Gamberale F. (1978) "Effects on Humans of Solvents in the Respiratory Air: a Method for Estimation of Uptake." *Environmental Research* 15(2), 1-4.
4. Breton, M., Frillici, P., Palmer, S., Spears, C., Arienti, M., Kravett, M., Shayer, a., and Suprenant, N. (1988) *Treatment Technologies for Solvent Containing Wastes*, Noyes Data Corporation, Park Ridge, New Jersey.
5. Hayes, M. E., Hood, C. C., Miller, R. E., and Sharpe, R. (1988) U.S. Patent 4,740,247.
6. Salisbury, F. B., and Ross, C. W. (1978) *Plant Physiology*, 2nd edition, Wadsworth Publishing Company, Inc., Belmont, California.
7. Comacho, R. (1989) Personal Communication, Industrial Waste Specialists, City of Austin, Texas, August.
8. Yokouchi, Y., and Ambre, Y. (1985) "Aerosols Formed From the Chemical Reaction of Monoterpenes and Ozone." *Atmospheric Environment*, 19(8), 1271-1276.
9. American Public Health Association. (1985) *Standard Methods for the Examination of Water and Wastewater*, 16th edition., Washington, D.C.
10. Cadena, F., Drohobyczer, A., Beach, M. I., and Barnes, D., (1988) "A Novel Approach to Simplified Respirometric Oxygen Demand Determinations." *Proceedings*, 43rd Annual Industrial Waste Conference, Lewis Publishers, 459-467.
11. Gaudy, A. F., Jr., Rozich, A. F., Garniewske, S., Moran, N. R., and Dkambaram. (1987) "Methodology for Utilizing Respirometric Data to Assess Biodegradation Kinetic." *Proceedings*, 42nd Annual Industrial Waste Conference, Lewis Publishers, 573-584.
12. Gaudy, A. F., Jr., and Gaudy, E. T. (1988) *Elements of Bioenvironmental Engineering*, Engineering Press, Inc., San Jose, California.
13. Dang, J. S., Harvey D. M., Jobbagy A. and Grady C. P. L. Jr. (1989) "Evaluation of biodegradation kinetics with respirometric data." Res. J. Wat. Pollut. Control Fed. 61, 1711-1721.
14. Jacquez, R. B., Cadena, F., Prabhakar, S., and Beach, M. I. (1989) "Gas Transfer Limitations in Environmental Respirometry." *Proceedings*, 44th Annual Industrial Waste Conference, Lewis Publishers, 425-433.

Section Four
BIOLOGICAL SYSTEMS – A. GENERAL

28 THE EFFECTS OF FREEZE-DRYING AND STORAGE ON A PHENOL DEGRADING BACTERIAL SUPPLEMENT

Roberta L. Gaiek, Engineer
Malcolm Pirnie, Inc.
Buffalo, New York 14219

Clifford R. Lange, Graduate Student
A. Scott Weber, Associate Professor
Department of Civil Engineering
State University of New York at Buffalo
Buffalo, New York 14260

INTRODUCTION

Commercial bacterial supplements have been promoted as an effective and economical way to enhance biological treatment processes. Addition of these specialized bacteria to the naturally occurring population of a bioreactor for enhancement of population diversity and/or activity is known as bioaugmentation. Bioaugmentation can be effective in increasing biodegradation by 1) adding bacteria with enzymatic systems capable of biodegrading previously persistent organics, and/or 2) adding bacteria with higher metabolic rates. In particular, bacterial supplements are often promoted as being capable of rapidly initiating biodegradation of many xenobiotic compounds including aromatics (such as phenol and chlorinated phenols). However, past studies have shown that lengthy acclimation periods may be required from these specialized bacteria prior to target compound biodegradation[1,2,3].

Typically, commercial supplements are freeze-dried to allow for ease of storage and application. Unfortunately, this preparation procedure often results in viability losses and reductions of metabolic activity upon supplement reconstitution which may lead to poor supplement performance. The purpose of this paper is to present and discuss the results of a laboratory investigation conducted to investigate the effects of freeze-drying and storage on bacterial supplement performance.

BACKGROUND

Factors identified as important for the successful employment of commercial bacterial supplements are discussed in this section. Lange[2] has outlined these parameters as including: supplement viability, presence and retention of necessary genetic coding, and activation of inducible enzyme systems (acclimation). The effect of these factors on supplement biodegradation potential is discussed below.

Supplemental Viability

To achieve enhanced biodegradation through supplementation, the biological treatment system must first be inoculated with the desired organisms. Subsequently, the bacterial supplement and/or their genetic coding for target compound biodegradation must be retained in the system. Therefore, supplement viability plays an important role in bacterial supplement performance. If supplement viability is lost at any time during the bioaugmentation process, no enhancement of biodegradation will occur.

There are two major factors which can influence the presence of viable supplement bacteria in the biological system. The first factor is the number of viable organisms delivered into the process being augmented. This initial viability of supplement bacteria is influenced by a number of factors including the freeze-drying and storage process. The second key factor influencing the number of viable supple-

45th Purdue Industrial Waste Conference Proceedings, © 1991 Lewis Publishers, Inc., Chelsea, Michigan 48118. Printed in U.S.A.

ment bacteria is the growth and proliferation of the desired organism once it has been introduced into the system.

Activation of Inducible Enzyme Systems

For a supplement to increase biodegradation of xenobiotic compounds it must have high levels of catabolic enzymes or be capable of rapidly inducing these enzymes upon introduction to the reactor. Ideally, the inducible enzyme systems are present in the supplement prior to freeze-drying and storage and are subsequently maintained throughout. This, however, is often not the case and an acclimation period is required before catabolic enzymes are induced. Lange[3], observed higher COD removal efficiency and higher oxygen uptake rates for waste acclimated commercial supplements than for non-acclimated supplements. It was hypothesized, based on the results of this study, that loss of enzymatic capability during freeze-drying and storage was the primary difference between acclimated and non-acclimated supplements.

Presence and Retention of Necessary Genetic Coding

Generally, the enzyme systems capable of xenobiotic chemical biodegradation are coded for on plasmids (extra-nuclear DNA) rather than in the nucleus itself[4]. Plasmids can replicate independently of the cell. That is, one cell can contain multiple copies of a plasmid. Consequently, the larger quantities of genetic coding may allow for amplified expression of enzyme systems. In addition, plasmids can distribute genetic information to indigenous populations through transposition and conjugation, thus leading to an improvement in xenobiotic compound biodegradation[5]. However, plasmid free strains of bacteria cannot synthesize enzyme systems capable of catabolizing such compounds. Therefore, supplements lacking the necessary catabolic plasmids cannot improve xenobiotic degradation. Consequently, plasmid loss occurring in a bacterial supplement will result in a reduction in the ability of the supplement to degrade xenobiotic compounds.

Plasmid loss during the preparation and employment of a supplement generally can result from: 1) plasmid destruction during storage and reconstitution, and/or 2) selection of organisms which lack plasmids. Nierman and Feldblyum[6], found that plasmids, such as those coding for xenobiotic biodegradation, often were lost in freeze-dried cultures of *B.subtilis* and *S.cerevisiae*, which were reconstituted after a nine month storage period. Although the authors had not concluded a direct cause for the plasmid loss, it is possible that the loss may be attributed to the attack of free radicals and other metabolic by-products which form and accumulate during freeze-drying and storage[7].

STUDY APPROACH

To assess the effects of freeze-drying and storage on supplement biodegradation a series of batch studies were conducted to assess each of the aforementioned factors. Phenol was used as the target compound. The lag period proceeding initiation and the rate and degree of phenol biodegradation were determined for a non-commercial supplement before and after freeze-drying and after various storage periods.

To determine if specific phenol degrading microorganisms were lost during freeze-drying and storage, differential plate counts were used to quantify groups of known phenol degraders. Additionally, the percent of phenol biodegrading microorganisms was determined using replicate plating on both rich media and media containing phenol as sole carbon and energy source.

To evaluate the supplement's ability to maintain inducible enzyme systems during the storage and reconstitution period, the cellular levels of catabolic enzymes required for catechol degradation were measured for the supplement during each batch study. Catechol degrading enzymes were chosen as the study enzyme because: 1) they are commonly studied inducible enzymes, 2) the meta pathway of catechol degradation is coded on plasmid DNA, and 3) catechol is a central intermediate in the bacterial metabolism of aromatic and hydroaromatic compounds[8].

Catechol can be degraded by two separate pathways (shown in Figure 1). The B-ketoadipate pathway (ortho pathway) employs the inducible enzyme catechol 1,2-oxygenase for catechol ring cleavage. In this pathway phenol is eventually transformed to the tricarboxylic acid cycle intermediates succinate and acetyl-CoA. The second pathway employed by bacteria to cleave catechol is the meta pathway which uses catechol 2,3-oxygenase to cleave the aromatic ring. This pathway gives rise to muconic semialdehydes which are eventually converted to pyruvate and acetaldehyde. The cellular levels of both enzymes were tracked during the batch studies.

Additionally, experiments were conducted to determine the presence of plasmid(s) containing the genetic coding necessary for target compound biodegradation. Presence of the plasmid(s) was

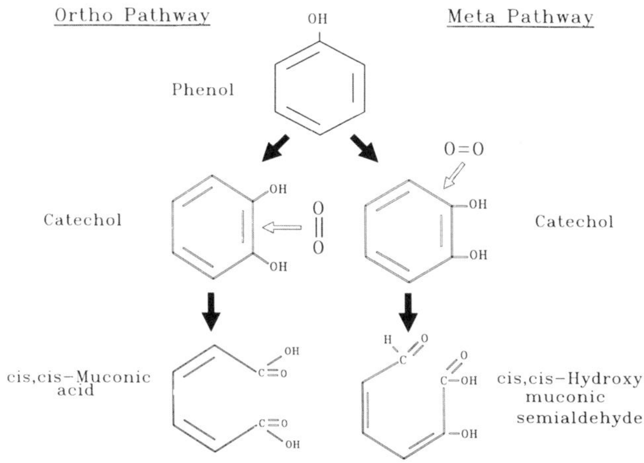

Figure 1. Pathways of bacterial phenol biodegradation.

detected using gel electrophoresis. These tests were performed on a strain of bacteria isolated from the supplement. Presence of the plasmid(s) was determined before and after freeze-drying and as a function of storage time.

STUDY METHODOLOGY

The purpose of this section is to describe the methods and materials used during this experimental investigation.

Supplement

A culture of bacteria capable of using phenol as a source of carbon and energy was developed and maintained at the University at Buffalo for use as a model supplement in this study.

Freeze-drying, Storage, and Reconstitution.

The supplement was centrifuged at 5000 rpm for 10 minutes, and washed one time with a phosphate buffer. Pellets of the supplement were transferred to 150 mL Labconco Freeze Dry Flasks. The freeze-drying apparatus employed consisted of a Labconco Drying Chamber, Model 75102, (able to accommodate up to 12 flasks) and a high vacuum pump (Edwards, Model EDM20) with a displacement of 20 m^3h^{-1}. The freeze-drying time for this experiment was three days. At the end of a freeze-drying run, samples were removed from the vacuum, quickly transferred to 125 mL Nalgene containers, and sealed. All freeze-dried supplements were stored at room temperature with minimal exposure to light.

The procedure for reconstitution of dried supplements was suspension of 0.2 grams of dry supplement in 100 mL of a sterile phosphate buffer. The resulting suspension was allowed to stand for approximately 15 minutes before testing.

Batch Degradation Studies.

Batch phenol degradation studies were conducted to assess the performance of the supplement before and after freeze-drying and as a function of storage time. The batch studies were conducted in glass 1-liter batch reactors. To initiate a study, 100 mL of the reconstituted supplement and 900 ml phenol substrate solution (see Table I) were added to the batch reactor. The initial phenol and MLVSS concentrations in the reactor were both 150 mg/L. Reactor contents were aerated with humidified diffused air as well as mixed with a gang magnetic stirrer (Wheaton Biostir 6, Model III). Parameters measured, as a function of time, during each experimental run included: reactor phenol concentration, ATP concentration, enzyme activity levels, and supplement viability and diversity.

Table I. Composition of Feed Solution Used in Batch Testing

Compound	Amount
Phenol	160 mg/L
K_2HPO_4	14 mg/L
KH_2PO_4	27 mg/L
$(NH_4)_2SO_4$	100 mg/L

Viability and Diversity Testing.

Viable cell counts were determined using heterotrophic plate counts. Heterotrophic plating as well as plating onto differential agars was performed as outlined in *Standard Methods*, Method 907A.[9] Presented in Table II are the differential agars which were employed to quantify groups of known phenol degraders during this study. All plates were incubated at 30°C for two days prior to counting.

Adenosine Triphosphate (ATP) analyses were conducted to assess the metabolic state of the supplement. ATP levels were determined using methods outlined by Turner Design.[10] ATP concentrations were assessed for the supplement using a Turner Fluorometer, Model 112. ATP levels are reported as ug ATP/mL.

Enzyme Assay.

Measurement of cellular levels of the inducible enzymes, catechol 1,2-oxygenase (ortho) and catechol 2,3-oxygenase (meta) were performed throughout the study. Twenty mL volumes were withdrawn from the bioreactor and washed 1 time with a 0.05M TRIS buffer (pH = 7.5). The suspension was again centrifuged and resuspended in 5 mL of TRIS buffer. The cells were disrupted using a sonic dismembrator (Fisher, Model 300) at high power for 10 minutes. The supernatant was separated from the residue by centrifugation at 8000 g for 10 minutes to obtain a cell-free extract. Measurement of enzyme activity was performed using methods outlined by Kojona, Y.,et al.,[11] for catechol 2,3-oxygenase and Hegeman, G. D., et al.,[12] for catechol 1,2 oxygenase.

Plasmid Detection.

Detection of plasmids was conducted on a strain of bacteria isolated from the supplement population which was capable of phenol degradation using the plasmid encoded meta cleavage pathway. The strain was assayed daily for activity of the meta enzyme. When significant levels of catechol 2,3-oxygenase were obtained, the presence of plasmids was confirmed using gel electrophoresis. The effect of freeze-drying on plasmid stability was assessed by comparing gels before and after freeze-drying.

Techniques employed for the extraction and electrophoresis of plasmids are outlined by Hackett et al.[13] A Davis System-gel unit (20 cm long gel) served as the gel electrophoresis tank. A 0.3% agarose gel with an applied voltage of 100 volts was used to resolve the DNA fragments. All gels were run for a period of 1 hour.

Sample Analyses.

Presented in Table III is a summary of the analytical tests and methods employed during the conduction of this study. Measurements of pH and temperature were performed using *Standard Methods*, method 423 and method 212, respectively. Total suspended solids (TSS) and volatile suspended solids (VSS) were determined using *Standard Methods*, methods 209 B and 209 C, respec-

Table II. Solid AGAR Medias Used for Population Study

Agar Medium	Type of Organisms Selected
Standard Methods agar	Heterotrophic Bacteria
Malt extract agar	Yeast
BBL Actinomyces agar	Actinomyces
Difco E.C. medium	Coliforms
Difco MacConkey agar	Enterics
Difco Pseudomonas Isolation agar	Pseudomonas

Table III. Analytical Tests and Methods Used in Experimental Studies

Test	Method
Temperature	*Standard Methods*, Method 212
pH	*Standard Methods*, Method 423
TSS	*Standard Methods*, Method 209 B
VSS	*Standard Methods*, Method 209 C
Plate Counts	*Standard Methods*, Method 907 A
Phenol	CHEMetrics, Inc.
ATP	Turner Design
Catechol 2,3-Oxygenase	Kojona, Y., et al.
Catechol 1,2-Oxygenase	Hegeman, G. D., et al.

tively. All plate counts were performed following method 907 C in Standard Methods. Phenol concentrations were determined using CHEMetrics, INC., Phenols Test Kit (Model p-12). Adenosine Triphosphate (ATP) analyses were conducted in accordance with Turner Design using a Turner Fluorometer, Model 112.

EXPERIMENTAL RESULTS

Presented in this section are the results of the laboratory studies conducted during the investigation.

General Effects of Freeze Drying on Supplement Performance

Based on the results presented in Figure 2, which depicts phenol degradation profiles for freeze-dried and non-freeze dried populations, it can be inferred that freeze-drying had an adverse effect on supplement performance. Before freeze-drying, phenol degradation was initiated immediately, while a lag period of 6 hours was required prior to rapid phenol degradation for the freeze-dried supplement. In both cases, the rate and degree of phenol biodegradation were similar once degradation was initiated.

The phenol degradation profiles obtained both before and after freeze-drying at time of 0, 17, and 45 days of supplement storage are presented in Figure 3. Once again, the most significant difference in these profiles is that the lag (or acclimation) time which is evident as the supplement storage time increases. Before freeze-drying, no lag period was observed prior to phenol degradation. However, as the storage time increased the acclimation period increased from about 6 hours after no storage to 24 hours after 45 days of storage. It should be noted, however, the phenol was still completely degraded regardless of storage time.

Supplement Viability and Diversity

Presented in Table IV are the results of the heterotrophic plating conducted to assess supplement viability as a function of freeze-drying and storage. The trend observed here was that there was a decrease in viability due to the actual freeze-drying process as well as a decrease in viability with increasing storage time. An overall loss in supplement viability of two orders of magnitude occurred over the 45 days of storage.

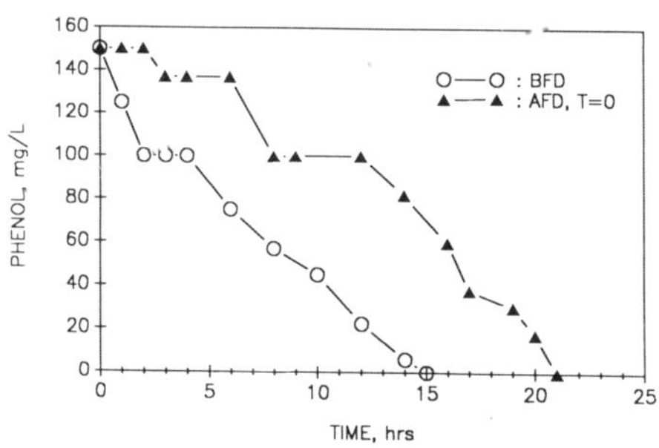

Figure 2. The effect of freeze-drying on supplement phenol biodegradation.

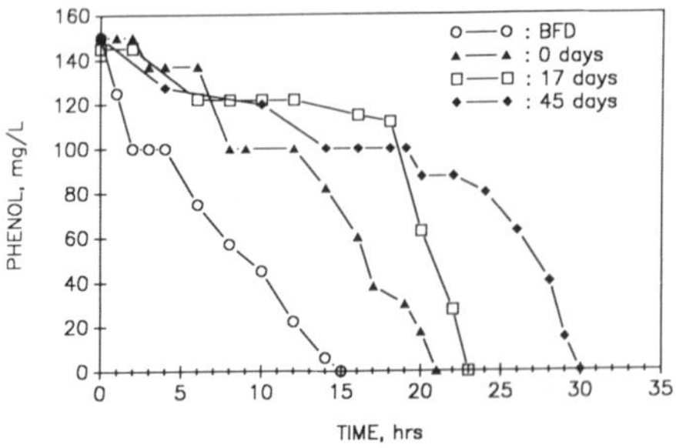

Figure 3. The effect of storage time on supplement phenol biodegradation.

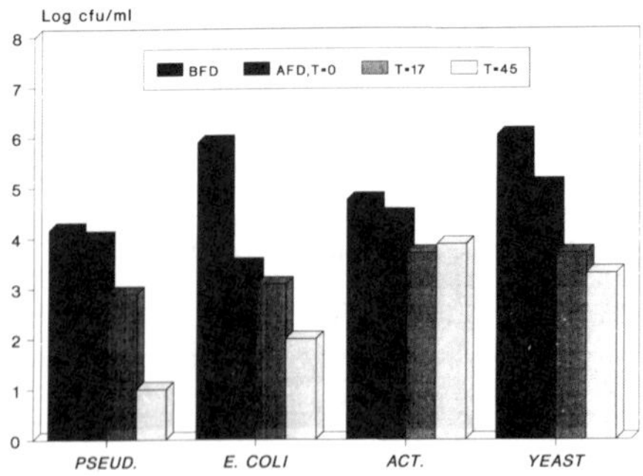

Figure 4. The effect of freeze-drying and storage on supplement population diversity.

Having observed a loss of supplement viability, it was desired to ascertain whether a general decrease in supplement viability occurred or if certain populations were being selectively lost during the freeze-drying and storage process. The selective loss of certain populations would in turn decrease the supplement diversity which could be detrimental to supplement performance. The results of selective plating techniques performed to assess the supplement diversity are presented in Figure 4. The various phenol biodegrading populations shown in Figure 4 include *Pseudomonas, E. Coli, Actinomycetes,* and *yeast*. The viability of each population is presented as a function of freeze-drying and storage. No great loss in any one population occurred. However, the same trend of decreasing viability with increasing storage time, which was evident for the heterotrophs, was observed. The *Pseudomonas* population suffered the greatest viability loss (approximately 3 orders of magnitude overall) during the freeze-drying and storage process while *Actinomycetes* appeared to be the most resistant.

To determine further if phenol degraders were being retained or lost, replicate plating was conducted to assess the percentage of phenol degraders present in the supplement. The results of these experiments are presented in Figure 5. Included in this figure are the number of viable heterotrophs as well as the percentage of phenol degraders present in the population before and after freeze-drying at 0, 17, and 45 days of storage.

Table IV. The Effect of Storage Time on Supplement Viability

Storage Time	Viability (CFU/mL)
BFD	2.0×10^5
AFD, T = 0	1.5×10^5
AFD, T = 17 days	8.2×10^3
AFD, T = 45 days	2.0×10^3

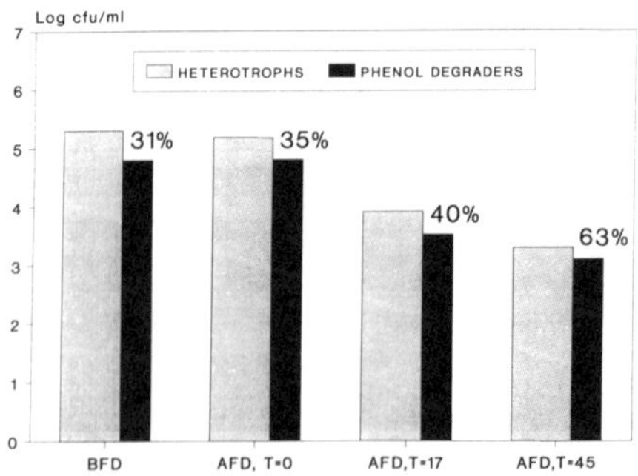

Figure 5. Retention of supplement phenol degraders after freeze-drying and storage.

The same trend was evident for the number of viable phenol degraders as was observed for the viable heterotrophs. That is, the viability of phenol degraders decreased with freeze-drying and increasing storage time. However, the percentage of phenol degraders increased from about 31% before freeze-drying to 63% after 45 days of storage. Therefore, although the number of viable phenol degraders was decreasing with storage time, the rate at which it was decreasing was most likely slower than for the non-phenol degrading population.

Enzyme Testing

The relative contributions of the two enzymatic pathways for phenol biodegradation were assessed before and after freeze-drying and are shown in Figure 6. Included in this figure are reactor levels of phenol, enzyme activities, and ATP concentrations. Before freeze-drying (BFD), phenol degradation occurred immediately with a moderate amount of enzyme activity present throughout the study period. Before freeze-drying, the meta enzyme appears to play a more significant role in phenol degradation than did the ortho enzyme.

After freeze-drying (AFD), an acclimation period of about 6 hours was required prior to the initiation of rapid phenol degradation. This acclimation period coincides with a lag period prior to enzyme induction. As the enzyme levels rise, there is a concurrent decrease in the phenol concentrations. Also significant, is that after freeze-drying, the ortho enzyme appears to be predominate. Thus, a switch in the primary pathways of phenol degradation occurred as a result of freeze-drying.

The reactor ATP levels measured as a function of time are presented in Figure 6. Before freeze-drying, there is an increase in ATP levels as the phenol is degraded. Because phenol was the sole substrate, this implies that phenol was biologically transformed into products which subsequently enter the TCA cycle and yield energy. A depression in cellular ATP levels was observed for the freeze-dried populations when exposed to phenol which was attributed to energy investment for enzyme synthesis. As enzyme induction occurred, the reactor phenol concentrations began to decrease and ATP levels rose. The ATP observations were consistent with other cellular tests and indicate a loss of phenol biodegradative capacity as a result of freeze-drying.

A summary of the phenol concentration and enzyme activity measured in the reactors after freeze-drying at storage times of 0, 3, 17, and 45 days is presented in Figure 7. There are two significant events which can be observed from these plots. First, the lag time prior to rapid phenol degradation increased with increasing storage time of the supplement. These lag periods were consistent with the periods required for significant buildup of the required enzymes. Thus, the lag periods experienced for phenol biodegradation are indeed periods of supplement acclimation to the target compound.

A second significant point which can be discerned from these plots is that a switch in the enzymatic pathways before and after freeze-drying occurred. As shown earlier (Figure 6), the meta pathway was the primary mechanism of phenol degradation prior to freeze-drying. As shown in Figure 7, after freeze-drying, the ortho pathway became the most significant pathway of phenol degradation. After only 3 days of supplement storage the meta enzyme activity was virtually undetectable.

Plasmid Testing

The large loss in the meta enzyme activity (as opposed to the ortho enzyme activity) observed after freeze-drying prompted further investigation. As noted earlier, the meta pathway generally is coded for on plasmid DNA, while the ortho pathway is typically found on chromosomal DNA. Therefore, further experiments were conducted to assess supplement plasmid retention during freeze-drying and storage.

A representative plasmid extraction sample run before and immediately after freeze-drying are shown in Figure 8. The two sets of samples shown in Figure 8 differ only in the volume of sample used in each run. Before freeze-drying, a distinct band was evident in the gel. After freeze-drying the plasmid band was no longer evident and spreading of the nuclear material throughout the gel was observed. This spreading may be indicative of lower molecular weight DNA due to plasmid break up or damage. Plasmid testing also was conducted after 4 days of supplement storage. Again, no plasmid band was evident in the gel and large amounts of spreading of the nuclear material was observed.

Cellular enzyme activity also was measured during the plasmid testing. The results of this testing is presented in Table V. Enzyme levels were measured for each pathway for before and immediately after freeze-drying and after 4 days of supplement storage. A loss in enzymatic activity was observed for both pathways due to the freeze-drying and storage process. However, a much greater reduction in activity was observed for the meta pathway. A loss of about 85% of activity occurred for the meta enzyme due to freeze-drying. Furthermore, after only 4 days of storage a loss of 99% of the original

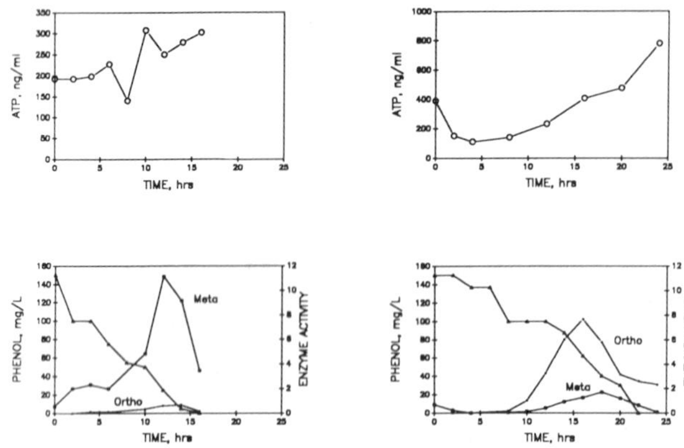

Figure 6. The effect of freeze-drying on factors influencing supplement performance.

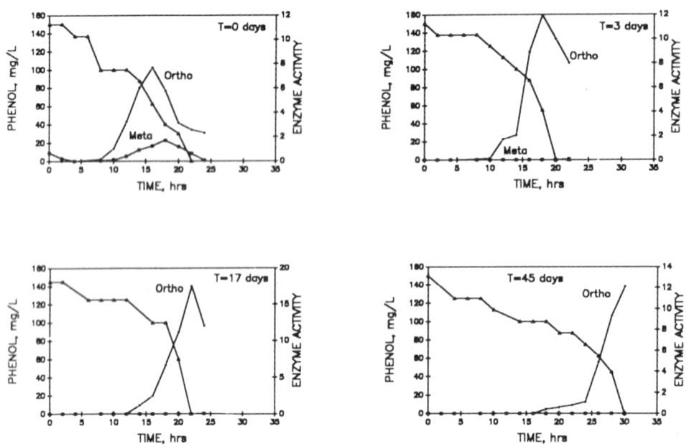

Figure 7. The effect of storage time on factors influencing supplement performance.

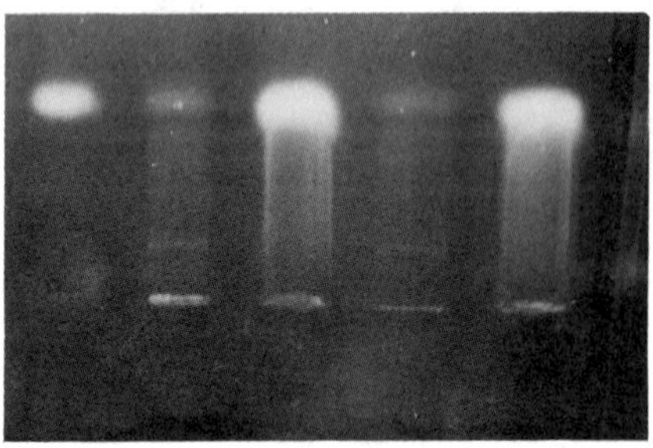

Figure 8. Agarose gel showing plasmic DNA before and after freeze-drying.

Table V. Enzyme Retention During Plasmid Testing

Storage Time	Catechol Oxygenase Activity* Ortho	Meta
BFD	12.7	73.5
AFD, T = 0 days	6.13	11.8
AFD, T = 4 days	1.8	0.8

*Enzyme activity expressed in units/mL.

meta activity present was evident. This observation is consistent with the enzymatic measurements presented earlier.

DISCUSSION OF RESULTS

Based on the results derived from this study freeze-drying and storage of bacterial supplements can be detrimental to supplement performance. Reductions in the supplement's biodegradative capabilities resulted both from the freeze-drying process and subsequent storage time. In particular, all three factors hypothesized as effecting biodegradative abilities were shown to be effected adversely by freeze-drying and storage.

An overall loss in supplement viability due to the freeze-drying and storage process was observed during viability testing. The phenol degrading portion of the supplement population appeared to be better able to survive during freeze-drying and storage than the non-phenol degrading organisms. This observation may be due to the fact that many yeast and *Actinomycetes* which are capable of phenol degradation generally are more resistant to harsh environmental conditions (such as freeze-drying) than are bacterial cells. This trend was exhibited by the population diversity experiments which showed that the *Actinomycetes* were better able to survive the freeze-drying and storage process. Consequently, in this particular case, the viability of the supplement may not have effected the performance of the supplement to as great a degree as the other parameters investigated.

A significant finding of this investigation was the switch in the major pathway of phenol degradation which occurred from before to after freeze-drying. Two factors were hypothesized to be the major reasons for the loss of the meta pathway. The first factor would be the near complete disappearance of the organism(s) from the mixed population which possess the meta pathway. The second factor is the destruction of the meta pathway genetic coding without the loss of the organism.

Differential plating experiments conducted to assess viability did not show a total loss of any one population. Of the populations present in the supplement, the *Pseudomonas* appeared to be most susceptible to freeze-drying and storage. However, significant numbers of these organisms remained viable after freeze-drying and storage. Accordingly, it is likely that the loss of the meta pathway was not due to organism loss.

Destruction of the plasmid DNA during freeze-drying and storage was indicated in the plasmid testing. In addition, a greater decrease in the plasmid coded meta enzyme activity was observed during both the plasmid testing and batch biodegradation testing. Based on these results, the loss or destruction of plasmids during freeze-drying and storage was an important factor affecting supplement performance.

In this study, supplement degradation of phenol after freeze-drying and storage was observed. This capability resulted from the presence of an alternative (non-plasmid coded) biodegradative pathway. However, because the ortho pathway was not the primary system operative prior to freeze-drying, significant induction periods were required by the supplement before phenol biodegradation. Although the target compound was still completely removed, induction periods such as these will severely minimize the effectiveness of the supplements. Furthermore, it can be hypothesized that freeze-drying and storage may be most detrimental to supplements relying solely on plasmid coded biodegradative abilities, which is the case for many xenobiotic degrading supplements.

CONCLUSIONS

A number of factors considered to be important in effecting supplement biodegradative abilities were investigated using a mixed population which was capable of phenol degradation. Based on this investigation a number of conclusions can be made. These include:

1. Acclimation periods were evident after freeze-drying and subsequently increased with supplement storage time.
2. A loss in supplement viability occurred after freeze drying; the viability continued to decrease as storage increased. In addition, no selective loss of the phenol degrading population was observed.
3. Two phenol biodegradative pathways were operative in the supplement population prior to freeze-drying. A loss of the plasmid coded meta pathway was observed after freeze-drying and storage.
4. There was evidence for plasmid destruction during the freeze-drying and storage period which was hypothesized as the primary reason for the dominance of the ortho pathway after freeze-drying.

REFERENCES

1. Lange, C.R., Hartman, J.R., Chong, N.M., Weber, A.S., and Matsumoto, M.R., "Constraints of Bioaugmentation in Enhancing Biological Treatment Process Performance", *Proc. 42nd Ind. Waste Conf. Purdue Univ.*, Lewis Publishers, Chelsea, MI, 275–284, (1988).
2. Lewandowski, G., Salerno, S., McMullen, N., Gneiding, L., and, Adamowitz, D., "Biodegradation of Toxic Chemicals Using Commercial Preparations", *Environmental Progress*, Vol. 5, 212–217, (1986).
3. Lange, C.R., *Constraints of Bioaugmentation for Enhanced Hazardous Waste Biodegradation*, M.S. Thesis, State University of New York at Buffalo, (1988).
4. Williams, P.A., "Catabolic Plasmids", *Trends Biochem. Sci.*, Vol 6, 23–26, (1981).
5. Johnston, J.B., and Robinson, S.G., *Genetic Engineering and New Pollution Control Technologies*, Hoyes Pub., New Jersey, (1984).
6. Nierman, W.C., and Feldblyum, "Cryopreservation of Cultures That Contain Plasmids", *Developments in Industrial Microbiology*, Vol. 26, 423–433, (1984).
7. Heckly, R.J., "Free Radicals in Dry Biological Systems", in *Free Radicals in Biology*, Vol. 2, W.S. Pryor, ed., Academic Press, 135–158, (1976).
8. Ornston, N.L., and Yeh, W., "Recurring Themes and Repeated Sequences in Metabolic Evolution", in *Biodegradation and Detoxification of Environmental Pollutants*, A.M. Chakrabarty, ed., CRC Press, Boca Raton, FL, 106–108, (1982).
9. APHA, *Standard Methods for the Examination of Water and Wastewater*, 16 edition, American Public Health Association, Washington D.C., (1985).
10. Turner Designs, *Photometric Facts: Live Biomass Measurements in the Water Pollution Control Plant*, (1980).
11. Kojona, Y., Itada, N., and Hayashi, O., "Metapyrocatechase: a New Catechol-cleaving Enzyme", *J. Biological Chemistry*, Vol. 236, No. 8, 2223-2228, (1961).
12. Hegeman, G.D., "Synthesis of the Enzymes of the Mandelate Pathway by *Pseudomonas Pudita", J. Bacteriology*, Vol. 91, No. 3, 1140–1154, (1966).
13. Hackett, P., Fuchs, J., and Messing, J., *An Introduction to Recombinant DNA Techniques*, Benjamin/Cummings Publishing Co., Redwood City, CA, (1988).

29 EFFECTS OF BIOAUGMENTATION ON NITRIFICATION BIOKINETICS

Stan A. Zachopoulos, Process Engineering Specialist
Havens and Emerson, Inc.
Cleveland, Ohio 44114

Yung-Tse Hung, Professor
Department of Civil Engineering
Cleveland State University
Cleveland, Ohio 44115

INTRODUCTION

As the effluent permit requirements become more stringent, higher degrees of wastewater treatment are required. Industrial and municipal effluents must be very low in ammonia-nitrogen concentrations in order to be allowed for discharge. The most effective and economic method of treatment for ammonia removal at the concentrations commonly found in municipal and industrial wastewaters is by biological means.

Biological oxidation of ammonia first to nitrite (NO_2) and subsequently to nitrate (NO_3), referred to as nitrification, is accomplished mainly by the action of two aerobic autotrophic bacterial species, *Nitrosomonas* and *Nitrobacter*. The nitrifying microorganisms exhibit a much slower growth rate than that of the heterotrophic microorganisms and are more sensitive to environmental conditions such as temperatures, pH and dissolved oxygen concentrations. In order to achieve the desired levels of nitrification, wastewater treatment facilities have to provide longer hydraulic and solids residence times or, in the case of high organic strength wastes, a separate nitrification stage following the organic removal stage.

In order to increase the organic matter and ammonia removal, many wastewater treatment facilities supplement specific bacterial strains to the biological system. This bacterial supplementation is referred to as bioaugmentation. Although many applications of a bioaugmentation have been reported with mixed results, there has been no single comprehensive study regarding the effects of bioaugmentation on the nitrification performance of wastewater treatment systems and the kinetic parameters of the process.

The objectives of this study were to examine the effects of bioaugmentation with heterotrophic and nitrifying microorganisms on the nitrification process in a combined carbon oxidation-nitrification aerobic suspended growth biological system under various operating conditions in terms of aeration time, sludge age and organic strength in the influent; to develop nitrogen removal kinetic rates; and to examine the effects of bioaugmentation on the nitrogen removal rates under the applied operating conditions.

EXPERIMENTAL PROCEDURES

Two identical bench-scale reactor systems, each with three continuous flow aerated reactors, were operated in parallel for a 30-month period. One system was operated as a control and the second was bioaugmented with a daily dose of 10 mg/L of a commercial bacterial product. The bacterial product was preserved in a liquid suspension and, according to the manufacturer, contained heterotrophic microorganisms along with the nitrifiers *Nitrobacter* and *Nitrosomonas*.

Primary effluents from municipal wastewater treatment plants were used as a feed in this study. Since the strength of this water was relatively weak, sodium acetate and glucose were added as an organic matter source to the raw sewage for certain reactor runs. Ammonium chloride and potassium biphosphate were also added to provide nitrogen and phosphorus nutrients for the microorganisms. The chemical doses were added to maintain a BOD:N:P ratio of 100:5:1.

45th Purdue Industrial Waste Conference Proceedings, © 1991 Lewis Publishers, Inc., Chelsea, Michigan 48118. Printed in U.S.A.

Table I. Experimental Protocol

A. *Constant Parameters*

Reactor No.	HDT (Hours)	Flow Rate (L/d)	Bacterial Dose ppm/d
1	6	2	10
2	12	2	10
3	24	2	10
4	6	2	0
5	12	2	0
6	24	2	0

B. *Variable Parameters*

Run No.	Sludge Age (Days)	Influent TOC (mg/L)	Influent NH_3-N (mg/L)	Influent Strength Characterization
L5A	5	60	20	Low
L10A	10	20	30	Low
L20A	20	60	40	Low
L5B	5	60	50	Low
L10B	10	30	50	Low
L20B	20	20	40	Low
M5	5	145	85	Medium
M10	10	90	50	Medium
M20	20	110	100	Medium
H5	5	280	125	High
H10	10	215	110	High
H20	20	170	105	High

The feed wastewater was introduced to the aeration tank and the effluent was collected after sedimentation. The settled sludge was returned to the aeration tank, while an amount of sludge was wasted daily to control the sludge age.

The three reactors in each system were operated at hydraulic detention times of 6, 12 and 24 hours, respectively. Twelve runs were conducted using different organic strengths in the influent and different sludge ages. Sludge ages of 5, 10 and 20 days were applied using low strength, medium strength and high strength influents. The low strength influent runs were conducted twice for each sludge age. The duration of each run was from 2 to 3 months. This period was sufficient to produce statistically significant results. The reactors were operated at a constant room temperature of about 22°C. The operating characteristics of the twelve runs conducted are shown in Table I.

Laboratory analyses of the influent, effluent, and mixed liquor of each reactor unit were conducted in order to assess the degree of treatment. The TOC concentrations were used to indicate the organic matter present in the wastewater.

RESULTS AND DISCUSSION

The results indicate that there was no significant difference in effluent quality in terms of TOC according to hydraulic detention time and according to bioaugmentation conditions for all of the runs conducted. The biodegradable organic matter was removed in detention times shorter than those examined. Further treatment, beyond the minimum detention time required, did not result in any additional improvement.

In all of the runs, a significant reduction in ammonia-nitrogen took place. Most of the ammonia was removed within the first 12 hours of aeration while little ammonia removal took place during longer detention times (Figure 1). Nitrate production also increased with detention time, but not linearly, due to the product inhibition process (Figure 2). Effluent ammonia concentrations were higher at the higher organic strength runs (Figure 3). As Figure 3 indicates, the ammonia removal was complete at a 5-day sludge age during the low strength runs, and at a 10-day sludge age during the medium strength runs. The higher degree of ammonia removal took place at a 20-day sludge age during the high strength runs.

Significant amounts of nitrate formed only in the low and medium organic strength runs. Only small amounts of nitrate formed in the high strength runs (Figure 4). Among the high strength runs,

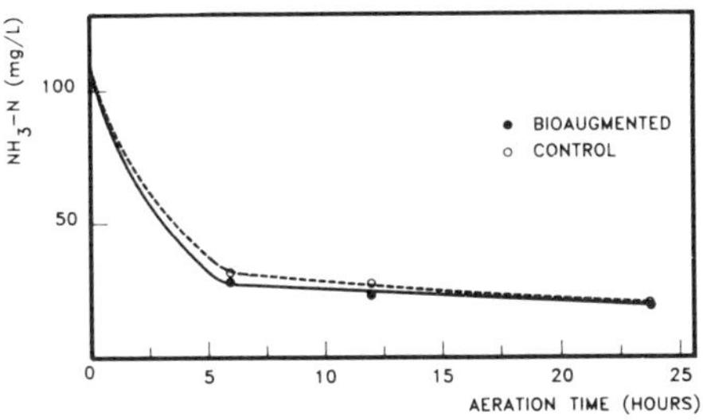

Figure 1. Effluent ammonia-nitrogen concentrations vs. aeration time (medium strength, 20-day sludge age).

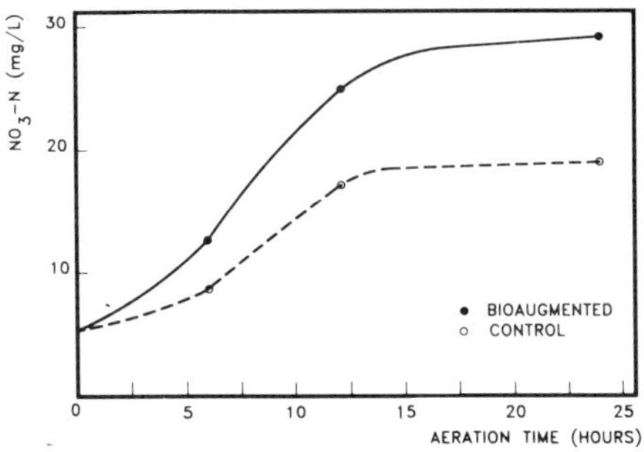

Figure 2. Effluent nitrate-nitrogen concentrations vs. aeration time (medium strength, 20-day sludge age).

less nitrate was formed at the 20-day sludge age run when less solids were produced. This indicated the ammonia-nitrogen was removed during the high strength feed runs was used for cell synthesis and was not oxidized to end products.

The bioaugmented reactors produced lower effluent ammonia concentrations in all reactor runs except for the 5-day sludge age, low and medium strength runs (Figure 3). This conclusion was substantiated by statistical analysis at a 0.05 level of significance. The bioaugmented reactors also produced higher nitrate concentrations compared to the control reactors in all the runs, except the low and medium strength, 5-day sludge age runs (Figure 4). Overall, the bioaugmented reactors removed 65% of the influent ammonia, while the control reactors removed 63%. The bioaugmented reactors also produced 12% more nitrate than the control reactors.

The ammonia removal rate, r which represents the amount of ammonia-nitrogen removed per unit of mixed liquor volatile suspended solids (MLVSS) per day was calculated for each reactor. The ammonia removal rates varied from 0.05 to 0.15 d^{-1} and generally increased with increasing sludge age. However, a calculation of the ammonia removal rates could not explain the impact of each individual parameter, such as hydraulic detention time, MLVSS concentration, sludge age, or level of decay on the process performance.

The ammonia removal model developed by Adams and Eckenfelder[1] was initially used to examine the performed reaction. This model assumes a first order nitrification reaction and is described by the equation:

$$Ne/No = e^{-kXt} \qquad (1)$$

where:
- Ne = effluent ammonia concentration, mg/L
- No = influent ammonia concentration, mg/L
- k = specific nitrification rate coefficient, L/mg.d
- X = MLVSS concentration, mg/L
- t = detention time, days

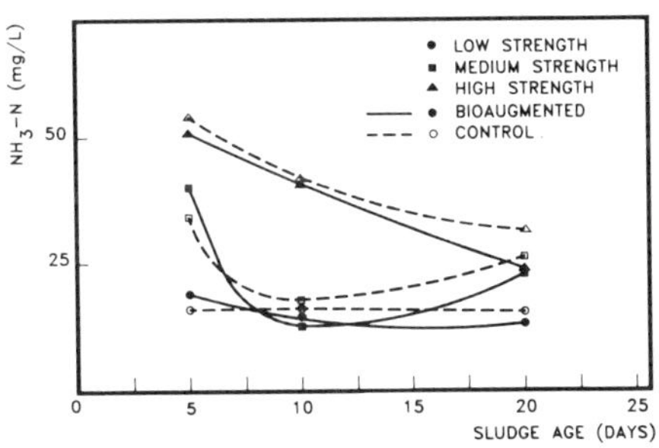

Figure 3. Effluent ammonia-nitrogen connectrations vs. sludge age.

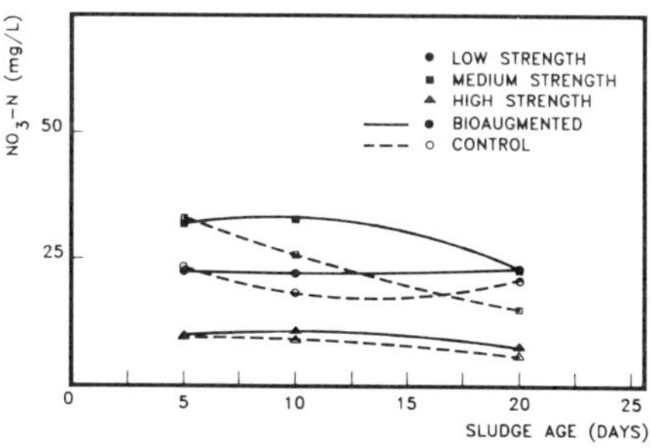

Figure 4. Effluent nitrate-nitrogen concentrations vs. sludge age.

Application of Equation (1) to the data obtained during the study revealed the model applies only to a small aeration time interval. The model also fails to describe the effects of the decay process. It was found the decay process was directly related to the ammonia concentration in the reactor and to the aeration time. The effects of the decay process on the nitrification performance could be described by the equation:

$$\frac{dN}{dt} = -K_nN + b_nNt \tag{2}$$

where: Kn = nitrification rate coefficient, day^{-1}
b_n = decay coefficient, day^{-2}

Integration of Equation (2), after substitution of K_n with kX and b_n with bX, gives the relationship:

$$\frac{-\ln(Ne/No)}{X} = kt - \frac{b}{2}t^2 \tag{3}$$

where: k = specific nitrification rate coefficient, L/mg.d
b = specific decay coefficient, $L/mg/d^2$

Equation (3) is a quadratic equation of the form:

$$Y = A + Bx + Cx^2 \tag{4}$$

with $Y = \frac{-\ln(Ne/No)}{X}$

B = k
C = -b/2
x = t, and
A = Y intercept at t = 0 (Yo)

The data obtained were statistically fitted to Equation (3) using a curvilinear regression analysis. Applications of the developed model are shown in Figures 5 and 6. The nitrification process followed four distinct stages (Figure 5). At very low aeration times nitrification did not occur (Stage I). As the aeration time increased, nitrifying microorganisms began to develop and nitrification occurred at its maximum rate (Stage II). When ammonia removal was complete, the nitrification rate increase became equal to zero (Stage III). Further aeration caused decay of the biomass and decline of the ammonia removal rate per unit of MLVSS (Stage IV). Application of the data to the proposed model resulted in very high correlation coefficients (up to 0.90). The correlation was significant with a 0.5 level of significance for all applications of the new model.

The coefficients obtained by the application of the developed model are shown in Tables II and III for the bioaugmented and control reactors, respectively. The average values of the parameters obtained during the two low strength runs for each sludge age are reported in these tables.

The nitrification rate coefficients obtained by the developed model were higher at a 5-day sludge age during the low strength runs, at a 10-day sludge age during the medium strength runs and at a 20-day sludge age during the high strength runs (Figure 7). The above conditions were optimum for nitrification for each influent strength category. The decay coefficients were also higher in the runs with the optimum nitrification rates (Figure 8). The X-intercept (Figure 5) represents the minimum

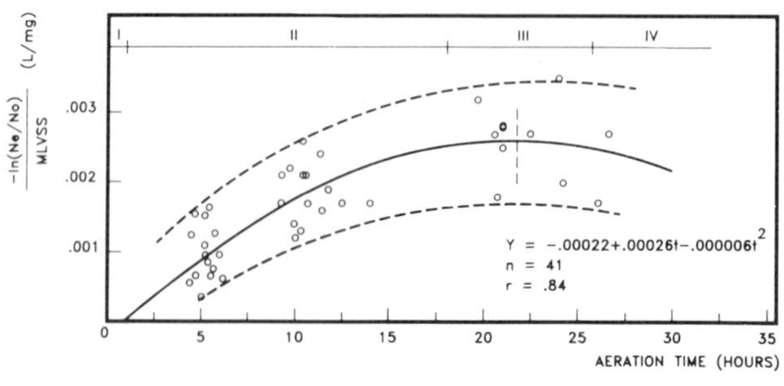

Figure 5. Determination of nitrification rates (low strength, 5-day sludge age, control reactors).

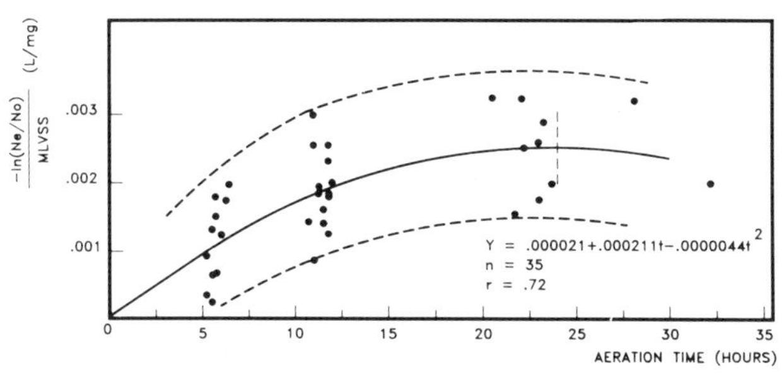

Figure 6. Determination of nitrification rates (medium strength, 10-day sludge age, bioaugmented reactors).

Table II. Summarized Results of Bioaugmented Reactors

Run No.	Y_o (L/mg)	Y_{max} (L/mg)	k (L/(mg.d))	b (L/(mg.d^{-2}))
L5	-0.00020	0.00280	0.00768	0.0105
L10	0.00081	0.00130	0.00139	0.0021
L20	0.00142	0.00246	0.00089	0.0008
M5	-0.00046	–	0.00242	0
M10	0.00002	0.00255	0.00510	0.0045
M20	0.00052	0.00110	0.00079	0.0006
H5	-0.00032	0.00092	0.00240	0.0023
H10	0.00011	0.00078	0.00156	0.0018
H20	0.00030	0.00160	0.00427	0.0070

aeration time required for initiation of the nitrification process. The Y-intercept (Yo) describes the ability of the system to perform nitrification immediately after initiation of aeration. The Yo values

Table III. Summarized Results of Control Reactors

Run No.	Y_o (L/mg)	Y_{max} (L/mg)	k (L/(mg.d))	b (L/(mg.d^{-2}))
L5	-0.00026	0.00364	0.00698	0.0064
L10	0.00037	0.00120	0.00175	0.0018
L20	0.00148	0.00210	0.00139	0.0020
M5	-0.00067	0.00207	0.00355	0.0023
M10	-0.00030	0.00271	0.00510	0.0044
M20	0.00031	0.00105	0.00182	0.0022
H5	-0.00010	0.00078	0.00187	0.0020
H10	-0.00004	0.00084	0.00194	0.0022
H20	-0.00004	0.00072	0.00197	0.0025

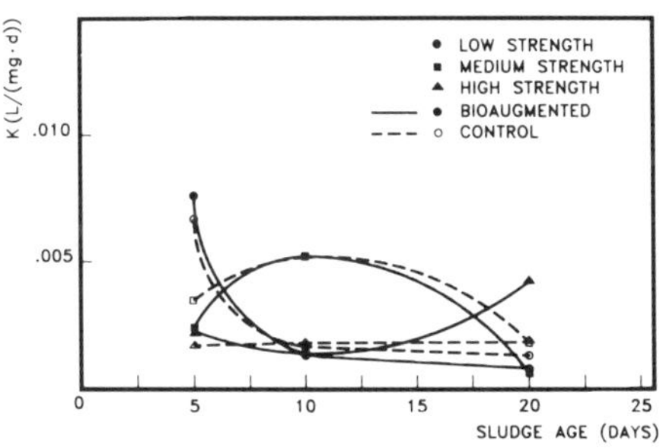

Figure 7. Nitrification rate coefficients vs. sludge age.

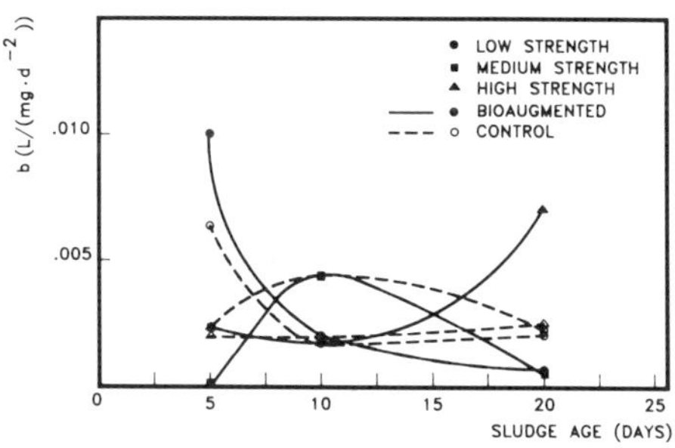

Figure 8. Decay coefficients vs. sludge age.

were negative for all runs with a 5-day sludge age. The Yo values continuously increased with increasing sludge age (Figure 9) and they decreased with increasing influent strength.

In general, bioaugmentation did not improve the specific nitrification rate coefficients nor the decay coefficients. The nitrification rates were higher in the bioaugmented reactors compared to the control reactors when the optimum sludge age was applied for each influent strength. The Yo values were higher in the bioaugmented reactors, indicating conditions proper for nitrification were developed earlier in the bioaugmented reactors than in the control reactors.

SUMMARY AND CONCLUSIONS

Based on results obtained from the bench-scale laboratory study using continuous flow activated sludge reactors, the following conclusions can be made concerning the nitrification process in a combined organic matter-nitrogen removal system and the application of bioaugmentation to the biological system:

1. The experiment did not demonstrate any significant difference in organic effluent quality according to bioaugmentation applied because the organic removal took place in detention times shorter than those examined.
2. The effluent ammonia-nitrogen was decreased and the effluent nitrate concentrations were increased with increasing hydraulic detention times.
3. The effluent ammonia concentrations were lower when lower organic strengths were present in the influent. The amount of nitrate produced by ammonia oxidation was lower at high organic strength runs.
4. The nitrification performance, as measured by the amounts of ammonia removed and nitrate produced, was better in the bioaugmented reactors compared to the non-bioaugmented reactors, except for the low feed strength, short sludge age runs.
5. The ammonia removal followed a first order reaction during the first period of aeration. By increasing the aeration time, the decay process started affecting the reaction reducing the nitrogen removal rate.
6. For each of the examined three influent organic strengths, the ammonia removal rate coefficients, calculated by the developed nitrogen removal biokinetic model, were higher for those sludge ages in which higher ammonia removal took place. These sludge ages were the 5-day for the low strength feed, the 10-day for the medium strength feed and the 20-day for the high strength feed. The decay coefficients were also higher in these runs.
7. According to the developed nitrogen removal model, bioaugmentation did not increase the nitrogen removal rates nor the decay coefficients. The improvement exhibited in some of the runs by the bioaugmented reactors were due to the establishment of conditions proper for nitrification earlier than in the control reactors.
8. The developed model could be useful for the analysis, operation, and design of combined carbon oxidation-nitrification systems. The model can reveal the minimum conditions required for nitrification, the optimum nitrification conditions, as well as the maximum conditions allowed to prevent a decrease in nitrogen removal rates.

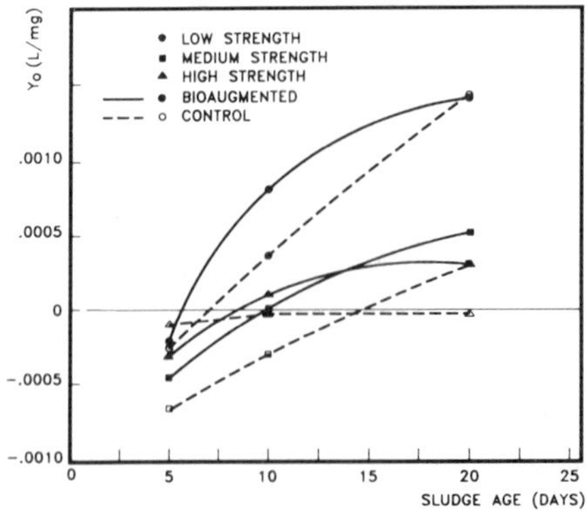

Figure 9. Yo values vs. sludge age.

In summary, bioaugmentation could enhance the nitrification process in certain cases when the proper conditions in terms of influent strength and sludge are provided.

REFERENCE

1. Adams, C. E. Jr., Eckenfelder, W. W. Jr., "Nitrification Design Approach for High Strength Ammonia Wastewaters," *Journal WPCF*, 49, 413 (1977).

30 THE EFFECT OF INTERSPECIES COMPETITION ON BACTERIAL SUPPLEMENTATION EFFICACY

Robert M. Cowan, Graduate Student
Kevin J. Shanahan, Graduate Student
A. Scott Weber, Associate Professor
Department of Civil Engineering
State University of New York at Buffalo
Buffalo, New York 14260

INTRODUCTION

Bacterial supplementation, or bioaugmentation, is a technique in which bacterial strains are added to a biological waste treatment process as a means of enhancing process performance. The bacterial culture used as the supplement may consist of specially selected naturally occurring strains, mutated strains, or even bioengineered organisms. These cultures usually are purchased from commercial supplement manufacturers but could be produced on site. Supplementation has been suggested for reducing process start-up time through the addition of large quantities of organisms, recovery from process upsets, enhancement of treatment rates, and/or for the addition of unique degradative abilities such as the breakdown of some of the more recalcitrant priority pollutants.

For bacterial supplementation to have a significant effect on a biological treatment process, the supplement must be capable of: 1) metabolizing the waste constituent of interest; and 2) effectively competing for a niche in the bioreactor. Often the supplements recommended by their manufacturers are not useful within the context of these two requirements. For example, Lange, et al[1] tested several commercial supplements suggested by their manufacturers' for treatment of a multicomponent hazardous waste and found most to be incapable of initiating degradation of the waste constituents without a period of acclimation. They also noted significant losses of the supplemented cultures from the reactors indicating competition effects.

For a supplement to maintain an effective niche in a biological reactor, it must compete with the indigenous population for limiting nutrients or be added to the reactor on a continuous basis. This competition is multifaceted; it is the combined effect of many interactions which occur between all the organisms in a reactor and cannot be explained solely on the basis of pure culture growth rates of the individual populations.

The interspecies interactions which can occur between two organisms have been classified by population biologists into the six categories listed in Table I [2]. Each axis of the table is labeled as the effect the presence of one organism has on the growth of the other. A positive sign (+) indicates that the presence of one organism benefits the other, a negative sign (-) indicates a harmful effect, and a zero (0) indicates no effect. Most mathematical models developed to date for describing bioaugmentation have been formulated assuming either neutralism or simple competition as the only interspecies interactions. Neutralism, where the presence of the indigenous population has no effect on the supplement and vise-versa is rare. Simple competition is the case where the two populations compete for the same limiting substrate but in no other way affect each others growth. The use of these simplifications may severely limit model utility.

The objective of this paper is to present the findings of a study conducted to demonstrate that interspecies interactions, other than simple competition, can affect the outcome of supplementation and should be considered in modeling efforts.

BACKGROUND

To date, few attempts have been made to mathematically model supplementation processes. Where these attempts have been made, the effects of interspecies interactions were largely ignored. The only study in which experimental data were directly compared to model simulations was performed by Kennedy et al[3]. The model used was based on an assumption of neutralism. The model provided

45th Purdue Industrial Waste Conference Proceedings, © 1991 Lewis Publishers, Inc., Chelsea, Michigan 48118. Printed in U.S.A.

Table I. Interspecies Interactions Between Two Populations

		Effect of Supplement on Indigenous Population		
		+	0	–
Effect of Indigenous Population on Supplement	+	(+,+) Mutualism	(0,+) Commensalism	(–,+) Predation
	0	(+,0) Commensalism	(0,0) Neutralism	(–,0) Ammensalism
	–	(+,–) Predation	(0,–) Ammensalism	(–,–) Competition

acceptable simulation of the experimental data for short term performance of a chemostat, but failed after about ten hours. The authors did recognize that the failure of the model over the long term was most likely due to the fact that it ignored changes in the growth kinetics of the supplement population in the reactor from those found in pure culture, in other words, the model failed because it ignored competition effects.

Two other investigations into modeling of supplementation, both of which use data from the literature for calibration of the models are those by Hull and Kapuscinski[4] and Rittman[5].

The model developed by Hull and Kapuscinski[4] is based on neutralism. The model is used to study the effect of product dosage rate on the efficacy of supplementation. The conclusion derived from the model is that dosage rates normally prescribed will not have a substantial impact on specific substrate removal. This conclusion is in part due to the fact that the model ignores the possibility that the supplement could use constituents of the waste other than the target compound for growth and therefore increase in concentration to a point such that a substantial impact is achieved. Conversely, if an interspecies interaction such as commensalism were to occur, the impact of even a large supplement dose might be negligible, or at best short lived.

Rittman[5] discussed competition as well as some of its potential impacts, but did not include an effective term for it in the formulated model. The model is loosely based on simple competition for the 'primary' or easily degraded substrate. The target organic is assumed to be at concentrations too low to contribute significantly to the growth of the supplement organisms. Due to these limitations the model is not generally useful but the results obtained by the author are useful. Rittman also came to the conclusion that supplementation dosages, as normally prescribed, are not likely to contribute any significant impact to process performance.

Davis and Parnham[6] performed a study in which a simple competition term was employed in the model. Competition with selective flocculation between two strains of *Saccharomyces cerevisiae* was the focus of the study. One strain had a lower specific growth rate and was very flocculent, whereas the faster growing strain did not flocculate well. The model used included terms for selective recycle of the flocculent culture. While model simulations from the study appeared to do a fair job at predicting the general trends found experimentally, some error was recognized by the researchers. The error was attributed to the selective recycle term but may have resulted from competition effects not accounted for by the model.

Model System

To demonstrate the importance of interspecies interactions on supplementation a model system consisting of two organisms and two substrates was utilized. The model system employed consisted of the organisms *Escherichia coli* and *Pseudomonas aeruginosa* and the substrates glucose and lactose. The substrates glucose and lactose were chosen as surrogate easily degraded and xenobiotic organic waste components respectively. Lactose was chosen as a surrogate xenobiotic for this work because it is degraded via an inducible enzyme pathway as are many hazardous organics of interest such as phenol and its chlorinated congeners, additionally its degradation products, glucose and galactose both shunt into common degradative pathways as occurs for most biodegradable xenobiotics.

The organisms *P. aeruginosa* and *E. coli* were selected based upon their degradative abilities for glucose and lactose. *P. aeruginosa* acted as the indigenous population and was only capable of degrading glucose. *E. coli*, the model supplement was capable of both glucose and lactose degradation.

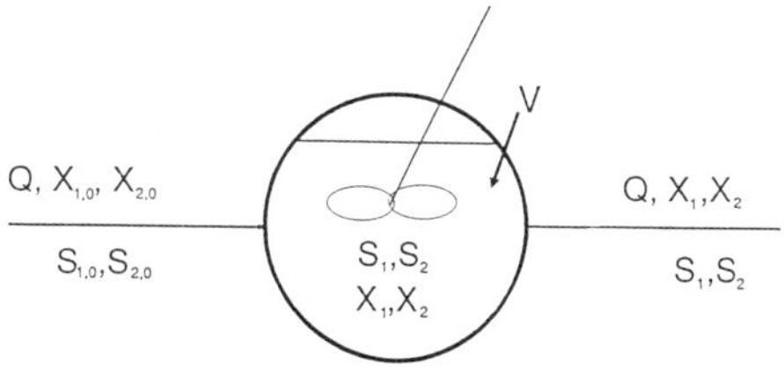

Figure 1. Schematic of model system with two organisms and two substrates.

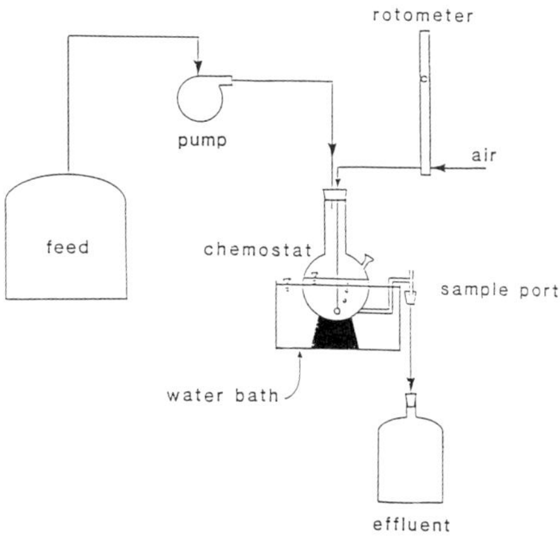

Figure 2. Schematic of chemostat used for continuous culture studies.

Model Formulation

As part of this investigation, mathematical models for simulating bacterial growth under a number of competitive situations were formulated. For Cases 2 and 3 where interactions occur between the supplement and indigenous populations, they were assumed to follow simple competition. Simple competition, although hypothesized to be inadequate for describing interactions between the two populations, was used because of its simplicity and because the extent of interactions between the two populations was unknown.

The assumption of simple competition as valid for describing the interspecies interactions was tested by comparing model simulations to data derived from the laboratory phase of this investigation. Mathematical models describing interactions between the two organisms in the model system were formulated for growth in a chemostat following Monod kinetics. Additional assumptions inherent in these models are: the chemostat reactor is completely mixed; only viable cells utilize substrate; no acclimation period is required; and diauxic growth is not considered.

The only interactions assumed to occur between the two populations was competition for the shared substrate, glucose, designated as S_1.

In this study, three cases were considered as presented below. Pure culture growth of a single compound is presented as Case 1. Case 2 represents a competitive situation where two organisms are grown on a common limiting substrate. For Case 3, the model is formulated for two organisms grown on two substrates. Because of the interest in supplementation in this work, one organism, modeling an indigenous population, was assumed to be capable of only utilizing one of the substrates. The second organism, modeling a supplement, was assumed to be capable of growth on both substrates. Competition occurs for the shared substrate. A schematic of a chemostat is given in Figure 1.

Case 1: Pure Culture Growth on One Substrate

Pure culture growth models were formulated for the purpose of assessing model validity under noncompetitive conditions. The governing equations were:

$$dX/dt = X(\mu - b - D)$$
$$ds/dt = D(S_o - S) - \mu X/Y_g$$

where:

$\mu = \mu_m S/(K_s + S)$
x = Cell concentration (cells/ml)
μ = Specific growth rate (hr^{-1})
b = specific decay rate (hr^{-1})
D = V/Q = Dilution rate (hr^{-1})
v = Volume of liquid in chemostat (L)
Q = Volumetric flow rate in and out of chemostat (L/hr)

S = Substrate concentration in reactor and effluent (mg/L)
S_o = Influent substrate concentration (mg/L)
Y_g = Yield = (cells/ml)produced/(mg/L)substrate consumed
μ_m = maximum specific growth rate (hr^{-1})
K_s = Monod half velocity constant (mg/L)

Case 2: Two Organisms Grown on One Substrate

This model was formulated for simulation of mixed culture growth on one substrate, either S_1, glucose, or S_2, lactose. The governing equations were:

$$dX_1/dt = X_1(\mu_1 - b_1 - D)$$
$$dX_2/dt = X_2(\mu_2 - b_2 - D)$$
$$dS/dt = D(S_o - S) - \mu_1 X_1/Y_{g1} - \mu_2 X_2/Y_{g2}$$

where subscript 1 and 2 denote given value is for organism 1 and organism 2, respectively.

Case 3: Two Organisms Grown on Two Substrates

The overall general model is presented here for mixed culture growth of both organisms on both substrates. This scenario would be considered most typical of supplementation processes. The governing equations are:

$$dX_1/dt = X_1(\mu_{11} + \mu_{12} - b_1 - D)$$
$$dX_2/dt = X_2(\mu_{21} + \mu_{22} - b_2 - D)$$
$$dS_1/dt = D(S_{o1} - S_1) - \mu_{11}X_1/Y_{g11} - \mu_{21}X_2/Y_{g21}$$
$$dS_2/dt = D(S_{o2} - S_2) - \mu_{12}X_1/Y_{g12} - \mu_{22}X_2/Y_{g22}$$

where the model formulations were developed based on concepts outlined in a paper by Smouse.[7] The growth rate expression for $\mu_{i/j}$ was,

$$\mu_{i,j} = \mu_{mi,j}S_j \,/\, [K_{si,j} + S_j + K_{si,j}S_{j\pm 1} \,/\, K_{si,j\pm 1}]$$

and subscripts i and j denote organism subscripts 1 and 2, and substrate subscripts 1 and 2, respectively.

MATERIALS AND METHODS

Bacterial Strains

Escherichia coli strain K-12 wild type (ATCC-e23716) and *Pseudomonas aeruginosa* (ATCC-e10145) obtained from American Type Culture Collection (ATCC) were used throughout this investigation. The cultures were maintained using lyophilization, storage on nutrient agar slants, and through monthly serial transfer to fresh full lawn working plates as suggested in the literature.[8]

Media and Cultivation

All experiments were performed at 37°C and media pH values were maintained between 6.8 and 7.2.

Standard basal medium. Standard basal medium[9] used for batch experiments had the following composition per liter: KC1, 1.0g; KH_2PO_4, 4.2g; $FeSO_4{\cdot}7H_2O$, 0.0005g; $CaC1_2$, 0.01 g; $MgC1_2{\cdot}6H_2O$, 0.2 g; NH_4C1, 0.8 g; Na_2SO_4, 0.2 g; K_2HPO_4, 8.5 g.

Alternate basal medium. An alternate basal medium was used for all continuous culture experiments because the formulation is based on a stable stock solution which permits easy generation of large media volumes. Batch kinetic tests confirmed that growth on both the standard basal media and alternate basal media was identical. The alternate basal media formulation per liter was: stock mineral solution, 10 mL; KH_2PO_4, 1.36 g; $Na_2HPO_4{\cdot}7H_2O$, 4.0 g; nitrogen stock solution, 0.5 mL; adjust pH to 7.0 using 6N NaOH. The stock mineral solution composition on a per liter basis was: $(NH_4)_6Mo_7O_{24}{\cdot}4H_2O$, 0.0095 g; EDTA, 0.12 g; $FeSO_4{\cdot}7H_2O$, 0.35 g; $Na_2B_4O_7{\cdot}10H_2O$, 0.01 g; $MgSO_4$,

10.00 g; CaO, 2.00 g; $CuSO_4 \cdot 5H_2O$, 0.02 g; $ZnSO_4 \cdot 7H_2O$, 0.55 g; $MnSO_4 \cdot 7H_2O$, 0.08 g; NH_4NO_3, 0.10 g; Nitrilotriacetic acid, 3.0 g; and 6.5 mL $CoSO_4$ solution (1000 mg/L). The nitrogen stock solution consisted of 10 mg/L N as NH_4Cl.

Nutrient broth. Nutrient broth was used for growth of inoculums for the continuous culture experiments. The nutrient broth consisted of 3.0 g beef extract (Difco) and 5.0 g peptone (Difco) per liter.

Stock glucose and lactose solutions. Glucose and lactose were added to appropriate feed solutions as concentrated stock solutions sterilized separately. All continuous culture experiments were operated with influent glucose and lactose concentrations of 200 mg/L. Mixed substrate chemostats were fed a media containing 100 mg/L of each substrate.

All growth media solutions, except for the glucose and lactose stocks for the batch experiments which were sterilized by filtration through 0.45μm filters, were sterilized by autoclaving at 121°C for 20 minutes for volumes up to 1 liter. The 20 liter carbouys containing 15 liters of feed for the continuous culture experiments were autoclaved for 80 minutes at 121°C.

Distilled and organic-free water. Batch kinetic experiments were performed according to the methods of Shehata and Marr[9]. This method requires the use of water with minimal organic contamination. The water utilized for this research was tested for the ability to support the growth of *E. coli* and *P. aeruginosa* on basal salts medium with no added substrate. Water purified using a 4-Module NANOpure System, Barnstead Model D3798, was found to be superior to both glass distilled and water from a stainless steel still when support of bacteria growth was considered. Organic free controls were run with each set of batch culture experiments to confirm the maintenance of acceptable water quality.

Batch Kinetics Experiments

Because both pure and mixed culture experiments were to be conducted as part of this investigation, and because it was desired to have uniformity between pure and mixed culture kinetic parameter determinations, batch procedures were used throughout the study.

Batch experiments were performed in 500mL Erlenmeyer flasks containing 100 or 200 mL of media. The flasks were mixed and surface aerated on a Labline Orbital Shaker (model 3520) placed in a temperature controlled incubator set at 37°C. The values for μ_{max} and K_s were determined using a constant substrate batch experiments similar to those developed by Shehata and Marr[9]. The flasks were inoculated with approximately $1x10^3$ cells/mL and allowed to grow until the cell concentration was no greater than $1x10^6$ cells/mL. Samples were taken at the time of inoculation and periodically for the next 8 to 12 hours and analyzed for cell concentration using viable counts on spread plates. Substrate concentration at the time of the final sample was determined for sterile filtered samples stored at 4°C until analyzed. It was assumed that substrate concentration was constant throughout the batch experiments where cell concentrations remained below $1x10^6$ cells/mL. Substrate concentrations found for the final sample were compared to the make up concentrations. Results of these comparisons supported the assumption of constant substrate concentration. This method of determining kinetic parameters would be especially useful where sensitive analytical methods are not available for substrate concentration determinations.

Continuous Culture Experiments

All continuous culture experiments were performed in 670 mL chemostats shown schematically in Figure 2. Sterile feed was supplied via peristaltic pumps, (Masterflex model 7553-30 with #13 pump heads). Reactor volume was controlled by overflow to the effluent line. Aeration and mixing were supplied using filtered, humidified, building air sparged through sintered glass aerators. Temperature control was achieved by immersing the reactors in water maintained at 37°C. Samples were taken from the effluent/sample port indicated in Figure 1 by restricting the flow of air out of the exhaust port, thus causing increased flow out the effluent line. The reactors were filled to capacity and the flow rate adjusted prior to inoculation.

Cell Concentration Measurements

Cell concentrations reported are viable cell counts as determined according to methods outlined in the *Manual of Methods of General Microbiology*[8]. For pure culture studies, both *E. coli* and *P. aeruginosa* concentrations were determined by total counts of colonies cultured on MacConky agar (Difco 0075-05-5). For mixed culture growth determination of the concentration of each species was possible because MacConky agar is a differential media with *E. coli* colonies appearing dark red and *P. aeruginosa* colonies appearing white. When differences in cell concentration between the two species were of two orders of magnitude or greater, simple plating of dilutions to MacConky agar was insufficient for determination of the concentration of the less abundant organism. In these cases, if *P. aeruginosa* was the more abundant species, *E. coli* concentrations were determined by placing MacConky agar plates inoculated with the appropriate dilution in a anaerobic environment. Under anaerobic conditions *E. coli* but not *P. aeruginosa* will grow on MacConky agar, thus creating a media selective for *E. coli*. When *E. coli* was the predominant species and greater sensitivity was needed for determining *P. aeruginosa* concentration, Pseudomonas Isolation Agar (Difco 0927-01-9) was used.

Determination of Substrate Concentrations

For this research it was required that an analytical method be used which was capable of accurately quantifying glucose and lactose concentrations down to 1 mg/L. To achieve this sensitivity, a modification of a glucose analysis method available as a diagnostic kit from Sigma Chemicals, St. Louis, MO, was required. Using the Glucose No. 16-UV diagnostic kit and an HPLC fluorescence monitor (Shimadzu RF-535), glucose and lactose concentrations could be quantified. The substrate concentrations were found based on the concentration of the protonated form of nicotinamide adenine dinucleotide, (NADH) produced enzymatically. NADH has a fluorescence emission spectrum with a peak at 400nm for excitation at 340nm. Determination of lactose concentrations were made possible by the addition of β-galactosidase to the enzymatic solution supplied by the kit. For mixed substrate analysis the glucose concentration of the sample was determined first by performing the standard glucose test. Lactose concentrations were then determined by the difference in the light emission between the test with β-galactosidase and the standard glucose analysis. The increased signal strength was attributed to the lactose in the sample.

RESULTS

Experimental results derived from this investigation are presented in three parts. First the kinetic data determined for use in the simulation model are presented. Pure culture growth of the two organisms independently on the two model substrates is presented second. Lastly, the results of mixed culture growth experiments are presented. Where appropriate model simulations are presented with experimental data for comparison purposes.

Determination of Kinetic Parameters

To investigate the effect of interspecies interactions on the growth kinetics for individual organisms grown in mixed culture, batch experiments must be utilized. In a batch reactor, provided that no interspecies interactions other than simple competition are occurring, each organism present will grow at its respective pure culture growth rate. If deviations from the pure culture kinetics are observed, these can be attributed to other interactions such as ammensalism or commensalism. Because mixed culture kinetic data is assessed as part of this study, all kinetic constants (both pure and mixed culture), other than yield, were derived from batch experiments.

Presented in Figure 3 are the experimental data for the three organism-substrate pairs in which growth occurred. No plot is provided for the growth of *P. aeruginosa* on lactose because *P. aeruginosa* can not metabolize lactose. To determine the values for μ_{max} and K_s, Hanes linearizations were performed on the experimental data. The kinetic constants were determined by a least squares fit of the linearized data. The resultant kinetic models are shown as the curves plotted on Figures 3a, b and c.

A summary of the kinetic data used for model calibration is given in Table II. The decay rate constants were found by monitoring the loss of viability of washed cells inoculated to organic free media. Yield constants were calculated by averaging yield values for all continuous culture experiments performed for a given organism-substrate pair because in the batch culture experiments the substrate concentration was constant, therefore no yield data could be obtained.

Based on an examination of the kinetic data presented in Table II, it is concluded that: 1) for growth on glucose, *P. aeruginosa* had higher μ_{max} and yield values, and a lower K_s value than did *E. coli*; 2) for the growth of *E. coli*, lower μ_{max} and K_s values and a higher yield were observed for lactose than for glucose; and 3) the decay rate of *P. aeruginosa* was slightly higher than that for *E. coli* independent of substrate.

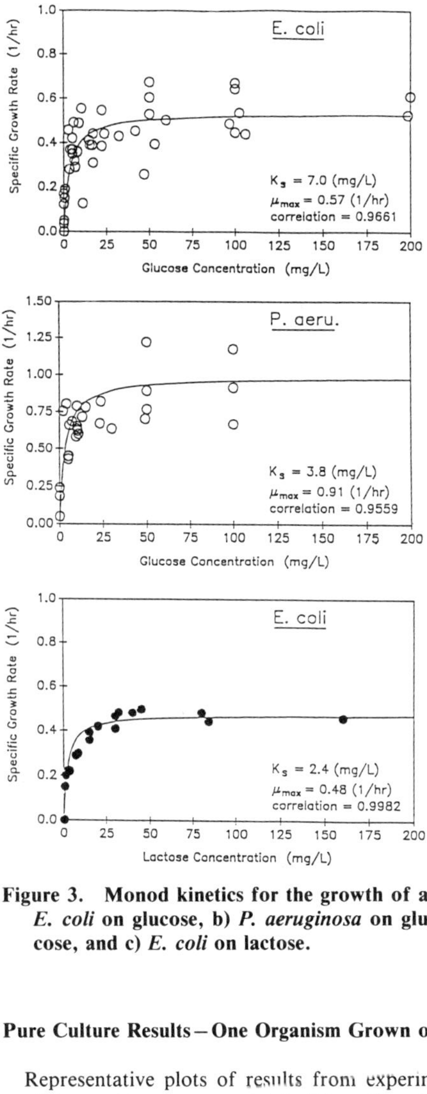

Figure 3. Monod kinetics for the growth of a) *E. coli* on glucose, b) *P. aeruginosa* on glucose, and c) *E. coli* on lactose.

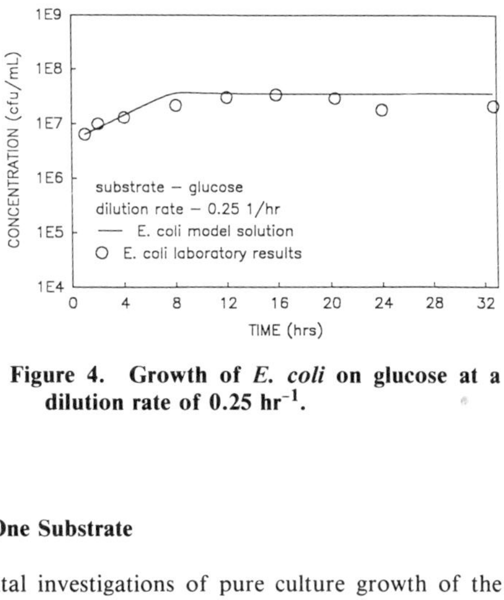

Figure 4. Growth of *E. coli* on glucose at a dilution rate of 0.25 hr^{-1}.

Pure Culture Results – One Organism Grown on One Substrate

Representative plots of results from experimental investigations of pure culture growth of the model organisms are presented in this section. In total ten experimental conditions were investigated. These were pure culture growth of *E. coli* on glucose, pure culture growth of *P. aeruginosa* on glucose, and pure culture growth of *E. coli* on lactose, each at dilution rates of 0.083, 0.167 and 0.25 hr^{-1}, and pure culture growth of *P. aeruginosa* on lactose at a dilution rate of 0.25 hr^{-1}. All experi-

Table II. Model Calibration Data[2]

	μ_{max} 1/hr	K_s mg/L	Yield CFU/mg	Decay 1/hr
E. coli				
glucose	0.57	7.0	1.95×10^5	0.02
lactose	0.48	2.4	3.88×10^5	0.02
P. aeruginosa				
glucose	0.91	3.8	9.22×10^5	0.03
lactose	0.00	0.0	0.00	0.03

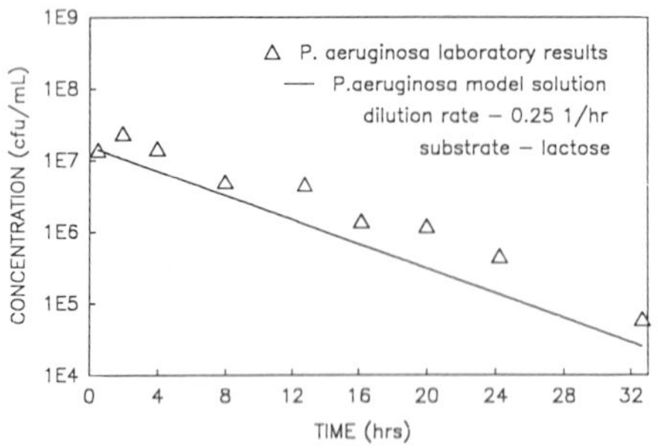

Figure 5. Growth of *P. aeruginosa* on lactose.

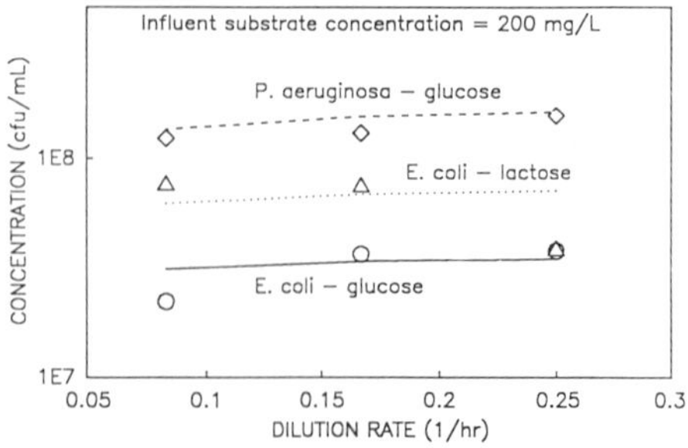

Figure 6. Cell concentrations at steady state for various dilution rates, experimental data and model predictions.

ments conducted at dilution rates of 0.25 and 0.083hr^{-1} were repeated twice with good reproduction of results.

Presented in Figure 4 are the experimental data and model simulations for pure culture growth of *E. coli* on glucose. In this particular experiment, *E. coli* was inoculated to a chemostat of 200 mg/L glucose at a dilution rate of 0.25 hr^{-1}. After inoculation, the concentration of *E. coli* increases to its steady state value of approximately 4×10^7 CFU/mL. Once this steady-state concentration had been achieved, performance of the reactor was consistent, with little deviation from the steady state concentration. The response in experimental data is in good agreement with the model prediction. This supports the use of the model for simulation of pure culture conditions, therefore if simple competition is a valid assumption the model should be useful for simulation of mixed culture conditions as well.

The ability of the mathematical model to simulate the loss of *P. aeruginosa* from a reactor fed only lactose is illustrated in Figure 5. Because *P. aeruginosa* is incapable of metabolizing lactose, no growth occurs and it washes out of the reactor. The slope of the washout predicted by the model simulation closely matches that found experimentally. The offset observed between the model and experimental data is possibly caused by the continued division of the cells inoculated to the reactor, as the inoculum cultures were grown on a nutrient rich media, Nutrient Broth.

A summary of the steady state viable cell concentrations found for each experimental pure culture case investigated are presented in Figure 6. The data points represent the average of the experimentally determined values and are compared against the line representing those determined using the mathematical model. Good agreement is apparent between the experimental and model predictions except for the growth of *E. coli* on lactose at the highest dilution rate, 0.25 hr^{-1}. This discrepancy of the experimental steady state cell concentration below that predicted by the model may have resulted from the drop off of cell concentration normally seen for growth in a chemostat as the dilution rate approaches the maximum specific growth rate. This experiment was repeated for confirmation of the observed result. Identical steady state cell concentrations were observed for both experiments.

Mixed Culture Results

Mixed culture experiments and model simulations were conducted in order to test the validity of the simple competition model for predicting the outcome of mixed culture growth. The experiments performed consisted of the concurrent growth of *E. coli* and *P. aeruginosa* on glucose alone, lactose alone and a mixture of glucose and lactose. Dilution rates of 0.083, 0.167, and 0.25 hr^{-1} were studied in this phase and matched those used in the pure culture experiments.

Two organisms grown on glucose. The experimental and model results for the three concurrent growth experiments with *E. coli* and *P. aeruginosa* cultured on glucose are shown in Figure 7. For the dilution rates of 0.083 and 0.167 hrs^{-1}, there is good correlation between the experimental data and model simulations based on pure culture growth kinetics. In both cases washout of *E. coli* occurs and a steady state concentration of *P. aeruginosa* similar to that predicted is achieved. At a dilution rate of 0.25 hrs^{-1}, model prediction for *E. coli* growth deviates significantly from experimental results. Although wash out of *E. coli* is predicted based on model simulations, it remains in the reactor coexisting with *P. aeruginosa* under what appears to be steady state conditions. In addition to the fact that *E. coli* maintains a presence in the system, the observed *P. aeruginosa* concentration at steady

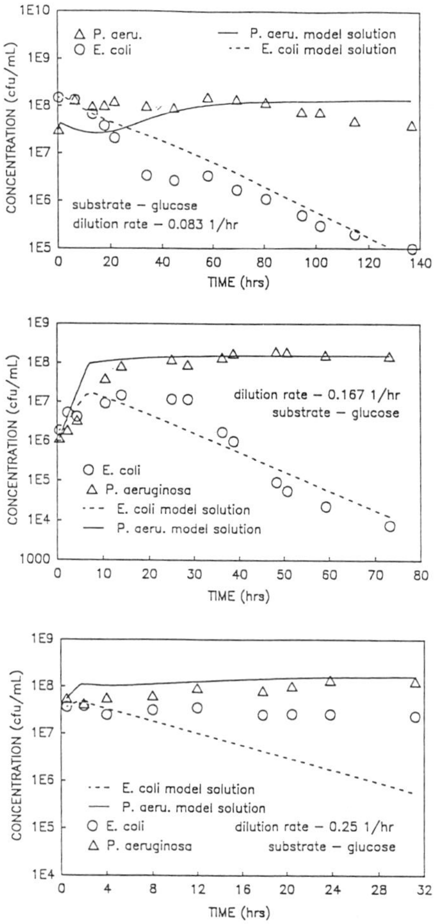

Figure 7. Mixed culture growth of *E. coli* and *P. aeruginosa* on glucose at dilution rates of a) 0.083 hr^{-1}, b) 0.167 hr^{-1}, and c) 0.25 hr^{-1}.

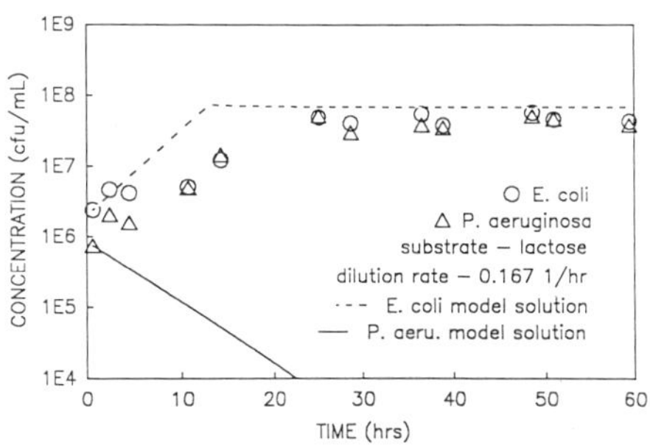

Figure 8. Mixed Culture growth of *E. coli* and *P. aeruginosa* on lactose at a dilution rate of 0.167 hr^{-1}.

state is lower than that predicted by the model. This occurs because, although not predicted, a significant amount of glucose is consumed by *E. coli*, thus reducing the amount available to *P. aeruginosa* and the concentration of *P. aeruginosa* achieved. Clearly at this growth rate, the simple competition assumption is not valid and other interspecies interactions are important.

Two organisms grown on lactose. In Figure 8, data derived from the concurrent growth on lactose experiments are presented for a dilution rate of 0.167 hr^{-1}. Failure of the model to simulate the events in the reactor indicate that other interspecies interactions besides simple competition are important. In addition to the data presented here for a dilution rate of 0.167 hr^{-1}, dilution rates of 0.083 and 0.25 hr^{-1} were investigated. Similar results were found for each dilution rate. For each growth rate investigated the simple competition model predicted the loss of *P. aeruginosa* due to its inability to initiate lactose degradation. However, in all cases the steady state *P. aeruginosa* concentration was close to that attained by the *E. coli* population. It is evident that *P. aeruginosa* benefits due to the presence of *E. coli*, apparently growing on the metabolic products of *E. coli's* degradation of lactose. The initial products of *E. coli* metabolism of lactose are known to be glucose and galactose and both these and the subsequent metabolic intermediates can be utilized by *P. aeruginosa* for growth. Further evidence to support the conclusion that *P.aeruginosa* utilizes *E. coli* metabolites is the fact that the *E.*

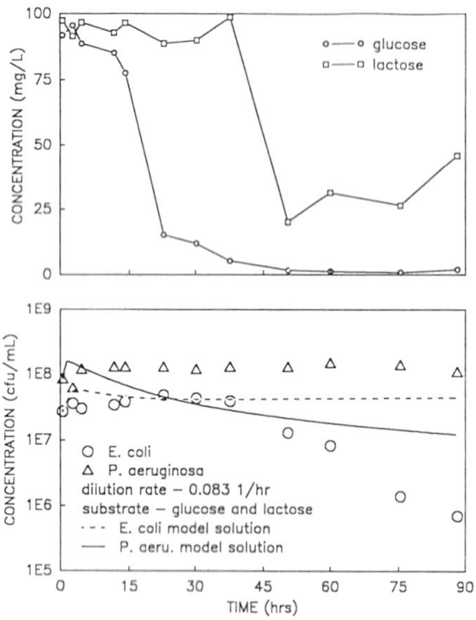

Figure 9. Mixed culture growth of *E. coli* and *P. aeruginosa* on glucose and lactose at a dilution rates of 0.083 hr^{-1}. a) Substrate concentrations observed. b) Organism concentrations observed and predicted.

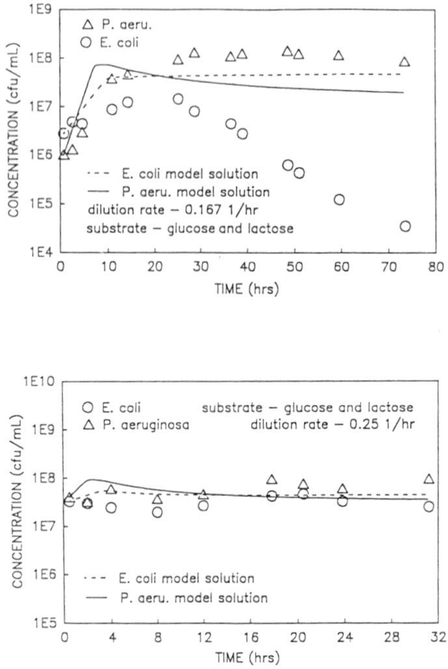

Figure 10. Mixed culture growth of *E. coli* and *P. aeruginosa* on glucose and lactose at dilution rates of a) 0.167 hr^{-1} and b) 0.25 hr^{-1}.

coli concentrations observed were lower than those predicted, indicating that some of the mass and energy supplied by the lactose is not being used for the growth of *E. coli*.

The dip in cell concentration of both species at the beginning of the experiment is most likely due to the need for induction of the lactose metabolic pathway. After *E. coli* acclimation to lactose, growth of both organisms proceeded.

Two organisms grown on two substrates. Presented in Figures 9 and 10 are the experimental results and model simulations for the three concurrent growth experiments performed with both model organisms and substrates. Depicted in Figure 9 are the predicted and measured growth of both organisms cultured at a dilution rate of 0.083 hrs^{-1}. The curves shown in the upper plot are the substrate data collected during this experiment. Based on measured substrate profiles diauxic growth is apparent during the initial portion of the experiment. The glucose concentration is reduced to below 10 mg/L prior to the induction of lactose degradation, which occurred after approximately 22 hours of culture time. Almost concurrent with the reduction in lactose concentration is a decrease in *E. coli* growth. This result is unexpected based on the simple competition assumption because only *E. coli* was presumed to be capable of lactose biodegradation. Apparently, as was noted earlier, *P. aeruginosa* out competes *E. coli* for the consumption of the metabolic products of the lactose degradation. The decline in the *E. coli* concentration continued throughout the remainder of the study with a subsequent increase in lactose concentration. It is likely that increases in lactose concentration may lead to a recovery of the *E. coli* concentration preventing total loss of the surrogate supplement from the reactor.

The experimentally derived concentration of *P. aeruginosa* in the reactor deviated from that predicted by the model. The concentration of *P. aeruginosa* obtained is significantly higher than predicted. This elevated concentration is presumed to result from *P. aeruginosa* utilizing the breakdown products of lactose. Accordingly for this case, it appears that the interspecies interactions not accounted for by the simple competition model do effect the efficacy of the treatment.

Cell concentration plots for the other growth rates studied using both substrates and organisms are presented in Figure 10. It is evident that the simple competition model also is not valid for the 0.167

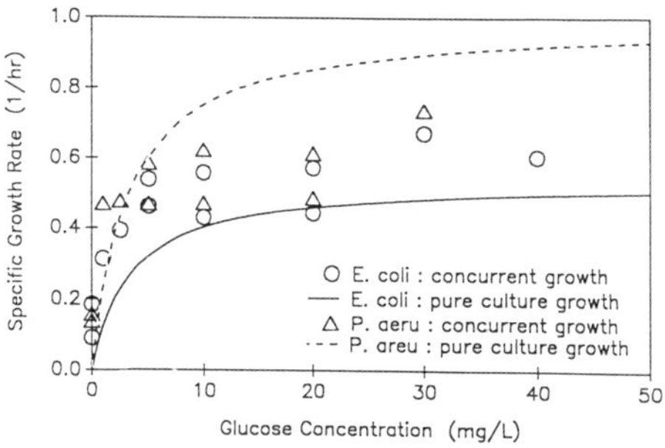

Figure 11. Mixed culture kinetics data for growth on glucose.

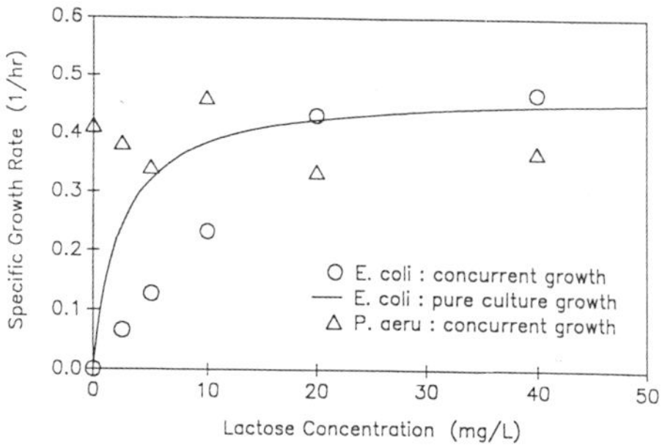

Figure 12. Mixed culture kinetics data for growth on lactose.

hrs^{-1} dilution rate. For concurrent growth of the model organisms at a dilution rate of 0.25 hrs^{-1} both organisms remain in the reactor throughout the experiment. Although the result was predicted by the simple competition model, it probably only occurs surreptitiously. It should be recalled that at a dilution rate of 0.25 hrs^{-1} *E. coli* remained in the concurrent growth chemostat fed only glucose (Figure 6(c)) which was not predicted by the model.

Summary of Mixed Culture Growth

In summary, a model formulated based on the assumption that pure culture kinetics are valid under mixed culture conditions, was capable of adequately simulating pure culture growth in a chemostat. However, when mixed culture conditions were considered, this simple competition model often failed to predict even the general outcome of the experiment. This failure suggests that other interspecies interactions may be important. It was hypothesized that interactions, other than simple competition affect the growth kinetics causing the mixed culture growth rates of the affected organisms to be different than those derived for pure culture. To provide evidence for this hypothesis, mixed culture growth kinetics were investigated.

Mixed Culture Growth Kinetics

Experiments were performed for the determination of the growth kinetics of *E. coli* and *P. aeruginosa* on glucose and on lactose when the other organism was present. The experiments were conducted using methods identical to those used to collect the pure culture data. Any change in the kinetics from those found for pure culture growth should be attributed to interspecies interactions other than simple competition.

Mixed culture growth on glucose. Figure 11 is a plot of the mixed culture kinetic data for the growth of *E. coli* and *P. aeruginosa* on glucose in comparison to the pure culture kinetic models presented in Figure 3. The mixed culture kinetics appear to deviate significantly from the pure culture models confirming the presence of interactions other than simple competition. It also is interesting to note that at lower substrate concentrations, corresponding to lower growth rates, or dilution rates in a chemostat, *P. aeruginosa* appears to have a kinetic advantage, while at higher glucose concentrations the kinetics for both organisms appear to converge. This trend was evident in the results of the mixed culture growth on glucose studies presented in Figure 7 where growth rate or dilution rate appeared to be a factor affecting the outcome of the experiment.

Mixed culture growth on lactose. Mixed culture growth kinetics on lactose are presented in Figure 12. *P. aeruginosa* attains significant growth rates in a solution containing lactose as the sole carbon source when *E. coli* is present, even though it can not initiate the metabolism of lactose. Again, as with the mixed culture kinetics for growth on glucose, at low substrate concentrations *P. aeruginosa* appears to have a kinetic advantage over *E. coli* which is lost at higher lactose concentrations. Clearly interactions which are not adequately described using a simple competition assumption are occurring.

SUMMARY

The purpose of this paper was to demonstrate that interspecies interactions occur which can have a significant impact on biological supplementation and therefore should be included in supplementation models. To achieve this objective, a model system consisting of two organisms and two substrates was used to simulate the supplementation process.

The results of experimental investigations conducted with the model system under continuous culture were compared with numerical simulations. The numerical simulations closely matched the experimental results for the pure culture experiments thereby corroborating the model. The numerical simulations for the mixed culture conditions often failed to predict the outcome of the experiments suggesting that interspecies interactions other than simple competition were occurring. This was confirmed by the mixed culture kinetic data. An important variable in the success of simple competition models to predict the outcome of mixed culture experiments was dilution rate.

ACKNOWLEDGMENTS

This research was funded jointly by the Occidental Chemical Corporation and the New York State Center for Hazardous Waste Management. The manuscript was not reviewed by either agency. Accordingly the opinions expressed in this article are those of the authors and not the funding agencies.

REFERENCES

1. Lange, C.R., J.R. Hartman, N.M. Chong, A.S. Weber, and M.R. Matsumoto, "Constraints of Bioaugmentation in Enhancing Biological Treatment Process Performance," *Proc. 42nd Ind. Waste Conf. Purdue Univ.*, Lewis Publishers, Chelsea, MI, 275–284, 1988.
2. Bailey, J.E. and D.F Ollis, *Biochemical Engineering Fundamentals* Second Edition, McGraw-Hill, New York, 1986.
3. Kennedy, M.S., J. Grammas, W.B. Arbuckle, "Parachlorophenol Degradation Using Bioaugmentation," *Research Journal WPCF*, Volume 62, Number 3, 227–233, 1990.
4. Hull, S.J. and R.B. Kapuscinski, "Efficacy of Bioaugmentation Products as Predicted by a Model of Steady-State Flocculent Cultures," *Proc. 42nd Ind Waste Conf Purdue Univ*, Lewis Publishers, Chelsea, MI, 297–308, 1988.
5. Rittman, B.E., "Reactor Considerations in the use of Bioaugmentation in Biological Treatment", *Bioaugmentation as a Means to Enhance Waste Treatment*, Lewis Publishers, Chelsea, MI, In Press, 1988.
6. Davis, R.H. and Parnham, C.S.III, "Competitive Yeast Fermentation with Selective Flocculation and Recycle," *Biotechnology & Bioengineering*, In Press, 1988.
7. Smouse, P., "Mathematical Models for Continuous Culture Growth Dynamics of Mixed Populations Subsisting on a Heterogeneous Resource Base: 1. Simple Competition," *Theoretical Population Biology, 17*, 16–36, 1980.
8. Manual of Methods for General Microbiology, American Society for Microbiology, 1981.
9. Shehata, T.E. and Marr, A.G., "Effect of Nutrient Concentration on the Growth of *Escherichia coli," Journal of Bacteriology*,107(1),210-216,1971.

31 BIODEGRADATION KINETICS OF A SELF-INHIBITORY SUBSTRATE BY STABLE STEADY-STATE BIOFILMS

Pablo B. Sáez, Graduate Research Assistant
Bruce E. Rittmann, Professor
University of Illinois at Urbana-Champaign

Qui-bo Zhang, Research Assistant Professor
Research Center for Eco-Environmental Sciences
Beijing, People's Republic of China

INTRODUCTION

In the treatment of industrial wastewaters and hazardous wastes, many organic contaminants fall into the category of self-inhibitory substrates, which are biodegradable electron donors that inhibit the rate of microbial metabolism when present at high concentrations. The utilization kinetics for self-inhibitory substrates usually are described with the Haldane modification of the Monod expression:[1-4]

$$(-r_{ut}) = \frac{q_m S X_a}{K + S + S^2/K_i} \tag{1}$$

in which $(-r_{ut})$ = the rate of substrate utilization $[M_sL^{-3}T^{-1}]$, S = the substrate concentration $[M_sL^{-3}]$, X_a = concentration of organisms active in degrading the substrate $[M_xL^{-3}]$, q_m = maximum specific rate of substrate utilization in the absence of any inhibition effects $[M_sM_x^{-1}T^{-1}]$, K = half-maximum-rate concentration (in the absence of inhibition effects) $[M_sL^{-3}]$, and K_i = self-inhibition constant $[M_sL^{-3}]$. When S increases initially from zero, $(-r_{ut})$ increases similarly to the Monod function. However, when S becomes greater than $S_{hi,dg} = \sqrt{KK_i}$, $(-r_{ut})$ is a decreasing function of S. Dispersed-growth process cannot sustain stable steady states at $S > S_{hi,dg}$.[3,5]

Sáez et al.[6] and Gantzer[7] recently showed that biofilm processes ought to be able to operate stably and at steady state for concentrations greater than $S_{hi,dg}$. Extension of the range of concentrations for stable operation is made possible by the "protection" afforded by diffusion-driven gradients in the biofilm. As is illustrated schematically in Figure 1, stable utilization of a self-inhibitory substrate is possible deep in the biofilm when diffusion through the outer (and inhibited) zone reduces the concentration to noninhibitory values for the deeper zone, which can have strong positive growth. [Recall that the net growth rate of cells in contact with S is

$$r_{gr} = Y(-r_{ut}) - bX_a \tag{2}$$

in which r_{gr} = cell net growth rate $[M_xL^{-3}T^{-1}]$, Y = true yield $[M_xM_s^{-1}]$, and b = specific maintenance-decay coefficient $[T^{-1}]$. When $(-r_{ut})$ is small, r_{gr} becomes negative, which is the situation for the zone of inhibition.]

Although Sáez et al.[6] and Gantzer[7] showed that stable steady-state biofilms are possible for $S > S_{hi,dg}$, they also showed that biofilms have an upper-limit concentration for stable operation. That concentration is defined as $S_{hi,bf}$, and it can be computed only by solution of the differential equations that constitute the steady-state model [e.g., equations 9–15 in Sáez et al.[6]]. $S_{hi,bf}$ occurs when the zone of positive growth occupies only a small portion of the biofilm; any reduction in the size of the positive-growth zone, by its being pushed relatively deeper into the film with a larger S, results in a net negative growth rate integrated over the entire film.[6,7]

Whereas $S_{hi,bf}$ defines the upper-limit concentration for maintaining a stable steady-state biofilm, S_{lo} defines the lower-limit concentration. S_{lo} is the lowest concentration to support a stable steady-state biofilm; for $S < S_{lo}$, r_{gr} is always negative[8]. Sáez et al.[6], Gantzer[7], and Sáez and Rittmann[9] showed that S_{lo} is computed for self-inhibitory substrates by

45th Purdue Industrial Waste Conference Proceedings,

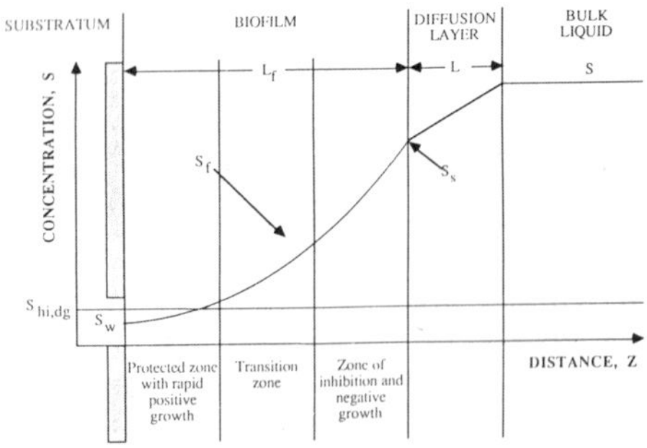

Figure 1. Schematic representation of how diffusion of a self-inhibitory substrate to "protected" layers allows stable steady-state utilization for $S > S_{hi,dg}$. Note that S = substrate concentration in the bulk liquid, S_s = substrate concentration at the outer biofilm surface, Sf = substrate concentration within the biofilm, S_f = substrate concentration at the attachment surface, L = diffusion-layer thickness, L_f = biofilm thickness, and z = distance normal to the attachment surface.

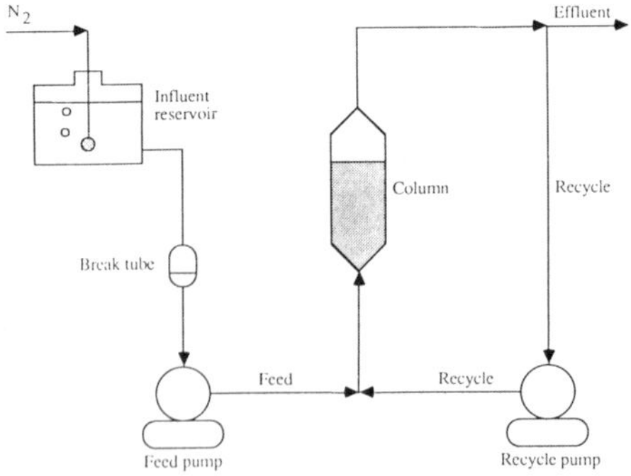

Figure 2. Schematic illustration of the anaerobic, completely-mixed fluidized-bed system.

$$S_{lo} = \frac{K_i - \sqrt{K_i^2 - 4K_iK\,S_{min}^{*2}}}{2S_{min}^*} \quad (3)$$

$$\text{in which } S_{min}^* = \frac{b'}{Yq_{min} - b'} \quad (4)$$

$$b' = b + b_s \quad (5)$$

and b_s = first-order biofilm-detachment coefficient $[T^{-1}]$, and b′ = overall first-order biofilm-loss coefficient $[T^{-1}]$.[10] S_{min}^* first introduced by Sáez and Rittmann,[11] is the fundamental biotic characteristic that describes the relative growth potential of the cells. As given in eq. (4), S_{min}^* is defined as the ratio between the specific loss rate (b′) and the maximum specific net growth rate in the absence of any inhibition effects (Yq_m - b′). For noninhibitory substrates, S_{min}^* is also identical to S_{lo}^* [= S_{lo}/K], as is demonstrated by taking the limit of eq. (3) when $K_i \to \infty$. S_{lo} and S_{min}^* differ between dispersed-growth and biofilm processes in that b_s = 0, for dispersed growth. Thus, when $b_s > 0$, S_{min}^* and S_{lo} are larger for biofilms than for dispersed growth.

This paper reports on experimental studies evaluating the kinetics of steady-state biofilms treating a self-inhibitory substrate, phenol. Anaerobic, nonmethanogenic conditions were maintained in order to observe only the response of the phenol degraders to a wide range of phenol concentrations. Particularly important are the values of S_{lo} and S_{hi}. If the "protection" mechanism works as illustrated in Figure 1 and modeled by Sáez et al.[6] and by Gantzer,[7] $S_{hi,bf}$ should be considerably larger than the highest substrate concentration capable of supporting a stable steady-state dispersed-growth process, $S_{hi,sg}$.

EXPERIMENTAL METHODS

The reactor was a bench-scale, completely mixed, fluidized-bed biofilm reactor operated under fermenting conditions. The system is illustrated diagrammatically in Figure 2. The fluidized bed was constructed of a 15.5-cm long by 1.6-cm diameter (I.D.) polyethylene tube. The attachment medium was 10 ml of anthracite, 20 by 30 mesh. Deoxygenated influent was fed continuously at 9 mL/hr by a peristaltic pump. Effluent recycle at 5820 mL/hr was used to control bed fluidization at approximately 50% and to provide completely mixed conditions within the bed.

Table I. Composition of the Influent Liquid

Compound	Concentration, mg/L
phenol	55 to 6,000, depending on the experiment
KH_2PO_4	3,400
K_2HPO_4	136
NH_4Cl	54
$FeCl_3 \cdot 6H_2O$	5
$CaCl_2 \cdot 2H_2O$	50
$ZnCl_2$	0.5
KOH	varied to control pH to 5.6
H_3PO_4	varied to control pH to 5.6

The influent liquid consisted of monohydric phenol in a solution of mineral salts and micronutrients, as presented in Table I. It was deoxygenated by continuous purging with N_2 gas. The pH was controlled at 5.6 to minimize development of methanogens.

The bacterial inoculum was effluent from a methanogenic fluidized-bed reactor in the author's laboratory.[12] Continuous feeding of phenol at 55 mg/L and at pH = 5.6 established an initial steady-state biofilm of phenol fermenters. Phenol concentrations in the influent were stepwise increased. A 9-day period was provided for an approach to the new steady state. Then, samples were analyzed for influent and effluent phenol concentrations. The phenol flux was computed by using a steady-state mass balance around the completely mixed reactor,

$$J_{exp} = \frac{S^\circ - S}{a\theta} \tag{6}$$

in which J_{exp} = the experimental phenol flux in mg phenol/cm^2-day, S° = influent phenol concentration in mg/L, S = effluent and reactor-liquid phenol concentration in mg/L, a = specific surface area for biofilm (67 cm^{-1} for the experimental conditions), and θ empty-bed detention time (0.144 days for the experimental conditions).

Phenol concentration was determined spectrophotometrically according to the 4-aminoantipyrine method.[13]

RESULTS AND DISCUSSION

Figure 3 presents influent concentrations and phenol fluxes for all runs with effluent phenol concentrations less than 1000 mg/L. Influent and effluent concentrations were virtually identical and fluxes were zero for all higher concentrations. The pattern of J versus S conforms generally with the predictions of biofilm modeling:[6,7] the flux rises rapidly as S increases above an S_{lo} value (near 25 mg/L in Fig. 3), reaches a maximum value near $S_{hi,bf}$ (about 15×10^{-3} mg/cm^2-day at 500–700 mg/L), and drops to zero for $S > S_{hi,bf}$. Modeling predicts an immediate drop to zero flux for $S > S_{hi,bf}$. However, the flux for S° = 730 mg/L (S = 700 mg/L) is low, but larger than zero. The non-zero value probably reflects residual biofilm from the prior steady state. In other words, the 9-day period was not sufficient to wash out the entire biofilm. Nevertheless, the complete inhibition of biofilm activity for $S \geq 950$ mg/L confirms that complete wash out occurred and that $S_{hi,bf}$ was between 400 and 730 mg/L.

A non-linear least-squares technique was used to find kinetic parameters that fit the experimental data well. The fitting procedure used the experimental points for S < 730 mg/L and the constraint that 400 mg/L < $S_{hi,bf}$ < 730 mg/L; hence, the "correct" flux for S° = 730 mg/L was assumed to be zero. The sum of squares of residuals (SSR) was computed as

$$SSR = \sum_{\text{all } i} (J_{exp} - J_m)^2 \tag{7}$$

in which J_m = the model-predicted flux for the ith experimental value of S and the trial kinetic parameters. The following parameters or parameter groups were varied to obtain good fits: $q_m X_f$, K, K_I, and b'/Yq_m, in which X_f = biomass density in the biofilm [$M_x L^{-3}$]. The model-predicted fluxes were obtained by using orthogonal collocation with Legendre polynomials in planar coordinates to discretize the governing ordinary second-order non-linear differential equations.[6] The resulting system of non-linear algebraic equations was solved with the packaged subroutine ZSPOW.[14] The physical and mass-transport parameters shown in Table II also were needed.

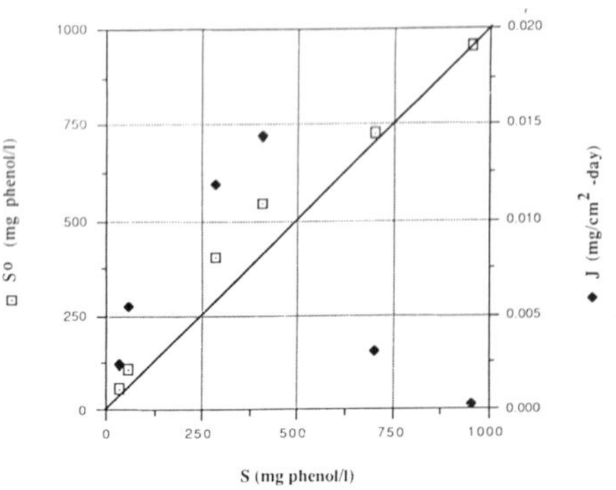

Figure 3. Performance of the steady-state fluidized-bed biofilm reactor fed phenol. The diagonal line gives a 1:1 slope for comparison of S and S° values.

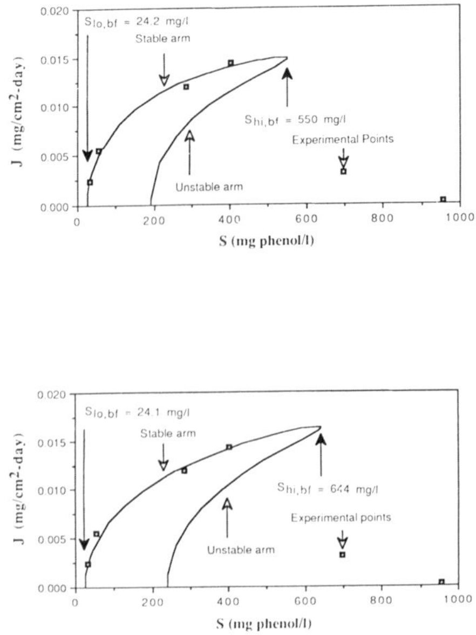

Figure 4. Model-predicted steady-state data for (a) example 1 and (b) example 2 are compared to the experimental data for non-zero fluxes. Physical, mass-transport, and kinetic parameters are given in Tables II and III.

Application of the non-linear least-squares method yielded a range of parameter combinations that could describe the experimental data with nearly the same SSR [approximately 9.1×10^{-7} (mg/cm^2-day)^2]. This lack of discrimination resulted largely from the small number of experimental points that had positive fluxes. Table III provides two divergent examples of parameter combinations giving nearly the same SSR. Example 1 describes a bacterium that can utilize phenol relatively rapidly ($q_mX_f = 4.6 \times 10^{-3}$ mg/cm^3-day), has relatively high growth potential ($S^*_{minbf} = 0.1$), but is strongly inhibited ($K_i = 21.5$ mg/L and $K_i^* = 0.1$). On the other hand, example 2 is for a bacterium with a relatively slow maximum utilization rate ($q_mX_f = 8.1 \times 10^{-4}$ mg/cm^3-day), with poor growth

Table II. Physical and Mass Transport Parameters Used to Obtain Model-Predicted Fluxes

Parameter	Value	Reference
mean medium diameter	0.068 cm	measured
medium density	1.55 g/cm^3	[15]
shape factor	0.63	[15]
water kinematic viscosity	775 cm^2/day	[16]
phenol diffusion coefficient		
in water	0.77 cm^2/day	[17]
in the biofilm	0.62 cm^2/day	[17]
porosity	0.52	computed from measured with no expansion
specific surface area	67 cm^{-1}	computed
empty-bed detention time	0.144 days	computed
Reynolds no.	6.1	computed
modified Reynolds no.	25.4	computed
Schmidt no.	1007	computed
thickness of effective diffusion layer	2.2×10^{-3} cm	computed [18]

Table III. Two Examples of Parameter Combinations Giving "Good" Fits to the Experimental Data

Parameter	Example 1	Example 2
$q_m X_f$, mg/cm^3-day	4.6×10^{-3}	8.1×10^{-4}
b'/Yq_m	0.091	0.5
K, mg/L	215	21.9
K_i, mg/L	21.5	262.8
K_i	0.1	12
S^*_{minbf}	0.1	1
$S_{lo,bf}$	0.113	1.101
$S_{lo,bf}$, mg/L	24.2	24.1
$S^*_{hi,bf}$	2.56	29.4
$S_{hi,bf}$, mg/L	550	644
SSR, (mg/cm^2-day)^2	9.1×10^{-7}	9.1×10^{-7}

potential ($S^*_{minbf} = 1.0$), but with relatively weak inhibition by phenol ($K_i = 262.8$ mg/L and $K^*_i = 12$).

Figure 4 shows how the two examples fit the experimental data for the non-zero fluxes. Neither the shapes of the curves nor the SSR values can discriminate which set is "better." Sáez et al.[6] evaluated S^*_{min} values (termed Θ^* in that paper) and found that the experimental values ranged from < 0.01 to 0.23 for self-inhibitory substrates in dispersed-growth reactors. This value range and common sense suggest that S^*_{minbgf} values less than 1.0 are likely to give stable steady states. Thus, example 1 seems to have a parameter set which is more reasonable and likely to have lead to stable steady states.

Table III and Figure 4 show that both parameter sets yielded very similar values of $S_{lo,bf}$ (approximately 24 mg/L) and $S_{hi,bf}$ (550–644 mg/L). This result emphasizes the fundamental nature of $S_{lo,bf}$ and $S_{hi,bf}$ as delimiting markers for biofilm-process performance: although the parameter sets differed extensively, the ranges of substrate concentrations over which process performance was possible were similar.

Table IV compares the $S_{lo,bf}$ and $S_{hi,bf}$ values for stable steady-state biofilm processes with the predicted $S_{lo,dg}$ and $S_{hi,dg}$ values for stable steady-state dispersed-growth reactors having the same kinetic parameters (given in Table III). Biofilm and dispersed-growth reactors have the same S_{lo} if detachment is negligible, although $S_{lo,dg}$ is less than $S_{lo,bf}$ if detachment is important. On the other hand, $S_{hi,bf}$ values are 8.1 to 8.5 times larger than $S_{hi,dg}$. This ratio demonstrates the value of diffusion resistance and concentration gradients in "protecting" the inner layers from inhibition. In this case, dispersed-growth systems could not sustain stable steady states for phenol concentrations greater than 68–76 mg/L; however, biofilm processes were stable for as high as 644 mg/L.

Finally, Figure 5 presents the model-predicted stable steady-state values of S_f and the net specific growth rate for various values of S and for the parameters of Example 1. The S_f curves show that the biofilm becomes deeper as S increases from $S_{lo,bf}$ [= 24.2 mg/L] to about 323 mg/L. Then, inhibition in the outer layers makes the biofilm more shallow. The net specific growth-rate curves demonstrate how the zone of positive specific growth rate moves from the outer layers (S near $S_{lo,bf}$) to the inner layers (S near $S_{hi,bf}$). Thus, the outer layers are in net negative growth, but they protect the inner layers through diffusional resistance. The biofilms become unstable for $S > 550$ mg/L, because the biofilm volume having a net negative growth rate always is larger than the maximum volume having a net positive growth rate.

Table IV. Comparison of S_{lo} and S_{hi} Values (mg/L) for Biofilm and Dispersed-Growth Systems When Biofilm Detachment is Negligible

S_{lo}	Example 1	Example 2
S_{lo}		
Biofilm, $S_{lo,bf}$	24.2	24.1
Dispersed Growth, $S_{lo,dg}$	24.2	24.1
S_{hi}		
Biofilm, $S_{hi,bf}$	550	644
Dispersed Growth, $S_{hi,dg}$	68	76

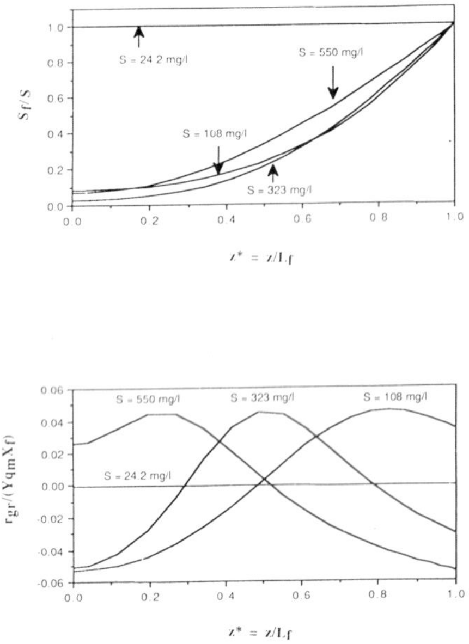

Figure 5. Normalized substrate concentration and net specific growth-rate profiles for selected S values show how the non-inhibitory concentrations were pushed deeper into the biofilm as S increased. Parameters for Example 1 were used.

CONCLUSIONS

Experiments using an anaerobic, completely-mixed, fluidized-bed biofilm reactor demonstrated that stable, steady-state removal of a self-inhibitory substrate (phenol) was possible for liquid concentrations of approximately 24 mg/L to as much as 644 mg/L. A steady-state-biofilm model employing the Haldane expression for the utilization rate of a self-inhibitory substrate was used to interpret the experimental results. Parameter estimation using a non-linear least-squares approach yielded a range of parameter sets that fit the experimental data nearly equally. Nonetheless, the fundamental characteristics $S_{lo,bf}$ and $S_{hi,bf}$ were very similar for each parameter set: $S_{lo,bf}$ was approximately 24 mg/L, while $S_{hi,bf}$ was 550–644 mg/L. According to the kinetic analysis, "protection" of the inner layers of the biofilm by diffusion through the outer layers increased the upper limit on stable steady-state operation to 550–644 mg/L from 68–76 mg/L for dispersed growth. Thus, the stable upper-concentration limit for steady-state biofilms was increased more than 8-fold, in comparison to dispersed-growth processes with the same microorganisms.

ACKNOWLEDGMENTS

This research was supported by funds provided by the U.S. National Science Foundation through grant ECE-83-51844 and from Chemical Waste Management, Inc. of Riverdale, Illinois.

REFERENCES

1. Rozich, A. F. and A. F. Gaudy, Jr. "Critical Point Analysis for Toxic Waste Treatment," *J. Environ. Engr. 110*, 562–572 (1984).
2. Gaudy, A. F. Jr., W. Lowe, A. Rozich, and R. Colvin. "Practical Methodology for Predicting Critical Operating Range of Biological Systems Treating Inhibitory Substrates," *J. Water Pollution Control Fedn. 60*, 77–85 (1988).
3. Godrej, A. N. and J. H. Sherrard. "Kinetics and Stoichiometry of Activated Sludge Treatment of a Toxic Organic Wastewater," *J. Water Pollution Control Federation 60*, 221–226 (1988).
4. Klecka, G. M. and W. J. Maier. "Kinetics of Microbial Growth on Pentachlorophenol," *Biotechnol. Bioengr. 49*, 46–53 (1985).
5. Suidan, M. T., I. N. Najm, J. T. Pfeffer, and Y. T. Wang. "Anaerobic Biodegradation of Phenol: Inhibition Kinetics and System Stability," *J. Environ. Engr. 114*, 1359–1376 (1988).
6. Sáez, P. B., H. T. Chang, and B. E. Rittmann. "Modeling Steady-State Substrate -Inhibited Biofilms," *Proceedings: International Conference on Physiochemical and Biological Detoxifica-*

tion of Hazardous Wastes, 1988, Y. C. Wu, Editor, (Lancaster, PA, Technomic Publishing Company), 1989, Vol. 2, pp. 567–586 (1988).
7. Gantzer, C. J. "Inhibitory Substrate Utilization by Steady-State Biofilms," *J. Environ. Engr. 115*, 302–319 (1989).
8. Rittmann, B. E. and P. L. McCarty. "Model of Steady-State Biofilm Kinetics," *Biotechnol. Bioengr. 22*, 2343–2357 (1980).
9. Sáez, P. B. and B. E. Rittmann. "Discussion of Kinetics and Stoichiometry of Activated Sludge Treatment of a Toxic Organic Wastewater," *J. Water Pollution Control Fedn. 61*, 357–358 (1989).
10. Rittmann, B. E. "The Effect of Shear Stress on Biofilm Loss Rate," *Biotechnol. Bioengr. 24*, 501–506 (1982).
11. Sáez, P. B. and B. E. Rittmann. "Improved Pseudoanalytical Solution for Steady-State Biofilm Kinetics," *Biotechnol. Bioengr.* 32, 379–385 (1988).
12. Wirtel, S. A. "Effects of Substrate Loading and Particle Filtration on the Performance of Methanogenic, Fluidized-Bed Biofilm Reactors," M.S. Thesis, Dept. Civil Engineering, University of Illinois, Urbana, Illinois (1988).
13. *Standard Methods for the Examination of Water and Wastewater*, 15th ed., (Washington, DC, Amer. Public Health Assn., 1980).
14. *Math/PC-Library*, Houston, Texas, (International Mathematics and Statistical Library, 1984).
15. Perry, R. H. and D. Green. *Perry's Chemical Engineers' Handbook*, 6th ed., (McGraw-Hill Book Co., New York, 1984).
16. Reynolds, T. D. *Unit Operations and Processes in Environmental Engineering*, (Brooks/Cole Engineering Div., Monterey, CA, 1982).
17. Namkung, E. and B. E. Rittmann. "Evaluation of Bisubstrate Secondary Utilization Kinetics by Biofilms," *Biotechnol. Bioengr. 29*, 335–342 (1987).
18. Rittmann, B. E. "Comparative Performance of Biofilm Reactor Types," *Biotechnol. Bioengr. 24*, 1341–1370 (1982).

32 HIGH TDS WASTEWATER TREATABILITY STUDY—DESIGN CONSIDERATIONS

Enos L. Stover, President
Alan W. Obayashi, Director of Process Development
Stover & Associates, Inc.
Stillwater, Oklahoma 74076

INTRODUCTION

The applicability of biologically treating a particular wastewater is a function of the biological degradability of the dissolved organics present in the wastewater. When considering the economics of a biological treatment system, the time required to biologically degrade the dissolved organics is of primary importance. The degradation rate of an organic compound is a function of the molecular structure of the compound, the prevailing environmental conditions, the genera and species of microorganisms utilizing it as a food source, and the time required for microorganisms to develop the enzymes necessary for substrate utilization.

The amenability or resistance of certain classifications of organic compounds to biological oxidation is well documented as follows:

1. Aliphatic or cyclic aliphatics are usually more susceptible to biological degradation than aromatics.
2. Unsaturated aliphatics, such as acrylics, vinyl, and carbonyl compounds, are generally biodegradable.
3. Molecular size is significant concerning the biodegradability of an organic. Polymeric and complex molecular substances have shown resistance to biological degradation, part of which is attributed to the inability of the necessary enzymes to approach and attack susceptible bonds within the compound structure.
4. Structural isomerisms in organic compounds affect the relative biodegradability of many compound classes. For example, primary and secondary alcohols are easily degraded, while tertiary alcohols may be resistant.
5. The addition or removal of a functional group affects the biological oxidation. A hydroxyl or amino substitution to a benzene ring renders the compound more degradable than the parent benzene, while a halogen substitution causes it to be less biodegradable.
6. Many organic compounds are easily biodegraded at low concentrations, but are biostatic or bio-toxic at higher concentrations.

Dissolved inorganic constituents can also affect the biological treatment processes. High levels of dissolved inorganics (high TDS) can have significant negative impacts on the activated sludge process. Some commonly observed problems include poor sludge settling characteristics, dispersed solids, and more sensitivity to temperature changes and variability in organic loadings. High TDS wastewaters up to 2% to 3% TDS can and are being successfully treated biologically. However, the variability in TDS concentrations at these levels is a critical operational parameter, and extreme care must be used in both adapting and acclimating the biological culture, as well as, taking extreme care to minimize the variability of the TDS concentrations.

The results of high TDS wastewater biological treatability studies on a chemical plant wastewater are discussed in this paper. Major considerations centered around biological treatment kinetics as developed by Stover and Kincannon for defining treatment capabilities in the activated sludge process. Other factors included sludge settling, thickening and dewatering characteristics, along with oxygen transfer considerations. Test results indicated activated sludge treatability to be feasible with major consideration being the variability of TDS levels.

45th Purdue Industrial Waste Conference Proceedings, © 1991 Lewis Publishers, Inc., Chelsea, Michigan 48118. Printed in U.S.A.

BIOLOGICAL TREATABILITY STUDIES

Operation of bench-scale activated sludge systems over varying operating conditions provided the performance data and biokinetic descriptive information (constants) required for design. The biological treatability investigations were conducted to provide the following information:

Descriptive biokinetic constants.
Organic removal rates.
Hydraulic retention times.
Oxygen requirements.
Oxygen transfer characteristics.
Sludge production.
Sludge settling and clarification characteristics.
Sludge thickening and dewatering characteristics.

All systems were started with biological sludge from an existing high TDS activated sludge system to speed up the acclimation and stabilization process. The systems were stabilized at the desired operating condition as soon as possible (two to three weeks). Then stabilized data collection was initiated for detailed evaluation and design purposes. During this time period the flow rates, recycle rates, and sludge wastage rates were maintained at a constant rate to simulate steady-state operation. All systems were monitored for pH, temperature, DO, and DO uptake rates. Suspended and volatile solids were run daily to control F/M and the growth rate or sludge retention time in the desired range. Removal of BOD, COD, and TOC were routinely monitored across each system on a predetermined schedule.

The systems were operated as continuous flow activated sludge systems with new feed prepared twice per week. Each wastewater was characterized in terms of pH, TSS, VSS, BOD, COD, TOC, NH_3-N, TKN, TDS, TP, and PO_4-P. Nutrients were added to each feed as appropriate for biological treatment. Representative effluents from each reactor were also characterized to determine treatment performance and develop the biokinetic constants required for the design evaluations.

GENERAL ASPECTS OF BIOLOGICAL TREATMENT

Biological removal rate kinetics can be measured in terms of BOD, COD, TC, TOC or any other parameter. The most commonly measured parameters are BOD, COD, or TOC removal rates. Removal rates reactions occurring in an aerobic biological system, although biochemically complex, are based on several fundamental processes. Biodegradable organic matter is used as a source of food by microorganisms comprising the biomass. In the process of assimilation, a portion of the organic matter is biochemically oxidized to obtain energy necessary for activity, growth, and cell maintenance. The remainder of the destructible organic material is assimilated and biochemically synthesized into new cell material (sludge production). Throughout these assimilation reactions, biomass undergoes basal metabolism reactions called endogenous respiration and oxygen is utilized in destruction and regeneration of cellular materials.

Since a biological treatment process is a viable system, its performance and reliability are dependent upon stability of the growth environment and absence of inhibitory factors. The possible inhibitory factors include extremes in pH, temperature, high TDS concentrations, and presence of toxic or inhibitory compounds. Optimum aerobic removal rate reactions occur in the pH range of 6.5 to 8.5, with temperatures between 10 and 40 degrees celsius, total dissolved solids concentrations of less than three percent, and absence of toxic substances, such as certain heavy metals or specific complex organics.

ACTIVATED SLUDGE PROCESS KINETICS

The one most critical engineering parameter for successful reliable design and operation of activated sludge systems is the F/M ratio. Relationships for F/M in terms of BOD, COD, or TOC can be developed for a particular plant and wastewater characteristics and used effectively for both design and operational process control. The fundamental basis for F/M design is that the mass substrate removed is a function of the mass substrate applied or the F/M ratio. Mathematical description of the substrate utilization rate is expressed as a function of the substrate loading rate (F/M) by monomolecular kinetics or changing order kinetics.

When considering an activated sludge reactor volume, a mass balance of substrate into and out of that reactor volume can be made as follows:

mass of substrate into the reactor	=	mass of substrate out of the reactor	+	mass of substrate consumed biologically	
FS_i	=	FS_e	+	$\frac{(dS)}{(dt)G} V$	(1)

where:
F = flow rate, MGD
S_i = influent substrate concentration, mg/L
S_e = effluent substrate concentration, mg/L
V = reactor volume in million gallons
$\frac{(dS)}{(dt)G}$ = substrate utilization rate, lb/lb·day

Mathematical description of this substrate utilization rate as a function of the substrate loading rate or F/M ratio based on monomolecular kinetics follows:

$$\frac{(dS)}{(dt)G} = \frac{U_{max}\, X \frac{FS_i}{XV}}{K_B + \frac{FS_i}{XV}} \tag{2}$$

where: $\frac{FS_i}{XV} = \frac{F}{M}$ = food to microorganism ratio, lb/lb·day
U_{max} = maximum specific substrate utilization rate, lb/lb·day
K_B = proportionality constant or substrate loading at which the rate of substrae utilization is one-half the maximum rate, lb/lb·day

By plugging this substrate utilization rate, equation 2, developed by Stover and Kincannon into equation 1, the resulting expression can be solved for the required aeration tank volume for design as follows (equation 3):

$$V = \frac{FS_i/X}{\frac{U_{max}\, S_i}{S_i - S_e} - K_B} \tag{3}$$

The resulting expression can also be solved for the effluent substrate concentration that would correspond to a specific F/M ratio and reactor volume available, as follows:

$$S_e = S_i - \frac{U_{max}\, S_i}{K_B + \frac{FS_i}{XV}} \tag{4}$$

Equation 4 can then be used to project plant performance with existing facilities at future loadings and design considerations. It can also be used for operational process control. Once the kinetic constants, U_{max} and K_B, have been determined the required F/M ratio at any given loading (F and S_i) can be determined to reliably meet the required effluent quality (S_e).

The kinetic constants, which represent the treatability characteristics of the wastewater, can be determined experimentally for any given activated sludge plant. These constants can be developed from pilot plant or laboratory scale studies, as well as, from the full-scale plant data when sufficient data is available to accurately determine these constants. These constants can be determined by linearizing equation 2 or taking the reciprocal as follows:

$$\frac{1}{\frac{(dS)}{(dt)G} \cdot \frac{1}{X}} = \frac{K_B}{U_{max}} \cdot \frac{1}{\frac{FS_i}{XV}} + \frac{1}{U_{max}} \tag{5}$$

where

$$\frac{(dS)}{(dt)G} \cdot \frac{1}{X} = \frac{F(S_i - S_e)}{XV}$$

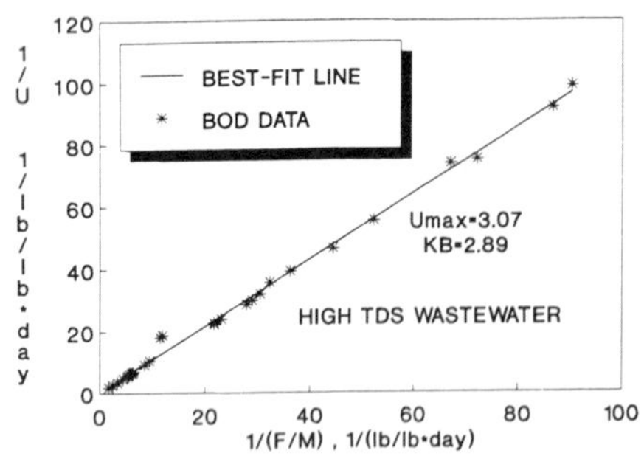

Figure 1. Reciprocal specific substrate utilization rate versus reciprocal specific substrate loading rate for BOD kinetics.

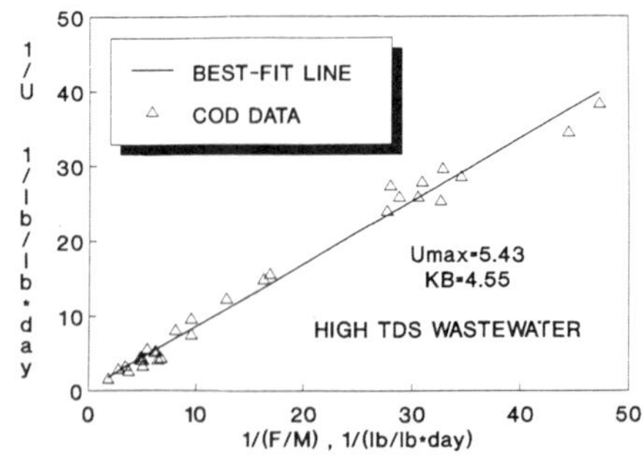

Figure 2. Reciprocal specific substrate utilization rate versus reciprocal specific substrate loading rate for COD kinetics.

Now the reciprocal of the specific substrate utilization rate, $F(S_i - S_e)/XV$ is plotted against the reciprocal of the specific substrate loading rate, FS_i/XV = F/M ratio. In this linearized form, the Y-axis intercept is equal to $1/U_{max}$, and the slope of the line is equal to K_B/U_{max}.

TEST RESULTS

The chemical plant wastewater treatability characteristics are presented in Figures 1, 2, and 3 in terms of BOD, COD, and TOC, respectively. In these figures the reciprocal of the specific substrate utilization rate is plotted as a function of the reciprocal of the specific substrate loading rate or F/M ratio. The substrate removal kinetic constants were developed from these figures by linear regression as follows:

Parameter	U_{max}	K_B
BOD	3.07	2.89
COD	5.43	4.55
TOC	1.94	2.20

These kinetics show a relatively low treatability rate compared to most wastewaters.

A comparison of the treatability kinetics of this wastewater is presented in Table I with two other industrial wastewaters. The maximum specific substrate utilization rate is significantly higher in terms of both BOD and COD for both of the other wastewaters. The kinetic constants in terms of TOC for all three wastewaters were fairly similar. Both U_{max} and K_B in terms of COD for the high TDS wastewater were higher than U_{max} and K_B in terms of BOD. The opposite relationship was observed for the other two wastewaters.

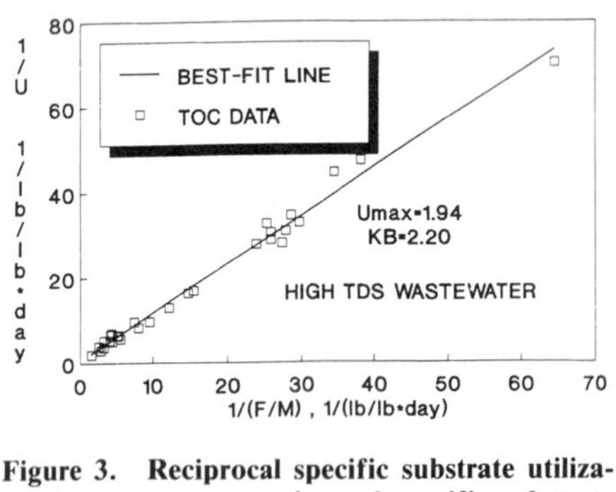

Figure 3. Reciprocal specific substrate utilization rate versus reciprocal specific substrate loading rate for TOC kinetics.

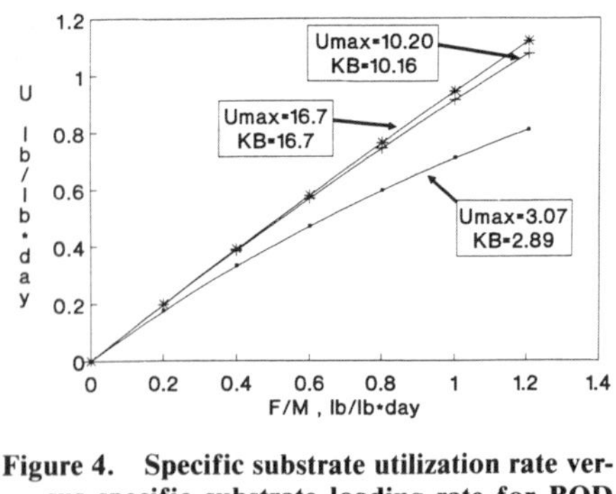

Figure 4. Specific substrate utilization rate versus specific substrate loading rate for BOD kinetics—comparison of three different wastewaters.

Table I. Comparison of Treatability Characteristics of Three Different Wastewaters

	BOD	COD	TOC
		High TDS Organic Chemical Wastewater (This Study)	
U_{max}	3.07	5.43	1.94
K_B	2.89	4.55	2.20
		Low TDS Organic Chemical Wastewater	
U_{max}	10.20	7.75	2.00
K_B	10.16	9.26	2.80
		High Strength Carbohydrate Wastewater	
U_{max}	16.70	12.30	2.30
K_B	16.70	12.30	2.20

A comparison of the treatability characteristics of the three wastewaters (including this study) in terms of BOD is presented graphically in Figure 4. The specific substrate utilization rate is plotted as a function of the F/M ratio in order to show a relative comparison of both the maximum substrate utilization rate and the rate of substrate utilization as a function of changing F/M ratio.

A detailed wastewater characterization program was conducted to develop mass balances of flows and pollutant loads from the production facilities. Sample collection consisted of both grab and composite samples as determined appropriate. The major parameters of concern were analyzed on both individual wastewater streams and the total wastewater stream. The test results from the characterization program and treatability studies were developed into the process design basis summarized in Table II.

Table II. Process Design Basis

Wastewater Characteristics	
Flow, gpd	110,000
BOD, mg/L	2,000 (1,000– 5,000)
COD, mg/L	11,000 (5,000–30,000)
TOC, mg/L	1,800 (700– 3,000)
TDS, mg/L	15,000 (5,000–25,500)
Treatability Characteristics	
U_{max} (BOD)	3.07
K_B (BOD)	2.89
U_{max} (COD)	5.43
K_B (COD)	4.55
U_{max} (TOC)	1.94
K_B (TOC)	2.20
Nutrient Requirements	
N	Addition Required
P	Addition Required
Oxygen Requirements	
Energy, lb/lb BOD removed	0.96
Endogenous, lb/day/lb MLVSS	0.31
Oxygen Transfer	
Alpha	0.81
Beta	0.86
Sludge Production, lb/lb BOD removed	0.75
Clarification Rate, gpd/ft^2*	410
Cake Solids, %**	21
Effluent Quality	
BOD, mg/L	30
TSS, mg/L	30
pH	6–9

*Chemical addition requirements of ferric chloride and polymer for control of dispersed solids. Flocculator clarifier recommended.

**Buchner funnel tests.

With the biokinetic data developed during the treatability study, the required aeration tank volume can be properly sized using equation 3, the required effluent quality, and the design wastewater flow rate and characteristics. With the other treatability characteristics, the proper aeration equipment, nutrient requirements, clarifier size, and dewatering equipment can be properly sized and defined. Because of the high effluent TSS observed during the treatability tests, chemical addition was required for controlling the TSS using ferric chloride and a high molecular weight cationic polymer.

SUMMARY AND CONCLUSIONS

When considering the applicability of treating a particular group of organic compounds the rate of degradation is a primary concern. The amenability or resistance to biodegradation is dependent on the type or class of organic compound. Furthermore, high levels of inorganic constituents can also effect the biological treatment processes. High TDS wastewaters of up to 2 to 3% can be successfully treated, however, the variability in TDS is a critical operating parameter.

The results of a high TDS wastewater biological treatabilty study on an organic chemical wastewater were presented in this paper. The biokinetic equations developed by Stover and Kincannon were used to estimate the kinetic constants, U_{max} and K_B. Good linear correlations were obtained for the kinetics in terms of BOD, COD, and TOC removal which were used to size the aeration tank volume. In comparing the treatability characteristics with two other wastewaters, i.e., a low TDS organic chemical wastewater and a high carbohydrate wastewater, the maximum specific substrate utilization rate of the high TDS wastewater was significantly lower than the other wastewaters.

Due to the high effluent TSS levels observed when treating the high TDS wastewater, a flocculator clarifier was recommended for design. Chemical addition capabilities were also provided. The high variabilities in both TDS and organic loading rates contributed to routinely high effluent dispersed solids levels. Operations were difficult under these highly variable conditions and modifications were recommended to both reduce the level of TDS and maintain more consistent concentrations.

Biological systems treating high TDS wastewaters are more sensitive to environmental changes and are more prone to process upsets. The higher the TDS levels the lower the biological stability to environmental changes. Major considerations include changes in temperature, pH, TDS levels, organic loading rates, and specific compounds that can be bio-static or bio-toxic. Longer start-up times are required for acclimation and adaption to high TDS levels. Typically, lower reaction rate kinetics are observed, as indicated in this study, for both the conventional parameters and specific organics. It can become critical to match the loading rates with the removal rates in order to prevent accumulation of specific organics levels in the reactors. When loading rates exceed removal rates serious problems can occur, such as increased volatile emission rates and/or inhibition and toxicity problems. When properly evaluated on a biological kinetic basis, these problems can be defined and avoided by proper engineering measures. Such engineering measures to increase stability can consist of various combinations of the following:

Lower F/M ratios and associated higher sludge retention times.
Higher hydraulic retention times.
Use of equalization and surge tanks for both organic loads and TDS controls.
Use of two-stage biological treatment systems.
Use of powdered activated carbon assists.

Section Four
BIOLOGICAL SYSTEMS—B. AEROBIC

33 BIOREMEDIATION OF ETHANOL AND PETROLEUM HYDROCARBON CONTAMINATED GROUNDWATER USING A ROTATING BIOLOGICAL CONTACTOR

Carole L. Engelder, Research Engineer
Sharon A. Matson, Technician
Amoco Oil Company
Amoco Research Center
Naperville, Illinois 60566

INTRODUCTION

Historical tank and piping leakage of petroleum product and recent spills of fuel-grade ethanol have triggered soil and groundwater remediation efforts at an Amoco Oil Company petroleum products marketing terminal in the midwest. Over thirty-five monitoring wells and one pumped well have been installed at the terminal by an engineering firm to determine the actual hydraulic capture zone and aquifer drawdown of the pumped well, and map the free product thickness. Location of the wells at the terminal are shown in Figure 1.

Currently the site is being remediated by pumping up to 200 gpm of the groundwater to a publicly-owned treatment works (POTW) sewer at a cost of $1.40 per 1000 gal. This amounts to over $12,000 per month for the groundwater disposal. In addition, the groundwater discharge to the sewer is limited to < 200 ppm ethanol and < 25 ppm total petroleum hydrocarbons (TPH).

Given the readily biodegradable nature of the ethanol and light petroleum products, biological treatment was chosen for study of the site clean-up. One remediation scenario would be to remove the ethanol and petroleum hydrocarbons from the pumped groundwater in above-ground treatment and discharge the treated stream to surface waters or recharge on-site to the aquifer.

Objectives of this study were to determine bioremediation efficiencies during treatment of the contaminated groundwater at the marketing terminal. Data acquired can also be applied indirectly to in-situ bioremediation options.

BACKGROUND

To investigate biological remediation of the groundwater, an above-ground process called the rotating biological contactor (RBC) was chosen. Also known as bio-disc or rotary disc, this technology utilizes an attached-growth of bacteria to degrade the organic contaminants in the groundwater. Studying the removal using the "biofilm" can develop information for both full-scale above-ground biotreatment methods, as well as in-situ bioremediation accomplished through soil-attached bacteria.

An RBC consists of circular sheets of media that revolve in groundwater pumped through a tank. Microbial biofilm grows on the media and is alternately exposed to oxygen in the air and organics in the water as the discs continuously rotate. The discs, mounted along a shaft, are typically divided by baffles into stages to enhance plug-type flow. Biofilm processes have a number of advantages over suspended growth biotreatment methods, such as activated sludge. The RBC does not require continual monitoring of the biomass and the subsequent adjusting of waste sludge and return sludge flows. Biomass control in an RBC is not necessary.

The amount and type of microorganisms in the biomass develops naturally along the RBC staging to best treat the available waste. Organic contaminants in the groundwater are gradually consumed as the influent flows through the four RBC stages, with the majority of biodegradation occurring in the first stage.

45th Purdue Industrial Waste Conference Proceedings, © 1991 Lewis Publishers, Inc., Chelsea, Michigan 48118. Printed in U.S.A.

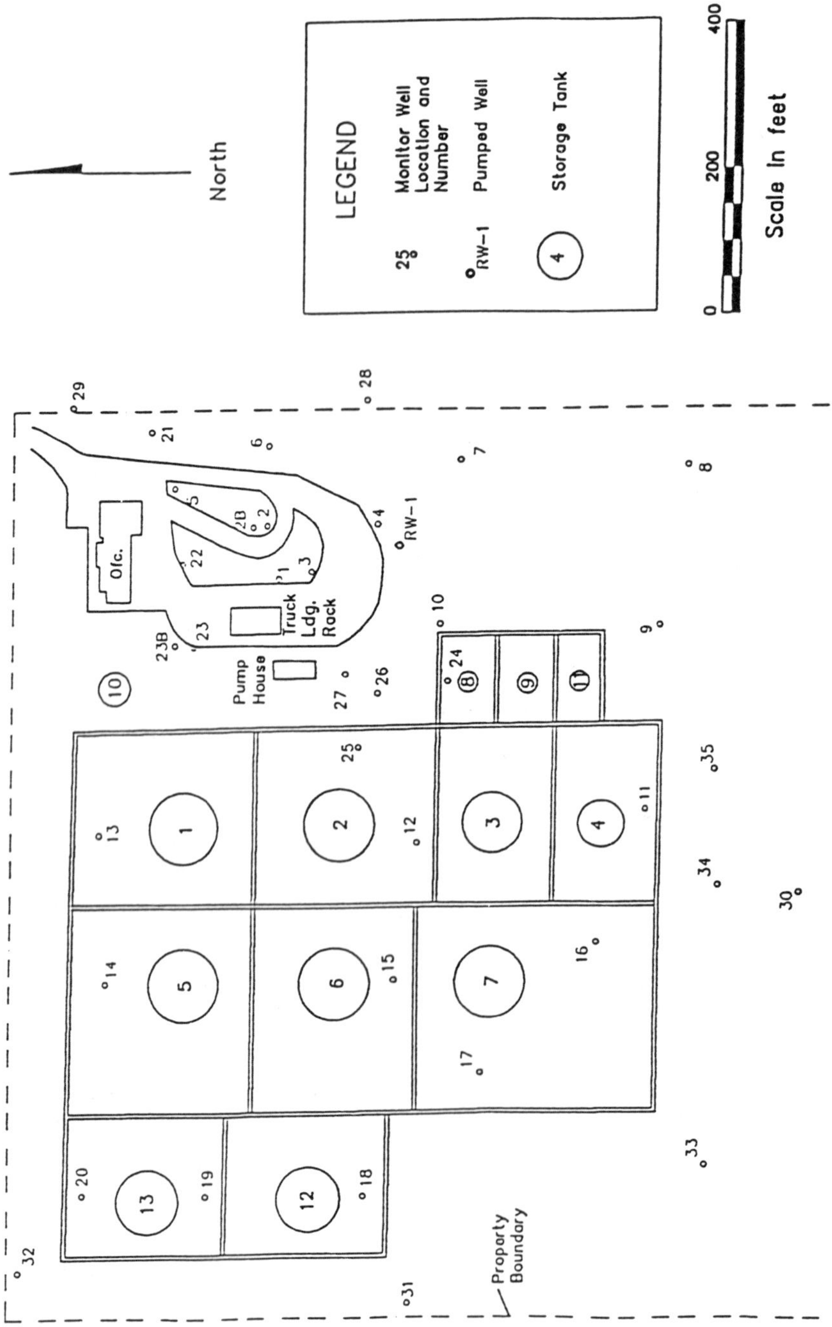

Figure 1. Well locations at the marketing terminal.

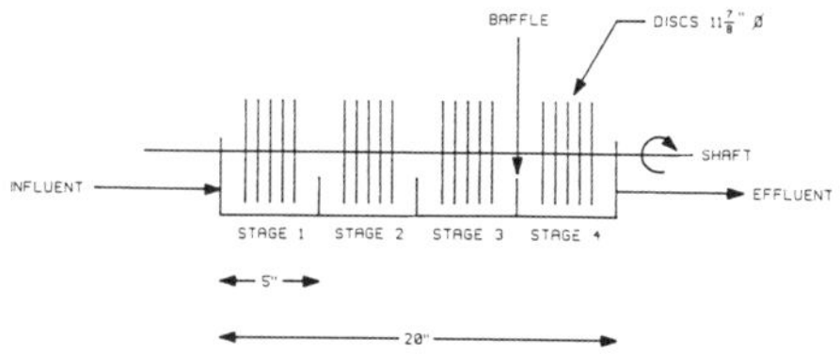

Figure 2. Bench-scale RBC design.

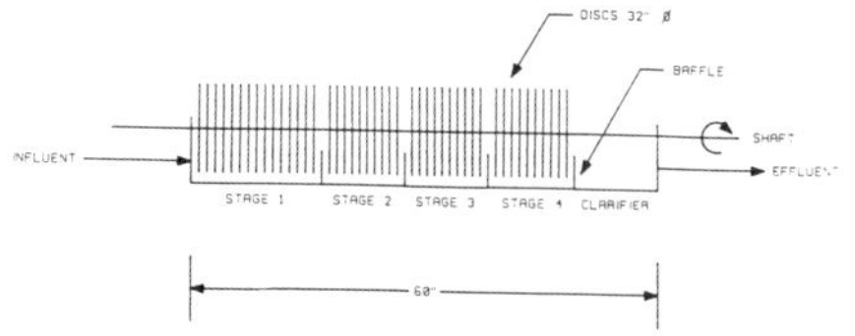

Figure 3. Pilot-scale RBC design.

This diversity of microorganisms, as well as the staged design, make the RBC process less susceptible to organic and hydraulic surges, compared to activated sludge systems. A wide range of flows and organic loadings can be treated by an RBC.

Excess bacteria, or sludge, from an RBC is readily settleable. Smaller secondary clarifiers are required for the RBC than for an activated sludge plant. The RBC sludge is also thicker than activated sludge, 2–4% total solids versus 1% solids generally found in activated sludge. Costs for sludge dewatering operations are therefore reduced.

RBCs are designed based on "decomposition curves" of percent influent organic removal versus hydraulic loading. Typically, from the hydraulic load in gpd/ft^2 and the waste stream in gpd, the required RBC media surface area is determined, ft^2.

EXPERIMENTAL METHODS

In order to assess the feasibility of remediating the contaminated groundwater at the marketing terminal biologically, preliminary lab treatability studies were conducted using a bench-scale RBC. Petroleum hydrocarbon and ethanol removal rates determined in the bench-scale unit were used for scale-up to a pilot RBC. Actual groundwater testing at the marketing terminal was performed in a pilot-scale vessel.

Bench-scale RBC

Based on initial characterization of the groundwater, a bench-scale RBC was designed for 230 ppm chemical oxygen demand (COD). Using RBC design curves, hydraulic loading (L) to the reactor was assigned a value of 1 gpd/ft^2. After constructing a conveniently sized tank and calculating the available media surface area (A), the flow through the RBC (Q) was determined as:

$$Q = A \times L$$

The bench reactor consisted of four stages with five 11 7/8" diameter acrylic discs/stage, see Figure 2. Based on the total disc surface area of 31.4 ft^2, the design flow through the RBC, containing ethanol and petroleum hydrocarbons as 230 ppm COD, was approximately 80 mL/min. Specifications of the bench-scale RBC are contained in Table I.

An artificial influent feed was devised to simulate the groundwater contamination found at the terminal. The primary constituent providing COD to the feed was fuel-grade ethanol. Prepared from ASTM spec D4806 ethanol (98%), unleaded gasoline was blended in to yield the fuel-grade product containing 93% ethanol.

Table I. Bench-Scale RBC Design Specifications

Organic content	230 ppm COD
Hydraulic load	1 gpd/ft^2
Number of Stages	4
Discs per stage	5
Disc diameter	11.875 inches
Media surface area	31.42 ft^2
Hydraulic flow	83 mL/min
Tank capacity	5.5 gallons
Hydraulic residence time	4.2 hours
Disc peripheral velocity	1 ft/sec
Disc rotation speed	20 rpm

Table II. Pilot-Scale RBC Design Specifications

Organic content	475 ppm COD
Number of Stages	4
Discs per stage	15, 10, 10, 10
Disc diameter	32 inches
Media surface area	503 ft^2
Tank capacity	128 gallons
Disc peripheral velocity	1 ft/sec
Disc rotation speed	7 rpm
Hydraulic flows	0.5 and 0.8 gpm
Hydraulic load	1.43 and 2.29 gpd/ft^2
Hydraulic residence time	4.3 and 2.7 hrs

In addition to ethanol, the groundwater and soil have been contaminated by leakage of petroleum products from tanks and piping. To supply a variety of these constituents to the artificial feed in the laboratory, wastewater obtained from an Amoco refinery was used. At the refinery, the wastewater flows from the process sewer through API-design oil/water separators and dissolved air flotation (DAF) units. Effluent obtained from the DAF was combined with diluted fuel-grade ethanol to produce the RBC influent stream, or "synthetic" groundwater.

Finally, nutrients were required for the biological groundwater treatability study. A COD:N:P ratio of 100:5:1 was used.[1] Nutrients nitrogen and phosphorus were added as the salts NH_4Cl and KH_2PO_4, dissolved in the ethanol feed. Sodium carbonate was supplied to the RBC as needed for buffering capacity to maintain a neutral pH.

During reactor start-up, the RBC was seeded with refinery wastewater and waste activated sludge bacteria. After six days in batch operation with daily additions of bacterial seed and wastewater, continuous flow was initiated.

Soluble organic carbon (SOC) removals through the bench-scale RBC were measured as Total Organic Carbon (TOC). Filtered samples were analyzed in an O.I.Corporation Model 700 TOC analyzer with autosampler. Influent and effluent SOC were compared to obtain the contaminant removal rate.

During this study, the comparative ratio between COD and TOC was approximately 4:1. In other words, for the design COD of 230 mg/L, a TOC of about 60 mg/L would be expected.

Pilot-scale RBC

After the feasibility of biotreating the contaminated groundwater was confirmed, the RBC process was scaled-up for a field study. Additional characterization at the terminal revealed a higher COD in the groundwater than previously assumed, consequently the pilot unit was designed for a 475 ppm COD. This was based on grab and composite averages from the pumped well at the terminal.

The pilot-scale RBC was comprised of four stages and a clarifier section, see Figure 3. The first stage contained 15 discs while the remaining three stages each housed 10 discs for a total of 45. The acrylic discs were 32" in diameter netting a media surface area of about 500 ft^2. At the terminal, two groundwater flow rates were investigated: 0.5 gpm and 0.8 gpm. Complete design specifications for the pilot reactor are given in Table II.

On-site, the pilot-scale RBC was placed next to the pumped well. A tap, after the POTW Neptune meter, was installed to feed the RBC, depicted in Figure 4. Effluent from the RBC flowed with the pumped groundwater down the sanitary sewer to the POTW.

The well, operated automatically by a level control, pumped intermittently. To maintain continuous flow to the RBC, an equalization tank was utilized. RBC influent level in the equalization tank was sustained by a ball-float and solenoid-control system. In addition to the equalization tank, a nutrient feed tank was established to supply nitrogen and phosphorus (from commercial fertilizer) as required. Figure 5 locates the RBC relative to the supplementary tanks and well.

To establish bacterial biofilm on the RBC media, seed was obtained from a local wastewater treatment facility. The POTW served as an excellent source of seed because of the use of "trickling filters" there, a wastewater treatment process using bacterial biofilm, like the RBC. Seed was obtained in 5-gallon buckets at the return activated sludge pump station.

Analysis of parameters during the pilot testing was performed both on-site and by a local commercial lab. Sampling frequencies and analysis are given in Table III.

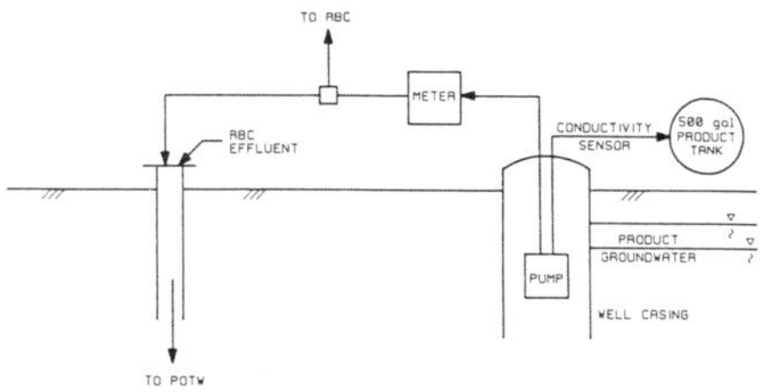

Figure 4. Pumped well system and pilot RBC feed.

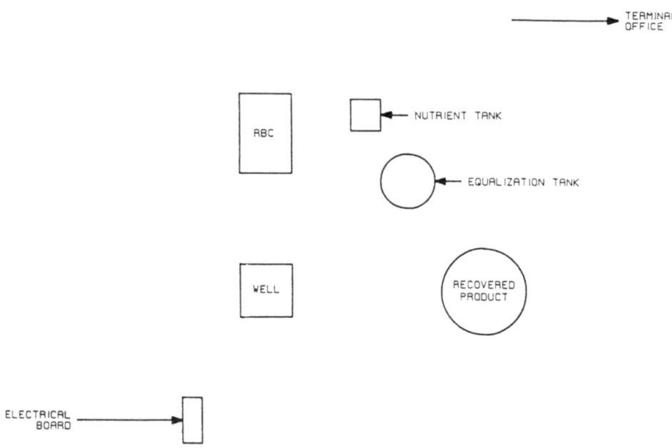

Figure 5. Location of pilot-scale RBC and auxiliary tankage.

RESULTS AND DISCUSSION

Objectives of this study were to investigate the biodegradability of the contaminated groundwater at the marketing terminal. In both the bench-scale unit treating a simulated groundwater and the pilot unit on-site, the ethanol and petroleum hydrocarbons were removed with a high level of efficiency.

Bench-scale RBC Results

In the lab, a bench-scale RBC was used primarily to study the toxicity of the ethanol contamination on biofilm bacteria. The microorganisms that thrive on the RBC media are similar to those that attach to soil particles in the aquifer. As a readily biodegradable compound, the ethanol did not prove toxic to the bacteria at the levels expected in the pumped groundwater.

Table III. Pilot Study Sampling and Analysis Program

Daily	pH	Terminal Personnel
	COD	
	NH_3-N	
	Dissolved O_2	
	Temperature	
Monthly/ Weekly	Oil and Grease	Commercial Lab
	Ethanol	
	Total Hydrocarbons	
	Phenols	
	BOD	
	Benzene	
	Toluene	
	Ethylbenzene	
	Xylene	
	Total Suspended Solids	

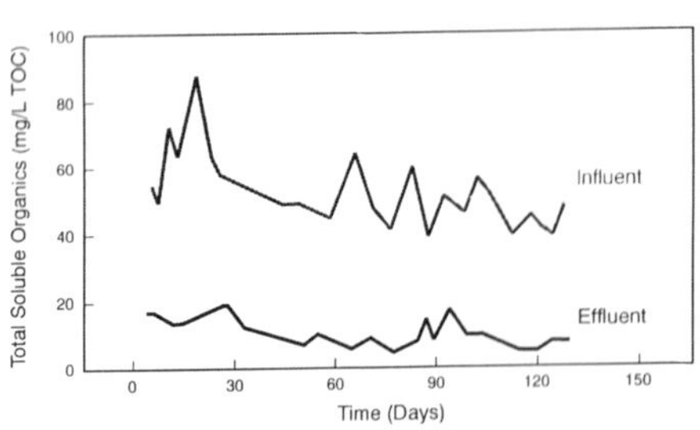

Figure 6. Bench-scale organics influent and effluent concentrations.

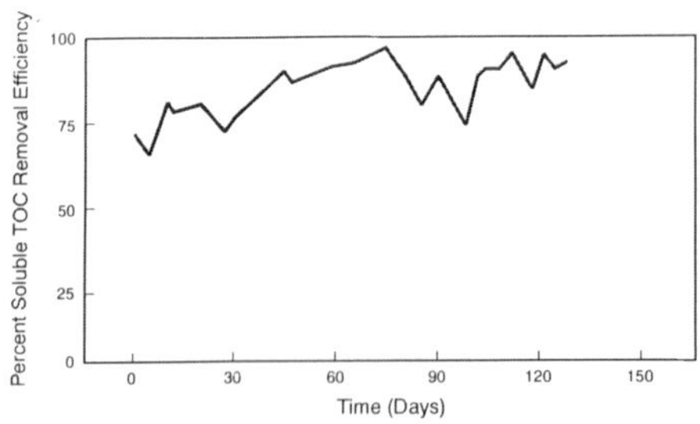

Figure 7. Bench-scale organics removal efficiency.

Organic contaminant removal rates were monitored using TOC. Figure 6 also depicts the SOC influent and effluent to the bench-scale RBC; percent removals over time are shown in Figure 7. After approximately 2–4 weeks to acclimate, the biofilm bacteria averaged 87% removal of the organics in the influent. The influent feed was typically 50 mg/L SOC, or 200 mg/L COD (using the 4:1 COD:TOC ratio previously discussed). The wastewater effluent averaged 6 mg/L SOC, after the biofilm reached steady-state.

Pilot-scale RBC Results

After fabrication and transport to the terminal, the pilot RBC was seeded and started in batch operation June 1989, with a continuous "start-up" flow rate of 0.2 gpm. After approximately 2 weeks, a steady-state bacterial biofilm was assumed and flow increased to the testing rate of 0.5 gpm. A second, higher flow rate of 0.8 gpm was investigated later in the fall.

Performance monitoring parameters were checked routinely by terminal personnel: pH, DO, NH_3-N, and COD. Table IV summarizes these parameters. pH varied seasonally with the groundwater influent from the equalization basin, averaging pH = 6.5. Pilot RBC pH values remained slightly above neutral, preempting the need for alkaline buffering at any time during the study. The groundwater has sufficient buffering capacity to satisfy the acidity produced during biological metabolism. (Carbonate buffering to raise the pH was required continually during bench-scale testing of the fabricated groundwater.)

Table IV. Pilot-Scale Monitoring Parameter Summaries

	pH		
Date	*Equalization Tank*	*Stage 1*	*Clarifier*
June-July	6.7	7.4	7.5
August	6.5	7.0	7.0
Sept-Oct	6.6	7.2	7.3
	Dissolved Oxygen [mg/L]		
Date	*Equalization Tank*	*Stage 1*	*Clarifier*
June		6–7	5–7
July-Aug	5	1–4	1–5
Sept-Oct		5–6	5–6
	Ammonia-Nitrogen [mg/L]		
Nutrient Tank *Flow rate*	*Nutrient Tank*	*Stage 1*	*Clarifier*
5–10 mL	500–1000	0.1–3	0.0–1
	Chemical Oxygen Demand		
Flow rate [gpm]	*Influent [mg/L]*	*Effluent [mg/L]*	*Percent Removal*
Start-up (0.2)	268	24	91.0%
0.5	314	29	90.9%
0.8	334	36	89.3%

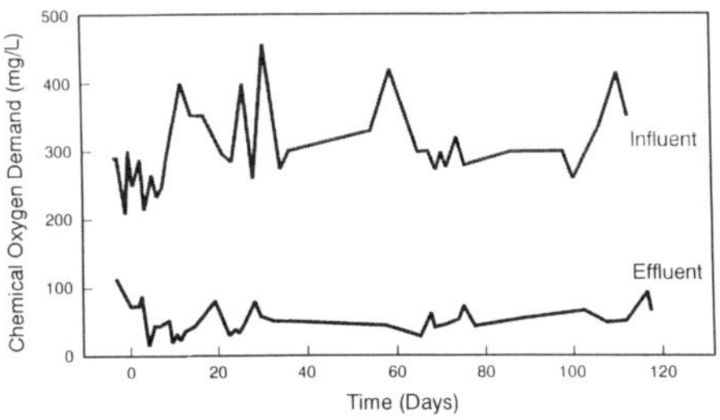

Figure 8. Pilot RBC organics influent and effluent concentrations.

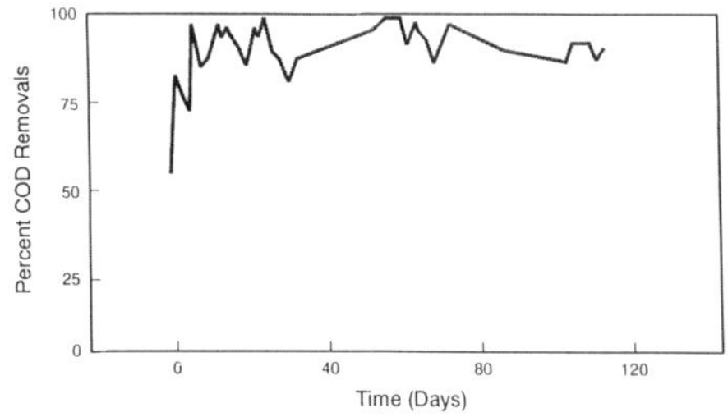

Figure 9. Pilot RBC organics removal efficiency.

Dissolved oxygen levels necessary for aerobic/facultative bacterial activity were maintained by the bio-disc rotation throughout the study, without additional aeration. Lower oxygen levels (1 mg/L DO) were recorded during the hotter summer months when the solubility of oxygen decreased.

Nutrients (primarily nitrogen and phosphorus) were supplied to the RBC by commercial fertilizer dissolved at a high concentration (500–1000 mg/L NH_3-N) in a "nutrient tank". Flow from the nutrient tank to the first stage of the RBC was regulated routinely to maintain detectable levels of ammonia in the clarifier. Nitrogen was assumed not limiting if ammonia was measured in the effluent.

Chemical oxygen demand was also monitored by terminal personnel to determine the status, or "health" of the biofilm. Percent COD removal by the pilot-scale RBC was slightly better than that measured in the bench reactor and was not effected significantly by the increase in flow rate.

Influent and effluent COD concentrations over the course of the on-site pilot study are shown in Figure 8, with the corresponding percent removals of the COD given in Figure 9.

Normal probability plots of the effluent COD levels at the two groundwater flow rates are contained in Figure 10. Slightly lower effluent quality is expected with an increased flow through the RBC. At 95% probability, the 0.5 gpm rate will achieve an effluent COD concentration of approximately 53 mg/L or less. At 0.8 gpm, the effluent groundwater will contain less than 67 mg/L of COD 95% of the time.

In addition, specific groundwater contaminants were measured through a commercial lab. These parameters and the pilot unit's removal performance at both tested flow rates are contained in Tables V and VI.

Total petroleum hydrocarbon (TPH) removals were excellent at both flow rates. Even though the concentration of TPH in the groundwater increased substantially toward the end of the study to almost 90 ppm, the organics were removed to well below 1 ppm.

During the first phase of testing, the volatile compounds (BTEX) were removed to below detectable limits (1 ppb). As the concentrations of these organics also increased in the groundwater, removals decreased slightly at the higher flow rate, but were still above 96% for benzene, xylene, and ethylbenzene.

Calculated phenol and O&G removal efficiencies were helped by the increase in organic concentrations in the groundwater. The removal efficiency improved to 86% for both at 0.8 gpm groundwater flow through the RBC.

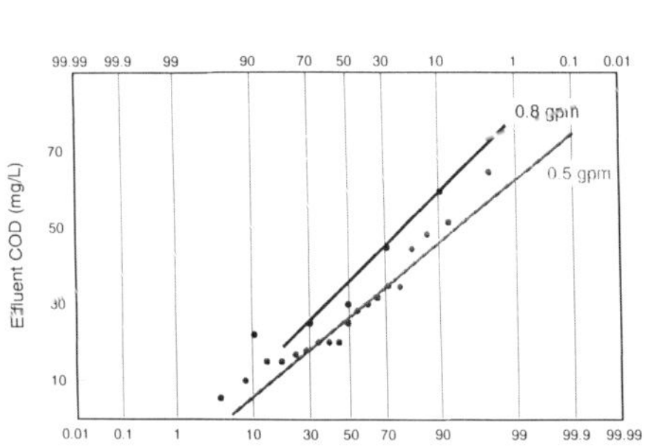

Figure 10. Probability of pilot RBC effluent quality.

Table V. Pilot Plant Performance at 0.5 GPM Flow

Constituent	Average Influent	Effluent	Removal
Total Petroleum Hydrocarbons	3400 ppb	15	99.6%
Benzene	703 ppb	<1	>99.9
Toluene	430 ppb	<1	>99.8
Xylenes	180 ppb	<1	>99.4
Ethylbenzene	52 ppb	<1	>98.1
Phenols	0.021 ppm	0.009	57.1
Ethanol	46 ppm	<1	>97.8
BOD	220 mg/L	21	90.5
O&G	2.9 mg/L	0.8	72.4

Table VI. Pilot Plant Performance at 0.8 GPM Flow

Constituent	Average Influent	Effluent	Removal
Total Petroleum Hydrocarbons	89,250 ppb	3650	95.9%
Benzene	3,800 ppb	108	97.2
Toluene	1,425 ppb	340	76.1
Xylenes	11,750 ppb	395	96.6
Ethylbenzene	3,325 ppb	97	97.1
Phenols	0.155 ppm	0.022	85.8
Ethanol	63 ppm	12	81.0
BOD	219 mg/L	46	79.0
O&G	5.9 mg/L	0.8	86.4

At the 0.5 gpm flow rate, or 4-hour residence time in the pilot-scale RBC, ethanol was degraded to below its analytical detection limit of 1 ppm. Removal efficiency deteriorated slightly to 81% with the increased flow rate and subsequent shorter hydraulic residence time.

Finally, BOD removal, an overall organic indicator, expectedly decreased with the higher flow rate as well. Although total BOD level in the groundwater did not increase during the investigation, the reduced residence time did not allow enough time for complete degradation of the organics tracked.

CONCLUSIONS

Above ground biological treatment of contaminated groundwater at a petroleum products marketing terminal proved successful for organics removals at the RBC design flow rate of 0.5 gpm. Total petroleum hydrocarbons, volatile organics, and ethanol were removed from the groundwater at efficiencies greater than 98%. Depending on discharge limits determined for the effluent stream, higher flow rates can be maintained with slight decreases in product removal efficiency.

REFERENCE

1. A. F. Gaudy, Jr. and E. T. Gaudy, *Microbiology for Environmental Scientists and Engineers*, McGraw-Hill Book Company: New York, 1980.

34 METHODS FOR UPGRADING AND RETROFITTING INDUSTRIAL ACTIVATED SLUDGE PLANTS: AN OVERVIEW AND CASE HISTORIES

Sam E. Shelby, Jr., Principal
Carl E. Adams, Jr., Principal
The ADVENT Group, Inc.
Brentwood, Tennessee 37027

INTRODUCTION

With increasingly stringent wastewater discharge regulations in the United States, particularly for the organic chemicals industry, and a new emphasis on control of effluent toxicity, many industries are faced with upgrading and retrofitting activated sludge plants. These upgrades may involve increased removals of specific organics (i.e., phenol), improved winter performance, and reduced effluent aquatic toxicity. This paper presents methods available to upgrade such systems, rationale for process selection, and several case histories demonstrating these methods.

BACKGROUND

Most activated sludge systems encountered in industry today consist of a single-stage complete mix aeration basin with final clarification system. The activated sludge system is often preceded by equalization and pH control and, perhaps, TSS and oil and grease removal. Figure 1 schematically presents the components of such a system. Single stage systems were selected in the past due to capital cost effectiveness, resistance to shock loads, and operational ease. However, problems encountered, especially with long detention time systems treating high-strength industrial wastes include the following:

- Poor winter kinetic performance for BOD and ammonia-nitrogen removal due to basin cooling (especially with surface aerators);
- The lack of selectivity or specificity for removal of priority pollutants and other toxic, but degradable organics;
- Development of poor settling sludge; and
- The presence of effluent aquatic toxicity in some cases.

Poor dissolved oxygen control has also been observed in such systems. In order to upgrade these systems to accommodate plant expansions or more stringent effluent limits, considerations should

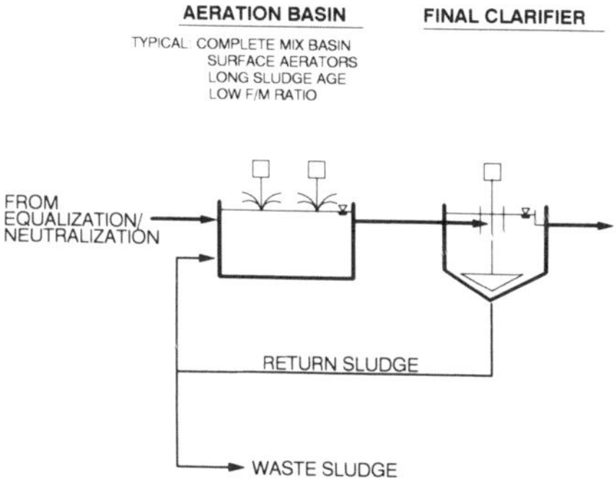

Figure 1. Typical industrial activated sludge system.

45th Purdue Industrial Waste Conference Proceedings, © 1991 Lewis Publishers, Inc., Chelsea, Michigan 48118. Printed in U.S.A.

now be given to process alternatives that address these problems, but include maximum utilization of existing facilities.

ALTERNATIVE UPGRADE METHODS

Alternative methods of upgrading industrial activated sludge systems discussed herein are:

- Addition of subsurface aeration systems;
- Addition of basin covers;
- Use of two-stage aeration systems for achieving necessary removal of specific organics. These systems include:
  - 2 stage, 1 sludge systems; and
  - 2 stage, 2 sludge systems.
- Addition of powdered activated carbon; and
- Nitrogen control, including nitrification and dentrification. Each is discussed below.

Addition of Subsurface Aeration Systems

As mentioned above, many industrial activated sludge systems have relatively long detention times that contribute to low winter basin temperatures that lead to decreased BOD, COD, or TOC removal. The performance of nitrifying systems decreases in cold winter conditions, and in extreme cold conditions (aeration basin temperatures below 10 to 15° C) biological nitrification will stop altogether. These cold weather problems are more pronounced in systems with surface aerators, where heat losses from the aerator spray are quite significant. One method to combat these cold performance problems is to replace or supplement the surface aerators with subsurface aerators. Both diffused and induction-type aerators can lead to warmer basin temperatures that can improve winter performance. Although induction-type aerators do not have heat losses from the aerator spray pattern (as with surface aerators), cold air is induced into the aeration basin mixed liquor in the winter, making these units less attractive from a heat conservation standpoint as compared to subsurface aerators, where compressed air is heated by the blowers before sparging or diffusion into the aeration basin.

Recent experience has shown that subsurface aerators will result in winter basin operating temperatures 10 to 15° C higher as compared to the use of surface aerators. The larger difference will be found for lower winter air temperature found in northern locations. This temperature difference can be sufficient to permit satisfactory nitrification and/or organics removal. Since organic removal rates approximately double for each 10° C increase in activated sludge system operating temperature, a 10 to 15° C temperature increase can insure more consistent permit compliance or significantly increase the capacity of an existing system. Also, due to the recent development of highly efficient fine and medium bubble diffusers, subsurface aerators can offer operating cost savings as compared to surface aerators even at liquid depths of 6 to 10 ft.

It should be mentioned that although aerators can be used to increase winter basin operating temperatures, their use can lead to excessively high temperatures in summer months. The maximum desired aeration basin operating temperature is 100 to 103° C, and at values consistently above this range problems can be encountered with poor sludge settling rates, reduced organic removals, and increased effluent TSS levels. If such problems are encountered or anticipated, wastewater cooling is recommended. Methods include the use of wastewater cooling towers, spray cooling systems, or other cooling methods.

Addition of Basin Covers

In addition to using subsurface aerators to maintain desired winter operating temperatures, basin covers can be added to further conserve heat losses. Covers are not recommended for aeration basins equipped with surface aerators due to high humidity levels and corrosion problems with the surface aerator motors that would be located under the cover. Also, maintenance costs will increase further due to higher labor costs associated with working in an enclosed environment, where a self contained breathing apparatus is required.

However, if coupled with subsurface aerators, basin covers can avoid the above described maintenance problems, and provide increased temperatures of 3 to 7° C above subsurface aerated basins without covers. Both structural and inflatable covers have successfully been used in industrial wastewater treatment applications. Basin covers can also provide a means for off-gas collection for odor control or for off-gas incinerator or scrubbing, if required at present or in the future. This will be increasingly important in the future as air toxics regulations are exacted and as air emissions from aeration basins are regulated by State and Federal agencies.

Other Temperature Control Methods

In addition to subsurface aerators and basin covers, other temperature control methods can be used to maintain higher winter aeration basin operating temperatures. These include insulation of above ground aeration tanks, live steam injection into the aeration basin, and possible routing of high temperature cooling water streams through the treatment plant. However, use of the later step should be carefully considered, since it will increase hydraulic loads and shorten detention times, factors that can lead to reduced treatment system performances.

Use of Two-Stage Aeration Systems

The use of two-stage aeration systems can provide several process advantages over single-stage systems. First, a two-stage system can achieve better organic removals than a single-stage system of the same total volume, due to the increased reaction rank kinetics of the two-stage system. Also, the first stage of a 2-stage system will generally be somewhat warmer, further improving biological removal rates. These factors can lead to improved removal of BOD, COD, and TOC and of specific organics, such as phenol. If organics are a significant contributor to effluent aquatic toxicity, a two-stage system should result in better organic removals that may lead to lower toxicity levels, as compared to a one-stage system.

Two-stage systems can be one-sludge or two-sludge systems depending on the use of intermediate clarifiers, as shown in Figures 2 and 3. This writer has experience with a one-sludge two-stage system that achieves nitrification on an organic chemical plant wastewater, whereas a one-sludge single-stage system was unable to nitrify. In this case, inhibitory materials were sufficiently reduced in the first stage to permit the growth of nitrifiers in the second stage.

Another advantage of a two-stage system is that it can lead to improved sludge settling as compared to one-stage system. Filamentous sludges often develop in one-stage systems due to low organic

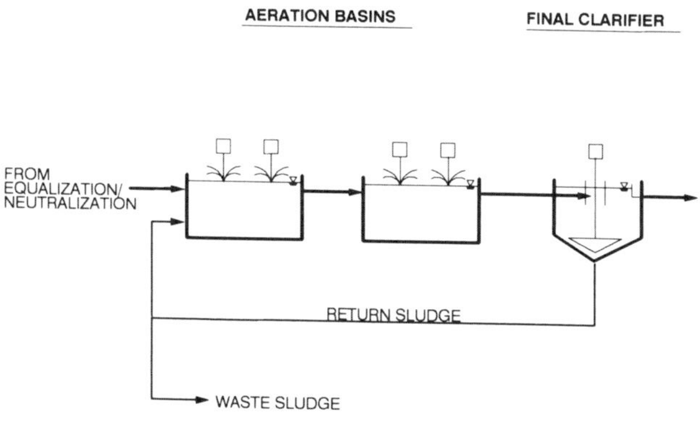

Figure 2. Two-stage one-sludge activated sludge system.

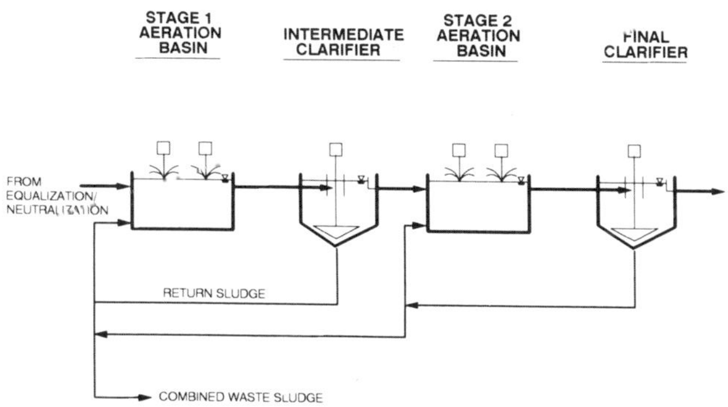

Figure 3. Two-stage two-sludge activated sludge system.

loading rates. This can be overcome, for instance, if two parallel aeration basins are operated in series. In this case the organic loading on the first stage basin is high enough to promote the growth of good settling floc formers, and promote good sludge settling rates.

Two-stage systems can be achieved by series operation of parallel aeration basins, addition of a new aeration basin, sectioning an existing, completely mixed basin, or by adding aeration equipment and returning sludge to an upstream tank or equalization basin. If a new first stage aeration basin is added or if aeration equipment is added to an existing basin, consideration should be given to using subsurface aerators for temperature control/heat conservation purposes.

In order to add operational flexibility, new plants are being designed with two aeration basins capable of series or parallel operation. For series operation, the first stage basin is equipped with additional aeration capacity since 60 to 80% of the overall oxygen requirement may be needed in the first stage. The aeration equipment in the second stage basin may be controlled by mixing requirements or by oxygen requirements for nitrification, if expected. Although this capability is slightly more expensive due to piping and aeration costs, it provides significant operating flexibility for the plant.

Powdered Activated Carbon Additions

The use of powered activated carbon (PAC) additions can be used to improve the performance and increase the capacity of existing aeration basins. Upon a brief acclimation period after addition, the powdered carbon is incorporated into the sludge biomass and retained in the system. Powdered carbon can adsorb organics and thereby improve performance, and in some cases with very inhibitory wastes, is essential for the basic viability of the activated sludge process. PAC additions can be used to reduce effluent aquatic toxicity, if organics or ammonia nitrogen are significant contributors to the toxicity. In wastewaters with high ammonia nitrogen levels, powdered carbon can adsorb toxic and inhibitory materials and promote the growth of more sensitive nitrifying organisms that would be unable to thrive without PAC additions. In addition to the above mentioned advantages, PAC additions can lead to improved sludge settling rates and lower effluent TSS levels, improved shock load resistance, and color reductions. Selected heavy metals will also adsorb to the carbon and lead to lower effluent levels, an important factor in complying with water quality criteria. PAC sludges also dewater well, another process benefit.

A recently developed modification of this process is a two-stage process with PAC additions to each stage or the second stage. This modification involves a new second stage aeration basin-clarification system following an existing activated sludge system. This two-stage process has the same benefits of a two-stage biological system mentioned above, but the staged addition of PAC to the two stages results in better carbon utilization and offers additional organic removal capability. In fact, this arrangement has been applied in a major U.S. industrial treatment plant in lieu of granular activated carbon adsorption columns. Short aeration basin detention times, on the order of a few hours, can be used if the process is to serve primarily as an adsorption step. Nevertheless, the aeration basin would be mixed and aerated to promote biological activity. If nitrification is desired, longer detention times consistent with the required sludge age for nitrification can be used. This process, which should achieve low effluent TSS levels eliminating the need for final filters, can be economically attractive as compared to filters and carbon columns. Virgin carbon additions and high mixed liquor carbon levels in the second stage allow the system to achieve low effluent organics levels and to control aquatic toxicity, also.

If PAC additions are initiated for an existing non-PAC system, increased aeration basin mixing capacity may be required. The heavier PAC sludge will settle rapidly and may not be appropriate for treatment plants equipped with hydraulic sludge take-offs on the final clarifiers. Also, royalty fees must be paid to the patent holder for the use of the process.

Nitrogen Control

Ammonia and nitrogen control is an increasingly important factor in industrial treatment plant operations, particularly due to aquatic toxicity and water quality considerations. Treatment methods for ammonia reduction include air stripping, steam stripping, ion exchange adsorption, breakpoint chlorination, and biological nitrification. Biological nitrification involves the conversion of ammonia to nitrites and ultimately to nitrates by biological organisms that can grow in activated sludge systems. In industrial plants with existing activated sludge systems, treatment plant upgrades to promote biological nitrification are generally the most attractive method to reduce ammonia, although source treatment by steam stripping may be applicable in some cases. The other treatment methods are available, but do not receive widespread use for various reasons.

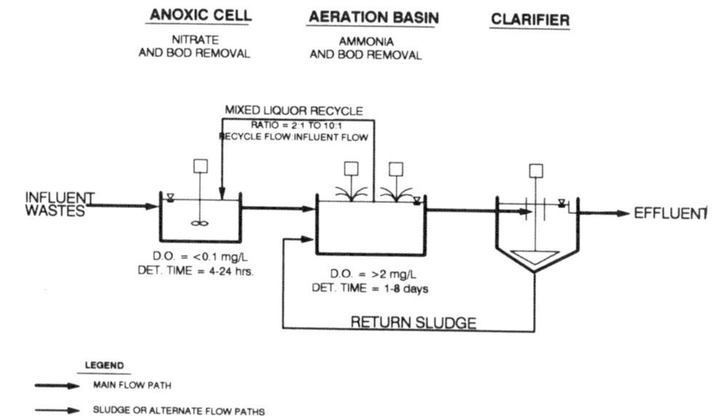

Figure 4. Nitrifying activated sludge system with first stage anoxic denitrification.

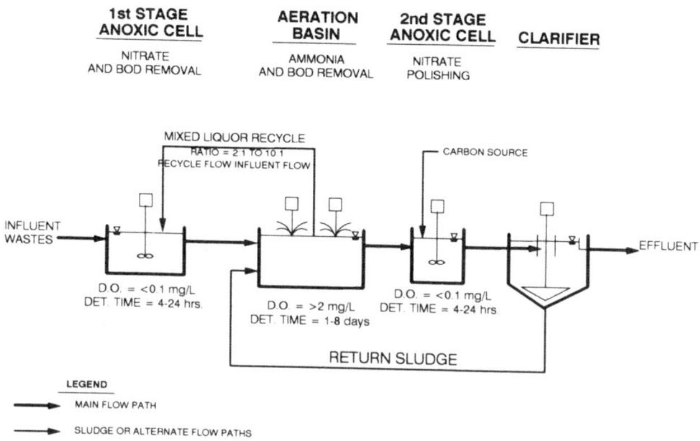

Figure 5. Nitrifying activated sludge system with first and second stage anoxic denitrification.

Biological Nitrification. All of the upgrade methods previously described can be used to promote biological nitrification. In design of these systems not only must influent ammonia-nitrogen be considered, but hydrolyzable influent organic-nitrogen must be considered, since it will be biologically converted to ammonia-nitrogen and present in the effluent as ammonia, unless converted to nitrates by nitrifying organisms. This is often observed in non-nitrifying activated sludge plants, where effluent ammonia levels are greater than influent levels.

It is well known that nitrifying organisms are more sensitive than BOD-removing organisms. Nitrifiers require higher dissolved oxygen levels and very low levels of toxic and inhibitory materials,

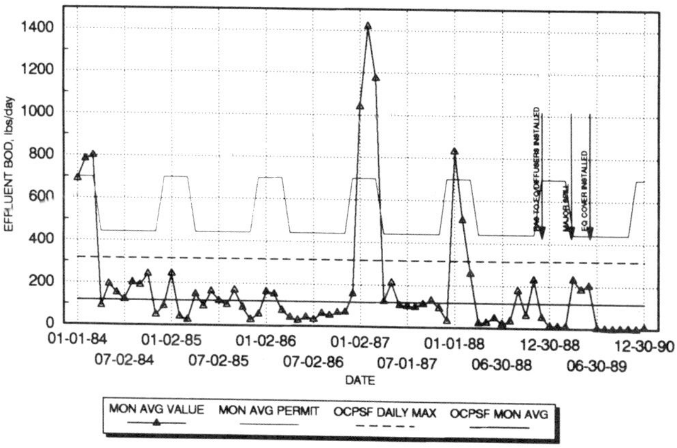

Figure 6. Discharge BOD vs. time monthly average value.

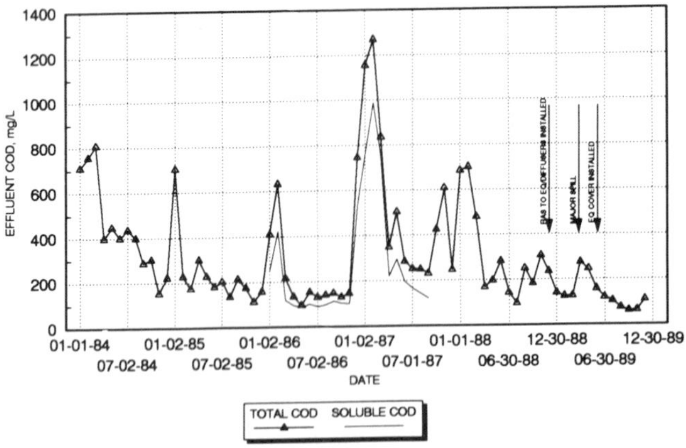

Figure 7. Discharge COD vs. time monthly average values.

such as certain specific organics and heavy metals, and do not function at basin temperatures below 10 to 15° C. Therefore, the temperature control methods, two-stage systems, or PAC additions can be used to upgrade existing activated sludge systems to achieve consistent biological nitrification.

Biological Dentrification. The presence of high nitrate-nitrogen levels in the aeration basin effluent can lead to denitrification in the final clarifiers, causing high effluent solids levels and/or final clarifier failure due to floating sludge blankets. This writer has experience where nitrate-nitrogen levels of 30 to 50 mg/L and higher caused final clarifier failure in an activated sludge system.

In order to reduce nitrate-nitrogen levels and avoid these problems, biological denitrification must be employed. In nitrifying systems this can be accomplished by the use of anoxic denitrification basins prior to or after the aeration basin, as indicated in Figures 4 and 5. The preferred approach is to locate the anoxic basin prior to the aeration basin and to recycle nitrified mixed liquor to contact with the influent waste in this basin. The anoxic denitrification basin is mixed but not aerated, and the raw waste serves as a carbon source for the denitrifying organisms. Here the organisms use chemical oxygen from nitrate, rather than dissolved oxygen, to "aerobically" degrade the raw waste organics. This will result in a reduced BOD load and oxygen requirements for the aeration basin. The recycle rate of mixed liquor can be adjusted to control the level of nitrate present in the aeration basin effluent and thereby prevent or minimize final clarifier denitrification operating problems.

If extremely low levels of nitrate-nitrogen are required in the plant discharge, a second anoxic denitrification basin prior to the final clarifier would be used as in Figure 5. This basin would be similar to the first-stage anoxic basin, except a supplemental carbon source would be provided to insure that adequate organics exist for rapid and complete denitrification. A high-strength raw waste source can be used if available. Otherwise, a slipstream of raw wastewater or a purchased carbon

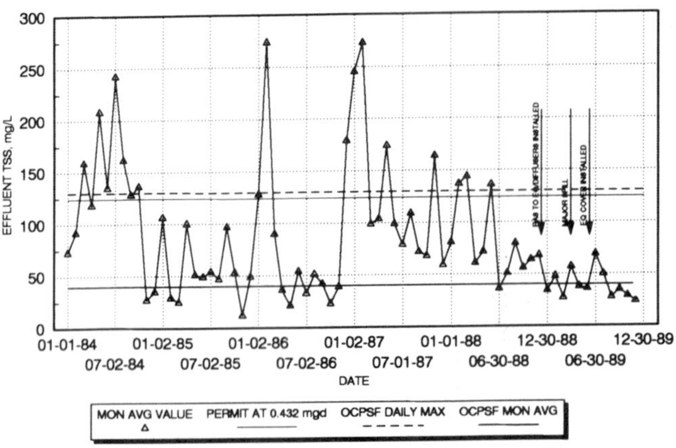

Figure 8. Discharge TSS vs. time monthly average values.

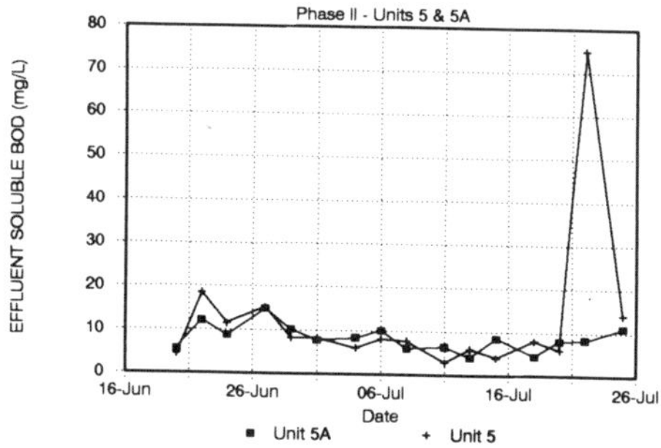

FIGURE 9A.

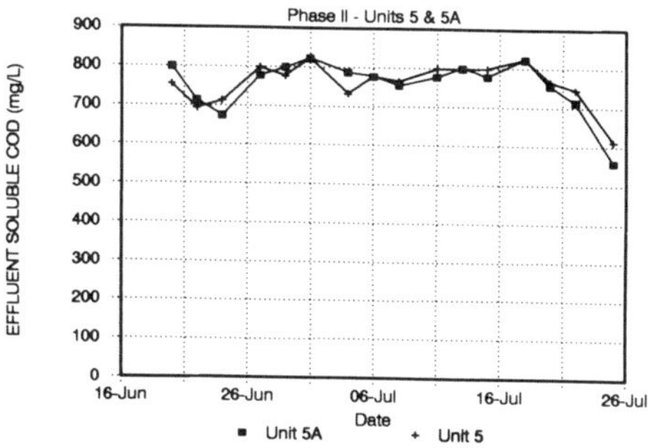

Figure 9. Effluent soluble BOD, a) effluent soluble COD.

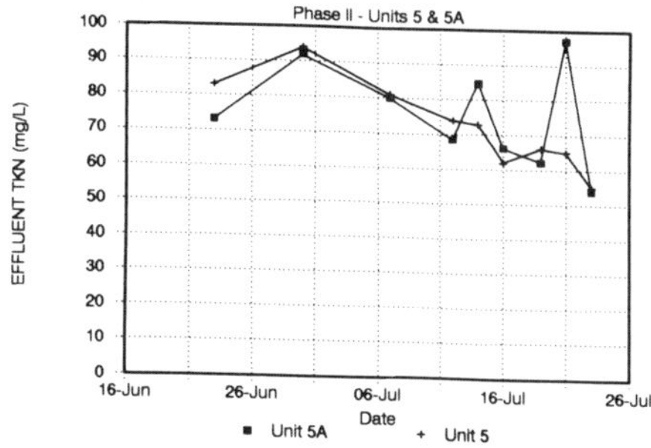

FIGURE 10A.

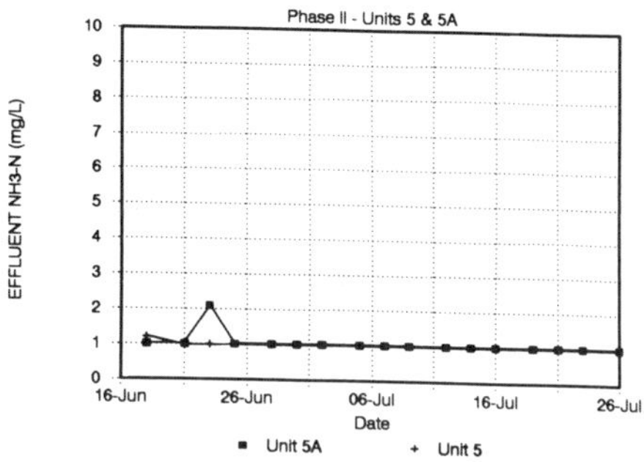

Figure 10. Effluent TKN, a) Effluent NH_3–N.

source can be provided. The organic source must be controlled and added carefully to prevent excessive additions and permit violations for BOD, COD, or TOC.

CASE HISTORIES

Several case histories are presented, demonstrating the methods discussed above. Each is discussed below.

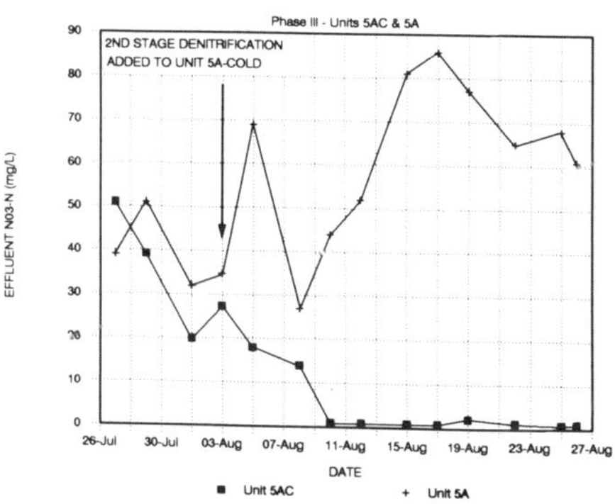

Figure 11. Effluent NO_3–N.

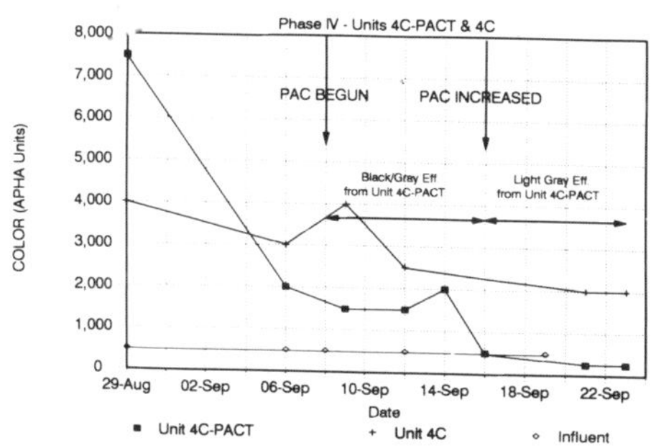

Figure 12. Influent and effluent color.

Table I. Case 3 Treatability Test Results

Effluent Parameters	Units	Phase I Complete Mix	Series	Phase II Complete Mix	Series
BOD-total	mg/L	50.0	23.5	21.0	19.0
BOD-Soluble	mg/L	9.4	8.7	15.0	10.0
TOC-Soluble	mg/L	64	52	56	51
COD-Soluble	mg/L	386	358	—	—
Total Phenolics	mg/L	0.041	0.034	0.041	0.030
Phenol-Soluble	mg/L	—	—	0.015	0.008
TSS	mg/L	31	34	26	17

Case 1

This case involved upgrading an existing equalization-activated sludge system at an organic chemicals and plastics plant by returning sludge and adding a subsurface aeration system to the equalization basin. Technical evaluations indicated this modification would be effective to improve winter performance and increase system capacity with minimal capital costs. After positive results were first obtained, a cover was added to the equalization basin to minimize heat loss and further improve performance. Results are presented in Figures 6, 7, and 8 showing the dramatic increase in performance with these improvements.

Case 2

This case involves a pharmaceutical plant treatment system upgrade to achieve ammonia-, organic-, and nitrate-nitrogen reductions and to control effluent color levels. This included adding aeration basin capacity with subsurface aerators to remove ammonia-nitrogen and organic-nitrogen, adding two anoxic denitrification basins, and powdered activated carbon additions to remove color, further reduce specific organics, and improve sludge settling, if necessary. A three-stage aeration basin-denitrification system will be employed to meet low effluent nitrate-nitrogen limits. A low-volume, high strength waste source from the plant will be added to the second anoxic basin as a carbon source for denitrification. Figures 9, 10, 11, and 12 show performance profiles for key parameters from treatability test data used to design the plant.

Case 3

A two-stage activated sludge system has been used for over two years to achieve effective phenol control, from over 100 mg/L in the raw waste to less than 0.020 mg/L in the final effluent at a large industrial plant, rather than use the more expensive alternate, activated carbon. Table I summarizes experimental test data used as a design basis for the expansion and Table II summarizes full-scale data from the plant after expansion.

Table I data shows the performance advantages of a two-stage system over a complete mix aeration basin. The plant expansion consisted of two new aeration basins to replace the existing activated sludge aeration system. The two basins were designed for parallel or series operation and equipped with subsurface aerators for temperature control purposes. The plant is operated in series mode and Table II data show average annual performance for 1989 with this mode of operation. These data indicate satisfactory compliance with the permit limits applicable to this discharge.

Table II. Case 3 1989 Full Scale Performance (Two-Stage System)

Parameters	Units	1989 Annual Average	Permit	
			Average	Maximum
BOD	mg/L	12	24	64
TOC	mg/L	24	80	238
COD	mg/L	88	350	1075
TSS	mg/L	23	40	130
Total Phenolics	mg/L	0.080	0.800	1.600
Phenol	mg/L	0.010	0.015	0.026

35 BIOLOGICAL NUTRIENT REMOVAL: A PILOT PLANT STUDY

Syed R. Qasim, Professor
Department of Civil Engineering
The University of Texas at Arlington
Arlington, Texas 76019

Clinton E. Parker, Professor and Chairman
Department of Civil Engineering
The University of Texas at Arlington
Arlington, Texas 76019

Robert T. McMillon, Superintendent
Wastewater Treatment Plant
City of Fort Worth, Texas 76101

INTRODUCTION

In 1980 a research program was initiated using bench-scale laboratory reactors to study an anoxic/anaerobic/aerobic treatment train for biological nutrient removal. In 1985 a pilot plant was constructed to field test the treatment process. These facilities were operated at varying flow and Qr/Q ratios, and at effective mixed liquor suspended solids (MLSS) concentrations. The results of 25-months of pilot-plant operational data indicated that the effluent concentrations of total BOD, TSS and nitrate nitrogen were less than 10 mg/L. Ammonia nitrogen was less than 2 mg/L. The solids settleability was excellent, and foaming due to *Nocardia*, was effectively controlled. Influent BOD concentrations of less than 89 mg/L significantly reduced the system's ability to remove phosphorus. A strong relationship between the amount of carbon source in the influent, phosphorus release in the anoxic and anaerobic tanks and phosphorus uptake in the aeration basin was established.

BACKGROUND INVESTIGATIONS

Biological Nutrient Removal

The importance of nitrogen and phosphorus in stimulating eutrophic conditions has been well documented and, as a result, over the last decade an increase emphasis has been placed on limiting these elements in wastewater effluents. Separate treatment processes capable of removing either nitrogen or phosphorus are well established. Both the theoretical and practical aspects on nitrogen removal are well understood. However, theories that explain the mechanism(s) responsible for the excess biological phosphorus uptake and the success of biological phosphorus removal has not been agreed upon.

Conditions for the separate removal of nitrogen and phosphorus by a biological system are well accepted. Anoxic conditions and a suitable carbon source are required for denitrification and the release of gaseous nitrogen. Conditions for nitrogen removal are shown in Figure 1.

Phosphorus removal conditions are shown in Figure 2. Phosphorus uptake is a temporary excess uptake by a biomass that has been stressed under anaerobic conditions. Timely separation of the biomass from the liquid phase after exposure to aerobic conditions is essential. Arvin has provided a good review of recent advances made in biological phosphorus removal.[1]

Bench scale results from a treatment train capable of significant nutrient removal and that can be easily added to an existing activated sludge treatment plant have been previously reported.[2-5] The purpose of the work reported here was to evaluate the treatment train under field conditions, to establish operational variables and environmental criteria for removal of phosphorus and/or nitrogen and to evaluate how maximizing the removal of one influences the removal of the other.

45th Purdue Industrial Waste Conference Proceedings, © 1991 Lewis Publishers, Inc., Chelsea, Michigan 48118. Printed in U.S.A.

Figure 1. Conditions for biological nitrogen removal.

Priority Processes

In the early to late 1970's three proprietary processes (PhoStrip, A/O, and Bardenpho) were developed to biologically remove phosphorus from wastewater.[1,6,7] All three processes depend upon a mechanism of cellular uptake under aerobic condition, triggered after exposure to anaerobic conditions. Since excess phosphorus uptake following anaerobic condition is temporary removal of biomass, solids separation must be timely practiced after excess uptake under aerobic conditions. The PhoStrip process employs chemical precipitation of phosphorus in a side stream where anaerobic conditions are used to release excess phosphorus. Neither the A/O nor the Bardenpho process incorporate chemical precipitation.

The A/O and Bardenpho processes incorporate biological removal of both nitrogen and phosphorus. The processes vary in process design and treatment trains used and the degree of nitrogen that can be removed. The A/O treatment process designed for combined nitrogen-phosphorus removal uses an anaerobic/anoxic/aerobic treatment train with internal recycle between the aerobic and anoxic reactors. The Bardenpho process uses a five stage treatment train consisting of a series anaerobic/anoxic/aerobic/anoxic/aerobic reactor sequence. Another process like the A/O process is the UCT process. This process requires an anaerobic/anoxic/aerobic reactor sequence and two internal recycle lines, one between the aerobic and anoxic and one between the anoxic and anaerobic. Based on a review of the PhoStrip, A/O and Bardenpho processes, it appears that these processes are capable of producing an effluent with 2 mg/L of total phosphorus.[7] Efforts to produce effluents with 1 to 2 mg/L of phosphorus have produced marginal and unpredictable results.[8] Evaluations of these processes indicate that before any process is selected to meet specific effluent limitations a pilot plant evaluation is necessary to minimize the risk of poor performance. Further research in predicting effluent quality based upon wastewater characteristics, process design and operational parameters, as well as the fundamental biochemical mechanism involved, is needed if a valid understanding of factors affecting performance is to be developed.

BENCH-SCALE STUDY

The need to minimize changes at existing activated sludge treatment plants so as to achieve significant nutrient removal is economically important to entities striving to produce environmentally acceptable effluents. In 1980 a program was initiated at The University of Texas at Arlington to

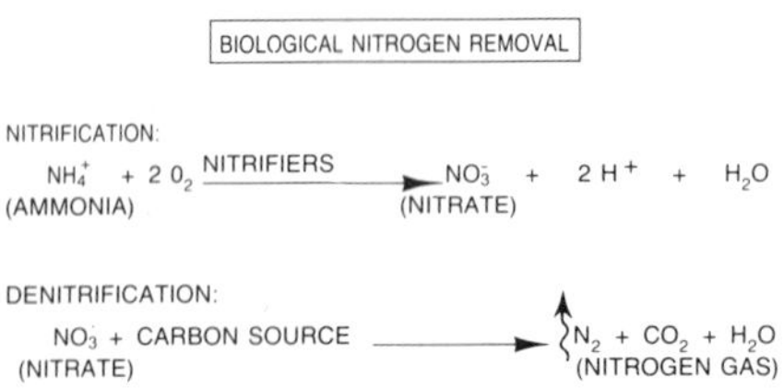

Figure 2. Conditions for phosphorus removal.

evaluate a simplified treatment regime that would require minimum unit process additions and use a single recycle line (common to most activated sludge designs). An extensive bench scale study was made using combined domestic and synthetic wastewater as feed to an anoxic/anaerobic/aerobic reactor sequence with recycle from the settling basin following the aerobic unit to the anoxic unit.

Results from the initial bench scale study were reported by Udomsinrot et al.[2,3] The system was operated at six mixed liquor suspended solids (MLSS) concentrations, ranging from 1200 mg/L to 5000 mg/L, and a return flow to influent ratio (Qr/Q) of 1.36. At low MLSS the settling characteristics of biomass deteriorated and overall results were affected. At MLSS concentrations above 2600 mg/L, 96% phosphorus removal was achieved, 65% of the total nitrogen was removed and ammonia nitrogen in the effluent was less than 1% of the influent.

In 1985 a cooperative program was initiated with the City of Fort Worth, Texas, to conduct a pilot plant study of the anoxic/anaerobic/aerobic treatment train using primary effluent. In order to develop design information a second bench scale study using primary effluent from the treatment plant at the pilot plant site was carried out.[5] The bench scale system was operated at Qr/Q ratios of 1.33 and 1.00 and MLSS was maintained between 2200 and 3000 mg/L. Nitrogen removal averaged 65% and almost complete nitrification was achieved.

Phosphorus removal significantly fluctuated. Reduced phosphorus removal because of nitrates in the anaerobic reactor in excess of 5 mg/L has been reported by Barnard.[8] Analyses of nitrogen being carried into the anaerobic reactor showed only traces of nitrate and was eliminated as a concern. A need for a minimum COD concentration in the anaerobic zone to accomplish phosphorus release and subsequent uptake in the aerobic reactor was reported by Davelaar et al.[9] Data from the bench study indicated a correlation of influent COD with phosphorus release and removal. An influent COD of at least 250 mg/L was required to maintain phosphorus removal above 80%. These data also indicated that the influent BOD should be greater than 100 mg/L.

PILOT PLANT STUDY

Pilot Plant Description

Construction of the pilot plant employing the anoxic/anaerobic/aerobic reactor sequence used in bench scale studies was completed at the City of Fort Worth, Texas, Village Creek Wastewater Treatment Plant in February 1986.

The pilot system was designed for maximum flexibility and an influent capacity of up to 68.1 m^3/d (18,000 gpd) while independently varying unit detention times, Qr/Q ratios between 1 and 3 and the air supply up to 16.5 L/s (35 cfm). It was fully instrumented to control and record flow and to alert main treatment plant operators of a malfunction at any time (day or night). Influent to the pilot plant was primary effluent. Figure 3 and Figure 4 show the arrangement of the reactors and the flow regime.

Pilot Plant Operation

The period of operation covered by the results presented here include 13 months during Phase I, and 12 months during Phase II (including start-up). Each of the operational phases are described below.

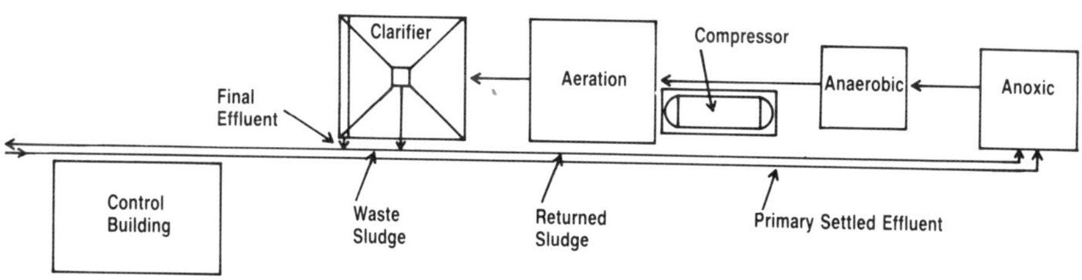

Figure 3. Schematic flow diagram of the pilot plant.

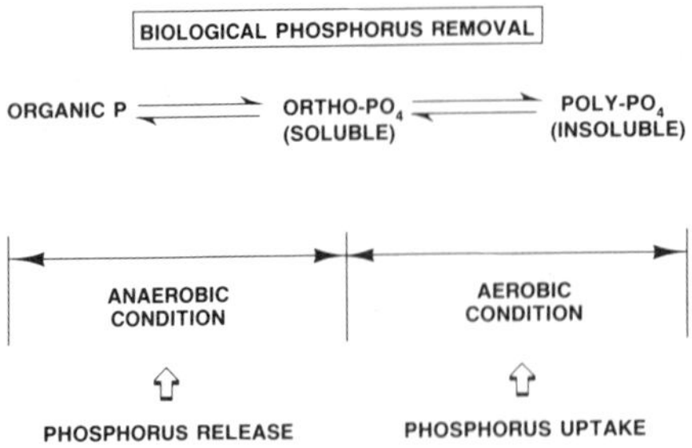

Figure 4. Photograph of pilot plant facility.

Phase I Operation

During Phase I operation the system was operated at three recycle ratios (Qr/Q) 2.0, 2.5, and 3.0. The average detention times for these Qr/Q ratios are given in Table I.

The recycle sequence during the period of operation was first at a Qr/Q of 2.5, then 3.0 and 2.0. Interruptions in operation due to malfunctions of flow meters, pumps, mixers or air compressor were minimum and did not significantly affect the operation of the facility. (Flow data for the influent was not obtained because of a meter malfunction for about 10 days in June-July 1986).

Laboratory analyses were conducted in accordance with *Standard Methods.*[10] Because of the time factor in the analyses of 5-day biochemical oxygen demand (BOD), a correlation between BOD and COD was made to estimate the BOD. Routine BOD analyses were made to check the correlation. Mixed liquor suspended solids were normalized across the anoxic, anaerobic and aerobic reactors to obtain an effective MLSS (EMLSS) and an effective (EMLVSS) for the system. The EMLSS concentration maintained in the system was 3100 mg/L. Table II shows the parameters measured and the analysis matrix used to evaluate the system.

Phase II Operation

Phase II operation was conducted over a 12-month period. During this phase the following conditions were maintained: EMLSS concentration was 2200 mg/L, and Qr/Q ratio was 0.5. The routine monitoring during this phase included all those parameters that were measured during the phase I operation. The average detention times for phase II operation are summarized in Table III.

RESULTS

Phase I Operation

The phase I operation data presented here covered a period from March 1986 through March 1987.[5,11] The dissolved oxygen in the aeration basin was maintained above 2.0 mg/L during the period of study. Morning DO values were consistently higher than afternoon values. Average monthly temperatures were the same in all reactor temperatures. The final clarifier temperature was one to two degrees lower than the reactors. Average temperature of all reactors over 13-month period was 22°C with maximum and minimum average monthly temperatures of 29 and 17°C.

Table I. Detention Times at Different Qr/Q Ratios During Phase I Operation[a]

Qr,(Q)[b,c] L/min	Qr/Q	Detention Times Based on Q + Qr, and (Q), Minutes: Anoxic Reactor	Anaerobic Reactor	Aerobic Reactor	Clarifier
99.8(39.9)	2.5	26(90)	13(47)	85(294)	74(256)
95.7(31.9)	3.0	46(176)	26(100)	93(366)	81(318)
61.6(30.8)	2.0	61(183)	35(104)	126(379)	110(330)

[a]After operation at Qr/Q of 2.5, liquid levels in the anoxic and anaerobic reactors were increased by 57 and 73%, respectively.
[b]Values in parentheses are based on influent flow only.
[c]L/min = 0.2642 gpm.

Table II. Data Collection Matrix

Parameter	Influent	Effluent	Aerobic Reactor	Anoxic Reactor	Anaerobic Reactor	Frequency[a]
Flow	×	×				C
Recycle flow			×	×	×	C
COD	×	×		×	×	W
BOD[b]	×	×	×	×	×	W
TSS		×	×	×	×	W
MLSS			×	×	×	W
MLVSS			×	×	×	W
Settleable solids			×	×	×	W
NO_3-N		×		×	×	W
NH_3N	×	×				W
TKN	×	×				W
Ortho-P(as P)	×	×		×	×	W
Total-P(as P)	×	×				W
DO			×			d
Temperature		×	×	×	×	d

[a]C-continous; W-4 to 5 times weekly; d-daily.
[b]Calculated from routine correlation with COD.

Removal of waste solids was based upon maintaining an EMLSS of 3100 mg/L. The MLVSS/MLSS ratio was essentially the same for all reactors at each Qr/Q, averaging 0.72. The sludge age based on observed increase in biological solids was 12 to 15 days. Average food to mass ratio was 0.12 per day.

Settleability of mixed liquor solids in the aerobic reactor was documented for Qr/Q of 2.0 and 3.0. Settleability was excellent, however, a significant change occurred in the sludge volume index (SVI) when Qr/Q was changed from 3.0 to 2.0. The SVI increased from 91 mL/g to 260 mL/g. This increase occurred during the winter months and is a characteristic that has been observed in the main plant in conjunction with a significant growth of *Nocardia*.

Overflow rates for the final clarifier varied between 24.4 and 36.7 m^3/m^2.d (600 and 900 gpd/ft^2) based upon (Q + Qr), and 8.2 and 10.6 m^3/m^2.d (200 to 260 gpd/ft^2) based upon Q. The average TSS concentration in effluent was 5 mg/L. A very small increase in the average effluent suspended solids was observed (from 4 to 7 mg/L) with an increase in the overflow rate. The influent and effluent data for BOD, COD, phosphorus and nitrogen are presented in Table IV.

Over the period of study for all three Qr/Q ratios the effluent COD averaged 39 mg/L with an average of 7 mg/L exerted by the effluent suspended solids. The influent and effluent summary data for BOD, COD, phosphorus and nitrogen are presented in Table IV. Soluble COD, phosphorus and nitrate concentrations in anoxic and anaerobic reactor are given in Table V.

Phase II operation

The Phase II operation data presented here covered a period from October 1987 through September 1988. During this period the dissolved oxygen in the aeration basin averaged 3.5 mg/L. As before, the DO in the morning was slightly higher than that in the afternoons. The average annual temperature of all reactors was 22°C with maximum and minimum average monthly temperatures of 27 and 13°C respectively. The EMLSS concentration and Qr/Q ratio maintained were 2200 mg/L and 0.5. The average MLVSS/MLSS ratio was 70%. The average sludge age and food to mass ratio were 13 days and 0.24 per day. The average SVI was 90, and a maximum of 235 was recorded during the winter months. The overflow rate for the final clarifier varied between 15.9 and 10.7 m^3/m^2.d (390 and 263

Table III. Detention Time at Qr/Q Ratio of 0.5 During Phase II Operation

QR,(Q)[a,b]	QR/Q	Detention Times Based on Q + Qr, and (Q), Minutes			
L/min		Anoxic	Anaerobic	Aerobic	Clarifier
20.3(40.7)	0.5	60(90)	30(45)	192(288)	168(252)

[a]Values in the parentheses are based on influent only.
[b]L/min = 0.2642 gpm.

Table IV. Influent and Effluent Summary Data for Phase I Operation

Operating Condition	Recycle Ratio, Qr/Q								
	2.0			2.5			3.0		
Parameter[a]	Max	Min	Avg	Max	Min	Avg	Max	Min	Avg
BOD									
Influent	182	64	114	233	32	122	381	47	136
Effluent soluble	3	2	2	3	2	2	3	2	2
COD									
Influent	413	145	257	527	71	275	864	106	308
Effluent soluble	83	20	33	95	19	32	143	16	31
Total phosphorus[b]									
Influent	8.2	3.0	5.9	—	—	—	9.0	5.8	7.6
Effluent	6.2	0.2	3.3	—	—	—	7.4	2.3	4.3
Orthophosphate[b]									
Influent	7.2	3.0	5.3	7.0	1.4	5.2	8.8	5.4	6.8
Effluent	5.5	0.1	3.3	6.2	0.7	3.0	7.3	1.0	4.0
Ammonia									
Influent	35.1	6.4	14.8	25.8	2.7	11.6	23.5	9.7	13.8
Effluent	1.5	<.05	0.1	0.7	<.05	0.1	0.5	<.05	0.1
Total nitrogen[c]									
Influent	41.4	12.7	21.1	32.1	9.0	17.9	40.6	16.0	19.9
Effluent	8.8	1.6	6.0	13.9	3.1	8.7	9.9	2.8	4.0
Nitrate[c]									
Effluent	7.6	2.1	4.8	12.7	3.1	—	8.7	1.6	2.9

[a]Parameters in mg/L; Removal in percent.
[b]As phosphorus.
[c]As nitrogen.

gpd/ft^2) based upon (Q + Qr) and Q respectively. The average suspended solids in the effluent was 11 mg/L. The average COD in the effluent was 51 mg/L with an average of 11 mg/L exerted by the effluent suspended solids. The influent and effluent summary data for BOD, COD, phosphorus, and

Table V. Phase I Summary Data of Soluble COD, Phosphorus and Nitrate Concentrations in Anoxic and Anaerobic Reactors.

	Values for Given Recycle Ratio, Qr/Q					
	COD[a]		Orthophosphate[a,b]		Nitrate[a,c]	
Qr/Q Range	Anoxic reactor	Anaerobic reactor	Anoxic reactor	Anaerobic reactor	Anoxic reactor	Anaerobic reactor
2.0						
Maximum	123	111	18.0	18.7	1.3	2.6
Minimum	29	33	3.0	3.3	0.1	0.1
Average	52	50	8.1	8.0	0.4	0.4
2.5						
Maximum	—	—	16.5	15.0	—	—
Minimum	—	—	3.0	2.7	—	—
Average	—	—	8.5	7.7	—	—
3.0						
Maximum	64	60	15.0	17.0	4.6	0.6
Minimum	25	29	4.0	4.0	0.1	0.1
Average	44	43	7.5	7.4	0.5	0.3

[a]Soluble fraction; mg/L.
[b]As phosphorus.
[c]As nitrogen.

Table VI. Influent and Effluent Summary Data for Phase II Operation

	Qr/Q = 0.5		
Parameter[a]	Max	Min	Avg
BOD			
Influent	179	71	130
Effluent	8	7	8
COD			
Influent	371	147	268
Effluent	190	10	40
Total Phosphorus[b]			
Influent	18	0.5	11
Effluent	10	0.2	2
Orthophosphate[b]			
Influent	10	0.5	5
Effluent	7	0.2	1
Ammonia[c]			
Influent	31	8.0	15
Effluent	29	<0.1	2
Total nitrogen[c]			
Influent	29	10	18
Effluent	18	3	11
Nitrate[c]			
Effluent	14	0.1	7

[a]Parameter in mg/L.
[b]As phosphorus.
[c]As nitrogen.

nitrogen are presented in Table VI. Soluble COD, phosphorus and nitrate concentrations in anoxic and anaerobic reactors are given in Table VII.

Nitrogen and Phosphorus Removal

Nitrogen

Average total nitrogen removal in Phase I operating at Qr/Q ratios of 2.0 and 3.0 was 72 and 80% respectively. During Phase II however total nitrogen removal was 39%. Nitrogen removal increased with an increase in recycle. This was expected since an increase in recycle increases the nitrate load on the anoxic reactor achieving greater denitrification.

Phosphorus

In Phase I, the results did not substantially change with Qr/Q. Also, phosphorus release in the anaerobic reactor and uptake in the aerobic reactor were low. In Phase II however, the phosphorus removal was significantly higher (average effluent concentration of 2.0 mg/L). Also, phosphorus release in anaerobic reactor and uptake in the aerobic reactor were significantly higher.

Others have reported that an available nitrate nitrogen concentration higher than 5.0 mg/L at the time the biomass is exposed to anaerobic conditions can prevent fermentation and affect phosphorus removal.[8] Nitrate data from the anoxic reactor indicate an average nitrate nitrogen concentration of 0.5 mg/L or less. This appears to rule out nitrates as responsible for the lower than expected

Table VII. Phase II Summary Data of Soluble COD, Phosphorus, and Nitrate Concentrations in Anoxic and Anaerobic Reactors

	COD		Orthophosphate		Nitrate	
	Anoxic	Anaerobic	Anoxic	Anaerobic	Anoxic	Anaerobic
Maximum	170	42	43	50	1	1
Minumum	34	28	small	small	small	small
Average	77	74	18	19	0.4	0.3

phosphorus removal in Phase I. The availability of carbon has also been shown to affect phosphorus removal[5,9]. Based on the average influent BOD, and BOD dilution due to recycle, the composite BOD feed to the anoxic reactor averaged 35 to 39 mg/L for the range of Qr/Q evaluated in Phase I. The composite BOD feed to the anoxic reactor in Phase II was 89 mg/L. It is evident from this study that biological phosphorus removal is influenced by the availability of biodegradable carbon in high enough concentration. Data from a bench scale study indicated that composite BOD feed must be above 65 mg/L to achieve 80% phosphorus removal.[5]

It was observed that during periods of high scum formation on the aeration basin of the main treatment plant essentially no scum formed in the pilot plant. Control of *Nocardia* scum is a concern in many activated sludge plants. Sezgin and Karr reported scum could be controlled by low sludge age or using a basin to mix the return sludge with influent to the aeration basin prior to aeration.[12] Glasser[13] was not successful in controlling scum by low sludge age, but was able to provide effective control by alternating aerobic and anoxic periods, maintaining an anoxic condition 12% of the exposure time. Experience with the anoxic/anaerobic/aerobic treatment train reported on here supports the use of periodic anoxia in the control of *Nocardia* scum. The system was free of scum and produced a sludge with excellent settleability.[5,11]

ACKNOWLEDGMENTS

Financial support for this work was provided by the City of Fort Worth. The pilot plant was located on the site of the City's Village Creek Wastewater Treatment Plant. Laboratory analyses were performed in the Environmental Engineering Laboratory of the Civil Engineering Department at The University of Texas at Arlington.

REFERENCES

1. Arvin, E., "Research Advances in the Understanding of Biological Phosphorus Removal," *Proceedings of the International Conference on New Directions and Research in Wastewater Treatment and Residuals Management*, University of British Columbia, Vancouver, Canada, June 23–28 (1985).
2. Udomsinrot, K., "Biological Nutrient Removal from Municipal and Industrial Wastewater," Ph.D. Dissertation, Department of Civil Engineering, The University of Texas at Arlington, Arlington, Texas (1985).
3. Udomsinrot, K., S. R. Qasim, C. E. Parker, W. Chiang, "Biological Nutrient Removal from Municipal and Industrial Wastewater," *Proceedings of the International Conference on New Directions and Research in Wastewater Treatment and Residuals Management*, University of British Columbia, Vancouver, Canada, 364, June (1985).
4. Qasim, S. R., and K. Udomsinrot, "Biological Nutrient Removal in Anoxic-Anaerobic-Aerobic Treatment Process," *International Journal of Environmental Studies*, 30, 257–270, (1987).
5. Qasim, S. R., C. E. Parker and O. Alsaadi, "Bench Scale Studies on Biological Nutrient Removal Using Primary Effluent from the Village Creek Wastewater Treatment Plant," report CE-FW-1-86, Department of Civil Engineering, The University of Texas at Arlington, Arlington, Texas (1986).
6. Levin, G. V., G. J. Topol and A. G. Tarnay, "Operation of Full-Scale Biological Phosphorus Removal Plant," *J. Water Pollution Control Federation*, 47, (3), 577–590 (1975).
7. U. S. Environmental Protection Agency, "Emerging Technology Assessment of PhoStrip, A/O and Bardenpho Processes for Biological Phosphorus Removal," EPA/600/S2-85/008, U.S. Environmental Protection Agency, Water Engineering Research Laboratory, Cincinnati, Ohio, (1985).
8. Barnard, J. L., "The Bardenpho Process," in Advanced Water and Wastewater Biological Nutrient Removal, (Edited by F. M. Wanielista and W. W. Eckenfedler, Jr.), *Ann Arbor Science Publishers*, 79–113, (1978).
9. Davelaar, D., R. T. Davis and G. S. Wiechers, "The Significance of an Anaerobic Zone for the Biological Removal of Phosphate from Wastewaters," *Water*, SA 4, April, 54 (1978).
10. "Standard Methods for the Examination of Water and Wastewater," 16th Ed., *American Public Health Assoc., American Water Works Assoc. and Water Pollution Control Federation*, Washington, D.C. (1985).
11. Parker, C. E., S. R. Qasim, R. T. McMillon, "Enhanced Nutrient Removal by Biological Treatment System," *International Journal of Environmental Studies*, 33, 275–284, (1989).
12. Sezgin, J., and P. R. Karr, "Control of Actinomycete Scum on Aeration Basins and Clarifiers," *J. Water Pollution Control Federation*, 58, (10), 972–976 (1986).
13. Glasser, J. A., "Control of Nocardia Scum in Activated Sludge by Periodic Anoxia," *J. Water Pollution Control Federation*, 59, (11), 972–977, (1987).

36 PERFORMANCE OF A PILOT-SCALE DEEP COLUMN BIOLOGICAL REACTOR ON SYNFUELS WASTEWATER

John R. Gallagher, Research Associate
Energy and Environmental Research Center
University of North Dakota
Grand Forks, ND 58202

INTRODUCTION

The anticipated need for a variety of wastewater treatment systems for synthetic fuels plants has spurred the testing of a novel bioreactor. Improved reactor configurations are needed that can attain high biomass concentrations and are resistant to toxic and other shocks. Limitations with respect to oxygen transfer and oxygen solubility prevent the maintenance of greater than 3,000 mg/L of biomass in conventional suspended growth biological systems.

Plug flow reactors have a gradient of substrate concentration through the reactor length, resulting in a mixture of substrate removal rates ranging from maximum to approximately the same as a complete mix system (CSTR). The combination of a high-dissolved oxygen reactor with a plug flow design provides a significantly greater removal rate potential than traditional biological systems. In addition to the kinetic advantage, plug flow reactors also are said to have fewer sludge settling problems during clarification than CSTRs. Poorly settling microbes, such as filamentous bacteria, are favored at low substrate concentrations, whereas flocculent bacteria are favored at high substrate concentrations.[1]

While biomass concentrations in conventional CSTRs may be in the range of 1 to 3 g/L, pure oxygen activated sludge (POAS) systems may attain MLSS in the range of 4 to 8 g/L at solids retention times of 3 to 15 days. POAS systems have not been widely used since they require: 1) a source of pure oxygen, and 2) high energy inputs for mixing.[2] Previous investigators have evaluated increased hydrostatic pressure for improving oxygen transfer, but did not incorporate both the use of pure oxygen or plug flow into the design.[3-8] The reactor used in this study took advantage of the potential of pure oxygen for increased biomass concentrations and plug flow for improved kinetics in a tall column (Deep Column) reactor.

METHODS

Reactor. In order to obtain the high-dissolved oxygen in the reactor and plug flow, a tall, narrow column reactor was envisioned. The reactor water column height would lead to improved oxygen solubility, due to increased hydrostatic pressure, and result in improved plug flow. In addition, since synfuels plants typically use pure oxygen in processing, pure oxygen was used for aeration of the reactor.

The theoretical oxygen saturation at 20°C with air at mean sea level is about 9 mg/L. Using 2 atmospheres of total pressure, one of standard pressure and one hydrostatic, increases the oxygen saturation to about 18 mg/L. The use of pure oxygen for aeration allow for about 87 mg/L of dissolved oxygen under 2 atmospheres. Reaction temperature under such a situation is a tradeoff; higher temperatures, to a point, result in faster reaction times and growth rates, but also result in reduced oxygen solubility. For simplicity, room temperature was used for testing; however, in commercial applications, oxygen solubility versus reaction rate considerations should be further evaluated.

For construction of the pilot reactor, a height sufficient to give 1 atmosphere of hydrostatic pressure was considered adequate to demonstrate the principle. A minimum reactor diameter of 4 inches was desired to reduce wall effects. The essential configuration of the reactor is that of a U, with the reactor and the oxygenation columns forming the legs of the U. In general, the flow pattern is that of an upflow suspended solids bioreactor. The pilot-scale reactor is shown in Figure 1 with the column lengths and widths and the clarifier roughly to scale. Design specifications for the system are given in Table I.

The liquid flow in the oxygenator is countercurrent to a flow of pure oxygen through a stainless steel fine-bubble diffuser and was pumped into the bottom of the reactor column. The added feed, sludge, and oxygenated water flowed in a plug through the column and then by gravity to the

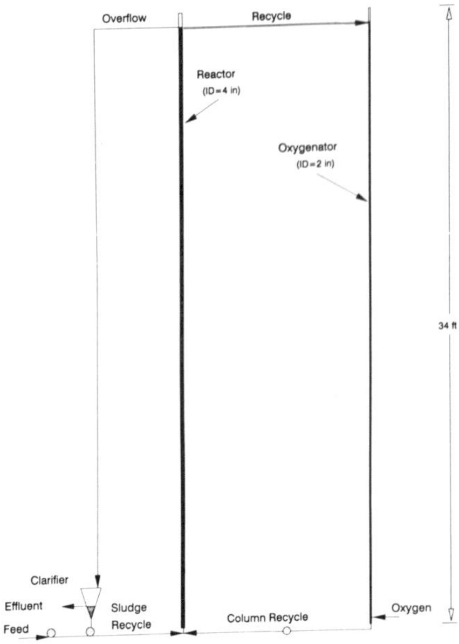

Figure 1. Process flow schematic of the pilot-scale Deep Column reactor. The reactor, oxygenator, and clarifier are drawn roughly to scale.

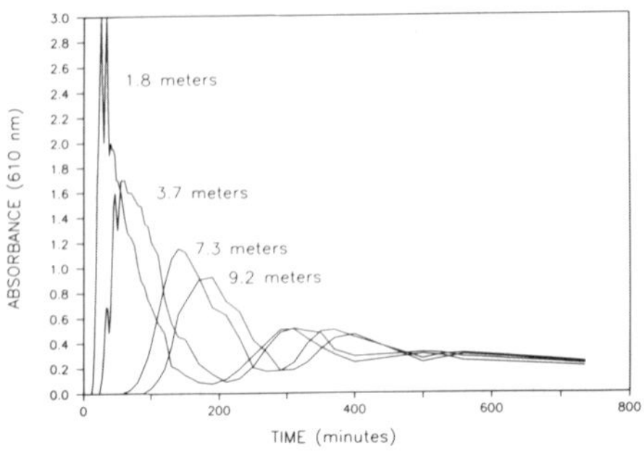

Figure 2. Plot of time versus absorbance at 610 nm for the dye injected into the Deep Column reactor as measured at the four reactor sample ports (indicated by distance from the bottom of the reactor) during the first tracer study.

oxygenator column. This arrangement ensured that the highest concentrations of biomass, feed, and dissolved oxygen entered the reactor column together. An effluent stream was drawn off the top of the reactor and flowed to the clarifier. Clarified effluent was removed by gravity, while the thickened sludge was returned to the reactor as described above. A YSI Model 56 dissolved oxygen (DO) probe

Table I. Design Characteristics of the Pilot-Scale Deep Column Reactor

Item	Specification
Reactor Column	
Material	Excelon R-4000 schedule 40 pipe
Total Height	10.97 m (36 ft)
Liquid Level	10.33 m (34 ft)
ID	10.226 cm (4.026 in.)
OD	11.430 cm (4.500 in.)
Liquid Volume	85.09 liters
Oxygenation Column	
Material	Excelon R-4000 schedule 40 pipe
Total Height	10.97 m (36 ft)
Liquid Level	10.33 m (34 ft)
ID	5.25 cm (2.067 in.)
OD	6.03 cm (2.375 in.)
Liquid Volume	22.36 liters
Clarifier	
Material	304-S with a plexiglas front
Size	rectangular inverted pyramid, 44.5 cm by 85 cm by 122 cm tall
Liquid Volume	variable

Table II. Composition of the Synthetic Wastewater

Constituent	mg/L	mg COD/L	mg TOC/L
Phenol	240	564	185
o-Cresol	30	76	23
Formate	80	28	21
Acetate	200	214	80
Ammonia	100	—	—
Thiocyanate	100	124	21
Phosphate	11	—	—
Total	—	1006	330

continuously monitored the DO at the top of the reactor. The quantity of oxygen applied to the water in the oxygenator was adjusted to maintain the DO at the top of the reactor around 3 to 4 mg/L. This control system ensured that excess oxygen was not added to the water, while preventing oxygen limitations in the reactor.

Wastewater Composition. The intended use of the Deep Column reactor is for the treatment of complex industrial wastewater, such as from the production of substitute natural gas. Although the use of actual gasification wastewater was desirable, the high variability of actual gasification wastewaters are often extremely difficult to analyze. Therefore, a synthetic wastewater was chosen for evaluation of the Deep Column process.

The synthetic wastewater was designed to be representative of a pretreated gasification wastewater.[9,10] Thus the primary organic component was phenol. Secondary organic components were cresol and the fatty acids, acetate and formate. Thiocyanate was also added, since this component is often formed from cyanide and sulfide after quenching of gasification condensates and is not removed in pretreatment. Ammonia was added to provide a nitrogen source and to simulate the background of gasification wastewater. Phosphorus was added in a ratio of 100:1, BOD to phosphorus, to meet biological requirements. The estimated COD of the synthetic wastewater was 1006. Since the organic components are all biodegradable, the BOD should approximate the COD. The composition of the synthetic wastewater is shown in Table II.

The pH of the synthetic water was adjusted to 7.0 with sodium hydroxide. Batches of the synthetic wastewater were made 20X concentration, then diluted with cold water into the feed storage tank, as needed. The feed tank had a volume of about 2040 liters (540 gal.) and was equipped with a nitrogen purge to reduce biological activity. The pH of the mixed liquor was maintained at 7.0 by the addition of sodium bicarbonate via an automatic pH control system.

Analytical. Standard methods of analysis for chemical constituents, oxygen uptake rates, and sludge settling were used.[11]

RESULTS AND DISCUSSION

Flow Analysis. Tracer studies were conducted to determine the flow pattern through the system in general and the reactor in particular. Since the reactor was designed to test the effects of both pure oxygen and plug flow, it is extremely important that the reactor flow pattern closely simulate plug flow. Two tracer studies were performed. In the first, dye concentrations from a slug addition were followed at ports located along the length of the reactor and at the effluent. These data, shown in Figure 2, indicated that the Deep Column system tended to be completely mixed after about 7 hours under the test conditions. In the second study, dye concentrations were followed in the effluent of the reactor. The data from that analysis were analyzed to determine the dispersion coefficient. The results showed that the reactor had a dispersion coefficient of 0.0217. This compares to an ideal plug flow reactor with a dispersion number of 0, or a complete mix reactor with a dispersion number of infinity.[2]

Process Performance. Operation of the pilot unit was begun by filling it with treated wastewater from acclimated bench-scale CSTRs and seeding from the same units. Initial feeding was in batches, with a switch to continuous mode when enough biomass had accumulated to accommodate the minimum feed pump rate. The feed rate was gradually increased over a period of time until the first planned loading rate, equal to about 2.6 days hydraulic retention time (HRT), was attained. Solids wasting was not performed; the unit was allowed to come to an equilibrium at the various loading rates.

The initial plan was to test four organic loadings to determine both performance and to make kinetic estimates. However, data at low loadings indicated that little differences could be observed in

Table III. Summary of Process Performance

Parameter	Range	Mean ± SD[a]
Temperature	13.0 top 33.2	23.0 ± 3.2 °C
pH	5.4 to 7.6	6.9 ± 0.2 units
HRT	0.08 to 2.63	1.4 ± 0.8 days
Fluid Velocity	0.495 to 15.35	2.1 ± 2.9 cm/s
Recycle Ratio	2.35 to 105	23.6 ± 28.0
Oxygen Uptake Rate	0.1 to 3.0	0.7 ± 0.5 mg/L-min.
Settleable Solids	100 to 897	467 ± 197 mL/L
SVI	59 to 292	134 ± 51 mL/g
Mixed Liquor VSS	1140 to 6690	3710 ± 1489 mg/L
SRT	1.6 to 130	31 ± 33 days
COD Removal	68.5 to 99.5	93.0 ± 7.8%

[a]SD = Standard deviation of the mean.

the reactor with height. Thus operations were changed to provide continued increases in loading to monitor performance, but to obtain kinetic estimates via batch methods.

Overall the performance of the system was excellent. A small number of upsets were experienced that were attributed to power failures and excursions of temperature and pH. Summary data for the reactor performance, including ranges, means, and standard deviations, for the period of operation are shown in Table III. In general, the poorest performance was only observed during an upset.

Figure 3 shows the change in the HRT as the feed rate was increased through the study period. Figure 4 shows the variation in the oxygen uptake rate through the period of operation. The rate of oxygen uptake is an indicator of biomass activity. This figure shows an increase in oxygen uptake that correlates with the increase in loading to the reactor. Figure 5 shows the variation in the concentration of biomass in the reactor, expressed as mixed liquor volatile suspended solids (MLVSS). Note that the highest observed biomass concentration was not at the highest loading. The Deep Column had only been at the higher loading for about a week before operations ceased. Undoubtedly, the steady-state biomass concentration would have been about 7 g/L. However, insufficient time remained in the project to continue operation.

The sludge settling was very good overall, especially when considering that for much of the test period, the loading rate was being changed. Figure 6 shows a sludge settling curve for the Deep Column at a time when the HRT was 0.74 days, the SVI was 200 mL/g, and the MLVSS was 3700 mg/L. This figure indicates a good settling sludge. Although gravity sludge separation was used for this research, other investigators have used dissolved air flotation for sludge separation.[6] This method should apply well to the Deep Column due to the higher solubilities of gases with increased pressure. Additionally, energy and capital costs may be reduced using dissolved air flotation.

Kinetics. The determination of kinetics in the continuous mode was desired; however, the change in substrate and biomass with reactor height were negligible at low loadings. Thus kinetics were estimated in the batch mode. Completely mixed, stirred batch reactors were used with food to microorganism ratios (F/M) ranging from 0.34 to 1.24 per day. The batch tests were performed on

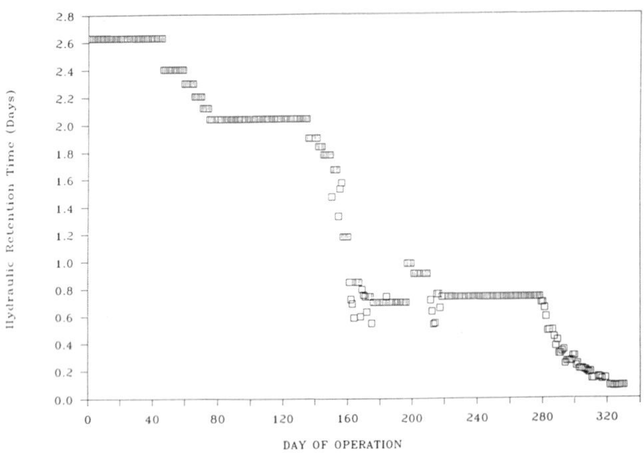

Figure 3. Variation in the hydraulic retention time (HRT) for the pilot-scale Deep Column reactor during the period of operations.

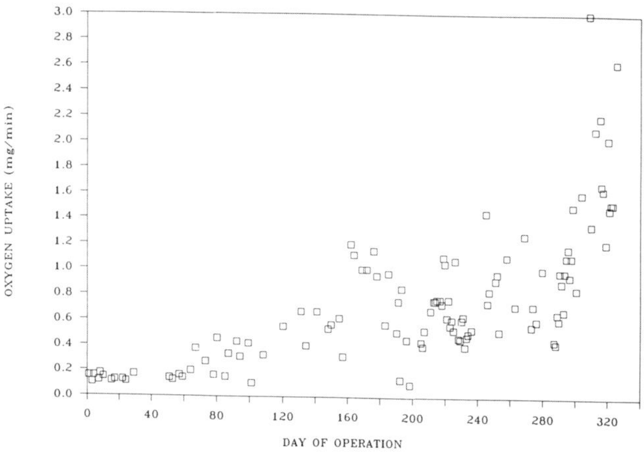

Figure 4. Variation in the oxygen uptake rate of the mixed liquor in the pilot-scale Deep Column reactor during the period of operations.

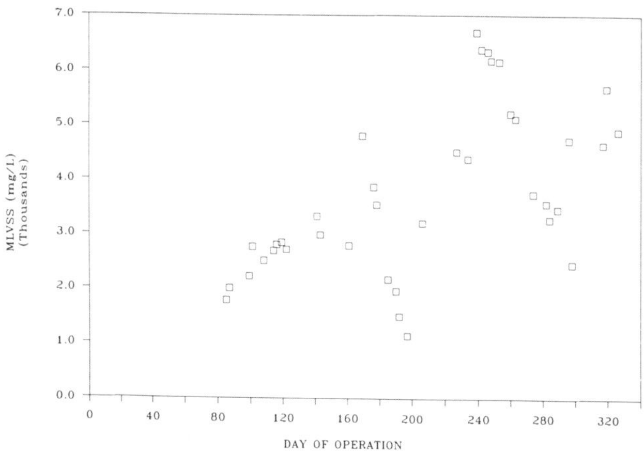

Figure 5. Variation in the mixed liquor volatile suspended solids (MLVSS) concentrations in the pilot-scale Deep Column reactor during the period of operations.

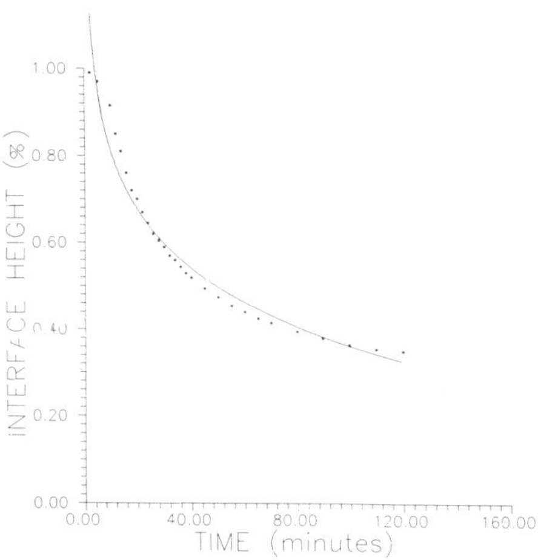

Figure 6. Sludge settling curve for the pilot-scale Deep Column reactor. Conditions were HRT ≈ 0.74 days, SVI = 200 mL/g, and MLVSS = 3700 mg/L.

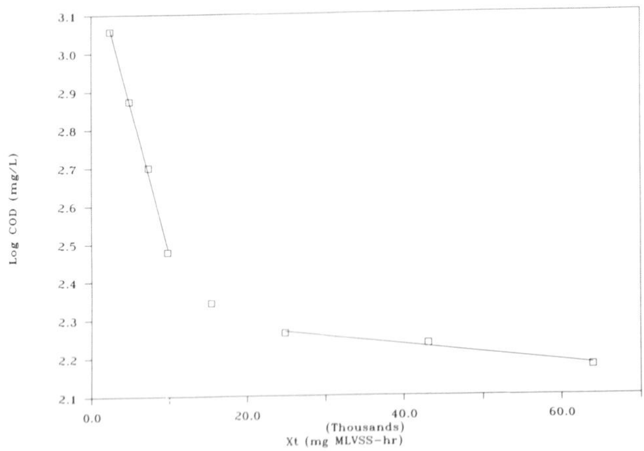

Figure 7. Plot used for determination of K, the reaction rate constant, for the batch reactor operated at a F/M of 0.67 per day.

two occasions with biomass from the Deep Column that had been operating at HRTs of 3.6 and 10 hours. The data from these two tests were combined and used to estimate the kinetic coefficients shown in Table IV according to the methods outlined by Irvine[12] and Reynolds.[13]

The pseudo-first order reaction rate constant, K, was determined by the plot shown in Figure 7.[13] This figure reveals a diphasic plot of K with a transition period between two linear areas between 10,000 and 24,000 mg MLVSS-hr. The initial, higher removal rate was 0.18 L/g MLVSS-hr (R^2 of 0.998), which occurred at substrate concentrations of 1600 to 300 mg/L as COD. The lower linear rate was 0.024 L/g MLVSS-hr (R^2 of 0.969) and occurred during substrate concentrations of 180 to 150 mg/L as COD. The average rate was 0.0056 L/g MLVSS-hr.

The diphasic nature of COD removal is due to the heterogenous nature of the feed. Some components of the feed are more easily degraded, or simply preferentially used, over others. Diphasic kinetics of COD removal of this waste complicate design, since it is not known which rates can be applied to which components of the waste stream. However, since it is known that there is an order of preference decreasing from fatty acids > phenol > thiocyanate > cyanide, it can be assumed that thiocyanate was the rate-limiting component of the COD.[9]

Table IV shows the kinetic values obtained from the analysis discussed above. All kinetic values obtained are within usual ranges reported for petroleum, coking, or other industrial wastewaters by other investigators.[2,13,14,15] Although the reaction rate was not significantly higher than those found in other, more conventional systems, the Deep Column system was capable of attaining biomass concentrations of over 6 g/L as volatile suspended solids. Even higher concentrations of biomass could have been attained with increased SRTs at the higher loadings. Additionally, it is not known to what extent batch kinetic estimates can predict the activities of a continuous plug flow system.

At the highest loading (lowest HRT), even though not at steady state, the COD removal was 90%, and no phenol was detected in the effluent (detection limit $\leq$0.1 mg/L).

Table IV. Kinetic Coefficients from Batch Testing

	Basis			R^2
Coefficient	COD	TOC	Units	(COD basis)
YG	0.47	2.01	mg MLVSS/mg S[a]	0.980
Kd	0.17	0.19	1/day	"
Y′	0.31	0.31	mg O_2/mg S	0.918
Ke′	0.25	1.08	mg O_2/mg MLVSS-day	"
K	0.18[b]	–[d]	1/g MLVSS-hr	0.948[c]
U_{max}	0.213	–	1/hour	0.882
Km	230	–	mg/L S	"

[a]S = substrate (COD or TOC).
[b]Highest observed value of the diphasic kinetics.
[c]Linear portion of higher rate.
[d]Not available.

CONCLUSIONS

Reliable estimates of reaction rate constants have not been obtained in a continuous mode, as a result of difficulties with obtaining substrate and biomass differences with reactor length in continuous operation. Rates obtained via batch methods were diphasic, indicating inhibition and/or diauxy. Implications for diphasic kinetics on actual operations are not known. Despite the fact that continuous kinetics could not be obtained, performance of the unit was excellent. The unit was very stable, required very little maintenance, and gave consistent removals. At moderately high HRTs (about 1 day) reasonable quantities of nitrification could be sustained.

The performance obtained at the 2.16 hour HRT, the lowest attained, revealed a nonsteady state MLVSS of 5000 to 5500 mg/L and COD removals of 90 to 96% with an inhibitory wastewater. This is excellent performance in any treatment system.

Overall, the concept of a plug flow, pure oxygen system has been tested on the pilot-scale; the unit was operated for over 400 days; preliminary kinetic evaluations have been obtained; and performance at HRTs of from 2.6 days to 2.16 hours has been demonstrated. The Deep Column reactor concept appears to be viable for applications that have pure oxygen available. Systems could, with reduced efficiency, be operated with air instead of oxygen. Pilot-scale comparisons with air should be conducted to demonstrate performance.

Although nitrification was not a goal of this research program, it was gratifying to find that it could be sustained at the 1-day HRT in the Deep Column. However, nitrification was lost in the Deep Column when the loading was increased to provide HRTs of less than 1 day. In an initial batch test, ammonia removal during the 24-hour reaction period averaged 98% at F/Ms up to 0.56 per day.

ACKNOWLEDGMENTS

This research was funded by Cooperative Agreement DE-FC21-86M10637 with the U.S. Department of Energy. The assistance of Jeff Hood, Amy Hartman, Brad Stevens, Michael Koehmstedt, Dan Toms, Dan Stepan, and Gale Mayer is acknowledged for construction, maintenance, data reduction, etc. of the Deep Column reactor.

REFERENCES

1. Rensink, J. H. 1974. New approach to preventing bulking sludge. *J. Water Poll. Control Fed.* 46:1888–1894.
2. Grady, C. P. L., Jr., and H. C. Lim. 1980. *Biological Wastewater Treatment*. New York: Marcel Dekker.
3. Urza, I. J., and M. L. Jackson. 1975. Pressure aeration in a 55-ft bubble column. *Ind. Eng. Chem., Process Des. Dev.* 14:106–113.
4. Jackson, M. L., D. R. James, and B. P. Leber, Jr. 1975. AIChE Symp. Series Water-1975 71:159–167.
5. Jackson, M. L., C. Shen, and C. Plopper. 1978. Deep tank aeration cuts process time. *Indust. Wastes* Nov/Dec. p. 31–36.
6. Jackson, M. L. 1981. Deep tank aeration/flotation for fermentation wastewater treatment. Purdue Indust. Waste Conf. 36:363–374.
7. Jackson, M. L., and C. Shen. 1978. Aeration and mixing in deep tank fermentation systems. *AIChE J.* 24:63–71.
8. Jackson, M. L. 1986. High capacity aeration in a very tall tank. *Biotechnol. Progress* 2:61–66.
9. Gallagher, J. R., and G. G. Mayer. 1985. Process performance of pilot-scale activated sludge treatment of Great Plains Gasification Plant's stripped gas liquor. Purdue Indust. Waste Conf. 40:1–12.
10. CH2M Hill and EMRC. 1988. Treatability testing of KILnGAS and Texaco coal gasification wastewaters. Electric Power Research Institute Report AP-5893.
11. APHA. 1985. *Standard Methods for the Examination of Water and Wastewater*. 16th Edition, Washington, D.C.
12. Irvine, R. L. 1980. *Activated Sludge: Stoichiometry, Kinetics, and Mass balances*, American Chemical Society, Washington, D.C.
13. Reynolds, T. D. 1982. *Unit Operations and Processes in Environmental Engineering*. Monterey, CA: Brooks/Cole Engineering.
14. Metcalf and Eddy, Inc. 1979. *Wastewater Engineering: Treatment, Disposal, Reuse*. New York: McGraw-Hill, Inc.
15. Benefield, L. D., and C. W. Randall. 1980. *Biological Process Design for Wastewater Treatment*. Englewood Cliffs, NJ: Prentice-Hall, Inc.

37 THE INFLUENCE OF RAW WASTES DOSING SYSTEM ON TECHNOLOGICAL PARAMETERS OF ROTATING BIOLOGICAL DISC (RBD) PROCESS

Wojciech Janczukowicz, Fulbright Scholar
Department of Civil and Environmental Engineering
New Jersey Institute of Technology
Newark, New Jersey 07102

Ewa Klimiuk, Lecturer
Maria Hull, Lecturer
Faculty of Water Protection and Inland Fishery
Academy of Agriculture and Technology
10-957 Olsztyn-Kortowo, Poland

INTRODUCTION

In biochemical processes, systems with step input of food to successive stages of biological reactors in series are commonly known and used. It allows to control[1] microorganisms growth in a more proper way, creates better conditions for organic substrate utilization and for more uniform biomass growth in all reactor stages.

In wastewater treatment technology, there are also known examples of step feed applying in activated sludge process[2], and in high-rate tower trickling filters.[3] However, there is little information available concerning the application of this system in RBCs.

In RBC, relations between amounts of biomass, organic substrate and dissolved oxygen concentration in every stage, are very important, because:

- the lack of oxygen in first stages limits the range of applied organic loading rate, and influences formation of anaerobic layers in biofilm
- excess organic loading rate at first stage causes intensive biomass growth, what in consequence can create some operating problems (with shafts and media)

Taking into consideration these phenomenon, the investigations on influence of wastes system feeding in RBD on technological conditions and parameters of wastewater treatment processes were carried out.

METHODOLOGY

The investigations were carried out using the experimental apparatus shown in Figure 1.

Synthetic wastes were used as the influent feed throughout this study. They were prepared by adding commercial dry milk powder to tap water. Highly concentrated milk solution was pumped to hydraulic mixer, where by mixing with tap water, wastes with required concentration were obtained. Prepared wastes flowed by feeders to two four-stages RBD units. Table I contains a summary of the detailed dimensions of RBD units used during two series.

In Control RBD unit, in both series, whole stream of wastes was conveyed to the first stage. For these conditions, the same hydraulic loading rate was applied to all stages.

In Step Feed RBD unit, according to examined systems, raw wastes were conveyed to several initial stages.

In series I, raw wastes flowed upon to the first three stages with the same flow rate (33/33/33 RBD). It was assumed that the wastes would not be conveyed to the fourth stage, because of necessity of creating proper conditions for nitrification organisms.[4,5]

45th Purdue Industrial Waste Conference Proceedings, © 1991 Lewis Publishers, Inc., Chelsea, Michigan 48118. Printed in U.S.A.

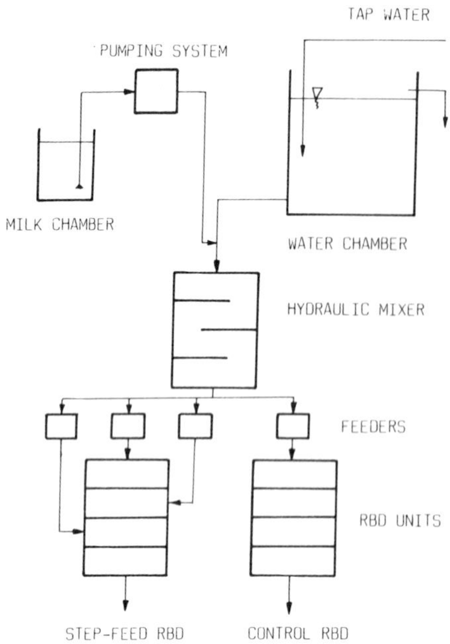

Figure 1. Schematic diagram of experimental apparatus.

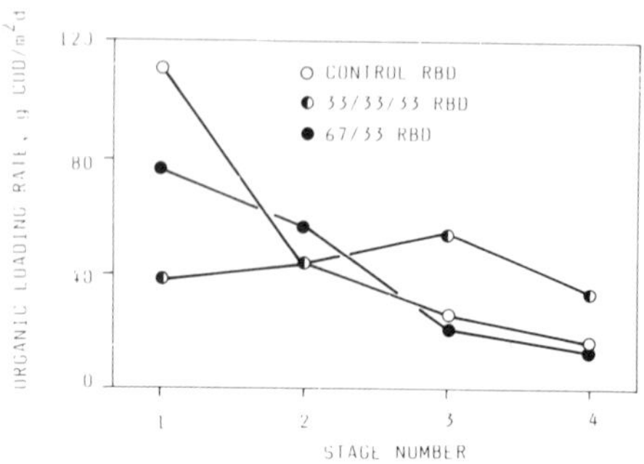

Figure 2. Organic loading rates for the four stages of the RBD units—series I and II.

In series II, raw wastes flowed up to the first stage (two thirds of flow rate) and to the second stage (one third of flow rate). This RBD unit was called 67/33 RBD.

In both series, the technological efficiency of Control and Step Feed units were compared at hydraulic loading equal 0.054 m^3/m^2d. The mean influent concentration was 510 mg COD/l (Table II). The investigations were carried out in temperature equal 19.3°C (series I) and 15.4°C (series II). The mean liquid temperatures in various stages of both units and in influent are contained in Table III.

Table I. Summary of the Dimensions of RBD Experimental Units

Parameter	Units	Values
Number of stages		4
Number of disks/stage		6
Disk diameter	m	0.8
Total surface area/stage	m^2	6.0
Total surface area/unit	m^2	24.0
Water volume/stage	L	30.0
Submergence	%	40.0
Rotational speed	r/min	7.5

Table II. Summary of Mean Steady-State Operating Conditions

Parameter	Series I	Series II
Influent COD conc., mg/L	517	512
Influent BOD conc., mg/L	383	318
Organic loading rate, g COD/m^2d	27.9	27.7
Organic loading rate, g BOD/m^2d	20.7	17.2
Hydraulic loading rate, m^3/m^2d	0.054	0.054

Table III. Summary of the Observed Mean Liquid Temperature (°C) in the Various Stages of the RBD Units

Series Unit Stage	I Control RBD	 33/33/33 RBD	II Control RBD	 67/33 RBD
Influent	13.0	13.0	10.1	10.1
1	13.7	14.1	10.9	11.2
2	13.9	14.2	11.4	11.4
3	14.0	14.3	11.7	11.6
4	14.1	14.5	12.0	11.9

ANALYTICAL CONTROL OF WASTES AND BIOFILM

The influent to the system and the effluent from every stage was monitored in grab samples.

Temperature, DO, pH, suspended solids and volatile suspended solids were measured once a day, whereas organic, ammonia, nitrite and nitrate nitrogen, twice a week. Samples for COD test were taken twice a day.

Values of all indicators were determined in filtered samples. Dissolved oxygen was measured using a DO probe meter. All analyses were carried out according to Hermanowicz's methodology[6].

Wet, dry attached biomass, volatile solids in biomass and the content of biofilm microorganisms were also determined during investigations. A sample from 20 cm^2 was taken for weight and microbiological analysis. The biofilm taken for weight and microbiological analysis. The biofilm was scraped from five different places. The results of wet and dry biofilm mass were related to the unit of disc area.

Microbiological investigations included total number of heterotrophic bacteria, ammonificational bacteria, fungi and microfauna composition of biofilm.

Biofilm samples used for bacteria and fungi examination were ground in a porcelain mortar adding physiologic saline (0.85% NaCl solution) to obtain uniform suspension diluted 1:10. Then many successive dilutions were prepared from this suspension and later quantitative inoculations were carried out from them on suitable culture medium.

Cultures of bacteria and fungi were made in dilution 10^{-4}, 10^{-5}, 10^{-6} and 10^{-7}, respectively.

Total number of heterotrophic bacteria was estimated by cultural plate method in Bacto Tryptone Glucose Yeast Extract Agar medium (Difco Supplementary Literature, 1968).

Number of ammonificational bacteria was estimated in bulion-agar medium with 3% bactopeptone.

Both kinds of cultures were incubated in 25°C for 48h and then the results were read counting new colonies.

Total number of fungi was determined in constant medium according to Lodder and Kregci-Van Rij[7]. Cultures were incubated in 25°C and after 7 days the results were read.

Examinations were carried out for 3 weeks, in each series with the frequency once a week.

Fungi examinations were made only for series II. Amount of bacteria and fungi was calculated for 1 g VS of biofilm.

Examinations of microfauna deposit were carried out for 3 weeks in each series with the frequency of twice a week.

Biofilm samples were taken to the measuring cylinder 20 cm^3 containing 2 ml of water. After having read the biofilm volume the sample was blown through in order to obtain uniform suspension.

Microscopic preparates were made from suspension volume 0.05 cm^3 and then animal species composition and species amount were estimated. If it was necessary the sample was diluted 2, 5, 10 or 15 times with water.

The number was given as number of individuals for 1 cm^2 of disc taking into the consideration the sample dilution.

Animal biomass was calculated on the basis of nomograms[8] expressed in mg of wet mass for 1 cm^2 of a disc.

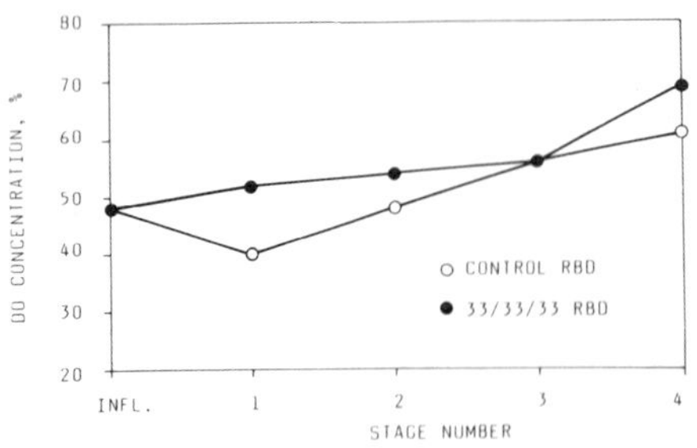

Figure 3. Mean steady-state dissolved oxygen concentration in the four stages of the RBD units – series I.

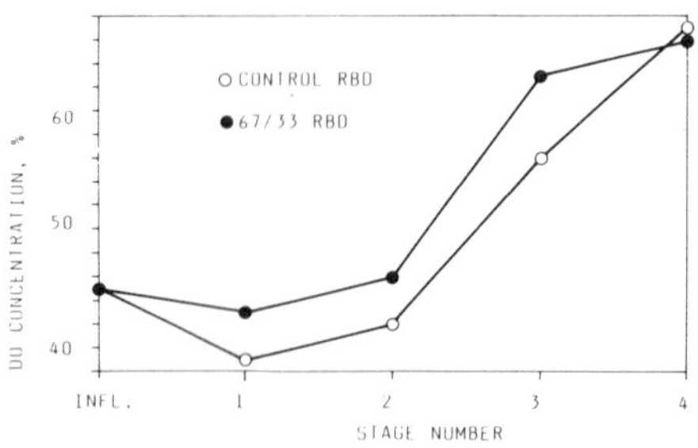

Figure 4. Mean steady-state dissolved oxygen concentration in the four stages of the RBD units – series II.

Analytical control of process as well as biofilm biocenosis examination were started after three weeks of the beginning of experiment.

RESULTS

A discussion of the experimental results were based on following parameters.

Organic loading rate

The results of investigations show that changes in feeding system effect the value of organic loading for the following stages significantly.

For Control RBD, in both series, the highest organic loading rate was obtained at the first stage. At the same time, first stage loading rate was six times higher than for fourth stages (respectively 112 and 19 g COD/m^2d – Figure 2).

In 33/33/33 RBD, organic loading rate of three initial stages increased from 37 to 54 g COD/m^2d.

In 67/33 RBD, it was stated that at significantly lower organic loading rate for the first stage, similar loading as in Control RBD for second stage was obtained.

Dissolved oxygen

In both series, in three stages, advantageous oxygen conditions were in RBDs with step feeding. In fourth stage, DO concentration in basin of Control RBD was higher than for 67/33 RBD and lower than for 33/33/33 RBD (Figure 3,4).

COD

Disadvantageous influence of step feeing on the effluent quality was observed in series I. In 33/33/33 RBD at third stage the increase of COD value was noticed. It was caused by conveying of raw wastes to this stage (Figure 5). In 67/33 RBD, lower COD values than in Control RBD were observed in all stages (Figure 6).

pH

Changes of pH values from 7.2 to 8.1 were observed during treatment process. However, significant differences in pH of effluent from Control and Step Feed RBDs were not stated (Figure 7,8). Similar tendencies of pH profiles in RBC were presented by[9,10,11].

Nitrogen compounds

In both series higher organic nitrogen utilization was observed in RBD with step feeding (Figure 9,10).

Conversion of nitrogen compounds mainly had a tendency to decompose organic nitrogen and utilization of its products to biofilm mass synthesis. However, accumulation of significant amounts of ammonia, nitrite as well as nitrate nitrogen was not observed on the following stages of Control and Step Feed RBDs (Figure 11–16).

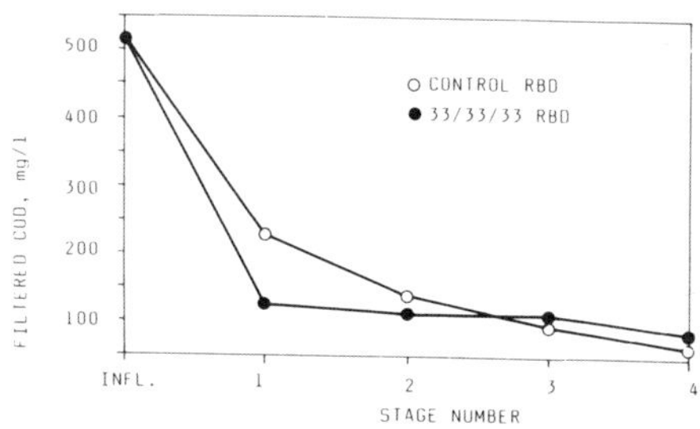

Figure 5. Mean steady-state mixed liquor filtered COD concentration in the four stages of the RBD units—series I.

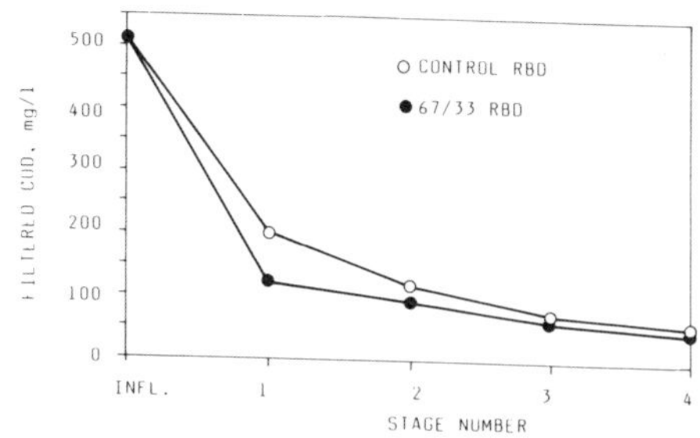

Figure 6. Mean steady-state mixed liquor filtered COD concentration in the four stages of the RBD units—series II.

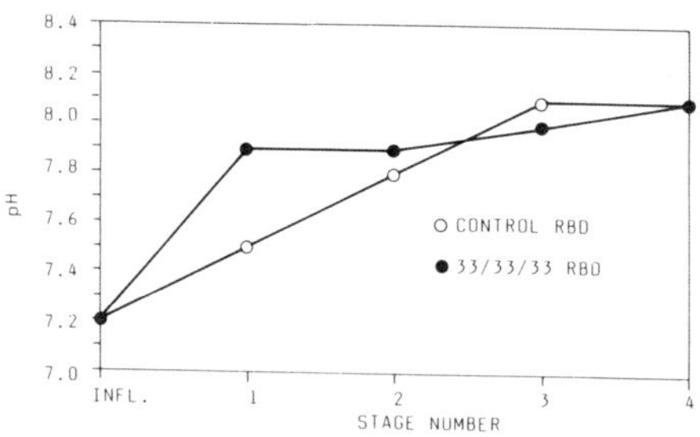

Figure 7. Mean pH values in the four stages of the RBD units—series I.

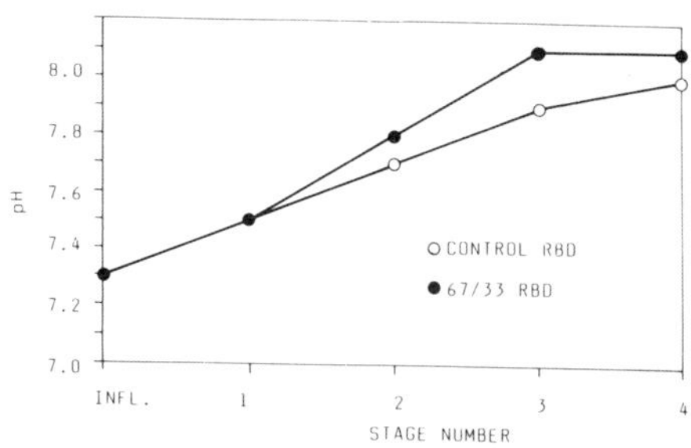

Figure 8. Mean pH values in the four stages of the RBD units—series II.

Attached biomass

Successive decrease of the amount of attached biomass on the following stages was stated in case of Control RBD in both series (Figure 17,18). These observations prove results of earlier investigations.[12,13] It is impossible to say that for all stages, there was a direct dependence between organic loading rate and biomass amount.

In 33/33/33, the insignificant increase of biofilm mass for following stages was observed in series I. For the first three stages, this increase can be attributed to amount and quality of conveyed organic substrate. The greatest amount of biofilm was observed on the disks of the fourth stage, although the organic loading rate for this stage was the lowest (Figure 17).

In series II, the biomass amount in the first and second stage of 67/33 RBD was on nearly the same level. The greatest amount of biomass was observed on third stage (Figure 18). Similarly, as in Control

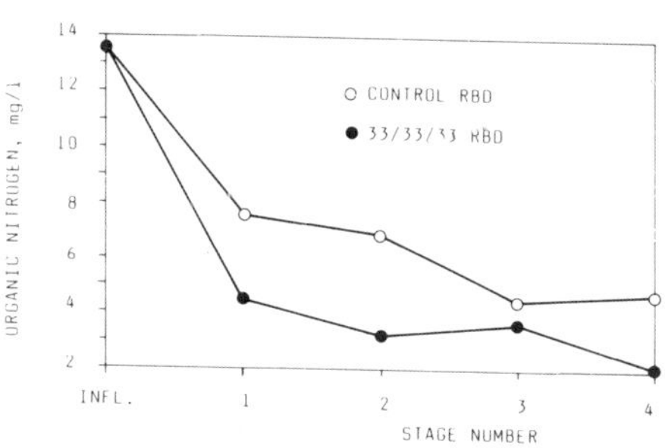

Figure 9. Mean steady-state organic nitrogen concentration in the four stages of the RBD units—series I.

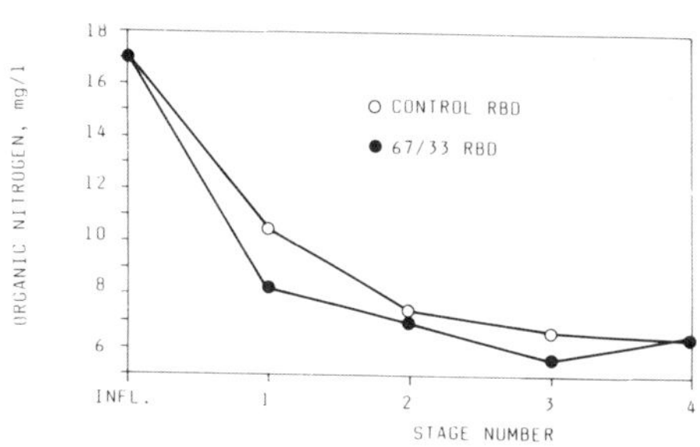

Figure 10. Mean steady-state organic nitrogen concentration in the four stages of the RBD units—series II.

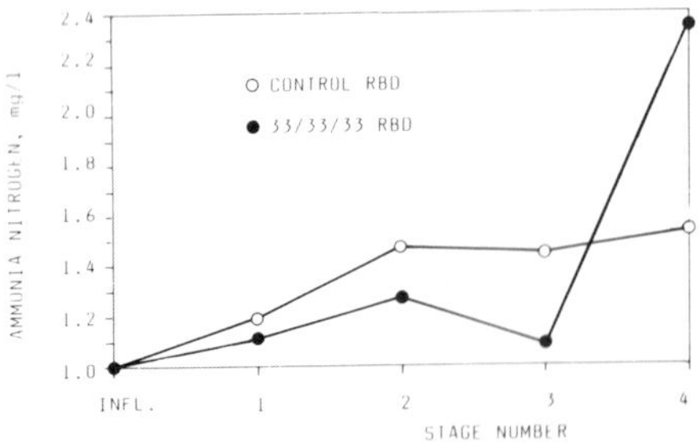

Figure 11. Mean steady-state ammonia nitrogen concentration in the four stages of the RBD units – series I.

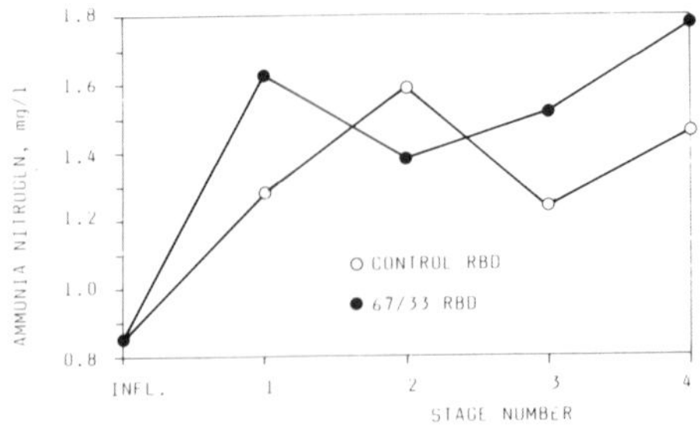

Figure 12. Mean steady-state ammonia nitrogen concentration in the four stages of the RBD units – series II.

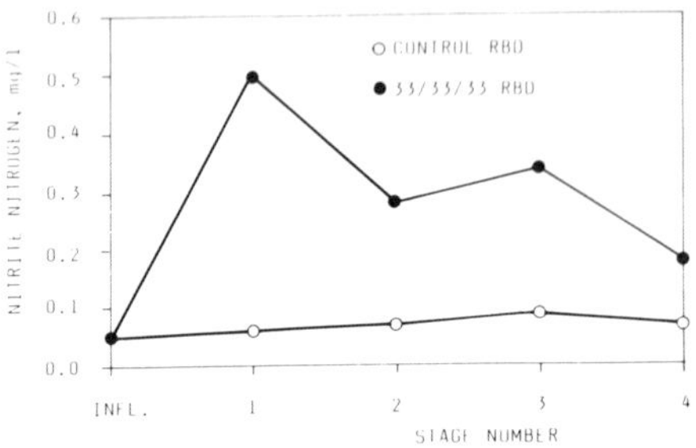

Figure 13. Mean steady-state nitrite nitrogen concentration in the four stages of the RBD units – series I.

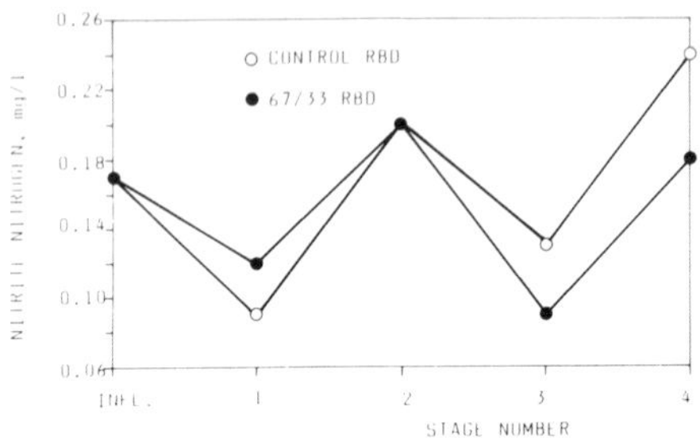

Figure 14. Mean steady-state nitrite nitrogen concentration in the four stages of the RBD units – series II.

RBD, there was not a close correlation between biomass amount and the value of organic loading rate.

The above mentioned data points out in many cases the lack of relationship between the organic loading applied, removed and biomass amount in following stages of RBDs. A characteristic feature is that a greater amount of biomass can be observed in RBD stages situated directly behind those where raw wastes were conveyed. It is the result that in those stages better oxygen conditions which together with access to substrate creates increased development of biofilm.

In both series of investigations, the amount of dried biomass of all stages in Control and Step Feed RBDs did not differ significantly. They were 1116 and 1043 g for series I, and 1416 and 1476 g for series II, respectively.

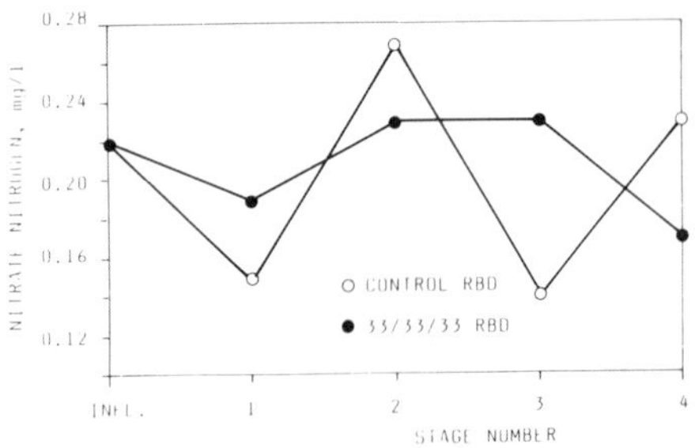

Figure 15. Mean steady-state nitrate nitrogen concentration in the four stages of the RBD units – series I.

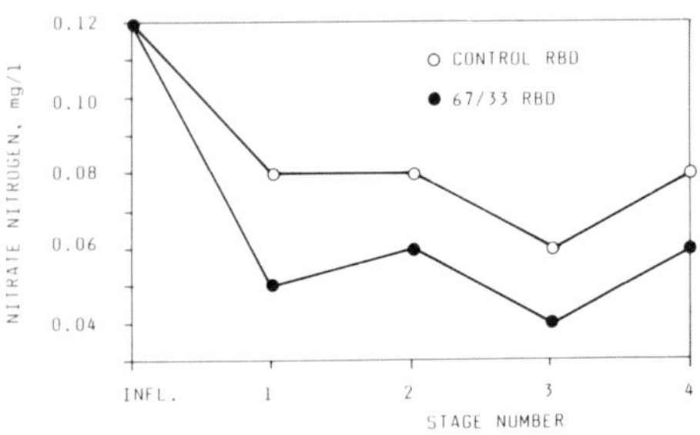

Figure 16. Mean steady-state nitrate nitrogen concentration in the four stages of the RBD units – series II.

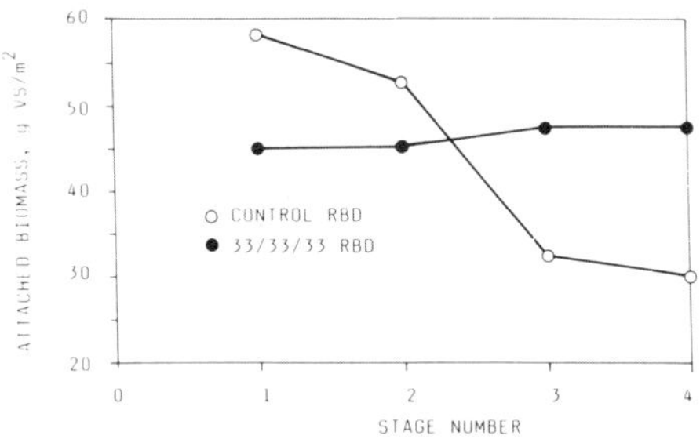

Figure 17. Attached biomass in the four stages of the RBD units – series I.

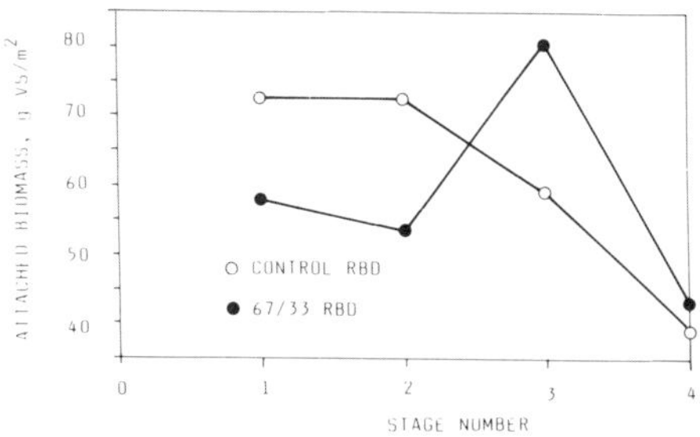

Figure 18. Attached biomass in the four stages of the RBD units – series II.

Total number of heterotrophic bacteria, ammonificational bacteria and fungi

On the Control RBD in I and II series of investigations, total number of heterotrophic bacteria, ammonificational bacteria and fungi (series II) decreased from 1 to 4 stages of RBD (Table IV).

On the Step Feed RBDs, the number of bacteria and fungi increased in the range of stages being directly conveyed by raw wastes: on 33/33/33 RBD from 1 to 3 stage and on 67/33 RBD from 1 to 2 stage.

On further stages, the number of microorganisms was estimated by used methods and it decreased significantly, especially the number of bacteria.

Total number of heterotrophic and ammonificational bacteria was several times higher in series I than in series II (Table IV) both on the Control RBD and Step Feed RBD.

RBD microfauna

During the time of investigations, only Zooflagellata, Rhizopoda and Ciliata appeared in biofilm. Zooflagellata were not taken into consideration in further examinations because of little biomass in comparison to other Protozoa.

Table IV. Total Number of Heterotrophic, Ammonificational Bacteria and Fungi * 10^{10}/g VS (mean values)

Unit/stage	Bacteria Heterotrophic	Bacteria Ammonificational	Fungi
33/33/33 RBD			
1	9.9	1.5	–
2	17.8	3.0	–
3	38.8	5.5	–
4	2.3	1.7	–
Control RBD			
1	10.5	2.7	–
2	7.3	1.90	–
3	1.1	0.77	–
4	0.2	0.36	–
67/33 RBD			
1	2.02	1.12	4.45
2	3.00	1.44	8.00
3	0.22	0.92	3.50
4	0.05	0.27	3.00
Control RBD			
1	4.70	1.80	250.0
2	2.55	0.74	150.0
3	0.29	0.54	38.0
4	0.16	0.29	10.0

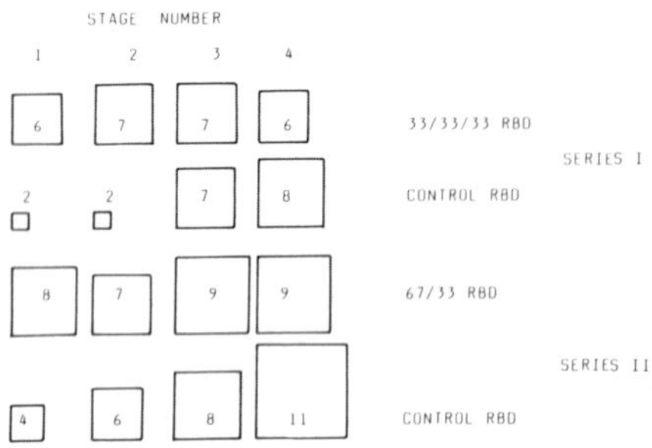

Figure 19. Animal species number in examined RBD units.

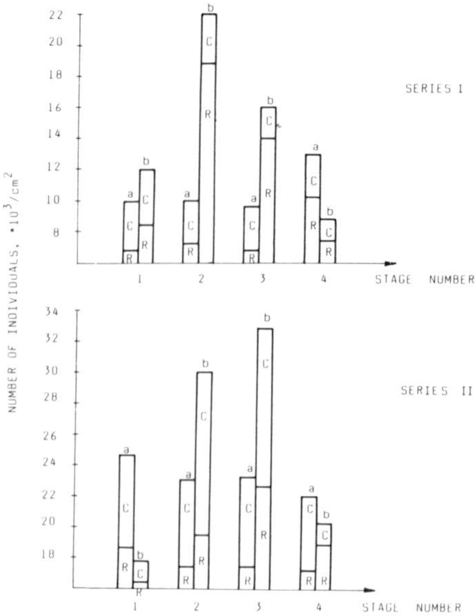

Figure 20. Total number of animals and Ciliata (C), Rhizopoda (R) participation in examined RBD units. (a – Step Feed RBD; b – Control RBD).

Number of species

From among Rhizopoda, there were only species of Amoeba which were not identified precisely. A total of 18 species from Ciliata were determined: 11 species in series I and 14 in series II. On the Control RBD, in both series, the number of taxons increased from 1 to 4 stage, especially on the last two stages (Figure 19). Significant equalization of species composition on all stages was observed on Step Feed RBDs. On 4 stage the number of taxons was lower in Step Feed RBDs than in Control RBDs.

Number of organisms

Number of organisms in biofilm of both RBDs in series I was 2 times lower than in series II (Figure 20) which corresponded to higher amount of bacteria and higher COD of the effluent. On both Control RBDs, due to high loading number of organisms, was lower on stage 1, increased on middle stages, and on 4 stage decreased again. Ciliata predominated on first stages and Rhizopoda on the final ones. There was significant equalization of organisms on all stages in Step Feed RBDs (Figure 20). The difference between RBDs concerned quantitative structure on stage 4. Ciliata dominated on 67/33 RBD which correspond to lower COD of treated waste. Rhizopoda dominated on 33/33/33 RBD and the quality of effluent was there.

Ciliata species were divided into several groups (Table V) according to their mobility, ability of feeding, way of feeding and kind of food they were living on.

In series I among the Ciliata, Colpidium group dominated on all stages of both RBDs, and their number decreased from 1 to 4 stage (Table VI). In series II, Colpidium dominated on 1 and 2 stages of 67/33 RBD or 1-3 stage of Control RBD. Cyclidium dominated on the final stages.

Organism biomass

Organism biomass was higher in series II and on both kinds of RBDs. It increased within 1-2 stage, reaching the same value on 2 stage (Figure 21). On 1 stage, the biomass was 2 times on 67/33 RBD, and on stages 3,4 on Control RBD. In series I organism biomass was higher in the whole system of 33/33/33 RBD (Figure 21). Significant species succession within the biomass was observed only in series

Table V. Species Composition of Indicator Ciliata Groups

Group name	Species Forming Group
Colpidium	Colpidium campylum, C. colpoda, Glaucoma scintillanus, Glaucoma sp. (chattoni?), Colpoda cucullus, Tetrahymena pyriformis, Dexiotricha colpidioides.
Cyclidium	Cyclidium citrullus, C. versatile
Chilodonella	Chilodonella uncinata, Thrithigmostoma cucullulus
Oxytricha	Oxytricha sp., Aspidisca lynceus, Euplotes affinis
Litonotus	Litonotus fasciola, L. lamella, Trachelophyllum pussilum, Loxophyllum sp.

Table VI. Percentage Participation of Cilata Group in Total Animal Number

	33/33/33 RBD				Control RBD			
Group	1	2	3	4	1	2	3	4
Colpidium	64.3	62.0	68.0	31.3	61.8	21.3	19.7	26.8
Cyclidium	0.01	1.5	0.4	1.4	–	–	0.28	17.5
Chilodonella	12.2	9.1	10.7	2.8	–	–	0.01	2.8
Oxytricha	–	–	–	–	–	–	–	–
Litonotus	–	–	0.16	0.77	–	–	–	–
	67/33 RBD				Control RBD			
Colpidium	44.7	67.9	13.3	2.0	81.5	73.0	30.3	1.5
Cyclidium	6.7	5.7	60.6	72.7	–	2.5	16.3	19.5
Chilodonella	17.4	11.4	6.0	1.6	0.1	0.2	15.4	4.7
Oxytricha	0.13	–	0.5	2.2	–	–	0.08	2.4
Litonotus	–	–	0.08	0.2	–	–	0.04	3.4

II and there was domination of Colpidium on the first stages, RBDs and Oxytricha (67/33 RBD) or Oxytricha and Litonotus (Control RBD) on the final stages (Table VII).

On 33/33/33 RBD there was no species succession and Colpidium dominated in the whole system. On Control RBD, the succession was from Colpidium to Rhizopoda.

SUMMARY

The results of investigations on the influence of feeding system four-stages RBD on conditions and technological parameters of treatment process, allows one to state, that changes in the system of raw wastes feeding cause the following effects:

- the increase of oxygen concentration in the initial stages of RBD
- biomass movement inside the system
- the change of treatment process efficiency expressed by COD removal
- equalization of species composition, the number and animal biomass on the following stages of the system

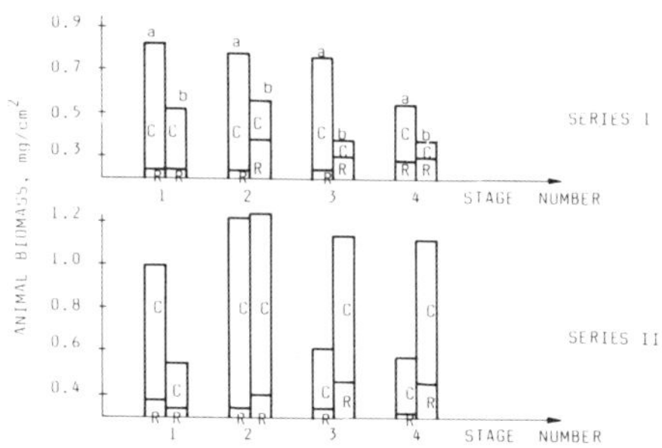

Figure 21. Total animal biomass and Ciliata (C), Rhizopoda (R) participation in examined RBD units. (a – Step Feed RBD; b – Control RBD).

Table VII. Percentage Participation of Ciliata Groups in Total Animal Biomass

	33/33/33 RBD				Control RBD			
Group	1	2	3	4	1	2	3	4
Colpidium	94.0	91.0	90.3	73.6	86.5	50.0	48.7	38.5
Cyclidium	0.03	0.09	0.03	0.02	–	–	0.05	2.0
Chilodonella	2.9	3.2	2.9	1.3	–	–	0.05	1.3
Oxytricha	–	–	–	–	–	–	–	–
Litonotus	–	–	0.4	0.4	–	–	–	–
	67/33 RBD				Control RBD			
Colpidium	75.2	90.9	37.7	5.2	89.1	87.7	66.0	3.5
Cyclidium	0.8	0.5	11.5	13.8	–	0.3	2.4	1.8
Chilodonella	8.5	4.3	4.6	12.1	1.8	2.5	9.0	0.2
Oxytricha	3.1	–	32.8	62.1	–	–	2.3	37.5
Litonotus	–	–	2.3	4.8	–	–	2.6	37.6

However, significant differences in the values of total biofilm mass amount and utilization of organic nitrogen were not observed.

After investigating two step feeding systems, the system where raw wastes were conveyed to two stages appeared more effective. Additionally, this system was more advantageous than Control RBD.

The highest number and biomass of Ciliata and lowest quantity of heterotrophic bacteria were observed on 67/33 RBD. It proves the significant role of Ciliata in clarifying treated wastes.[14] At the same time, significant succession of species was observed: from the domination of species requiring high concentration of dispersed bacteria (Colpidium group) on the first stages, to the domination of species with decreased feeding needs, but with higher feeding abilities (Oxytrichia, Cyclidium groups) on the final stages.

The rate of biochemical removal of organic substrate in the reactors with biofilm depends on the amount and quality of dissolved organic compounds and oxygen concentration. These factors had a significant influence on the growth and thickness of biofilm. If oxygen does not penetrate the deepest layers of biofilm, then in the biofilm mass division into oxygen and anaerobic layers takes place. The last one practically does not take part in the process of substrate removal. In anaerobic layer, fermentation mechanisms dominate and gaseous products of this process together with shearing forces may be the cause of temporary sloughing of biomass.[15] These phenomena are the reason of biofilm overdevelopment and cause operating problems.

According to the experiment assumptions, step feeding of two or three initial stages influences the decrease of organic loading rate on the first stage and increase on the second and third ones. Therefore, in the first stage the oxygen conditions are improved and in further stages, the amount of available organic substrate increases. It may cause the advantageous relationships between the amount of available oxygen and substrate. The decrease of oxygen limitation in the first stage, and the substrate in the following, may influence the global increase of the amount of active biomass in the system and intensification of treatment process. The relationship between the amount of active biomass and the depth of oxygen and substrate penetration was stated earlier.[16,17,18] The balance of species composition and amount of microorganisms on all the biofilm stages are the result and proof of advantageous movement in the system of oxygen and substratum.

From above mentioned results, it comes out, that conditions giving the increase participation of active biomass in total biofilm and better treatment efficiency were only obtained for 67/33 RBD. It results that the change of feeding system to four-stages RBD may give better technological effects. Moreover, it was stated, that this procedure improves operating conditions coming out from more uniform biomass growth on all stages. It is of significant importance for the first stage, where biofilm overdevelopment and its intensive sloughing cause irregularities in the mechanical loading of shafts and drive system. It leads to installation failures.

CONCLUSIONS

The change of wastes feeding system to four-stages RBD from the system with the same hydraulic loading with more uniform organic loading for all stages, causes the biomass movement inside the system and improvement of oxygen conditions.

It was stated, that for four-stages RBD, higher effectiveness of COD removal can be obtained by feeding wastes to the first and the second stage. Conveying of raw wastes to the third stage decreases efficiency.

The carried out experiment concerned the RBD with high organic loading. Therefore, the results can be applied in plants where high organic loading of the first stage causes biomass overdevelopment and the presence of anaerobic layers in the biofilm. The change of feeding system in this case may improve operating conditions without any significant investments.

REFERENCES

1. "Ecological Aspects of Used-Water Treatment," vol. 3, The Processes and Their Ecology, Ed.: Curds, C. R., H. A. Hawkes, Academic Press Inc, London, 1983.
2. Gould, R. H., "Talmans Island Opens for Worlds Fair," *Munic. Sanit.*, 10,185 (1939).
3. Ingram, W. T., "A new approach to trickling filters design," Proceedings of the American Society of Civil Engineers, Journal of Sanit. Eng. Div., 83 (3) 1956.
4. Antonie, R. L., et al., "Evaluation of a rotating disk wastewater treatment plant," *Journal Water Poll. Control Fed.*, 46, 498 (1974).
5. Wend, C. N., A. H. Molof," Nitrification in the biological fixed-film rotating disc system," Journal Water Poll. Control Fed., 46, 1674 (1974).
6. Hermanowicz, W., et al. "Fizyczno-chemiczne badanie wody," Arkady, Warszawa, 1976.
7. Rodina, A., "Mikrobiologiczne metody badania wod," PWRiL, Warszawa 1968.
8. Cislenko, L. L., "Monogrammy dla opredielenija viesa vodnych organismov po razmieram i formie tiela," Nauka, Leningrad, 1968.
9. Antonie, R. L., F. M. Welch, "Preliminary results of a novel biological process for treatment diary wastes," Proceedings: 24th Industrial Waste Conf., Purdue University, 115 (1969).
10. Pajak, A. P., R. C. Loehr, "Treatment of poultry manure wastewater using a rotating biological contractor," *Water Research*, 10, 399 (1976).
11. Pescod, M. B., et al., "Biological disc filtration for tropical waste treatment," *Water Research*, 6, 1509 (1972).
12. Sack, W. A., et al., "Evaluation of the Bio-Disc treatment process for summer camp application," EPA Project No S- 800707, National Environmental Research Center, Cincinnati, OH, 1973.
13. Pano, A., E. J. Middlebrooks, "Kinetics of carbon and ammonia nitrogen removal in RBC's," *Journal Water Poll. Control Fed.*, 55, 956 (1983).
14. *Environmental Biotechnology*, Ed.: Rehm, H. J., Reed, G., Vol 8, 1986.
15. Huang, J. C., et al., "Growths and activities of fixed-films in treating sugar waste," Proceedings 38th Purdue Ind. Waste Conf., Purdue University, 817 (1983).
16. Owen, D. T., K. J. Williamson, "Oxygen limitation in heterotrophic biofilm," Proceedings 31st Purdue Ind. Waste Conf., Purdue University, 267 (1976).
17. Harris, N. P., G. S. Hansford, "A study of substrate removal in microbial film reactor," *Water Research*, 12, 855 (1976).
18. Famularo, J., et al., "Application of mass transfer to rotating biological contactors," *Journal Water Poll.* Control Fed., 50, 653 (1978).

38 COMPARATIVE STUDIES OF ACTIVATED SLUDGE AND ROTATING BIOLOGICAL DISC PROCESS (I).

Ewa Klimiuk, Lecturer
Maria Hull, Lecturer
Faculty of Water Protection and Inland Fishery
Academy of Agriculture and Technology
10-957 Olsztyn-Kortowo, Poland

Wojciech Janczukowicz, Fulbright Scholar
Department of Civil and Environmental Engineering
New Jersey Institute of Technology
Newark, New Jersey 07102

INTRODUCTION

Rotating Biological Discs process (RBD), as a relatively new system, required checking of utilization possibilities in treating different kinds of wastes. It was natural and necessary to compare this process with other methods, especially with the activated sludge process. The investigations concerned mainly economical aspects and technical conditions and were less concentrated on kinetic growth of microorganism or creation of biotic factors.[1,2,3,4,5]

Although RBD is similar to the activated sludge process, it is considered as completely mixed reactors; comparison of their technological parameters is very difficult. For the activated sludge process, substrate removal rate is refered to biomass of microorganisms determined as volatile suspended solids in the dried biomass (VSS) or protein content. It is commonly accepted that for steady-state conditions, the value of VSS is constant. In fixed-film reactors only part of the biomass is responsible for substrate removal. This part is called an active layer. The depth of this layer was determined by using very sensitive and specific laboratory methods.[6,7] It is almost impossible to apply these methods to a technical scale. Effectiveness of the RBD process is mainly calculated in relation to the area of biodiscs, neglecting the mass of biofilm which is responsible for the biodegradation process. These simplifications cause the knowledge of biofilm growth parameters to be insufficient. Consequently operating problems caused by biofilm overdevelopment occur in many wastewater treatment plants.

Since in engineering practice we have a large knowledge about growth kinetics of activated sludge, it is very useful to assume, that activated sludge biomass can be a pattern for comparing the efficiencies of both processes.

The aim of this paper is the estimation of biofilm mass which is able to metabolize the same amount of substrate as activated sludge biomass. In this case, an activated sludge process as a well known system could serve as the basis for technological evaluation of RBD.

MATERIALS AND METHODS

Description of experimental unit

The experimental unit consisted of sewage chamber, RBD and activated sludge reactors and a tank for treated wastes (Figure 1).

An activated sludge tank with an inner settling chamber had a volume of 2.1. RBD basins with a capacity of 2 l contained eight of 0.25 m diameter discs. The total surface area of media was 0,8 m^2. The discs were rotated with speed of 8 rpm.

45th Purdue Industrial Waste Conference Proceedings, © 1991 Lewis Publishers, Inc., Chelsea, Michigan 48118. Printed in U.S.A.

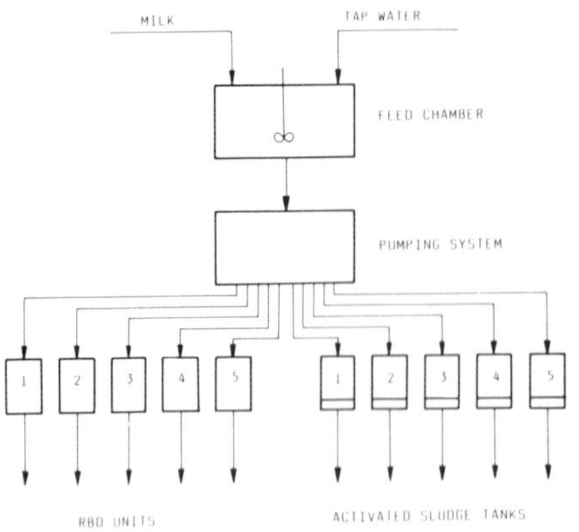

Figure 1. Schematic diagram of experimental apparatus.

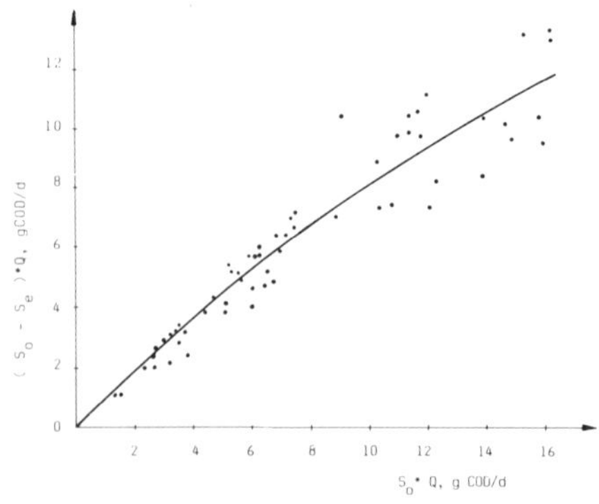

Figure 2. Relationship between organic load applied and removed for activated sludge process; $S_0 = 1200$ g COD/m^3.

Raw wastes

The synthetic milk wastes were prepared by mixing water with commercial milk. The final concentration was 1200 mgCOD/l (I series) and 2050 mgCOD/l (II series).

Analytical control of the process

The organic content of influent and effluent was determined in filtered samples by COD test (method with potassium dichromate).[8] Biomass of activated sludge and biofilm were measured by volatile solids in the dried biomass (VSS and VS, respectively). The results were determined for a sludge in mg VSS/l and for a biofilm in mg VS/cm^2.

The mass of attached growth was determined by calculating the dry weight of biomass scraped from a known area of each disc.

Technological parameters of the process

The investigations were carried out in two series with different influent concentrations. RBD and activated sludge reactors were operated in parallel. In both systems, typical ranges of hydraulic and organic loadings were applied. The mean values of COD were used for calculating the technological parameters of both processes. The results for the activated sludge process are presented in Table I and for RBD process in Table II.

RESULTS AND DISCUSSION

Comparative studies of COD removal rate in RBD and activated sludge processes

The COD removal rate was analyzed in relation to influent load. According to experimental assumptions, the influent load increase in each examined series was caused by the change of hydraulic loading (Table I,II). Experimental data suggested that the relationship between the amounts of COD removed and applied could be described as a first order equation. The resulting curves after simulation and experimental data are presented in Figure 2 and 3 for activated sludge and in Figure 4 and 5 for RBD. Coefficient values of k obtained from simulation are in Table III.

In activated sludge process concentration, influence on COD removal kinetic was not observed. It is proved by similar coefficient values of k (0.04 and 0.035). The range of obtained values of k may suggest that global COD removal rate for the examined organic load was more limited by mass transport than the rate of enzymatic reactions.

In the RBD process, the influence of influent concentration on kinetic COD removal was significant.

Table I. Technological Parameters of Activated Sludge Process (Average Values)

		Series I (S_o = 1200 mg COD/1) Unit number					Series II (S_o = 2050 mg COD/1) Unit number				
Technological Parameters	Units	1	2	3	4	5	1	2	3	4	5
Flow rate	1/d	2.0	3.5	4.2	6.7	9.1	2.2	4.2	5.6	6.1	10.0
Biomass concentration in reactor	g VSS/1	3.15	4.05	4.20	3.80	3.10	2.85	3.00	3.40	3.40	4.37
Organic loading	g COD/g VSS · d	0.38	0.52	0.60	1.06	1.76	0.79	1.44	1.69	1.94	2.35
Aeration time	h	24	13.7	11.4	7.2	5.3	21.8	11.4	8.6	7.9	4.8

Table II. Technological Parameters RBD Process (Average Values)

		Series I (S_o = 1200 mg COD/1) Unit number					Series II (S_o = 2050 mg COD/1) Unit number					
Technological Parameters	Units	1	2	3	4	5	1	2	3	4	5	6
Flow rate	1/d	3.7	6.5	10.1	27.7	50.4	3.7	6.5	10.6	27.7	37.4	50.4
Organic loading rate	g COD/m^2d	5.6	9.80	15.2	41.6	75.6	9.5	16.7	27.2	71.0	95.8	129.2
Hydraulic loading rate	m^3/m^2d	0.005	0.008	0.013	0.035	0.063	0.005	0.008	0.013	0.035	0.047	0.063

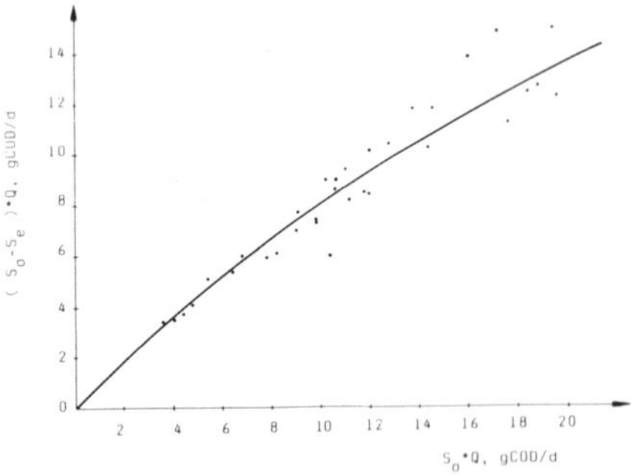

Figure 3. Relationship between organic load applied and removed for activated sludge process; S_0 = 2050 g COD/m^3.

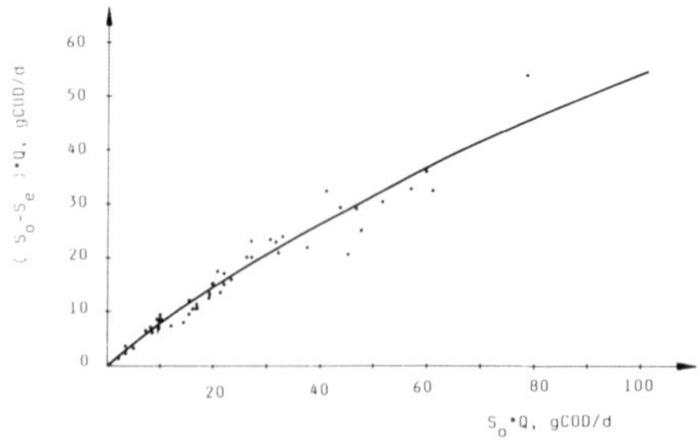

Figure 4. Relationship between organic load applied and removed for RBD; S_0 = 1200 g COD/m^3.

Low values of k coefficient in RBD, six times lower for series I and twice lower for series II, point out that the kinetic rate of COD removal for lower influent concentration is influenced by diffusion limitations, while for higher influent concentration, by the rate of enzymatic reactions.

Substrate removal rate is not only dependent on its composition (the same for both methods), but on quantity and activity of microorganisms able to remove substrate as well. For both systems, the reactors volume was 2 l. In the whole tank, the total active sludge biomass was 6–8 g (VSS), whereas in RBD system biofilm mass changed from 30 to 90 g (VS) relative to the organic loading. Higher values of biofilm mass do not prove lower biofilm ability for removing substrate, because hydraulic loading for RBD was much higher than applied in activated sludge process. The only possibility for comparing both systems is if the mass of biofilm equivalent to activated sludge (taking into consideration obtained treatment effects) can be estimated.

Comparative studies of COD removal rate for activated sludge and RBD process took into consideration the biomass responsible for biodegradation process.

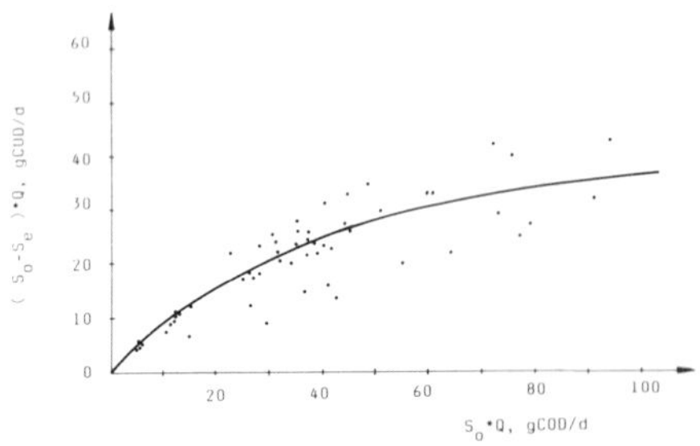

Figure 5. Relationship between organic load applied and removed for RBD; S_0 = 2050 g COD/m^3.

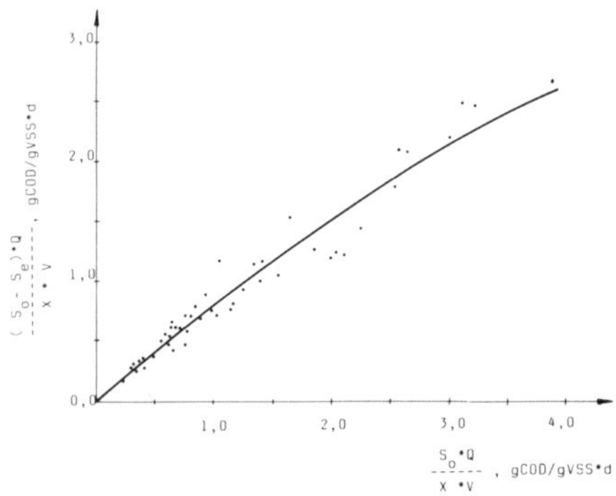

Figure 6. Relationship between organic load applied and removed related to the sludge biomass; S_0 = 1200 g COD/m^3.

Table III. Coefficient Values of K Obtained for First Order Reaction

Parameter	K	
Number of series	I	II
Activated Sludge Process	0.04	0.035
RBD Process	0.007	0.024

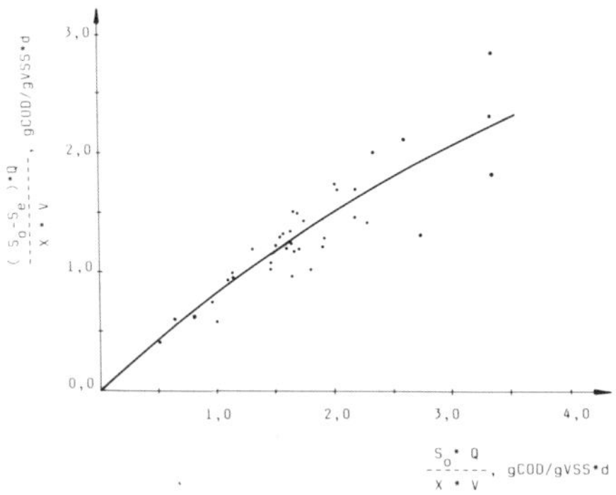

Figure 7. Relationship between organic load applied and removed related to the sludge biomass; $S_o = 2050$ g COD/m^3.

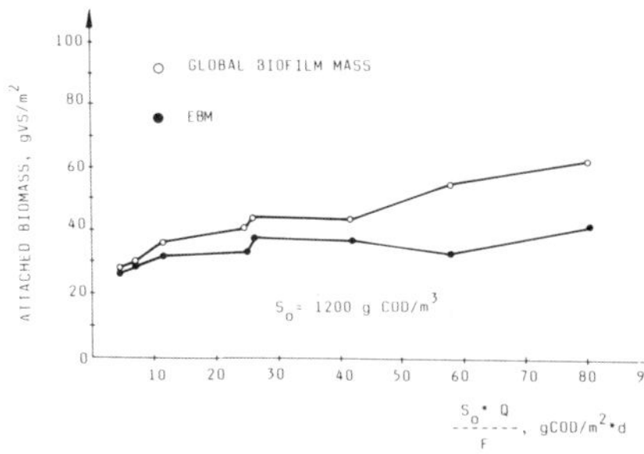

Figure 8. Global biofilm mass and EBM with respect to COD loading; $S_o = 1200$ g COD/m^3.

As it was stated earlier, the relationship between removed and the applied COD load followed the first order reaction. In activated sludge process, the value of microorganisms concentration in the reactor was constant. So, it was assumed that the relationship between the COD load removed and applied as related to the sludge biomass (VSS) would also follow the first order reaction. Figures 6 and 7 present experimental data and curves for this relationship.

For RBD, it was assumed that the global biofilm mass consisted of active and nonactive parts:

$$X_B = X_{BA} + X_{BN} \tag{1}$$

If substrate removal processes in both methods had the same rate, then the amount of substrate load removed by biofilm would equal the amount of load removed by activated sludge:

$$\frac{R_{AS}}{V^*X} = \frac{R_B}{A^*X_{BA}} \tag{2}$$

It results from Equation 2, that the ratios of substrate removal rates and microorganism quantities are equal.

Experimental data enables one to estimate the values of R_{AS}, R_S, V, X, X_B using Equation 1 and 2. The parameters A and X_{BA} have to be estimated. The value of biofilm surface (A) was assumed to be equal to the media surface (F). Moreover, it was assumed that the activated sludge biomass unit could be treated as an activity metabolic standard. In this case $X_{BA} = X_{BE}$. After using all these assumptions Equation 2 becomes:

$$\frac{R_{AS}}{V^*X} = \frac{R_B}{F^*X_{BE}} \tag{3}$$

In order to calculate X_{BE} value, the right side of Equation 3 was multiplied and divided by X_B. Then:

$$\frac{R_{AS}}{V^*X} = \frac{R_B}{X_B{}^*F} * \frac{X_B}{X_{BE}} \tag{4}$$

After defining:

$$\frac{R_{AS}}{X^*V} \text{ as } R_{ASX} \text{ and } \frac{R_B}{X_B{}^*F} \text{ as } R_{BX}$$

Equation (4) can be written as:

$$X_{BE} = \frac{R_{BX}}{R_{ASX}} * X_B \tag{5}$$

In Equation 5, R_{BX} and R_{ASX} represents the amounts of COD load removed per biofilm or activated sludge biomass, respectively.

Calculations of biofilm mass equivalent to activated sludge biomass.

The relationship from Equation 5 was used for calculating the amount of biofilm mass equivalent to the amount of activated sludge biomass (Equivalent Biofilm Mass). For this purpose, based on experimental results, the amounts of COD load applied (S_o*Q/X_B*F) and COD removed ($(S_O-S_e)*Q/X_B*F$) per all biofilm mass were calculated. Next, using relationships from Figures 6 and 7, for every value of COD load applied to RBD, a referring value of COD removed load for activated sludge was estimated.

According to Equation 5, the ration R_{BX}/R_{ASX} multiplied by the value of global biofilm mass per area (X_B) give searching values of equivalent biofilm mass (EBM). In Figures 8 and 9, the global values of biomass amount calculated on the basis of average values from particular series and biofilm mass from equivalent to activated sludge biomass calculated from Equation 5 were presented.

Taking into consideration the above presented data for wastes, with COD concentrations equal to 1200 mg/L, the EBM ranged from 95 to 62.2% of total amount of biofilm mass and it depended on organic loading. For wastes with COD concentration equal to 2050 mg/L, these values were between 96.7 and 51.5%.

SUMMARY AND CONCLUSIONS

The above presented investigations on the comparison of treatment effectiveness by activated sludge and RBD processes enabled us to determine the parameter which can be helpful in choosing one of the methods given specific technological conditions. This parameter was defined as the amount of biofilm mass which could remove the same amount of substrate as unit biomass of activated sludge.

As it was pointed by Hartman,[9] the rate of substrate removal in outer layer of biofilm was very high and reached much higher values than in activated sludge flocs. Hartman explains it as the increase of enzymatic activity due to significant rates of substrate inflow and product outflow from biofilm.

La Motta and all[10] showed that dependence between the waste velocity flow through fixed film reactors and effectiveness of treatment existed. For proper flow velocities, effectiveness is mainly limited by the rate of enzymatic process in biofilm.

It is also known that there are some parts of biofilm mass which have been neglected because they were not responsible for treatment effectiveness. The increase of the waste velocity flow may cause the rapid growth of this part of biofilm. Their thickness can be much higher than the thickness of the active layer. Rotating Biological Discs are very advantageous but there is one very important disadvantage concerning problems with shafts caused by great and uneven biofilm overdevelopment. Only a small part of this biofilm takes an active role in treatment processes.[7,11,12] There must be a compromise between possibilities of applying high hydraulic loading (which is an advantage) and the amount of biofilm mass produced in these conditions.

The determination of Equivalent Biofilm Mass (EBM) allows the comparison of both systems. It also enables one to show which part of global biomass is active as much as activated sludge.

We can distinguish three cases:

- Biofilm is very active and thin. It was theoretically determined EBM could be higher than its real value. In that case RBD would be more advantageous than activated sludge
- The global amount of biofilm equals the value of EBM. In that case (from the techno-

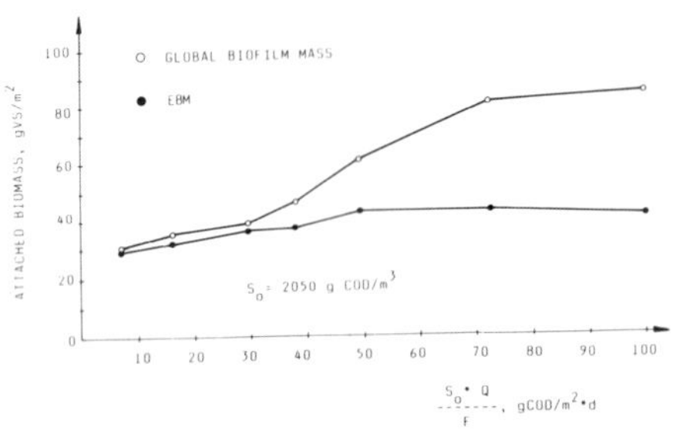

Figure 9. Global biofilm mass and EBM with respect to COD loading; S_o = 2050 g COD/m^3.

logical point of view), both systems have the same effectiveness. Economical aspects should decide choice of the method.
- The amount of global biofilm mass is higher than the value of EBM. In this case, high activity of outer layer is an insufficient criterion because RBD produces too much useless biofilm. For these conditions, the activated sludge process is the right method.

The above presented considerations led to the following conclusions:

1. Calculated values of EBM were lower than the amount of total biofilm mass in the whole range of applied organic loading. It means that the activated sludge method is more effective for this type of sewage.
2. For organic loading between 4 to 40 g COD/m^{2}d (influent concentration 1200 mg/L) and for 6 to 30 g COD/m^{2}d (influent COD concentration 2050 mg/L), the amount of global biofilm mass and value EBM are very similar. In this range of organic loading both methods have similar effectiveness.
3. With the increase of organic loading above 40 g COD/m^{2}d, intensive biofilm growth occurs and in these conditions the application of activated sludge process is more advantageous. It mainly concerns wastes with influent COD concentration equal to 2050 mg/L.

NOMENCLATURE

X_B – global biofilm mass, g/m^2
X_{BN} – nonactive part of biofilm, g/m^2
X_{BA} – active part of biofilm, g/m^2
X_{BE} – Equivalent Biofilm Mass, g/m^2
A – biofilm surface area, m^2
F – media surface area, m^2
Q – influent flow rate, m^3/d
R_{AS} – organic load removed in activated sludge process, g COD/d
F_B – organic load removed in RBD process, g COD/d
R_{ASX} – organic load removed in activated sludge process per biomass of sludge (VSS), g COD/d*g (VSS)
R_{BX} – organic load removed in RBD process per biofilm mass (VS), g COD/d*g (VS)
S_o – influent COD concentration, g/m^3
S_e – effluent COD concentration, gm^3
V – liquid volume in the reactor, m^3

REFERENCES

1. Botuk, B. O., et al., "Pogruzennyje diskowyje wraszczapiszczijesa biologiczeskije filtry," Wodosnabzenije i sanitarna technika, 1,13 (1975).
2. Clark, J. H., et al., "Performance of a rotating biological contactor under varying wastewater flow," *Journal Water Poll. Control Fed.*, 50, 896 (1978).
3. Antonie, R. L., et al., "Evaluation of a rotating disk wastewater treatment plant," *Journal Water Poll. Control Fed.*, 46, 498 (1974).
4. Poon, C. P. C., et al., "Factors controlling rotating biological contactor performance," *Journal Water Poll. Control Fed.*, 51, 601 (1979).
5. Paolini, A. E., G. Variali, "Kinetic considerations on the performance of activated sludge reactors and rotating biological contactors," *Water Research*, 16, 155 (1982).
6. Characklis, W. G., et al., "Dynamics of biofilm processes: Methods," *Water Research*, 16, 1207 (1982).
7. Hoehn, R. C., A. D. Ray, "Effect of thickness on bacterial films," *Journal Water Poll. Control Fed.*, 45, 2302 (1973).
8. Hermanowicz, W. et al., "Fizyczno-chemiczne badanie wody i sciekow," Arkady, Warszawa 1976.
9. Hartman, L., "Biologische Abwasserreigung," Springer-Verlag, Berlin, Heidelberg, New York, 1983.
10. La Motta, E. I., "Extermal mass transfer in a biological film reactor," *Biotechnol. and Bioeng.*, 18,1359 (1976).
11. Kornegay, B. H., J. F. Andrews, "Kinetics of fixed film biological reactors," Proceedings: 22nd Purdue Ind. Wastes Conf., Purdue University, 620 (1967).
12. Characklis, W. G., "Fouling biofilm development. A process analysis," *Biotechnol. and Bioeng.*, 23, 1923 (1981).

39 COMPARATIVE STUDIES OF ACTIVATED SLUDGE AND ROTATING BIOLOGICAL DISC PROCESS (II – MICROFAUNA)

Maria Hull, Lecturer
Ewa Klimiuk, Lecturer
Faculty of Water Protection and Inland Fishery
Academy of Agriculture and Technology
10-957 Olsztyn-Kortowo, Poland

Wojciech Janczukowicz, Fulbright Scholar
Department of Civil and Environmental Engineering
New Jersey Institute of Technology
Newark, New Jersey 07102

INTRODUCTION

Comparative microfauna investigations of different types of wastewater treatment plants lead to the recognition of its specific development. Utility of this kind of investigation is greater if ecology of one of the comparative systems is well-known and activated sludge can be regarded as such.[1] Recently, the knowledge concerning rules of growth RBD biocenosis has enlarged,[2] but is still not complete and requires further experiments. Earlier comparative studies of activated sludge and RBD microfauna were mainly qualitative.[3] Trials of carrying out quantitative analogies probably encountered limitations in the methodolgy. RBD microfauna is not distributed within the whole volume of the film but mainly in the thin, oxygen layer which is technically impossible to separate during the time of sample taking. In this case, total film biomass cannot be the base for qualitative comparison with activated sludge.

In the present paper, the amount of biofilm mass equivalent (EBM) to activated sludge biomass in relation to organic load removed was the base (see part I of this article) and quantitative parameters of disc microfauna were referred to it.

Comparative studies of activated sludge and RBD biofilm included: species composition, frequency, number and microfauna biomass.

METHOD OF MICROFAUNA EXAMINATION.

The investigations were carried out on a laboratory scale by using five tanks with activated sludge and five (one stage) RBD reactors. The experimental units and technological conditions of experiment are described in the first part of this paper.

Microscopic observations were started after twenty days of experiment and carried out for one month with frequency 2–3 days. The biofilm was scraped of the first disc of RBD from 16 cm^2 and placed in 20 mL cylinder with 2 cm^3 of water. After measuring the volume, the sample was blown through by pipettes to obtain homogeneous suspension. The samples of activated sludge were taken directly from aeration tanks.

Microscopic sections prepared from the volume of 0.05 cm^3 of biofilm and activated sludge were examined to describe microfauna species composition. The number of microorganisms was estimated using 640x magnification for Zooflagellata and Rhizopoda, and 256x magnification for Ciliata and Metazoa. According to the number of microfauna, the sample was diluted two, three, ten and fifteen times. The average number from three repetitions was finally related to one milligram of dried activated sludge biomass or to one milligram of dried biofilm mass equivalent to activated sludge (for RBD).

Zooflagellata, Rhizopoda, Rotatoria and Nematoda were treated as complete systematic units, however, Ciliata was examined as species.

45th Purdue Industrial Waste Conference Proceedings, © 1991 Lewis Publishers, Inc., Chelsea, Michigan 48118. Printed in U.S.A.

Table I. Microfauna Occurence Frequency in Activated Sludge in Relation to Organic Load Removed Rate

Taxons	Organic load removed, g COD/g VSS d				
	< 0.3	0.3–0.6	0.7–1.0	1.1–1.4	1.5–1.9
Zooflagellata	100	100	100	100	100
Rhizopoda	100	100	100	100	100
Opercularia coarctata	100	91	92	64	66
Aspidisca costata	100	83	78	57	50
Oxytricha fallax	50	50	50	14	33
Rotatoria	100	100	78	71	66
Vorticella convallaria	25	25	21	14	33
Trachelophyllum pussilum	25	50	35	57	33
Histriculus similis	50	58	42	28	16
Opercularia phryganea		25	14	35	—
Nematoda		45	28	50	—
Chilodonella uncinata		16	7	64	33
Paramecium caudatum		8	7	7	—
Urostyla sp.		8	—	—	—
Opercularia microdiscum		8	—	9	—
Vorticella extensa			7	—	—
Trimyema compressum			7	14	16
Opercularia minimum			7	—	—
Colpidium campylum				21	16
Glaucoma scintillans				7	—
Cyclidium glaucoma				7	—
Number of samples	4	12	14	14	6
Number of taxons	9(43%)	15(71%)	16(76%)	18(85%)	12(57%)

Microfauna biomass was determined by comparison with geometric bodies,[4] assuming that dried biomass is the tenth part of the wet biomass.

Biocenotic parameters (species number, frequency, number, biomass) were related to the value of organic load removal rate, calculated per unit of activated sludge biomass (R_{ASX}), and in case of discs – per unit of biofilm biomass equivalent to activated sludge (R_{BEX}). The number and microfauna biomass corresponding to the assumed load ranges were summed up and represented as arithmetical averages for a given load range.

RESULTS

Species composition and species frequency

Twenty-one taxons were identified in 50 samples of activated sludge: 17 species from Ciliata and Zooflagellata, Rhizopoda, Rotatoria and Nematoda (Table I). The number of microfauna species increased with organic load removal ranges (R_{ASX}) from 43 to 85%, and it was the greatest in the range 1.1 – 1.4. At higher load, the number of species was reduced to 57%. Zooflagellata and Rhizopoda were found in each examined sample of activated sludge, whereas the frequency of other organisms depended on the load rate. Attached and crawling Ciliata and Rotatoria were mainly found in the load which did not exceed 0.6 R_{ASX}. At higher load, free-swimming Ciliata, including predatory, and Nematoda increased their frequency. Besides Zooflagellata and Rhizopoda the following species: Opercularia coarctata, Aspidisca costata, Rotatoria, Oxytricha fallax, Trachelophyllum pussilum, Histriculus similis and Vorticella convallaria were characterized by the highest frequency.

Zooflagellata, Rhizopoda, Rotatoria, Nematoda and 15 species of Ciliata were observed in 72 samples of biofilm (Table II). The number of species decreased with load from 83 to 40 % and the highest was in the range < 0.3–0.6 R_{BEX}. Zooflagellata and Rhizopoda were found in each film sample. Seven species from Ciliata were observed most often and they appeared in a broad range of load, but with different frequency. Opercularia coarctata, Cyclidium glaucoma, Chilodonella uncinata and Colpidium colpoda were the most numerous or as the only ones below 0.6 R_{BEX}. Colpidium compylum, Glaucoma scinillans and Tetrahymena pyriformis preferred a higher load. Colpidium campylum, Glaucoma scintillans and Chilodonella uncinata were the most common in the

Table II. Microfauna Occurrence Frequency in Biofilm in Relation to Organic Load Removed Rate

Taxons	Organic Load Removed, g COD/g VSS d < 0.3	0.3–0.6	0.7–1.0	1.1–1.4	1.5–1.9
Zooflagellata	100	100	100	100	100
Rhizopoda	100	100	100	100	100
Colpidium campylum	66	94	95	100	100
Tetrahymena pyriformis	16	16	12	25	25
Glaucoma scintillans	66	72	70	75	100
Chilodonella uncinata	88	94	75	75	50
Colpidium colpoda	66	61	16	12	25
Opercularia coarctata	50	16	12	12	–
Rotatoria	33	5	8	12	–
Cyclidium glaucoma	66	77	–	–	–
Cyclidium versatile	22	5	–	–	–
Dexiotricha colpidiopsis	5	5	–	–	–
Drepanomonas revoluta	22	11	–	–	–
Platyophrya sp.	16	–	–	–	–
Thrithigmostoma cucullus	16	–	–	–	–
Colpoda cucullus	–	5	4	–	–
Paramecium caudatum	–	5	4	–	–
Podophrya fixa	–	–	8	–	–
Opercularia microdiscum	–	–	–	12	–
Nematoda	–	–	–	12	25
Number of samples	18	18	24	8	4
Number of taxons	15(83%)	15(83%)	12(60%)	11(55%)	8(40%)

whole load range. During the examination, Metazoa appeared rarely on the disc: Rotatoria mainly at the lowest and Nematoda at the highest load.

Microfauna number

In activated sludge, total microfauna number increased from about 2000 individuals in 1 mg of dry matter at the lowest load to about 27000 in the load range 0.7 – 1.9 (Figure 1). Very weak development of all animals with some predominance of Ciliata over the other groups corresponded with the lowest load (Table III). The number of animals increased significantly and Zooflagellata and Rhizopoda were in abundance at the load above 0.3. Rotatoria developed the best in the load range 0.3 – 1.0;

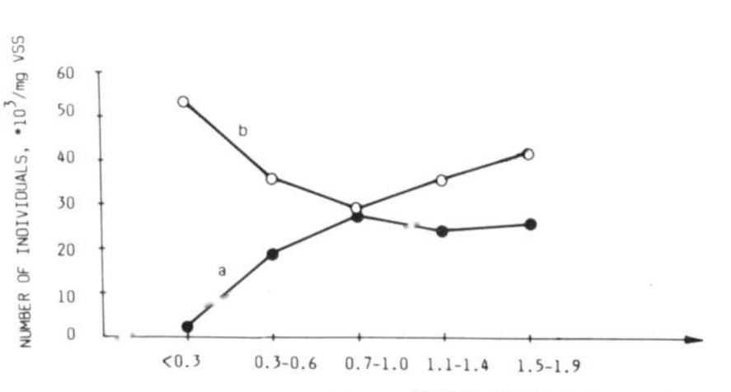

Figure 1. The changes of animals number in activated sludge(a) and biofilm(b) in relation to organic load removed.

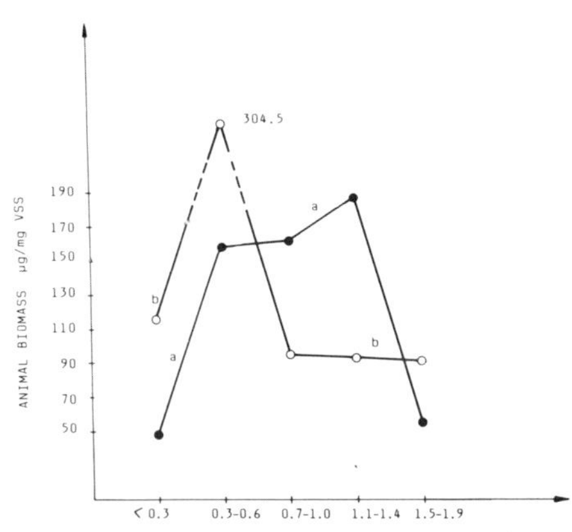

Figure 2. The changes of total animal biomass in activated sludge(a) and biofilm(b) in relation to organic load removed.

Table III. Distribution of Microfauna of Activated Sludge and RBD in Relation to Organic Load Removal Rate. Number in Thousand Individuals/mg VSS of Activated Sludge and Biofilm Mass Equivalent to Activated Sludge. Biomass in μg/mg VSS of Activated Sludge and Biofilm Equivalent to Activated Sludge.

Organic Load Removed,	Number					Biomass				
g COD/g VSS d	< 0.3	0.3–0.6	0.7–1.0	1.1–1.4	1.5–1.9	< 0.3	0.3–0.6	0.7–1.0	1.1–1.4	1.5–1.9
Activated sludge:										
Zooflagellata	0.16	7.6	7.9	8.7	9.2	0.003	0.13	0.14	0.15	0.16
Rhizopoda	0.10	4.6	12.9	9.4	16.2	0.02	7.3	19.8	15.0	26.0
Ciliata	1.9	5.4	6.2	6.2	1.1	44.9	94.2	80.0	143.2	19.8
Rotatoria	0.1	1.4	1.5	0.5	0.2	5.0	56.0	60.0	25.0	9.0
Nematoda	–	0.001	0.001	0.004	–	–	1.1	1.1	4.4	–
Biofilm:										
Zooflagellata	21.0	12.2	14.1	17.8	17.3	0.28	0.22	0.25	0.31	0.30
Rhizopoda	11.7	6.4	3.3	9.5	16.5	19.0	10.22	5.3	15.2	26.4
Ciliata	21.5	17.9	11.6	10.0	9.0	95.9	294.0	89.2	67.8	62.1
Rotatoria	0.008	0.002	0.003	0.005	–	0.4	0.1	0.05	0.1	–
Nematoda	–	–	–	0.008	0.002	–	–	–	8.8	2.2

Ciliata 0.3–1.4 and Rhizopoda 1.5–1.9 R_{ASX}. Zooflagelata were numerous in the range 0.3–1.9, showing the tendency to increase.

Opelcularia coarctata, O. phryganeae, Aspidisca costata and Histriculus similis and Trachelophyllum pussilum determined the Ciliata number in activated sludge. Aspidisca group (all Hypotricha were included here) predominated at load below 0.3 and Opercularia group (all Peritricha) in the rest of the load range (Table IV). Participation of Trachelophyllum predator did not exceed 5.7% and free-swimming bacteriovrous Ciliata–2%.

Microfauna number in RBD changed from 54000 individuals (in 1 mg of biofilm dry matter equivalent to activated sludge as treatment efficiency is concerned) at load < 0.3 to 29000 in 0.7–1.0 R_{BEX} (Figure 1). The number increased again to 42000 individuals at higher load. Protozoa mainly influenced the total number. At load < 0.3–0.6 there was predominance of Ciliata and above this range-Zooflagellata and Rhizopoda (Table III). The number of Metazoa was not high.

Ciliata occurring in RBD were put together into groups as in activated sludge. Opelcularia coarctata constituted Opercularia group developing numerously only at the load below 0.3 R_{BEX} (Table IV). Cyclidium glaucoma and C. versatile were included into Cyclidium group and Ch.uncinata and Thrithigmostoma cucullulus into Chilodonella group. Both groups developed numerously at the load below 0.6. Colpidium campylum, C. colpoda, Glaucoma scintillans, Parameciun candatum and Tetrahymena pyriformis were included into Colpidium. The group, with Colpidium domination, maintained a significant predominance in the range 0.3–1.9 R_{BEX}, giving way to Cyclidium group at the low range. Other species (Platyophrya, Drepanomonas, Colpoda, Dexiotricha, Podophrya) constituted a small percentage of Ciliata number.

Animal biomass

Total animal biomass in activated sludge increased from 49.9 to 187.8 μg of dry matter in 1 mg of sludge dry matter in the range < 0.3–1.4 R_{ASX} (Figure 2). Ciliata biomass mainly decided about the amount of biomass and Rotatoria biomass in the range 0.3–1.0 (Table III). In the range, 1.5–1.9 total biomass decreased to 55 g, and Rhizopoda demonstrated the highest participation. Zooflagellata biomass was not very large in spite of its high number.

Crawling Ciliata (Aspidisca group) were the most numerous below 0.3 R_{ASX} load and attached Ciliata (Opercularia group) were the most abundant in other ranges in the biomass of Ciliata group (Table IV). High biomass in the range 1.1–1.4 was influenced by high participation of Opercularia phryganeae.

Animal biomass in RBD increased from 115.7 to 304.5 g of dry matter (in 1 mg of biofilm dry mass equivalent to activated sludge as efficiency is concerned) in the range < 0.3–0.6 R_{BEX} (Figure 2). As higher load, biomass decreased and maintained in the range 91–95 μg. Ciliata biomass (Table III), mainly Colpidium group (Table IV) determined animal biomass in the biofilm. High biomass in the range 0.3–0.6 R_{BEX} was due to high Colpidium colpoda participation. Rhizopoda contribution did not exceed 25 % (1.5–1.9) and Nematoda 10 % (1.1–1.4). Zooflagellata and Rotatoria were of little significance in the biofilm (Table III).

SUMMARY AND DISCUSSION

Organic load applied is one of the most significant factors determining micro and macrofauna development in wastewater treatment plants.[5,6,7] The amount of organic load removed being the function (first order) of organic load applied and simultaneously treatment efficiency factor was applied as a comparative parameter in our studies.

In comparative ranges of organic load removed of activated sludge and biofilm, one expects certain similarities, microfauna showed many significant differentiations. In both systems, Protozoa were the most essential components of microfauna. Zooflagellata and Rhizopoda developed as the most numerous, especially at the highest load, and Ciliata created the largest biomass. Zooflagellata participation in animal biomass was very little. Rotatoria developed well, but only in activated sludge. In the range of lower load, they created biomass not much smaller than Ciliata biomass. Nematoda developed rather weakly in both systems. Weak development of Metazoa in biofilm can be explained by one stage of RBD and rather short period of investigations, because this group appears numerously in RBD, especially Nematoda.[8]

The influence of organic load removal was expressed by quantitative–qualitative changes of microfauna which significantly differentiated both systems.

Similar number of taxons was found in both systems. The following species were common among Ciliata: Colpidium campylum, Glaucoma scintillans, Chilodonela uncinata, Cyclidium glaucoma,

Table IV. Percentage Participation of Particular Groups in Ciliata Number and Biomass in Relation to Organic Load Removed Rate

Organic Load Removed,	% of Number					% of Biomass				
g COD/g VSS d	< 0.3	0.3–0.6	0.7–1.0	1.1–1.4	1.5–1.9	< 0.3	0.3–0.6	0.7–1.0	1.1–1.4	1.5–1.9
Activated sludge:										
Aspidisca	50.3	18.5	15.9	8.4	34.0	89.5	25.3	39.0	6.9	15.1
Opercularia	48.9	78.0	80.6	87.5	58.3	10.5	74.3	60.2	92.2	84.1
Trachelophyllum	0.8	3.4	3.1	3.2	5.7	0.03	0.2	0.2	0.14	0.3
Others	–	0.2	0.4	0.9	2.0	–	0.2	0.6	0.7	0.5
Biofilm:										
Opercularia	12.0	0.05	0.03	0.2	–	10.8	0.14	0.2	1.2	–
Cyclidium	42.0	13.4	–	–	–	4.7	0.4	–	–	–
Chilodonella	21.7	20.0	5.1	0.9	4.0	9.8	2.4	1.3	0.1	1.1
Colpidium	21.7	66.0	93.2	98.9	96.0	74.0	97.0	98.4	98.7	98.9
Others	2.6	0.5	0.7	–	–	0.7	0.06	0.07	–	–

Opercularia coarctata and Paramecium caudatum, but their participation in both biocenosis was different. Attached Ciliata and crawling Ciliata (mainly Opercularia, Aspidisca, Histriculus) dominated in activated sludge with little participation of free-swimming and predatory species. Free-swimming species dominated exclusively in RBD. Cyclidium spp. with quite numerous Opercularia coarctata, Chilodonella uncinata and Colpidium colpoda prevailed at the lowest load. Colpidium campylum accompanied by Glaucoma scintillans, Tetrahymena pyriformis and Colpidium colpoda dominated in the other range.

Ciliata from Colpidium, Glaucoma, Tetrahymena are very demanding bacteriovorous, numerously developing at bacteria concentrations at least 10^7 in cubic centimeter of water; whereas Chilodonella uncinata, Aspidisca costata and Opercularia coarctata accept concentration of 10^6 bacterial cells.[9] Food needs are connected with species mobility and feeding ability determined by the degree of specialization of the mouth structure.[2] In all cases a decrease in loading (i.e. decrease of bacteria concentration) we observe the following species succession: Colpidium (or Glaucoma)–Opercularia–Vorticella (or Carchesium, Zoothamnium)–crawling Ciliata.[5]

Cyclidium spp. make a particular group. They are characterized by great mobility with well developed mouth apparatus in comparison to its small body size. From our observations, it comes out that Cyclidium spp. develop in great quantities, mainly in the last stages of RBD. It proves great feeding ability, which can be maintained at decreased bacterial concentration. Cyclidium type can be a useful indicator of low RBD loading.

The number of taxons and animal biomass in activated sludge increased with the loading to the value 1.1–1.4 COD/ mg VSS (R_{ASX}). At the load below 0.3 and above 1.4, the number of species and biomass was on a similar low level.

In RBD the highest species number and animal biomass were observed at much lower load, not exceeding 0.6 COD/mg VS relative biofilm mass equivalent to activated sludge (R_{BEX}).

In this range (0.3–0.6) animal biomass was twice as high than activated sludge. At higher load (0.7–1.4) proportions were reversed, and animal biomass in RBD was nearly twice lower than in activated sludge. It was also characteristic that in the range 0.7–1.9 R_{BEX}, load did not influence animal biomass. The relationship between the number of total animals and load was not very clear, especially in RBD. Perhaps, it was caused by the difficulties of exact estimation of numbers of Zooflagellata and small Rhizopoda.

Species composition and animal biomass as indicators of bacteria concentration and food utilization prove that living conditions are potentially more advantageous in RBD than in activated sludge. Mainly, food conditions and washing out intensity of the system determine it. In RBD, dispersed bacteria concentrate in thin layer of wastes washing the film; whereas in the sludge they are spread in the whole volume of the tank. Free-swimming organisms in the activated sludge are easily washed out[11], which cannot be practically observed in RBD.[10] The factor, however, which can strongly limit RBD microfauna development, contrary to activated sludge, is the oxygen conditions in the biofilm worsening with the loading increase. In the present investigations, the upper limit of dynamic animal development was determined by the load 0.6 mg COD/ mg VS of the biofilm equivalent to the activated sludge. It corresponded to the influent loading about 1 mg COD/mg VS or nearly 40 g COD/m^2d (see first part of article). Up to this loading level, total biofilm mass was approximate to EBM. At higher loading, biofilm mass in disc strongly attached reaching the size far exceeding activated sludge biomass. As it is known in the thick film, anaerobic processes are intensified and their products limit the quantitative and species microfauna development especially Ciliata. In these conditions usefulness of animals as loading indicators decreases and bacterial indicators (e.g. Zooglea, Beggiatoa, Thiothrix) become more important.[7]

The main function of sedimentators in aerobic processes is to clarify effluent wastes.[11] The present investigations show that the role of animals in waste clarification, in the load range 0.3-0.6, is much more significant in RBD than in activated sludge. In the load range 0.7-1.4, it is reversed.

REFERENCES

1. Ecological Aspects of Use-Water Treatment, Ed: Curds, C.R., H.A. Hawkes, Vol.1, Academic Press, London, 1975.
2. Bishop, P.L., N.E. Kinner, *Aerobic Fixed-Film Processes, Biotechnology,* Eds: Rehm, H.J., G.Reed, Vol. 8, chapter 3, 1986.
3. Madoni, P., F.P. Ghetti, " The structure of ciliated Protozoa communities in biological sewage treatment plants," *Hydrobiologia,* 83, 207 (1981).
4. Cislenko, L.L., " Nomogrammy dla opriedielenija viesa vodnych organizmov po razmieram i formie tiela," Nauka, Leningrad, 1968.

5. Sladka, H., V.Sladecek, " A guide of organisms from waste water plants," Prace a studie 162, Vyz. Ustav Vodohosp., Praha, 1985.
6. Kinner, N.E., C.R.Curds," Development of Protozoa and Metazoa communities in Rotating Biological Contactor biofilms," *Water Research*, 21,481 (1987).
7. Sudo, R., M.Okada, T. Mori, " Rotating Biological Contactor microbe control in RBC," *J. Wat. Wast.*, 19,1 (1977).
8. Hul, M., M.Krzemieniewski, J.Zakrzewski, " The effect of technical paramaters of a moving tape-bed biofilter upon formation of its biocenosis," *Acta Acad. Agricult. Techn. Olst., 14, 115 (1985).*
9. Bick, H., S.Kunze, " Eine Zusammenstellung von autokologischen und saprobiologischen Befuden an Susswasserciliaten," *Int. Rev. ges. Hydrobiol.*, 56, 337 (1971).
10. Kinner, N.E., " An evaluation of the feasibility of using Protozoa and Metazoa as indicators of RBC effluent quality," Proceedings: Second International Conf. on Fixed – Film Biol. Processes, 74 (1984).
11. Curds, C.R., " The ecology and role of Protozoa in aerobic sewage treatment processes," *Ann. Rev. Microbiol.*, 36, 27 (1982).

40 DENITRIFICATION OF HIGH-STRENGTH INDUSTRIAL WASTEWATERS

William W. Clarkson, Assistant Professor
Ben J. B. Ross, Graduate Research Assistant
Srikanth Krishnamachari, Graduate Research Assistant
School of Civil Engineering
Oklahoma State University
Stillwater, Oklahoma 74078

INTRODUCTION

Industries producing highly concentrated nitrogenous waste streams include chemical fertilizers, munitions, nuclear fuel processing, and semi-conductor manufacturing, among others. Regenerant streams from ion exchangers used to remove contaminants from drinking or process waters may also present problems in treatment and disposal of strong nitrogenous wastes. Biological nitrification and denitrification have been applied to tertiary treatment of low-strength municipal wastewaters, with nitrogen concentrations typically < 50 mg/L. Limited studies have shown the potential of these processes to treat nitrogen concentrations $> 1{,}000$ mg/L efficiently, particularly in the absence of significant amounts of biodegradable organic matter.

Previous studies[1,2] have shown that fixed-biofilm reactors were capable of completely nitrifying between 500 and 1,000 mg NH_4^+-N/L in a semiconductor waste, providing an economically viable alternative to air stripping. Denitrification could be used as a second stage process in a complete removal system for wastes of this type, or alone for treatment of wastes containing nitrate and/or nitrite-N. Denitrification may be carried out heterotrophically by common facultative bacteria, for example species of *Pseudomonas, Alcaligenes, Paracoccus, Bacillus, Propionibacterium*, etc. These organisms metabolize compounds for carbon and energy. Much less well-known are the various autotrophic denitrifying bacteria. Different species are capable of deriving energy from oxidation of hydrogen, reduced iron and sulfur ions, etc. for thee incorporation of inorganic carbon. Examples of sulfur oxidizers are *Thiosphaera pantotropha, Thiomicrospira denitrificans*, and *Thiobacillus denitrificans*[3]. Both heterotrophic and autotrophic denitrifying bacteria use nitrate as a terminal electron acceptor to facilitate the oxidation of substrate. The intermediate and final products of this process are given in the following general sequence from more oxidized to more reduced:

$$NO_3^- \dashrightarrow NO_2^- \dashrightarrow NO \dashrightarrow N_2O \dashrightarrow N_2$$

Certain species of bacteria may be more or less able to use each of the intermediates to oxidize substrate. Environmental conditions may also affect how completely the reduction may be carried out, resulting in accumulation of certain intermediate compounds.

Comparison of the stoichiometry of heterotrophic and autotrophic denitrification reveals that, whereas the heterotrophs are net alkalinity producers, autotrophic denitrifiers consume alkalinity (are net producers of acidity) in much the same way as nitrifying bacteria. Following are calculated stoichiometries for the two systems:

Autotrophic[3]:

$$NO_3^- + 0.79\ S_2O_3^= + 0.27\ HCO_3^- + 0.20\ H_2O\ =\ 0.05\ C_5H_7O_2N + 0.47N_2 + 1.56\ SO_4^= + 0.28\ H^+$$

Heterotrophic[4]:

$$NO_3^- + 1.08\ CH_3OH + H^+\ =\ 0.065\ C_5H_7O_2N + 0.47\ N_2 + 0.76\ Co_2 + 2.44\ H_2O$$

The purpose of this study was to determine the upper concentration and loading rate limitations of both heterotrophic denitrification using methanol as carbon and energy source, and autotrophic denitrification using bicarbonate carbon and thiosulfate as electron donor. Studies were carried out in bench-scale attached film expanded bed (AFEB) and upflow sludge blanket (USB) reactors. There-

45th Purdue Industrial Waste Conference Proceedings,

fore, comparisons were made between different microbial populations and between two high-rate biological reactor configurations.

RESEARCH OBJECTIVES

The overall goal of the study was to demonstrate the feasibility of denitrification of concentrated waste streams having NO_3^--N concentrations in excess of 500 mg/L. Specific objectives were as follows:

- Determine the maximum nitrate concentration which could be treated without significant loss of efficiency.
- Determine the maximum volumetric loading and removal rates achievable.
- Establish the stoichiometric chemical requirements for high-rate denitrification under the experimental conditions.
- Compare results of heterotrophic denitrification with methanol to autotrophic denitrification with thiosulfate.
- Compare the performance of two high-rate reactors, the attached film expanded bed (AFEB) and the upflow sludge blanket (USB).

MATERIALS AND METHODS

Experimental Apparatus

Bench-scale AFEB and USB reactors were constructed for these experiments. One AFEB and one USB were tested simultaneously under heterotrophic and autotrophic conditions; thus a total of four reactors were used. Each had a total volume of approximately 2.5 L, approximately 30 to 40 percent of which was occupied by denitrifying biomass under the operating conditions of the study. These reactor configurations were chosen due to their demonstrated capacity for high loading and conversion rates in a variety of applications. Extremely high biomass densities, and high cell residence times coupled with low hydraulic residence times make them ideal candidates for any anaerobic or anoxic treatment system. Energy requirements of these systems are relatively low, and solids may be handled efficiently, perhaps obviating the need for secondary clarification.

AFEB systems have been fully described elsewhere.[5] The essential elements of this system are that it is a fixed biofilm process employing small, lightweight, inert support media which are suspended in an upward flowing recycle stream taken from the reactor above the bed. Diatomaceous earth particles (~ 0.2 - 0.6 mm diameter) were used in this research. The bed was expanded approximately 15 to 20% by the recycle flow. The light support media are easy to fluidize slightly in this manner, resulting in greater biomass densities than a typical fluidized bed reactor. A schematic diagram of the experimental AFEB system is shown in Figure 1. The reactors were constructed from an acrylonitrile plastic Imhoff settling cone attached to an upper segment of plexiglass tubing. A funnel was placed in the reactor to allow for gas capture and to provide a water seal to the level of the effluent overflow port.

The USB reactors do not employ any biofilm support media, nor in this case was there any provision for effluent recycle. The biomass forms a "blanket" of sludge which will exhibit flocculant or granular properties depending on the operating conditions of the system. Influent was introduced into the reactor at the bottom, and mixing was accomplished with a magnetic stirrer set at low speed. A stir bar in a cage was balanced in the rounded bottom portion of the reactor vessel. The mixing system was intended to slowly and continuously agitate the sludge blanket just enough to enhance granulation and contact of biomass with substrate, but did not always perform satisfactorily. The USB reactors were made of glass and as indicated in the process schematic diagram, Figure 2, were fitted with an outer collar to collect effluent overflowing the center section. A funnel in the inner part served the same purpose as in the AFEB.

Nitrogen Substrates

This study was meant to determine limits on the technology in terms of nitrate concentrations, volumetric conversion rates, treatment efficiencies, and actual stoichiometry in the absence of any other inhibitions or interferences. Thus, synthetic substrates were made up to provide for the growth needs of heterotrophic and autotrophic bacteria. Recipes for the synthetic wastes are given in Table I for a nitrate-N concentration of 500 mg/L. This concentration was adjusted as desired.

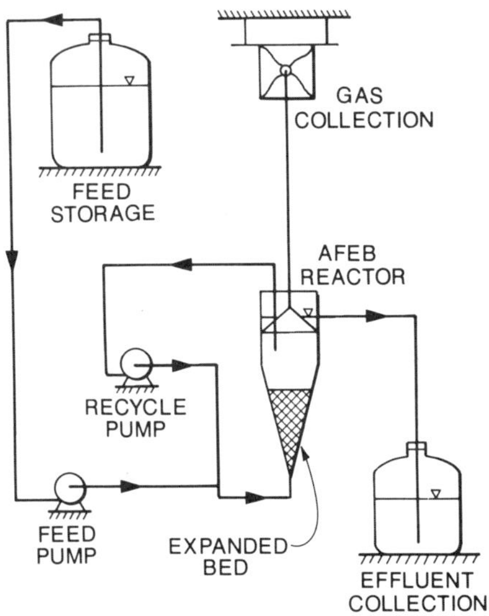

Figure 1. Schematic of the bench-scale experimental AFEB systems.

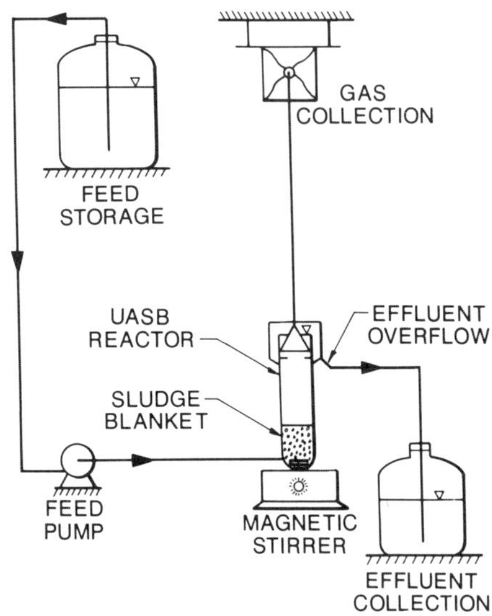

Figure 2. Schematic of the bench-scale experimental USB systems.

Analytical Procedures

Total and volatile suspended solids, pH, alkalinity, and COD (closed reflux culture tube procedure) were analyzed according to Standard Methods.[6] Nitrate, nitrite, and sulfate were measured with a Dionex 2000 i/SP ion chromatograph, following Standard Methods. Thiosulfate was determined by titration with iodine.[7] Attached biomass was quantified as volatile solids by procedures described by Clarkson.[8]

Innoculation, Start-up, and Reporting of Results

Inoculum for both sets of experiments was obtained from return sludge of the Ponca City, Oklahoma activated sludge municipal wastewater treatment plant. The sample was divided and fed methanol or thiosulfate once per day for two weeks to encourage the growth of heterotrophic and autotrophic denitrifying bacteria, respectively. After two weeks of inoculum development, the AFEB and USB reactors were seeded by filling to about two-thirds volume and initiating continuous feeding. Initial conditions were NO_3^--N concentrations between 250 and 500 mg/L and hydraulic residence times (HRT) of 6 hours. As washout of biomass occurred during the initial start-up period, small

Table I. Synthetic Industrial Wastewater Recipes (made up with tap water)

Autotrophic Denitrification (500 mg NO^-_3-N/L)	
Ingredients	*Concentration, g/L*
Potassium Nitrate, KNO_3	3.60
Sodium Thiosulfate, $Na_2S_2O_3$	7.20
Sodium Bicarbonate, $NaHCO_3$	3.60
Mono Potassium Phosphate, KH_2PO_4	0.10
Ferrous Sulfate, $FeSO_4$	0.01
Heterotrophic Denitrification (500 mg NO^-_3-N/L)	
Ingredients	*Concentration, g/L*
Potassim Nitrate, KNO_3	3.60
Methanol, CH_3OH	1.50
Mono Potassium Phosphate, KH_2PO_4	0.035
Magnesium Sulfate, $MgSO_4$	0.004
Ferrous Sulfate, $FeSO_4$	0.002

amounts of fresh inoculum from the semi-continuous seed reactors were added to replace the loss. The autotrophic USB developed a granular appearance, in place of the light, flocculant initial growth, after about six weeks. The autotrophic AFEB also became fully coated with biofilm and increased in bed volume between 350 to 450 mL within the same time period. The heterotrophic AFEB volume grew from 420 to 700 mL in seven to eight weeks. Due in part to uneven performance of the mixing system, the heterotrophic USB required about 12 weeks for development of a distinct sludge blanket with a clear blanket/supernatant interface. Truly granular sludge aws not observed in this reactor; however the appearance of the blanket was characterized by tight floc formation.

Start-up was deemed complete when the reactors exhibited significant biomass development and a stable response in terms of treatment efficiency. Experimental data were taken from that point onward. Each data point represents the mean of at least three individual samples taken after daily monitoring indicated that a steady-state condition had been reached. In this study, steady-state conditions were considered to have been achieved when no significant change in removal efficiency or other key parameters could be observed over a span of at least 10 hydraulic residence times.

All feed concentrations, loading rates, and conversion rates are expressed in terms of "nitrogen equivalent" (*e.g.* mg N_e/L). This represents the total of nitrate-N and nitrite-N, the soluble oxidized nitrogen forms present in a given sample. Small amounts of nitrite were found in most influent and effluent samples. At the highest loading rates, particularly in the heterotrophic systems, high concentrations of nitrite accumulated in the effluent. Thus, conversion efficiency in terms of N_e concentration may be relatively low when very little nitrate remained after treatment. It is felt that this was due to particular bacterial species' inability to reduce nitrite under the experimental conditions.

RESULTS AND DISCUSSION

The experimental strategy for each of the reactor systems consisted of two series of tests conducted in sequence. The first series was designed to determine an upper limit on substrate nitrogen concentration, while the second series was intended to determine the maximum loading and removal rates achievable by increasing loading rates (decreasing HRT) at substrate concentrations somewhat below the previously determined upper limit.

Constant HRT/Variable Concentration Experiments

At the conclusion of the start-up phase, both AFEB reactors were operated at approximately 3.5 hr HRT, and both USB reactors at approximately 6 hr HRT. These feed rates were maintained throughout phase one of the experiments, while feed concentrations were increased stepwise until a significant decrease in treatment efficiency was observed. Initial target concentrations were 750 mg N_e/L in the autotrophic AFEB, 700 mg N_e/L in the heterotrophic AFEB, 275 mg N_e/L in the autotrophic USB, and 200 mg N_e/L in the heterotrophic USB. Results of these experiments are summarized in Tables II and III for the AEFB and USB reactors, respectively.

AFEB reactors. Data from Table II are plotted in Figures 3 and 4 to show the effect of influent nitrogen concentration on treatment efficiency and N_e reduction in the AFEB reactors. The plots reveal a clear break in the response curve for autotrophic denitrification at about 1,000 mg N_e/L. This breakdown in efficiency probably comes about as a result of produce formation (*e.g.* sulfate buildup) rather than as a direct function of influent nitrate loading. The break point for heterotrophic denitrification is not quite as well defined, but Figure 4 shows the upper concentration limit to be in the range of 1,400 to 1,500 mg N_e/L for near complete efficiency. In terms of volumetric loading rate, these upper limits equate to about 7 and 11 kg $N_e/m^3 \cdot d$ for the autotrophic and heterotrophic AFEB systems, respectively.

USB reactors. Results of the USB experiments from Table III are plotted in Figures 5 and 6 and show a similar response. The autotrophic system again begins to experience difficulty at about 1,000 mg N_e/L feed concentration. The heterotrophic USB, while lagging noticeably behind the autotrophic reactor throughout the range of values tested, again breaks down at influent concentrations close to 1,400 mg N_e/L.

Constant Concentration/Variable HRT Experiments

Based on the results of the first series of tests, feed concentrations were selected for each reactor in order to test the effects of increasing loading rate on system performance. This was accomplished by increasing the pumping rate of substrate through the reactors, thus decreasing the HRT incrementally until the pattern of response could be ascertained. The autotrophic AFEB reactor was operated at 750

Table II. Summary of AFEB Reactors at Constant HRT and Variable Feed Concentration

Infl. Conc. mg N_e/L	N_e Consumed mg/L	N_e Consumed %	Exp. Bed Vol. mL	HRT hour	Attached Biomass kg VS/m^3	Gas Collected L/d	Loading Rate kg $N_e/m^3 \cdot d$	Removal Rate kg $N_e/m^3 \cdot d$
Heterotrophic Reactor:								
689	688	99.9	900	3.1	55	3.1	5.36	5.35
986	926	93.9	1000	3.4	—	—	6.90	6.48
1004	999	99.5	1000	3.4	82	6.1	7.03	6.99
1495	1412	94.4	1000	3.4	—	—	10.47	9.88
1671	1516	90.7	1000	3.4	83	9.4	11.70	10.61
1748	1211	69.3	1000	3.4	80	—	12.24	8.48
Autotrophic Reactor:								
741	738	99.6	875	3.4	23	5.8	5.42	5.40
882	879	99.7	1000	3.8	27	—	5.65	5.63
1041	1026	98.6	1000	3.8	21	6.9	6.66	6.57
1242	828	66.7	1000	3.8	—	—	7.95	5.30

Table III. Summary of USB Reactors at Constant 6 Hour HRT and Variable Feed Concentration

Infl. Conc. mg N_e/L	N_e Consumed mg/L	N_e Consumed %	Gas Collected L/d	Loading Rate kg $N_e/m^3 \cdot d$	Removal Rate kg $N_e/m^3 \cdot d$
Heterotrophic Reactor:					
199	199	100	–	0.80	0.80
291	290	99.7	–	1.16	1.16
402	401	99.8	–	1.61	1.61
592	409	69.1	–	2.37	1.64
782	597	76.3	–	3.13	2.39
1030	792	76.9	2.4	4.12	3.17
1308	885	67.7	3.1	5.23	3.54
1485	664	44.7	1.6	5.94	2.66
Autotrophic Reactor:					
277	265	95.3	3.4	1.11	1.05
496	487	97.9	6.9	1.98	1.96
758	756	99.7	–	3.03	3.02
801	761	95.0	7.5	3.20	3.05
939	923	98.3	7.8	3.76	3.69
1086	864	79.6	–	4.34	3.46

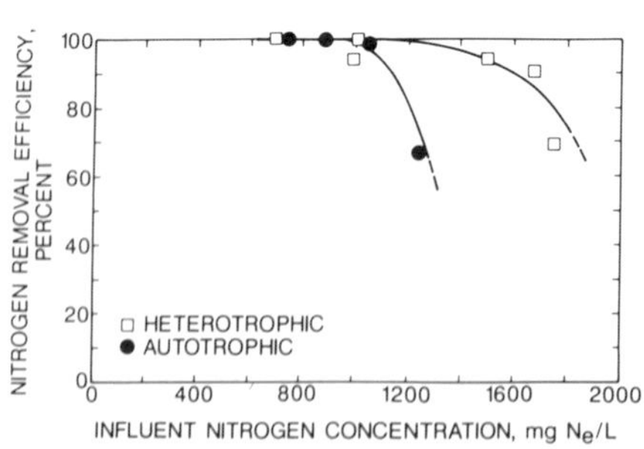

Figure 3. Treatment efficiency as a function of influent nitrogen concentration in AFEB reactors operated at constant 3.5 hour HRT.

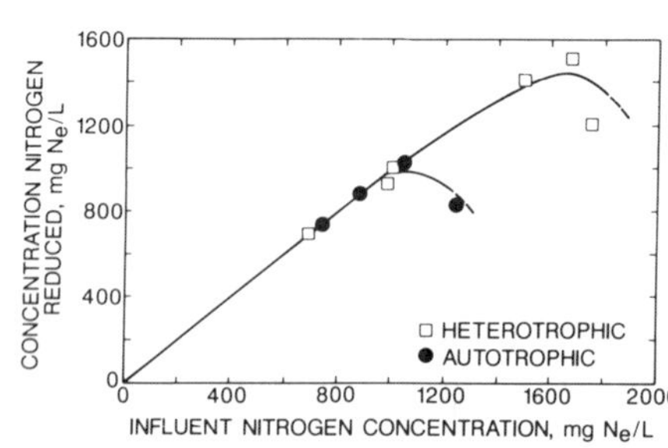

Figure 4. Influent nitrogen concentration limits on treatment efficiency in AFEB reactors operated at constant 3.5 hour HRT.

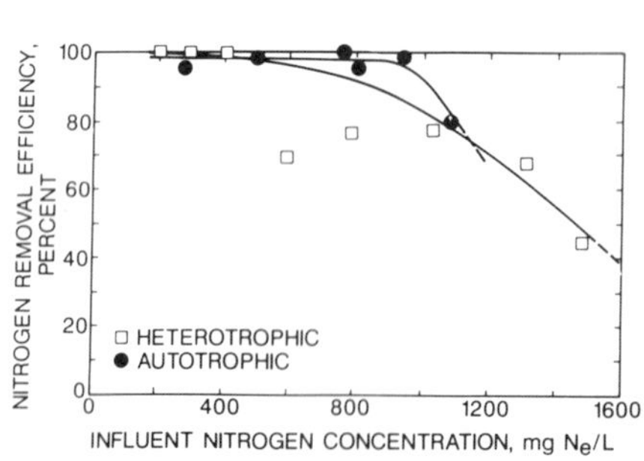

Figure 5. Treatment efficiency as a function of influent nitrogen concentration in USB reactors operated at constant 6 hour HRT.

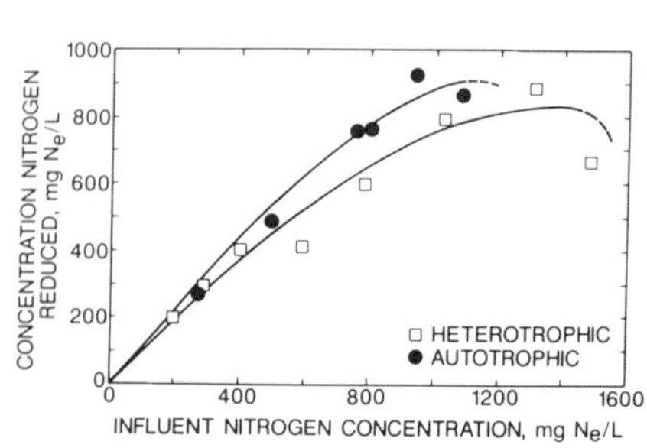

Figure 6. Influent nitrogen concentration limits on treatment efficiency in USB reactors operated at constant 6 hour HRT.

mg N_e/L, while the heterotrophic AFEB received 925 mg N_e/L due to the higher limiting value found in phase one experiments. Both USB reactors were fed at 500 mg N_e/L. Even though the pattern of limiting concentration values was similar in the USB tests to the AFEB values, all feed concentrations over 500 mg N_e/L resulted in treatment efficiencies less than 80% in the heterotrophic USB. Results of this second phase of testing are summarized in Tables IV and V for the AFEB and USB reactors, respectively.

AFEB reactors. Data from Table IV are plotted in Figure 7. The heterotrophic reactor shows a linear relationship of decreasing treatment efficiency with increasing loading rate. Conversion efficiency dropped below 90% at all loading rates over 10 kg $N_e/m^3 \cdot d$. However the loss of efficiency was gradual with increasing loading, such that 50% removal efficiency was still observed at a volumetric loading in excess of 40 kg $N_e/m^3 \cdot d$. At this operating condition, the heterotrophic AFEB was converting methanol substrate at a rate of 69 kg COD/$m^3 \cdot d$, a rate comparable to the highest reported for biological treatment processes[8].

The autotrophic AFEB demonstrated near 100% conversion efficiency throughout the entire range of loadings tested. Failure actually occurred as a result of physical upset of the expanded bed from excessive biogas production, which tended to carry media to the surface of the reactor. The biological capacity of the reactor had not been reached at 17.5 kg/$m^3 \cdot d$, the highest loading rate test completed.

USB reactors. Data from Table V, plotted in Figure 8, again indicate a dramatic difference in response between the two bacterial populations. The heterotrophic system shows a much lower treatment efficiency throughout the range of experimental loadings than the autotrophic system. The heterotrophic response is linear, as in the AFEB, but efficiencies are much lower. Response of the autotrophic USB is very similar to the AFEB, and in this case the system was pushed far enough to observe the loading rate limitation on efficiency. This break point clearly occurred between influent loadings of 15 and 20 kg $N_e/m^3 \cdot d$.

General Results

Stoichiometric parameters for chemical consumption and product formation were in accord with predicted values for all of the systems except that the autotrophic bacteria exhibited a lower than expected ratio of N_e reduced to alkalinity consumed, in comparison with the calculated stoichiometry presented earlier. Specific removal rates were calculated for the AFEB reactors, based on the attached biofilm volatile solids in each reactor. Attached autotrophic biomass averaged 25 kg VS/m^3 of settled bed volume, and the specific removal rate in this case ranged from 0.21 to 0.67 kg N_e/kg VS · d. In the heterotrophic system, attached biomass averaged 82 kg VS/m^3. With this extremely high biofilm concentration, specific activity may be assumed to be low, and in fact was measured as 0.09 to 0.14 kg N_e/kg VS · d. On the basis of organic removal, the rate for the heterotrophic AFEB was between 0.31 and 0.43 kg COD/kg VS · d. Low specific activities are generally associated with long cell residence times in biological systems characterized by high biomass concentrations in the reactor and low effluent volatile suspended solids. Effluent VSS in these experiments averaged 47 mg/L in the autotrophic AFEB, 59 mg/L in the autotrophic USB, 46 mg/L in the heterotrophic AFEB, and 94 mg/L in the heterotrophic USB.

Discussion

In order to view the results of this research in proper perspective, they must be assessed in comparison to other studies conducted under comparable conditions. Table VI is a summary of results obtained by other investigators. It should be noted that most experiments were conducted with feed concentrations ranging from 500 to 1,500 mg N_e/L. One autotrophic study is reported, and certain heterotrophic systems used organic substrates other than methanol. Reactor types represented include static filters or packed bed reactors (PBR), upflow sludge blankets (USB), one complete-mix continuous stirred tank reactor (CSTR), and fluidized bed reactors (FBR).

In the temperature range of this study (~ 20–25°C) the highest documented volumetric removal rate is 12.1 kg $N_e/m^3 \cdot d$, comparable to results obtained in this study with the heterotrophic AFEB. However, both the autotrophic AFEB and USB reactor systems significantly exceeded this rate. Part of the effect may be attributable to the fact that most other studies were conducted on real process waste streams containing impurities which could inhibit the denitrification process. The extremely positive effect of elevated temperature on reaction rates can be seen by comparing the results of Bosman et al.[15] at 20 and 38°C. Denitrification rates could thus be enhanced significantly when treating warm process wastes.

Table IV. Summary of AFEB Reactors at Constant Feed Concentration and Variable HRT

Infl. Conc. mg N_e/L	N_e Consumed mg/L	N_e Consumed %	Exp. Bed Vol. mL	HRT hour	Attached Biomass kg VS/m^3	Gas Collected L/d	Loading Rate kg $N_e/m^3 \cdot d$	Removal Rate kg $N_e/m^3 \cdot d$
Heterotrophic Reactor:								
934	880	94.2	1000	3.4	—	—	6.54	6.16
943	796	84.4	1000	2.3	81	7.1	9.90	8.36
912	787	86.3	1000	1.7	82	—	12.77	11.02
941	675	71.7	900	1.0	—	14.0	21.96	15.74
904	472	52.2	600	0.5	—	—	42.19	22.02
Autotrophic Reactor:								
741	738	99.6	875	3.3	23	5.3	5.42	5.40
771	767	99.5	900	1.7	21	8.3	10.96	10.91
747	740	99.1	850	1.3	20	11.7	13.53	13.41
773	763	98.7	850	1.1	—	—	17.46	17.23

Table V. Summary of USB Reactors at Constant Feed Concentration and Variable HRT

Infl. Conc. mg N_e/L	N_e Consumed mg/L	N_e Consumed %	HRT hour	Gas Collected L/d	Loading Rate kg $N_e/m^3 \cdot d$	Removal Rate kg $N_e/m^3 \cdot d$
Heterotrophic Reactor:						
513	373	72.7	6	—	2.05	1.49
504	363	72.0	4	1.8	3.02	2.18
520	286	55.0	3	—	4.16	2.29
522	281	53.8	1.5	4.0	8.35	4.49
498	133	26.7	1	—	11.95	3.19
Autotrophic Reactor:						
510	491	96.3	6	—	2.04	1.96
476	470	98.7	3	7.5	3.81	3.76
486	469	96.5	2	—	5.83	5.62
490	468	95.5	1.25	24.6	9.41	8.98
497	476	96.2	0.8	40.9	14.30	13.72
492	157	36.2	0.6	18.5	20.10	6.40

CONCLUSIONS

These experiments have demonstrated the feasibility of applying high-rate biological treatment technology to denitrification of solutions over a concentration range of 500 to > 1,500 mg N_e/L. Specific conclusions which can be drawn from this study include the following:

- The autotrophic AFEB achieved 99% nitrogen removal at loadings up to at least 17.5 kg/$m^3 \cdot d$, feed concentration = 760 mg N_e/L, and HRT = 1.1 hr.

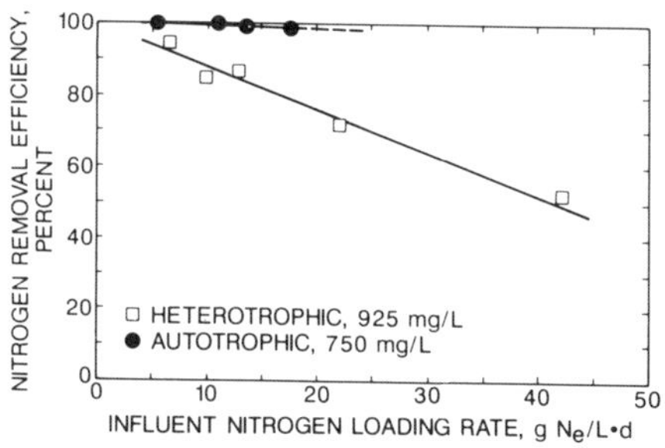

Figure 7. Treatment efficiency as a function of loading rate in AFEB reactors operated at constant feed concentrations (Heterotrophic = 925 mg N_e/L, Autotrophic = 750 mg N_e/L).

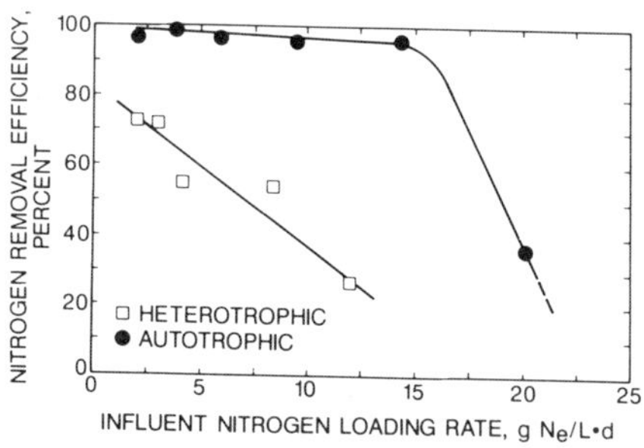

Figure 8. Treatment efficiency as a function of loading rate in USB reactors operated at constant feed concentration of 500 mg N_e/L.

Table VI. Concentrated Nitrate Removal Rates in Previous Studies

Reactor	HRT hr	Concentration mg N/L	Removal Rate kg N/$m^3 \cdot d$	Substrate	Reference
PBR	0.24	3500	5.6	Methanol	(9)
PBR	—	970	5.6	Thiosulfate	(10)
PBR	2.4	1000	12.1	Methanol	(11)
USB	—	500	12.0	Fusel Oil	(12)
USB	—	900	7.2	Methanol	(13)
CSTR	—	1220	14.7 (32°C)	Methanol	(14)
FBR	—	1450	10.6 (20°C)	Molasses	(15)
FBR	—	1450	38.4 (38°C)	Molasses	(15)

- Upper limit of autotrophic AFEB loading was not reached due to physical upset from excessive biogas production.
- The autotrophic USB demonstrated that granular sludge development is achievable with this bacterial population.
- The autotrophic USB achieved 96% nitrogen removal efficiency at loadings up to 14.3 kg $N_e/m^3 \cdot d$, feed concentration = 500 mg N_e/L, and HRT = 0.8 hr.
- Upper limit of autotrophic USB loading was encountered at between 15 and 20 kg $N_e/m^3 \cdot d$, before treatment efficiency dropped below 99%.
- The autotrophic bacteria showed a definite limiting feed concentration for complete conversion efficiency at about 1,000 mg N_e/L.
- The heterotrophic AFEB achieved 91% removal efficiency at a loading of 11.7 kg $N_e/m^3 \cdot d$, feed concentration = 1670 mg N_e/L, and HRT = 3.4 hr.
- The maximum removal rate in the heterotrophic AFEB was noted at a loading of 42 kg $N_e/m^3 \cdot d$. Conversion rate at this condition was 22 kg $N_e/m^3 \cdot d$, which corresponds to an organic conversion rate of 69 kg $COD/m^3 \cdot d$.
- Heterotrophic bacteria tolerated higher feed concentrations of nitrogen than the autotrophs before significant reductions in treatment efficiency occurred.
- Both the heterotrophic AFEB and USB reactors, in contrast to the autotrophic units, showed a linear response of treatment efficiency to loading rate.
- Accumulation of nitrite was noted when treatment efficiency dropped at higher loading rates, particularly in the heterotrophic systems.
- The heterotrophic USB performed relatively poorly, achieving a maximum conversion rate of only 4.5 kg $N_e/m^3 \cdot d$. This may have been related to poor sludge granulation or to problems with the mixing system in addition to any biological factors.

Three of the four systems tested therefore demonstrated superior potential for use in a number of industrial waste treatment applications. It is felt that the autotrophic systems are particularly well suited to industrial situations, provided that dilution of very strong wastes can be accomplished to bring the feed concentrations into the completely treatable range.

ACKNOWLEDGMENTS

The experiments reported in this paper constituted the MS thesis research of the two junior authors. Support for this work was provided through a grant from the Oklahoma State University Center for Water Research.

REFERENCES

1. Collins, A. G., W. W. Clarkson, and M. Vrona, "Fixed-film Biological Nitrification of a Strong Industrial Waste," *J. Water Pollut. Control. Fed.*, 60, 499–504 (1988).
2. Collins, A. G., W. W. Clarkson, and S. A. Florio, "Nitrification of a Strong Industrial Waste Using Aerobic Expanded Bed Reactors," *Proc. Industr. Wastes Symposia*, Session 15, 61st Ann. Conf. of the WPCF, Dallas (1988).
3. Ross, B. J. B., "High-rate Autotrophic Denitrification of Simulated Industrial Wastewater," M.S. Thesis, Oklahoma State University (1989).
4. McCarty, P. L., L. Beck, and P. S. Amant, "Biological Denitrification of Wastewaters by Addition of Organic Materials," *Proc. 24th Industr. Waste Conf.*, Purdue Univ. Ext. Serv. 135, 1271–1285 (1969).
5. Schraa, G. and W. J. Jewell, "High Rate Conversions of Soluble Organics with a Thermophilic Anaerobic Attached Film Expanded Bed," *J. Water Pollut. Control Fed.*, 56, 226–232 (1984).
6. APHA, AWWA, and WPCF, *Standard Methods for the Examination of Water and Wastewater*, 16th edition (1985).
7. Pierce, W. C. and E. W. Hanisch, *Quantitative Analysis*, Wiley, New York (1948).
8. Clarkson, W. W., "Fermentation of Particulate Organic Matter to Methane in a Thermophilic Anaerobic Attached Film Expanded Bed Reactor," Ph.D. Thesis, Cornell University (1986).
9. Jewell, W. J. and R. J. Cummings, "Denitrification of Concentrated Nitrate Wastewaters," *J. Water Pollut. Control Fed.*, 47, 2281–2291 (1975).
10. Claus, G. and H. J. Kutzner, "Physiology and Kinetics of Autotrophic Denitrification by *Thiobacillus denitrificans, Appl. Microbiol. Biotechnol.*, 22, 283–288 (1985).
11. Blaszczyk, M., E. Galka, E. Sakowicz, and R. Mycielski, "Denitrification of High Concentrations of Nitrites and Nitrates in Synthetic Medium with Different Sources of Organic Carbon, III: Methanol," *Acta Microbiol. Polon.*, 34, 195–206 (1985).

12. Klapwijk, A., J. C. M. van der Hoeven, and G. Lettinga, "Biological Denitrification in an Upflow Sludge Blanket Reactor," *Water Research*, 15, 1-6 (1981).
13. Miyaji, Y. and K. Kato, "Biological Treatment of Industrial Wastes Water by Using Nitrate as an Oxygen Source," *Water Research*, 9, 95–101 (1975).
14. Bode, H., C. F. Seyfried, and A. Kraft, "High-rate Denitrification of Concentrated Nitrate Wastewater, *Water Sci. Technol.*, 19, 163–174 (1987).
15. Bosman, J., A. A. Eberhard, and C. I. Baskir, "Denitrification of a Concentrated Nitrogenous Industrial Effluent Using Packed Column and Fluidized Bed Reactors," *Prog. Water Technol.*, 12, 667-680 (1978).

41 PERFORMANCE CHARACTERISTICS OF AN ACTIVATED SLUDGE SYSTEM WHEN NITRATE IS THE SOLE SOURCE OF NITROGEN

Nannette Corey, Engineer
James M. Montgomery, Consulting Engineer
Metarie, Louisiana 70002

Larry Benefield, Professor
Department of Civil Engineering
Auburn University, Alabama 36849

INTRODUCTION

On a weight basis most of the organic matter of a bacterial cell consists of macromolecules which belong to four classes: nucleic acids, proteins, polysaccharides, and complex lipids. Although the percentages will vary depending on the actual growth conditions, a rapidly growing cell will contain approximately 60% protein, 19% nucleic acid, 15% lipid and 6% carbohydrate.

Organic removal in an activated sludge process is accomplished by aerobic heterotrophic microorganisms which utilize a portion of the organic material as a carbon source for synthesis of new biomass and oxidize the remainder to provide energy for the synthesis function as well as for the other nongrowth functions. In order for the synthesis function to proceed, the wastewater must contain an adequate supply of all the elements which are found in the cytoplasmic material of a cell.

Nitrogen is a component of proteins, nucleic acids, amino sugars, etc., and is, therefore, essential for microbial growth. Generally, the nitrogen requirement is easily met with municipal wastewaters, but many industrial wastewaters do not contain sufficient nitrogen and require its addition as a supplement.

When nitrogen addition is required, it is normally added in the ammonium form. For wastewaters totally deficient in nitrogen, the cost for supplemental nitrogen can be quite high. Therefore, it is desirable in such cases to minimize, to the greatest extent possible, the overall operational costs for the activated sludge process. One way this might be accomplished is to use nitrate as a source of nitrogen. It is possible that the use of this nitrogen source could alter certain system operating parameters such as sludge production in such a manner that the overall effect of nitrogen supplementation on operating costs would be minimal. However, information in this area is very limited. The objective of this research was to, therefore, determine how the operating characteristics of the activated sludge proces are affected when nitrate is used as the sole source of nitrogen.

LITERATURE REVIEW

Ammonia is the only form of inorganic nitrogen that can be directly assimilated into amino acids. The ability of an organism to utilize other forms of inorganic nitrogen depends on the presence of enzymes or enzyme systems that are able to convert these compounds to ammonia. Some of the major reactions which have been proposed to describe the interconversion of various inorganic nitrogen compounds are presented in Figure 1.

Certain microorganisms have the ability to utilize nitrate as a terminal electron acceptor for anaerobic respiration through a process known as dissimilatory nitrate reduction. Nitrate can also be used by some microorganisms as a source of nitrogen through a process known as assimilatory nitrate reduction.

Two types of dissimilative nitrate reduction processes have been identified in microorganisms.[2] These are fermentative nitrate reduction and respiratory nitrate reduction. Fermentative nitrate reduction is the term used by Broda[3] to describe the process when no membrane-bound enzymes, cytochromes or electron transport phosphorylations are involved. In this case nitrate acts as an electron

45th Purdue Industrial Waste Conference Proceedings,

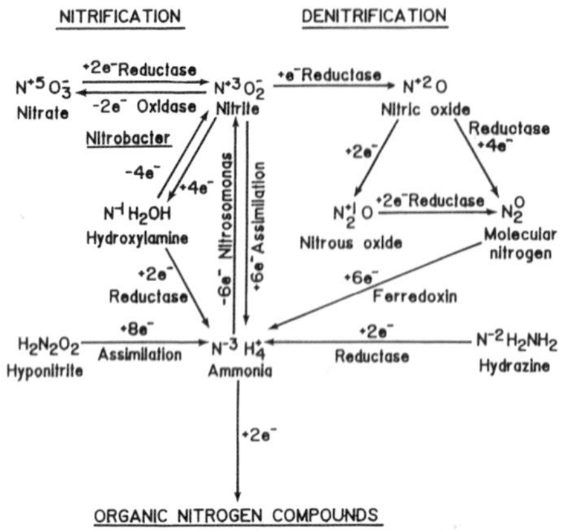

Figure 1. Pathways of inorganic nitrogen metabolism[1].

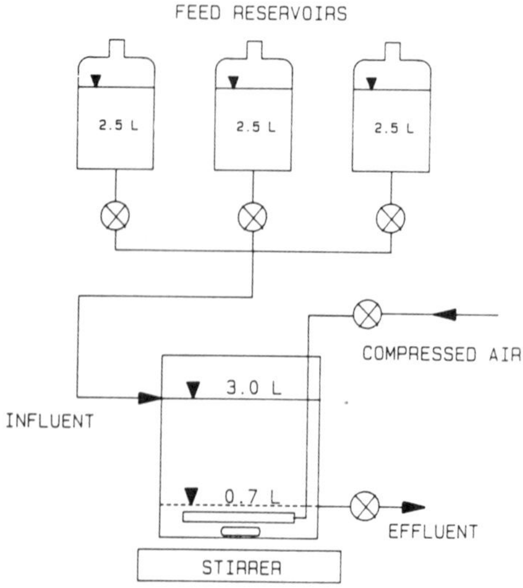

Figure 2. Schematic of SBR system.

acceptor to reoxidize reduced NAD, thus converting substrate which would otherwise be involved in the regeneration of NAD.

Nitrate respiration employs membrane-bound enzymes, cytochromes, and electron transport phosphorylations. Most of the bacteria capable of nitrate respiration are aerobic respirers which can anoxically use nitrate instead of oxygen as a terminal electron acceptor. Sometimes the reduction is only to nitrite, but often it proceeds to nitrous oxide or dinitrogen and is then termed denitrification.

The key enzyme in nitrate respiration is nitrate reductase which reduces nitrate to nitrite. Actually two different nitrate reductase enzymes have been found to be involved in nitrate respiration.[4] These enzymes have been classified as nitrate reductase A and nitrate reductase B. Nitrate reductase A seems to always play a role in nitrate respiration, but the function of nitrate reductase B is different in different bacteria and little information is available on the basic properties of this particular enzyme. In general, nitrate reductase A is an inducible enzyme which is formed only in the absence of oxygen and in the presence of nitrate. Under aerobic conditions the formation of both nitrate reductase A and nitrate reductase B is repressed.

The exact pathway of assimilatory nitrate reduction has not been clearly established. Certain workers feel that organisms assimilate nitrate by first reducing it to nitrite through the action of assimilatory nitrate reductase and then reducing nitrite by nitrite reductase to ammonia.[5] Other workers feel that nitrate assimilation occurs by reduction to nitrite, hydroxylamine, and finally to ammonia.[1] Whatever the pathway, the first step involves the action of a nitrate reductase enzyme to reduce nitrate to nitrite. The nitrate reductase enzyme involved in assimilatory nitrate reduction is somewhat different from the nitrate reductase enzyme involved in dissimilatory nitrate reduction. In general, the assimilatory nitrate reductase enzyme is formed during aerobic growth with nitrate as a nitrogen source but is repressed by the presence of ammonium ions. Some studies on assimilatory nitrate reductases show the existence of an active form and an inactive form that depends on the redox state of the environment. Under reducing conditions, the enzyme is converted into the inactive form.[4]

Few studies have been conducted to determine how activated sludge systems respond to the use of nitrate as the sole source of nitrogen. In one of the earliest studies in this area, Symons and McKinney[6] reported that nitrite and nitrate nitrogen were less available than ammonia nitrogen. They found that the composition of the activated sludge varied according to the type and amount of nitrogen supplied with the type of wastewater treated. In their study, the nitrogen content of the sludge was observed to fluctuate between 1.5 and 11.3% while the carbon to nitrogen ratio of the sludge varied between 3.8 to 30. These workers also concluded from their studies that even in the absence of nitrogen, organic removal would continue for three to four weeks before a system failed. However, the solids accumulation would be greater than in a nitrogen supplemented system, and the final products of cell synthesis would no longer be predominately proteins but rather polysaccharides.

Table I. Average SBR Operating Data

Reactor	Ammonia	Nitrate
Original feed in reservoir (L/cycle)	2.50	2.50
Remaining feed in reservoir (L/cycle)	0.13	0.12
Volume fed (L/cycle)	2.37	2.37
Residual reactor volume (L/cycle)	0.70	0.61
Total reactor volume (L/cycle)	3.07	2.99
Volumes wastes (L/cycle):		
5 day solids retention time	0.20	0.21
10 day solids retention time	0.10	0.11
15 day solids retention time	0.07	0.07

In a much later study Emerson and Sherrard[7] reported that when ammonia is the supplied form of nitrogen, compared with nitrate, a higher mixed liquor volatile suspended solids (MLVSS) concentration results in the aeration basin. This high MLVSS concentration associated with the addition of nitrogen in the ammonium form produces a higher volumetric oxygen demand and results in higher waste sludge production. They also reported that at similar biological solids retention times the addition of ammonia results in better effluent quality as measured by chemical oxygen demand (COD).

MATERIALS AND METHODS

The purpose of this investigation was to determine how the operating characteristics of the activated sludge process are affected when nitrate is used as the sole nitrogen source. This was accomplished using two aerated sequencing batch reactors with identical feed compositions except for the nitrogen source. One reactor was supplied with ammonia as its sole nitrogen source and the other was supplied with nitrate. Each system was operated at biological solids retention times of 5, 10, and 15 days. Sufficient time was allowed for the systems to stabilize at each operating point and then the appropriate systems parameters were measured for a minimum of five consecutive days. Each system was characterized in terms of dissolved nitrogen and phosphorus concentration, mixed liquor nitrogen and phosphorus content, and chemical oxygen demand. A time series profile of COD was also established at each solids retention time.

Bench scale sequencing batch reactors were constructed from 1/4″ plexiglass and the systems were arranged as shown in Figure 2. The feed reservoirs were placed above the reactors to facilitate gravity flow. Influent and effluent flows were controlled by solenoid valves. Sludge wasting was accomplished periodically by pumping from the mixed liquor. Aeration, which was also controlled by a solenoid valve, was supplied using laboratory compressed air and a 4″ wall-type aquarium aerator. A 4″ stick-type portable aerator was used in the nitrate reactor at the 15 day retention time. Mixing was accomplished using a teflon coated magnetic stir bar controlled by a magnetic stirrer below the reactor. All operations were controlled with bench top programmable electronic timers. The reactors were kept covered at all times to prevent algae growth. Solids were collected from the effluent containers daily in two-liter cylinders and returned to the reactors. The volumes shown in Figure 2 are approximate. Actual operating volumes are given in Table I.

The seed sludge was obtained from the aeration basin of the Southside Treatment Plant at Auburn, Alabama which treats primarily domestic wastewater. Six weeks were allowed for the initial acclimation period at a biological solids retention time of 15 days. Four weeks were allowed for system stabilization for the next two changes to retention times of 10 and 5 days. The sequencing batch reactors were operated on 8 hour cycles resulting in 3 cycles per day. These cycles will be subsequently referred to by number: cycle one starting at 7:15, cycle two starting at 15:15, and cycle three starting at 23:15. At the beginning of each cycle the feed was allowed to flow as rapidly as possible into the reactor. This feeding period required about 5 minutes. The mixed liquor was stirred and aerated for a total feed and mix time of 6 hours. One-third of the daily total wasting volume was pumped out of the reactor just prior to the end of the aeration period. The mixed liquor was then allowed to settle for 1 hour. At the end of this settling period the supernatant was drained from the reactor within a period of about 10 minutes. The system remained idle for the remainder of the cycle resulting in a total draw and idle time of 1 hour.

The composition of the feed used in this research is given in Table II. A concentration base feed was made weekly and stored at 4°C. This concentrated stock was diluted daily with tap water and the

Table II. Feed Composition

Material	Concentration (mg/L)
Glucose	100 as COD
Acetate	100 as COD
Glycerol	100 as COD
$NaHCO_3$	200
K_2HPO_4	5 as P
KH_2PO_4	5 as P
$FeCl_3 \cdot 6H_2O$	1 as Fe
$MgSO_4 \cdot 7H_2O$	1 as Mg
$MnSO_4 \cdot H_2O$	1 as Mn
Nitrification Inhibitor	3
In addition:	
Ammonia Reactor = NH_4Cl - 20 mg/L as N	
Nitrate Reactor = KNO_3 - 20 mg/L as N	

nitrogen sources were added to each feed at this time. The feed was dispensed into the feed reservoirs prior to cycle two.

RESULTS AND DISCUSSION

BSRT Profiles

The biological solids retention time profiles (BSRT) show the average values at each retention time along with the range over which the average was taken. Figure 3 shows the BSRT profiles for the mixed liquor volatile suspended solids (MLVSS). The ammonia reactor MLVSS was always about 200 to 400 mg/L greater than the level in the nitrate reactor and the concentration of MLVSS increased with an increase in retention time in each reactor.

The nitrogen present at the end of the aeration period of one cycle was determined and extrapolated over a 24-hour period to obtain a daily nitrogen balance. No data are presented for the 5 day solids retention time because of a problem with nitrite measurements. However, this problem was soon

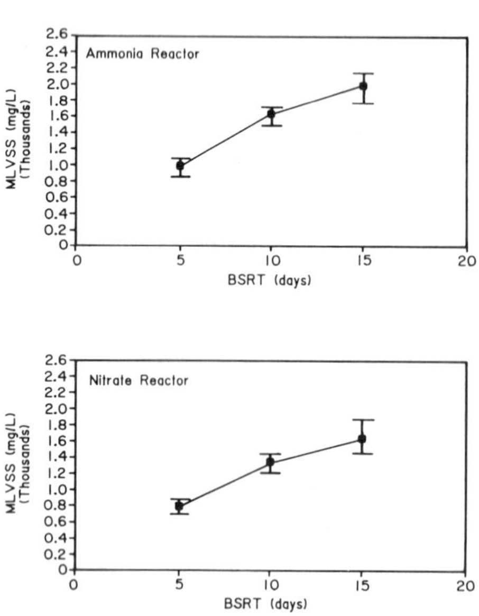

Figure 3. Relationship between average MLVSS concentration and BSRT.

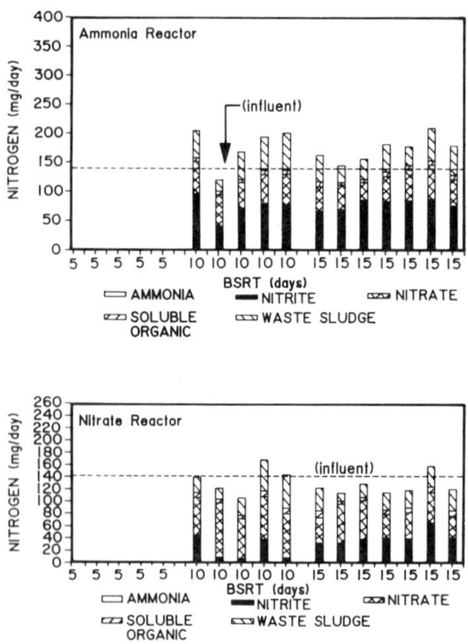

Figure 4. Nitrogen balance histograms.

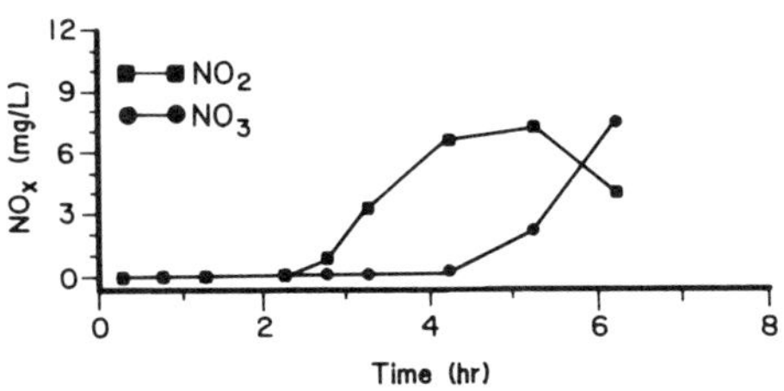

Figure 5. Nitrite and Nitrate concentration profiles produced during nitrification in a sequencing batch reactor system[8].

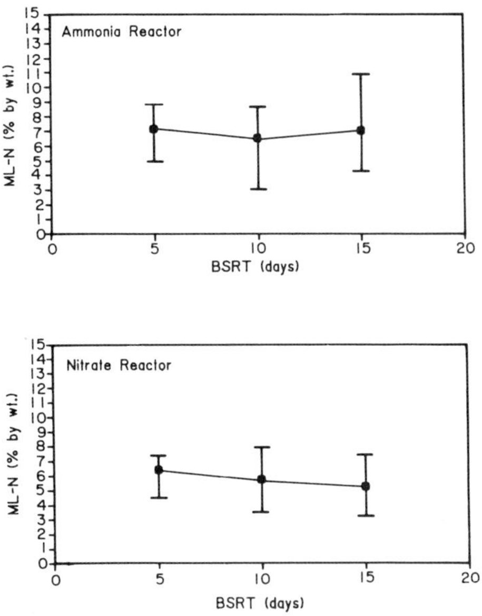

Figure 6. Relationship between mixed liquor nitrogen content and BSRT.

corrected and nitrite measurements were made during the testing period for the 10 day and 15 day biological solids retention time. At the 10 and 15 day retention times the percent recovered was generally within 20% of the 100% balance point.

The nitrogen balance histograms depicted in Figure 4 show the amount and forms of nitrogen recovered daily on a mass basis. The total influent nitrogen during each 24-hour period averaged 142.6 milligrams. In general no ammonia was present in either reactor system. Occasionally a small amount of ammonia was detected but not in an amount significant enough to appear in the histograms. In the ammonia reactor all the influent ammonia was converted to to other forms of nitrogen while in the nitrate reactor, which had no influent ammonia, it is suspected that ammonia was formed but was utilized immediately.

The ammonia reactor showed a build up of nitrite along with nitrate indicating that nitrification was occurring even though a nitrification inhibitor was used. The nitrate reactor showed more nitrate than nitrite indicating incomplete reduction of nitrate. Nitrite and nitrate data presented in Figure 4 illustrate the usefulness of sequencing batch reactors as a research tool because they allow the researcher to observe the time course of reactions. During the process of nitrate reduction, nitrite accumulates until the nitrate has almost completely disappeared because the nitrite reductase enzyme is partially repressed in the presence of nitrate. Although for different reasons, the sequential nature of the nitrification reaction will also express itself in an SBR system. Data presented in Figure 5 were taken from Nicol[8] who studied biological phosphorus removal in an SBR system with a 2 hour anaerobic period at the beginning of the feed cycle. In his system, nitrite and nitrate remained at zero throughout the anaerobic period. About one hour into the aerobic period nitrite began to appear (the lag in nitrite appearance was probably due to the formation of intermediate compounds following ammonia oxidation). It reached a maximum near 9 mg/L after about 2 hours of aeration and decreased thereafter to near 3 mg/L by the end of the aerobic period. The increase in nitrate was delayed with respect to that of nitrite but once it appeared it continued to increase throughout the aerobic period reaching a concentration of near 10 mg/L.

The organic nitrogen content was only a small percentage of the total nitrogen present and the amount remained relatively constant in each reactor. The ammonia reactor always had a larger total amount of nitrogen in the waste sludge because the MLVSS concentration was always higher in this reactor. Figure 6 shows the mixed liquor nitrogen expressed as a percent by weight. The mixed liquor nitrogen (ML-N) remained relatively constant in each reactor with the ammonia reactor having a higher percentage at 6.9% ML-N and the nitrate reactor having about 5.9% ML-N. These averages

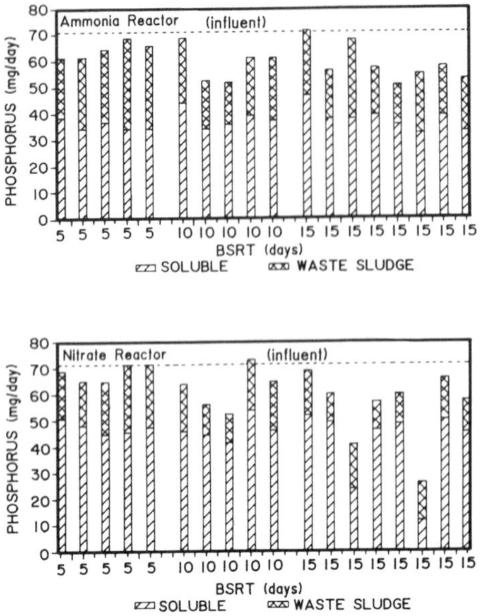

Figure 7. Phosphorus balance histograms.

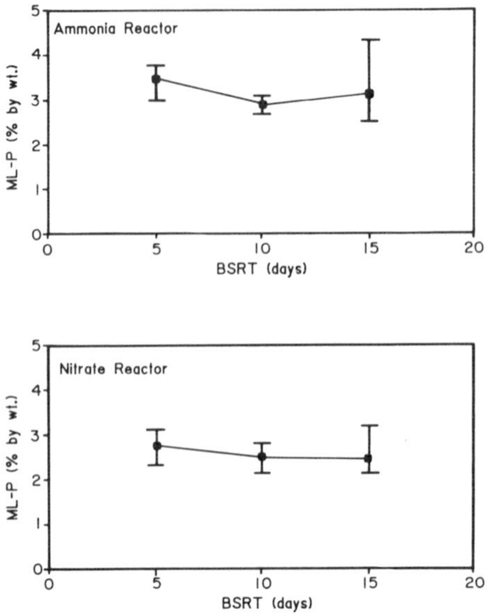

Figure 8. Relationship between mixed liquor phosphorus content and BSRT.

are much lower than the 12.2% indicated by McCarty.[9] This difference may be real or it may be the result of problems encountered in the TKN digestion process for the mixed liquor. The solids present during the digestion process caused the liquid to become foamy resulting in overflow at times and a possible loss of nitrogen. Because of this problem, it was difficult to obtain accurate results which is evident in the wide range of values depicted in Figure 6.

Phosphorus balances were determined daily by the same procedure used for the nitrogen balances. The percent phosphorus recovered was generally within 20% of the 100% balance point. Inaccuracies are most probably due to the digestion and dilution procedure involved in the total-phosphorus test.

The phosphorus histograms which are presented in Figure 7 show the amount and forms of phosphorus recovered daily. The influent phosphorus over each 24 hour period averaged about 71.3 milligrams. The ammonia reactor had a larger total amount of phosphorus in the waste sludge because of the higher MLVSS concentration in this reactor.

Figure 8 shows the mixed liquor phosphorus (ML-P) expressed as a percent by weight. This figure shows that the phosphorus contained in the biomass remained relatively constant in both reactors with the ammonia reactor having the higher percentage at 3.2% ML-P while the nitrate reactor percentage was lower with 2.6% ML-P. These percentages are near the 2.3% value indicated by McCarty.[9]

It is important to note that the mixed liquor nitrogen and the mixed liquor phosphorus percentages were both higher in the ammonia reactor. This is significant because it indicates metabolic differences in the microorganisms of each reactor.

COD Time Series Profiles

Figure 9 shows the soluble COD time series profile for the 10 day biological solids retention time. The COD time series profile is typical of the COD reductions observed at the three biological solids retention times observed in this study. Initial COD concentrations were calculated by adding the COD remaining in the reactor after the idle period with the COD concentration of the feed. Typically the profiles show a rapid decrease in COD concentration over the first 15 to 30 minutes with a slower decrease over the next 1 or 2 hours in both reactor systems. Once reaching a low level, the COD concentration fluctuates—probably because of cell lysis and resynthesis.[10]

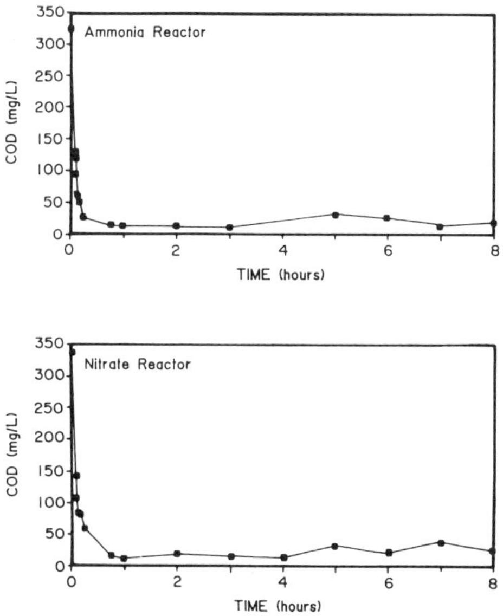

Figure 9. Time series COD at BSRT of 10 days.

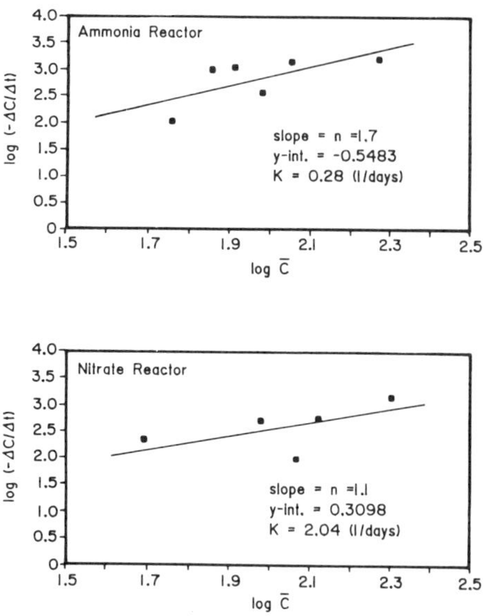

Figure 10. Logarithm of rate vs. logarithm of average COD concentration at BSRT of 5 days.

Order of Reaction and Biokinetic Constants

The data from the linear part of the time series COD plots were used to determine the order of the reaction of substrate removal using the method presented by Swinbourne.[11] A typical plot from the study is shown in Figure 10. The order of reaction and the rate of substrate utilization were found to vary at the different BSRT values. The order of reaction for the ammonia reactor averaged about 2.0 while the nitrate reactor average was 1.5. The rate of substrate utilization for the ammonia reactor averaged about 0.17 $days^{-1}$ with the nitrate reactor averaging about 1.17 $days^{-1}$.

Figure 11 shows the plots of specific substrate utilization versus specific growth rate which were used to determine the microorganisms yield coefficient (Y) and the decay coefficient (K_d) using the method presented by Benefield and Randall.[12] The yield coefficient for the ammonia reactor was 0.94 while the value for the nitrate reactor was lower at 0.66. The decay coefficient for the ammonia reactor was determined to be 0.16 $days^{-1}$ while the nitrate reactor decay coefficient was 0.09 $days^{-1}$.

Sludge Settling Characteristics

Figure 12 depicts the BSRT profiles for the sludge volume index (SVI). The SVI values obtained ranged from 30 to 50 indicating excellent sludge settling characteristics. At the 10 day BSRT the settling velocity in the ammonia reactor was 27.8 ft/hr and in the nitrate reactor was 20.2 ft/hr. The settling velocities at the 15 day retention times were 21.1 and 20.2 ft/hr., respectively, in the ammonia and nitrate reactors. Compared to the results presented by Bisogni and Lawrence,[13] these settling velocities are quite rapid. The settling velocity was not measured at the 5 day BSRT because the low solids level in the reactor made it extremely difficult to identify the location of the sludge interface.

CONCLUSIONS

Treatment of nutrient deficient wastewaters often requires the addition of ammonia or nitrate as a source of nitrogen for the microbial population to survive and effectively treat the wastewater. The addition of ammonia can cause ammonia to be discharged in the effluent which often violates the effluent permit limits. When faced with this problem, it is often advantageous to use nitrate instead of ammonia as the sole source of nitrogen.

This study has shown that wastes can be treated effectively by biological treatment processes when nitrate is supplied as the sole source of nitrogen. The microbial population developed in this study was able to survive and effectively remove organic material when supplied with nitrate as its sole source of

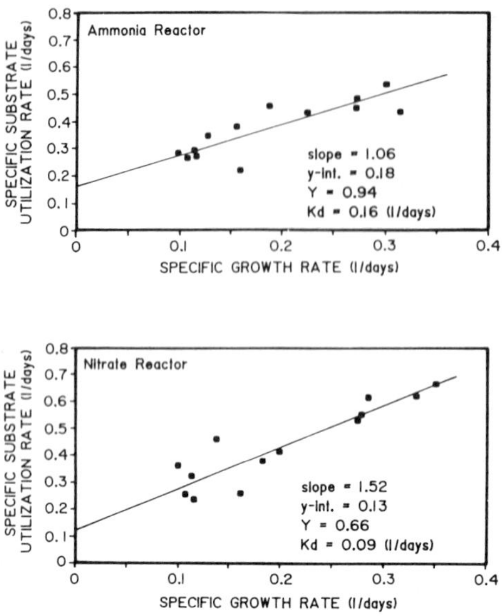

Figure 11. Specific substrate utilization rate vs. specific growth rate.

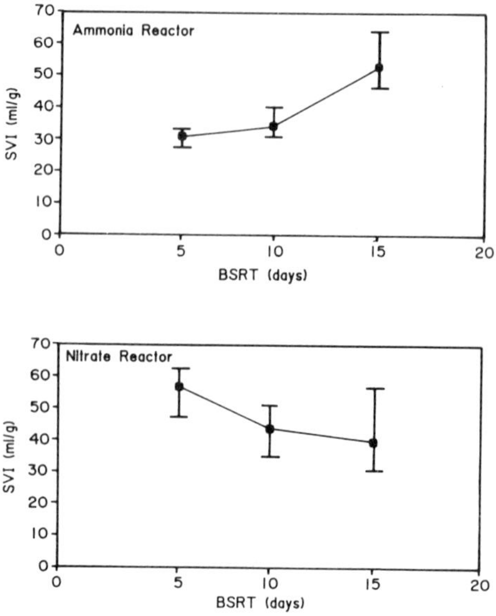

Figure 12. Relationship between sludge volume index and biological solids reention time.

nitrogen. The percentages of nitrogen and phosphorus required per unit of biomass were lower for the microorganisms developed with nitrate as compared with ammonia. This is significant as it indicates metabolic differences in the two microbial populations.

At any of the biological solids retention times employed in this study, the mixed liquor volatile suspended solids (MLVSS) content was lower in the nitrate reactor and was associated with a lower microbial yield coefficient when compared to the ammonia reactor. As noted in Emerson and Sherrard's study, this lower MLVSS concentration would require less oxygen and less nitrogen resulting in a more cost-effective treatment.

The level of treatment obtained was basically the same for each nitrogen source studied. In each case essentially all of the carbonaceous oxygen demand was removed within the first hour of treatment and each sludge had very good settling properties.

REFERENCES

1. Moat, A. G. and Foster, J. W., *Microbial Physiology*, Wiley-Interscience, New York, NY (1988).
2. Fenchel, T. and Blackburn, T. H., *Bacteria and Mineral Cycling*, Academic Press, New York, NY (1979).
3. Broda, E., "The History of Inorganic Nitrogen in the Biosphere," *Journal of Molecular Evaluation, 7*, 87 (1975).
4. Stouthamer, A. H., "Biochemistry and Genetics of Nitrate Reductase in Bacteria," *Advances in Microbial Physiology, 14*, 315 (1976).
5. Ingraham, J. L., Maaloe, O., and Neidhardt, F. C., *Growth of the Bacterial Cell*, Sinauer Associates, Inc., Sunderland, Massachusetts (1983).
6. Symons, J. M. and McKinney, R. E., "The Biochemistry of Nitrogen in the Synthesis of Activated Sludge," *Sewage and Industrial Waste, 30*, 875 (1958).
7. Emerson, S. L. and Sherrard, J. H., "Aerobic Biological Treatment of Nitrogen-Deficient Organic Wastewaters," *Journal of the Water Pollution Control Federation, 55*, 467 (1983).
8. Nicol, J. P., "The Effect of Feed Variability in a Laboratory Sequencing Batch Reactor," Master of Science Thesis, Department of Civil Engineering, Auburn University, Auburn, AL (1988).
9. McCarty, P. L., "Phosphorus and Nitrogen Removal by Biological Systems," *Proceedings*, Wastewater Reclamation and Reuse Workshop, Lake Tahoe, California, June 25–27 (1970).
10. Viessman, W., Jr. and Hammer, M. J., *Water Supply and Pollution Control*, 4th Edition, Harper and Row, New York (1971).
11. Swinbourne, E. S., *Analysis of Kinetic Data*, Appleton-Century-Crofts, New York (1971).

12. Benefield, L. D. and Randall, C. W., *Biological Process Design for Wastewater Treatment*, Prentice-Hall, Inc., New Jersey (1980).
13. Bisogni, J. J., Jr. and Lawrence, A. W., "Relationship Between Biological Solids Retention Time and Settling Characteristics of Activated Sludge," *Water Research, 5*, 753 (1971).

42 ATP TOXICITY TEST IN ACTIVATED SLUDGE SYSTEMS

PR. Senthilnathan, Senior Environmental Engineer
Merck & Company
Elkston, Virginia 22827

J. J. Ganczarczyk, Professor
Department of Civil Engineering
University of Toronto
Toronto, Ontario, M5S 1A4, Canada

INTRODUCTION

The microbial population in activated sludge plants are vulnerable to toxic substances in the influent. Particularly, the municipal sewage treatment plants experience sometimes unexpected upsets due to toxins discharged by industries. In order to prevent such upsets, toxicity screening tests of industrial discharges are recommended.

Among various methods available for the short term toxicity screening, adenosine triphosphate (ATP) assay has gained a broad popularity. At present, many commercial testing equipment and protocols are available to measure the level of ATP in activated sludge. Generally, the short term toxicity screening test is carried out as follows: the suspected influent is added to a sample of the activated sludge and aerated while another sample of the activated sludge is aerated under endogenous conditions as a control. At selected time intervals (e.g. 30, 60 and 120 minutes), the level of ATP in the test and control activated sludge is determined. A substantial decrease in ATP level is considered to indicate the toxicity of the tested wastewater.

However, some precautions are necessary for a correct application of this method. This paper identifies some measures which have to be considered to avoid wrong conclusions.

BACKGROUND

Toxicity testing methods using microorganisms has an excellent potential in predicting the toxicity of wastewaters. Various methods for this purpose are described elsewhere.[1,2] Among various methods, respirometry, ATP measurement, and dehydrogenase activity measurement (for e.g. by 2, 3, 5-triphenyl tetrazolium chloride [TTC]) are broadly used for measuring the toxicity of wastewaters to activated sludge.

Several investigators reported ATP measurement as a useful tool for measuring viable biomass concentration[3-20] applied the ATP measurement for predicting toxicity to activated sludge. For toxic substances causing uncoupling of oxidative phosphorylation, ATP measurement was more sensitive than respirometry in predicting the toxicity.[21]

Adenosine triphosphate (ATP), a universal constituent of living cells, plays an important role in bacterial metabolism. It is also known as the "universal energy currency". Hydrolysis of ATP to ADP (adenosine diphosphate) yields about 7 kcal/mol of ATP. ATP decomposes rapidly upon cell death. ATP is usually measured by a bioluminescent enzymatic reaction as shown below.

$$\text{Luciferin} + \text{Luciferase} + \text{ATP} \xrightarrow{Mg^{++}} \text{Luciferin-Luciferase-AMP} + \text{Pyrophosphate}$$

$$\text{Luciferin-Luciferase-AMP} \xrightarrow{O_2} \text{Oxyluciferin} + \text{Luciferase} + \text{AMP} + CO_2 + \text{Light}$$

45th Purdue Industrial Waste Conference Proceedings, © 1991 Lewis Publishers, Inc., Chelsea, Michigan 48118. Printed in U.S.A.

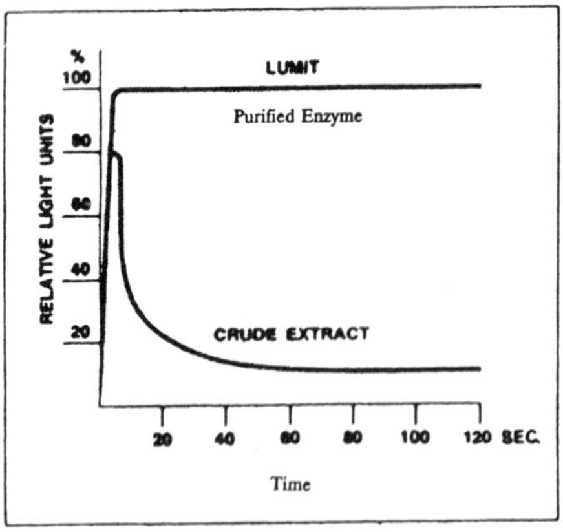

Figure 1. Effect of enzyme purification on ATP assay (Source—3M Co., Minneapolis, U.S.A.).

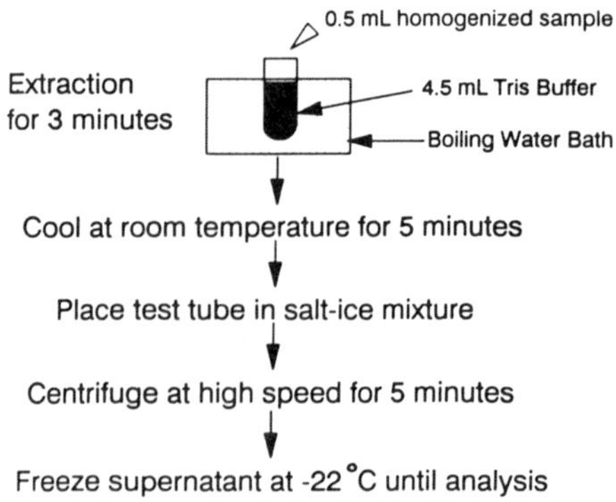

Figure 2. Extraction of ATP.

ATP reacts with enzyme luciferin in the presence of luciferase and magnesium ions (catalyst) to form luciferin-luciferase-AMP (AMP—adenosine monophosphate) which in turn reacts with oxygen to produce a quantum of light for each mole of ATP. The light emitted is measured by luminometers in terms of relative light units (RLU). Firefly extracts are used as a source of the required enzymes for these reactions. The intensity of emitted light increases rapidly within few seconds and declines due to activity of other enzymes utilizing ATP. The nature of light emission depends on the purity of enzymes and the concentration of inhibiting compounds. Recent advances in purification of firefly extracts and the addition of stabilizing agents have resulted in almost a stable emission of light (Figure 1). At present, sensitive instruments are available to measure ATP in picogram level (equivalent of 10^3 colony forming units/mL).

The main purpose of this study was to check the validity of the short term toxicity testing protocols for testing the toxicity of wastewaters to activated sludge by means of ATP assay.

MATERIAL AND METHODS

Adenosine triphosphate (ATP) Assay

Adenosine triphosphate (ATP) content of the activated sludge was measured with a Lumac Luminometer—Model 1070 (Lumac/3M, Minnesota, U.S.A.). The details of the method are given below. The steps for extracting ATP from activated sludge are shown in Figure 2.

A 20 mL sample of activated sludge was homogenized at a high speed for 30 seconds in a Warring blender. From this suspension, 500 μL of sample was pipetted by a 1 mL wide-mouthed pipette into 4.5 mL of boiling Tris-EDTA buffer (0.02 M Tris, 2 mM EDTA, pH 7.75). After boiling at 100°C for 3 minutes, the sample was removed from the water bath, cooled for 5 minutes at room temperature, and then placed in a mixture of salt and ice. One mL of this solution was centrifuged at 10,000 rpm for 5 minutes to remove the cellular solid debris. The supernatant was decanted into another vial and frozen at -22°C until analysis.

The frozen extracts were thawed at room temperature (20°C). These extracts were diluted when necessary with Tris-EDTA buffer. A 100 μL of this solution was placed in a cuvette. To this cuvette, an equal volume of Lumit HS reagent (Lumac/3M, Minnesota, U.S.A.) containing purified Luciferene-Luciferase enzyme and stabilizing agents was added and the Relative Light Units (RLU) emitted during the bioluminescent reaction were measured. The readings were integrated for 10 seconds. Internal standardization technique (Figure 3) was used to quantify the actual amount of ATP present in the sludge. For this technique, 10 μL of a known concentration of pure ATP solution was added to 100 μL of sludge extract and the relative light units were measured as described above. From these readings, the amount of ATP per RLU and subsequently the quantity of ATP in the sludge samples were calculated. The instrument was tested with standard ATP and the resulted calibration curve is shown in Figure 4.

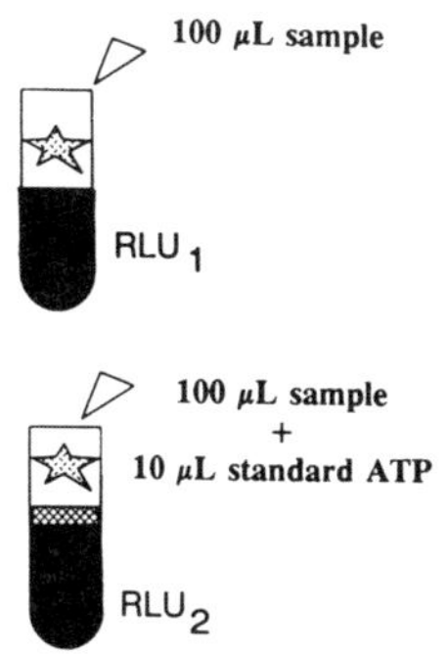

$$ATP\ (sample) = \frac{Amount\ of\ Standard\ ATP}{(RLU_2 - RLU_1)}\ RLU_1$$

Figure 3. Internal standardization.

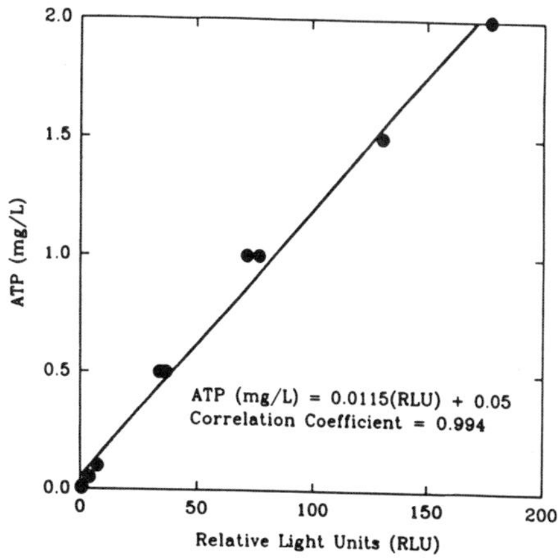

Figure 4. ATP calibration curve.

Toxicity Testing

Batch reactor studies were conducted to determine the effect of various substrates on the ATP levels in different types of activated sludges. These studies were conducted in 4 liter aspirator bottles. Three liters of test sludge (municipal activated sludge or activated sludge acclimated to phenol) were aerated in each bottle. One reactor served as a control without addition of any substrate. To the other reactors, known amount of substrates (glucose or phenol) were added. At selected time intervals, samples were taken from the reactors for ATP analysis. The initial concentration of mixed liquor suspended solids (MLSS) and mixed liqour volatile suspended solids (MLVSS) in the reactors were measured according to the Standard Methods.[22]

RESULTS AND DISCUSSIONS

Toxicity Due to Phenol

In order to test the toxicity of phenol to municipal activated sludge, two 3 liter portions of this sludge were aerated with and without phenol addition, respectively. At the same time, in order to study the effect of phenol on phenol acclimated sludge, 3 liters of activated sludge acclimated to phenol was also aerated with phenol addition. The initial concentration of phenol in the above experiments was 500 mg/L. Figure 5 depicts the results of these experiments. Endogenously respiring municipal activated sludge lost only about 28% of its ATP content in 5 hours. Whereas, the ATP content of municipal activated sludge fed with phenol decreased rapidly from 3.2 mg/L to 1.4 mg/L in 20 minutes, to 1.2 mg/L in 60 minutes, and reached almost a constant value of 0.95 mg/L (30% of initial ATP) after two hours. This observation indicated that phenol is toxic to municipal activated sludge. Interestingly, phenol acclimated sludge also lost 50% of its ATP content in 20 minutes and reached a minimum value of 1.67 mg/L (45% of initial ATP) in an hour. But, during prolonged aeration for a period of five hours, its ATP level increased gradually to a level greater than that of the initial level. It appeared that initially phenol was also toxic to activated sludge acclimated to phenol. If a judgment regarding the toxicity of phenol to phenol acclimated sludge would be made within an hour or two since the start of the experiment, it would have been erroneous.

Experiments with Glucose as a Substrate

In order to understand whether the reason for the initial reduction in ATP level in the above experiments was due to inhibitory nature of phenol, further tests were conducted with municipal activated sludge fed with a non-inhibitory substrate (glucose). During these tests, one portion of sludge was fed with 1000 mg glucose/L and another portion was fed with 500 mg phenol/L. The results presented in Figure 6 were very similar to that of the previous experiments. The ATP level decreased rapidly also for non-inhibitory substrate (glucose).

An analysis of previous work conducted by other researchers on the effect of different components on ATP pool revealed that two different types of observations were made. The major difference

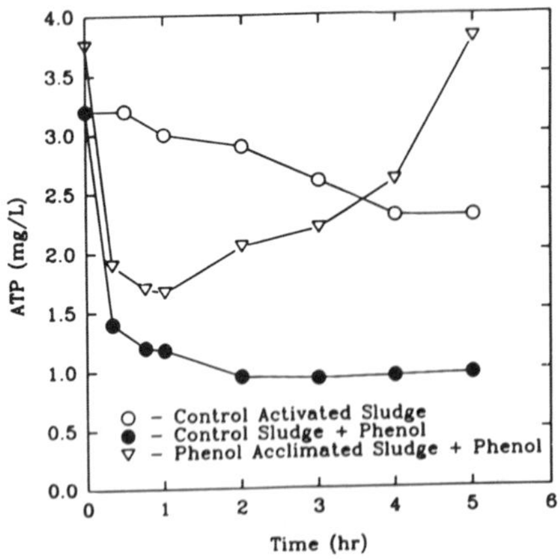

Figure 5. Response of activated sludge acclimated to phenol.

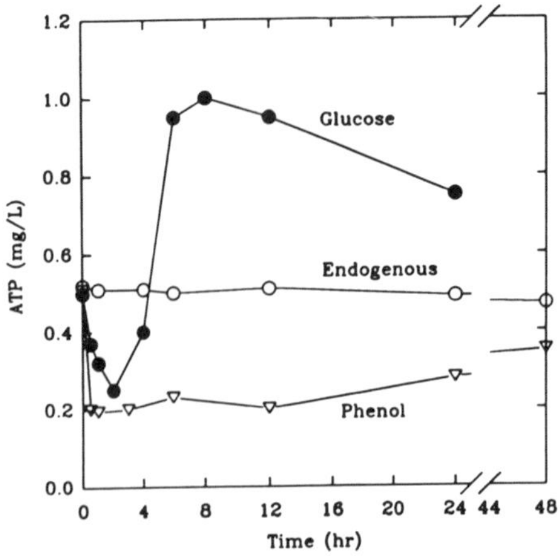

Figure 6. Response of activated sludge from sewage treatment.

between the experiments resulting in different responses was the initial amount of microbial inoculum. Following section analyzes the two types of ATP toxicity testing experiments in detail.

Studies with a Small Amount of Inoculum

Many studies on ATP toxicity testing utilized a small amount of inoculum.[30] The actual amount of inoculum varied from few transfer loops to 10 mL, and the initial number of viable organisms was in the range of 10^3 to 10^6 CFU (Colony Forming Units)/mL. Typical results from this type of study (Figure 7) indicated that the ATP pool increased continuously due to non-inhibitory substrates and decreased due to inhibitory substrates. Similar observations were reported by D'Eustachio and Levin[23] for pure cultures of *Pseudomonas fluorescens* and *Bacillus subtillis* and by Forrest[24] for *Streptococcus faecalis*. Initial reduction in ATP pool due to noninhibitory substrate was not observed in these studies. It appears that it might be due to the small amount of inoculum. However, there is no direct experimental evidence available to support this conclusion.

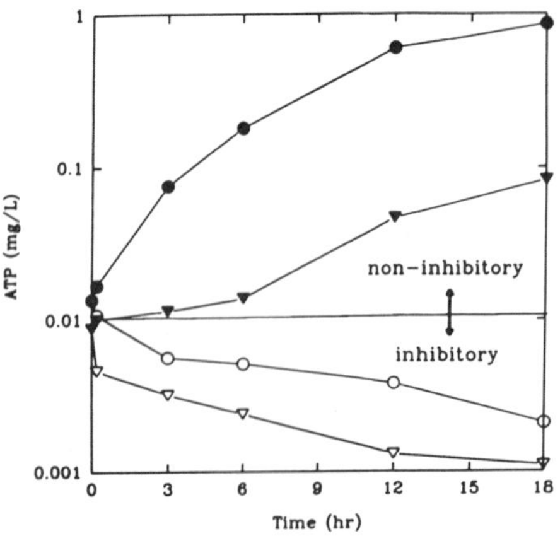

Figure 7. ATP tests with a small amount of inoculum.[30]

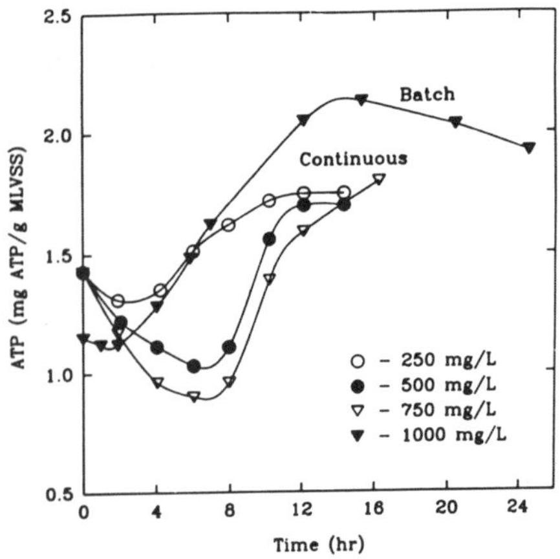

Figure 8. ATP tests with skim milk as a substrate.[6]

Studies with a Large Amount of Inoculum

Patterson et al.,[4] Brezonik and Patterson,[6] Chu et al.,[7] Schneider,[21] Goronoszy and Eckenfelder,[25] and this study, utilized a large amount of activated sludge as an inoculum in the ATP tests. In all these tests an initial reduction in ATP pool due to non-inhibitory substrates was observed. Figure 8 depicts the results from the study by Brezonik and Patterson.[6] The results presented in Figure 8 were obtained with skim milk - a non-inhibitory substrate. For the activated sludge obtained from continuous reactors, the magnitude and duration of ATP pool reduction varied with the initial concentration of substrate added. But, the maximum increase in ATP pool over the initial value was approximately the same (about 20%) for all the studied levels of substrate loading. For the activated sludge obtained from a fill and draw reactor (acclimated to batch feeding), there was only a slight decrease in ATP level during the first 2 hours after feeding skim milk. However, 14 hours after substrate addition, the ATP level reached a maximum value that was 57% above the initial ATP level.

Table I summarizes the ATP responses of different activated sludges, namely, municipal activated sludge, activated sludge acclimated to skim milk, and activated sludge acclimated to phenol to glucose, skim milk, and phenol respectively. In all these cases, the ATP level decreased initially and reached a minimum value. The time required to reach the minimum value of ATP ranged from 1.0 to 7.5 hours. After reaching a minimum value, the ATP level increased gradually. The time required to build the ATP pool to a value equivalent to that of the initial ATP level ranged from 4 to 11 hours. After prolonged aeration, ranging from 8.0 to 16 hours, the ATP level reached a maximum value and subsequently decreased.

It is important to note that in all of the above experiments, although the substrates were non-inhibitory and were the sole carbon source, if the conclusions were made within a short period ranging from 4 to 11 hours, the tested substrates would have been considered toxic. Hence, caution is needed in judging the toxicity of wastewaters which usually contain unknown quantities of multiple substrates. The duration of decreased ATP level due to non-inhibitory substrate might vary with the type of sludge (microbial population and history of acclimation), the metabolic activity of sludge (sludge age), the initial amount of sludge, the initial concentration of substrate and the characteristics of substrate. The actual reasons for this reduced ATP level due to non-inhibitory substrates are not well understood. However, some possible reasons are analyzed later in this paper.

ATP Levels During Endogenous Respiration

The ATP levels of endogenously respiring sludge were almost even for a period of 48 hours. Similar observations were noted by Patterson et al.[3,4] for activated sludge and Strange et al.[26] for *Aerobacter aerogenes*. This occurrence might be due to the starvation survival of organisms[27,28] and/or due to the growth of new organisms that utilize the lysed organics, and/or the growth of nitrifying organisms, and/or the presence/growth of eukaryotic organisms in the sludge.

Possible Reasons for the Reduction in ATP Pool Due to Non-inhibitory Substrates

The initial reduction in ATP pool due to non-inhibitory substrates might be attributed to metabolic reactions. Microorganisms have to spend energy during the initial steps of substrate level phosphorylation. For instance, glucose is initially converted to pyruvate by means of glycolysis (Figure 9). During the initiation of glycolysis, 2 moles of ATP are spent by the microorganisms. At the end of glycolysis, the energy released by oxidizing the substrate is used to phosphorylate adenosine diphosphate (ADP) to ATP and there is a net production of 2 moles of ATP. Each mole of pyruvate is further oxidized to carbon dioxide via Krebs cycle to yield four pairs of electrons carried by nicotinamide adenine dinucleotide (NAD), one pair of electrons carried by flavin adenine dinucleotide (FAD), one mole of ATP, and three moles of carbon dioxide.[32] The NADH and $FADH_2$ molecules produced through Krebs cycle are oxidized by oxidative phosphorylation to yield 34 moles of ATP (Figure 10). Theoretically, one glucose molecule may yield 38 moles of ATP at maximum metabolic efficiency. However, in reality, many bacteria lack some of the electron transport steps and are less efficient. Many intermediate compounds produced during the above reactions might also be used by the cells for purposes other than producing ATP. The metabolic pathways also would differ according to the environmental conditions around the bacteria and may result in decreased ATP production.

In addition to the above reason, ATP drop could occur due to the death of eukaryotic organisms (possibly present in the inoculum) such as protozoa, rotifers, and nematodes. ATP reduction could also occur due to oxygen depletion during the initial reaction period when the substrate concentration is high. Brezonik and Patterson (1971) observed a temporary decrease in ATP pool due to a decrease in dissolved oxygen level. Some organisms, such as acenetobacter (possessing the characteristics of luxury uptake of phosphorus) might store the phosphorus in the forms other than that of ATP (as

Table I. Summary of ATP Responses to Different Substrates

Initial Substrate Concentration (mg/L)	Time to Reach Minimum ATP Value (hr)	Time to Reach Maximum ATP Value (hr)	Time to Break-even with initial Level of ATP (hr)
Municipal Activated Sludge + Glucose			
1000	2.0	8.0	5.0
Skim Milk Acclimated Activated Sludge + Skim Milk (as BOD_5) (Adapted from Brezonik and Patterson, 1971)			
250	3.0	12.0	5.5
500	5.5	13.5	10.0
750	7.5	16.0	11.0
1000	2.0	16.0	4.0
Phenol Acclimated Activated Sludge + Phenol			
500	1.0	–	5.0

ADP, AMP, GDP [guanosine diphosphate], GTP [guanosine triphosphate]). Soluble substrate uptake and intracellular storage product formation during the initial phase is an energy consuming process. If aneraobic conditions exist during the initial phase, the final acceptence of electrons is delayed to a later anoxic or aerobic stage. These conditions can promote enhanced biological phosphorus removal.[29]

Other explanations for this initial reduction of ATP pool are also possible. Measurement of ATP does not account for energy stored in other forms such as NADH, $FADH_2$, and GTP. To date, the actual reasons for this initial reduction of ATP pool due to non-inhibitory substrates are not yet conclusively understood.

SUMMARY AND CONCLUSIONS

1. Two types of ATP toxicity testings can be recognized; those with a small inoculum and those typical for activated sludge systems.
2. Although toxic compounds reduce the ATP pool, the reduction of ATP level during the short-term test need not necessarily be due to toxicity. Such reduction also can occur due to non-inhibitory substrates.
3. However, after a prolonged aeration, the level of ATP ultimately increases beyond the

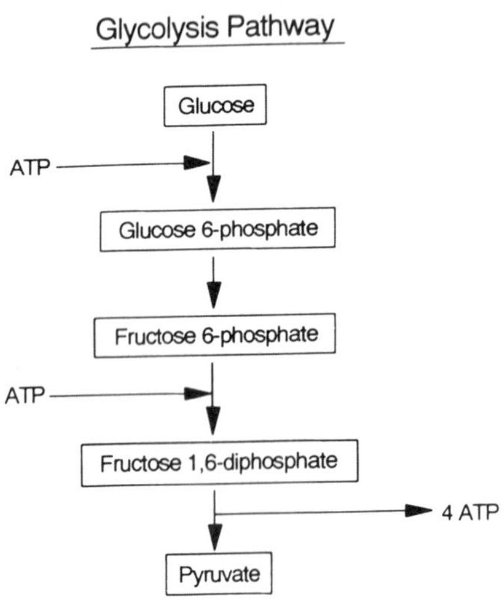

Figure 9. Glycolysis pathway.

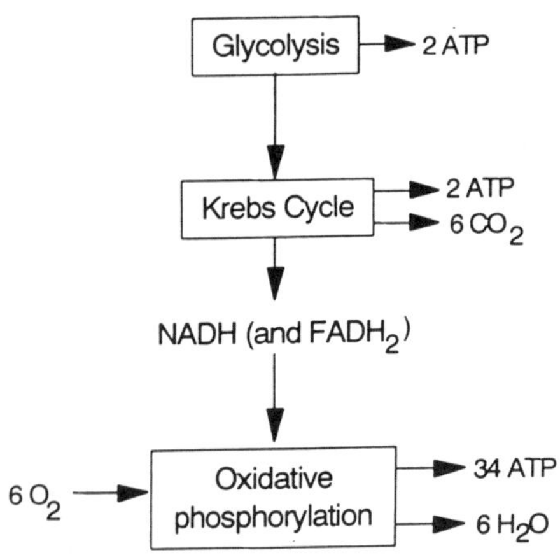

Figure 10. Summary of glucose metabolism.[31]

initial level for non-inhibitory substrates. Whereas, for toxic or inhibitory substances, the ATP level does not recover even after a prolonged aeration.

4. The reduced level of ATP can be sustained even for a period of about 10 hours depending on the initial concentration of biomass and non-inhibitory substrate. Hence, in order to obtain reliable results, ATP toxicity screening tests should be carried out for a longer duration, preferably for 12 hours or longer or at least for the time equivalent to one hydraulic retention time in the aeration basin.
5. The ATP level of endogenously respiring activated sludge was almost constant even after 48 hours of aeration. This might be due to starvation survival of bacteria, or due to the growth of new population of bacteria feeding on lysed products, or the growth of nitrifying bacteria.
6. The presence of eucaryotic organisms, such as protozoa, rotifers, and nematodes may also affect the ATP level in activated sludge. The exact effect of their presence is still unknown.

REFERENCES

1. Liu, D., and Dutka, B. J., Eds., (1984) *Toxicity Screening Procedures using Bacterial Systems*, Marcel Dekker, Inc., New York.
2. Bitton, G., and Dutka, B. J., Eds., (1986) *Toxicity Testing Using Microorganisms,* Vol. I & II, CRC Press, Boca Raton.
3. Patterson, J. W., Brezonik, P. L., and Putnam, H. D., (1969) Sludge activity parameters and their application to toxicity measurements and activated sludge, *Proc. 24th Purdue Ind. Waste Conf.*, 127–154.
4. Patterson, J. W., Brezonik, P. L., and Putnam, H. D., (1970) Measurement and significance of adenosine triphosphate in activated sludge, *Environ. Sci. Tech.*, 4, 569–575.
5. Weddle, C. L., and Jenkins, D., (1970) The viability and activity of activated sludge, *Water Res.*, 5, 621–640.
6. Brezonik, P. L., and Patterson, J. W., (1971) Activated sludge ATP: effects of environmental stress, *J. San. Engg. Div., ASCE.*, 97, 813–824.
7. Chu, G. C. Y., Erickson, L. E., and Fan, L. T., (1973) Dynamic behavior of a complete-mixing activated sludge system, *Biotechnol. Bioengg.*, 15, 1101–1121.
8. Upadhyaya, A. K., and Eckenfelder, W. W. Jr., (9175) Biodegradable fraction as an activity parameter of activated sludge, *Water Res.*, 9, 691–695.
9. Tifft, E. C., Jr., and Spiegel, S. J., (1976) Use of adenosine triphosphate assay in disinfection control, *Environ. Sci. Technol.*, 10, 1268–1272.
10. Levin, G. V., Schrot, J. R., and Hess, W. C., (1975) methodology for application of adenosine triphosphate determination in wastewater treatment, *Environ. Sci. Tech.*, 9, 961–965.
11. Afghan, B. K., Tobin, R. S., and Ryan, J. F., (1977) Improved method for quantitative measurement of ATP and its application to measure microbial activity in natural waters, activated sludges, and sediments, Presented at 2nd ATP Methodology Symp., San Diego, CA.
12. Kennicutt, M. C., (1980) ATP as an indicator of toxicity, *Water Res.*, 14, 325–328.
13. Burns, R. G., and Slater, J. H., Eds., (1982) *Experimental Microbial Ecology*, Blackwell Scientific Publications, London.
14. Parker, C. E., (1982) Surrogate parameter analysis for organic priority pollutants, *J. Wat. Pollut. Control Fed.*, 54, 77–86.
15. Roe, P. C., and Bhagat, S. K., (1982) Adenosine triphosphate as a control parameter for activated sludge process, *J. Wat. Pollut. Control Fed.*, 54, 244–254.
16. Kang, S. J., Pugh, L. B., and Borchardt, J. A., (1983) ATP as a measure of active biomass concentration and inhibition in biological wastewater treatment processes, *Proc. 38th Purdue Ind. Waste Conf.*, 751–759.
17. Johnson, I. R., and Stafford, D. A., (1984) Control of the activated-sludge process using adenosine triphosphate measurements, in *Microbiological methods for Environmental Biotechnology,* Grainger, J. M., and Lynch, J. M., Eds., Academic Press, New York, N.Y., pp. 183–195.
18. Bitton, G. B., and Koopman, B., (1986) Biochemical tests for toxicity screening, in *Toxicity Testing Using Microorganisms,* Bitton, G., and Dutka, B. J., Eds., CRC Press, Boca Raton, p. 27–56.
19. Nouvion, N., Block, J. C., and Faup, G. M., (1987) Effect of biomass quantity and activity on TOC removal in a fixed-bed reactor, *Water Res.,* 21, 35–40.
20. Jensen, B. K., Eriksen, T., and Jorgenson, P. E., (1988) Determination of active biomass, Presented at Int. Conf. on Water and Wastewater Microbiology, Irvine, California, Feb. 8–11.
21. Schneider, C. G., (1987) Screening wastewater for toxicity to activated sludge, Ph. D. Thesis, Vanderbilt University, TN.

22. APHA, (1985) *Standard Methods for the Examination of Water and Wastewater,* 16th edition, American Public Health Association, Washington D.C.
23. D'Eustachio, A. J., and Levin, E. I., (1967) Levels of adenosine triphosphate during bacterial growth, *Bact. Proc.*, 20, 121–122.
24. Forrest, W. W., (1965) Adenosine triphosphate pool during the growth cycle in *Streptococcus faecalis, J. Bacteriol.*, 90, 1013–1016.
25. Goronoszy, M. C., and Eckenfelder, W. W. Jr., (1988) Floc-load as it relates to enzymatic transfer of soluble substrate and sludge bulking control, Presented at Int. Conf. on Water and Wastewater Microbiology, Irvine, California, Feb. 8–11.
26. Strange, R. E., Wade, H. E., and Dark, F. A., (1967) Effect of starvation on ATP concentration in *Aerobacter aerogenes, Nature,* 199, 55–67.
27. Morita, R. Y., (1982) Starvation survival of heterotrophs in the marine environment, *Adv. Microbial Ecology*, 6, 171–193.
28. Henis, Y., (1987) Survival and dormancy of bacteria, in *Survival and Dormancy of Microorganisms*, Henis, Y., Ed., John Wiley & Sons, New York, pp. 1–108.
29. Jenkins, D., (1987) Discussion of feature article on bulking control, *J. Wat. Pollut. Control Fed.*, 59, 733–734.
30. Seyfried, P. L., and Horgan, C. B. L., (1983) Effect of cadmium on lake water bacteria as determined by the luciferase assay of adenosine triphosphate, in *Aquatic Toxicology and Hazard Assessment,* Bishop, W. E., Cardwell, R. D., and Heidolph, B. B., Eds., ASTM Publication STP-802, pp. 425–441.
31. Freifelder, D., (1987) *Molecular Biology,* Jones and Bartlett Publishers, Inc., Boston, U.S.A.
32. Norton, C. F., (1986) *Microbiology*, Addison-Wesley Publishing Co., Donmills, Ontario, Canada.

Section Four
BIOLOGICAL SYSTEMS – C. ANAEROBIC

43 HIGH pH EFFECTS IN ANAEROBIC TREATMENT OF LIQUID INDUSTRIAL BYPRODUCTS

Eric A. Seagren, Graduate Research Assistant
University of Illinois
Urbana, Illinois 61801

Audrey D. Levine, Assistant Professor
Civil and Construction Engineering
Iowa State University
Ames, Iowa 50011-3232

Richard R. Dague, Professor
Civil and Construction Engineering
Iowa State University
Ames, Iowa 50011-3232

INTRODUCTION

The success of anaerobic digestion of industrial wastes is dependent on environmental conditions such as pH, the absence of toxic substances, and many other factors. The majority of research on pH in anerobic digestion has been focused on low pH effects.

For optimum operation of anaerobic reactors, the hydrogen ion concentration must be maintained within a narrow range near pH 7.0.[1-5] However, limited research has been published on the effects of high pH on the performance of anaerobic systems. Typically, industrial waste streams that have high pH levels are otherwise ideal candidates for anaerobic treatment. For example, grain processing wastes and egg breaking wastes may have pH levels above 10. Another case where a high pH may be imposed on anaerobic systems is in the application of chemical coagulation to separate anaerobic biomass in the anaerobic activated sludge process.[6] A side benefit of anaerobic treatment of appropriate high pH wastes is the methane enrichment of the gas phase which occurs as pH increases.

The objective of this research was to investigate high pH effects in completely-mixed, suspended growth, anaerobic reactors, in particular the upper pH limit that is possible while simultaneously avoiding toxic inhibition from the alkaline earth and alkali metals (Na^+, K^+, Ca^{++}, and Mg^{++}) and ammonia (NH_3/NH_4^+). In addition, the effect of high pH on the different trophic groups of the anaerobic mixed culture was evaluated.

BACKGROUND

Optimum pH for Anaerobic Treatment

The three groups of bacteria involved in the anaerobic digestion of organic material (i.e., the hydrolytic/fermentative bacteria, the acetogens, and the methanogens) have different optimum values for pH in pure culture. When all three groups of bacteria involved in the anaerobic process are functioning together as an integrated system, the optimum pH range has to accomodate all of the species which are involved. A summary of the results of several laboratory studies that investigated the optimum pH range for anaerobic mixed cultures is presented in Table I. One point to note about the results in Table I is that two of the studies used anaerobic filters. These findings may not be comparable to suspended biomass anaerobic systems since, as pointed out by Speece and Parkin[8] (1983), with an anaerobic filter the methanogens located in the interior of the biofilm are not subject to the same conditions (e.g., pH) as exist in the bulk solution. Also, as pointed out by Porter[7], the cessation of methane production that Clark and Speece[4] found above pH 9 could have been due to

45th Purdue Industrial Waste Conference Proceedings, © 1991 Lewis Publishers, Inc., Chelsea, Michigan 48118. Printed in U.S.A.

Table I. Optimum pH Ranges for Growth of Anaerobic Mixed Cultures

Seed	Substrate	Reactor Type	Optimum pH	Reference
Digesting sewage solids	Calcium acetate; calcium butyrate; ethyl alcohol plus inorganic salts/yeast	Test tubes with inverted vials for gas collection	7.0	[1]
Primary digesters	Primary sludge	Lab scale digesters	7.1	[2]
Enriched culture; pear peeling wastes	Acetic acid, plus alkalinity and salts	Erlenmeyer flasks (continuous fed)	6.5–7.1	[5]
Digested sludge	Sodium acetate plus inorganic salts	Anaerobic filter	6–8	[4]
Digester supernatant	Glucose plus alkalinity and nutrients	Anaerobic filter	8 (max)	[7]

ammonia inhibition. Clark and Speece based their conclusion that methane production essentially stopped above pH 9 on results gathered when the pH in the filters was 9.3 and 9.4. With an influent ammonia concentration of 236 mg/L, as N, the free ammonia concentration at pH 9.3 would be 160 mg/L, a toxic concentration.

Others have recommended pH ranges for anaerobic treatment. McCarty[3] recommended a pH range of 6.6 to 7.6 with an optimum of approximately 7.0 to 7.2. On the basis of field observations, Albertson[2] reported on optimum pH range of 6.8 to 7.2 for anaerobic digestion, but noted that digesters operated successfully outside this range.

A concept that is useful in reflecting inhibition of methane production as a result of pH is the pH Inhibition Factor (pH-IF) introduced by Clark and Speece.[4] The pH-IF is defined as the ratio of the methane production rate at nonoptimum pH to the methane production rate at optimum pH. If no inhibition is present in the reactor, the pH-IF will be near unity. Likewise, if the pH is inhibitory or stimulatory, the pH-IF will be below or above unity, respectively. The pH-IF concept is used later in this paper as a reflection of the degree of methane production inhibition as a result of high pH.

Effect of pH on Equilibrium Chemistry

The pH in anaerobic processes controls the equilibrium concentration of several important chemical species. It is well known that, in the near neutral pH range, the buffering capacity of the system is dominated by the carbonate system, with carbonate acid and bicarbonate species being the most important. Two other chemical systems of importance to anaerobic processes, and whose form depends on the solution pH, are the ammonia system (NH_3 or NH_4^+) and the sulfide system (insoluble, soluble, or gaseous). Parkin and Owen (1986) provide a good review of the toxicity associated with ammonia and sulfides.

The pH of the solution influences metal precipitation. The anionic species that are primarily responsible for metal precipitation in anaerobic digesters are sulfide (S^{-2}), carbonate (CO_3^{-2}), and phosphate (PO_4^{-3}).[9] Equilibrium and solubility relationships for sulfide, carbonate, and phosphate can be used to calculate the concentrations of soluble species present at a given pH. Below pH 8, the relative fractions of the soluble species which are metal precipitating (S^{-2}, CO_3^{-2}, and PO_4^{-3}) are small.

Cellular Responses to pH

Enzymes are generally active over a specific pH range. Typically, an optimum pH exists for each enzyme mediated reaction. The pH influences the ionization state of the free enzyme, the enzyme-substrate complex, and the substrate. Enzymes are proteins that have many ionizable groups and can exist in different states of ionization. The distribution between the different ionic forms depends on the pH and the ionization constants of the functional groups. It has been postulated that only one of the ionic forms of the enzyme is biologically active.[10] Since protein configuration and activity are pH dependent, it follows that cellular transport processes, reactions, and, in turn, growth rates, are pH dependent.[11]

Intracellular pH is close to neutrality for all micoorganisms. In order to protect important internal cellular reactions from changes in external pH, microorganisms have evolved cell membranes that have limited permeability to protons.[12] According to Daniels et al,[13] methanogenic bacteria have the ability to maintain a transmembrane hydrogen ion gradient. However, their intracellular pH is some-

Table II. Composition of Synthetic Feed Used in pH Inhibition Studies

Characteristic, units	Value
Sodium Bicarbonate, g/L	0.6
Minerals:[a]	
Iron, mg/L as Fe	2.48
Zinc, mg/L as Zn	0.25
Nickel, mg/L as Ni	0.25
Cobalt, mg/L as Co	0.25
Manganese, mg/L as Mn	0.25
Nonfat dry milk, g/L	12.4
Fat, g/L	0.12
Lactose, g/L	6.32
Total Kjeldahl Nitrogen, g/L	0.67
Phosphate, g/L	0.27
Protein, g/L	4.50
Alkalinity, mg/L as $CaCO_3$[b]	350
Chemical Oxygen Demand, mg/L	14,000
pH[c]	7.2–7.3

[a]Added from stock solutions of chloride salts.
[b]Alkalinity of control feed (Reactor 3).
[c]pH of control feed (Reactor 3).

times influenced more by the external pH than is the case for some eubacteria. The iso-pH (i.e., internal pH = external pH) was found to be approximately 6.7 for *Methanobacterium thermoautotrophicum,* spheroplasts of *Methanospirillum hungatei,* and *Methanobacterium bryantii.* In a medium of pH 8.9, the internal pH of *Mb. thermoautotrophicum* was estimated to be 6.6.

Research indicates[13] that for *Mb. thermoautogtrophicum, Mb. bryantii,* and spheroplasts of *Msp. hungatei,* the membrane potential makes up the major portion of the proton-motive force. In addition, the contribution of the pH gradient to the proton-motive force decreases as the pH of the media rises. For *Mb. thermoautotrophicum* and *Mb. bryantii,* no large pH-dependent change in membrane potential was evident. Raising the external pH, and therefore decreasing the chemical potential, caused a decrease in the proton-motive force. This would seem to indicate that any cellular functions that are directly dependent on the proton-motive force may be favored when the media is more acidic than pH 7.0.[13]

The transport of solute molecules across the semipermeable cell membrane is also affected by pH.[12] Because cell membranes have low permeability to charged compounds, only nonionized substrate molecules enter the cell by passive diffusion. Substrates supporting growth of methanogens, such as formate, acetate, and methylamines, are weak acids and bases and therefore distribute across the cytoplasmic membrane by diffusion, depending on pH.[13] In addition, Huster and Thauer[14] have hypothesized that *Mb. thermoautotrophicum* can passively diffuse pyruvic acid through the lipid membrane in the undissociated form. Most solute molecules cross the semipermeable cell membrane through active transport which may be pH dependent. pH dependent transport systems have been described by Balch and Wolfe[15] for the coenzyme 2-mercaptoethanesulfonic acid (HS-CoM) in a strain of *Methanobrevibactor ruminantium* (optimum pH range 6.2 to 8.0 or 6.9 to 7.8, depending on experimental procedure) and by Jarrell and Sprott[16] for nickel in *Mb. bryantii* (sharp pH optimum of approximately 4.9 at 35°C). Substrate uptake may be dependent on extracellular enzyme activity, either free in solution or in the periplasmic space. In this situation, the external pH exerts a direct influence on substrate uptake.[12]

EXPERIMENTAL PROCEDURE

Reactor Operation

In this study, three completely-mixed anaerobic reactors, each with a liquid volume of 4.23 L, were operated in parallel at a hydraulic retention time of 14 days (feed rate of 0.21 mL/min). The substrate fed to each reactor was identical and consisted of animal-grade, low heat, nonfat dry milk (NFDM), sodium bicarbonate alkalinity, minerals, and tap water. The bicarbonate alkalinity was added in an amount of 0.043 g $NaHCO_3$ per g COD fed. The composition of the substrate is shown in Table II.

There were two components of the laboratory study. During Phase 1 (233 days) the quasi-steady-state operating conditions and treatment performance of the reactors were characterized. The condi-

Table III. Modifications to Feed Solution During Phase 2 of Study

	Reactor 1 Feed Modification[b]			Reactor 2 Feed Modification[b]		
Time, days[a]	Chemnical Added	Conc g/L	pH	Chemical Added	Conc g/L	pH
234–237	NaOH	2.0	11.6–11.8	NaOH	2.0	11.6–11.8
238–256	NaOH	3.1	11.8–12.0	NaOH	2.0	11.6–11.8
257–263	NaOH	3.8	12.0–12.2	NaOH	2.0	11.6–11.8
264–277	NaOH	4.5	12.2–12.3	NaOH	2.0	11.6–11.8
278–284	NaOH	4.5	12.2–12.3	NaOH	2.0	11.6–11.8
	KOH	0.4				
285–291	NaOH	4.5	12.3–12.4	NaOH	4.5	12.2–12.4
	KOH	1.1				
292–301	NaOH	4.5	12.2–12.5	NaOH	4.5	12.2–12.3
	KOH	2.2				
302	37% HCl	64.9	1.2			
303–308	Regular Feed		7.3–7.7			
309–312	Regular Feed		7.3			

[a]From initiation of study (phase 1 includes days 0 through 233).
[b]Additional components added to feed solution defined in Table II.

tions in all three reactors were essentially identical. During Phase 2, the pH levels in Reactors 1 and 2 were increased to evaluate the effects of high pH on methanogenisis. The modifications to the feed during Phase 2 are shown in Table III. The pH in Reactor 3 was not changed, it serving as a control unit.

A flow diagram for one reactor with its associated gas scrubbing, collection and sampling system configurations, is shown in Figure 1. The three identical reactors were each seeded with 4.2 L of liquid from a pilot-scale submerged media anaerobic reactor and were maintained at 35°C. The reactors were mixed continuously at about 1,500 rpm using an electric laboratory mixer with a stainless steel three-blade marine-type propeller (4 cm diameter) attached. The seal around the mixer shaft was the reactor liquid itself. The influent feed line was inserted through the stainless steel tube connected to the reactor lid. Synthetic waste was pumped from refrigerated, one-liter glass bottles into the reactors using a peristaltic pump (1 to 100 rpm) that was connected to a timer and operated intermittently, initially for 15 seconds every one minute and later for 30 seconds every two minutes.

Monitoring Program

The monitoring program used for this study is outlined in Table IV. In addition to the parameters listed, the average total gas production rate (L/day) and hydraulic retention time (HRT) were monitored daily. Because the reactors were completely-mixed without recycle, the solids retention time (SRT) was equal to the HRT.

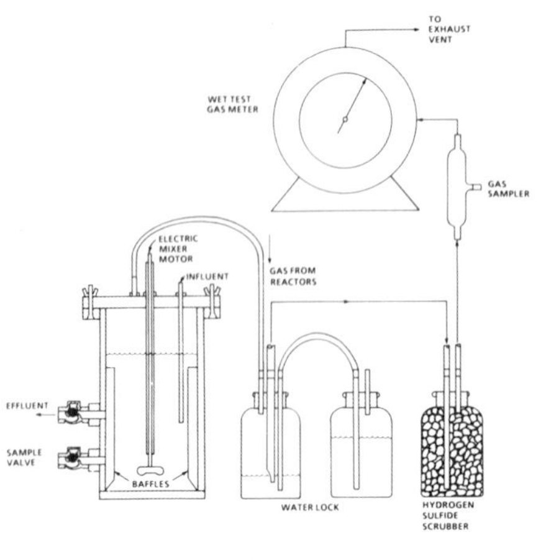

Figure 1. Schematic illustration of typical completely stirred tank reactor (CSTR) and gas handling system.

Table IV. Summary of Parameters Monitored in this Study

Parameter	Analytical Method	Frequency
pH	pH meter	Daily
Total alkalinity and acids	Direct titration[a]	Weekly
Total organic acids (C2-C5)	Gas chromatograph	Weekly
Gas composition	Gas chromatograph	Weekly or Bi-weekly
Chemical oxygen demand	Standard Methods (508B)[b]	Weekly
Total suspended solids	Standard Methods (209C)[b]	Weekly
Volatile suspended solids	Standard Methods (209D)[b]	Weekly
Ammonia	Standard Methods (417E)[b]	Weekly

[a]Method from DiLallo and Albertson.[17]
[b]From *Standard Methods*.[18]

RESULTS AND DISCUSSION

Phase 1: Quasi-Steady-State Conditions and Performance

It was assumed that a quasi-steady-state had been achieved if the gas production rate was relatively constant for at least three hydraulic retention times under constant operating conditions. During Phase I (days 1 through 233) the quasi-steady-state operating conditions for Reactors 1, 2, and 3 were, respectively, average influent COD, g/L, 14.37, 14.37, and 14.37; average HRT, days, 14.1, 14.1, and 14.4; and, corresponding average organic loading rates, g/L/day, 1.02, 1.02, and 1.00. A summary of the average quasi-steady-state treatment performance for each reactor is presented in Table V.

Methane recoveries for all three reactors were close to the theoretical value of 0.35 L CH_4(STP)/g COD removed. The pH of each reactor was near neutral, and the alkalinity levels were around 2500 mg/L, as $CaCO_3$. The concentrations of the C2 to C5 volatile organic acids were low, further evidence of efficient treatment.

Phase 2: High pH Effects

During Phase 2 the operating conditions for Reactors 1, 2, and 3 were, respectively, average influent COD, g/L, 13.87, 13.72, and 13.68; average HRT, days, 14.0, 14.1, and 14.8; and corresponding average organic loading rates, g/L/day, 0.99, 0.97, and 0.92. These values were quite similar to the conditions during Phase 1. A direct consequence of excess hydroxide addition to the feed was a pH increase in Reactors 1 and 2. These changes in pH and pH-IF for Reactors 1, 2, and 3 (the control) are shown in Figures 2a,b, 3a,b, and 4a,b, respectively.

Table V. Quasi-Steady-State Treatment Performance[a]

Parameter, units	Reactor		
	1	2	3
Total Alkalinity, mg/L as $CaCO_3$	2504	2503	2491
Chemical Oxygen Demand:			
Total COD, g/L	3.09	3.16	3.08
Filtered COD,[b] g/L	1.65	1.32	1.59
Gas production, L/day			
Total Gas Production	1.64	1.71	1.49
Methane Production	1.10	1.15	1.04
Organic acids,[c] mg/L as acetic	37	30	39
pH	7.08	7.08	7.12
Suspended solids, mg/L:			
Total	949	1350	993
Volatile	871	1268	930

[a]Values reported are average values for each parameter calculated from monitoring data collected during the pseudo-steady-state period through day 233.
[b]Determined by filtering the sample through a Whatman GF/C glass fiber filter.
[c]Total organic acids represents the sum of C2 through C5 volatile organic acids, mg/L as acetic (acetic + propionic + iso-butyric + 2-methyl-butyric + n-butyric + iso-valeric + n-valeric).

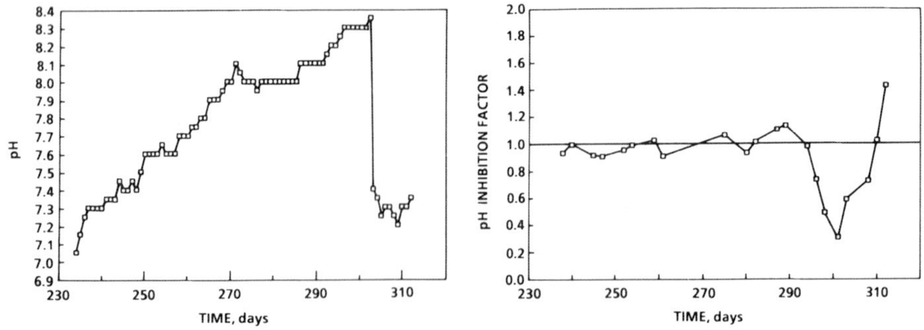

Figure 2. Variation in pH and pH-IF for Reactor No. 1 during phase II.

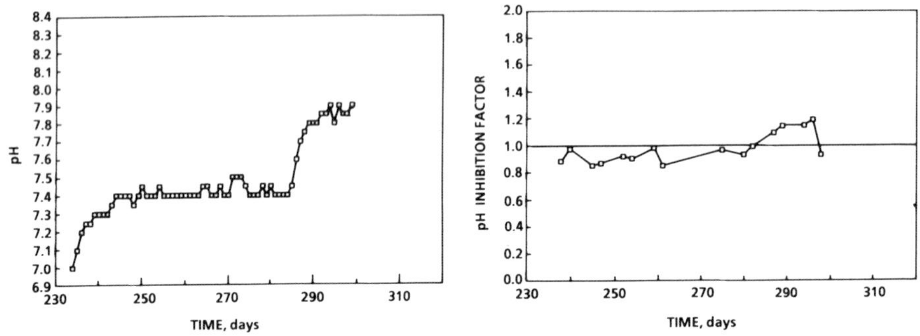

Figure 3. Variation in pH and pH-IF for Reactor No. 2 during phase II.

The hydroxide additions to Reactor 2 (Figure 3a) were conducted in two steps. First the pH was increased to about 7.4 to approximate the upper pH limit recommended for optimum anaerobic treatment. Then, starting on day 284, the pH in Reactor 2 was increased rapidly to approximately 8.0.

In Reactor 1, the pH was increased gradually, as shown in Figure 2a. Thus reactor responses to rapid increases in pH could be compared to responses from gradual pH changes.

No excess hydroxide was added to the control reactor (No.3) and no significant variation in pH was observed for this unit, as shown in Figure 4a.

As the pH in the reactors increased, the minor dissolved CO_2 species, H_2CO_3, was converted to bicarbonate (Figure 5), which made the hydration of aqueous CO_2 more favorable, allowing more

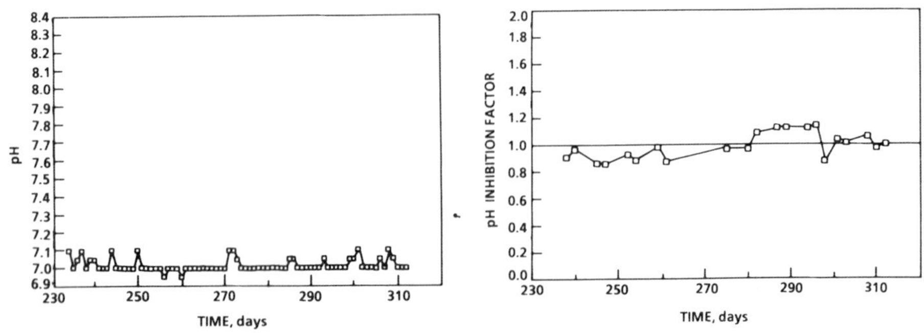

Figure 4. Variation in pH and pH-IF for Reactor No. 3 (the control) during phase II.

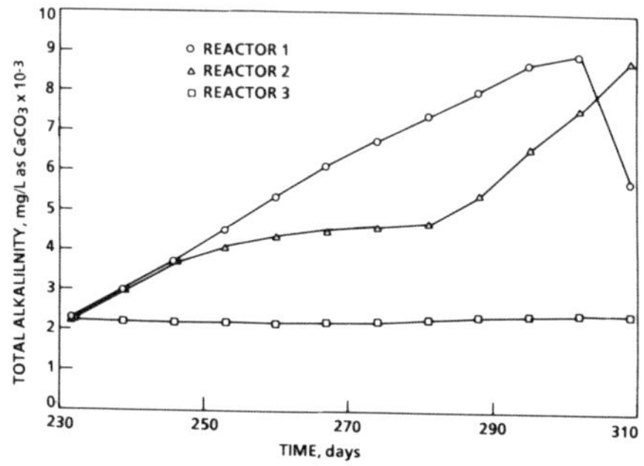

Figure 5. Variation in total alkalinity (5a), methane percentage (5b) and volatile acids (5c) during phase II.

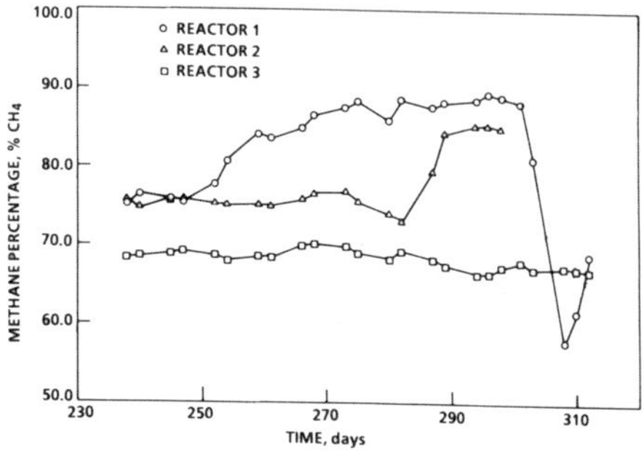

Figure 6. Variation in percentage methane during phase II.

gaseous CO_2 to dissolve in the liquid phase and, in turn, increasing the relative methane content of the gas phase (Figure 6). It can be seen that although the gas produced is not pipeline quality (>97% CH_4), methane enrichment is significant. For example, between days 292 and 301, the influent pH to Reactor 1 ranged from 12.3 to 12.5 and the gas produced averaged 89% CH_4. In comparison, the pH of the influent to Reactor 2 ranged from 7.2 to 7.6 and the gas produced averaged 67% to CH_4.

Inhibition of Methane Production

The methane production rate was used to identify potential system inhibition as a result of high pH. Methane production is an excellent monitoring parameter because it is a measure of the health of the methanogenic population, the most sensitive bacterial group in anaerobic systems. Thus, methane production rates can be used as diagnostic tools to detect the onset of inhibition. Because the quasi-steady-state methane production was different for each reactor (as shown in Table V), methane production data was normalized using the pH-IF (Clark and Speece, 1971).

As expected, no inhibition due to pH was observed in Reactor 3 (the control), as shown in Figure 4b, and the pH-IF remained near 1.0. Likewise, no evidence of pH inhibition was observed in Reactor 2 (Figure 3b). When Reactor 2 was operated at a pH of 7.4 (days 244 to 285), the pH-IF remained near 1.0. For the experimental operating conditions used in this study, a pH of 7.4 did not represent an upper limit for optimum treatment. Between days 285 and 294, the pH of Reactor 2 was raised to approximately 7.9 and held there until day 299. Again, no inhibition was evident.

Reactor 1 was the only unit in which inhibition occurred, as indicated by the pH-IF (Figure 2b). Clearly, as long as the reactor pH was equal to or less than 8.1 (up to day 291), no inhibition was evident. Further, when the reactor pH was between 7.0 and 8.2, the pH-IF was 1.0 ± 0.2. However, a dramatic decrease in pH-IF occurred at pH 8.3 (days 296 to 302) and by day 301 (after 6 days at pH 8.3), the pH-IF had decreased to 0.31.

To determine if the extensive inhibition that resulted from 11 days of exposure to high pH levels was reversible, the pH in Reactor 1 was reduced rapidly to near 7.0. On day 302, the feed solution was acidified with hydrochloric acid to a pH of 1.2. This change in the feed solution pH was designed to neutralize the reactor pH rapidly and thereby prevent significant washout of organisms. Within one day the reactor pH decreased to 7.40, the pH-IF increased to 0.59, and the methane production rate nearly doubled. This rapid recovery is an indication that the methanogenic organisms were still viable and the changes resulting from adverse pH levels were inhibitory but not irreversible.

After the rapid decrease in reactor pH, the acidic feed for Reactor 1 was replaced with neutral feed at pH 7.3 (Table III). During this time period (days 303 to 312), Reactors 1 and 3 (control) received identical feed solutions. However, even with a neutral feed solution, the methane production rate was depressed. The rate of system recovery is reflected in the parabolic shape of the pH-IF versus time curve for days 303 to 310 (Figure 2b). After seven days of using the control feed solution, the pH-IF continued to increase beyond 1.0. This increase in pH-IF beyond unity could be due to the consumption of excess substrate accumulated during the inhibition period.

The lag in recovery of methane production after day 303 was probably due to regrowth of the methanogens. Assuming that the high pH was bacteriostatic and the methanogens were not growing during the seven days when the pH was 8.3, approximately 40% of the methanogens would have washed out by day 302 (ignoring cell decay). The high pH may have been bacteriocidal to some degree, but this cannot be determined from these results. It is interesting to compare these findings for recovery from high pH inhibition with those of Clark and Speece[4] for an anaerobic filter. Clark and Speece evaluated pH levels of 8.4, 8.6, 9.3, and 9.4 for periods of 3 to 4 days. In each case, when the pH was restored to optimum, the pH-IF promptly recovered to 1.00. It is likely that the difference in recovery pattern between the CSTR and the biofilm reactor is due to the fact that while the methanogens may have been reversibly inhibited in both cases, the nongrowing cells would tend to be retained in the biofilm reactor whereas in the CSTR they would be washed out. Similarly, the discrepancy between the pH-IF found here for pH 8.3, which steadily decreased with time to 0.31 after seven days, and the pH-IF of 0.78 found by Clark and Speece[4] with pH 8.4 for four days can (at least in part) be explained by the combination of inhibition and washout in the CSTR.

pH Effects on Various Trophic Groups of Organisms

Variations in the C2 to C5 organic volatile acids provide further insight into the effect of pH inhibition on various groups of anaerobic organisms. The concentrations of the individual organic acids in Reactor 1 are shown as a function of time in Figure 7. The concentrations of acetic, propionic, and n-butyric acids were consistent through day 281, although the iso-butyric concentrations increased and low levels of 2-methyl-butyric and isovaleric acid were detected for the first time. By day 286, the pH in Reactor 1 was 8.1 and by day 288, further changes were evident in the concentrations of organic acids. During this period, no inhibition of methane production was evident, but significant increases in the levels of acetic, iso-butyric, 2-methyl-butyric and iso-valeric acid occurred.

When methane production was inhibited (after day 294), further changes occurred in the organic acid concentrations, with the acetic acid concentration increasing 60 fold between days 288 and 309. The other analyzed organic acids also increased during this period, although more gradually (except for n-butyric acid which remained at about the same level).

These effects can be explained by looking at the effects of stressing the acetoclastic and CO_2-reducing methanogens on two trophic levels, the hydrolytic/fermentive bacteria and the acetogenic bacteria. As the CO_2-reducing methanogens became stressed by the high pH levels, there would have been an increase in the H_2 concentration. In this situation, there is a tendency for the electrons of NADH, which are generated in the catabolism of carbohydrates by hydrolytic/fermentative bacteria, to be utilized in catabolyzing pyruvate to reduced products, especially propionate, and possibly butyric, valeric and caproic acid, rather than acetate, CO_2 and H_2.[19] An increase in H_2 partial pressure would also inhibit butyrate acetogenesis and propionate acetogenesis by H_2-producing acetogens,[19] resulting in accumulation of butyrate and propionate. It is interesting to note that propionate acetogenesis is more sensitive to H_2 partial pressure than butyrate acetogenesis[20] and the levels of propionic acid were observed to increase while the levels of n-butyrate remained low. As the acetoclastic methanogens became stressed by the high pH, there was an accumulation of acetate. It is not surprising that the increases of acetic acid were much greater than the other acids since approximately 70% of the methane produced from organic matter in nature is formed by means of the methyl group of acetate.[19]

Potential Causes of Inhibition Above pH 8.1

Why were pH levels above 8.1 inhibitory? One potential cause of inhibition is salt toxicity. Bicarbonate salts and volatile acid salts were present in the reactor before and during the inhibition period. In general, salt toxicity is associated with the cationic portion of the salt rather than the anionic portion[21]. The two major sources of cations were: 1) sodium and potassium added with the hydroxide used in to raise the pH, and 2) ammonium ions formed during degradation of substrate protein. The feed solution with the highest pH had a sodium concentration of 2.59 g/L, as Na^+, and a potassium concentration of 1.53 g/L, as K^+. Compared to the range of stimulatory and inhibitory values suggested by McCarty and McKinney,[21] the feed concentration was below inhibitory levels. Also, it has been reported that an antagonistic cation, such as potassium, can significantly reduce the toxicity of an inhibitory cation such as sodium.[22] Considering that the same levels of Na^+ were fed to Reactor 2 without ill effects and that potassium is an antagonistic cation for Na^+, inhibition was probably not caused by salt toxicity.

Ammonia is another potential source of toxicity. The total ammonia concentrations measured during this study did not exceed the range that McCarty[22] reported as causing no adverse effect (200 to 1,000 mg/L). The highest value of free ammonia calculated for Reactor 1 was 96.2 mg/L, as N (when

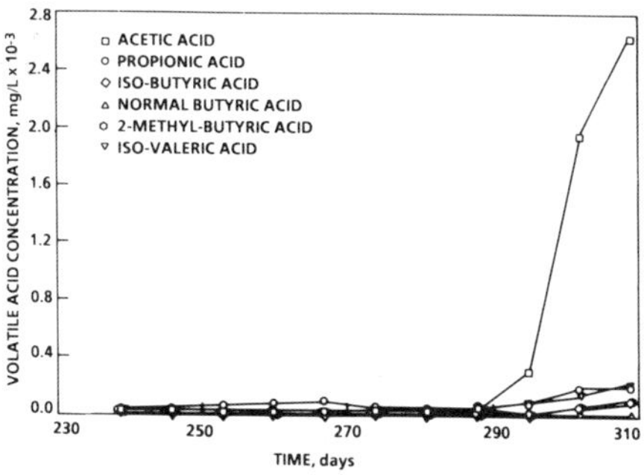

Figure 7. Variation in volatile acids during phase II.

Reactor 1 was already inhibited), which is below the level of 100 mg/L, as N, above which free ammonia toxicity can be a problem.[23] It should also be noted that McCarty and McKinney[21] reported free ammonia concentrations of 130 mg/L and 150 mg/L, as N, as causing organism death. On Day 291, immediately before Reactor 1 began to show signs of inhibition, the free ammonia concentration was only 62 mg/L, as N. Therefore, it appears that ammonia was not a cause of inhibition.

Sulfides in the reactor liquid are another potential source of toxicity. However, no significant source of sulfides was present in the feed solution. The availability of metals as nutrients is pH dependent, but at pH values near 8, the anionic species that are primarily responsible for precipitating metals in anaerobic digesters (i.e., S^{-2}, CO_3^{-2}, and PO_4^{-3}) would still not be the dominant species of sulfides, carbonates, and phosphates present.

At the cellular level, several factors can be affected by a high pH and lead to inhibition of the methanogens. The data gathered in this research are not appropriate for evaluating cellular level responses to high pH. However, in the absence of other causes of inhibition, as discussed above, it must be concluded that the inhibition of the methanogens that was observed at pH levels above 8.1 was related to pH effects at the cellular level, such as enzyme activity, membrane transport, or the proton-motive force.

CONCLUSIONS

Steady-state methane production was not inhibited at pH levels up to 8.1. As long as the reactor pH can be maintained below 8.1, treatment of high pH industrial wastes without neutralization may be possible in well-buffered anaerobic systems. The most basic substrate treated in this study had a pH of about 12. However, for wastes containing higher concentrations of organic or ammonia nitrogen, ammonia toxicity, not high pH per se, could become the primary inhibitory factor.

Based on methane production and C2 to C5 volatile organic acid data, it appears that methanogenic bacteria were inhibited when the pH in the reactor was above 8.1. Acetogenic, hydrolytic and fermentative bacteria were functioning when the reactor pH was above 8.1.

Based on the recovery from high pH inhibition found in this research for a CSTR and by Clark and Speece (1971) for a biofilm reactor, the effect on methanogens of high pH appears to be bacteriostatic; hence, reactors that can better retain the inhibited, non-growing bacteria in the system will recover more quickly from excessively high pH episodes.

Further research is needed to determine the actual cause for inhibition of the methanogenic bacteria at pH levels over 8.1 in the absence of other inhibitory factors. Important avenues of investigation are biochemical reactions, trace metal interactions, and cellular level responses such as pH effects on substrate uptake or membrane functions.

ACKNOWLEDGEMENTS

This research was supported, in part, by a grant from the United States Department of Agriculture, contract number 89-34188-4370 (Melvin J. Schlattman, Administrative Contact, and David R. MacKenzie, Technical Contact) through the Iowa Biotechnology Byproducts Constortium, and by the Engineering Research Institute at Iowa State University.

AUTHORS

Eric Seagren is a graduate research assistant at the University of Illinois at Urbana. At the time this research was conducted, he was a graduate research assistant in the Department of Civil and Construction Engineering at Iowa State University. Audrey D. Levine is an assistant professor, and Richard R. Dague is professor and chairman in the Department of Civil and Construction Engineering at Iowa State University. Correspondence should be directed to Richard R. Dague, Department of Civil and Construction Engineering, 394 Town Engineering Building, Iowa State University, Ames, IA 50011.

REFERENCES

1. Heukelekian, H., and Heinemann, B. (1939). "Studies on the methane-producing bacteria. I. Development of a method for enumeration." *Sewage Works Journal*, 11, 3, 426–435.
2. Albertson, O. E. (1961), "Ammonia nitrogen and the anaerobic environment." *Journal Water Pollution Control Federation*, 33, 9, 978–995.
3. McCarty, P. L. (1964a). "Anaerobic waste treatment fundamentals: Part II: Environmental Requirements and Control." *Public Works*, 95, 123–126.
4. Clark, R. H., and Speece, R. E. (1971), "The pH tolerance of anaerobic digestion." *Advances in Water Pollution Research*, Volume 1, S. H. Jenkins (Ed.), Pergamon Press, Oxford.
5. Van Den Berg, L., Patel, G. B., Clark, D. S., and Lentz, C. P. (1976). "Factors affecting the rate of methane formation from acetic acid by enriched methanogenic cultures." *Canadian Journal of Microbiology*, 22, 1312–1319.
6. Dague, R. R., McKinney, R. E., and Pfeffer, J. T. (1970), "Solids Retention in Anaerobic Waste Treatment Systems." *Journal Water Pollution Control Federation*, 42, (2), Pt. 2, R29.
7. Porter, J. R., Jr. (1981). "Effect of high pH on anaerobic waste treatment in anaerobic filters." M.S. Thesis, Engineering Library, University of Iowa, Iowa City, Iowa, 52242.
8. Speece, R. E. and Parkin, G. F. (1983), "The response of methane bacteria to toxicity." Proceedings of the Third International Symposium on Anaerobic Digestion. Boston, Massachusetts.
9. Callander, I. J., Barford, J. P. (1983), "Precipitation, chelation, and the availability of metals as nutrients in anaerobic digestion. I. methodology." *Biotechnology and Bioengineering*, XXV, 8, 1947–1957.
10. Dixon, M. and Webb, E. C. (1979). *Enzymes*, 3rd Edition, Longman Group Limited, London.
11. Bailey, J. E., and Ollis, D. F. (1986), *Biochemical Engineering Fundamentals*, 2nd Edition, McGraw-Hill Book Co., New York.
12. Forage, R. G., Harrison, D. E. F., and Pitt, D. E. (1985). "Effect of environment on microbial activity." *Comprehensive Biotechnology*, Volume 1, M. Moo-Young (Ed), Pergamon Press, Oxford.
13. Daniels, L., Sparling, R., and Sprott, G. D. (1984). "The bioenergetics of methanogenesis." *Biochimica et Biophysics Acta*, 768, 2, 113–163.
14. Huster, R. and Thauer, R. H. (1983). "Pyruvate assimilation by *Methanobacterium thermoautotrophicum*." *FEMS Microbiology Letters*, 19, 2/3, 207–209.
15. Balch, W. E. and Wolfe, R. S. (1979), "Transport of coenzyme M (2-mercaptoethanesulfonic acid) in *Methanobacterium ruminantium*." *Journal of Bacteriology*, 137, 1, 264–273.
16. Jarrell, K. F. and Sprott, G. D. (1982). "Nickel transport in *Methanobacterium bryantii*." *Journal of Bacteriology*, 151, 3, 1195–1203.
17. DiLallo, R., and Albertson, O. E. (1961). "Volatile acids by direct titration." *Journal Water Pollution Control Federation*, 33, 4, 356–365.
18. *Standard Methods for the Examination of Water and Wastewater* (1986). 16th Edition, American Public Health Association, Washington, D.C.
19. Bryant, M. P. (1979) "Microbial methane production—theoretical aspects." *Journal of Animal Science*, 48, 1, 193–201.
20. Daniels, L. (1984). "Biological methanogenesis: Physiological and practical aspects." *Trends in Biotechnology*, 2, 4, 98–98.
21. McCarty, P. L., and McKinney, R. E. (1961). "Salt toxicity in anaerobic digestion." *Journal Water Pollution Control Federation*, 33, 4, 399–415.
22. McCarty, P. L. (1964b). "Anaerobic waste treatment fundamentals: Part III: Toxic materials and their control." *Public Works*, 95, 91–94.
23. Parkin, G. F. and Owen, W. F. (1986). "Fundamentals of Anaerobic Digestion of Wastewater Sludges." *Journal of the Environmental Engineering Division*, ASCE, 112 (EE 5), 867–920.

44 ANAEROBIC BIODEGRADATION OF INDUSTRIAL SOLVENTS – THE CASE OF ACETONE AND ISO-PROPANOL

E. Terzis, Experimental Officer
J. Diaper, Student
Dept. of Environmental Science
University of Bradford
Bradford, W. Yorkshire, BD7 1DP
England

INTRODUCTION

Increased industrial water usage in recent years has resulted in higher volumes of waste water, rich in organic pollutants. Waste waters from pharmaceutical, cosmetic, textile, rubber and other industries will inevitably contain solvents, often in high concentrations. The presence of solvents is undesirable in an industrial effluent. Direct disposal in a river or to the sewage system is prohibited in U.K. because of their effect in living organisms,[1] on the sewage reclamation system itself and also because of their flammability. Iso-propanol and acetone are two commonly used solvents and easily found in an industrial effluent.

Anaerobic treatment of organic waste has long been incorporated in the biological treatment of waste waters. Anaerobic digestion of waste organic matter is a very complex process. Any description can only be a simplified account of the real and complex processes taking place in an anaerobic system. Such an account is given in Fig. 1. Experience in the U.K. is primarily associated with the anaerobic digestion of sewage sludge but the field of application is by no means restricted to that process and it is increasingly applied to industrial wastes.[2] To investigate the applicability of the anaerobic digestion of iso-propanol and acetone the Monod kinetic model was employed.[3,4] The estimation of the kinetic parameters of the Monod model provides a quantitative description of the anaerobic degradation of iso-propanol and acetone and also offers useful design information for such process.

THE KINETICS OF THE ANAEROBIC PROCESS

In a completely mixed continuous flow system with solids recirculation (Fig. 2, see also list of symbols) the rate of change of the bacterial mass and substrate concentration are described by two differential equations.

$$(dX/dt) = \mu.X.V - K_d.X.V - [W.X+(Q - W)].X_e \qquad (1)$$

$$(ds/dt) = [Q.(S_o - S)] - (\mu/Y).X.V \qquad (2)$$

Assuming that anaerobic digestion exhibits Monod type kinetics the bacterial growth rate can be linked with the concentration of the limiting substrate (S).

$$\mu = \mu_m.S/(K_s+S) \qquad (3)$$

Under steady-state conditions Equations 1 and 2 yield;

$$\Theta_s^{-1} = \mu - K_d \qquad (4)$$

$$\mu = Y.[Q.(S_o - S)]/(X.V) \qquad (5)$$

The growth (Y) and decay (K_d) parameters can be determined as the slope and the intercept, respectively, of the plot of Equation 4 as $1/\theta_s$ vs. μ. The maximum specific growth rate (μ_m) and half-velocity coefficient (K_s) can be determined by rearranging Equation 3 to obtain linearity and plotting $1/\mu$ vs. $1/S$. Such plots are shown in Fig. 3.

45th Purdue Industrial Waste Conference Proceedings,

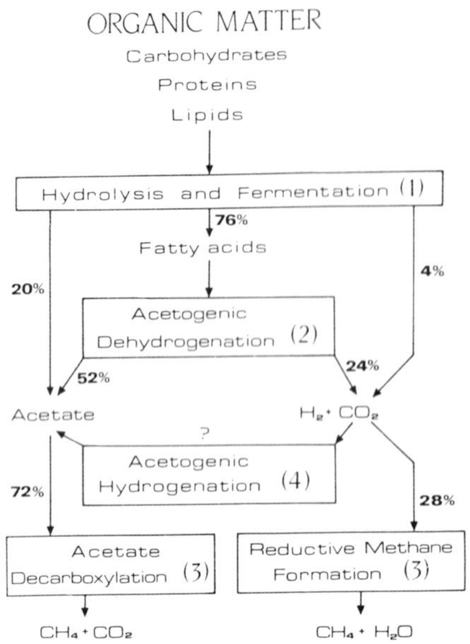

Figure 1. The stages of anaerobic digestion.

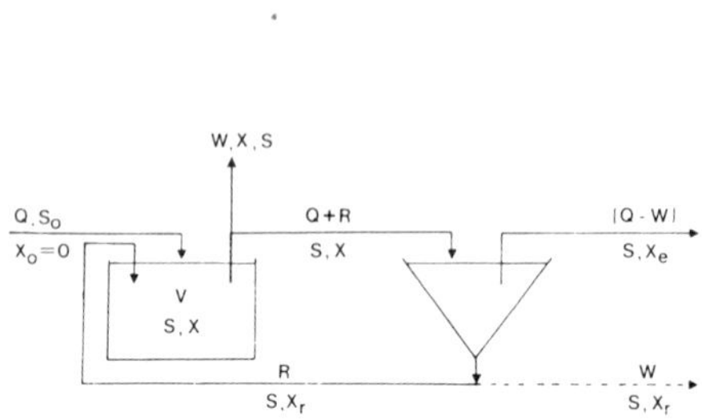

Figure 2. Flow scheme for continuous flow stirred reactor.

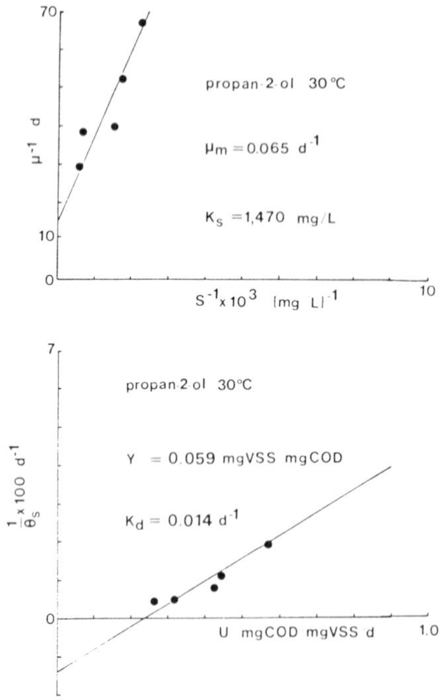

Figure 3. Linear plot for kinetic and growth coefficients.

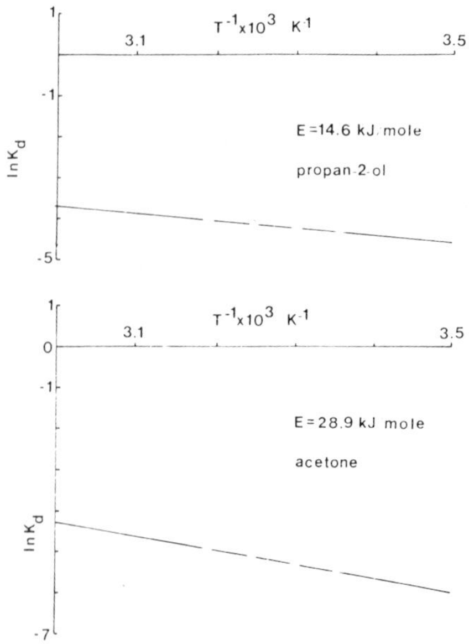

Figure 4. Arrhenius plot for the decay coefficient.

Expressions for effluent substrate concentration (S) and bacterial mass (X) can be derived as a function of θ_s and these parameters are of chief interest to the plant operator of a biological treatment plant;

$$S = [K_s.(1+\Theta_s.K_d)] / [\Theta_s.(\mu_m - K_d) - 1] \quad (6)$$

$$X = (\Theta_s/\Theta).[Y.(S_o - S)] / (1+\Theta_s.K_d) \quad (7)$$

In order to sustain steady state bacterial growth a minimum substrate concentration is needed. This must be enough as to support a growth rate of the bacterial population greater than its loss rate due to lysis and endogenous decay mechanisms ($\mu \geq K_d$). We can implement this if we substitute in the Monod equation the endogenous decay constant (K_d) for the specific growth rate (μ):

$$K_d = \mu_m.S / (K_s+S)$$

solving for S we have,

$$S_{(min)} = K_s.K_d / (\mu_m - K_d) \quad (8)$$

but,

$$\mu_m - K_d = \Theta_{s(min)}^{-1},$$

and so,

$$S_{(min)} = K_s.K_d.\Theta_{s(min)} \quad (9)$$

The minimum substrate concentration is defined as $S_{(min)}$ and it can be associated with $\theta_{s(min)}$. The $S_{(min)}$ represents the lower limit of the effluent substrate concentration, because at substrate concentration lower than $S_{(min)}$ the bacterial growth is always going to be less than the bacterial loss and the process cannot function. There is no way to achieve lower effluent substrate concentration than $S_{(min)}$ under any process loading at fixed environmental conditions.

THE EFFECTS OF TEMPERATURE

Anaerobic digestion is subject to temperature influences just as any other biological or chemical process. Since bacteria have no internal capacity to maintain a favourable temperature, they are subject to any modifications of the environmental temperature and they must react accordingly. The Arrhenius equation in the logarithmic form is used to describe the effect of increasing temperature on the rate limiting reaction as follows.

$$\ln(k) = \ln(A) - E/R.(1/T) \quad (10)$$

where k is the rate constant at a given temperature T (°K), R is the gas constant, 8.314 J/K/mole, E is the activation energy (kJ/mole), and A is a constant. Thus the activation energy E can be obtained from the slope of the linear plot of ln(k) against the reciprocal of the absolute temperature. Such plot for the decay coefficient is shown in Fig. 4. The Arrhenius law is applicable in biological systems within a limited temperature range, because enzymes present in these systems are inactivated at high temperatures, due to the denaturation of the protein[5].

EXPERIMENTAL METHODS

The main experimental apparatus used in these studies consisted of four bench scale anaerobic digesters. These units constructed from five litre quickfit bottles, fitted with the necessary tubing for feeding the digester and withdrawing the effluent. Complete mixing of the contents of the digesters was achieved by using a magnetic stirrer. Heating of the digesters was provided by two one meter long heat belts, surrounding each digester and controlled using a thermocouple detector placed inside the digester and connected with a temperature controller, set at the required temperature. In this way the desirable temperature was maintained within ±1°C. The generated gas was collected using a inverted measuring cylinder, by displacement of acidified water to prevent the carbon dioxide dissolving. The volume of the gas collected this way was dependent on the pressure and ambient temperature. Both these variables were monitored and the gas volume was corrected to S.T.P. Fig. 5 represents a graphical representation of the entire system.

The operation described here was designed to provide determination of the kinetic coefficients, at various temperatures and solids retention times. The anaerobic systems were seeded with actively digesting sewage sludge to provide an active microbial population. The digesters were then purged of

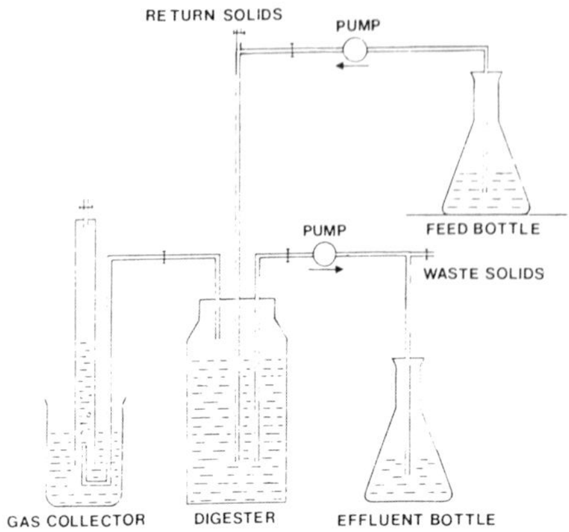

Figure 5. Bench scale anaerobic digester.

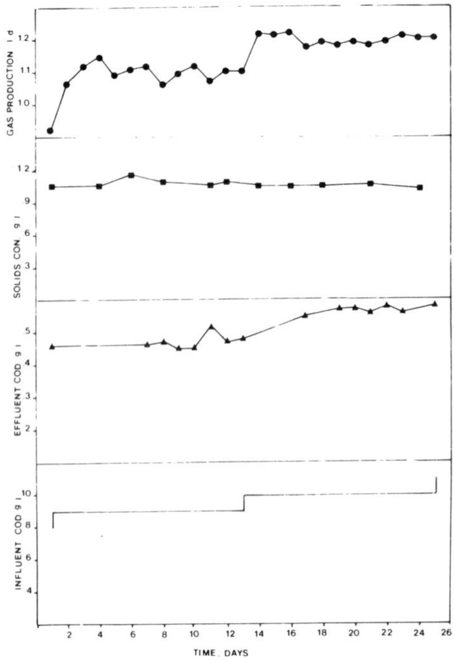

Figure 6. Change of system variables from one steady state to another for acetone at 30°C.

oxygen with nitrogen gas and set at the desirable temperatures. The anaerobic units were left without feeding until no gas was produced from undigested solids. The feeding of the anaerobic units started with a synthetic waste water containing an inorganic nutrient salt solution and 1 g/L iso-propanol. No solids were wasted, but for sampling, until a stable acclimatised microbial population established. Indication for this, was the values of gas production and composition, and the effluent COD. When after two weeks such a population was established,the steady state programme started.

The basic requirement for the determination of the kinetic coefficients from the equations outlined earlier, is that the digester should be operated at various solids retention times and effluent substrate concentration. This was achieved by operating the system at various biological loading rates. The digesters were operated at different loading rates by altering the feed strength, and adjusting the solids waste rate, hence, varying the solids retention time, to keep the concentration of the mixed liquor volatile suspended solids (representing biomass concentration) in the reactor constant. By varying the solids retention time the effluent substrate concentration was inevitably varied.

At each steady state achieved of such operation of the digesters accurate measurements of the bacterial population and effluent substrate concentration was necessary. It was felt that the volatile suspended solids (VSS) would be a reasonable approximation of the biomass concentration as the feeding substrate was soluble and no solids was contained in the feed. The chemical oxygen demand (COD) of the waste water was used as a measure of the synthetic waste water strength as it is commonly used in practice to determine the strength of industrial waste waters.

The criteria used to assess the steady state conditions were volatile solids concentration, effluent C.O.D., gas production and gas composition. The volume of the waste solids was determined the first 3-5 days after a change from one feed concentration to another and the digesters reached steady state the next 7-10 days. When the above four parameters had become constant i.e., variation within 5%, it was apparent that a new steady state had been reached (Fig. 6). The measurements of these parameters was carried on for 7 days and the mean values were used to determine the kinetic coefficients.

The effluent C.O.D., volatile suspended solids were routinely measured every three days and at the steady state daily. The following analyses were performed at least weekly on mixed liquor of each

Table I. Kinetic Coefficients of Iso-Propanol for Substrate Utilization and Biological Growth

Temperature °C	K_s (mg COD/1)	μ_m (day $^{-1}$)	K_d (day $^{-1}$)	Y (mg VSS/mg COD)
25	670	0.047	0.013	0.066
30	1470	0.065	0.014	0.059
35	1050	0.095	0.016	0.061
40	4000	0.125	0.017	0.074

digester: pH, alkalinity, weight of suspended solids and gas composition. The gas production was measured daily.

The C.O.D., volatile suspended solids, suspended solids, pH and alkalinity were determined as described in Yorkshire Water Authority's Methods of Analysis.[6] The gas composition analysis using the Orsat analyser was made according to the *Standard Methods for the Examination of Water and Wastewater*, 13th Edition.[7]

RESULTS AND DISCUSSION

The experiment demonstrates that once a suitable bacterial population is established, anaerobic digestion proceeds equally well at all temperatures from 25° to 40°C, inclusively, provided that the population is maintained at the temperature to which it was adjusted. An optimum temperature for the anaerobic digestion of iso-propanol and acetone was not possible to be positively determined but the indications are that it should lie around the 40°C.

The interest in rate models is associated with predicting operating conditions and effluent quality. In order of any prediction to be made, the four basic coefficients Y, K_d, μ_m and K_s must be estimated. Scarcity of data for the kinetic constants and specially for K_s makes such predictions very difficult. Values for Y, K_d and μ_m have been widely researched and have been reported for various environmental conditions. The Monod half-velocity coefficient, however, has not been determined as frequently and less is known either of its absolute value for specific bacterial species or of its dependence on environmental conditions as temperature and pH. The low rate coefficient values are a major reason for the paucity of data and can be used in determining applicability of the process. Because the rate coefficients are so low, obtaining steady state values takes a great deal of time.

The kinetic and growth parameters of the anaerobic digestion of iso-propanol and acetone have been determined at 25°C–40°C (Tables I and II). The estimated growth and kinetic constants reported agree reasonably well with the values reported in the literature[8].

It was proposed previously that the effect of temperature on the various coefficients would follow the Arrhenius law. The Arrhenius equation has been successfully used to describe the effect of increasing temperature on the rate-limiting reaction in biological systems.

The effect of temperature on methane fermentation is best illustrated considering the temperature variation of the maximum specific growth rate. The maximum specific growth rate seems to be a very strong function of temperature, approximately doubles with 10°C increase in temperature, with activation energies for iso-propanol and acetone of +51.4 kJ/mole and +58.9 kJ/mole respectively.

The significance of this is that more stable and fast fermentation occurs at the higher temperature compared with the lower temperatures. It is possible to operate an anaerobic system at a lower temperature with the same efficiency as one at higher temperature by increasing the solids retention time (SRT), e.g. by doubling the solids retention time in the system at 25°C, it should operate at a rate equivalent to one at 35°C with a shorter SRT.

Table II. Kinetic Coefficients of Acetone for Substrate Utilisation and Biological Growth

Temperature °C	K_s (mg COD/1)	μ_m (day $^{-1}$)	K_d (day $^{-1}$)	Y (mg VSS/mg COD)
25	1330	0.046	0.004	0.038
30	2000	0.071	0.005	0.040
35	2380	0.111	0.006	0.043
40	2860	0.140	0.007	0.051

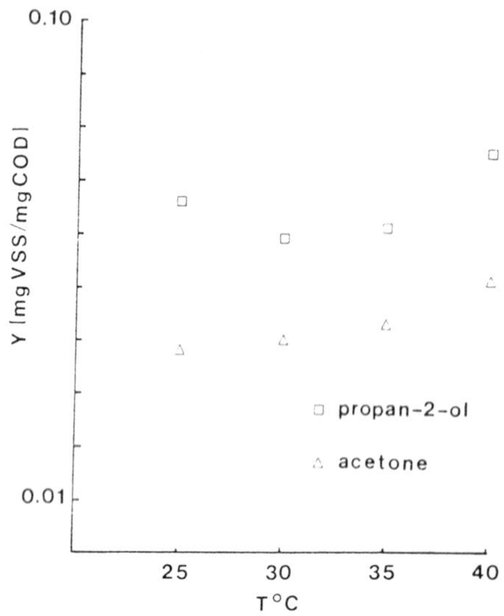

Figure 7. The temperature variation of the yield coefficient.

Very few values of the half-velocity coefficient have been published[8], and very rarely at different temperatures. Two studies have shown that, K_s for acetic acid and propionic acid[9] and also lipids[10] decreases with increasing temperature.

The information gained from the half velocity coefficient cannot be considered to be anymore than that which can be gleaned from the analytical geometry of the hyperbolic curve generated by the Monod equation. The half-velocity coefficient can be considered as some quantitative expression of the sensitivity of μ (or of the bacterial population, expressed through the observed μ) to the concentration of the limiting substrate. The greater the K_s the less affinity of the bacterial population for the specific substrate. In that case, K_s may increase or decrease with increase in temperature depending on the specific substrate and bacterial species. As bacterial species have different temperature optima, it seems reasonable to assume that K_s may decrease or increase with increasing temperature depending on the specific reaction. It is also noteworthy that Golueke[11] reported increased volatile fatty acid (VFA) concentrations in anaerobic digestion above 40°C. The increased VFA concentration indicates that the affinity of the bacterial population to the substrate is decreased with the increase in temperature. This would result in higher value of K_s at the higher temperature. For both iso-propanol and acetone K_s increases with increasing temperature, with activation energies +77.6 kJ/mole and +38.4 kJ/mole, respectively.

Temperature is expected to influence the yield coefficient but not to a great extent, provided that no significant change took part in the metabolism of the bacterial population. This seems to be the case for iso-propanol and acetone as it is shown in Fig. 7. However, the absolute values of the yield for iso-propanol and acetone differ so markedly, that it may be that different bacterial populations were present in each case, especially with their different values for K_s and different values for the energy of activation for the respective K_d values.

The coefficient of endogenous decay, is however, a function of temperature, even not a very strong one. The activation energy for K_d was determined to be +14.6 kJ/mole for iso-propanol and +28.9 kJ/mole for acetone.

Table III. Reaction Mechanism for Methanogenesis from Iso-Propanol and Acetone

	Reactants	Products	$\triangle G^{\circ\prime}$ kJ/reaction
1.	$4CH_3CHOHCH_3$	$4CH_3COCH_3 + 4H_2$	+98.8
2.	$HCO_3^- + H^+ + 4H_2$	$CH_4 + 3H_2O$	-135.6
3.	$4CH_3COCH_3 + 32H_2O$	$12HCO_3^- + 12H^+ + 32H_2$	+713.2
4.	$8HCO_3 + 8H^+ + 32H_2$	$8CH_4 + 24H_2$	-1084.8
SUM 5.	$4CH_3CHOHCH_3 + 5H_2O$	$9CH_4 + 3HCO_3^- + 3H^+$	-408.4

Comparison of the maximum specific growth rates for propan-2-ol and acetone reveals that those are essentially equal, with variations in the range of -2% to +17%. In a series of consecutive steps there will often be one that proceeds much more slowly than all the others because its rate constant is much smaller than those relating to previous or subsequent steps. This one reaction thus determines the rate of conversion of the initial reactants into the final products, and is the rate-limiting step of the whole sequence. Since the overall reaction must proceed at the rate of its slowest stage, the rate of formation of product will the rate of conversion of this limiting step.

The maximum specific growth rates of iso-propanol and acetone are almost identical, considering of course that we are referring to biological systems. The determining step of the anaerobic digestion of iso-propanol seems to be the anaerobic degradation of acetone as the final reaction in a series of steps. Iso-propanol is firstly oxidised to acetone and hydrogen and afterwards acetone is fermented to methane and carbon dioxide. The overall mechanism is presented in Table III, thermodynamic values obtained from reference.[12] It seems that acetone is firstly oxidised to H_2 and CO_2 and subsequently converted to methane, as no intermediate products were detected during its fermentation.

LIST OF SYMBOLS

V = reactor volume (liters)

Q = inlet flowrate (liters/day)

R = recycle biomass flowrate (liters/day)

W = waste biomass flowrate (liters/day)

S_o = influent substrate concentration (mgCOD/l)

S = effluent substrate concentration (mgCOD/l)

X = concentration of biomass in the reactor (mgVSS/l)

X_e = concentration of biomass in effluent (mgVSS/l)

X_o = concentration of biomass in influent (mgVSS/l)

X_r = concentration of biomass in the recycle line (mgVSS/l)

θ = hydraulics retention time (V/Q, days)

θ_s = solids retention time ($X.V/[W.X-(Q-W).X_e]$, days)

μ = specific growth rate (d^{-1})

μ_m = maximum specific growth rate (d^{-1})

K_s = half velocity coefficient (mgCOD/l)

K_d = decay coefficient (d^{-1})

Y = growth yield coefficient (mgVSS/mgCOD)

REFERENCES

1. Browing, E., *Toxicity of Industrial Solvents*, Elsevier Publishing Company, London, 1965.
2. Lettinga, L. G., Van Velsen, L., De Zeeuw, W. and Hobma, S. W., The application of anaerobic digestion to industrial pollution control. In *Anaerobic Digestion*, ed. D. A. Stafford, B. I. Wheatley and D. E. Hughes, Applied Science Publishers, London, 1980, pp. 167–186.
3. Lawrence, A. W., Application of process kinetics to design of anaerobic processes. In *Anaerobic Biological Treatment Processes*, Advances in chemistry series, *105*, ACS, Washington, D.C., 1971, pp. 163–189.
4. Lawrence, A. W. and McCarty, P. L., Unified basis for biological treatment design and operation. *J. of the Sanitary Eng. Division*, ASCE, 1970, *96*, pp. 757–777.
5. Laidler, K., *Physical Chemistry with Biological Applications*, Addison Wesley, 1980, pp. 386–446, pp. 213–221.
6. Yorkshire Water Authority, *Methods of Analysis*, June 1981.
7. American Public Health Association, *Standard Methods for the Examination of Water and Wastewater*, 13th edition, Washington D.C., 1971, pp. 541–546.
8. Henze, M. and Harremoes, P., Anaerobic Treatment of Wastewater in Fixed Film Reactors—A Literature Review. In *Anaerobic Treatment of Wastewater in Fixed Film Reactors*, ed. M. Henze, Pergamon Press, Exeter, 1983, pp. 1–101.
9. Lawrence, A. W. and McCarty, P. L., Kinetics of methane fermentation in anaerobic treatment. *J. of Water Pollut. Control Fed.*, 1969, *41*, pp. R1–R17

10. O'Rourke, J. T., *Kinetics of anaerobic treatment at reduced temperatures*, Ph.D. Thesis, Dept. Civil Eng., Stafford University, March 1968.
11. Golueke, C. G., Temperature effects on anaerobic digestion of raw sewage sludge. *Sewage and Industrial wastes*, 1958, *30*, pp. 1225–1232.
12. Thauer, R. K., Jungermann, K. and Decker, K., Energy conservation in chemotrophic anaerobic bacteria. *Bacteriol. Rev.*, 1977, *41*, pp. 100–190.

45 ANAEROBIC TREATMENT OF CANNERY INDUSTRY WASTEWATERS

Eugenio Foresti, Associate Professor
University of São Paulo
São Carlos School of Engineering
Av. Dr. Carlos Botelho, 1465
13560 – São Carlos – SP – Brazil

INTRODUCTION

The nature of the organic matter of cannery industry wastewater makes them suitable for biological treatment, specially because they rarely present toxicant or inhibitory compounds in their composition. In fact, preparation and operations for processing animal and vegetable raw materials into edible products scarcely make use of chemicals. Thus, wastewater characteristics are mainly related to the nature of the organic matter processed. Normally, the only operations which necessarily use chemicals are the cleaning and sterilization of equipment, floors, and walls. The use of disinfectants and caustic soda at the end of each working cycle results in short-time concentrated discharges, and may lead to operational problems for those wastewater treatment plants which are not protected against shock-loads. However, flow equalization and influent pH control normally have enough diluting and neutralizing effect to permit the use of biological processes for cannery wastewater treatment.

Until recently, aerobic processes have been considered the most convenient method to remove the organic matter from cannery wastewaters. This trend has gradually changed since Young and McCarty[1] successfully made use of anaerobic filters for the treatment of soluble wastes. Following their work, studies concerning the feasibility of anaerobic processes have been initiated all over the world. According to Speece,[2] there are several different anaerobic reactor configurations employed to treat a large variety of industrial wastewaters. Nevertheless, most of the available data refer to lab-scale plants, and an evident lack of information about scale-up criteria, and full-scale reactor performance still persists. This is particularly true in Brazil, where climate conditions are largely favorable to the use of anaerobic processes.

This paper presents some performance data of anaerobic reactors used in the study of cannery wastewater treatment carried out at the São Carlos School of Engineering (Brazil) since 1977.

PREVIOUS STUDIES

The first pilot anaerobic filter for industrial wastewater treatment in Brazil was installed at a meat-packing industry – Frigorifico Bordon – São Paulo.[3] After five months of continuous operation, BOD_5 removal efficiency higher than 70% was achieved for the hydraulic retention time (HRT) of 12 hours (HRT) is always referred to the void volume in this paper). Based on the performance of the pilot filter, full-scale plants were constructed to treat low-strength wastewater from soft-drink and meat-packing industries. Campos et al.[4] reported on five years operation of two full-scale anaerobic filters constructed at a meat-packing industry. COD removal efficiency has been higher than 75% (unfiltered samples) for HRT of 13h. In spite of the fact that the filters are not covered, no aggressive odors have been detected in the treatment area.

Pivelli[5] operated two lab-scale filters used for the treatment of potato processing wastewaters. For organic loading rate (OLR) of 1,75 kg COD $m^{-3}d^{-1}$, and HRT of 24h, BOD_5 removal efficiency higher than 80% was achieved.

In 1983, an intensive research program on wastewater treatment was proposed to a cannery industry – Conservas Alimenticias Hero S.A., in São Carlos, which mainly processes vegetables. It included the design, construction, and operation of three anaerobic reactors at that industry: an anaerobic filter (AF), a fluidized bed reactor (FBAR), and an up-flow anaerobic sludge blanket reactor associated to an anaerobic filter (UASB/AF). For the AF, COD removal efficiency higher

45th Purdue Industrial Waste Conference Proceedings,

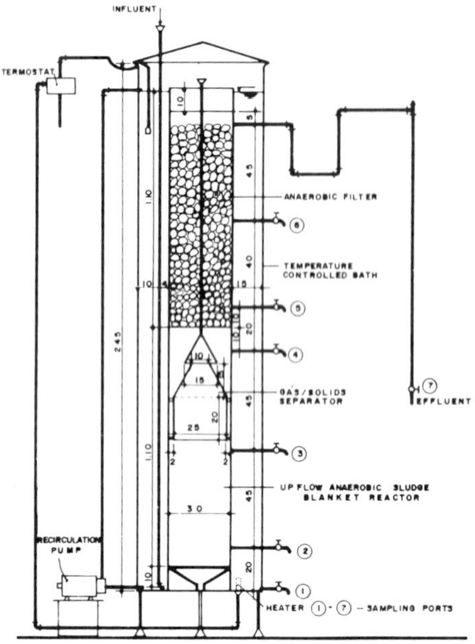

Figure 1. Pilot plant. UASB reactor followed by anaerobic filter.

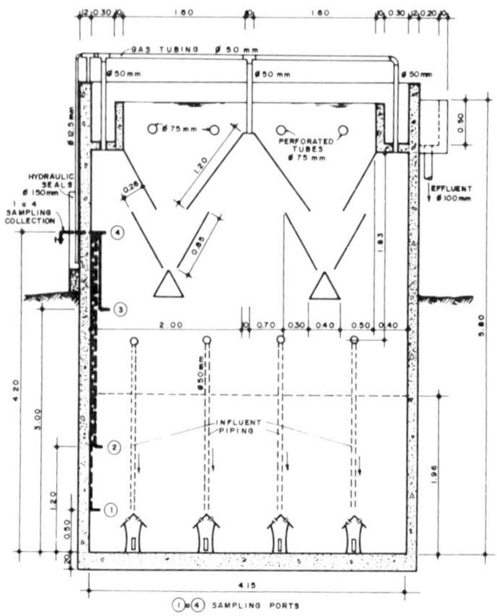

Figure 2. UASB prototype plant. Cross section A.

than 80% was achieved for HRT of 16h, at temperatures ranging from 20 to 31°C.[6] The FBAR operated at HRT of 2h with mean COD removal efficiency of 63%.[7] The UASB/AF achieved mean COD removal efficiency of 80% at HRT of 6h.[8]

Based on the above experience, it was decided to construct an UASB reactor prototype plant, and the start-up of this reactor was fully reported by Camacho.[9]

Some results from the pilot plant and from the prototype plant are presented as follows.

CHARACTERISTICS AND PERFORMANCE

Characteristics of the Reactors

The scheme of the pilot reactor is shown in Fig. 1. It was composed of two reactors sequentially fitted in a vertical position, immersed in a temperature controlled bath (35 ± 2°C). The first unit was an UASB reactor with volume of approximately 80l, while the other was an AF. The total measured volume was 125l.

Fig. 2 shows the schematic drawing of one cross section of the prototype plant designed according to conventional criteria for UASB reactors[10] with a calculated volume of $88m^3$.

Both the reactors were fed with industry wastewaters coming from a primary sedimentation tank. For the pilot plant, the influent was stored in a 200l reservoir where received $NaHCO_3$ for alkalinity, after being pumped to the reactor. The prototype was fed directly by pumping the effluent from the primary tank to the inlet chamber where Na_2CO_3 was added.

The gas/solids separation device of the pilot plant did not function properly since the start-up period, when the gas pressure in the out-let tube was supposed to maintain a free surface within the separator. Poor solids separation and clogging of the out-let tube led to the decision of opening it to the atmosphere. Therefore, a large fraction of the biogas was driven to the stone packing bed. On the other hand, losses of gas in the prototype plant have appeared due to construction imperfections. However, no adverse effect beyond the loss of gas have been verified on the performance of the reactor.

Table I. Operation Phases According to Fernandes[9] and Camacho[10]

Reactor	Phase	HRT (h)	Duration (days)
	1	24.0 (15.4)	75
Pilot	2	12.0 (7.70)	136
	3	6.0 (3.80)	106
			Total 307
	Intermittent feeding	60.0	25
	1	48.0	63
Prototype	2	24.0	38
	Transition	18.0	4
	3	12.0	66
	4	7.5	59
			Total 225

Performance

For both the reactors, operation phases were defined by the HRT applied due to seasonal characteristics of the industry which processes different raw materials at different seasons. Tomatoes, fruits, and grains are used during their harvest period, resulting in significant changes in the wastewater quality. Sharp fluctuations in COD may occur not only from one season to another, but also from one day to the other, depending on the availability of raw material, thus making OLR a useless parameter for the definition of operation phases.

Table I shows the operation phases for the pilot and prototype plants, including the values of HRT, and the duration (in days) of each phase. Figures in parenthesis represent the estimated values of HRT for the UASB reactor of the pilot plant, thus excluding the volume of the AF.

Table I shows that the UASB reactor of the pilot plant was operated at very low HRT in the phase 3 (approximately 3, 8h). For the prototype plant the lower HRT was 7,5h.

Table II shows the main results obtained in each phase. COD values refer to average values from raw influent and effluent samples. OLR values were obtained by the product of influent COD and the corresponding volumetric loading rate (VLR), expressed in volume of wastewater treated per day per unit volume of reactor ($m^3.m^{-3}d^{-1}$).

The temperature of the pilot reactor was maintained at $35 \pm 2°C$ during the experimental period, except in the last 30 days of the phase 3, when the heater was switched-off. No evident effects on the reactor performance was verified after the change in the temperature of operation. On the other hand, the prototype reactor was operated at ambient temperature, and the figures in the Table II refer to average values obtained in each phase. Other results (pH, Total Alkalinity, and Volatile Suspended Solids) are plotted in Fig. 3 against VLR.

There were marked differences in the behaviour of the two units as shown in Table II and Fig. 3. For the pilot plant, COD removal efficiency gradually decreases as VLR increased, while the prototype has tended to improve its performance as the time of operation increased. Diverse value was obtained during phase 2, when the highest mean COD removal efficiency was achieved. However, the

Table II. Main Results Obtained in Each Phase

Reactor	VLR ($m3{\cdot}^{m-3}{\cdot}d^{-1}$)	COD (mg/L) Inf.	COD (mg/L) Eff.	E (%)	OLR (kg COD m^{-3} d^{-1})	T (°C)	
	1.0	287	37	87	0.3	35.0 ±	2.0
Pilot	2.0	342	47	85	0.7	35.0 ±	2.0
	4.0	298	57	80	1.2	*	
						Inf.	Eff.
	0.5	495	184	58	0.2	27.5	22.0
	1.0	872	177	76	0.9	29.0	27.0
Prototype	2.0	394	135	66	0.8	30.0	28.5
	3.2	438	126	69	1.4	29.5	29.0

*35.0 ± 2.0 except during the last 30 days.

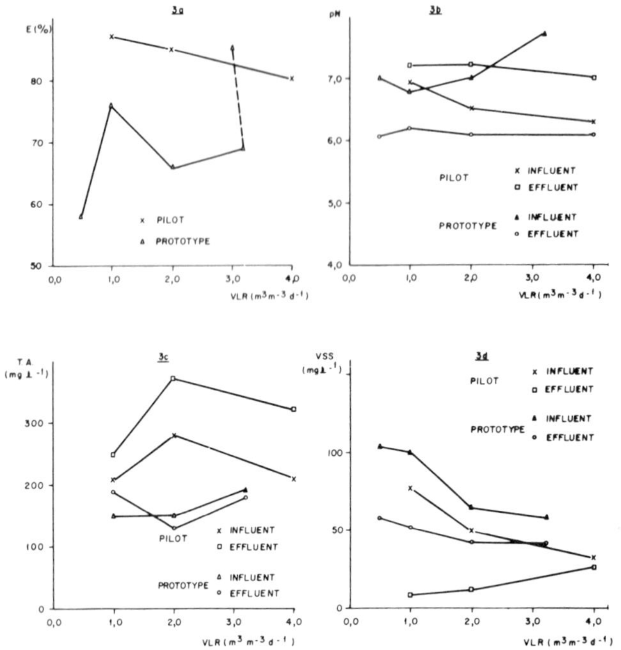

Figure 3. Effect of VLR on the performance of the reactors. Parameters: Fig. 3a. – E (COD removal efficiency); Fig.3b. – pH; Fig. 3c. – Total Alkalinity; Fig. 3d. – VSS.

influent COD was also the highest in that phase. Moreover, while there was no definite pattern for the influent COD, the effluent mean value tended to decrease from one phase to the next during the prototype operation period. It is likely that biomass accumulation and acclimatization can be accounted as the main reason for that performance. On the other hand, the decrease of the mean COD removal efficiency by the pilot plant as VLR increased, was probably caused by losses of solids from the UASB, due to the poor performance of the gas-solids separator. Part of the solids were retained in the AF, and the rest exited increasing the effluent COD.

Mean influent COD differed significantly in both the experiments probably due to previous degradation in the holding tank of the pilot plant. The mean HTR in this tank was estimated in 12h, since it was filled once a day to maintain the system operating 24h. For the pilot plant, low values of influent COD were surely related to pre-acidification, while low values of effluent COD actually reflected the high degree of treatment achieved. As a result, the assimilating capacity of the prototype reactor seemed to be lower than that showed by the pilot reactor. Fig. 3a also shows a recent result of the prototype reactor confirming the gradual increase of the COD removal efficiency, after the start-up period, for VLR of 3.0 $m^3.m^{-3}.d^{-1}$ (HRT = 8h) with COD removal efficiency higher than 80%. Therefore, it can be concluded that the reactor did not achieve the upper limit of its capacity during the experimental period. This shows that start-up of UASB reactors may be very time consuming, particularly when treating low strength wastewaters.

The pH responses of the reactors seem to reflect the different stages of acidification of the influent. For the pilot plant, the addition of $NaHCO_3$ was effective for improving the buffer capacity of the system. This effectiveness could also reflect the high stage of acidification of the influent. On the other hand, the higher influent pH (after the addition of Na_2CO_3) was not maintained in the prototype, probably due to the acidification process within the reactor. Therefore, although the influent pH was maintained close to 7.0, the effluent pH was mostly out of the range considered the best for anaerobic processes. The increase of the wastewater buffer capacity seems to be the main problem to be solved for a reliable operation at the moment. Although the UASB reactor has presented a good performance in general terms, the operation at low pH represents a potential risk of collapse.

Fig. 3c shows that the pilot reactor gained alkalinity during all the phases, while the prototype consumed some of the initial alkalinity. It can be concluded that the prototype did not achieve the best performance during the experimental period.

For the pilot reactor Fig. 3d shows an increase of VSS concentration when VLR increased from 2.0 to 4.0 $m^3.m^3m^{-3}.d^{-1}$, although the influent VSS had decreased from phase 2 to phase 3. Since the VSS content in the unfiltered samples affected the COD values, the decrease of the COD removal efficiency in the last phase partially resulted from the increase of VSS losses. Therefore, since the gas/solids separator did not function properly, the general performance of the reactor was affected. On the contrary, the prototype plant showed a tendency to improve its VSS holding capacity with time, despite the increase of VLR. Thus, it can be concluded that the gas/solids separator was properly designed in that case.

Profiles of the Reactors

Fig. 4 and Fig. 5 show the COD profiles of the pilot and prototype reactors respectively, in different operation phases. The figures in parenthesis are the results of VSS concentration in each sampling port. For the pilot reactor, there are two profiles of phase 3. The first (day 272) resulted from sampling occurred 23 days after the beginning of this phase, and the other (day 314) was obtained 25 days after the heater had been switched-off. For the prototype reactor, the profiles refer to the last sampling of each phase.

For both the reactors, the highest COD removal was achieved in the sludge layer nearer the bottom, where VSS concentration was normally higher than in any other region. In general, the AF of the pilot plant showed a poor performance in the COD removal efficiency, except at the end of the experiment. It can be explained by the low values of VSS concentration within the filter, and by noting that no significant biomass attachment growth occurred. This hypothesis can be admitted since the filter was fed with very low strength influent.

The pilot plant profiles show a progressive loss of VSS from the UASB reactor with the increase of VLR, indicating problems with the gas/solids separator. From the COD profile, it can be admitted that the remaining biomass presented a high specific activity since the reactor performed well even under low VSS concentration. Fig. 4 also shows that the AF had important role in the retention of VSS at high VLR. It probably was responsible for the high COD removal efficiency after the heater had been switched-off (phase 3, second profile).

The prototype reactor profiles (Fig. 5) show that negligible COD removal occurred above the sludge layer, since the VSS concentrations were very low. They also show a very low biomass growth during the start-up as indicated by the values of VSS concentrations from the samples of intermediate ports. From these results it can be inferred that low mixing was probably the main problem during the start-up, since VSS concentrations seemed to be adequate for achievement of a better performance if compared with those found for the pilot reactor.

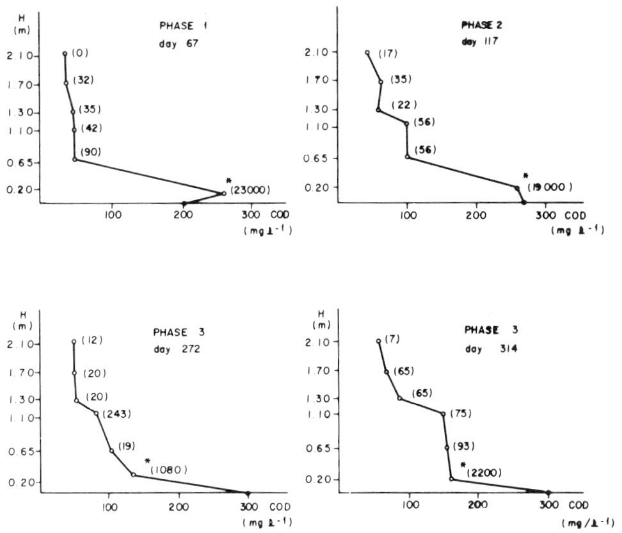

Figure 4. Pilot plant profiles. The numbers in parentheses indicate the values of VSS (mg/L). (0) – not detected in the VSS analysis. (*) – estimated values from the results of TVS.

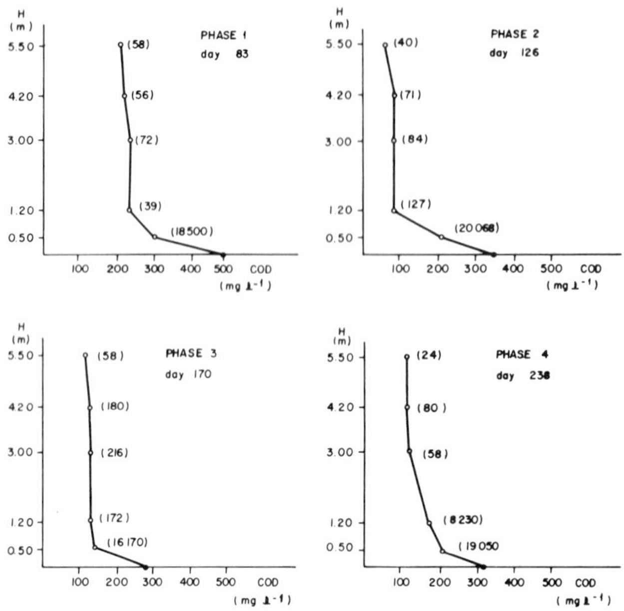

Figure 5. Prototype profiles. The numbers in parentheses indicate the values of VSS (mg/L).

The gas/solids separator of the prototype reactor did function well, and has been responsible for the increase of the sludge height and density. Samples recently collected have shown that the sludge bed is completely granulated along its height.

CONCLUSIONS

UASB reactors are reliable for the treatment of very low strength cannery wastewaters, at temperatures ranging from 20 to 30°C.

In pilot plant studies using a hybrid reactor (UASB/AF), the mean COD removal efficiencies of 87 and 80% were achieved for HRT values of 24 and 12h, respectively. The USAB reactor of the pilot plant presented high efficiency even operating at a calculated HTR of 7.7h (≥80%). Except during the last phase, the AF presented a poor performance probably due to the low substrate concentration resulted from the UASB reactor.

The pilot reactor was operated at 30°C most of time. However, the heater switching-off appeared to have no effect on the system performance.

Studies with the UASB prototype reactor operated at ambient temperature (20 to 30°C) confirmed the results of the pilot plant, although COD removal efficiencies had been smaller during the start-up period. Nevertheless, recent results have shown an increase in the COD removal efficiency (>80%) at HRT of 8 hours.

Significant differences in the behaviour of the pilot and prototype reactors can be attributed to on the conception and modes of operation. Among others, distinct mixing patterns and stages of influent acidification were probably the main factors affecting the performance of the reactors.

The results from sludge layer growth and VSS concentration increase showed that a long start-up period would be necessary for the UASB to achieve its best performance. This is particularly true for full-scale plants. Studies on cannery wastewater nutritional deficiencies have been proposed in order to reduce the start-up period.

ACKNOWLEDGMENTS

The author acknowledges the FAPESP—Fundação de Amparo a Pesquisa de São Paulo for the financial support, and the industry Conservas Alimenticias Hero S.A. for the construction of the reactors, as well as for the facilities provided.

REFERENCES

1. Young, J. C.; McCarty, P. L. Anaerobic filter for waste treatment. *JWPCF*, vol. 41, (5), (1969).
2. Speece, R. E. Anaerobic biotechnology for industrial wastewater treatment. *Env. Sci. Technol.* 17, (9), (1983).
3. Foresti, E. et al. Pilot anaerobic filter for meat-packing wastewater treatment (in Portuguese). Presented at the XVI AIDIS Congress. Dominican Republic (1978).
4. Campos, J. R. et al. Wastewater anaerobic treatment of food processing industry: two study cases. *Wat. Sci. Technol.* 18, (12), (1986).
5. Pivelli, R. P. Studies on the applicability of pilot anaerobic filter for the treatment of potato processing wastewaters. (in Portuguese) MSc. Thesis. São Carlos School of Engineering. The University of São Paulo. Brazil, (1983).
6. Paula Jr., D. R. Performance of a pilot anaerobic filter in the treatment of cannery industry wastewaters. (in Portuguese) MSc. Thesis. São Carlos School of Engineering. The University of São Paulo. Brazil, (1984).
7. Akutsu, J. Biological treatment of cannery industry wastewater using fluidized bed reactor. (in Portuguese) MSc. Thesis. São Carlos School of Engineering. The University of São Paulo. Brazil, (1984).
8. Fernandes, J. A. Applicability of a UASB/AF reactor in the treatment of cannery industry wastewaters. (in Portuguese) MSc. Thesis. São Carlos School of Engineering. The University of São Paulo. Brazil, (1984).
9. Camacho, R. D. P. Performance of a prototype UASB reactor in the treatment of cannery industry wastewaters. MSc. Thesis. São Carlos School of Engineering. The University of São Paulo. Brazil, (1986).
10. Lettinga, G. Design, operation, and economy of anaerobic treatment. *Wat. Sci. Technol.* 15, (8/9), (1985).

46 TREATMENT OF TANNERY WASTEWATER BY UPFLOW ANAEROBIC SLUDGE BLANKET REACTOR

Anwar Khursheed, Research Associate
Rashid H. Siddiqi, Professor
Department of Civil Engineering
Aligarh Muslim University
U.P. India

INTRODUCTION

Kanpur city located on the banks of River Ganga is expected to generate 355 MLD municipal wastewater by the year 2020. The 25 km stretch of the river along the city has been identified as the most polluted reach due to discharge of untreated domestic and industrial wastes under a national "Ganga Action Plan" of management of river water quality. The city has a cluster of about 150 tanneries on the south bank of the river discharging a total of 5.8 MLD wastewater. Under the action plan it is proposed to intercept the waste from the tanneries by an industrial sewer and take it to a common treatment facility where it may be mixed with 130 MLD municipal wastewater and treated through a proposed conventional treatment plant or treated separately after suitable dilution by an Upflow Anaerobic Sludge Blanket (UASB) reactor. The present paper describes laboratory studies conducted to assess the anaerobic treatability of the waste using a bench scale 3.2 L reactor, which paralleled the construction and operation of a 10 m^3 UASB pilot plant.

TANNING PROCESS AND WASTEWATER CHARACTERISTICS

The tanneries at Kanpur process skins to finished or semifinished chrome and vegetable tanned leathers. Some units process only semifinished leather to finished product. The flow from individual units varies from 10 to more than 100 m^3/d and 10 to 30 L/kg of hide processed depending upon the capacity and process used.

An extensive survey was conducted in which discharges from each individual tannery were determined both in terms of quality and quantity. Based on this survey, a composite for the 150 tanneries was prepared by mixing predetermined volumes of waste flows from specific operations of a larger size tannery [Table I].[1]

The characteristics of the composite wastewater prepared according to Table I are given in Table II.

MATERIALS AND METHODS

The laboratory UASB reactor set up is shown in Figure 1. It consisted of a transparent rigid plastic pipe (85 mm i.d. x 630 mm length) with a settling funnel fitted at the top. The gas phase was maintained at 175 mm water pressure by providing a water seal. The feed was pumped in at the bottom of the tank using a peristaltic pump from a feed reservoir which was filled every day. Methane gas was measured by displacement of a 5% NaOH carbon dioxide scrubbing solution in a Marriott flask.

Wastewater composite sample was collected according to Table I and transported to the laboratory. The feed was prepared by settling the waste for two hours and mixing the supernatant with local domestic wastewater to obtain required dilutions.

The reactor was seeded with sludge obtained by settling raw municipal wastewater. Earlier studies had indicated that the sludge obtained in this manner had adequate methanogenic activity. During the start up the reactor was fed an acetate solution of 500 mg/L COD prepared in domestic wastewater at 24 h HRT for about 60 days. After this period when the COD of the reactor effluent had come down to about 30 mg/L, the acetic acid was replaced by settled tannery waste.

45th Purdue Industrial Waste Conference Proceedings,

Table I. Relative Quantities of Pioneer Tannery Waste Streams Collected for Projected Composite from 150 Tanneries[1]

No.	Type of Liquor	Percentage
1.	Soaking and liming (chrome tanning)	14
2.	Soaking and liming (vegetable tanning)	14
3.	Deliming	7
4.	Chrome and rechrome	8
5.	Vegetable tanning	3
6.	Wet finishing	6
7.	Washings	48

Table II. Characteristics of Projected Composite Waste from 150 Tanneries,[1] mg/L Except pH

No.	Parameter	Value
1.	pH	8.4–9.4
2.	Alkalinity (as $CaCo_3$)	2000–2900
3.	BOD, 5 days 20°C	2100–3200
4.	COD	4800–6200
5.	Total solids	26400–36000
6.	Dissolved solids	23000–30000
7.	Suspended solids	3400–6000
8.	Chloride (as Cl^-)	11000–13000
9.	Sulfate (as SO_4)	1700–2750
10.	Chromium (as Cr)	170–225
11.	Sulfide	80–150

Methanogenic activity tests were made in serum bottles containing about 1.5 g VSS (microoganism), 1000 mg acetate COD and other inorganic nutrients in a total of 500 mL reaction mixture. Unacclimated methanogenic organisms were obtained from a laboratory digester maintained on a synthetic medium containing acetic acid as carbon source.

All analyses were performed according to *Standard Methods*.[2] Chromium was determined by AAS.

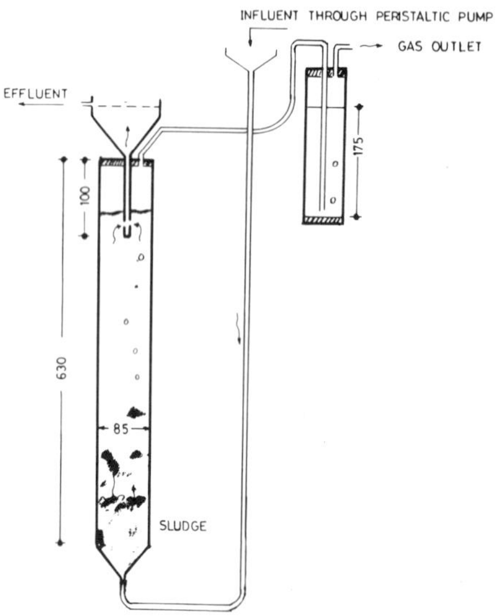

Figure 1. Laboratory UASB reactor (all dimensions in mm).

RESULTS AND DISCUSSION

Reactor Performance

The reactor was operated for a total period of 340 days. After 250 days of operation it had to be stopped due to non-availability of waste. It was started again after six months. Once the feed was re-started, it could be revived in 4 days indicating that the biomass had maintained its viability during the long period of shutdown.

The proportion of tannery waste in the feed was increased in three intermediate steps from 1 in 20 to 1 in 2 resulting in influent COD variation from 450 to 2650 mg/L. The reactor was operated at HRT of 24 h for all dilutions except for the last one when it was operated at HRT of 12 h. The proportion of the waste in the feed was stepped up once the reactor had reached steady operation as evidenced by consistent effluent COD concentration. It is realized that the system may not have reached steady state as there was no measurement of biomass accumulation in the reactor and that expelled with the effluent.

Figures 2 and 3 show the influent and effluent COD values for 1 in 6.7 and 1 in 3 dilutions of the waste at 24 h HRT and for 1 in 2 dilution of the waste at 12 h HRT, respectively. The periods during which the reactor was considered to be operating at steady state are also indicated. It is seen that when the input COD concentration was increased from 780 to 1375 mg/L at the same HRT of 24 h, it took about 30 days for the reactor to adjust to a steady performance [Figure 2]. When the input COD concentration was increased to 2650 mg/L, simultaneously with a decrease in HRT from 24 h to 12 h, the period required for adjustment to steady performance was about 60 days [Figure 3]. This longer duration could also be due to the fact that the system was not operating for 6 months prior to this period.

The steady state COD reductions at various dilutions of the waste against influent COD concentration and volumetric COD load on the reactor are shown in Figures 4 and 5, respectively. The fact that at the lowest loading a maximum of 80% reduction in COD was obtained, and that the COD reductions were about 80% both at low and high loadings indicates that there was no toxic effect. The reactor could be operated satisfactorily at a volumetric loading of 5.25 kg COD/m^3.d. Further, it is concluded that 80% of the COD of the settled tannery waste is biodegradable through anaerobic reactions.

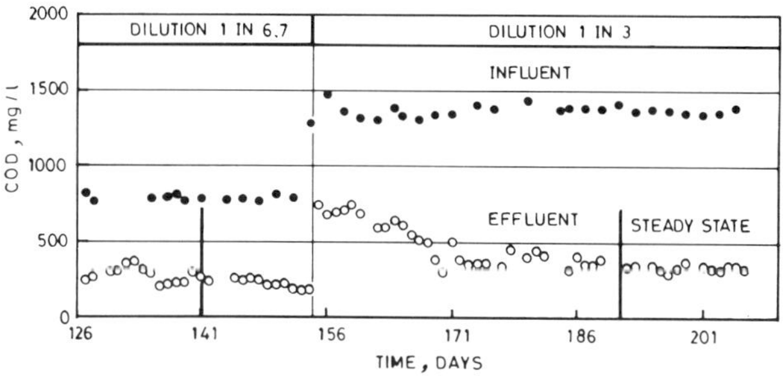

Figure 2. Response of bench scale UASB to 1 in 6.7 and 1 in 3 tannery waste dilution, HRT 24 hr.

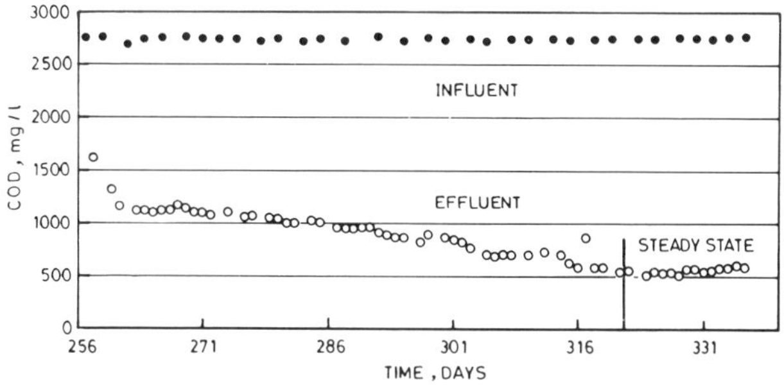

Figure 3. Response of bench scale UASB to 1 in 2 tannery waste dilution, HRT 12 hr.

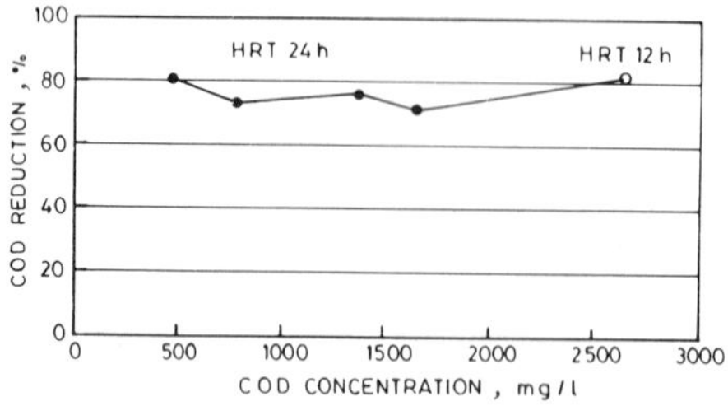

Figure 4. Response of UASB to changing influent concentration.

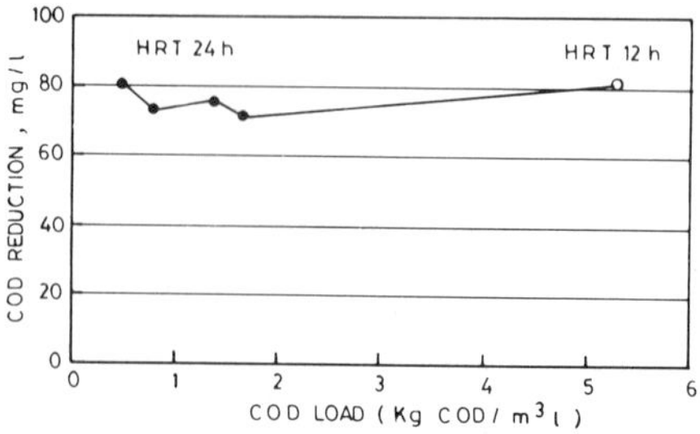

Figure 5. Response of UASB to changing load.

Toxicity of Tannery Wastewater

Table III gives average concentrations of constituents of influent and effluent of the reactor during various steady state periods, which may individually or in combination be toxic to methanogenesis.

Tannins: Arora and Chattopadhya,[3] while studying treatment of vegetable tanning liquor containing 4,500 mg/L tannins by anaerobic filter operated at 12–24 h HRT, concluded that 70 to 90% tannin reduction was possible. Field and Lettinga,[4] using pure gallotannic acid, showed that 700 mg/L concentration of the acid resulted in 50% inhibition of methanogenesis. Data of Table I show that 72 to 87% reduction of the influent tannins in the combined waste was possible. Toxicity studies with unacclimated methanogenic culture and combined tannery waste as a source of tannins showed 50% decrease in activity when the tannins concentration was 325 mg/L [Figure 6]. It is therefore concluded that the tannins in waste would not cause toxicity to anaerobic treatment systems and can be degraded as long as it contains an acclimated biomass.

Chromium: Jackson-Moss, *et al.*,[5] using synthetic feed, showed that methanogenic organisms can be acclimated to trivalent chromium up to a concentration of 5000 mg/L. In the present case the concentration of chromium was much lower and no toxicity was observed. Data of Table I show that at higher influent concentrations 40 to 50% of the chromium was retained in the reactor, probably adsorbed on the sludge or as hydroxide precipitate. Thus with controlled sludge discharge, the reactor would also serve as a chromium separation or concentrating device.

Table III. Influent and Effluent Characteristics During Steady State Periods

		COD, mg/L		Tannins, mg/L		Chromium mg/L		Total Sulfide mg/L		
Dilution	HRT,	Inf.	Eff.	Inf.	Eff.	Inf.	Eff.	Inf.	Eff.	pH Eff.
1 in 20	24	462	87	25	6	26	12	–	–	7.5–8.1
1 in 6.7	24	787	212	52	22	56	39	6	148	7.4–7.9
1 in 3.0	24	1375	325	145	65	84	42	23	152	7.5–7.8
1 in 2.5	24	1662	475	190	116	122	70	7	335	7.3–7.7
1 in 2	12	2625	500	225	138	148	88	29	356	7.0–7.6

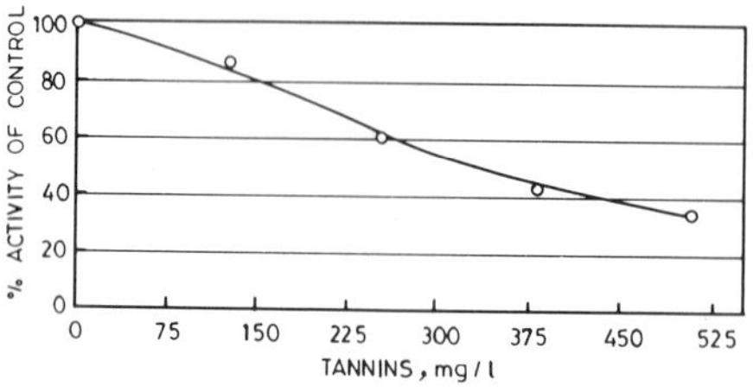

Figure 6. Toxicity of vegetable tanning waste to unacclimated methanogenic organisms pH 7.0.

Table IV. Mass Balance for COD Metabolized

No.	Distribution	mg/d	percent
1.	COD removed	10240	100
2.	Methane COD collected	1028	10.0
3.	Methane COD dissolved in effluent	454	4.4
4.	COD used in $SO_4^=$ reduction	6003	58.6
5.	Unaccounted COD	2755	27.0

Sulfide: Toxicity of sulfides in anaerobic digesters is related to hydrogen sulfide gas pressure in the head space and pH of the liquor.[6] The pH not only influences the concentration of unionized hydrogen sulfide specie, considered responsible for toxicity, but a high pH has a synergestic toxic effect also. Koster *et al.*,[7] showed that while 250 mg/L unionized hydrogen sulfide resulted in 50% inhibition of acetoclastic activity in pH range of 6.4–7.2, only 90 mg/L had similar effect in pH range of 7.8–8.0. The sulfide data of Table III show an increase in sulfides as the waste is treated anaerobically. This is attributed to sulfate reduction. In the case of 1 in 2 dilution of the feed the total effluent sulfide concentration was 356 mg/L. For the pH range of the effluent only about 10-20% of this would be in molecular specie form[7] and therefore would not cause toxicity.

Gas Production and COD Balance

Gas production measurements were made only for a period of 20 days when the reactor was operating at 12 h HRT. Table IV shows a mass balance for COD metabolized during this period. The COD used in sulfate reduction is based on influent and effluent sulfate concentrations during this period.

A part of unaccounted COD may be attributed to net synthesis of biomass and its retention in the reactor. However, this fraction would indeed be very small because of large sludge retention time. There was no intentional wasting of biomass from the reactor.

It may also be noted that more than half of the COD removed was used in sulfate reduction and only about 15% was channelled through mathanogenesis. There is a possibility that if sulfate concentration in the influent increases further without a proportional increase in COD, mathane production may cease completely. This may result in upset of reactor performance as one of the functions of methane in UASB reactor is to gently stir its contents.

Methanogenic Activity of Sludge

The methanogenic activity of the reactor sludge is given in Table V. The decrease in the activity with time may be attributed to a large population of acid producers and sulfate reducing bacteria at higher loadings.

SUMMARY AND CONCLUSIONS

Tannery wastewater which contains high concentrations of organic matter, tannins, chromium and sulfates, needs careful management. Conventionally aerobic systems requiring large power inputs have been proposed for its treatment. Present study shows that UASB reactor which requires little or no energy is a viable option.

By carefully selecting a proper dilution of the waste the system can handle the toxic constituents in the waste. In the present case no toxicity was observed when the reactor was operated at 12 h HRT and 5.25 kg $COD/m^3 \cdot d$, load, giving 81% reduction in COD. The reactor not only stabilizes organic matter but also concentrates chromium in the sludge.

Table V. Methanogenic Activity of Reactor Sludge

Sludge	g COD methane/g VSS.d
Seed	0.24
After 150 days	0.127
After 340 days	0.119

Because of presence of sulfates a large fraction of input COD, as high as 58% in the present case,was metabolized in its reduction. However, there was sufficient methanogenic activity which kept the system well stirred and effective.

The present study was carried out with settled waste. It is envisaged that in a full scale plant, waste with suspended material can also be treated, provided sufficient storage volume for the sludge is provided in the reactor.

ACKNOWLEDGMENT

The investigation reported in this paper was sponsored by M/S Haskoning, Royal Dutch Consulting Engineers and Architects, Indo-Dutch Environmental and Sanitary Engineering Project, Kanpur Mirzapur under Ganga Action Plan.

REFERENCES

1. *Tannery Waste Management Volume II*, Project report, Indo-Dutch Environmental and Sanitary Engineering Project, Kanpur-Mirzapur, under Ganga Action Plan, November, 1988.
2. *Standard Methods for the Examination of Water and Wastewater*, 16th edition, APHA, 1985.
3. Arora, H.C. and Chatopadhya, S.N., "Anaerobic Contact Filter Process: A Suitable Method for the Treatment of Vegetable Tanning Effluents," *Water Pollution Control*, 79, 501, 1980.
4. Field, J.A. and Lettinga, G.,"The Methanogenic Toxicity and Anaerobic Degradability of a Hydrolyzable Tannin," *Water Research 21*, 367, 1987.
5. Jackson-Moss, C.A., *et al.* "The Effect of Trivalent Chromium on Anaerobic Digestion," *JALCA*, 389, *83*, 1988.
6. Lawrence, A.W.,et al, "The Effect of Sulfides on Anaerobic Treatment," *Proceedings 19th Industrial Waste Conference* 343, Purdue University, 1964.
7. Koster, I.W., *et al.*,"Sulfide Inhibition of the Methanogenic Activity of Granular Sludge at Various pH Levels," *Water Research 20*, 1561, 1986.

47 A KINETIC MODEL OF LOW LEVEL ALKALINE SOLUBILIZATION FOR ENHANCED ANAEROBIC DIGESTION

Jih-Gaw Lin, Visiting Assistant Professor
Department of Civil and Environmental Engineering
Washington State University
Pullman, Washington 99164-2910

Bill T. Ray, Assistant Professor

Bruce A. DeVantier, Associate Professor
Department of Civil Engineering and Mechanics
Southern Illinois University at Carbondale
Carbondale, Illinois 62901

INTRODUCTION

Sludge processing accounts for approximately 30 to 40% of the capital costs and 50% of the operating costs in a typical municipal wastewater treatment plant.[1] Both primary and waste activated sludges must be concentrated and stabilized before ultimate disposal. Anaerobic digestion is the most common stabilization method. Since waste activated sludge (WAS) represents about two-thirds of the total sludge volume, any process for improving WAS digestion will significantly reduce overall wastewater treatment plant costs.

It has been recognized that anaerobic digestion occurs in a three step model,[2] shown in Figure 1. In this model, the first stage is acidogenesis, which involves the solubilization of polymers such as cellulose, starch, or proteins, to oligomers. Solubilization is primarily accomplished by extracellular, hydrolytic enzymes produced and excreted by the bacterial population for this specific purpose. Essentially no organic waste stabilization occurs during solubilization. The organic matter is only converted into a soluble form that can pass through bacterial cell walls where it can be utilized for energy or synthesis by bacteria.[2] Once complex organics are solubilized, they are fermented to organic acids such as propionic, butyric, and the longer chain organic acids. For the most part the organic material merely changes form during this first stage. A relatively small fraction is actually consumed as energy. The second stage is acetogenesis, through which acetogenic bacteria convert the longer-chain fatty acids (3C and up) into acetate, molecular hydrogen (H_2), and CO_2, which are the true substrates for the methane formers. In the third stage, acetoclastic methanogens break apart the acetate to form CH_4 and CO_2.

Alkaline solubilization of WAS has been shown to be a potential process for improving the rate of anaerobic sludge digestion.[3] WAS solubilization is also recognized as a rate-limiting step in the overall

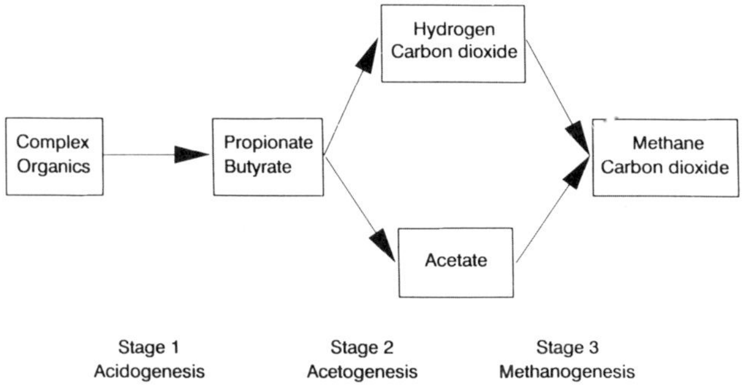

Figure 1. Biochemical model of methane production.

45th Purdue Industrial Waste Conference Proceedings, © 1991 Lewis Publishers, Inc., Chelsea, Michigan 48118. Printed in U.S.A.

digestion process.[4,5,6] The purpose of this study was to develop a simple model which describes the reaction kinetics and also adequately explains the rate-limiting step in the digestion process.

LITERATURE REVIEW

Digestion of WAS, in which most of the substrate is enclosed within cell membranes, requires conversion of cell bound substrate into more readily accessible forms before it can be utilized by viable anaerobes. Autolysis is the predominant means of release for these substances in a natural system. An alternative is to accelerate this process by physical or chemical treatment. The alkaline solubilization process has been investigated for its ability to enhance the anaerobic digestion of WAS.[3,7]

It is generally accepted that only a portion of the sludge added to a digester is biodegradable and that the degradable portion of WAS is comprised primarily of active bacterial cells. Gossett and Belser[4] found that the ultimate anaerobic biodegradability of WAS, measured by volatile solids destruction, is dependent on the solid retention time (SRT) of the activated sludge process. At an SRT of 5 days, the WAS was 60% biodegradable but at an SRT of 30 days the sludge was only 40% biodegradable. The ultimate biodegradability was determined by using an extended digestion time of 128 days. The authors estimated that the ultimate biodegradability of the sludge was 68% as the SRT approached zero.

Stuckey and McCarty[8] estimated the ultimate anaerobic biodegradability of WAS to be 48 to 53% of the COD under mesophilic conditions. Pavlostathis and Gossett[9] estimated that the ultimate WAS biodegradability ranged from 51.7 to 61.5% at a SRT of 10 days. Acetic acids were found to be only 90%degradable at a temperature of 35°C and a hydraulic retention time (HRT) of 30 days.[10]

Several investigators have studied the kinetics of anaerobic digestion of WAS. Gossett and Belser[4] conducted a laboratory study of the kinetics applicable to anaerobic digestion of WAS. They proposed a model which considers the rate of conversion of potential substrate to available substrate as being first-order with respect to the concentration of viable WAS organisms in the anaerobic digester. This specific availability rate was determined to be 0.22 day^{-1}.

Based on the conceptual model of biological sludge digestion, Pavlostathis and Gossett[9] developed and evaluated a comprehensive kinetic model for the continuous-flow stirred reactor without recycle, at steady state. The model includes preliminary conversion mechanisms, acidogenesis, and methanogenesis. The preliminary conversion mechanisms include death/lysis, followed by solubilization to available substrate for the anaerobic digester microflora. Both the death/lysis rate of viable cells and the cellular solubilization rate are considered to be first-order. The death/lysis rate constant was found to be 2 day^{-1}, and the solubilization rate constant was 0.15 day^{-1}. Microbial growth rate and substrate utilization rate in acidogenesis and methanogenesis are based on the Monod equation. In addition, Pavlostathis and Gosset[9] derived a set of steady-state equations which can be used to describe the performance of a continuous-flow completely-mixed anaerobic digester without recycle and to predict its effluent quality.

The results of the studies[4,5,7,9] dealing with the anaerobic digestion of WAS have indicated that solubilization of particulates is the slowest step and therefore controls the overall anaerobic digestion process. In contrast, the rate limiting step for primary waste sludge is the conversion of volatile acids to methane.[11]

Kinetic models have been developed for traditional anaerobic digestion of WAS by Gossett and Belser,[4] and Pavlostathis and Gossett.[9] There is a need to develop similar models for anaerobic processes which utilize solubilized WAS.

REACTION MODEL

The model presented here is intended to describe, in part, the kinetics of a high-rate digester fed solubilized WAS. It is hoped that the model will also provide some insight into the limitations of such systems. It is not intended to be used as a design tool for the development of a full scale high-rate reactor.

Three different feed sludges were used in this project. The control sludge (1% TS), without pretreatment, was fed into the first reactor. The second reactor received a sludge solubilized with 20 meq/L of calcium hydroxide, and a third reactor received the sludge solubilized with 20 meq/L of sodium hydroxide. The reactors were operated at temperature at 35°C and at HRTs of 20, 13, 10, 7.5, and 5 days. The performance of these reactors has been presented elsewhere.[3]

The reactors receiving the pre-solubilized sludges yielded higher organic removals than the reactor receiving the control sludge. This implies that the availability of soluble substrate is the rate limiting

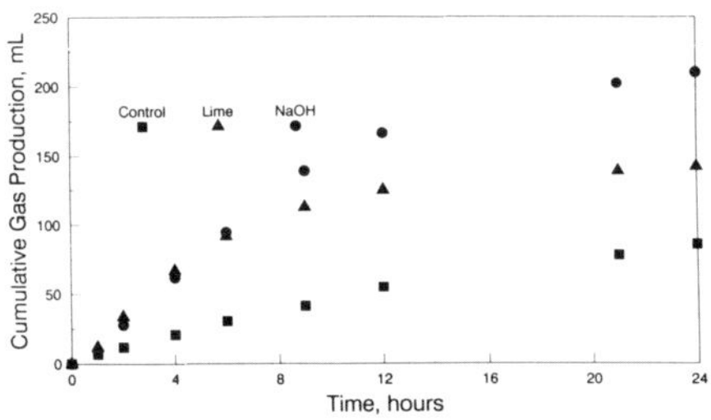

Figure 2. Cumulative gas production versus time during feeding cycle for 10 day HRT.

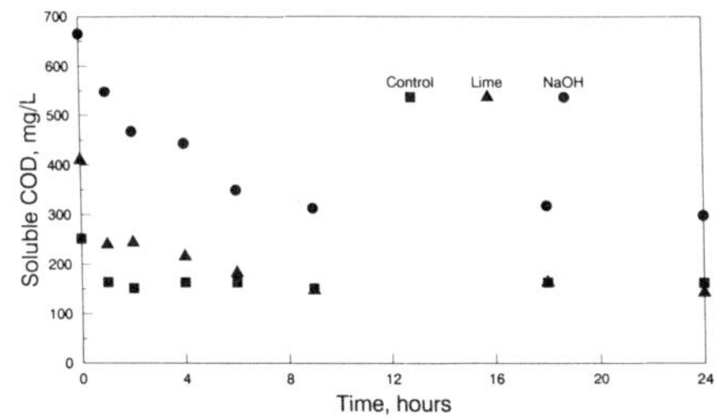

Figure 3. Soluble COD versus time during feeding cycle for 10 day HRT.

step in this digestion process. In addition, the nature of the soluble COD and gas production curves during a feed cycle (see Figures 2 and 3) indicates that beyond approximately the first six hours after feeding, there is no substantial further reduction of soluble COD while the gas production continues to proceed at some slow, but steady rate. This suggests that biodegradable soluble COD is continuously produced through sludge solubilization at a rate equal to the biodegradable soluble COD being converted to methane gas.

Model Generation

As shown in Figure 1, anaerobic digestion is generally described as a three-step process. This process can be represented by the following lumped reactions:

$$A \xrightarrow{k_1} B \xrightarrow{k_2} R \xrightarrow{k_3} S \qquad (1)$$

where: A = complex organics
B = acetate precursors
R = acetate
S = methane

Using the concept of this three-steep process, a model can be developed to predict the gas production as a function of time. This model represents the major complex microbiological reactions, although conversion of complex organics to methane is a multistep process. Also, this model accounts only for the methane produced by the acetate path. However, approximately 70% of the methane formed in an anaerobic digester comes from this pathway.[12]

It is further assumed that the conversion of acetate to methane is very fast in relation to the conversion of the acetate precursors to acetate. This assumption can be supported by two observations. First, the organic acid concentration in the reactor was usually below 100 mg/L as acetate, indicating that organic acids were being consumed as rapidly as they were formed. Second, O'Rourke[11] found that the maximum specific rate of substrate utilization for acetate was 6.7 g BOD/g VSS-day. With VSS concentrations in the reactors in the range of 5 to 10 g VSS/L, this would suggest that k_3 is much greater than either k_1 or k_2. Thus, the model can be simplified to the form:

$$A \xrightarrow{k_1} B \xrightarrow{k_2} S \qquad (2)$$

Several simplifications and assumptions were made in developing this model. The model assumes first order, irreversible, series reactions. It is also assumed that all variables are expressed on a COD unit basis. The complex organics, A, can be approximated by the particulate (nonfilterable) COD, and that the acetate precursor, B, can then be approximated by the soluble (filterable) COD. Thus, the concentrations of the reactants and products can be approximated by:

where: A = particulate (nonfilterable) COD, mg/L
B = soluble (filterable) COD, mg/L
S = methane COD, mg/L

The reaction rates are given as follows:

$$\text{where:} \quad r_A = k_1C_A \tag{3}$$
$$r_B = k_1C_A - k_2C_B \tag{4}$$
$$r_S = k_2C_B \tag{5}$$

The initial conditions are

$$\text{where:} \quad C_A = C_{A0} \tag{6}$$
$$C_B = C_{B0} \tag{7}$$
$$C_S = 0 \tag{8}$$

For a batch reactor of the type used the reaction rate varies with time as follows:

$$r_A = -\frac{dC_A}{dt} \tag{9}$$

$$r_B = \frac{dC_B}{dt} \tag{10}$$

$$r_S = \frac{dC_S}{dt} \tag{11}$$

Elimination of the reaction rate, r_A, from Equation 3 and 6 yields a differential equation, the solution of which is:

$$C_A = C_{A0}\exp(-k_1t) \tag{12}$$

Similarly, the concentration of B is:

$$C_B = [k_1C_{A0}/(k_2-k_1)]*[\exp(-k_1t)-\exp(-k_2t)] + C_{B0}\exp(-k_2t) \tag{13}$$

Finally, using the mass balance equation, the concentration of S is given by:

$$C_S = C_{A0}[1-\exp(-k_1t)] - [k_1C_{A0}/(k_2-k_1)]*[\exp(-k_1t)-\exp(-k_2t)] + C_{B0}[1-\exp(-k_2t)] \tag{14}$$

Comparison with Experimental Data

Using the equations and assumptions discussed previously, the model was fit to the gas production data for each of the three reactors at each of the four HRTs. In the first step, it was assumed that 70% of COD (both soluble and particulate) added to the reactor was anaerobically biodegradable,[2] and that the soluble COD in the reactor prior to feeding was inert. A non-linear regression was used as a numerical analysis technique to determine the "best fit" values of the kinetic constants. Thus, by using a non-linear regression technique, the kinetic parameters k_1 and k_2 for each reactor at each retention time were determined from the batch gas production data. Unfortunately, the model using these k values could not adequately describe the gas production, as can be seen in Figure 4. Physically, the gas production (C_S) in Equation 14 at infinite time should be the sum of the initial particulate COD (C_{AO}) and soluble COD (C_{BO}). For Reactor 2 at the 20 day retention time the prediction of gas production (C_S) would be 4290 mg/L COD, e.g., from the sum of C_{AO} and C_{BO}, when time is infinite. However,

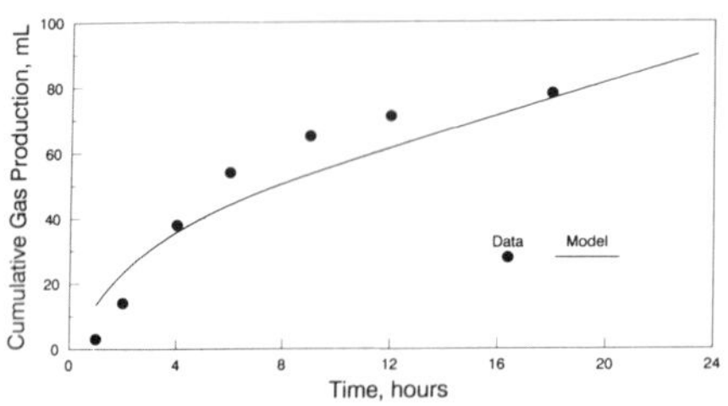

Figure 4. Comparison of model with 20 day HRT data with lime for C_{AO} determined by chemical analysis.

this value is clearly much higher than the gas production measured during the 24 hour test (see Figure 4). Similar predictions were obtained for the other retention times and also for the other two reactors.

From this failure came the concept that the C_{AO} could be divided into a long term and a short term fraction. This is illustrated by Figure 5. As can be seen, the short term C_{AO} is obviously lower than the long term C_{AO} which was measured as 70 percent of the influent total COD. Since there is no physical test which can determine the short term C_{AO}, it must be treated as a fitting coefficient (just as k_1 and k_2) in the regression analysis.

It might be assumed that C_{BO} could also be similarly divided into long term and short term fractions. This was not done for several reasons. First, it is the sum of initial components A and B:

$$C_{UO} = C_{AO} + C_{BO}$$

which defines the ultimate gas production. Thus increasing C_{BO} will force C_{AO} to be smaller for a given C_{UO}. The mathematical consequence of this fact is that if a non-linear regression method is required to determine the best fits values of both C_{AO} and C_{BO}, it will fail to converge due to the large number of C_{AO}, C_{BO} pairs which produce the same C_{UO}. Secondly, from a more physical perspective, it should not be expected that chemically solubilized substrate (with lime or NaOH) should exhibit long term degradation kinetics characteristic of the production of intermediate forms. Following this argument, then, the control reactor should exhibit a different kinetic response (in terms of k_1 and k_2), when the model is fitted with C_{BO} as an experimentally determined value and C_{AO} determined as a fitting parameter.

The batch feeding cycle used in this project was 24 hours for the 20 day, 13 day, 10 day, and 7.5 day HRTs, and 12 hours for the 5 day HRT. The tendency toward a plateau in gas production on the order of days indicates that C_{AO} should be a short term value. This second step was similar to the first step except that C_{AO} was treated as an unknown. The kinetic parameters k_1 and k_2, as well as C_{AO}, were determined. This model yielded better results (see Figure 6) than the first step (see Figure 4). For this second step, the values of k_1, k_2, and C_{AO} are shown in Table I.

In a third step, the values obtained for C_{AO} were used to find the universal k_1 and k_2 values for each reactor. It should be expected that the variation of k_1 and k_2 seen in Table I is due to statistical

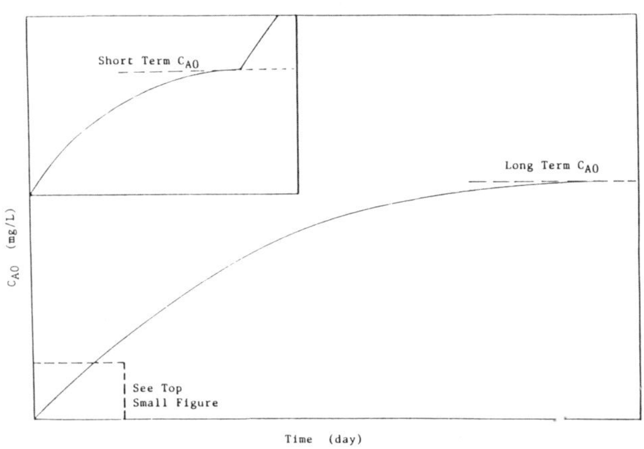

Figure 5. A concept of long term and short term C_{AO}.

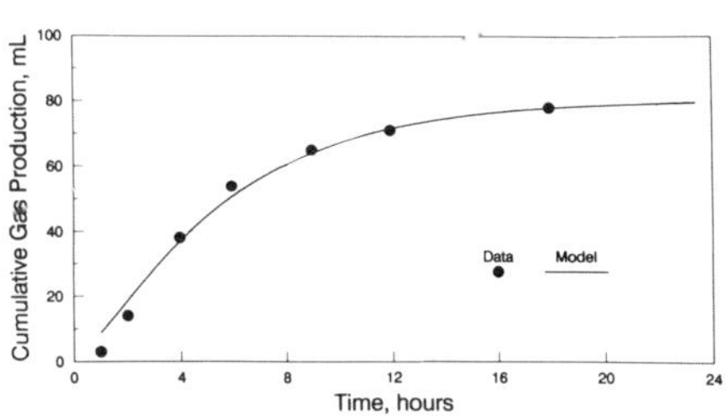

Figure 6. Comparison of model with 20 day HRT data with lime for C_{AO} determined by numerical analysis.

Table I. The Values of k_1, k_2, and C_{A0} from the Second Step

	Reactor 1 Control			Reactor 2 Lime			Reactor 3 NaOH		
HRT days	k_1 hr^{-1}	k_2 hr^{-1}	C_{A0} mg/L	k_1 hr^{-1}	k_2 hr^{-1}	C_{A0} mg/L	k_1 hr^{-1}	k_2 hr^{-1}	C_{A0} mg/L
20.0	0.039	1.37	370	0.62	0.20	94	0.88	0.15	114
13.0	0.034	20.3	358	1.10	0.21	79	0.50	0.14	146
10.0	0.041	24.6	278	0.92	0.19	195	0.38	0.13	212
7.5	0.031	1.80	281	0.42	0.12	133	0.35	0.06	271
5.0				1.01	0.11	10	0.90	0.08	423
Avg.	0.036	12.0		0.81	0.17		0.60	0.11	

Table II. The Values of Kinetic Parameters from the Third Step

Kinetic Parameters	Reactor 1 Control	Reactor 2 Lime	Reactor 3 NaOH
k_1 (hr^{-1})	0.036	1.64	1.15
k_2 (hr^{-1})	3.35	0.14	0.08

variation. Thus if all the data sets for a given reactor are combined, the best fit estimates of k_1 and k_2 can be made for the reactor type. The average values for k_1 and k_2 found in step 2 were used as initial values. However, in order to obtain convergence, the k_2 value used for Reactor 1 was 1.60 hr^{-1} instead of the 12 hr^{-1} found in step 2. The values determined for k_1 and k_2 for each of the three reactors are shown in Table II. The predicted gas production is compared with the observed data in Figure 7 (a) and (d) for four HRTs.

By comparing Table I and Table II, it is noted that the value of k_2 (3.35 hr^{-1}) for Reactor 1 is approximately twice as high as the initially estimated value of k_2 (1.6 hr^{-1}), but k_1 (0.036 hr^{-1}) does not change. In contrast, the values of k_2 (0.14 hr^{-1} for Reactor 2 and 0.08 hr^{-1} for Reactor 3) are similar to the initial estimates of k_2 (0.17 hr^{-1} for Reactor 2 and 0.11 hr^{-1} for Reactor 3), but the values of k_1

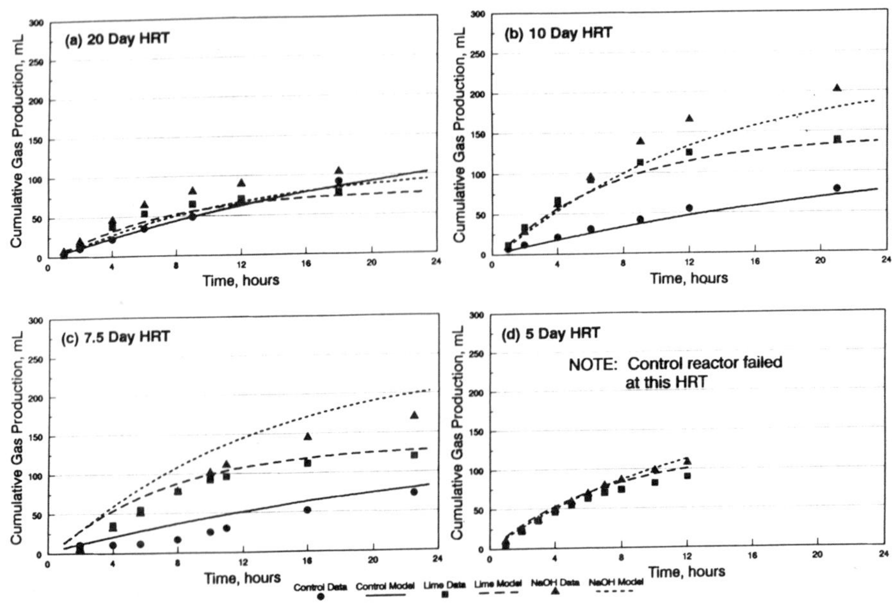

Figure 7. Comparison of model with actual gas production.

(1.64 hr^{-1} for Reactor 2 and 1.15 hr^{-1} for Reactor 3) are approximately twice as high as the initial estimated values of k_1 (0.81 hr^{-1} for Reactor 2 and 0.60 hr^{-1} for Reactor 3). This indicates that the solubilized sludges are different kinetically from the sludge which is not chemically pretreated, and that k_1 is more reliably predicted for the control as compared to k_2 being more reliably predicted for the solubilized sludges.

The k_1 value 0.036 hr^{-1} for Reactor 1 (control) is lower than the k_2 value for 3.35 hr^{-1} (Table II). A plausible physical explanation is that the conversion of soluble substrate to methane precursors and those precursors to methane are rate limited by biomass if sufficient substrate is available. For the reactors fed the presolubilized waste activated sludges, the k_1 values are higher than the k_2 values. This would indicate that the pretreatment of waste activated sludge actually improves the first step in the reaction and further stimulates the biological conversion which is occurring in the anaerobic reactor. Thus, if the model is correct for Reactor 1, the rate limiting step is not solubilization but acetogenesis (the conversion of acetate precursors to acetate).

CONCLUSIONS

The following are some of the conclusions that can be inferred from this research project:

A kinetic model has been developed which provides some insight into the limiting reactions within the single-stage high-rate reactor during digestion of solubilized WAS. Although the model is not sufficiently verified to be used for design purposes, it correlates well with the laboratory data observed in this study.

The complex organics are approximated by the particulate COD and that the acetate precursors are approximated by the soluble COD. It is observed that the initial particulate COD (C_{AO}) can be further divided into a long term and a short term fraction. The short term C_{AO} is lower than the long term C_{AO} which was measured as 70% of the influent total COD.

The model substantiates the fact that the rate limiting step in the digestion of untreated waste activated sludge is the conversion of a particulate biomass into a soluble substrate, while the rate limiting step in the digestion of solubilized waste activated sludge is the conversion of soluble organics to acetate.

REFERENCES

1. Vesilind, P. A., *Treatment and Disposal of Wastewater Sludges*, Ann Arbor Science Publishers, Ann Arbor, Michigan, 1974.
2. Parkin, Gene F. and Owen, William F., "Fundamentals of Anaerobic Digestion of Wastewater Sludges," *Journal of the Environmental Engineering Division*, ASCE, Vol. 112, No. 5, pp. 867–920, 1986.
3. Ray, B. T., J.-G. Lin, and R. V. Rajan, "Low Level Alkaline Solubilization for Enhanced Anaerobic Digestion," *Journal of the Water Pollution Control Federation*, Vol. 62, No. 1, pp. 81–87, 1990.
4. Gossett, J. M. and Belser, R. L., "Anaerobic Digestion of Waste Activated Sludge," *Journal of the Environmental Engineering Division*, ASCE, Vol. 108, No. EE6, pp. 1101–1120, 1982.
5. Ray, B. T., J. C. Huang, and B. A. Dempsey, "Sludge Digestion by Anaerobic Fluidized Beds: II. Process Kinetic Model," *Journal of the Environmental Engineering Division*, ASCE, Vol. 115, No. 6, pp. 1156–1170, 1989.
6. Eastman, J. A. and Ferguson, J. F., "Solubilization of Particulate Organic Carbon During the Acid Phase of Anaerobic Digestion," *Journal of the Water Pollution Control Federation*, Vol. 53, No. 3, pp. 352–366, 1981.
7. Rajan, R. V., Lin, J.-G., and Ray, B. T., "Improved Anaerobic Digestion with Low Level Chemical Pretreatment," *Proceedings of the 43rd Industrial Waste Conference*, Purdue University, 1989.
8. Stuckey, D. C. and McCarty, P. L., "Thermochemical Pretreatment of Nitrogenous Materials to Increase Methane Yield," *Biotechnology and Bioengineering Symposium*, No. 8, pp. 219–233, 1978.
9. Pavlostathis, S. G. and Gossett, J. M., "A Kinetic Model for Anaerobic Digestion of Biological Sludge," *Biotechnology and Bioengineering*, Vol. 28, No. 10, pp. 1519–1530, 1986.
10. McCarty, P. L., "Anaerobic Waste Treatment Fundamentals: IV, Process Design," *Public Works*, pp. 95–99, December, 1964.
11. O'Rourke, J. T., "Kinetics of Anaerobic Waste Treatment at Reduced Temperatures," Ph.D. Dissertation, Stanford University, Stanford, California, 1968.
12. Lawrence, A. W. and McCarty, P. L., "Kinetics of Methane Fermentation in Anaerobic Treatment," *Journal of the Water Pollution Control Federation*, Vol. 41, No. 2, pp. R1–R17, 1969.

48 UPGRADING THE ANAEROBIC CONTACT PROCESS CAPACITY USING POLYURETHANE CARRIERS

Veysel Eroglu, Associate Professor
Izzet Ozturk, Associate Professor
Department of Environmental Engineering
Istanbul Technical University
80626 Maslak, Istanbul—Turkey

Erol Bircan, Environmental Engineer
Sahan Kimya A.S., Inonu Caddesi No. 70
80090, Gumussuyu, Istanbul—Turkey

INTRODUCTION

In recent years extensive investigations have been carried out all over the world on industrial wastewater treatment. The organic conventional parameters such as BOD, COD and TOC which are significant indicators in wastewater pollution are usually removed by biological methods. Among these methods aerobic biological ones are preferred over the anaerobic systems due to simplicity in their operation and control. However, for strong organic wastewaters anaerobic treatment systems become increasingly preferable over the aerobic methods since anaerobic processes are bioenergy producers. Consequently, anaerobic biological treatment methods are being continuously developed.

The most important points in the anaerobic processes are the preservation of sufficient biomass within the system and the maintenance of necessary contact between the substrate as well as biomass. For this purpose, in recent years it has been practiced to keep the micro-organisms at fixed surfaces by putting carriers into the main reactors.

In this study, the effect of the polyurethane (PUR) carriers was investigated on the anaerobic treatment of wastes from a citric acid plant in Turkey. Since the company has planned to double the capacity of the factory without any extension in the existing treatment plant, two model anaerobic digesters were used to test the feasibility of the PUR carriers for upgrading the existing treatment system.

PRODUCTION OF CITRIC ACID

Citric acid is produced by surface fermentation method from the sugar-beet molasses at the FURSAN factory which is the subject of this research. The yearly production of this plant is 3500 tons composed of citric acid monohydrate, citric acid anhydrate and trisodium citrate dehydrate. Under the suitable conditions the molasses is transformed to citric acid and oxalic acid. With the treatment of lime the oxalic acid is settled as calcium oxalate and hence it is separated from the system. Subsequently, calcium citrate is also settled and filtered. Sulphuric acid is added and the calcium sulphide precipitates are removed from the system.The citric acid solution passes through an active carbon bed, cationic and anionic ion exchangers. Then it is evaporated and crystallized. Finally, the crystals are dried and packed for commercial purposes.

SOURCES AND CHARACTERISTICS OF WASTEWATERS

The wastewaters produced during the citric acid production the wastewater can be classified five groups.

1. Wastewater from calcium citrate filter.
2. Wastewater from the washing of molasses tanks and connections.
3. Wastewater from the activated carbon filter, anionic and cationic ion-exchangers.
4. Cooling waters, and
5. Domestic wastewaters

45th Purdue Industrial Waste Conference Proceedings, © 1991 Lewis Publishers, Inc., Chelsea, Michigan 48118. Printed in U.S.A.

Table I. Flowrates and Characteristics of Wastewaters

Parameters	Sources of Wastewaters 1	2	3	4	5
Flowrate, m^3/d	370	75	75	75	75
COD, mg/L	35,000	—	—	—	—
BOD_5, mg/L	25,000	1,000	1,800	710	150
TKN, mg/L	1,800	90	50–260	30	110
Total P, mg/L	40	—	—	—	—
pH	5.5–7	—	—	—	—
$SO_4^{=}$, mg/L	1,300	—	—	—	—
Ca^{++}, mg/L	1,000	—	—	—	—
T, °C mg/L	45–60	—	—	—	—

The quantities and properties of these waters are presented in Table I.

The specific wastewater flow rate from the factory is 70 m^3 per ton of product. At present, in the plant, concentrated wastewaters are purified in a two stage anaerobic treatment plant. This treatment system is composed of an acid reactor, a neutralization unit and a methane reactor. The last unit is a lamella sedimentation tank. The settled flocs are recycled to the methane reactor . The volumes of acid and methane reactors are 700 m^3 and 2100 m^3, respectively. Presently, the organic load of the treatment plant is about 4 kg COD/m^3 per day. The removal efficiency of COD is 80% . The effluent COD concentration is about 5000 mg/L. The flow diagram for the system is given in Figure I.

The design and performance data for the existing treatment system are presented in Table II.

MATERIAL AND METHODS

The design organic loading rate is about 6 kg COD/m^3.d for the existing two-stage anaerobic treatment plant. However, at present, the treatment system has been operated on an organic loading rate of 4 kg COD/m^3.d. At present conditions, 80% of COD removal can be achieved. Since the company has planned to double the production of citric acid without any extension in the existing anaerobic treatment plant, a treatability study was conducted to test the feasibility of the PUR carriers for upgrading the existing methane digester.

Two model anaerobic methane digesters were used in this study. The volume of each reactor was 10 liters with corresponding diameter of 19 cm and a height of 39 cm . One of the reactors (Reactor A) was filled with 1 liter of PUR carriers which corresponded to 10% of the reactor volume with particle sizes of 3 to 7 mm. The second reactor (Reactor B) was operated as a completely mixed contact digester. Both reactors were kept in a climate room maintaining the temperature at 35 ± 2°C . The flow diagram of the model treatment system is shown in Figure 2.

The effluent from the existing full-scale acid reactor was used as the feed. The seed sludge was supplied from the existing methane reactor. The VSS content and sludge volume index (SVI) are 11

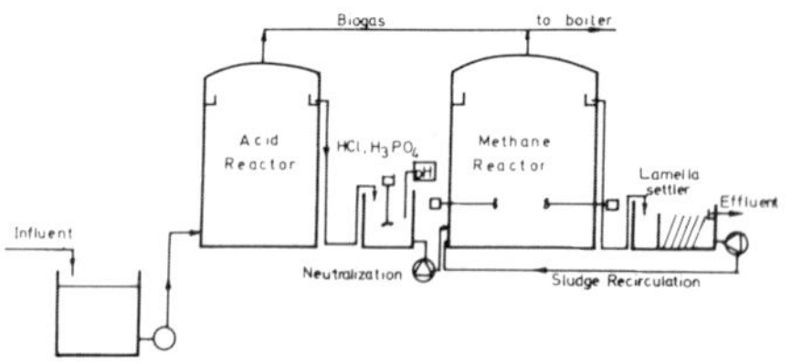

Figure 1. Flow diagram of existing treatment system.

Table II. Design and Performance Parameters for Existing Treatment System

Reactor	Design Parameters HRT (day)	Design Parameters OLR (kg COD/m^3.d)	Performance Parameters HRT (day)	Performance Parameters OLR (kg COD/m^3.d)	Performance Parameters COD Removal (%)
Acid reactor	1.7	21	5.8–4.0	5–7.3	30–20
Methane reactor	5.3	2–6 (6)	17.4–12.0	1.7–2.4	81–78

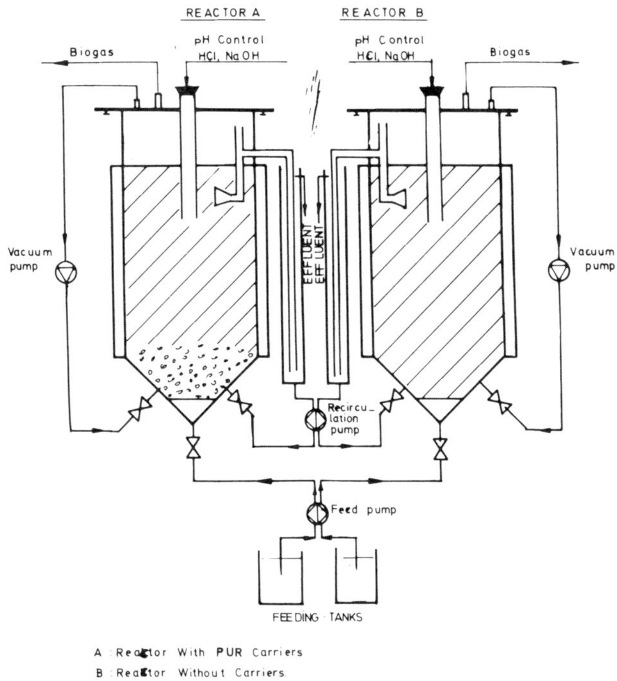

Figure 2. Flow scheme of model reactors.

kg/m^3 and 27 mL/gram, respectively.[2] Since the volumetric loading rate (VLR) was about 4 kg COD/m^3.d in the existing anaerobic treatment system, the model reactors were started up with this organic loading rate (OLR). Diluted HCL or NaOH was added to the reactor directly, in order to maintain a pH of 7.1-7.8. Important operation parameters including pH, COD, volatile acid, suspended solids concentration were analyzed in the influent and effluent of the reactors according to *Standard Methods*.[1] Biogas volumes from the digesters were measured by using water displacement method.

The effluent from the existing acid reactor was diluted with tap water to maintain a COD of about 17.5 gr/L for the start up period. During two weeks of the start up, hydraulic retention time (HRT) and OLR were 2.89–3.75 days and 4.6–5.5 kg COD/m^3.d, respectively. The start up period ended on the 15th day model reactors were fed with the effluent from the full-scale acid reactor without any dilution.

The experimental procedure involved in this study was summarized in Table III. The samples were collected and analyzed at least 3 times per week. Both reactors were operated in parallel untill the 62nd day, after which only Reactor A was continued to reach the maximum OLR.

CHARACTERISTICS OF CARRIER

Causing the organisms to attach on certain carrier materialsis a suitable way of overcoming the problems resulting from the washing-out of organisms that do not readily flocculate. Carrier materials also raise the biomass concentration in the reactor. Polymer based porous carriers have been found as being an interesting solution for retention of the anaerobic sludge in the reactor in the last decade. One of the most popular carriers is polyurethane (PUR) havingactivated carbon inside . The main properties of modified polvurethane carriers are summarized in Table IV.[3]

Table III. Experimental Procedure for Model Reactors

Parameters	Reactor A (With PUR Carrier)	Reactor B (Without Carriers)
HRT, days	1.9– 5.0	2.60– 5.0
Influent COD, g/l	25.3–35.5	25.5 -33.0
OLR, kg/m^3.d	6.6–13.04	6.6 -10.25

Table IV. Properties of Modified PUR Carriers

Variable Property	from–to	necessary for
1. *Physico-chemical*		*Biokinetics*
Ionic charging		Process stability
Hydrophily	(+),(+),(-)	Adherence of the organisms
	-	
Surface activity	adsorbent	Transportation process
2. *Physical*		*Process Technology*
Density (g/cm^3)	1.03–1.10	Circulation and
Particle size (mm)	3–10	Retention of the carrier
Sedimentation velocity (m/h)	40–150	OLR

RESULTS AND DISCUSSION

Substrate Utilization

The anaerobic treatability studies have been carried out for ll5 days including 15 days of start up period. Table V summarizes influent and effluent characteristics including COD, volatile acid and suspended solid concentrations together with parameters such as influent flow rates (Q_f),organic loading rates (OLR), gas production rates (Q_g), COD removal efficiencies (E) and specific gas production yields(Y_g).

The major advantage claimed for PUR carriers is their ability to retain higher biomass concentrations in the reactor which considerably improves the reaction rate per unit volume.Evidence of this may be seen in Figure 3 in which COD removal rates and efficiencies were plotted as a function of COD loadings.

The results were obtained using two parallel reactors under identical operating conditions. In Reactor A, which contains one liter PUR carriers occupying 10% of its volume, a COD removal efficiency of 83% was achieved at an OLR of 12 kg COD/m^3.d which corresponds to a hydraulic retention time (HRT) of about 3.5 days. The corresponding soluble COD removal rate is 8.3 kg COD/m^3.d. A COD removal efficiency of 70% was observed at an OLR of 9 kg COD/m^3.d in Reactor B which contains no carrier. The corresponding HRT and COD removal rates are 3.5 days and 7.0 kg COD/m^3.d, respectively for Reactor B.

Results from this study have shown that using PUR carriers improves the treatment capacity of the anaerobic contact process significantly. Design organic loading rate of existing anaerobic contact reactor can easily be doubled by upgrading the system with PUR carriers.

Biogas Production

One of the most important features of anaerobic treatment is the production of biogas. Biogas production was measured daily during the experimental study. A very strong relationship was observed between the daily gas production rate and the COD loadings as illustrated in Figure 4. The relationship between gas production rates (Q_g) and organic loading rates is linear up to an OLR of about 12 kg COD/m^3.d for Reactor A. A similar relationship exists up to an OLR of about 8 kg COD/m^3.d for Reactor B. Specific gas production yields (Y g) were about 0.5 liter per kg COD removed for both reactors at the steady-state operating conditions. CO_2 contents of the gas produced were also determined from time to time for both experimental plants. Average methane contents of the biogas were in the range of 75 to 80% in both reactors.

System Efficiency

Considering the n-th order substrate (COD) removal kinetics,anaerobic treatment process can be expressed by the following equation:

$$-\frac{dS}{dt} = k_{(n,s)}.X_o \left(\frac{S}{S_o}\right)^n \quad (1)$$

where: $K_{(n,s)}$ = the n.th order substrate removal kinetic constant (1/d)
X_o = average biomass concentration in the reactor (mg/L)
S_o = influent COD concentration (mg/L), and

Table IV. Summary of Waste Characteristics and Operating Parameters

Time (day)	Q± (L/d)	HRT (d)	Inf. COD S_o gCOD/L	OLR kg COD/m^3.d	Volatile Acid (VA) gHAC/L	Reactor A COD(S) g/L	Reactor A VA g/L	Reactor A SS g/L	Reactor A COD %	Reactor A Q_g L/d	Reactor A Y_g L/gCODr	Reactor B S g/L	Reactor B VA g/L	Reactor B SS g/L	Reactor B COD %	Reactor B Q_g L/d	Reactor B Y_g L/g
1–7	2.66	3.75	17.25	4.59	—	6.36	4.02	0.84	63	—	—	8.16	4.10	0.77	53	—	—
8–14	3.45	2.89	16.0	5.52	5.7	3.65	1.76	0.39	77	—	—	5.65	2.4	0.55	65	—	—
15–22	2.0	5.0	33.0	6.60	11.0	4.56	1.37	0.35	86	26	0.46	10.75	5.33	0.70	67.4	20	0.45
23–39	2.83	3.53	29.06	8.22	8.8	6.03	1.60	0.84	79	35	0.54	9.50	4.0	0.75	67.3	30	0.54
40–52	3.27	3.06	25.5	8.34	10.8	6.0	1.50	—	76.5	32	0.50	6.15	1.5	0.80	76	30	0.47
53–61	3.84	2.60	26.7	10.25	11.4	4.73	1.20	1.0	82	42	0.50	7.3	3.4	1.2	73	23	0.31
62–77	4.0	2.47	28.5	11.4	11.8	4.8	1.40	—	83	44	0.47						
78–87	3.6	2.78	28.0	10.1	11.4	4.65	1.60	0.85	83	—	—						
88–100	5.0	2.0	25.3	12.65	11.1	4.9	1.6	0.6	81	52	0.51						
101–111	5.3	1.9	24.6	13.04	—	6.3	3.3	—	74	47	0.49						
112–115	3.6	2.78	35.5	12.78	—	5.5	2.8	—	~85	—	—						

S = Effluent COD, Q_g = Gas flow rate, Y_g = Specific gas yield, OLR = organic loading rate, HRT = Hydraulic retention time.

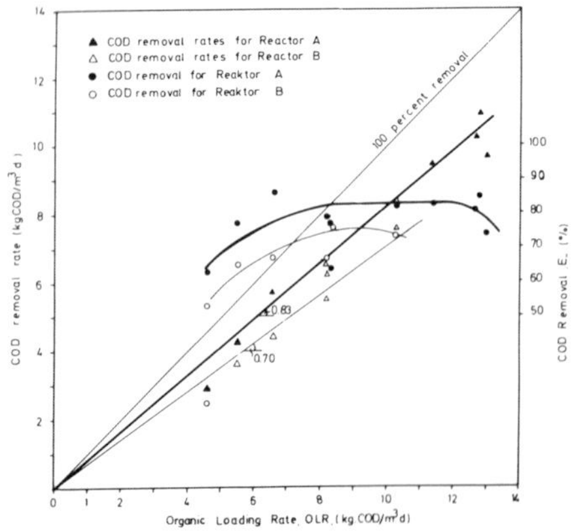

Figure 3. OLR vs. COD removal rate and system efficiency (E).

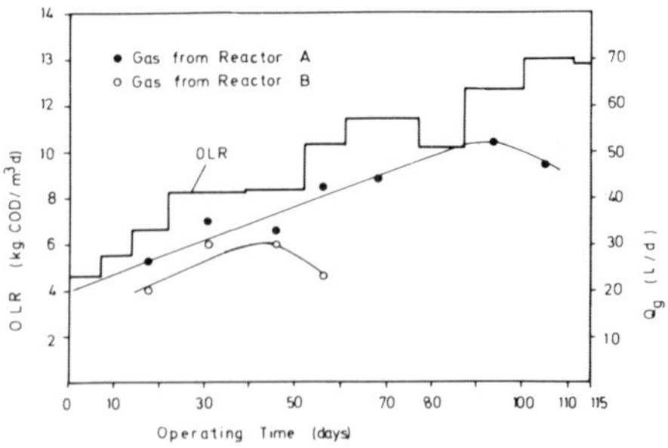

Figure 4. Operating time vs. OLR and gas production rate.

S = effluent COD concentration (mg/L)

Integrating Eq.1 for n = 2 yields the following linear expressions:

$$\frac{\Theta}{E} = a + b.\Theta \tag{2}$$

or

$$\frac{S_o}{L_{OH4}} = a + b.\Theta \tag{3}$$

where: E = the system efficiency, $(S_o - S)/S_o$
Θ = the hydraulic retention time (d)
L_{OH4} = CH_4conversion efficiency in terms of COD equivalent

The results presented in Table V for Reactor A were plotted in Figure 5. This figure shows that COD removal efficiencies can be expressed by the following linear equation for OLR's of less than 13 kg COD/m^3.d:

$$\frac{E}{\Theta} = 1.143\ \Theta + 0.25 \tag{4}$$

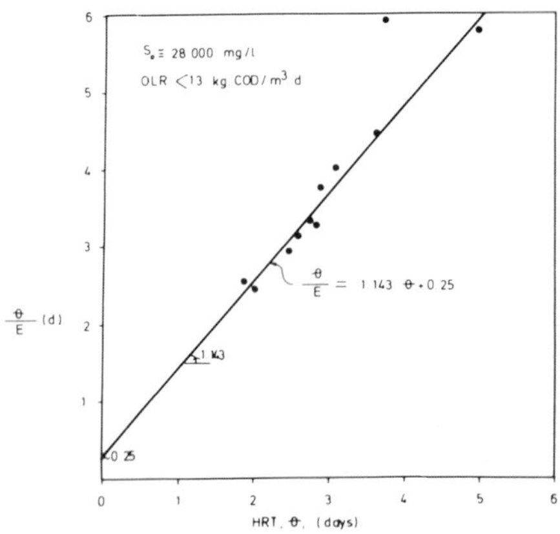

Figure 5. HRT vs. HRT/E for reactor A.

Equation 4 can be used to predict the process performances for PUR carrier assisted anaerobic digester (Reactor A) at various operating conditions.

CONCLUSIONS

Using PUR carriers for microorganism attachment is a suitable way to overcome the problems resulting from the washing-out of the biomass and for upgrading the existing anaerobic contact digesters. Results from this study have indicated that design organic loading rate of existing anaerobic contact reactor can easily be doubled by upgrading the system with PUR carriers. A linear relationship exists between gas production rates and organic loading rates up to 12 kg COD/m^3.d. Specific gas production rates are about 0.5 liter per g COD removed. COD removal kinetics can be expressed by the second order multiple substrate kinetics. In a carrier assisted anaerobic reactor containing 10% of PUR carriers by its volume, a COD removal efficiency of 83% was achieved at an OLR of 12 kg COD/m^3.d which corresponds to a HRT of about 3.5 days. The experimental study, which was lasted more than 3 months, has clearly showed that using PUR carriers is an appropriate and feasible alternative for upgrading the existing anaerobic digester.

REFERENCES

1. AWWA, APHA and WPCF, *Standard Methods for the Examination of Water and Wastewaters*, 15th Edition (1983).
2. Bircan, E., "Upgrading of Anaerobic Contact Processes Using PUR Carriers," MSc. Thesis (In Turkish), Department of Environmental Engineering, Istanbul Technical University (1989).
3. Pascik, I. and Baumgarten, I., "Anaerobic Municipal Wastewater Treatment Trial With Immobilised Microorganisms," Paper Presented at Specialized Conference in Klagenfurt, Germany (1984).
4. Grau, P. Dohanyos, M. and Chudoba, J. "Kinetics of Multicomponent Substrate Removal by Activated Sludge," Water Research, Vol. 9, pp. 637–642, (1975).
5. Öztürk, I., Demir, I., and Ubay, G., "A Unified Model for Anaerobic Treatment Systems," Proceedings of the 5th International Workshop on Instrumentation, Automation and Control of Wastewaters and Transport Systems, IAWPRC, Kyoto (1990).

49 ANAEROBIC TREATMENT OF FOOD ADDITIVE WASTEWATER IN A FLUIDIZED-BED REACTOR

Kenneth F. Potochnik, Senior Development Engineer
Air Products and Chemicals, Inc.
Allentown, PA 18195–1501

Mauro Caringi, Environmental/Safety Manager
Nestle Foods Corporation
Purchase, NY 10577

Ralph Sgammato, Manager of Engineering
FIDCO, Inc.
New Milford, CT 06776

INTRODUCTION

FIDCO, Inc., a subsidiary of Nestle Foods Corporation, operates a manufacturing plant and is host to a research laboratory in New Milford, Connecticut. The manufacturing plant produces a variety of food products and additives. The bulk of the manufacturing activities involve batch processes which generate a process wastewater stream highly variable in both flow and concentration. Operation of the research laboratory on an eight hour per day basis and the nature of the laboratory activities, contribute further to the variability of the combined (manufacturing plant plus research lab) wastewater stream. The waste stream is currently treated in an aerobic biological treatment system owned and operated by FIDCO.

Although the existing aerobic system is capable of meeting the current effluent regulatory requirements, anticipated future increases in production and more stringent effluent quality requirements will necessitate an increase in treatment capacity and performance. Ammonia removal is of particular concern. The influent ammonia concentration is high and the existing treatment facility lacks the aeration capacity for both carbonaceous BOD removal and nitrification.

The normally high concentration of biodegradable organic compounds in the combined wastewater stream favors the use of an anaerobic biological process as a pretreatment step prior to aerobic treatment. Installation of an anaerobic pretreatment step would reduce the influent load on the aerobic system making existing aeration capacity available for nitrification. Anaerobic pretreatment would also reduce the total waste sludge production at the treatment plant.

A fixed-film biological fluidized-bed reactor was evaluated for the anaerobic pretreatment of the New Milford plant and research laboratory wastewaters. The fluidized-bed process involves passing wastewater upward through a bed of media, such as sand, at a velocity sufficient to expand and fluidize the media. Once fluidized, the media provides a very large surface area for biological growth, in part leading to the development of reactor biomass concentrations five to ten times greater than those achieved in suspended growth anaerobic systems. This high biomass concentration makes possible the use of smaller reactors than those required for suspended growth processes or other fixed-film processes. Also, the high biomass concentration enhances the ability of the fluidized-bed to handle inhibitory feed components and variations in organic load.[1,2,3]

Fluidized-bed systems also have the capacity to handle wide variations in hydraulic loading. The majority of the flow required to fluidize the support media is provided by recirculated effluent; therefore, even large changes in the wastewater flowrate will not upset the system. The flowrate through the fluidized bed remains constant. The high recycle flowrate required for fluidization also reduces the variability of the wastewater composition.

A pilot test of the fluidized-bed system was conducted at FIDCO between January and May of 1989. The principal objectives of the study were to assess the feasibility of anaerobic fluidized-bed treatment of a highly variable food additive wastewater and to obtain information for process design

45th Purdue Industrial Waste Conference Proceedings, © 1991 Lewis Publishers, Inc., Chelsea, Michigan 48118. Printed in U.S.A.

of a full-scale fluidized-bed system for the FIDCO application. The specific objectives of the study were:

1. Start up the pilot unit by gradually acclimating an anaerobic seed material in the fluidized-bed reactor to the FIDCO wastewater and subsequently accumulating a large inventory of attached anaerobic biomass.
2. Operate the pilot unit at several pseudo steady-state organic loading rates (constant wastewater feed rates with normal variations in wastewater concentrations) to determine the optimum average organic loading.
3. Determine process design parameters.
4. Evaluate the effect of influent TKN (total Kjeldahl nitrogen) and NH_3 on the performance of the process.

PILOT PLANT EQUIPMENT AND OPERATION

Equipment

A transportable pilot-scale system was installed at FIDCO, Inc.'s wastewater treatment plant in New Milford, CT. The pilot unit employed at FIDCO (Figure 1) consisted of a clear plastic reactor column six inches in diameter and 10-feet tall. The reactor contained a bed of silica sand media upon which active biological films could attach and grow.

The sand bed was fluidized by recycle stream circulated by a positive displacement recycle pump. Wastewater feed was stored in a 200-liter feed tank and injected into the recycle stream using a variable speed peristaltic pump. Acid and caustic solutions for pH control, were also injected at this point using similar pumps. The addition of the acid and caustic solutions was regulated by an automatic pH controller. The pH sensing electrode was located in the recycle line downstream of the recycle pump.

Following influent injection, the combined recycle/feed stream passed through the recycle pump, across the pH probe and through an electric recycle heater. The heater automatically controlled the temperature of the recycle stream to maintain the desired reactor temperature. In the FIDCO study, the reactor was operated in the mesophilic temperature regime and controlled at 35°C ± 1.5°C.

A rotameter, mounted in the recycle line in parallel with the pH probe, was used to measure the combined recycle flowrate. To prevent fouling, the flowmeter was normally off-line. Recycle flow was directed through the meter only when a flow measurement was required.

After exiting the heater, the combined recycle stream entered a cone-shaped liquid distributor at the bottom of the reactor and flowed upward through the fluidized bed. At the top of the reactor the majority of the treated wastewater was withdrawn and recirculated. A four-inch diameter sand trap located in the effluent recycle line was provided to protect the recycle pump in the event a process upset caused sand to carry out of the reactor.

Treated effluent overflowed the reactor and entered a gas/liquid separator which allowed any entrained biogas to disengage prior to effluent collection. The effluent from the separator flowed through a trap designed to prevent a air backflow into the separator and reactor and then emptied into an effluent collection tank.

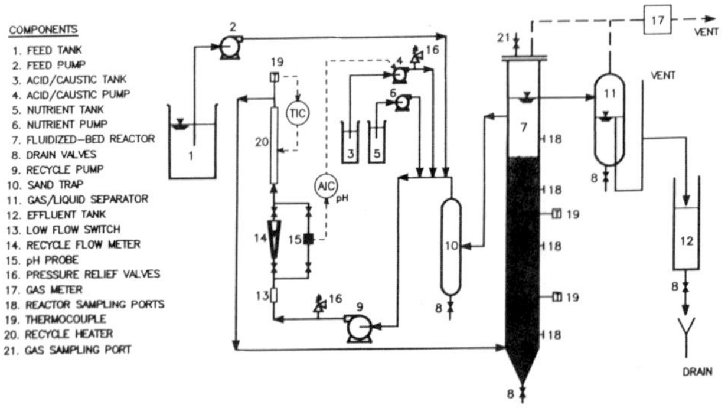

Figure 1. Flow diagram of the anaerobic fluidized-bed pilot plant.

Biogas was collected from the head space of both the reactor and the gas/liquid separator. During the start-up phase of the study, when the gas production rate was low, produced gas was collected in an 80-liter sample bag. The bag was emptied daily through a wet-test gas meter to measure the volumetric rate. Later in the study the gas meter was connected directly to the reactor/separator to continuously measure gas production.

Sampling and Analyses

The pilot plant was attended to eight hours per day, five days per week by an operator employed by FIDCO. Composite influent and effluent samples were collected daily and analyzed on-site for total and soluble chemical oxygen demand (COD) alkalinity (as $CaCO_3$), total volatile fatty acids (VFA), total suspended solids (TSS), volatile suspended solids (VSS), temperature and pH. Once each week composite influent and effluent samples were submitted to an environmental testing laboratory for analysis of total biochemical oxygen demand (BOD), TKN, and ammonia nitrogen (NH_3-N).

Analysis for influent and effluent nitrate (NO_3-N) and orthophosphate (PO_4-P) were conducted at Air Products' environmental laboratories in Allentown, PA.

Samples of the biogas produced by the fluidized-bed reactor were collected weekly in evacuated steel bombs and sent to Air Products' laboratories for GC determination of CO_2, CH_4, O_2, and N_2 content. The rate of gas production was measured using a wet-test gas meter as discussed in the previous section.

The feed rate to the pilot unit was calculated daily from the liquid volume collected in the effluent tank. This volume was checked by allowing the feed pump to discharge into a graduated cylinder for a known period of time. The total flowrate entering the bottom of the reactor (recycle plus feed) was measured using an in-line rotameter.

The fluidized-bed could be observed directly through the clear reactor sidewalls. The bed height was measured and recorded daily. The values for COD loading (g COD/L-day) and influent HRT for the reactor were calculated based on the volume occupied by the fluidized-bed.

The temperature in the fluidized-bed reactor was measured using two thermocouples inserted through the reactor sidewall. This temperature was automatically controlled at $35°C \pm 1.5°C$.

The pH in the reactor was monitored continuously using an automatic pH controller equipped with a strip-chart recorder. Acid and caustic usage were also monitored to allow estimation of the chemical requirements for the design of a full-scale system.

Summary of Operation

The test program conducted at FIDCO, Inc. consisted of an initial biomass accumulation and growth period and two pseudo-steady-state operating (constant HRT) periods.

At start-up the reactor was filled with media taken from an operating commercial-scale aerobic fluidized-bed system. The media consisted of 30–50 mesh silica sand coated with aerobic biofilm. Seed material, consisting of anaerobic digester sludge and belt filter press supernatant, was obtained at the local municipal wastewater treatment plant and added to the reactor. Three liters of FIDCO wastewater (COD = 20,000 mg/L) were also added. The reactor was then topped off with warm tap water and sealed. The recycle pump was adjusted to provide a superficial liquid velocity of approximately 24.4 m/hr in the reactor. The system was operated on complete recycle for one week before beginning continuous wastewater feed.

The initial biomass acclimation and growth phase involved gradually increasing the influence COD loading to the reactor to develop a stable, balanced population of anaerobic microorganisms. Wastewater feed was supplied to the pilot plant from a 200-liter day tank. Manufacturing plant wastewater was pumped into the day tank each morning using a submersible pump located in the treatment plant's influent wetwell. The contents of the day tank were sampled daily and analyzed for COD. Based on the results of the analysis, the feed rate to the pilot unit was adjusted to provide the desired COD loading.

The influent to the pilot plant did contain wastewater from the research laboratories on the FIDCO site because of difficulties in obtaining representative samples of the combined wastestream.

Over a period of 55 days the organic loading to the pilot reactor was increased from 2 g COD/L-day (0.13 lb COD/ft^3-day) to approximately 8 g COD/L-day (0.5 lb COD/ft^3-day). The influent hydraulic residence time (HRT) was decreased from approximately 7.5 days to 36 hours over the same period. Plots of the COD loading and influent HRT as a function of time are presented in Figures 2 and 3, respectively. Figure 4 is a plot of influent COD concentration over the course of the study.

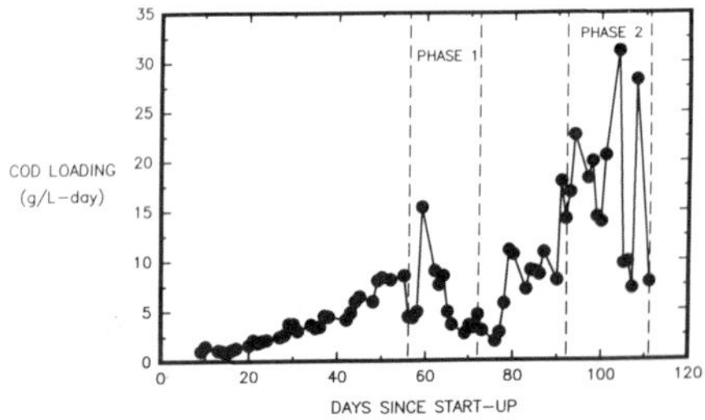

Figure 2. COD loading history of the fluidized-bed pilot plant.

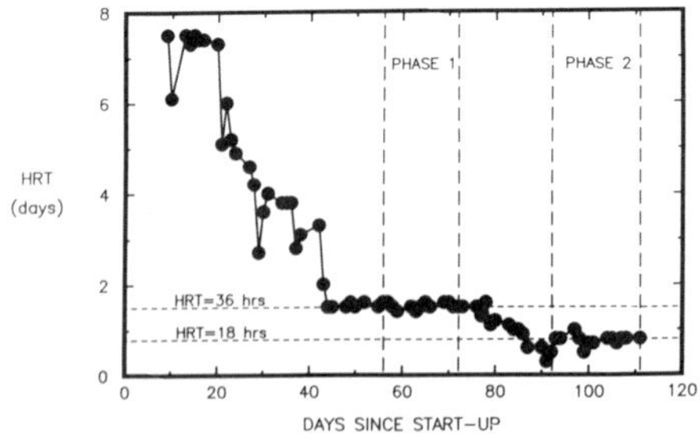

Figure 3. Influent HRT history of the fluidized-bed pilot plant.

On day 56 the wastewater feed supply was switched from the day tank to an overflow tank to begin the first pseudo-steady-state test. Untreated manufacturing plant wastewater was continuously pumped from the wetwell to a 200-liter tank located near the pilot plant. Wastewater flowed through the tank at a rate of 30 to 50 L/min and overflowed back to the wetwell. Feed for the pilot unit was withdrawn from the tank at a rate of 0.63 L/hr, thus exposing the fluidized-bed system to the normal variations in the manufacturing plant wastewater composition and pH, while maintaining a constant influent HRT of approximately 36 hours. (See Table I for a summary of operating conditions).

During the 16-day test, the influent characteristics varied widely. Influent total COD ranged from 4,250 mg/L to 21,500 mg/L with an average value of 8,835 mg/L. Influent pH varied from a high of 8.5 to a low of 6.0. Upon completion of the 36-hour HRT test, the pilot unit was placed back on feed from the day tank and the COD loading was gradually increased to 16 g/L-day (1 lb/ft^3-day).

On day 92 the pilot plant was again fed from the overflow tank at an influent HRT of 18 hours (influent flowrate or 1.29 L/hr) to begin a 19-day test. As in the previous test, influent conditions varied widely. The large swings in the influent COD concentration during the test period (5,500 mg/L to 24,000 mg/L) produced COD loading variations from 7.2 g/L-day (0.45 lb/ft^3-day) to over 31 g/L-day (1.94 lb/ft^3-day).

Through most of the 111-day study, the volatile acid (VA) concentrations in the fluidized-bed reactor remained within acceptable limits (below 1,000 mg/L). On two occasions, at the start of each constant HRT test phase, the VA concentration rose briefly to approximately 1,600 mg/L in response to high acids in the influent. The VA concentration returned to normal within a few days and gas production remained unaffected.

Several brief mechanical upsets occurred during the study; air injection into the recycle line, a short power outage and several incidents of feed line plugging. None of these events had a lasting impact on the performance of the fluidized-bed system.

RESULTS AND DISCUSSION

The pilot tests were conducted under constant flowrate conditions; so the study did not evaluate the response to the full-scale plant's weekly flowrate cycle which varied from 15.1 to 30.3 m^3/day (4,000

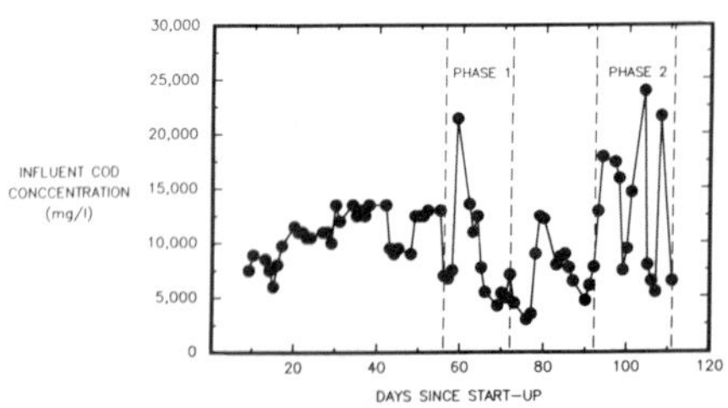

Figure 4. Daily composite influent COD to the pilot plant.

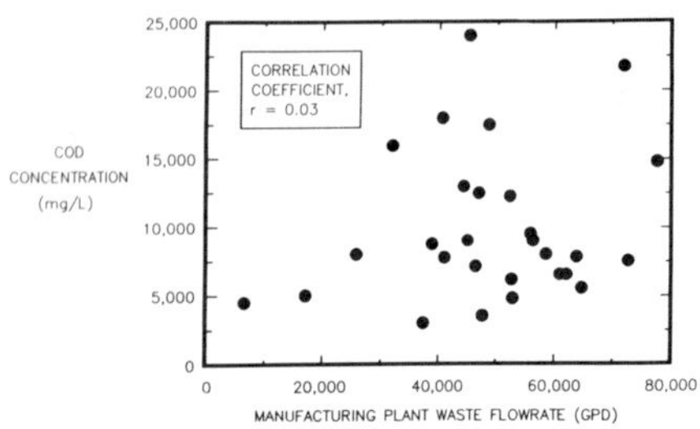

Figure 5. Relationship between wastewater flowrate and COD concentration.

Table I. Pilot Plant Operating Conditions

	36 Hr. HRT Test			18 Hr. HRT Test		
	Max.	Min.	Avg.	Max.	Min.	Avg.
Feedrate (mL/min)	11.3	9.9	10.4	29.7	16.7	21.5
COD-t loading (g/L-day)	15.4	2.7	5.9	31.1	7.2	16.8
Influent HRT (days)	1.61	1.40	1.53	0.95	0.52	0.74
Reactor temp. (deg F)	95.9	92.8	95.3	96.2	92.2	94.7
Reactor pH	7.5	7.0	7.2	7.4	7.0	7.2
Influent Composition						
COD-t (mg/L)	21500	4250	8835	24000	5500	12589
COD-s (mg/L)	19500	3000	7492	15000	4500	8583
Volitile acids (mg/L)	3450	960	1993	4515	1665	2681
TSS (mg/L)	2506	301	938	3740	324	1237
VSS (mg/L)	1880	228	691	1858	296	815
Alkalinity (mg/L)	3820	1760	2540	9200	1900	3593
NH_3-N (mg/L)	998	566	728	1031	512	772
TKN (mg/L)	870	669	802	**	**	703
NO_3-N (mg/L)	0.201	0.021	0.111	0.178	0.115	0.147
PO_4-P (mg/L)	47	19	33	**	**	97
pH	8.5	6.0	7.7	10.9	6.0	7.3

**Single data point.

to 8,000 gal/day) on weekends to 151 to 303 m^3/day (40,000 to 80,000 gal/day) on weekdays. The absence of this test data, however, does not detract from the results and conclusions drawn from the study. Comparison of plant flow and wastewater composition data collected during the test periods suggests that there is no "typical" waste stream for the FIDCO plant. No correlation was found between either the influent COD concentration (Figure 5) or the influent ammonia concentration and the total daily wastewater flow to the treatment plant ($r = 6.03$ and $r = 0.08$, respectively). These results indicate that the influent wastewater strength is independent of the wastewater flowrate. Therefore, the variations in loading experienced by the pilot plant during the test periods can be considered representative of those experienced by the full-scale plant, even in the absence of a variable feedrate.

The FIDCO pilot study demonstrated the fluidized-bed system's ability to treat a continuously changing waste stream. Examination of the influent COD loading and concentration ranges presented in Table I reveals significant variations around the mean. Although the influent flowrate remained approximately constant during each test period, the COD loading to the reactor varied by more than a factor of five. Substantial variations in the concentrations of influent volatile acids (VA), TSS, alkalinity, ammonia and pH were also experienced.

The process performed well during the two constant-HRT tests. As expected with a highly variable wastewater feed stream, variation in effluent quality and gas production was observed. The average influent and effluent compositions for the two test phases are presented in Table II, the table also gives the average percent removal of the components listed.

COD and BOD Removal

Removal of COD in the reactor was determined daily from analysis of influent and effluent composite samples. The data show a higher COD removal efficiency during the 18-hr HRT test period than during the 36-hr HRT test. Figure 6, a graph of COD removal efficiency as a function of time since the start of the given test period, shows the daily variability in removal efficiency. During both test periods the COD removal was variable but stable, trending neither up or down over the course of the test period. Removal of COD ranged between 41 and 77% during the 36-hr HRT test and between 53 and 87% during the 18-hr HRT test.

The difference in removal efficiency between the two HRT conditions is shown in Figure 7, a log versus probability plot of the removal efficiency. From Figure 7 the median values for COD removal are 59.5% at an HRT of 36-hr and 68.5% at an 18-hr HRT. The mean values given in Table II are slightly higher at 61.3 and 71.0%, respectively. The shallower slope of the 18-hr HRT test data suggests slightly greater stability. The difference, however, is small.

Table II. Pilot Plant Performance Summary

	36 Hr. HRT Test			18 Hr. HRT Test		
	Avg. Influent	Avg. Effluent	Change (%)	Avg. Influent	Avg. Effluent	Change (%)
COD-t (mg/L)	8835	3421	61.3%	12589	3657	71.0%
COD-s (mg/L)	7492	2812	62.5%	8583	2470	71.2%
Volitile acids (mg/L)	1993	874	56.2%	2681	881	67.1%
TSS (mg/L)	938	413	56.0%	1237	599	51.6%
VSS (mg/L)	691	269	61.1%	815	336	58.8%
Alkalinity (mg/L)	2540	4472	-76.1%	3593	4248	-18.2%
NH_3-N (mg/L)	728	859	-18.0%	772	874	-13.3%
TKN (mg/L)	802	1050	-30.9%	703	834	-18.7%
NO_3-N (mg/L)	0.11	0.12	-11.3%	0.15	0.13	10.2%
PO_4-P (mg/L)	33	51	-56.4%	97	24	75.3%
pH	7.7	7.2		7.3	7.2	

The observed relationship between COD removal and organic loading is contrary to what is normally expected. The average influent flowrate during the 18-hr HRT test was twice as high as that during the 36-hr HRT test (21.5 mL/min versus 10.4 mL/min) and the COD loading was higher by a factor of 2.85 (16.8 g/L-day versus 5.9 g/L-day). Surprisingly, the COD removal efficiency under the higher flow, higher road conditions was 15% greater.

The results suggest that the fluidized-bed process had not yet achieved a steady-state biomass concentration by the end of the study. Anaerobic biofilms, which grow slowly, continued to accumulate within the reactor over the length of the study. Thus, the food to microorganism ratio (F/M) during the second test period (18-hr HRT) may have been equal to or lower than the F/M during the earlier test period, even though the average organic loading, on a volumetric basis, was nearly three times higher. Unfortunately, data on the attached biofilm concentration within the reactor is not available to confirm this hypothesis. It is likely, based on the performance data however, that the process would have provided effective treatment at higher COD loadings and shorter influent detention times had the study been continued.

Effluent COD concentrations ranged from approximately 1,200 mg/L to over 8,000 mg/L. As shown in Figure 8, a log versus probability plot of effluent COD during the two test periods, there is little difference in effluent quality between the two loading conditions. The median effluent COD concentration for the 36-hr HRT test was 3,017 mg/L. The median was slightly higher, at 3,300 mg/L during the higher rate 18-hr HRT test. The scatter in the effluent concentration data is primarily a function of the highly variable influent wastewater.

The data base for BOD removal is substantially smaller than that for COD removal. Twenty-four-hour composite influent and effluent samples were sent out once each week for determination of total BOD. On two occasions the reported influent BOD concentration exceeded the COD concentration. These data were considered erroneous, and were discarded. The remaining influent BOD values ranged from a high of 6,300 mg/L to a low of 3,400 mg/L. BOD removal averaged 71.7%, a significant reduction for a pretreatment step. Based on the limited data available, the BOD removal efficiency appears to be independent of both feedrate and loading over the range tested.

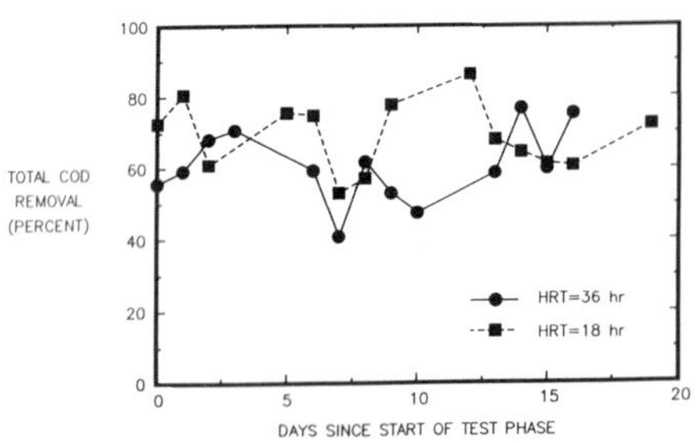

Figure 6. COD removal during each pseudo-steady-state test.

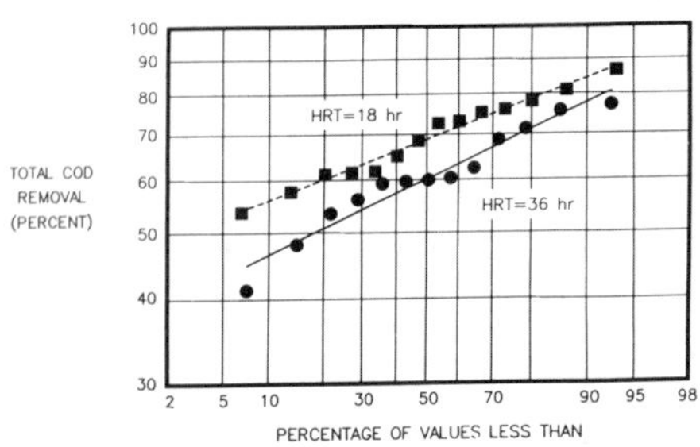

Figure 7. Log-normal frequency distribution of COD removal efficiency.

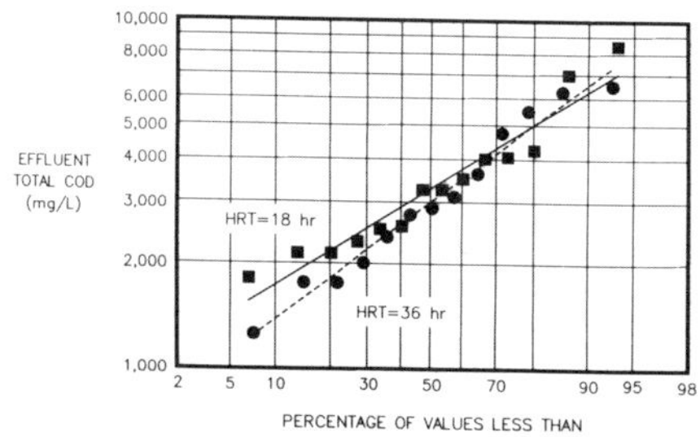

Figure 8. Log-normal frequency distribution of effluent total COD.

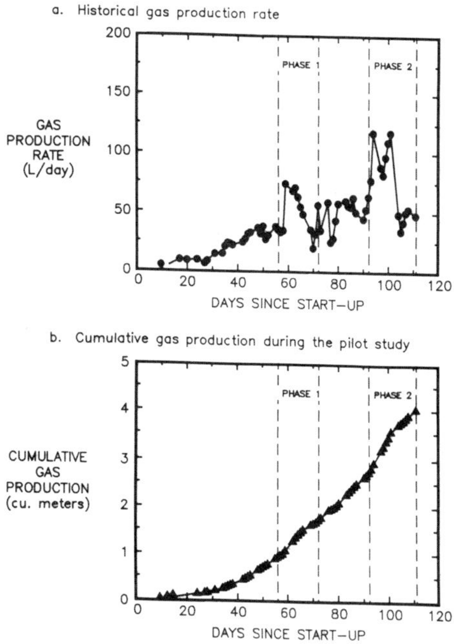

Figure 9. Gas production history of the fluidized-bed pilot plant.

Gas Production

The rate of gas production in an anaerobic system, and the methane content of the gas depends upon the type of wastewater being treated and the treatment rate (the organic loading).[4] Therefore, the gas production rate and the methane yield (m^3 CH_4 produced per kg COD removed) measured in pilot plant varied widely. A plot of the gas production rate and the cumulative gas produced over the course of the study is presented in Figure 9. As expected, the rate of gas production correlates roughly with organic loading, with the highest gas rates achieved at times of highest load. The average gas yield for the FIDCO wastewater was calculated to be 0.34 m^3/kg COD removed. At an average measured methane content of 71.3% the methane yield was 0.24 m^3 CH_4/kg COD removed.

pH Control and Alkalinity

The ability of the fluidized-bed process to handle wide swings in the influent pH is confirmed by the pilot-plant results. Figure 10 is a graph of the influent pH and the fluidizied-bed reactor pH over the course of the study. Feed pH normally ranged between 6.0 and 9.0, although in one instance, the pH

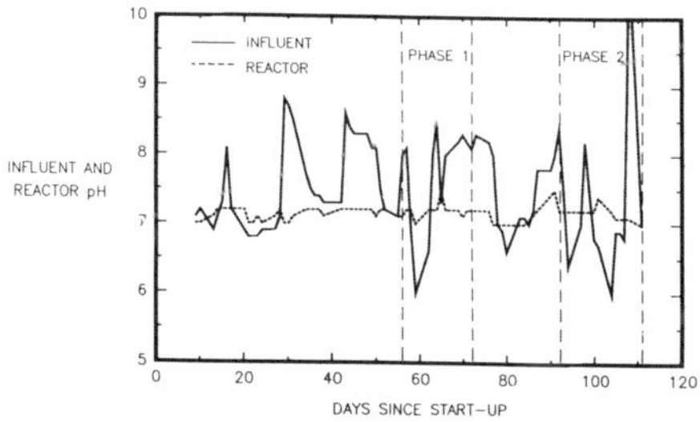

Figure 10. Historical pH values for the pilot plant reactor and influent.

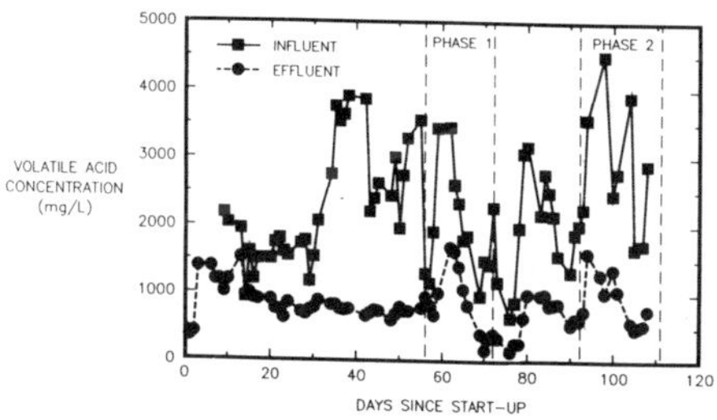

Figure 11. Historical volatile acid concentrations in the pilot plant influent and effluent.

spiked to 10.9. As Figure 10 indicates, the pH within the reactor remained stable at a pH of 7.2 ± 0.2.

The stable reactor pH control results primarily from the large dampening effect of the effluent recycle, which reduces the demands on the acid and caustic addition systems to respond to rapid changes in influent pH. Regular addition of small amounts of 1% sulfuric acid (H_2SO_4) were required to maintain the pH around 7.0.

The influent alkalinity, expressed as mg/L $CaCO_3$, was quite high, normally ranging between 2,000 mg/L and 4,5000 mg/L. This explains in part the minimal caustic requirement for pH control. The alkalinity of the treated effluent was between 20% and 50% greater than that of the influent even with the addition of acid to the recycle stream. The increase in alkalinity is most likely the result of volatile acid metabolism within the fluidized-bed reactor. Volatile fatty acids (VFA) primarily acetic, butyric and propionic acids, are the main metabolic substrates for methanogenic anaerobic bacteria. Normally, these acids are produced by other anaerobics in the reactor. At FIDCO, the influent wastewater also contained high concentrations of volatile acids. A plot of influent and effluent volatile acid concentrations measured during the pilot study is shown in Figure 11. Even with the production of additional volatile acids inside the reactor, a net acid reduction of 50 to 70% was observed across the system.

Solids Removal & Biomass Yield

Influent Total Suspended Solids (TSS) concentrations during the two test periods ranged from 300 mg/L to over 3,700 mg/L, of which approximately 70% were volatile suspended solids (VSS). On average, the fluidized-bed reactor achieved 52% TSS removal during the test periods. Effluent total and volatile suspended solids concentrations averaged 500 mg/L and 300 mg/L, respectively.

Biomass was never wasted from the fluidized-bed. It is likely that anaerobic biofilms were accumulating in the fluidized-bed throughout the study, making determination of the steady-state biomass yield difficult. However, on the basis of effluent VSS during the test periods, the average total biomass yield was calculated to be 0.05 kg VSS/kg COD removed. Typical yields for similarly loaded fluidized-bed reactors range between 0.04 and 0.08 kg VSS/kg COD removed.[5]

Nutrient Requirements

The addition of supplemental nutrients to the influent wastewater was not required during the study. The influent contained adequate quantities of the primary nutrients nitrogen (as ammonia) and phosphorus (as orthophosphate).

Influent ammonia concentrations ranged from 500 mg/L to 1,050 mg/L (as N). These concentrations are high, but below the concentration considered to be toxic to anaerobic bacteria (1,500 mg/L).[6] The excellent COD removal performance indicates that the fluidized-bed reactor was not adversely effected by the high ammonia concentrations. The decomposition of organic nitrogen compounds in the fluidized-bed resulted in a net increase in ammonia of approximately 15% across the system.

FULL-SCALE PROCESS DESIGN

A design for a full-scale fluidized-bed system for FIDCO, Inc. was based on the pilot data. The design parameters developed from pilot testing include design COD loading, COD removal, BOD removal, TSS removal, gas yield, biomass yield, nutrient requirements and pH control requirements (Table III). In addition to the parameters obtained from pilot testing, a design feedrate and a design influent composition were determined from data collected at the existing FIDCO treatment plant.

Probability analysis was employed to arrive at a design concentration using values measured during the pilot study. The design total COD concentration was taken to be 13,040 mg/L, the 80 percentile value. That is, 80% of the measured influent total COD concentrations were less than or equal to 13,040 mg/L. For soluble COD, an 80% design value of 9,200 mg/L was used.

A similar approach was employed to obtain a design flow basis. Total plant influent flowrate was measured by FIDCO during the later half of the pilot study. The data show a weekly cycle in plant flowrate. Flow is low on the weekends (15.1 to 30.3 m^3/day) when the manufacturing plant operates at reduced capacity and the research laboratory is closed. The wastewater flowrate increases dramatically on Monday when manufacturing and research activities resume. The weekday flowrate at the treatment plant was observed to peak at 341 m^3/day (90,000 gal/day) on several occasions. Because the influent COD concentration is independent of the wastewater flowrate, the design flowrate for a full-scale system was again taken to be the 80-percentile value, 318 m^3/day (84,000 gal/day).

Table III. Summary of Results

	36 hr HRT	18 hr HRT	Design
COD-t loading (g/L-day)	5.9	16.8	16.0
HRT (hgrs)	36.7	17.7	18.0
COD-t removal (%)	59.7	68.5	70.0
BOD-t removal (%)	67.0 *	76.0 *	70.0
Methane yield (L CH_4/gCOD rem)	0.43	0.22	0.24
Biomass yield (gVSS/gCOD rem)	0.06	0.04	0.05
Nutrients required	NR	NR	NR
Acid required (gal/1000 gal) (10% H_2SO_4)	4.6	2.4	4.5
Caustic required (gal/1000 gal) (50% NaOH)	MIN	MIN	MIN

*Single data point.

Based on the pilot data, a design total COD loading of 16 kg COD/m^3-day was assumed with an expected COD removal efficiency of 70%.

At the design conditions given above, a reactor 6.71 meters (22 ft) in diameter with a fluidized-bed depth of seven meters (23 ft) would be required. The total reactor height including liquid freeboard and dome cover would be approximately 11.9 meters (39 ft). The size of the reactor could be reduced by installing upstream equalization to reduce the variations in flow and concentration.

The design presented here is considered to be conservative. The pilot plant results suggest that the system would have provided effective treatment at higher rates than were actually tested.

CONCLUSIONS

The following conclusions can be drawn from the results of pilot testing the anaerobic fluidized-bed wastewater treatment system at FIDCO, Inc.

1. The anaerobic fluidized-bed system is capable of effectively pretreating a highly variable waste stream. An average COD removal efficiency of 70% was achieved at an average COD loading of 16 g/L-Day. Effective treatment at higher loadings is expected but cannot be confirmed.
2. The pilot plant achieved an average BOD removal efficiency of over 70%. BOD removal was independent of loading over the range tested.
3. The high influent ammonia concentrations typical of the FIDCO wastewater have no adverse effect on treatability. Decomposition of organic nitrogen compounds results in a 15% increase in ammonia concentration across the fluidized-bed reactor.
4. Supplemental nutrient addition is not required. The FIDCO wastewater contains sufficient concentrations of both nitrogen and phosphorus to support biological growth in both an anaerobic pretreatment system and an aerobic secondary treatment system. The influent ammonia-nitrogen and phosphate concentrations average 500 mg/L and 35 mg/L, respectively.
5. The fluidized-bed reactor achieved approximately 50% reduction in influent total suspended solids.
6. The total gas yield averaged 0.34 m^3/kg COD removed with an average methane content of 71.3% on an air-free basis. The average methane yield was 0.24 m^3CH_4/kg COD removed.
7. The fluidized-bed system provides excellent pH control. Acid addition to the recycle stream is required. The influent wastewater normally contains sufficient alkalinity (3,000 mg/L as $CaCO_3$) to allow stable pH control. Caustic addition for pH control was not normally required.

ACKNOWLEDGMENTS

The operating and engineering staff of the FIDCO, Inc. wastewater treatment facility provided valuable assistance throughout the pilot study.

REFERENCES

1. Switzenbaum, M. S. and Jewell, W. J., "Anaerobic Attached-Film Expanded-Bed Reactor Treatment," *Journal Water Pollution Control Federation*, 52, 1952 (1980).
2. Jeris, J. S., "Industrial Wastewater Treatment Using Anaerobic Fluidized-Bed Reactors," *Water Science and Technology*, 15, 169 (1983).
3. Sutton, P. M. and Huss, D. A., "Anaerobic Fluidized-Bed Treatment: Pilot to Full-Scale Demonstration," Presented at the 56th Annual Water Pollution Control Federation Conference, New Orleans, LA (1984).
4. Kennedy, K. J., Van Den Berg, L. and Murray, W. D., "Advanced Fixed Film Reactors for Microbial Production of Methane From Waste," Proceeding of the 35th Purdue Industrial Waste Conference (1981).
5. Sutton, P. M., and Li, A., "Single Phase and Two Phase Anaerobic Stabilization in Fluidized Bed Reactors," *Water Science Technology*, 15, 333 (1983).
6. McCarty, P. L., "Anaerobic Waste Treatment Fundamentals, 2," *Public Works*, pp. 109, Sept. 1964.

50 ANAEROBIC TOXICITY OF RESIN ACIDS IN CHEMITHERMOMECHANICAL PULP WASTEWATER

P. J. McCarthy, Environmental Scientist
R. C. Landine, Vice President
ADI Limited
Fredericton, N.B.
Canada E3B 4Y2

K. J. Kennedy, Research Associate
National Research Council
Ottawa, Ontario,
Canada K1A 0R6

R. L. Droste, Associate Professor
Department of Civil Engineering
University of Ottawa
Ottawa, Ontario
Canada K1N 6N5

INTRODUCTION

Resin acids are present in a number of wood processing wastewaters. These wastewaters are being treated anaerobically although, at present, little is known about the effects that resin acids have on anaerobic bacteria or of their fate in an anaerobic environment. Andersson and Welander[1] reported an abietic acid/oleic acid mixture to be inhibitory at 1,178 mg/L. However Koster and Cramer[2] reported inhibition to 50% of control activity by oleic acid at 1,235 mg/L which left some doubt as to the effect of abietic acid. Field et al.[3] found 50% of control activity for an abietic acid concentration of 114 mg/L. McFarlane and Clark[4] reported that resin acids are degraded in an upflow anaerobic sludge blanket (UASB) reactor with the exception of dehydroabietic acid. It was also concluded by the same authors that anaerobic bacteria could acclimatize to high levels of dehydroabietic acid.

It has been stated by Welander[5] and Hall and Cornacchio[6] that chemithermomechanical pulp (CTMP) wastewater can be toxic to anaerobic bacteria. The objective of this study was to characterize the toxicity of BCTMP wastewater for fiber and nonfiber components and to compare BCTMP wastewater anaerobic toxicity to toxicity caused by pure resin acids.

MATERIALS AND METHODS

Materials

Pure resin acids were purchased from Helix Biotech. Scientific Ltd., Richmond, B.C. Resin acid analysis was carried out by Econotech Laboratory, New Westminister, British Columbia. Anaerobic granular sludge was provided by Paques Lavalin and originated in Roermond, Holland where it was used for the treatment of recycled paper wastewater and therefore was not acclimatized to CTMP wastewater. The sludge was stored at 4°C until used. BCTMP-N wastewater, with a resin acid concentration of 36 m/L, was provided by Quesnel River Pulp Company, Quesnel, British Columbia. The mill has an alternating thermomechanical/BCTMP process with peroxide bleaching. The "N" in BCTMP-N indicates that the wastewater has normal fines/fiber content. Quesnel River Pulp Company also produces a high fines/fiber wastewater when producing certain special grades of pulp. Wastewater was stored upon receipt at -20°C until it was used.

Anaerobic Toxicity Assay

A slightly modified anaerobic toxicity assay (ATA) described by Cornacchio et al.[7] was used for all assays. The modifications were:

45th Purdue Industrial Waste Conference Proceedings,

1. Methanol was used to dissolve the resin acids and therefore the first spike for these samples also includes methanol degradation. The volume of methanol used was 200 microliters which provided an additional carbon source of 6,000 mg COD/L.
2. After gas production ceased in all the serum bottles a second spike of acetate/propionate was given which provided a final serum bottle soluble COD of 3,300 mg/L.

Each study was run in duplicate with controls. The same volume of methanol was added to the controls. Serum bottles (150 mL) were mixed throughout the experiment in a room controlled at 35°C. The desired final concentration of volatile suspended solids (VSS) in the serum bottles was 5 g/L. Gas volume was measured by water displacement and gas composition was measured by gas chromatography according to the method reported by Kennedy and van den Berg.[8]

Activity was measured as the initial rate of methane production. The initial period was defined as the period of constant methane production that occurred directly after spike injection. The initial period can be as short as one or two days, and as long as a week and is dependent on the available carbon source, the state of the biomass, and the presence or absence of toxins.

Activity can be normalized to the VSS present in the serum bottles becoming specific activity. Normalization of activity is desirable when the amount of VSS in an experimental set of serum bottles is variable.

BCTMP Preparation

One-half of the BCTMP-N waste was thawed and was then centrifuged at a relative centrifugal force (RCF) of 12,100 g for 30 minutes. Supernatant was stored and labelled BCTMP-NC (normal fines centrifuged) and was assumed to be fiber-free. (Not all of the material in the fines is fiber but fines and fiber will be used interchangeably in this paper). The centrifugate, raw fiber, was collected and labelled fiber. Fiber in BCTMP-N wastewater was 0.6% (w/v). The second half of the wastewater was stored as BCTMP-N.

Toxicity of BCTMP-N vs. BCTMP-NC

For both BCTMP-N and BCTMP-NC wastewaters, dilutions of 20, 40, and 60%wastewater were tested for inhibition of anaerobic bacterial activity. The percent dilution of the wastewater includes the volume of sludge present in the serum bottle. For example, in the 60% dilution, the following are present in the serum bottle: 30 mL of wastewater; 10 mL of defined medium; 8 mL of sludge and 2 mL of soluble carbon source.

Fiber Preparation

A known weight of wet fiber (centrifugate) was washed four times in HPLC grade methanol pouring off the supernatant each time. The pellet was saved and labelled methanol insoluble fiber (MIF). The combined supernatant, collected from washing the fiber, was again centrifuged at a RCF of 12,100g for 30 minutes. The pellet was combined with the MIF. The supernatant was labelled methanol soluble fiber (MSF). Methanol was aspirated off and the dried MIF and MSF were weighed. Solutions were made up on a proportionality basis with respect to the weight of the original wet fiber.

For example, if the separation started with 2.4 g of wet fiber, which is the equivalent of 400 mL ($0.4\ L \times 6\ g/L = 2.4\ g$) BCTMP-N, then to make 100 mL of MIF or MSF solutions that have a fiber content equivalent to BCTMP-N wastewater, one fourth of the MIF or MSF is required.

It should be noted that any volatile compounds would be aspirated off in the process so that compounds such as terpenes would not exist in significant concentrations in either MSF or MIF. The toxic compounds remaining in MSF and MIF are therefore non-volatile.

Toxicity of MSF vs. MIF

Single and double strength solutions of MIF and MSF were tested. Single strength represents that amount of MIF or MSF that would be present if there was 100% wastewater in the serum bottle. To test the toxic effect of a raw fiber, a parallel study was run using raw fiber at single strength.

Toxicity of Resin Acids

A resin acid mixture was prepared that matched as closely as possible the mean resin acid composition reported in BCTMP-N wastewater from the Quesnel River Pulp Company on five separate occasions. Seven resin acids were used and their percent composition in the final mixture is indicated in Table I. A series of resin acid concentrations ranging from 20 to 700 mg/L were assayed using the ATA.

Table I. Resin Acid Concentration in BCTMP-N Wastewater

Resin Acid	Percentage
Pimaric	7
Sandarocopimaric	6
Isopimaric	12
Levopimaric	21
Dehydroabietic	22
Abietic	26
Neoabietic	6
	100

RESULTS AND DISCUSSION

Activity and Duplicate Variability

In each set of experiments the VSS concentrations in the serum bottles was relatively constant and normalization on the data to specific activity was not done. Variability in the duplicate serum bottles was assessed. One measure of variability is the percent deviance of the duplicate serum bottles from their mean rate of methane production. First spike variation was 5.0% with a standard deviation of 4.2%. Spike two variation was 4.8 plus or minus 5.2%. The low variability is indicative of good duplication.

Toxicity of BCTMP-N vs. BCTMP-NC Wastewaters

The results of the anaerobic toxicity assays for all analyses are summarized in Table II. For the centrifuged wastewater, BCTMP-NC, only the 40% dilution displayed slight inhibition in spike 1. Inhibition increased with wastewater concentration for spike 2 for the same wastewater. Soluble toxic substances which inhibit anaerobic bacteria were therefore masked in spike 1 degradation and their effects were evident in the spike 2 results. The 20% BCTMP-NC wastewater never displayed inhibition which suggested that dilution to 20% and/or removal of the fiber rendered the wastewater free of toxic effects. In fact there may be a net beneficial effect at this dilution since 20% BCTMP-NC waste performed better than the control for both spikes 1 and 2.

In contrast to BCTMP-NC, the raw wastewater (BCTMP-N) toxicity increased with concentration for spike 1. The difference can be attributed to the presence of the fiber in BCTMP-N. All concentrations of BCTMP-N inhibited methane production in spike 2. Since 20% BCTMP-N wastewater was toxic in the second spike, then it can be concluded that to remove the toxic effects, not only is dilution to 20% wastewater necessary, but the fiber must also be removed.

Table II. Summary of Anaerobic Toxicity Assays

Toxicant	Spike 1*	Spike 2*
BCTMP-NC 20%	150	109
40%	91	70
60%	118	54
BCTMP-N 20%	117	65
40%	78	70
60%	48	65
MSF (single strength)	33	73
MSF (double strength)	13	56
MIF (single strength)	95	96
MIF (double strength)	94	100
Raw fiber (100%)	39	41
Resin acids		
20 mg/L	102	100
40 mg/L	119	84
80 mg/L	97	84
160 mg/L	63	65
320 mg/L	41	59
700 mg/L	NA**	38

*All numbers under spike 1 and 2 are based on percentage of control activity.
**Acetate/propionate was not given to these bottles in spike 1.

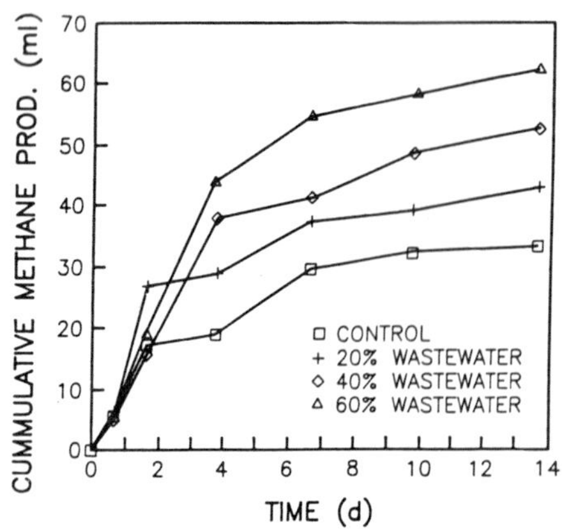

Figure 1. Cumulative methane production from the degradation of BCTMP-N wastewater and acetate/propionate; Spike 1.

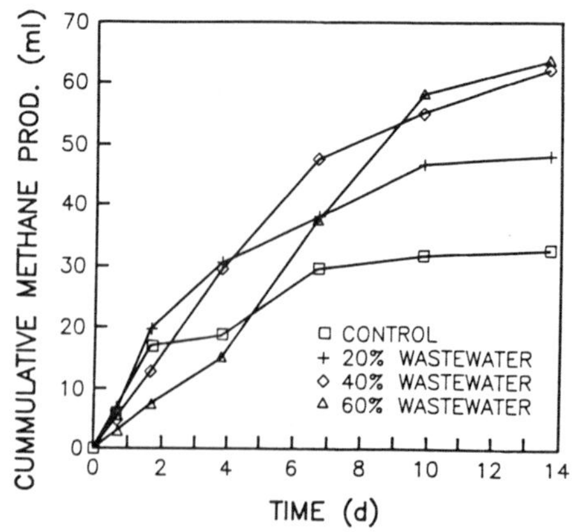

Figure 2. Cumulative methane production from the degradation of BCTMP-NC wastewater and acetate/propionate; Spike 1.

It must be noted that for spike 1 of both BCTMP-NC and BCTMP-N, wastewater was being degraded as well as the acetate/propionate spike (control). Higher wastewater concentration increases the organic loading which may have a beneficial or detrimental effect on the anaerobic microbial activity depending on the organic loading and the consequent increase in toxicant concentration. This may explain the higher than control activity of BCTMP-NC at two of the three wastewater concentrations in spike 1. It is not clear why BCTMP-NC at 40% in spike 1 exhibited slightly lower activity than that of the 60% wastewater but it does suggest that the interaction of organic loading and toxicant concentration is complex.

BCTMP-NC wastewater in the second spike neither exhibited greater inhibition with concentration increase nor a greater inhibition at the 60% level than 60% BCTMP-NC wastewater. This was unexpected since the presence of the fiber proved to add toxicity in 20% wastewater. It can be seen in Figures 1 and 2 that 60% BCTMP-N wastewater degradation took much longer than either the control or BCTMP-NC 60% wastewater degradation. As a result the BCTMP-NC bottles were inactive longer without an easily degradable carbon source before the introduction of the second spike. It is thought that the length of time the serum bottles remain in the presence of toxic substances without an easily degradable carbon source may have an effect on the apparent toxicity in the second spike. Results (not included here) on the shutdown and re-startup of laboratory scale continuous reactors fed with BCTMP-N waste also suggested that the length of time the sludge is without an easily degradable carbon source in the presence of BCTMP-N wastewater has an adverse effect on anaerobic activity.[9] Continuous experiments are underway to investigate this hypothesis.

Toxicity of MSF vs. MIF

Figures 3 and 4 demonstrate that most of the inhibitory substances contained in the fibers are soluble in methanol. Results listed in Table II indicate that MIF had no inhibition in either spike 1 or 2 whereas MSF percent control activity is 73% at single strength concentration in spike 2 and 33% in spike 1. Double strength MSF is even more inhibitory. It is possible that the methanol wash dissolved material that would normally not be dissolved in BCTMP wastewater or in the defined medium plus water. This was tested with raw fiber at 0.6 g/100 mL or the same proportionate strength as single strength raw fiber content is similar to that caused by MSF single strength. In spike 2, the raw fiber inhibition was even greater than both single and double strength MSF. Two factors may explain this response. It may indeed take longer for the toxicants to dissolve from the raw fiber in the serum bottle than in methanol which would both help to decrease the toxicity in spike 1 and increase the toxicity in spike 2. Secondly, it was stated earlier that MSF was void of volatile substances. These substances may play a role in increasing the toxicity in spike 1 and to a lesser extent, as gasses escape, in spike 2.

Resin Acid Toxicity

Resin acid toxicity is also summarized in Table II and spike 2 toxicity is displayed graphically in Figure 5. Generally, in both spike 1 and 2, inhibition increases as the concentration of resin acids increases. At a resin acid concentration of 40 mg/L there is an improvement in activity with respect to the control in spike 1 and only 16% inhibition over the control in spike 2. Spike 1 of the resin acids exhibits a classical inhibition curve as discussed by McCarty.[10] Inhibition decreases as concentration

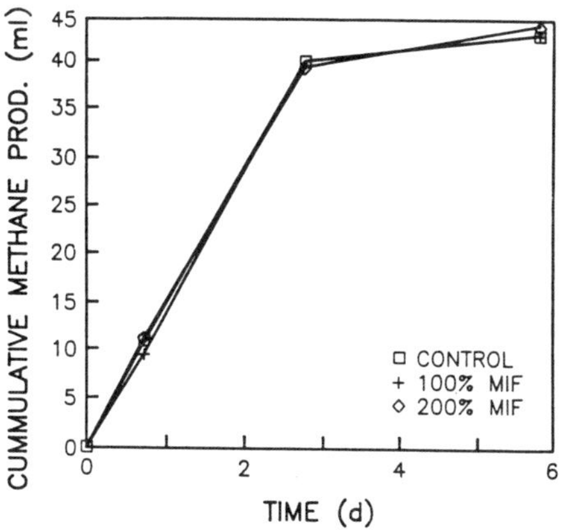

Figure 3. Cumulative methane production from the degradation of acetate/propionate in the presence of the methanol insoluble fiber fraction.

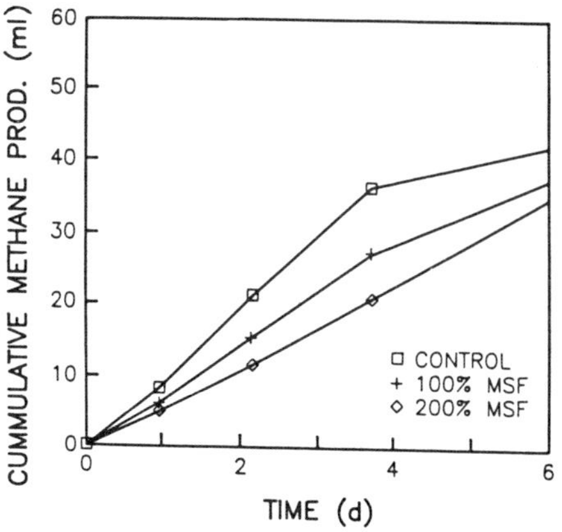

Figure 4. Cumulative methane production from the degradation of acetate/propionate in the presence of the methanol insoluble fiber fraction.

increases until a certain level of toxicant is attained at which point an actual beneficial effect is seen. After this point, as the concentration of the toxicant increases, the inhibition increases.

Toxicity of Resin Acids vs. BCTMP Wastewater

For this comparison only spike 2 will be considered since in the second place spike all serum bottles received identical organic loadings (spike 1 had varying concentrations of the wastewater which caused the difference in organic loading). From Table II, 40% BCTMP-NC exhibits 70% control activity. If Figure 5 is entered on the abscissa at 70% control activity, then 40% BCTMP-NC wastewater corresponds to 140 mg/L resin acids. The actual resin acid concentration in the BCTMP-N wastewater was 36 mg/L. There is at most 14 mg/L resin acid in 40% raw wastewater and even less in centrifuged wastewater (estimate based on the maximum possible resin acid concentration if there was no resin acid partitioned into the fiber). Similarly, 60% BCTMP-NC corresponds to 320 mg/L resin acids. Barring a synergistic effect of resin acids with a non-toxic compound in the wastewater it can be concluded that not all of the toxicity of BCTMP-N wastewater is caused by resin acids.

Toxicity of Methanol Soluble Fiber vs. Resin Acids

Analyses have been done on the partition of resin acids into the soluble and non-soluble (fiber) phases of the wastewater.[11] At the very least 35% of the resin acids remain in the wastewater. More generally however, filtered wastewater contains 60% or more of the resin acid concentration present in the raw wastewater. Since the MSF was prepared from wastewater that had a resin acid concentra-

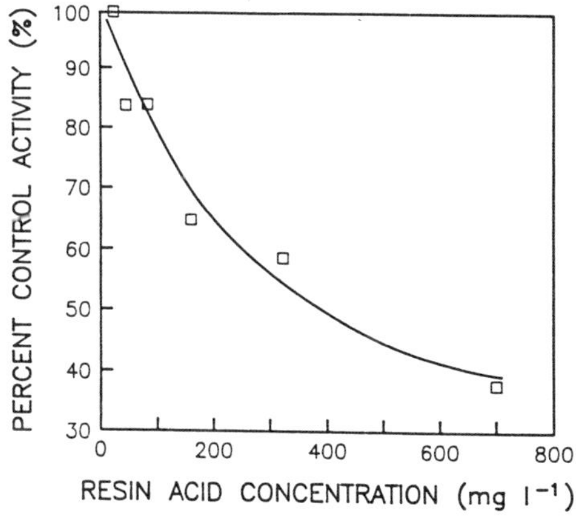

Figure 5. Resin acid caused inhibition in Spike 2.

tion of 36 mg/L, 65% (or 23 mg/L) is the maximum possible concentration of resin acids in the reconstituted fiber. In order to achieve the same inhibition attained by MSF (single strength) a resin acid concentration of 140 mg/L would have had to be present (Figure 5). It can be concluded that resin acids were not responsible for all the inhibition caused by MSF.

CONCLUSIONS

The following conclusions are based on a sludge that was not acclimatized to CTMP wastewater. Anaerobic toxicity in BCTMP-N wastewater is partly soluble and partly fiber associated. The anaerobic toxicity that is removed with the fiber is methanol soluble. Resin acids inhibit anaerobic activity but BCTMP wastewater toxicity to anaerobic bacteria can only be partially explained by the presence of resin acids. The possible synergistic effects that other compounds may have in conjunction with the resin acids in BCTMP-N wastewater toxicity have not been excluded.

ACKNOWLEDGMENTS

This research was partially funded by NSERC. Gratitude is extended to the National Research Council for the use of facilities. The granular sludge was provided by Paques Lavalin and the wastewater was supplied by Quesnel River Pulp Company, Quesnel, B.C. We would also express appreciation for the technical advice of Eva Andras, Debbie Richardson and C. M. Elliott, employees of NRC.

REFERENCES

1. Andersson P. and Welander, T. G. (1985) Wastewater Treatment Method. European Patent Application. 0159972.
2. Koster, J. J. and Cramer A. (1987) Inhibition of methanogenesis in granular sludge by long-chain fatty acids. *Appl. Envir. Microbiol. 53*, 403–409.
3. Field, J. A., Leyendeckers, M. J. H., Sierra Alvarez, R., Lettinga, G. and Habets, L. H. A. (1988) The methanogenic toxicity of bark tannins and the anaerobic biodegradability of water soluble bark matter. *Wat. Sci. Techol. 20*, 219–240.
4. McFarlane, P. and Clark, T. (1988) Metabolism of resin acids in anaerobic systems. *Wat. Sci. Techol. 1*, 273–276.
5. Welander, T. (1988) An anaerobic process for treatment of CTMP effluent. *Wat. Sci. Technol. 20*, 143–147.
6. Hall, E. and Cornacchio, L. A. (1988) Anaerobic treatability of Canadian pulp and paper mill wastewaters. *Pulp & Paper Can. 89*, T188–192.
7. Cornacchio, L., Hall, E. R. and Trevors, J. T. (1986) Modified serum bottle procedures for industrial wastewaters. Wastewater Technology Centre, *Technology Transfer Workshop on Laboratory Scale Anaerobic Treatability Testing Techniques*. Environment Canada, EPS 6-EP-83-1.
8. Kennedy, K. J. and van den Berg, L. (1982) Stability and performance of anaerobic fixed film reactors during hydraulic overloading at 10–35°C. *Wat. Res. 16*, 1391–1398.
9. Kennedy, K. J. (1989) Personal communication.
10. McCarty, P. (1964) Anaerobic waste treatment fundamentals: Part three; Toxic materials and their control. *Public Works*. November, 91–94.
11. de Vegt, A. (1988) Internal Paques Lavalin report.

51 POISONING OF SULFATE REDUCTION WITH MOLYBDENUM IN ANAEROBIC REACTORS FED GLUCOSE

Zablon Oonge, Graduate Research Assistant
Gene F. Parkin, Professor
Department of Civil and Environmental Engineering
The University of Iowa
Iowa City, Iowa 52242

INTRODUCTION

The presence of sulfate in wastewater can reduce the amount of methane that can be produced from the degradation of wastewater organics. Sulfate reducing bacteria (SRB) use various electron donors partially and completely oxidizing them while reducing SO_4^{2-} to H_2S. It is well documented that both sulfate reduction and methane production can occur in the same environment. It is also well documented that SRB may outcompete the consortium of bacteria responsible for methane production. Several explanations for the reported predominance of SRB include energetic advantages, kinetic advantages, and toxicity. There is considerable interest in preventing sulfate reduction to H_2S from occuring. One possibility to add molybdenum (Mo) since some literature reports indicate that Mo is toxic to SRB at lower concentrations than those reported to be toxic to methane-producing organisms.[1-3]

The microbial community involved in the anaerobic processing of complex organic matter such as glucose has generally been divided into three groups. Fermentative bacteria break down complex substrates into fatty acids, H_2, and CO_2. Fatty acids larger than acetic acid, such as butyric and propionic acids, are oxidized to acetate by a second group of bacteria known as acetogenic bacteria. H_2 and CO_2 are also formed by this group of bacteria. Finally, waste stabilization takes place when the methane forming bacteria convert acetate and H_2/CO_2 into CH_4.

Sulfides produced from sulfate reduction are known to be toxic to methane producing bacteria. Varying levels of dissolved sulfides ($DS = H_2S + HS^- + S^{2-}$), from 100–800 mg/L, have reportedly been associated with toxicity to methane production.[4-16] It is widely believed that H_2S is the toxic sulfide form. As such, when levels of sulfides are known, pH data are also required to use in determining H_2S levels. Kroiss and Wabnegg[17] have related methanogenesis to the level of free H_2S in solution. They found that a free H_2S level of around 50 mg/L inhibits acetogens and methanogens by about 50% while complete inhibition occurred at a free H_2S level of around 200 mg/L.

It has also been reported that sulfides are toxic to SRB.[18,19] Recent work by Hilton and Oleszkiewicz has shown that it is DS and not H_2S that inhibits sulfate reduction. They propose that a situation can be created where DS levels are high and H_2S levels are low, such as in the case with high pH, causing more of the substrate to be converted to methane by inhibiting the SRB.

The interaction between sulfides and heavy metals is an important phenomenon in anaerobic waste treatment. Both sulfides and heavy metals are nutrients at low concentrations and toxic at high concentrations. Even if both sulfide and heavy metals are added at sufficient concentrations, precipitation of the metal sulfide may make them both unavailable to the bacteria.

Molybdenum has been used to inhibit the activity of SRB in the studies of interaction of SRB with methane producing consortia.[1,2,3,20] found that 20 mM of MoO_4^{2-} completely inhibited sulfate reduction in intertidal sediments and inhibited acetate oxidation to CO_2 by 98.5%. They concluded that unless Mo has unknown inhibitory effects on other respiratory processes, the degree to which it inhibits acetate oxidation indicates the amount of acetate oxidized via sulfate reduction.

The purpose of the study described herein is to attempt to assess the feasibility of selectively poisoning sulfate reduction by Mo. At the same time the effect of COD/S ratio on interaction between SRB and methane producing consortia was studied for systems fed glucose.

45th Purdue Industrial Waste Conference Proceedings,

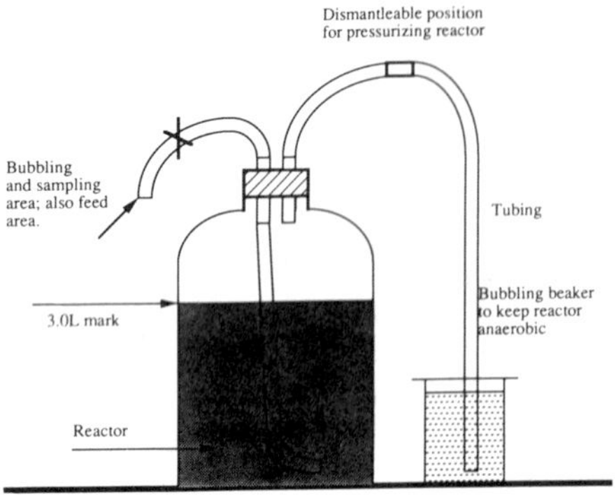

Figure 1. Schematic diagram of an anaerobic reactor.

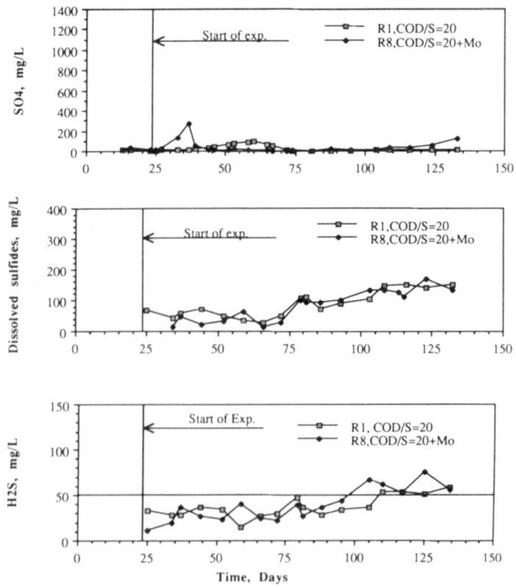

Figure 2. Effect of Molybdenum on SO_4 at COD/S = 20.

MATERIALS AND METHODS

Enrichment Cultures

Three types of stock cultures were used as seed for the experiments: acetate, propionate and lactate enrichment cultures. The acetate and propionate enrichment cultures were operated at a solids retention time (SRT) of 25 days while the lactate enrichment culture was kept at 40 days SRT. All stock cultures were kept at a temperature of 35°C. The organic loading rate (OLR) is maintained at 0.5 g COD/L-d. The stock cultures were methane producing and have sulfate reducers maintained by feeding magnesium sulfate at a COD/S ratio of 40. The methane-producing organisms are primarily acetate-utilizing methanogens, the primary species being *Methanosarcina mazei* and *Methanothrix soehngenii*. The sulfate reducers are primarily *Desulfovibrio*.[21] After seeding with enrichment culture, reactors were maintained at an SRT of 25 days by wasting 120 mL from each reactor daily after good mixing and replacing with an equal amount made up of glucose as the organic feed plus nutrients and sulfate.

Reactor Design

Eight bench-scale, anaerobic, fill-and-draw reactors were used (Figure 1). The reactor volume was four liters and was filled to the three liter level with stock cultures. Two tubings were connected to the reactor through the rubber stopper. One tube ended at the bottom of the reactor and was used for both feeding and mixing contents of the reactor by bubbling using nitrogen gas. The second tubing ended at the rubber stopper. It was used for venting gas from the reactor and for increasing pressure in the reactor to force the effluent out of the reactor using nitrogen gas.

Feed Composition

Feed to the reactors consisted of nutrient solution of the composition given in Table I and an OLR of 0.375 g COD/L-d using glucose. An initial COD/S ratio of 20 was used until the reactors reached a quasi-steady state.

The organisms were acclimated to the new substrate, glucose, by starting the organic load at low levels and gradually increasing it to the level of 0.375 gCOD/L-d at which experiments were done. Reactor startup lasted 66 days. The reactors were then divided into four pairs with COD/S of 20, 15, 10, and 7.5. One reactor from each pair was fed molybdenum in the form of $Na_2MoO_4{\cdot}2H_2O$ at 0.2 mM of Mo. The conditions for each reactor were:

Table I. Nutrient Feed Solution to the Reactors

Constituent	Concentration mg/L
NH_4Cl	400
$MgCl_2$	400
KCl	400
$FeCl_2 \cdot 4H_2O$	40
$CaCl_2 \cdot 2H_2O$	25
$(NH_4)_2 \cdot HPO_4$	80
$CoCl_2 \cdot 6H_2O$	2.5
KI	2.5
$(NaPO_3)_6$	10
NH_4VO_3	0.5
$MnCl_2 \cdot 4H_2O$	0.5
$ZnCl_2$	0.5
$Na_2MoO_4 \cdot 2H_2O$	0.5
HBO_3	0.5
$NiCl_2 \cdot 6H_2O$	0.5
Cysteine	0.5

Reactor 1 (R1): COD/S = 20 which is 18.75 mg SO_4-S/L-day and no Mo;
Reactor 2 (R2): COD/S = 15 which is 25 mg SO_4-S/L-day and no Mo;
Reactor 3 (R3): COD/S = 10 which is 37.5 mg SO_4-S/L-day and no Mo;
Reactor 4 (R4): COD/S = 7.5 which is 50 mg SO_4-S/L-day and no Mo;
Reactor 5 (R5): COD/S = 7.5 with 0.2 mM of Mo which is 19.2 mg/L MoO_4-Mo;
Reactor 6 (R6): COD/S = 10 with 0.2 mM Mo;
Reactor 7 (R7): COD/S = 15 with 0.2 mM Mo;
Reactor 8 (R8): COD/S = 20 with 0.2 mM Mo.

Analytical Methods

COD, VSS, sulfide (total and dissolved) and total alkalinity were measured according to *Standard Methods*.[22] Volatile acids were analyzed using a Hewlett-Packard model 5830A gas chromatograph with a glass column packed with 60/80 carbopack C/0.3% carbowax 20M/0.1% H_3PO_4 (Supelco). The column was operated at 120°C and the FID temperature was 250°C with helium as the carrier gas. Sulfates were measured using a Dionex ion chromatograph model 2000i connected to a programmable Dionex 4270 integrator. Glucose was measured using an OVT-direct glucose reagent, a solution of 6% ortho-toluidine in glacial acetic acid stabilized with thiocarbamide, supplied by Fisher. Instructions come with the reagent when ordered. Effluent COD values reported were calculated by subtracting the COD equivalent of the sulfide content of effluent (32 mg as sulfide equivalent to 64 mg COD).

Un-ionized hydrogen sulfide levels were calculated from pH and dissolved sulfide data using the reactions and equilibrium constants taken from Sawyer and McCarty[23] and listed below. Equilibrium constants were corrected to 35°C. Ionic strength corrections were not made.

$H_2S <=> HS^- + H^+ \qquad K_1 = 1.21 \times 10^{-7}$
$HS^- <=> S^{2-} + H^+ \qquad K_2 = 2.83 \times 10^{-13}$
$[H_2S] = DS[1 + (K_1/[H^+]) + (K_1K_2/[H^+])]^{-1}$
DS is total dissolved sulfides

RESULTS AND DISCUSSION

Attainment of Steady State

Table II summarizes the quasi-steady-state condition in the reactors with a COD/S of 20 and OLR of 0.375 g COD/L-day. The quasi-steady-state is defined arbitrarily as variation of biomass (VSS) and volatile acids concentrations within 10 and 20%, respectively. Once quasi-steady-state was achieved, feed COD/S was changed.

Table II. Quasi-Steady-State Conditions for All the Reactors

Reactor		HAc mg/L	HPr mg/L	COD mg/L	VSS mg/L	Tot. S^{2-} mg/L
R1;	X	22	123	457	528	94
	S	12	53	100	242	15
R2;	X	24	47	324	600	86
	S	19	46	100	188	23
R3;	X	28	164	564	720	103
	S	18	28	105	193	22
R4;	X	44	118	483	468	72
	S	50	56	109	151	16
R5;	X	37	123	519	656	88
	S	30	60	179	138	22
R6;	X	54	50	363	696	77
	S	46	38	91	176	15
R7;	X	23	46	338	692	80
	S	16	10	56	135	23
R8;	X	40	92	518	676	77
	S	26	43	161	149	22
All	X	34	95	446	629	85
	S	31	60	136	182	21
	n1	10	10	4	5	6
	n2	80	80	32	40	48

X = mean; n1 = number of samples analyzed in each reactor.
S = standard deviation; n2 = number of samples analyzed in all reactors.

Effect of Changing COD/S

Figures 2–10 show the effects of changing COD/S ratios at an OLR of 375 mg COD/L-d. COD/S was varied from 20 to 15, 10, and 7.5, increasing the sulfur loading rate (SLR) from 18.75 mg S/L-day to 25, 37.5 and 50 mg S/L-day, respectively, for each pair of reactors. Reactors fed with high SLR were observed to have high levels of dissolved sulfides (Figures 2–5). There was a general increase in dissolved sulfides and H_2S in the reactors during the whole experimental period. Reactors with higher

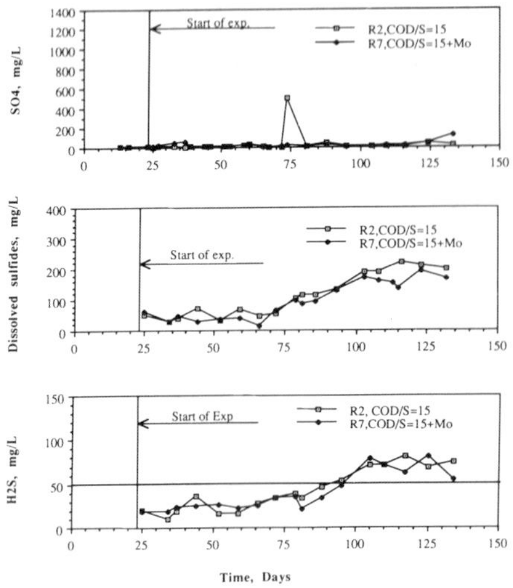

Figure 3. Effect of Molybdenum on SO_4 at COD/S = 15.

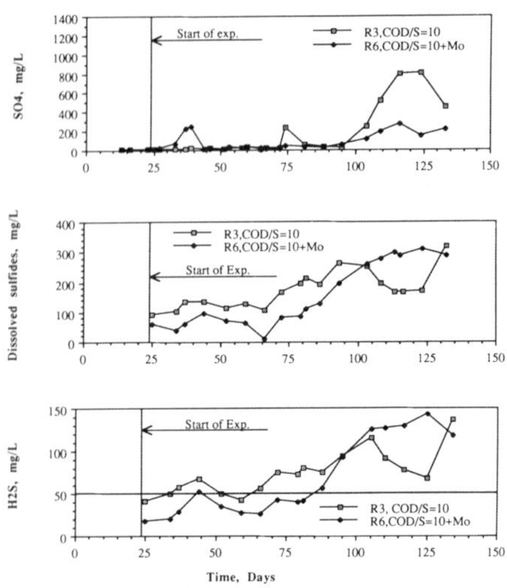

Figure 4. Effect of Molybdenum on SO_4 at COD/S = 10.

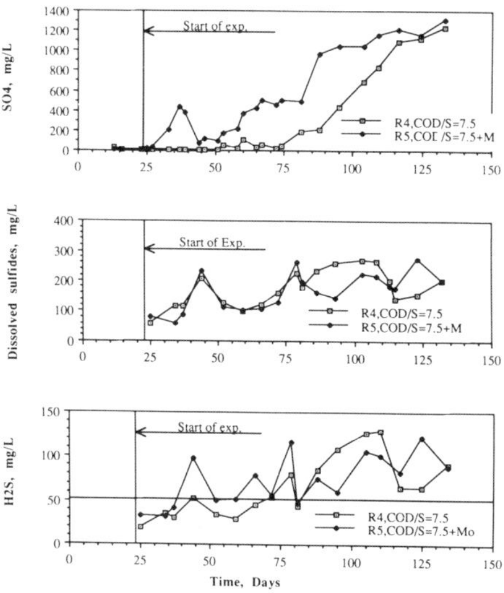

Figure 5. Effect of Molybdenum on SO_4 at COD/S = 7.5.

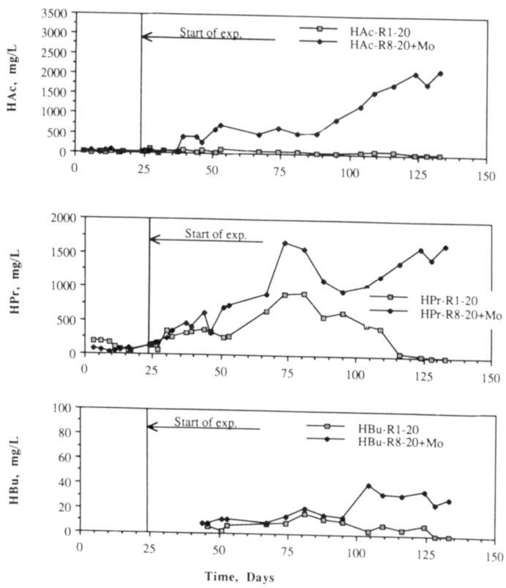

Figure 6. Effect of Molybdenum on Volatile Acids at COD/S = 20.

SLR also showed significant SO_4 in the effluent indicating that SRB were adversely affected. Reactor R4 (Figure 5) show this increased effluent SO_4 starting around day 75. As can be seen the level of H_2S was above 50 mg S/L and DS above 100 mg S/L.

As noted in the introduction, the literature suggests that sulfide levels of 100–800 mg S/L are associated with decreased process performance. These represent H_2S levels of 50–430 mg S/L, with most reports indicating that toxicity is first manifested at H_2S levels near 50 mg S/L. This level is noted as a point of reference in Figures 2–5.

Accumulation of volatile acids was most noticeable in reactors with high SLR (Figures 6–9). Reactors R3 and R4 (Figures 8 and 9) demonstrate the effect of COD/S on volatile acids. In reactor R3, HPr levels continued to rise until a peak of about 1750 mg/L was attained at around day 120 after which HPr started to decrease. HAc on the other hand was stable throughout this period. During this

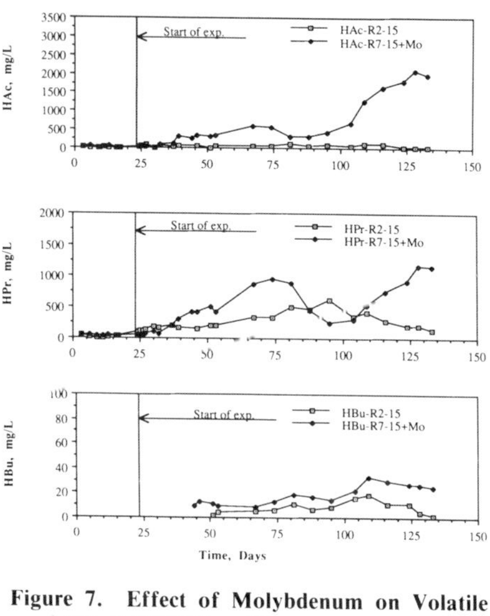

Figure 7. Effect of Molybdenum on Volatile Acids at COD/S = 15.

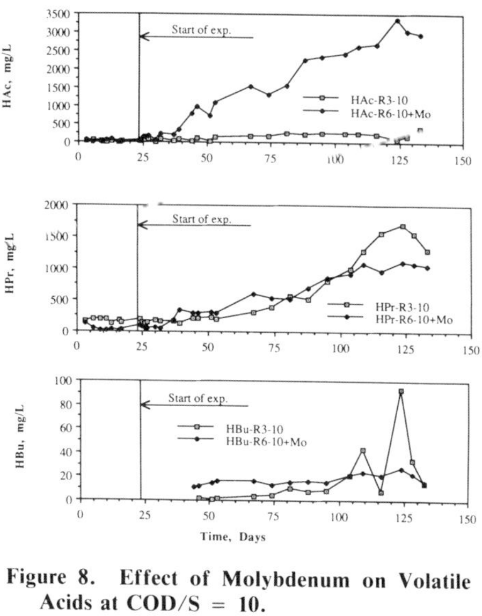

Figure 8. Effect of Molybdenum on Volatile Acids at COD/S = 10.

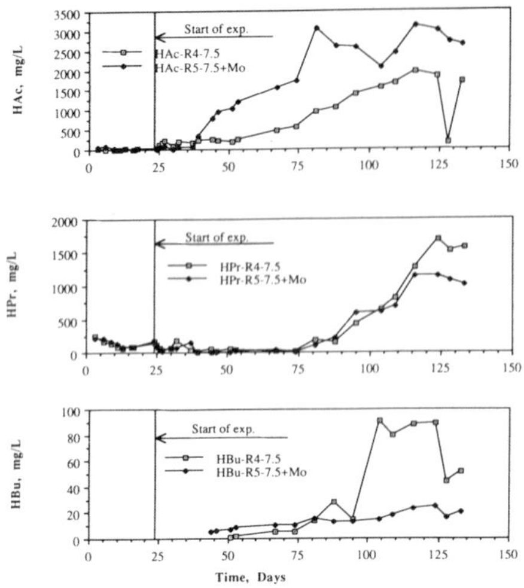

Figure 9. Effect of Molybdenum on Volatile Acids at COD/S = 7.5.

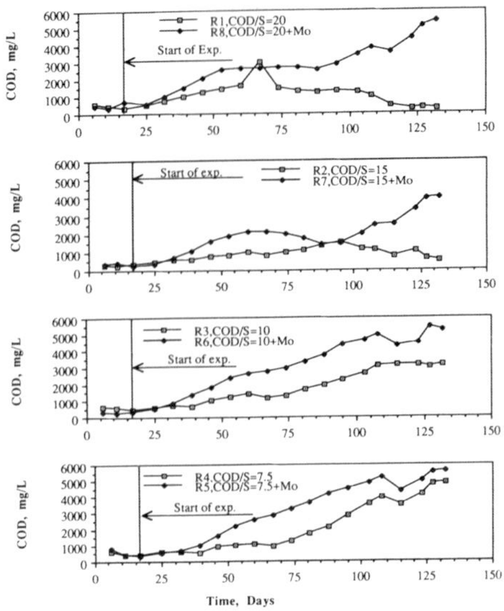

Figure 10. Effect of Molybdenum on COD.

same period H_2S levels were above 50 mg S/L and DS was above 100 mg S/L (Figure 4). Reactor R4 had both HAc and HPr rise from day 75, except for one day (day 127). Here, H_2S and DS were also above 50 and 100 mg S/L, respectively.

COD continued to increase in most reactors except for reactors R1 and R2 where it stabilized (Figure 10). With an OLR of 375 mg COD/L-d and an SRT of 25 days, the maximum theoretical COD is 9375 mg COD/L. Reactors R5, R6, and R8 had CODs greater than 5000 mg COD/L.

Effects of Molybdenum

Molybdenum was introduced to the reactors as slug doses over three days up to a level of 0.2 mM. Thereafter the concentration level was maintained by addition of a quantity equal to the wasting rate. In studying the effects of molybdenum it is important to investigate the toxicity of sulfides at the same time to differentiate the toxicity due to molybdenum alone from that due to sulfides and/or a combination of sulfides and molybdenum.

An effect of molybdenum on sulfate reducers should translate itself in decreased formation of sulfides and an increase of SO_4 in the effluent. Reactors R1, R2, R7, and R8 did not show a significant effect on sulfate reduction (Figures 2 and 3). In these reactors the levels of H_2S and DS were not significantly different within each pair though they continued to increase throughout the research period. The level of H_2S was below 50 mg S/L during most of the period. Reactors R3, R4, R5, and R6 (Figures 4 and 5) exhibited similar behavior. There was no significant difference in H_2S and DS levels between the pairs. However, H_2S was above 50 mg S/L for most of the experimental period for these reactors.

All reactors fed Mo exhibited small SO_4 peaks starting around day 30 and ending around day 45. During the same period, the H_2S level was below 50 mg S/L except for reactor R5 (Figure 5). It is during this period that one can postulate that sulfate reducers were affected by the Mo toxicity. The organisms appeared to acclimate to the Mo toxicity and quickly recovered. Significant levels of SO_4 in the effluent started to appear after day 95 in reactors R3 and R6. For reactor R4 the increase started after day 75, while for reactor R5 the recovery from the initial slug dose lasted for a short period and on day 52 increases started which lasted until the end of the research. In all cases the level of H_2S was above 50 mg S/L when the increase occured. DS levels were above 100 mg/L. It was inferred that Mo was perhaps initially toxic to SO_4 reducers and that the SO_4 reducers recovered from the Mo toxicity quickly. Reactors with COS/S of 10 and 7.5 were affected by sulfide toxicity mainly.

One of the goals of the research was to study whether the effects of Mo toxicity are more pronounced on SO_4 reducers as compared to the methane-producing consortia. Butyric acid (HBu), propionic acid (HPr), and acetic acid (HAc) are products of anaerobic degradation of glucose. Levy et al.[24] found that acetate can also be formed from CO_2 and H_2 when methanogenesis is inhibited. Thus, if Mo inhibits methanogenesis then it is possible to have substantial accumulation of HAc as a result of failure of acetate utilizing methanogens and from formation from CO_2 and H_2. Figures 6–9 indicate that the response to Mo for HAc comes 10 to 15 days after dosing. All reactors fed with Mo started accumulated HAc and this accumulation continued to the end of the research. In contrast, for the reactors not fed Mo, substantial accumulation is seen only in reactor R4. This accumulation coincides with the period when the level of H_2S is above 50 mg S/L. Considering the set of reactors R1 and R8 which had a COD/S ratio of 20 (Figure 6), the level of H_2S did not rise above 50 mg S/L until after day 100 for both reactors. Reactor R8 had an accumulation of HAc throughout the experimental period while there was no accumulation on reactor R1. It is reasonable to assume that the acetate utilizers were affected by Mo toxicity from these data.

HPr utilizers were also affected by the Mo toxicity. Considering reactors R1 and R8 (Figure 6), HPr started accumulating in both reactors immediately after R8 was dosed with Mo. There was a lot more accumulation in reactor R8 than in reactor R1. it is not clear what is causing the stress in reactor R1, but this environmental condition appeared to peak around day 75 and thereafter HPr levels decreased to zero. In contrast, the HPr accumulation in reactor R8 continued to rise until the end of the research. It was concluded that reactor R8 initially responded to Mo toxicity and later to a combination of Mo and sulfide toxicity after day 100. The H_2S in this pair of reactors was above 50 mg S/L after day 100.

The systems with higher SLR (COD/S = 10 and 7.5) also show evidence of Mo toxicity to HAc utilizers. Figures 8 and 9 show significant differences in levels of HAc between reactors fed Mo (R5 and R6) and those without Mo (R3 and R4). There was no significant difference in levels of HPr in these pairs of reactors. It would appear that HPr was responding to sulfide toxicity as the level of H_2S was above 50 mg/L in these reactors and that of DS was above 100 mg S/L. Since all reactors fed Mo received equal amounts, and that only reactors with high SLR did not show significant effects on Mo on higher acids (HPr and HBu), it is possible that Mo was being removed out of the solution by either being precipitated as a sulfide or by being chelated. The reduced Mo in the solution would not be sufficient to affect all acid utilizers. It is reasonable to conclude that these higher acids accumulations were mostly due to sulfide toxicity.

In summarizing the effects of Mo on volatile acids, the following observations are noted:

- Mo definitely affected HAc utilizers.
- The effect of Mo on other acids, HPr and HBu, was worse at high COD/S ratios (20 and 15) than at low COD/S ratios (10 and 7.5). This is possibly due to removal of Mo from active participation in the solution by reaction with produced sulfides.
- In all systems there was very little difference in H_2S and DS levels between reactors fed Mo and those without Mo. Thus, with a large difference in levels of acids between reactors in a pair of reactors, this difference is probably attributable to Mo toxicity.

Figure 10 illustrates the effect of Mo on effluent COD. It is important to emphasize the direct relationship between the levels of volatile acids in the reactors and the COD of the effluent. In all cases, the reactors fed Mo had higher COD than their twin with similar COD/S ratio. This is also evident from Figure 11a.

Data for glucose (Figure 12) show there was very little effect on glucose conversion until after day 95. The data shown is typical for all reactors. The fermentative bacteria which are responsible for converting glucose to HBu, HPr and HAc remained unaffected for most of the experimental period.

SUMMARY AND CONCLUSIONS

Table III contains a summary of performance of the reactors towards the end of the investigation. An average of results for the last three weeks was used.

When feeding Mo it had been expected that the sulfate reducers would be selectively poisoned leaving the methane-producing consortia to dominate. This did not happen. Reactor R5 with the lowest COD/S and fed Mo show only 43% of COD removal. Reactor R8 showed 47% COD removal and 94% SO_4 removal. Whereas the low sulfate feed to R8 may not have generated sufficient sulfides to remove Mo out of active participation in the reactors, the methane-producing consortium was perhaps more severely inhibited by the available Mo in this particular reactor than in the reactors with low COD/S ratios (R6 and R7). Reactors R4 and R5, the two reactors with the lowest COD/S, had poor SO_4 removals compared with the others. It is the effect of sulfide toxicity on SRB that was dominant here; the effect of Mo was not evident. Glucose removals were very high in all systems.

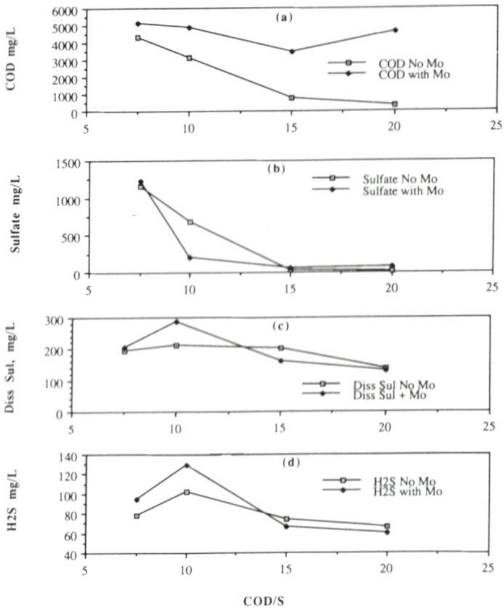

Figure 11. Effect of COD/S: a) COD vs. COD/s, b) Sulfate vs. COD/S, c) Dissolved sulfide vs. COD/S, d) H_2S vs. COD/S.

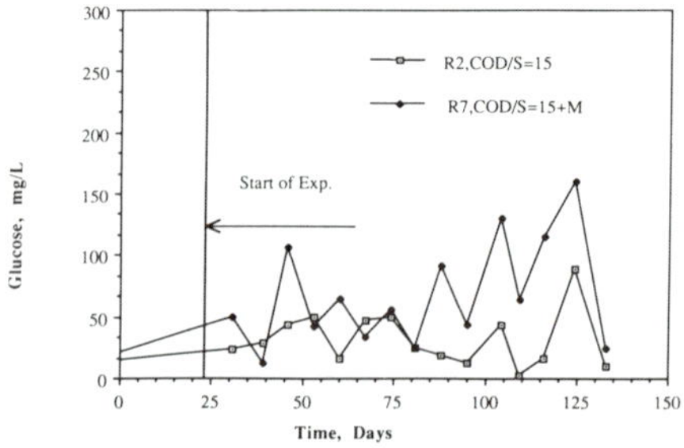

Figure 12. Typical effect of Molybdenum on glucose.

The following conclusions are drawn from the investigation:

- Reactors with a high COD/S ratio had good sulfate removals and/or conversions. Those reactors showing any significant SO_4 accumulation were thought to be affected by sulfide toxicity. The effect of Mo on SRB at the concentrations used in the reactors was not significant. There was an initial response when Mo was dosed but the organisms appeared to acclimate to it.
- Glucose removals/conversions were very high in all systems. Fermentative bacteria were therefore very slightly affected by the form of toxicity in the reactors.
- Stress appeared in the form of the accumulated volatile acids HAc, HPr, and HBu. Mo was toxic to acid utilizers. Methane-producing bacteria were the most affected as indicated by significant accumulation of HAc in all systems fed Mo. Mo affected HPr utilizers more at high COD/S than it did at low COD/S. The same pattern was seen for HPu utilizers. It is suggested that at low COD/S ratios, Mo was removed from active participation by produced sulfides. The combination of sulfide toxicity and Mo that remained caused the effect seen in these reactors. At high COD/S there was a definite effect of Mo on acid utilizers.
- The research showed that it is not feasible to selectively poison sulfate reducers using Mo under the conditions tested.

Table III. Summary of Performance During Last Three Weeks (Average of 4 Samples)

Reactor	Glucose Removal %	COD Removal %	SO_4 Removal %	H_2S (mg S/L)	DS (mg S/L)
R1,COD/S = 20	>99	97	>99	55	138
R2,COD/S = 15	>99	92	98	71	202
R3,COD/S = 10	>99	67	78	102	213
R4,COD/S = 7.5	>99	51	69	78	195
R5,COD/S = 7.5 + Mo	98	43	67	105	206
R6,COD/S = 10 + Mo	98	46	94	130	285
R7,COD/S = 15 + Mo	99	60	96	68	160
R8,COD/S = 20 + Mo	98	47	94	85	130

ACKNOWLEDGMENTS

The research was supported by funds provided by African American Institute (AAI) which runs the African Graduate program (AFGRAD) for the United States Aid for International Development (USAID), and by the Biotechnology Byproducts Consortium at The University of Iowa.

REFERENCES

1. Balba, M. T., and Nedwell, D. B., "Microbial Metabolism of Acetate, Propionate and Butyrate in Anoxic Sediment from the Colne Point Saltmarsh, Essex, U.K.," *Journal of General Microbiology, 128*, 1415–1422, 1982.
2. Banat, I. M., Nedwall, D. B. and Balba, M. T., "Stimulation of Methanogenesis by Slurries of Saltmarsh Sediment after the Addition of Molybdate to Inhibit Sulphate-reducing Bacteria," *Journal of General Microbiology, 129*, 123–129, 1983.
3. Winfrey, M. R. and Ward, D. M., "Substrates for Sulfate Reduction and Methane Production in Intertidal Sediments," *Applied and Environmental Microbiology, 1*, 193–199, 1983.
4. Rudolfs, W. C., and Amberg, H. R., "White Water Treatment II. Effects of Sulfides on Digestion." *Sewage and Ind. Wastes*, 24, 1278 (1952).
5. Aulenbach, D. B., and Huekelekian, H., "Transformation and Effects of Reduced Sulfur Compounds in Sludge Digestion." *Sewage Ind. Wastes*, 27, 1147 (1955).
6. Lawrence, A. W. and McCarty, P. L., "The Role of Sulfide in Preventing Heavy Metal Toxicity in Anaerobic Treatment," *Journal of Water Pollution Control Fed., 37*, 392–406, 1966.
7. Cappenberg, T. E., "A Study of Mixed Continuous Cultures of Sulfate-reducing and Methane-producing bacteria," *Microbial Ecology, 2*, 60–72, 1975.
8. Bryant, M. P., Campbell, L. L., Reddy, C. A. and Crabill, M. R., "Growth of *Desulfovibrio* in Lactate or Ethanol Media in Sulfate in Association with H_2-utilizing Methanogenic Bacteria," *Applied and Environmental Microbiology, 33*, 1162, 1977.
9. Khan, A. W. and Trottier, T. M., "Effect of Sulfur-Containing Compounds on Anaerobic Degradation of Cellulose to Methane by Mixed Cultures obtained from Sewage Sludge," *Applied Environmental Microbiology, 35*, 1027, 1978.
10. Winfrey, M. R., and Zeikus, J. G., "Effect of Sulfate on Carbon and Electron Flow during Microbial Methanogenesis in Freshwater Sediments," *Applied and Environmental Microbiology, 33*, 275–281, 1977.
11. Mountfort, D. O. and Asher, "Effect of Inorganic Sulfide on the Growth and Metabolism of *Methanosarcina barkeri strain DM*," *Applied Environmental Microbiology, 37*, 670, 1979.
12. Parkin, G. F., Speece, R. E., Yang, C. H. J. and Kocher, W. M., "The Response of Methane Fermentation Systems to Industrial Toxicants," *Journal of Water Pollution Control Federation, 55*, 44, 1983.
13. Parkin, G. F., Sneve, M. A., Loos, H., "Anaerobic Filter Treatment of Sulfate-Containing Wastewaters" (1990) Accepted for publication in *Water Science and Technology*.
14. Parkin, G. F., Lynch, N. A., Kuo, W. C., Van Keuren, E. L., Bhattacharya, S. K. "Interaction Between Sulfate Reducers and Methanogens Fed Acetate and Propionate" (1990) Accepted for publication in the *Journal Water Pollution Control Federation*.
15. Koster, I. W., et al., "Sulfide Inhibition of the Methanogenic Activity of Granular Sludge at Various pH-Levels." *Water Research*, 20, 1561 (1986).
16. Hilton, B. L. and Oleszkiewicz, J. A. "Sulfide-induced Inhibition of Anaerobic Digestion". (1988) *J. Env. Eng. Div., ASCE, 114*, pp. 1377–1391.
17. Kroiss, H., and Plahl-Wahnegg, F., "Sulfide Toxicity with Anaerobic Waste Treatment." In *Advanced Anaerobic Waste Treat. Conf.*, Amsterdam, AWWT Symposium Secret., The Hague, pp. 72–85 (1983).
18. Gunnarson, L. A. H., and Ronnow, P. H., "Interrelationships Between Sulfate Reducing and Methane Producing Bacteria in Coastal Sediments with Intense Sulfide Production." *Marine Biology*, 69, 121 (1982).
19. Postgate, J. R., *The Sulfate Reducing Bacteria*, Cambridge: Cambridge Univ. Press (1979).
20. Banat, I. M., Lindstrom, E. B., Nedwell, D. B. and Balba, M. T., "Evidence for the coexistence of two distinct functional groups of sulfate-reducing bacteria in salt marsh sediment," *Applied and Environmental Microbiology, 42*, 985–992., 1981.
21. Carlson, J. M., "Propionate Production from Acetate in Acetate Enrichment Cultures," M.S. Thesis, University of Iowa, Iowa City, IA, 1988.
22. American Public Health Assn., Water Pollution Control Federation, *Standard Methods for the Examination of Water and Wastewater, 15th Ed.*, American Public Health Assn., New York, N.Y., 1985.

23. Sawyer, C. N. and McCarty, P. L., "*Chemistry for Environmental Engineering*," 3rd Ed.; McGraw-Hill, 1978.
24. Levy, P. F., Bernard, G. W., Garcia-Martinez, D. V., Sanderson, J. E., and Wise, D. L., "Organic Acid Production from CO_2/H_2 and CO/H_2 by Mixed Culture Anaerobes," *Biotech and Bioengineering, 23*, 2293, 1981.

52 A STUDY INTO THE EFFECT OF A SERIES OF SHOCKS ON A PILOT-SCALE ANAEROBIC FILTER

M. E. Caine, Senior Research Associate

G. K. Anderson, Senior Lecturer

T. Donnelly, Research Associate
Department of Civil Engineering
University of Newcastle upon Tyne
Newcastle upon Tyne
United Kingdom NE1 7RU

INTRODUCTION

The imposition of more stringent industrial consents to discharge led to companies examining the potential of in-house treatment.[1] Of the options available, high rate anaerobic digestion processes such as the Anaerobic Filter[2] and the UASB[3] have inherent advantages over both aerobic and physical/chemical techniques. These include: low operating costs, low sludge yield and the production of a high grade fuel gas.

Scepticism has resulted in inertia within process industries when considering anaerobic digestion. Doubts have been based on several factors: (i) the poor performance of many full-scale UASBs of crude design treating non-saccharide waste, (ii) the apparent lack of comprehensive process control, (iii) the time needed to develop the required concentration of biomass at a startup and (iv), the resistance of a "sensitive" biological process to accidental shocks.

Careful process selection and the employment of skilled operators will overcome (i) and (ii). However the consequence of reactor failure following severe shock is usually reseeding and restart. This is very expensive in terms of time and seed sludge. The possibility of litigation also arises.

This study was designed to show the effect of a series of severe shocks (pH, temperature, organic and hydraulic) on a pilot-scale Anaerobic Filter, (AF), operating on a dairy effluent. Effluent quality, gas production and process stability were monitored throughout an eight hour shock period and for a further sixteen hours. The relationship between the above were discussed.

MATERIALS AND METHODS

Description of Process

The study was carried out at the SERC anaerobic digestion facility based on-site at Birds-Eye Walls in Gloucester. The facility comprised a 13 m^3 balance tank supplying four reactors: Anaerobic Filter, UASB, Contact Process (all nominally 5m^3) and an Expanded Bed (0.5m^3). The actual liquid volume of the Anaerobic Filter was 5.28 m^3. The media (a variety of Pall ring with A:V ratio 160m^2:m^3) were contained in two 1.64 m^3 compartments 1.5 m above the base of the reactor. The balance tank and Anaerobic Filter are illustrated in Figure 1.

Composition of Wastewater

The wastewater was produced by mixing ice cream and water-ice products to a target COD of 6000 mg l^{-1}. The two main CIP (cleaning in place) detergents were also added at the concentrations found in the factory wastewater. This accurately represented an ice-cream factory effluent. The balance tank pH was maintained automatically between 6.60 and 6.80 using sodium hydroxide. Nutrients, N and PO_4 (dosed at COD:N:PO_4, 100:2:1) and chelated trace elements (Fe,Mn,Co,Ni dosed at 0.02 mg l^{-1} metal) were added manually. The physical and chemical characteristics of the raw wastewater during steady state prior to the shock study are given in Table I.

45th Purdue Industrial Waste Conference Proceedings,

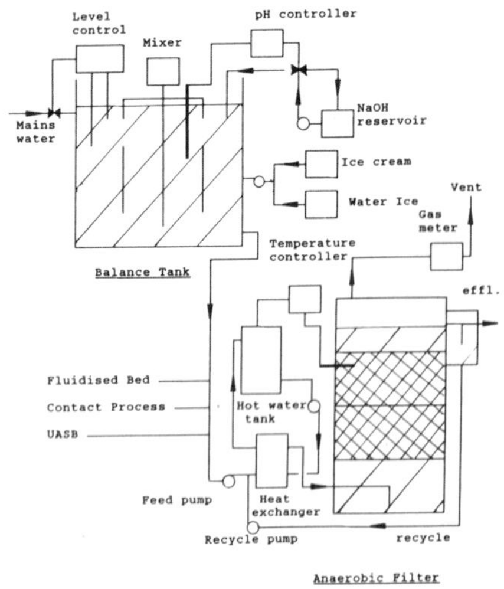

Figure 1. Balance tank and pilot scale anaerobic filter.

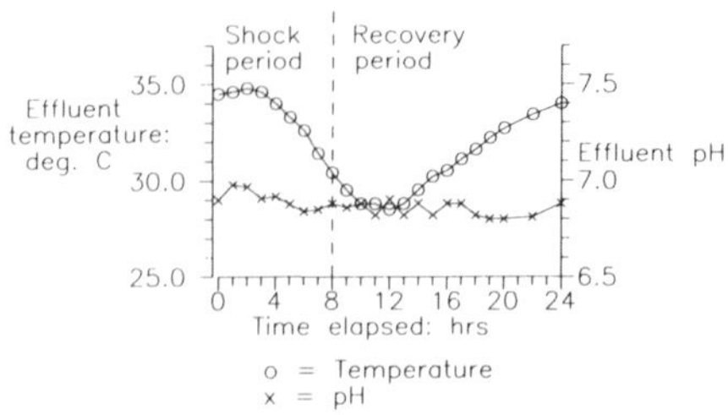

Figure 2. Variation in effluent temperature and pH during low temperature shock.

Mixing Characteristics

A sophisticated mathematical model was used to simulate the results of a series of exit age distribution experiments. From this the volumes for the mixed zone, plug flow zone and dead zones within the reactor were defined. The reactor was predominately mixed ($4.8m^3$) with the remainder of the volume in the reactor ($0.5m^3$) plug flow in character. There was negligible dead zone. This confirmed the observation of Young and Young.[4] At the steady state described above, when gas production consistently exceeded $1m^3$ m^{-3} reactor·day, recycle (and upflow velocity) was found to have little effect on mixing or performance. At lower loading rates (applied during start-up or as a result of a diurnal variation in feedstrength) the resultant drop in gas production was seen to effect mixing. This was manifest in rapid channelling of cold, untreated wastewater, although the introduction of recycle prevented this.

Steady State Performance

The AF was in continuous 7-day operation for a period of 10 weeks prior to the shock study. The OLR was increased from 2.5–3.0 to 6.0–8.0 kg COD m^{-3} day^{-1} over a 3 week period. This was followed by a 5 week period during which the effect of upflowvelocity/recycle on performance and

Table I. Wastewater Characteristics

Parameter	Units	Value	Sample Frequency	Sample Type
Average COD	mg l^{-1}	5770	daily	composite
COD range	mg l^{-1}	4020–7550	daily	composite
Soluble COD	mg l^{-1}	2220	weekly	composite
Suspended solids	mg l^{-1}	1180	daily	composite
Volatile suspended solids	mg l^{-1}	1120	daily	composite
Suspended oil and grease	mg l^{-1}	610	weekly	composite
Total VFA	mg l^{-1}	630	twice weekly	grab
TKN	mg l^{-1}	164	weekly	composite
NH_3-N	mg l^{-1}	58	weekly	composite
PO_4-P	mg l^{-1}	18	weekly	composite
Total Alkalinity	mg l^{-1}	1190	weekly	grab
Bicarbonate alkalinity	mg l^{-1}	400	weekly	grab
VFA Alkalinity	mg l^{-1}	790	weekly	grab

Table II. Steady State Effluent Characteristics and Reactor Performance

Parameter	Units	Value	Sample Frequency	Sample Type
OLR	kg COD m^{-3} day^{-1}	7.8	–	–
Gas production	l $hr.^{-1}$	600	–	–
Applied CH_4 yield	m^3CH_4 kg COD^{-1}	0.28	–	–
Methane	%	75	Twice weekly	grab
Settled COD	mg l^{-1}	810	Daily	composite
Total COD	mg l^{-1}	1500	Daily	composite
Set. COD removal	%	86	–	–
Suspended solids	mg l^{-1}	640	Daily	grab
Volatile suspended solids	mg l^{-1}	550	Daily	grab
Total VFA	mg l^{-1}	58	Twice weekly	grab
TKN	mg l^{-1}	91	Weekly	composite filtered
NH_3-N	mg l^{-1}	68	Weekly	composite filtered
PO_4-P	mg l^{-1}	18	Weekly	composite filtered
Total alkalinity	mg l^{-1} $CaCO_3$	1600	Weekly	grab
VFA alkalinity	mg l^{-1} $CaCO_3$	150	Weekly	grab
Bicarbonate alkalinity	mg l^{-1} $CaCO_3$	1450	Weekly	grab

mixing was examined. A quasisteady state was then established over a 2 week period. Details of this are given in Tables II and III. At this point the HRT was between 16 and 17 hours. The degree of recycle was set at 130% to give an upflow velocity of 17m day^{-1}. The shock experiments were carried out over a 4 week period.

Analyses

Analyses for SS,VSS,COD and Alkalinity were carried out according to *Standard Methods*.[5] Settled COD refers to the COD of effluent supernatant which was allowed to settle in a 100ml measuring cylinder for 30 minutes. Settleable solids refer to those solids which settled in 30 minutes and was given by the difference between the unsettled effluent SS and the settled effluent supernatant SS. Volatile Fatty Acids and gas composition were analysed by gas chromatography. Specific methanogenic activity was measured using a Warburg respirometer.[6]

Shock Studies

The shock studies were designed to indicate the effect of operator error on the process, the error being rectified after 8 hours, equivalent to one working shift. Analyses were carried out throughout the shock period and for a further sixteen hours, approximately one hydraulic residence time. The programme of analysis was modified slightly according to the shock that was applied. The standard framework of tests is given in Table IV.

Analyses of liquor (profile) suspended solids, volatile suspended solids, volatile fatty acids, filtered COD, pH and temperature were carried out at 0, 12 and 24 hours.

RESULTS AND DISCUSSION

Low Temperature Shock

The shock was applied by switching the main heating elements off for eight hours. Following an initial increase in effluent temperature a steady fall was noted as shown in Figure 2. Temperature profiles (Figure 3) revealed that the temperature in the bed of the reactor fell to 25.5°C at the end of

Table III. Steady State Reactor Profile Data

Sample Port	Reactor Height:m	ss:mg l^{-1}	vss:mf $^{-1}$	Total VFA:mg l^{-1}
1	0.5	2160	1910	380
2	1.95	1850	1680	160
3	3.4	1000	870	65
Effluent	4.4	620	570	58

Table IV. Programme of Analysis

Test	Frequency hr.$^{-1}$	Test	Frequency hr.$^{-1}$
Feed pH	1	Feed COD	2
Feed temperature	1	Effluent COD	2
Effluent pH	1	Feed ss,vss	2
Effluent temperature	1	Effluent SS,VSS	2
Gas production	1	Gas composition	4
Feed VFA	2	Eff. VFA	2

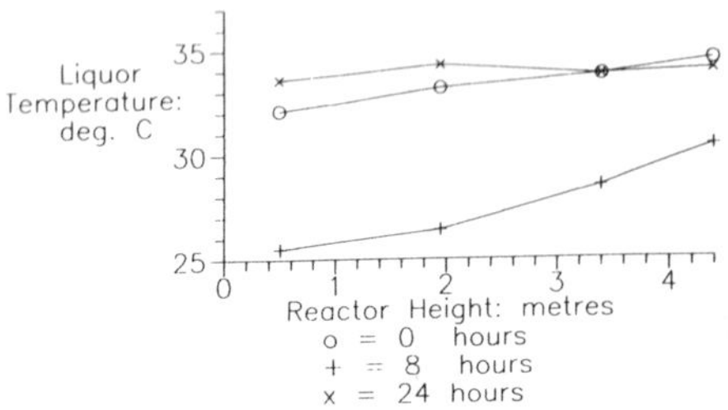

Figure 3. Variation in rector profile temperature during low temperature shock.

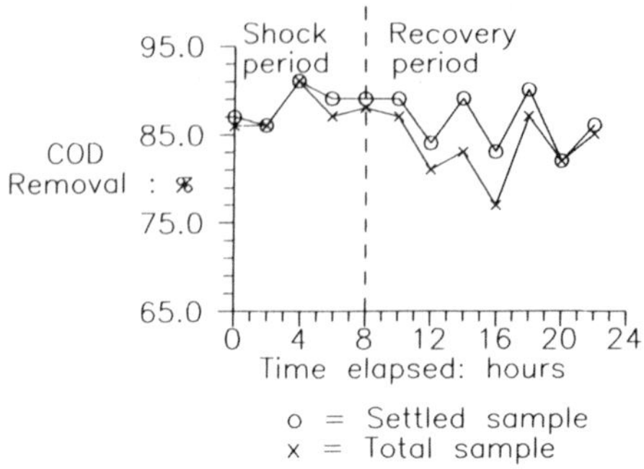

Figure 4. Variation in COD removal during low temperature shock.

the shock period. Sixteen hours after the power had been restored to the heaters the temperature throughout the reactor was still slightly below the operating range of 35–75°C.

Performance

The effluent quality remained high throughout the study. Figure 4 illustrates the consistently high level of COD removal. The small disparity between settled and total COD removal shown in Figure 5 reflected the low settleable solids content of the effluent which averaged 70 mg l^{-1}. Diurnal variations in feed strength were reflected in effluent COD as shown in Figure 6.

Methane production decreased marginally relative to the organic load applied during the shock period (Figure 7). Gas composition also showed little variation over 24 hours ranging between 66% and 73% methane. Gas production increased once power was returned to the heaters revealing the increasing activity of biomass within the reactor. Methane yield also climbed during the recovery period. This complemented the phase of slightly reduced membrane yield noted between hours 4–10.

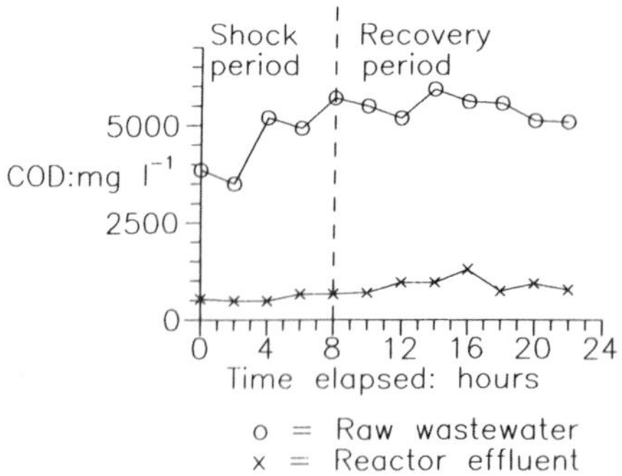

Figure 5. Wastewater and effluent COD during low temperature shock.

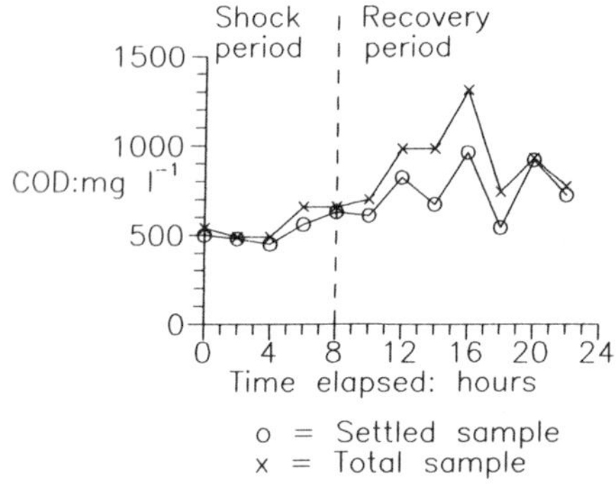

Figure 6. Settled and total effluent COD during low temperature shock.

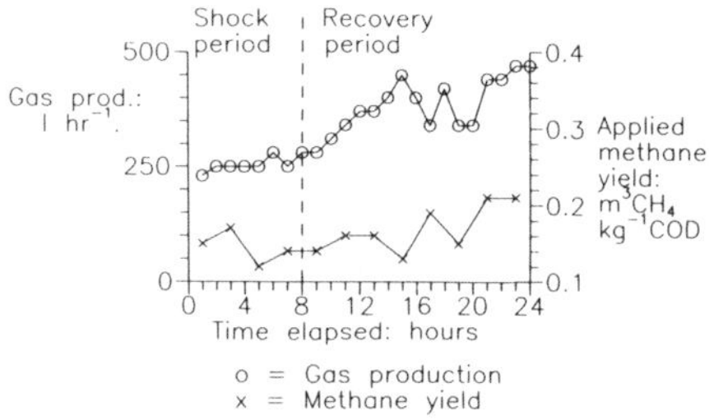

Figure 7. Gas production and methane yield during low temperature shock.

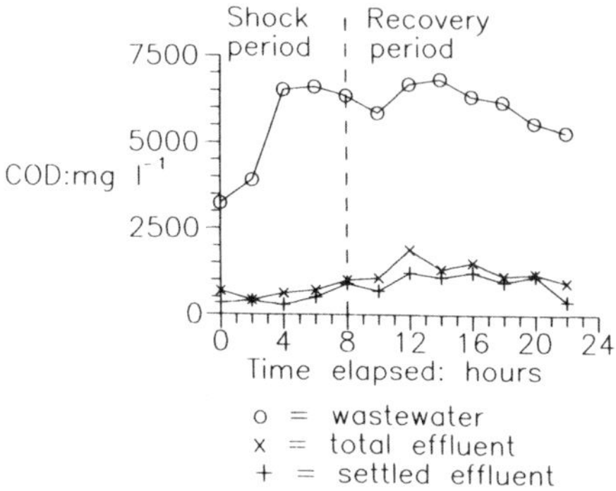

Figure 8. Wastewater COD, settled and total effluent COD during pH shock.

This effect cannot be ascribed to the increase in feed strength seen between 4–6 as such increases characteristically produced as immediate increase in gas production.

The stability of the process was not disturbed to any degree as shown by the effluent volatile acids remaining below 50 mg/l throughout the study. No marked variation was observed in effluent pH which lay between 6.60–6.80 over 24 hours.

The reason that the process withstood this shock was the use of recycle. Recycling hot liquor prevented the temperature at the bed of the reactor falling immediately to that of the raw wastewater which was between 19.4 and 20.4 °C during the experiment. The activity of the sludge bed was not severely reduced and sufficient gas production maintained to retain the completely mixed character of the process. The cooling load was therefore spread quite uniformly throughout the reactor.

Without recycle, a cold slug of untreated of wastewater would enter the reactor. Activity would decrease leading to a fall in gas production. On restoring power the partially treated slug would be displaced as a very poor effluent although the long term stability of the process would not be affected. This was found to be the case in an earlier low temperature shock carried out at a different steady state.

Low pH Shock

The low pH shock was caused by turning off caustic recirculation for 8 hours. Hydrolytic and preacidogenic activity reduced the balance tank pH to 5.29 at the end of the shock period. Once pH control was returned the balance tank pH returned to the set point (6.70) only 3 hours.

Performance

The effluent quality remained satisfactory throughout the study as illustrated in Figure 8. Small variations in both effluent quality and COD removal reflected diurnal changes in feed strength (Figures 8 and 9).

Gas production was not significantly affected by the decrease in feed pH and increased broadly in the line with organic load as shown in Figure 10. Any decrease in gas production due to low wastewater pH may have masked this increase while the consistent methane yield suggested that this was not the case.

Resistance to low pH can be ascribed to the alkalinity reserve within the liquor which ranged between 1500 and 2000 mg l^{-1} $CaCO_3$ (total alkalinity) at steady state. This was sufficient to buffer the effect of unneutralised volatile acids introduced into the reactor.

Organic Shock

Concentrated ice-cream and water-ice products were pumped into the balance tank to raise the wastewater COD to approximately 12,000 mg l^{-1} for 8 hours. Some difficulty was experienced in increasing the COD quickly and four hours elapsed before the COD was raised to above 11000 mg l^{-1}. Once a further 8 hours had elapsed, the influent COD was allowed to fall to 6000 mg l^{-1}.

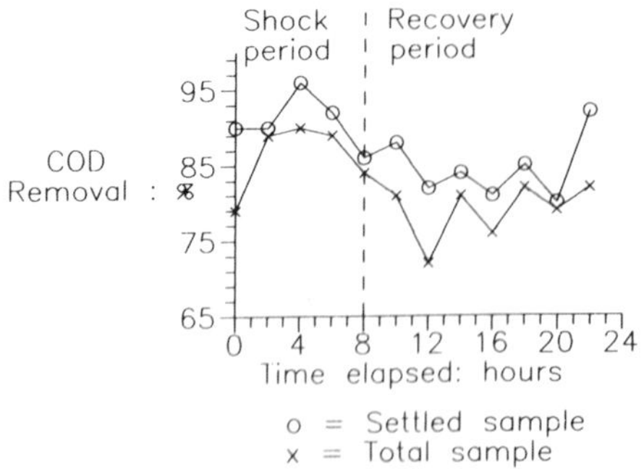

Figure 9. Variation in COD removal during pH shock.

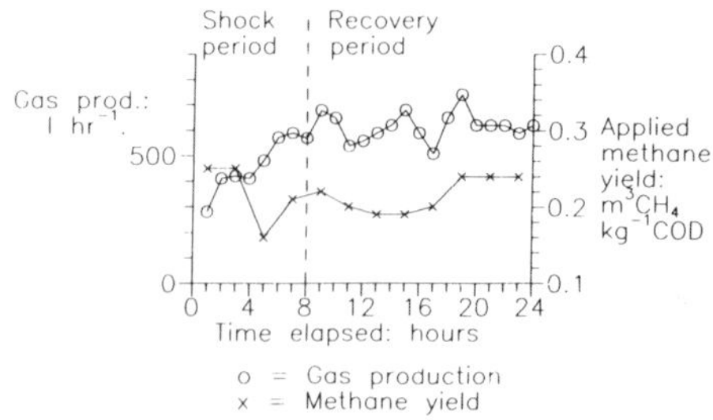

Figure 10. Gas production and methane yield during pH shock.

Performance

The effluent quality decreased gradually throughout the period of the organic shock with the increase in effluent COD disproportionate to that of the wastewater. Effluent COD and COD removal are shown in Figures 11 and 12.

The difference between settled and total COD removal was clearly exaggerated at some points due to the loss of settleable solids. Figure 13 indicates the degree of settleable solids loss over the 24 hours. These had the appearance of milk solids rather than biomass. Reactor suspended solids profiles indicated that the level of solids in the bed of the reactor had remained steady, lying between 23000 and 25000 mg l^{-1} suspended solids over 24 hours. Analysis of the suspended oil and grease content of

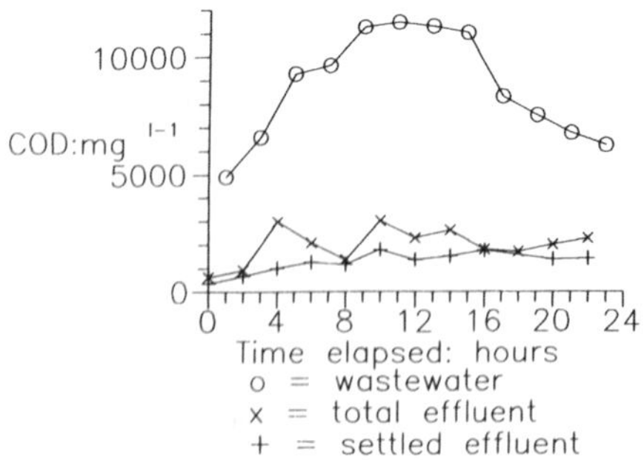

Figure 11. Wastewater COD, settled and total effluent COD during organic shock.

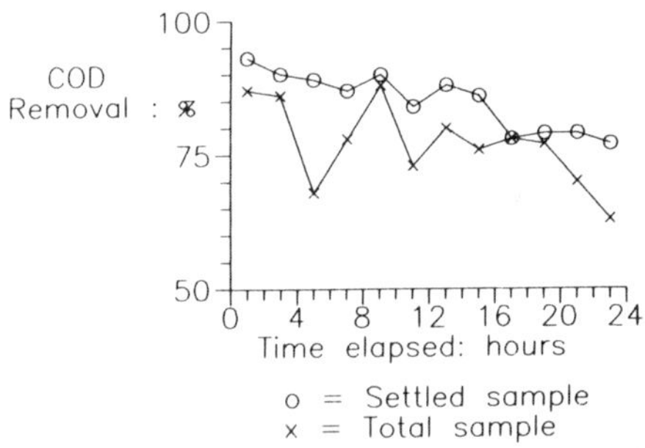

Figure 12. Variation in COD removal during organic shock.

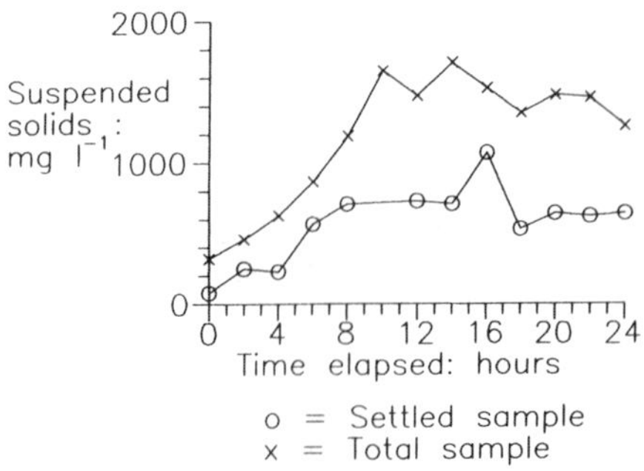

Figure 13. Effluent suspended solids during organic shock.

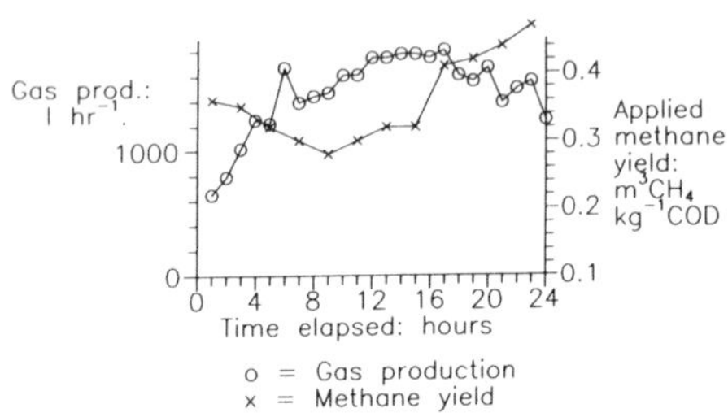

Figure 14. Gas production and methane yield during organic shock.

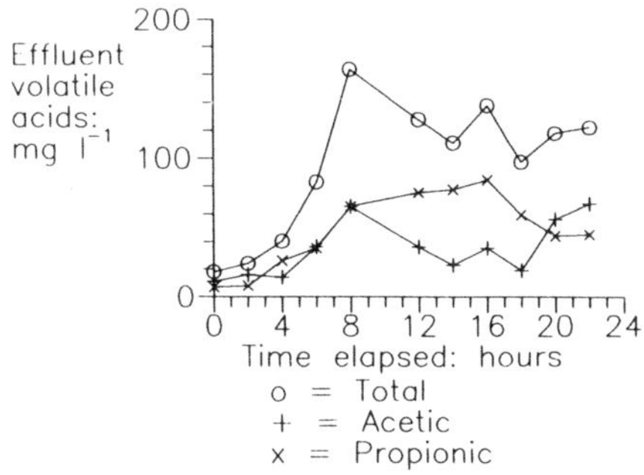

Figure 15. Effluent volatile acids duirng organic shock.

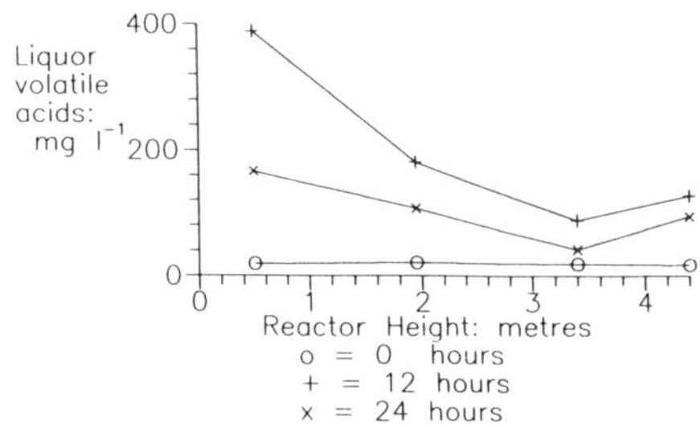

Figure 16. Variation in reactor profile volatile acids during organic shock.

the effluent showed that this had decreased in line with COD removal falling from 90% to 78% during 24 hours. The increase in settleable solids was therefore attributed to insoluble milk proteins and small quantities of bacterial flocs and fragments washed from both the sludge bed and the packing as a result of increased gas production.

The increase in gas production broadly reflected the increase in feed strength. The lower COD removal was confirmed by a decrease in methane yield during the shock period. As the shock was stopped the methane yield rose to above the theoretical maximum as residual volatile acids in the liquor were consumed. Both gas production and methane yield are illustrated in Figure 14.

The methane content fell immediately from 81% to 67% on application of the shock. This cannot be attributed to changes in pH displacing more CO_2 from the liquor. At steady state the increase in loading rate will not produce an intrinsic change in the biogas composition given that the ratio of components (fat:carbohydrate:sugar) remains constant.[7] Analysis of the wastewater composition during the shock showed the proportion of fat to have increased to 60% of the increased wastewater COD. At steady state this would have increased the proportion of methane in the off-gas.

Stability was maintained throughout the shock period although there was evidence that the system was under stress. In addition to reductions in methane content, methane yield and COD removal total effluent volatile acids increased from below 50 mg/l to 164 mg l^{-1} after 8 hours (Figure 15). The ratio of propionic to acetic acid, which was close to unity at steady state, increased to between 2-3:1 in the effluent and almost 5:1 in the bed of the reactor. The increase and recovery in reactor VFA concentration is shown in Figure 16. No significant variation in reactor or effluent pH was noted despite the increases in volatile acids.

The increase in volatile fatty acids may be explained as complex responses by the microbial ecology to the shock applied. The activity of individual groups of bacteria within a mixed anaerobic population is regulated by the partial pressure of hydrogen in the liquor.[8] Hydrogen is produced as a result of many fermentative reactions which ultimately generate the methanogenic precursors, namely, acetic acid, carbon dioxide and hydrogen.[9] However only autotrophic methanogens consume hydrogen. This obligate H_2-utilizing group of methanogens is regarded as the most sensitive to changes in the environment[10] (e.g. in pH and redox potential) induced as a result of shock. On application of the shock, the partial pressure of hydrogen increased as the autotrophic methanogens were inhibited. This is manifest indirectly by the change in gas composition as both methane production and carbon dioxide assimilation decrease. Acetogenic bacteria, responsible for the conversion of propionic acid to acetic acid are rendered inactive by the increase in hydrogen partial pressure. A synchronous change in the acidogenic metabolism of glucose may have also occurred, reducing the hydrogen partial pressure and generating propionic acid, i.e. . . .

$$C_6H_{12}O_6 + 2H_2 \text{ -----> } 2CH_3COOH + 2H_2O \quad (1)$$

Both result in an increase in the propionic acid concentration.

No evidence of acetic acid overload[11] was found prior to the increase in propionic acid. The production of propionic acid as a primary response to shock may be characteristic of a bacterial population developed from fat based wastewater. Such cultures may have relatively high concentrations of acetoclastic methanogens, since primary product of triglyceride fermentation is acetic acid.

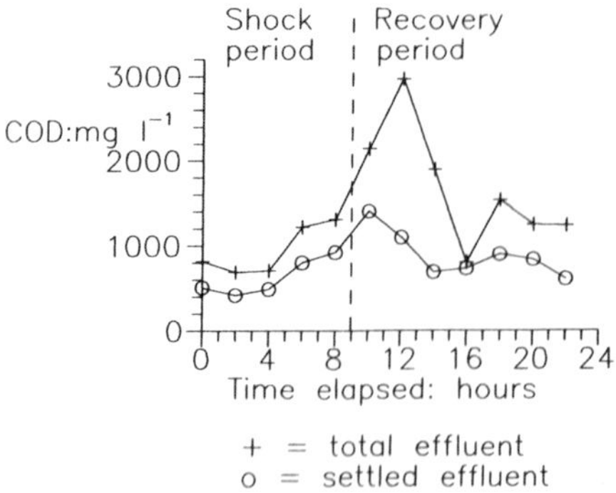

Figure 17. Settled and total effluent COD during hydraulic shock.

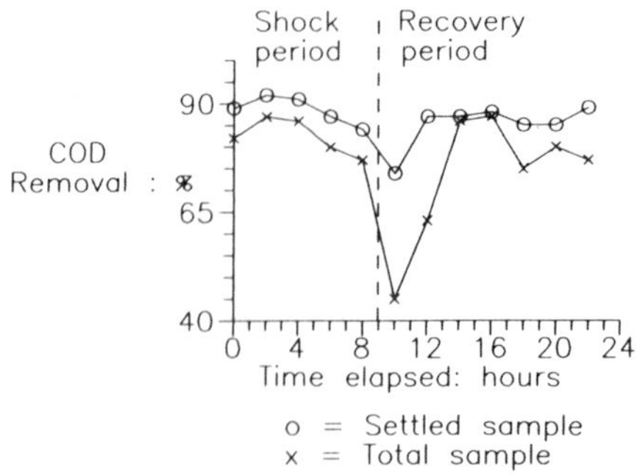

Figure 18. Variation in COD removal during hydraulic shock.

Hydraulic Shock

The shock was applied by increasing the feed rate to the process from 330 1 hr.$^{-1}$ to 565 1 hr.$^{-1}$ for eight hours. As a result upflow velocity of the liquor was increased from 16 m day^{-1} to 21 m day^{-1} over the shock period. The hydraulic retention time was reduced from 15.6 to 9.3 hours. The increase was implemented over 1 hour and recovery was monitored for 15 hours. No attempt was made to reduce the feedstrength and maintain the organic loading rate at a steady state as this was considered unrealistic. The shock was therefore a combined organic/hydraulic one.

Performance

The effluent quality deteriorated throughout the shock period, while total effluent COD increased from steady state of 1200 mg l^{-1} to a maximum of 2970 mg l^{-1}. Settled COD removal remained high however, falling below 80% only once over 24 hours. The variation in effluent COD and COD removal are shown in Figures 17 and 18 respectively. The deterioration in effluent quality was due to the loss of significant amounts of sludge from the bed of the reactor. The increase in effluent suspended solids is shown in Figure 19. The extent of sludge loss is shown from suspended solids profiles in Figure 20. Effluent quality returned to normal after 20 hours.

At the start of the shock, solids in the bed of the reactor had increased to in excess of 60000 mg^{-1}. The solids were not flocculent but discrete particles measuring up to 3mm in diameter[12]. They were rapid settling and highly active; the maximum SMA was measured at 0.30 1 CH_4 g vss^{-1} day^{-1}. Two factors were coincident with the start of sludge loss from the reactor: (i) Gas production increased to above 1000 1 hr.$^{-1}$. This increased the gas velocity to approximately 20 m day^{-1} and the combined fluid velocity to 41 m day^{-1}. (ii) Effluent volatile acid concentrations increased to 180 mg l^{-1}. Both gas production and effluent volatile acid concentrations are illustrated in Figures 21 and 22, respectively.

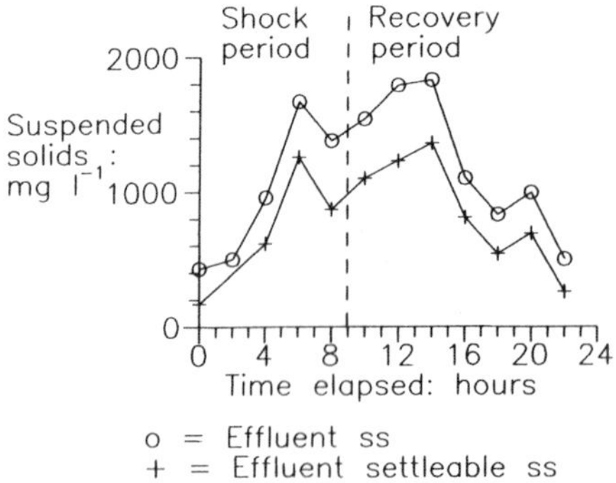

Figure 19. Effluent total and settleable solids during hydraulic shock.

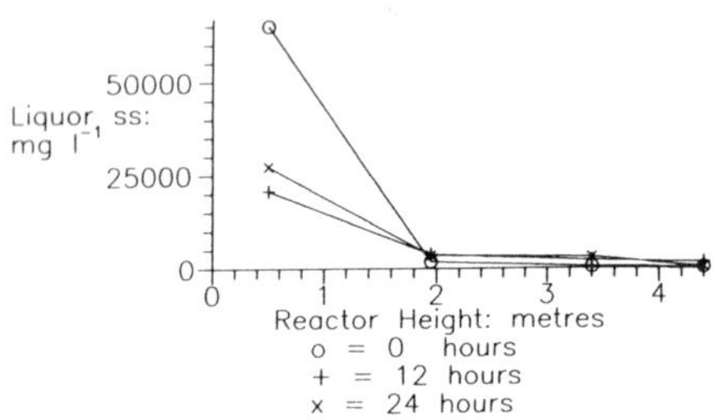

Figure 20. Variation in reactor profile suspended solids during hydraulic shock.

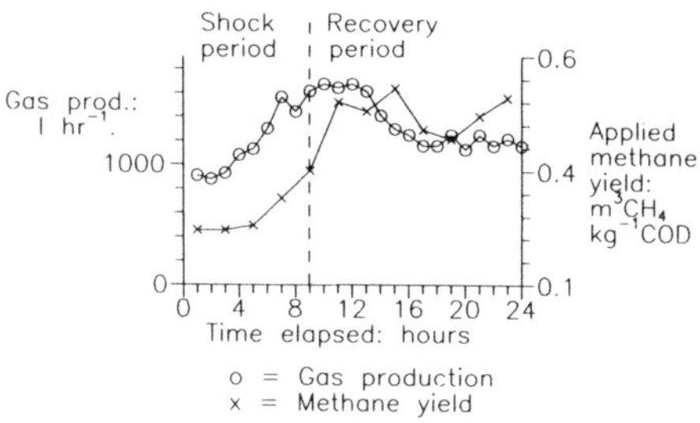

Figure 21. Gas production and methane yield during hydraulic shock.

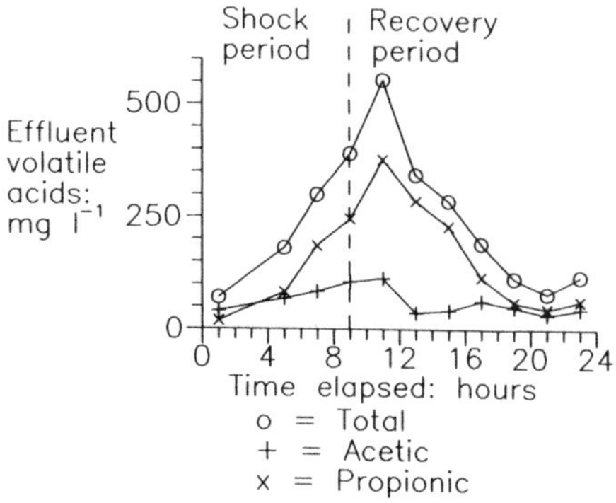

Figure 22. Effluent volatile acids during hydraulic shock.

Gas and fluid velocities were not felt to be primarily responsible for the solids loss however. Higher gas velocities were induced in the organic shock with no resultant loss of sludge. Also during a study into upflow and recycle, combined fluid velocities of 40 m day^{-1} were applied to the process with no increase in solids loss. The reduction in feed rate did not effect a rapid reduction in solids loss.

The presence of increased concentrations of volatile acids in the packed region of the reactor was thought to be significant. This was thought to alter the dynamics of the solid/liquid separation occurring within the media. At steady state the biomass particles, containing gas, rise into the packing where they collide with other discrete particles or pieces of media coated with biomass. The gas may be disentrained from the particle as a microbubble, depending on the size of that bubble and the nature of the collision. The biomass is then free to settle to the sludge bed. Several collisions may be required to effect this. Once the level of substrate increases in the packed region the potential arises for the particle to generate sufficient gas to lower the density of the particle to a point where it resists settling. The probability of the particle being washed through the packing therefore increases. The increased fluid velocities in the reactor will enhance this effect, increasing the density to which a particle of biomass falls before it resists settling.

Gas production and composition revealed the increasing organic load and developing instability. As in the organic shock, gas production rapidly increased to above 1000 l hr.$^{-1}$. A decrease in CH_4 yield followed by an increase to above the theoretical maximum was also noted (Figure 21) as residual volatile acids were consumed. The methane content fell to 56% from a steady state of 72% before climbing to 78% after 24 hours. Overall process stability was maintained despite the loss of sludge from the bed of the reactor. Effluent pH fell from 6.70 to 6.47 after 12 hours and recovered quickly to 6.79 after 15 hours. There was however clear evidence of stress to the system and the process may have been near its operational limits. High levels of volatile acids were found throughout the reactor with the increase in propionic relative to acetic again characteristic. This led to concentrations of volatile acids in the effluent increasing to 555 mg l^{-1} after 10 hours. The variation in reactor and effluent volatile acids is given in Figures 22 and 23, respectively.

Closer examination of the effluent volatile acids and gas composition confirmed the effect noted during the organic shock. Here, as instability developed, acetic acid levels also increased as heterotrophic methanogenic species became inhibited. Once the steady state hydraulic and organic condi-

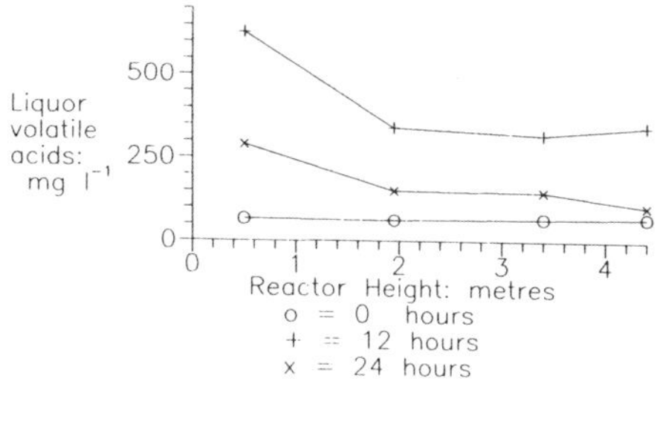

Figure 23. Variation in reactor profile volatile acids during hydraulic shock.

tions were restored, the volatile acids fell. The sequence in which the volatile acids were removed was also characteristic of recovery in anaerobic systems[11]. Acetic acid levels fell first, to concentrations approaching steady state (approximately 50 mgl^{-1}) before significant reductions in propionic acid were noted. This indicates that the partial pressure of hydrogen had to be lowered before acetogenic activity could be initiated to reduce the concentration of propionic acid.

The increase in effluent volatile acids was a result of the sudden reduction of hydraulic retention time. The consequence of this increase was the loss of biomass from the bed of the reactor. It was not clear whether the limiting HRT for the process (treating this wastewater) had been breached. Earlier periods of operation at lower wastewater COD (4500 mg l^{-1}) gave consistently high performance at HRT's of 8–9 hours however.

Design Implications

(a) Recycle.

The response of the AF to pH and in particular low temperature shocks advocated the use of recycle. It effectively spread the effect of any change in the wastewater composition through both the reactor and the bacterial population contained within it. This maintained gas production and the mixed character of the process. However, recycle has two undesirable consequences: (i) increased upflow velocity and (ii) increased pumping costs. These could both be minimised by linking recycle with gas production. Thus recycle would be initiated at low loads or periods of low activity experienced during certain shocks. Above certain levels of gas production eg $2m^3$ m^{-3} of reactor day^{-1} (equivalent to 440 1 $hr.^{-1}$ on the reactor described above) recycle would cease. This would reduce pumping costs and lower solids loss during conditions of severe shock.

(b) Settler

The use of a settler downstream of the AF is advocated during both steady state and shock conditions. During steady state operation up to 200 mg l^{-1} of settleable organic material is produced. This is a consequence of fat and protein emulsions being broken during the digestion process. The amount of settleable material increased during shock periods. In extreme cases, large amounts of biomass were lost from the reactor. The use of a settler would increase COD removal during steady state by 5–10%. During periods of shock this improvement may rise to 40%. The option of returning viable biomass to the reactor is also available.

(c) Process Control

The response of the AF to shocks highlighted parameters that could be incorporated in a process control system. Gas production/composition reflected both the load applied and the balance of digestion within the reactor. The response to changes was almost immediate, offering the potential for a very short control loop. Increasing volatile fatty acids in the reactor was also a primary indicator of instability.

Conclusion

The Anaerobic Filter was found to be resistant to a series of severe shocks. Low temperature (minimum reactor temperature 25.5°C) and low pH (minimum wastewater pH 5.29) resulted in small transient changes in reactor performance and negligible changes in reactor stability. Organic shock produced a 24% reduction in COD removal, reflecting the increased amount of suspended organic material in the effluent. Evidence of developing instability was noted. Settled COD removal was maintained above 75% however and steady state performance was reestablished within 24 hours of the shock period. Total COD removal fell to a minimum of 45% during hydraulic shock as sludge from the bed of the reactor was washed out with the effluent. This was primarily a result of high volatile acids throughout the reactor disturbing the settling dynamics of the packed region. Settled effluent quality remained above 70% however and steady state was also rapidly reestablished.

The use of recycle related to gas production is suggested to improve performance and resistance to certain shocks. The use of a settler increased effluent quality during steady state and markedly during shock. Temporary infringements of consent to discharge during shock may be prevented by the use of a settler. Changes in gas composition/production and volatile acids induced during shock indicated their potential incorporation in a system of process control.

ACKNOWLEDGMENTS

The authors gratefully acknowledge the contributions of C. Nichols and L. Smith and the support of Unilever Research Ltd. and Birds-Eye Walls Ltd. (Gloucester).

REFERENCES

1. Anderson, G. K., Donnelly, T., "Anaerobic Digestion of High Strength Industrial Wastewaters." *J. Inst. Publ. Hlth. Engrs.*, 1977, 3, 5.
2. Young, J. C., McCarty, P. L., "The anaerobic filter for wastewater treatment," in Ref. 25 *J. Water Pollut. Control Fed., 41*, R160 1969.
3. Lettinga, G., van Nelsen, A. F. M., Hobma S. W., de Zeeuw, W. Klapwijk A., "Use of the upflow sludge blanket (USB) reactor concept for biological wastewater treatment especially for anaerobic treatment." *Biotechnol. Bioeng. Vol. 12*, No. 4, April 1980, pp. 669–734.
4. Young, H. W., Young, J. C., "Hydraulic characteristics of upflow anaerobic filters." *J. Environ. Eng.*, 17, 416A, 1983.
5. Standard Methods for the analysis of water and wastewater, 16th ed.. New York: American Public Health Association, 1985.
6. James, A., Chernicharo, C. A. L., Campos, C. M. M., "Development of a new methodology for the analysis of specific methanogenic activity." *Wat. Res.*, accepted and in press 1989.
7. "Anaerobic treatment processes and methane production." Notes on water pollution No. 64, D.O.E. 1964.
8. Bryant, M. P., Wolin, M. J., Wolf R. S., "Methanobacillus omelianski, a symbiotic association of two species of bacteria." *Arch. Microbiol.*, 59, 20–31, 1967, ecology.
9. Mosey, F. E., "Anaerobic Biological Treatment." *The Inst. of Wat. Pollut. Cont.*, Symposium on the treatment of wastes from the food and drinks industry, 10th Jan. 1974.
10. Forday, W. and Greenfield, P. F., "Anaerobic Digestion." *Effluent and Water Treatment Journal*, Oct. 1983, pp. 405–412.
11. Mosey, F. E., "New development in the anaerobic treatment of industrial wastes." *Effluent and Water Treatment Journal*, Mar. 1983, pp. 85–93.
12. Kennedy, K. J., Gorur, S. S., Elliot, C. A., Andras, E., Guiot, S. R., "Media effects on performance of anaerobic reactors." *Wat. Res.* Vol. 23, No. 11, pp. 1397–1405, 1989.

Section Four
BIOLOGICAL SYSTEMS – D. SLUDGE TREATMENT

53 IMPROVED SLUDGE DEWATERING

Edwin R. Hendricks, Senior Engineer
BBS Corporation
Columbus, Ohio 43229

INTRODUCTION

Overview of Solids Disposal

Ultimate disposal of wastewater sludge is and has been an increasingly complex and expensive proposition for the City of Columbus, Ohio.

Sludge incinerators have been in operation for over thirty years. The incineration capacity is initially limited by the amount of water contained in the sludge. Historically, the two multihearth incinerators at the Southerly WWTP have been limited to about 5.5 wt/hr. Two new incinerators have recently come on line and they provide about 7.0 wt/hr each. Total incineration capacity at the SoWWTP is about 25 wt/hr and, allowing 3 of 4 units on line continuously, the reliable capacity is about 430 wt/d.

When it is possible to obtain sludge cake dry enough such that minimal auxiliary fuel is required to sustain burning, the condition is described as autogenous combustion. Analysis of the incinerators indicates that autogenous combustion will be obtained when the sludge cake is about 24% to 28% solids. At that point, fuel requirements necessary to maintain control of the incinerators should fall to less than 0.10 gallon, per wt. of sludge burned.

The projection of changes in incinerator performance as feed sludge cake dryness is increased from 18% to 30% solids is inexact due to the number of critical variables. However, the projection of near autogenous combustion is reliable. Recent operations where 26–28% cake was burned with greatly reduced fuel indicates verification of the projections.

Contrarily, projection of quantitative changes in composting performance as feed sludge cake is varied from 18% to 30% is not practical, given the limited scope of studies and time allotted. However, initial surveys of the composting situation clearly indicate that the impact of improved cake solids would be even greater on composting than on incineration. For instance, it is projected that doubling the cake solids would result in using about 70% less wood chips. The combined effect of decreased wood chip addition, improved mixing and reduced handling are expected to combine synergistically to provide significant improvements at composting.

There does not seem to be any overwhelming advantages or disadvantages to ongoing land application programs vis-a-vis cake solids improvement. The City currently practices surface application of cake, and injection of thickened sludge is being piloted on a full scale basis. Analysis to date indicates similar costs for both methods. In other works, the reduced cost of hauling sludge cake is offset by dewatering costs. However, there may be some advantage to the drier filter press sludge in that handling and storing the 28% cake is expected to be much easier than the 18% cake.

Systematic Analysis

In order to provide guidance for selection of new solids handling facilities and/or expansion of existing facilities, a systematic cost comparison of selected solids handling models was undertaken. Key assumptions of that analysis were as follows:

1. Incinerator capital costs was treated as sunk cost (capital cost not included in the analysis).

45th Purdue Industrial Waste Conference Proceedings,

2. Similarly, other previously expended costs or sunk capital costs were not included in the alternative present worth analysis, including the following:
   a) Thickening centrifuges—4 ea.
   b) Anaerobic digesters—6 ea.
   c) Dewatering centrifuges—6 ea.
   d) Miscellaneous buildings, tankage, piping, and equipment.
3. Thickening of primary and waste activated sludge was included in all models.
4. The first analysis was based upon cake solids being improved to 21% (from 18%) through improvements to existing centrifuges and purchase of additional machines.
5. All costs necessary to upgrade or maintain existing facilities were included in the analysis. Key findings of the systematic analysis were as follows:
   a) Land application and incineration were nearly equal in cost.
   b) Composting was projected in gain efficiency based upon expected mixing improvements. Costs were higher than incineration, but relatively equal within the limits of error of the analysis.
   c) Anaerobic digestion provides some improvement in efficiency, due to the reduction in the weight of water ultimately disposed of. The key advantage of anaerobic digestion is the low cost of rehabilitating the existing digesters.
   d) None of the alternatives considered had a serious impact on manpower needs.
   e) None of the alternatives had a serious impact on maintenance costs. (Annual and 27 year cycle).
   f) None of the alternatives had a serious impact on energy costs.

However, it was very obvious at this point that a drier sludge cake would have significant impact on energy costs, and potential secondary impacts of reducing future expenditures for capital improvements. In other words, as incineration and composting capacity are limited by cake solids, their capacity to dispose of solids increase as cake dryness increases.

Pilot Studies

A series of pilot studies was undertaken from November 1985 to August 1986. The general goals of the studies were to compare belt filter presses, centrifuges, and membrane filter presses. Due to severe time limitations and other constraints, the test conditions were less than ideal. Interpolation and engineering judgement were applied to the test work, and the results were applied to further cost analysis.

Constraint on testing include the following:

1. Existing centrifugal dewatering machines were somewhat outdated, and only one of six machines had what was considered the proper gear ratio. Side by side testing was not possible due to scheduling constraints and plant limitations.
2. Significant variations in dewaterability occurred during the test period. (November 1985 to August 1986).
   - Primary sludge characteristics (PS) ranged from good with low volatiles during wet weather to poor with high volatiles and septicity during hot dry weather.
   - PS thickness ranged from 2-1/2% to 6% plus.
   - WAS varied in septicity and thickness. It was, at that time, partially aerobically digested before dewatering.
3. Thickening of primary sludge (PS) during the testing was conducted in the clarifiers by practicing intermittent withdrawal which often appeared to contribute to septicity, and lack of thickening.
4. The mix of PS:WAS normally varied widely. One MFP test was conducted with sludge samples carefully mixed for that test.

The test program included the following:

1. Two MFP bench scale series tests.
2. One trailer scale centrifuge test.
3. Two full scale centrifuge test.
4. One trailer scale belt filter press test.

In general, the worst dewatering conditions encountered were during the MFP tests. Indicators of these worst case conditions were that the PS was not as thick as other tests and had higher volatiles. Attempts at normalizing the data, based upon broad assumptions, were not conclusive, but did tend to support the final conclusions. While it is not considered appropriate to compare the test data without significant explanation and normalization, it can be stated that the filter presses tended to yield sludge cakes with precoat about 5% drier than the centrifuges and belt presses. As will be explained below, this difference combined with other factors was enough to yield significant cost-effective advantage to the filter presses. (Current operations indicate the original assumptions underrated the presses, and they are in fact more cost effective than was first thought).

Additional Analysis

The reliability of the equipment was a significant factor in this analysis. Historically, the SoWWTP reportedly maintained about 2/3 of its centrifuge dewatering equipment on a ready status. It was felt that this was due in part to the fact that there was surplus equipment. A factor of 5/6 availability was used for analysis. It was projected that an improved preventive maintenance program for the dewatering centrifuges would also justify the 5/6 availability.

In order to obtain information about operational and maintenance aspects of MFP, visitations to municipal facilities were undertaken. It was found that, in general at existing facilities, 100% availability of machinery is the norm, with about 10% scheduled down time for preventive maintenance, lubrication and cleaning.

Considering numerous judgmental factors, it was decided that about 70% overall availability would be allowed for the MFP. This conservative position was based mainly on the premise that new machinery must be viewed with some skepticism against actually operating equipment.

In summary, key input to the cost effective systematic analysis was as follows:

1. Capital cost, adjusted for replacement.
2. Dewatering equipment capacity.
3. Operational costs of dewatering devices.
4. Incineration costs dependent on consistency of feed cake solids.

Sensitivity analysis was carried out for a range of cake solids. Based judgmentally upon the most likely cake solids output, 20% for centrifuges and 25% for MFP it was found that centrifuges were about 24% more costly then MFP.

The basis for choosing MFP was finally based upon judgmental factors. The key factors regarding machine capacity and availability were intentionally weighted towards a worst case for the MFP while some allowances were made for expectations of improved performance from the centrifuges. Conversely, expected improvements in sludge thickening were factored into the analysis and expected to have greater favorable impact on the MFP than the centrifuges. Generally, the centrifuges do not benefit greatly from prethickening as they tend to operate based on solids output limitations and not hydraulic throughput.

However, the MFP capacity is significantly improved by thickened sludge in as much as cycle time is reduced. The squeeze time is related to the flow rate of filtrate out of the cake. This phenomenon is predictable and can be factored into the analysis, such that it was possible to predict squeeze time reductions. Predicted MFP capacity should increase about 25% due to prethickening the sludge

However, the Southerly Wastewater Treatment Plant has not previously had experience with thickening processes and there was concern that the necessity of a synergistic combination of thickening and dewatering might be somewhat liberal. (Recent results have shown otherwise).

Therefore, the decision to choose MFP was made based on the estimated cake solids output of 25% and 35,000 dry pounds per day per machine.

It is interesting to note that the analysis included the projection that savings in incineration costs would reach as much as $1,500,000 per year. This relates to a projected savings on the order of $700,000 in the near future for current flows and loadings. The original projection was based on converting to one City plant, instead of the current plan to operate two City plants.

For various reasons, and especially because the decision to operate one plant was changing to a two-plant operation, the decision to install only four (4) MFP was made, although the original analysis indicated nine machines for a one-plant scenario.

The rated capacity is about 105,000 dry pound per day against actual demand of 200,000 to 270,000 dry pound per day dependent on various unresolved questions such as plant flow rating and anaerobic digestion disposition. Therefore, the plan, as detailed design was initiated in late 1986, was to run two or three of the existing centrifuges in conjunction with the MFP.

Recent Developments

As construction approached substantial completion on about July 1, 1988, start up testing was accomplished with the test press.

The feed sludge consistency increased from the 2% to 3% seen in the testing phase to about 6.5%. It is also notably less septic due to the new thickening operations.

Five runs of the test press confirmed previous optimism about the positive effect of preconditioning the sludge. Average results of those runs were 39.2%. Full scale production for a fourteen day commissioning run resulted in 29% cakes.

Limitations on MFP at Columbus dewatering are likely to be due to external causes, relative to labor and maintenance availability.

Full-scale operation and subsequent optimization are expected to be very fascinating and ultimately yield a significantly more cost effective sludge solids handling process for the City of Columbus. Furthermore, the MFP produce about 98% capture of solids while the centrifuges produce about 85% capture. This may turn out to be as critical as the savings in fuel costs inasmuch as the reduction in recycle is projected to have significant benefits to the nitrification process.

The reduction in recycle will reduce the MLSS inventory necessary in aeration by as much as 10% to 20%. Initial estimates indicate two to four future aeration tanks may not be needed due to the reduced recycle.

In summary, the successful operation of the filter presses can save the City of Columbus a net value of one-half million dollars per year, plus the future construction of two to four aeration tanks.

Design Summary

The key components of the Solids Handling System, their essential dimensions, capacities and input and output parameters, are as follows:

Filter Press System Components:

1. Process Wells
2. Willett Pumps
3. Filter Presses
4. Screw Conveyors
5. Process Air System
6. Precoat System
7. Ferric Chloride System
8. Polymer System

Sludge Dewatering Press Details:

1. Type: Recessed Plate with Membrane Filter Press
2. Number: 4
3. Solids Loading: 35,000 dry lb/d per press
4. Influent Solid: 4 to 6% dws
5. Capture Eff.: 99%
6. Dewatered Solids: 28% dws (w/precoat)
7. Destination: Incineration and Composting
8. Operating Pressure: 150 psig.

Operational Filter Press System Components and Control Alternatives

The following abbreviated notes are representative of the "walk through" type presentation and discussion of the press operation.

1. A press cycle time is about 110 minutes.
2. Cycle Components:
   a. Prefill .... 10 min.
   b. Precoat .... 10 min.
   c. Filtration .... 25 min.
   d. Membrane Squeeze .... 15 min.
   e. Membrane Exhaust .... 5 min.
   f. Core Blow .... 5 min.
   g. Cake Discharge .... 40 min.

Total .... 110 min.

3. Precoat—a timed cycle of the circulating equipment is the key control, as the precoat suspension is circulated through the presses.
4. Filtration—Willett pumps fill up the presses.
   a. Stall pressure—primary control set to shut off Willetts.
   b. Elapsed time—secondary control that shuts off the Willetts if elapsed time expires before stall pressure is reached.
   c. Stall—80–100 psi is the normal range.
   d. DCC reads the pressure gage and shuts off filtration.
   e. If the feed sludge is thin, then stall pressure is not likely to control.
   f. Stall is reached in 20–25 min. now, but elapsed time will cause shutdown in about 30 min. if stall is not reached.
5. Membrane Squeeze:
   a. Air receivers are the source which feeds into a manifold along the top of the press and causes the membrane to inflate. (100–150 psi is available there).
   b. The filtrate drain is a good indication of what is happening. A steady stream of flows occurs during the filtration cycle. Flow will fall off at the end of the cycle. Visual inspection of filtrate clarity will help verify proper operation.
   c. Membrane squeeze shortens cycle time and improves cake release. Time set for membrane squeeze is about 15 minutes, but can be adjusted with a thumb wheel of the DCC panel.
6. Membrane Exhaust:
   a. Valves in the air manifold piping switch position and the air is exhausted to the atmosphere as the membrane deflates.
   b. Manual operation of most sections of the press is possible with local control through H-O-A switches.
7. Core Blow:
   a. A valve in the pipe from the air receiver opens and forces the sludge core from movable end to the fixed end. Core sludge is displaced through core blow piping to process wells.
   b. Following timed delay, an alarm will sound indicating operator attention required.
   c. Dripping occurs, indicating hydraulic pressure has been relieved and the press can be opened.
   d. The Bombay doors then open.
   e. An alarm sounds while Bombay doors open and continues until doors are in full open position.
   f. If a fault occurred such as a hydraulic line failure, then alarm would indicate doors are not open.
8. Cake discharge then occurs. There is a nine second travel time for the plate shifter. The operator assists the cake release with a canoe paddle.

There are key peripheral systems that are required for the press operation. They are as follows:

- Process Air
- Plant Air
- Plate Washer
- Precoat
- Chemical Feed
- Control System

A brief discussion of those units is as follows:

1. The process air compressors can provide up to 250 psi to the air receiver. Currently set for about 175 psi. Air dryers operate in pairs, are on all the time. They are regenerative type units. They alternate towers automatically. One tower is drying air while the other tower is regenerating. Cycle time is currently set at 10 minutes. Air receivers with auto moisture trap are included.
2. Plant air compressors are tied to other plant air systems. This system also has dryers. These units power the air controlled valves on the membrane air manifold and exhaust lines.
3. The presses include an automatic washer system. The cloth wash pumps operate at high pressures (1200 psi).

4. The precoat silo equipment is having a dusting problem, which will be mitigated by future improvements.
5. Chemical Feed System:
   a. The polymer bag loader has an automatic dust filter and exhaust that starts when the bag loader door is opened.
   b. The polymer feeder control panel has timer settings for screw feeder, mixing and aging time. Essentially the operation is automatic on a timed function basis. Control is through the programmable controller.
   c. A ferric chloride system was provided but is not required and is not used.
6. The control of the press cycle is essentially automatic. Two "computer units" operate in tandum in order to provide operational backup. There is no alarm printer on CRT output used for normal operation. This may be modified in the future.

Recent data from the Startup Operation indicates the following:

- Cycle Time: About 10 per day per press versus 11 projected.
- Operational Labor: Two operators per shift versus three projected.
- Cake solids output for a 50/50 mixture by weight of PS/WAS is about 28%.
- Over all dry weight production capacity to date is about 90% of projected and may increase to 110% to 120% of projection as debugging continues.
- Centrifuge dewatering produces about 20% cake solids with a 50/50 mixture.

Summary

- The current net results of utilization of the sludge presses is that the daily production of wet tons of solids requiring disposal can be reduced by about 15% to 20%. This means that if necessary all of the sludge produced can be incinerated with three of four incinerators in operation. Therefore the reliability of sludge disposal at the SoWWTP is significantly increased. This analysis assumes three presses and two or three centrifuges, will be operated.
- The cost of fuel to burn sludge is reduced and will be reduced further when the incinerators are upgraded to handle drier cake solids.

54 SIDE OUTLET RECYCLE FOR MULTIPLE HEARTH SEWAGE SLUDGE INCINERATORS

Charles A. Richmond, Incinerator Specialist
BBS Corporation
Columbus, Ohio 43229

CONVENTIONAL MULTIPLE HEARTH FURNACE

Traditionally, Multiple Hearth Furnace manufacturers have written about the virtues of the Multiple Hearth Furnaces as follows (Figure 1).

"The Multiple Hearth Furnace consists of a circular steel shell surrounding a number of vertically stacked chambers (hearths) and a central rotating shaft to which rabble arms are attached. The rabble arms have rabble teeth attached to their bottom surfaces. The furnaces are furnished in a number of diameters 6′ OD to 25′-9″ OD and number of hearths (from four to sixteen).

In a MHF with a ODD number of hearths, the top hearth is an outfeed hearth, that is the sludge enters the center and moves outward to a series of peripheral holes where it falls through to the next hearth which is an infeed hearth.

In a MHF with an EVEN number of hearths, the sludge enters the top hearth near the wall of the hearth. The rabble arms and teeth move the sludge inward to the center drop hole where it falls to hearth #2 where it is moved outward to the peripheral drop holes.

The number of rabble arms per hearth, the size, spacing and number of rabble teeth on each arm move the sludge into an archimedes spiral (Figure 2) creating a number of rows of material. The surface area of the rows is dependent upon the angle of repose of the sludge, thus area may be more than 125% of the flat area of the hearth.

Sludge ignited in the center section of the MHF generates hot gases which travel counter-current to the downward flow of the solids. In so doing, that heat energy contained in the hot gases evaporate the moisture in the sludge, and in giving up the heat, the gases cool down. Because of the moisture content, the sludge stays at a temperature between 160–200°F.

The exit temperature of the hot gases at the top of the MHF will generally be 700–900° F. Because of the moisture content of the sludge, the low solids temperatures of 160–200 will insure that no odorous materials exit the gas outlet of the MHF.

The MHF has three basic zones:

1. Two or more upper hearths on which a major portion of the free moisture is evaporated.
2. Two or more intermediate hearths on which sludge burns, producing temperatures in excess of 1,400°F.
3. A bottom hearth that serves as an ash cooling hearth and in which preheated air from the air cooling hearth and in which preheated air from the air cooling of the center shaft is admitted to form a portion of the required combustion air. The hot ash is cooled and the preheated air is further heated from the hot ash.

The instantaneous position of the zones within the MHF vary with the moisture content of the sludge and the combustible content, along with the higher heating value of the combustible. As these factors vary during operation, the zones naturally and automatically adjust themselves for the variation and the fire moves up or down. This inherent ability makes for simple and efficient design.

The center shaft consists of two concentric shafts. Cool air is blown up the center tube out into the rabble arms to cool them and returned to the annulus between the outer and inner tubes. This provides cooling to the rabble arms and the heat removed is returned to the MHF as preheated sludge combustion air.

45th Purdue Industrial Waste Conference Proceedings, © 1991 Lewis Publishers, Inc., Chelsea, Michigan 48118. Printed in U.S.A.

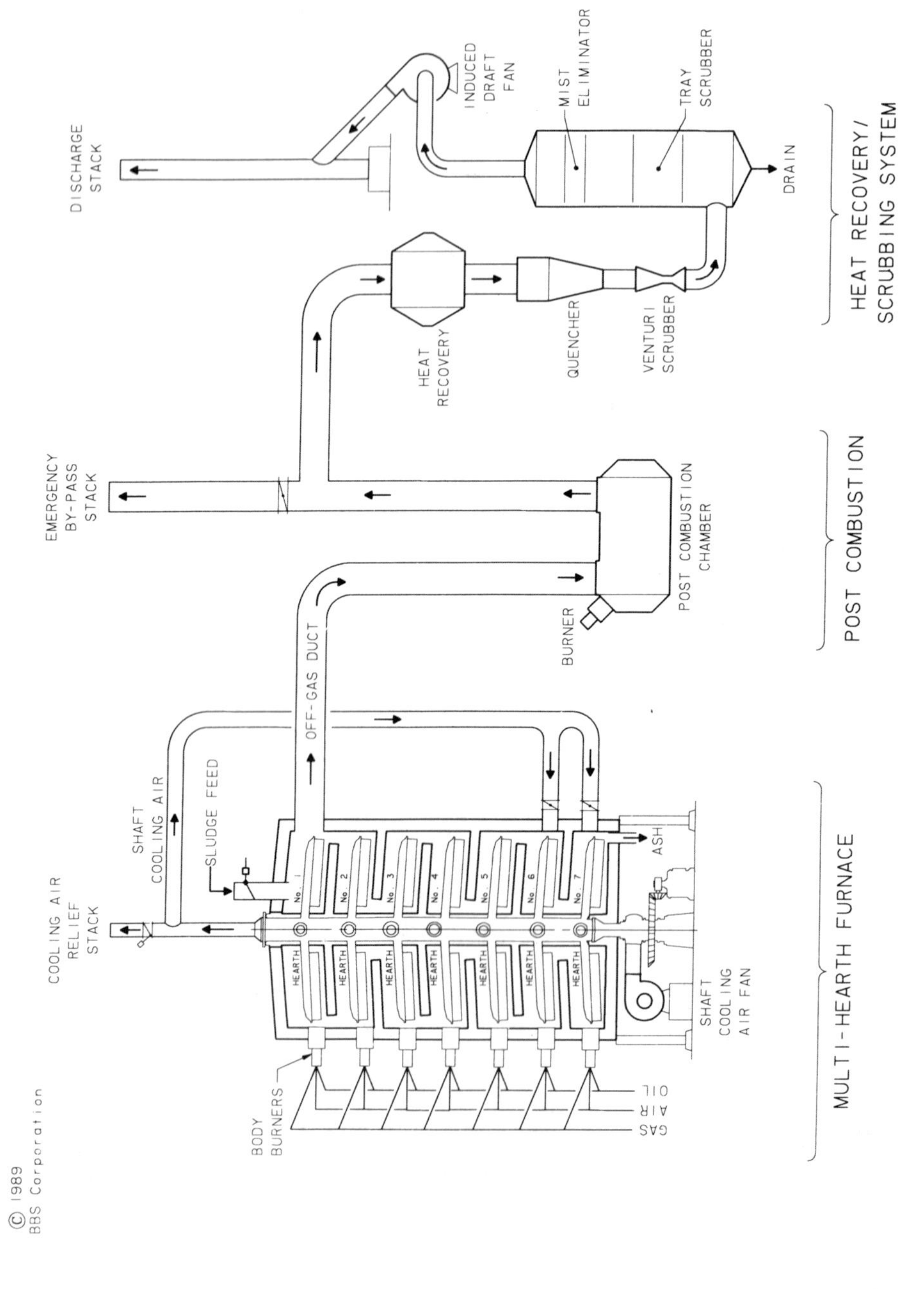

Figure 1. Conventional design flowsheet.

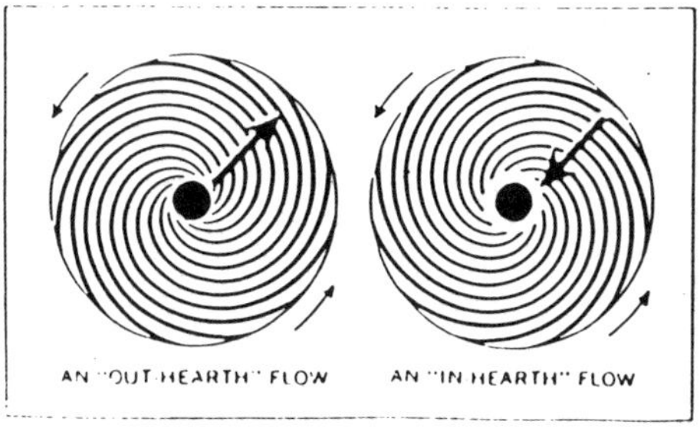

Figure 2. Archimedes spiral path.

The above explanation of how the MHF works was satisfactory for many years. As the use of activated sludge increased, the moisture content increased to 78–82%. Polymer was used as a dewatering aid.

With the increase in fuel costs in 1974, the need for better dewatering methods was obvious. Several solutions were offered:

A. Pressure filterpresses with plates and frames or hollow chambers, yielding wet cake of 40% or more dry solids. The major problems were too dry a cake with fires on the top hearth of the MHF, and the non-continuous nature of the method, which resulted in troublesome storage hoppers.
B. Thermal conditioning processes which made the older coil filters effective again. The disadvantage of this system was the odors generated in the treatment process, and the fact that the sludge now was so dry off the dewatering device that control of the MHF became almost impossible. Fires on the top hearth, odors from the MHF, excessive temperatures within the MHF resulted in early failures of refractory and castings.

Solutions utilized to control the temperature in the MHF were:

A. Add a lot of air at the top of the MHF to cool it down. This resulted in larger scrubbers, larger ducts, larger Induced Draft Fans, operating at higher speeds, using more horsepower.
B. Put the sludge into the second or third hearth so that fire at the top hearth would not occur, and possibly the combustion would be completed in hearth #1 or #2. This solution did not change the higher operating temperatures present within the combustion zone (1,800°F), which caused refractory and casting failures.
C. Water was added to the sludge to rewet it so that the MHF would be more controllable. This solved the temperature problem but did not solve the odor generation in the drying zone of the MHF.

EMPIRICAL DESIGN CRITERIA

In designing a Multiple Hearth Furnace, several empirical factors are considered:

A. The GROSS HEAT RELEASED within the MHF. Each MHF size has a maximum number of BTU's per hour capacity, above which refractory failure occurs, and alloy casting life is shortened.
B. The MAXIMUM WATER EVAPORATION RATE, this is an amount of water evaporated per sq. ft. of MHF area, which when exceeded, has led to unburned sludge exiting the MHF.
C. The MAXIMUM SOLIDS CAPACITY PER HOUR PER SQUARE FOOT OF HEARTH AREA. This is a mechanical limitation, controlled by the number of rabble arms, the size and number of teeth, and the angle of the teeth. Speed of shaft rotation has little effect on the operation.

   It may offer a short term solution to a momentary problem, but due to the nature of the MHF it is not a long term solution. The reason shaft speed is of limited significance is the process of drying during the "Constant Rate Drying Period" (90% to 40% water) is a "Diffusion Rate Controlled" process.
D. FIRE POINT—This is the temperature at which the partially dried sludge will ignite. It generally is 1,250 F at 40% moisture.
E. The WATER COEFFICIENT—This is the number of BTU's which are available for each pound of water evaporated. It is obtained by multiplying the % combustible by the

higher heating value of the combustible, and dividing that product by the % water. It is a known fact that if that product falls below 2,250 BTU per pound of water to be evaporated, then auxiliary fuel is required. It is known that if the coefficient is between 2,500 and 3,000 BTU available per pound of water evaporated, the MHF runs autogenously. It is also known that if the water coefficient is over 3,500 BTU's per pound of water evaporated, the sludge will probably burn on the top hearth (Figure 3).

Heat Transfer

Heat transfer within MHF is primarily by radiation from the two diatomic molecules formed in combustion, namely carbon dioxide and water.

The amount of heat transferred is described by the equation:

Q = U × A *T.
Q = the total amount of heat transferred per hour.
U = an overall heat transfer coefficient which is a function of the gas phase temperature, i.e.—as the gas phase temperature goes up, U increases, as the gas phase temperature goes down, U decreases.
A = the total surface area exposed to the hot gases.
*T = the difference in temperature between the hot gases and the material on the bed.

The empirical values commonly used for U are:

*T 600 F U = 6
*T 800 F U = 8
*T 1000 F U = 10
*T 1200 F U = 12
*T 1400 F U = 14
*T 1600 F U = 18

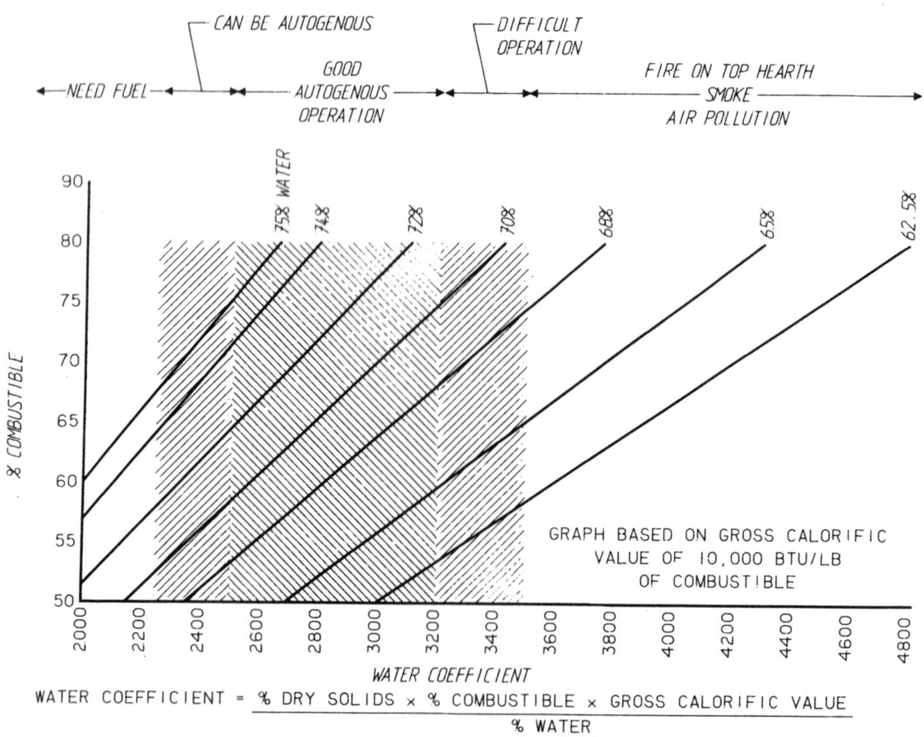

Figure 3.

Thus, it can be seen that as the gas temperature decreases, the one variable in the equation, U falls off sharply, but so does *T, resulting in significantly smaller amount of heat transferred per hearth, as the gases travel upward out of the MHF.

At a 1,400°F gas phase temperature, the amount of heat transferred on an 18′-9″ OUT hearth is 3,074,400 BTU/hr.

At a 600°F gas phase temperature, the amount of heat transferred on an 18′-19″ OUT hearth is 439,000 BTU/hr.

This equation is the most important relationship within the MHF and controls much of its behavior.

Desirable Design Criteria

The desirable design criteria then is to design a MHF such that:

A. The location and temperature of the combustion zone is pre-selected and maintained in one hearth or area, so that the OPERATOR is running the furnace and not the sludge running the operator.
B. Fuel economy is greatest, this means *LOW* Carbon Monoxide in the flue gases and high Carbon Dioxide.
C. No odors are emitted from the system, this means *NO* low molecular weight compounds in the flue gases.
D. Wet cakes with *HIGH* water coefficients are easily processed *WITHOUT* top hearth fires, and without lowering the number of hearths available when more area is desired due to an increase of moisture in the wet cake feed.
E. Wet cakes with low water coefficients are processed WITHOUT generating odors, and WITHOUT the risk of some UNBURNED sludge escaping out the calcine outlet.
F. Clinkers are eliminated, unless grease or skimmings are fed.

Side Outlet Recycle Process

In 1977 a different method of operating a MHF was designed and sold to the City of San Mateo, California. This method met ALL of the desirable design criteria listed above. The system went on line in 1980 and has been in continuous operation since, except for a brief down-time to repair the earthquake damage from October 1989 (Figure 4).

This process provides *STABILITY OF OPERATION* unattainable with a conventional flowsheet. The arrangement utilizes the recycle of furnace gases from the top hearth to below the burning hearth to control the fire location and to avoid direct flame burners on the furnace sidewall; it provides for combustion gases to exit the furnace at a side location on the burning hearth to provide a hot, low-odor,low-hydrocarbon, low-particulate exit gas. This process can be retrofitted to existing MHF's to comply with the new proposed regulations (40 CFR part 503).

The Side Outlet Recycle is an arrangement of ductwork, fans and controls around a multiple hearth sludge incinerator to provide simple, easy control of combustion and a clean exit gas at pressure drops of approximately 20 inches wc through the scrubbing system.

Combustion Control

Recycle allows instrumental control of the combustion process with minimal need for operator attention. Control of the combustion process is the ability to hold the fire at the designated burning hearth regardless of changes in sludge feed rate or character; control is accomplished by varying the quantity of combustion gases which pass over the drying sludge in the top hearths of the furnace; more gas quantity recycled, more water evaporation on the drying hearths; less gas quantity recycled, less water evaporation on the top hearths. It is desirable to evaporate just enough water so that the sludge is at its "FIRE POINT" upon entering the combustion hearth. A thermocouple on the hearth above the burning hearth sends a signal to an indicator/controller which controls an inlet damper to the recycle fan.

If the sludge is very dry as charged (35–40% dry solids) then the recycle fan circuit is disabled and the damper at the top to the post combustion chamber is opened slightly, controlled by the temperature on the top hearth.

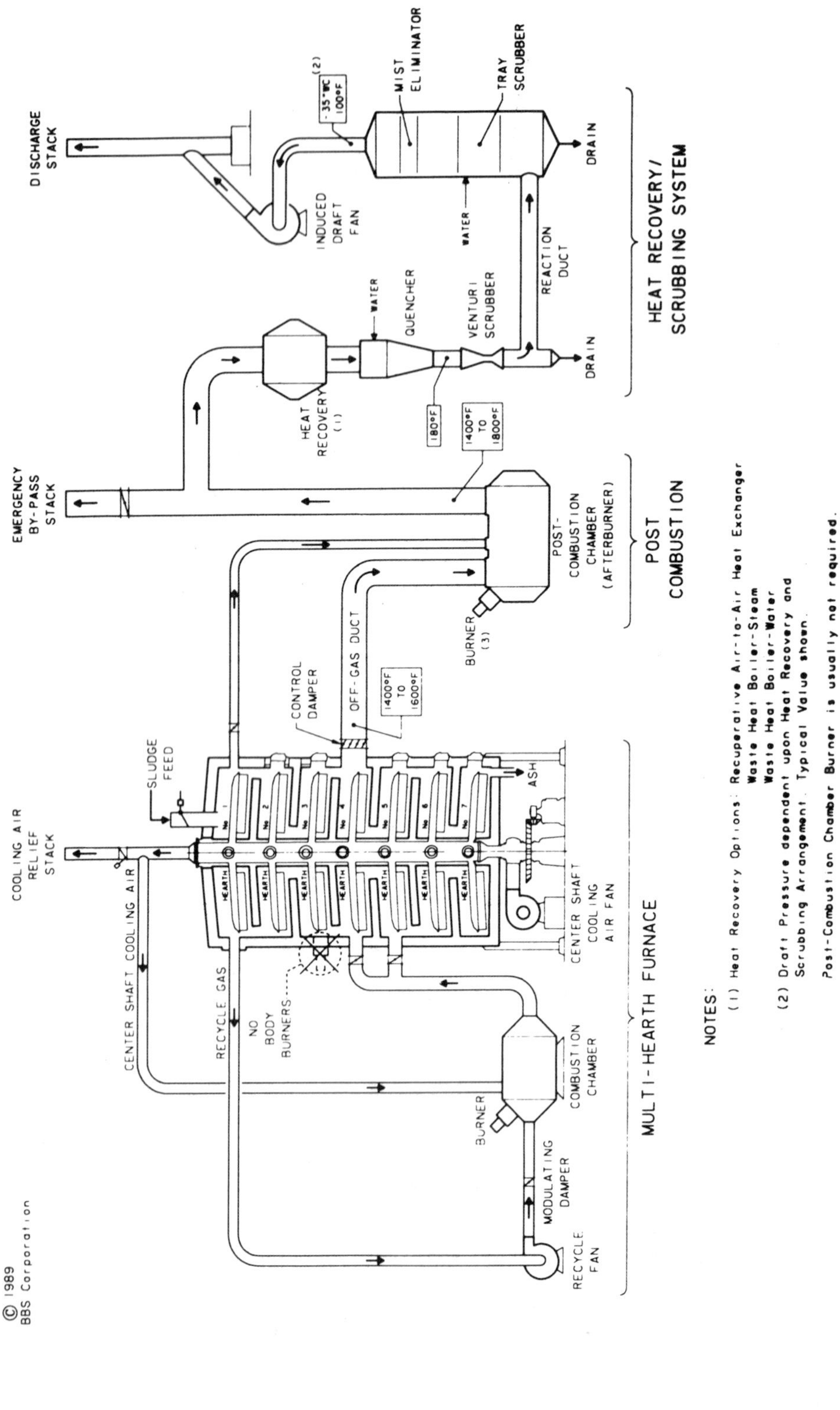

Figure 4. Side outlet recycle flowsheet.

Combustion Completion

In an up-draft MHF, combustion in the gas phase has little or no time for the reaction to go to completion. The hot gases from the combustion are immediately cooled on the hearth above, where the gas temperature falls below 1,250°F, stopping further reaction. With the side outlet recycle process, the flue gases discharged from the side of the MHF enter a baffled refractory lined chamber which provides both time and turbulence for the combustion reactions to go to completion. At 1,400°F, 0.75 seconds residence time appears to be adequate.

Furnace Draft

Furnace draft is controlled by the position of a multi-plate high temperature alloy damper at the side outlet. A draft controller on the combustion hearth actuates an alloy multi-louver damper at the side outlet. This is a single loop control.

Scrubber Pressure Drop

The scrubber total pressure drop is maintained by a differential pressure controller across the total scrubbing system. This is an independent control loop, and can be set to the pressure drop required for the desired particulate removal. Most particulate from an MHF is larger than 2 microns, therefore 99+% removal can be attained at 20″ wc. If emission test work indicates a different particulate size prevails, then a different total pressure drop must be used.

Temperature Control

Temperature control is by means of a thermocouple on the burning hearth sending a signal to a temperature indicator-controller. Auxiliary fuel is introduced into one external combustion chamber, in response to the demand for more heat. At 1,400°F (or lower) the fuel is shut off. This is a single control loop.

Excess Air

The excess air is controlled by an oxygen analyzer. As more air is required it is introduced either with an independent combustion air fan or with the centershaft cooling air return. The excess air can be operated routinely at 50% thus saving fuel, over the conventional multiple hearth which frequently runs at 250% excess air.

Air Emission Test Results

The average air emissions quality to be expected is:

Particulate	0.25–0.65 Lbs/Ton of Dry Solids
Hydrocarbons	10–15 ppm
Carbon Monoxide	150–200 ppm
Oxygen Content	7–8%

Section Five
RESPIROMETRY

55 USE OF RESPIROMETRY TO EVALUATE THE BIODEGRADABILITY OF EMULSIFIED SPECIALTY CHEMICAL PRODUCTS

Richard J. Colvin, Project Engineer
Alan F. Rozich, President
Bioprocess Engineering, Inc.
Wilmington, Delaware 19808

Bruce J. Hough, Principal Environmental Engineer
Hercules Incorporated
Wilmington, Delaware 19894

Anthony F. Gaudy, Jr., Professor Emeritus
Department of Civil Engineering
University of Delaware
Newark, Delaware 19716

INTRODUCTION

In recent years, the demand for higher quality effluents in the pollution control field have led to the need to quantify the removal capability of biological treatment processes. One tool that has recently seen resurgence in the environmental field is the use of respirometry for the design and operation of biological wastewater treatment plants. Respirometry or oxygen uptake measurements are key because they can be utilized to quantify the ability of a particular biomass system to degrade target wastes. Several papers have been recently published describing the use of respirometric techniques to determine the values of the biokinetic constants for biological systems treating toxic and nontoxic organic compounds.[1,2,3,4] Work has also been done utilizing these techniques on actual municipal[5,6,7] and industrial[8] waste systems. Other results showed that respirometric techniques can be used to predict plant effluent quality for a municipal activated sludge process.[5] For the current work, Hercules Incorporated desired to evaluate the biodegradability of five of its paper chemical products in anticipation of the startup of their HKC Korea Waste Plant which has to treat these production facility wastewaters. The goals of this work were twofold. The first objective was to estimate the effluent quality that will be produced by the activated sludge process as measured by the Korean COD method. The second goal of this work was to identify any troublesome or recalcitrant components which may contribute to the effluent COD by passing through the biotreatment process.

MATERIALS AND METHODS

The general testing approach involved the operation of a bench-scale activated sludge unit in conjunction with separate batch testing. A waste stream containing equal COD portions of the target components was developed and fed to a bench-scale activated sludge reactor which was operated at a long mean cell residence time (MCRT) in order to simulate extended aeration conditions. After a three-week acclimation period, biomass was harvested from the bench-scale reactor and used in separate batch respirometry tests to identify the biodegradability of the individual component wastes. Bench-scale reactor effluent samples were also collected toward the end of the work to determine COD and BOD removal capability.

A 1.5 liter glass reactor was used as the bench scale activated sludge reactor. This reactor was charged with 1 liter of tap water and 500 ml of biomass seed that was obtained from a Hercules Incorporated plant treating similar wastes. Mixing and dissolved oxygen were provided using an

45th Purdue Industrial Waste Conference Proceedings, © 1991 Lewis Publishers, Inc., Chelsea, Michigan 48118. Printed in U.S.A.

Table I. Nutrient Concentrations Used In Each Respirometry Study

Nutrient	Concentration/Respirometer Flask
Sodium Molybdoate	10 mg/L
Ferric Chloride ($FeCl_3 \cdot 6H_2O$)	0.25 mg/L
Magnesium Sulfate ($MgSO_4 \cdot 7H_2O$)	22.5 mg/L
Calcium Chloride ($CaCl_3 \cdot 2H_2O$)	20 mg/L
Di-Ammonium Phosphate	30 mg/L
Urea	30 mg/L
BOD Phosphate Buffer (No Nitrogen)	10 mg/L
Tap Water	100 mg/L

overhead mixer and diffused air, respectively. Initially, reactor feed rates were set at 1000 mg COD/L/day but were later reduced to 500 mg COD/L/day in order to attenuate inhibition effects.

After the three-week acclimation period in the bench-scale reactor, biomass was harvested on a weekly basis and used in respirometry tests. These tests were designed to examine the biodegradability of the five individual waste components to determine which waste or wastes are difficult to treat and contribute significantly to the effluent COD.

A total of four respirometry tests were performed using a six unit, electrolytic respirometer*. Each test was run using six, 1-liter reactor flasks with each flask containing the desired amount of waste sample, nutrients and biomass seed for each study. Cumulative oxygen uptake was monitored on each flask for a period of 48 hours at sample intervals between 0.1 and 2.0 hours. The nutrient mixture used in each flask for all four studies is listed in Table I. The waste samples added to each flask were taken from stock solutions made from the initial concentrated waste samples diluted from their initial COD values using distilled water. Initial waste strengths were targeted at 200 mg/L but varied from 50 to 500 mg/L COD. Reasons for the variation in initial COD values were largely attributed to the difficulty encountered with obtaining homogenous mixtures of the emulsified products. The remaining 1-liter flask volume was composed of biomass seed and distilled water. In order to obtain the target biomass concentration in each flask, a suspended solids sample was taken from the bench-scale reactor the day before performing a respirometer test. To concentrate the biomass to the desired value, 1 liter of mixed liquor was taken from the continuous flow reactor and centrifuged for 5 minutes at 5000 rpm. The centrate was discarded and the biomass was resuspended in 50 mL of distilled water. Initial seed concentrations of between 30 and 50 mg/L were used for each respirometry test.

The overall strategy for the individual batch tests was as follows:

Test 1

All five products were individually tested for biodegradation using seed from the bench-scale reactor; a blank (no waste, only seed) reactor was also set-up to check for endogenous oxygen uptake rates. The target initial COD for all five components was 200 mg/L COD.

Test 2

This was a repeat of Test 1, with the substitution of an "All Waste" reactor (equi-COD mixture of all five components) for the endogenous reactor. The rationale was to compare the degradation ability of the biomass on the waste mixture and on the individual components in that mixture.

Test 3

Same approach as Test 2.

Test 4

This test focussed on the bench-scale reactor effluent for BOD evaluation. The other components were tested to verify the biodegradation difficulty which was observed for waste samples B, C and D; Sample A was deleted because it consistently showed better biodegradation rates than the other components.

*ER-100 Electrolytic Respirometer marketed by Bioscience Management, Inc. (Bethlehem, PA).

During each respirometry test, samples were collected to determine the extent of biodegradation that occurred in each flask. Analyses were performed during studies 2, 3, and 4 in which initial and final, total and soluble COD samples were taken along with suspended solids readings on each flask. This information was utilized to check the cumulative oxygen uptake or BOD data obtained from the respirometer by performing a COD mass balance on the system. This approach was employed because it provides a methodology to check the biodegradation data. All waste (substrate) COD in a given aerobic system must be channeled to cell mass or oxidized to carbon dioxide (assuming that the COD is not chemically oxidized or physically removed by mechanisms such as stripping). Carbon dioxide production coincides with oxygen consumption (uptake). The respirometric testing technique allows the analyst to cross-check biodegradation data since COD depletion data is convertible to oxygen uptake data and vice-versa.

Initial soluble COD samples were taken from each flask before the addition of biomass seed. Final soluble COD samples were filtered through 0.45 μm Whatman Filter papers. All COD samples were preserved using 0.5 mL of concentrated sulfuric acid and refrigerated before analysis. The COD method used during this investigation was the colorimetric technique developed by Hach.[9]

Finally, the Korean COD (KCOD) analysis was performed on the bench-scale reactor effluent to obtain an indication of the expected COD levels in the effluent of a treatment plant handling these wastes. This test differs from the conventional COD because the KCOD test utilizes potassium permanganate as an oxidizing agent. This procedure which was obtained from Hercules Incorporated is given in the Appendix.

RESULTS AND DISCUSSION

As previously stated, the purpose of this work was twofold. First, a bench-scale reactor was operated to evaluate the COD (Korean COD) and BOD removal capability of an activated sludge reactor being fed 500 mg/L COD consisting of an equi-COD mixture of the five target wastes. This reactor was also used to provide a source of biomass seed for batch respirometry tests. The objective of the respirometry tests was to identify "troublesome" individual waste components which would pass through the biotreatment system and contribute to the effluent soluble COD.

BENCH-SCALE REACTOR

The effluent from the reactor was analyzed after seven weeks of operation for BOD and COD (both the U.S. and Korean method). Soluble effluent COD using the U.S. COD method was measured to be 134 mg/L; this represents a soluble COD removal efficiency of 73% ((500–134)/500). This number is in the range observed by other Hercules biotreatment facilities which are treating similar wastes. Soluble effluent BOD was observed to be 2.3 mg/L.

The Korean COD method yielded a value of 29 mg/L KCOD (Korean COD) for the same soluble effluent sample. The ratio of COD to KCOD is 4.6 (134/29). The total KCOD of the effluent sample (soluble portion plus 30 mg/L of suspended solids was 46 mg/L). Current estimates suggest that the HKC Korea's waste plant will receive an influent of approximately 3,000 mg/L COD. Using the observed COD removal efficiency of 73%, the soluble effluent COD will be 810 mg/L ((0.27) (3000)) or 176 mg/L KCOD (810/4.6). It should be noted that previous work on landfill leachate treatment[10] validates the extrapolation of COD performance data for estimating effluent COD characteristics at higher COD influent concentrations (Note: Bench-scale reactor was fed 500 mg/L COD).

RESPIROMETRY TESTS

The results of the batch respirometry tests are presented in Figures 1 through 4 and Tables II through V. Analysis of the data in Tables II through V indicates that approximately half of the batch test results represent good COD depletion and oxygen uptake balances. This is stated because the COD depletion can be accounted for in the mass of oxygen utilized and in the increase in cell mass. It is believed that a higher percentage of the results would yield good balances if more consistent COD data were obtainable (The difficulties with the COD analysis which were attributed to the physicalities and handling of the highly concentrated, emulsified waste samples are discussed in the Materials and Methods). The purpose of the balance analysis was to check the biodegradation data in order to eliminate any anomalous results. Detailed discussions on the balance procedure are given elsewhere.[1,2]

The results of the first respirometry test are presented in Figure 1 and tabulated in Table II. COD data judged to be anamolus via the COD/oxygen uptake balance check were deleted from the table. After the review of the respirometry results, it was determined that Waste A was most biodegradable,

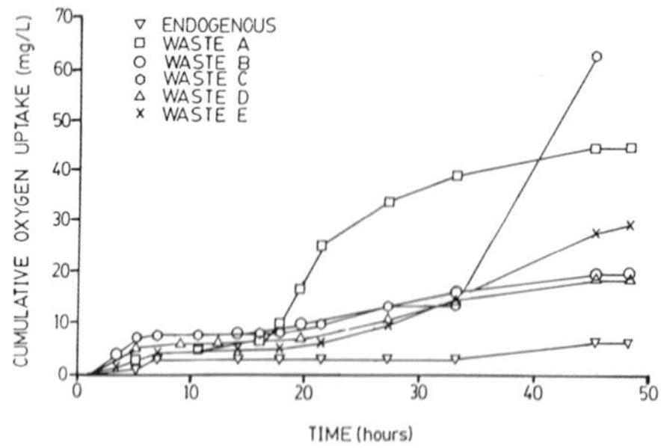

Figure 1. Respirometric Test #1: Individual components plus endogenous seed reactor

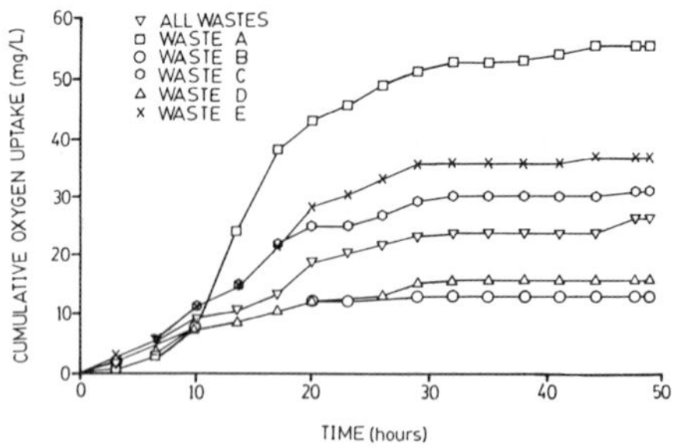

Figure 2. Respirometric Test #2: Individual components plus "All Waste" reactor

while Waste B and Waste D were problematic. For this test, the oxygen uptake (BOD) results for Waste C were erroneous (inordinately high) due to faulty equipment (subsequent tests identified the biodegradability ranking of this component).

Figure 2 depicts the results of Test 2 which is tabulated in Table III. The endogenous reactor was replaced by a reactor which received an equal COD mixture of each of the target wastes; that is, the same mixture which was fed to the bench-scale system. In Figure 2, it is interesting to note that respirometry results for the "All Waste" reactor fall in the middle along with those results for Waste C. Waste A is well above the "All Waste" system which indicates that this component is degrading relatively well. Conversely, the other products show lower results which means again that these components are more resistant to biodegradation by the seed biomass. Similar results were also obtained in Test 3.

The goal of Test 4, shown in Figure 4, was to verify the difficulty regarding the biodegradability of Wastes B and D and to assess the BOD of the bench-scale reactor effluent. These results are interest-

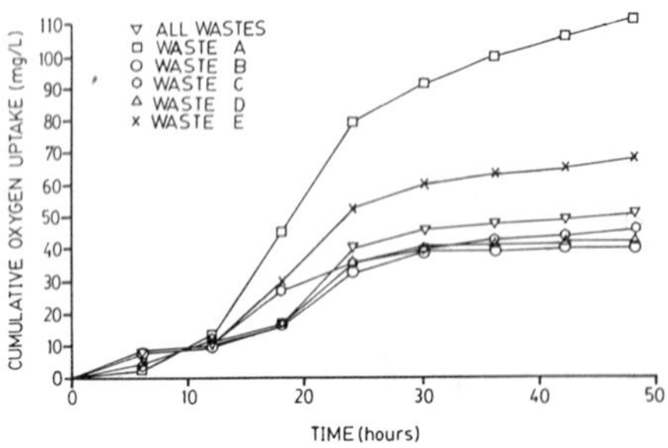

Figure 3. Respirometric Test #3: Individual components plus "All Waste" reactor

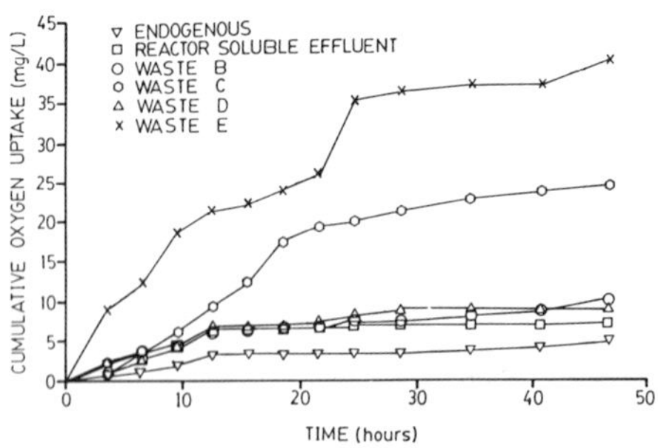

Figure 4. Respirometric Test #4: Components B, C, D, and E; effluent BOD; and endogenous reactor

Table II. Soluble COD Results for Hercules Respirometer Study #1

Flask Component	Soluble COD Initial (mg COD/L)	Soluble COD Final (mg COD/L)	COD Removal (mg COD/L)	BOD 48 hours (mg 02/L)
Endogenous	–	–	–	6.1
Waste A	76	–	–	44.3
Waste B	118	80	38	19.2
Waste C	122	78	44	118.7*
Waste D	76	61	15	18.3
Waste E	80	–	–	28.9

*This value is disregarded because the Electroylsis Cell was not working properly.

Table III. BOD, COD, and Solids Results for Hercules Respirometer Study #2

Flask Component	Total COD Initial (mg COD/L)	Soluble COD Initial (mg COD/L)	Soluble COD Final (mg COD/L)	COD Removed (mg COD/L)	Solids Initial (mg/L)	Solids Final (mg/L)	Solids Produced (mg/L)	BOD 48 Hours (mg O_2/L)
Waste A	153	–	72	–	50	98	48	55.5
Waste B	122	64	11	53	50	76	26	12.8
Waste C	284	–	102	–	50	88	38	31
Waste D	294	–	173	–	50	80	30	15.7
Waste E	264	–	–	–	50	64	14	36.9
All Wastes	213	94	82	12	50	54	4	26.1

Table IV. BOD, COD, and Solids Results for Hercules Respirometer Study #3

Flask Component	Total COD Initial (mg COD/L)	Initial (mg COD/L)	Soluble COD Final (mg COD/L)	COD Removed (mg COD/L)	Initial (mg/L)	Solids Final (mg/L)	Solids Produced (mg/L)	BOD 48 Hours (mg O_2/L)
Waste A	–	104	38	66	50	58	8	110
Waste B	–	87	68	19	50	50	–	39.6
Waste C	–	160	79	81	50	56	6	45.2
Waste D	129	87	18	69	50	74	24	41.7
Waste E	261	–	170	–	50	60	10	67.4
All Wastes	211	152	119	33	50	62	12	50.1

Table V. BOD, COD, and Solids Results for Hercules Respirometer Study #4

Flask Component	Total COD Initial (mg COD/L)	Soluble COD Initial (mg COD/L)	Soluble COD Final (mg COD/L)	COD Removed (mg COD/L)	Initial (mg/L)	Solids Final (mg/L)	Solids Produced (mg/L)	BOD 48 Hours (mg O_2/L)
Endogenous	73	–	–	–	30	24	–	4.8
Reactor Effluent (Soluble)	164	134	130*	4	30	32	2	7.1
Waste B	57	40	28	12	30	28	–	10
Waste C	91	62	40	22	30	34	4	24.6
Waste D	79	45	–	–	30	38	8	9.1
Waste E	48	–	–	–	30	16	–	41.8

*Korean COD performed on this sample.

Table VI. ΔCOD/Initial COD Values Calculated for Each Waste From the First Three Batch Respirometry Tests

Test No.	Waste A	Waste B	Waste C	Waste D	Waste E
1	— +	0.322	0.361*	0.373	0.713
2	0.561	0.420	0.332	0.139	0.378
3	0.635*	0.220	0.510*	0.790*	0.532
Avg.	0.598	0.321	0.401	0.434	0.541

+ΔCOD/Initial COD Value was greater than 1.0.
*Used actual final ΔCOD for this computation.
Note: The results of Test 4 are not included because Waste A was not analyzed in this test.

ing in that the BOD of the reactor effluent is essentially negligible. Furthermore, as was the trend in the other three tests, Waste B and D showed relative resistance to biodegradation which Waste C appeared to be more amenable to biotreatment.

Biodegradability Ranking

At the request of Hercules, we were asked to take this analysis one step further and formulate a methodology to quantify the biodegradation ranking of each waste component and to develop an algorithm for estimating the effluent soluble COD contribution of each waste. It was decided to utilize the data in the batch tests to rank the components by the products of △COD (COD removed) divided by initial COD. This is essentially the percentage or fraction of initial COD removed. △COD was computed using the respirometric data (BOD) according to the following expression:

$$\Delta COD = \frac{\text{oxygen uptake}}{1 - (OX)\,(YT)} \quad (1)$$

where: XO = cell COD value (1.42); and
YT = cell yield (0.35, computed from data)

Missing initial COD values (caused by analytical difficulties) were estimated in one of two ways. One way was to use the total COD value and subtract the COD contribution attributed to suspended solids. The other way was to utilize the COD and oxygen uptake balance principle and the final soluble COD value to compute initial COD's.

Table VI lists the △COD/Initial COD values for the five Hercules' wastes tested. Based on the information given in Table VI, the biodegradability ranking for each waste is given in Table VII. In order to predict the impact of the influent waste mixture on effluent COD characteristics, the batch results listed in Table VII need to be "scaled-up" to the bench-scale system. This is stated because bench-scale reactors which simulate full-scale systems generally have greater COD removal than batch reactors. The bench-scale reactor which simulated a continuous flow treatment situation achieved a 73% COD removal rate for the waste mixture. A ranking procedure for individual wastes can be formulated if it is assumed that each waste is characterized by its own COD removal ratio which results in an aggregate △COD for the waste mixture:

△COD Waste A + △COD Waste B + etc. . . . = △COD Waste mixture.

Adding up the average △COD/Initial COD numbers from Table VII and dividing by 5 (the bench-scale waste mixture consisted of equal COD contributions of the five products), yields 0.459. It was noted that the bench-scale unit had a COD removal rate of 0.73; the scale-up factor is calculated as

Table VII. Summary of ΔCOD/Initial COD Data and Biodegradability

Waste	Average Δ COD/Initial COD*
A	0.598
E	0.541
D	0.434
C	0.401
B	0.321

*Test 4 not included in calculations.

Table VIII. Adjusted ΔCOD/Initial COD Factors for Each Waste

Waste	Adjusted Factor
A	0.951
E	0.860
D	0.690
C	0.638
B	0.510
Combined Waste	0.730

0.73/0.459 or 1.59. The adjusted △COD/Initial COD numbers for the products are given in Table VIII. The information in Table VIII can be utilized by inserting each removal factor into a formula for predicting the impact of waste stream composition on effluent COD; however a more rigorous bench-scale evaluation of this methodology and algorithm is required to determine the overall utility of this approach.

SUMMARY AND CONCLUSIONS

The results of this work demonstrate another useful application of respirometry for analyzing biotreatment situations. In this case, the goal was to determine the potential contributions which various waste components would make to the effluent COD of the wastewater treatment plant. The general approach consisted of feeding a simulated waste stream which was composed of five target waste products to a bench-scale activated sludge system. After a three-week acclimation period, biomass was harvested from the reactor and used in separate batch respirometric tests to evaluate the biodegradability of the individual waste products.

Four sets of batch tests were performed over a seven-week period. These tests were able to distinguish the biodegradation characteristics of the individual products. Additionally, because the wastes were emulsions, some difficulty was encountered in interpreting COD data. However, since a respirometric methodology was utilized for the batch tests it was easier to determine which COD data were anamolous. This is stated because, for each batch test, COD and oxygen balances can be performed which allows for an extra check on the data.

Finally, an attempt was made to utilize the data from batch respirometric tests to develop biodegradability rankings for the individual waste components. The goal is to try and predict the contribution to effluent COD that each component will make. Although the rankings methodology and associated algorithm require more rigorous verification, this initial effort indicates some potential as a relatively convenient technique for analyzing complex multicomponent biodegradation situations.

REFERENCES

1. Gaudy, A. F. Jr., Rozich, A. F., Garniewski, S., Moran, N. R., and Ekambaram, A., "Methodology for Utilizing Respirometric Data to Access Biodegradation Kinetics", *Proceedings*, 42nd Industrial Waste Conference, Purdue University, p. 573 (1988).
2. Gaudy, A. F., Jr., Ekambaram, A., and Rozich, A. F., "A Respirometric Method for Biokinetic Characterization of Toxic Waste," *Proceedings*, 43rd Industrial Waste Conference, Purdue University, p. 35 (1989).
3. Gaudy, A. F., Jr. Ekambaram, A., Rozich, A. F., "Comparison of Respirometric Methods for Determination of Biokinetic Constants for Toxic and Nontoxic Wastes," *Proceedings*, 44th Industrial Waste Conference, Purdue University (1990).
4. Grady, C. P. L., Jr., Dang, J. S., Harvey, D. M., Jolobagy, A., and Wang, L. L., "Determination of Biodegradation Kinetics Through Use of Electrolytic Respirometry". *Water Science and Technology*, 21:957 (1989).
5. Rozich, A. F., Colvin, R. J., Gaudy, A. F., Jr., and Martin, J., "Application of a Process Model Calibrated with Respirometry to Predict Full-Scale Activated Sludge Performance," *Proceedings*, 45th Industrial Waste Conference, Purdue University (1990).
6. Rozich, A. F., and Colvin, R. J., "Use of Respirometry for Operating Activated Sludge Plants Treating Toxic Wastewaters" *Presented*, Water Pollution Control Speciality Conference on Analytical Techniques and Residuals Management, Atlanta, GA (1988).
7. Rozich, A. F., Colvin, R. J., Gaudy, A. F., Jr., and Martin, J., "Application of a Process Model Calibrated with Respirometry to Predict Full-Scale Activated Sludge Performance" (with A. F.

Rozich, A. F. Gaudy, Jr., and J. Martin), *Submitted*, 45th Industrial Waste Conference, Purdue University, Lafayette, IN (1990).
8. Gaudy, A. F., Jr., and Rozich, A. F., "Development of a Practical Test for Assessing Effects of Industrial Effluents on Patapsco Activated Sludge," Progress Report Prepared for Department of Public Works, Bureau of Water and Wastewater, City of Baltimore, Baltimore, Maryland, January (1988).
9. Hach Chemical Company, "Hach Water Analysis Handbook", Loveland, CO, 2-144-146 (1980).
10. Gaudy, A. F., Jr., Rozich, A. F., and Garniewski, S., "Treatability Study of High Strength Landfill Leachate", *Proceedings*, 41st Industrial Waste Conference, Purdue University, p. 627 (1987).

APPENDIX

KOREAN COD PROCEDURE

This procedure was provided by Hercules Incorporated and involves adding excess permanganate and digesting the sample at 100°C for 1 hour under alkaline conditions. After digestion, the sample is then acidified using a 30% sulfuric acid solution. The excess permanganate is titrated using thiosulfate with a starch, potassium iodine color indicator. The exact procedure used is listed below along with the formula for calculating the Korean COD of the sample.

Procedure

1. Add the following into a 300 mL digestion flask with condenser.

   - 1 mL sample (1 mL deionized water for blank)
   - 20 mL 0.25N $KMnO_4$
   - 0.5 mL 45% KOH
   - 25 mL Deionized Water

2. Heat flask in water bath at 100°C for 60 minutes
3. Cool sample and add the following:

   - 1 mL 10% KI
   - 2 mL 30% H_2SO_4
   - 2 mL 1% Soluble Starch

4. Titrate sample with 0.025 N Na_2SO_3 to clean end point.
5. Calculate Korean COD using the following formula.

   Korean COD (mg KCOD/L) = (a - b) (f) (100/v) (0.2)

   a – 0.025 N sodium thiosulfate consumed while measuring the blank sample (mL).

   b – 0.025 N sodium thiosulfate consumed while measuring the test sample (mL)

   f – sodium thiosulfate concentration factor (0.025).

   v – amount of sample (mL).

56 A COMPUTERIZED RESPIROMETRIC METHOD FOR DETERMINING INHIBITION POTENTIAL OF WASTEWATERS

Martha I. Beach, President
N-CON Systems Co.
Larchmont, New York 10530

Ricardo B. Jacquez, Professsor
Fernando Cadena, Assoc. Professor
Atul Shah, Graduate Research Assistant
New Mexico State University
Las Cruces, New Mexico 88003

INTRODUCTION

Twenty years ago, with the passage of the Clean Water Act, the federal government embarked on a program to clean up the navigable waters of the United States. Along with restrictions, guidelines and discharge permits, the construction grants program provided funds for the design and construction of thousands of Publicly Owned Treatment Works (POTW). In some instances the failure of new and sophisticated waste treatment plants to meet their effluent permit limits has been traced to toxic wastes from industrial operations. In 1984, the Environmental Protection Agency (EPA) issued its national "Policy for Development of Water Quality-Based Permit Limitations for Toxic Pollutants." This Policy is being implemented by the National Pollutant Discharge Elimination System (NPDES) and represents a major change in Federal and State pollution control strategies. Permit discharge compliance is no longer based on specified end-of-pipe technology or specific single chemical analyses. The Environmental Protection Agency (EPA) has incorporated whole effluent toxicity tests to their control strategy because there is now a general consensus that an evaluation of effluent toxicity, when adequately related to instream conditions can provide a valid indication of receiving water impacts.[1]

In an effort to limit the entry of toxic materials into the POTW, stringent pre-treatment regulations have been promulgated. This in turn has resulted in increased interest in the effects of various materials on biological treatment systems, as well as on the receiving waters.

In spite of the fact that most engineers generally refer to activated sludge or biomass in the singular, microbiologists and bio-process engineers are quick to point out that activated sludge consists of an innumerable and diverse population of aerobic microorganisms. These microorganisms consume the carbonaceous material in the wastewater if they are healthy and have sufficient oxygen available.

WHAT IS TOXICITY?

Toxicity can not truly be defined as an absolute. We should think more in terms of functional toxicity. Functional toxicity must take into account the form and concentration of the substance, exposure time, type and condition of the exposed specie(s). Functional toxicity occurs when a substance or substances cause reduced metabolism in the subject organism(s) resulting in reduced activity and/or food intake, inability to reproduce, or death.

In this context, a substance is considered toxic if, when taken into a POTW, it causes an upset to the biological process(es) of the plant and allows the discharge of the toxic substance, as well as other substances, in partially or untreated form in excess of discharge permit limits. This discharge will, in turn, be toxic to the biota of the receiving waters.

BIOMONITORING TECHNIQUES

A wide array of biomonitoring techniques have evolved for determining the effect of treatment plant effluent on receiving waters. The best known are those involving various species of fish and

---

45th Purdue Industrial Waste Conference Proceedings,

invertebrates. However, such tests are not appropriate for determining the toxic or inhibitory effects of incoming materials on the waste treatment plants biological processes.

For this purpose researchers have found that respirometry, using seed or inoculum from an activated sludge process is one of the most sensitive for determining the effect of potential toxic materials on the biological processes of a wastewater treatment plant.

BACKGROUND OF WASTEWATER RESPIROMETRY

The use of respirometry to study the oxygen consumption of waste treatment processes dates back to the middle 1920s with the Warburg and Sierp manual respirometers. Don Bloodgood, who founded the Purdue Industrial Waste Conference, developed and worked with a modified Warburg respirometer back in 1938. Montgomery describes a variety of respirometers developed in England, the United States, Europe and South Africa during the 1950s and 60s to aid in the determination of biochemical oxygen demand (BOD/5).[2] In 1960 Jenkins observed that: "Use of manometric (respirometric) methods for determination of BOD would eliminate any errors introduced by dilution, would give a quicker answer, and if a standard apparatus to study large volumes (500–1000 ml) of respiring liquid, could be developed, the course of oxidation of a representative sample of the liquid could be followed by the changes in the chemical nature of the samples, as well as by observing directly the exact course of oxygen-uptake."[3] Observation of oxygen uptake rate is also of prime concern for activated sludge biokinetics and determining inhibition potential or toxicity.

SOURCES AND CORRECTION OF ERRORS

Because respirometers depend on measurement of minute changes in vacuum, in a closed reactor, caused by the respiration of bacteria, they are necessarily sensitive to all types of pressure changes. Temperature and atmospheric changes, as well as leaks, are the prime causes of error. Errors caused by temperature changes are relatively simple to control with constant temperature water baths or incubators. Such errors may also be compensated for mathematically. However, because of their dynamic nature over time, it is virtually impossible to compensate for atmospheric pressure variations, whether caused by weather, operation of lab vents or opening and closing doors. It is therefore, essential that the reference side of the sensor be isolated from changes in atmospheric pressure, particularly for tests lasting more than a few hours. In 1984 Drohoybczer and Cadena developed a new respirometer with a pressure reference that could be sealed at the start of each test. This system eliminated errors caused by atmospheric pressure changes.

Oxygen is supplied by a precisely regulated source and delivered to each reactor by its own individually calibrated oxygen metering device (OMD). Oxygen delivery is controlled and monitored by the system computer.

COMPUTER CONTROL OF RESPIROMETER FUNCTIONS

The combining of a personal computer with the pressure compensated respirometer creates a unique system in which the computer controls the delivery of oxygen to the reactors, collects, stores and produces numeric and graphic displays of the data generated. During the test(s) accumulated oxygen demand, as well as oxygen uptake rate, are displayed on monitor. The instrument, called a PCBOD, was reported on by Cadena at the Purdue Industrial Waste Conference in 1988.[4] This respirometer, is now available from N-CON Systems as the "Comput-OX" wastewater respirometer in 2, 4 or 8 reactor models.

BENEFITS OF A STANDARD BIOCULTURE

Reduction of instrument errors has increased attention on the need for standardized biological test proceedures. The organisms used in respiration measurements range from fish to microorganisms. Regardless of size, the presence of toxic materials disrupts normal metabolism, hinders respiration or kills the test organisms. Single species tests are often too sensitive provide repeatable results when exposed to a wide variety of potentially toxic substances. If respirometry is to take advantage of the sensitivity of microorganisms, it is essential to maintain a reliable source of healthy, broad spectrum population. For comparative purposes it is also essential that the culture be representative of microorganisms found in both industrial and domestic wastewater treatment plants and provide reproducible results.

The following are the results of tests using an 8 reactor Comput-OX respirometer tests with a commercially available broad based bioculture as a reference microbial population.

REFERENCES

1. Grothe, Donald R., Kimerle, Richard A., Charles D. Malloch "A Perspective on Biological Assessments," *Water Environment & Technology*, Vol. 2, Number 4 (April 1990) pp. 62–67.
2. Montgomery, H. A. C., "the Determination of Biochemical Oxygen Demand by Respirometric Methods," *Water Research*, Vol. 1, No. 1, Page 632, 1967.
3. Jenkins, David "The Use of Manometric Methods in the Study of Sewage and Trade Wastes" *Waste Treatment*, pp. 99–125. Pergamon Press, New York 1960.
4. Cadena, Fernando; Drohobyczer, Aleksander; Beach, Martha "A Novel Approach to Simplified Respirometric Oxygen Demand Determinations" 43rd Purdue Industrial Waste Conference, Purdue University 1988.

57 ADVANCED MEASUREMENT CELLS FOR AUTOMATED QUANTIFICATION OF BIODEGRADATIVE MICROBIAL OXYGEN UPTAKE IN FLUID, SEMI-SOLID AND SOLID MEDIA

J. B. Hill, Research Chemist
R. D. Bleam, Manager of Technical Services
T. G. Zitrides, President
Bioscience Management, Inc.
Bethlehem, Pennsylvania 18017
C. S. McDowell, President
Tri Bio, Inc.
Allentown, Pennsylvania 18013

INTRODUCTION

In the late nineteenth century, scientists first noticed that microorganisms consumed oxygen while decomposing organic wastes. Since then, oxygen uptake measurements have been used as a measure of the treatment of wastewaters.

One of the most popular oxygen uptake measurement techniques was the biochemical oxygen demand (BOD) test, from which has been derived the standardized five-day BOD test.[1] Although many newer and more sophisticated experimental procedures have been developed, including chemical oxygen demand (COD) and total organic carbon (TOC) analyses, the BOD test maintains its usefulness today. An increasingly popular experimental procedure used to measure oxygen uptake, which is closely related to the BOD test, is respirometry. This method is designed to monitor the progression of biodegradation reactions in terms of oxygen uptake.[2,3] There are two types of respirometry, electrolytic and manometric. Manometric respirometry measures oxygen uptake by monitoring pressure reduction in a closed system. Commercial manometric systems use either air or oxygen to provide the oxygen requirements to the system. The principle of both manometric oxygen and electrolytic respirometry systems, is to maintain the dissolved oxygen level in the biodegrading sample during oxygen demand by continuous replacement of the oxygen consumed with oxygen from the headspace of the sample reactor. In electrolytic systems, oxygen is generated by electrolytic dissociation of water to oxygen and hydrogen. In manometric oxygen systems, oxygen is supplied as a pure gas from cylinders. Electrolytic instruments hydrolyze either copper sulfate or sulfuric acid. The sample oxygen demand is determined by converting the power required (to generate oxygen and reestablish the initial oxygen composition in the headspace) into oxygen equivalents.

The remaining discussion and results will focus on electrolytic respirometric instruments. Electrolytic respirometry was initially developed to test wastewaters. It automates the labor intensive five-day BOD test.[4] It can measure the biodegradability of a wastewater without dilution of the sample, unlike the standard BOD test. The technique was designed to more closely simulate the environment of a wastewater treatment plant, and therefore give a better indication or measurement of its state of operation. Commercial instruments were originally developed in the 1970's primarily for wastewater samples of moderate strength.

However, due to its versatility, electrolytic respirometry found its way into an increasingly wide range of applications for water borne contaminants of both high and low strength. Larson and Perry[3] have shown that the rate and extent of biodegradation of organic compounds in low strength natural waters can be assessed with electrolytic respirometry. Other researchers have used electrolytic respirometry to determine biodegradation kinetics. Grady, Dang et al. used this technique to determine kinetics for single organic compounds in dilute aqueous solutions in batch reactors, and found it to be much less labor intensive than other techniques.[5] Still others demonstrated that respirometry and chemical oxygen demand analyses are sufficient to characterize kinetic constants for wastewaters,

45th Purdue Industrial Waste Conference Proceedings, © 1991 Lewis Publishers, Inc., Chelsea, Michigan 48118. Printed in U.S.A.

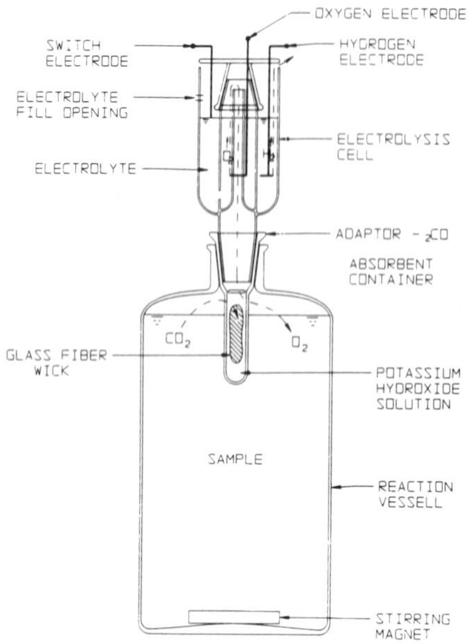

Figure 1. Schematic diagram of the standard reactor and electrolysis cell.

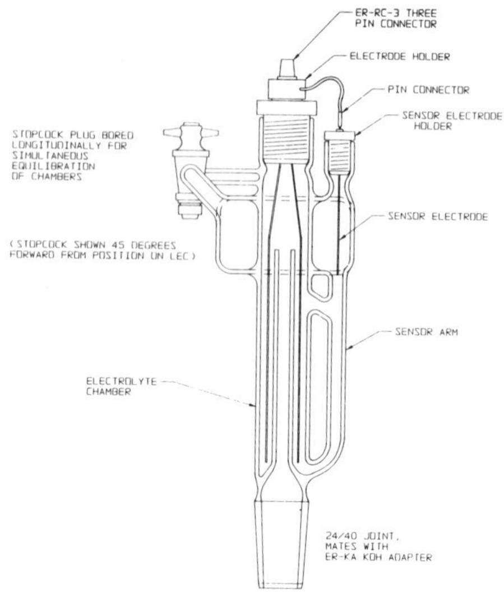

Figure 2. Schematic diagram of the low oxygen electrolytic cell.

replacing more complex chemical analyses.[6] Respirometric data can also be used to obtain biodegradation information for toxic wastewaters. This can consist of treatability data as well as the determination of biokinetic information. A practical methodology, incorporating respirometric and COD data into a process control model in order to define guidelines for operation of activated sludge plants treating toxic wastes,[7] and other types of biotreatment processes has been reported.[8]

More recently attention has been drawn to the cleanup of sites containing contamination in soils, sludges, surface waters and groundwaters. However, in many sites the contaminants are insoluble, inhibitory, highly dilute or highly concentrated. Some reporters and many practitioners have turned to respirometry as a means to characterize the biodegradation kinetics of these liquid, solid and semi-solid wastes. Unfortunately, in many cases, respirometry was found to be lacking. Specifically, conventional respirometers dealing with high strength wastes demonstrated inadequate gas to liquid oxygen transfer rates, inadequate oxygen generation rates, and inadequate carbon dioxide absorption rates. When analyzing low strength wastes, respirometers showed systematic errors yielding significant experimental error. These were caused by initial gas composition errors, atmospheric pressure errors, and errors in oxygen transfer/gas diffusion. Mixing problems were also encountered with slurries, sludges and insolubles. This paper will discuss modifications in reactor design and in oxygen transfer cell design to circumvent these limitations and to expand the range of sample types and waste strengths that can be analyzed using respirometric instruments. Applications of these modifications are microbial toxicity testing, reactor design studies, in-situ biodegradation simulations, and biological treatability testing of industrial sludges.

MATERIALS AND METHODS

Modifications of the standard electrolytic respirometry equipment (Figure 1) have been developed to expand the capabilities of this technique to allow for testing of high strength wastes, groundwaters, soils and sludges. All designs discussed are compatible with existing respirometric equipment. They all have been prototyped and tested.

Low Oxygen Demand Electrolytic Respirometer Cell

Standard respirometers (Figure 1) measuring uptakes in low strength wastes show experimental error caused by initial gas phase composition, atmospheric pressure changes, and limitations in oxygen transfer.

Potential oxygen transfer errors can occur because: small amounts of oxygen consumption will not displace enough electrolyte to cause the electrolysis cell to produce oxygen. Atmospheric pressure error is introduced when electrolyzing sulfuric acid which produces hydrogen and oxygen. While oxygen is added to the reactor, the hydrogen must be allowed to escape to the atmosphere through a hole in the outer wall of the electrolysis cell. This opening may allow atmospheric pressure changes to affect the pressure within the electrolysis cell.

The low oxygen electrolysis cell was derived from a modified electrolytic cell which was used successfully to determine BOD kinetics of receiving waters with low levels of BOD[9] (Figure 2). A small change in the electrolyte level in the chamber of the low oxygen cell causes a larger change in the electrolyte level in the sensor tube, which is a much narrower tube than in the standard cell. This increases the sensitivity of the low oxygen cell to smaller increments of oxygen demand. The large glass envelope acts as a pressure buffer in which these volume changes can occur with minimum interference by any of the concomitant pressure changes within the system. A stopcock serves to equilibrate the two compartments with the ambient atmosphere before the experiment is initiated. When the stopcock is closed and the respirometer is operating, the instrument electrolyzes copper sulfate, producing oxygen in response to demand in the reactor bottle, in order to maintain equilibrium. This eliminates atmospheric pressure as a variable.

A respirometric experiment was designed to test the sensitivity of the low oxygen cell. Two reactors with 100 mg of a mixed BOD inocula, containing equal portions of B. subtilus, B. cereus, and P. fluorescens, and a casamino acid solution of 20 mg/L were used to compare the cumulative oxygen uptake with a low oxygen electrolysis cell against a standard electrolysis cell.

Enhanced Transfer Reactor

Limits with the conventional respirometer reactor when analyzing high strength wastes include inadequate oxygen diffusion from the generation point to the use point, inadequate oxygen transfer and inefficient carbon dioxide absorption.

Oxygen is produced in the electrolysis cell located remotely from the biodegrading sample. Oxygen is pulled into the headspace of the reactor by air swirling induced by the liquid mixing action caused by a stirring bar. Thus, gas composition gradients can occur with high oxygen concentrations located away from the high demand locations. Carbon dioxide buildup can occur at the gas/liquid interface causing inefficient oxygen transfer, since oxygen transfer is a direct function of its gas store concentration.

Also, oxygen in the headspace of the reactor passively diffuses into the liquid. Carbon dioxide, produced during biodegradation, is also in the headspace of the reactor. It is absorbed by the potassium hydroxide trap through passive transfer. The potassium hydroxide trap is located remotely from the carbon dioxide source. Thus, carbon dioxide must diffuse to the absorption point. This creates carbon dioxide gradients with carbon dioxide concentration highest at the gas/liquid interface.

The enhanced transfer cell (Figure 3) is designed to enhance the transfer efficiency of both oxygen and carbon dioxide during biodegradation. When in operation, the enhanced transfer reactor is set on the respirometer like the standard reactor, and mixing is attained identically. However, the enhanced transfer reactor attains active gas/liquid interaction, while the standard reactor relies on passive interaction.

As oxygen is consumed within the reactor, pressure is reduced in the headspace. This causes the electrolyte to break contact with the switch electrode in the electrolysis cell, and oxygen is produced. The vacuum created by mixing pulls the oxygen from the cell into the reactor. The internal aeration tower forces the oxygen through the carbon dioxide scrubber and then down through the tower to the bottom of the reactor. The oxygen is dispersed throughout by mixing, and gases rise to the top of the reactor. At the surface, gases are exchanged actively and the vacuum forces passage of headspace gases through the carbon dioxide scrubber. This active bubbling of the gases through the potassium hydroxide causes a more efficient removal of the carbon dioxide. Therefore, the response of the electrolysis cell to oxygen consumption is positively affected.

The enhanced transfer reactor is best suited for oxygen uptake analyses of high oxygen demand wastes such as landfill leachates and for reactor configurations where high oxygen demands can be anticipated, such as batch reactors.

Two experiments were designed to test the transfer efficiency of this reactor against the standard reactor. Excessive amounts of BOD and inocula were added to ensure high oxygen demand. In the first experiment, each reactor contained one liter of a stock feed solution containing 7 g/L dextrose, 8 g/L peptone, 0.3 g/L of yeast extract, 1 g/L of butyric acid, 0.17 g/L $MgSO_4$, 0.26 g/L NH_3Cl, 0.06 g

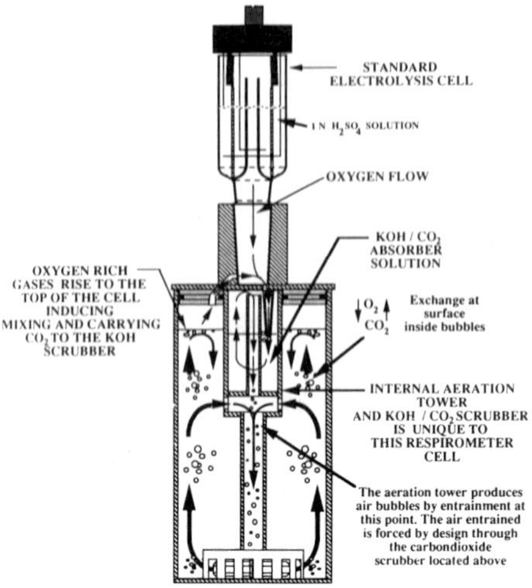

Figure 3. Schematic diagram of the enhanced transfer reactor.

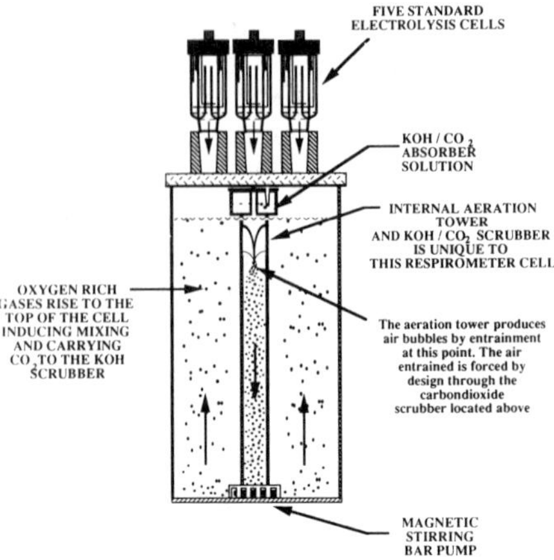

Figure 4. Schematic diagram of the large volume/high transfer reactor.

$FeCl_3$, and 0.8 g/L of $NaHPO_4$ with an ultimate BOD of approximately 20,000 mg/L, and 10 grams of a commercially available microbial inoculum. One milliliter of BOD buffer solutions[10] was also added.

In the second experiment, each reactor contained one liter of a feed solution containing 10 g/L glutamate, 10 g/L yeast extract and 20 g/L of glucose, along with 10 g of a commercially available BOD inoculum. Again, 1.0 milliliter of BOD buffer solutions[10] was added.

Large Volume/High Transfer Rate Reactor

The maximum reactor size of commercial electrolytic respirometers is two liters. Therefore, the maximum sample volume which can be analyzed is two liters. This introduces sampling variations as an error because of limited sample volume.

Also, larger sample volumes require increased oxygen production because of the increased oxygen demand exerted by more substrate and microorganisms.

A standard commercial electrolysis cell can produce 64 mg/L/hr. However, large sample volumes with high oxygen demand may exceed this capacity. Oxygen and carbon dioxide transfer will also be limited by larger sample volumes, if passive transfer occurs. A standard electrolysis cell can produce 64 mg/L/hr of oxygen. However, the actual transfer rate of oxygen in the standard reactor is typically 25–30 mg/L/hr. Increased sample volumes will reduce the transfer rate more because of limitations with mixing, which drives the passive oxygen transfer.

The large volume/high transfer rate reactor (Figure 4) functions respirometrically like the smaller version.

Large sample volume capability is not the only modification to this reactor. The top of this reactor is designed to hold five standard electrolysis cells simultaneously. Because of larger sample volumes, the oxygen demand exerted within the reactor may exceed the limitations for the instrument with one electrolysis cell. Therefore, additional electrolysis cells may be added in order to meet the oxygen requirements. Cells will produce oxygen in series to meet the oxygen demand exerted during biodegradation. By adding the oxygen uptake data for each electrolysis cell on the reactor, the total cumulative oxygen uptake may be attained.

Oxygen transfer and carbon dioxide absorption occur in this reactor, as in the enhanced transfer reactor.

As oxygen is consumed within the reactor, pressure is reduced in the headspace. This causes the electrolyte to break contact with the switch electrode in the electrolysis cell, and oxygen is produced.

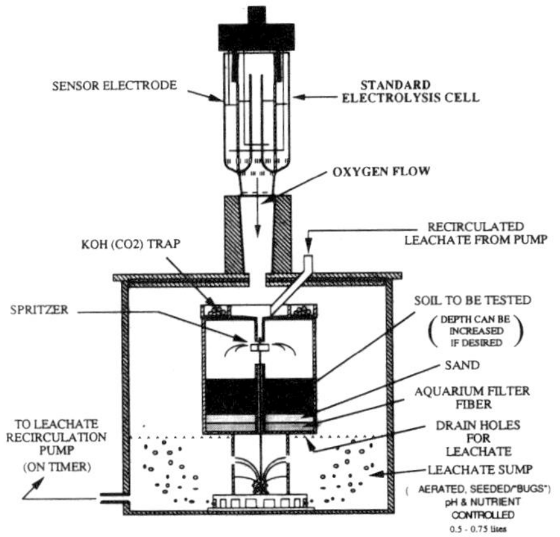

Figure 5. Schematic diagram of the leachate recirculation reactor.

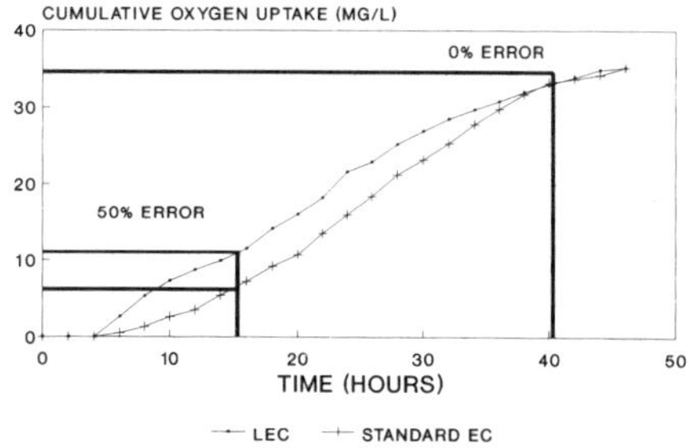

Figure 6. Cumulative oxygen uptake data comparing the low oxygen cell and the standard cell.

The vacuum created by mixing pulls the oxygen from the cell into the reactor. The internal aeration tower forces the oxygen through the carbon dioxide scrubber and then down through the tower to the bottom of the reactor. The oxygen is dispersed throughout by mixing, and gases rise to the top of the reactor. At the surface, gases are exchanged and the vacuum forces passage of headspace gases through the carbon dioxide scrubber. This active bubbling of the gases through the potassium hydroxide causes a more efficient removal of carbon dioxide, while the internal aeration tower increases the oxygen transfer into the sample.

A respirometric experiment designed to acclimate a microbial population to a substrate and measure oxygen uptake was used to test this reactor. Nine liters of a solution containing 9 grams of a commercial microbial inoculum, 1.83 g/L K_2 (HPO_4), 0.73 g/L KH_2PO_4 and 0.2 g/L NH_4Cl, was added to the reactor. A coking wastewater with 240 mg/L of cyanide, 835 mg/L of thiocyanate and a COD of 2500 mg/L was the substrate. Increasing substrate additions of 10, 20, 40, 80, 160, 360, and 640 milliliters were made to the reactor at various times. Three standard electrolysis cells were mounted on the reactor.

Leachate Recirculation Reactor

Existing respirometric instruments are designed for wastewater samples of moderate strength. Soils and sludges must be slurried with water to enable oxygen uptake measurement. Soil slurries are difficult to mix at solids concentrations greater than 10% with a stirbar. Slurries can cause uncoupling of the magnetic stir bar and reduced mixing capability. This mixing limitation is detrimental in two ways. Organic compounds found in soils may be only slightly soluble and mixing in the slurry is needed to solubilize them. If mixing is not efficient, solubilization decreases, and the compound is not easily available for the microorganisms to degrade. Also, mixing is needed for oxygen transfer to occur. The mixing of the liquid causes the air in the headspace to swirl and pulls oxygen, which is produced in the electrolysis cell, into the reactor. If mixing is inefficient, oxygen transfer is negatively affected.

The leachate recirculation reactor (Figure 5) is designed to measure oxygen uptake in solid and semi-solid samples. The soil to be tested is placed in the center well of the reactor. Underneath the soil is a layer of sand, followed by a layer of aquarium filter fiber. Drain holes for leachate are on the bottom of the center well. Leachate from the soil is collected in the bottom of the reactor and is aerated by mixing. A pump recirculates the leachate to the top of the center well, where it passes through a carbon dioxide scrubber. After passing through the scrubber, the leachate is redistributed over the soil, which is driven by the stirbars at the bottom of the reactor. This provides adequate moisture for biodegradation in the soil. This reactor is designed for use with a standard electrolysis cell.

An experiment was designed to use potting soil, contaminated with mineral spirits, to test the function of this reactor. Approximately 300 grams of potting soil was placed in the reactor and 750

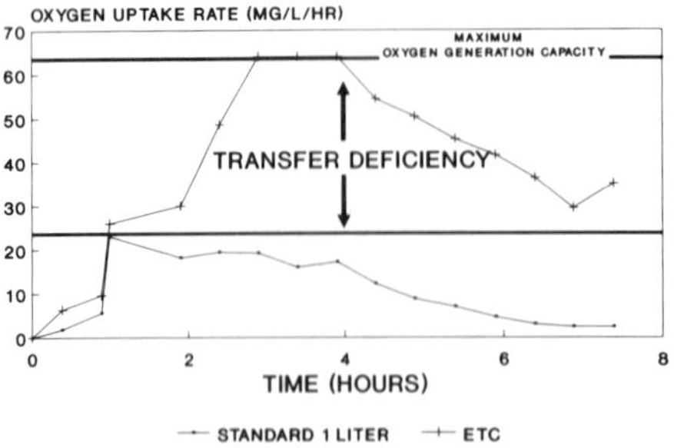

Figure 7. Oxygen uptake rate data comparing the enhanced transfer reactor and the standard reactor.

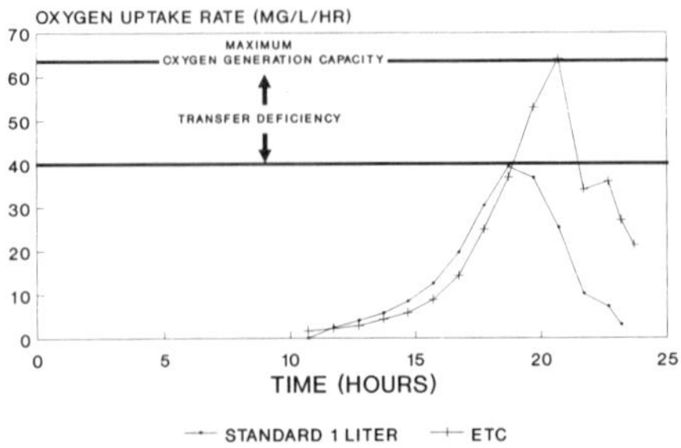

Figure 8. Oxygen uptake rate data comparing the enhanced transfer rector and the standard reactor.

milliliters of tap water was placed in the leachate sump. This experiment was designed only to determine if the reactor could measure the respiration of the soil.

RESULTS AND DISCUSSIONS

A series of experiments was developed to test the design and function of each of the modifications to the standard electrolytic respirometry equipment which have been discussed.

Low Oxygen Electrolysis Cell

In comparing the oxygen uptake rates exerted by a low BOD wastewater in a reactor with a standard electrolysis cell and a reactor with a low oxygen electrolysis (Figure 6), the sensitivity to oxygen demand of the low oxygen electrolytic cell is demonstrated. After 40 hours of this experiment, the oxygen uptake of each reactor is 35 mg/L. However, at 15 hours the oxygen uptake of the reactor with the low oxygen cell is 10 mg/L, and the oxygen uptake with the standard cell is 5 mg/L. This is a 50% differential. The improved accuracy can be attributed to the better sensitivity of the low oxygen cell due to its ability to sense smaller increments of oxygen demand because less electrolyte must be displaced in order to break contact with the switch electrode. The improved accuracy may also be attributed to the reduced impact of atmospheric pressure variation due to the resulting from a closed electrolysis cell. This demonstrates the effectiveness of the low oxygen cell at low BOD concentrations. At 35 mg/L of oxygen uptake or greater, the sensitivity of the low oxygen cell is diminished. But at lesser concentrations, the sensitivity can reduce experimental error. This allows for better sensitivity and reduced error when analyzing low BOD concentration wastes such as groundwaters and surface waters.

Enhanced Transfer Reactor

In comparing the standard one liter reactor with the enhanced transfer reactor, the transfer deficiency of the standard reactor with high strength wastes becomes apparent (Figures 7 and 8). In Experiment 1 (Figure 7) after an initial lag phase, the oxygen uptake rates increased in each reactor. However, the oxygen uptake rate of the standard reactor reached a maximum of only 24 mg/L/hr, while the oxygen uptake rate of the enhanced transfer reactor attained 64 mg/L/hr, which is the maximum oxygen generation capacity of this instrument. This demonstrates the increased oxygen transfer with the enhanced transfer reactor.

The actual oxygen uptake rate in the enhanced transfer reactor probably exceeds 64 mg/L/hr. The oxygen transfer defficiency in both the standard reactor, which is caused by transfer inability, and the enhanced transfer reactor, which is caused by instrument limitations probably causes the microorganisms to switch to other metabolic mechanisms for degradation. Substrate utilization may also be a factor. This explains the consistent drop in oxygen uptake rates after the maximum uptake rate in each reactor has been attained. However, this does not diminish the effectiveness of the enhanced transfer reactor to transfer significantly more oxygen to the microorganisms than the standard reactor.

In Experiment 2 (Figure 8), a similar situation to that in Experiment 1 occurs. After an initial lag phase, the oxygen uptake rates in each reactor increase similarly. However, the oxygen uptake rate of the standard reactor reached a maximum of 39 mg/L/hr, while the maximum oxygen uptake rate in

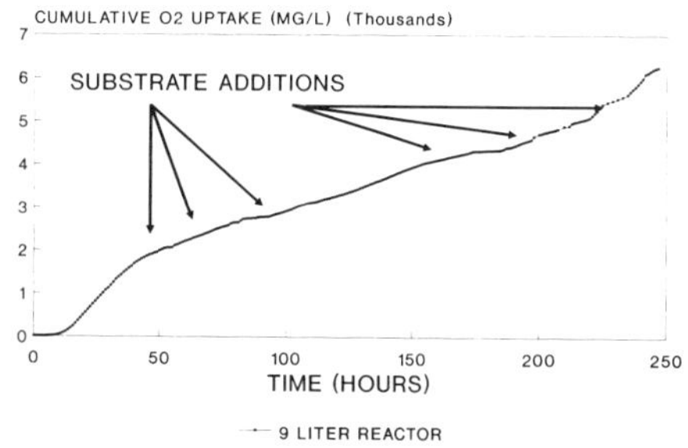

Figure 9. Oxygen uptake rate data obtained with the nine liter reactor.

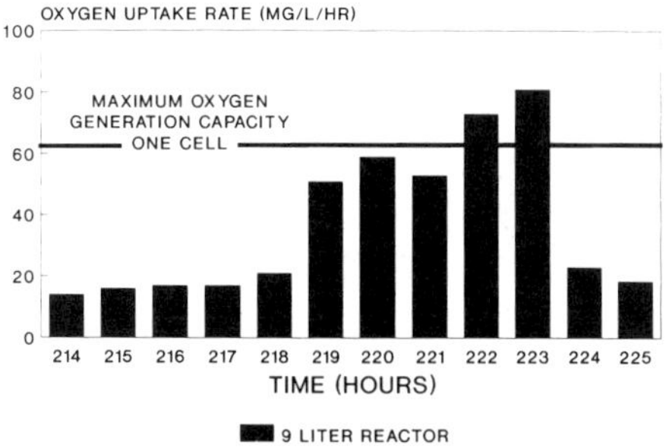

Figure 10. Oxygen uptake rate data obtained with the nine liter reactor.

the enhanced transfer reactor reached 64 mg/L/hr, the limit of the instrument. This reinforces the oxygen transfer efficiency of the modified reactor.

The actual oxygen uptake rate in the enhanced transfer reactor may exceed 64 mg/L/hr. Again, the inefficient ability of both reactors to supply the microorganisms with the required oxygen for biodegradation probably causes the microorganism to switch to other metabolic mechanisms for degradation. However, the effectiveness of the modified reactor to transfer more oxygen to the microorganisms than the standard reactor is apparent.

These two experiments demonstrate the increased oxygen transfer ability with the enhanced transfer reactor to allow for direct monitoring of high strength wastes such as landfill leachates and batch reactors.

Large Volume/High Transfer Rate Reactor

A respirometry test was used to acclimate a microbial population to a possibly inhibitory substrate. Increasing substrate additions were made to the reactor to minimize the inhibition effect of the substrate to the microorganisms, and to incremently increase oxygen uptake rates (Figure 9).

After each substrate addition, oxygen uptake rates increased. Initial substrate additions of 10, 20, 40 and 80 mLs caused small increases in oxygen uptake. However as acclimation and substrate addition increased, so did oxygen uptake. Subsequent substrate additions of 160, 320 and 640 milliliters produced dramatic increases in oxygen uptake rates. After the addition of 360 milliliters of substrate at 214 hours, the oxygen uptake rate exceeded 64 mg/L/hr, (Figure 10) which is the maximum oxygen uptake rate ability of this instrument with one electrolysis cell. However, this reactor was able to supply the increased oxygen demand exerted by the microorganisms because of the additional electrolysis cells. Two standard electrolysis cells produced oxygen simultaneously in order to meet the maximum oxygen requirements of the microorganisms, which was 81 mg/L/hr. With the three standard electrolysis cells, there was a reserve oxygen production capacity of 111 mg/L/hr.

As discussed previously, the transfer efficiency of a reactor used to analyze a high strength waste is very important. The oxygen production rate may be adequate but the transfer of the oxygen and carbon dioxide is what drives the experiment. When acclimated microorganisms receive the required amount of oxygen to metabolize a waste and carbon dioxide is sufficiently absorbed, the biodegradation will continue as long as the substrate is available. This is the condition of the experiment using the large volume/high transfer rate reactor.

This reactor allows for testing of large samples of wastewater with high oxygen demand, or entire pilot scale chemostats and batch reactors, without dilution.

Leachate Recirculation Reactor

An experiment using the leachate recirculation reactor was designed to test the function of this modified reactor. Figure 11 shows oxygen uptake data which was obtained. Oxygen uptake progressed during the first twenty hours of the experiment. During the initial recirculation of leachate, the leachate caused flooding in the soil, which depressed the activity of the microorganisms and slowed oxygen uptake probably because it stopped oxygen transfer. When recirculation was stopped, and the liquid percolated through the soil, oxygen uptake returned to its initial rate. During the second

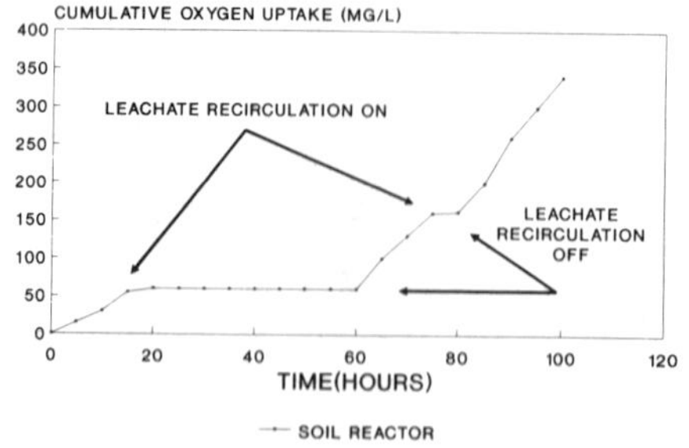

Figure 11. Cumulative oxygen uptake data obtained with the leachate recirculation reactor.

recirculation, oxygen uptake slowed again. However, the recirculation was on for only a short time. When recirculation was stopped, oxygen uptake returned and actually exceeded its initial rate.

The leachate recirculation reactor has the ability to measure oxygen uptake in solid and semi-solid matrices. However, the leachate recirculation must be carefully monitored so that the leachate only serves to provide moisture and oxygen to the biodegrading soil, and does not flood the soil and depress activity of the microorganisms.

Slurries which were previously required in order to analyze these sample types, and caused excessive mixing problems at higher solids concentrations, may now be eliminated. This modified reactor will allow for the direct analyses of solid and semi-solid matrices, such as soils and sludges.

CONCLUSION

Existing commercial respirometric equipment has several functional limitations. These limitations derive from the original design objective to measure biochemical oxygen demand in moderately contaminated liquids. The limitations include oxygen transfer rate, oxygen production rate, carbon dioxide absorption rate, mixing of mixed media slurries, and relatively low sensitivity. We have reviewed these limitations in one commercial electrolytic respirometer and developed modifications to the standard design to overcome them.

Oxygen transfer efficiency can be increased by changing from passive to active transport of oxygen between the headspace and liquid. Carbon dioxide absorption can be enhanced by actively bubbling the headspace gases through the carbon dioxide scrubber. High oxygen demanding wastes which require high oxygen production and transfer rate, can be analyzed using a large reactor with multiple electrolysis cells for increased oxygen generation. Solids, such as soils and sludges, can be analyzed in a reactor which recirculates leachate back into a biodegrading solid matrix in an enclosed reactor. Finally, electrolysis cell design can be altered to increase sensitivity and eliminate atmospheric pressure errors when measuring low oxygen demanding wastes.

Each of these modifications alleviates limitations of existing commercial equipment for use in an expanded array of operations to include high strength fluid, solid and semi-solid matrices. These modifications can be used to adapt commercial respirometry equipment for use on a much wider range of practical biodegradation problems.

REFERENCES

1. Young, J. C. (1977) "Biochemical Oxygen Demand: Measurement and Application" Chapter 1 pp.1.
2. Young, J. C. and Baumann, E. R. (1976) "The Electrolytic Respirometer I. Factors affecting oxygen uptake measurements." *Wat. Res.* 10, 1031–1040.
3. Larson, R. J. and Perry, R. L. (1980) "Use of the electrolytic respirometer to measure biodegradation in natural waters." *Wat. Res.* 15, 697–702.
4. Young, J. C., Garner, W., Clark, J. W. (1965) "An Improved Apparatus for Biochemical Oxygen Demand. *Analytic. Chem.* 37, 784.
5. Grady, C. P. L., Jr.., Dang, J. S., Harvey, D. M., Jobbagy, A. and Wang, X. L. (1988) "Determination of biodegradation kinetics through use of electrolytic respirometry." *Wat. Sci. Technol.* 21, 957–968.
6. Gaudy, A. F., Jr., Rozich, A. F., Garniewski, S., Moran, H. R., and Ekambaram, A., "Methodology for utilizing respirometric data to assess biodegradation kinetics." presented at the 42nd Annual Purdue Industrial Waste Conference, Purdue University, West Lafayette, IN 1987.

7. Rozich, A. F., and Colvin, R. J. 1988. "Use of Respirometry for Operating Activated Sludge Plants Treating Toxic Wastewaters." presented at the Water Pollution Control Federation Specialty Conference, Atlanta, Georgia, April 1988.
8. Rozich, A. F. and Zitrides, T. G. "Practical Approach for the Biological Remediation of Hazardous Waste Sites." presented at the HMCRI RCRA Superfund Conference, Las Vegas, Nevada, 1988.
9. Hickey, C. W. and Nagels, J. W. (1985) "Modifications to electrolytic respirometer systems for precise determination of BOD exertion kinetics in receiving waters. *Wat. Res.* 19, 463–470.
10. *Standard Methods for the Examination of Water and Wastewater*. (1985) The 16th Edition. Part 507.4 a, b, c and d. p. 527.

58 APPLICATION OF A PROCESS MODEL CALIBRATED WITH RESPIROMETRY TO PREDICT FULL-SCALE ACTIVATED SLUDGE PERFORMANCE

Richard J. Colvin, Project Engineer
Alan F. Rozich, President
Bioprocess Engineering, Inc.
Wilmington, DE 19808
Anthony F. Gaudy, Jr., Chairprofessor of Civil Engineering
University of Delaware
Newark, DE 19716
John Martin, Operations Engineer
City of Baltimore
Baltimore, MD 21202

INTRODUCTION

Biological systems, such as activated sludge, often comprise the backbone of the process technology which is utilized for treating municipal and industrial wastewaters. With the increasing regulatory constraints and requirements that pollution control managers must face, activated sludge systems in both the public and private sectors will be expected to deliver better effluent quality on a more consistent basis. This will require a control strategy which can react to variations in influent waste characteristics such as changes in waste components, in strength of the waste, or in flow rate. Also, pollution control professionals need to be increasingly concerned with controlling the levels of toxicity and organics in effluent streams. Toxic components of influent waste must be considered from two viewpoints: 1) their effect on the treatment process and its primary function, removal of the influent COD; and 2) pass-through of toxic materials into the plant effluent.

As is the case with any process, in order to realize optimal performance, it is necessary to utilize process control models for delineating guidelines for both design and operation. Although several models[1,2,3,4] are available for designing fluidized biological processes such as activated sludge, it is rare for these models to be utilized for making specific recommendations for operation or to be used on a frequent basis as a predictive tool by plant management. Part of the reason for this may be the fact that some of the models are primarily suited for use in design. However, another explanation for this is the fact that it is difficult and time-consuming to evaluate the biological constants that must be inserted into the process control models to make them applicable for a particular treatment situation.

This paper describes the application of a respirometric protocol that enables practitioners to determine specific operating strategies which are based on character and strength of the influent wastewater and on the biodegradation capability of the plant biomass. The procedure utilizes a proven process control model which is calibrated using biodegradation measurements made with an electrolytic respirometer. Data collection as well as process analysis are computerized and coordinated via a software package. A series of verification studies indicate that the protocol accurately predicts process performance for a full scale activated sludge treatment plant.

BACKGROUND

Process Control Models

Most process control models for activated sludge systems are based on the fact that net specific microbial growth rate, μ_n, (reciprocal of θ_c) can be described as a function of the concentration, S, of exogenous waste, or substrate. Since engineering controls can be applied to control μ_n, and if the relationship between μ and S is defined, then it is feasible to use the model to select the values of these controls that are needed to maintain effluent quality, S. An example of such a predictive modelling

45th Purdue Industrial Waste Conference Proceedings, © 1991 Lewis Publishers, Inc., Chelsea, Michigan 48118. Printed in U.S.A.

approach that can be applied for both noninhibitory and inhibitory, or toxic, wastes has been given in the literature[4]. It should be noted that the mathematical relationship between μ and S (see Appendix for details) quantifies the biodegradative capability of the biomass on the target waste.

Respirometric Calibration of Process Control Model

The key for the successful application of a model is to insure that it is properly calibrated. For activated sludge models, this entails the quantification of the biokinetic constants contained in the equations which relate μ and S; i.e., the evaluation of the biodegradative capability of a biomass on a particular waste.

Two methods have been previously utilized for evaluating the biokinetic growth constants: these are the batch growth study technique[4] and the substrate utilization method[5]. Although both techniques have been shown to produce reasonable estimates of the constants for both noninhibitory and inhibitory wastes, they are rarely, if ever, used by plant managers for control of the treatment process because of the level of analytical effort that is associated with these procedures. With a simpler and less labor-intensive method which is capable of yielding reliable kinetic results is available, then the routine evaluation of the constants and the associated use of that model can be incorporated into the normal decision-making regimen that comprises the operation of a plant's activate sludge system.

It has been demonstrated that respirometric measurements can be utilized to obtain values of the biokinetic constants[6]. The approach consists of utilizing oxygen uptake readings to "construct" calculated biomass growth or substrate utilization curves; the curves are analyzed to compute specific growth rates, i.e., μ values, for different substrate concentrations, S[6]. The μ and S data are then fit to the appropriate growth rate function (equations (1) and (2), given in the Appendix) to determine the values of the biokinetic constants.

It should be stressed that oxygen uptake measurements cannot be used directly as a substitute for growth or substrate utilization data. Rather, the respirometric readings must be converted into equivalent growth or substrate utilization data before it is feasible to compute values of μ for corresponding S values. Equations 3 and 4 are used to convert oxygen uptake readings into predictive biomass and substrate utilization curves, respectively.

$$X_{(t)} = X_o + \frac{O_{2(t)}}{(1/Y_t - O_x)} \tag{3}$$

$$S_{(t)} = S_o - \frac{O_{2(t)}}{1 - O_x Y_t} \tag{4}$$

where:
- X_o = initial biomass concentration, mg/L;
- S_o = initial soluble COD concentration, mg/L;
- O_2 = cumulative oxygen uptake, mg/L;
- O_x = unit COD of cell mass, mg COD/mg X;
- Y_t = cell yield, mg/mg.

Values of $O_{2(t)}$ are taken from the accumulated O_2 uptake curves obtained using the respirometer while values of Y_t, the cell yield, and O_x, the unit cell COD, are known or calculated from data obtained from a separate batch unit. If it is desirable to convert O_2 uptake data into growth data, equation (3) is used to produce a biomass growth curve as a function of time. A growth rate can be determined for each reactor system and for each corresponding set of respirometric data by plotting the calculated X values on arithmetic paper. The slope of the straight line portion of these plots represents the growth rate, μ. This produces a set of μ and S data since each reactor receives a different initial concentration of the waste.

Substrate utilization data are generated by utilizing Equation 4 which produced a substrate "drawdown" curve; this curve is analyzed to yield a set of μ and S data[5].

An example of respirometric measurements which have been converted into growth data is given in Figure 1. These curves illustrate the fact that O_2 uptake cannot be used directly as a substitute for measurement of growth. Although the curves are plotted using the natural logarithms of the measured O_2 uptake and calculated biomass, respectively, only the biomass curve yields a straight-line (exponential) phase which can be used to estimate μ, the specific growth rate. The calculated X curve was generated using the oxygen uptake curve and Equation 3; values for Y_t, O_x and X_o were 0.44, 1.23, and 190 mg/L, respectively. The value of 0.053 h^{-1} for μ, is computed from the slope of the straightline portion of the calculated biomass curve.

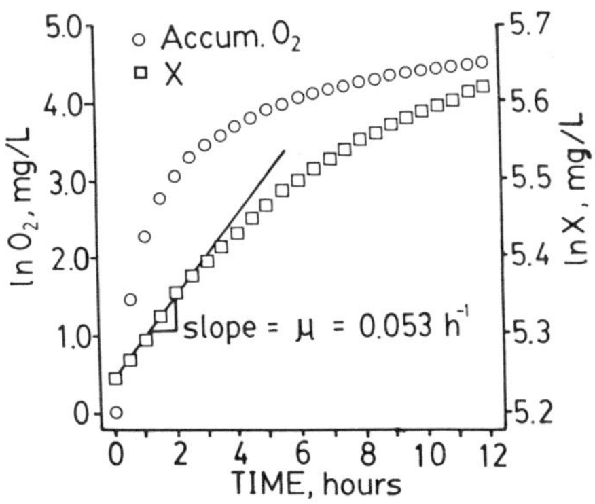

Figure 1. Plot Showing Conversion of Oxygen Uptake Data into Growth Data

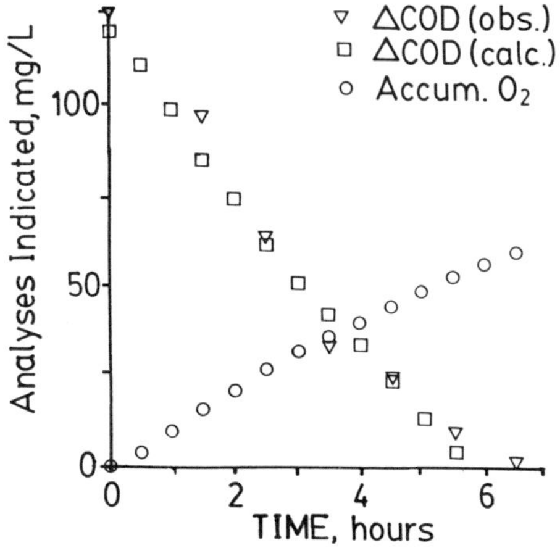

Figure 2. Plot Showing Conversion of Oxygen Uptake Data into Usable COD (substrate) Data

Figure 2 depicts a set of substrate utilization data which were calculated from O_2 uptake readings. In this figure, the S values (calculated as mg/L of metabolizable COD) were computed using Equation 4 and the same Y_t and O_x values which were used in Figure 1. The value of S_o for this analysis is the readily biodegradable fraction of the initial COD (i.e., $\triangle$COD) which is calculated as the difference of the initial soluble COD and residual soluble COD. In Figure 4, it is evident that the predicted substrate utilization curve compares favorably with the measured readily biodegradable COD values (residual subtracted from actual COD readings). A set of μ and S data is generated by analyzing the substrate utilization drawdown curve[5].

APPLICATION AND MODEL TESTING

The biokinetic constants for the primary effluent of the Patapsco Wastewater Treatment Plant (Baltimore, MD), a municipal oxygen activated sludge plant which receives a large industrial contribution from chemical manufacturers, were evaluated over a six-month period using an electrolytic respirometer. Waste strength and usable substrate were both expressed as $\triangle$COD[7,8], this produces a better assessment of the amount of influent carbon that is readily amenable to biodegradation in the biotreatment system. Details regarding the measurement of $\triangle$COD and other applications are available in the literature[7,8].

Earlier work[9,10] on the same waste stream indicated that it was very difficult to utilize the batch growth study and substrate utilization techniques for generating the growth data which are needed for

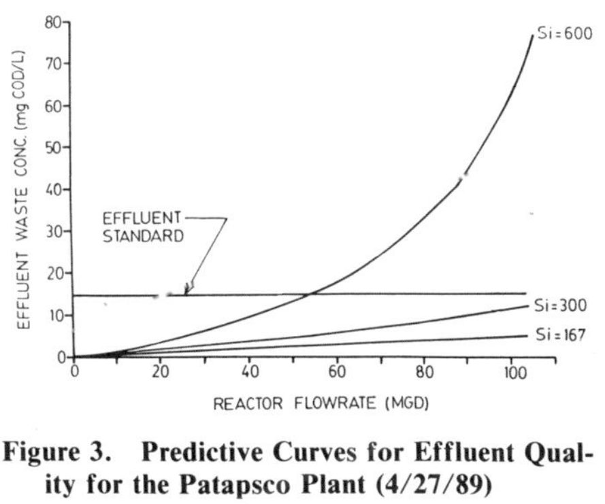

Figure 3. Predictive Curves for Effluent Quality for the Patapsco Plant (4/27/89)

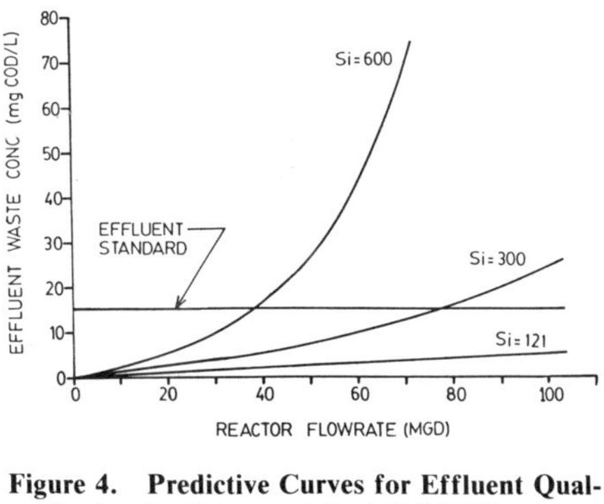

Figure 4. Predictive Curves for Effluent Quality for the Patapsco Plant (8/17/89)

Table I. Values of the Biokinetic Constants for the Patapsco Plant Biomass

Date	μ_{max} h^{-1}	K_s mg/L ΔCOD	K_i mg/L ΔCOD	Y_t mg/mg
4/25/89	0.109	15	—	0.41
4/27/89	0.090	15	—	0.41
7/24/89	0.064	28	108	0.29
7/31/89	0.027	17	—	0.29
8/9/89	0.053	28	108	0.31
8/17/89	0.036	16	—	0.27
8/29/89	0.155	1.4	204	0.55
9/5/90	0.090	1	—	0.46
9/11/89	0.037	15	—	0.27

Note: $K_d = 0.005\ d^{-1}$ was assumed for each study.

determining the values of the biokinetic constants. A number of procedural difficulties were encountered which significantly increased the labor needed for collecting the raw data. For example, the presence of influent suspended solids and low soluble COD concentrations necessitated modification of the waste prior to kinetic testing. The waste had to be filtered and, in several instances, concentrated in order to provide a sufficient COD concentration to permit the measurement of growth rates over a range of initial COD concentrations for the batch growth study technique or to provide a sufficiently high initial concentration for the substrate utilization technique. Additionally, it was observed that the waste was occasionally colored, which interfered with optical density readings. This interference was occasionally exacerbated by the fact that the waste was decolorized during the course of a batch growth test. These facts and the fact that the waste had to be modified tend to compromise the values of the biokinetic constants which were obtained from these data.

The application of respirometric methods for producing the data needed for evaluating the biokinetic constants was found to be relatively easy and convenient in comparison to the older procedures. For instance, the waste did not require any preparation or pretreatment; it was used in its original form and concentration. Optical density interference from suspended solids or the color of the waste die not present a problem since only O_2 uptake was being monitored. If values for the cell yield and cell COD (Y_t and O_x, respectively) and the initial cell concentration, X_o, are known, the O_2 measurements can be transformed into growth or substrate utilization data. Finally, although it was observed that the solids contained in the primary effluent exerted an oxygen demand, it was also found that this demand was not sufficient to disrupt the transformation of O_2 readings into growth or substrate utilization data. That is, the O_2 demand of these solids was so small that it could be neglected because it did not upset the △COD and oxygen balance of the system.

The values of the biokinetic constants which were determined using an electrolytic respirometer for the Patapsco primary effluent are given in Table I. These numbers were taken from nine runs which were performed from April through September, 1989. This information is inserted into process control models to provide predictions for effluent quality as △COD for various values of the engineering control parameters and influent loadings.

Table II lists the values of engineering parameters, recycle sludge concentration, X_R; recycle flow ratio, α; primary effluent △COD (computed from soluble BOD_5 data), S_i; and primary effluent flow rate, F. Actual and predicted values for effluent quality are also provided. The predictions were made using the information in Tables 1 and 2 and the model equations which have been presented elsewhere[4].

Table II shows that, despite the complex nature of the treatment process and the waste, the model, calibrated with respirometric measurements of plant biomass, provides a reasonable prediction for effluent quality. It needs to be emphasized that it is often difficult to obtain accurate values for the input engineering parameters (e.g., flow rates, recycle flows, recycle sludge concentrations and influent waste strengths) for such a large facility. This is stated because it is often difficult to obtain representative composite samples or to acquire accurate measurements of process flow rates and other operational parameters. However, since the staff at the Patapsco plant were exceptional in their record-keeping efforts, it facilitated the modelling efforts. The success of the modelling exercise also underscores the utility of competent record-keeping and monitoring activities.

Figures 3 and 4 illustrate examples of the predictive utility of the modelling approach for determining the plant capacity or, in other words, for defining the "performance envelope" for different influent waste strengths, S_i. The predictive curves in Figure 3 were constructed utilizing the kinetic

Table II. Values of Engineering Parameters and Actual and Predicted Values for Effluent Quality for the Patapsco Plant

Date	X_R mg/L VSS	α	S_i mg/L ΔCOD	F MGD	Soluble Effluent COD mg/L Actual	Pred.
4/25/89	14,035	0.37	161	45.7	5	1
4/27/89	9,485	0.40	167	45.3	11	2
7/24/89	14,600	0.38	143	55.4	6	3
7/31/89	13,000	0.34	159	73.9	5	9
8/9/89	14,130	0.44	115	44.0	8	2
8/16/89	10,730	0.44	121	44.6	3	2
8/29/89	9,310	0.37	169	41.5	5	0.1
9/5/89	8,900	0.35	121	40.0	3	0.1
9/11/89	10,360	0.37	192	44.4	6	4

constants and engineering parameters which were relevant for 4/27/89 while those curves for Figure 4 were generated using the appropriate information from 8/17/89. The figures show that, at the actual S_i values (167 mg/L and 121 mg/L for 4/27/89 and 8/17/89, respectively), the plant has a wide operating range before experiencing performance problems. (In this case, plant capacity is more likely to be limited by the capacity of the secondary clarifiers.) As S_i increases to 600 mg/L COD, it is seen that both figures indicate susceptibility to effluent deterioration via COD detailed explanation), the plant capacity at an S_i of 600 mg/L is 55 MGD for 4/27/78 and 38 MGD for 8/17/89.

It should be noted that the primary difference in conditions between the predictions in Figures 3 and 4 is the difference in the values of the biokinetic constant, μ_{max}. The engineering parameters used for the predictions are relatively close in value. The point is that the model predictions are sensitive to the values of the biokinetic constants which are determined using the respirometer. The biokinetic constants, in turn, are sensitive to various environmental and plant conditions (e.g., waste quality, temperature, pH, etc.) which effect the ecology of the activated sludge biomass. Thus, as conditions change which effect the biomass and, consequently, plant capacity, the respirometric modelling procedure has the ability to quantify the impact of these changes in predicting plant capacity.

The application of process control models calibrated with respirometry will generally follow the scheme depicted in Figure 5. Initially, one would test the incoming waste to determine the kinetic response of the plant biomass as quantified by the values of the biokinetic constants which are evaluated via respirometry. This information can then be used in concert with a bioprocess model to allow the plant operator to make a more informed decision regarding operational alternatives. For example, some plants periodically have to contend with treating a high strength waste (e.g., production campaigns in industrial treatment plants) and it would be prudent to know the loading rate that can be tolerated without adversely affecting effluent quality. In this instance, the waste is tested and the model is used to determine allowable loading rates, the necessary process control adjustments to maintain effluent quality, or both. Additionally, the algorithm can also be applied to determine if the plant has the appropriate aeration capacity to handle the loading. Because the bulk of the analytical

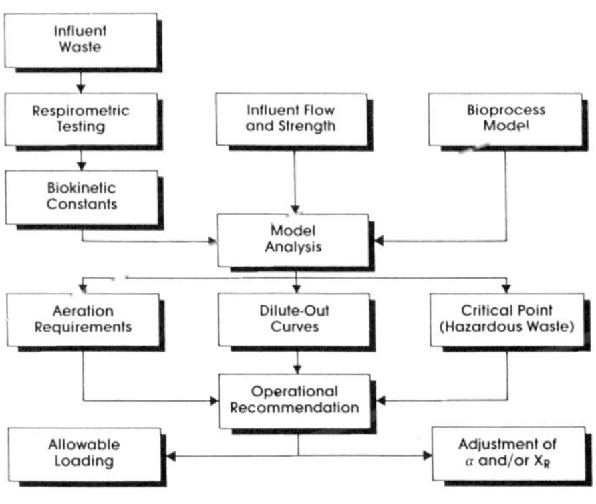

Figure 5. Flow diagram showing application of respirometrically calibrated models for activated sludge process control.

effort associated with the kinetic evaluation is automated, if need be it can be performed on a daily basis, which would allow plant management to respond rapidly to changing influent conditions.

It should be pointed out that the scheme given in Figure 5 is also suitable for operating activated sludge processes which are treating toxic, or hazardous, wastes. For this application, one includes a critical point analysis in order to determine the operational location of the reactor's critical point, i.e., that combination of the loading and engineering control parameters that cause systems which are treating toxic wastes to be prone to sudden effluent deterioration and failure[4,5,11]. The critical point analysis is used to provide information to avoid operation near the failure point, which will allow management to optimize process loadings without risking effluent deterioration. Additional information regarding the significance and importance of the critical point analysis for toxic waste treatment has been given elsewhere[5,11].

SUMMARY

This paper presented an application of utilizing respirometric measurements for a treatment plant's biomass and influent waste stream to calibrate process control models for defining operating criteria for activated sludge systems. It was pointed out that one of the major drawbacks regarding the application of biokinetic models for process control has been the inherent level of labor that is normally associated with determining the values of the biokinetic constants which are used in the models. The advent of new respirometric techniques have now eliminated much of the analytical effort that previously characterized the evaluation of the constants. This makes the use of process control models for making specific operational recommendations convenient and attractive.

In this paper, an example was presented of the application of the respirometric method for a municipal oxygen-activated sludge plant which receives a large industrial contribution from chemical manufacturers. The values of the biokinetic constants which were determined respirometrically were inserted into process-control models to make predictions for effluent quality. The predicted and actual values for effluent quality compared favorably. Predictive curves for effluent quality for different influent waste strengths were also presented. These curves are especially useful in defining the treatment capacity of a facility.

Finally, a flow chart was given in Figure 5 which outlines a procedural plan for implementing the use of process control models in concert with respirometric testing. This chart provides general guidelines for implementing and incorporating the biokinetic methodology into the normal decision-making regimen for plant operations. It was also noted that, if the bioreactor had to contend with toxic or hazardous wastes which are inhibitory, it would be appropriate to generate critical point curves in order to identify and avoid the operating condition(s) that produce critical points in reactors which are treating these wastes, since critical point violations result in severe effluent deterioration and wash-out of the biomass.

REFERENCES

1. Adams, C. E., Jr. and Eckenfelder, W. W., Jr. (eds), *Process Design for Industrial Waste Treatment*, Enviro, Nashville, Tenn. (1974).
2. Lawrence, A. W. and McCarty, P. L., "A Unified Basis for Biological Treatment Design and Operation," *J. San. Eng., Div., ASCE, 96*, 757 (1970).
3. Ramanathan, M. and Gaudy, A. F., Jr., "Steady-state Model for Activated Sludge with Constant Recycle Sludge Concentration" *Biotechnology and Bioengineering, 13*, 125 (1971).
4. Rozich, A. F., Gaudy, A. F., Jr., and D'Adamo, P. C., "Predictive Model for Treatment of Phenolic Wastes by Activated Sludge," *Water Research, 17*, 1453 (1983).
5. Gaudy, A. F., Jr., Lowe, W. L., Rozich, A. F., and Colvin, R. J., "Practical Methodology for Predicting Critical Operating Range of Biological Systems Treating Inhibitory Substrates," *Presented*, 59th Annual Water Pollution Control Federation Conference, Los Angeles, CA (October 1986); *J. Water Pollution Control Fed.*, In Press, (1987).
6. Gaudy, A. F., Jr., Rozich, A. F., Garniewski, S., Moran, R., and Eckambaram, A., "Methodology for Utilizing Respirometric Data to Assess Biodegradation Kinetics," *Presented*, 42nd Industrial Waste Conference, Purdue University, West Lafayete, IN (May, 1987).
7. Gaudy, A. F., Jr. and Gaudy, E. T., *Microbiology for Environmental Scientists and Engineers*, McGraw-Hill Co., New York, (1980).
8. Gaudy, A. F., Jr. and Gaudy, E. T., "COD Gets Nod Over BOD Test," *Ind. Water Eng., 9*, 30 (1972).
9. Rozich & Gaudy, Inc., "Determination of the Numerical Values of the Biokinetic Constants and Implications to the Design of the Expanded Facilities of the Patapsco Wastewater Treatment

Plant," Engineering Report to the Wastewater Facilities Division, City of Baltimore, MD (August, 1986).
10. Gaudy, A. F., Jr., Rozich, A. F., and Lowe, W. L., "Design Criteria for Treatment of Combined Wastes," *Presented*, Annual Conference of the Environ. Eng. Div., ASCE, Orlando, FL (July, 1987).
11. Rozich, A. F. and Gaudy, A. F., Jr., "Critical Point Analysis for Toxic Waste Treatment," *J. Environ. Eng. Div., ASCE, 110*, 562 (1984).

APPENDIX

Growth Rate Functions

The biokinetic growth constants are utilized in growth rate expressions to relate μ to S. If the waste is noninhibitory, equation (1), the Monod equation, is employed for this purpose.

$$\mu = \frac{\mu_{max} S}{K_s + S} \tag{1}$$

where: μ_{max} = the maximum specific growth rate of which a biomass is capable under specified conditions of temperature and pH in the absence of any restriction on growth rate by limiting nutrient concentration, e.g., oxygen, nitrogen or carbon source, day^{-1}

K_S = the saturation constant: the shape factor in the rectangular hyperbola form of the Monod relationship; it is defined as the concentration of the limiting substrate at which $\mu = 0.5\ \mu_{max}$, mgl^{-1}

S = soluble substrate concentration, mgL^{-1}

However, if the waste is inhibitory, as would be the case for a hazardous or toxic waste, it is appropriate to apply equation (2), the Haldane Function.

$$\mu = \frac{\mu_{max} S}{K_s + S + S^2/K_i} \tag{2}$$

where: K_i = the inhibitor constant: used in the Haldane and other model equations for inhibitory substrates to account for the effect of concentration of the inhibitory substrate on the specific growth rate, mgL^{-1}

If a toxic waste is being treated, then it is especially crucial to assess the biokinetic constants in order to maintain treatment efficiency and avoid severe effluent deterioration; reasons for this have been detailed in the literature[4,11].

Determination of Effluent Criteria in Terms of $\triangle$COD

Effluent standards for secondary treatment are 30 mg/L for BOD_5 and 30 mg/L for suspended solids. However, the predictions of the kinetic model are made in terms of soluble substrate ($\triangle$COD) in the effluent. Thus, it is essential to estimate or calculate the allowable soluble substrate levels from the effluent standard. One approach is to estimate the allowable soluble BOD_5 in the effluent by subtracting out the suspended solids contribution to the total effluent BOD_5. By assuming that 70% of the effluent suspended solids are biological with a composition equivalent to $C_5H_7NO_2$, and assuming that 70% of the ultimate BOD_5 is exerted during the 5-day incubation period, the BOD_5 exerted by the suspended solids can be calculated and subtracted from the 30 mg/L BOD_5 allowed. This yields the allowable soluble BOD_5 in the effluent.

$$BOD_5\ (\text{soluble}) = 30 - (30)(1.4)(0.7)(0.7) = 10\ \text{mg/L}$$

The concentration of soluble substrate which can be expected to cause a BOD_5 of 10 mg/L is estimated by assuming that 90% of the $\triangle$COD will register as ultimate BOD (carbonaceous) and that 70% of this will be exerted in a 5-day incubation period. Then, allowable $\triangle$COD can be computed.

$$\triangle COD = 10/((0.9)(0.7)) = 16\ \text{mg/L}$$

Thus, when using dilute-out curves such as those given in Figure 3, one desires to achieve something under 16 on the effluent quality scale in order to identify the operating parameters that will prevent the system from experiencing an effluent violation caused by a biochemical failure. If a safety factor is to be added, then it might be appropriate to use 10 mg/L as the criterion for effluent $\triangle$COD. However, 20 mg/L should certainly be taken as the upper limit for effluent quality.

Section Six
NEW WASTE PROCESSES

59 USE OF TAILORED ZEOLITES FOR REMOVAL OF BENZENE AND TOLUENE FROM WATER

Fang Gao, Environmental Engineer
JOAB Inc.,
Las Cruces, New Mexico 88004

Fernando Cadena, Associate Professor
Department of Civil Engineering
New Mexico State University
Las Cruces, New Mexico 88003

Robert W. Peters, Environmental Systems Engineer
Energy Systems Division
Argonne National Laboratory
Argonne, Illinois 60439

INTRODUCTION

Cases of groundwater contamination by leaking underground storage tanks (USTs) have been increasing throughout the United States since early 1970s.[1,2,3] The most common cases are the contamination caused by leaking gasoline USTs. It is estimated that in the United States the total number of leaking gasoline USTs could be about 100,000 at the present time, and this figure may increase to about 350,000 by 1995.[4] Groundwater contamination by benzene, toluene and xylenes (BTXs) is coincident with the leaking USTs since gasoline contains approximately 3% benzene, 9 to 11% toluene, or 10 to 12% xylenes as antiknock additives.

BTXs are toxic to humans. These chemicals are leukemogenic, mutagenic, teratogenic, and carcinogenic.[5,6,7,8,9] This problem is compounded by potential exposure to the groundwater contaminated by BTXs. A large portion of the U.S. population (about 50%) uses groundwater as a source of drinking water. About 75% of the U.S. cities obtain all of part of their water supplies from groundwater.[10] In the southwest, such as New Mexico, more than 95% of the local population is dependent upon groundwater.[11]

Several investigators are working on the removal of BTXs from the contaminated water through adsorption using tailored clays as adsorbents.[12,13,14,15,16] Tetramethylammonium (TMA +) is used in their studies to tailor the clay content of the natural soils, and to improve the adsorption capacity of the soils toward BTXs. Their results from batch isotherm studies have shown significant improvement, especially in the studies of benzene adsorption.

Under aqueous condition, the sodium ions (Na+) sitting between the clay platelets become hydrated, causing interplatelet expansion.[15] During the tailoring process, the tetrahedral TMA^+ ions enter the clay's interlamellar space and replace the sodium ions. The replacement of the hydrated sodium ions results in some shrinkage of the interlamellar space. However, such shrinkage is limited due to the strong holding of the rigid tetrahedrons of TMA^+ ions. The interlamellar space between the platelets and between the TMA^+ tetrahedrons remains stable and ample enough for benzene molecules to enter and become adsorbed.

Previous studies on TMA^+ tailored clays and clay-containing soils have shown significant improvement of BTX adsorption over their natural state.[15,16,17] For example, a batch adsorption isotherm test for benzene versus LA soil (a natural soil with 36% clay content obtained from the Los Angeles area) shows that the linear adsorption coefficient (K) of the TMA^+ tailored LA soil is about 40 times greater than that of the natural (non-tailored) LA soil.[17]

45th Purdue Industrial Waste Conference Proceedings, © 1991 Lewis Publishers, Inc., Chelsea, Michigan 48118. Printed in U.S.A.

Although clays possess a significant potential as adsorbent for removal of BTXs, as shown previously, their practical application is much limited by their low permeability (i.e., hydraulic conductivity). When water passes through a clay medium, the clay particles expand due to the hydration of the interplatelet metal cations (most commonly monovalent cations like Na^+). Such hydration reduces the interparticle space, thus lowering the permeability of the clay medium. When a tailoring agent, such as TMA^+, is added, interlamellar shrinkage may occur due to the replacement of the hydrated cations by TMA^+ ions. As pointed out previously, the shrinkage is limited by the rigid tetrahedral structure of TMA^+ ions. Some TMA^+ ions, due to their hydrophobic nature, may become attached onto the outer surface of the clay particles. The coagulation effect among these TMA-attached clay particles may become so significant due to the interparticle closeness that the permeability of the clay medium is further reduced.

The purpose of this project was to study the feasibility of using natural zeolites, instead of clays, as base media for removal of BTXs because zeolites have excellent hydraulic conductivity and relatively high CEC (cation exchange capacity). The adsorption capacity and permeability of three zeolites were tested and analyzed under both natural and tailored conditions.

PHYSICAL AND CHEMICAL CHARACTERISTICS OF ZEOLITES

Zeolites are natural minerals usually found in volcanic sedimentary deposits. They are built of aluminum- and silicon-based pyramid-shaped or tetrahedral molecules assembled in a large network of open channelways, similar to a sponge with uniform holes.[18] The 3-dimensional lattice of zeolite molecules consists of $(SiO_4)^{-4}$ and $(AlO_4)^{-5}$ tetrahedrons strongly bonded at all corners.[19] The framework of the open cavities and inner channels of the zeolite tetrahedrons have net negative charges which are balanced by exchangeable cations. Sodium ions (Na^+) are the most common exchangeable cations in zeolites. These metal cations are loosely bound within the innerchannels and are feasibly and readily replaced by cations with higher valence or cations having greater affinity to the zeolite innersurface. These characteristics impart to zeolites a relatively great CEC and a great potential to be used as chemical barrier agents.

The tetrahedral framework of zeolite features an outstanding stable structure. Such stability makes the zeolite particles behave quite differently from the clay particles under aqueous condition. When water is introduced into a zeolite medium, the natural exchangeable cations (Na^+) become hydrated. However, the hydration will be limited by the stable innerchannel space defined by the rigid tetrahedral structure. As a result, no expansion of zeolite particles is induced, and the interparticle porous space and the permeability of the zeolite medium remain relatively unchanged. By comparison, under aqueous condition, the clay platelets expand with the hydration of the exchangeable cations due to the absence of bonding at the edge of the clay platelets, resulting in reduction of interparticle porous space and permeability.

Under tailoring condition, the tailoring agent (TMA^+) can easily replace the hydrated cations (Na^+) which are very loosely attached to the innersurface of the zeolite particles due to hydration. The replacement is also favored by the hydrophobic (or lipophilic) nature of TMA^+ ions and their greater affinity to the zeolite innersurface. Some TMA^+ ions become attached to the outer surface of the zeolite particles, which may produce a coagulating effect. Such effect, however, will be limited as long as the zeolite particles are coarse enough to maintain interparticle pores within the medium.

MATERIALS USED AND EXPERIMENTAL PROCEDURES

This project was divided into two phases: First, batch adsorption isotherm study, and second, column adsorption breakthrough study. Three zeolites, SC-, XY- and CH-Zeolites, were tested in the first phase and CH-Zeolite was tested as a representative medium in the second phase. The 20×50 mesh granular zeolite samples were provided by Teague Mineral Products, a mineral production company located in southeast Oregon. These zeolite samples were analyzed at the Soil and Water Laboratory (College of Agriculture, New Mexico State University). The results show that the CECs of SC-, XY-, and CH-Zeolites are 50.9, 72.7, and 71.6 meq/100g, respectively.

First Phase: Batch Adsorption Isotherm Study

A batch adsorption technique was applied to all three zeolites. The target pollutants were benzene for all three zeolites and toluene for only CH-Zeolite which was used as a representative solid medium. The adsorption characteristics were modeled using a linear isotherm ($\frac{x}{m} = K C_O$) to simplify the analysis.

The experimental steps were:

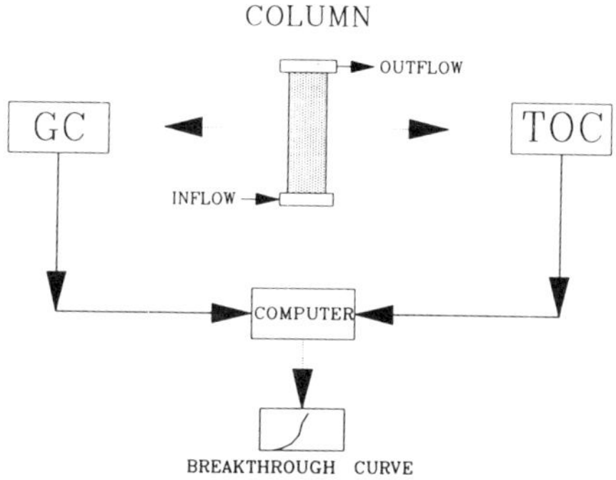

Figure 1. Column adsorption experimental setup.

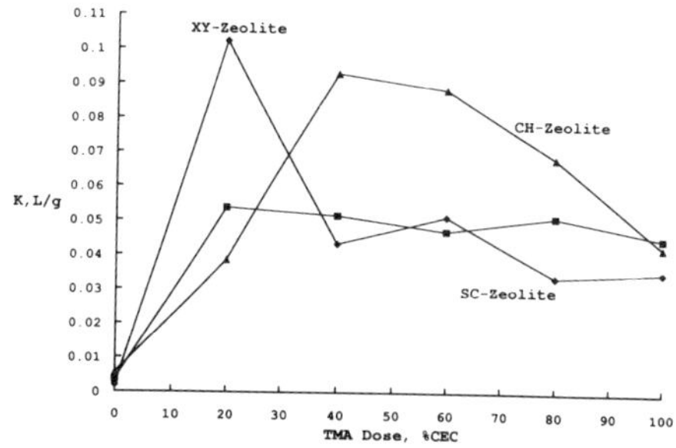

Figure 2. Summary of batch adsorption tests for benzene.

1. Six homogenous zeolite suspensions of 25 g/L were prepared by magnetic stirring for 20 minutes.
2. Tetramethylammonium chloride (TMA-Cl) was added to each suspension at 0, 20, 40, 60, 80, and 100% CEC respectively to tailor the zeolite in each suspension.
3. The tailoring process was run under stirred condition for 15 minutes (preliminary study showed that the exchange process was complete in about 5 minutes).
4. Each suspension was split into 5 pairs of 14-mL test tubes (reactors).
5. Various amounts of pollutant solution (benzene or toluene solution) were injected into corresponding pairs of reactors, yielding 5 different pollutant concentrations from 0 mg/L to 120 mg/L in 5 pairs of reactors. The reactors were covered with Teflon-sealed caps immediately to minimize volatilization.
6. The adsorption process was performed on a head-over-head mixer for 6 minutes (preliminary tests showed that the adsorption process was complete in about 3 minutes).
7. Solid-liquid phase separation was done on a centrifuge at 3500 rpm for 5 to 8 minutes to obtain clear supernatants in the reactors.
8. 5 μL of the clear supernatant in each pair of reactors was injected with microsyrenge into a gas chromatograph (GC) equipped with flame ionization detector (FID) to determine the pollutant concentration in the liquid phase (supernatant).
9. The amount of pollutant adsorbed on the zeolite particles was obtained from the difference between the initial concentration and the concentration in the liquid phase after adsorption.
10. Results obtained in Step 9 were used to establish six linear isotherms for the six tailoring agent concentrations, and six linear adsorption coefficients (K) were obtained. Among those K values, the maximum one corresponded to the optimum dose of TMA^+.
11. Steps 1 through 10 were performed for each zeolite, and six linear adsorption coefficients were obtained for each zeolite.

Second Phase: Column Adsorption Breakthrough Study

A short glass column was used in this study to further explore the feasibility and practicality of using zeolites as chemical barrier media. The dimension of the column was 7.0 cm I.D. and 30 cm in length. The CH-Zeolite was used as a representative medium and tailored with the optimum TMA^+ dose determined in the batch isotherm study. The target pollutants were benzene and toluene. The experimental setup is shown in Figure 1.

Experimental steps were:

1. 1100 grams of CH-Zeolite were packed in the column, yielding a pore volume of 670 mL.
2. The column was flushed with tap water in an upflow pattern until the effluent became clear.
3. An upflow of the polluted water (benzene or toluene solution) was injected continuously into the column. Samples of influent and effluent were taken at 1/4 to 1/2 pore-volume intervals and analyzed on GC-FID to determine the pollutant concentrations.

Table I. Summary of Batch Adsorption Study

Type of Zeolite (and CEC)	Dose of TMA (% CEC)	Linear Adsorption Coefficient (K,L/g)	
		For Benzene	For Toluene*
SC-Zeolite	0	0.0036	–
50.89 meq/100 g	20	0.0538	–
	40	0.0516	–
	60	0.0472	–
	80	0.0514	–
	100	0.0454	–
XY-Zeolite	0	0.0020	–
72.72 meq/100 g	20	0.1023	–
	40	0.0434	–
	60	0.0514	–
	80	0.0340	–
	100	0.0354	–
CH-Zeolite	0	0.0056	0.0076
71.57 meq/100 g	20	0.0385	–
	40	0.0930	0.0171
	50	–	0.0271
	60	0.0884	0.0242
	80	0.0686	0.0140
	100	0.0428	0.0156

*Only CH-Zeolite was tested for toluene adsorption.

4. After the breakthrough of the pollutant (i.e., the effluent concentration became constant and equal to the influent concentration), the column was re-flushed with tap water until no pollutant was detected in the effluent.
5. A TMA-Cl solution was injected continuously in an upflow fashion to tailor the zeolite medium in the column. Samples of both influent and effluent were taken at 1/4 pore-volume intervals and analyzed with a TOC (total organic carbon) analyzer to determine the concentrations of TMA^+.
6. After the breakthrough of the TMA^+, the column was flushed with tap water again to remove the excessive or dissolved TMA^+ in the column. Data obtained in Step 5 and in this step were used to perform TMA^+ mass balance calculations.
7. Steps 3 and 4 were repeated to assess the adsorption behavior of the tailored zeolite column. Data collected were used to develop the breakthrough curves.

RESULTS AND DISCUSSION

Batch Adsorption Isotherm Study

The batch adsorption characteristics of three zeolites were modeled using a linear isotherm ($x/m = K\ C_o$). The adsorption capacities of the same zeolite samples but tailored with different doses of TMA-Cl can be estimated simply by comparison of their linear adsorption coefficients (K). A greater K value represents a greater adsorption capacity. The summary of the K values is presented in Table I. Graphical summary of the same results is shown in Figures 2 and 3.

Based on the results shown in Table I and Figures 2 and 3, the following findings are drawn:

(1) All three zeolites show very low adsorptions for benzene in their natural forms. The K values for SC-, XY-, and CH-Zeolites are 0.0036, 0.0020, and 0.0056 L/g, respectively. The low adsorption is likely due to the unavailability and inaccessibility of the zeolite innerchannels, which are occupied by the hydrated spheres of the naturally existing cations, such as Na^+. Very few benzene molecules may attach to the outer surface of the zeolite particles under these conditions. Such attachment is also minimized due to the absence of strong bonding sites.

(2) After tailored with TMA^+, all three zeolite samples show significant improvement in their adsorption toward benzene. The maximum K values for SC-, XY-, and CH-Zeolites are 0.0538, 0.1023, and 0.0930 L/g, respectively. These values are 14, 50, and 16 times greater than those of the corresponding non-tailored zeolites. The TMA^+ cation has an overall size of approximately 4.5 Å,[20] which allows itself to enter easily the zeolite innerchannel whose size is approximately 5.4 Å.[19] After

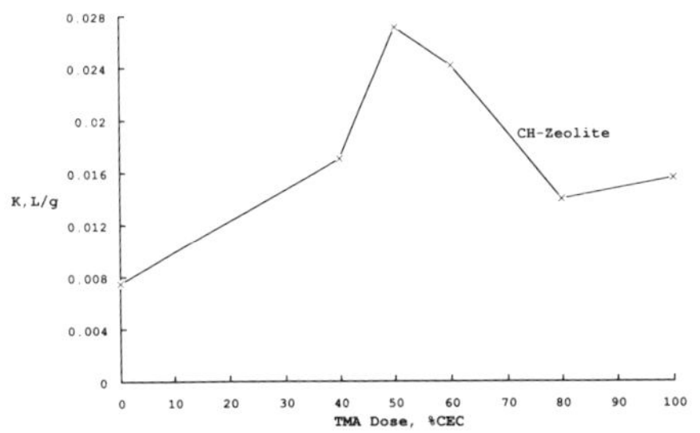

Figure 3. Summary of batch adsorption tests for toluene.

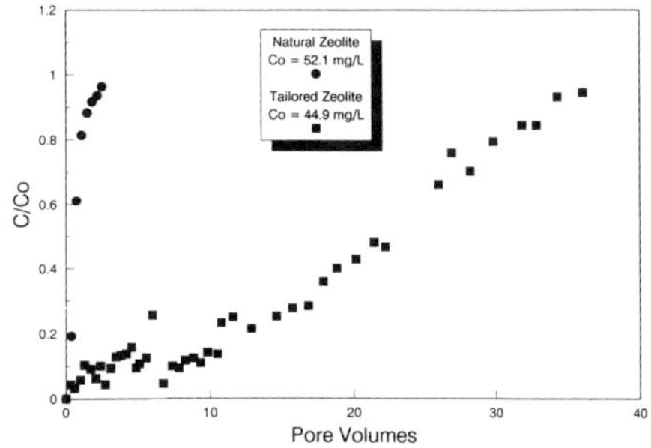

Figure 4. Column adsorption breakthrough curves: CH-Zeolite for benzene adsorption.

replacement of the hydrated Na^+ by TMA^+, the cavities of the zeolite innerchannels become available and are then accessible to benzene molecules which have an edge-to-edge size of about 4.3 Å.[21] By entering the zeolite innerchannels, benzene molecules become strongly trapped within the zeolite particles.

(3) The doses of TMA^+ which produce the maximum adsorptions for benzene are not at 100% CEC of all three tested zeolites. The optimum TMA^+ doses for SC-, XY-, and CH-Zeolites are 20, 20, and 40% CEC of the corresponding zeolites. Applying additional amount of TMA^+ in the tailoring process produces adverse effects on the adsorption of benzene in all three zeolite tests. A reasonable explanation for this phenomenon is that the exchange of the hydrated Na^+ by TMA^+ and the intrusion of benzene molecules into the zeolite innerchannels are limited due to the small difference between the size of TMA^+/benzene and the size of the zeolite innerchannel. During the tailoring process, the exchanged TMA^+ ions may hinder soluble TMA^+ ions from entering the innerchannels and exchanging with the natural cations. Additional TMA^+ ions will stay in the liquid phase (in the bulk solution), instead of exchanging and becoming trapped in the zeolite innerchannels. During the adsorption process, some benzene molecules enter the zeolite innerchannels first. These adsorbed benzene molecules and the exchanged TMA^+ ions inside may obstruct the access to the innerchannels. On the other hand, soluble TMA^+ ions in the bulk solution may compete for the benzene molecules, attracting them away from the zeolite particles.

(4) The tailoring process is less dramatic for the adsorption of toluene. This is mainly due to the greater size of the toluene molecules since the CH_3-group adds an approximate 1.10 Å tail structure to the toluene molecule. Although TMA^+ tailoring can also open some cavity space in the zeolite innerchannels, the toluene molecules appear to be mainly trapped at the edge portions of the channel openings, instead of entering the innerchannels.

Column Adsorption Breakthrough Study

The column adsorption characteristics in this study were modeled using breakthrough curves. The breakthrough curves for the natural and the TMA^+ tailored CH-Zeolites using benzene as the target pollutant are shown in Figure 4. The breakthrough curves for CH-Zeolites using toluene as the target pollutant are shown in Figure 5. In both figures, C and C_0 represent the pollutant (benzene or toluene) concentrations in the column effluent and influent, respectively. The unit pore volume for CH-Zeolite column was 670 mL. During the tailoring process, mass balance of TMA^+ was performed and the results are presented in Table II. Flow rates through the natural and tailored columns were also monitored, and the results are presented in Table III.

The following observations are drawn based on the results shown in Figures 4 and 5, and in Tables II and III:

(1) In the tests for benzene, natural CH-Zeolite showed almost no adsorption. It can be seen in Figure 6 that $C = 1/2\ C_0$ (or $C/C_0 = 0.5$) occurred when less than 1 pore volume of the polluted water had eluted through the column. The occurrence of $C/C_0 = 0.5$ at pore volume less than 1 is commonly explained as lack of adsorption in the short column breakthrough studies. On the other hand, the TMA^+ tailored column exhibited significant adsorption of benzene since the occurrence of $C/C_0 = 0.5$ was retarded until about 22 pore volumes of the political water had passed through the column.

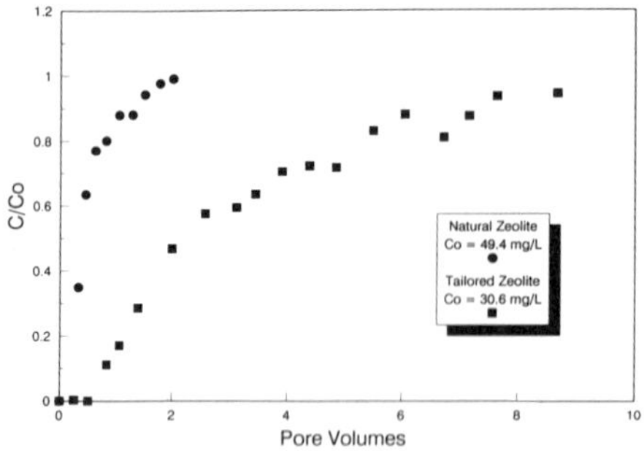

Figure 5. Column adsorption breakthrough curves: CH-Zeolite for toluene adsorption.

(2) In the tests of toluene, no adsorption was observed by the natural CH-Zeolite. Figure 7 shows that $C/C_0 = 0.5$ happened at about pore-volume = 0.5. After tailored with TMA^+, CH-Zeolite showed obvious adsorption for toluene. The improvement, however, was much smaller than that observed in the benzene tests. Since $C/C_0 = 0.5$ occurred at about 2.5 pore volumes, the retardation for toluene was 2 pore volumes, which is small as compared to the 21 pore volume retardation for benzene. These results are compatible with those obtained in the batch isotherm study.

(3) Table II shows the results of TMA^+ mass balance performed during the tailoring process. In the tests for benzene, the TMA-Cl solution used for tailoring was 51.45 g/L which was equal to 34.48 g TMA-Cl per pore volume (670 mL). This solution made the dose of TMA^+ ions in 670 mL solution equal to 40% CEC of 1100 g CH-Zeolite, which was the optimum dose determined in the batch isotherm study. In the test for toluene, TMA-Cl solution was 64.33 g/L or 43.10 g per 670 mL, the optimum dose of 50% CEC of 1100 g CH-Zeolite. Results of the mass balance show that the actual doses of TMA^+ ions in the benzene and toluene tests were 38% and 56% CEC respectively for 1100 g CH-Zeolite. Therefore, the adsorption for both benzene and toluene occurred within small ranges of the optimum tailoring conditions found in the batch studies. Another point to be noted is that the greater bulk concentration of TMA^+ ions favored the mass transfer of TMA^+ from the tailoring solution to the ion exchange site on zeolite particles. This explanation is consistent with mass transfer theory.

(4) Results from the flowrate monitoring (Table III) show that tailoring with TMA^+ did not alter the permeability of CH-Zeolite when it was used as adsorption medium. These results demonstrate favorable chemical and hydraulic characteristics of CH-Zeolite for being used as chemical barrier in field conditions.

Table II. Mass Balance for TMA in CH-Zeolite Column Tailoring

	Benzene Test	Toluene Test
Total amount of TMA in tailoring solution	89.88 g	119.95
Eluted TMA	29.76 g	32.57
Dissolved TMA in column	27.40 g	38.86
Exchanged TMA	32.72 g	48.52
Actual dose of TMA (% CEC)	37.96	56.28
Optimum dose (% CEC) from batch study	40	50

Table III. Flowrates in CH-Zeolite Column Tests

	Flowrate (mL/min)*	
	Benzene test	Toluene test
Natural CH-Zeolite column	23.74	17.18
Tailored CH-Zeolite column	26.39	25.46

*Flowrate varied slightly due to the small hydraulic head differences made during the tests.

CONCLUSION

(1) All three zeolites tested in this project have shown good potentials as adsorbents for removal of benzene from water. The adsorption capacity of zeolites is significantly improved by tailoring with TMA^+.

(2) Both batch isotherm and column breakthrough tests have shown that TMA^+ tailoring also improves the adsorption for toluene. The improvement, however, is less dramatic than for benzene adsorption. This is likely due to the incompatibility of the steric properties of the toluene molecules with respect to the innerchannels of the zeolite particles.

(3) Tailoring with TMA^+ does not alter the excellent permeability of the zeolite medium.

ACKNOWLEDGMENT

Thanks are given to Argonne National Laboratory, New Mexico Water Resources Research Institute and Teague Minerals Products for their consistent cooperation and cordial support which were very important for the completion of this project. Thanks also go to Francine Vigil and Enrique Cazares for their help in completion of the experimental phase of this project.

REFERENCES

1. McKee, J. E., F. B. Laverty, and R. H. Hertel, 1972, "Gasoline in Groundwater," *J. Water Pollut. Control Fed.*, 44:293–302.
2. Jercinovic, D. E., 1984, "Petroleum Products Contamination of Soil and Water in New Mexico", Research Project Report, NM Environmental Improvement Division, Santa Fe, NM.
3. Kohlhaas Corporation, 1989, "Information Booklet: MUSTs for USTs in New Mexico", Albuquerque, NM.
4. Manko, J. M., 1985, "Opinion: Bhopal Tragedy Illuminates in Dangers in Leaking Tanks", *Manag. World Wastes*, 28:117–119
5. Parmeggiani, L., 1983, editor, *Encyclopedia of Occupational Health and Safety*, 3rd ed., International Labor Office, Geneva, Switzerland.
6. Sax, N. I., 1984, *Dangerous Properties of Industrial Materials*, 6th ed., Van Nostrand Reinhold Company, Inc., New York, NY.
7. Sittig, M., 1985, *Handbook of Toxic and Hazardous Chemicals and Carcinogens*, 2nd ed., Noyes Publications, Park Didge, NJ.
8. Rinsky, R. A., A. B. Smith, and R. Hornung, 1987, "Benzene and Leukemia: an Epidemiologic Risk Assessment", *New England J. Med.*, 35:87–98.
9. Fawell, J. K., and S. Hunt, 1988, *Environmental Toxicology: Organic Pollutants*, Ellis Horwood Ltd., Chichester, England.
10. Collins, A. G., and A. I. Johnson, 1988, editors, *Groundwater Contamination: Field Methods*, ASTM Symposium, 1986.
11. McQuillan, J. E., 1984, "Organic Water Contaminants in New Mexico", Research Project Report, Environmental Improvement Division, NM.
12. Wolfe, T. A., D. Turgut, and E. R. Baumann, 1986, "Adsorption of Organic Pollutants on Montmorillonite Treated with Amines", *J. Water Pollut. Control Fed.*, 58:68–75.
13. Cadena, F., and S. W. Jeffers, 1987, "Use of Tailored Clays for Selective Adsorption for Hazardous Pollutants", *Proc. 42nd Ind. Waste Conf.*, Purdue Univ., 42:113–119.
14. Boyd, S. A., M. M. Mortland, and C. T. Chiou, 1988, "Sorption Characteristics of Organic Compounds in Hexadecyltrimethyl-ammonium-smectite", *Soil Sci. Soc of American J.*, 52:652–659.
15. Cadena, F., 1989, "Use of Tailored Bentonite For Selective Removal of Organic Pollutants", *ASCE J. Environ. Engineering*, 115:757–767.
16. Peters, R. W., F. Cadena, and M. W. Page, 1989, "Containment of Organic Leachates from Leaking Underground Storage Tanks Using Tailored Soils as Liners", paper presented at Symposium on Hazardous Waste and Environmental Management in the Gas Technology, sponsored by Inst. Gas Tech., Chicago, IL, May 1–2, 1989.
17. Peters, R. W., K. Banks, F. Cadena, and F. Gao, 1990, "Adsorption Characteristics of BTX Compounds onto Tailored Soils", paper presented at 1990 AIChE Winter National Meeting, Orlando, FL, March 18–22, 1990.
18. Schmidt, R. R., 1987, "Zeolites: Their Market Potential and Occurrence in New Mexico", report of investigation, New Mexico Energy and Minerals Department, Santa Fe, NM.
19. Teague Mineral Products, 1989, "Bentonite and Zeolite", produce information brochure, Teague Mineral Products, Inc., South Adrian, OR.

20. Weast, R. C., M. J. Astle, and W. H. Beyer, 1989, editors, *Handbook of Chemistry and Physics*, 70th ed., CRC Press, Inc. Boca Raton, FL.
21. Wingrove, A. S., and R. L. Caret, 1981, *Organic Chemistry*, Harper & Row Publishers, Inc., New York, NY.

60 REMOVAL OF CADMIUM AND CHROMIUM FROM WASTEWATER USING FLY ASH

T. Viraraghavan, Professor
Faculty of Engineering
Univeristy of Regina
Regina, Saskatchewan
Canada S4S OA2

Ganesh A. K. Rao, Engineer
Tata Chemicals Limited
Bombay, India

INTRODUCTION

The heavy metals discharged in the wastewaters can be toxic to aquatic life and cause natural waters to be unsuitable as potable water sources. The ubiquitous nature of heavy metals, their extreme toxicity even in trace quantities, their tendency for bioaccumulation in the food chain, and the stricter environmental regulations related to heavy metal discharges make it necessary to develop schemes for the removal of heavy metals from wastewaters.

In spite of the effectiveness of activated carbon in the removal of heavy metals from wastewaters, the high cost of activated carbon has restricted its more widespread use. Since adsorption is one of the more effective methods for removing heavy metals from wastewaters, an inexpensive and more easily available adsorbent would make the removal of heavy metals from wastewaters by adsorption an economically viable alternative. It has been demonstrated that inexpensive adsorbents such as tree bark, discarded automotive tyres, human hair, sawdust, corncobs, peanut hulls, straw, coffee grounds, coconut shells, and rice husks are effective in removing heavy metals from wastewaters. The abundance of fly ash and its easy availability makes it an economical adsorbent for removing heavy metals from wastewaters.

Fly ash is defined as the finely divided residue that results from the combustion of ground coal. Fly ash is a particulate substance ranging from 0.5 to 100 microns in size and composed primarily of silica (SiO_2) and alumina (Al_2O_3). Other macroelements present to a greater or lesser degree include carbon, calcium (as lime or gypsum), magnesium, sulfur (as SO_4^{-2} and SO_3^{-2}), sodium and potassium.[1]

The use of fly ash in wastewater treatment has been studied in recent years and experimental results have shown fly ash to be a suitable adsorbent for the removal of heavy metals from wastewaters. Gangoli et al.[2] examined the suitability of fly ash as an adsorbent for the removal of chromium from aqueous solutions. They concluded that the removal of hexavalent chromium using fly ash involved a chemisorption mechanism associated with the bonding between active alumina sites and the chromate anion. The experimental evidence from their study indicated that chromate ions are partially removed from aqueous solutions by fly ash. The removal of hexavalent chromium by adsorption on fly ash was also studied by Grover and Narayanaswamy.[3] They concluded that fly ash with a fairly good adsorbent for the removal of hexavalent chromium from aqueous solutions and that the adsorption was probably due to pore diffusion. Yadava et al.[4] investigated the removal of cadmium by fly ash at different conditions by varying the contact time, temperature and pH. Their results showed that an increase in temperature from 20°C to 40°C decreased the adsorption of cadmium on fly ash from 96.07% to 83.78%. The removal of cadmium by fly ash increased from 11.85% to 89.82% on increasing the pH of the cadmium solution from 4.0 to 8.5.

This study was therefore initiated to evaluate the effectiveness of locally available fly ash in the removal of cadmium and chromium from municipal wastewater. Laboratory batch kinetic and isotherm studies were conducted to evaluate the adsorption capacity of fly ash in respect of cadmium and chromium in wastewater. The effects of contact time, pH and temperature on the adsorption capacity of fly ash were also investigated.

45th Purdue Industrial Waste Conference Proceedings,

MATERIALS AND METHODS

Wastewater

The wastewater used in the study was collected from the City of Regina Municipal Wastewater Treatment Plant located 20 km west of the City of Regina. The treatment plant receives mostly residential wastewater and the effluent from a few industries in and around the city, of which an oil refinery and a steel mill are the major contributors. The wastewater used in this study was the raw wastewater collected at the bar screens before the wastewater enters the primary sedimentation tanks. All the wastewater samples collected were grab samples. The wastewater was collected and stored in plastic containers thoroughly washed with 1 M HCl and deionized water before sampling to avoid metal contamination. The wastewater samples were acidified with 1 M HCl to pH below 2.0 immediately upon collection to preserve the soluble metals. The wastewater samples were stored in a cold storage room (4°C) before being used for the batch experiments. The total chromium in the wastewater will be referred to as chromium in this study. The wastewater was spiked with stock solutions of cadmium and chromium before being used in the batch studies to raise the cadmium and chromium concentrations to approximately 1 mg/L.

Fly Ash

The fly ash used in the study was obtained from Unit #2 of the Poplar River Power Station of the Saskatchewan Power Corporation. The fly ash generated from the combustion of lignite coal was collected from the first bunker of the Poplar River's electrostatic precipitators. The fly ash samples were analyzed for cadmium, chromium, copper, nickel and zinc.

Batch Kinetic Studies

The batch studies were conducted using a Jar Test Apparatus (manufactured by Phipps and Bird Inc.).

The kinetics of cadmium and chromium adsorption by fly ash was studied in batch experiments. The wastewater in each of the jars was spiked with cadmium and chromium stock solutions used for atomic absorption spectroscopy containing cadmium metal in 2% nitric acid and potassium dichromate in distilled water. 0.5 mL each of cadmium and chromium stock solutions were added to 500 mL of the wastewater in each of the jars raising the cadmium and chromium concentrations to just above 1 mg/L. The wastewater was spiked with the stock solutions to make it representative of wastewaters with industrial effluents which contain approximately 1 mg/L each of cadmium and chromium. 10 g of fly ash was added to 500 mL of the wastewater and a control without fly ash was also carried through to check if any cadmium of chromium was adsorbed onto the walls of the glass beakers. 10 g of fly ash was added to 500 mL of deionized water to check for any leaching of cadmium or chromium during the study period. The fly ash and wastewater were mixed at 100 rpm and maintained at the room temperature 21° ± 1°C. The wastewater from each jar was filtered and analyzed for cadmium and chromium at the end of each one-hour period during the 10-hour study period to determine the equilibrium time necessary for the maximum adsorption of cadmium and chromium on peat. Equilibrium time is defined as the contact time required for the concentrations of cadmium and chromium in the wastewater to reach an equilibrium value. The wastewater in each of the jars was filtered through a 0.45 micrometer Millipore membrane filter using a vacuum pump and analyzed for cadmium and chromium by atomic absorption spectroscopy. The wastewater in the control jar and the deionized water in the jar (to study leaching) were also analyzed for cadmium and chromium by atomic absorption spectroscopy. All batch kinetic studies were replicated.

Batch Adsorption Studies at Various pH

The effect of pH on the adsorption of cadmium and chromium on fly ash was studied by the addition of 10 g of fly ash to 500 mL of the wastewater. The pH of the wastewater was adjusted to be in the range 2.0 to 7.5 by the addition of either 1 M HCl or 1 M NaOH. The fly ash and wastewater were mixed at 100 rpm for 3 hours and the cadmium and chromium concentrations at the end of the study period were determined by atomic absorption spectroscopy. The pH of the wastewater at the end of the 3 hour study period was also determined.

Batch Adsorption Studies at Various Temperatures

The removal of cadmium and chromium from wastewater by fly ash at various temperatures was studied by conducting batch experiments at 5° ± 1°C, 10° ± 1°, 15° ± 1°C, and 21° ± 1°C. Representative weights (1.0g, 2.0g, 3.0g, 4.0g, 5.0g, 6.0g, 7.0g, 8.0g, 9.0g and 10.0g) of fly ash were mixed with 500 mL of the spiked wastewater at 100 rpm for 3 hours. The pH of the wastewater was adjusted with either 1 M HCl or 1 M NaOH for maximum adsorption of cadmium and chromium. The cadmium and chromium concentrations at the end of the study period were determined by atomic absorption spectroscopy. The removal of cadmium and chromium by fly ash from the actual wastewater was studied by conducting batch experiments with the wastewater without any acidification of the wastewater after collection. All the batch adsorption studies were replicated.

Desorption Studies

Desorption studies were carried out at 21°C ± 1°C. These were conducted by mixing representative weights (1.0g, 2.0g, 3.0g, 4.0g, 5.0g, 6.0g, 7.0g, 8.0g, 9.0g, and 10.0g) of fly ash with 500 mL of the spiked wastewater for 3 hours. After 3 hours the fly ash was separated from the wastewater by filtration and mixed with deionized water for 3 hours. The cadmium and chromium concentrations in the deionized water at the end of the 3 hour study period was determined by atomic absorption spectroscopy after filtering the deionized water through a 0.45 micrometer membrane filter. The desorption studies were replicated.

Analytical Procedures

Methods outlined in *Standard Methods*[5] were followed for the analyses. The study involved the analytical demonstration of the cadmium and chromium concentrations in the wastewater, and the pH of the wastewater. For fly ash, the analysis involved the determination of the cadmium, chromium, copper, nickel, and zinc content of the fly ash sample. The specific method followed in the determination of each parameter is indicated below.

The pH of the wastewater was determined electrometrically using a Accumet Selective Ion Analyzer (Model 750) manufactured by the Fisher Scientific Ltd. Cadmium and chromium concentrations in the wastewater were determined by atomic absorption spectroscopy using a Varian AA10 Atomic Absorption Spectrometer (Section 300 APHA 1985). A set of standards (0.50 mg/L, 1.00 mg/L, 1.50 mg/L, and 2.00 mg/L for cadmium, and 2.00 mg/L, 4.00 mg/L, 6.00 mg/L, and 8.00 mg/L for chromium) were prepared for cadmium and chromium using atomic absorption spectroscopy stock solutions manufactured by Fisher Scientific Ltd. The standards were prepared using deionized water and calibration curves were plotted for the cadmium and chromium concentration range found in the wastewater. Single element hollow cathode lamps of cadmium and chromium were used as light sources. The wastewater was filtered through a 0.45 micrometer Millipore membrane filter before being analyzed for cadmium and chromium. The elements cadmium and chromium were analyzed at a wavelength of 228.8 nm and 357.9 nm respectively. Three replicate readings of the same sample were taken with an integration time of 3 seconds and a delay time of 10 seconds. The soluble metal concentrations in the wastewater were determined using an air-acetylene flame. The atomic absorption spectrometer had a detection limit of 0.002 mg/L for cadmium and 0.02 mg/L for chromium, and a sensitivity of 0.025 mg/L for cadmium and 0.10 mg/L for chromium.

The metals in the fly ash sample were determined by digesting the samples and analyzing by inductively coupled argon plasma—atomic emission spectroscopy (ICP-AES). These analyses were carried out at the Saskatchewan Research Council Laboratory at Saskatoon, Saskatchewan.

RESULTS AND DISCUSSION

Wastewater Characteristics

The characteristics of the wastewater used in the study are shown in Table I. The concentrations in the wastewater were found to range from 0.077 mg/L to 0.400 mg/L for cadmium and from 0.009 mg/L to 0.259 mg/L for chromium. Since the cadmium and chromium concentrations in the Regina municipal wastewater were lower than the cadmium and chromium concentrations in wastewaters of other Canadian cities, the wastewater was spiked with cadmium and chromium stock solutions to raise the cadmium and chromium concentrations to above approximately 1 mg/L and thus making the test samples comparable to those of other Canadian cities.

Table I. Wastewater Characteristics

Characteristic	No. of Samples	Range	Mean	Std.Dev.
pH	30	7.08–7.69	–	–
BOD	11	45–239	156	71
TOC	9	21–81	53	20.2
COD	4	323–508	405	68
SS	25	72–1106	409	255
VSS	25	54–588	246	133
TKN	4	37–40.9	39.2	1.58
NH_3-N	4	20.2–29.7	23.5	3.65
PO_4-P	125	3.5–24.2	8.4	4.65
Cadmium	4	0.077–0.400	0.297	0.122
Chromium	6	0.009–0.250	0.116	0.074
Copper	6	0.146–1.372	0.073	0.419
Nickel	7	0.185–1.095	0.539	0.299
Zinc	7	0.162–1.228	0.612	0.362

Note: 1) All values are in mg/L except pH.
2) Values for Cd, Cr, Cu, Ni, and Zn are based on analyses conducted at the University of Regina; all other values were obtained from the City of Regina Wastewater Treatment Plant.

Fly Ash

Results of typical analyses of the fly ash carried out by the Technical Services and Research Division of the Saskatchewan Power Corporation, Saskatchewan are presented in Table II. The fly ash had a surface area of 1.5 to 1.7 m^2/g and 50 to 60% of the fly ash was retained on a 325 mesh sieve. The cadmium, chromium, copper, nickel, and zinc contents of the fly ash are shown in Table III.

Batch Kinetic Studies

The kinetics of cadmium and chromium adsorption by fly ash was studied by mixing for 10 hours 10 g of fly ash in 500 mL of wastewater containing 1.116 mg/L of cadmium and 1.145 mg/L chromium. The plots for cadmium concentration versus time (Figure 1) and chromium concentration versus time (Figure 2) showed that equilibrium was attained in three hours for the adsorption of both cadmium and chromium by fly ash. It was found that 97% of the cadmium and 32% of the chromium

Table II. Typical Analysis of Fly Ash (Source: Saskatchewan Power Corporation, Regina, Saskatchewan)

Constituent	Value
CaO, % weight/weight ignited	11.5
Na_2O, % weight/weight ignited	0.60
K_2O, % weight/weight ignited	2.10
P_2O_5, % weight/weight ignited	<0.1
Al_2O_3, % weight/weight ignited	21.80
MgO, % weight/weight ignited	4.30
TiO_2, % weight/weight ignited	0.80
SiO_2, % weight/weight ignited	50.70
Fe_2O_3, % weight/weight ignited	4.50
BaO, % weight/weight ignited	0.20
SrO, % weight/weight ignited	0.20
SO_3, % weight/weight ignited	0.53
Loss on ignition, %	0.84
Copper, ppm	67
Chromium, ppm	33
Nickel, ppm	30
Cadmium, ppm	0.1
Zinc, ppm	57

Table III. Metal Concentrations in Fly Ash

Constituent	Concentration, ppm
Cadmium	< 1
Chromium	80
Copper	81
Nickel	34
Zinc	81

was adsorbed on to fly ash in the initial three hours and the adsorption of cadmium and chromium did not increase with time after the initial three hours.

The three hours equilibrium time for cadmium adsorption on fly ash in this study as compared to the two hour equilibrium time obtained by Yadava et al.[4] suggests that the adsorption of cadmium on fly ash used in this study is a slower process as compared to the fly ash studied by them. This may be due to the higher surface area (5.77 m^2/g) of the fly ash used by Yadava et al.[4] as compared to the surface area of 1.7 m^2/g for the fly ash used in this study. The 56% silica and 26% alumina in the fly ash used by them as compared to the 50% silica and 21% alumina in the fly ash used in this study could also contribute to the difference in the equilibrium time for the adsorption of cadmium on fly ash. The 24-hour equilibrium time obtained by Grover and Narayanaswamy[3] for chromium adsorption on fly ash suggests that the adsorption of chromium on fly ash is also very much dependent on the type of fly ash studied.

The batch kinetic data for the adsorption of cadmium and chromium on peat was tested for a first order reaction. The rate equation for a first order reaction is given by (6):

$$-\ln C_a/C_{ao} = kt \tag{1}$$

where: c_a = concentration of solute after time "t", mg/L
c_{ao} = initial concentration of solute, mg/L
k = reaction - rate constant, $time^{-1}$.

The reaction rate constants for the adsorption of cadmium and chromium were determined from the plots of $-\ln (C_a/C_{ao})$ versus time and were found to be 0.405 $hour^{-1}$ and 0.0134 $hour^{-1}$ respectively. This indicates a preferential adsorption of cadmium on fly ash as compared to chromium. The rate constant of 1.398 $hour^{-1}$ obtained by Yadava et al.[4] for cadmium adsorption on fly ash shows that the rate of adsorption of cadmium on fly ash is higher, possibly due to higher surface area (5.77 m^2/g) of the fly ash used by them.

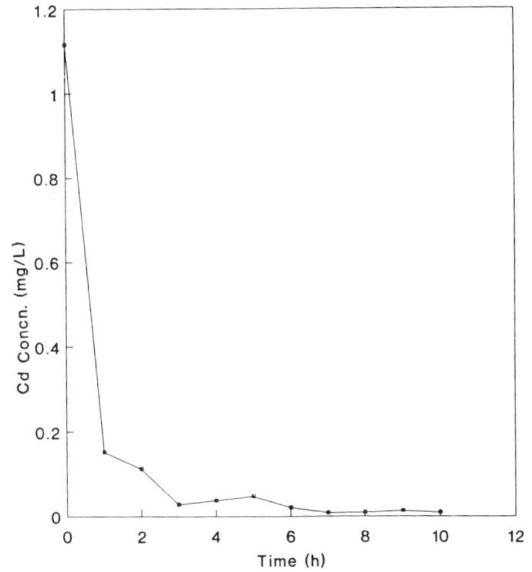

Figure 1. Removal of cadmium by adsorption on fly ash.

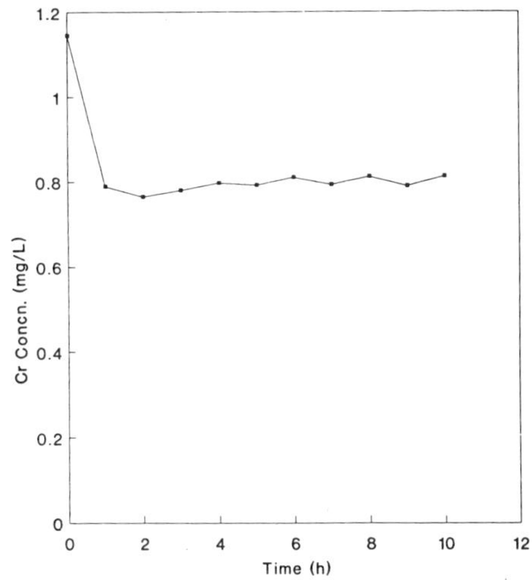

Figure 2. Removal of chromium by adsorption on fly ash.

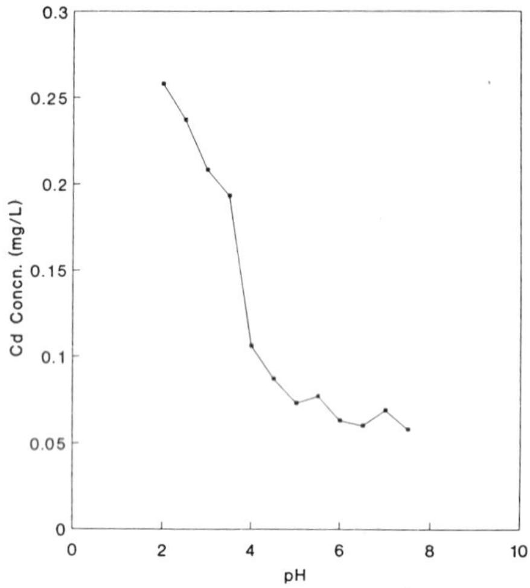

Figure 3. Effect of pH on cadmium adsorption on fly ash.

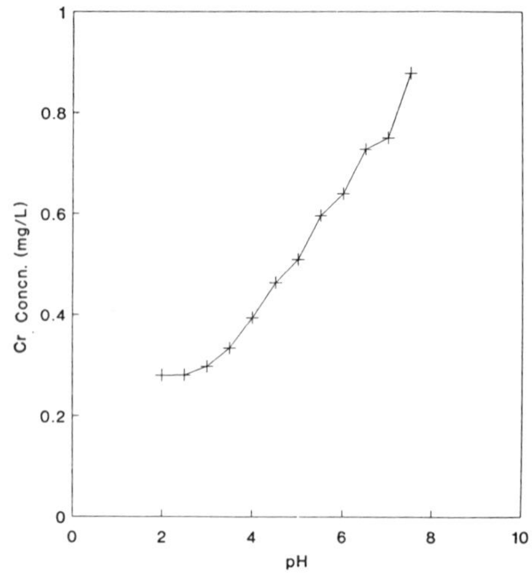

Figure 4. Effect of pH on chromium adsorption on fly ash.

Adsorption at Various pH

The effect of the pH on the adsorption of cadmium and chromium on fly ash were studied by varying the pH of the wastewater in the pH range of 2.0 to 7.5. The removal of cadmium by adsorption on fly ash was found to increase with an increase in the pH of the wastewater from 2.0 to 7.5 at 21° ± 1°C (Figure 3). The effect of pH on the adsorption of cadmium on fly ash may be explained on the basis of aqua complex formation on the oxides present in the fly ash. A positive charge develops on the surface of the oxides of fly ash in an acidic medium as follows:

$$MO + H\text{-}OH \xrightarrow{H+} M\text{-}OH_2^+ + OH^- \tag{2}$$

A lowering on the extent of adsorption results as the surface charge thus developed is not suitable for cadmium adsorption which results in the chloride ion getting readily associated with the surface of the fly ash. In an alkaline medium the surface of the fly ash becomes negatively charged and thus favours the adsorption of cadmium. These results agree well with the results obtained by Yadava et al.[4].

The effect of pH on the adsorption of chromium on to fly ash is shown in Figure 4. With an increase in pH from 2.0 to 7.5, the removal of chromium by adsorption on fly ash was found to decrease at 21° ± 1°C. The variation in the adsorption of chromium with varying pH can be explained on the basis of the role silica plays on the adsorption of chromium on fly ash. The fly ash used in these studies contained 50.70% silica, and hence influenced the adsorption of chromium on fly ash to a great extent. Gangoli et al.[2] reported that the surface of silica may exhibit considerable affinity towards metal ions. The central ions of silicates have a very strong electron affinity and therefore the oxygen atoms bound to the silicon ions have a low basicity, rendering the silica surface as a weak acid. The oxygen atoms on the silica surface are free to react with water, forming surface silanol (-Si-OH) groups. The acidity of the silanol groups determines the dependence of the charge silica surface results, and at high pH values a negatively charged surface prevails. Therefore, the adsorption of anions like chromium is greater in acidic pH in alkaline pH. Gangoli et al.[2] have indicated that chromium is perhaps adsorbed on the alumina surface in fly ash by a strong tendency towards chemical bonding between the chromate group and the alumina. Grover and Narayanaswamy[3] obtained similar results of improved chromium adsorption at pH below 3.0 using fly ash.

Adsorption at Various Temperatures

The adsorption data can be interpreted using several mathematical relationships which describe the distribution of the solute between the liquid phase and the solid phase.

The Langmuir, the Freundlich, and the Branaur-Emmett-Teller (BET) isotherms are shown below in Equations 3, 4, and 5, respectively[7].

Table IV. Langmuir and Freundlich Equations for Cadmium Adsorption on Fly Ash

Condition	Equation	Corr. Coeff.
Langmuir		
5° ± 1°C	1/(x/m) = 43.47 + 229.07 1/C	0.957
10° ± 1°C	1/(x/m) = 108.69 + 322.06 1/C	0.944
15° ± 1°C	1/(x/m) = 227.27 + 450.04 1/C	0.834
21° ± 1°C (pH adjusted)	1/(x/m) = 344.82 + 102.02 1/C	0.906
21° ± 1°C (no pH adjustment)	1/(x/m) = 28.90 + 138.02 1/C	0.984
Freundlich		
5° ± 1°C	log(x/m) = −1.13 + 2.92 logC	0.827
10° ± 1°C	log(x/m) = −1.55 + 2.87 logC	0.726
15° ± 1°C	log(x/m) = −1.701 + 3.54 logC	0.682
21° ± 1°C (pH adjusted)	log(x/m) = −1.15 + 2.14 logC	0.826
21° ± 1°C (no pH adjustment)	log(x/m) = −0.925 + 2.41 logC	0.876

The Langmuir equation is of the form shown below.

$$\frac{x}{m} = \frac{abC}{1 + aC} \tag{3}$$

The Freundlich isotherm is shown below.

$$\frac{x}{m} = kC^{(1/n)} \tag{4}$$

The BET isotherm is of the form presented below.

$$x/m = ACx_m / (Cs - C)\,(1 + (A-1)C/Cs) \tag{5}$$

where: x = amount of solute adsorbed (mg)
m = weight of adsorbent (mg)
C = concentration of solute in solution (mg/L), and A, a, b, k, n, x_m are constants.

Adsorption studies were conducted at 5° ± 1°C, 10° ± 1°C, 15°C ± 1°C, and 21° ± 1°C to determine the influence of temperature on the adsorption of cadmium and chromium on to fly ash. Adsorption data from studies on cadmium adsorption on fly ash were found to fit well the Langmuir isotherm for the weight range of 0.25 to 2.5 g of fly ash in 500 mL of the wastewater. Table IV shows the Langmuir and Freundlich isotherm equations for the adsorption of cadmium on fly ash at various temperatures. Detailed analysis of the variance showed that the Langmuir model described the adsorption data better. The "t" test showed that the coefficient of X was significant for the 95% confidence level and hence X and Y were related. The values of F (at 5° ± 1°C, 10° ± 1°C, 15° ± 1°C, and 21° ± 1°C) in the analysis of variance were significant at 95% confidence level; hence the linear model is well correlated. An analysis of the results showed that the cadmium adsorption capacity of fly ash decreased with an increase in temperature indicating the adsorption process to be exothermic. These results are similar to the results obtained by Yadava et al.[4] The decreasing trend of adsorption with temperature may be due to the weakening of the adsorptive forces between the active sites of fly ash and cadmium. Since physical adsorption is an exothermic process, an increase in the wastewater temperature resulting in a decrease in the adsorption capacity indicates physical adsorption to be the rate-determining step.

In the case of adsorption of chromium on fly ash the resulting adsorption data were found to fit the Langmuir and Freundlich isotherms (Table V). The "t" test (for equations with relatively higher correlation) showed that the coefficient of X was significant for the 95% confidence level except for the adsorption of chromium on fly ash at 5° ± 1°C, and hence X and Y were related. The values of F (at 5° ± 1°, 10° ± 1°, 15° ± 1°, and 21° ± 1°) in the analysis of variance (for equations with the best correlation) were significant at 95% confidence level except for adsorption of chromium at 5° ± 1°C;

Table V. Langmuir and Freundlich Equations for Chromium Adsorption on Fly Ash

Condition	Equation	Corr. Coeff.
Langmuir		
$5° \pm 1°C$	$1/(x/m) = -232.55 + 1661.13\ 1/C$	0.579
$10° \pm 1°C$	$1/(x/m) = -192.30 + 1265.18\ 1/C$	0.582
$15° \pm 1°C$	$1/(x/m) = -434.78 + 952.0\ 1/C$	0.698
$21° \pm 1°C$ (pH adjusted)	$1/(x/m) = -22.93 + 1295.80\ 1/C$	0.510
$21° \pm 1°C$ (no pH adjustment)	$1/(x/m) = 454.54 + 2673.79\ 1/C$	0.727
Freundlich		
$5° \pm 1°C$	$\log(x/m) = 0.022 + 21.41 \log C$	0.991
$10° \pm 1°C$	$\log(x/m) = 0.024 + 20.20 \log C$	0.992
$15° \pm 1°C$	$\log(x/m) = -0.022 + 9.13 \log C$	0.991
$21° \pm 1°C$ (pH adjusted)	$\log(x/m) = -2.537 + 1.66 \log C$	0.379
$21° \pm 1°C$ (no pH adjustment)	$\log(x/m) = -2.744 + 5.78 \log C$	0.384

hence the linear model is well correlated. The Langmuir and Freundlich constants indicating adsorption capacity showed that the chromium adsorption capacity of fly ash decreased with increasing temperatures. The decrease in adsorption capacity with an increase in temperature showed the exothermic nature of chromium adsorption of fly ash. As in the case of cadmium adsorption on fly ash, a decrease in the adsorptive forces between the active sites of silica and chromium may lower the chromium adsorptive capacity of fly ash at higher temperatures.

The temperature coefficients for the Langmuir and Freundlich constants for the adsorption of cadmium and chromium on peat were obtained from the van't Hoff-Arrhenius equation.[7] The van't Hoff-Arrhenius equation is of the form shown below.

$$K_T = K_{21}\, \Theta^{(T-21)} \tag{6}$$

where: K_T, K_{21} = constants at temperatures T°C and 21°C respectively, and
Θ = temperature coefficient.

The temperature coefficients for the Langmuir and Freundlich constants for adsorption of cadmium and chromium on fly ash were obtained from the plots of the Langmuir and Freundlich constants versus the temperature. The temperature coefficients for the Langmuir and Freundlich constants for cadmium and chromium adsorption on fly ash are shown in Table VI. For the temperature range of 5°C to 21°C, the variation in the value of Θ for cadmium adsorption on fly ash ranged between 1.003 and 1.235 for K, between 1.012 and 1.066 for n, between 1.197 and 1.372 for a, and between 1.072 and 1.138 for b. For the same temperature range, the Θ values for chromium adsorption on fly ash ranged between 1.445 and 2.625 for K, between 1.173 and 1.328 for n, between 1.010

Table VI. Temperature Coefficients for the Langmuir and Freundlich Constants for Cadmium and Chromium Adsorption on Fly Ash

	Langmuir Constant	Temperature Coefficient	Temperature Range
Cadmium	a	$\Theta = 1.189$	5°C to 21°C
	b	$\Theta = 1.138$	5°C to 21°C
Chromium	a	$\Theta = 1.029$	5°C to 21°C
	b	$\Theta = 1.020$	5°C to 21°C
	Freundlich Constant	**Temperature Coefficient**	**Temperature Range**
Cadmium	K	$\Theta = 1.010$	5°C to 21°C
	n	$\Theta = 1.007$	5°C to 21°C
Chromium	K	$\Theta = 1.428$	5°C to 21°C
	n	$\Theta = 1.184$	5°C to 21°C

Table VII. Equilibrium Parameter "R" for Cadmium and Chromium Adsorption on Fly Ash

Condition	R values	
	Cadmium	Chromium
5° ± 1°C	0.800	1.204
10° ± 1°C	0.693	1.258
15° ± 1°C	0.605	2.258
21° ± 1°C (pH adjusted)	0.797	1.306
21° ± 1°C (no pH adjustment)	0.217	1.020

and 1.179 for a, and between 1.155 and 1.633 for b. A decrease in the temperature coefficient, Θ, for cadmium and chromium adsorption on fly ash indicates an increase in the rate of cadmium and chromium adsorption on fly ash with an increase in temperature. The negative slope of the plots of the Langmuir and Freundlich constants for cadmium and chromium adsorption on fly ash versus temperature indicate a decrease in the cadmium and chromium adsorption capacity of fly ash with an increase in temperature. This indicates that the adsorption of cadmium and chromium on fly ash to be an exothermic process.

The essential characteristics of a Langmuir isotherm can be expressed in terms of a dimensionless constant separation factor or equilibrium parameter which is defined as:[8]

$$R = 1/(1 + aC_o) \tag{7}$$

where: a = Langmuir constant, and
C_o = initial solute concentration (mg/L).

This parameter indicates the shape of the isotherm as follows:

Value of R	Type of Isotherm
$R > 1$	Unfavorable
$R = 1$	Linear
$0 < R < 1$	Favorable
$R = 0$	Irreversible

Table VII shows the R values for the adsorption isotherms of cadmium and chromium adsorption on fly ash. For cadmium adsorption on fly ash, the R values were found to be less than 1 and greater than 0 indicating favorable adsorption of cadmium on fly ash in the temperature range 5° ± 1°C to 21° ± 1°C. The adsorption of chromium on fly ash was found to be unfavorable in the temperature range of 5° ± 1°C to 21° ± 1°C, based on the R values in Table VII.

Desorption Studies

Desorption studies of fly ash in deionized water showed that neither cadmium nor chromium was desorbed from fly ash during the three-hour study period.

General Discussion

Batch adsorption experiments have shown that fly ash is an effective adsorbent for the removal of cadmium and chromium from wastewaters. The adsorption of cadmium and chromium on fly ash was found to be an exothermic process.

Fly ash was found to have a higher adsorption capacity for cadmium as compared to chromium. While both cadmium and chromium needed an equilibrium time of three hours, the adsorption of chromium was higher (52.80%) in the pH range 2 to 3, while cadmium was adsorbed to a greater extent (94.80%) in the pH range 7 to 8. This indicates that fly ash would be a suitable adsorbent for the removal of cadmium from municipal wastewaters as the pH of the municipal wastewater was found to be in the pH range 7 to 8. This would not limit the use of fly ash to treatment of municipal wastewaters as the calcium oxide content of fly ash (11.50%) would raise the pH of an acidic wastewater appreciably for fly ash to be an effective adsorbent for the removal of cadmium from acidic industrial effluents. Activated carbon was found to have a much higher adsorption capacity for cadmium and chromium; however, fly ash which is easily available at a fraction of the cost of activated carbon, will be suitable for small-scale wastewater carbon, will be suitable for small-scale wastewater treatment plants. Results of the three-hour desorption study should be interpreted with caution in respect of possible leaching from land disposal of the used fly ash.

Application to Practice

The results of this study can be used to design batch adsorption systems for the treatment of wastewaters contaminated with cadmium and chromium. Batch adsorption studies have shown that fly ash adsorbed 93% of the cadmium and 44% of the chromium in the wastewater.

For systems where the total volume of wastewater is small and generated over a period of 8 to 12 hours, one reaction vessel may be adequate for the treatment of the wastewater. In cases where a continuous system is required, two reaction vessels may be used by alternatively using each of the reaction vessel for the treatment of the wastewater. Fly ash can be added to the reaction vessel containing the wastewater and the mixture may be agitated for 3 hours (equilibrium time) such that the maximum amount of cadmium and chromium are adsorbed on fly ash. The wastewater and fly ash can then be separated after agitation to prepare the reaction vessel for another batch of the wastewater. Since the Langmuir and Freundlich isotherms modeled the adsorption of cadmium and chromium on fly ash, respectively, the amount of fly ash required to decrease their concentration in the wastewater can be calculated. The design capacity of the reaction vessel will reflect the minimum reaction time (equilibrium time) required and the volume of the wastewater to be treated.

Fly ash can be disposed of for use in the cement and concrete industry, or can be used as a fill material for roads and construction sites and as a filler in asphalt mix.

CONCLUSIONS

The following conclusions can be drawn from this study.

1. The kinetic studies indicated that equilibrium in the adsorption of cadmium and chromium on fly ash was reached in three hours of contact between fly ash and the wastewater.
2. The reaction rate constants for cadmium and chromium adsorption on fly ash were found to be 0.405 $hour^{-1}$ and 0.0134 $hour^{-1}$.
3. The reaction rate for adsorption of cadmium and chromium on fly ash was found to increase with temperature as indicated by the temperature coefficients for the Langmuir and Freundlich constants.
4. Fly ash was found to adsorb a maximum of 93% of the cadmium in the wastewater at $5° \pm 1°$ in the pH range of 7 to 8 and a maximum of 44% of the chromium in the wastewater at $5° \pm 1°$C in the pH range of 2 to 3.
5. The adsorption of cadmium and chromium on fly ash were found to be exothermic.
6. The adsorption of cadmium on fly ash was described by the Langmuir isotherm at $5° \pm 1°$C, $10° \pm 1°$C, $15° \pm 1°$C, and $21° \pm 1°$C.
7. The adsorption of chromium on fly ash generally followed the Freundlich isotherm at $5° \pm 1°$C, $10° \pm 1°$C, and $15° \pm 1°$C, and the Langmuir isotherm at $21° \pm 1°$C.

ACKNOWLEDGMENTS

The study was supported by an operating grant to the first author by the Natural Sciences and Engineering Research Council (NSERC) of Canada. The second author would like to thank the Province of Saskatchewan for providing a scholarship through the Faculty of Graduate Studies and Research for the period September to December, 1989.

REFERENCES

1. Torrey, S. (1978). *Coal Ash Utilization—Fly Ash, Bottom Ash and Slag*, Noyes Data Corporation, Park Ridge, New Jersey.
2. Gangoli, N., Markey, D. C., and Thodos, G. (1975). "Removal of heavy metal ions from aqueous solutions with fly ash", *Proceedings of the 2nd National Conference on Complete Water Reuse: Water's Interface with Energy, Air and Solids*, Chicago, Illinois.
3. Grover, M., and Narayanaswamy, M. S. (1982). "Removal of hexavalent chromium by adsorption of fly ash", *J. Environ. Engrg.* Div., Institution of Engineers (India) 63, 36–39.
4. Yadava, K. P., Tyagi, B. S., Panday, K. K., and Singh, V. N. (1987). "Fly ash for the treatment of cadmium (II) rich effluents", *Environ. Tech. Lett.*, 8, 225–234.

5. *Standard Methods for the Examination of Water and Wastewater*. (1985) 16th Edition, American Public Health Association, Washington, D.C.
6. Levenspiel, O., (1974). *Chemical Reaction Engineering*, Wiley Eastern Private Ltd., New Delhi, India.
7. Benefield, L. D., Judkins, J. F., and Weand, B. L. (1982). "*Process Chemistry for Water and Wastewater Treatment*", Prentice-Hall, Inc., Englewood Cliffs, New Jersey, 07632.
8. Weber, T. N., and Chakravorti, R. K. (1974). "Pore and solid diffusion models for fixed-bed adsorbers", *AIChE Journal*, 20(2), 228–238.

61 PACKED-ENTRAPPED-MIXED MICROBIAL CELL PROCESS FOR REMOVAL OF PHENOL AND ITS COMPOUNDS

P. Y. Yang, Professor
T. S. See, Graduate Research Assistant
Department of Agricultural Engineering
University of Hawaii at Manoa
Honolulu, HI 96822

INTRODUCTION

The biodegradation of hazardous wastes by conventional-biological treatment systems has been reviewed by Grady.[1] It was observed that most of the conventional-biological systems were capable of degrading toxic-organic materials. The removal of toxic pollutants by primary treatment, chemical clarification, trickling filter, activated sludge, aerated lagoon, and facultative lagoon have been evaluated by Hannah, et al.[2] It was concluded that the activated sludge process provided the best removal of conventional pollutants as well as toxic priority pollutants. Regarding the response of phenol-acclimated activated sludge process to quantitative shock loading, it was concluded by Rozich and Gaudy[3] that design of a slow specific growth rates (higher solid retention time, SRT) are essential for systems that undergo shock loadings. Similarly, anaerobic suspended cell systems have potential to degrade toxic organics if acclimation of microbial cells to the organics and longer SRT's are provided.[4]

Immobilized microbial cell systems have been increasingly used in the bioconversion processes, including the production of food, beverages, pharmaceuticals, macromolecules such as organic acids and amino acids, other industrial chemicals, and fuels such as alcohol, methane and hydrogen. Biological wastewater treatment is within the category of bioconversion processes. Immobilized cell systems, such as biofilms, have been widely used for decontamination and methane production. This type of process is particularly suitable for the treatment of non-conventional or toxic wastewater, which requires a longer cell or solid residence time. Under conditions that would have caused washout of the free cells, immobilized cells could still be retained in the reactor;[5] i.e., SRT could be increased by immobilization of cells.

Phenol and phenolic compounds have been considered as toxic organics and tend to be prevalent in the industrial wastewater streams. Studies on the application of immobilized pure culture and systems for synthetic phenolic wastewater treatment were reported.[6,7,8,9,10] The application of immobilized pure culture characteristic in the actual wastewater treatment process. Application of immobilized (entrapped) mixed culture cell for phenol and phenol related compounds was previously reported.[11,12,13] These compounds could be successfully treated. However, the effectiveness of absorbed and entrapped mixed microbial cells for the treatment of phenol was not thoroughly compared[14] and required further investigation.

Protection of groundwater and clean-up for hazardous wastes/wastewater including pesticide wastes and pesticide contaminated water have become two of the major concerns of the USEPA and the states; certainly Hawaii is no exception.

Carbaryl (phenol related compound) is the most extensively used carbamate pesticide.[15] It has been found in groundwater and is on the EPA National Groundwater Survey List.[16] The detoxification of carbaryl using pure culture isolated from soil to describe the differences in each of the decomposition process was investigated.[17] It was concluded that the mixed cultures of soil microbes were more effective in transforming carbaryl than in pure cultures. Thus, the packed-entrapped-mixed microbial cell process was proposed to investigate its removal. This treatment process attempts to provide an effective degradation of phenol and its related compound (such as carbaryl) from water and wastewater systems.

45th Purdue Industrial Waste Conference Proceedings, © 1991 Lewis Publishers, Inc., Chelsea, Michigan 48118. Printed in U.S.A.

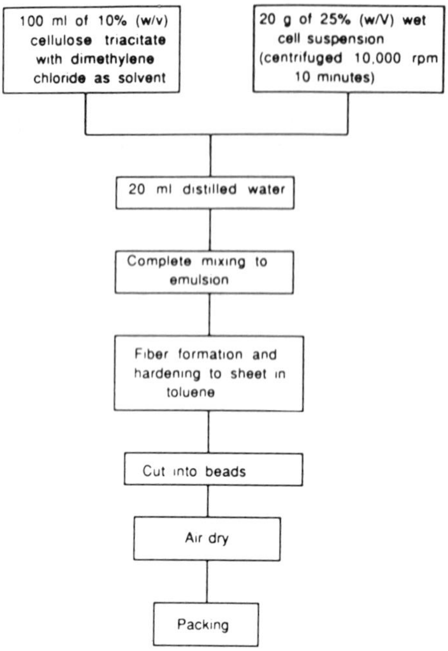

Figure 1. Procedure for preparation of entrapped microbial cell.

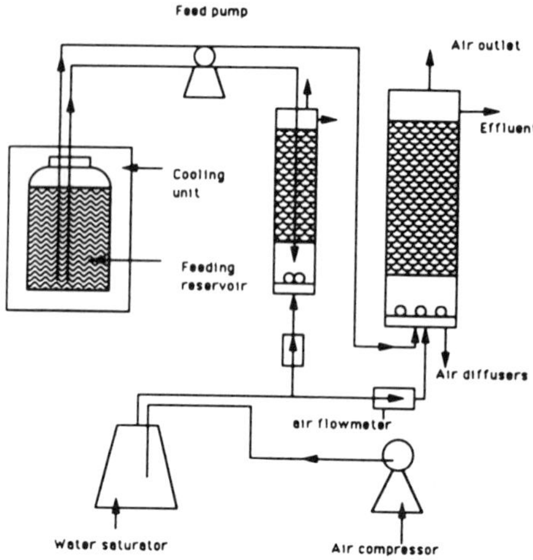

Figure 2. Experimental set-up for entrapped-mixed microbial cell process.

MATERIALS AND METHODS

Microbial Acclimation

Mixed populations of microorganisms which are capable of decomposing phenol and carbaryl (1-naphytl-N-methyl-carbamate) were isolated from soil by enrichment culture technique. The phenol-acclimated microbial cells were used for both the treatment of phenol and carbaryl. The carbaryl-acclimated microbial cells were used for the treatment of carbaryl only.

Immobilization of Microbial Cells

The method of immobilization of microbial cells was a modified technique of Kolaric et al.[18] As shown in Figure 1, wet cell suspension was added to cellulose triacetate dissolved in methylene chloride. After mixing, it was poured in a container with toluene. As soon as it hardened, the fibers were cut into desired sizes and shapes. For adsorbed cell process (without entrapment), the acclimatized cells were allowed to grow on the surface of the carriers of cellulose triacetate.

Experimental Equipment

Two types of carries were used in this study; one is for the phenol acclimated cell column, the other is for the carbaryl acclimated cell column. For the phenol acclimated unit, the carriers were placed into two columns with a total effective volume of 970 mL and a void volume of 300 mL (packing ratio of 69%). For the carbaryl acclimated unit, the carriers were placed into a column with a total effective volume of 5.0 liters and a void volume of 4.0 liter and 3.0 liter (packing ratio of 20% and 40%, respectively). The void volume was measured as the actual volume passed through the column packed with the carrier. As shown in Figure 2, the columns were equipped with an air diffuser, an air compressor, feed tube, pump and reservoir with a cooling unit. The substrate (as shown in Table I) was pumped in directly from a refrigeration tank (at 4°C) to the column with a peristaltic pump. Mixing and aeration were accomplished by pumping compressed air through the columns at a rate of 6ℓ/ℓ/min. The temperature at the reactors was maintained at 25°C.

Table I. Synthetic Organic Wastewater

Phenol	250	mg/L
or		
Carbaryl (Sevin)	100	mg/L
$(NH_4)_2 \cdot SO_4$	250	mg/L
$MgSO_4 \cdot 7H_2O$	50	mg/L
$FeCl_3 \cdot 6H_2O$	0.25	mg/L
$CaCl_2$	3.75	mg/L
KH_2PO_4	263.5	mg/L
K_2HPO_4	535	mg/L
$MnSO_4 \cdot H_2O$	5	mg/L
Tap Water	50	mg/L

Experimental Procedure

For adsorbed and entrapped microbial cell treatment of phenol, the COD loading rate was maintained less than 2.0 g/L/day. For phenol acclimated cell column and carbaryl acclimated cell column of treating carbaryl, the COD loading rates were stepping increased from 0.2 to 0.9 g/L/day and from 0.3 to 1.2 g/L/day, respectively. The volume of calculating loading rate was based on void volume. The process performance was evaluated by measuring effluent concentration COD, suspended solid, pH and NH_4-N.

Analytical Methods

Concentrations of COD and suspended solids were analyzed according to procedures in *Standard Methods.*[19] Ammonium-nitrogen was determined by the procedure given by Ecker and Lockard.[20] pH was measured with an Orion 501 pH analyzer.

RESULT AND DISCUSSION

Performance of Treating Phenol With and Without Entrapped Cell

Based on our previous study,[14] it was concluded that the application of entrapped microbial cell for low organic loading rate (<2.0 gCOD/L/day) might not be necessary when it was compared to those without entrapped microbial cell process. However, study of a long term of process operational behavior was not investigated at the COD loading rate of less than 2.0 g/L/day.[14] As shown in Figures 3 and 4, the effluent TCOD (Total COD) and SCOD (soluble COD) were monitored for 55 days of operation. It can be seen that the start-up period of 10 days is required for both processes. Also, both effluent TCOD and SCOD concentrations in the process with microbial cell entrapment are lower and stable than those in the process without microbial cell entrapment. Regarding TCOD removal efficiency as shown in Figure 5, the entrapment process maintains a higher range of 92–95% TCOD removal efficiency after starting up period than the process without cell entrapment (85–95%). The comparison is shown in Table II. Therefore, it can be concluded that the microbial cell entrapment is

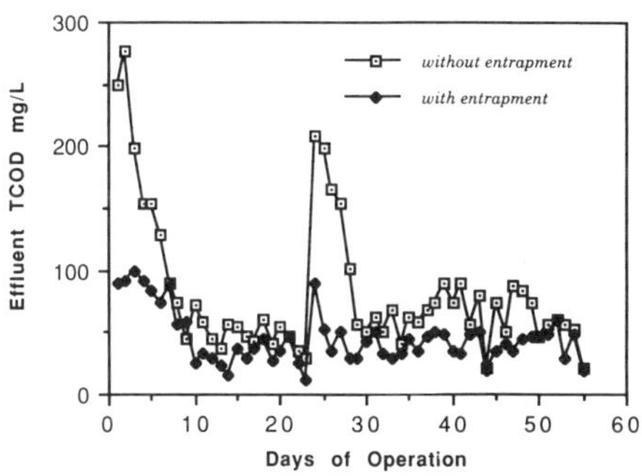

Figure 3. Operational performance of with & without cell entrapment (Effluent TCOD).

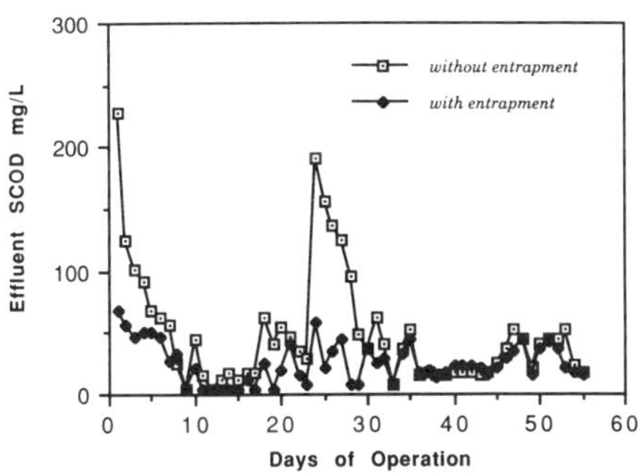

Figure 4. Operational performance of with & without cell entrapment (Effluent SCOD).

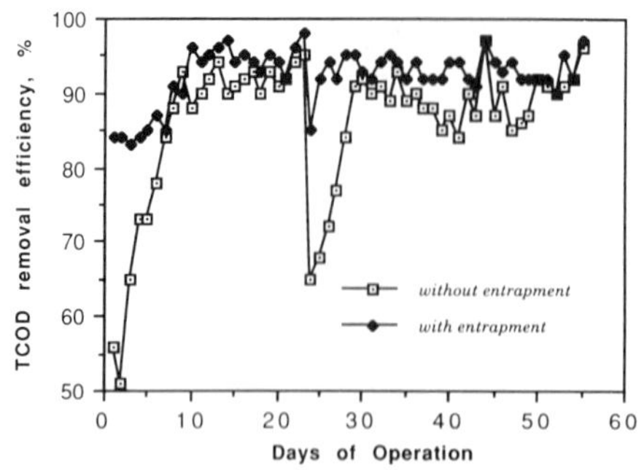

Figure 5. TCOD removal efficiency for with & without cell entrapment processes.

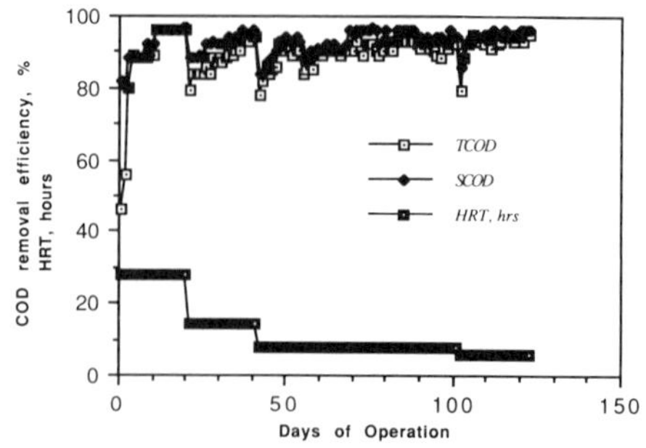

Figure 6. COD removal efficiency at 28, 14, 8 & 6 HRTs.

still required in order to achieve a stable and high TCOD removal efficiency. It also can offer a better way in preparation of a package plant operation when the entrapment of microbial cell is applied.

Process Performance of Treating Carbaryl

Using Phenol Acclimated Microbial Cell

Because of the similar structure of benzene for both phenol and carbaryl, the acclimated phenol column was used to treat carbaryl. After the termination of using phenol acclimated microbial cells in treating phenol, the same column was used for treating carbaryl. The experiment was started with an operation HRT (Hydraulic Retention Time) of 28 hours, then gradually decreased to 14 and 8 hours. For a loading rate of 0.2–0.3 gCOD/day (nominal HRT = 28 hrs), the system was able to attain TCOD and SCOD removal efficiency of 96% after a a start up period of 10 days (Figures 6 and 7). When the loading rate was increased to 0.5 gCOD/L/day and at 0.9 gCOD/L/day (nominal HRT = 8 hrs), removal efficiency dropped to 80% but gradually increased to greater than 90% after 6 days. The system has a long term operational stability. For a period of about 125 days, the unit is still very stable in terms of COD removal efficiency. Also, the suspended solid concentrations in the effluent, as shown in Figure 8, were able to be maintained under 10 mg/L after 10 days of starting up period. The comparison is summarized in Table III. The high process performance in terms of high COD

Table II. Comparison Between W/ and W/O Entrapment Processes

	% COD Removal Efficiency		Effluent SCOD
	TCOD	SCOD	(Mg/L)
With entrapment	92	96	25
Without entrapment	86	91	49

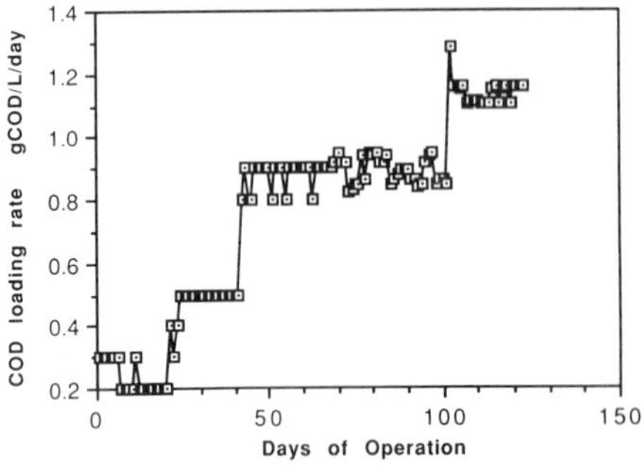

Figure 7. Operational condition for COD loading rate for treatment of carbaryl.

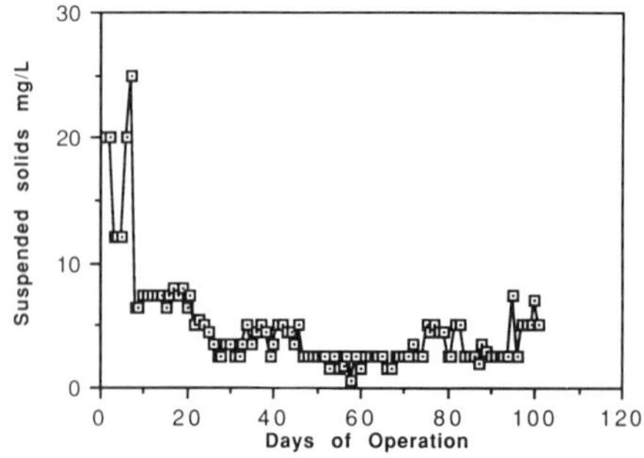

Figure 8. Operational performance of treating carbaryl (suspended solids).

Table III. Aerobic Biodegradation of Carbaryl (1-Naphtyl N-Methyl Carbamate) Using Phenol-Acclimatized Immobilized Cells

HRT	28	14	8
%TCOD removal efficiency	96	90–94	89–93
%SCOD removal efficiency	96–97	95–96	88–96
Total SS (mg/L)	<10	<10	<10
SRT (days)	10–49	10–49	12–33

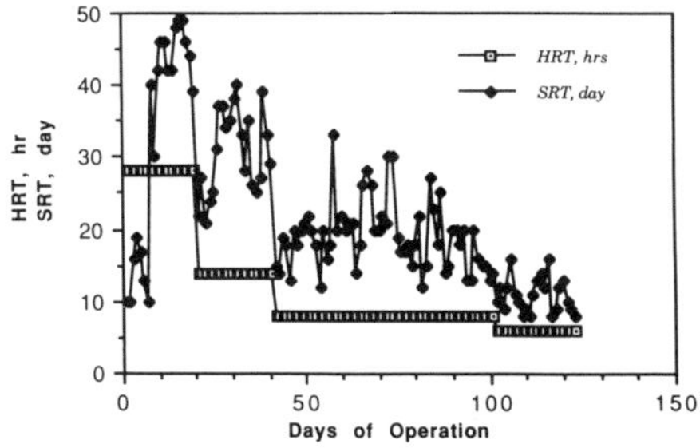

Figure 9. SRT & HRT performance vs. days of operation.

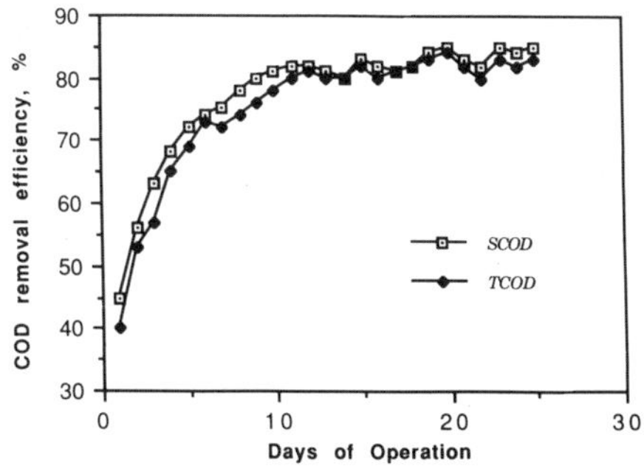

Figure 10. COD removal efficiency at a loading rate of 0.3 gCOD/L/day (packing ratio of 20%).

removal efficiency and low concentration of suspended solid in the effluent are due to the high SRT's (Solid Retention Time) of 10–50 days which could be maintained (Figure 9). The estimation of SRT was reported previously.[11]

Using Carbaryl Acclimated Microbial Cell

After the beginning of this experiment, the carrier packing ratio was 20%. After 25 days of operation, the packing ratio was increased to 40%. During the operation of a packing ratio of 20%, a 0.3 gCOD/L/day loading rate (HRT = 24 hours) was maintained and removal efficiencies of TCOD and SCOD were about 85% after a start up period of 12 days (Figure 10). Because of the lower COD removal efficiency compared to the phenol acclimated column (as shown in Figure 6), the packing ratio was increased to 40%. The process performances in terms of solid retention time, COD removal efficiency and concentrations of effluent COD and suspended solid are shown in Figures 11, 12, 13, 14 and Table IV. Under the COD loading rate of 1.32 g/L/day (HRT of 6 hours), the TCOD and SCOD removal efficiencies can be achieved in the range of 90–96% and 90–98%, respectively. The concentration of effluent SCOD and suspended solids can be maintained under 20 mg/L and 15 mg/L, respectively. The required period of attaining the steady state is about 4–7 days. The SRT's is generally higher in the packing ratio of 40% than 20%.

For practical application, the treatment unit may not be continuously used. Therefore, the entrapped mixed microbial cell column was purposely stored in a refrigerator (5°C) without aeration

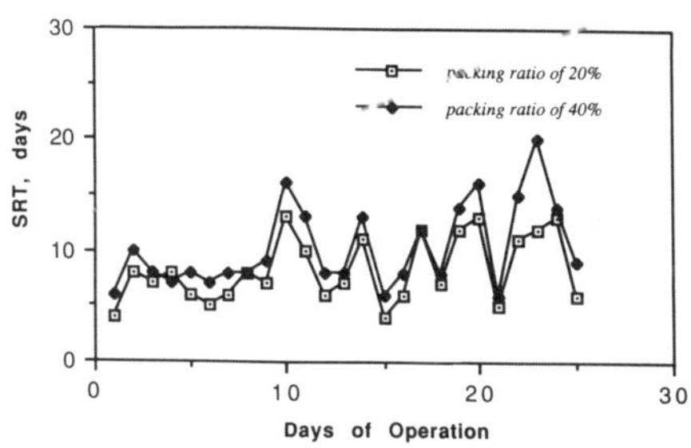

Figure 11. Solid retention time vs. days of operation.

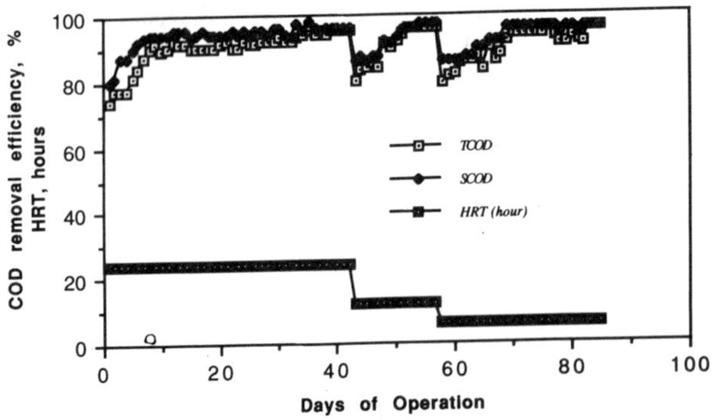

Figure 12. COD removal efficiency at 24, 12, & 6 HRTs (carbaryl at 5.0 li bioreactor & 40% packing).

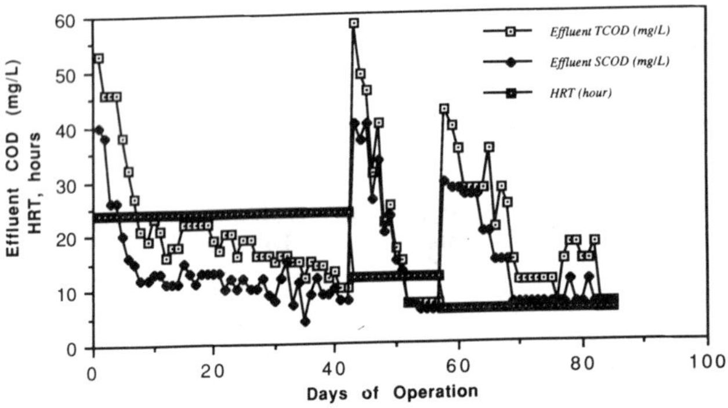

Figure 13. Effluent COD at 24, 12 & 6 HRTs (carbaryl at 5.0 li bioreactor & 40% packing).

for 45 days and was reused for its capability in removal of carbaryl. As shown in Figure 15, the COD removal efficiency can be maintained in the range of 90–96% at the HRT of 6 hours after 5 days of recovering period. Compared to the original treatment efficiency (i.e., prior to the storage), the stored immobilized cell unit still provides an active treatment efficiency in removing the carbaryl. This is very important in the application of small quantity of wastewater production because of its weakness of continuing operation.

The advantage of using biological treatment of toxic organics is the ultimate oxidation or destruction of toxic organic into innocuous substances, such as carbon dioxide and water. Thus, a treated effluent from the operation of HRT of 12 hours was compared with an untreated influent subjected to the high performance liquid chromatography. As shown in Figure 16, both the carbaryl, and α-

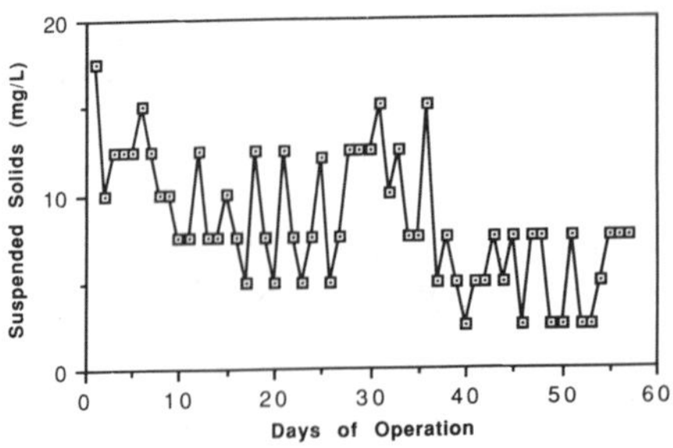

Figure 14. Operational performance of 40% packing (Effluent suspended solid).

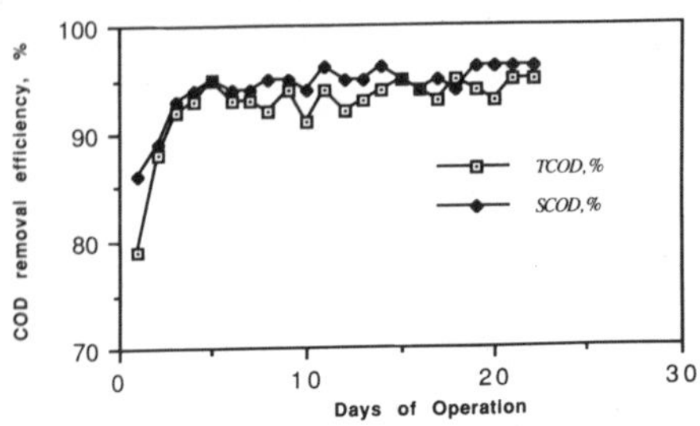

Figure 15. COD removal efficiency of entrapped-phenol-acclimatized cells after 45 days of refrigeration (influent: carbaryl; HRT: 6 hrs).

Table IV. Aerobic Biodegradation of Carbaryl Using Carbaryl-Acclimatized Immobilized Cells

HRT, hrs	24	12	6
% mean TCOD	90	90	90
% mean SCOD	94	92	93
% TCOD range at steady state	90–95	90–96	90–96
% SCOD range at steady state	90–98	91–97	90–96
Number of days to attain steady state at TCOD removal efficiency >90% (initial time: start-up)	7	5	9
Number of days to attain steady state at SCOD removal efficiency >90% (initial time: start-up)	4	5	6
COD loading rate (gCOD/L/day)	0.36	0.65	1.32

naphthol contained in the effluent were completely oxidized. This demonstrates that the entrapped mixed microbial cell process is very effective in destruction of the toxic organics into innocuous substances.

Operational Criteria and Application

Based on the previous work on the treatment of phenol using the entrapped mixed microbial cell process[11,12,13,14] and the present work for the treatment of phenol and carbaryl, the operational criteria and application of using the entrapped mixed microbial cell process are developed.

For different types of wastewater, acclimation prior to the immobilization of mixed microbial cells is required. Although phenol acclimated cells could degrade certain substituted phenols, specific acclimation is still recommended, especially for wastewater containing more than one carbon source. Mono-carrier (cellulose triacetate) is more preferable than bi-carrier (cellulose triacetate and calcium alginate), since it is easier to prepare and the process performance is better because of its stronger mechanical strength when the aeration is applied.[11]

For high phenol content,[13] wastewater with a COD concentration of 500–600 mg/L, loading rate should be operated under 9.0 gCOD/L/day in order to maintain the COD removal efficiency higher than 90%, and to maintain high effluent quality in terms of low COD and SS concentration (< 30

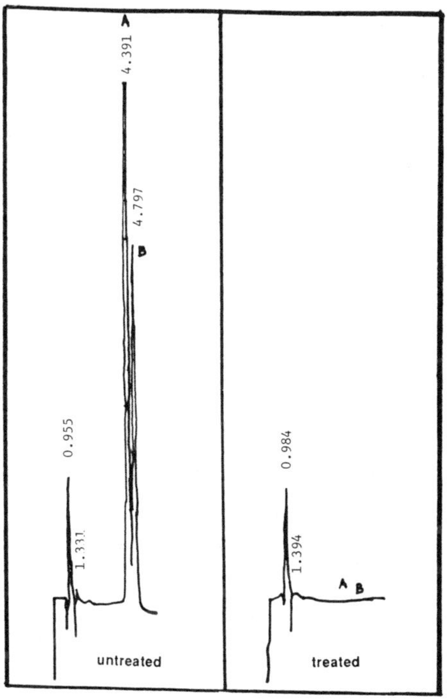

Figure 16. High performance liquid chromatography of untreated and treated samples (A = carbaryl; B = α-naphthol).

mg/L). For operating a loading rate under 2.0 gCOD/L/day, as presented in the study, entrapped mixed microbial cell process is still required because high COD removal efficiency and high quality of effluent SS and COD can be maintained.

Carriers should be rinsed with water regularly in order to remove the biomass build up on the surface. The frequency of washing depends on the loading rate applied. At low loading rate (under 2 gCOD/L/day), carriers can be operated without washing for more than a month due to the low microbial growth rate. At loading rate of 8.5g/COD/L/day, weekly washing is recommended. For a higher strength of wastewater, a series of columns might be required to maintain the high effluent quality.

Equalization and secondary settling tanks are not required as the loading rate is operated under 9.0 gCOD/L/day. Primary settling facility is required since high influent solids might result in clogging problems.

A selected pesticide, carbaryl, can be treated by entrapped-acclimated-microbial cell process under the loading rate of 1.32 gCOD/L/day (based on the present result). Rinse of the column may not be required under the loading rate of 1.32 gCOD/L/day. A small and mobile treatment unit can be designed as a package plant for the biological treatment of pesticide water/waste by using the acclimated-entrapped-mixed-microbial cell process. Pesticide water/waste usually generates in a small amount which is often difficult to design and operate because of the great variation in flow and pollution load and lack of continuation of treatment operation. The application of entrapped microbial cell process for treatment of this type of wastewater with toxic characteristics and small quantity of wastewater production holds great potential because of its powerful and reliable capability compared to the conventional-biological treatment process.

CONCLUSION

Entrapped mixed microbial cell process for treatment of phenol and phenol related compounds (carbaryl) is feasible. Entrapment of mixed microbial cell process provides better and stable process performance compared to the process without entrapment of microbial cells. Under COD the loading rate of 1.2 g/L/day, carbaryl can be treated by the entrapped-acclimated-mixed-microbial cell process by achieving a COD removal efficiency of more than 90% and SS concentration of lower than 15 mg/L. The process can effectively maintain the SRT's in the range of 30–50 days; even the HRT's are operated in the range of 6–24 hours. The application of entrapped microbial cell process for the treatment of toxic organics, such as phenol and its related compounds, can be developed into a package treatment plant which must be efficient and practical in application.

ACKNOWLEDGMENT

The funding for this research was provided by the USDA Special Grant: Tropical/Subtropical Agricultural Research (89-34135-4612) and the Hawaii Institute of Tropical Agriculture and Human Resources of the University of Hawaii. Also, the assistance of Dr. Carl J. Miles, Agricultural Biochemistry Department, University of Hawaii, in the analysis of pesticide of using HPLC is gratefully acknowledged.

REFERENCES

1. Grady, C. P. L., Jr., "Biodegradation of Hazardous Wastes by Conventional Biological Treatment". *Hazardous Wastes and Hazardous Materials*, 3, 333–365 (1986).
2. Hannah, S. A., et al., "Comparative Removal of Toxic Pollutants by Six Wastewater Treatment Processes". *Journal of Water Pollution Control Federation*, 58, 27–34 (1986).
3. Rozich, A. F. and Gaudy, A. F., Jr., "Response of Phenol-Acclimated Activated Sludge Process to Quantitative Shock Loadings". *Journal of Water Pollution Control Federation*, 57, 795–804 (1985).
4. Young, L. Y. and Rivera, M. D., "Methanogenic Degradation of Four Phenolic Compounds". *Water Research*, 19, 1325–1332 (1985).
5. Cheethan, P. S. J. and Bucke, C., "Immobilization of Microbial Cells and Their Use in Wastewater Treatment". *Microbial Methods for Environmental Biotechnology*, Grainer, J. M. and Lynch, J. M., eds., Academic Press, Inc., 219–234 (1984).
6. Bettman, H. and Rehm, H. J., "Degradation of Phenol by Polymar Entrapped Microorganisms". *Applied Microbiology and Biotechnology*, 20, 285–290 (1983).
7. Bettman, H. and Rehm, H. J., "Continuous Degradation of Phenol(s) by Pseudomonas putida P8 Entrapped in Polyacryamidehydrazide". *Applied Microbiology and Biotechnology*, 22, 389–393 (1985).

8. Westmeier, F. and Rehm, H. J., "Biodegradation of 4-Chlorophenol by Entrapped Alcaligene sp. A 7-2". *Applied Microbiology and Biotechnology*, 22, 301–305 (1985).
9. Takahashi, S. et al., "Treatment of Phenolic Wastewater by Aureobasidium pollulans Adhered to the Fibrous Supports". *Applied Microbiology and Biotechnology*, 13, 175–178 (1981).
10. Enrhardt, H. M. and Rehn, H. J., "Phenol Degradation by Microorganisms Adsorbed on Activated Carbon". *Applied Microbiology and Biotechnology*, 21, 32–36 (1985).
11. Yang, P. Y., et al., "Immobilized Mixed Microbial Cells for Wastewater Treatment". *Biological Wastes*, 23, 295–312 (1985).
12. Yang, P. Y. and Wang, M. L., "Entrapped Mixed Microbial Cell Process for Treatment of Phenol and its Related Compounds". Proceedings of Water Pollution Control in Asia, Bangkok, Thailand, 457–463 (1988).
13. Yang, P. Y. and Wang, M. L., "Immobilized Mixed Microbial Cell Process for Toxic-Organic Water and Wastewater Treatment." Proceedings, International Conference on Physiochemical and Biological Detoxification of Hazardous Wastes, Atlantic City, New Jersey, 531–551 (1988).
14. Yang, P. Y. and Wang, M. L., "Packed-Entrapped-Mixed Microbial Cells for Small Wastewater Treatment". Proceedings, International Specialized Conference on Design and Operation of Small Wastewater Treatment Plants, Trondheim, Norway, 349–356 (1989).
15. Kiene, R. P. and Capone, D. G., "Stimulation of Methanogenis by Aldicarb and Several Other N-methy Carbamate Pesticides". *Applied and Environmental Microbiology*, 51, 1247–1251 (1986).
16. U.S. Environmental Protection Agency, "Health Advisories for 50 Pesticides". NTIS Report, PB-113543 (1987).
17. Bollag, J. M. and Liu, S. Y., "Degradation of Sevin by Soil Microorganisms". *Soil Biology and Biochemistry*, 3, 337–345 (1971).
18. Kolarik, M. J., et al., "Glucose Isomerase Cells Entrapped in Cellulose Acetates". *In Immobilized Enzymes in Food and Microbial Process*, eds, Colson and Cooney, New York (1974).
19. *Standard Methods for the Examination of Water and Wastewater*, APHA, AWWA and WPCF, 16th Edition (1985).
20. Ecker, R. E. and Lockard, W. R., "Specific Effect on Limiting Nutrient of Physiological Events During Culture Growth". *Journal of Bacteriology*, 82, 511–516. (1961).

62 FULL-SCALE ELECTRON BEAM TREATMENT OF HAZARDOUS WASTES–EFFECTIVENESS AND COSTS

Charles N. Kurucz, Associate Professor
Department of Management Science

Thomas D. Waite, Professor
Department of Civil Engineering
University of Miami
Coral Gables, Florida 33124

William J. Cooper, Director
Michael G. Nickelsen, Research Associate
Drinking Water Research Center
Florida International University
Miami, Florida 33199

INTRODUCTION

Extensive research regarding the effectiveness of high energy electrons in treating aqueous based waste streams has been conducted over the last year and one half. The full scale research facility at the Virginia Key Wastewater Treatment Plant in Miami, Florida includes a 1.5 MeV, 50 ma electron accelerator and is plumbed to handle waste flow at 120 gpm. Source streams connected directly to the plant are potable water, chlorinated oxygenated secondary effluent, and sludge containing 2 to 8% solids. Six thousand gallon tank trucks can also be connected to the plant influent lines for large batch experiments.

Bacterial disinfection experiments have been conducted as well as experiments on halogenated and aromatic organic compounds such as 1,1,1 trichloroethane, trichloroethylene(TCE), chloroform, benzene and toluene.[1,2,3] Recent efforts have concentrated on these and other hazardous organic compounds regulated by the USEPA and identified as frequently occuring wastewater and groundwater contaminants.

This paper will present a description of electron beam technology, summarize the removal effectiveness of the process based on experiments conducted to date, and provide first approximation cost estimates of full scale treatment.

ELECTRON BEAM TECHNOLOGY

Electron beams have been in commercial use since the 1950s. Early applications involved the crosslinking of polyethylene film and wire insulation. The number of applications has since grown to include sterilization of medical supplies, rubber vulcanization, disinfection of wastewater, food preservation, curing of coatings, etc. Today there are several hundred electron processing systems installed for industrial applications in over 25 countries.

Electron beam processing involves exposing the material to be irradiated to a stream of high energy (fast) electrons. These electrons interact with the material in less than 10^{-12} seconds to produce electrons of lower and lower energy. Eventually a large number of slow electrons with energies less than 50 eV is produced and these electrons interact with molecules to produce excited states of these molecules, positive ions and electrons. Eventually the electrons slow to thermal energies and get trapped. In materials of low dielectric constant most electrons do not escape the pull of the positive ions formed when they were produced. The electrons are attracted back to the positive ions causing a chemical reaction. This is termed direct radiolysis. In high dielectric materials such as water and aqueous solutions, most electrons escape the pull thus leaving both the positive ions and electrons free to react with the water or waste components in it. This is referred to as indirect radiolysis. Since the ratio of direct to indirect radiolysis of a waste is approximately the weight fractions of waste to water[4] the radiolysis of water is the primary mechanism of destruction using high energy electrons. The

45th Purdue Industrial Waste Conference Proceedings, © 1991 Lewis Publishers, Inc., Chelsea, Michigan 48118. Printed in U.S.A.

efficiency of ionizing radiation initiating chemical reactions is measured by G values, where G is defined as the number of radicals, excited states or other products, formed (or lost) in a system absorbing 100 eV of energy.

Irradiation of pure water has been studied extensively with numerous excellent reviews on the subject.[5-7] It is thought that the initial radiation process (10^{-16}–10^{-14} sec) results in the formation of excited molecules H_2O, H_2O^+, e^- and, after 10^{-7} sec the products are:

	products	G value	
H_2O -/\/\/\–>	e^-_{aq}	2.6	(1)
	H·	0.55	
	OH·	2.7	
	H_2	0.45	
	H_2O_2	0.71	
	H_3O^+	2.7	

The three radicals formed are the most reactive species. The e^-_{aq} and H· are reducing radicals, and OH· is an oxidizing radical.[5-7] These reactive species then interact with any organic molecules present in the water. Since the aqueous electron and the hydroxyl radical OH account for over 90% of these reactive species, their chemistry is of primary interest for the electron beam process. Reaction rate constants for selected organic compounds have been tabulated[8] and it can be determined which species is most reactive with a particular organic compound.

High energy electrons for industrial applications are generally produced by electron accelerators or generated in situ by gamma ray sources such as $Cobalt^{60}$. In an electron accelerator a current is used to produce a stream of electrons which are accelerated by applying an electric field generated at a given voltage. The applied voltage determines the speed and thus the energy of the accelerated electrons. For example, an applied voltage of 2.0 MV will accelerate electrons to an energy level of 2.0 MeV. The number of electrons generated per unit time is proportional to the beam current and thus the beam power given by the product of beam current and the accelerating voltage. For example, a 2.0 MeV accelerator operating at 50 ma will produce 100 KW of beam power. Commercial machines up to 3.0 MeV and 40 ma are readily available and machines of much higher voltage or current can be obtained.

The amount of energy from the high energy electron beam that is absorbed by an irradiated material per unit mass is called dose. A common unit of dose is the rad, defined as the energy absorption of 100 ergs per gram of material. The amount of energy that is transfered to the irradiated material is not constant as the electrons penetrate deeper into the material. This phenomenon is illustrated in Figure 1 where it can be seen that energy transfer near the surface is less than the maximum transfer, which occurs at approximately one half of the maximum depth of penetration. The maximum depth of penetration is directly proportional to the energy of the incident electrons and inversely proportional to the density of the material being irradiated. Thus absorbed dose depends on the type and thickness of the material, accelerating voltage, beam power, and the length of time the material is exposed to the electron beam.

THE ELECTRON BEAM RESEARCH FACILITY

Wastewater streams at the Miami Electron Beam Research Facility are presented to a scanned 1.5 MeV, 50ma beam in a falling stream approximately 48″ wide and .15″ thick. A cross section of this stream produced by a 120 gpm flow over a 48″ wide weir is shown in Figure 2. Since the maximum penetration in water is approximately .29″ for 1.5 MeV electrons, some electrons pass through the stream and thus not all of the beam energy is transfered to the water. With the addition of overscanning the waste stream to insure that the edges of the stream are irradiated, more energy is lost with the result that the efficiency of energy transfer is approximately 65%. Thus when the electron beam is operating at 50 ma (75 KW) and 120 gpm the waste stream is receiving an average dose of approximately 650 kilorads. Total power consumption, including pumps, chillers and other auxilliary equipment is about 120 KW. Varying the beam current and holding the flow constant changes the absorbed dose in a linear fashion, allowing for experimentation at doses lower than 650 kilorads.

Figure 3 shows a schematic overview of the facility. The facility was completed in 1983 at a cost of $1,729,000. At present, several influent streams are connected to the plant. They are potable water, chlorinated secondary effluent and anaerobically digested sewage sludge. The secondary wastewater treatment is extended (pure oxygen) aeration and the effluent is chlorinated prior to the electron beam. The sludge contains 2–3% solids. In addition, we have the capacity to connect 6000 gallon tank trucks to the influent lines. Thus large scale batch experiments can be conducted on groundwater or any other source of contaminated water for which treatability studies are desired.

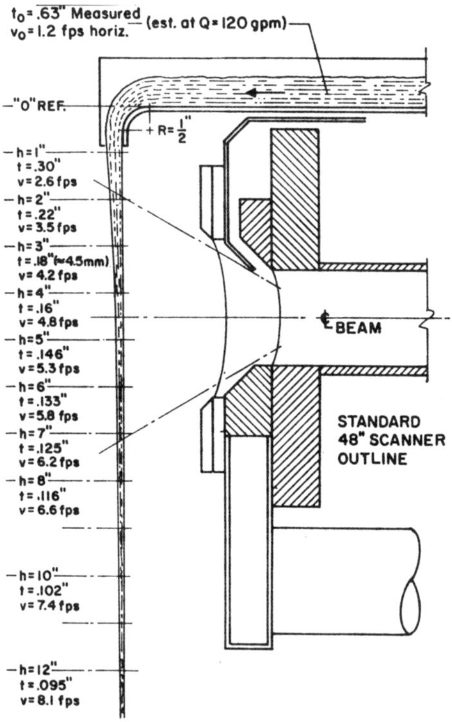

Figure 1. Cross section of the waste stream presented to the electron beam with a Weir delivery system.

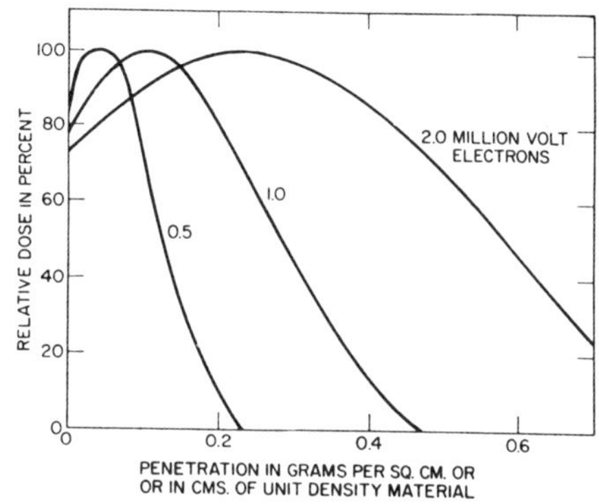

Figure 2. Distribution of relative dose at varying depth in water for electrons of several energies.

The electron beam has been instrumented to continuously monitor and record various operating parameters (voltage, current, flow) and absorbed dose. The absorbed dose is measured using resistance temperature devices (RTDs) mounted in the waste stream immediately before and after the beam. They are connected via an analog interface to a portable computer which continuously reads and records temperatures. The temperature difference between influent and effluent streams is converted to absorbed dose by the relationship that a change of 1°C in water is equivalent to a dose of approximately 418.6 Krads.

TREATMENT EFFECTIVENESS

Recent experiments have focused on establishing the relationship between electron beam dose and the destruction of various hazardous organic compounds. In a typical experiment chemicals were injected into the influent stream at controlled rates and samples taken before and after the waste stream passed through the beam. The absorbed dose was varied by changing the beam current and the order of the current settings was randomized to average out any systematic variation which might have been present from variables not being specifically controlled. The analyses were performed at the Drinking Water Research Center located at Florida International University. The toxic organics were determined using gas chromatography(Hewlett-Packard Model 5890 GAs Chromatograph equipped with a DB-5 Bonded Phase Megabore with a Superscript™ column and a Model PI-52-02A Photoionization Detector).

The results of experiments with organic compounds that are frequently found at Superfund sites are summarized in Table I. Although experiments have been run with varying water qualities (potable water, chlorinated secondary, and raw watewater) and initial concentrations,[3] the data presented in this paper were primarily obtained using chlorinated secondary wastewater and initial concentrations of approximately 100μg L^{-1}. As can be seen, doses of 50 to 800 kilorads effectively remove most of the frequently occuring organics. A typical removal curve is presented in Figure 4.

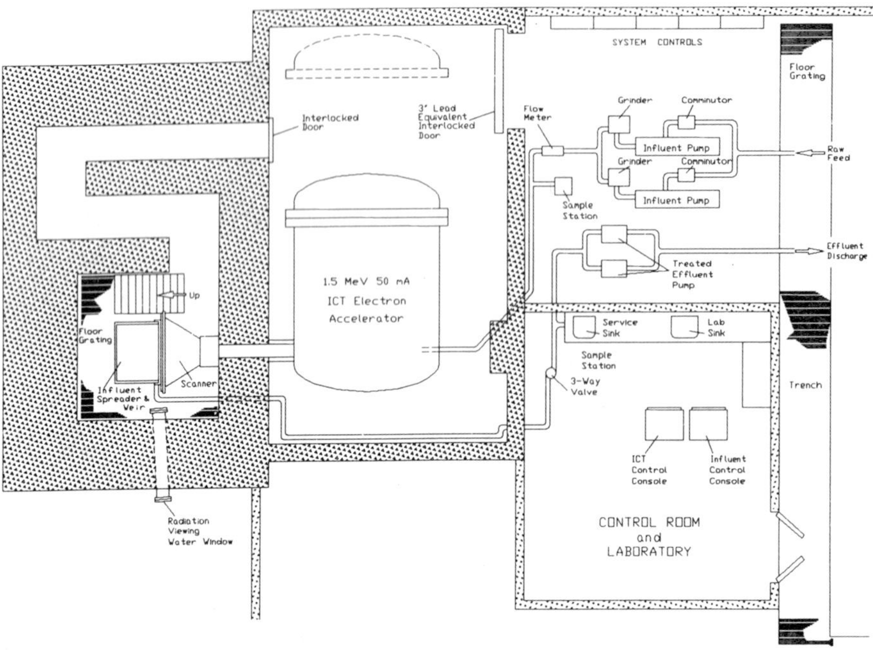

Figure 3. Plan view of the Miami electron beam research facility.

A summary of the removal percentages for all organic compounds that have been investigated to date is presented in Table II. Note that in many cases when the removal percentage is given as >99, the actual removal was to below detection limits (i.e. >99.99%). Although recent experiments with concentrations up to 1,500μg L^{-1} indicate concentration effects for some compounds, it is still clear that the process is very effective over a wide range of conditions. Additional experiments are being planned to further define concentration and water quality effects, and to identify reaction by- products. To date we have not identified any reaction by-products other than sub-μM concentrations of aldehydes when destruction is near complete. When complete destruction does occur the compounds are reduced to CO_2, H_2O and salts.

ESTIMATED TREATMENT COSTS

The cost of treatment using the electron beam technology depends on many factors such as the dose required to obtain the desired detoxification, the volume of waste to be treated, the size of the treatment facility, the time utilization of the facility, and the manner in which capital recovery is handled. For example, the estimated current cost of the permanent 1.5 MeV facility at Miami is given in Table III. The estimated capital requirements represent an approximate 5% annual inflation of the total price actually paid for the Miami facility. Changes in the amortization period or interest rate can significantly affect annual capital charges. Also note that no indirect costs such as supervision, overhead, etc. are included.

Estimated total hourly costs for the current system and a modified system operating at various flow rates for 6,000 hours per year are given in Table IV. The current system can be operated over a range of flow rates from 60 to 160 gpm. Since the maximum absorbed dose delivered by the electron beam depends on flow rate, an application requiring a dose of 1290 Krads would limit flow to 60 gpm. A dose requirement of approximately 500 Krads such as that required for TCE would allow treatment at approximately 160 gpm. Modifications to the current waste stream delivery system would allow operation at much higher flow rates. Also changes to the beam operating system would allow treatment to take place without the presence of a full time operator. These changes yield the cost and flow

Table I. Removal Percentages for the 25 Most Frequenty Identified Substances at 546 Superfund Sites

Rank	Substance	Percent of Sites[a]	Percent Removal	Required Dose (Krads)
1	Trichloroethylene (TCE)	33%	>99%	500
2	Lead and compounds	30	–	–
3	Toluene	28	97	650
4	Benzene	26	>99	650
5	PCBs	22	–	–
6	Chloroform	20	83	650
7	Tetrachloroethylene (PCE)	16	>99	500
8	Phenol	15	88	800
9	Arsenic and compounds	15	–	–
10	Cadmium and compounds	15	–	–
11	Chromium and compounds	15	–	–
12	1,1,1-Trichloroethane	14	89	650
13	Zinc and compounds	14	–	–
14	Ethylbenzene	13	92	650
15	Xylene	13	91	650
16	Methylene chloride	12	77	800
17	trans-1,2-Dichloroethylene	11	93	800
18	Mercury	10	–	–
19	Copper and compounds	9	–	–
20	Cyanides (soluble salts)	8	–	–
21	Vinyl chloride	8	–	–
22	1,2-Dichloroethane	8	60	800
23	Chlorobenzene	8	97	650
24	1,1-Dichloroethane	8	–	–
25	Carbon tetrachloride	8	>99	50

[a]Adapted from McCoy & Assoc., *Haz. Waste. Consult.* 3:2(1985).

estimates given for the modified system. Thus a modified system would allow carbon tetrachloride to be treated with 50 Krads at a flow of 2,100 gpm.

The examples above indicate the range of treatment possibilities. The 160 gpm flow yields operating costs of approximately \$2.5/1000g while the 2,100 gpm flow gives approximately \$.25/1000g. In either case, the costs compare favorably to a reported operating cost of \$2.59/1000g for a UV peroxidation process[9] and total costs of approximately \$200/1000g for a mobile wet oxidation process.[10] In the latter case the evaluation was done for very high concentrations (.1 to 4.0 wt%). For these same concentrations carbon/incineration was estimated to cost \$76/1000g to \$517/1000g. Research is currently underway to provide the basis for cost estimates for transportable electron beam systems and systems of different accelerating voltages and power. The costs of operating transportable units will be significantly higher than permanent facilities due to transportation costs, increased maintenance requirements, and shorter useful lives.

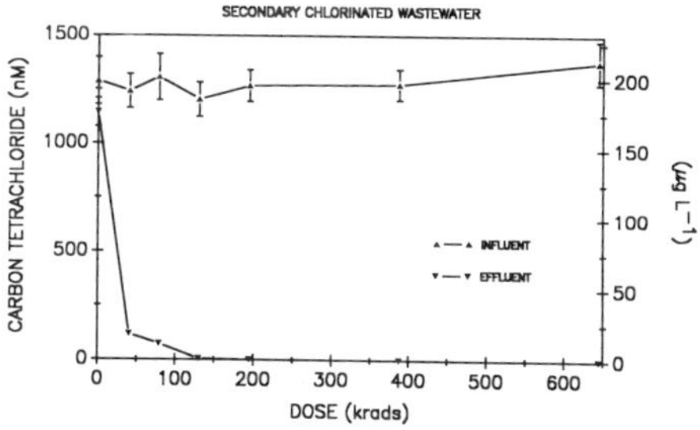

Figure 4. Removal of carbon tetrachloride at several irradiation doses in secondary chlorinated wastewater (error bars indicate one standard deviation from the mean; where no error bars are seen the error is within the data point).

Table II. Summary of Percent Removal of Various Organic Compounds by Treatment Application Area

	Percent Removal	Required Dose (Krads)
Drinking Water		
Chloroform	83%	650
Bromodichloromethane	>99	80
Dibromochloromethane	>99	80
Bromoform	>99	80
Wastewater/Groundwater Treatment		
Carbon tetrachloride	>99	50
Trichloroethylene (TCE)	>99	500
Tetrachlorethylene (PCE)	>99	500
trans-1,2-dichloroethene	93	800
cis-1,2-dichloroethene	98	800
1,1-dichloroethene	>99	800
1,2-dichloroethane	60	800
hexachloroethane	>99	800
1,1,1-trichloroethane	89	650
1,1,2,2-tetrachloroethane	88	650
hexachloro-1,3-butadiene	98	800
Methylene chloride	77	800
Groundwater Treatment		
Benzene	>99	650
Toluene	97	650
Chlorobenzene	97	650
Ethylbenzene	92	650
1,2-Dichlorobenzene	88	650
1,3-Dichlorobenzene	86	650
1,4-Dichlorobenzene	84	650
m-Xylene	91	650
o-Xylene	92	650
Dieldrin	>99	800
Total phenol	88	800

Table III. Estimated Current Costs for the Miami Permanent Electron Beam Facility Costs

System: 1.5 MeV, 50 ma Electron Beam

Capital Costs($1000s):

Installed Beam	1,850	
Support Facility[a]	500	
	2,350	
Amortization	(10 years @ 15%) -	$466/year
	(20 years @ 15%)	$374/year

Operating Costs (hourly):

Operator	$20.0
Power[b]	10.5
Water[c]	2.5
Maintenance	8.0
	$41.0

[a]Shielding, delivery system, etc.
[b]150 KW @ $.07/KWH.
[c]2K gph @ $1.25/1000g.

Table IV. Total Hourly Cost vs. Flow and System Modifications

Flow (gpm)	Dose (Krad)	Current System	Modified System
60	1290	98	na
120	645	103	86
160	484	106	na
	670	na	89
2100	50	na	96

Amortization: 20 years @ 15%.
na = not applicable.

CONCLUSIONS

High energy electrons produced by electron beam accelerators effectively remove hazardous organic compounds from aqueous solutions. Although the treatment costs for this process are highly dependent on the required dose and thus the allowable flow rate, they are competetive with alternative treatment methods. Additional research is required to determine whether high concentration waste streams are best treated with higher doses or by recirculation with smaller accelerators.

REFERENCES

1. Cooper, W.J., Waite, T.D., Kurucz, C.N., Narbaitz, R.M., and Greenfield, J.H. Electron Beam Irradiation: An Emerging Treatment of Water and Wastewater. Joint AWWA-WPCF Symposium on Emerging Technologies, San Diego, CA, August 1989.
2. Waite, T.D., Kurucz, C.N., Cooper, W.J., Narbaitz, R.M., and Greenfield, J.H. Dissinfection of Wastewater Effluents with Electron Irradiation. Proceedings of National ASCE/NCEE Conference. Houston, Texas, July 1989.
3. Cooper, W.J., Nickelsen, M.G., Meacham, D.E., Waite, T.D., and Kurucz, C.N. Treatment of Industrial Hazardous Wastes with High Energy Electrons. Proceedings of the 7th National RCRA/SUPERFUND Conference and Exhibition, St. Louis, Missouri, May 1990.
4. Singh, A., Sagert, N.H., Borsa, J., Singh, H., and Bennett, G.S. The Use of High-Energy Radiation for the Treatment of Wastewater: A Review. Proceedings of the 8th Symposium on Wastewater Treatment, Montreal, Canada, 1985.
5. Pikaev, A.K. *Pulse Radiolysis of Water and Aqueous Solutions*. Indiana University Press. Bloomington, IN. 295 pp., 1967.
6. Bielski, B.H.J. and J.M. Gebicke. Species in Irradiated Oxygenated Water. *Adv. Radiation Chemistry*, 2:177–279, 1970.
7. Draganic, I.G. and Z.D. Draganic. *The Radiation Chemistry of Water*. Academic Press, N.Y., N.Y. 242 pp. 1971.
8. Buxton, George V., Clive L. Greenstock, W. Phillip Helman and Alberta B. Ross. Critical Review of Rate Constants for Reactions of Hydrated Electrons, Hydrogen Atoms and Hydroxyl Radicals (OH/O-) in Aqueous Solution. *Journal of Physical and Chemical Reference Data*, 17: 513–886, 1988
9. Yost, K.W. UV Peroxidation: An Alternative Treatment Method for Organic Contamination Destruction in Aqueous Waste Streams. Proceedings of the 43rd Purdue Industrial Waste Conference, West Lafayette, IN, 1988.
10. Eckert, C.A., Leman, G.W., and Yang, H.H. Homogeneous Catalysis for Wet Oxidation: Design and Economic Feasibility of a Mobile Detoxification Unit. Hazardous Materials Control, Vol. 3, No. 2, 1990. ©Table 1. Removal Percentages For The 25 Most Frequently Identified Substances At 546 Superfund Sites

Section Seven
NEW EQUIPMENT APPLICATION

63 MEMBRANE AIR STRIPPING: OPERATING PROBLEMS AND COST ANALYSIS

Sylvia E. Schwarz, Graduate Research Assistant
Michael J. Semmens, Professor
Civil and Mineral Engineering Department
University of Minnesota
Minneapolis, Minnesota 55455
Karl J. Froelich, Technical Manager
Hoechst Celanese
Charlotte, North Carolina 28217

INTRODUCTION

Volatile organic compounds (VOCs) traditionally have been removed from water by packed tower air stripping. Substantial literature is available concerning this method.[1,2,3]

In packed tower air stripping, VOC laden water flows down from the top of a tower packed with small ceramic, metal or plastic packing material. Air flows counter-currently from the bottom of the tower. As the water flows around the packing elements, it is broken into small streams and droplets thus creating a large surface area. VOCs transfer easily from the water phase into the air phase and are carried out the top of the tower. The efficiency of this method of treatment depends on the value of the Henry's constant, H, for the compound to be stripped, the air to water ratio, and the specific surface area for mass transfer provided by the packing.

Microporous hollow fiber membranes recently have been investigated as an alternative to packed tower air stripping.[4] These fibers are made of non-wetting microporous polypropylene with pores covering 40 percent of their surface area. Because the pores cover so much of the fiber surface area, and because the liquid phase will not wet the pores, gases may transfer from one side of the membrane through the pores to the other side with negligible membrane resistance. In addition the fibers provide a very high surface area for mass transfer.[5]

In laboratory-scale air stripping studies using microporous hollow fiber membranes it was shown that overall mass transfer coefficients (K_La values) for VOC removal were 3 to 7 times higher than those commonly observed in packed tower air strippers.[5] In addition the mass transfer rate was observed to be liquid film diffusion controlled.

Zander et al.[6] compared the VOC removal performance of a small commercial hollow fiber membrane module with a pilot-scale air stripping tower. The membrane system compared favorably for the six compounds studied. It was shown that equivalent removals could be achieved in the membrane module at lower air flow rates. The study was of short term duration however, and did not address aspects of long-term fiber usage such as fouling of membrane surfaces.

Zander et al.[6] also observed that some compounds were removed more poorly than they should be based on theoretical predictions. These poor removals corresponded to conditions when the stripping constant, R (= QgH/Qw) was less than 2. The authors speculated that the high packing density of the fibers in the commercially-made modules may have led to air flow channeling and poor flow characteristics on the shell side of the module.

An important advantage of the hollow fiber membrane modules is that they are smaller than packed towers and can be mounted either vertically or horizontally. If mounted horizontally, the need to pump water 30 feet to the top of a tower is eliminated. Water pressure drops across a single module are therefore much lower than those required of packed towers. However, very large air pressure

45th Purdue Industrial Waste Conference Proceedings, © 1991 Lewis Publishers, Inc., Chelsea, Michigan 48118. Printed in U.S.A.

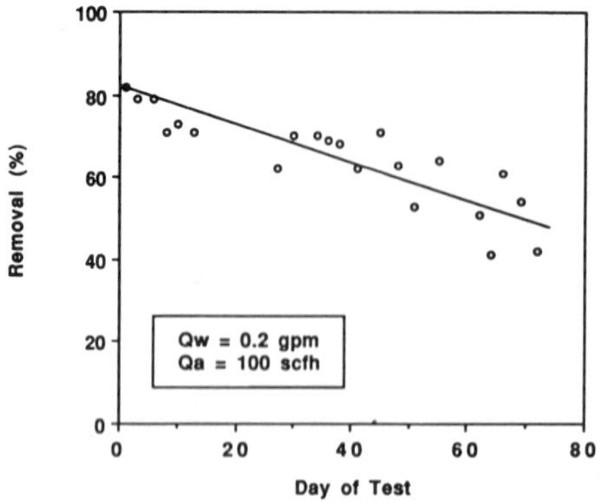

Figure 1. Change in TCE removal.

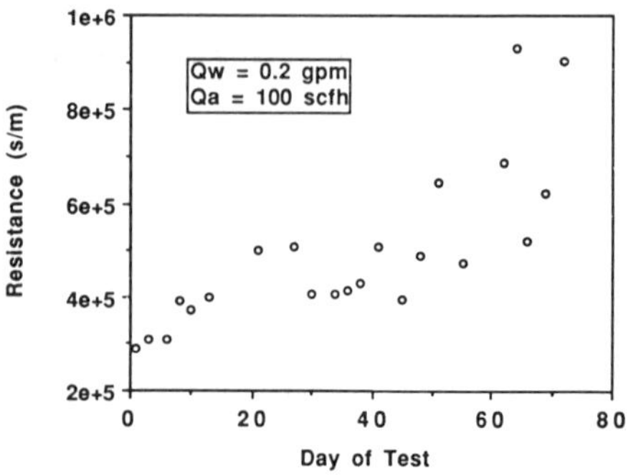

Figure 2. Change in resistance to TCE transfer.

drops were observed across the module.[6] The modules have been designed primarily for medical purposes and have not been optimized for industrial use where the cost of pumping large volumes of air is a concern.

This paper attempts to assess the relative cost of the membrane air stripping process and address some of the issues raised in the earlier work of Zander et al.[6] specifically: 1) the impact of fouling and long-term industrial usage of the fibers, 2) excessive air pressure drops across the shell side of the modules, and 3) the influence of fiber packing density.

Fouling of Hollow Fiber Membrane

Two studies were conducted to test fouling effects on hollow fiber membranes. The first involved a long-term field study of fiber performance, and the second study, conducted in the lab, characterized the kinetics of iron fouling on the membrane fiber surfaces.

Pilot Study

A small pilot hollow fiber membrane air stripping unit was set up to treat ground water contaminated with a high concentration (1500 μg/L) of trichloroethylene (TCE). This unit consisted of a commercially-made hollow fiber membrane module* with 5400 fibers tightly packed in a polycarbonate shell and potted at the ends with polyurethane. The fibers had an effective length of 54.6 cm, a total surface area of 3.4 m^2 and 40% of this surface was covered by 0.03 μm pores. The contaminated ground water was fed through the lumen of the fibers. The flow rates of air and water were measured with rotameters and inlet and outlet pressures of air and water were measured with pressure gages. Air was supplied by a 1-hp air compressor** and fed counter-currently through the shell side of the module. A water flow rate of 0.2 gallons per minute and an air flow rate of 100 standard cubic feet per minute were maintained throughout the testing period of 75 days.

Inlet and outlet water samples were taken approximately 3 times a week for a variety of operating conditions and after the samples were taken, the flows were set back to 0.2 gpm and 100 scfh. Samples were collected in 35mL sample bottles with no headspace, and brought back to the lab for analysis. Samples were extracted using pentane[7] and analyzed on a gas chromatograph+ equipped with a capillary column and an electron capture detector.

Results

Figure 1 shows the change in removal rate as a function of time. For the 75 days of testing there was an overall reduction in removal efficiency of approximately 30%. Removal is related to mass transfer by the following equation:[5]

*Hoechst Celanese, Inc., Charlotte, N.C.
**Dayton Speedaire, Chicago, IL.
+Hewlett Packard 589-0A, St. Paul, MN.

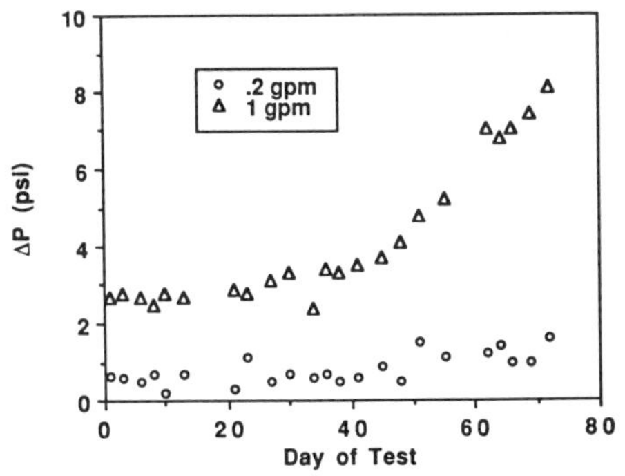

Figure 3. Change in water headloss across module.

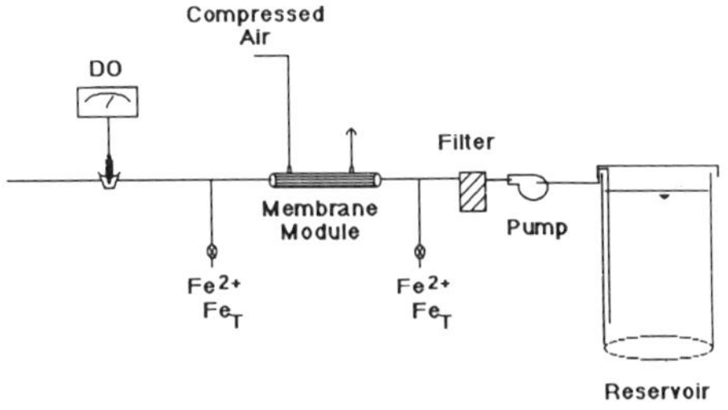

Figure 4. Laboratory apparatus for iron fouling experiments.

$$\frac{C_i}{C_o} = \frac{\exp(K_L a \frac{L}{v}(1 - \frac{1}{R})) - \frac{1}{R}}{1 - \frac{1}{R}} \tag{1}$$

where C_i is the influent concentration, C_o is the effluent concentration, K_L is the mass transfer coefficient, a is the specific surface area, L is the length, v is the velocity and R is the stripping factor as defined above.

The inverse of the mass transfer coefficient represents the resistance to mass transfer and in Figure 2 this resistance is plotted as a function of time. The initial resistance to mass transfer is that for the clean membrane. Semmens et al.[5] showed that in air stripping of volatile compounds the membrane resistance is negligible and the overall resistance is governed by liquid film diffusion. However as the membrane becomes fouled, its resistance becomes significant and the overall resistance to mass transfer is observed to increase. This increase is apparent in Figure 2.

The fouling was accompanied by a concomitant increase in pressure drop across the module, as can be seen in Figure 3.

Following the testing period, the membranes were brought back to the lab and cleaned with acid. This procedure was found to restore their initial mass transfer characteristics.

Four types of fouling may be important when hollow fiber membranes are used in water treatment applications:

1) *suspended solids accumulation*
Suspended solids in the water can quickly plug the membrane modules. The problem is quite different from the usual problems observed with ultrafiltration and reverse osmosis. In the air stripping process no solids accumulate at the membrane wall since no water crosses the membrane, however the ends of the hollow fibers can become plugged by influent solids and pre-filtration may be essential.
2) *biological fouling and slime development*
Bacteria will grow on almost any surface, and polypropylene hollow fibers are no exception. Depending on the water and the type of material being stripped, this may or may not be a problem.
3) *calcium carbonate deposition*
$CaCO_3$ deposition on the fibers may occur during air stripping when carbon dioxide is stripped from a groundwater. The increase in pH at the membrane wall can cause a local supersaturation of calcium carbonate. The amount of $CaCO_3$ which may be expected to precipitate can be calculated based on chemical equilibria.
4) *iron oxidation*
Reduced iron, which is frequently found in groundwaters, is oxidized as oxygen transfers through the membrane to form a precipitate of Fe $(OH)_3$. Some of the ferric hydroxide formed may deposit on the walls of the membrane and some will be carried out with the water stream. The rate of iron oxidation is strongly dependent on the pH and the concentration of iron as shown in Equation 2:

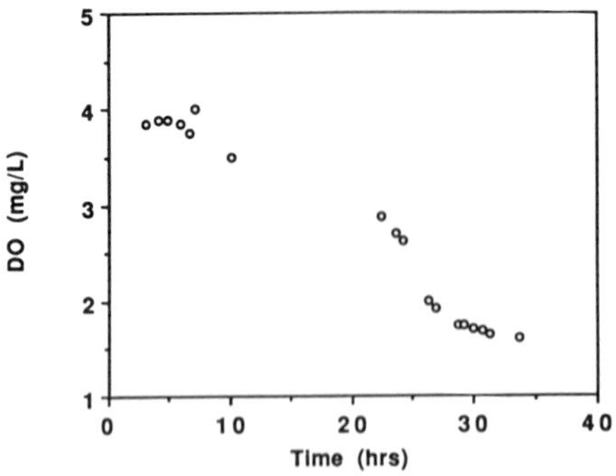

Figure 5. Typical effluent dissolved oxygen profile during an iron oxidation experiment.

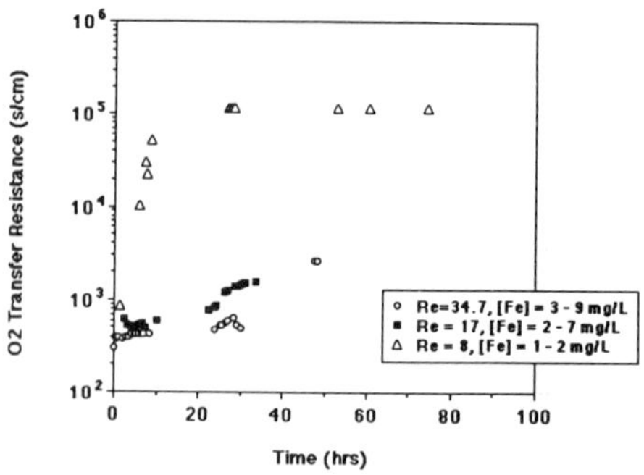

Figure 6. Changes in the overall resistance to oxygen transfer as function of time.

$$\frac{d[Fe^{2+}]}{dt} = -k[Fe^{2+}][OH^-]^2 pO_2 \qquad (2)$$

The reaction occurs quite rapidly; the rate constant in Equation 2 is greater than $2 \times 10^{13}\ M^{-2}\ atm^{-1}\ min^{-1}$.[8]

Laboratory Studies on Iron Oxidation

In the field study iron fouling appeared to be the most severe problem in the test module and high concentrations of iron were released from the membrane when it was acid-cleaned after 75 days of operation. The other forms of fouling were not apparently important during the test period. As a result, a second series of experiments were conducted in the laboratory to evaluate the factors influencing the rate of iron fouling of polypropylene fibers under controlled conditions. Figure 4 shows a schematic of the apparatus used. A 55-gallon tank was filled with uncontaminated water that had been deoxygenated by the the addition of sodium sulfite and a trace of cobalt nitrate as a catalyst. The pH was controlled using a bicarbonate-CO_2 buffer[9] and ferrous sulfate ($FeSO_4$) was added to provide concentrations from 5 to 20 mg/L. This feed solution was prefiltered using a cartridge-type water filter and pumped through the lumen of the fibers of a lab-made module containing 180 fibers. Air was blown counter-currently through the shell side of the module. Inlet and outlet samples of water were taken and analyzed for Fe^{2+} and total Fe using the 1,10-phenanthroline method.[10] Effluent dissolved oxygen was measured using a dissolved oxygen electrode.

Figure 5 shows a typical dissolved oxygen curve collected during these tests. When no iron is present in the water, the change in concentration of oxygen in the water can be described as:

$$\frac{dC}{dt} = K_L a\ (C^* - C) \qquad (3)$$

where C is the concentration of oxygen transferred, C* is the equilibrium concentration of oxygen (which is dependent on temperature), K_L is the mass transfer coefficient and a is the specific surface area. This can be integrated to:

$$-\ln\left(\frac{C^* - C}{C^* - C_o}\right) = K_L a\ \frac{L}{v} \qquad (3)$$

where C_o is the initial dissolved oxygen concentration in the water, L is the length of the fibers and v is the velocity of flow through the fibers ($\frac{L}{v}$ is the detention time in the module).

The overall mass transfer coefficient for oxygen transfer, K_L, can be estimated from equation 4 when no iron is present in the water. When a chemical reaction takes place with oxygen (such as the oxidation of iron) the rate of oxygen transfer is increased.[11] The amount of dissolved oxygen in the effluent will be lower since some of the oxygen is used in the reaction. In such a case the total amount of oxygen transferred may be estimated as the sum of the measured increase in dissolved oxygen plus the amount of oxygen required to oxidize the iron.

$$4Fe^{++} + 2H_2O + O_2 \longrightarrow 4Fe^{+++} + 4OH^- \qquad (5)$$

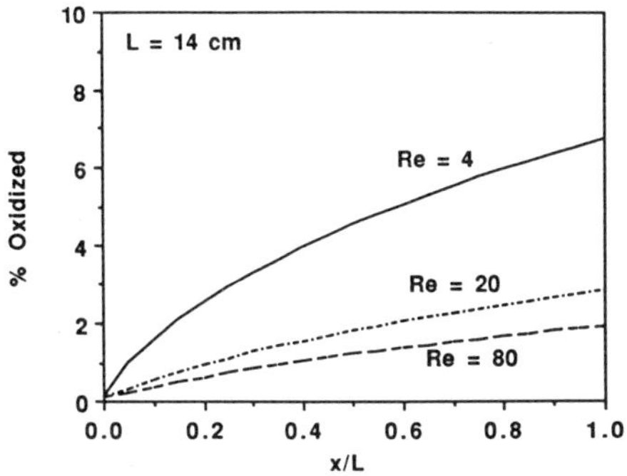

Figure 7. Calculated percent iron oxidation along fiber length at pH 7.

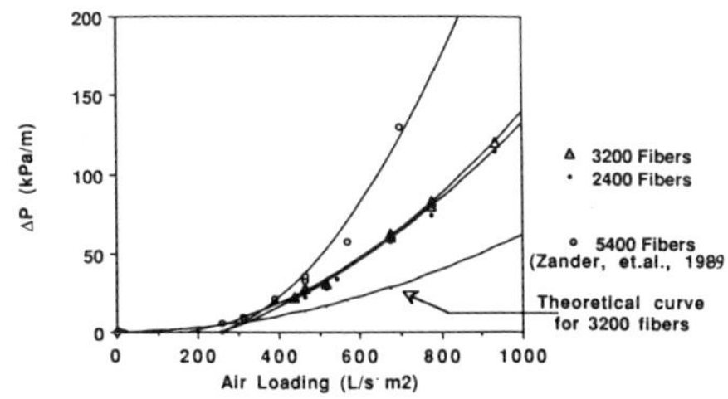

Figure 8. Pressure drop vs. air loading for 3 hollow fiber modules with different packing densities.

An "effective" mass transfer rate can be experimentally found by calculating the amount of oxygen used for oxidation plus the amount of oxygen in the effluent water and using this as the value for C in Equation 4. Values of the effective mass transfer coefficient were estimated in this way during the test period.

The inverse of the mass transfer coefficient reflects the resistance to mass transfer. Resistance values were calculated with time from the oxygen transfer and the iron oxidation data, and three sample runs are shown in Figure 6. The initial resistance is approximately equal to the liquid film resistance for oxygen transfer, and the membrane resistance is negligible. As the fibers are fouled by ferric hydroxide deposition the resistance increases. In some cases it appears to reach an asymptotic value.

Many similarities exist between the increase in mass transfer resistance and that of heat transfer resistances. These heat transfer fouling resistances have been studied extensively by engineers. The general equation for heat transfer fouling resistance as a function of time is:[12]

$$R_f = R_f^\circ * (1 - e^{-\Theta/\Theta_c}) \qquad (6)$$

where R_f is the resistance due to fouling, R_f° is the asymptotic value of R_f, θ is time and θ_c is the time it would take to achieve the asymptotic value if fouling proceeded linearly at the initial rate. In iron fouling of microporous membranes, R_f° and θ_c are functions of the oxidation kinetics and the velocity of flow through the fibers. Equation 6 assumes that the rate of deposition is fastest initially and declines as fouling continues, however our data did not always follow this trend. Our mass transfer data did agree with the fouling behavior observed in heat transfer in that an asymptotic resistance was reached.

Figure 6 also shows the effect of velocity on the value of the asymptotic resistance. Although heat transfer fluids always flow in the turbulent range, and the flows in hollow fibers are always laminar, both show the pattern of higher fouling rates at lower velocities. The asymptotic resistance in iron fouling of hollow fibers seems to be more dependent on the velocity than on the influent iron concentration. An asymptotic fouling resistance may be expected when the higher velocities and increased shear within the fibers prevent the accumulation of further ferric hydroxide deposits.

In addition the reduced detention time at higher velocities allows less oxidation to occur, as the contact time within the module is reduced. This is demonstrated in Figure 7 which shows the calculated percent oxidation profile of ferrous ion along the length of the fibers at different water Reynolds numbers.

The concentration of iron(III) which attaches to and fouls the fibers is much lower than the concentration which is actually oxidized. Following each run, the fibers were acid cleaned. This cleaning restored the fibers to their original mass transfer characteristics and the acid was then analyzed for iron concentration. Table I shows the average difference in ferrous iron concentration measured before and after the membrane module (the amount of iron oxidized). The amount oxidized was approximately constant in three runs, yet the amount of iron recovered in the acid cleaning decreases for increasing velocities.

If the mass of recovered iron is divided by the volume treated, an average concentration of deposition is obtained. This value is always less than the average value of difference in ferrous

Table I. Amount of Fe^{3+} Deposited on Fibers as a Function of Velocity

Run #	ΔFe^{2+} ($\frac{mg}{L}$) (avg)	Fe^{3+} Recovered ($\frac{mg}{cm^2}$)	Velocity (cm/s)	Volume Treated (L)
Run 4	.58	.39	2.0	166.2
Run 5	.42	.26	4.3	166.2
Run 6	.47	.21	8.7	218.7

concentration. This means that a substantial amount of iron which is oxidized does not deposit on the surfaces of the fibers but passes out of the module.

Since a large fraction of iron(III) is carried out of the module by the water stream, and since an industrial application for membrane air stripping might use modules in series, second stage and subsequent modules would be more subject to particulate fouling and fiber plugging than the first of a series of modules. Some sort of treatment must be planned for filtration or settling of suspended solids before the water can be used for any purpose following air stripping.

Module Pressure Drops

Figure 3 shows an increase in water pressure drop across the hollow fiber module due to fouling. Although the initial pressure drop through the module is not excessive, the increase due to membrane fouling can significantly increase operating costs.

In their paper, Zander et al.[6] observed that the shell-side pressure drops across hollow fiber membrane modules were excessive and much higher than the air pressure drops found in packed towers. The high pressure drop may be attributable to the high fiber packing density in the commercial module, and tests were made to evaluate the influence of lower packing densities. Figure 8 shows data obtained in three modules with different packing densities (the module used by Zander et al. contained 5400 fibers; we tested modules containing 3200 and 2400 fibers packed into the same shell diameter). It is clear that the air pressure drops remain very high. Notice also in Figure 8 that the module with 2400 fibers has approximately the same shell-side pressure drop as that with 3200 fibers. The manufacturer has suggested that this is due to loosely potting the fibers in the module which causes some fibers to be pushed up into the exit air port. This was observed in the module with 2400 fibers. Newer designs for these modules include a mesh surrounding the fiber bundle which will prevent fiber blockage of the exit port.[13] The cost analysis developed below is based upon the pressure drop data for the module with 3200 fibers.

The calculated pressure drop across the membrane module for 3200 fibers is also shown in Figure 8, and it shows that the pressure drop can be reduced significantly if there is not a major headloss across the air ports.

Influence of Packing Density on Performance

Studies were conducted to analyze the effect of various fiber packing densities on the removal of volatile compounds from water. The two modules containing 3200 and 2400 fibers were tested. Performance tests were run on these modules to determine the removal efficiency of three compounds (1,1,2-trichloroethane, trichloroethylene and carbon tetrachloride). Water containing the compounds was pumped through the lumen and air was blown counter-currently over the outside of the fibers. Water inlet and outlet samples were taken at different water flow rates and analyzed on the gas chromatograph. The observed removals were compared with the theoretical removal rates predicted using mass transfer correlations.

The results of this analysis are inconclusive, as differences in removal rates for the three different packing densities were within experimental error. The poor packing of the fibers in the modules may be responsible for the scatter and lack of difference. A module with a nominal packing density of 30% may have clusters of fibers packed much more densely and other areas where the packing is much looser. It is likely that the air will channel preferentially through areas of low packing density; densely packed areas will therefore provide poor contact between the fibers and the air, and this will lead to poorer removals. A photograph of three modules containing different numbers of fibers is shown in Figure 9 to illustrate the non-uniform packing. In addition Figure 10 displays side views of two of the modules. The module with 2400 fibers has large gaps between clusters of fibers. Figures 9 and 10 show

Figure 9. Distribution of fibers around the module diameter. Left to right: 5400 fibers, 3200 fibers, 2400 fibers.

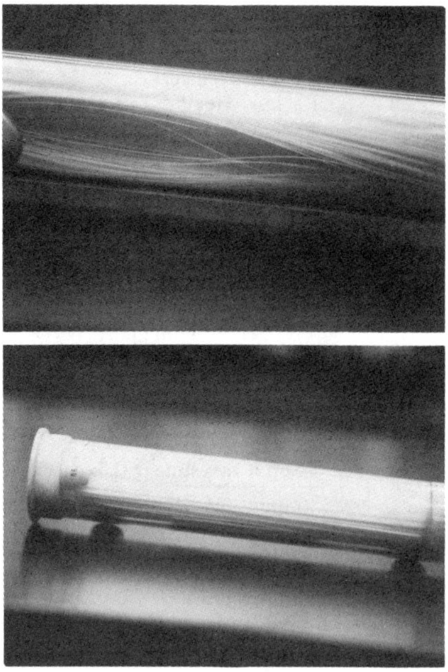

Figure 10. Two hollow fiber modules. Top: 2400 fibers, bottom: 3200 fibers.

the difficulty in potting the fibers so that they are held somewhat rigidly and are uniformly distributed within the module.

Cost Analysis

Figure 11 shows a comparison of the cost of air stripping using hollow fiber membranes with that of packed towers. Table II shows the assumptions that were made in this cost analysis, and in Table III the costs are broken down into operating and capital costs. The packed tower method of air stripping remains the least expensive option at this time costing only \$0.98 per thousand gallons treated compared to \$1.37 for the hollow fiber membrane method. Although the packed tower is less expen-

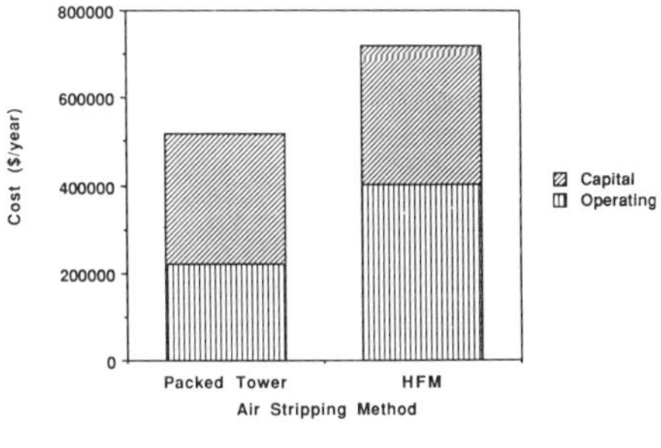

Figure 11. Comparison of costs of air stripping.

Table II. Assumptions

Influent water:	1500 μg/L TCE
Removal required:	99.9%
Water flow rate:	1000 gpm
Operating hours:	24 hr/day, 365 day/yr

Table III. Manufacturing Cost Estimate

Packed Tower

2 Packed Towers
Packing Height: 26.5 ft
Packing Material: Ceramic Berl Saddles, 1 inch
2 Air Compressors
3 Water Pumps
Operating Hours: 24 hr/day, 365 days/year
M&S Index, 1989: 895.1[17]

Operating Costs:	
Air Pumping costs (/yr):	$103940
Water Pumping costs (/yr):	14880
Labor costs (1 hr/shift):	26280
Maintenance, repairs (4% equipment cost):	74320
Total Direct Operating Costs Per Year:	**$219420**
Capital Investment (Direct Costs)	
(Straight line depreciation assumed, 10 years)	
Purchased Equipment (60% Cap. Inv.)	$1858044
Installation (16% Cap. Inv.)	495500
Instrumentation & Controls (5% Cap. Inv.)	154800
Piping (Installed) (8% Cap. Inv.)	247750
Electrical (Installed)(8% Cap. Inv.)	247750
Total Direct Capital Costs:	**$3003844**
Total Yearly Capital Costs:	**$ 297381**
Total Cost per 1000 gallons:	**$ .98**

Hollow Fiber Membrane System

1 Bank of Modules, 94 Modules per Bank
Each module is 1.8m (6 ft) long, 0.3048 m (1 ft) in diameter
1 Air Compressors
2 Water Pumps
Operating Hours: 24 hr/day, 365 days/year

Operating Costs:	
Air Pumping costs (/yr):	$357011
Water Pumping costs (/yr):	7225
Labor costs (1 hr/shift):	26280
Maintenance, repairs (6% equipment cost):	12027
Total Direct Operating Costs Per Year:	**$402543**
Capital Investment (Direct Costs)	
(Straight line depreciation assumed, 10 years)	
Purchased Equipment (60% Cap. Inv.)	$2000450
Installation (16% Cap. Inv.)	500115
Instrumentation & Controls (5% Cap. Inv.)	166705
Piping (Installed) (8% Cap. Inv.)	266730
Electrical (Installed) (8% Cap. Inv.)	266730
Total Direct Capital Costs:	**$3200730**
Total Yearly Capital Costs:	**$ 316872**
Total Cost per 1000 gallons:	**$ 1.37**

sive, the closeness of these numbers in the cost analysis is encouraging. Table III also describes the designs for the hollow fiber membrane and the packed tower air stripping systems which were used in this cost analysis. The design for the hollow fiber membrane system was chosen to minimize the cost of pumping air and water, plus the cost of membranes. A module diameter of 0.305 m (1 foot) was assumed for all modules. The packed tower design was based on the mass transfer correlations developed by Onda.[14] Capital cost information for the packed tower system came from Peters and Timmerhaus[15] and for the hollow fiber membrane system from Rutledge.[16]

The analysis is somewhat unfair as the packed tower technology has been around for many years and the process has been optimized. By comparison the hollow fiber membrane technology is in its infancy and the design of the membrane modules for this type of application has not been optimized. For example, the development of a commercial crossflow module could dramatically modify the economic analysis. A crossflow module configuration could increase the mass transfer rates by an order of magnitude while simultaneously reducing the pressure drop across the module. Such developments are likely to make membranes cost effective.

The modular design of the hollow fiber system lends itself to smaller scale and point-of-use applications such as in-house water treatment systems. In such applications the hollow fiber membrane modules are superior to packed towers. In large scale applications the modular design offers little economy of scale and packed tower air stripping is still the least expensive option.

A large portion of the cost of air stripping using hollow fiber membranes is due to pumping costs. The pressure drop across a module decreases as the length of the module decreases, but if small modules are used, the number of modules required for a treatment objective is increased and the headlosses due to fittings, contractions, and expansions becomes a major factor in the cost of water pumping. The use of large diameter modules will reduce this cost, however, the contact between the air and the fibers may be inadequate in a large fiber bundle. New designs for large modules are underway to ensure good phase contact and to reduce pressure drops.[13] Capital costs for hollow fiber modules are also expected to decrease as more industrial uses are found for them.

CONCLUSIONS

Operational aspects of hollow fiber membrane air stripping were explored in both lab and long term field tests. Problems such as fouling of the fibers, excessive pressure drops across the modules and poor fiber packing were described. Each of these problems will influence the design for membrane modules for practical application.

Chemical fouling was observed to occur at the membrane wall as a result of oxygen transfer from the air to the water. Iron fouling was shown to be responsible for a deterioration in the overall mass transfer coefficient over time, however, the fibers were easily cleaned with acid. The acid-cleaned fibers were as effective in VOC removal as the virgin fibers. By comparison, a packed tower is difficult to clean when its packing elements have become fouled.

The pressure drop across existing commercial shell and tube membrane modules is high and increases if fouling occurs. The pressure drop could be reduced if a good crossflow module design could be developed. In addition, such a design would resolve fiber packing problems that reduce the effective contact between the air and the fibers, and provide higher mass transfer coefficients.

At the present time, hollow fiber membrane air stripping only appears feasible when used in small scale applications.

REFERENCES

1. Ball, W. P., Jones, M. D., & Kavanaugh, M. C., "Mass Transfer of Volatile Compounds in Packed-Tower Aeration". *Jour. WPCF* 56:2 (1984) p 127–136.
2. Amy, G. L., Cooper, W. J., "Air Stripping of Volatile Organic Compounds Using Structured Media", *Journal of Environmental Engineering-Div. ASCE*, 112:4, August, (1986) pp. 729–743.
3. Clark, R. M., Eilers, R. G., & Goodrich, J. A., "VOCs in Drinking Water: Cost of Removal", *Journal of Environmental Engineering-Div. ASCE*, 110:6, (December, 1984) pp. 1146–1162.
4. Qi, Zhang, Cussler, E. L., "Microporous Hollow Fibers For Gas Absorption" I and II., *Journal of Membrane Science* 23 (1985) pp. 321–332 and 333–345.
5. Semmens, M. J., Qin, R. & Zander, A. K., "Using a Microporous Hollow-Fiber Membrane to Separate VOCs From Water", *Journal of the American Water Works Association*, 81:4, (1989) pp. 162–167.
6. Zander, A. K., Semmens, M. J., & Narbaitz, R. M., "Removing VOCs by Membrane Stripping", *Journal of the American Water Works Association*, 81:11, (1989) pp. 76–81.
7. Mehran, M. F., Slifker, R. A. & Cooper, W. J., "A Simplified Liquid-Liquid Extraction Method

for Analysis of Trihalomethanes in Drinking Water", *Journal of Chromatographic Science*, 22, (June, 1984) pp. 241-243.
8. Davison, W., and Seed, G., "The Kinetics of the Oxidation of Ferrous Iron in Synthetic and Natural Waters", *Geochim Cosmochim Acta* 47, (1983) pp. 67-79.
9. Millero, F. J., Sotolongo, S. & Izaguirre, M. "The Oxidation Kinetics of Fe(II) in Seawater", *Geochim Cosmochim Acta* 51, (1987) pp. 793-801.
10. *Standard Methods for the Examination of Water and Wastewater*, 16th Edition, Franson, M. A. H. (Managing Editor), American Public Health Association, (1985), pp. 215-219.
11. Cussler, E. L., *Diffusion: Mass Transfer in Fluid Systems*, Cambridge University Press, London, (1984) pp. 325-376.
12. Epstein, N. "Fouling: Technical Aspects (Afterword to Fouling in Heat Exchangers)" in *Fouling of Heat Transfer Equipment*, Somerscales, E. F. C., and Knudsen, J. G. (eds.) Hemisphere Publishing Corp., Washington,(1981), pp. 31-53.
13. Bailey, M., Personal communication (April, 1990).
14. Onda, K. et al. "Mass Transfer Coefficients Between Gas and Liquid Phases in Packed Columns", *Journal of Chemical Engineering Japan*, 1:56 (1968).
15. Peters, M. S. and Timmerhaus, K. D., *Plant Design and Economics for Chemical Engineers*, 3rd Edition, McGraw-Hill Book Company, New York, (1980).
16. Rutledge, F., Personal communication (April, 1990).
17. "Economic Indicators", *Chemical Engineering*, 97:4 (April, 1990), p. 244.

64 NANOFILTRATION MOTIVATES FUTURE WASTEWATER RECLAMATION

Mao-Yuan Tur, Group Leader
Ming-Shean Chou, Group Leader
Hung-Yuan Fang, Department Manager
Environmental Engineering Department
Refining & Manufacturing Research Center
Chinese Petroleum Corporation
Taiwan 60036, R.O.C.

INTRODUCTION

To meet stricter environmental regulations, tertiary treatment is generally required to polish the secondary effluent at industry. The polished wastewater, after removing silica and divalent ions such as calcium, magnesium, and sulfate, can be recovered as cooling-tower make-up water. Membrane processes are generally considered as cost-effective for removing these ions. Conventional reverse osmosis (RO) process for desalination is operated in a pressure range of 200–800 psi. Power consumption is a main factor in the operation and maintenance cost. Developments of the membrane technology in the last decade have resulted in a type of innovated membrane which operated in a typical pressure range of 50–200 psi while still exhibiting ion-rejecting capability.[1,2] The type of membrane was developed by FilmTech Co. (USA) and named as "nanofiltration (NF)" membrane owing to its characteristic of rejecting molecules with sizes in the order of one nanometer. The membrane is made of cross-linked aromatic polyamides and has negative surface charges. Thus, the anion repulsion mainly determines the solute rejection. It exhibits typical rejections of 40–70% for monovalent ions and 70–98% for di- or multivalent ions, respectively. Organics with molecular weight above 300 can also be separated from the lower ones. Due to these characteristics, nanofiltration has been applied to partial water desalination, water softening, groundwater treatment, sulfate removal from saline in oil fields, and separating monovalent salts from organics in the MW range of 300–1,000.

This article will be focused on the applicability of this nanofiltration technique to wastewater reclamation in petroleum refining industry. Demineralization efficiency of nanofiltration membrane NF-70 (FilmTech Co.) was tested. Operating and maintenance costs for this process were then estimated and compared with those of reverse osmosis.

EXPERIMENTAL

Feed

Feed water was prepared by filtrating secondary effluent from CPC's refinery through a mixed media followed by activated carbon adsorption to remove suspended solids, colloids, and trace organics contained therein. Characteristics of the feed water are shown in Table I. For preventing the membrane from scaling upon a recovery of 80%, the feed was brought to a Langelier Saturation Index (LSI) of -1.70 by adding 85 mg/L hydrochloric acid and 4 mg/L scale inhibitor (Flocon-100, Pfizer Co.) following the suggestions of ASTM: D-3739-83.[3]

Test Unit and Test Method

Schematic diagram of the nanofiltration test unit is shown in Figure 1. The whole NF system is equipped with a 1.5 gpm procon pump capable of maintaining a liquid system pressure of up to 150 psig. The system contains a chamber which holds an element of spiral-wound membrane (NF 70-2514, FilmTec) with an effective area of 5 sq. ft. System pressure and concentrate flowrate were regulated with a needle valve. Cooling system is not required for the concentrated flow because of neglecting heat dissipation at a system pressure of less than 130 psig and a total operation time of less than 90

45th Purdue Industrial Waste Conference Proceedings,

Table I. Characteristics of Feed Water

	Unacidified	Acidified
pH	8.1	5.8
Cation, mg/L		
$[Na]^+$	240.5	240.5
$[Mg]^{2+}$	31.3	31.3
$[Ca]^{2+}$	75.9	75.9
Anion, mg/L		
$[Cl]^-$	292.1	356.2
$[HCO_3]^-$	146.0	35.8
$[NO_3]^-$	16.8	16.8
$[SO_4]^{2-}$	354.6	354.6
TDS, mg/L	1157.2	1111.1
SiO_2, mg/L	45.0	45.0
TOC, mg/L	2.0	2.0
LSI[a]	+ 0.5	-1.7

[a]LSI: Langelier Saturation Index.

min. Batch tests were made at a series of system pressure of 40, 70, 100, and 130 psig. Samples were collected at recoveries of 25, 50, and 80% and the operation time recorded.

Analytical

Cations (Na, Mg, and Ca) and anions (chloride, nitrate, and sulfate) were analyzed by ion chromatography (Model 2010i, Dionex Co., USA). Total dissolved solid (TDS) was taken as sum of total ions. Total alkalinity was determined by titrating the water sample to pH = 4.0 with 0.02 N sulfuric acid. Silica was analyzed according to *Standard Methods*[4] by a colorimeter (Model 5101, Skalar Co., Holland). Total organic carbon (TOC) was analyzed by persulfate/UV-oxidation method (Model DC-80, Dohrmann Co., USA).

RESULTS AND DISCUSSION

Definitions of Recovery and Rejection

Recovery for batch test is defined as the percentage of feed which passes through the membrane and is recovered as permeate. It can be expressed as:

$$\text{Recovery} = Qp/Qf \times 100\% \tag{1}$$

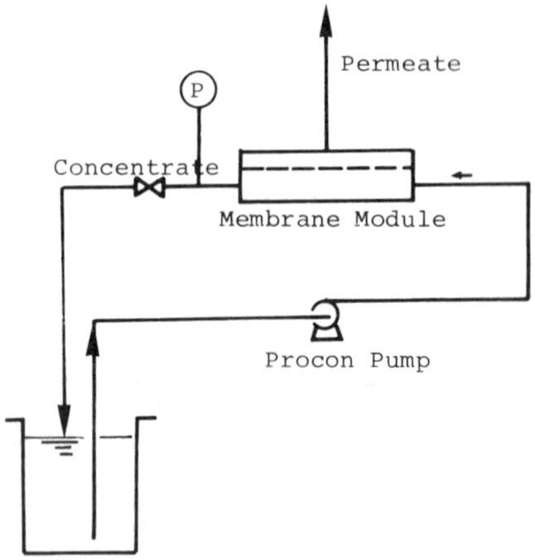

Figure 1. Schematic diagram of bench scale NF test unit.

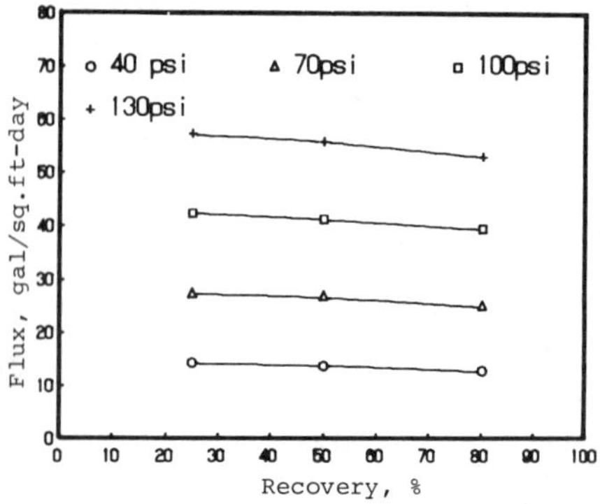

Figure 2. Effect of recovery and operating pressure on the flux.

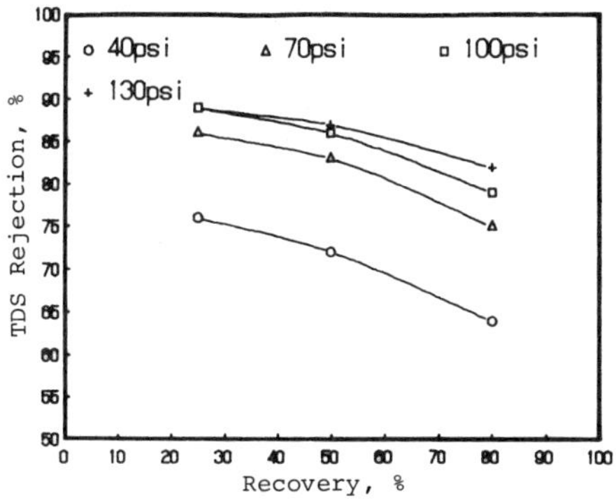

Figure 3. Effect of recovery and operating pressure on the rejection of total dissolved solid.

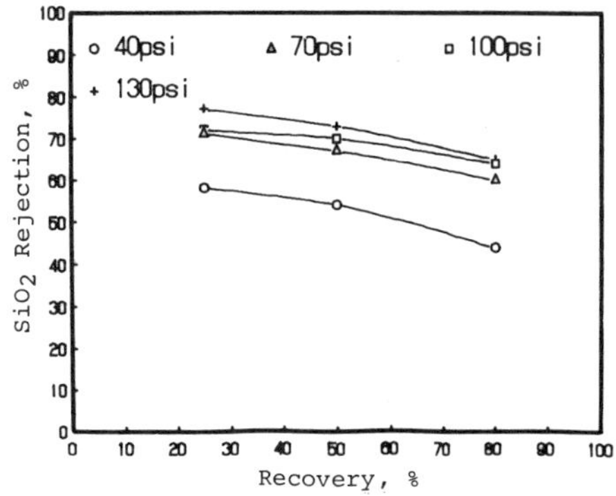

Figure 4. Effect of recovery and operating pressure on the rejection of silica.

where: Qp = mass or volume of permeate accumulated
Qf = mass of volume of initial feed water

Rejection is a measure of membrane effectiveness for molecule separation and is defined as:

$$\text{Rejection} = (1 - Cp/Cf) \times 100\% \tag{2}$$

where: Cp = solute concentration in permeate
Cf = solute concentration in the fresh feed

Fluxes

For a given membrane, permeate water flux depends mainly on the pressure gradient and the solute flux depends mainly on the concentration difference between concentrated and permeate fluid similar to that depicted by the Fick's law. In this study, as shown in Figure 2, water flux through the membrane increases proportionally with increasing net driving pressure (pressure difference minus osmotic pressure) and decreases slightly with increasing recovery. For example, water flux is 27.2 gal/sq. ft-day (gfd) at 70 psig and 25% recovery. This flux makes energy saving more significant because the equivalent result can be obtained only at a pressure as high as 250 psig for reverse osmosis membrane.

Rejection

Figure 3 shows that the rejection of TDS decreases with increasing recovery and decreasing operating pressure. As recovery becomes higher, the circulating stream becomes more concentrated with dissolved solids which result in the increase of osmotic pressure and solute concentration gradient. Therefore, permeate degrades at higher recoveries. As can also be seen from Figure 3, operations at higher pressure result in higher TDS rejections. However, the higher solute flux coupled with increasing water flux at higher pressure makes the enhancement of TDS rejection not proportional to the system pressure. Variation of silica rejection with system pressure and recovery exhibits similar trends to that of TDS as shown in Figure 4.

Another major concern of the NF performance is the rejection of individual ion at the desired recovery of, for example, 80% in this study. As shown in Figure 5, rejection of ions is insensitive to operating pressure in the practical range of 70–130 psig except for nitrate. The rejection pattern shown in Figure 5 implied that the number of charges on the ions may be significant for their separation. Therefore, higher rejection (95%) was obtained for divalent ions such as sulfate, magnesium, and calcium, and medium rejection (65%) for monovalent ions such as chloride, sodium, bicarbonate (alkalinity), and silica. A rejection as low as 30% was observed for nitrate. It was also observed that nitrate exhibits lower rejection compared with other anions in reverse osmosis. As described in the previous section, the NF70 membrane was constructed of crosslinked aromatic polyamides with negative surface charges. Thus, the anion repulsion is the most important control factor for solute rejection as observed in this study and reported by Eriksson.[1,2]

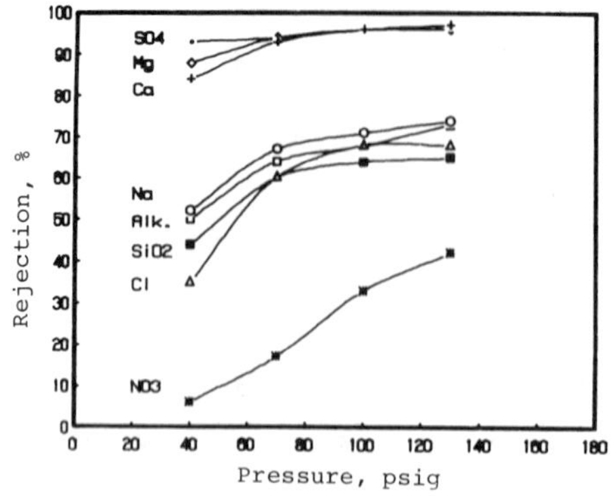

Figure 5. Effect of operating pressure on the rejection of ions and silica at recovery 80%.

Table II indicates that the quality of permeate at the 80% recovery and the 70 psig pressure is better than used for cooling-tower make-up in the aspect of TOC and scaling tendency.

ECONOMICS

In order to compare the cost of RO and NF system on an equal basis, a plant with 840 gpm product flow is considered. Despite some difference in product quality, due to inherently unequal rejection of ions and silica for both cases, the total number of element and its arrangement are identical. It consists of six 140-gpm racks, each with a 4 + 2 array of six spiral-element pressure vessels. Design basis of the plant is shown in Table III. The lower recovery in the RO system is attributable to its high silica rejection and the saturation limitation in the concentrate stream. By using identical fluxes and nearly the same recovery, the fouling index in both cases is assumed to be identical. In addition, due to the type of thin film composite, there is no compaction effect for both cases.

Capital costs for the RO and NF systems are shown in Table IV.[5] Each includes pumping system (pumps, valves, and piping), electrical control devices, membrane and pressure vessel. The costs were estimated by assuming that: 1) pumping system is less expensive by $20,000 in the NF because of its lower pressure; 2) NF membrane element is 20% more expensive than RO membrane element; and 3) NF pressure vessel is 50% cheaper than ROs. The cost of electrical control devices is, of course, identical. As shown in the table, capital cost of NF is only 5% more expensive than that of RO.

Table V shows operating and maintenance (O&M), chemicals, electricity cartridge replacement, labor, depreciation, and interest costs for the two plants. Both have the same O&M costs except for membrane replacement and electricity. The membrane cost of NF is 20% higher than that of RO. Electrical energy was estimated to be 3.56 kwh/kgal-product for the RO plant by assuming that the system is operated at 70% recovery and 240 psig with combined energy efficiency of pump and motor of 0.7. That for the NF plant was estimated to be 1.30 kwh/kgal at a recovery of 80% and a feed pressure of 80 psig and a boosting pressure of 40 psig after 50% recovery. Energy efficiency was taken

Table II. Characteristics of Permeate[a] and Cooling-Tower (C/T) Make-Up Water

	Permeate	C/T make-up
pH	5.6	7.0
Cation, mg/L		
[Na]+	80.1	26.0
[Mg]2+	2.0	40.0
[Ca]2+	4.7	16.0
Anion, mg/L		
[Cl]-	144.3	20.0
[HCO_3]-	13.1	146.0
[NO_3]-	13.9	trace
[SO_4]2-	14.1	70.0
TDS, mg/L	272.2	318.0
SiO_2, mg/L	18.0	38.0
TOC, mg/L	1.0	3.0

[a]Permeate: Obtained at 70 psig and 80% recovery.

Table III. Design Bases for RO and NF Plants

	RO	NF
Product flow, gpm	840	840
Feed temp., °C	25	25
Membrane area, sq.ft/element	330	330
Total number of element	216	216
Flux, gpd/element	5600	5600
Recovery, %	70	80
Fouling index (3 years)	0.85	0.85
Compaction factor (3 years)	1.0	1.0
Pressure, psig		
Initial	240	80
3 Year	280	100
Membrane life, year	3	3
Availability, %	90	90
Rejection, %		
$[Na]^{+}$	98	70
$[Mg]^{2+}$	98	95
$[Ca]^{2+}$	98	95
$[Cl]^{-}$	98	60
$[HCO_3]^{-}$	95	65
$[NO_3]^{-}$	90	20
$[SO_4]^{2-}$	98	95
SiO_2	95	60

Table IV. Capital Cost (US$) for the RO and NF Plants[5]

	RO	NF
Feed Pump and Interstage Booster Pump with Valves and Piping	109,200	89,200
Electrical Control Devices	45,000	45,000
8″ Membrane Element	330,000	396,000
Pressure Vessel	64,800	43,200
Total	549,000	573,400

Table V. Operating & Maintenance Cost (US$/Kgal) for the RO and NF Plants

	RO	NF
Chemicals		
HCl, 85 mg/L (33% HCl @ $0.06 /lb)	0.18	0.18
Scale inhibitor, 4 mg/L (Flocon-100 @ $1.00 /lb)	0.04	0.04
Cleaning	0.04	0.04
Electricity (@ $0.08 /lb)	0.28	0.10
Cartridge replacement		
Membrane replacement	0.27	0.33
Prefilter	0.04	0.04
Labor	0.08	0.08
Depreciation and interest (@ 20 %/yr)	0.27	0.28
Total	1.20	1.09

at the same value as RO. Difference in energy consumption between RO and NF is 2.26 kwh/kgal-product, which is $0.18 /kgal-product at an electricity cost of $0.08 /kwh. Therefore, as seen in Table V, O&M cost saving is about $0.11 /kgal-product by using NF70 membrane instead of RO (FT30) membrane.

CONCLUSION

NF70 membrane with negative surface charges opens new areas for membrane separation. Nanofiltration is also a low energy consumption alternative to low pressure reverse osmosis where water softening is required. The shortage of water resource in this decade makes nanofiltration potential for wastewater reclamation in industry.

REFERENCES

1. Eriksson, P., "Nanofiltration Extends the Range of Membrane Filtration," *Environmental Progress*, Vol. 7, pp. 58–62 (1988).
2. Eriksson, P., "Nanofiltration—What It Is and Its Applications," Presented at the 6th Annual Conference Technology/Planning Conf., Cambridge, MA., USA., Nov. 1–3 (1988).
3. ASTM: D 3739-83, "Standard Practice for Calculation and Adjustment of the Langlier Saturation Index for Reverse Osmosis."
4. APHA, AWWA, WPCF, *Standard Methods for the Examination of Water and Wastewater*, 16th Ed., 1985.
5. Crowdus, F., "System Economic Advantages of a Low Pressure Spiral RO System Using Thin Film Composite Membranes," *Ultrapure Water*, July/August, 1984.

65 CONSIDERATIONS FOR SELECTION, DESIGN, AND OPERATION OF INCLINED PLATE SETTLERS FOR INDUSTRIAL WASTEWATER TREATMENT

Greg J. Humpal, Environmental Engineer
Robert Chiesa, Group Leader
Facility and Process Engineering
RMT, Inc.
Madison, Wisconsin 53708

INTRODUCTION

The popularity of the inclined plate settler (IPS) can be largely attributed to the fact that they use significantly less space and cost less than a conventional clarifier. Most units also do not have any sludge rake mechanism that would require repairing, thus it is a simpler clarifier. The IPS normally also includes the rapid mix tank and the flocculation tank that are attached to the clarifier making a convenient package.

The IPS units have been used in the metal finishing industry, electroplating industry, steel mills, foundries, aggregate processing power plant bottom ash, mining mineral processing, coal wash water, municipal wastewater, water treatment, and the paper industry.

The purpose of this paper is to summarize important factors to be considered by consulting engineers and industries when incorporating an IPS into an industrial wastewater treatment system. The discussion will cover the basic IPS unit and associated equipment necessary for a complete clarification system. Operational conditions that can affect performance will also be addressed.

In this paper, the term "IPS unit" and IPS refer to just the basic inclined plate settler, while the term "IPS system" refers to the complete system, which includes the associated components typically required for treating industrial wastewaters.

IPS SYSTEM

The IPS system is made up of more than just the inclined plate separator. There are several auxiliary pieces of equipment that are vital to the success of the IPS, see Figure 1. They include the following:

- pH adjustment tank
- Chemical feed system

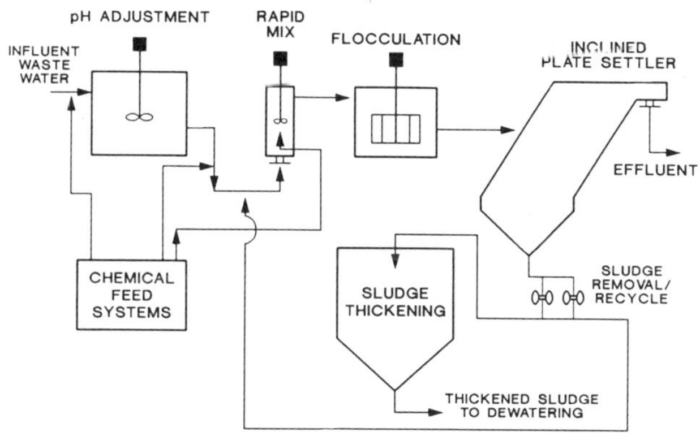

Figure 1. Typical IPS System.

45th Purdue Industrial Waste Conference Proceedings, © 1991 Lewis Publishers, Inc., Chelsea, Michigan 48118. Printed in U.S.A.

- Rapid mix tank
- Flocculation tank
- Sludge removal/recycle
- Sludge thickening

Not all of these are required for each installation, but, when required, they can lead to the success or failure of the IPS system. This section of the paper will discuss the importance of each IPS system component and the design considerations for each.

Inclined Plate Settler

The IPS is the heart of the system. The IPS provides solids removal and some sludge storage. The IPS unit consists of a pack or packs of parallel plates. The plates are typically spaced at 2-inch intervals and are inclined at 45° and 55° angles from the horizontal.

Each manufacturer uses different approaches to ensure even flow distribution through the plate packs. Without some positive means to distribute the flow through the plates, a short circuit occurs reducing the performance of the settler. A short-circuit can occur between the plates when the inlet and outlet do not promote even distribution across the plate. A short-circuit also occurs when the space between one set of plates receives more flow than other spaces. This can be caused by inlet or outlet conditions in the IPS. Several manufacturers prevent this type of a short circuit by having industrial orifices for each plate spacing and by raising the top of the plates above the water level.

pH Adjustment

The pH adjustment system, when required, can directly lead to the failure of the IPS system if it is not designed properly. The most common failure mechanism is that the pH is not adjusted to the required setpoint to precipitate the heavy metals in the waste stream and thus the dissolved metals flow through the IPS. The inability to maintain the required pH setpoint can be caused by inadequate size of the chemical feed system, inadequate size of the mixer in the pH adjustment tank, inadequate detention time in the pH adjustment tank, or a short circuit of wastewater through the tank. Another failure mechanism occurs if the pH adjustment tank mixer is too large or if the baffles are not adequate to prevent vortexing from occurring in the tank. This can cause air to be entrained in the wastewater and to be trapped in the forming floc particles in the flocculation tank. The floc then floats to the top of the IPS and over the weirs or out the orifices. Baffles should be provided to limit vortexing, and the mixer should be sized to pump the wastewater to achieve an effective tank volume turnover rate of 2.5 to 3.0 times per minute.[1] The pH adjustment tank must also have a long enough detention time for a complete chemical reaction, particularly with lime, which might cause scaling in the IPS if the reaction if not complete.

Chemical Feed System

A chemical feed system is required when polymer or other chemicals are needed to remove the constituents of concern from the waste stream. If the chemicals cannot be fed at the required dosage or are ineffective in performing their function, the IPS system will fail. Care needs to be taken in selecting the chemicals before designing the IPS by using jar tests and by adequately sizing the equipment. The equipment need to be sized to allow feeding the chemicals both at several times the dosage estimated and at fractional amounts of the dosage estimated. Polymer storage can also affect performance, since, once activated, polymer has a limited shelf life.

Rapid Mix Tank

The rapid mix tank provides the dispersion of the chemicals being fed into the wastewater. The rapid mix tank is usually provided as part of or as an attachment to the IPS. The rapid mix tank has a chemical injection port that injects the chemical into the center of the waste stream directly below the mixer. Typically, the mixers used are high-intensity, single-speed mixers. The rapid mix tank is normally sized for a 10- to 30-second detention time.

Flocculation Tank

The flocculation tank provides gentle mixing, which results in the floc particles colliding and growing to larger sizes that will settle more rapidly. If there is too much energy, the particles will be sheared to smaller sizes again, and, if there is too little energy, the particles will not collide. The flocculation tank is also normally provided as part of the IPS system.

Sludge Removal/Recycle

The sludge removal is crucial to the success of the IPS system. Without it, the clarifier quickly fills with sludge and ceases to function. Typically, some type of positive displacement pump (such as an air-operated diaphragm pump) is used to remove the solids from the clarifier. The piping also must be sized to prevent plugging, as well as to maintain relatively high velocities during the pumping operation. Two- to four-inch pipe is usually the minimum used. Water flush of the sludge piping can also be provided to clean the sludge pipe.

Sludge recycle is required when the possibility exists for an inadequate amount of solids in the incoming waste stream to promote good floc formation.

Sludge Thickening

Sludge thickening can be accomplished in a separate tank or in a tank that replaces the bottom hopper of the IPS. Usually, if the thickener is part of the IPS, a sludge thickener rake is used to assist in the thickening process.

DESIGN CRITERIA

This section will discuss design criteria used to select the IPS, including design variables, equipment variables, and the need for pilot-scale and bench-scale testing.

Design Variables

Design variables are those variables that set the size of the IPS and the accessories required. Table I lists both the dependent and independent IPS design variables. The independent variables need to be set before the dependent variables are determined.

Wastewater characteristics are the most important independent variable to determine. Time should be taken to collect representative data on at least the flow rate, influent suspended solids, type of solids, and if the waste stream is corrosive. The type of solids includes basic characteristics (e.g., metal hydroxide, precipitate, ash, sand), such as whether the solids are abrasive, the density of the solids, the size distribution, and the "stickiness", (i.e., whether they will adhere to the IPS plates). For most waste streams, information on the type of solids is needed only in general terms; however, if the designer of the manufacturer of the IPS has no experience with the solids, specific data will be required. Normally, the temperature of the waste need only be known in general terms, but, if a high or low temperature is expected, then temperature data should be collected. Tests of wastewater settling characteristics can be performed, but care should be taken to duplicate expected operating conditions, including treatment chemicals used and temperature.

The independent variables of the rapid mix tank are detention time and mixing energy factor. A typical detention time is 10 to 30 seconds (see Table II).[2] The manufacturer's usually have rounded the sizes of rapid mix compartments, so the actual detention time will vary. The mixing energy factor is also an independent variable. Typical mixing energy factors are in the range of 1–2 hp/ft^3/sec.

The flocculation tank has the detention time and the mixing energy factor as independent variables. A detention time of 1 to 4 minutes is typical for a flocculation tank (see Table II).[2] It is best to confirm the detention time through jar testing. The power required for flocculation is very hard to set exactly, because it is very dependent on the floc formed by the polymer used. Thus, typically, the flocculator has an adjustable speed motor to allow for optimizing the energy input while in operation. A reliable manufacturer (based on experience) should be consulted on recommended sizes. Different manufacturers use different types of flocculators. As mentioned previously, which type of flocculator is better is dependent on the floc being formed and the wastewater solids characteristics.

The independent variables of the inclined plate settler are sludge thickening rate, hydraulic loading rate, and sludge storage detention time. The sludge thickening rate is only important to the IPS if a sludge thickener section is being purchased with the IPS. The sludge thickening rate can be determined by bench-scale studies or by manufacturer's data for similar sludges. The manufacturer, however, often provide only one size of thickener for each size of IPS. If it is not large enough, a separate thickener should be considered. The required sludge storage hopper section detention time is usually based on the disposal method of the sludge. Typically, one size of the sludge storage section is available from the manufacturer of the IPS per the size of the IPS. If more volume is required than is available in the IPS sludge storage hopper compartment, a separate tank should be provided. The sludge should never be stored such that it builds up in the plate pack of the settler, because this would reduce performance of the settler.

Table I. Inclined Plate Settler System Design Variables

Independent Variable[1]	Dependent Design Variable: Parameter	Dependent Design Variable: Affected by Independent Variables
A. Polymer Storage, Handling, and Feed	a. Polymer storage requirements	A.1, A.2, A.3, F.1
*1. Polymer type (e.g., cationic, anionic, non-ionic, manufacturer, and make)	b. Polymer equipment required	A.1, A.2
*2. Polymer form (e.g., dry, liquid, emulsion)	c. Sizing of polymer equipment	A.1, A.2, A.3, F.1
*3. Polymer dosage (mg/L)	d. Materials of construction for polymer equipment	A.1, A.2
*4. Point of polymer injection	e. Design of polymer injection location	A.4
B. Rapid Mix Tank	f. Rapid mix tank volume (gal)	B.1, F.1
*1. Rapid mix tank detention time (sec)	g. Rapid mixer motor horsepower range (HP)	B.1, B.2, F.1
*2. Rapid mix tank energy input	h. Flocculation tank volume (gal)	C.1, F.1
C. Flocculation Tank	i. Flocculation motor horsepower range (HP)	C.1, C.2, F.2
*1. Flocculation tank detention time (min)	j. Flocculator design	F.3, F.4
*2. Flocculation tank energy input	k. Materials of construction for rapid mix tank, rapid mixer, flocculation tank, and flocculator	F.6, F.7, F.8, F.9
D. Inclined Plate Settler	l. Inclined Plate Settler settling area (ft^2)	D.1, F.1, F.5
*1. Hydraulic loading rate (gpm/ft^2)	m. Effluent TSS concentration (mg/L)	A.1, A.2, A.3, D.1, E.1, F.1, F.3, F.4, F.5
*2. Sludge thickening rate (lb dry solids/hr-ft^2)	n. Angle of inclined plate settler plates (degrees)	F.3, F.4, F.7
3. Sludge storage section detention time (hours)	o. Material of construction for plates in settling zone	F.5, F.7, F.8, F.9
E. Sludge Recycling and Sludge Wasting	p. Underflow sludge solids concentration (% by weight)	A.1, A.2, D.2, E.1, F.1, F.2, F.3, F.4
*1. Sludge recycling rate (lb/hr)	q. Sludge recycling pump capacity (gpm)	E.1[2]
2. Point of recycling sludge injection	r. Sludge wasting pump capacity (gpm)	E.2, E.3, F.1, F.2[2]
3. Sludge wasting frequency (occurrences per day)	s. Clearances for sludge wasting and sludge recycling pumps and valves (inch)	F.3
4. Sludge wasting duration (min/occurrence)	t. Materials of construction for sludge wasting and sludge recycling pumps and piping	F.5, F.7
F. Wastewater Characteristics		
1. Wastewater flow rate (gpm)		
*2. Influent total suspended solids (TSS) concentration (mg/L)		
**3. Influent TSS particle size distribution and maximum particle size		
4. Influent TSS particle density		
5. Wastewater settling characteristics		
6. Abrasiveness of TSS particles		
7. "Stickiness" of TSS particles		
8. Influent wastewater temperature (°F)		
9. Wastewater corrosiveness		

1. Items marked with an asterisk (*) should be determined from bench or pilot testing; items marked with a double asterisk (**) are optional for testing.
2. Also depends on underflow sludge solids concentration (i.e., Dependent Design Variable Item p).

Table II. Recommended Design Variables

Wastewater	Overflow Rate (gpm/ft^2)	Plate Angle (degrees)	Rapid Mix Time (seconds)	Flocculation Time (minutes)
Metal Finishing				
Plating/Rinses	0.5–0.6	55	10	1
Sulfide	0.5–0.6	45/55	10	1
Precipitation Anodizing	0.1–0.6	55	10	1
Steel				
BOF Scrubber	0.8	45	None	1
Pickle Rinse	0.5–0.6	55	10	1
Mill Scale	1.0	45	None	1
Power				
Bottom Ash	0.8	45/55	None	1
Fly Ash	0.6–0.8	45	10	1
Coal Pile	0.1–1.0	45/55	10–15	0–1
CPI				
Cooling Tower Blowdown	0.5–0.6	55	10	1
Pigment	<0.5	45/55	0–30	0–2
Brine	0.4–0.5	55	10	2

The hydraulic loading rate is a very important variable and is based on the wastewater type. Table II lists hydraulic loading rates for various wastewaters.[2] The numbers are based on data from manufacturers, so, typically, the engineer should decide if a design safety factor needs to be applied. The loading rate can also be determined from bench or pilot studies. Since all IPSs have some parts of the plates that are not fully utilized for settling, some manufacturers recommend using an effective area of 80% of the total projected plate areas for design sizing. The angle of incline for the plates is based on an angle that will allow for a free discharge of the solids from the plates. The typical angles available from IPS manufacturers are 45° and 55° angles from the base horizontal line. The slope depends on the nature of the solids to be separated. For heavy solids, such as sand, etc., the plates are usually placed at a 45° angle; while for light solids, such as metal hydroxide sludge, the plates are inclined at a 55° angle. The angle is set to allow for continuous removal of the solids by gravity from the plate pack. The 45° incline provides more settling area, 24% per plate of equal size, than does the 55° incline. The most important consideration for selection of the angle is whether the plate pack will be self cleaning for the sludge particles generated. If the designer is unsure, the manufacturer might be able to give a recommended incline for the plates. If the sludge is border line, the designer should select the steeper slope. Table II lists various wastewaters versus plate angles.[2] If you have a special or "non-standard" waste, the manufacturer should be consulted or a pilot test conducted.

The independent variables of the polymer storage, handling, and feed are polymer type, polymer form, polymer dosage, and point of injection. Polymer type(s) needed are dependent on wastewater characteristics. The polymer types are cationic, anionic, and non-ionic. Each manufacturer usually has several formulations of each type that have been developed for different wastewaters. Jar testing should be performed to identify the polymers that are most effective and the appropriate optimum polymer dosage range. The polymer form should also be considered when selecting a polymer. Considerations are performance, owner preference, capital costs for polymer handling equipment, polymer cost, and ease of use. The polymer forms include the following: dry, which is in a granular form; emulsion, which is typically 50% polymer and 50% oil; and solution type, which is 2 to 8% polymer and which is very viscous. The polymer dosage rate is typically determined initially through jar testing, and is then adjusted during operation to optimize performance. The point of polymer injection is typically in the bottom of the rapid mix compartment just below the rapid mixer impeller. Care must be taken not to shear the polymer with the rapid mixer. In two polymer systems, one polymer is usually injected into the pipe line and dispersed upstream of the rapid mix compartment.

The independent variables of sludge wasting and sludge recycling are sludge wasting duration, sludge wasting frequency, and sludge recycling rate. The sludge wasting duration, expected sludge concentration, dry weight of generated sludge and sludge wasting frequency need to be set initially in order to size the sludge pumps. The pump sizing also has to include sludge pipe velocity considerations. The sludge recycling rate only has to be considered if the solids in the wastewater might be of a concentration that is too low to promote flocculation or for certain types of wastewater (e.g., fly ash, bottom ash, BOF scrubber, mill scale). Often, the sludge recycling pump is the same size as the waste pump, thereby allowing it to be used as a standby for the waste pump. The sludge pumps usually have variable pumping rate capacity.

Equipment Variables

Each manufacturer of IPS units designs them slightly differently with many of these differences patented. The design professional's job is to sort through the various manufacturers, designs, and claims, and to select acceptable IPS units. This section will discuss some of these differences and how they may affect performance or selection.

Materials of construction are standard for all manufacturers. Typically, the tank is carbon steel with the plates constructed of fiber reinforced plastic (FRP). Several manufacturers offer optional designs of stainless steel tanks, while others offer FRP tanks. For highly corrosive environments or abrasive solids, the material of construction of the tanks, and related piping values etc. should be evaluated. Under normal industrial wastewater treatment conditions, the coated carbon steel is sufficient, but it will require periodic painting.

The only difference between rapid mix tanks from various manufacturers are detention time and mixer sizing. The detention time should be above a minimum, but, if it is too large, extra mixing horsepower will be required. Mixer horsepower should be large enough to mix the wastewater and the chemicals. Too large of a mixer will increase costs, but, it will also cost the manufacturer more, it is unlikely that they will supply an oversized mixer.

The flocculation tanks will vary in detention time, with a minimum being required, but the main concern is the flocculator mechanism. Several manufacturers use the paddle-type flocculator mechanism that is similar to the type that has been used in water treatment plants for years. For light floc, a rake flocculator is preferred. Another standard type is the propeller mixer flocculator. The propeller-type flocculator tends to shear the particles that come close to the propeller. The propeller type though has been used effectively and is preferred with heavy solids that might settle in the flocculation tank. The normal detention time is 1 to 4 minutes in the flocculation tank, but this time is dependent on the solids and the floc being formed.

The IPS unit has many differences that are unique to different manufacturers, including plate spacing, flow distribution devices, sludge removal, and feed zone. The majority of the manufacturers use a standard 2-inch plate spacing. Some manufacturers use a smaller plate spacing (e.g., under 1.5 inches) to achieve a higher effective service area per unit volume. The main problems with the smaller plate spacing is the potential for plugging of the plate. In addition the smaller spacing causes interferences between the sludge sliding down the plate with the water flowing up the plates. The flow distribution devices are the means of distributing the flow through the plates. This is the most unique item to each manufacturer. The most effective means to distribute the flow is the use of an orifice per plate spacing with the plates extending above the water level, thus preventing short-circuiting of flow up one spacing and over top of the plates to other orifices. This design guarantees that each spacing receives equal amounts of flow. Another design is with a launder positioned above the plates with the orifice spaced above the plate spacing. A gap of several inches is left between the orifices and the top of the plates. At one installation, it was noticed that flow was coming up the plate pack and carrying solids across the top of the plates with this design, during a buoyant solids problem. Thus, flow through all the plate spacing can be uneven. Some manufacturers try to distribute the flow using devices on the influent to the plates. These devices will break up the floc particles if they are designed to be hydraulically effective. Many of the manufacturers rely on sludge hoppers to facilitate removal of the sludge with some units having multiple hopper design. These units are taller units. Some manufacturers use sludge scraper or screw mechanisms to assist in removal of the sludge. Thus, the IPS units have flat or nearly flat bottoms. These units usually use up more area, but are shorter. If the facility can accommodate the taller units with the non-mechanical sludge hopper, they are the best selection because they do not have moving parts under the water to repair.

The IPS feed zone varies with different manufacturers. Several manufacturers use a common feed zone in the middle of the unit with the water entering the plate packs through an opening in the side of the plate pack near the bottom. Other manufacturers allow the water to enter the bottom of the plate pack. It has been postulated by others, that the bottom feed tends to resuspend the solids as they fall from the bottom of the plates. The side entrance feed causes the lower portion of the plates to be dysfunctional for settling because the feed enters above it. With these units, the manufacturer, for the most part, recommends using an effective settling area of 80% of the total plate pack projected area for design. It is probably justified to use 80% effectiveness for the bottom feed units also to take into account any resuspension of solids.

Table III. Start-Up Versus Full Operation of an IPS System[a]

IPS System Status	Duration (months)	Average Effluent TSS (mg/L)[b]
• Temporary system during construction	6	31
• Start-up of permanent system	1	19
• Initial long-term system performance	18	8
• Recent system performance	5 years	< 5

[a]See Reference 3 for a description of the IPS system at this plant.
[b]Based on daily 24-hour composite samples over the duration shown.

Other items to consider in the selection of IPS units are platforms, extension legs, sludge draw-off ports, and vibrators. The platforms provide access to the rapid mix and flocculator drive motor and also allow inspection of performance. The platforms typically are provided with ladders. If possible, some type of stairs should be provided for easier access. Extension legs are required if the IPS is followed by a taller gravity filter that hydraulically requires the IPS unit to be higher. The IPS unit should be provided with sludge draw-off ports to allow spot checking of the sludge in the sludge hopper. A vibrator is sometimes used to vibrate the sludge hopper. This may help thicken the sludge as well as prevent rat holing of the sludge while it is being pumped out.

PILOT-/BENCH-SCALE TESTING

The need to pilot- or bench-scale test varies with each application. If a site's wastewater is sufficiently similar to another manufacturing plant, no testing is needed. If the type of wastewater is generally known, then bench-scale testing will suffice in providing the needed information. A bench test should be conducted to test the effectiveness of the system, especially for metal plating wastewater if chelating agents may be present. Tests should also be conducted to determine the best polymer types to assist in sizing and selecting the polymer equipment. If the general characteristics of the wastewater are unknown, especially settleability, the angle of incline required, or the potential for particles to stick, over time, to the plates and plug them, pilot testing is recommended subsequent to bench testing to narrow the unknown parameters. For some cases, it may be possible to send a sample of the wastewater to an IPS manufacturer to run the test, but, if the wastewater may degrade or change over time, the test should be conducted on-site.

OPERATIONS CONSIDERATIONS

All too often after well-designed IPS systems are installed, they do not achieve the desired performance. This can occur when the IPS system is considered by plant operating personnel to be a "black box" which is given no operating attention. However, major improvements in IPS system performance can sometimes be made through simple operating changes which do not require significant design modifications. An example of this is a coal-fired power plant which uses two IPS units in parallel to treat a varying mixture of bottom ash wastewaters, coal pile runoff, metal-cleaning wastewaters, boiler blowdown, and other suspended solids-laden wastewaters at the plant.[3] One such plant installed half of the IPS treatment system on a temporary basis to achieve a regulatory compliance deadline, while the remainder of the system was under construction. The IPS system was designed to achieve an effluent TSS of 25 mg/L. Table III shows how the plant operating personnel have been able to dramatically reduce the effluent TSS concentration over a period of several years by improving IPS system operation.

Table IV shows the operating variables for an IPS system. Although this table shows that there are a large number of operating variables, only three simple day-to-day operating variables (i.e., polymer dosage, sludge recycling rate, and sludge wasting rate) need to be controlled by the operator. Therefore, operation of an IPS system is relatively simple once initial start-up operational settings have been established. Typical operator time for IPS system operation is 1–2 hrs/day for an 8-hr/day operation and 3–5 hrs/day for a 24-hr/day operation. These time estimates include collecting wastewater and sludge samples, performing simple operation and maintenance activities, and completing standard operating logs.

Table IV. Inclined Plate Settler System Operating Variables

Independent Variables	Dependent Variables
1. Influent Wastewater Flow Rate and Temperature[a]	a. Effluent TSS concentration
2. Wastewater Solids Characteristics[a]	b. Underflow sludge solids concentration
• Influent TSS particle size	
• Influent TSS particle density	
• Influent TSS particle surface charge	
3. Primary Start-up Variables:[b]	
• Polymer type and form	
• Rapid mixer speed	
• Flocculator speed	
4. Secondary Start-up Variables:[b]	
• Point of polymer injection	
• Point of sludge recycle injection	
5. Day-To-Day Operating Variables:[c]	
• Polymer dosage	
• Sludge recycle rate	
• Sludge wasting rate	

[a]Wastewater characteristic variable—not subject to operator control.
[b]These variables are normally set during initial start-up and not normally changed during subsequent operation.
[c]These variables are normally controlled during daily operation to optimize IPS system performance.

The following are recommendations for initial start-up and ongoing operation to help achieve optimum IPS performance.

Initial Start-Up

During the initial IPS system start-up, the primary start-up variables shown in Table IV should be optimized for the system. Typically, this optimization is performed under the direction of a competent professional (such as the IPS supplier's start-up personnel or the design consultant's start-up engineer). Once these variables are set during initial start-up, they do not normally need to be changed by plant personnel. During start-up, initial settings are also established for the polymer dosage and sludge recirculation rate. However, at this time, it is more important to optimize the primary start-up variables, as the operators will be able to later optimize these day-to-day variables as they become familiar with the operation of the IPS system.

However, should the wastewater flow rate, temperature, or solids characteristics subsequently change significantly, the settings for these primary start-up variables should be re-evaluated based on the changed conditions. Typically, the operator would notice that the solids characteristics have changed either directly (through knowledge of modifications to the processes generating the wastewater being treated or visual observation of the wastewater) or indirectly (through a significant long-term increase in the effluent TSS concentration and/or decrease in the underflow sludge solids concentrations).

Sometimes even when the primary start-up variables are optimized, the desired system performance is still not achieved. In such cases, it is then necessary to evaluate the secondary start-up variables listed in Table IV. Sometimes, seemingly minor modifications to the locations of polymer and/or recycling sludge injection location(s) can significantly improve IPS performance. Strictly speaking, these secondary start-up variables are design variables (as shown in Table IV). However, temporary changes to the polymer and/or recycling sludge injection location(s) can generally be made quite easily, so that the effect of changes in these variables can be investigated during start-up, if necessary.

In some instances, even adjusting these secondary start-up variables does not result in achieving the desired IPS system performance. In such cases, it may be necessary to re-evaluate the key design variables upon which the IPS system was based, particularly the hydraulic loading rate. The problems may be due to changed wastewater characteristics since the original settling tests were conducted.

Ongoing IPS Operation

Once the settings for the initial start-up variables have been established, ongoing IPS system operation consists of controlling the day-to-day operating variables to maintain peak system operation. These day-to-day variables are also listed in Table IV. Many IPS systems are designed with automatic control of the polymer feed and sludge recycling rates. However, even for these systems, the operator should frequently check the effectiveness of the polymer feed and sludge recycling rates through simple jar tests similar to those used during initial start-up. The operator should also measure the sludge level in the sludge storage compartment of the IPS, and to control the amount of sludge wasting so as to maintain the sludge level within reasonable limits.

During day-to-day operation, the operator should collect wastewater samples from the flocculation tank and visually check the samples for signs of a well-flocculated wastewater (i.e., large and strong flocs, rapid floc settling rate, and clear supernatant). The operator should also collect samples from the sludge hopper and note the sludge volume of the sample after a fixed amount of time (typically 15 to 30 minutes) for settling. Significant deviations in floc quality or sludge volume from a previous days' results should signal a change in IPS system performance, and should trigger an investigation by the operator into the cause(s). Plant operating personnel should also consider periodically performing additional simple operational tests (such as jar settling tests at various polymer dosages and sludge recirculation rates) to verify that these variables are being maintained at their optimum for peak IPS performance. By building up a database of IPS system performance over an extended period of time, the operating personnel will develop a perspective in which to place day-to-day variations in treatment performance, and will be able to use this database to guide daily system operation as well as to provide a source of information to assist operators in diagnosing and correcting problems during periods of system upsets.

CONCLUSION

The IPS unit does not exist by itself but is part of an overall system. Care needs to be taken in design of the IPS unit and the IPS system for the IPS unit to work properly. The engineer needs to take into account the design variable presented in Table I when designing an IPS to achieve the optimal design and performance. The unit also needs to be operated correctly and the operating variables adjusted over time as needed to optimize performance.

REFERENCES

1. Johannes, R. D., et al., "Electroplating/Metal Finishing Wastewater Treatment Practical Design Guidelines" *Proceedings of the 43rd Purdue Industrial Waste Conference*, May 1988.
2. Brett Boyd, Parkson Corporation, Personal Correspondence.
3. Chiesa, R. and Bangert S., "Achieving the EPA Effluent Limitations for Power Plants: Approaches Taken by Utilities in Wisconsin", *Proceedings of the 35th Purdue Industrial Waste Conference*, May 1980.

66 SEMIVOLATILES SCREENING AT CERCLA SITES

Roy O. Ball, Principal
Mary Jo Anzia, Project Engineer
Environmental Resources Management
North Central, Inc.
Deerfield, Illinois 60015

INTRODUCTION

The Comprehensive Environmental Response, Compensation and Liability Act of 1980 (CERCLA) and the Superfund Amendments and Reauthorization Act of 1986 (SARA) require the investigation and subsequent cleanup of hazardous waste sites listed on the National Priorities List (Superfund sites). The history of Superfund site activities is frequently uncertain. In addition, disposal and/or discharge records for Superfund sites are often inaccurate and incomplete. Thus, the nature and extent of contamination is unknown and is usually highly variable over short distances. As a result, thorough investigations of the soil and ground water at Superfund sites are generally required as part of the site Remedial Investigation (RI).

The collection and analysis of investigative soil samples utilizing the Contract Laboratory Program (CLP) protocol, which is required by the United States Environmental Protection Agency (USEPA) for Superfund sites, is tedious and expensive. The soil investigation can be accelerated by conducting a soil screening program to delineate areas of suspected contamination, as well as areas of little or no contamination. By implementing a screening program, the number of soil samples required for CLP analyses can be significantly reduced. Generally accepted soil screening methods exist for volatile organic compounds (HNu headspace technique), metals (plasma AA), PCBs (Clor-N-Soil™), asbestos containing materials and petroleum hydrocarbons (Total Petroleum Hydrocarbons). However, a literature review did not identify any generally accepted soil screening tools for semivolatile organic compounds. This paper presents a discussion of the theory and use of fluorescence as an indicator of soil contamination due to the presence of fluoresceable semivolatile organic compounds.

THEORY OF FLUORESCENCE

Fluorescence is the process in which radiation is emitted by molecules or atoms that have been excited by absorption of radiation.[1] Energy levels of an atom can be represented by a series of horizontal lines. The difference in energy between energy levels is represented by the vertical distance between two lines. If level N represents the ground state, or the state in which the atom is normally found, by absorption of light of frequency ν_{FN}, the atom is raised to level F and, if no other energy levels exist between N and F, can return to N only by reemission of light of the same frequency (ν_{VN}). This type of fluorescent emission in which the wavelengths of the emission and absorption lines coincide is called resonance radiation.[2] Several levels (C,D, etc.) can exist between N and F. Under these conditions, other transitions can occur, which result in the emission of frequencies smaller than ν_{FN}. The empirical law stating that the wavelength of light emitted by a fluorescent material is longer (lower energy) than that of the radiation used (to excite the fluorescence was formulated by Stokes[3].

Atoms and molecules that fluoresce have well-defined excitation and emission spectra. If an organic compound is fluorescent, its emission band corresponds to the same electric transition as its absorption band of the greatest wavelength (i.e., its first absorption band).[4] There is an approximate mirror-image relationship between the shape of the absorption spectrum and the shape of the fluorescence spectrum.[5] In general, the fluorescence yield is independent of the wavelength of the exciting light.[6] The quantum efficiency of a molecule is the ratio of the total emitted light to the total absorbed light. Therefore, the quantum efficiency is also independent of the wavelength of the exciting light. If the exciting light used is of a wavelength that is different from that of the absorption peak, a smaller portion of the light will be absorbed and proportionally less light will be emitted-illustrating the constancy of the quantum efficiency. However, the shape and location of the emission spectrum will not change.[7]

45th Purdue Industrial Waste Conference Proceedings, © 1991 Lewis Publishers, Inc., Chelsea, Michigan 48118. Printed in U.S.A.

Table I. Representative Excitation and Emission Frequencies[a]

Compound	Solvent	Excitation Frequency (nm)	Emission Frequency (nm)
Acenapthene	pentane	291	341
Anthracene	pentane	420	430
Benzo(a)anthracene	pentane	284	382
Benzoic acid	70% H_2SO_4	285	385
Chrysene	pentane	264	381
Fluoranthene	pentane	354	464
Fluorene	pentane	300	321
Phenanthrene	pentane	252	362
Pyrene	pentane	330	382

[a]Schulman, pp. 213–229.

Instrumentation for the measurement of fluorescence generally consists of a light source (to electronically excite the sample), a monochromator (to separate the light of desired energy from the source), a sample compartment, a second monochromator (to isolate the sample's fluorescence energy from the excitation energy), a photodetector (to translate the fluorescent light into an electrical signal), and a readout system (to determine the intensity of fluorescent light emitted).[8]

FLUORESCENCE SCREENING METHODOLOGY

Prior to beginning the field screening, several determinations must be made. First, the fluorescence screening procedure requires an approximate definition of the known or suspected contamination so that excitation and emission frequencies of the compounds can be tabulated and used to design necessary filters for the fluorometer. Excitation and emission frequencies for several semivolatile organic compounds are presented in Table I. The choice of excitation and emission wavelengths can affect the sensitivity of a fluorescence test. The correct choice of wavelength filters minimizes the fluorescence of blanks and compounds of little concern while retaining high sensitivity for the compound of interest. Once the wavelength is determined, the appropriate filters can be obtained from the fluorometer manufacturer.

Secondly, the extracting solvent must be selected. In general, the solvent must be capable of rapidly dissolving the compounds of interest. The solvent should not readily emulsify in the presence of water or soil and should have little or no fluorescence at the wavelengths of interest. Finally, the solvent should have as low a toxicity and as high a flash point as practical. Under most circumstances, hexane or iso-octane are suitable choices.

Finally, the approximate densities of the predominant soils at the site must be determined. This can usually be accomplished without laboratory analysis of the soils by visual inspection and the selection of standard densities based on soil type.

Upon completion of the preliminary determinations, soil collection can begin. The fluorescence screening procedure is best accomplished in a protected area, such as a field trailer. This is necessary for protection of the fluorometer and will likely provide more accurate measurements. The soil samples can be collected in an inexpensive, sealable glass jar, such as a Mason jar. The soil samples should be screened for fluoresceable materials as rapidly as possible after collection within a uniform time interval. The following procedure is used to determine the fluorescence of the soil:

1. Approximately 40 to 50 grams of the sample is placed in a 125 milliliter (mL) Erlenmeyer flask.
2. An equal volume of extracting solvent (soil weight/density) is placed in the flask, and the flask is agitated for a period of at least 15 minutes.
3. The contents of the flask are then decanted to produce solid-free extract. Alternatively, for soils with a very high fines content, the flask contents are centrifuged with the same objective.
4. The solid-free extract is used to prepare a serial dilution set using the extracting solvent as the diluent. In most cases, either the first extract or one serial dilution will be sufficient to obtain a relative fluorescence reading.
5. If desired, known amounts of reagent contaminants similar to those expected at the site can be added to the native soil matrix and a response curve developed.

The actual fluorescence value is then calculated using the relative fluorescence reading, the gain setting of the fluorometer, the dilution factor, and the fluorometer aperture required, as shown in the equation below.

$$FL_A = \log_{10} \frac{FL_R \times DF}{GAIN \times APER} \tag{1}$$

where: FL_A = actual fluorescence
FL_R = relative fluorescence
DF = dilution factor
GAIN = fluorometer gain setting
APER = fluorometer aperture required

The $\log_{10}$ (two significant digits) of the actual fluorescence value is generally presented to avoid overstating the precision of the field test.

PROBLEMS AND INTERFERENCES

There are several circumstances that can cause interferences in the fluorescence technique and corresponding variability in fluorescence readings. For instance, if the sample extract opacity is poor, lower fluorescence readings may be obtained due to reflection of the excitation and emission light. The sample extract may be centrifuged to improve the opacity and minimize interference. Another problem that may be encountered in field screening is the presence of precipitation in the sample extract. If precipitation is present in the sample extract, the extract can be allowed to stabilize before obtaining a fluorescence reading.

A sample may contain hardened tars which may not dissolve in the selected solvent with shaking. If this is the case, a Soxhlet apparatus and the extraction procedure described in Method 5520D of *Standard Methods*[9] may be used to produce a useful sample extract.

A final problem is that of reproducibility. The fluorescence screening results may exhibit significant variability. However, as a field screening technique, the procedure provides results which satisfy the screening objectives. As previously stated, the results should be reported as $\log^{10}$ values (with 2 significant digits) and applied on this basis.

The fluorescence tests can be performed at the rate of 14 samples/hour, with one analyst using a single shaker and fluorometer (assuming the sample extracts do not have to be centrifuged). By increasing the number of shakers, up to 28 samples/hour can be processed. The test equipment cost (1989) is approximately \$5500 for fixed equipment (excluding the centrifuge) and \$200 for disposable supplies per hundred samples.

SEMIVOLATILE SCREENING APPLICATIONS

Soil screening for fluoresceable semivolatile organic compounds has been conducted at several Superfund sites. The following section discusses the site-specific methodology, screening results, and correlation of the screening results with CLP semivolatile organic compound analytical results at one of the sites.

Indiana Superfund Site

The Indiana Site is an approximately 32-acre site on the National Priorities List, which was operated as an unpermitted landfill for 17 years. A variety of wastes, including foundry sand, household waste, slag, and industrial sludge were accepted, and no apparent landfill segregation of the wastes was performed. To minimize the number of soil samples for CLP analyses necessary to characterize the site, an extensive soil screening program was proposed and approved by the USEPA, Region V.

As a basis for soil sample locations, a 100-foot grid was established over the site. Soil screening samples were collected from each grid intersection at depth intervals of two feet, to a minimum total depth of ten feet. The samples were field screened for several possible contaminant fractions, including semivolatile organic compounds. The methodology described above was used to obtain fluorescence readings for 442 soil samples. Upon completion of the soil screening activities, locations for ten soil samples were selected. These soil samples were analyzed for the Target Compound List (TCL) semivolatile organic compounds, among others.

Table II. Semivolatile Screening and Analytical Results

Sample	Total TCL Semivolatiles (ppb)	Actual Fluorescence (LOG_{10})	Soil Type
A	25045	-1.02	foundry sand
B	0	0.60	brown sand
C	5704	3.35	foundry sand
D	3995	3.27	foundry sand
E[a]	5134	3.27	foundry sand
F	1282	2.59	foundry sand
G	4020	2.22	foundry sand
H	690	2.27	foundry sand
I	490	0.83	olive sand
J	137	1.71	olive sand
K	240	2.65	clay cap

[a]Field replicate sample of D.

In addition to aiding in the selection of analytical soil sample locations, the extensive screening program also provided a clear description of the composition of the shallow soils and fill at the site. The site consists of six predominant geologic facies: 1) fill material, 2) clay cap material, 3) foundry sand, 4) brown sand, 5) olive gray sand, and 6) olive gray clay. The first three geologic facies are materials that were placed on the site during landfilling operations, while the last three are naturally-occurring. The soil screening actual fluorescence values ($\log_{10}$), TCL total semivolatiles concentrations and soil type for each of the ten soil samples collected are shown in Table II.

The fluorescence screening and semivolatile results were tested for correlation using Olmstead and Tukey's corner test for association.[10] A graph of the results is shown on Figure 1. As shown on Table II, Sample A contains the highest total semivolatiles concentration, yet has an uncharacteristically low actual fluorescence value. The data set for the entire boring at that location indicates that the contamination at that point is highly variable over small distances. Therefore, this sample point has been eliminated for the statistical analyses. Olmstead and Tukey's corner test shows positive correlation of the fluorescence screening data and the total semivolatile organic concentration at the 97% confidence level. The data shown in Figure 1 were also analyzed by linear regression and showed a correlation coefficient of 0.74. This correlation coefficient for eight degrees of freedom is significant at the 99% confidence level.

The actual fluorescence values were also grouped by geologic facies, as shown on Figure 2. The distinct grouping of the fluorescence data by soil type and the higher mean values associated with foundry sand and fill materials demonstrate the objective basis of the fluorescence screening. The higher actual fluorescence values obtained from the fill and foundry sand are indicative of the elevated concentration of semivolatile organic compounds found in these two facies.

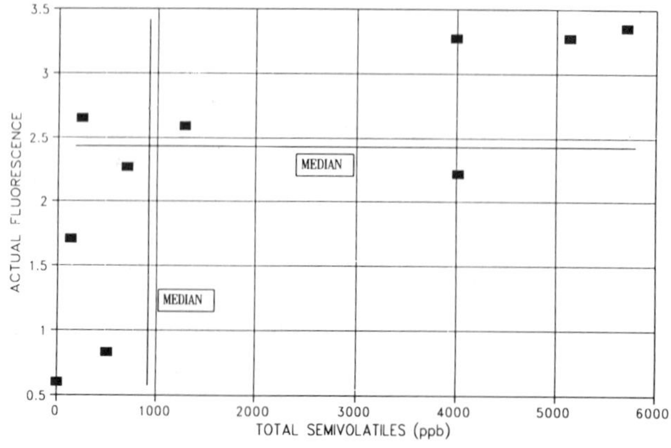

Figure 1. Semivolatile correlation.

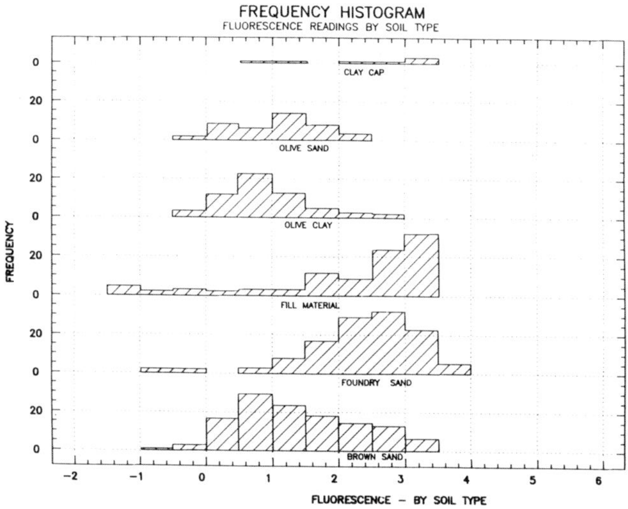

Figure 2. Frequency histograms

SUMMARY

The collection and analyses of soil samples is a requirement of nearly every investigation of CERCLA sites. Soil screening is often conducted prior to collection of soil samples for TCL analyses to characterize the site and aid in the selection of soil sample locations. Generally accepted screening techniques exist for volatile organic compounds, metals, PCBs and asbestos-containing materials. However, there are no generally accepted screening techniques for semivolatile organic compounds. The fluorescence screening technique is a useful procedure for determining the relative presence of semivolatile organic compounds in soil.

The fluorescence screening technique results, like other field screening results, have inherent variability. However, as an initial indicator of the extent and location of semivolatile organic compound contamination, the technique provides satisfactory data and can provide reliable direction for the appropriate location and number of soil samples for TCL analyses.

ACKNOWLEDGMENT

The use of fluorescence as a screening tool for semivolatiles was initially developed by Roger A. Minear, now Director of the Institute for Environmental Studies at the University of Illinois at Urbana-Champaign, and one of the authors (Ball) at the University of Tennessee in Knoxville in 1977.

REFERENCES

1. Ralph S. Becker, *Theory and Interpretation of Fluorescence and Phosphorescence* (New York: Wiley Interscience, 1969), p. 1.
2. Ibid., p. 87.
3. Robert C. Weast, ed., *Handbook of Chemistry and Physics*, 55th ed. (1918; rpt. Cleveland, Ohio: CRC Press, 1974), p. F-110.
4. Peter Pringsheim, *Fluorescence and Phosphorescence* (New York: Interscience Publishers, Inc., 1949), p. 394.
5. Becker, p. 89.
6. Pringsheim, p. 307.
7. Sequoia-Turner Corporation, *Operator's Reference Manual—Model 450 Digital Fluorometer* (Mountain View, California, September 1987), p. B.1.
8. Stephen C. Schulman, *Fluorescence and Phosphorescence Spectroscopy: Physiochemical Principles and Practice* (1977; rpt. Elmsford, new York: Pergamon Press, 1979), p. 134.
9. Lenore S. Clesceri, Arnold E. Greenberg and R. Rhodes Trussel, eds., *Standard Methods for the Examination of Water and Wastewater,* 17th ed. (1917; rpt. Baltimore, Maryland: Port City Press, 1989), p. 5-45.
10. Robert R. Sokal and F. James Rohlf, *Biometry* (San Francisco: W. H. Freeman and Company, 1969), p. 538.

Section Eight
LANDFILLS – A. GENERAL

67 SANITARY LANDFILL DESIGN: INWARD HYDRAULIC GRADIENT VS. OUTWARD HYDRAULIC GRADIENT

Richard W. Eldredge, President
Eldredge Engineering Associates, Inc.
Naperville, Illinois 60563

Timothy A. Bannister, Senior Hydrogeologist
Wehran Engineering Corporation
Naperville, Illinois 60540

This paper compares the significant differences between landfills which are sited and designed based on inward hydraulic gradient as opposed to those with an outward gradient. Specific examples of both type of landfills will be discussed. Current design trends favor the outward gradient and the exclusion of all moisture sources for an extended period of time. The authors believe that these design requirements are simply postponing the normal biodegradation of the organic waste within a municipal waste landfill and suggest that inward gradient sites may prove to be a better design alternative.

An alternative method of siting and design based on inward hydraulic gradients would stimulate digestion at a controlled rate or with controlled collection and disposal of biodegradation by-products.

The authors believe that:

1. If all liners leak, the site designer should control the direction of the flow for the period of time that leaching characteristics have environmental consequence.
2. It is better to remove a minor amount of water which infiltrates to establish an inward gradient than to be required to remove almost all of the infiltration and still have an outward gradient condition.

INTRODUCTION

Recent attitudes in sanitary landfill design (particularly in the regulatory community) have resulted in the impression that sanitary landfills sited well above groundwater levels are always preferable to those sited below groundwater levels. At first glance, siting landfills above groundwater levels seems to be a preferable design strategy, especially if the concept of total isolation of wastes from the environment is considered paramount. However, it has been the experience of the authors that a design strategy based primarily on "isolation" from (i.e. distance above) the surrounding groundwater regime is perhaps ignoring possible advantages to siting a landfill below groundwater level.

The purpose of this paper is to raise the issue of outward hydraulic gradient design versus inward hydraulic gradient design. The following sections present what we mean by outward gradient and inward gradient designs, three major aspects to this issue, and a discussion of examples of each type of design.

OUTWARD AND INWARD HYDRAULIC GRADIENT DESIGNS: DEFINITION

Schematics of outward and inward hydraulic gradient designs are presented in Figures 1 and 2. It is assumed for this discussion that both landfill designs have a properly designed clay and/or membrane liner overlain by a leachate collection system. Note that regardless of how low the leachate level becomes within the site in Figure 1, the hydraulic gradient is always outward. In order to maintain the inward hydraulic gradient, shown in Figure 2, the level of leachate must be lowered to a level equal to or below the outside groundwater level.

45th Purdue Industrial Waste Conference Proceedings, © 1991 Lewis Publishers, Inc., Chelsea, Michigan 48118. Printed in U.S.A.

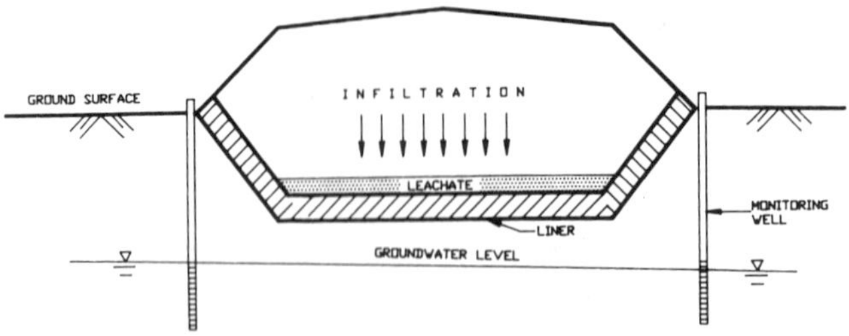

Figure 1. Outward hydraulic gradient.

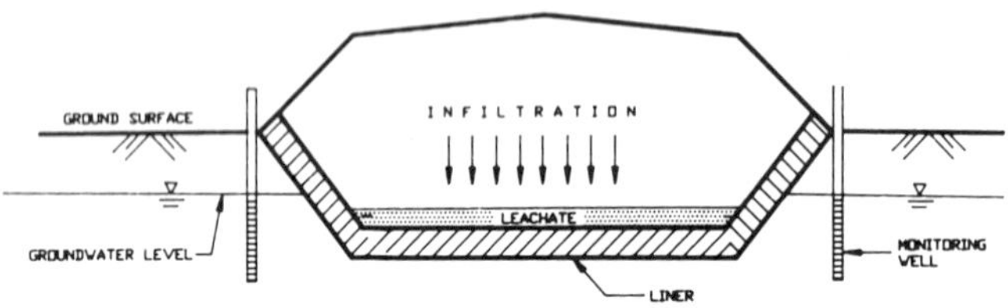

Figure 2. Inward hydraulic gradient.

With these rather simple, yet clear definitions, we can now examine three major aspects of each siting design.

LINER LEAKAGE

It has been said that "all landfills leak". This statement, though seemingly rather cynical, is probably accurate. The critical question, however, is "*how much* does a landfill leak and what will be the impact on the environment?"

Given the large surface area under landfills, the variablity of naturally occurring underlying soils, construction variables, and other factors, it is safe to conclude that leakage through landfill liners will occur.[2,3] Skyscrapers built in earthquake country will undoubtedly sway; the goal is not to keep them from swaying, but to control the swaying in such a way to prevent a catastophe. Likewise, because landfill liners (due to many different factors) will undoubtedly leak, the goal should be to control the leakage in such a way as to reduce the impact upon the environment, specifically groundwater.

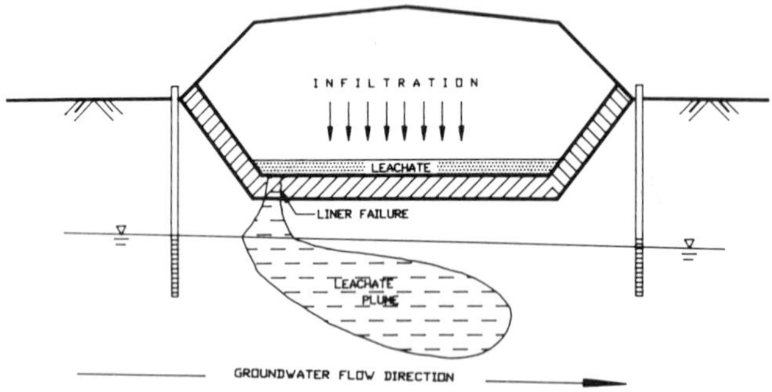

Figure 3. Liner leak under outward gradient scenario.

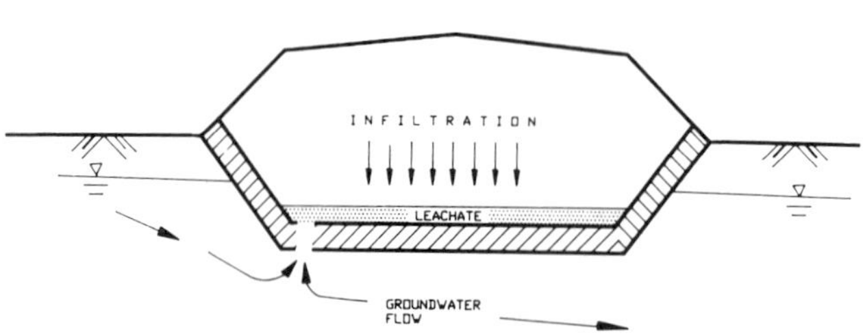

Figure 4. Liner leak under inward gradient scenario.

The anticipated liner leak in an outward gradient site as shown in Figure 3 will make its way due to gravity, through the unsaturated zone where it then joins the saturated zone and moves along with the prevailing groundwater flow direction. Properly placed monitoring wells (vertically as well as horizontally) identify the plume before adjacent properties are affected.

With a similar size leak in Figure 4, we see that due to the inward hydraulic gradient, surrounding groundwater flows into the site, becomes mixed with leachate, and outside groundwater remains unpolluted. A leak has occurred but has been controlled in a way that protects the environment. One may argue that a negative side to this is that a greater volume of leachate must be collected and treated. However, depending on the size of the leak, there is a good chance that the cost of handling and treating the additional leakage into the site of Figure 4 may be considerably less than investigating, handling and treating the leachate plume in Figure 3.

Perhaps another obvious benefit to the site design of Figure 4 is that the landfill owner is rewarded for the installation of a better liner, knowing that the less secure the liner, the more leachate must be treated, the higher the operating costs and the fund that must be set aside for post-closure maintenance. On the other hand, in Figure 3, it may be decades before a significant plume is discovered, perhaps even beyond the currently regulated post-closure period, at which time the landfill owner (and the post-closure funding) may well be gone.

BIODEGRADATION

There is a considerable amount of literature on anaerobic biodegradation in sanitary landfills. Though there may be disagreement on the length of time of primary methane generation in a landfill[1] it is widely accepted that one of the major factors favoring enhanced biodegradation is increased moisture content. Little, if any, anaerobic digestion takes place in the absence of moisture. Not surprisingly, a saturated fill will provide the best environment for anaerobic digestion, waste degradation, and early settlement.

Figure 5 shows that under outward hydraulic gradient conditions, the goal is try and keep the fill as dry as possible so that a minimal head of leachate is maintained resulting in minimal infiltration from any source. Unfortunately, this counters anaerobic digestion and simply prolongs the potential for biodegradation. Under inward hydraulic gradient conditions, Figure 6, the volume of saturated fill undergoing anaerobic digestion is potentially larger because the goal is no longer to completely remove all leachate to reduce pressure gradients, but only to remove enough leachate to assure that an inward gradient is consistently maintained.

It is not suggested here that outward gradient sites do not undergo anaerobic digestion. There are potentially many locally saturated or high moisture zones above the landfill base where anaerobic digestion occurs. There is the potential, however, of even greater anaerobic biodegradation if site design permits a significant accumulation of water within the landfill.

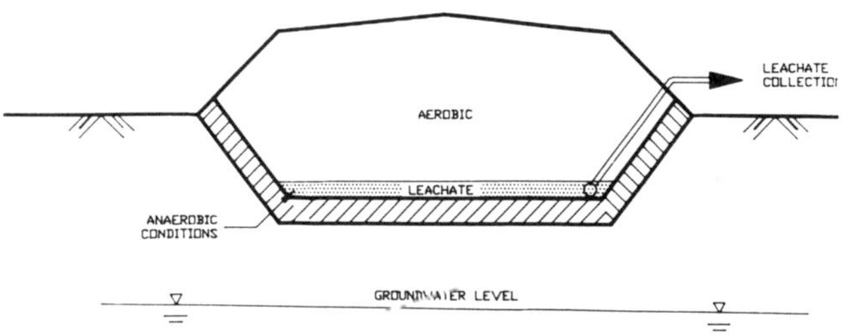

Figure 5. Biodegradation under outward gradient conditions.

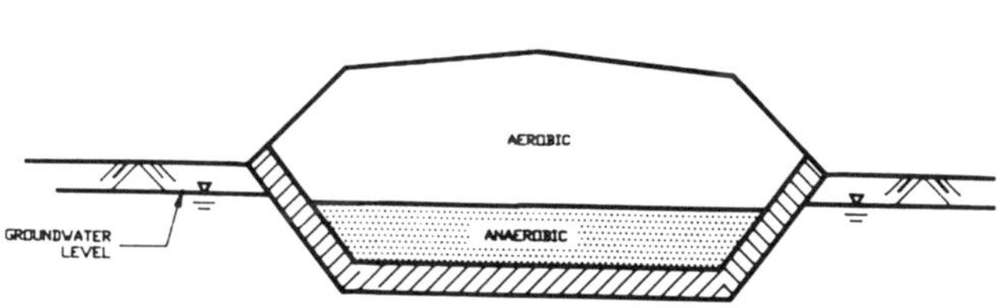

Figure 6. Biodegradation under inward gradient flow.

MONITORING

The third major aspect of inward vs outward hydraulic gradient design is that of monitoring. The very fact that groundwater monitoring plays such an important role in landfill design and operation is testimony to our concern for our ability to accomplish the "total isolation principle".

Referring again to Figure 3, proper placement of monitoring installations, both vertically and horizontally, cannot be overestimated. Proper placement of monitoring installations requires accurate information on groundwater flow directions. The monitoring system must be able to identify changes in flow directions which may result from other area activities throughout the life of the site (pumping, etc.).

While the focus of groundwater monitoring in an outward gradient site is on groundwater flow direction and quality, the focus of monitoring for an inward gradient site is on groundwater levels inside and outside the site as represented in Figure 7. As long as monitoring installations adequately represent head relationships vertically and horizontally in and outside the landfill, we can be confident that, should a leak occur, leakage will be inward, and thus leachate will not escape the site.

Intuitively, it seems that accurate representation of head values (water levels) inside and outside a landfill utilizing piezometers has a greater chance of success than representation of groundwater quality by sampling of monitoring wells outside the contaminated fill. Consider a hypothetical example where 750 feet of downgradient landfill perimeter is monitored every 250 feet. Assuming the wells are 4 inches in diameter and the wells are purged for three well volumes each time they are sampled, the diameter of sampling influence for each well is about 13 inches; thus, sampling of wells spaced 250 feet apart represents 0.4 percent of the groundwater leaving the downgradient perimeter of the landfill (assuming uniform flow across the perimeter). In contrast, piezometers may fairly represent surrounding groundwater levels hundreds of feet away from the monitoring site, depending on groundwater gradients and the level of accuracy needed for verifying inward gradient conditions.

The next section provides examples of each type of site which further illustrate the aspects described thus far.

EXAMPLE OF INWARD HYDRAULIC GRADIENT SITE

The landfill shown in Figure 8 consists of about 240 acres and is situated on a major two-layer aquifer system. The landfill does not have a liner nor did it originally have a leachate collection system. To upgrade the site to current design standards, a soil-bentonite backfill wall with an average permeability of 5×10^{-8} cm/sec was installed around the 3.2 mile perimeter of the landfill. The cut-off wall extends through both aquifer layers and is keyed into an underlying shale unit at an average depth of 91 feet.

In order to insure that leachate does not migrate away from the site, the landfill operator must meet a performance specification which requires that inside groundwater or leachate levels be maintained a minimum of six inches below outside groundwater levels. Interestingly enough, the regulatory agency did not require a prescriptive standard on the cut-off wall permeability. The cut-off wall could have a permeability of 1×10^{-5} cm/sec as long as inside levels are maintained below outside groundwater levels. However, knowing that increased leakage into the site would require greater operating costs, the landfill operator was careful to see that the lowest possible permeability was achieved for the cut-off wall. A leachate collection drain was installed just inside and concentric with the cut-off wall in the upper aquifer layer as seen in cross-section view in Figure 9.

Currently, methane gas from biodegradation in the landfill is being collected from the leachate collection system and serves the heating needs of an adjacent commercial wholesale flower nursery.

Each circle shown along the cut-off wall in Figure 8 represents a group of four monitoring wells in which groundwater levels inside and outside the cut-off wall and in upper and lower aquifer layers are tracked throughout the year. Several of these wells also serve as water quality monitoring wells. Figure

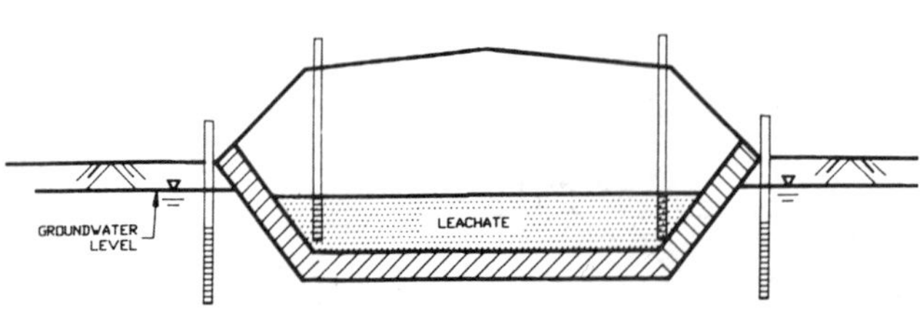

Figure 7. Monitoring for inward gradient.

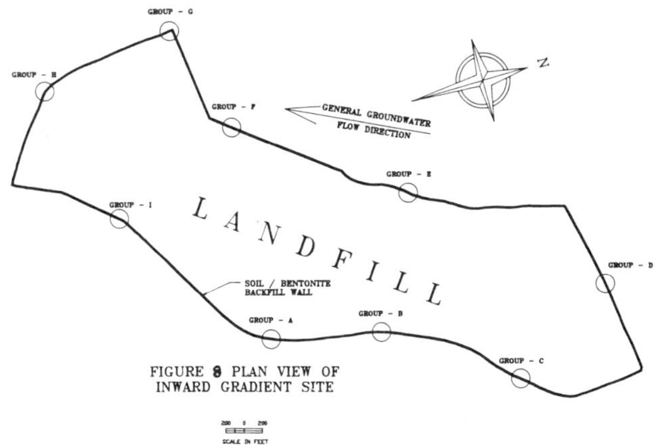

Figure 8. Plan view of inward gradient site.

9 shows a cross-section at Monitoring Group C which displays average groundwater levels in each aquifer for the year 1989. Figures 10 and 11 are hydrographs from Monitoring Group C showing how the inward hydraulic gradient condition has been maintained. Note from Figure 9 that inside levels are kept far below outside levels; there is plenty of room for the landfill operator to allow inside levels to rise if desired. Currently, since methane gas is being collected in the leachate collection system, the preference is to keep inside levels at a minimum. This will likely change as the method of methane gas collection is altered.

EXAMPLE OF OUTWARD HYDRAULIC GRADIENT SITE

The focus of this example is on the monitoring problem involved in outward gradient conditions discussed earlier. Figure 12 shows a leachate plume emanating from a 12-acre landfill with a relatively dense monitoring network. The average groundwater seepage velocity is about 110 feet per year. Longitudinal dispersivity is about 100 feet and transverse dispersivity is estimated to be 20 feet. The source of the leachate leak is about 25 feet square and the leak has been active for about 10 years.

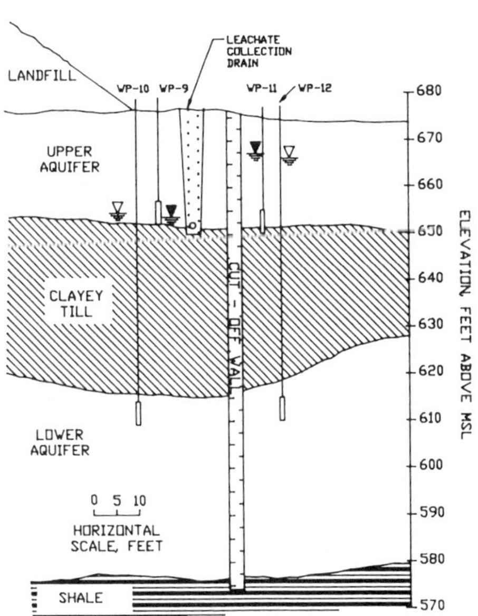

Figure 9. Cross-section group C.

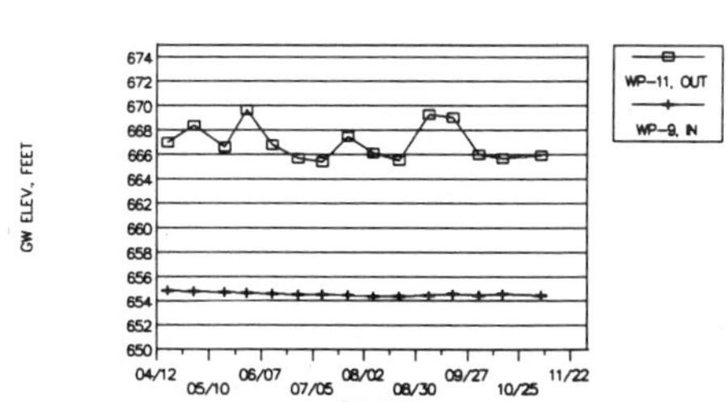

Figure 10. Performance monitoring wells group C, upper aquifer.

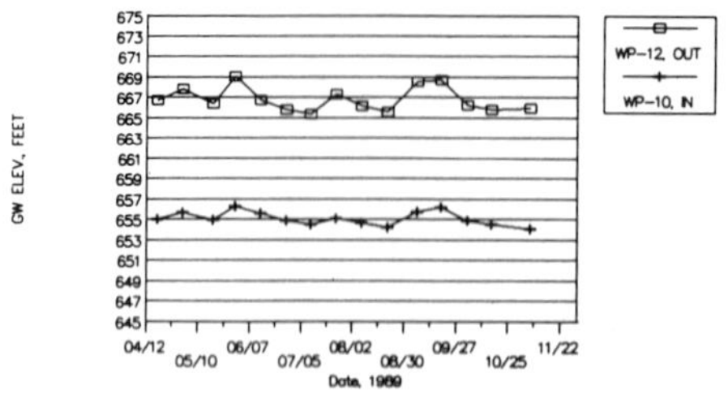

Figure 11. Performance monitoring wells group C, lower aquifer.

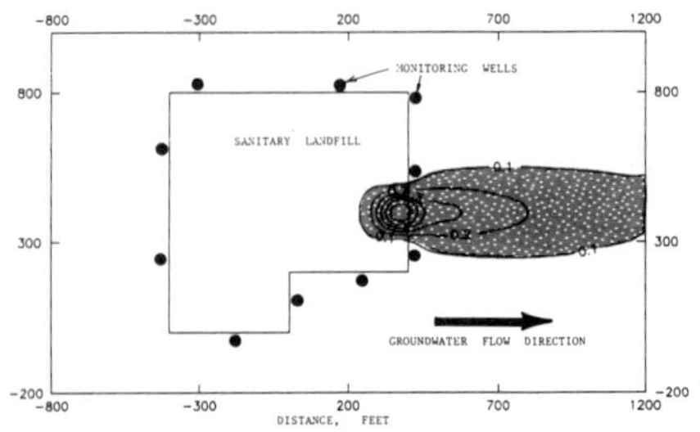

Figure 12. Plume from outward gradient site.

The scenario in Figure 12 shows that even under a relatively stringent monitoring network, leachate migration from a leak could go undetected for a significant length of time.

SUMMARY

Though current trends in sanitary landfill design seem to favor outward hydraulic gradient conditions, there are several advantages to inward hydraulic gradient design that should be considered. Inward hydraulic gradient conditions provide a safety factor in the event of liner leakage. Depending upon site conditions, an inward gradient design could allow a greater volume of waste to remain saturated, thus enhancing anaerobic biodegradation. Inward hydraulic gradient sites have particularly less sensitive monitoring requirements than outward gradient sites.

It is the authors' belief that inward hydraulic gradient design should be considered a viable and advantageous approach to sanitary landfill design. Specific site conditions may often dictate that an outward hydraulic gradient approach is preferable; however, the outward hydraulic gradient design approach should not necessarily be considered the de facto standard for environmentally sound sanitary landfill design.

REFERENCES

1. Gunnerson, Charles G., and David C. Stuckey, "Integrated Resource Recovery; Anaerobic Digestion; Principles and Practices for Biogas Systems", *World Bank Technical Paper Number 49*, The World Bank, (1986).
2. Lu, James C.S., Bert Eichenberger, and Robert J. Stearns, "Leachate from Municipal Landfills; Production and Management", *Pollution Technology Review No. 119*, Noyes Publications, Park Ridge, New Jersey, (1985).
3. Massmann, Joel, and R. Allan Freeze, "Groundwater Contamination from Waste Management Sites: The Interaction Between Risk-Based Engineering Design and Regulatory Policy", *Water Resources Research*, Vol 23, No. 2, pp 351-367, (February, 1987).

68 CASE HISTORY: USE OF REMOTE SENSING TO DETERMINE POTENTIAL SITES OF OIL DISPOSAL AT A MUNICIPAL LANDFILL

Carol D. Rugge, Research Assistant
Dr. R. C. Ahlert, Professor II
Rutgers University
Department of Chemical and Biochemical Engineering
Piscataway, NJ 08855-0909

INTRODUCTION

Fountain Avenue Landfill (FAL) and adjoining Pennsylvania Avenue Landfill (PAL) are unique landfills in that each is bounded on three sides by water and on the fourth by a highway and a residential area. Oil disposal is suspected at both fills. PAL, which opened in 1956, was closed in 1979 due to spontaneous fires. In 1983, PCB-contaminated oil was found to be seeping from PAL into the surrounding waters. An oil lens up to eight feet thick was later found to be floating on PAL's water table.[1]

FAL was in operation from 1962 to 1985. Figure 1 shows Fountain Avenue Landfill as it appeared in 1985.[2] Due to its proximity to PAL and the appearance of oily seeps, FAL may also have received oil. Since FAL is much larger than PAL—197 acres, 130 feet maximum elevation compared to PAL's 110 acres, 80 feet elevation—FAL may present a PCB problem of much greater magnitude.[1] This paper describes the development of FAL over its forty-year history, the use of remote sensing to delineate potential oil disposal sites, and past and present surveys of contamination. This information will be used to guide future remediation efforts.

HYDROGEOLOGY OF THE REGION

FAL is bounded by Jamaica Bay to the south, Old Mill Creek to the east and Hendrix Creek to the west. PAL, which lies west of FAL, is bounded by Fresh Creek to the west and Hendrix Creek to the east. Jamaica Bay, a brackish, shallow bay covering 39 square miles, has an estimated 1% exchange rate with the Atlantic Ocean, and is a sink for contaminants released via discharges of municipal wastewater treatment plants and landfill leachate.

FAL and PAL overlie the Upper Glacial Aquifer, which rests on a layer of Gardiners clay. The clay is believed to be impermeable. However, under Jamaica Bay, the thickness of clay is negligible in some areas and contaminants can be transported to deeper aquifers. Recharge percolating through the fills moves horizontally towards Jamaica Bay, i.e. radially outward from leachate mounds. However, some recharge continues downward to the Upper Glacial aquifer, where it can eventually disperse towards the Bay. The amount of recharge is estimated to be about 0.18 MGD at Pennsylvania Avenue, with 1% flowing to the Upper Glacial Aquifer, and 0.47 MGD at Fountain Avenue, with 20% moving to the Upper Glacial Aquifer.[1]

POTENTIAL SOURCES OF CONTAMINATION

At Pennsylvania Avenue Landfill, there were 19 recorded deliveries of oil from 1974 to 1979, reportedly used for dust control. Extensive drilling indicated presence of an estimated 155,000 gallons of oil floating on the water table, at depths up to eight feet thick. Between six and twelve million gallons of potentially mobile oil were estimated to reside in soil pores, while an additional one to two million gallons were considered to be unrecoverable. Chemical analysis of this oil showed the presence of PCBs, lead and solvents. At Fountain Avenue Landfill, there were 232 oil deliveries over the same time period. Oily seeps have since been apparent at FAL, but no estimate of an oil lens has been made.[1]

45th Purdue Industrial Waste Conference Proceedings, © 1991 Lewis Publishers, Inc., Chelsea, Michigan 48118. Printed in U.S.A.

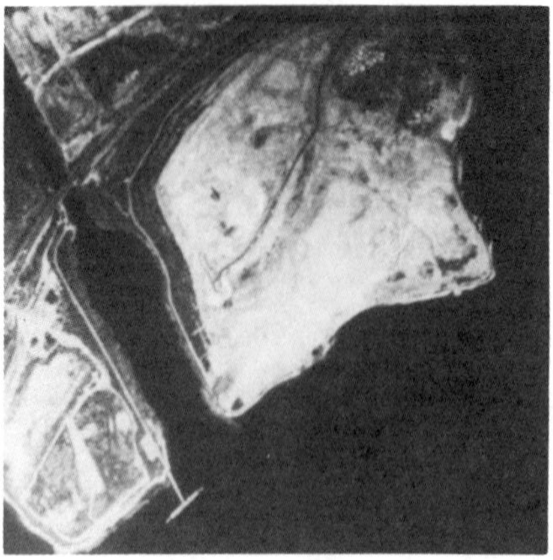

Figure 1. FAL as it appeared in 1985.

Figure 2. Aerial of FAL taken in 1954.

USE OF REMOTE SENSING IN FOUNTAIN AVENUE LANDFILL INVESTIGATIONS

Potential oil disposal sites are apparent on historical aerial photographs of Fountain Avenue Landfill. High altitude photography was available for this site from 1947 to 1985 as a result of mapping missions flown for unrelated projects. The photography, available as color, color infrared or black-and-white photography at scales ranging from 1:10,000 to 1:120,000, was obtained from the National Ocean Survey (NOS), U. S. Geological Survey (USGS) or National Aeronautics Space Agency (NASA). Current aerial coverage documents the present state and extent of the landfill. Historical coverage shows what features lie buried beneath the present land surface.

To accurately relate mapped features to a ground survey, it was necessary to digitally process the images to remove distortion introduced by the position of the plane and the camera. The photograph or color transparency was scanned into digital format, and several road intersections in existence over the time period of interest were chosen as reference points. The Universal Transverse Mercator (UTM) coordinates of these points were determined by digitizing a USGS 7.5' Quadrangle. A regression routine in the ERDAS image processing system was used to align the image to the map, as described by Airola and Kosson.[3] The images, once corrected, could be used to accurately locate present and past features.

HISTORICAL AERIAL PHOTOGRAPHY AVAILABLE FOR FAL

The historical aerial photographs of FAL provide insight into past activities at the site:[4]

pre-1947	The future site of FAL is two peninsulas separated by a creek of roughly 100 acres.
1947	The eastern-most peninsula displays a figure eight configuration, believed to be composed of dredge spoils.[2]
1954	The western peninsula has been extended in preparation for landfill operations (see Figure 2).
1961	The eastern peninsula has been expanded to twice the original size; all streams have been covered. Dredge material appears to have been spread over the entire existing landmass, as well as over the new extensions.
1963	A circular liquids disposal pit, accessed by an oil-stained road, is visible on the eastern peninsula (see Figure 3).
1966	FAL has an area believed to be an automobile junkyard northwest of the circular pit, and the circular pit has been enlarged. Roads are also more extensive, indicating increased landfilling activity.[2] This view is shown in Figure 4.
1969	FAL reaches its present size of 297 acres. The junkyard has swelled to three times its original size, over the former site of the circular pit; a circular anomaly appears in the northwest corner, and the liquid pit to the east is no longer visible.
1972	Dark regions appear along the entire eastern part of the fill, as well as the western corner. No oil disposal basin is visible in the southeastern corner.

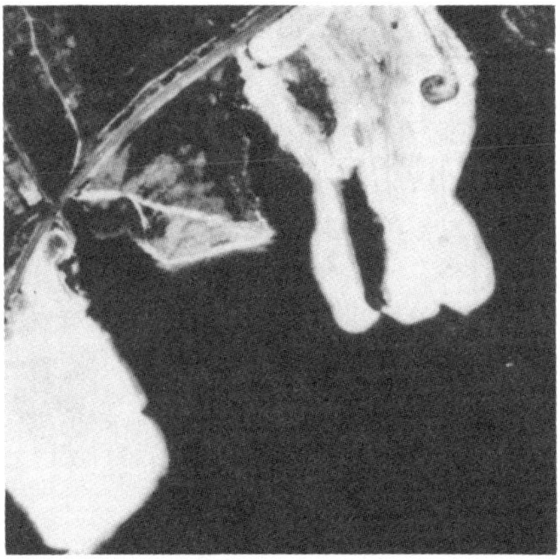

Figure 3. Aerial of FAL taken in 1963.

Figure 4. Aerial of FAL taken in 1966.

1973	Traces of vegetation appear in former dark areas.
1976	Rectangular areas are apparent along the highway and in the northern and eastern corners of the fill.
1985	The fill is showing signs of vegetation along the highway.

The historical photography shows two potential containment basin locations at Fountain Avenue. The large, dark circular area apparent in the northeast corner of the fill, seen in the 1963 and 1966 photographs, may have contained liquids. In 1976, the stained, rectangular areas are visible.

PREVIOUS CONTAMINATION SURVEYS

Previous surveys of contamination of groundwater, surface water, sediment and soil have been made at Fountain Avenue and Pennsylvania Avenue by various consultants at the request of the site owners. The most recent results are discussed here.

Groundwater. Nineteen wells at PAL and twenty-one wells at FAL were sampled during 1983 and 1984. In the FAL leachate mound, high concentrations of toluene, naphthalene, dimethylphenol and zinc were detected. Substantial amounts of toluene, n-nitrosodiphenylamine, PCBs and antimony were found in the leachate mound at PAL. The upper region of the Upper Glacial Aquifer contained naphthalene and antimony under FAL, and benzene, phthalate and antimony under PAL. In the lower portion of the Upper Glacial Aquifer, thallium was found under FAL, and antimony under PAL.[1]

Surface Water. Surface water was sampled around the two landfills during 1982 and 1984. Table I shows heavy metal concentrations in surface water in the bay far from the landfills, near the landfills, and in leachate. High levels of antimony, lead and nickel were found in water near both landfills, mirroring leachate content.[1]

Table I. Heavy Metal Concentrations in Surface Water Near PAL and FAL, Sampled in 1982 and 1984

Metal	Heavy Metal Concentrations (μg/L)			
	Bay	PAL	FAL	Leachate
Antimony	NT	240	350	350
Cadmium	4	10	30	30
Copper	10	ND	40	17
Lead	15	40	160	170
Nickel	60	80	140	140

Table II. Compounds Detected in PAL Sediments in 1984

Substance	Concentration (ppb)
Total PCBs	80
Lead	18,000–30,000
Volatile Organics:	
Toluene	230–840
Ethylbenzene	29–160
Chlorobenzene	15
Xylenes	52–154
Base/Neutral Extractables:	
Naphthalene	280–2,100
Phthalates	230–1,600
Dichlorobenzene	290

Sediments. Sediments were sampled around PAL during 1984. Table II shows PCBs, lead, volatile organics and base/neutral extractables detected in this study.[1]

Soil. Soil at PAL was also sampled during 1984. PCBs and lead were detected, as well as various volatile organics and base/neutral extractables (see Table III). In addition, PCBs were detected at levels up to 4.7 ppm, while lead was found in concentrations ranging from 114 ppm to 0.3%. [1]

These studies show that waste oil and industrial solvents at PAL have significantly contaminated surrounding surface water, ground water and sediments. The oil problem at FAL may be much worse due to a greater recorded number of oil deliveries.

RUTGERS UNIVERSITY SOIL SAMPLING

A map of the landfill is shown in Figure 5. Twelve soil samples were taken around the perimeter in 1988. Four soil samples were taken on the Old Mill Creek shoreline (8S-1, 1, 2A, and 2B), one at the south end of the fill on the Bay side (3A), six on Hendrix Creek coast (8S-6, 8S-7A, 8S-7C, 8S-8, 8S-9, and 8S-11), and one in a marsh in the corner of the fill near Hendrix Creek and the Belt Parkway (4). Four additional samples were taken in 1989 in an active seep about 100 feet above sea level (J-1A, J-2A, J-1B, J-2B).

SOIL ANALYSIS

Soil samples were analyzed for polychlorinated biphenyls (PCBs), phthalates, polyaromatic hydrocarbons and long-chain hydrocarbons. All soils were extracted with field moisture present; this has been shown to aid in the desorption of PCBs.[5] Excess moisture was removed with sodium sulfate, and cyclohexane was selected as the solvent.

High resolution gas chromatography with an electron capture detector (ECD) was selected for PCB analysis. The high resolution afforded by a capillary column allowed analysis of concentrated extracts without clean-up. The ECD was used in this analysis because it is extremely sensitive to chlorine, easy to operate, and gives reproducible results.

Table III. Compounds in PAL Soil Samples in 1984

Substance	Concentration (ppb)
Volatile Organics:	
Benzene	10–160
Toluene	22–26,000
Ethylbenzene	58–4,800
Chlorobenzene	17–280
Xylenes	2,120
Propylbenzene	2,860
Base/Neutral Extractables:	
Naphthalene	260–36,000
Phthalates	2,300–28,000
Dichlorobenzene	27,000
Phenanthene	220–17,000

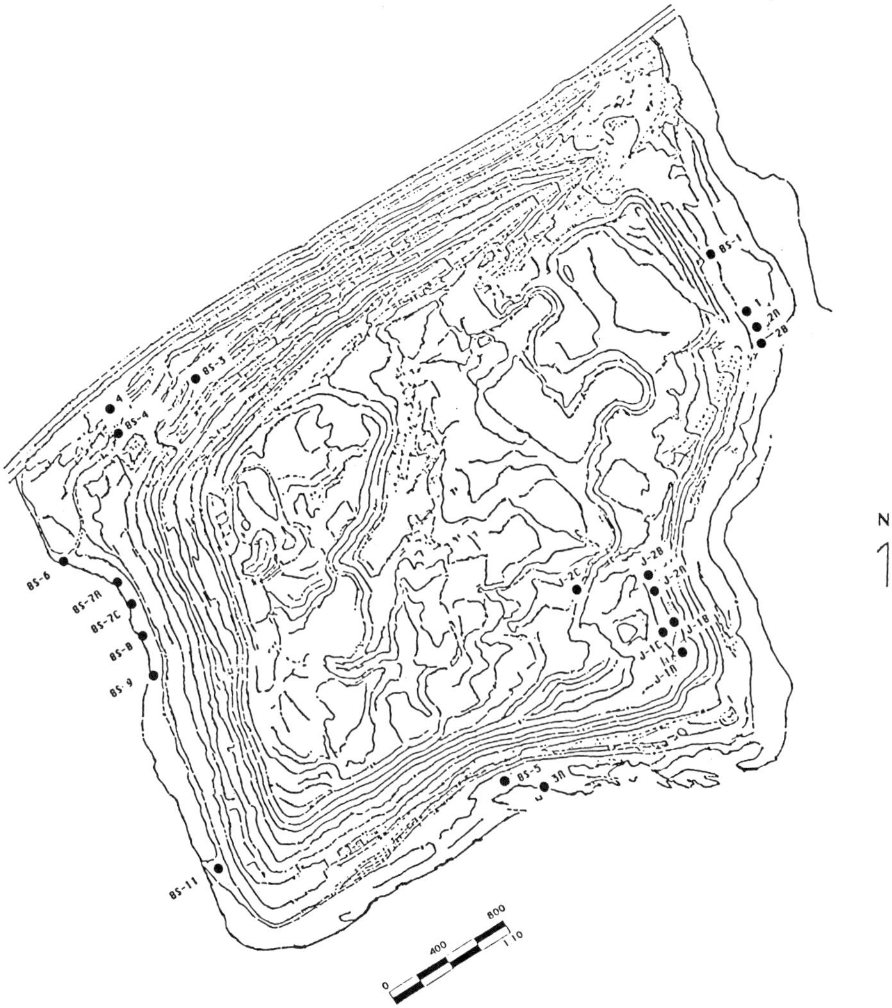

Figure 5. Rutgers University sample locations at FAL.

typically found in waste oil, like hydrocarbons, polychlorinated biphenyls and bis (2-ethylhexyl) phthalate are summarized in Table V.

A mass selective detector equipped with a capillary column confirmed hydrocarbon analyses and provided an overview of sample composition[4].

RESULTS

Polynuclear Aromatic Hydrocarbons. As shown in Table IV, 8S-1, near the entrance to the fill, southern-most sample 3A, and western coast sample 8S-9 contained fluoranthene. The soil in the running seep showed low concentrations of anthracene (40–60 μg/kg dry soil) and fluoranthene (110–290 μg/kg dry soil). One sample from the seep also showed the presence of indeno (1,2,3-cd) pyrene.

Table IV. Polynuclear Aromatic Hydrocarbons in FAL Soil

Location	Anthracene (μg/kg dry soil)	Fluoranthene (μg/kg dry soil)
8S-1	ND	130
J-1A[a]	60	180
J-1B	40	290
J-2A	ND	110
J-2B	40	200
3A[b]	ND	200
8S-9	ND	210

[a]910 μg/kg dry soil Indeno(1,2,3-cd)Pyrene.
[b]140 μg/kg dry soil Chrysene.

Oil Indicators. Polychlorinated biphenyls, bis(2-ethylhexyl) phthalate and long-chain hydrocarbons were used as oil indicators. Concentrations or presence of these substances in FAL soil samples are shown in Table V. All samples showed evidence of PCBs, predominantly Aroclor 1242 and 1254, with high concentrations detected near the northwest pit and the running seep. The phthalate was detected at concentrations ranging from 180 to 16,000 μg/kg dry soil in all samples except the northwest corner. Hydrocarbons containing from 20 to 33 carbons were detected in both the seep and sample 8S-1. Mass spectrometry showed that the long-chain hydrocarbon pattern found in 8S-1 and the new seep were very similar.[4]

CONCLUSIONS

Soil sampling at Fountain Avenue Landfill shows that there appears to be substantial contamination of coastal soil. This could be due to disposal of industrial waste in the large circular pit near the northeast corner, the circular basin to the west, or the rectangular bermed areas. To investigate these sources, there are two courses of action presently planned. Firstly, the ground coordinates of the round basin will be obtained from geocorrected images. A ground survey of the basin perimeter will outline possible drilling sites. Secondly, the location of the active seep will be related to buried features by drilling in the middle of the seep, surveying the sites of the borings, and relating these locations to geo-referenced images.

Remote sensing and soil analysis at this landfill has provided a hypothesis for the presence of PCB-contaminated oil. Oil-containment structures are visible in historical photographs, and traces of oil were detected in the soil. Borings taken at accurately-known locations via aerial photography will confirm the relationship of containment basins in the photos to the presence of oil.

Table V. Oil Indicators in FAL Soil

Site	PCBs (μg/kg dry soil)	Phthalate (μg/kg dry soil)	Hydrocarbons (Present/ND)
4	760	ND	ND
1	20	ND	ND
2A	30	ND	ND
2B	20	180	ND
8S-1	130	ND	Present
3A	150	2,100	Present
J-1A	730	7,040	Present
J-1B	500	16,200	Present
J-2A	380	3,820	Present
J-2B	400	5,340	Present
8S-6	110	ND	ND
8S-7A	80	ND	ND
8S-7C	50	420	Present
8S-8	30	ND	ND
8S-9	190	1,260	ND
8S-11	140	1,090	ND

ACKNOWLEDGMENTS

A special note of thanks goes to W. V. Black for his development of the PCB analytical method, as well as retrieval of coastal samples in 1987. Thanks also goes to D. S. Kosson and A. T. Andrews for their part in sampling the new seep in 1989.

REFERENCES

1. Ahlert, R. C., Chepiga, M. M., Enzminger, J. D., Goldfarb, W., Kosson, D. S., Motto, H. L., Rauch, A., and Witmer, C. M. (1985) *Evaluation of Landfill Closure Requirements*. Rutgers University unpublished report.
2. Pederson, L. K. (1989) *Integrated Monitoring Techniques for Evaluation of a Hazardous Waste Landfill*. Master's Thesis, Rutgers University.
3. Airola, T. M. and Kosson, D. S. (1989) "Digital Analysis of Hazardous Waste Site Aerial Photographs," *Journal of Water Pollution Control Federation*, 61(2), 180–183.
4. Rugge, C. D. (1989) *Polychlorinated Biphenyl Analytical Methodology and Application to a Contaminated Municipal Landfill*, Master's Thesis, Rutgers University.
5. Johnson, R. E. and Starr, R. I. (1972) "Ultrarapid Extraction of Insecticides from Soil Using New Ultrasonic Technique", *Journal of Agriculture and Food Chemistry*, 20(1) 48–51.

69 THREE DIFFERENT GENERIC LANDFILL TOPOGRAPHIES AND THEIR MANAGEMENT

I. Ponsoye, Environmental Engineer
FRANCE-DECHETS-BP 29-78440
Gargenville, France

C. Fraissinet, Doctor in Geology
F.D. Conseil-BP 29-78440
Gargenville, France

INTRODUCTION

FRANCE-DECHETS is a subsidiary company of SITA and Lyonnaise des Eaux. Its speciality is hazardous industrial waste landfilling.

FRANCE-DECHETS operates 20 landfills in France and has several projects abroad (SITA operates 80 municipal waste landfills with 5 Mt/yr).

When considering only industrial wastes, FRANCE has 11 sites. Among them 10 are operated by SITA of which 8 are operated by FRANCE-DECHETS.

Our current authorized airspace capacity is 3 Mm3. FRANCE-DECHETS treats 370,000 T/year of industrial wastes. Looking back over our 15 years of experience we can divide the waste treated into 5 main categories:

- waste from incineration plants (fly ash),
- waste from water treatment plants (solid sludges),
- paint wastes (solids) after phase separation,
- waste from metallurgical industries (slags, sands from smelting works),
- contaminated materials (soils).

For each of these 5 categories we developed specific methodologies for each of the following steps of the process:

- waste identification and analysis before acceptance,
- control analysis before delivery,
- pretreatment and packaging for transport and storage,
- cell preparation and operation.

The landfilling process is based on 3 key words = security, control and transparency. Designing a landfill includes many engineering disciplines with a large variety of specialities = geology, hydrogeology, chemistry, geotechnology, environment. This paper describes part of the process according to the different types of initial site topographical configurations: above ground containment, quarry, valley. How these factors affect the cost per cubic meter of waste deposited will be considered. We will not detail general constraints such as geological context, accessibility, existence of a natural outlet for treated water, factors which are determinants for the choice of a site, but we will emphasize those directly related to the design of the site.

WATER IS ENEMY NUMBER ONE

Throughout the site life one wants to avoid as much as possible the contact of wastes with water.

This is why some people thought of sites in deserts and others of salt mines.

As far as surface storage is concerned, what can be done?

- during the construction phase, one makes sure that the groundwater table will be protected and that no infiltration into the waste can occur.

M = 1 million.

45th Purdue Industrial Waste Conference Proceedings, © 1991 Lewis Publishers, Inc., Chelsea, Michigan 48118. Printed in U.S.A.

- during the operating phase, one can think of a roof over the part being filled or over an intermediate temporary storage zone.
- during the closure phase, wastes will be protected by an impervious and repairable cover layer, over which surveillance can be maintained.

THE HYDROLOGICAL CONTEXT: HOW IT CAN AFFECT THE CONSTRUCTION COST

Assuming groundwater is to be drained all around the site in order to avoid any infiltration into the wastes, the technique used will depend on the depth involved. We will consider three cases, where:

L = the half side of the squared base perimeter
H = the depth of the vertical barrier for containment.
Therefore the vertical area of the barrier is $8\ L \times H$
Horizontal area of the site is $A = 4L^2$

If $0 < H < 6$m, a trench 1.5 m large will be excavated and filled with:

- a 0.5 m sand layer on the bottom
- a 1 m top soil layer on top
- a gravel layer in between, equipped with drainage pipes.

The corresponding costs can be estimated as:

$$C = \underbrace{\$2.5/m^2 \times 8 \times L \times H}_{\text{earthworks}} + \underbrace{\$7.5/ml \times 8 \times L}_{\text{0.5 m sand layer}} + \underbrace{\$27.5/m^2 \times 8 \times L \times (H-1.5)}_{\text{gravel layer}} + \underbrace{\$4/ml \times 8 \times L}_{\text{drain pipes}}$$

If $6 < H < 12$ m, a cut off wall is to be installed.

The corresponding cost can be estimated as:

$$C = \underbrace{\$142{,}000}_{\text{fixed}} + 8\ LH \times \$\ 74/m^2$$

If $12 < H < 50$ m, a cut off wall will be installed with very specific equipment.

The corresponding cost can be estimated as:

$$C = \underbrace{\$167{,}000}_{\text{fixed}} + 8\ LH \times \$225/m^2$$

In the three cases the construction cost related to the hydrogeological context increases with the length (L) and the depth (H).

The value of this cost per cubic meter of wastes will depend on the configuration of the rehabilitated landfill.

To illustrate this estimation of costs we examine four theoretical configurations.

case 1a	average cover slope 5:1	average side slope 3:1
case 1b	average cover slope 5:1	average side slope 1:1
case 2a	average cover slope 3:1	average side slope 3:1
case 2b	average cover slope 3:1	average side slope 1:1

Figures 1 to 4 show the variation of construction costs in $\$/m^3$ as functions of H and L.

Calculations are based on a vertical barrier of (H + 1) meters.

Unit costs decrease as L increases. Case 2b offers the largest volume for wastes when H and L are fixed, and therefore the lowest cost unit.

As an example we give in Tables I and II, respectively, typical unit costs when H = 5 m and H = 30 m.

GEOMETRICAL ASPECTS: HOW THEY CAN AFFECT CONSTRUCTION AND CLOSURE COSTS

The first engineering configuration to be studied on an initial topographic basis is the one which would characterize the final aspect of the landfill.

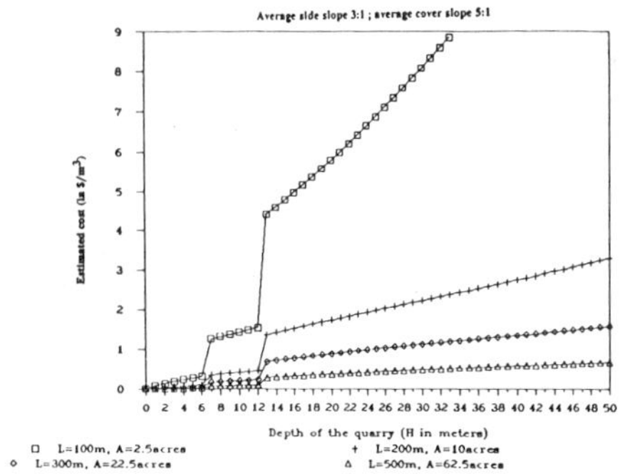

Figure 1. Estimation of the cost due to hydrogeological context of a quarry.

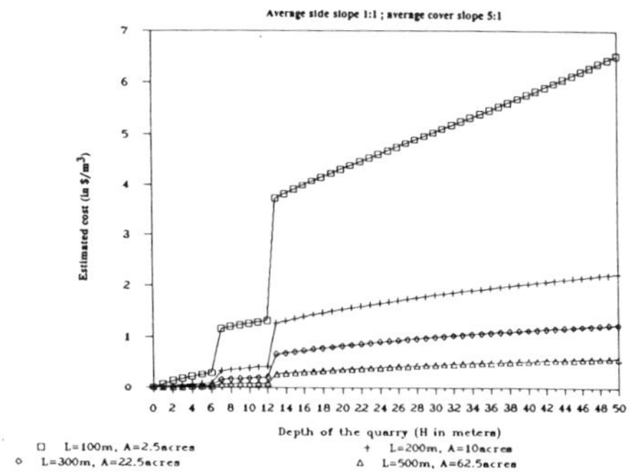

Figure 2. Estimation of the cost due to hydrogeological context of a quarry.

This will depend on the maximum acceptable height, the shape required and the future use of the land.

From the aesthetic point of view, one wants the final aspect of the site to be fully integrated in the surroundings.

From the technical point of view we consider that the final cover should have:

- a minimum slope of 5:1 (20%), because of settlement expected, and the effectiveness of rainfall water drainage over the impervious layer.
- a maximum slope of 3:1 (33%) because of friction angles and the desirable shear stress limits between the layers of the geocomposite cover.

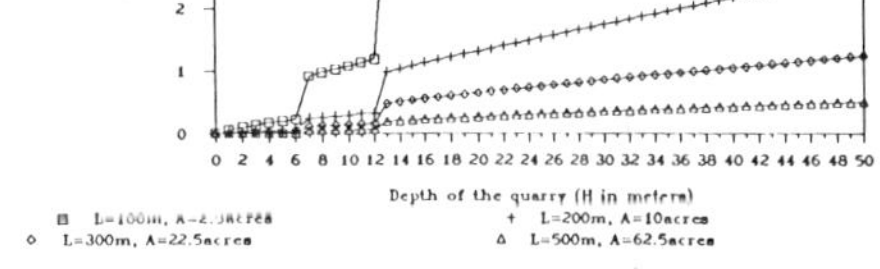

Figure 3. Estimation of the cost due to hydrogeological context of a quarry.

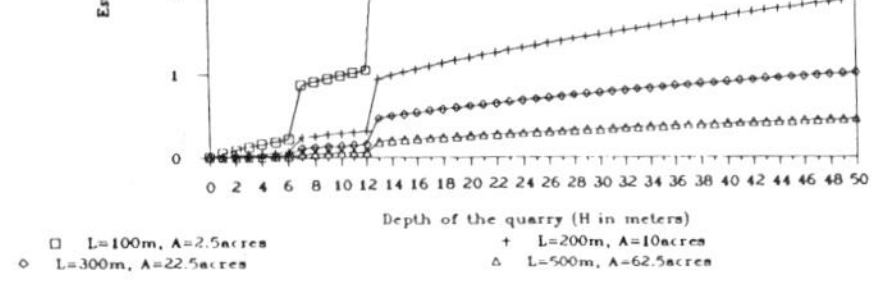

Figure 4. Estimation of the cost due to hydrogeological context of a quarry.

Table I. Typical Costs for H = 5m

H = 5m		L = 100 m C = $120,000	L = 200 m C = $235,000	L = 300 m C = $385,000	L = 500 m C = $585,000
case 1a	V(m^3)	0.44×10^6	2.876×10^6	8.92×10^6	38.2×10^6
	C($/$m^3$)	0.27	0.08	0.043	0.015
case 1b	V(m^3)	0.456×10^6	2.912×10^6	8.96×10^6	38.28×10^6
	C($/$m^3$)	0.26	0.08	0.043	0.015
case 2a	V(m^3)	0.616×10^6	4.28×10^6	13.72×10^6	60.4×10^6
	C($/$m^3$)	0.19	0.055	0.028	0.0097
case 2b	V(m^3)	0.636×10^6	4.32×10^6	13.76×10^6	60.4×10^6
	C($/$m^3$)	0.19	0.054	0.028	0.0097

Table II. Typical Costs for H = 30 m

H = 30m		L = 100 m C = M$5.75	L = 200 m C = M$11.3	L = 300 m C = M$16.9	L = 500 m C = M$28.0
case 1a	V(m^3)	0.712×10^6	5.08×10^6	15.08×10^6	58.4×10^6
	C($/$m^3$)	8.075	2.22	1.12	0.479
case 1b	V(m^3)	1.144×10^6	6.24×10^6	16.96×10^6	61.6×10^6
	C($/$m^3$)	5.026	1.81	0.996	0.454
case 2a	V(m^3)	0.888×10^6	6.52×10^6	19.88×10^6	80.4×10^6
	C($/$m^3$)	6.47	1.73	0.85	0.348
case 2b	V(m^3)	1.32×10^6	7.68×10^6	21.76×10^6	83.6×10^6
	C($/$m^3$)	4.356	1.47	0.776	0.335

If dikes are to be constructed, an average external slope of 2:1 (50%) is currently practised.

From this final configuration, the total airspace volume available for wastes can be deduced. We assume that most of the construction materials needed can be supplied from the initial site, and we initially neglect those which will be hauled from outside of the site.

From the final area to be covered, and the thickness of the cover, the volume of materials to be stockpiled is calculated. In order to minimize earthworks while excavating this volume, one wants to profile the bottom of the landfill with minimum slopes of 3% for the bottom itself and maximum slopes of 3:1 (33%) for the slopes in the case of a quarry.

From the profiles of the bottom, the side slopes and the cover, the total area A' for a geocomposite lining system GLS is deduced.

Therefore the ratio of the area of the GLS to the volume of wastes A'/V can be calculated. The average cost of the water infiltration protection and drainage works per cubic meter of wastes is directly related to this ratio. The higher it will be, the higher the cost will be.

When designing a quarry, the depth and slopes are primary determinants. One wants to minimize earthworks, to optimize A'/V and have stabilized slopes including protection and drainage layers during the filling phase.

Another determining factor will be whether the quarry is exploited concurrently with landfilling. If such is the case, the constraints are more strict: the landfill should stay completely independent from the hydrological point of view, and the landfilling rate must be adjusted to the evolution of the quarry site.

Typical slopes are 3:1 (33%). Sometimes steeper slopes are employed up to 1:1 (100%), but then the questions of slope stability and erosion protection layers is critical and the use of geotextile reinforcement should be considered.

Figures 5 to 8 illustrate the estimation of GLS cost related to the depth with the same four theoretical profiles. Calculations are based on two different average costs of the GLS of $ 30/m^2 and $ 40/m^2.

One can see that costs decrease rapidly as H and L increase. Let us have a look at typical values in $/$m^3$ (See Table III).

In the case of a valley the cost of the GLS will be between the one corresponding to an aboveground containment and the one corresponding to a quarry.

When designing a valley landfill, one of the most important factors during the construction phase is the volume of the dike to be constructed at the base of the valley, and the eventual embankments. The ratio of the volume of materials to the volume of waste will affect the cost. This ratio can vary in the range from 1:5 up to 1:2. If materials are to be hauled, the cost can be from $ 3 to $8 per cubic meter of wastes.

GEOMETRICAL ASPECTS: HOW THEY AFFECT OPERATING COSTS

Designing the profile of the bottom, and of the different cells, should be carefully studied. One wants to minimize earthworks and to have the maximum airspace volume. However, one also wants, assuming there is no roof, to minimize as much as possible the surface of waste exposed to rainfall and to separately drain away clean surface waters and leachates.

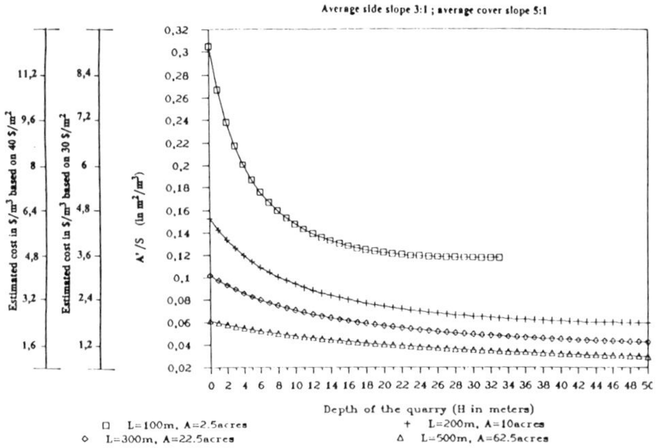

Figure 5. Estimation of G.L.S. cost.

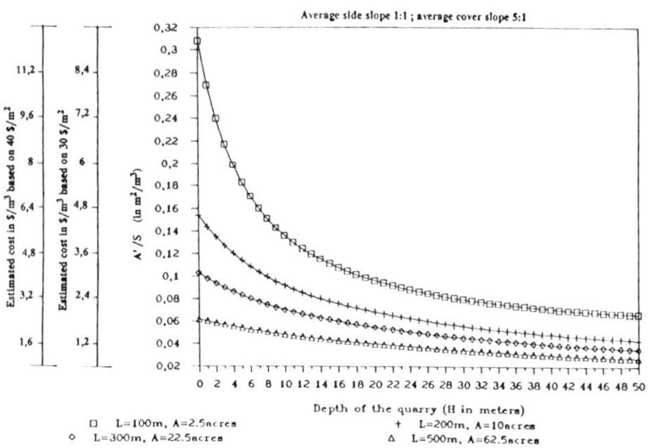

Figure 6. Estimation of G.L.S. cost.

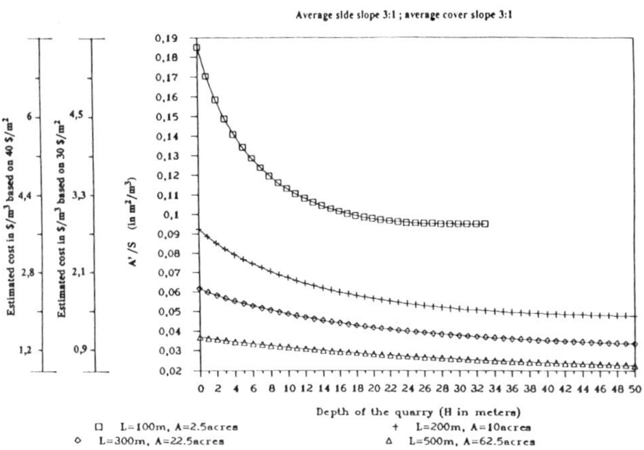

Figure 7. Estimation of G.L.S. cost.

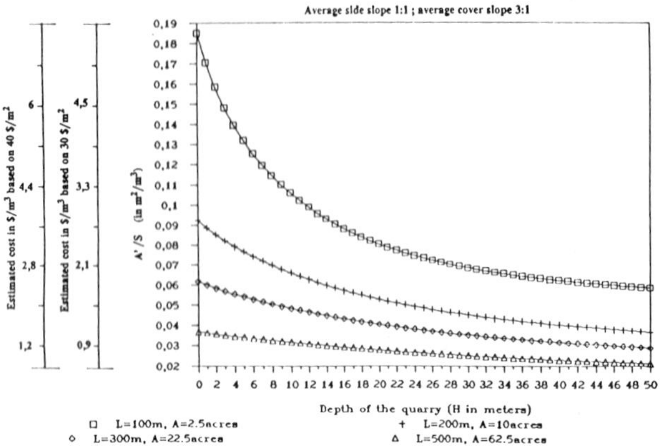

Figure 8. Estimation of G.L.S. cost.

Table III. Typical Costs for Various Values of H and L

		L = 100 m		L = 200 m		L = 300 m		L = 500 m	
		\$30/m²	\$40/m²	\$30/m²	\$40/m²	\$30/m²	\$40/m²	\$30/m²	\$40/m²
Above (H = 0)	case 1	9.15	12.20	4.56	6.08	3.06	4.08	1.83	2.44
ground containment	case 2	5.99	7.40	2.77	3.70	1.85	2.46	1.11	1.48
Quarry (H = 30 m)	case 1a	3.51	4.68	1.95	2.60	1.48	1.97	1.06	1.41
	case 1b	2.38	3.17	1.66	2.22	1.36	1.81	1.02	1.36
	case 2a	2.85	3.80	1.54	2.05	1.13	1.51	0.77	1.03
	case 2b	2.09	2.78	1.38	1.84	1.07	1.43	0.76	1.01

Minimum bottom slopes are to be around 3% for drainage effectiveness. Each cell should be separately drained to a specific point.

During operations, cells will be filled and covered (with an initial cover) as soon as possible.

Several leachate pumping wells, if possible outside of the pound are preferable.

The depth of the quarry will directly affect operating costs of leachate pumping.

CONCLUSIONS

Some theoritical profiles of three different generic landfills have been studied. The initial topography of a site, is, of course, the principal determinant in the choice of configuration.

In spite of the high initial investment cost, construction works need to be considered on a large scale, so that the average cost per cubic meter of wastes can be optimized. Further research and development is needed in order to employ steeper slope angles and geocomposite layers.

ACKNOWLEDGMENTS

The authors wish to thank their colleagues in FRANCE-DÉCHETS who contributed substantially to this paper through their extensive operating experience in waste landfilling.

REFERENCES

1. Geotechnical Practice for Waste Disposal 87, *ASCE*, Michigan (June 15–17, 1987).
2. Waste Alternatives Landfill of the 90s, (December 1988).
3. *Waste Management and Research*, Number 7 (1989).
4. Second International Landfill Symposium, Sardinia (II) (October 9–13, 1989).
5. Environment Ministry and National Agency for Waste Recovery and Disposal = Guide for Industrial Wastes Disposal and Recovery (1989) (in French).

Section Eight
LANDFILLS – B. LEACHATES

70 EVALUATION OF A FULL-SCALE PHYSICO-CHEMICAL TREATMENT PLANT FOR REMOVAL OF BDAT CONSTITUENTS FROM HAZARDOUS WASTE LANDFILL LEACHATE

Bill Y. M. Liu, Project Engineer

Alan Li, Project Manager

Michael P. Henry, Project Manager
Chemical Waste Management, Inc.
Research and Development
Geneva, Illinois 60134

Bruce Kruckel, Operational Manager
Chemical Waste Management, Inc.
Model City, New York 14107

INTRODUCTION

Leachate, generated by water percolating through landfills, can be classified into the two broad categories: non-hazardous and hazardous leachates. Basically, hazardous leachate is generated from hazardous waste or co-disposal landfills, whereas non-hazardous leachate is generated from hazardous waste or co-disposal landfills, whereas non-hazardous leachate is defined as leachate from non-hazardous solid waste landfills. Because of the high concentrations of contaminants in leachate, leachate is generally treated before it can be discharged or disposed of.[1]

Both types of leachate can be treated with conventional wastewater treatment technologies, such as biological treatment for organic removal, metal precipitation for metal removal and granular activated carbon (GAC) adsorption for effluent polishing.[1] The effluent from these leachate treatment plants usually discharges into Publically Owned Treatment Works (POTWs) or to surface waters under state or federal National Pollution Discharge Elimination System (NPDES) permits. The treatment residues are then disposed of in a hazardous landfill for hazardous treatment residues and a solid waste landfill for non-hazardous residues. However, the treatment practice for hazardous waste leachate will be significantly impacted and might have to change when the EPA promulgates the Third-Third landban regulation regarding multi-source leachate on May 8, 1990.

The EPA has set on overall schedule for the land disposal program, which requires all land disposal restriction regulations to be in place by May 8, 1990.[2] The EPA divided listed hazardous wastes into three groups (the "Third"), to be regulated in successive stages over a period of 66 months. The multi-source hazardous leachates, regulated by the Third Third, will have to meet the treatment standards for wastewaters and non-wastewaters before land disposal.[3] The EPA developed the standards representing the best demonstrated available technology (BDAT).

The BDAT constituents, totalling 231 compounds, include volatile organics (63 constituents) and semi-volatile organics (107 constituents), metals (16 constituents), inorganics other than metals (3 constituents), organochlorine pesticides (20 constituents), phenoxyacetic acid herbicides (3 constituents), organophosphorus insecticides (5 constituents), polychlorinated biphenyls (7 constituents) and dioxins and furans (7 constituents).

In the proposed rule for the Third Third landban[2], leachate has to meet the wastewater standard, consisting of 228 BDAT compounds, before land disposal in the surface impoundment. Treatment residues, such as metal and biological sludges from the leachate treatment plant, will have to meet the non-wastewater standard before disposal in a landfill. The standard for non-wastewater includes 205

45th Purdue Industrial Waste Conference Proceedings,

BDAT compounds based on total constituent analysis (TCA) and 11 BDAT metals based on toxicity characteristic leaching procedure (TCLP).

The Third Third regulations will not affect the effluents of those facilities discharging to POTWs or into surface water under state or federal NPDES permits. However, for those facilities which use surface impoundment for their final effluents, the Third Third landban will significantly affect their practice. In addition, leachate treatment residues, such as sludge from metal precipitation processes or spent GAC from adsorption processes, are subject to BDAT treatment standards.

The performance of a full-scale physico-chemical leachate treatment plant for BDAT constituents in leachate was evaluated and the results are presented in this paper. The impacts of the proposed rule on the treatment practice of a physico-chemical treatment plant is also discussed. The full-scale treatment plant in this study consists of metal precipitation, sand filtration and GAC adsorption.

LEACHATE TREATMENT FACILITY

Chemical Waste Management, Inc. (CWM) operates the leachate treatment facility used in this study. The facility accepts leachates from off-site generators along with leachate generated from the site's hazardous waste landfill. The flow diagram of this leachate treatment facility, is given in Figure 1.

The leachate is pretreated to separate oils from the leachate with the treated aqueous phase being sent to a leachate holding tank. The leachate is then adjusted to an acidic pH and ferrous sulfate is added to reduce chromium. Lime is then added to the alkalization tank to raise the leachate pH to 9 or higher for metal precipitation. The resulting slurry is then dewatered by a filter press and the filtrate is neutralized to pH 7 with hydrochloric acid before it passes through a sand filter. The filtered effluent goes to a two-stage GAC unit for organic removal. The effluent from the GAC columns is further polished by aeration in facultative ponds prior to discharge under a State Pollution Discharge Elimination System (SPDES) permit. Filter cakes are collected and disposed of on-site in a hazardous waste landfill. Spent GAC is regenerated off-site and reused.

SAMPLING AND ANALYSIS

Three sets of samples were taken from the locations labelled in Figure 1 at a weekly interval in accordance with the procedures described in the Generic Quality Assurance Project Plan for the Land disposal Restriction Program.[4] Samples of the metal precipitation feed, filter press filtrate and dewatered cake, GAC unit influent and effluent and the spent GAC from the lead unit were collected. All samples were analyzed for the 231 BDAT compounds according to EPA SW-846 procedures.[5] In addition, the metal sludge and spent GAC were also stabilized with four different formulae to study the effects of chemical stabilization. TCLP and TCA tests were performed on all stabilized samples.

Since hexachlorophene and aramite cannot be determined under EPA methodologies, all samples were analyzed for the full list of BDAT compounds except for these two compounds. All samples were preserved at the time of collection and sent to an off-site laboratory for analysis.

RESULTS AND DISCUSSION

Of the 231 BDAT compounds, only 77 were found in detectable concentrations in at least one of the process streams sampled. Consequently, only these compounds are addressed in the remainder of the paper.

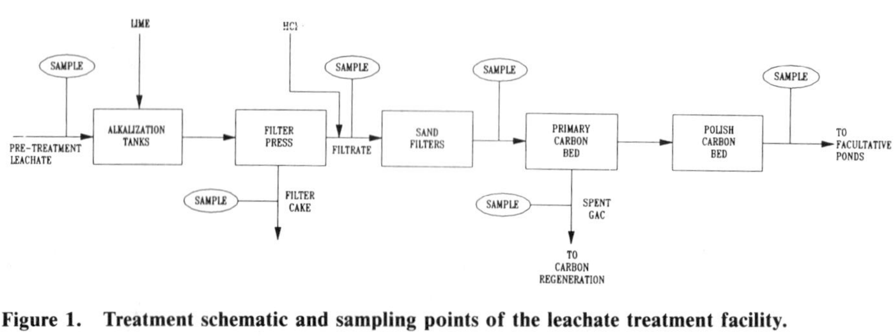

Figure 1. Treatment schematic and sampling points of the leachate treatment facility.

Table I. Optimum pH for Lime Precipitation

Metal	Optimum pH	Major Precipitate
Arsenic	12	AsO_4^{-3}
Barium	10.5	$BaCO_3$
Cadmium	10.8	$Cd(OH)_2$,$CdCO_3$
Chromium (+3)	>7.0	$Cr(OH)_3$
Copper	>8.0	$Cu(OH)_2$
Lead	10	$Pb(OH)_2$,$PbCO_3$
Nickel	9.9	$Ni(OH)_2$
Silver	>11.1	AgOH
Zinc	>11	$Zn(OH)_2$

Metal Precipitation Process

In the metal precipitation process, most metals form hydroxide, sulfide or carbonate precipitates at an alkaline pH.[6] The optimal pH and the major form of precipitate during lime precipitation are tabulated in Table I.[6] The data in this table indicates that it is not possible to select a single pH that is optimal for removal of all metals.

Table II lists the average metal concentrations of the metal precipitation process influent, the filter press filtrate and sand filter effluent for each metal. Data presented in Table II are the average values of the three samples. The removal is poor in the filter press filtrate, suggesting that some metal precipitate may be carried over in the filtrate. However, sand filtration helped to remove a significant portion of the metal solids resulting in a lower metal concentrations in the sand filter effluent. This indicates that sand filtration not only prevents solids from plugging the GAC filter but also helps polish the metal concentration in the filter press effluent.

The concentrations of the volatile and semivolatile organic compounds are listed in Tables III and IV, respectively. About 30% of the volatile organics were removed in the overall metal precipitation process (Table III). Like metals, poor organic removals were observed in the pressure filtration alone. About 70% of semivolatile organics was removed by the overall process, while only 45% by the filter press. It is possible that some organics are enmeshed in precipitate and carried over in the filtrate. Those flocs were filtered out in the sand filter, giving the overall process good performance on metal and semivolatile organic removal. The lower removal of volatile organics, versus semivolatile organics, is expected because volatile compounds are generally smaller and simpler molecules which are not enmeshed by metal precipitates as easily as the larger semivolatile molecules.

The concentrations of the inorganic compounds throughout the metal precipitation process are given in Table V. Inorganics, namely fluoride and sulfide, were also removed by the metal precipitation process. Since the solubilities of metal sulfides are significantly lower than those of the metal hydroxides,[7] most metals will form sulfide precipitates before forming hydroxide precipitates. For example, the solubility of lead hydroxide is 10^8 times greater than that of lead sulfide. Therefore,

Table II. Average Metal Concentrations in Metal Precipitation

Metal	Influent, μg/L	Filtrate, μg/L	Sand Filter Effluent, μg/L	Removal Efficiency %
Arsenic	89	48	26	70.8
Barium	143	200	72	49.7
Cadmium	68	23	ND	100
Chromium (hexavalent)	8	ND	ND	100
Chromium (total)	58	221	41	29.3
Copper	ND	650	100	–
Lead	17	25	12	29.4
Mercury	2.4	0.3	0.4	80
Nickel	650	643	76	88.3
Selenium	7.9	3.7	ND	100
Vanadium	ND	35	ND	–
Zinc	297	2400	30	87.9
Total Concentration	1339.9	5952.3	363.4	73.3

ND = Nondetectable.

Table III. Performance of Metal Precipitation on Removal of BDAT Organics

Constituent	Influent, μg/L	Effluent μg/L	Sand Filter Effluent, μg/L	Removal Efficiency %
Acetone	12233	12800	11967	2.2
Benzene	28	15	ND	100
n-Butyl alcohol	2200	4700	ND	100
Chlorobenzene	29	17	ND	100
Chloroform	195	112	75.5	61.3
1,2-Dibromoethane	ND	30	ND	–
Dibromomethane	ND	144	78	–
1,1-Dichloroethane	150	27.5	ND	100
1,2-Dichloroethane	90	74	59	34.4
Ethyl benzene	39	12	ND	100
Isobutyl alcohol	4833	3250	2600	46.2
Methanol	24667	22000	18833	23.7
Methyl ethyl ketone	9033	10067	7667	15.1
Methyl isobutyl ketone	8200	7000	5033	38.6
Methylene chloride	3067	1657	600	80.4
Pyridine	440	440	330	25
1,1,2,2-Tetrachloroethane	817	260	68	91.7
Tetrachloroethene	69	28	ND	100
Toluene	473	163	69	85.4
1,1,1-Trichloroethane	ND	16	ND	–
1,1,2-Trichloroethane	ND	99	55	–
Trichloroethene	383	263	116	69.7
1,2-Xylene	86	29	ND	100
1,3-Xylene and 1,4-Xylene	180	38	ND	100
1,4-Xylene	ND	52	ND	100
Total Concentration	67200	63300	47550.5	29.2

ND = Nondetectable.

sulfide was most likely removed in the form of metal precipitate. Addition of lime has been reported as the standard technique for precipitation of high concentrations of fluoride ion.[5] The lime reacts with fluoride in the wastewater to produce calcium fluoride precipitate.

Table IV. Performance of Metal Precipitation on Removal of BDAT Semivolatiles

Constituent	Influent, μg/L	Effluent μg/L	Sand Filter Effluent, μg/L	Removal Efficiency %
p-Chloro-m-cresol	ND	ND	140	–
para-Cresol	1350	1085	540	60
Dibenzo(a,i)pyrene	0.035	ND	ND	100
Dibenzo(a,e)pyrene	0.53	ND	ND	100
2,4-Dimethylphenol	275	ND	ND	100
Phenol	10600	9933	6800	35.8
Phthalic acid	32000	13000	4400	86.3
Total Concentration	44225	24018	11880	73.1

Table V. Performance of Metal Precipitation on Removal of BDAT Inorganics

Constituent	Influent, μg/L	Effluent μg/L	Sand Filter Effluent, μg/L	Removal Efficiency %
Cynide	237	160	150	36.7
Fluoride	2200	1067	867	60.6
Sulfide	8800	1500	3000	65.9
Total Concentration	11237	2727	4017	64.3

Table VI. Average Volatile Organic Concentrations in GAC Adsorption

Constituent	Influent μg/L	Effluent μg/L	Removal Efficiency %
Acetone	11967	4657	61.1
Chloroform	75.5	ND	100
Dibromoethane	78	ND	100
1,2-Dichloroethane	59	ND	100
Isobutyl alcohol	2600	2000	30
Methanol	18833	16500	12.4
Methyl ethyl ketone	7667	1110	85.5
Methyl isobutyl ketone	5033	ND	100
Methylene chloride	600	404.5	32.6
Pyridine	330	ND	100
1,1,2,2-Tetrachloroethane	68	ND	100
Toluene	69	ND	100
1,1,2-Trichloroethane	55	ND	100
Trichloroethene	116	ND	100
Total Concentration	47550.5	24671.5	48.1

ND = Nondetectable.

GAC Adsorption Process

Table VI lists the average concentrations of volatile organics in the influent and effluent of the GAC unit. About two-thirds of the volatile organics detected in the GAC influent were completely removed. The effluent still had five organic compounds with significant concentrations. They are acetone, isobutyl alcohol, methanol, methyl ethyl ketone and methylene chloride. The reason for these compounds appearing in the effluent is because they are not very adsorbable. Acetone and methanol are not adsorbable because their high solubility and polarity in water act to oppose the attraction of the adsorbate to the carbon.[8,9] Methanol and acetone both are small molecules and tend to be less adsorbable because fewer chemical bonds will form.[8] Isobutyl alcohol and methyl ethyl ketone are more adsorbable than methanol and acetone because they are less polar and less soluble molecules.[8,9] However, research has shown that isobutyl alcohol and methyl ethyl ketone are not very adsorbable.[8] Although methylene chloride is highly insoluble in water, it also displays a poor adsorbability at low influent concentration.[10]

The average concentrations of semivolatile organics are given in Table VII. Of the five semivolatile organics found in the influent, only phthalic acid and dibenzo(a,e)pyrene were found in detectable concentrations in the GAC effluent. Phthalic acid is soluble in water at pH greater than 6 and is expected to be less adsorbable.[11] This is due to the dissociation of the two carboxyl groups at pH greater than 6.

Average metal concentrations, given in Table VIII, reveal the poor adsorption of metals by GAC. Five metals have higher effluent concentration than influent concentrations, also the effluent has a higher concentration of total metals. These results agree with those of other researchers.[1] Results from previous studies show that there is no reduction by activated carbon for arsenic and barium at low concentrations.[1] The reported removal efficiency for nickel was 4%, while lead and mercury were reported to have higher effluent concentrations than the influent concentrations.[1]

Table VII. Average Semivolatile Organic Concentrations in GAC Adsorbption

Constituent	Influent μg/L	Effluent μg/L	Removal Efficiency %
P-Chloro-m-cresol	140	ND	100
para-Cresol	540	ND	100
Phenol	6800	ND	100
Phthalic acid	4400	5050	–
Dibenzo(a,e)pyrene	ND	0.039	–
Total Concentration	11880	5050.039	57.5

ND = Nondetectable.

Table VIII. Average Metal Concentrations in GAC Adsorption

Constituent	Influent μg/L	Effluent μg/L	Removal Efficiency %
Arsenic	26	11.3	56.5
Barium	72	83.5	–
Chromium (total)	41	62	–
Chromium (hexavalent)	ND	220	–
Copper	100	130	–
Lead	12	8.2	31.7
Mercury	0.4	ND	100
Nickel	76	112.7	–
Zinc	30	ND	100
Total Concentration	357.4	407.7	–

ND = Nondetectable.

The average concentrations of inorganics, shown in Table IX, indicate no removal of cyanide and fluoride. Sulfide displays a modest degree of adsorption.

Overall Treatment Plant Performance

Only sixteen BDAT compounds were detected in the GAC effluent. The concentrations of all compounds in the influent and effluent, their removal efficiencies, and the concentrations for the proposed rule for wastewaters are listed in Table X. Concentrations of all metals and inorganics were reduced to levels meeting the proposed standards, whereas five volatile and one semivolatile organic compounds could not meet the proposed rule. Since these compounds are poorly adsorbable to GAC, but are readily biodegradable, the addition of a biological treatment unit to the treatment train should better treat these organic compounds. Table XI lists the removal efficiencies of each treatment unit for BDAT groups, showing that semivolatile organics, metals and inorganics are mostly removed in metal precipitation. The volatile organics are removed about equally in the metal precipitation and GAC adsorption.

In summary, GAC can polish most organic compounds to a non-detectable level except for those poorly adsorbable compounds. Since the GAC effluent could meet the proposed treatment standards for those compounds, a biological treatment unit should be added to treatment facilities handling leachates with high concentrations of those non-adsorbable organic compounds.

Treatment Residues

The proposed rules for the Third Third landban requires metals in treatment residues to pass standards based on TCLP tests and non-metals to pass standards based on TCA.[7] In order to ensure that BDAT compounds were properly retained in the metal sludge, the sludge was stabilized by four different formulae to compare the effect of chemical stabilization.

Table XII tabulates the TCA results of the unstabilized and four stabilized metal sludges along with the proposed TCA standards for non-wastewaters.[3] The corresponding TCLP results of those five materials as well as the proposed rules are shown in Table XIII.

Table XII shows that the stabilized sludges have lower organic concentrations than the unstabilized sludge. Four compounds display significant concentrations in all four stabilized samples, namely: methanol, bis(2-ethylhexyl)phthalate, phenol and 1,2,4 trichlorobenzene. Comparison with the proposed rule shows that acetone is the only volatile organic which will not meet the proposed standard in

Table IX. Average Inorganic Concentrations in GAC Adsorption

Constituent	Influent μg/L	Effluent μg/L	Removal Efficiency %
Cyanide (total)	150	270	–
Fluoride	867	1000	–
Sulfide	3000	1600	46.7
Total Concentration	4017	2870	28.6

ND = Nondetectable.

Table X. Comparison of the GAC Effluent and the Proposed BDAT Standards for Wastewaters

BDAT Compound	Influent (μg/L)	Effluent (μg/L)	% Removal	Proposed Rule (μg/L)
Volatile Organics				
Acetone	12233	4657	61.9	162
Benzene	28	ND	100	136
n-Butyl alcohol	2200	ND	100	137
Chlorobenzene	29	ND	100	32
Chloroform	195	ND	100	46
1,1-Dichloroethane	150	ND	100	59
1,2-Dichloroethane	90	ND	100	211
Ethyl benzene	39	ND	100	32
Isobutyl alcohol	4833	2000	58.6	125
Methanol	24667	16500	33.1	33
Methyl ethyl ketone	9033	1110	87.7	16
Methyl isobutyl ketone	8200	ND	100	32
Methylene Chloride	3067	404.5	86.8	89
Pyridine	440	ND	100	8
1,1,2,2,-Tetrachloroethane	817	ND	100	32
Tetrachloroethene	69	ND	100	56
Toluene	473	ND	100	80
Trichloroethene	383	ND	100	54
1,2-Xylene	86	ND	100	182
1,3 and 1,4-Xylene	180	ND	100	182
Semivolatile Organics				
para-Cresol	1350	ND	100	1315
Dibenzo(a,i)pyrene	0.035	ND	100	41
Dibenzo(a,e)pyrene	0.53	0.039	92.6	41
2,4-Dimethylphenol	275	ND	100	36
Phenol	10600	ND	100	26
Phthalic Acid	32000	5050	84.2	20
Metals				
Arsenic	89	11.3	87.3	1390
Barium	143	83.5	41.6	1150
Cadmium	68	ND	100	200
Chromium (Total)	58	220	–	370
Copper	ND	130	–	1280
Lead	17	8.2	51.8	280
Mercury	2	ND	100	150
Nickel	650	112.7	82.7	550
Zinc	297	ND	100	1020
Inorganics				
Cyanide	237	270	–	1900
Fluoride	2200	1000	54.5	35000
Sulfide	8800	1600	81.8	14000

Table XI. Treatment Efficiency of BDAT Groups in Each Unit

BDAT Group	% Removal Efficiency		
	Metal Precipitation	GAC Adsorption	Total
Volatile Organic	29.2	34.1	63.3
Semivolatile Organic	73.1	15.4	88.5
Metal	73.3	-3.8	69.5
Inorganic	64.3	10.2	74.5

Table XII. TCA Results of Metal Sludge (μg/L)

Constituent	Unstabilized Sludge	Stabilized Sludge 1	2	3	4	Proposed Rule
Volatile Organics						
Acetone	3500	1200	ND	ND	ND	140
Methanol	9100	11000	9500	8500	4900	140000
Methyl Isobutyl ketone	2100	ND	ND	ND	ND	200000
Methylene chloride	ND	170	ND	ND	330	31000
Tetrachloroethene	190	ND	ND	ND	ND	6200
Toluene	160	ND	ND	ND	ND	28000
Trichloroethene	370	ND	ND	ND	ND	5600
Semivolatile Organics						
Bix(2-ethylhexyl)phthalate	36000	22000	15000	15000	18000	28000
Di-n-butyl phthalate	1200	ND	ND	ND	580	28000
Phenol	9700	3200	2400	2600	2300	6200
1,2,4-Trichlorobenzene	1300	1100	560	730	430	19000

ND = Nondetectable.

unstabilized and formula 1 stabilized sludge. Other volatile organics will have no difficulty meeting the proposed standards. All semivolatile organics will meet the standards in the stabilized sludge, while phenol and bis(2-ethylhexyl)phthalate will not without stabilization. Stabilization was shown to be effective in limiting the leaching of metals from the metal sludges to below the TCLP standards.

Spent GAC usually goes to regeneration and reuse. The final disposal of GAC can be difficult to handle because of the high concentrations of compounds adsorbed by it. The spent GAC sample was also stabilized by four formula to test the effect of fixation. The TCA and TCLP results for the unstabilized and four stabilized spent GAC are listed in Tables XIV and XV, respectively.

The TCA results show that stabilization cannot completely immobilize organic compounds in spent GAC. Despite the high concentrations in TCA tests, only acetone, methylene chloride, pyridine and phenol exceed the proposed standards. Therefore, either incineration or other technologies should be investigated for the final disposal of spent GAC.

The TCLP results of the spent GAC shows that arsenic, total chromium, lead and nickel will not meet the standards even after the stabilization. The high metal concentration might explain why the GAC is ineffective to remove metals because the high metal concentrations remain after regeneration.

CONCLUSIONS

The effectiveness of metal precipitation depended on the efficiency of separating precipitate from the stream for the facility evaluated in this study. With sand filtration, the metal precipitation process can reduce BDAT metals by about 60%.

The metal precipitation unit, followed by sand filtration, removed about 30% of volatile and about 73% of semivolatile organics.

The GAC adsorption could remove most organics, but was ineffective for the removal of certain organics, such as acetone and methanol. A physico-chemical treatment train is therefore insufficient

Table XIII. TCLP Results for Metal Sludge (μg/L)

Constituent	Unstabilized Sludge	Stabilized Sludge 1	2	3	4	Proposed Rule
Metals						
Arsenic	51	8.9	8	5	39	5600
Cadmium	220	ND	ND	ND	ND	66
Chromium (total)	83	48	61	28	ND	5000
Mercury	0.2	ND	ND	0.2	0.2	200.0
Nickel	1300	23	ND	ND	76	320

ND = Nondetectable.

Table XIV. TCA Results for Spent GAC (μg/L)

Constituent	Unstabilized Sludge	Stabilized 1	Stabilized 2	Stabilized 3	Stabilized 4	Proposed Rule
Acetone	140000	5000	18000	39000	ND	140
Chloroform	4500	3900	4700	3900	4000	6200
Dibromomethane	2200	1600	1500	1300	1600	16000
1,1-Dichloroethane	2200	1700	2000	1800	1800	6200
1,2-Dichloroethane	4800	3900	4700	4200	4200	140000
1,1-Dichloroethylene	ND	1100	13000	1100	1100	140000
Ethyl ether	ND	4300	4200	4100	4200	2000000
Methanol	5100	11000	6500	7600	9300	2000000
Methyl ethyl ketone	590000	ND	ND	32000	ND	31000
Methyl isobutyl ketone	340000	110000	120000	100000	60000	16000
Methylene chloride	39000	45000	35000	28000	33000	6200
Pyridine	ND	40000	ND	ND	65000	6200
1,1,1-Trichloroethane	1900	1500	2000	1500	1600	5600
1,1,2-Trichloroethane	1400	ND	ND	ND	ND	85
Trichloroethane	4400	4000	5000	3500	3300	3200
Vinyl chloride	1000	ND	ND	ND	ND	6200

ND = Nondetectable.

Table XV. TCLP Results for Spent GAC (μg/L)

Constituent	Unstabilized Sludge	Stabilized 1	Stabilized 2	Stabilized 3	Stabilized 4	Proposed Rule
Metals						
Arsenic	12000	5100	8400	6700	3500	5600
Barium	34000	42000	320000	59000	48000	100000
Beryllium	830	680	800	670	710	–
Cadmium	640	ND	ND	ND	ND	66
Chromium (total)	19000	20000	25000	13000	20000	5000
Copper	170000	120000	87000	70000	110000	–
Lead	6500	5600	4700	12000	16000	510
Nickel	37000	27000	23000	18000	24000	320
Selenium	2900	ND	1200	ND	ND	5600
Vanadium	12000	13000	23000	9400	14000	–
Zinc	23000	79000	180000	77000	80000	–

ND = Nondetectable.

to meet the proposed rules of Third Third for the facility evaluated. A biological treatment unit is needed to make a complete BDAT treatment train for leachates containing non-adsorbable organics.

The observed removal of metals by GAC was low, as would be expected.

Stabilization of metal sludges with cement-based reagents was demonstrated to be an effective method for the fixation of organics and metals. The stabilized metal sludges can pass the proposed rules for TCA and TCLP with no difficulties.

Spent GAC will not meet the standards for TCA and TCLP even after stabilization and should be treated by other technologies, such as incineration.

REFERENCES

1. Shuckrow, A. J., Pajak, A. P. and Touhill, C. J., *Hazardous Waste Leachate Management Manual*, Noyes Data Corporation, Park Ridge, NJ (1982).
2. Berlow, J. R. and April, R., Methodology for Developing Best Demonstrated Available (BDAT) Treatment Standards, *EPA/530-SW-89-017L*, USEPA, Washington, D.C. (December, 1988).
3. *Federal Register* (November 18, 1989).
4. *Generic Quality Assurance Project Plan for Land Disposal Restrictions Program (BDAT)*, USEPA, Washington, DC (March 1987).

5. USEPA, *Test Methods for Evaluation Solid Waste, SW-846*, EPA Office of Solid Waste and Emergency Response, Washington, DC (1986).
6. Patterson, J. W., *Industrial Wastewater Treatment Technology*, 2nd ed., Butterworths, Stoneham, MA (1985).
7. Conner, J. R., *Chemical Fixation and Solidification of Hazardous Wastes*, Van Nostrand Reinhold, New York, NY (1990).
8. Perrich, J. R., *Activated Carbon Adsorption for Wastewater Treatment*, CRC Press, Boca Raton, FL (1981).
9. Godfrey, N. B., "Solvent Selection via Miscibility Number", *Chemtech*, p. 359–363 (1972).
10. Dobbs, R. A. and Cohen, J. M., *Carbon Adsorption Isotherm for Toxic Organics*, EPA-600/8-80/023, Cincinnati, OH (1980).
11. *The Merck Index*, 9th ed., Merck & Co., Rahway, N.J. (1976).

71 TREATMENT OF CONTAMINATED GROUNDWATER AND LEACHATE FROM SUPERFUND SITES AT POTWs – A CASE STUDY

Vasiliki Keramida, President
Ontario Environmental, Inc.
Indianapolis, Indiana 46204

Robert F. Holm, Administrator
Water and Land Protection Division
Indianapolis Department of Public Works
Indianapolis, Indiana 46221

INTRODUCTION AND BACKGROUND

In the Comprehensive Environmental Response, Compensation and Liability Act (CERCLA) of 1980, commonly called Superfund, Congress set up a trust fund (Superfund) to be managed by the U.S. EPA and help pay for the study and cleanup of hazardous waste sites that threaten public health and the environment. Sites eligible for Superfund action are listed on the National Priorities List (NPL) by the U.S. EPA.

In 1984 the U.S. EPA listed the Northside Sanitary Landfill (NSL) site on the NPL, as a potential source of onsite and offsite contamination, and authorized a remedial investigation/feasibility study (RI/FS) of the site. The NSL site is located in Boone County, Indiana, about 10 miles northwest of Indianapolis (Figure 1). The site is near the intersection of State Road 32 and U.S. 421 and is bounded on the east and south by Finley Creek and farmlands, on the north by uncultivated farmland, and on the west by an unnamed ditch which separates NSL from another Superfund site, the Environmental Conservation and Chemical Corporation (Enviro-Chem) site (Figure 2). The unnamed ditch joins Finley Creek at the southwestern corner of the landfill and Finley Creek flows into Eagle Creek about one-half mile downstream from the sites (Figure 2). Eagle Creek flows south from its confluence with Finley Creek for approximately ten miles before it empties into Eagle Creek Reservoir, one of the sources of drinking water for the City of Indianapolis (Figure 3).

The Northside landfill occupies approximately 168 acres. Approximately 70 acres of the 168-acre site has been operated as a landfill. The site have been active since at least 1962 and hazardous waste was co-disposed of with solid waste at the NSL facility since 1967. By 1981, NSL had accepted at least 16 million gallons of hazardous substances.[1] Today, NSL continues to operate as a solid waste landfill. The facility had recurring operational deficiencies as reported by the Indiana State Board of Health, and preliminary investigations by the U.S. EPA prior to the site's listing on the NPL had revealed environmental problems at the site, including groundwater contamination.[1] The RI/FS work at the NSL site was started by EPA in 1984 and the Record of Decision (ROD) was issued in 1987.

The selected remedy presented by EPA in 1987 was a combined remedy for both the NSL and the adjacent Enviro-Chem sites. An integral component of that remedy was the collection of leachate and groundwater and on the on-site treatment of these wastes.[2] EPA had concluded that both inorganic and organic toxic contaminants in the leachate/groundwater mixture needed to be treated in order to meet water quality based NPDES permit limits for discharge into Finley Creek, and had proposed an onsite treatment system consisting primarily of a metal precipitation component and a powdered activated carbon treatment (PACT) component.[2] The actual treatment system was to be developed through bench and pilot scale testing during the predesign investigations for the final remedial action. EPA subsequently added ammonia and chloride to the list of conventional pollutants (BOD_5, TSS, etc.) to be limited in the NPDES permit, thus complicating even further the required treatment process sequence.

Persuasive comments received by the EPA on the ROD convinced the agency that it was preferable to implement separate, complementary remedies for the Enviro-Chem and NSL sites, rather than applying the combined remedy originally selected. The idea, furthermore, of treating the NSL

45th Purdue Industrial Waste Conference Proceedings, © 1991 Lewis Publishers, Inc., Chelsea, Michigan 48118. Printed in U.S.A.

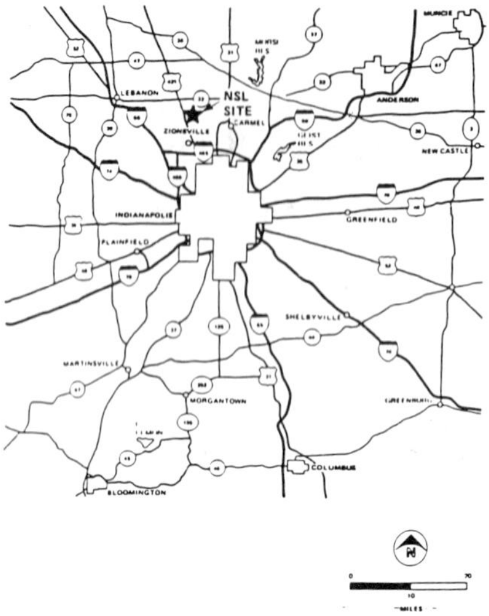

Figure 1. NSL site general location.

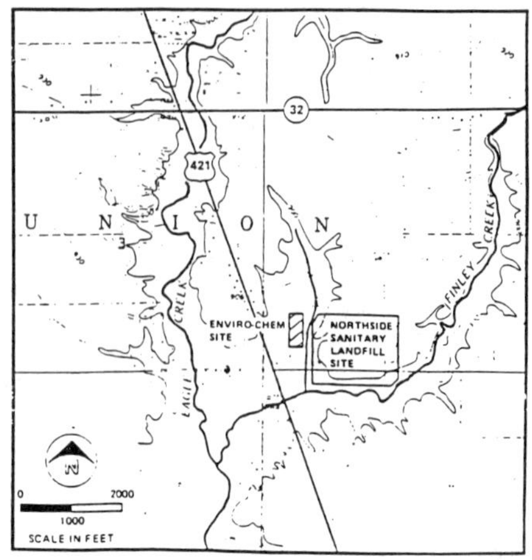

Figure 2. NSL site and surroundings.

leachate/groundwater waste at the Indianapolis POTW rather than at an on-site treatment plant started gaining momentum, and it was eventually adopted by the EPA as the preferred remedial component for the NSL site. The incorporation of such a scheme into the remedial plan pivoted on the

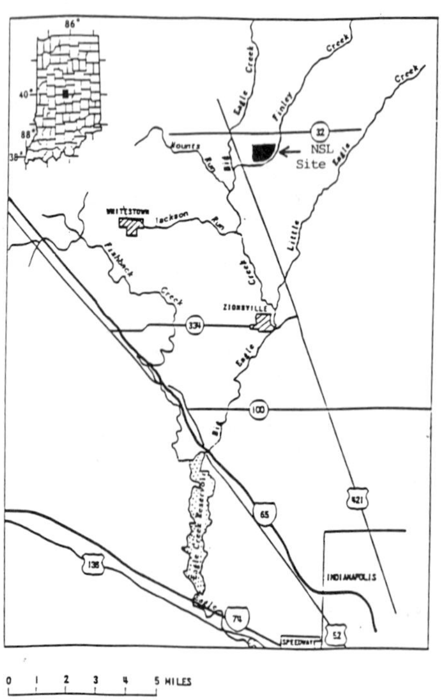

Figure 3. NSL site map.

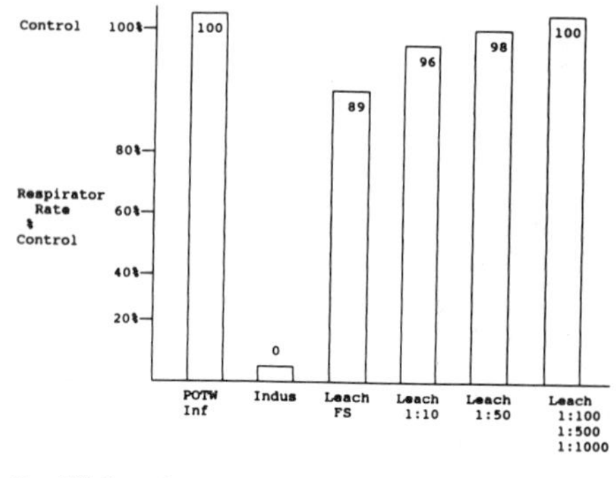

Figure 4. NSL Leachate inhibition evaluation.

compatibility evaluation performed by the Indianapolis Department of Public Works of the NSL leachate, for determining the acceptability of this wastestream into the City's sewage system.

Under the POTW treatment scheme the NSL waste would be transported from the NSL site to an appropriate entry point into the Indianapolis sanitary sewer system at the north part of Indianapolis by means of a pumping station and a forcemain, approximately 8 miles long. Preliminary evaluations of the construction and operation-maintenance costs associated with the POTW treatment option, including monthly compensation to the POTW, suggest that it is a much more economical alternative to on-site treatment. The POTW treatment option, furthermore, carries more long term effectiveness than the on-site treatment alternative and a reduced liability. The long term effectiveness of a complex on-site treatment plant in such a case could be questionable, due to the anticipated changes in the volume and consistency of the leachate/groundwater waste as the site goes through its post-closure cycle. The opportunities, therefore, for a malfunctioning of an on-site treatment plant could be numerous, and any such malfunctioning would have the possibility of seriously impacting the receiving stream, Finley Creek, which is part of a potential migration pathway from the NSL site to the Eagle Creek Reservoir.[1]

COMPATIBILITY EVALUATION

Landfill leachates are often comparable to complex industrial waste streams which contain both toxic inorganic and organic contaminants. Chapter 27 of the Code of Indianapolis and Marion County, Indiana, "Sewers and Sewage Disposal",[3] allows the conveyance by the sewers and treatment by the POTW of wastewaters such as the NSL leachate and groundwater, provided that such wastewaters meet all applicable requirements of the Code. In the case of the NSL waste it was decided that the applicable requirements of the code translated to a compatibility evaluation which included three components: the inhibition evaluation, the chemical analysis evaluation, and the sewer surcharge evaluation.

The inhibition testing, the first step in the evaluation process, was designed to determine whether the waste is compatible with the biological reactions employed by the Indianapolis POTW, namely the nitrification process, and therefore will not cause interference to the plant's operations. The chemical analysis evaluation, on the other hand, was based on a comparison of the leachate analytical data to the Code limits and the city's guidelines for industrial discharges.

Finally, the third component addressed concerns about sewer system surcharging and combined sewer overflows, and the possibility of the NSL waste being carried by such overflows to receiving streams.

TESTING PROTOCOL

Because of the limited number of monitoring wells on the NSL site it was assumed that groundwater samples from those wells would not be representative of the whole volume of groundwater which would eventually have to be collected. Furthermore, leachate generally contains a wider range of contaminants, and at higher concentrations, than the groundwater at landfill sites. The compatibility evaluation, therefore, addressed only the NSL leachate, assuming that the leachate would represent the "worst case scenario" of possible NSL waste streams.

In 1988, staff from the Indiana Department of Environmental Management (DEM), collected samples from each of the three leachate tanks presently used at NSL, and delivered them to the IDPW Laboratory. A composite sample was prepared by the laboratory staff, using equal volumes from each of the leachate samples, LT_1, LT_2 and LT_3, and was tested for inhibition. These samples were also tested for volatile organic compounds (VOCs), semivolatile organic compounds (SVOCs), pesticides/PCBs, cyanide, antimony, arsenic, beryllium, cadmium, chromium, copper, lead, mercury, nickel, selenium, silver, thallium, and zinc. The identification of the samples sent to an outside laboratory for analysis is as follows:

LT_1 – Leachate Tank #1
LT_2 – Leachate Tank #2
LT_3 – Leachate Tank #3
$LT_{1,2,3}$ – Composite
LT_4 – Leachate Tank #2 Duplicate
LT_5 – Trip Blank

Inhibition Test Protocol

The NSL leachate compatibility evaluation performed by the IDPW consisted of two distinct nitrification inhibition tests. Both of the Indianapolis Advanced Wastewater Treatment (AWT) plants include an Oxygen Nitrification process where nitrifying bacteria in the Activated Sludge oxidize the NH_3-N. Compatibility of a waste with a POTW such as the Indianapolis AWT plant has to be determined on the basis of nitrification inhibition, because of the extreme sensitivity of the nitrifying bacteria to toxic substances, in comparison to the rest of the biomass in the activated sludge. Two nitrification inhibition tests have been developed by the IDPW in order to determine inhibition and handle nitrification interference problems that certain industrial wastewaters could impose on the AWT plants.

The first test, called the Oxygen Utilization Rate (OUR) test, uses an enriched *Nitrosomonas* biomass which was progressively cultured from the POTW's nitrifying activated sludge over a period of several months and was subsequently maintained in the laboratory under strict growth conditions. The respiration rate exhibited by the pure *Nitrosomonas* culture when mixed with a particular waste is employed as the bench-mark parameter for determining inhibition. This test is conducted with a Gilson Model 5/6 Oxygraph. The OUR test is performed routinely by IDPW staff on composite samples collected from interceptor sewers and influent lines to the POTW, as well as on selective industrial wastestreams. The OUR test's speed and accuracy make it an excellent screening tool for early determination of toxicity problems in a specific sewer interceptor line, as well as a valuable device for evaluation of compatibility of new wastestreams. In summary, the test is performed by mixing a certain volume of a specific waste with a certain volume of *Nitrosomonas* biomass. Ammonia is added to the mixture at specified levels, if needed, to serve as food for the nitrifiers, and the mixture is allowed to react, while aerated, for approximately one half hour, prior to measuring the respiration rate of the biomass in the mixture. Controls are run simultaneously with the waste samples (biomass plus food, with no waste present). The compatibility of a waste is determined by comparing the respiration rate shown by the waste mixture to the respiration rate of the control. For simplicity of comparison purposes, the respiration rates are translated to percentages of the control (an average of all the control respiration rate reading is used as the 100% basis). Based on the test's variability, wastes exhibiting respiration rates of 85% and above are considered compatible with the nitrification process.

The second type of nitrification inhibition test used was the Ammonia Depletion test. This procedure is a verification of the OUR test results and is designed to check the compatibility of a waste under POTW "field" conditions. In the Ammonia Depletion test, predetermined volumes of the waste in question are mixed in an Erlenmeyer flask with an appropriate volume of actively nitrifying Return Activated Sludge (RAS), freshly collected from the Indianapolis POTW, along with ammonia solution for food. The mixtures are placed on a shaker table and are aerated. Initial determination of the NH_3-N concentration in each flash is made at time zero and then the NH_3-N readings are taken hourly during aeration, until nitrification is nominally complete. Usually, three hourly readings of NH_3-N concentration are recorded for each sample in an effort to parallel the ammonia depletion rates seen through the stages of the POTW's nitrification reactor system. Nitrification rate data for the waste mixtures are plotted and compared to the rates obtained from the controls run along with the waste samples. The idea behind this test is that the nitrifying bacteria present in the activated sludge from the POTW's reactor should deplete the ammonia present in a sample at similar rates as those observed in the plant's reactor. If the ammonia concentration in a sample remains steady or is reduced at a low rate over the 3-hour period, the assumption is made that the nitrifiers are not consuming the ammonia because the waste is inhibitory to them. Such waste is in turn classified incompatible with the POTW's processes.

RESULTS AND DISCUSSION

Inhibition Evaluation

In the Oxygen Utilization Rate (OUR) test the NSL leachate composite sample was tested at full strength as well as at dilutions of 1:10, 1:50, 1:100, 1:500 and 1:1000. Each dilution was run in duplicate. For comparison purposes an 8-hour composite sample from the Belmont POTW influent wastewater was also included in the test as well as a 24-hour composite sample from an industrial discharge to the City's sewers. The purpose of the inclusion of these two wastestreams was to illustrate how a non-inhibitive waste (Belmont) and an inhibitive waste at full strength (industrial) behave under the OUR test. The results of the OUR test are summarized in Figure 4. The respiration rates shown are the averages of the rates obtained from the duplicate samples. The full strength NSL leachate duplicates showed a respiration of 88% and 91% of the control, while the various dilutions of the leachate

Table I. NSL Leachate Samples Inhibition Evaluation Ammonia Depletion Test Results

Sample ID	Dilution	NH_3-N (ppm) Hourly Readings Time			
		0 hr.	1 hr.	2 hr.	3 hr.
Control 1		17	7	0	0
Control 2		17	6	0	0
Industrial	1:1	55	55	55	55
Duplicate	1:1	55	50	45	45
Belmont Influent	1:1	18	10	2	0
Duplicate	1:1	17	10	1	0
NSL Leachate	1:1	80	50	50	50
	1:1	80	65	60	60
	1:10	17	7	0.5	0
	1:10	18	10	0.5	0
	1:50	17	6	0	0
	1:50	18	8	0	0
	1:100	18	8	0	0
	1:100	18	7	0	0
	1:500	18	7	0	0
	1:500	17	6	0	0
	1:000	16	7	0	0
	1:000	17	6	0	0

recorded respirations ranging from 96% to 100% of the control. The Belmont influent sample's respiration was at 100% and the industrial sample at full strength showed a zero respiration rate.

The Ammonia Depletion Rate test was performed with NSL leachate at full strength as well as the same dilution ratios used in the OUR test. All samples were run in duplicate and the Belmont influent and industrial waste were also tested. There was no ammonia solution (food) added to the full strength leachate samples and the industrial samples because of the high ammonia concentrations found in these wastes during the initial sample ammonia determination. The results of the Ammonia Depletion test are illustrated in Table I, as well as in Figures 5 and 6. The NSL leachate samples at every dilution level, except at full strength, performed equally well as the control and better than the Belmont influent wastewater. The NH_3-N concentration in the NSL leachate samples dropped from approximately 18 mg/L to zero in 2 hours, indicating that the nitrifying activated sludge performed well in the presence of the leachate. It should be noted that in the case of the full strength leachate sample, even though there was substantial ammonia reduction observed, the ammonia concentration did not go to zero at the end of the 3-hour test period, due to the high initial ammonia content in the sample. By comparison, the industrial sample at full strength was inhibitory because, even though its initial ammonia concentration was also high, it did not show practically any ammonia reduction, suggesting that the nitrifying bacteria present were inhibited by the waste's constituents.

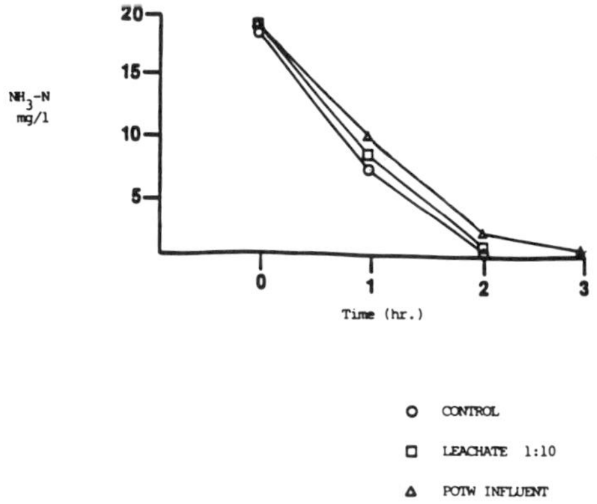

Figure 5. NSL leachate inhibition evaluation.

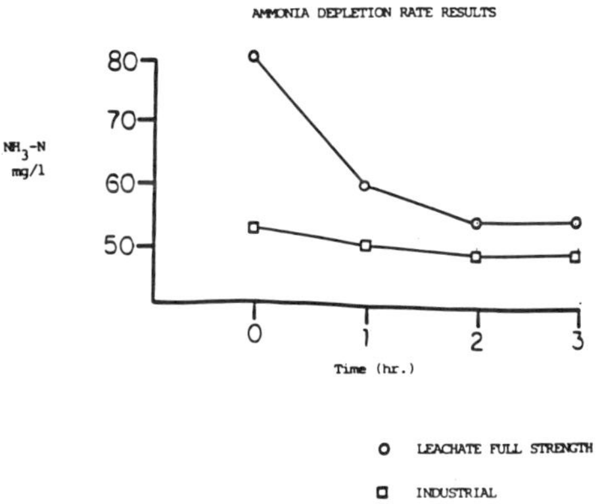

Figure 6. NSL leachate inhibition evaluation.

Table II. NSL Leachate Samples Chemical Analyses

Sample ID	
LT 1 - Leachate Tank 1	VOCs
LT 2 - Leachate Tank 2	VOCs
LT 3 - Leachate Tank 3	VOCs
LT 4 - Leachate Tank 2 Duplicate	VOCs
LT 1, 2, 3 - Composite	Semi Vol., Pesticides/PCBs, CN, Metals
LT 5 - Trip Blank	VOCs, Semi Vol., Pesticides/PCBs, CN, Metals

Note: Method References
VOCs - SW846-8240
Semi Vol. - SW846-8270
Pesticides/PCBs - SW846-8080

Based on the results of the two tests performed, we concluded that the NSL leachate was not inhibitory to the nitrification process of the Indianapolis POTW and, therefore, was compatible with the plant's operations, as far as treatability is concerned.

Chemical Analysis Evaluation

The NSL leachate samples were analyzed for the constituents and by the methods stated in Table II. The various compounds found in each sample are shown in Table III. There were no VOCs found in the sample from the leachate tank #3 (LT3). Seven VOCs were found in the sample from leachate tank #3 (LT3). Seven VOCs were found in the sample from leachate tank #1 (LT1), four in LT2 and three in its duplicate sample LT4. In addition, two VOCs appeared in the trip blank. One semivolatile compound was found in the composite of the three leachate tank samples (LT 1, 2, 3), as well as an estimated concentration of 700 ppb of unknown tentatively identified semivolatile compounds. No pesticides or PCBs were found in any of the samples and no cyanide. With regard to metals, only chromium, nickel and zinc are shown in the composite sample from the three leachate tanks, while chromium also appeared in the trip blank.

A comparison of the compounds found in the leachate samples to the discharge limits stated in the Indianapolis Sewer Use Ordinance is shown in Table IV. The majority of the compounds addressed in the Ordinance were not detected in the leachate samples, with the exception of chromium, nickel and zinc which were found at concentrations far below the allowable limits. It should be noted that for pentachlorophenol the Ordinance limit is 0.012 mg/L. Although no pentachlorophenol was detected, the laboratory could not achieve a detection limit below 0.05 mg/L due to the complexity of the

Table III. NSL Leachate Samples Chemical Analysis Results Compounds Detected

	Sample ID					
	LT1	LT2	LT4 (Dupl)	LT3	LT 1,2,3	Trip Blank
VOCs (ppb)						
Methylene Chloride	80	BDL	BDL	BDL	–	5
Acetone	220	BDL	BDL	BDL	–	BDL
Chloroform	15	136	BDL	BDL	–	BDL
Trichloroethene	10	BDL	BDL	BDL	–	21
Toluene	33	125	125	BDL	–	BDL
Ethyl Benzene	325	775	1270	BDL	–	BDL
Xylenes	510	160	320	BDL	–	BDL
SVOCs (ppb)						
Bis(2-ethylhexyl)phthalate	–	–	–	–	200	BDL
Unknowns	–	–	–	–	700	BDL
Metals (ppm)						
Chromium	–	–	–	–	0.16	0.13
Nickel	–	–	–	–	0.12	BDL
Zinc	–	–	–	–	0.09	BDL

BDL = Below detection limit.
– = Not tested for.

Table IV. NSL Leachate – Compounds Detected Comparison to the Indianapolis Sewer Discharge Limits (ppm)

Parameter	Limit	Leachate Sample
As	4.0	<0.002
Cd	1.2	<0.01
Cr	24.0	0.16
Cr (hex)	3.4	–
Cu	2.2	<0.02
CN(a)	0.4	–
CN(T)	8.0	<0.01
Pb	4.7	<0.1
Ni	7.3	0.12
Hg	0.025	<0.002
Ag	4.2	<0.05
Zn	38.0	0.09
Phenol	46.0	<0.05
Penta-Chlorophenol	0.012	<0.05

leachate sample matrix. Pentachlorophenol historically has not been found in any samples related to the NSL site. It is expected that the leachate-groundwater mixture will be generated by NSL in the future will be of lesser complexity than the leachate due to its dilute nature and, therefore, more amenable to the standard lower detection limit of 0.010 mg/L, so that compliance with the city's ordinance limit for pentachlorophenol can be verified.

There were certain organic compounds found in the leachate samples for which there was no limits specified in the city's Sewer Use Ordinance. The concentrations of these compounds were compared to numerical guidelines for discharges of organic compounds of concern established by the City of Indianapolis during its pretreatment study. These guidelines were set for organics present in the city's sewer system at a concentration within ten percent of a concentration reported to have an impact on treatment processes, sludge or aquatic life, but whose low levels in the POTW influent did not warrant the establishment of limits. Table V shows the comparison of the concentrations of the organics in the NSL leachate samples to the city's guidelines. The organics found in the NSL samples were at concentrations ranging from 11 to 65 times less than the city's guidelines for industrial discharges for the corresponding compounds. There were no guidelines for four of the compounds detected in the leachate, namely acetone, trichloroethene, xylenes, and bis (2-ethylhexyl) phthalate. These four compounds were among the organics for which there were no guidelines developed by the City of Indianapolis due to their extremely low concentrations reported for all four compounds in the NSL leachate do not suggest that it is possible for the compounds to impact the City's sewage system. The data generated from these specific leachate samples, therefore, indicated that there is no need for pretreatment of the NSL leachate prior to its discharge into the Indianapolis sewer system.

Table V. NSL Leachate Samples Comparison of Concentrations of Organic Compounds Found in the Leachate to Municipal Guidelines for Industrial Discharges to Sewers

Compound	Leachate Analysis (ppb) LT1	LT2	LT4 (Dupl)	LT3	Trip Blank	Maximum Municipal Guideline (ppb)
Methylene Chloride	80	BDL	BDL	BDL	5	5250
Acetone	220	BDL	BDL	BDL	BDL	NA
Chloroform	15	136	BDL	BDL	BDL	3650
Trichloroethene	10	BDL	BDL	BDL	21	NA
Toluene	33	125	125	BDL	BDL	7800
Ethyl Benzene	225	775	1270	BDL	BDL	15000
Xylenes	510	160	320	BDL	BDL	NA
Bis(2-ethylhexyl)phthalate	–	–	–	200	BDL	NA

BDL = Below detection limit.
– = Not tested for.
NA = Not available. No guidelines were established for these compounds due to their extremely low concentration in or absence from the municipal wastewaters.

Sewer Surcharge Evaluation

The preliminary figures for the leachate-groundwater effluent volume from the NSL site have been estimated at 15,000 gal/day (avg) and 30,000 gal/day (max). The NSL effluent will be pumped through a pipeline, approximately 8-miles long, to an appropriate entrance point to the Indianapolis sewer system in the city's North Belmont sewer basin. The North Belmont basin contains the North Belmont Interceptor which will transport the NSL effluent to the Indianapolis Belmont POTW. Review of available information on the sewer lines in the Indianapolis North Belmont basin indicated that there were adequate grades and slopes to allow large quantities of flow to be transported before the pipes could become surcharged. This was verified by ADS Services, Inc. in their "Mass Flow Monitoring Study", performed for the City of Indianapolis in 1986.[4]

The North Belmont Interceptor is a recently built line which was designed to provide maximum separation of storm water flow and sewage flow. There are no Combined Sewer Overflows (CSOs), therefore, in the path to be followed by the NSL effluent until the old combined sewers are encountered, approximately six miles from the point the NSL effluent enters the city's sewer system. There are three CSOs at that old combined sewer section. These three CSO sites, however, were among the CSOs eliminated from further consideration during the Indianapolis 1981–83 CSO study,[5] because of few or no overflow events observed. The ADS study[4] had determined that the average dry weather flow at a location in the North Belmont Interceptor upflow from the CSO area was 5.5 mgd during the study's June–July 1986 period. The peak flow at the same location was 14.5 mgd, measured shortly after a 1.06 inch rainfall event. No discharges from the three CSOs were observed during that rain event. These figures suggest that in the unlikely event a discharge occurred out of one of these CSOs, the NSL effluent will be so diluted as to have no impact on the stream receiving the CSO discharge. Indeed, the dilution ratio of sewage to NSL waste in the North Belmont Interceptor at that ADS study location upflow from the CSOs referenced above, under the 14.5 mgd sewage flow and 30,000 gal/day NSL waste flow conditions, is expected to be almost 500 to 1.

The possibility for exfiltration is not present in the North Belmont sewer basin due to the fact that infiltration rather than exfiltration occurs throughout the North Belmont basin. This was well documented in the Indianapolis 1981–82 infiltration/inflow study,[5] as well as in the ADS study.[4]

We can, therefore, conclude that there is no basis for concern regarding overflows and exfiltration problems along the city's sewer lines which will carry the NSL effluent to the POTW.

CONCLUSIONS

The results of this work have demonstrated that the leachate from a certain Superfund landfill site can be accepted by the Indianapolis sewers for treatment at the POTW, without need for pretreatment. Bioassays run on leachate samples from the NSL Superfund landfill site for toxicity and inhibition testing showed no characteristics for inhibiting the POTW nitrification process. Chemical analysis of the leachate samples indicated that the concentrations of the contaminants present in the leachate fall well below the city's sewer discharge limits and guidelines. Finally, all the available information on the sewer lines through which the NSL effluent will travel suggest no problems with CSOs or exfiltration.

REFERENCES

1. "Final Remedial Investigation Report, Northside Sanitary Landfill, Indiana,"prepared by CH2M-HILL for the U.S. EPA 5 (March 27, 1986).
2. "Combined Alternative Analysis Report for Northside Sanitary Landfill and Environmental Conservation and Chemical Corporation Site," prepared by CH2M-HILL for the U.S. EPA 5 (December 6, 1989).
3. Chapter 27, Code of Indianapolis and Marion County, Indiana, "Sewers and Sewage Disposal."
4. ADS Services, Inc. "Mass Flow Monitoring Report," Final Report for City of Indianapolis (1986).
5. Howard Needles Tammen & Bergendoff, "The Indianapolis Combined Sewer Overflow Study Report," (1983).

72 REVERSE OSMOSIS TREATMENT OF LANDFILL LEACHATE

Riley N. Kinman, Professor
Civil & Environmental Engineering
University of Cincinnati
Cincinnati, Ohio 45221

David L. Nutini, General Manager
RNK Environmental, Inc.
Covington, Kentucky 41017

INTRODUCTION

The feasibility of using reverse osmosis (RO) technology to treat landfill leachate was studied. The project was conducted on a pilot-scale in the field. The specific objectives of the study were to: 1) evaluate the RO treatment processing of landfill leachate; and 2) determine any other requirements necessary to handle the landfill successfully.

METHODS AND MATERIALS

A pilot-scale reverse osmosis (RO) treatment unit was set-up in the field in July, 1989. The RO was capable of treating 10,000 gallons of leachate per day. The unit itself consisted of a 5 micron prefilter, a pump, 4 × 50 square feet of polyamide membranes (total = 200 square feet), pressure valves, flow meters and an instrument panel, which included a conductivity meter. Operating pressure for the RO unit was between 200–600 psi. The RO unit is shown in Figure 1.

A flow diagram of the complete treatment system is presented in Figure 2. It consisted of two 20,000 gallon holding tanks. The first tank (Tank #1) was used as a settling tank. The leachate was pumped from this tank through a 10 micron filter into Tank #2. From Tank #2, it was pumped to a 55 gallon drum and then to the RO unit. The permeate (clean water) from the process was piped to one of two holding tanks. The concentrate (leachate) was piped back to Tank #1. Therefore, the leachate was concentrated during processing.

A flow diagram of the treatment process for fresh leachate (treated on October 24, 1989) is presented in Figure 3. The only change from the above was that a 4000 gallon tanker truck was hooked up directly to the RO. The concentrate stream was returned directly to the truck. Therefore, the feed was concentrated during processing.

Several trial runs were made during the study. Samples of the treated leachate (permeate) were taken during and at the end of each run. Analyses of these samples included the following parameters:

Figure 1. RO unit.

45th Purdue Industrial Waste Conference Proceedings,

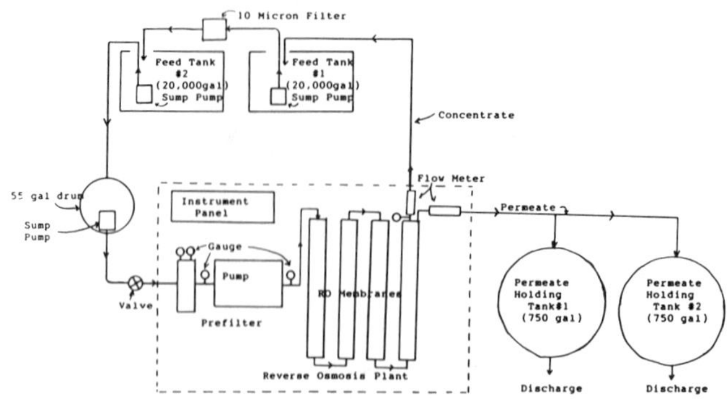

Figure 2. Flow diagram of RO treatment process.

pH, conductivity, Chemical Oxygen Demand (COD), alkalinity, total solids, volatile solids, ammonia and Total Kjeldahl Nitrogen (TKN).

A composite sample was also collected and analyzed. Analysis of this sample included a determination of the following additional parameters to the ones mentioned above: metals (Ar, Cd, Cr, Cu, Pb, Ni and Zn). These were measured by Atomic Absorption Spectrophotometry (AA). Volatile organic compounds (VOCs) were analyzed from this composite sample as well. This was performed by Gas Chromatography/Mass Spectrometry (GC/MS).

During the RO runs, additional observations and measurements were made. These included: pressure readings of all RO gauges, flow rates for permeate and concentrate streams, conductivity readings, visual observations of treated leachate, odors and other problems. Photographs were taken to document the results.

RESULTS AND DISCUSSION

Treatment runs of 2 to 6 hours in length were made during operation of the RO unit. Membrane pressures were operated in a range of 200 to 400 psi. Flow rates in the permeate and concentrate streams varied according to pressure settings. Typically at 200 psi, the permeate and concentrate flow rate were 0.8 gallon per minute and 10 gallons per minute, respectively. At the higher pressure (400 psi), the permeate and concentrate flow rates were 2 gallons per minute and 8.5 gallons per minute, respectively. Feed temperature was at 25°C except for the fresh leachate run on October 24, 1989, when the temperature was 22°C. A least squares plot of permeate flux versus pressure is shown in Figure 4. This graphically demonstrates the correlation of the two parameters. A high concentrate flow to permeate flow was maintained to prevent early fouling of the membrane. A brief five minute flush was adequate to restore permeate flux that may have declined. Permeate flux was best maintained at the lower pressure (200–250 psi), but tended to decrease slightly over time at the higher pressure.

RO treatment of landfill leachate resulted in a permeate stream of extremely high quality and a concentrated waste stream. Appearance of the permeate was like that of clean water (Figure 5).

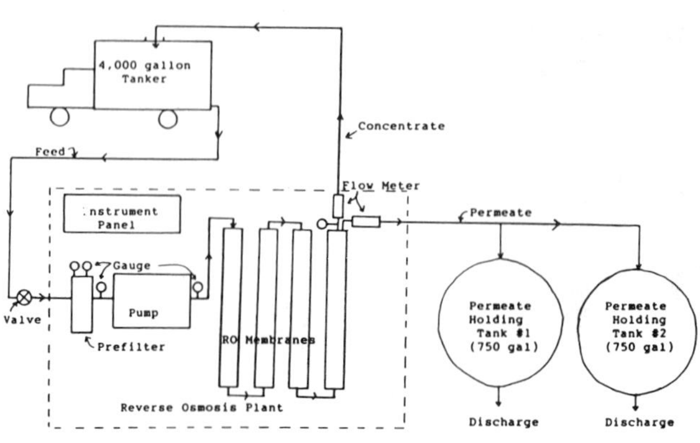

Figure 3. Flow diagram of RO treatment process of fresh leachate on October 24, 1989.

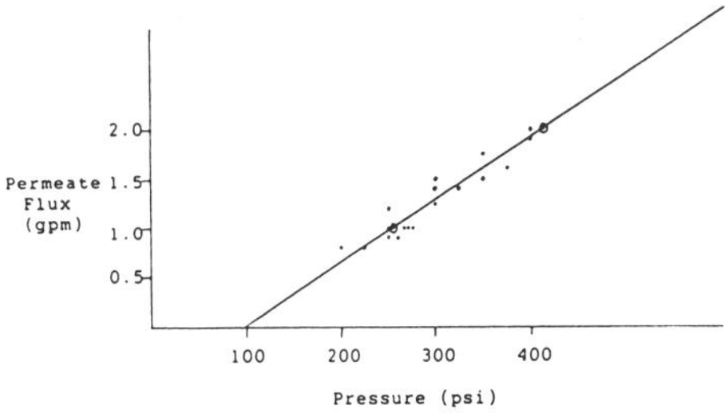

Figure 4. Composite of all runs – permeate flux vs. pressure when feed temperature was 25°C.

Figure 5. Appearance of permeate (clear) and concentrate (black) from the RO treatment unit.

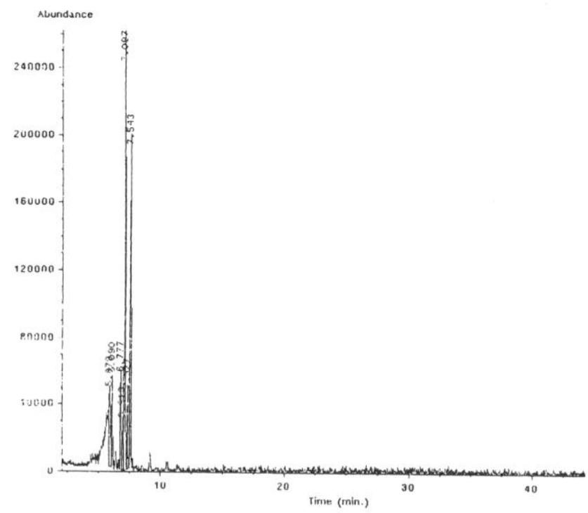

Figure 6. GC/MS analysis of August 23, 1989, permeate sample.

Conductivity, read on the RO machine meter, was at infinity in all runs. The results comparing the parameters analyzed for the leachate (feed) and permeate (clean water) are presented in Tables I and II. The values given in Table I were average values for all runs. The metal values given in Table II were analyzed from a composite sample.

The percent removals for the parameters are provided in Table III. Removals were all in the high 90's. Total solids were reduced by 97%. Organics were reduced by 97%. Metals were reduced to below detectable limits.

GC/MS Analysis was performed on various permeate samples to look for trace toxic compounds in those samples. Table IV presents the results of those analyses. Very few compounds were identified. No quantification of those identified compounds was determined. Levels are generally in the μg/L range. An example of one of the GC/MS permeate scans is presented in Figure 6. Only two trace compounds were identified (tetrahydrofuran and 2-butanone). These results were significantly different to the GC/MS scan of the feed samples (leachate). They graphically demonstrate the organic removals by the RO unit.

No major problems were encountered in the operation of the RO unit. As mentioned earlier, the higher operating pressures resulted in a slight decrease in permeate flux. This was overcome by

Table I. Analysis of Leachate (Feed) and Permeate During Various Runs (Average Values)

Parameter[a]	Leachate (Feed)	Permeate
pH	7.5	6.6
Alkalinity	2581	142
COD	888	27
Total Solids	4107	123
Volatile Solids	541	43
Ammonia	776	26
TKN	845	26

[a]All units in mg/L except for pH which has no units.

Table II. Analysis of Metals in Leachate (Feed) and Permeate

Metal	Leachate (Feed)[a]	Permeate
Ar	8	BDL
Cd	1	BDL
Cr	96	BDL
Cu	31	BDL
Pb	23	BDL
Ni	60	BDL
Zn	860	BDL

[a]All units in $\mu g/L$ except for pH which has no units.
BDL = Below detectable limits.

Table III. Percent Removals for Each Parameter

Parameter	Percent Removal (%)
Alkalinity	94.5
COD	97.0
Total Solids	97.0
Volatile Solids	92.1
Ammonia	96.6
TKN	96.9

operating at the lower pressures (200–250 psi). This resulted in a concentrate flux which provided sufficient self-cleaning of the membrane surface.

A slight hydrogen sulfide odor was noted in the permeate holding tank. This was noted mainly in permeates from treated leachate that had been in the feed tanks for a few days before treatment; thus, more biological activity had taken place in the tanks. Treatment of fresh leachate resulted in little odor. Therefore, it seems that leachate treatment by RO would provide better results if treated soon after collection. In the final design, the RO would be sized to handle one-day leachate accumulation to eliminate odor build-up.

The build-up of gas in the older leachate as a result of biological activity also presented a problem in priming the RO in start-up. Again, this problem could be eliminated by treating the leachate soon after collection.

Table IV. Results of GC/MS Analysis for Trace Toxic Compounds in the Permeate Samples from Various Runs

Date	Compounds Identified
8–16–89	Dimethoxymethane
	Tetrahydrofuran
8–23–89	Tetrahydrofuran
(12 noon)	2-Butanone
8–23–89	Dimethoxymethane
(2 pm)	Tetrahydrofuran
8–30–89	Tetrahydrofuran
	3-Methoxy-1-propene
	Trimethylhydrazine
8–31–89	(1-chloro-3-methoxypropyl)-benzene
	4-[(Trifluoroacetyl)oxy-benzoic acid
	Trimethylhydrazine
	3-methyl-butanal
9–7–89	Dimethoxymethane
	Tetrahydrofuran
	Heptachloro-hexafluoro-hexane
	6-methyl-1-heptane
	Hexane

CONCLUSIONS

1) RO treatment of landfill leachate is technically feasible.
2) Total solids levels were reduced from an average value of 4107 mg/L to a little over 100 mg/L, providing 97% removal of the solids.
3) Organics were reduced by 97%. Trace organics could be found in low μg/L range by GC/MS.
4) Metals as high as 860 μg/L were reduced below detectable limits.
5) Water was produced from landfill leachate with no pretreatment to a quality better than drinking water quality.
6) An optimum pressure for RO operation would probably be at 250 psi.
7) No major problems were encountered in the operation of the RO Treatment Unit.

RECOMMENDATIONS

Based on the results of the study, we have recommended the design of a mobile RO treatment system. Since the landfill owner operates several landfills, a mobile unit could serve all the landfills. We recommended that the RO treatment unit treat peak flow volumes of leachate. That is, 50% of the total peak volume would be treated by single pass through the RO. The other 50% plus the concentrate would be recycled back to the landfill (Figure 7).

During low volume production of leachate, the RO would operate in a multiple pass mode, thereby treating whatever volume of leachate is necessary and returning a minimal amount of recycle back to the landfill after concentrating the original leachate feed volume. This is demonstrated in Figure 8.

The ultimate goal of this system would be not to recycle more than 50% of the leachate produced. That is, treatment and reduction of more than 50% of the leachate would occur while gaining the control and benefits of leachate recycle. Those benefits include:

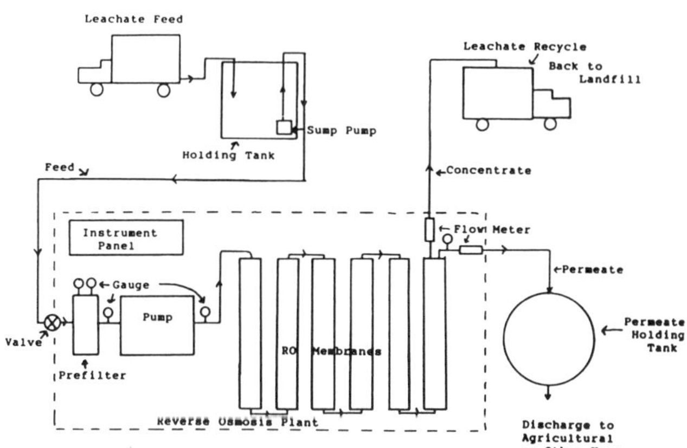

Figure 7. Single pass through reverse osmosis treatment system during high volume leachate production.

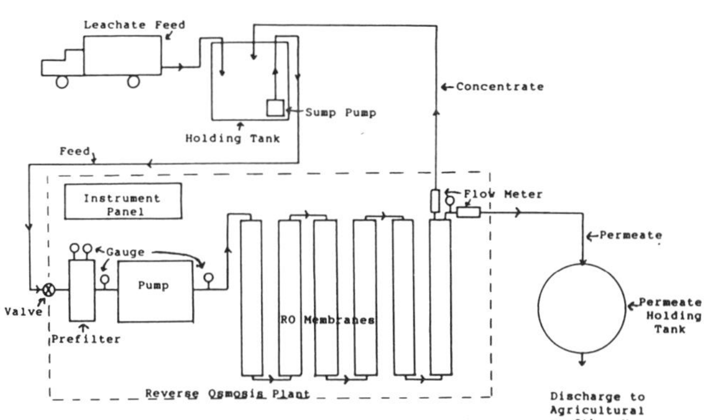

Figure 8. Multiple passes through reverse osmosis treatment system during low volume leachate production.

(1) Moisture returned to the landfill to enhance degradation.
(2) Nutrients returned to the landfill to enhance degradation.
(3) Microorganisms returned to the landfill to enhance degradation.
(4) Trace metals returned to control the release of these materials from the landfill.
(5) Sulfides are returned to the landfill which would aid in the control of the trace metals.

The Mobile RO Unit system would be designed to be contained on an enclosed truck bed. Besides the unit itself, the truck would contain its own power source or provisions could be made for power at each site. Garden hoses can quickly be hooked-up to the feed, permeate, and concentrate lines. A tanker truck can be used to collect the leachate and serve as the feed tank. The concentrate line may be recycled back to the feed tank truck during low flow volumes, as mentioned above, or a second tank truck may be required to collect and recycle the leachate back to the landfill. The permeate may be discharged directly to a sewer, spread on agricultural lands, or collected in a 500 to 1000 gallon storage tank as part of the mobile system and discharged later.

Section Nine
INDUSTRIAL WASTES – A. PETROLEUM WASTES

73 BIOLOGICAL TREATMENT STUDY OF PETROLEUM MARKETING TERMINAL EFFLUENT

Bruce D. Douglas, Project Manager
Patrick M. Maroney, Vice President
Brown and Caldwell Consultants
Pleasant Hill, California 94523-4342

David W. Pierce, Senior Process Engineer
Chevron Corporation
Richmond, California 94802-0627

INTRODUCTION

Petroleum marketing terminals (PMTs) receive bulk shipments of gasolines, middle distillates, lube oil, and other petroleum products from pipelines, tankers, barges, railcars, and trucks. The products are stored in tanks and warehouses and distributed to gas stations, wholesale outlets, and industrial facilities.

PMTs generate liquid effluents consisting primarily of tank bottom water and stormwater runoff that has come into contact with product. Historically, this effluent has been processed through an oil/water separator, which effectively removes oil and grease and suspended solids. The principal dissolved organic constituents of concern in PMT effluent are benzene, toluene, xylene, and phenolic compounds, with typical concentrations totaling about 25 mg/L. Heavy metals may be present, but their combined concentrations are usually less than 1 mg/L.

PMT effluent discharges are intermittent; sometimes days or weeks may pass without a discharge. On a long-term average basis, typical PMT effluent volumes can range from 1,000 to 10,000 gpd. Volumes may be as low as 100 gpd or as high as 20,000 gpd, depending upon rainfall patterns and tank water draw schedules.

The purpose of this project was to assess the treatability of PMT effluent by municipal wastewater treatment systems, which most commonly use the activated sludge process. A municipal treatment plant with a continuous influent flow could readily assimilate the intermittent discharges of biodegradable PMT effluent. The small volume of PMT effluent would have a minor impact on the total organic loading to the municipal plant. The project was conducted under contract with the American Petroleum Institute (API) by Brown and Caldwell Consultants.

The bench-scale, continuous-flow treatability study investigated two issues:

1. Can the activated sludge process remove PMT effluent contaminants?
2. Does PMT effluent interfere with the activated sludge process?

Researchers conducted the tests by operating parallel treatment units. Both units were operated as conventional activated sludge systems. The Control Reactor treated only domestic sanitary wastewater. The Test Reactor received a blend of 2% PMT effluent and 98% sanitary wastewater. Mechanical mixing and a low air-to-liquid flow rate ratio reduced volatile organic emissions so that removal results could be primarily attributed to biodegradation. The results indicate that activated sludge removes PMT effluent contaminants and that PMT effluent does not interfere with the activated sludge process.

45th Purdue Industrial Waste Conference Proceedings,

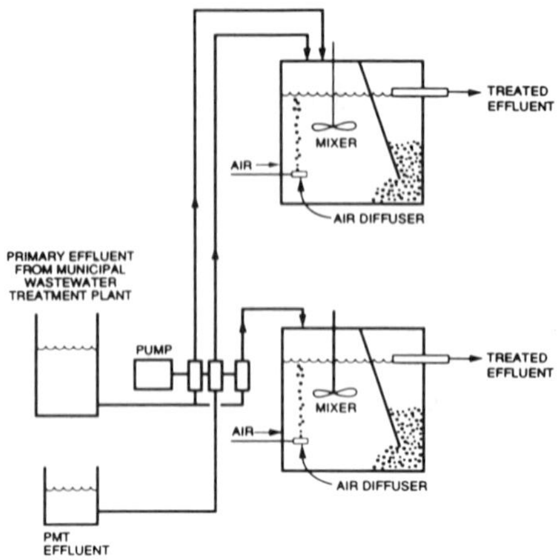

Figure 1. Process schematic.

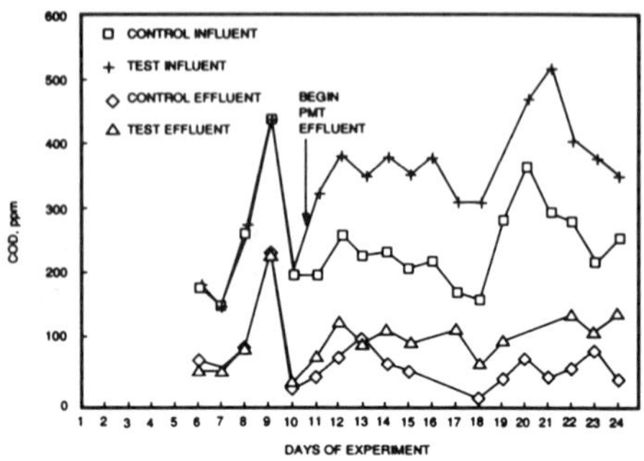

Figure 2. Chemical oxygen demand results.

MATERIALS AND METHODS

The bench-scale treatment systems were continuous-flow, complete mix, activated sludge bioreactors (Figure 1). Researchers operated the two reactors in parallel for 24 days. Each reactor was an acrylic plastic tank with a capacity of 24.8 liters. A removable divider apportioned the reactor into an aeration zone and a settling zone. The 19.6-liter aeration zone was mixed by a gang-stirrer paddle to minimize air stripping. The mixed liquor was aerated by distributing air, supplied by an air pump, through two diffuser stones. The mixed liquor flowed under the divider into the clarification zone, where the solids settled and were internally recycled back into the aeration zone.

This section describes the wastewater supplies, reactor operation, and start-up. The approach to minimizing the stripping of volatile compounds is then discussed.

Wastewater Supplies

The municipal wastewater was primary clarifier effluent from the Dublin/San Ramon wastewater treatment plant (DSR) in Dublin, California. DSR is a conventional activated sludge plant, serving approximately 100,000 people in the communities of Dublin, San Ramon, and Pleasanton, California. The industrial and commercial contribution to DSR was less than 5% of influent flow. Thus, the biomass of the DSR system was not acclimated to the contaminants in PMT effluent. This primarily domestic system was deliberately selected to create difficult test conditions to establish the viability of a municipal plant treating PMT effluent.

PMT effluent was collected from a PMT in the San Francisco Bay Area. It consisted of tank bottom water draws from leaded and unleaded gasoline storage tanks blended in equal proportions. No stormwater runoff was included. Using tank bottom water created a worst-case test condition, because tank bottom water is normally much stronger than typical PMT effluent, which usually has a stormwater component.

Researchers analyzed grab samples of the municipal wastewater and the PMT effluent for pH, chemical oxygen demand (COD), total suspended solids (TSS), metals, base/neutral and acid extractable priority pollutants (EPA Method 625), purgeable priority pollutants (EPA Method 624), and phenolic compounds before the start of the test. The results are presented in Table I.

The principal organic PMT effluent contaminants were benzene, ethylbenzene, toluene, xylene, phenol, naphthalene, and phenolic compounds, with concentrations ranging from 0.8 mg/L (naphthalene) to 53 mg/L (benzene). The COD was 6,700 mg/L. No heavy metal concentration exceeded 0.2 mg/L.

The municipal wastewater did not contain significant amounts of PMT effluent contaminants. The COD level was 190 mg/L. Naphthalene, benzene, ethylbenzene, and xylene were all below the detection limit. Phenol, phenolic compounds, and toluene were detected, but at concentrations two to three orders of magnitude less than PMT effluent concentrations. Phenolic compounds had the highest concentration at 0.08 mg/L, compared with 35 mg/L in the PMT effluent. No heavy metal concentra-

Table I. Compositions of Municipal Wastewater and PMT Effluent

Parameter	Municipal Wastewater Concentration, mg/L	PMT Effluent Concentration, mg/L
pH	7.2	8.5
COD	190	6,700
TSS	40	25
Phenolic compounds	0.08	35
Phenol	0.014	2.2
Naphthalene	<0.002	0.80
Benzene	<0.001	53
Ethylbenzene	<0.001	4
Toluene	0.003	38
Xylene isomers	<0.001	16
Arsenic	0.002	0.18
Chromium	0.0068	0.0128
Lead	0.017	0.065
Nickel	<0.001	0.075
Zinc	0.08	0.17

tion exceeded 0.1 mg/L. The comparison of the municipal wastewater with the PMT effluent emphasizes how this work presents a worst-case test of the treatability of PMT effluent: tank bottom water, undiluted with stormwater, was added at a high percentage to a biomass acclimated strictly to domestic wastewater.

Reactor Operation

Table II lists the operating parameters for this test. The influent flow rate was 0.01 gpm, which established a hydraulic detention time (HDT) of 8.6 hours in the aeration zone. The proportion of PMT effluent in the Test Reactor influent was maintained at 2%. The solids retention time (SRT) was 5.5 days. At the wasting rate required to establish this SRT, the mixed liquor suspended solids (MLSS) remained between 800 and 1,000 mg/L. Return activated sludge from DSR was used to charge each reactor during start-up.

Start-up

The purposes of start-up were to confirm that the reactors functioned as activated sludge units and to develop procedures for steady operation. Return activated sludge from DSR was used to seed each reactor. The reactors treated the same feed—DSR sanitary wastewater—for the first 10 days. Then PMT effluent (tank bottom water) was dosed to the Test Reactor.

During the first week of operation, researchers wasted one-third of the biomass daily to obtain a 3-day SRT. This proved to be too much stress on the biomass and prohibited the development of a stable MLSS concentration. Therefore, researchers re-seeded both reactors on Day 8 and, using a reduced wasting rate, operated with an SRT of 5.5 days.

Minimizing Air Stripping

Volatile organic emissions were not measured in the reactor offgas, but the experiment incorporated design and operational parameters to minimize volatiles stripping. The use of mechanical mixing by a gang-stirrer paddle allowed the air-to-liquid flow rate ratio to be held to only 60-to-1. Further-

Table II. Operating Parameters

Parameter	Actual
Feed flow rate	0.091 gpm
Hydraulic detention time	8.6 hr
PMT effluent percentage in Test Reactor feed	2%
Solids retention time	5.5 days
Mixed liquor suspended solids	800 to 1,000 mg/L

more, the biomass acclimated rapidly to the PMT effluent, promoting biodegradation over stripping as a removal mechanism.

The literature indicates that an estimate of 15% volatiles stripping is conservative. The Environmental Protection Agency estimates that only 8 to 12% of benzene (the most volatile of the principal PMT effluent organic compounds) is stripped during activated sludge treatment.[1] A bench-scale study reported 16 to 25% stripping of benzene, toluene, xylene, and ethylbenzene, using an air-to-liquid flow rate of 250 to 350,[2] much higher than the ratio in this experiment. Aerated lagoon studies by an API member company measured less than 1% volatiles stripping from a refinery wastewater feed. Thus, although stripping was not directly monitored, it is not considered to be a significant removal mechanism.

RESULTS

Conventional analyses conducted during this experiment were COD, MLSS, mixed liquor volatile suspended solids (MLVSS), dissolved oxygen (DO), dissolved oxygen uptake rate (DOUR), microbial evaluation, effluent solids concentrations, and 5-day biochemical oxygen demand (BOD). Data pertaining to these analyses are presented below, followed by a discussion of PMT effluent contaminant removals.

Chemical Oxygen Demand

Figure 2 presents the COD concentrations for the influent and effluent of each reactor. The COD data show that the two reactors were performing similarly while they were treating the same feed. Influent and effluent COD values were nearly identical between Day 6 (the first day of COD measurements) and Day 10 (the last day before the Test Reactor received PMT effluent).

The influent COD of the Control Reactor (the reactor treating domestic wastewater only) averaged 245 mg/L during the test, usually ranging between 200 and 250 mg/L until the final week, when several concentrations near 300 mg/L were measured. The effluent COD from the Control Reactor averaged 60 mg/L, or 75% removal.

The Test Reactor treated a blend of municipal wastewater and PMT effluent from Day 11 through Day 24. During this period, the average influent COD was 385 mg/L, a 60% increase over the average Control Reactor COD, caused by the high loading of PMT effluent organic compounds. This organic loading greatly exaggerated the impact of a typical PMT effluent discharged to a POTW. The COD values rose toward the end of the experiment, reflecting the higher COD of the municipal wastewater during those days. The influent COD of the Test Reactor was consistently 140 mg/L higher than that of the Control Reactor. This increment can be considered the PMT effluent contribution to the influent COD in this conservative test.

The Test Reactor's effluent COD averaged 110 mg/L between Day 11 and Day 24, representing 71% COD removal. This effluent value was achieved consistently from the first day of PMT effluent introduction into the feed. There was no period of poor removals. Thus, the biomass acclimated easily and rapidly to the PMT effluent contaminants.

Mixed Liquor Suspended Solids

The average MLSS concentrations in the Control Reactor and Test Reactor were 825 mg/L and 950 mg/L, respectively (see Figure 3). These values are lower than typical MLSS levels in full-scale activated sludge plants, but they reflect the realities of bench-scale experiments. The low values made the approach to treatment evaluation even more conservative, because the impact of PMT effluent on the process would be more easily noticed in a low-MLSS system.

MLVSS concentrations (the biomass) closely followed MLSS trends at about 85% of MLSS concentrations for each reactor, or 700 and 800 mg/L in the Control and Test Reactors, respectively. The reactor treating PMT effluent developed a biomass of comparable (or slightly greater) concentration. Again, the low solids concentrations created a very conservative indicator to evaluate the impact of PMT effluent on the biomass, as demonstrated dramatically by the food-to-microorganism (F:M) ratios (based on applied COD), of 0.95 versus 1.34 mg COD/mg MLVSS/day for the Control and Test Reactors, respectively.

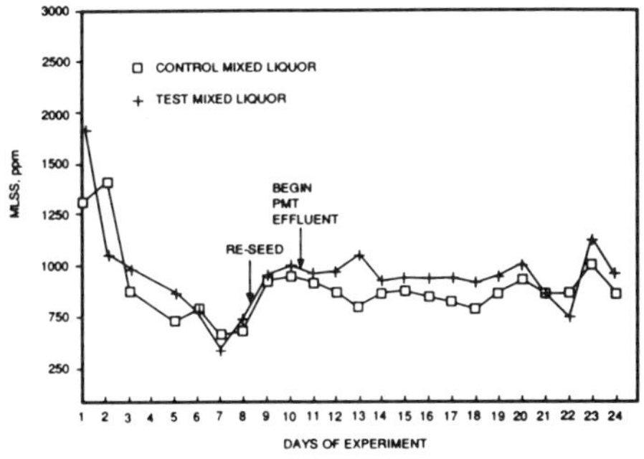

Figure 3. Mixed liquor suspended solids results.

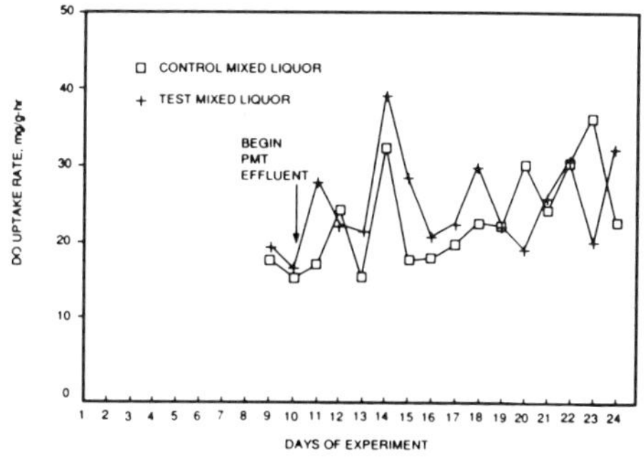

Figure 4. Dissolved oxygen uptake rate results.

Dissolved Oxygen

The average DO values were 2.0 mg/L and 1.6 mg/L for the Control Reactor and Test Reactor, respectively. Values are lower in the Test Reactor because the high organic loading caused greater oxygen use, depressing the DO. Figure 4 shows the DOUR results for both systems. The average DOUR in the Control Reactor from Day 11 through Day 24 was 23 mg O_2 per g biomass per hour (mg/g-hr), and the average DOUR in the Test Reactor over the same period was 25 mg/g-hr. The comparable oxygen uptake results indicate that PMT effluent did not affect microbial health negatively.

Microbial Composition

The mixed liquor from each reactor was scanned twice during the final week of the test. By performing scans near the end of the experiment, it was possible to determine if extended exposure to PMT effluent contaminants had caused the microbial population in the Test Reactor to change in comparison to the population in the Control Reactor. The scans described the type and abundance of filamentous bacteria and protozoa, as well as other microorganisms.

The results of the microbial scans indicated that the biomass populations were very similar. Both experienced filamentous growth problems. This was not of major concern, because DSR also experienced bulking during the time it supplied the test with domestic wastewater and seed sludge. The observations of protozoa were significant, since protozoa indicate an advanced microbial population and a stable biomass. Protozoa comprised 12% of the biomass in the Test Reactor in both scans. This was equal to or greater than the abundance of protozoa in the Control Reactor, which declined from 12 to 8% between the scans. The PMT effluent contaminants, therefore, had no debilatory effect on the protozoa.

Effluent Suspended Solids

The effluent suspended solids concentrations were low. This condition can occur in activated sludge systems experiencing bulking from filamentous growth, because the filaments tend to filter out suspended particles. The average suspended solids concentration in the Control Reactor effluent was 9 mg/L. In the Test Reactor effluent, the average suspended solids concentration was slightly higher at 13 mg/L, but in the same low range.

A bulking sludge may cause the sludge blanket to periodically rise too high within the clarification zone, carrying solids out of the reactor. Once in the Control Reactor, and twice in the Test Reactor, the sludge blanket did rise up to the level of the discharge pipe, producing an effluent suspended solids measurement in the range of 60 to 70 mg/L. The additional day that solids carried over and the slightly higher average effluent suspended solids concentration in the Test Reactor generally reflect the higher mass of solids in the Test Reactor mixed liquor and, thus, the clarification zone.

Biochemical Oxygen Demand

During the last week of the experiment, four BOD measurements were conducted for each reactor's influent and effluent. As with COD, BOD results demonstrate the organic overloading of the Test Reactor caused by the large proportion of PMT effluent comprised entirely of tank bottom water.

Table III. Organic PMT Compounds Results

Compound	Influent concentration, μg/L	Effluent concentration, μg/L	Percent removal
Benzene	385	30	92
Ethylbenzene	17	2	88
Toluene	255	18	93
Xylene isomers	66	9	86
Phenol	280	10	96
Phenolic compounds	940	130	86

The average BOD values in the Control Reactor influent and effluent were 135 and 15 mg/L, respectively. This represents 90 percent BOD removal, or 120 mg/L BOD removed. In the Test Reactor, influent BOD was 50 percent higher at 200 mg/L. The biomass responded well to this increased load, degrading more BOD than the Control Reactor (150 versus 120 mg/L). The Test Reactor's higher influent BOD load did affect removal efficiency (75 versus 90%), but this was expected. A higher biomass concentration (more typical of full-scale operation) would be needed in the Test Reactor to attain the lower effluent BOD concentrations observed in the Control Reactor.

PMT Effluent Parameters

The PMT effluent contaminant concentrations in the influent and effluent of the Test Reactor are presented in Table III. Typical removals were 85 to 95% for benzene, ethylbenzene, toluene, xylene, phenols, and phenolic compounds. For benzene, toluene, and phenol, removals were over 90 percent. The influent concentrations of these compounds were 200 to 400 μg/L, with effluents of 10 to 30 μg/L. For ethylbenzene and xylene, influent concentrations were 17 and 66 μg/L, respectively, and removals were still between 85 and 90%, with effluent concentrations of 2 and 9 μg/L, respectively. For the general category of phenolic compounds, the influent concentration was 940 μg/L, with removal down to 130 μg/L, or 86%.

DISCUSSION OF RESULTS

The objectives of the experiment were to determine 1) if the activated sludge process can remove PMT effluent contaminants, and 2) if the contaminants impair the process. These objectives were achieved by creating very conservative conditions with respect to PMT effluent contribution to an activated sludge system. Thus, a 2 percent blend of undiluted tank bottom water was fed to a bench-scale system supporting a biomass of less than 1,000 mg/L. Influent COD and BOD levels were 50 to 60% higher in the Test Reactor than in the Control Reactor.

This section discusses the ability of activated sludge to remove PMT effluent contaminants. It then examines whether PMT effluent interfered with the activated sludge process.

Activated Sludge Treatment of PMT Effluent Contaminants

The PMT effluent organics which were in measurable concentrations in the Test Reactor were benzene, ethylbenzene, toluene, xylene, phenol, and phenolic compounds. None of these compounds were detected in the Control Reactor. The compounds were monitored, and the results, presented in Table III, showed 85 to 95% removal occurred in the Test Reactor. Clearly, activated sludge can remove these contaminants.

Biodegradation is considered to be the most significant removal mechanism, because stripping was minimized by the design and operational parameters described earlier. If a conservative estimate of 15% stripping is subtracted from the overall removals of benzene, ethylbenzene, toluene, and xylene, the biodegradation of these volatiles is still between 70 and 80%, which is consistent with the BOD and COD removals in the Test Reactor. Hence, biodegradation of PMT effluent contaminants is comparable to the biodegradation of domestic organics in the same feed.

Analysis of Interference Indicators

Adverse impacts on the growth rate of the biomass, the DOUR, and the microbial composition indicate interference with the activated sludge process. These factors are discussed below.

Biomass Growth Rate. The biomass growth rate is evaluated in terms of the specific growth rate in the aeration zone (mg biomass (VSS) growth per hour per g VSS (mg/g-hr)). Since each reactor maintained a stable MLVSS concentration during the period when the Test Reactor received PMT effluent, the biomass growth is equivalent to the biomass wasted. The average biomass growth in the Control Reactor was 2,230 mg per day, or 90 mg/L per day. For the Test Reactor, the average biomass growth was 2,720 mg per day, or 110 mg/L per day. Dividing these values by the respective average MLVSS concentrations, the specific growth rates were 5.4 mg/g-hr for the Control Reactor and 5.7 mg/g-hr for the Test Reactor.

These comparable growth rates show that PMT effluent did not interfere with the biomass. In fact, the reason the Test Reactor specific growth rate is slightly higher is that the PMT effluent contaminants were readily biodegradable. Hence, the additional organic input from the PMT effluent caused higher results not only in specific growth rate, but also in all other parameters related to biomass growth. These include MLSS, MLVSS, and overall growth rate. Therefore, the PMT effluent contaminants should not have a negative impact on the growth rate of the biomass in an activated sludge system.

Dissolved Oxygen Uptake Rate. The average DOURs were 23 and 25 mg/g-hr for the Control Reactor and Test Reactor, respectively. These values are considered virtually identical within the accuracy of the measurement technique. Since a comparable DOUR was sustained in the Test Reactor throughout the test period, accompanied by consistent removal of PMT effluent contaminants, it is concluded that the PMT effluent did not impair the microbial activity in the activated sludge system.

Microbial Composition. Microbial scans performed during the last week of the experiment indicated that the biomass in the Test Reactor had not been impaired by the PMT effluent. Test Reactor microorganisms were similar to Control Reactor microorganisms. Where they were different, the Test Reactor biomass was in better condition. For example, the Test Reactor biomass showed a higher proportion of protozoa than the Control Reactor biomass.

CONCLUSIONS

Bench-scale reactors treating municipal wastewater and PMT effluent with the activated sludge process performed identically when treating the same influent. The response of the Test Reactor to influent containing PMT effluent indicated the effect of PMT effluent on the activated sludge process.

The Test Reactor was intentionally overloaded with PMT effluent organic compounds to test a worst-case situation. The reactor treated a 2 percent blend of undiluted tank bottom water. The biomass of the Test Reactor responded well, removing a greater mass of organic compounds than the Control Reactor removed.

Examination of process performance indicated that the activated sludge process can remove PMT effluent contaminants. All measurable organics—benzene, ethylbenzene, toluene, xylene, phenol, and phenolic compounds—had 85 to 95 percent removal.

Measurements of biomass viability indicated that PMT effluent constituents did not interfere with the process. The specific growth rate of the biomass was higher in the Test Reactor than in the Control Reactor. The DOUR was comparable in the two reactors. The microbial composition was substantially the same in the two reactors, with all differences indicating superior results for the Test Reactor.

The data presented in this report show that PMT effluent can be treated and should be easily accommodated by municipal wastewater treatment plants using activated sludge or an equivalent biological treatment process. This approach may be the most environmentally acceptable way to dispose of PMT effluent.

REFERENCES

1. *55 Federal Register 8292*, (March 7, 1990).
2. Weber, W. J., Jr., and B. E. Jones, *Toxic Substance Removal in Activated Sludge and PAC Treatment Systems*. Prepared for the U.S. Environmental Protection Agency. EPA-600/2-86-045 (1986).

74 DISPOSAL OF HAZARDOUS WASTES FROM PETROLEUM REFINERIES

Jeffrey S. Bryant, Sr. Environmental Supervisor
Charles W. Moores, V. P., Environmental Marketing & Sales
Badger Engineers, Inc.
Cambridge, Massachusetts 02142

INTRODUCTION

The petroleum refining industry is required to operate under increasingly stringent environmental regulations. In the past decade, refineries have had to conform to new environmental regulations dealing with effluent quality, solids disposal, vapor emissions, hazardous materials containment, and remediation of contaminated areas. Addressing these requirements invariably generates an increased quantity of solids requiring disposal. In many cases these solids fall under the "hazardous waste" definition. Disposing of these hazardous wastes is becoming a critical issue for the refining industry.

This paper reviews the sources and characteristics of hazardous and solid wastes in a refinery, discusses the available disposal alternatives, and presents examples of disposal scenarios which conform to current and proposed regulations.

PROBLEM DEFINITION

Regulations Impacting Refineries

A refinery may be constrained by environmental regulations issued from the federal, state, or local level. It is often the case, however, that state and local environmental agencies either defer to the federal environmental regulations or promulgate environmental regulations that are consistent with those issued by the U.S. Environmental Protection Agency (EPA). This paper addresses only the EPA environmental regulations impacting refinery solid wastes.

Because the monitoring and record keeping requirements as well as liabilities for improper disposal are significantly greater for hazardous wastes, it is important to identify which wastes in a refinery are hazardous. Under current regulations, the EPA uses two procedures to define wastes as hazardous: "listing" and "hazardous characteristics". The listing procedure involves identifying specific wastes that pose hazards to human health and the environment. For the refining industry, the EPA has "listed" the wastes shown on Table I.

The second procedure involves identifying properties or "characteristics" that, if exhibited by a waste, indicate a potential hazard if the waste is not properly controlled. The four characteristics that must be considered are toxicity, ignitability, reactivity, and corrosivity.

- Toxicity: A waste is hazardous due to the toxicity characteristic when the leachate, resulting from subjecting the waste to the "toxicity characteristic leaching procedure" (TCLP) test[2], exceeds specified concentrations for specific components. In this regard, EPA has finalized a new toxicity characteristic rule which, as indicated on Table II, adds 25 new substances to the Toxicity Characteristic list. This is in addition to the 14 constituents (eight of which are heavy metals) to be met as part of the previous law[4]. Refineries will have to comply with these new regulation by the fall of 1990.
- Ignitability: This characteristic refers to a waste that has an organic layer or one which exerts a high vapor pressure and a flash point less than 140°F.
- Reactivity: Wastes which can generate toxic gases when exposed to extreme pH conditions are hazardous due to the reactivity characteristic. Wastes containing sulfides or cyanides are examples of wastes demonstrating this characteristic.
- Corrosivity: These are wastes having a pH less than 2 or greater than 12.5.

45th Purdue Industrial Waste Conference Proceedings,

Table I. "Listed" Hazardous Wastes from Refineries[1]

Description	Waste Designation Code
Dissolved Air Flotation Float	K048
Slop Oil Emulsion Solids	K049
Heat Exchanger Bundle Cleaning Sludge	K050
API Separator Sludge	K051
Leaded Tank Bottoms	K052

Except for a relatively small number of solid waste streams, such as raw water treating sludge, it is possible that any of the solid waste streams from a refinery can conform to at least one of the criteria for hazardous waste. Therefore, all sources of solid waste generic to a refinery are briefly discussed.

Sources and Characteristics

This section identifies the sources and characteristics of solid wastes commonly generated by petroleum refining operations and associated wastewater treatment. These wastes are categorized according to those that are (a) continuously generated, (b) intermittently generated, or (c) waste repositories and contaminated soils requiring remediation. Table III summarizes these sources of "solid wastes". In this context "solid wastes" are considered to be any garbage, refuse, sludge, or other discarded material generated within the refinery or associated with pollution control facilities. A refinery block flow diagram indicating typical sources of solid wastes is shown in Figure 1.

Table II. New Constituents Added to the Toxicity Characteristic List[3]

New Constituent	Regulatory Level
Benzene	0.50 mg/L
Carbon tetrachloride	0.50 mg/L
Chlordane	0.03 mg/L
Chlorobenzene	100.0 mg/L
Chloroform	6.0 mg/L
m-Cresol	200.0 mg/L*
o-Cresol	200.0 mg/L
p-Cresol	200.0 mg/L
1,4-Dichlorobenzene	7.5 mg/L
1,2-Dichloroethane	0.50 mg/L
1,1-Dichloroethylene	0.70 mg/L
2,4-Dinitrotoluene	0.13 mg/L
Heptachlor	0.008 mg/L
Hexachloro-1,3-butadiene	0.5 mg/L
Hexachlorobenzene	0.13 mg/L**
Hexachloroethane	3.0 mg/L
Methyl ethyl ketone	200.0 mg/L
Nitrobenzene	2.0 mg/L
Pentachlorophenol	100.0 mg/L***
Pyridine	5.0 mg/L**
Tetrachloroethylene	0.7 mg/L
Trichloroethylene	0.5 mg/L
2,4,5-Trichlorophenol	400.0 mg/L
2,4,6-Trichlorophenol	2.0 mg/L
Vinyl chloride	0.20 mg/L

*If o-, m-, and p-Cresol concentrations cannot be differentiated, the total cresol concentration is used. The regulatory level for total cresol is 200.0 mg/L.

**Quantitation limit is greater than the calculated regulatory level. The quantitation limit, therefore, becomes the regulatory level.

***The EPA proposes to set a new regulatory level for this constituent, based on the latest toxicity information.

Table III. Summary of Refinery Solid Waste Sources and Characteristics

	Relative Volumes	High Oil/Orgs Content	High Solids Content	High/Low pH	High Metals Content	"Listed" Waste	Potentially Hazardous Waste	Non-Hazardous Waste
Continuous								
API/CPI Sludge	High	X	X		X	X		
DAF Sludge	High	X			X	X		
Catalyst Fines	High		X		X		X	
Coke Fines	Low		X					
HF Alky Sludge	Low			X			X	
Lime Sftng Sludge	Medium			X				X
Merox Blowdown	Low	X					X	
Raw Wtr Trt Sludge	Medium		X					X
Slops	Medium	X			X	X		
Spent Acids	Low			X			X	
Spent Amines	Low	X					X	
Spent Caustic	Medium	X		X			X	
Spent Clays	High	X	X				X	
Spent Glycols	Low	X					X	
Wst. Act. Sludge	Low	X	X					X
Intermittent								
BS&W	[3]	X	X				X	
Construction Wastes	[3]		X				X	
Contam. Soils	[3]	X	X				X	
Leaded Tank Bottoms	Low	X	X		X	X		
Mntc. Wastes	Low	X	X			X[1]		
Pond Cleaning	Low		X				X	
Unleaded Tank Bottoms	Low	X					X	
Vac. Truck Sludge	[3]	X	X		X[2]		X	
Remediation								
Remed. Wastes	[3]	X	X	?	?		X	

[1]Heat exchanger bundle cleanings sludge is a "listed" waste (K050).
[2]For example cooling tower sludges may contain metals (e.g. chromates).
[3]Varies.

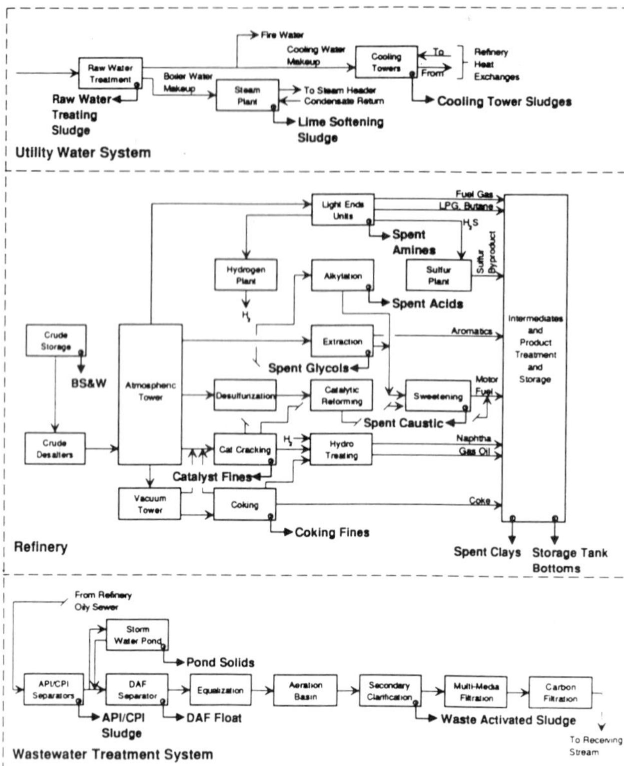

Other Solid Waste Sources:
- Remediation Wastes
- Construction/Maintenance Wastes
- Laboratory Wastes

Figure 1. Refinery schematic diagram indicating representative sources of solid waste.

Due to the lack of consistent data the total quantities of oils and solids generated by a refinery are difficult to estimate. By extrapolating 1978 data[5], it is suggested that the industrial average for total refinery solids production may exceed 0.32 tons of (oil + solids) per thousand barrels of crude oil processed. Furthermore, this is a significantly greater waste production rate than indicated by a similar 1974 study[6], suggesting that improved waste treatment and a greater focus on managing hazardous wastes has resulted in an increased quantity of material identified as refinery solid wastes.

Continuous Wastes

API Sludge—API sludge is a "listed" hazardous waste (K051). This waste stream, consisting of a mixture of solids, heavy hydrocarbons and water, is pumped from the bottom of the API oil/water separator. Due to the presence of sulfides in refinery wastewater, a number of metals can precipitate as insoluble metal sulfides to contribute to the solids fraction of the API sludge. Quantities of API sludge generated by a refinery can also vary widely, depending primarily on the degree of stormwater segregation from the process sewers. API sludge quantities can range from 0.0015 to 0.02 wt. percent (water free) of crude oil processed[7].

CPI Sludge—Corrugated plate interceptor (CPI) separators are also used to separate insolubles from refinery wastewater and are used in conjunction with (and sometimes in place of) API separators. Oils and other floatable material are removed from the top of CPI units and typically are pumped to the slops system. Bottom sludges are similar in composition to the API sludges, but quantities are generally lower with solids particles of smaller size. Since the corrugated plates often present a favorable environment for biological growth they require periodic cleaning for optimum performance. This can result in the CPI sludge containing some biological solids as well.

DAF Sludge—Dissolved air flotation (DAF) separators generally follow API separators in the refinery wastewater treatment flow scheme, and are used to remove residual quantities of oils and solids from the wastewater. Typically flocculants (e.g. aluminum or iron salts) and polymers are added to promote the removal of oils and solids. These are floated to the surface through the use of air bubbles introduced with the feed to the DAF unit. DAF sludges generally contain 1 to 5% solids, and the total sludge flow often exceeds that of the API sludge, depending on the quantity of the coagulants used and the removal efficiency of the upstream equipment.

Catalyst Fines—While there are several process units in refineries which use catalysts to promote reactions, the largest consumer of catalyst is the hydrocarbon cracking unit. The catalysts used in the cracking unit are usually composed of silica alumina or silica magnesia in microspherical shapes. The catalyst may be in a fixed bed, moving bed, or fluid bed arrangement, with the ease of regenerating catalyst favoring the fluid catalytic cracking (FCC) unit. Due to the high catalyst impingement energies associated with catalyst fluidization, there is a high rate of catalyst fines generation. These are often slurried to a pond or tank, where they are settled and periodically removed. While the catalysts are generally nonmetallic, there may be cyanides and organics associated with this catalyst fines waste.

Coke Fines—Coke is a nearly pure carbon product. The coke is typically removed from the coking vessel by means of drilling or hydroblasting. The cuttings from the coke removal operation are often slurried to the sewer as coke fines.

HF Alkylation Sludge—In alkylation systems using hydrogen fluoride as the catalyst for alkylation of unsaturated hydrocarbons, it is necessary to limit the escape of fluorides to the wastewater to meet regulatory limits. For this reason, any discharges from the HF alkylation unit which contain fluoride are treated at the unit with lime to promote calcium fluoride precipitation. The resulting calcium fluoride sludge will contain traces of alkylation hydrocarbons, which may become a source of volatile organic carbon (VOC) emissions if improperly handled.

Laboratory Wastes—Within a refinery samples of crude, intermediates, and products are taken and analyzed in the laboratory as part of a quality control program. Laboratory wastes can consist of broken glassware, disposable pipettes, spent filter media, absorbents, solvents, reactants, etc. Since the quantity of laboratory wastes is comparatively small with a highly variable composition, in many refineries these wastes are segregated and disposed of separately.

Lime Softening Sludge—Lime softening of clarified raw water is often practiced to prepare boiler feedwater. The addition of lime and soda ash increases the pH and precipitates calcium as a calcium carbonate sludge. This calcium carbonate sludge generally has a rather high solids concentration of perhaps 6% as it is pumped from the gravity clarifiers. This sludge is typically nonhazardous, and directed to land disposal.

Merox Blowdown—This process separates mercaptans from gasoline and other fuel grade hydrocarbons by contacting the liquid hydrocarbons with a metal catalyst, which is often dissolved in a caustic solution. The catalyst acts to convert the mercaptans to disulfides which are separated from the hydrocarbon stream in the circulating caustic solution. The disulfides are typically removed and directed to the sulfur recovery plant. A purge stream is taken of the circulating caustic/catalyst solution to remove impurities. Due to the presence of a high pH, ignitable organics, and heavy metals, this stream is typically hazardous.

Raw Water Treatment Wastes—Refineries consume water in their cooling water system, steam boiler system, for reagent makeup, and in process units. Often refineries draw their raw water from rivers and treat it to remove silts and other foreign materials. Gravity clarification is most often practiced, where coagulants are added to facilitate gravity settling. The resulting sludges are nonhazardous, and will have a solids content ranging from a few percent up to 6% from the bottom of the clarifier. While this sludge stream was commonly returned to the river in the past, in most cases this practice is no longer permissible.

Slops—Refinery slops systems consist of a number of tanks which receive the floating oils from separators (e.g. API, CPI, and DAF separators), oily sump wastes collected by vacuum trucks, oily wastes from equipment decommissioning, etc. The Slops tanks provide the long holdup times necessary to allow emulsified oils to agglomerate. Often heat and chemicals are added to promote the emulsion breaking. Periodically the oil is decanted from the slops tanks and returned to the crude charge while the water is drained from the tank bottoms and directed to the oily sewer. Solids are periodically removed for disposal.

Spent Acids—While currently few refineries utilize sulfuric acid as an alkylation "catalyst", it was a frequently used process in the past. In the sulfuric acid catalyzed alkylation process the acid acts to polymerize unsaturated compounds, to produce products which are added to the gasoline pool to increase the octane rating. Spent acid solutions, containing tarry polymer compounds, were frequently dumped into lagoons, or "acid pits", or alternatively, were discharged to the sewer. For this reason, in older refineries spent acid lagoons often exist which require remediation. The low pH generally causes these wastes to be hazardous.

Spent Amines—Amines (primarily monoethanolamine and diethanolamine) are used in gas sweetening operations to remove hydrogen sulfide. The sulfur rich amine solution is stripped to remove reduced sulfur compounds while the stripped (lean) amine solution is recycled to the absorber to remove additional reduced sulfur compounds from the refinery gas stream. Reduced sulfur compounds stripped from the rich amine stream are usually converted to elemental sulfur in a downstream Claus plant.

Periodically a purge stream is drawn from the circulating amine solution to remove inerts and impurities. Usually the sulfide concentrations are sufficiently high to cause this stream to be characterized as a hazardous waste. Discharging these spent amine solutions to the sewer system can cause upsets in the wastewater treatment plant due to the increased loading of chemically reduced sulfur and nitrogen compounds.

Spent Caustic—In many refineries aqueous solutions of caustic soda (NaOH) are used in "sweetening" units to extract acidic compounds from refinery feedstocks. Typical substances which are extracted using caustic soda are phenolics, naphthenics, sulfides, mercaptans, and cyanides. The resulting spent caustic solutions are typically directed to a spent caustic holdup tank, where the residual hydrocarbons collect at the surface to be decanted. The high pH and the high concentrations of sulfides cause this stream to be a hazardous waste.

In some cases the spent caustic is trucked to an offsite processor for recovery of the organic acids and sodium sulfhydrate (NaSH) as salable products.

Discharging the spent caustic to the sewer system is still practiced at some refineries, often to provide alkalinity to the biological treatment system. This disposal practice can place severe stress on the biological system if there are excessive concentrations of reduced sulfur compounds, cyanides, and organic acids.

Spent Clays—Clay treating is used in refineries as a finishing process to stabilize kerosine and jet fuels as well as lube oils. As these hydrocarbons are pumped through the clay bed, polymeric materials are absorbed on the fine clay particles. When the clay becomes saturated with adsorbed material, it must be replaced. Since land disposal was often a common practice in the past, in older refineries this spent clay can often be found on refinery property where it was dumped. The absorbed materials can slowly desorb as the spent clay comes in contact with water, thus posing a threat to groundwater. If, however, the spent clay is discharged to the sewer, the fine clay particles become nucleation points for hydrophobic materials, resulting in greatly increased quantities of oily sludge.

Spent Glycols—Glycol solutions are often used in a refinery as a solvent in liquid/liquid solvent extraction systems to extract various hydrocarbon fractions. To prevent the accumulation of impurities which reduce the selectivity of the recirculated solvent, a blowdown stream is taken and the solvent is replenished. The high aromatics content of this stream typically causes it to be hazardous.

Waste Activated Sludge—WAS consists of biological solids resulting from the biological treatment of refinery waste water. In addition to microbial cells, WAS contains adsorbed oil and insoluble metals. Sludge from the secondary clarifiers will range in consistency from 0.5 to 1.5% solids, and, after dewatering, typically have a solids content of 18 to 25%. To achieve greater sludge cake dryness, coagulants such as lime, metal salts (e.g. aluminum sulfate, ferric chloride, etc.), and polymers can be added. The estimated range of WAS sludge produced is 30 to 50 pounds of dry solids per thousand barrels of crude oil processed. Usually the hydrocarbon or metals content are not sufficiently high to cause this stream to be hazardous.

Intermittent Wastes

The following brief descriptions refer to intermittent sources of petroleum refinery solid wastes.

BS&W—Bottom (or basic) sediment and water (BS&W) refers to materials which accumulate in the bottom of crude oil storage tanks. BS&W consists of anything that is heavier than crude oil and separates from the crude oil by gravity. BS&W can result from poor separation operations at the point of crude oil production or may be entrained during transportation. Generally the solids consist of sand, clay, soil and corrosion products. The water in BS&W is usually salty unless the crude oil has been desalted at the point of production. In the latter case the quantity of BS&W is generally smaller.

BS&W is handled in two ways; (a) it is allowed to settle and accumulate in the crude oil tanks with periodic removal, or (b) the contents of the crude oil tanks is continuously mixed to prevent BS&W from separating. In the latter case, BS&W is removed in the crude oil desalter prior to introduction into the crude column.

Depending on the quality of the crude, BS&W can represent as much as 1% of the crude and can range in consistency from relatively fluid mixture of crude oil, solids and water to a relatively dense clay/sand/oil/water sediment.

Construction Wastes—Most refineries periodically execute construction projects to increase capacity, change unit operations, or to replace equipment. These projects generate construction wastes such as lumber, packaging materials, painting wastes, insulation wastes, etc. Due to the heterogeneous nature of these wastes, sorting may be required to segregate hazardous wastes.

Contaminated Soil—Soils can become contaminated due to spills during material loading and unloading and from equipment leaks. The cleanup process often results in a quantity of contaminated soils requiring remediation or disposal. The composition of these "soils" can range from loamy material to crushed stone or marine shells.

Leaded Tank Bottoms—The tank bottoms from the storage of leaded gasoline is a listed hazardous waste (K052). These wastes consist of sediments, water, and emulsions which are periodically pumped or drained from the bottom of the storage tank. Because this is generally a manual operation, the hydrocarbon content in the leaded tank bottoms is largely dependent on the vigilance of the operator.

Maintenance and Miscellaneous Wastes—Maintenance and miscellaneous wastes can consist of such things as absorbents, sand blasting grit, spent cleaning solvents, hydroblasting wastes, clothing, boots, gloves, painting wastes, rags, etc. Heat exchanger bundle cleaning sludge, being a "listed" waste (K050) is an example of a hazardous maintenance waste. Although these wastes are produced throughout the year, the largest quantities are generated during plant "turnarounds". While having diverse physical characteristics, they all are contaminated with oil.

Pond Cleanings—Most refineries, particularly in older refineries, have lagoons or ponds which are (or have been in the past) used for equalization to smooth the flow and concentration variations in the wastewater feed to the treatment plant. Such ponds may also have been used to store "off spec" wastewater not conforming to the effluent water quality requirements. Typically, the contents of these "off spec" ponds are returned to the wastewater treatment plant. Oily solids which settle to the bottom of these ponds must be periodically removed. Pond cleaning occurs relatively infrequently and can result in significant quantities of oily sludge ranging from a relatively fluid material with a high oil content to a relatively dense oily soil. This material can be very nonhomogeneous, containing such things as tires, gloves, steel containers, etc.

Unleaded Tank Bottoms—As with leaded tank bottoms, the sediments in storage tanks are periodically pumped from the bottom of intermediate and product storage tanks. While these wastes are not a listed hazardous waste, these may contain components which can impart a hazardous characteristic (e.g. ignitability, toxicity).

Remediation Wastes

The 1984 Hazardous and Solid Waste Amendments to the RCRA regulations prevent refineries from the direct land disposal of hazardous wastes. These regulations also stipulate that facilities serving as disposal sites for hazardous waste be restored to an environmentally viable condition, through the submittal of a closure plan as part of the "Part B Permit" approval process. Executing these closure plans often generates large quantities of remediation wastes for disposal.

This summary of sources of refinery solid wastes indicates that:

- the sources of solid waste within a refinery are numerous.
- the quantities and characteristics of solid wastes depend largely on the unit operations employed by the refinery and the operating practices within the refinery.

In this context, the next section discusses approaches to treatment and disposal of refinery solid wastes.

TREATMENT/DISPOSAL ALTERNATIVES

In the decade of the 1980's the following trends were observed:

- Reaction to environmental regulations and a desire to minimize liability exposure has encouraged refineries to focus on waste treatment issues, resulting in a greater quantity of solid wastes requiring processing and disposal.
- More rigorous design criteria and longer permitting times for new hazardous waste disposal facilities has caused the supply of new disposal facilities to lag far behind demand. The net effect has been a significant increase in disposal costs.

Furthermore, these trends are likely to continue in the future. Due to the large number of variables involved, the optimum solid waste management strategy is likely to be refinery-specific, depending on the refinery unit processes, crude and product slates, quantities and types of solids generated, availability of various disposal options (e.g. landfarming), etc. Therefore, a "systems approach" must be taken to determine the optimum disposal strategy that minimizes both cost and pollution liability risk.

Minimization/Abatement

Prolonging the usable life of a landfill and/or reducing disposal costs can be effectively realized through minimizing or abating the production of solid waste and/or treating the wastes prior to deposition in the landfill. Therefore, identifying methods of minimization/ abatement is a key aspect of any strategy. The following are examples of potential minimization/abatement opportunities within a refinery.

Organization/Management—The administrative organization of a refinery should be such that one person is responsible for meeting both refinery production quotas as well as environmental regulations. A common set of priorities must be developed and communicated to refinery operators, maintenance, and environmental personnel. For example, providing the proper organization, management, and incentives, in most cases will prevent, or at least mitigate, the effects of the "3 am emulsion dump" which facilitates a maintenance turnaround at the expense of environmental control.

Process Changes—Consideration of the following may be of value in minimizing and/or abating solid waste in various refinery process units:

- Optimize the operations of the refinery slops system to minimize the production of emulsions and the cycling of oils from slops to the sewer, to the API separators, and back to the slops system again.
- Minimize or eliminate the use of caustic to remove acidic compounds from gasoline. Where caustic washing cannot be eliminated, provide facilities to make an efficient phase separation rather than making phase separations in storage tanks.
- Eliminate the use of adsorbents such as clay where possible through the use of alternative processes (e.g. apply mild hydrogen treating to react with impurities, to make them amenable to subsequent removal).
- Minimize the loss of coke fines from coke cutting and handling operations through installing catch basins and improved housekeeping practices.
- Optimize asphalt blowing operations to eliminate emulsion formation.
- Consider substitution of regenerable catalysts for alkylation, rather than using inorganic acids.

Recycle/Reuse—Numerous solid wastes in a refinery contain crude oil, intermediates, and products. The wasting and disposal of these materials can be prevented by:

- Recovery of phenolics and sodium sulfides as byproducts from spent caustics.
- Process wastes with a high organic content (e.g. API/CPI sludges, DAF float, etc.) in the coker unit, if available, and if the product coke quality permits.
- Use solvent extraction to remove the hydrocarbon fraction of contaminated solid wastes (e.g. with processes using propane or TEA as a solvent). The extracted oils can be

recycled to the crude charge, while it is likely that the residue can be disposed of as a nonhazardous waste.

Wastewater Treatment Sludge—Wastewater treatment sludge production is the result of the interaction of solids, oil, and water. Minimizing the quantities of these materials entering the sewer system minimizes the quantity of resultant sludge.

The following are suggestions to prevent the introduction of solids into the wastewater system:

- Segregate uncontaminated stormwater and oily sewers to reduce the quantity of sand, soil fines, and inerts entering the sewer.
- Maintain clean, paved, and curbed operating areas.
- Prevent discharge of dry solids (e.g. coke fines, sand blasting fines, furnace decoking solids, etc.) into sewer.
- Avoid the overuse of wastewater flocculants (e.g. aluminum and iron salts, etc.).

Practices which can significantly reduce the quantity of oil discharges to the sewer include the following:

- Minimize sample purges to the sewer.
- Minimize oil discharges from equipment decommissioning.
- Maximize the use of sewer manhole inlet seals to minimize oil retention in the sewer system. This will prevent oil slugs during high flow conditions.
- Discharge tank bottoms directly to the sewer (with appropriate monitoring and controls to prevent oil entrainment), rather than discharging to the diked areas first.
- Optimize the operation of the crude oil desalters to minimize the quantity of oil in the desalter wastewater.

Since minimizing the wastewater discharged to the oily sewer often pays dividends by reducing the quantity of resultant sludges, common methods for reducing the wastewater quantities discharged to the sewer are listed:

- Segregate uncontaminated wastewater streams and direct them to surface discharge.
- Maximize the use of air cooling to minimize cooling tower blowdown.
- Eliminate the use of direct contact cooling and barometric condensers by installing surface condensers, etc.
- Maximize the cooling tower cycles of concentration.
- Institute spill cleanup procedures which minimize the use of water and maximize recovery.
- Optimize sour water stripping and maximize the reuse of stripped sour water.
- Maximize the use of nonsalty water such as rain water and fresh ballast water for refinery raw water makeup.
- Maximize the return and reuse of steam condensates.

Treatment

The following are brief descriptions of unit operations or systems which may be used to reduce the volume of waste and/or detoxify waste prior to landfill. Table IV shows the capabilities of these system as they relate to various refinery waste sources.

Sludge Conditioning/Dewatering—The objective of sludge conditioning processes is to thicken the sludge as much as possible by removing liquid. Often coagulants and polyelectrolytes are effective in destabilizing charge balances in the sludge to facilitate oil and particle solid agglomeration, thus releasing water.

Common mechanical sludge dewatering methods[8] are (approximately in the order of decreasing sludge cake dryness: a) plate and frame filters, b) belt filter presses, c) vacuum filters, and d) centrifuges.

Often plate and frame filters and vacuum filters require a precoat material to aid in separating the sludge cake from the filter cloth. Spent catalyst fines, and/or raw water treatment sludges, in some cases may serve as an effective precoat material.

High speed centrifuges generally are not a good application for sludges containing oils due to the tendency to form oil/water emulsions. These emulsions, when returned to the refinery waste treatment system, are found to be particularly difficult to treat.

Nonmechanical means of sludge drying (e.g. sludge drying beds) are not generally used because they are labor intensive, require an arid climate, and tend to emit odors from VOCs.

Table IV. Refinery Solid Wastes vs. Potential Treatment and Disposal Methods

	Stripping DWTRG/ Landfill	Recycle to Coker	Solvent Xtraction Landfill	DWTRG. RKI/ Landfill	DWTRG. FBI/ Landfill	Chem. Fixation/ Landfill	DWTRG/ Landfill	Bio. Processes (Landfarm)
Continuous								
API/CPI Sludge	X	X	X	X	X	X		X
DAF Sludge	X	X	X	X	X	X		X
Catalyst Fines							X	
Coke Fines		X		X	X	X	X	X
HF Alky Sludge	X			X	X			
Lime Sftng Sludge							X	
Merox Blowdown	X			X	X			
Raw Wtr Trt Sludge							X	
Slops	X	X	X	X	X	X	X	X
Spent Acids	X			X	X			
Spent Amines				X	X			
Spent Caustic	X			X	X			
Spent Clays			X	X	X			
Spent Glycols				X	X			X
Wst. Act. Sludge				X	X	X	X	X
Intermittent								
BS&W	X		X	X	X	X	X	X
Construction Wastes				X				
Contam. Soils	X		X	X	X	X	X	X
Leaded Tank Bottoms	X		X	X	X	X	X	X
Mntc. Wastes			X	X				
Pond Cleaning	X		X	X	X	X	X	X
Unleaded Tank Bottoms	X	X	X	X	X	X	X	X
Vac. Truck Sludge	X		X	X	X	X	X	X
Remediation								
Remed. Wastes	X		X	X	X	X	X	X

RKI = Rotary kiln incinerator.
FBI = Fluid bed incinerator.

Rotary Kiln Incineration—The objective of a rotary kiln incineration (RKI) system is to convert the carbon-containing compounds in the waste to carbon dioxide and water. This not only reduces the solid waste volume, but also generates heat and can destroy organic toxicants. These objectives are realized by tumbling the solid wastes in the reactor section of the RKI at relatively high temperatures (1400 to 2500°F) in the presence of air. A RKI system usually consist of: feed systems (e.g. for liquid, solid, and vapors), a rotary kiln reactor, secondary combustion chamber, waste heat boiler (optional), vapor quench, air pollution control system, induced draft fan, stack, ash handling system, and ancillary systems and controls.

Incinerating wastes in a RKI system accomplishes the following:

- Combustion of carbon compounds, thus "stabilizing" the waste.
- Destruction of toxic organics.
- Wastes having a high volatile content will experience a large volume reduction.

Usually any heavy metals present in the feed are also oxidized to metal oxides. These oxidized forms can be both more soluble and more toxic than chemically reduced forms of the metals. Subjecting the ash residue to the TCLP test will determine whether the ash is hazardous due to the presence of excessive concentrations of heavy metals in the leachate. If so, a slagging incinerator may offer the advantage of vitrifying the ash to encapsulate the heavy metals, thus making them unavailable for leaching. Chemical fixation processes, where chemicals are added causing the sludge material to solidify, also may be capable of minimizing the leachability of various compounds, including metals.

Fluid Bed Incineration—While conceptually the rotary kiln and fluid bed incineration systems are similar, the wastes fed to the fluid bed incinerator must be relatively homogenous with respect to fluidization properties, while a rotary kiln incinerator can accommodate a broader range of feed materials. The fluid bed incinerator attains mixing of wastes at combustion temperatures by fluidizing the waste particles with sand while combustion air and reaction gases provide the fluidization energy. The unit operations in a fluid bed incineration system are conceptually similar to those of a rotary kiln system.

Care must be taken in selecting the appropriate feed mixture, since low melting point eutectics can cause the bed material to become sticky and agglomerate, thus preventing fluidization. Despite this potential problem, Amoco has been successfully incinerating solid wastes at their Whiting Refinery for a number of years[9].

Chemical Fixation—Chemical fixation is a process where pozzolonic materials are mixed with the wastes to react and convert the wastes into a chemically and mechanically stable material. Typically, however, using this process to stabilize oily sludges is not a good application.

Stripping—Stripping is a separation process used to remove materials having a high vapor pressure from the waste material. Stripping can be accomplished in trayed or packed columns for liquids or in heated screw conveyors or rotary kilns for sludges.

Biological Processes—The primary objective of biological processes is to "stabilize" the waste by converting organic materials into inerts such as carbon dioxide and water. Since a prerequisite of most biological processes is that the waste organics be soluble in water, the low solubility of hydrocarbons generally leads to relatively low reaction rates for most oily sludges and refinery solids.

Despite this, landfarming is a biological process which is currently widely practiced for the disposal of refinery sludges due to the low capital and operating costs. Waste is spread on the land in the form of thickened sludge (thus avoiding the conditioning and dewatering steps) and, if the landfarm is on refinery owned land, generally the transportation costs are minimal. Due to the increasingly stringent restrictions on land disposal of hazardous wastes, there is some question whether this disposal method will be permitted indefinitely.

Landfilling—With greater scarcity of available landfill space, disposing of hazardous waste in Class I landfills is becoming increasingly expensive. Associated future pollution liabilities are also proving to be a deterrent to landfill disposal of refinery hazardous wastes.

Solvent Extraction—Extracting the hydrocarbon fraction of refinery solid wastes through solvent extraction has the advantage of recovering the hydrocarbons for reuse, while the solids are often rendered nonhazardous. Using a light hydrocarbon such as propane or butane (with an inherently high purity) has the advantages of reducing the introduction of hazardous constituents with the solvent. The process is also familiar to refiners in the form of the propane deasphalting process.

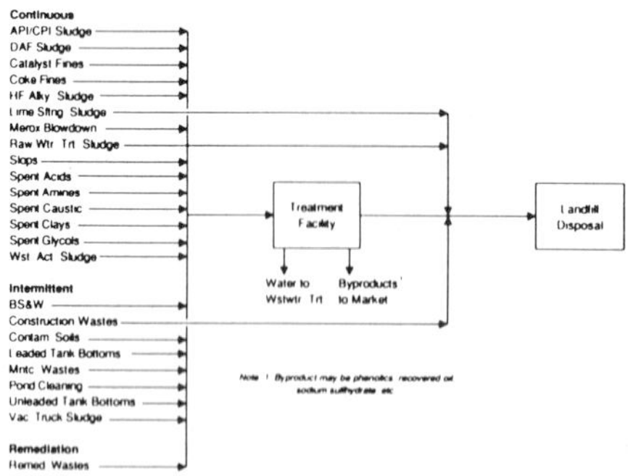

Figure 2. Minimum capital investment refinery solids disposal system.

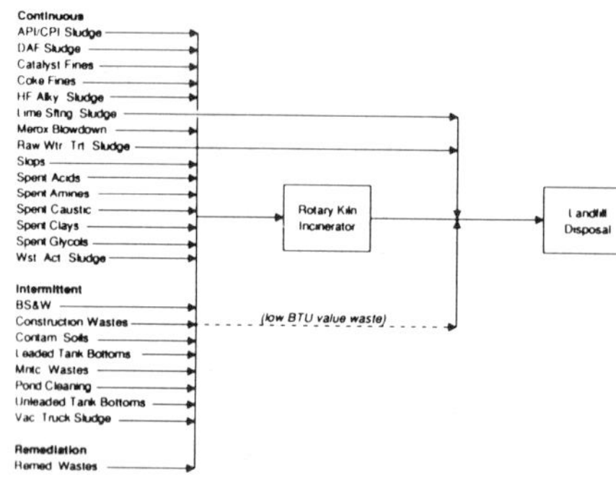

Figure 3. Minimum risk with low complexity refinery solids disposal system.

POTENTIAL DISPOSAL SCENARIOS

The following are three cases which present disposal scenarios in the context of alternative management strategies. The management strategies can be characterized as follows:

- Case I – Minimize capital investment.
- Case II – Minimize risk by reducing ultimate disposal volume; minimize system complexity.
- Case III – Minimize risk while optimizing efficiency and flexibility through application of "appropriate" technologies.

The correct disposal strategy for a given refinery is that which considers all important factors such as quantity, type and characteristics of solid wastes, capital and operating costs, risk of future environmental liability, safety, level of appropriate technology, and site-specific factors (e.g. plot space, opportunities for minimization/abatement, etc.).

Because the availability of landfarming as a long-term disposal option is open to question, landfarming has not been included in the disposal scenarios described below.

Case I

Case I represents a situation where the use of landfill disposal is maximized. As indicated schematically in Figure 2, all refinery "solid" wastes are collected and shipped to a solid waste facility for treatment. The objectives of the treatment facility are to:

- Reduce the volume of solid wastes going to landfill disposal.
- Reclaim the solvent and/or metals content for resale.
- Detoxify the wastes prior to landfill where possible.

The treatment facility is often operated by an offsite contractor. Most contract treatment facilities have a number of unit operations which separate and/or react with waste components. Some of the wastes such as raw water treatment wastes, nonhazardous refuse and construction wastes typically go directly to landfill disposal without intermediate treatment.

Material which remains following treatment is directed either to a hazardous waste landfill or a sanitary landfill, depending on whether the residual waste is hazardous.

From the refinery standpoint, this scenario has the following advantages:

- It involves a minimum of capital investment for the refinery.
- The costs associated with solid waste disposal are directly associated with the quantity and type of solid wastes generated.
- If properly managed this approach can provide incentives for waste abatement/minimization.

When utilizing the services of a disposal contractor for offsite treatment and disposal, however, the refiner can still be exposed to pollution liability for the improper disposal of the hazardous waste residues by the contractor.

Case II

Case II, shown schematically in Figure 3, depicts a treatment system featuring one rotary kiln incinerator to handle all incinerable wastes, with the ash residue going to landfill disposal. While a RKI is often not the optimum treatment system for all wastes, a properly designed and operated RKI will accommodate a wide range of feeds (e.g. spent caustics/acids, biological solids, remediation wastes, etc.).

If the RKI is equipped with a wet ash handling and a wet air pollution control system, soluble inorganics contained in the incinerator feed may be returned to the wastewater treatment system and ultimately escape in the refinery effluent. For incinerator feeds containing compounds which form nonhazardous soluble salts (e.g. spent caustics, sodium chloride, spent acids, etc.) this may be advantageous. However, consideration should be given to the escape of pollutants, such as soluble heavy metals, by this route. An alternative may be an incinerator with a dry ash and dry air pollution control system. However, this may merely transfer these problems to the landfill.

Incinerators which burn wastes containing alkaline earth metals such as sodium and potassium can experience problems associated with high erosion rates of the incinerator refractory due to the formation of low melting eutectics. Since refractories often contain heavy metals such as chromium, this can further aggravate the problem of heavy metals in the ash.

Incineration of nonhazardous refuse can be treated in the rotary kiln or transported to the landfill directly. For example, water treatment sludges and construction wastes can often be discharged directly to the landfill.

The Case II approach requires that the refiner address the following issues:

- The capital costs for a single technology treatment facility and landfill, particularly if the wastes are hazardous, can be high.
- Often significant periods of time are required to obtain environmental permits for treatment systems and landfills, particularly if there is significant local opposition.
- Often a single technology treatment facility, although flexible, is not optimum for the range of waste characteristics encountered.
- As long as there is capacity in the treatment facility and landfill there is generally no incentive to abate and/or minimize waste generation within the refinery.
- This approach allows the refiner to retain control of the waste treatment and disposal activities, and minimize risk associated with potential future liability.
- Using a single solid waste treatment unit to treat the majority of the sources can represent a potential reliability problem.

Case III

This approach, shown schematically in Figure 4, involves treatment of solid wastes in multiple treatment units, each chosen to best accommodate the characteristics of a particular type of solid waste. To manage environmental risk, ultimate disposal of residues in this scenario is at a captive

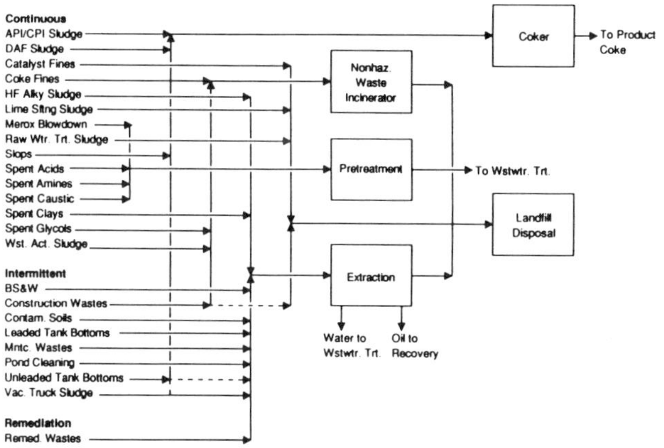

Figure 4. Application of "appropriate" technology refinery solids disposal system.

landfill owned and operated by the refinery. To implement this scenario, the solid wastes are generally segregated according to the following characteristics:

- High oil/organics content.
- Inorganic solids containing oil.
- Nonhazardous high-BTU containing wastes.
- Aqueous solutions of acids and alkali.
- Inorganic solids.

Once segregated, the solid wastes can be treated in unit operation which are suited to their characteristics. For example wastes with relatively high oil/organic and low inorganic content can often be used as feed to the refinery coker, thus converting a waste into a product.

If coking is not possible, wastes high in oil/organic can be combined with inorganic wastes which are contaminated with oils such as tank bottoms (e.g. BS&W), pond cleanings, contaminated soil, spent clays, etc. and pretreated in an extraction unit. Through the use of extraction units, oils are removed and recovered, while the oil-free solids can be discharged to the landfill.

Nonhazardous wastes such as maintenance wastes, office refuse, lab wastes and other selected streams can be incinerated in a relatively small nonhazardous waste incinerator. The objective of this unit operation is to reduce the volume of these wastes prior to disposal in a sanitary landfill.

Spent acids and caustic streams, because of their high water content, can be neutralized and pretreated to remove specific compounds such as oils, phenols, reduced sulfur compounds, etc., before they are discharged to the refinery wastewater treatment system. In selected cases these streams can also be recycled to a contract reclaimer.

As in Cases I and II, wastes having a low organic content, such as construction wastes and raw water treatment sludges can generally go directly to the landfill.

The Case III approach has the following features:

- Wastes with similar characteristics can be treated in unit operations which are optimized for a narrower range of waste characteristics.
- Pretreatment unit operations can be selected which facilitate recycle and/or recovery as opposed to disposal.
- Multiple solid wastes treatment facilities generally improve operating flexibility and reliability.
- Multiple waste treatment unit operations, however, increase operator and maintenance requirements.
- As in Case II, once the capital is expended to construct treatment facilities, there is little incentive for waste minimization/source abatement.
- By operating the waste treatment and landfill the refiner retains control of the treatment and disposal operations, thus reducing future environmental liabilities.

CONCLUSIONS

From this survey of the sources of refinery solid wastes and the available disposal alternatives, the following conclusions can be drawn:

- Increasingly stringent environmental regulations on effluent quality and a greater awareness of environmental risks is resulting in an increased rate of solid waste generation by the petroleum refining industry. These solid wastes not only come from a large number of sources but the solid wastes have diverse characteristics.
- The closing of existing hazardous waste landfills and the delays associated with permitting new landfills is creating a critical solids waste disposal problem for refineries.
- While solid wastes may be transferred to a contract disposer, the associated liabilities remain with the refiner.
- The cost effective solution for a particular refinery is one where a "systems approach" is taken to achieve an optimal mix of in-process minimization/abatement, recycle/byproduct recovery, and treatment/disposal approaches to the solution.

REFERENCES

1. 40 CFR Part 264, "Land Disposal Restrictions for First Third Scheduled Wastes; Final Rule", (August 17, 1988).
2. 40 CFR 268, "EPA Regulations on Land Disposal Restrictions", Appendix I, 51 FR 40636.
3. EPA, "Environmental Fact Sheet; Toxicity Characteristic Rule Finalized", Office of Solid Waste (March 1990).
4. 40 CFR 261.24, "Table 1 Maximum Concentration of Contaminants for Characteristic of EP Toxicity", 52 FR 29850.
5. Pojasek, R. B., *Toxic and Hazardous Waste Disposal: Volume Four*, Ann Arbor Science, 1980, pg. 204, (Jacobs Engineering Co. 1978).
6. *Manual on Disposal of Refinery Wastes: Volume on Solid Wastes*, Am. Petrol. Inst. (1980).
7. Beychok, M.R., *Aqueous Wastes From Petroleum and Petrochemical Plants*, John Wiley & Sons (1967).
8. *Process Design Manual for Sludge Treatment and Disposal*, EPA 625/179011 (1979).
9. Becker, K. P. and Wall, C. J. "Incinerate Refinery Waste In A Fluid Bed", Hydrocarbon Processing (October 1975).

75 COMBINED TREATMENT BY AERATED PONDS AND CHEMICAL CLARIFICATION COMPLETED BY RECIRCULATED COOLING SYSTEM

N. Galil, Senior Lecturer

M. Rebhun, Professor
Environmental and Water Resources Engineering
Technion—Israel Institute of Technology
Haifa, Israel

Biotreatment of wastewater from an integrated oil refinery by activated sludge and by RBC was studied and reported by the authors in previous Purdue Industrial Waste Conferences. The objective of this study is to develop a low cost process combination by utilizing facilities, which exist in most industries, as treatment means as well as recipient for the renovated effluent. The idea is to share the task of the bioprocess between: a) a partial primary stage biotreatment by two aeration ponds (AP); b) chemical clarification of the primary stage effluent for removal of MLSS and colloids synthesized in the first process, by utilizing the existing lime clarification-softening system; c) completion of the bioprocess by a recirculated cooling system (RCS) which has a high biooxidation capacity and also reuses the clarified effluent as make-up. The paper analyzes experimental data obtained from a pilot plant including two aerated tanks in series with a total hydraulic residence time of 2 days, followed by a lime clarifier. The main characteristics of the AP process is that hydraulic and sludge residence times are identical; two days are considered as very short SRT, especially in petrochemical wastewater, while in activated sludge SRT is usually above 15 days. The AP removed 60% of the COD—15% by stripping—and 80% of the BOD; the oil removal was up to 50%. The chemical lime clarification improved the AP effluent reducing: suspended solids from 160 to 8 mg/l; oil from 34 to 9 mg/l; total COD from 290 to 170 mg/l. The study includes data on volumes and characteristics of the chemical sludge produced. The organic matter residual from previous treatment stages, most of it in soluble form, will be biodegraded in the RCS. The characteristics of RCS working as biological reactor were studied by Rebhun et al. and reported elsewhere. This study emphasizes economy in capital investment and beneficial reuse of industrial wastewater. The combined system includes two biological barriers, AP and RCS, combined with the chemical clarification barrier, providing reliable and a low cost process combination.

INTRODUCTION

Refinery wastewater biological treatment data has been reported for many years. The data demonstrates efficient stabilization in the activated sludge process in which the biota metabolize the refractory components after long acclimation periods[1]. Adding adapted mutant bacteria has been suggested by others[2,3]. The presence of parafinic and aromatic hydrocarbons affects biofloc settling adversely and their biodegradability rate is much lower than that of municipal type organic material[4].

Phenols are a major component of refinery wastewaters, being produced in the cracking process and special gasoline washeries. While many report that phenol is a controlling toxicant in biological treatment[5,6], others report efficient phenol removal at influent concentrations up to 200 mg/l, after acclimation, which requires, however, a long reactor residence time to achieve good biofloc settleability[7].

The wastewater of the integrated refinery reported in this study is pretreated by gravity oil-water separation, followed by flocculation and dissolved air flotation. The effluent contains 60 to 280 mg/l BOD, 260 to 710 mg/l COD, 20 to 170 mg/l hydrocarbons; the phenols concentration was 2 to 6 mg/l, after taking concentrated phenolic streams, accumulated in the refinery, to separate treatment.

The severe regional water shortage, imposed to reuse treated effluent as make up in the cooling water system, minimized effluent discharge to the river. Biological treatments by activated sludge or

45th Purdue Industrial Waste Conference Proceedings, © 1991 Lewis Publishers, Inc., Chelsea, Michigan 48118. Printed in U.S.A.

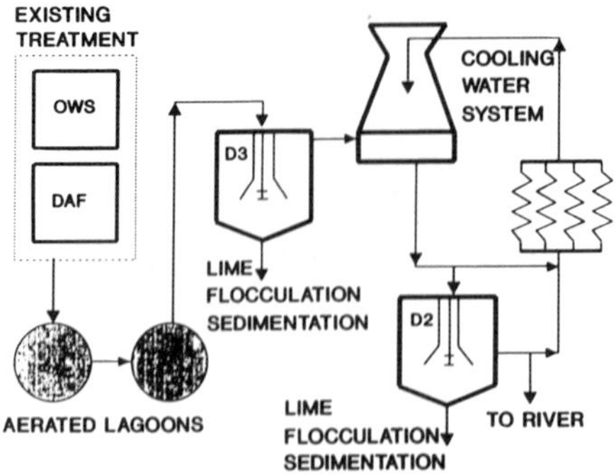

Figure 1. Refinery wastewater systems.

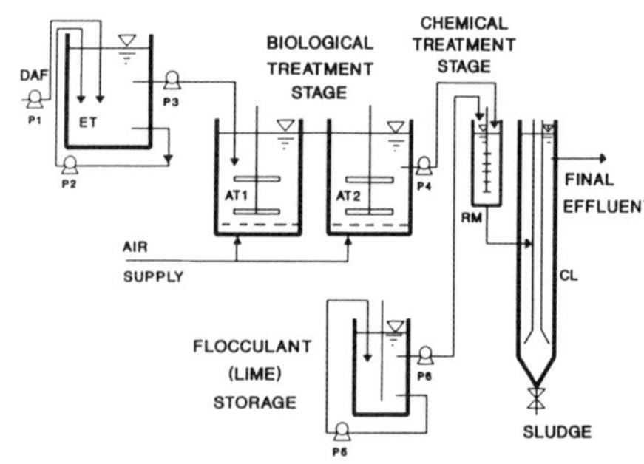

Figure 2. General description of the pilot plant.

by rotating biological reactor were considered and studied separately[8,9], but not adopted because of their relatively high investment cost.

The treatment sequence studied here is based on sharing the tasks of the bioprocess between: (a) a partial primary stage biotreatment by aeration ponds (AP); (b) chemical clarification of the primary stage effluent for removal of MLSS and colloids synthesized in the first process, by utilizing the existing lime clarification-softening system; (c) completion of the bioprocess by a recirculated cooling system (RCS) which has a high bio-oxidation capacity and also reuses the clarified effluent as make up. Aerated ponds are considered as the simplest system in terms of construction and operation[10,11]. Equalization of the feed is unnecessary since the ponds provide large, well-mixed volumes to dilute influent slugs of contaminants[12]. Two concrete storage tanks were suggested to be transformed into aerated ponds, however all the treatment sequence is based on existing facilities. A schematic description of the existing and the proposed wastewater treatment stages is shown in Figure 1.

The objectives of the study were to develop a low cost biotreatment process capable of producing a suitable and minimilly disturbing make up to the water RCS, achieving also a blow-down quality which fulfills requirements of discharge to the river.

EXPERIMENTAL DESIGN

The on-site pilot plant (Figure 2) consisted of an equalization tank of 1,000 l, two aeration tanks of 220 litres each, a rapid mixing cell, a clarifier and a flocculant (hydrated lime) preparation unit; all the necessary piping, pumps, flow measurements and regulation devices were included. Composite samples of the pilot plant feed and effluent were collected, stored and refrigerated for laboratory analysis; dissolved oxygen, temperatures, pH, and flow rates were determined at the pilot plant.

Routine analyses were performed according to *Standard Methods*[13].

RESIDENCE TIME DISTRIBUTION

An important feature of the aerated ponds process is the relatively long detention time. With detention times in order of days, any change occuring at one end of the process may only be felt much later at another end of the process. According to residence time distribution for two completely mixed reactors in series, the concentration of any impulse into the first reactor (AT1) will be felt according to the function

$$E_1(t) = C(t)/Co = e^{-t/\Theta} \quad (1)$$

where: c = represents the concentration of an impulse (mg/l);
t = real time (day);
Θ = hydraulic residence time per reactor (day).

The same impulse will be felt by the second reactor (AT2) according to the function:

$$E_2(t) = (t/\Theta)e^{-t/\Theta} \quad (2)$$

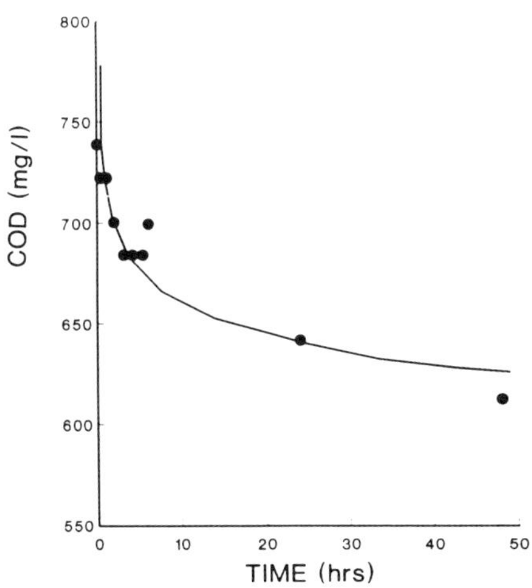

Figure 3. COD removal in stripping experiments.

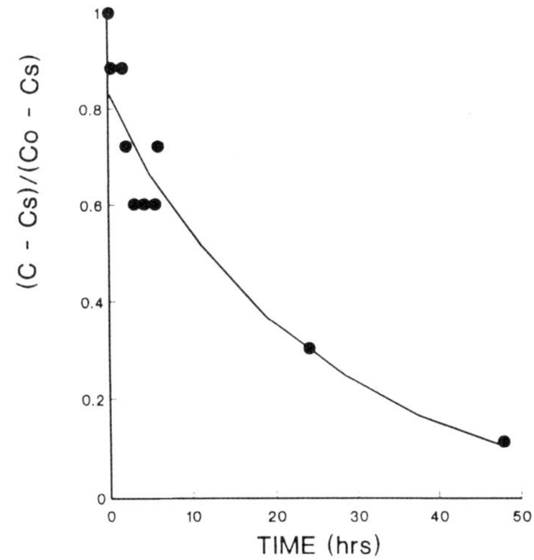

Figure 4. COD removal—rate experiments.

While the functions in Equations 1 and 2 describe the residence time distribution for reactors AT1 and AT2, in series, the fraction of an impulse having passed through the reactors, by time t, will be given by the integrals:

$$\int_o^t E_1(t)dt \qquad \text{for AT1} \tag{3}$$

$$\int_o^t E_2(t)dt \qquad \text{for AT2} \tag{4}$$

The calculations of reactor efficiencies and kinetic coefficients were based on residence time distribution, which approximates the true average influent concentrations, felt by each reactor during a given time period.

STRIPPING EXPERIMENTS

During the course of biological treatment, part of the dissolved organic matter was lost to the atmosphere through stripping, resulting from the aeration. In order to calculate process removal efficiencies and various kinetic parameters it was desired to differentiate between the organic matter lost to the atmosphere to that degraded by the biological treatment.

A fresh 2 liter dissolved air flotation effluent sample was filtered to remove suspended solids and was aerated in the laboratory for a period of 3 days, using a bubble porous stone aerator. Samples were withdrawn every hour, for the first 8 hours, for COD determination and sampling continued at 24, 48 and 72 hours. As shown in Figure 3, plots of COD versus time were made. A power function regression was performed on the experimental data to determine the equation of the curve, which was:

$$COD_t = 717t^{-0.0355} \tag{5}$$

where: COD_t = COD concentration after t hours (mg/l)
t = aeration (stripping) time (hours)

The curve was extrapolated to seven days to determine COD_7, which was taken to approximate the saturation concentration, C_s.

Knowing C_s, the overall COD gas transfer coefficient, K_{LA}, could be determined; therefore, the organic matter lost to the atmosphere could be approximated by the gas transfer equation (Figure 4):

Table I. Results of Stripping Experiments Conducted on Dissolved Air Flotation Effluent

Date	Co (mg/L COD)	Cs (mg/L COD)	Cs/Co	K_{LA} (1/he)
21 Dec 88	970	815	0.84	0.0638
24 May 89	295	260	0.88	0.1140
19 June 89	382	275	0.71	0.0341
21 June 89	626	410	0.65	0.0216
26 June 89	597	424	0.71	0.1200
3 July 89	738	597	0.81	0.0425
6 July 89	493	411	0.83	0.0544
Average:			0.78	0.0643
Standard deviation:			+0.08	+0.0356

$$(C-C_S)/(Co-C_S) = e^{-K_{LA}t} \quad (6)$$

where: C = final COD concentration after t hours (mg/l);
Co = initial COD concentration (mg/l);
K_{LA} = gas transfer coefficient (1/hr);
t = aeration time for stripping (hr).

Table I summarizes experimental results obtained for the ratio Cs/Co and for the gas transfer coefficient K_{LA}.

Though the stripping experiment was performed on filtered dissolved air flotation effluent, there was concern that bacterial growth during the stripping experiment could exert an oxygen demand thereby causing overestimation of the amount of COD stripped to the air. Calculations, based on kinetic data, showed that even without adding bactericide, the expected oxygen demand from bacteria present in the wastewater was always less than 5% of the total COD lost to the atmosphere; therefore, the stripping experiments were performed without the use of bactericides.

AERATED PONDS

During the experimental work the biomass level in bioreactors was most of the time 80–200 mg/l of MLVSS, with hydraulic detention times between 1/2 to 2 days.

The two stages bioprocess removed total organic matter by 60% as COD and by 80% as BOD (see Figure 5 and Table II). 92% of the phenols were removed already after the first aeration stage, AT1.

During the experimental work the dissolved air flotation unit of the refinery performed poorly, therefore high concentrations of oil entered to the bioprocess, 72 mg/l instead of 20–25 mg/l which is the usual level; in these conditions of oil "overloading", the aerated ponds removed 50% of it.

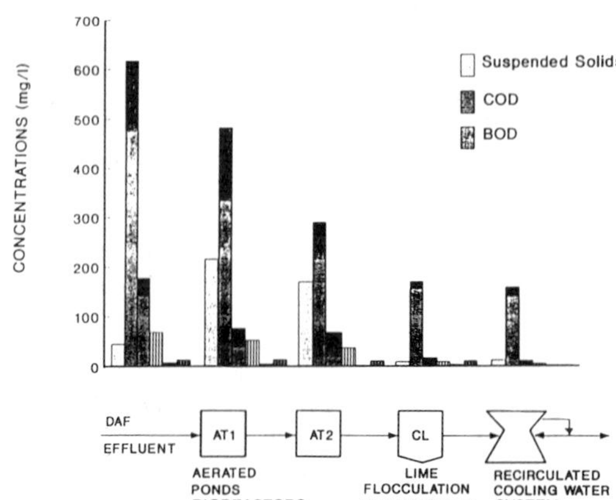

Figure 5. Effluent qualities after different treatment stages.

Table II. Effluent Qualities After Different Treatment Stages

Constituent	Influent	AT1	AT2	Lime	RCS[a]
pH	6.7–7.9 (7.2)[b]	6.7–7.5 (7.0)	6.7–7.5 (7.0)	7.4–11.7 (10.6)	7.8–8.4 (8.1)
Suspended solids	22–84 (46)	52–297 (217)	48–207 (168)	1–30 (8.3)	10
COD total	257–707 (617)	212–747 (480)	191–480 (290)	79–316 (167)	155
soluble	186–698 (476)	138–453 (333)	101–370 (216)	72–281 (155)	140
BOD total	59–283 (177)	21–146 (74)	18–93 (67)	3–47 (18)	10
soluble	43–229 (142)	6–88 (39)	2–54 (21)	1–36 (14)	5
Oil	17–176 (71)	14–93 (52)	10–60 (34)	3–19 (8.7)	3
Phenols	1.6–5.6 (5.1)	0.1–5.6 (1.4)	0–2.1 (0.4)	0–1.1 (0.3)	0.2
NH_4-N	4–24 (14.4)	2.6–20 (13.6)	3.2–15 (12)	3–14 (9)	0.5

[a]Appreciated values, based on reference 15, after additional treatment by lime flocculation-sedimentation.
[b]80% of the results achieved the value stated in parenthesis, or less. All concentration values are in mg/L.

Ammonia concentrations were not influenced by the bioprocess. This can be explained by the short biomass residence times, which in the case of AP have the same values as the hydraulic detention times, therefore the nitrification condition could not be satisfied.

A kinetic analysis was made, in terms of COD and BOD, based on calculated residence time distribution and eliminating the organic matter removed to the atmosphere by stripping. The results regressed to the first order reaction (Fig. 6A, Fig. 6B), given by the mathematical expression for CSTR:

$$\frac{Se}{So} = \frac{1}{1 + K\Theta} \qquad (7)$$

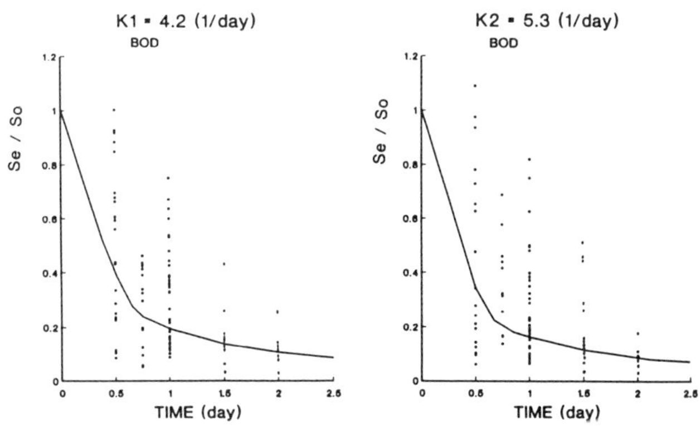

Figure 6A. Effect of superficial upflow velocity on biomass profile in single phase AFBR.

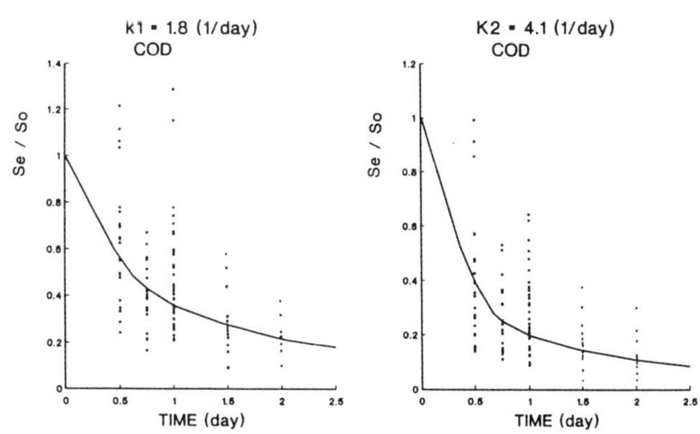

Figure 6B. Kinetics of COD removal.

Table III. Experimental Values of Bioprocess Rate Constant, K (1/day)

Reactor	AT1	AT2
BOD	4.2	5.3
COD	1.8	4.1

where: So = organic matter, COD or BOD, concentration in influent (mg/l);
Se = organic matter, COD or BOD, concentration in effluent (mg/l);
θ = hydraulic residence time in the reactor (day);
K = bioprocess rate constant (1/day).

The experimental values of K are summarized in Table III and discussed below.

LIME FLOCCULATION-SEDIMENTATION

The pilot unit CL simulated an existing process, treating the make up to the RCS by chemical flocculation with $Ca(OH)_2$ and by sedimentation. During the past years, 1/3 of this make up was fresh water and 2/3 was biological effluent from the municipal wastewater treatment plant of Haifa, Israel, while the process tasks were to achieve efficient clarification and softening by magnesium and calcium removal.

According to this study, all the make up to RCS will consist in the future on treated refinery wastewater, therefore the pilot unit CL worked on the AP effluent. In this case clarification has a special importance, because it must capture the solids produced by the biological process before the make up enters to RCS.

Lime flocculation-sedimentation was characterized in the experiments by high efficiency removal of suspended solids, always more than 90% and significant reduction of 75% of residual oil. The sludge accumulated contained 20–25% of organic solids originated by the aerated ponds and had good thickening and cake filtration characteristics.

DISCUSSION

The substrate removal rate, q, can be obtained from the mass balance on the bioreactor, resulting in the expression:

$$q = \frac{So - Se}{X\Theta} \tag{8}$$

where X is the MLVSS concentration in the reactor.

The removal rate, q, can be obtained also by using the Michaelis-Menten rate model:

$$q = \frac{\hat{q}\ Se}{K_S + S_e} \tag{9}$$

where: $\hat{q}$ = the maximal substrate removal rate (1/day);
K_s = saturation coefficient (mg/l).

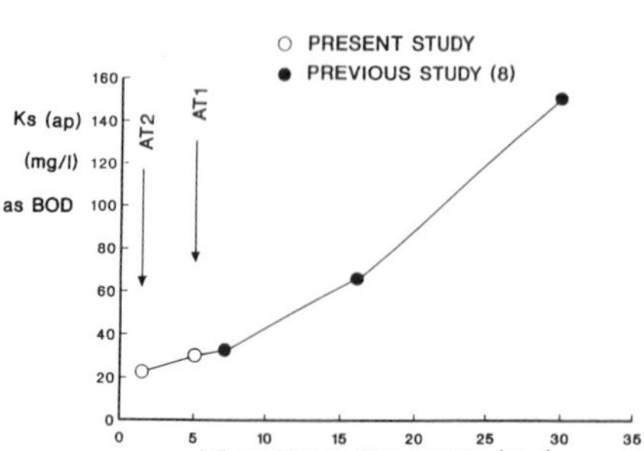

Figure 7. The apparent saturation coefficient vs. concentrations of phenols in the influent.

From Equations 8 and 9, the substrate concentration ratio can be obtained as:

$$\frac{Se}{So} = \frac{1}{1 + \frac{\hat{q}}{K_S + S_e} X\Theta} \qquad (10)$$

By comparing Equations 7 and 10, one can see that the bioprocess rate constant, K, obtained in the experimental work, is given by:

$$K = \frac{\hat{q}}{K_S + S_e} X \qquad (11)$$

In a previous work, Rebhun and Galil[8] studied the biokinetics of an activated sludge process, treating oil refinery wastewater. It was found that in the presence of different phenol concentrations, $\hat{q}$ was not affected and reached the value of 1.35. The variations in phenol concentrations affected only the K_s values, indicating a competitive inhibition[13,14]. By substituting values for X and S_e in equation[11], the apparent saturation coefficient, K_s(ap), could be calculated for AT1 and AT2. Fig. 7 shows values of K_s(ap) versus different phenol concentrations in the influent in the present as well as in the previous work[8]. The higher bioprocess rate constant, K, obtained in the second stage, AT2, is probably a result of the lower phenol concentration (1.4 vs. 5.1 mg/l) exerting a lower inhibition influence on the second stage bioprocess.

Although AP are not considered to be an intensive biotreatment, because of their low biomass concentrations and short biosludge residence time, this study shows that an effective biotreatment could be achieved by the two-stage AP in series, having detention times of one day per stage. Up to 30% of the total organics were removed by stripping; the biodegradable fraction was removed by 80–85%; and the biosolids were removed by the lime flocculation-clarification, which also reduced alkalinity (bicarbonate), calcium and magnesium, which reduced major scale formers.

The behavior of the RCS and its influence on the effluent quality in completing the previous treatment stages could be appreciated on the basis of the results obtained by Rebhun and Engel[15] and Rebhun and Grynberg[16]. Their studies on reuse of treated wastewater for industrial cooling systems show that RCS could operate as a bioreactor. Ammonia was completely bio-oxidized to nitrates; acid produced by bio-nitrification served for neutralization of excess lime supplied with the lime precipitation effluent and no external acid addition was necessary. Biogrowth may be controlled by mechanical cleaning or side stream treatment. The blow down from RCS, using reclaimed effluent was generally of better quality than the effluent before reuse, except for chlorides and mineral salinity.

CONCLUSIONS

The study has developed a combined treatment system which includes two biological barriers AP and RCS, combined with a chemical clarification barrier, providing a reliable and good quality effluent.

The production capacity of slimes in RCS can be substantially reduced, due to the efficient primary AP biotreatment as well as the lime flocculation-sedimentation.

The biosludge produced can be separated by lime clarifications, and easily treated and disposed together with the lime sludge.

The developed solution is based mainly on existing facilities, therefore the investment cost is appreciated as 35 percent of new activated sludge or RBC system.

ACKNOWLEDGMENTS

Credits: The project was supported by the Oil Refineries, Haifa, Ltd., and carried out at the Industrial Wastewater Treatment and Waste Renovation Laboratory at the Faculty of Civil Engineering, Technion-Israel Institute of Technology.

The invaluable assistance of Mr. Y. Brayer, Environmental Quality Engineer of Haifa Oil Refinery is greatly appreciated.

REFERENCES

1. Dickenson, L. R., and Giboney, T. J., "Stabilization of Refinery Wastewater with the Activated Sludge Process", Proc. 25th Waste Conf. Purdue Univ., West Lafayette, Ind. (1970).
2. Mahmud, Z., and Thanh, V. N., "Biological Treatment of Refinery Wastes", Proc. 33rd Waste Conf., Purdue Univ., West Lafayette, Ind. (1978).
3. McDowell, C. S., and Zitrides, G. T., "Accelerating the Dynamic Response of Bacterial Populations in Activated Sludge Systems", Proc. 34th Waste Conf., Purdue Univ., West Lafayette, Ind. (1979).
4. Groenewold, C. J., et al., "Comparison of BOD Relationships for Typical Edible and Petroleum Oils", Jour. Water Pollution Control Fed., 54, 389, (1982).
5. Christiansen, A. G., and Spraker, W. P., "Improving Effluent Quality of Petrochemical Wastewaters with Mutant Bacterial Cultures", Proc. 37th Waste Conf., Purdue Univ., West Lafayette, Ind. (1982).
6. Reynolds, J. H. et al., "Temperature Toxicity Model for Oil Refinery Wastes", Jour. Env. Eng. Div., Proc. Am. Soc. Civ. Eng., 100, 557, (1975).
7. McKinney, R. E., "Microbiology for Sanitary Engineers", McGraw Hill Book Company Inc., New York, N.Y., (1972).
8. Rebhun, M., Galil, N., "Inhibition by Hazardous Compounds in an Integrated Oil Refinery", Jour. Water Pollution Control Fed, 60, 1953 (1988).
9. Galil, N., Rebhun, M., "A Comparative Study of RBC and Activated Sludge in Biotreatment of Wastewater from an Integrated Oil Refinery", Proc. 44th Waste Conf. Purdue Univ., West Lafayette, Ind. (1989).
10. "Manual on Disposal of Refinery Wastes", Volume on Liquid Wastes, Chapter 13, 53, American Petroleum Institute, (1969).
11. Lamkin, J. C., Sorg, L. V., "Treatment of Refinery Wastes in an Aerated Lagoon", Proc. API 44 (III), 13–20, (1964).
12. Davies, B. T., Vose, R. W., "Custom Designs Cut Effluent Treatment Costs Case Histories at Chevron U.S.A., Inc.", Proc. 32nd Waste Conf., Purdue Univ., West Lafayette, Ind. (1977).
13. Aiba, S., et al., "Biochemical Engineering", 2nd Ed., Academic Press, Inc., New York, N.Y., (1973).
14. Laidler, K. J., "Introduction to the Chemistry of Enzymes", McGraw-Hill, New York, N.Y., (1954).
15. Rebhun, M. Engel, G., "Reuse of Wastewater for Industrial Cooling Systems", Journal Water Pollut. Control Fed., 60, 237 (1988).
16. Rebhun, M., Grynberg, H., "Use of Wastewater for Cooling Systems", Annual Report No. 2., Bronfman Fund., Technion Research and Development Foundation Ltd., Haifa, Israel, (1985).

76 PILOT TESTING AND DESIGN OF AN ACTIVATED SLUDGE SYSTEM FOR REFINERY WASTEWATER

Edward C. Copeland, Senior Project Engineer

James D. Cole, Senior Project Manager
Heritage Remediation/Engineering, Inc.
Indianapolis, Indiana 46251

INTRODUCTION

In 1989, a small Midwestern refinery determined that its existing lagoon-based wastewater treatment system was incapable of consistently complying with the requirements of its anticipated NPDES permit renewal. In addition, proposed stringent water quality standards, and other factors had made it clear that either an upgrade or replacement of the existing wastewater treatment system was required. This paper describes the investigation undertaken to determine the most effective system to treat the refinery wastewater, in terms of both cost and performance. The operations performed by the refinery put it in the Topping Subcategory for purposes of USEPA effluent guidelines limitations (40 CFR Part 419)[1].

WASTEWATER SOURCES

The refinery process operations generate the following wastestreams. The volatilized material from the preflash, crude tower, quench stack, and stabilizer all contain a mixture of water and product in vapor form. As this vapor is condensed, product separates from the water and is returned to the process. The wastewater is split into two streams, one of which is discharged directly to the wastewater treatment system. The remainder is diverted to the desalter. The desalter mixes oil and water intimately to remove dissolved salts and other inorganics from the oil. Water is removed by phase separation and split, with part of it being recycled to the head of the desalter, while the remainder is discharged to the wastewater treatment system.

The vacuum fractionation process is an additional source of wastewater. A mix of petroleum fractions and condensed steam is discharged to the "hot well", then to an oil water separator. The effluent from the separator passes through a cooling tower and is recycled to the vacuum process. There is a periodic blowdown from this operation to the wastewater treatment system.

In addition to these process sources, there exist numerous non-process sources of wastewater to the treatment system. Various sample ports on tanks and miscellaneous lines drain directly into the process sewers. Of particular concern are the tank water draws from the caustic treater and merox unit, where pure product such as jet fuel could be released directly into the sewer. Water drawn from remote tanks is hauled via portable tanker and discharged directly to the front end of the treatment system. Also, backwashes from the process water supply ion exchange system and the rock filter units are discharged to the wastewater treatment system. Finally, stormwater runoff from the refinery unit operations area, as well as water from the various tank farm diked areas, are diverted through the process wastewater treatment system. It should also be noted that sanitary wastes are discharged to an on-site septic system, not to the wastewater treatment system.

EXISTING WASTEWATER TREATMENT SYSTEMS

Noncontact cooling water is currently discharged directly through two outfalls, Outfall 002 and Outfall 003. Both Outfalls discharge to a lake.

In the process wastewater treatment system, process water and part of the stormwater flow into a two-train, in-series, API separator system (Figure 1). The wastewater is pretreated with a polymer to improve phase and gravity separation. The initial API separator, the North API, is 15 feet wide by 47 feet long by 5 feet deep, and is divided into three sections by concrete "under and over" baffle walls. Oil skimming is conducted manually in all three sections. Sludge removal is conducted biannually,

45th Purdue Industrial Waste Conference Proceedings, © 1991 Lewis Publishers, Inc., Chelsea, Michigan 48118. Printed in U.S.A.

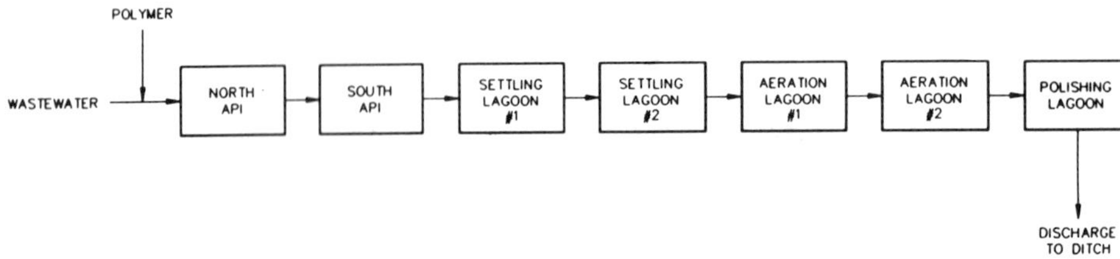

Figure 1. Existing wastewater treatment flow diagram.

which requires taking the unit off-line. The effluent from the North API flows into the South API (12 feet wide by 35 feet long by 12 feet deep). The South API is equipped with moving surface skimmers and sludge rakes; however, gross oil removal is still performed manually and sludge removal is practiced on an intermittent basis. Dry weather effluent flow rate from the South API is estimated at 150,000 gallons per day. Effluent from the South API enters a wet sump, where a battery of three level-controlled centrifugal pumps transfer the wastewater to the lagoon treatment system. Stormwater from the north tank farm is pumped separately to the lagoons, as is boiler blowdown, softener regeneration waste, and rock filter backwash.

The lagoon treatment system consists of five lagoons. The initial lagoon, with an estimated volume of 108,000 gallons, serves as a separation and equalization basin and is equipped with a mechanical oil skimmer. The second basin also serves as a settling as well as a compartmentalized biological (facultative) lagoon, subdivided by rubber baffles. The gross volume of this basin is estimated at 232,000 gallons. The third basin in the series is an aerated lagoon, equipped with two five horsepower mechanical surface aerators. The estimated volume of this basin is 500,000 gallons. Excess plant steam is blown off to this basin to maintain temperature in winter. The fourth lagoon, an aeration basin, contains one five horsepower blower and has an estimated volume of 460,000 gallons. The final treatment unit in this system is a polishing lagoon, estimated at 108,000 gallons. The effluent from the system flows through a Parshall flume and is discharged through Outfall 001 to a small stream. The Q7,10 low flow characteristics of the receiving stream is 0.0 cfs. The Q7,10 is described as the lowest observed seven-day consecutive average flow rate of the stream in a given ten-year period. The average daily flow from the lagoon system is 169,000 gallons per day.

TREATMENT SYSTEM EVALUATION

Based upon review of available literature[2,3,4,5], it was believed that biological treatment, in spite of the poor performance of the existing biological lagoons, was the best available means of treatment. Review of plant data, however, created concern about the ability of the wastewater to maintain a viable culture. COD and BOD_5 values in the raw wastewater averaged only 170 mg/L and 70 mg/L, respectively. Interestingly, data from the individual wastestreams showed the potential for a relatively high strength wastewater. Much of this data was suspect, however, so it was determined that on-site pilot testing of biological treatment was required. On-site analytical testing was conducted alongside the pilot tests to maintain adequate feedback as to the operation of the pilot system. Concurrently, the entire refinery was reviewed both for possible wastestream reduction as well as reasons for the apparent analytical anomaly.

Several unit operations normally found in the treatment of refinery wastewater were concluded not to be applicable to this site[6]. Dissolved Air Flotation (DAF) of the wastewater was determined to be inappropriate both because of the generation of a listed hazardous waste and the reduction of biologically assimilable material to the biotreatment units. A trickling filter, while excellent in reducing downstream shock loads, was also determined to be infeasible as either a stand–alone unit or as a pretreatment operation. Removal efficiencies of the trickling filter alone would not be sufficient to meet the anticipated water quality standards. As a pretreatment operation, the trickling filter was considered impractical as a result of the low strength of the wastewater. Downstream treatment units would not have sufficient assimilable material to sustain a viable biological system. Rotating biological contactors merited serious consideration based upon literature[7], but it was determined that maintenance requirements as well as the low loadings would render them unusable in this facility.

PILOT PLANT RESULTS AND DISCUSSION

To evaluate the treatability of the refinery's waste streams, a pilot- scale activated sludge evaluation was undertaken. The focus of the pilot-scale study was to establish the process design parameters necessary to design a full-scale treatment system that would enable the facility to meet future NPDES limits. The pilot operation was also necessary to predict effluent quality compliance with the proposed State Water Quality Standards. Pilot unit process parameters evaluated included: optimal mixed liquor suspended solids (MLSS) level as a function of settleability and temperature; optimal solids retention time (SRT) necessary for nitrification and carbonaceous BOD removal; net sludge production based on biological yield, SRT, and temperature; biokinetics such as Food to Microorganism ratios (F/M), endogenous decay, and wastage rates necessary to maintain MLSS; and oxygen and mixing requirements for the activated sludge. Additionally, the effects of batch dumps were simulated to determine the inhibitory and/or toxic effects of upstream unit process upsets on a biological treatment system. By varying process parameters and analyzing the performance data, baseline process kinetics were determined that could be used in designing a full-scale treatment system.

SETUP & PROCEDURE

The pilot study was conducted using six 5-gallon aquariums, which were modified with an adjustable baffle at one end (Figure 2). The baffle served to provide an area for quiescent settling of the mixed liquor, thus simulating a clarifier. The dimensions of the aquarium reactors were 16.4 inches long by 8 inches wide by 10 inches deep, with a freeboard of 2.25 inches. The addition of an adjustable baffle divided the reactor into two sections; the activated sludge section had a volume of 3.8 gallons, while the clarifier section had a volume of 0.6 gallons. The baffles provided a quiescent area necessary for solids subsidence of the activated sludge, and were installed at an angle of approximately 30° from the perpendicular. Surface area of the clarifier section was approximately 0.36 ft^2.

F/M ratios were determined by varying MLSS of the reactors while maintaining the same mass loading. The feed for the reactors, taken from the discharge side of the full-scale API sump pumps, was transferred to an eighty-five gallon agitated feed holding tank. See Figure 3 for reactor process flow diagram. Minimum hydraulic retention times (HRT) for optimum treatability were not evaluated, but were arbitrarily established at 1 day. Initial reactor setpoints are depicted in Table I. Additionally, one of the control reactors provided effluent for a bioassay evaluation and for a complete priority pollutant scan to determine the aquatic toxicity and relative effectiveness of the biological treatment system.

The reactors were initially seeded using activated sludge from a local Publically Owned Treatment Works (POTW) and purchased biofloc that the refinery was using to supplement their existing lagoons. The individual reactors were fed raw wastewater by peristaltic pumps and were allowed to acclimate for approximately three weeks. The initial alkalinity was measured and supplemented using calcium carbonate. After the acclimation period, MLSS levels were determined and the appropriate quantity of biomass was wasted to obtain the respective MLSS setpoints. Reactor effluent was used to replenish volume lost to wastage. Throughout the study, reactor MLSS were maintained at the design setpoints by periodic wasting.

Pilot units 1, 2, 3, 4, and 5 were operated for ten weeks. Reactor 6 was operated for only 3 weeks. Data from reactors 1, 2, and 3 were used to determine the baseline biological kinetics for the design of the wastewater system[8]. Reactors 4 and 5 evaluated the effects of process upsets on the pilot system by simulating batch dumps of various process units. The process units from which batch dumps were simulated included: desalter, pre-flash condenser, salt tank caustic waste from the jet fuel caustic scrubber, and API skimmings. Batch dumps were fed to the reactors by peristaltic pumps and were hydraulically loaded at a rate considerably greater than would be expected during a full-scale upset event.

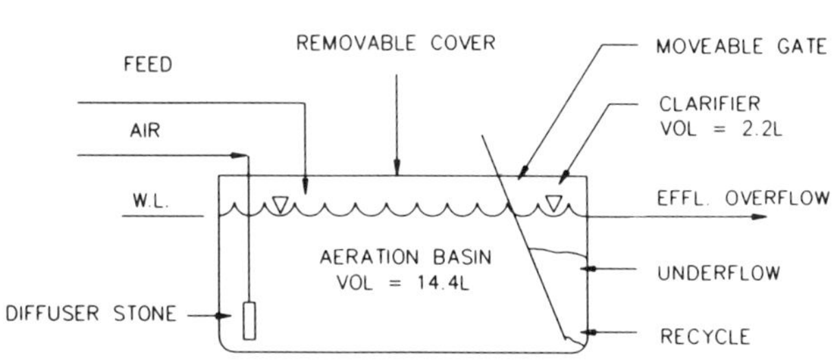

Figure 2. Typical pilot reactor.

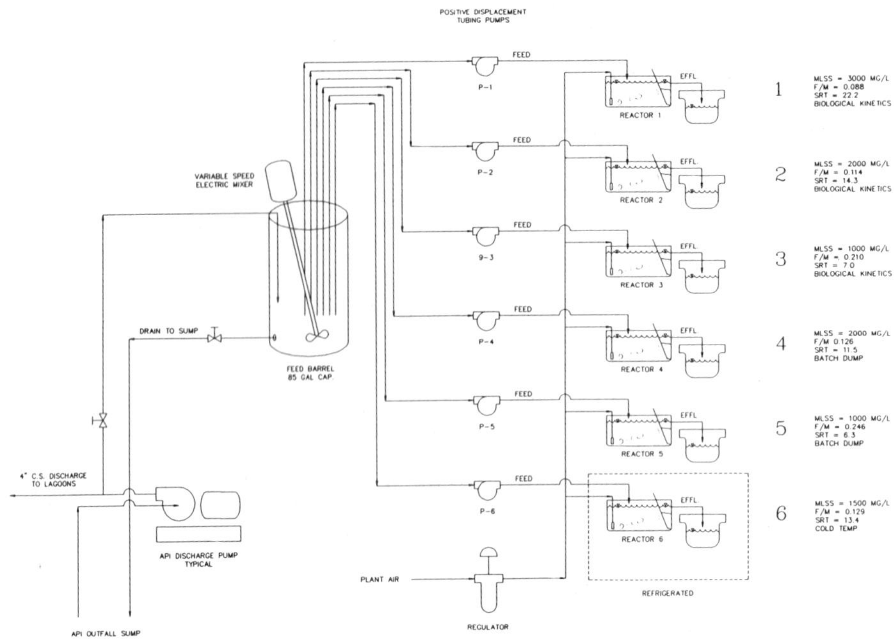

Figure 3. Pilot study reactor setup.

RESULTS

Reactors 1, 2, and 3 were operated at three different mixed liquor levels to determine biological kinetics such as F/M ratios, cell yield, and SRT. Overall, the F/M ratios for the bench-scale reactors were significantly lower than the literature values for biological treatment of refinery wastewater.[9] Loading rates based on Chemical Oxygen Demand (COD) and five day Biochemical Oxygen Demand (BOD_5) values were significantly lower than expected. Biological yield and endogenous decay coefficients were determined to be 0.74 lb MLSS/lb of COD removed and 0.016 day^{-1} respectively (Figure 4). BOD_5 removal efficiencies averaged 94.5% for the ambient temperature control reactors (reactors 1-3). COD removal for these reactors averaged 79.7%. All analyses were performed per *Standard Methods*[10]. Refer to Figures 5 and 6 for BOD_5 and COD performance graphs, respectively. All of the ambient temperature reactors exhibited nitrification activity for ammonia-nitrogen (NH_4-N) removal (Figures 7 & 8). Net solids production for a full-scale treatment system was calculated to be 165 lb/day.

The batch dump evaluations revealed that the process upsets exhibited an inhibitory effect on the biological system but no toxic effects were observed. In each case, the reactors recovered quickly from the batch dumps (usually within two to three HRT cycles). See Figure 9 for batch dump COD performance summary for reactors 4 and 5.

Reactor 6, the cold temperature unit, was evaluated for winter performance at a MLSS of 1500 mg/L and a basin temperature of 5-7°C. Overall reactor performance observed indicated a marked

Table I. Bench-scale Reactor Setpoints

Reactor MLSS	HRT	Parameter	Temp	
1	3000	1 day	Low F/M	Summer
2	2000	1 day	Control	Summer
3	1000	1 day	High F/M	Summer
4	2000	1 day	Batch Dumps	Summer
5	1000	1 day	Batch Dumps	Summer
6	2000	1 day	Cold Temp	Winter

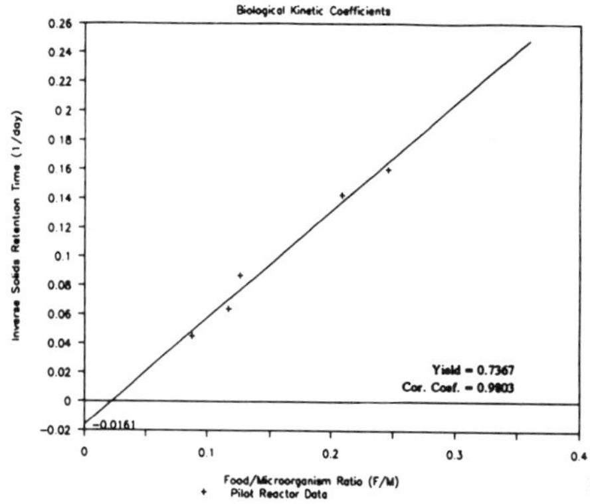

Pilot Plant Data

UNIT	CODr (gm/d)	SRT day	MLSS (mg/l)	COD F/M day	1/SRT (1/day)
1	3.539	22.24	2806	0.088	0.0450
2	3.483	15.78	2067	0.117	0.0637
3	3.366	7.02	1117	0.209	0.1425
4	3.946	11.51	2171	0.126	0.0869
5	4.069	6.25	1150	0.246	0.1601

Figure 4. Pilot data summary and regression analysis.

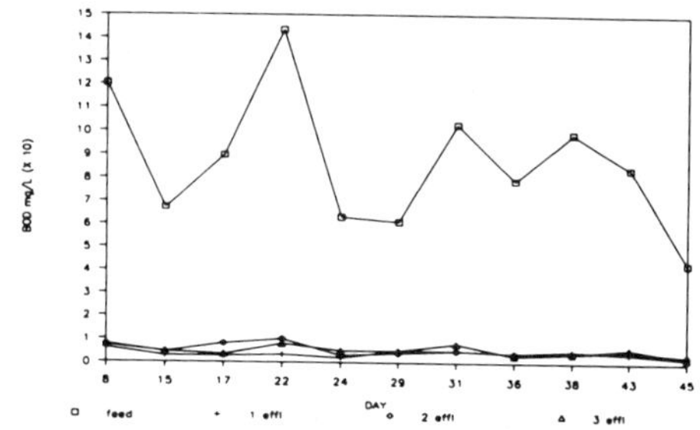

Figure 5. Pilot BOD performance data.

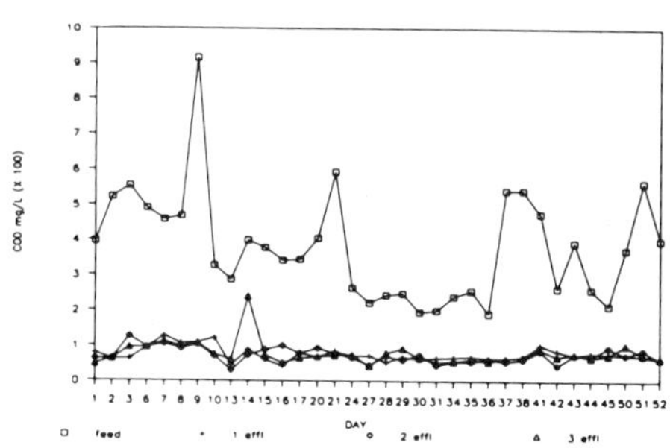

Figure 6. Pilot COD performance data.

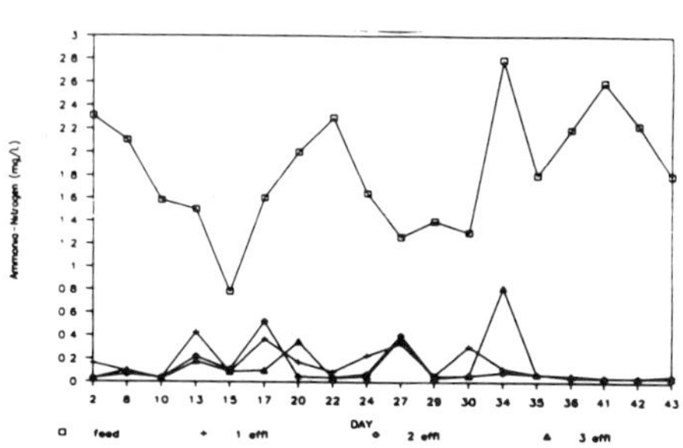

Figure 7. Pilot ammonia-nitrogen performance data.

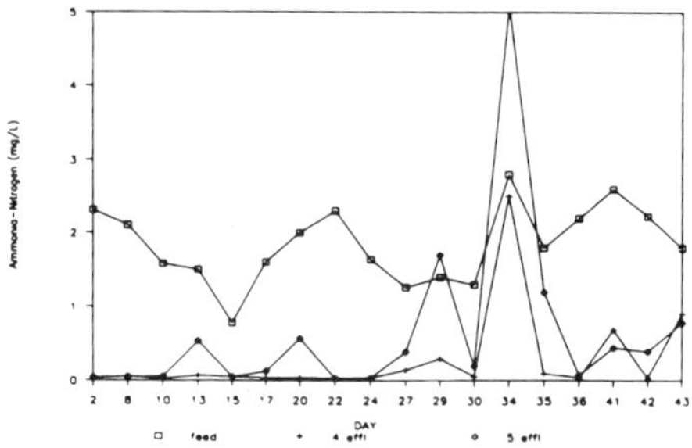

Figure 8. Pilot batch-dump reactors ammonia-nitrogen performance data.

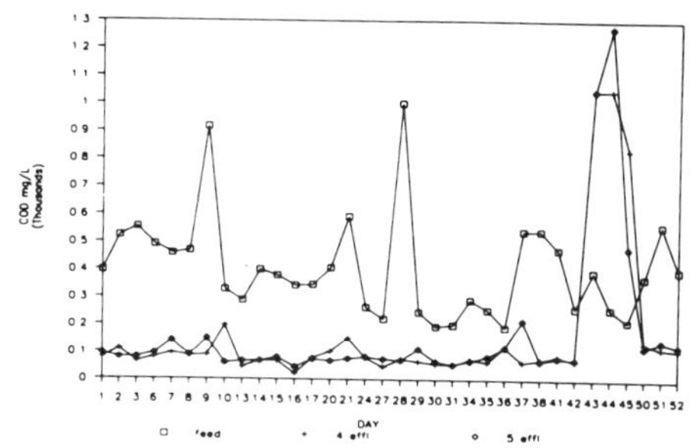

Figure 9. Pilot batch-dump COD performance data.

decrease in removal efficiencies for BOD_5, COD, and NH_4-N. There was no ammonia-nitrogen removal observed in the cold temperature reactor. This was expected given the limited temperature range in which the nitrifying microorganisms are able to propagate. Cold temperature removal efficiency for BOD_5 was 73.0%. This was an overall reduction in removal efficiency of 22.8% from the ambient temperature reactors. Cold temperature COD removal was 65.5%, which was a 17.8% reduction compared to the ambient units. This would strongly suggest that the design of the full-scale treatment system should include provisions for maintaining temperature in the biological reactor above 20°C in order for the refinery to meet their discharge ammonia-nitrogen criteria, and desired BOD_5 and COD effluent quality.

Oxygen uptake rate (OUR) analyses were conducted routinely on the activated sludge of each reactor to determine the relative oxygen demand. OUR's for the control reactors averaged 6.5, 5.7, and 5.5 mg/L O2/hr for reactors 1, 2, and 3 respectively. Calculation of full scale mixing horsepower, based upon oxygen uptake rates, determined that the system was mixing limited, not oxygen limited.

As a result of the low content of oxygen demand-inducing material in the wastewater, there was concern about solids bulking in the reactors. Bulking was not observed in either the control or the batch dump reactors throughout the pilot study. However, provisions were made in the full-scale design to address this concern.

Settleability of the sludge in the pilot system was monitored in the clarifiers daily, and in the aeration section of the reactor, twice weekly. In the aeration basin, the air supply was shut off and the solids/liquid interface subsidence over a five minute interval was noted. This observance was limited in its application to full-scale clarification design. The settling observed in the reactors accounted only for flocculent settling. Reactor depth was limited and, as such, the compression factor for hindered settling, which is the most crucial portion in a solids flux analysis, could not be evaluated. Full-scale clarifier design was undertaken using a conservative solids flux value of 10 $lbs/ft^2/d$.

A microbiologic exam was conducted on the activated sludge of the reactors to determine the predominant species of microorganism. The success of using a biological treatment system depends greatly upon the propagation of a microbial population capable of utilizing the waste constituents of the process water, and which are amenable to, and not adversely affected by, the subsequent treatment operations, i.e. settling, recycle, concentration. Using a microscope and slides, activated sludge from each of the reactors was observed and the various types and relative numbers of microorganisms were identified. The reactors exhibited normal floc (large agglomerations of bacteria) with little or no filamentous bacteria present. Additionally, substantial numbers of protozoa and multicellular animals, such as stalked and flagellated cilates and rotifers, were observed in the sludge. According to the literature,[11] this is indicative of a stable activated sludge system with long aeration times and low loading rates, as was the case for this pilot study.

Reactor 2 effluent was subjected to laboratory toxicity testing using *daphnia magna* and to a full priority pollutant scan to determine if the effluent exhibited any toxic effects either from conventional or nonconventional pollutants, and to evaluate if any of the 126 priority pollutants were present in the effluent[12]. The No Observed Effect Level (NOEL) from the daphnia 48-hour bioassay was 100 per cent for the reactor effluent. The considerable numbers of protozoans observed in the sludge supports the bioassay findings since protozoa are more susceptible to toxics than are bacteria. Likewise, the priority pollutant scan revealed no constituents above the detection limits for any of the priority pollutants.

WASTESTREAM SEGREGATION

It was believed that wastestream segregation and other modifications could also have a significant beneficial impact on the operation of any wastewater treatment system. The most obvious solution was isolation of stormwater. The facility sewer system was such that all stormwater, including that from the tank farms, was diverted through the wastewater treatment system. By isolating and controlling the stormwater, hydraulic overload could be greatly reduced. Therefore, it was recommended that a new outfall be permitted specifically for stormwater, and that all possible stormwater be diverted to it. It should be noted that sewers in the main process area consisted of open ditches; therefore, it was impossible to isolate stormwater that was generated in the process area.

Another modification that was explored was a change of outfall point. As noted previously, Outfall 001 discharged to a stream (i.e. a drainage ditch) with a Q7,10 of 0.0 cfs. This meant that the refinery effluent, which statistically can comprise the entire stream flow, had to meet very strict water quality standards. Additionally, more stringent water quality standards were envisioned, based upon proposed State regulations. However, a river with a Q7,10 of 38 cfs was only 0.8 miles away. Changing the discharge point of the process wastewater treatment system from the ditch to the river would provide a greater mixing zone. Thus, the effluent did not have to meet the water quality standard alone, but only after combination with the river flow. A computer analysis using USEPA-designed software was used to model the impact of the design effluent on the river. Relevant data from the United States Geological Survey (USGS) and the State regulatory agency were obtained and used in the model. Given the relative flows of the two streams and the data developed in the model, it became apparent that water quality limits would no longer be an issue, and that the effluent would be limited only by USEPA effluent guidelines.

A third evaluation conducted in the refinery was upstream capture of batch dumps or spills prior to their entering the wastewater treatment system. The four sources modelled as batch dumps in the pilot system were the most obvious to be evaluated. In particular, the desalter was suspected as a major contributor to upsets in the waste treatment system. Individual analysis of effluent from the desalter showed elevated levels of phenols, oil, BTEX, and other inhibitory and toxic compounds.

An oil/water/solids separator on the effluent from the desalter was evaluated to determine if it would mitigate the effects of the desalter on the treatment system. Based upon preliminary evidence, it appeared that the separator would reduce the loading to the treatment system, and therefore it would be advantageous to install one. Preliminary investigations on using oil/water separators on other wastestreams indicated them to not be economically justifiable.

It was noted that several non-process wastewaters were discharged to the process wastewater treatment system, including the rock filter backwash, softener backwash, and boiler blowdown. It was determined that all three of these streams could be diverted to either Outfall 002 or Outfall 003 without degradation of the quality of effluent from these outfalls. Simultaneously, the reduction in hydraulic loading to the process wastewater treatment system would reduce sizes of required units, increase concentration of biologically assimilable material, and generally result in a more stable and consistent wastestream to be treated.

Several other process modifications were recommended to improve treatability. Condensed steam blowoffs were diverted to the non-contact outfalls instead of to the process sewer. Laboratory wastes were collected and lab packed for off-site disposal, and numerous product samples taken to measure refinery process operations were collected and re-run, rather than being poured into the sewer.

DESIGN OF AND MODIFICATIONS TO THE WASTEWATER TREATMENT SYSTEM

Based upon the results of the site evaluation and the pilot plant, the conceptual design of the plant was undertaken[12]. It was determined that isolation of stormwater from the tank farms was feasible and advantageous. In order to control release of water from the tank farm dike areas, a pumping system and valve and piping network was installed in each diked area. Normally, all valves are closed and no flow is discharged. Rainwater collects in the diked area during a rainfall event and is held until it can be inspected by an operator. The operator will make a general determination whether there is possible contamination of the stormwater sufficient to cause the effluent to violate effluent standards. Should the stormwater exhibit gross contamination that renders it unsatisfactory for direct discharge, it will be pumped to the process wastewater treatment system. Because there may still be oil and solids in the water pumped from the diked areas, the wastewater will be diverted through an oil/water separator before discharge to the stream through new Outfall 004. Calculation of flow was made using standard rainfall and runoff curves to size the oil/water separator, and to determine the general rate of pumping required from the diked areas.

Non-process wastewater from boiler blowdown was diverted to Outfall 003. Analysis of the concentration of contaminants, flow, and temperature of these new streams determined that they would exhibit an insignificant impact on the overall discharge. The rock filter backwash and softener backwashes were diverted to Outfall 002. Previous analyses of the existing cooling water wastestream demonstrated that the backwash wastestream can be added without adverse effects, and that compliance with permit conditions can be maintained.

The process wastewater treatment system design is as follows. Design parameters are shown in Table II. A general process flow diagram is shown in Figure 10. Wastewater and stormwater from the unit will flow as before to a manhole upstream of the existing API separators. An anionic polymer is added to improve separation of oil, solids, and water in the API's. Mixing is achieved by using sparged air, which also improves the "floatability" of the lighter material and improves skimming efficiency. The wastewater then flows into the existing API separators. Sludge handling operations will be added to the South API to improve operation of the skimming and scraping system. Sludge from the API's will be dewatered in a recessed plate filter press prior to off-site disposal as a hazardous waste. Oil skimmings collected in the API's will be treated and reprocessed through the refinery. Overflow from the South API will then be pumped to a 125,000 gallon aerated equalization

Table II. Full-scale Design Parameters

Average Flow	0.125	MGD
F/M	0.117	(lb $COD_R d^{-1}$) lb MLSS
MLSS	2000	mg/L
COD_R	292	lb COD_R/day
k_d	0.016	day^{-1}
Effluent TSS	12	mg/L
Temperature	20	°C
Aeration Volume	0.90	Million Gallons
HRT	0.5	day
BOD_R	83	lb BOD_R/day
BOD Loading	0.004	(lb $BOD_R d^{-1}$)/1000ft^3
Solids Production	165	lb MLSS/day
SRT	15.8	day
Wastage	0.0039	MGD

tank, to provide storm surge protection to biological units downstream. From the equalization tank, the wastewater will be pumped to two activated sludge tanks. Capability for hydrogen peroxide addition prior to the wastewater entering the activated sludge tanks is provided to mitigate the effects of batch dumps. Under normal conditions, one tank would be used for contact stabilization in the activated sludge process, while the other tank would be used for sludge reaeration. Under storm conditions, both tanks could be used in parallel. Under spill conditions (such as high phenol or cyanide levels[13]), the sludge reaeration tank would be used to reduce impact and maintain a healthy biomass. The tanks will be side by side, separated by a "breech wall" that will equalize under storm conditions. The contact stabilization tank will be designed to have a normal hydraulic detention time of 12 hours, and MLSS of 2000 mg/L. The sludge reaeration tank will be the same physical size as the contact stabilization tank, but will be maintained at a MLSS level of 8,000–10,000 mg/L.

Effluent from the activated sludge system will flow to a clarifier. Sludge drawn from the bottom of the clarifier can be recycled to either the contact stabilization tank, the sludge reaeration tank, or an aerobic digester. Similarly, sludge can be wasted from either the contact stabilization tank or the sludge reaeration tank to the aerobic digester. Solids content of 1.5%–3% total solids will be maintained in this digester. Additionally, if the refinery is able to isolate a wastestream in tanks or process units that would cause harm to the activated sludge system (such as a high phenol or cyanide waste), this wastestream will be pumped directly to the aerobic digester for pretreatment. Decant from the digester is returned to the equalization tank at the head end of the treatment system. One preliminary design feature that merits further investigation involves the piping of all the blowers for each of the tanks requiring aeration (equalization basin, contact tank, reaeration tank, and digester) such that the offgas from the tanks can be diverted directly into the suction of the blowers rather than

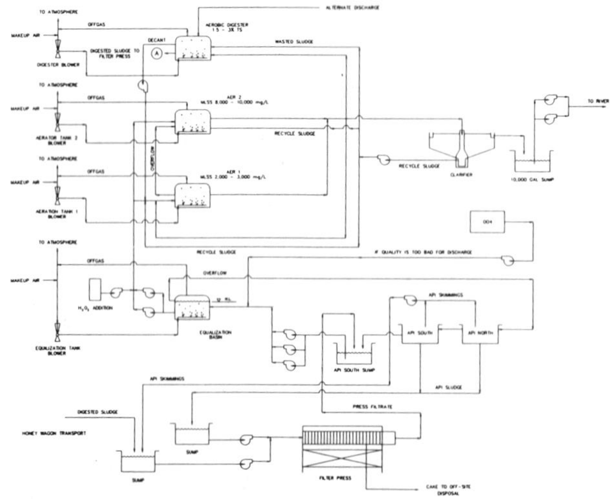

Figure 10. Proposed wastewater treatment system process flow diagram.

being released to the atmosphere. In this way, heat can be preserved in cold weather, or if air regulations require control of emissions, the offgas is collected and contained. Digested solids will be dewatered and disposed of, or land applied. Heat balance calculations confirmed that a closed top steel activated sludge tank will maintain a temperature sufficient under winter conditions that will prevent loss of nitrification activity, yet will not overheat under summer conditions.

The overflow from the clarifier goes to a 10,000 gallon sump, where it will be sampled prior to being pumped via a 0.8 mile pipeline to the river. An outfall structure meeting Corps of Engineers guidelines will be constructed at the river to ensure no erosion or other deleterious effects of discharge will occur.

CONCLUSION

The results of the pilot plant study demonstrated that the wastestream can successfully be treated using biological processes. Although biological kinetics and parameters such as organic loading and F/M were low compared to published literature values, the reactors exhibited excellent constituent removal efficiencies and biomass propagation under normal operating conditions. This is borne out by results of the toxicity testing, which showed a NOEL of 100%. Because of the relatively low F/M, the reactors were susceptible to batch dumps of process wastes. Although batch dumps significantly reduced the treatment effectiveness over a period of several days, the reactors recovered rapidly (2–3 HRT cycles), showing that the dumps were inhibitory to the biological process but not toxic to the biomass. Full-scale design addressed the potential effects of batch dumps by including an equalization tank upstream of the biological unit, provisions for peroxide addition, and sludge reaeration to provide unstressed biomass as makeup to the activated sludge. Concern for reduction in treatment efficiency during cold weather was addressed by enclosing the tanks to minimize heat losses. Modifications made upstream of the treatment system to reduce hydraulic loading should improve the treatability of the wastestream. Based upon the results of this study, it appears that a similarly configured full-scale system would enable the refinery to meet all permit and water quality standards as currently proposed.

REFERENCES

1. *Development Document for Effluent Limitation Guidelines New Source Performance Standards and Pretreatment Standards for the Petroleum Refining Point Source Category*, United States Environmental Protection Agency, National Technical Information Service, Washington, D.C. (October 1982).
2. Galil, Noah, and M. Rebhun. "Biotreatment Inhibition by Hazardous Compounds in an Integrated Oil Refinery", *Proceedings of the 42nd Purdue University Industrial Waste Conference*, pp. 163–173, Lewis Publishers, Chelsea, Michigan (1988).
3. LaGraff, M.B., D.R. Marrs, and P.W. McCallum. "Meeting BAT and Water Quality Standards at a Petroleum Refinery/Petrochemical Complex," *Proceedings of the 36th Purdue University Industrial Waste Conference*, pp. 320–328, Ann Arbor Science, Ann Arbor, Michigan (1982).
4. Meiners, Henry, and G. Mazewski. "Design, Start-up and Operation of a Refinery Treatment System with Muscle," *Proceedings of the 34th Purdue University Industrial Waste Conference*, pp. 710–718, Ann Arbor Science, Ann Arbor, Michigan (1980).
5. Mueller, James A., K.-B. Wu, and S. Kaczmarek. "Nitrification in Refinery Wastewater Treatment," *Proceedings of the 40th Purdue University Industrial Waste Conference*, pp. 507 522, Ann Arbor Science, Ann Arbor, Michigan (1985).
6. Hammer, Mark J., *Water and Wastewater Technology*, John Wiley & Sons, New York (1986).
7. Galil, Noah, and M. Rebhun. "A Comparative Study of RBC and Activated Sludge in Biotreatment of Wastewater From an Integrated Oil Refinery", *Proceedings of the 44th Purdue University Industrial Waste Conference,* Lewis Publishers, Chelsea, Michigan (1989).
8. Metcalf & Eddy, Inc. *Wastewater Engineering: Treatment, Disposal, Reuse*, McGraw-Hill Book Company, New York (1979).
9. Reitano, A. J. Jr. "Startup and Operation of a Refinery Activated Sludge Plant", *Proceedings of the 36th Purdue University Industrial Waste Conference*, pp. 310–319, Ann Arbor Science, Ann Arbor, Michigan (1982).
10. *Standard Methods for the Examination of Water and Wastewater,* 16th ed., American Public Health Association, New York (1985).
11. *Wastewater Treatment Plant Design,* WPCF Manual of Practice No. 8, Water Pollution Control Federation, Washington, D.C. (1977).
12. *Quality Criteria for Water, 1986 (the Gold Book)*, United States Environmental Protection Agency, National Technical Information Service, Washington, D.C. (1986).
13. Wong, Joseph M., and P. M. Maroney. "Cyanide Control in Petroleum Refineries" *Proceedings of the 44th Purdue University Industrial Waste Conference*, Lewis Publishers, Chelsea, Michigan (1989).

77 VOLATILE ORGANIC COMPOUND EMISSIONS FROM AIR FLOTATION BASINS IN REFINERY AND PETROCHEMICAL WASTEWATER TREATMENT

Dr. Saleh Al-Muzaini, Manager
Environmental Sciences Department
Kuwait Institute for Scientific Research
Safat, Kuwait

Dr. Fathy Hamoda, Professor
Civil Engineering Department
Kuwait University
Safat, Kuwait

Dr. Hosny Khordagui, Associate Research Scientist
Environmental Sciences Department
Kuwait Institute for Scientific Research
Safat, Kuwait

Dr. Eid Al-Khatib, Lecturer
Health Sciences Institution
The Public Authority for Applied Education and Training
Safat, Kuwait

INTRODUCTION

The Shuaiba Industrial Area (SIA) in Kuwait is one of the largest industrial complexes in the Arabian Gulf region. It generates some 23,000 m^3/d of industrial wastewater, which is discharged to the sea after partial treatment or with no treatment. Additional 3000 m^3/d of sanitary wastewater is also discharged without treatment. The flow for the year 2000 is projected to average 35000 m^3/d and to peak at 50,000 m^3/day. Presently, there are almost 12 major industries in SIA including petrochemical companies, three refineries, two power plants and an industrial gas corporation. Figure 1 presents the SIA with respect to Kuwait City and to other Gulf States.

The Shuaiba Area Authority (SAA) has taken steps to establish a central wastewater treatment facility to improve the wastewater to a quality adequate for landscape irrigation. The proposed treatment may include an oil removal system, such as dissolved air floatation, and an activated sludge system. Depending on the treatment used, the treated wastewater may be used for such purposes as: 1) irrigating food and nonfood crops; 2) creating recreational areas; 3) recycling for further use by industry; and 4) recharging brackish groundwater acquifers. Volatile organic compounds (VOCs) in industrial wastewater should be removed to avoid impairing the performance of the activated sludge system and to improve the water quality.

Wastewater samples were collected from existing source points that contain appreciable amount of purgeable VOCs. Each sample was analyzed for VOCs at the laboratories of the Kuwait Institute for Scientific Research (KISR). The identified VOCs were grouped according to chemical structure, and toxic properties.

The VOCs are defined as organic materials that are emitted into atmosphere either from using liquid fuels or from industrial wastewaters. VOCs discharged in wastewater effluents have toxic effects on the marine ecosystem and biological activities, and they are considered harmful to human beings[1,2,3]. Many of these compounds showed evidence of animal and human carcinogenicity and mutagenicity. VOCs are widely used in industrial applications such as in degreasing operations, cleaning and painting. They are also produced by industrial processes.

World protection agencies, such as the United States Environmental Protection Agency (USEPA), took steps to develop effluent standards for VOCs[4]. Others required industries to identify toxic VOCs;

45th Purdue Industrial Waste Conference Proceedings, © 1991 Lewis Publishers, Inc., Chelsea, Michigan 48118. Printed in U.S.A.

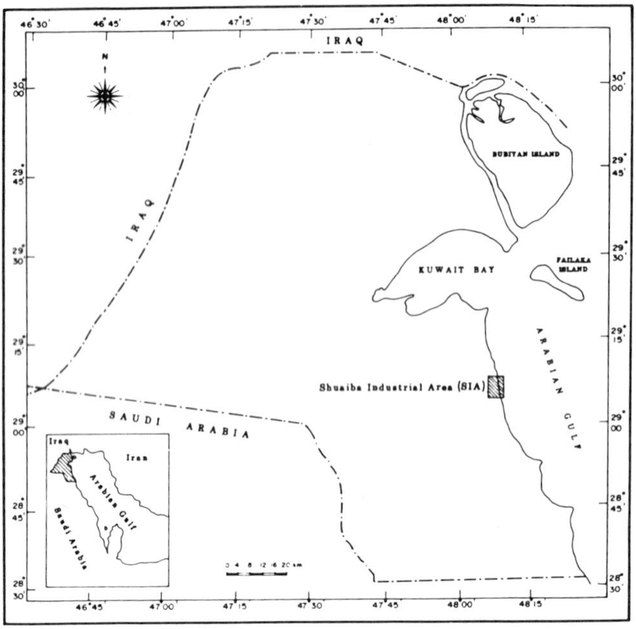

Figure 1. Location map of the Shuaiba industrial area (SIA).

therefore, steps were taken by world protection agencies to develop a control program for VOCs by establishing effluent protection agencies to develop a control program for VOCs by establishing effluent limitations which eventually required studying each VOC separately. Chapman et al[5] suggested that selection of individual priority pollutants for a monitoring program should be based on their anticipated fate and their effects on the environment. Keith and Teilliard[4] summarized the steps taken by the USEPA to establish effluent limitations for VOCs. Each compound was studied individually to determine a monitoring level in industrial discharges.

Al-Muzani[6] and Shunbo et al[7] investigated VOCs in the SIA. They concluded that benzene, chloroform, ethylbenzene, toluene, m-xylene, and o-xylene are found in refinery wastewater. They suggested screening industrial wastewater effluents to define which VOCs are present. Removal of VOCs from industrial wastewater was examined by Kincannon et al[8] who concluded that aromatic compounds were removed by a combination of stripping and biodegradation. Gaudy et al[9] suggested that joint mechanisms (air stripping and biological processes) are acceptable to separate VOCs. Removal by air stripping or biological methods were entirely dependent on process operation.

This study was initiated to gather information on VOCs in the industrial wastewater effluents and to recommend technical specifications that will help SAA to reduce VOC quantities during wastewater treatment. The information obtained from this study will be useful for SAA to design a monitoring system to protect workers in SIA.

EXPERIMENTAL PROCEDURE

Field sampling included the collection of liquid samples at the influent and effluent of each of the dissolved air flotation (DAF) and the induced air flotation (IAF) units at SIA were analyzed for VOCs. In addition, off-gas samples were collected at the top of each unit by aspirating air at 50 ml/min through sampling tubes containing activated charcoal which absorbed the vapors. Air sampling lasted for two hours at each location. The air pumps were calibrated before being used. Due to the nature of VOCs, composite sampling cannot be performed. Gas chromatography coupled with mass spectrometry (GC/MS) was used to identify the VOCs in both air and liquid samples.

The schematic diagram of the pilot plant is shown in Figure 2. This plant consisted of a holding tank, which is air-sealed and has a capacity of 4546 L. Two pumps were connected to the tank. The first pump was used to recirculate wastewater to the tank through a diffuser. The second pump was used to pump the holding tank effluent to the DAF unit. A 0.592 m^3 DAF basin used a pressurizing tank, flow meters and a rotating skimmer. The DAF basin and feed wastewater were kept at 24°C and the pH was not changed for all experiments.

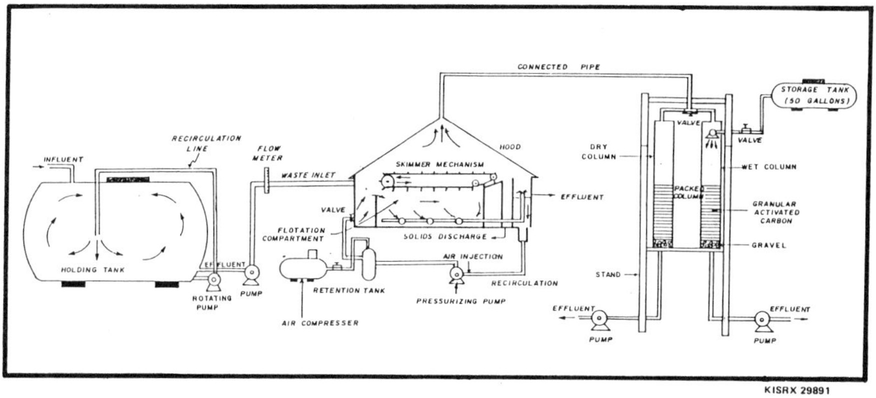

Figure 2. Schematic diagram of pilot plant.

Two packed columns of 99 cm diameter each were filled with granular activated carbon (GAC) to approximately 15 cm weight above a 10-cm deep gravel layer with an air-tight cover (Figure 3).

The hydraulic retention time in the DAF tank was designed to be about one hour. The DAF effluent flow rate was approximately 10 L/min and the recycling flow rate was about 7.5 L/min. The pressure level in the pressurization tank was kept at about 4.5 bar.

Sample preparation procedure and analyses were conducted according to the Federal Register.[10,11] All other analyses were performed according to *Standard Methods for the Examination of Water and Wastewater*[12].

Four series of experiments were performed using a continuous steady-state condition.

RESULTS

Field Sampling

Table I presents the VOC concentrations found in the influent wastewater and their removal efficiencies in both the DAF and the IAF field units, as well as the levels of organic vapors in the vicinity of these units. The results obtained indicate the common existence of major VOCs such as benzene, toluene, chlorobenzene, ethylbenzene and xylene in appreciable levels. Within the treatment units that apply

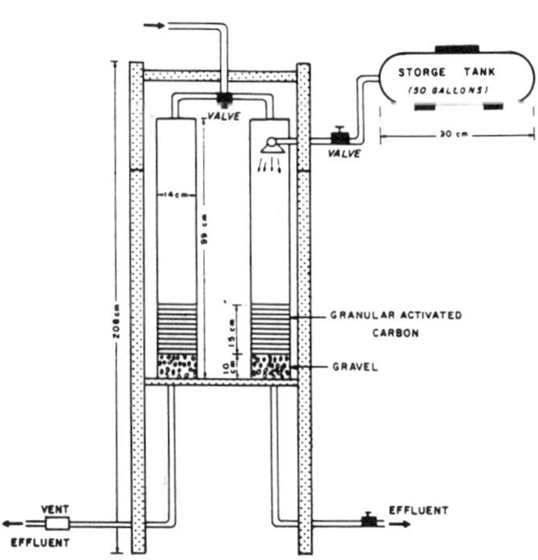

Figure 3. Details of activated-carbon columns.

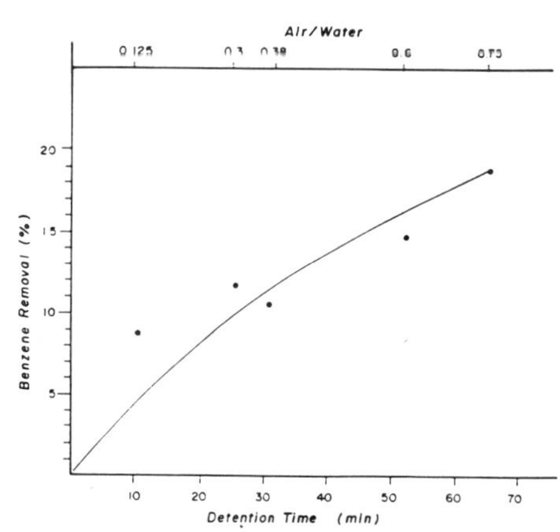

Figure 4. Effect of detention time and air-to-water ratio on the removal of benzene.

Table I. Analysis of Industrial Wastewater and Organic Vapors in the DAF and IAF Field Units[a]

System	Parameter	Compound				
		Benzene	Toluene	Chlorobenzene	Ethylbenzene	Xylene
Diffused Air Flotation Unit	Influent (μg/L)	2938	4445	60	1138	1153
	% Removal	17.8	28.0	ND[b]	24.1	19.5
	Air at Top of Unit (mg/m^3)	3.1	3.2	ND	0.4	0.4
Induced Air Flotation Unit	Influent (μg/L)	865	964	34	260	206
	% Removal	17.3	19.5	5.9	23.8	ND
	Air at Top of Unit (mg/m^3)	1.8	1.5	ND	0.1	ND

[a]Average of values obtained during four days of wastewater sampling and two days of air sampling.
[b]Not determined.

either diffused or induced air flotation for oil separation, the results show that these systems are not as efficient as it was originally thought for VOC emissions, although the DAF system showed comparatively higher VOC removal than the IAF system. The low VOC removal efficiencies observed could be attributed to the low detention time within flotation tanks, low air-to-water ratio, and/or accumulation of a spreading oil film on water surface that hinders gas exchange. Since volatilization of VOCs can be greatly affected by temperature this variable cannot be overlooked in the interpretation of results.

As far as air quality concern within the vicinity of the DAF and the IAF units, the reported levels are generally lower than the total weighted average (TWA) for eight hours occupational exposure specified by regulatory agency such as the US Occupational Safety and Health Administration (OSHA). However, in few occasions the levels of benzene, a known carcinogen, was exceeded. The levels of organic vapors within the vicinity of stripping units such as DAF and IAF should give reasons for concern but not alarm or panic.

Effects of Detention Time on Stripping of VOCs

A solution containing 50 μg/l of each of five VOCs (benzene, toluene, ethylbenzene, xylene and bromoform) was prepared by weighing a known amount of each pure compound, about 4 g per tank. The pure compounds were added to a methanol-water solution. This concentrated solution was then added to the tap water in the holding tank. The tank contents were then mixed thoroughly by a recirculation pump.

Liquid and air samples were collected every hour for 4 h per day and analyzed for a continuous steady-state condition. Liquid samples were taken from the influent and effluent of the pilot unit. Air samples were collected from the surrounding area (uncovered basin) as well as from above the DAF unit (covered basin). The contents of the holding tank and the DAF basin separately were mixed uniformly before sampling. Additional samples were collected from the wet and dry activated-carbon columns.

The liquid samples were collected in clear borosilicate glass vials of 40 ml capacity. Sampling was conducted with minimum disturbance to minimize losses. All samples were analyzed within 2 d of the sampling event. Figures 4–8 illustrates the effect of detention time and air-to-water ratio on the percentage removal of benzene, toluene, p-xylene, o-xylene and bromoform. The results indicate a significant increase in percentage VOC removal as a function of the air-to-water ratio. The rate of VOC removal was generally highest at a 0.5 air-to-water ratio and lower at higher ratios. The 0.5 air-to-water ratio, corresponding to 45 min detention time in the field, seems to be optimum for oil and grease removal. However, in the pilot studies, such retention time removed a maximum of 16% benzene, 14.5% toluene, 15% p-xylene and 14.5% o-xylene.

An air-to-water ratio of about 50 is needed to remove about 60% of the VOCs, whereas the applied ratio intended for oil and grease flotation is only 1/100 of the ratio used in stripping processes. An air-to-water ratio greater than 70 has been reported to remove 90% of similar VOCs. The pilot-unit data are in agreement with data collected from the field studies earlier in this investigation.

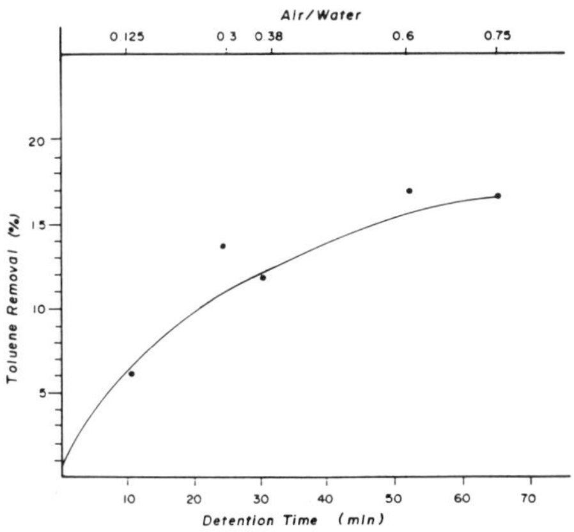

Figure 5. Effect of detention time and air-to-water ratio on the removal of toluene.

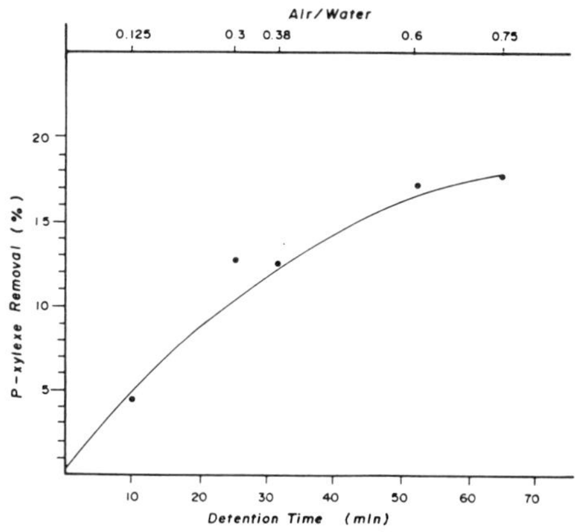

Figure 6. Effect of detention time and air-to-water ratio on the removal of p-xylene.

Effect of VOC Load on Stripping Efficiency

The purpose of this experiment was to determine the magnitude of the VOC load's effect on stripping efficiencies in the DAF pilot unit and its impact of the volatilization rates of VOCs.

Field operating conditions were simulated in the pilot unit. The effect of substrate concentration was studied by spiking the holding tank with model compounds (benzene, toluene, o-xylene, p-xylene and bromoform) at concentrations ranging from 0.25 to 6.25 mg/l.

The effect of VOC concentration on the volatilization rate (mg/l/h) is presented in Figures 9–12. The percentage removal (stripping efficiency) of the tested VOCs is illustrated in Table II for benzene, toluene, p-xylene, o-xylene and bromoform.

The trend observed was a decrease in the percentage VOC removal with increasing concentration (Table II). However, the volatilization rate increased for benzene, toluene, p-xylene and o-xylene. The observed increase in the volatilization rate was almost linear but leveled off at concentrations higher than 2.5 mg/l. The leveling off phenomenon can be attributed to the saturation of the constant volume of stripped air flowing through the water column.

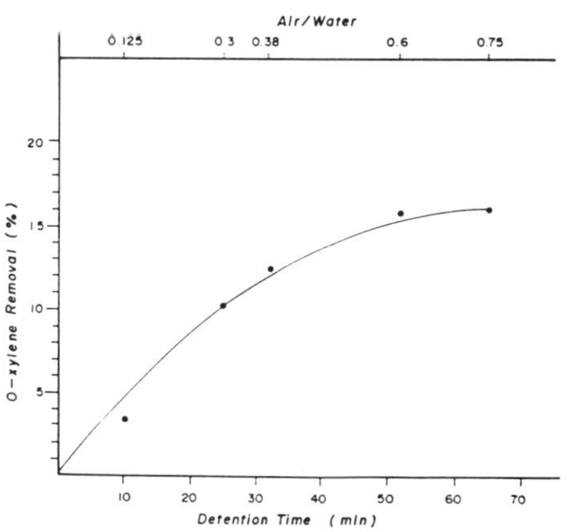

Figure 7. Effect of detention time and air-to-water ratio on the removal of o-xylene.

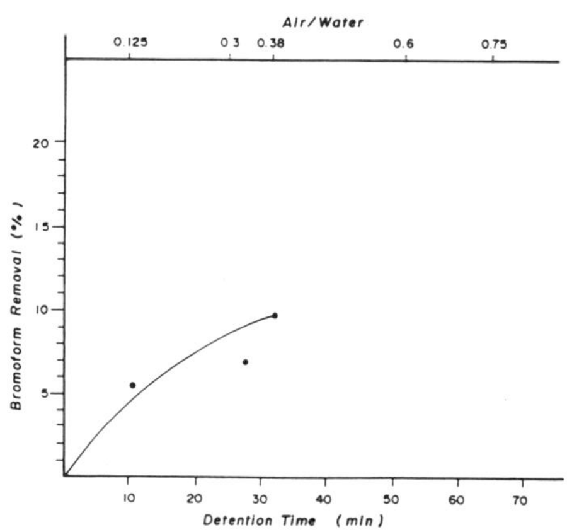

Figure 8. Effect of detention time and air-to-water ratio on the removal of bromoform.

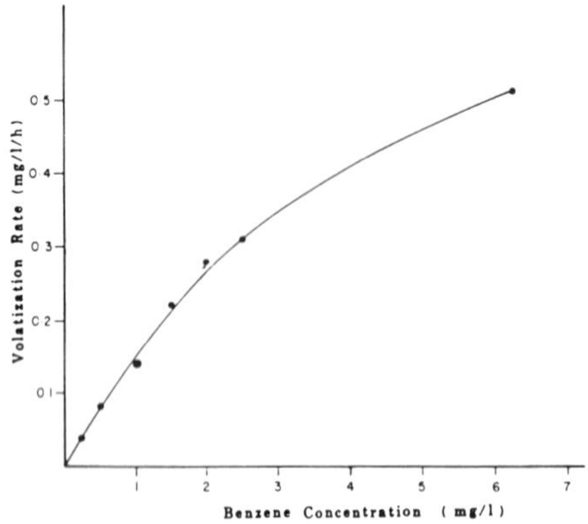

Figure 9. Effect of VOC concentration on the volatilization rate of benzene.

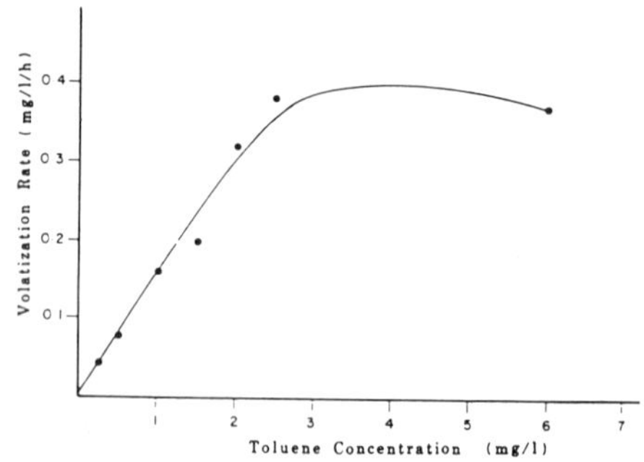

Figure 10. Effect of VOC concentration on the volatilization rate of toluene.

VOC Treatment and Removal Using Wet Activated-Carbon Column

The purpose of this investigation was to explore the possibility of using wet activated carbon for VOC control in the work place.

The emitted VOCs were determined, and a wet feeding wet system was spiked with VOCs in concentrations covering the detected range. The feed system consisted of a 200 L feed tank connected to the inlet of a packed granular activated-carbon column. The influent flow was gravitational and controlled by a valve at 1.6 L/min. The outlet for the column's effluent was controlled to provide a similar flow rate. Samples were collected at the column's inlet and outlet to determine the level of organics removed. The

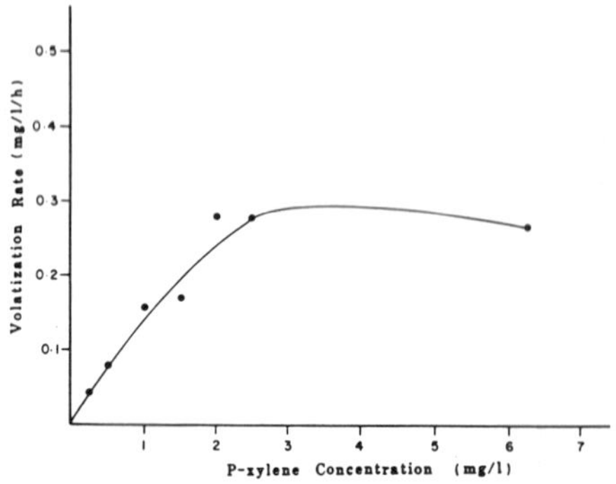

Figure 11. Effect of VOC concentration on the volatilization rate of p-xylene.

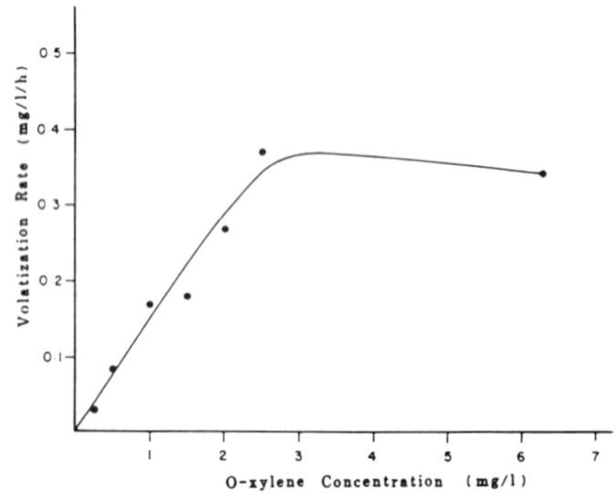

Figure 12. Effect of VOC concentration on the volatilization of o-xylene.

Table II. Effect of VOC Concentration on Stripping Efficiency of VOCs in Pilot Plant

	VOC Concentration (mg/L)						
Compound	0.25	0.50	1.0	1.5	2.0	2.5	6.25
Benzene	16.4	17.5	14.5	14.6	13.9	12.4	8.2
Toluene	17.5	16.6	16.5	13.6	16.1	15.4	5.8
O-xylene	13.8	17.5	17.2	11.9	13.7	15.1	5.5
P-xylene	18.5	13.7	15.9	7.9	14.3	11.3	4.2
Bromoform	—	—	13.1	—	5.3	—	4.0

— = No data.

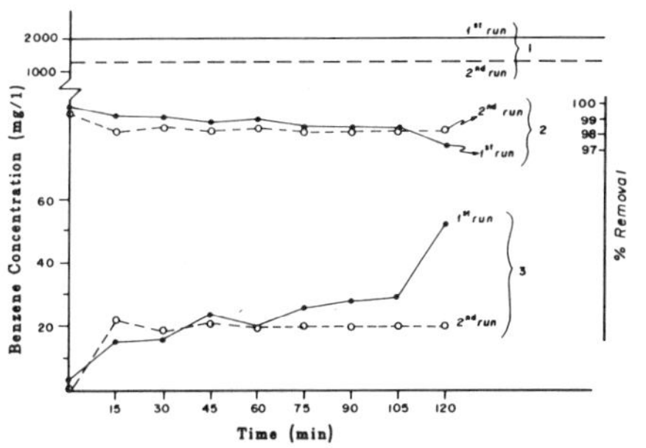

Figure 13. Removal of benzene by the wet granular activated-carbon process.

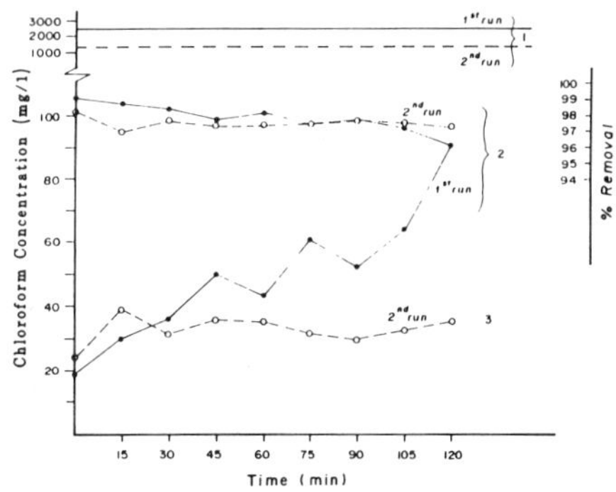

Figure 14. Removal of chloroform by the wet granular activated-carbon process.

model compounds used in the wet adsorption studies were benzene, toluene, chlorobenzene, bromoform, and o-xylene.

The percentage removals and the VOC levels in effluent for benzene, toluene, chlorobenzene, bromoform and o-xylene are shown in Figures 13–17. An adsorption isotherm could not be drawn because of the high activity of the newly packed carbon column. VOC breakthrough at the specified flow rate was minimal, and the adsorption process was efficient enough to remove VOCs at 96 to 99% for all compounds. Variation in VOC concentrations at the column inlet had an insignificant effect on the percentage removal.

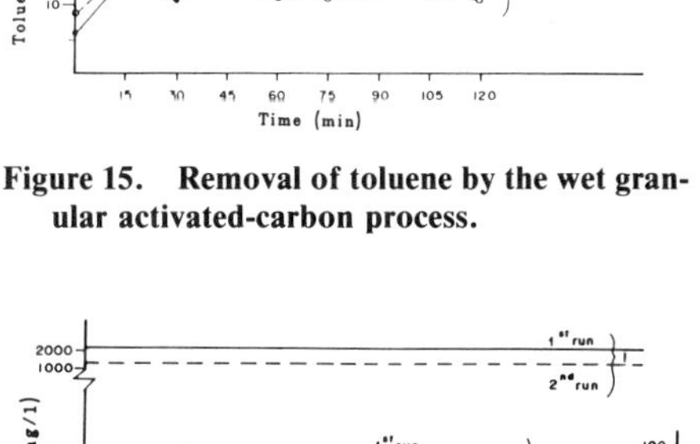

Figure 15. Removal of toluene by the wet granular activated-carbon process.

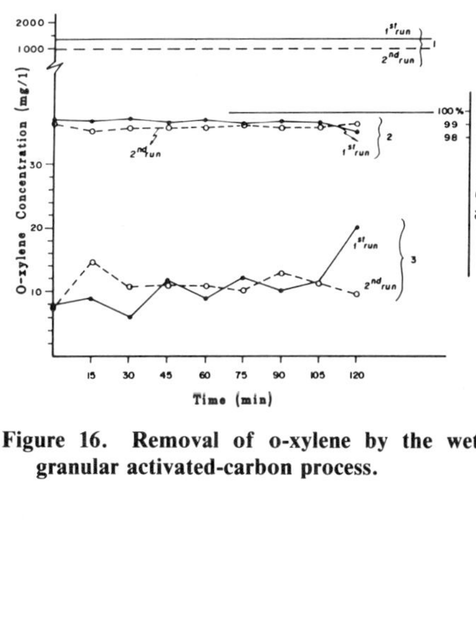

Figure 16. Removal of o-xylene by the wet granular activated-carbon process.

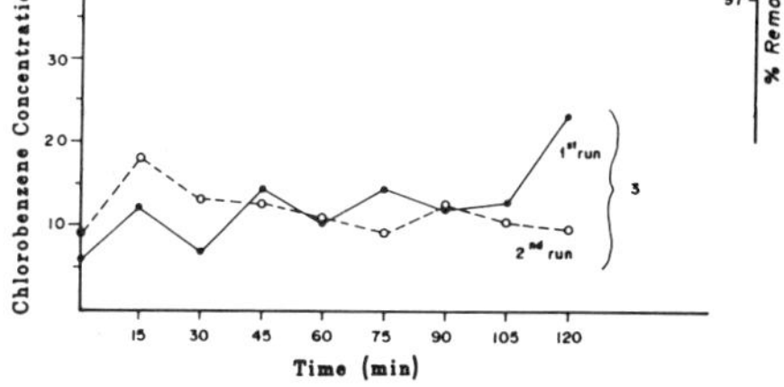

Figure 17. Removal of chlorobenzene by the wet granular activated-carbon process.

VOC Treatment and Removal Using Dry Activated-Carbon Column

This investigation was concerned with the removal of VOCs emitted from the DAF pilot unit by scrubbing the evolved gases through a dry granular activated-carbon column.

The system configuration is shown in Figure 3. The idea was to cover the DAF tank, connect it to the carbon column and then apply a negative pressure at the column outlet; the contaminated air would flow through the carbon column and then through the pores and voids of the carbon granules, where it would be deprived of its load of organics. Air samples were collected twice: the first from a confined room space using adsorption column bypass, and the second from the same confined room space and passing through the adsorption column. The difference in the concentrations of air pollutants per unit volume of the confined chamber was attributed to adsorption losses.

The percentage removal of VOC air contaminants using dry activated carbon ranged from 81.7% for benzene to 99% for the other compounds of interest. Compared with wet adsorption systems, the dry activated-carbon air scrubber is a competitive and effective system for the removal of VOCs emitted from petrochemical industries.

CONCLUSIONS

The percentage VOC removal was a function of the air-to-water ratio applied to the DAF pilot unit.

An air-to-water ratio greater than 50 has capability of removing 60% of VOCs.

The rate of VOC removal in mg/l/h was generally higher at 0.5 air-to-water ratio and lower at higher ratios.

The adsorption process in the wet activated carbon process was efficient enough to remove VOCs at 96% to 99% for all studied compounds.

The dry activated carbon was more effective for removing VOCs than the wet activated carbon process.

ACKNOWLEDGEMENT

The authors acknowledge the financial support for this work of both Kuwait Institute for Scientific Research and Shuaiba Area Authority. Thanks are extended to Kuwait Foundation for Advancement of Sciences for making it possible for the author to present this paper.

REFERENCES

1. Lucas, A. D., Health hazard evaluation report, Metropolitan Sewer District, National Institute for Occupational Safety and Health, Health Environmental Testing Association (HETA), *51-240-1031*, Cincinnati, Ohio (1981).
2. McGlothlin, J. D., and J. E. Cone, Health hazard evaluation report, Metropolitan Sewer District, National Institute for Occupational Safety and Health, Health Environmental Testing Association (HETA), *81-207-945*, Cincinnati, Ohio (1981).
3. Luker, P. A., and J. Lee, "Worker exposure to chlorinated organic compounds from the activated-sludge wastewater treatment process," *Am. Ind. Hyg. Assoc.*, 44: 109 (1983).
4. Keith, L., and W. Teilliard, "Priority pollutants I-A perspective view," *Environmental Science and Technology*, 13(4): 416–423 (1979).
5. Chapman, P., G. Romberg and A. Gary, "Design of monitoring studies for priority pollutants," *Journal of WPCF*, 54(1):292–297 (1982).
6. Al-Muzaini, S., "Ultimate treatment of SIA wastewaters for reuse: Treatment of refinery wastewaters using various modified activated sludge processes," *Ph.D. Dissertation*, School of Civil Engineering, University of Rhode Island, Rhode Island (1987).
7. Shunbo, F., P. Literathy, F. Abdally, and M. Al-Ali, "Study of micropollutants in industrial effluent discharges in the Shuaiba Industrial Area," Kuwait Institute for Scientific Research, *Report No. KISR1127A*, Kuwait (1983).
8. Kincannon, D. F. and A. Fazel, "Volatilization of Organics in Activated Sludge Reactors," Proc. 41st Purdue Industrial Waste Conf., Purdue University, Lewis Publishers, Chelsea, MI (1984).
9. Gaudy, A., B. Turner and S. Pusztaszer, "Biological treatment of volatile waste components," *Journal WPCF*, 35(1): 75–93 (1963).
10. Federal Register, "Method 602-purgeable halocarbons," *Federal Register*, 49(209):29–39 (1979).
11. Federal Register, "Method 602-purgeable halocarbons," *Federal Register* 49(209):40–48 (1979).
12. APHA, *Standard methods for the examination of water and wastewater*, American Public Health Association, Washington, D.C., U.S.A. (1986).

Section Nine
INDUSTRIAL WASTES – B. BREWERY AND DAIRY WASTES

78 TREATMENT OF BREWERY WASTEWATER USING ANAEROBIC FLUIDIZED BED REACTOR IN SINGLE-PHASE AND TWO-PHASE PROCESS

G. K. Anderson, Senior Lecturer
T. F. Chan, Croucher Foundation Scholar
Department of Civil Engineering
University of Newcastle
Newcastle-upon-Tyne, United Kingdom NE2 7RU

INTRODUCTION

With the present worldwide outcry to move towards a 'greener' future and the continuous tightening of legislation in relation to industrial waste disposal, anaerobic treatment of industrial wastewater addresses the environmental issue in two aspects. Firstly, anaerobic treatment reduces the amount of organic matter being discharged into the natural environment. Secondly, methane, one of the 'greenhouse' gases can be produced under controlled conditions and can be utilized to provide valuable energy. Therefore, there has been an ongoing effort to improve anaerobic treatment technology in order to achieve higher organic removal efficiency to minimize both capital and operating costs. In addition, there is an incentive to enhance the methane content in the off-gas produced thus lowering the subsequent cost in energy recovery. Moreover, and most important, the process has to demonstrate its stability and reliability during long term operation.

Since the first major study of an anaerobic filter,[1] the fixed-film concept was further developed, one of the results being the introduction of an anaerobic fluidized bed reactor (AFBR) in the early eighties.[2] In the AFBR, the support medium for the growth of microorganisms is fluidized by the up-flowing wastewater, which thus provides a large surface area for biomass attachment. This enables a high biomass concentration to be achieved thus reducing reactor size and the cost of reactor construction.[3] However, the AFBR has the potential disadvantage of a higher energy requirement due to the high recycling rate necessary to maintain the media in a fluidized state. The second inherent problem, as experienced by many other processes, is the need for uniform flow distribution at the inlet. Over the past ten years, the AFBR has been used to treat a wide range of industrial wastewaters successfully at laboratory to full-scale.[4,5,6]

A more fundamental approach towards the improvement of anaerobic treatment technology was that of phase separation, first proposed by Pohland and Ghosh.[7] The objective of phase separation is to separate the acid-forming microorganisms and the methane-forming microorganisms into two reactors in series. Since the optimum growth conditions for these two groups of microorganisms are different, consequently the environmental conditions in each reactor may be controlled accordingly. The yield coefficient data indicate that the total reactor volume can be reduced when phase separation is practised.[5] Evidence has shown that improvement in treatment efficiency and process stability under shock load can be attained by the two-phase process.[8,9,10] Notable results have been obtained in full-scale applications of the two-phase process.[11,12,13,14]

The objectives of this investigation were to study the treatment of wastewater from a brewery using a pilot-scale AFBR. The AFBR was first operated as a single-phase process (SPAFBR) treating the wastewater directly under various superficial up-flow velocities (SUV). In the next stage, the reactor was operated as a methane (second) phase process (MPAFBR) treating post-acidified brewery wastewater.

45th Purdue Industrial Waste Conference Proceedings, © 1991 Lewis Publishers, Inc., Chelsea, Michigan 48118. Printed in U.S.A.

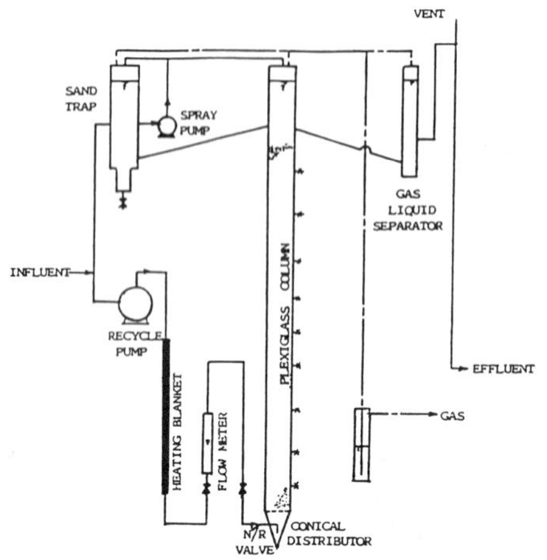

Figure 1. Schematic diagram of the AFBR.

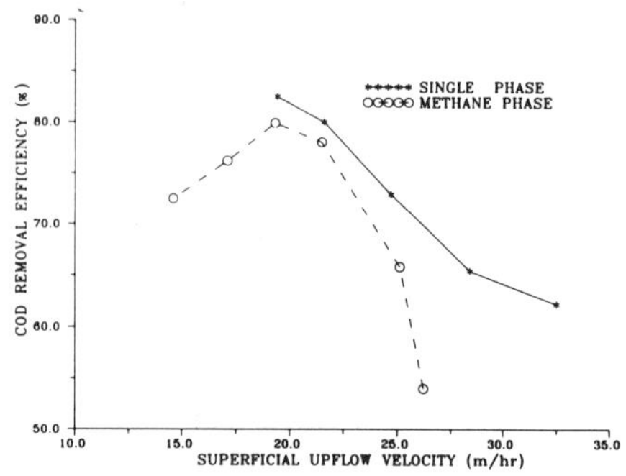

Figure 2. Effect of superficial up-flow velocity on COD removal efficiency by AFBR.

EQUIPMENT, MATERIALS AND METHODS

The 60 liter pilot-scale AFBR (Figure 1) in this study consisted of a Plexiglass column, 3 meters in height with an internal diameter of 0.16 metre. Quartz sand obtained by sieving and collecting the fraction which passed a 600 micron sieve but retained by a 425 micron sieve was used as the reactor medium. Approximately 30 kg of sand was employed to fill the reactor column to an unexpanded height of 1.1 meter.

Fluidization of the media bed was achieved by effluent recycle. The SUV was controlled by a variable speed Monopump and monitored by an in-line flow meter. The temperature of the reactor was maintained at 35°C by means of a heating blanket in the recycle line.

The feed to the SPAFBR was prepared by diluting spoilt beer (COD 85,000 to 95,000 mg/L) collected from the local brewery with tap water, in an attempt to produce wastewater resembling that discharged from a brewery. Ammonium bicarbonate (NH_4HCO_3) and potassium dihydrogen orthophosphate (KH_2PO_4) were added as nutrient. When the AFBR was operated as a single-phase process, feed was pumped directly into the recycle line from the feed tank. Under the methane-phase regime, the same feed was first acidified by a 90 litre contact process which was operated at 35°C and a pH of 5.25 before being pumped into the AFBR.

Single-Phase Operation

Start-up of the SPAFBR was achieved by seeding the reactor with 2 liters of anaerobic filter sludge which had been treating brewery waste and batch feeding for the first 6 days with concentrated spoilt beer. The SUV was 24 m/hr which resulted in an initial bed expansion of 25%. Continuous feeding began on the seventh day with a COD concentration of 1,000 mg/L at a hydraulic retention time (HRT) of 23 hours. The volumetric organic loading rate (OLR) was increased stepwise by increasing the COD concentration of the feed. Within 50 days, the SPAFBR had achieved a 94% COD removal efficiency at an OLR of 4.6 kg/m^3/day. The OLR was further increased to 16.6 kg/m^3/day by decreasing the HRT to 7.3 hour after 200 days and a COD removal efficiency of 87% was maintained. However the process eventually became unstable which resulted in a decrease in the treatment efficiency. The OLR was reduced to about 10 kg/m^3/day to regain treatment stability. The investigation into the effect of SUV took place after the SPAFBR had returned to stable operation.

During this study the COD of the feed and the OLR were kept at a very stable level (Table I). The SUV was increased to a maximum of 32.5 m/hr in run S1 and decreased stepwise until clogging occurred at the 0.85 to 0.95 m level. Effluent and gas samples were collected for COD and methane content analysis. Biomass samples were collected at the end of each run from sampling ports along the height of the AFBR column and the volatile solid (VS) content measured.

Table I. Loading Characteristics of Single-Phase AFBR

Run	SUV (m/hr)	OLR ($kgCOD/m^3/day$)	HRT (hr)	Influent COD (mg/L)
S1	32.5	9.46	9.7	3840
S2	28.4	9.10	10.1	3840
S3	24.7	9.38	9.7	3800
S4	21.6	9.81	9.8	4000
S5	19.4	9.74	10.1	4080

Methane-Phase Operation

The contact process was seeded with 500 mL of the sludge referred to and fed with concentrated spoilt beer without pH control for one week before continuous feeding started. The acidified effluent from the contact process was in turn pumped into the AFBR at an initial OLR of 3.6 $kg/m^3/day$. However, due to numerous mechanical failures during the first two months of operation, the effect of SUV was first investigated before other parameters were considered. In this case, the SUV was decreased from 26.2 m/hr to 14.6 m/hr without any sign of clogging. On the other hand, the treatment efficiency was observed to be deteriorating continuously when the SUV was below 17.1 m/hr so that the investigation was discontinued.

The post-acidified brewery wastewater characteristics are shown in Table II. The variation in COD concentration and degree of acidification made the task of maintaining a constant OLR to the MPAFBR very difficult.

RESULTS AND DISCUSSION

Process Efficiency

As illustrated in Figure 2, the SPAFBR attained a 62% COD removal efficiency at less than the maximum SUV in run S1. The removal efficiency increased to 82% at the minimum SUV of 19.4 m/hr during which partial clogging of the SPAFBR took place at the 0.85 to 0.95 meter level and channelling of flow was observed through the Plexiglass column. The improvement in COD removal was particularly significant between run S2 and S4. The MPAFBR under a maximum SUV of 26.2 m/hr could attain only 54% COD reduction despite the fact that the SUV was considerably lower than that in run S1. When the SUV was decreased to 21.5 m/hr, the removal efficiency increased rapidly to 78%. The highest removal efficiency of 80% was reached when the SUV was decreased further to 19.3 m/hr. However, the efficiency began to decrease as the SUV was decreased further in runs M5 and M6. In this case, no sign of clogging could be observed as was the case in the SPAFBR.

During single-phase operation, the methane content of the off-gas was consistently around the 70% level as long as the SUV was kept below 28 m/hr (Figure 3). An increase in the SUV resulted in a reduction of the methane content to 65%. A similar decrease in methane content took place when the SUV of the MPAFBR was increased to greater than 22 m/hr. However, the off-gas still contained over 80% methane throughout the methane-phase operation. Off-gas with over 87% methane content was collected if the MPAFBR was operated at a SUV of below 21 m/hr. This exceptionally high level of methane is extremely beneficial in subsequent energy reclamation.

When the SUV was decreased from its maximum to the minimum$_s$ the amount of organic matter utilized by the SPAFBR increased by 130 gCOD/day (36.9%). In the case of the MPAFBR, an additional 85 gCOD/day (58% increase) was removed when the SUV was decreased from 26.2 to 14.6 m/hr. However, Figure 4 shows that there was little improvement in terms of organic matter utiliza-

Table II. Loading Characteristics of Methane-Phase AFBR

Run	SUV (m/hr)	OLR ($kgCOD/m^3/day$)	HRT (hr)	Influent COD (mg/L)	Acidification %
M1	26.2	4.51	12.1	2270	94
M2	25.1	4.15	12.6	2177	94
M3	21.6	4.86	12.1	2447	94
M4	19.3	4.51	12.0	2248	100
M5	17.1	5.10	12.1	2558	100
M6	14.6	5.30	11.2	2464	90

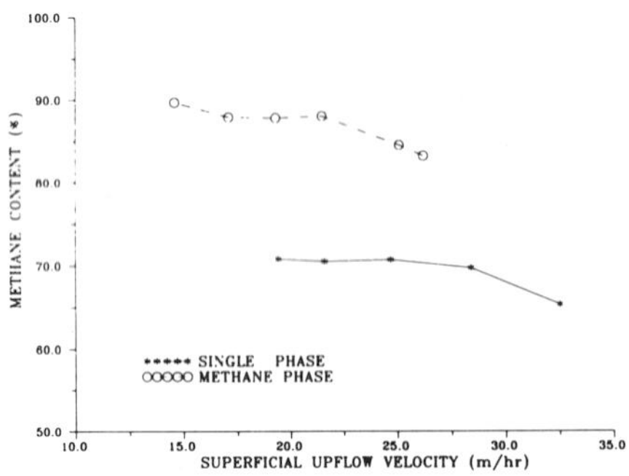

Figure 3. Effect of superficial upflow velocity on methane content of off-gas in AFBR.

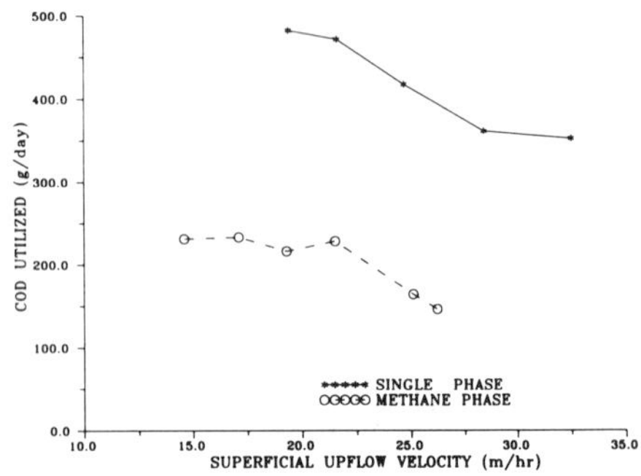

Figure 4. Effect of superficial upflow velocity on utilization of organic matter by AFBR.

tion below a SUV of 20 m/hr despite an increase in biomass hold-up within both the SPAFBR and the MPAFBR.

Biomass Hold-up

In the SPAFBR, the biomass accumulated within the reactor increased by about 250% when the SUV was decreased from run S1 to S5 (Figure 5). Similarly, there was about a 230% increase in biomass hold-up in MPAFBR between runs M1 and M6. The maximum quantity of biomass which accumulated in the MPAFBR was only 164 gVS compared with 1,190 gVS (725%) in the SPAFBR. This can be partly explained by the fact that the OLR applied to the MPAFBR was about half of that applied to the SPAFBR. Secondly, it is known that the biomass in a single-phase anaerobic process consists of all groups of microorganisms, namely the acid formers and the methane formers, converting organic matter to methane. However, in a methane phase process, such as the MPAFBR which was treating over 90% acidified wastewater, there is little need for the presence of acid-forming microorganisms. This observation is in agreement with the yield coefficient data presented by Henze

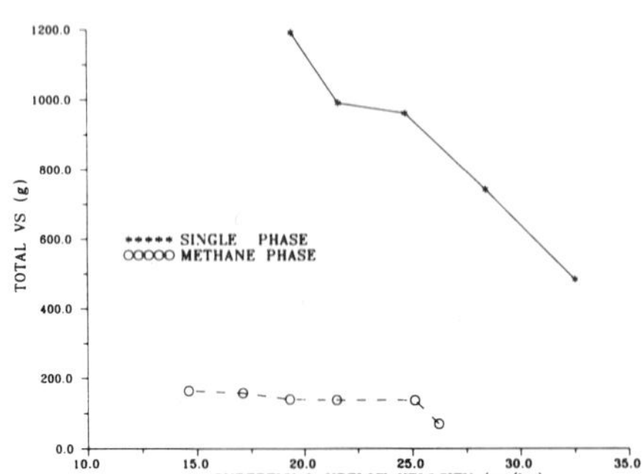

Figure 5. Effect of superficial upflow velocity on biomass attachment in AFBR.

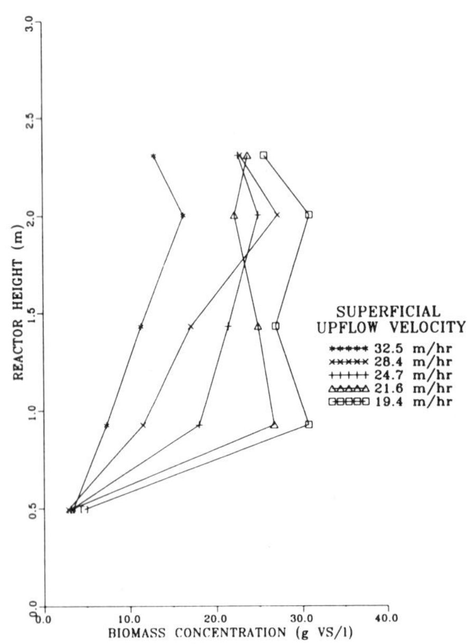

Figure 6. Effect of superficial upflow velocity on biomass profile in single-phase AFBR.

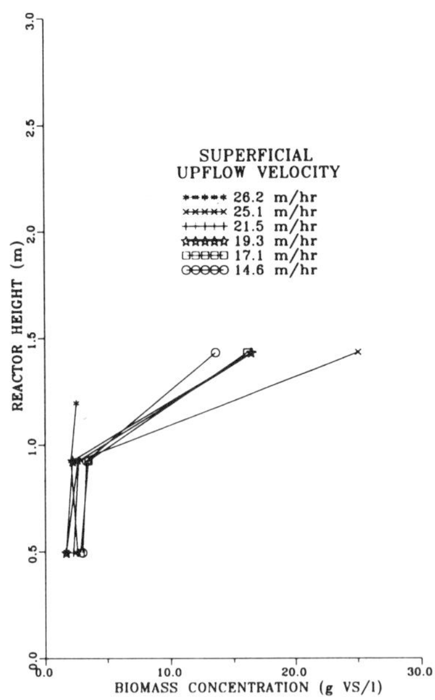

Figure 7. Effect of superficial upflow velocity on biomass profile in methane-phase AFBR.

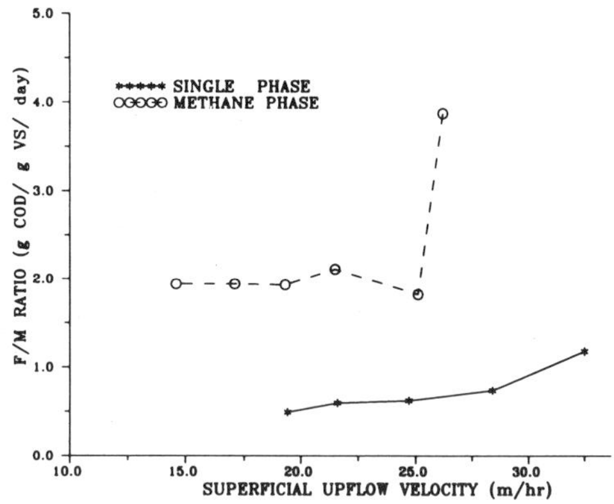

Figure 8. Effect of superficial upflow velocity on F/M ratio of AFBR.

and Harremöes,[5] this confirms the potential advantage of reactor size reduction in the design of a methane-phase process.

The biomass concentration profiles for the AFBR during both the single-and methane-phase operation are shown in Figures 6 and 7. In both operations, the biomass concentration in the AFBR varied along the reactor height. In the SPAFBR, it was never greater than 5 $kgVS/m^3$ below the 0.5 m level. Above the 0.5 m level, it increased throughout the whole reactor at all SUV and reached a maximum at the 2.0 m level except in run S4. The maximum biomass concentration measured was 30.8 $kgVS/m^3$ at the end of run S5, when clogging of the bed developed. Above the 2.0 m level, the biomass concentration began to decrease under all conditions.

The effect of SUV on the MPAFBR was not so apparent. This may be due to the lower OLR being applied and the lower biomass concentration expected in a methane process as discussed above. According to Figure 7, a maximum biomass concentration of 25 $kgVS/m^3$ was attained in run M2 but decreased to a constant level of between 12 and 15 $kgVS/m^3$. It is clear that little biomass (less than 4 $kgVS/m^3$) developed in the lower part of the MPAFBR below 0.92 m level.

The low biomass concentration in the lower part of the AFBR can be explained by the much higher shear forces the media particles experienced. First, there was an excessive amount of turbulent drag on the media caused by uneven distribution at the inlet of the reactor. This turbulent stream was observed visually through the Plexiglass column. The top of this turbulent zone appeared to be at the 1.2 m level. Secondly, there was the shear created when particles collided with each another, which was also induced by the strong turbulence.

The much lower biomass concentration between the 0.5 to 0.9 metre level in the MPAFBR indicated that the biofilm in this case was more susceptible to shearing. This observation is consistent with the suggestion that polysaccharides, which bind the microorganism to the biofilm, are mainly excreted by the acid forming bacteria.[15] Therefore, in the MPAFBR, where much fewer acidogens were present, the same degree of shearing would cause much greater damage on the biofilm, hence reducing the biomass concentration. On the other hand, this may have been the same reason which prevented clogging from occurring in the MPAFBR even at a much lower SUV.

The decrease in biomass concentration at the top of the SPAFBR was due not only to the maximum up-flow fluid flux (liquid and gas) present, but also to the much thicker biofilm[16] which decreased the bulk density of the bioparticles, in turn resulting in a much higher relative localized bed expansion. The same phenomenon took place in the MPAFBR since the bioparticles at the top of the fluidized bed were distinctly larger than those at the lower part of the bed when examined visually. However, direct measurement could not be made due to the absence of sampling points in the appropriate positions.

Food-Microorganism Ratio

By maintaining a stable OLR, the F/M ratio of the SPAFBR approached a constant value of 0.6 gCOD/gVS/day before clogging of the fluidized bed occured (Figure 8). This ratio is low compared with the proposed design value of 1 gCOD/gVS/day.[5,17] The decrease in F/M ratio in run S5 could be due to the development of an active biomass to replace the non-viable biomass trapped in the clogged section of the bed. However, the effect of a lower SUV was not investigated since further clogging of the bed would mean total channelling of flow and the fluidized regime would no longer exist. The MPAFBR showed a higher and very constant F/M ratio of 1.9 compared to the proposed design value of 2 gCOD/gVS/day despite the continued lowering of SUV. Therefore, it is apparent that the biomass within the AFBR tends to maintain an optimal F/M ratio between the extremes in SUV. The three-fold increase in F/M ratio for a MPAFBR was once again due to the different composition of microorganisms present in the biomass as discussed earlier.

Figure 9 indicates that it is more advantageous and economical in terms of COD removal efficiency to operate an SPAFBR at as low a SUV as possible, even at the expense of clogging part of the bed. On the other hand, there appears to be an optimum SUV of 20 m/hr for operating a MPAFBR. Operating the MPAFBR at a SUV higher or lower than this optimal results in deterioration in treatment efficiency. However, it is comparatively more efficient to operate a MPAFBR at SUV lower than the optimum SUV. Not only has the energy input to be considered, but also the process efficiency and stability, which is much more sensitive to changes in SUV especially at higher SUVs.

CONCLUSIONS

Variation in SUV affected the process efficiency of both the single-phase AFBR and the methane-phase AFBR, while the MPAFBR was more sensitive to changes in SUV.

The SPAFBR performed best at the lowest SUV (19.4 m/hr) when clogging of the fluidized bed had just began. The MPAFBR attained the highest efficiency when operated at a SUV of 20 m/hr.

Despite the sensitivity of the MPAFBR to SUV, it was able to achieve a removal efficiency comparable to the SPAFBR while providing off-gas with a methane content of 87%.

The higher value of F/M ratio for the MPAFBR implies a reduction in reactor volume compared to that of the SPAFBR in order to operate at the same OLR. This will indicate a reduction in construction and operating costs of the AFBR.

When designing an AFBR, it is necessary to account for the volume of the turbulent zone which is inefficient at removing organic matter due to the low biomass concentration.

ACKNOWLEDGMENT

T.F. Chan would like to express his gratitude to the Croucher Foundation, Hong Kong for supporting part of this research.

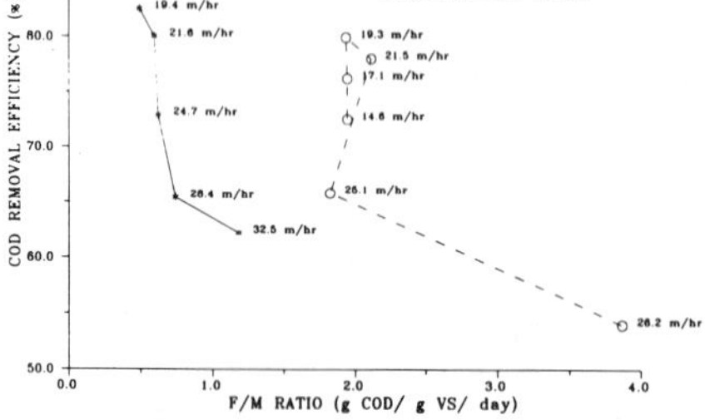

Figure 9. Effect of F/M ratio on COD removal efficiency.

REFERENCES

1. Young, J.C. and P.L. McCarty (1969) The anaerobic filter for waste treatment, *J. Water Pollut. Contr. Fed.* , 41, 5, R160.
2. Jewell, W.J., M.S. Switzenbaum and J.W. Morris (1981) Municipal wastewater treatment with the anaerobic attached microbial film expanded bed process, *J. Water Pollut. Contr. Fed.*, 53, 4, pp 482–490.
3. Cooper, P.F. (1981) The use of biological fluidised bed for the treatment of domestic and industrial wastewaters, *The Chemical Engineer*, Aug/Sept, pp 373–376.
4. Anderson, G.K., I. Ozturk and C.B. Saw (1989) Pilot scale experiences on anaerobic fluidized bed treatment of brewery wastes, Proc. Int. Symp. on Waste Management Problems in Agro-industries , Istanbul, Turkey, pp 201–211.
5. Henze, M. and P. Harremöes (1983) Anaerobic treatment of wastewater in fixed film reactors – a literature review, *Wat. Sci. Technol.* , 15, pp. 1–101.
6. Sutton, P.M. (1987) Suspended and supported growth anaerobic reactors: pilot to full-scale results, Aquatech '86, Anaerobic Treatment , pp 259–268.
7. Pohland, F.G. and S. Ghosh (1971) Developments in anaerobic treatment process, *Biotechnol. Bioeng Symp.* , No.2, pp. 85–106.
8. Cohen, A., A.M. Breure, J.G. van Andel and A. van Deursen (1980), Influence of phase separation on the anaerobic digestion of glucose – I. Maximum COD-turnover rate during continuous operation, *Water Res.*, 14, pp 1439–1448.
9. Cohen, A., A.M. Breure, J.G. van Andel and A. van Deursen (1982), Influence of phase separation on the anaerobic digestion of glucose – II. Stability and kinetic responses to shock loadings, *Water Res.*, 14, pp 1439–1448.
10. Bull, M.A., R.M. Sterritt and J.N. Lester (1984) An evaluation of single-and separated-phase anaerobic industrial wastewater treatment in fluidized bed reactors, *Biotechnol. Bioeng.*, 26, pp 1054–1065.
11. Pfeiffer W., U. Temper, A. Steiner, A. Carozzi, J. von Mücke (1986) Anaerobic wastewater treatment – results of a literature review, Aquatech '86, Anaerobic Treatment , pp 219–231.
12. Enger, W.A., W.M.A. van Gils, J.J. Heijnen, W.A.A. Koevoets (1986) Full-scale performance of a fluidized bed in a two-stage anaerobic waste water treatment at Gist-brocades, Aquatech '86, Anaerobic Treatment , pp 297–306.
13. Gönenç, İ.E. and M. Kerestecioglu (1989) Starting-up of a full scale two-phase contact process treating citric acid wastewater, Proc. Int. Symp. on Waste Management Problems in Agro-industries , Istanbul, Turkey, pp 221–228.
14. Oliva, E., J.C. Jacquart and C. Prevot (1990) Treatment of waste water at the El Aguila brewery (Madrid, Spain) Methanization in fluidized bed reactor, *Wat. Sci. Technol.* , 22, pp 483–490.
15. Ehlinger, F., J.M. Audic, D. Verrier and G.M. Faup (1987) The influence of the carbon source on the microbiological clogging in an anaerobic filter, *Wat. Sci. Technol.* , 19, pp 261–273.
16. Zhang, H. (1987) Investigation of a pilot scale anaerobic fluidized bed biological reactor (AFBBR) for brewery wastewater stabilization, unpublished report, Dept. of Civil Eng., University of Newcastle-upon-Tyne.
17. Heijinen, J.J., A. Mulder, W.Enger and F. Hoeks (1986) Review on the application of anaerobic fluidized bed reactors in waste-water treatment, Aquatech '86, Anaerobic Treatment , pp 159–171.

79 TREATMENT OF DAIRY WASTEWATER BY CHEMICAL COAGULATION

Frederic C. Blanc, Professor
Rita Navia, Graduate Student
Department of Civil Engineering
Northeastern University
Boston, Massachusetts, 02115

INTRODUCTION

Industrial dairy wastewaters are generally high strength wastes containing soluble, colloidal and suspended solids with a high concentration of BOD. The more than 5,000 industrial dairy plants in this country have a large waste load to treat and dispose of. If discharged into the municipal sewer system, such wastewaters often produce a high pollution load which is unacceptable to the Publically Owned Treatment Works [POTW]. This then results in a requirement for industrial pretreatment of such wastes before discharge to the collection system. Chemical coagulation-flocculation followed by sedimentation, which plays an important role in many solid liquid separation schemes and is widely used in water and wastewater treatment may be used in the pretreatment of dairy wastewaters in some instances. This method can remove BOD, COD and suspended solids from the waste stream and may be economically feasible for some of the industrial dairy product facilities which are located in urban areas.

In this study a laboratory investigation of chemical coagulation treatment of dairy processing wastewater using alum, ferric chloride, or ferrous sulfide together with lime was performed. Extensive testing to determine chemical requirements and combinations for various removal efficiencies was conducted. Following the laboratory studies, a cost analysis was performed evaluating treatment benefits and pretreatment process economics based on costs of local sewer user surcharges.

BACKGROUND

Dairy wastes are generated by disposing of residues, wasting of by-product, process equipment cleaning and leakage or spills. The main organic constituents in dairy wastewaters consist of fats, proteins and carbohydrates in various proportions which are associated with milk. Other organics found in some dairy industry wastewaters may come from processing companion products such as orange juice, other fruit juices, or ice cream and sherbet additives such as fruits, nuts, coloring substances or sugar. The BOD strength of dairy wastewaters varies depending on the type of production process. This parameter has been reported to vary from as low as 40 mg/L to as high as 10,000 mg/L, with a mean of 2,300 mg/L.[1] Suspended solids, which are mostly organic, are typically present in concentrations up to 2,000 mg/L.[1] COD values range from 80 mg/L to 95,000 mg/L with a typical mean of 4,500 mg/L for various types of dairy plants.[2] Phosphorus, which comes from milk and cleaning compounds is typically present in a mean concentration of 49 mg/L.[2] The reported values of pH range from 4.5 to 9.2.[3]

The treatment objectives for dairy wastewaters are described by the Environmental Protection Agency in Effluent Guidelines and Standards for Dairy Products.[4] Treatment or pretreatment processes for dairy wastewaters are usually the biological type with the most common being: activated sludge, trickling filters, rotating biological contactors, lagooning, land treatment, and anaerobic reactors. Anaerobic processes as described by Sobkowicz[5] for the treatment of cheese production wastewaters or Backman[6] for treating dairy wastewater have the capability of handling high organic loadings in a pretreatment application. Table I depicts the range of treatment efficiency for various biological treatment processes treating dairy wastewaters.

Chemical treatment or pretreatment of dairy wastewaters is not practiced to a great extent. There are not many reports of chemical treatment in the wastewater treatment literature. A physical-chemical pretreatment system was described by Bough and Swientek with reported cost savings.[8]

45th Purdue Industrial Waste Conference Proceedings,

Table I. Efficiency of Biological Treatment Processes for Industrial Dairy Effluents[7]

Process	Percent Reduction in BOD_5 Range
Activated Sludge	24–99
Aerated Lagoons	70–99
Oxidation Ditches	74–99
Septic Tanks	50–87
Trickling Filters	21–99

Tabakov and Karpati reportedly described the design of a pretreatment system consisting of equalization, neutralization, addition of alum, lime and polyelectrolyte followed by electroflotation.[9]

MILK COLLOIDS AND COAGULATION

Milk is made up mainly of four substances in addition to water: fat, casein, albumin, and lactose. Water comprises approximately 87% of milk by weight. Lactose makes up 5%, butter fat 3.7% and casein 3% of the milk. Lactose is a sugar accounting for 40% of the total milk solids and it is in solution. Casein is a protein which occurs in milk as a hydrophilic colloidal calcium phosphate complex containing approximately 7% of the inorganic material in milk. Casein molecules tend to join together to form extensive interwoven chains which are large and behave like colloidal particals.[10] Butter fat globules in milk have varying diameters. Most are above the colloidal size range but some are as small as 0.1 microns in diameter. Milk proteins coat the dispersed globules preventing cohesion. Rapid agitation of milk breaks this protective coating of protein permitting the butter fat to aggregate and eventually form a mass of butter.The addition of electrolytes of opposite charge to milk colloids such as casein reduces the electrical potential on the casein particles bringing the particles closer to the isoelectric point. The addition of acid causes dissociation of the casein calcium phosphate complex as well. At a pH in the 3 to 5 range, casein begins to precipitate out of the milk waste. An isoelectric point at pH 4.6 has been reported for casein.[11]

EXPERIMENTAL METHODS AND MATERIALS

To determine the effectiveness of chemical coagulation, flocculation, and sedimentation as a pretreatment measure for dairy wastewaters, a synthetic dairy wastewater was utilized. This wastewater was created by diluting fresh whole milk in a ratio of 1:15. Wastewater data from a local dairy processing facility was utilized in arriving at the dilution factor. Table II summarizes some of the characteristics of whole milk and the synthetic dairy wastewater used.The measured COD of the synthetic dairy waste ranged from 12,016 to 14,371 mg/L during the studies depending upon the brand of milk used and variability in dilution and analysis.

Numerous experiments were set up using a standard jar test apparatus with 500 mL beakers. The jar test procedure included a rapid mix at 80 rpm which continued for 1 minute after chemical addition followed by 20 minutes of slow stirring at 20 rpm and 30 minutes of settling. After settling, supernatant was withdrawn from the jars for analysis of BOD, COD, and suspended solids following procedures in *Standard Methods.*[12] A milk-fed batch activated sludge seed was maintained for seeding the BOD dilution water. Jar tests were conducted using a wide range of chemical dosages and combinations. Sample pH adjustment prior to chemical addition was not practiced. However, acid

Table II. Average Composition of Whole Milk and Synthetic Dairy Waste

Characteristics	Concentration	
	Whole Milk	Synthetic Dairy Waste
Total BOD, 5-day (mg/L)	102,500	7,421
Total COD (mg/L)	150,000	12,926
Total Solids (mg/L)	125,000	9,000
Volatile Solids (mg/L)	117,000	8,100
pH	7	7
Fat (mg/L)	36,000	–
Suspended solids (mg/L)	–	870
Alkalinity as $CaCO_2$	198	14
Phosphorus (mg/L)	1,000	70

addition to determine the effect of pH adjustment alone as a coagulation mechanism was investigated. The reagent grade coagulants used were:alum, ferric chloride, ferric sulfate and hydrated lime. All coagulants other than lime were fed from stock solutions in which 1 mL of solution contained 10 mg of the metal ion. In the preliminary investigations supernatant quality based on visual observation was used to select optimum dosage ranges for futher study. Batch settling tests were conducted to determine floc settling velocities for optimum chemical dosages.

EXPERIMENTAL RESULTS AND DISCUSSION

Tables III, IV, and V present a summary of the averaged results of batch test supernatant quality and percentage removals for various doses of alum, ferric chloride, and ferric sulfate, with and without lime in the practial optimum range which might be used on a milk waste. For this range in each case the floc settled well and the settled supernatant was not turbid on visual inspection. The results indicate that reductions in COD from 13,000 mg/L to 3,500 mg/L and BOD reductions from 8,400 mg/L to 1,500 mg/L are possible using chemical coagulation, flocculation and sedimentation.In terms of BOD and COD removal all coagulants performed well; however, the ferric chloride lime combinations achieved slightly better BOD and COD removal than the alum and lime or ferric sulfate and lime combinations. Suspended solids removals greater than 98% are possible with a wide range of coagulant applications reducing the suspended solids concentration in the wastewater from about 1,000 mg/L to concentrations less than 20 mg/L.All coagulants provided good removal of suspended solids with the ferric salts providing slightly better performance than the alum and lime. Table VI summarizes the best performance removals and final pH values at the optimum coagulant dose for each of the three coagulants.

The final pH values listed in Tables III, IV, V, and VI, are due to the coagulant addition and represent changes from the initial waste pH of 7. Use of acid alone to change the pH was investigated in the range from pH 7 to pH 2. Flocculation was observed at a pH of about 4.5 but the floc formed was too small in size and did not settle. In general the best BOD removal was observed for coagulant doses which resulted in a final supernatant pH between 4 and 5. Figure 1, which shows the BOD removal efficiency for various final supernatant pH values resulting from different applications of alum, ferric chloride, or ferric sulfate, illustrates that the pH 4 to 5 range is the area where best chemical treatment occurs.

Settling characteristics of the various flocs for optimum dose for each coagulant type listed in Table VI are presented in the settling curves from graduated cylinder bench tests shown in Figures 2, 3, and 4. The settling velocity values derived from the settling curve data show overflow rate values in the range of less than 100 gallons per square foot day. Such overflow rates are normally associated with sludge thickening in municipal wastewater treatment but are not unusual in the case of concentrated solids suspensions often associated with industrial wastewaters. The ferric coagulant flocs had higher settling velocities than the alum floc.

ECONOMIC CONSIDERATIONS

For a process to be utilized it must be economically feasible as well as technically feasible. In the coagulation studies performed in this research, a wide range of coagulant dosages provided acceptable BOD, COD and suspended solids removal efficiencies. In order to select the most appropriate treatment process, process economics must be investigated. The approach taken in this case was to perform a cost analyses based on chemical costs and sewer user surcharge costs in which costs were expressed in dollars per million gallons treated. The total cost with treatment was compared with the surcharge costs without treatment to obtain the savings which would be available for annualized capital cost recovery and other annual operation and maintainance costs. This cost comparison was computed for all coagulants, dosages and performance levels tested in the study.

The chemical costs used are based on prices in the New England area in January 1989. Sewer user surcharges were based on 1989 surcharges for the Charles River Water Pollution Control Facility, a POTW in eastern Massachusetts that provides secondary treatment. The sewer user surcharges for excess quantities are $ 0.115 per pound of five day BOD and $ 0.07 per pound of suspended solids at that facility. Although surcharges exist for other parameters such parameters were not included in the cost analysis. Included in the cost analysis was a cost for neutralizing the chemical treatment effluent from a pH of 4.5 to 7.0 based on titration curves for the optimum supernatants. This neutralization cost was based on the use of sodium hydroxide which has a higher chemical cost than lime neutralization.

Table VII summarizes the results of the economic analysis by illustrating the costs, sewer surcharges and differential for the optimum dose of each of the coagulant combinations. All costs in the table are

Table III. Chemical Treatment Effluent Quality and BOD, COD, Suspended Solids Removal

Aluminum Sulfate ($Al_2(SO_4)_3 \cdot 18\ H_2O$) + Lime ($Ca(OH)_2$)

Chemical Addition				Average Settled Water Quality					
$Al_2(SO_4)_3$ (mg/L)	Al (III) (mg/L)	$Ca(OH)_2$ (mg/L)	pH	SS (mg/L)	%SS Removal	COD (mg/L)	%COD Removal	BOD (mg/L)	%BOD Removal
371	30	0	5.4	75	88.24	5,558	53.54	3,300	52.00
494	40	0	5.1	18	97.12	5,117	57.37	2,683	63.10
618	50	0	4.5	24	97.25	4,870	59.49	2,125	70.43
741	60	0	3.9	46	93.83	4,715	61.00	2,000	73.51
371	30	15	5.5	66	89.62	5,548	53.62	3,400	50.55
494	40	15	5.2	45	92.83	5,365	55.15	3,250	52.73
618	50	15	4.6	38	94.06	5,779	51.94	2,380	63.00
741	60	15	4.0	31	94.01	4,843	57.00	2,231	70.00
371	30	30	5.8	44	95.01	5,418	55.00	3,285	65.00
494	40	30	5.5	13	96.23	4,715	61.00	2,125	71.86
618	50	30	5.0	14	97.77	4,247	67.36	1,975	73.84
741	60	30	4.1	20	96.86	4,747	60.73	2,100	72.19

Table IV. Results of Chemical Treatment Applied to Synthetic Dairy Waste

Ferric Chloride ($FeCl_3$) + Lime ($Ca(OH)_2$)

Chemical Addition				Average Settled Water Quality					
$FeCl_3$ (mg/L)	Fe(III) (mg/L)	$Ca(OH)_2$ (mg/L)	pH	SS (mg/L)	%SS Removal	COD (mg/L)	%COD Removal	BOD (mg/L)	%BOD Removal
387	80	0	5.1	142	83.84	4,604	67.23	2,535	71.73
436	90	0	4.9	41	95.23	4,174	67.81	2,000	73.31
484	100	0	4.7	24	96.64	4,149	69.29	1,864	76.75
536	110	0	4.1	16	98.10	3,679	72.90	1,700	79.17
387	80	15	5.4	29	96.57	4,122	67.95	2,100	75.00
436	90	15	5.1	13	98.48	4,160	67.66	2,500	70.23
484	100	15	5.0	11	98.69	3,924	69.13	1,875	76.23
536	110	15	4.6	14	88.36	3,969	69.14	2,000	76.19
387	80	30	5.5	60	87.53	5,304	59.20	2,400	71.43
436	90	30	5.4	56	93.98	4,463	65.60	3,000	60.10
484	100	30	5.2	33	96.06	4,270	66.65	2,150	74.40
536	110	30	4.9	14	98.33	3,817	70.33	2,100	75.00

Table V. Results of Chemical Treatment Applied to Synthetic Dairy Waste

Ferric Sulfate ($Fe_2(SO_4)_3$) + Lime ($Ca(OH)_2$)

Chemical Addition			Average Settled Water Quality						
$Fe_2(SO_4)_3$ (mg/L)	Fe(III) (mg/L)	$Ca(OH)_2$ (mg/L)	pH	SS (mg/L)	%SS Removal	COD (mg/L)	%COD Removal	BOD (mg/L)	%BOD Removal
286	80	0	4.5	9	98.96	3,709	72.17	2,000	74.03
322	90	0	3.5	17	98.35	3,969	69.76	2,200	71.43
358	100	0	3.3	16	98.41	4,047	69.16	2,200	71.43
394	110	0	3.0	36	96.43	4,126	68.56	3,650	52.60
286	80	15	4.9	41	95.31	4,291	66.80	2,597	65.00
322	90	15	4.3	15	98.55	4,126	68.31	2,400	67.42
358	100	15	3.9	39	96.11	4,008	69.46	2,250	70.78
394	110	15	3.7	12	98.60	4,098	70.00	3,212	72.00
286	80	30	5.0	19	98.15	4,283	67.37	2,250	70.78
322	90	30	4.4	21	97.96	3,969	69.76	2,350	69.48
358	100	30	4.0	13	98.75	3,930	70.06	2,050	73.38
394	110	30	3.8	11	98.80	4,013	71.00	2,921	73.50

Table VI. Optimum Dose and pH for the Removal of Contaminants from the Synthetic Dairy Waste

Chemical	pH	Optimum Dose (mg/L)[a]	Percent Removal BOD	Percent Removal COD	Percent Removal SS
Alum + Lime	5.0	50 + 30	73.84	67.36	97.77
Ferric Chloride	4.1	110	79.17	72.90	98.10
Ferric Sulfate	4.5	80	74.03	72.17	98.96

[a]Dosages shown are in mg/L as the metal ion Al^{3+} for alum and Fe^{3+} for Ferric Chloride and Ferric Sulfate.

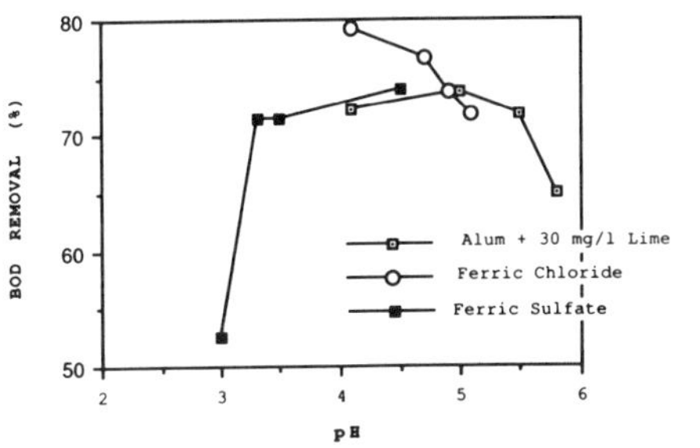

Figure 1. BOD removal vs. pH for coagulants.

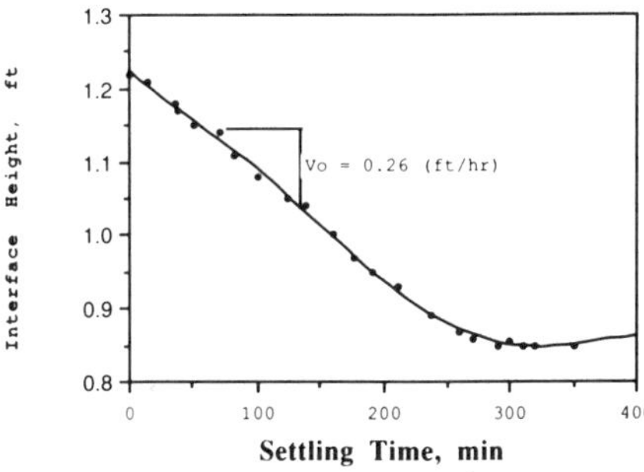

Figure 2. Settling curve for alum (50 mg/L as Al) and lime (30 mg/L).

presented in terms of dollars per million gallons treated. In each case the initial BOD of the untreated waste sample used in the jar test series for that chemical was used in the computation. Thus the initial BOD for the alum treated waste was 7,550 mg/L, the initial BOD for the ferric chloride treated waste was 8,400 mg/L and the initial BOD for the ferric sulfate treated waste was 7,370 mg/L. It turned out that the optimum chemical dose in each case also coincided with the maximum cost differential for that chemical. In the case of alum only the 40 mg/L Al-30 mg/L Ca dosage showed an equal cost differential while there were more alternative dosages of the ferric salts which would provide a cost differential close to the differential illustrated. The solids surcharge savings in each case was estimated to be approximately $400 per million gallons treated. The ferric salts provided better cost savings than the alum. It is felt that the total differential illustrated in Table VII represents a realistic number against which annualized capital cost and other pretreatment operational costs might be compared.

SUMMARY AND CONCLUSIONS

The results of this study have indicated that chemical coagulation is a viable treatment process for treating wastewaters from dairy products facilities. Treatment of the milk waste in this study pro-

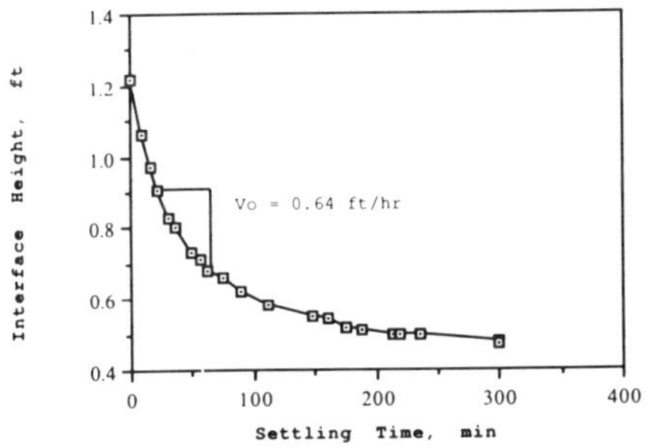

Figure 3. Settling curve for ferric chloride (110 mg/L as Fe).

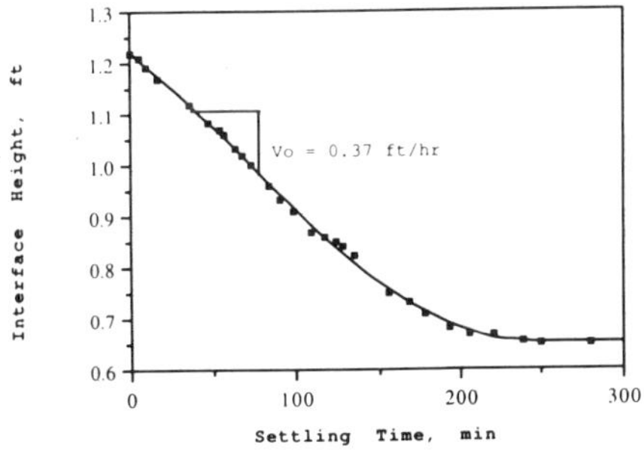

Figure 4. Settling curve for ferric sulfate (80 mg/L as Fe).

Table VII. Summarized Cost for the Coagulants Used at the Optimum Dose with Cost Savings

Coagulants	Chem. Cost ($/MG)	Neutralization NaOH ($/MG)	Neutralization $Ca(OH)_2$ ($/MG)	Sewer Surcharge with Chemical Treatment[a] ($/MG)	Total ($/MG)	Sewer Surcharge without Treatment[c] ($/MG)	Cost Differential ($/MG)	Cost Differential with SS Surcharge ($/MG)[b]
Alum + lime	1,841	215	159	1,640	3696	6935	3239	3649
Ferric Chloride	1,160	295	178	1,426	2881	7743	4862	5262
Ferric Sulfate	953	240	167	1,664	2857	7492	4635	5035

[a]Cost calculations based on use of NaOH for neutralization.
[b]A sewer surcharge savings for solids discharge was estimated at $400 per MG for all cases.
[c]Based on BOD only.

duced BOD removal efficiencies in the 70 to 75% range. COD removal efficiencies were slightly lower in the 65 to 75% range. Ferric salts appear to be the best coagulants from both an economic and cost standpoint. Economic viability of chemical coagulation was demonstrated for cases where the objective is to reduce sewer user surcharge costs in pretreatment applications. In some instances where better effluent quality is required chemical coagulation can be used before biological treatment to control milk solids and reduce the size of the biological treatment units required. In general, chemical treatment should be considered in dairy wastewater treatment process designs.

REFERENCES

1. Development Document for Effluent Limitations Guidelines and New Source Performance Standards for Dairy Product Processing, *EPA-440/1-74-021-a.*
2. Brown, H.B., Pico, R.F., "Characterization and Treatment of Dairy Wastes in the Municipal Treatment System", *Proc. 34 th Industrial Waste Conference*, Purdue Univ., 326–334, [1979].
3. Harper, W.J., *Upgrading Dairy Production and Treatment Facilities to Reduce Pollutants: Waste Characteristics and in Plant Control*, EPA Technology Transfer Seminar, Philadelphia, PA. August 20-21, [1973].
4. *Effluent Guidelines and Standards for Dairy Products,* EPA [Amended by 39FR 32993, September 13, 1974, 51FR 24996, July 9, 1986].
5. Sobkowicz, A.M., "Celrobic Anaerobic Treatment of Dairy Processing Wastewater", *Proc. 41 st Industrial Waste Conference*, Purdue Univ., 458–464, [1986].
6. Backman, R.C., Blanc, F.C., and O'Shaughnessy, J.C., "The Treatment of Dairy Wastewater by Anaerobic Up-Flow Packed Bed Reactor", *Proc. 40th Industrial Waste Conference*, Purdue Univ., 361–372, [1985].
7. Boyle,B.C. and Polkowski, L.B., *Alternate Methods of Treating or Pretreating Dairy Plant Wastes*, in Watson, S.K., "The Treatment of Dairy Plant Wastes", EPA Technology Transfer Seminar, Madison, WI, March 20-21, [1973].
8. Bough, W.A., and Swientek, R.J., " Dairy Installs Pretreatment System vs. Paying 75% of $1.5 Million City Treatment Plant–Saves $450,000", *Food Process.*, 42, 10,150 [1980].
9. Tabakov, D., and Karpati, A., "Treatment of Effluent from the Dairy Industry," Khranitelna Promishlenost (Hung.), 33, 7, 22 (1984): *Dairy Sci. Abstr.*(G.B.), 48, 519 [1986].
10. Li, A.Y., and Corrado, J.J., "Scale Up of Membrane Anaerobic Reactor System", *Proc. 40 th Industrial Waste Conference*, Purdue Univ., 399–404, [1985].
11. Webb, B.H., *By Products from Milk*, 1st ed., AVI Publ., Westport, Connecticut, 2 [1970].
12. *Standard Methods for the Examination of Water and Wastewater*, 16 th ed., APHA-AWWA-WPCF, Washington, D.C., [1985].

Section Nine
INDUSTRIAL WASTES—C. METAL PROCESSING WASTES

80 SELECTIVE PRECIPITATION OF MIXED METAL HYDROXIDES

Clyde S. Brooks, President
Recycle Metals
Glastonbury, Connecticut 06033

INTRODUCTION

Selective precipitation is examined here as a technically feasible approach for improving the efficiency for recovering non-ferrous metals from industrial waste effluents in the presence of contaminants such as aluminum, chromium or iron. Precipitation provides probably the lowest cost alternative for the non-selective removal of metals from industrial waste waters to meet EPA standards for effluents. Consideration of the pH dependence of hydroxide precipitation of mixed metal systems such as copper or nickel in the presence of aluminum, chromium or iron would seem to provide an opportunity to conduct staged hydroxide precipitation of the contaminants, aluminum, chromium or iron below about pH 5–6 delaying precipitation of the copper or nickel to a pH above 6. Unfortunately freshly precipitated aluminum, chromium and iron hydroxides provide considerable capacity for adsorption/ion exchange leading to mixed coprecipitates of the trivalent contaminants and the divalent copper or nickel frustrating efficient separations. Recent experimentation with iron precipitated in the presence of selected anionic surfactants in a moderate pH range 3.5–4.5 permitted efficient separation from copper or nickel which remained in solution pending a secondary separation process.[1]

The formation of mixed precipitates called jarosites occurs by hydroxide reaction with NaOH of KOH from sulfuric acid solutions of mixed metals.[2,3] A typical composition for a jarosite might be $Na(FeCr)_3(OH)_6(SO_4)_2$ with inclusion of anions such as phosphates or chromates. Investigations have shown a preferential ordering of metal cations coprecipitated in jarosite system is of the order of $Fe^{3+} >> Cu^{2+} > Zn^{2+} > Co^{2+} > Ni^{2+} > Cd^{2+}$. Advantage has been taken of the collection action of freshly precipitated aluminum or ferric hydroxide to remove by adsorption colloid flotation low concentrations of metal cations such as Cd^{2+}, Cu^{2+}, Cr(III), Ni^{2+} and Zn^{2+} in waste effluents in the presence of anionic flotation surfactants such as sodium lauryl sulfate.[4,5] Co-precipitation of ferrous and ferric cations to form ferrites with soluble metal cations such as Cd^{2+}, Cu^{2+}, Cr(III), Ni^{2+} or Pb^{2+} has been applied successfully to removal of metal contaminants from waste effluents.[6] More recent EPA funded investigations[7] have evaluated co-precipitation effects for iron and other heavy metals such as Cr(III), Cu, Ni, Zn and Pb in efforts to develop metal recovery processes from industrial waste streams.

The approach in the present investigation is quite different than in any of the studies cited above. In this investigation the objective has been to take advantage of the competitive adsorption of anionic surfactants in a moderately acid pH regime on freshly precipitated ferric hydroxide to effect separations from copper or nickel in solution. Furthermore, an evaluation of an extension of this approach to separations of copper or nickel form aluminum or chromium (III) hydroxides has been undertaken.

In this study, model experiments have been conducted in which ferric iron in solution with copper or nickel plus selected anionic surfactants, primarily carboxylic acid types, is adjusted in pH with NaOH to precipitate the iron as ferric hydroxide leaving the copper or nickel in solution. Screening tests were conducted with a series of anionic surfactants, lauric acid, oleic acid, ascorbic acid, acetic acid, citric acid, and ethylene diamine tetra acetic acid (EDTA) to establish the anionic providing the most effective competition with nickel for adsorption/ion exchange on freshly precipitated ferric

45th Purdue Industrial Waste Conference Proceedings, © 1991 Lewis Publishers, Inc., Chelsea, Michigan 48118. Printed in U.S.A.

hydroxide. Additional tests were conducted with acetic acid and EDTA, two of the most effective competitive surfactants, to optimize the concentration range for separation of copper and nickel from freshly precipitated ferric hydroxide. Additional evaluations were conducted with actual industrial waste systems such as electrochemical milling waste filter cake, and spent nickel hydrogenation catalyst, solubilized in sulfuric acid to determine the efficiencies for separation of nickel.

EXPERIMENTAL PROCEDURES

Model Tests

Simulated waste systems were prepared for the model tests with iron, 1000 mg/L ($FeCl_3 \cdot 6H_2O$ Mallinckrodt 5376) as the contaminant and varying amounts of a second metal such as chromium (III) ($Cr(NO_3)_3 \cdot 9H_2O$ Aldrich 23,925-9), copper ($CuSO_4 \cdot 5H_2O$ Mallinckrodt 4744), nickel ($Ni(NO_3)_2 \cdot 6H_2O$ Aldrich 24,407-4), or zinc ($ZnCl_2$ Aldrich 20,808-6) for the adsorption experiments. The systems prepared for the competitive experiments consisted of iron at 1000 mg/L and a second metal, nickel or copper in varying amounts (0.44–1.0 atomic ratio), relative to the iron and a number of anionic surfactants such as acetic acid (Mallinckrodt 2488), oleic acid (Mallinckrodt 2744), lauric acid (MCB LX0060-3), sodium lauryl sulfate (Aldrich 86,201-0), citric acid (MCB CX1723-1), ascorbic acid (Aldrich A9290-2), oxalic acid (MCB CB558), sodium dodecane sulfonate (MCB AX2490), and ethylene diamine tetraacetic acid (EDTA Matheson Coleman Bell CB1068) in amounts ranging from 30 to 120 mg/L.

The adsorption measurements to establish the coadsorption-ion exchange of the anionic surfactants were conducted by mixing the two metal salts in the desired ratios in 100 mL of aqueous solution and adding NaOH to adjust the pH to the range of 3.5 to 4.5 to completely precipitate the iron as $Fe(OH)_3$. The mixture was thoroughly mixed, the precipitate was separated by filtration and the concentration of the second metal, copper or nickel, was determined in the filtrate.

The competitive adsorption measurements in the presence of the anionic surfactants was conducted as described above for the binary metal systems except that the anionic surfactant in the concentration of interest, 250–1000 PPM, was introduced to be present when the pH adjustment with NaOH was made to precipitate the iron as ferric hydroxide.

Waste Systems

The industrial wastes examined consisted of an electrochemical milling waste filter cake (ECM), and a spent nickel/alumina hydrogenation catalyst. The waste sludges, as received, in an amount of 10 grams were solubilized with 20 wt.% H_2SO_4 and the insolubles separated by filtration. The acid solubilized wastes were analyzed for the metals of interest. Anionic surfactants of interest, primarily acetic acid and EDTA, were added in concentrations of 100 to 1000 mg/L and the pH adjusted with NaOH with vigorous mixing to completely precipitate the contaminant metal iron as ferric hydroxide or aluminum as aluminum hydroxide. The precipitates were separated by filtration and the concentrations of the metals of interest determined in the filtrate.

The metal concentrations for these experiments were analyzed by atomic absorption by Griswold and Fuss Environmental Laboratories, Inc., 360 Center St., Manchester, CT or Vargo Laboratories, Inc., 587 E. Middle Turnpike, Manchester, CT.

RESULTS AND DISCUSSION

Adsorption/ion Exchange of Metal Cations on Ferric Hydroxide

An indication of the extent of coprecipitation of metal cations such as Cr(III), Cu, Ni, and Zn on ferric hydroxide in hydroxide precipitations with NaOH in a moderately acid pH range of 4–5 is shown in Figures 1 and 2. The initial iron concentration was 1000 PPM and the second metal was present in amounts ranging from 0.1 to 1.0 weight ratio relative to the iron. The coprecipitation is extensive, being of the order of 20 to 80% for the divalent cations, Cu^{2+}, Ni^{2+}, or Zn^{2+}, and essentially quantitative for the Cr^{3+} which is in a pH range bordering where chromium hydroxide precipitates.

Competitive Adsorption of Metal Cations and Anionic Surfactants on Ferric Hydroxide

Earlier experimentation[1] established that for precipitation of ferric hydroxide in the presence of Ni^{2+} in a pH range of 3.5 to 5.5 that anionics such as sodium dodecane sulfonate and citric acid and notably lauric acid inhibit the coprecipitation of the nickel.

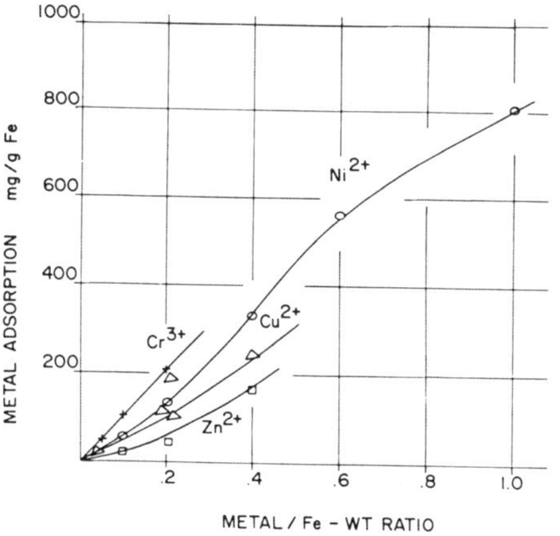

Figure 1. Adsorption of metal cations on ferric hydroxide at pH 4.5.

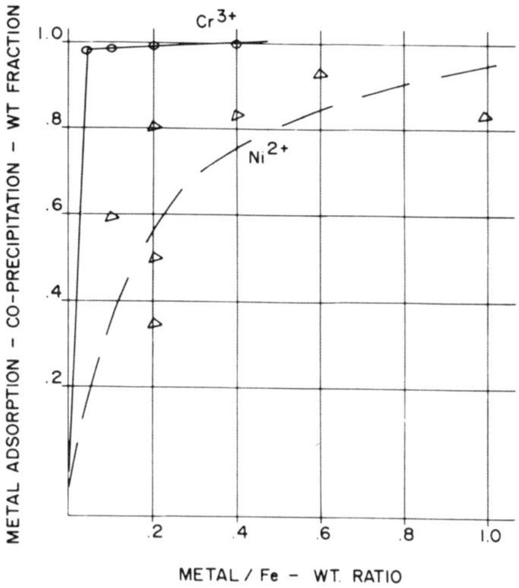

Figure 2. Adsorption—co-precipitation of Cr^{3+} and Ni^{2+} on $Fe(OH)_3$ at pH 4.5.

Screening tests were conducted for competitive adsorption at pH 4.5 for 200 mg/L of 7 carboxylic acid type surfactants, oleic acid, ascorbic acid, citric acid, acetic acid, lauric acid and EDTA with Ni^{2+} (1.0 Ni/Fe atomic ratio) present with precipitation of $Fe(OH)_3$ (Fe 1000 PPM). These experiments (Figure 3) indicated that the anionics, acetic acid, lauric acid and EDTA provided the most significant inhibition of the nickel coprecipitation with ferric hydroxide. Reductions of the order of 40–50% of the nickel coprecipitation were observed.

The inhibition of the incorporation of the second metal into the ferric hydroxide precipitate under these experimental conditions can be attributed to a combination of competitive adsorption and sequestration action. The present study did not include direct measurement of the adsorption of the anionics on the ferric hydroxide so that the relative contributions of adsorption versus sequestration is not established. In selecting agents to provide selective precipitation consideration must be given to the sequestration action on the iron. An assessment was made of the sequestration of ferric hydroxide precipitation by the anionic surfactants acetic acid, citric acid and EDTA for 1000 PPM Fe^{3+} for a Ni/Fe 0.48 atomic ratio. In Figure 4 a plot is shown of the pH required for complete $Fe(OH)_3$

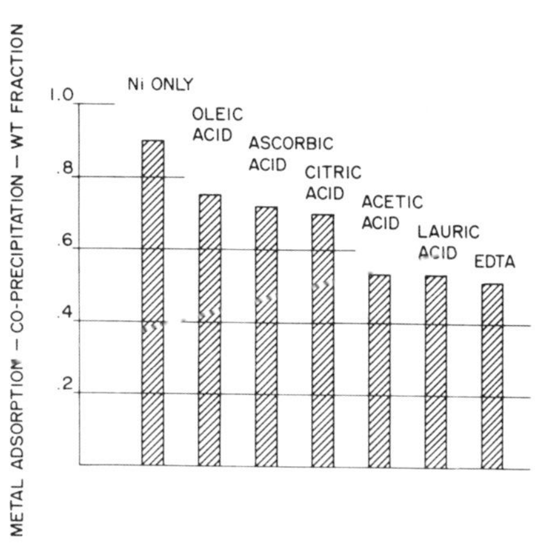

Figure 3. Competitive adsorption at pH 4.5 of carboxylic acids (200 PPM) and Ni^{2+} (1.0 Ni/Fe) on $Fe(OH)_3$(Fe 1000 PPM).

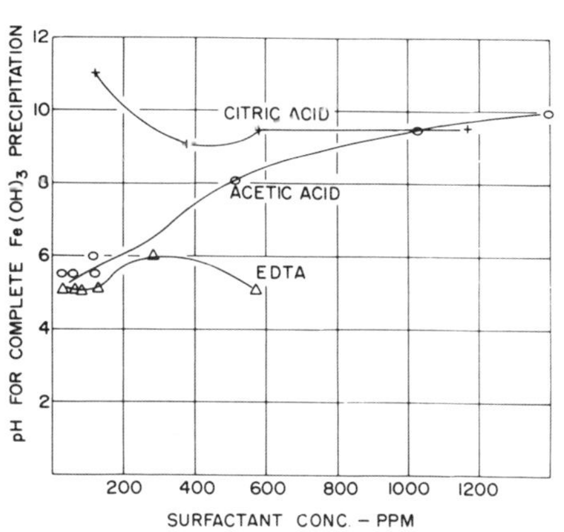

Figure 4. Sequestration of ferric hydroxide precipitation by anionic surfactants Ni/Fe 0.48 atomic ratio.

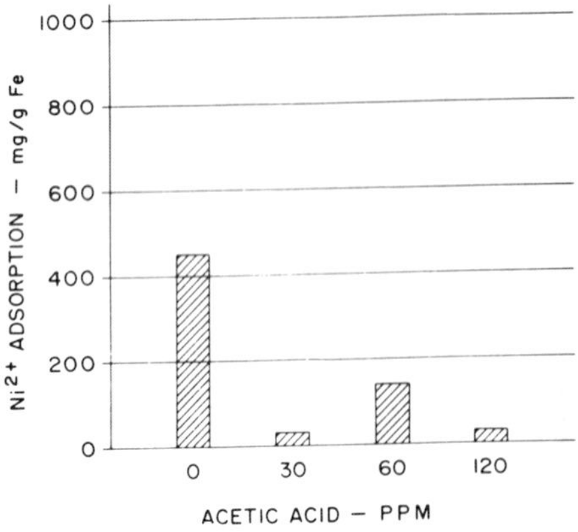

Figure 5. Ni^{2+} **Adsorption of** $Fe(OH)_3$ **in the presence of acetic acid at pH 5.5 Ni/Fe 0.48 atomic ratio.**

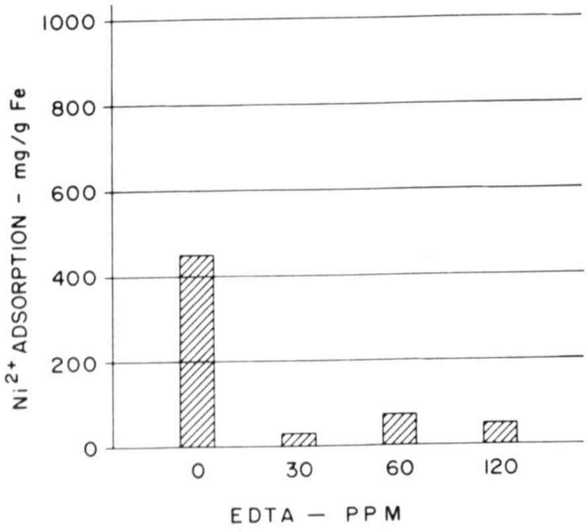

Figure 6. Ni^{2+} **Adsorption on** $Fe(OH)_3$ **in the presence of EDTA at pH 5.0 Ni/Fe 0.48 atomic ratio.**

precipitation versus concentration of the anionic surfactant. Acetic acid and EDTA at concentrations of 200 mg/L and below are seen to not affect significantly the sequestration of the iron. Citric acid at concentrations ranging from 100 to 1400 mg/L strongly sequesters the iron to precipitation above pH 9 precluding application for selective precipitation to a system like Ni/Fe.

Additional experiments were conducted with the systems iron (1000 PPM) in the presence of nickel (Ni/Fe 0.48 atomic ratio) for precipitation at pH 5.0–5.5 in the presence of acetic acid and EDTA in the concentration range 30–120 mg/L (Figures 5 and 6). Inhibitions of the nickel adsorption on the $Fe(OH)_3$ of the order of 80% are obtained for both of these agents in the concentrations used.

Similar experiments were conducted also with the systems iron (1000 mg/L) in the presence of copper (Cu/Fe 0.44 atomic ratio) for precipitation at pH 4.0 in the presence of acetic acid and EDTA in the concentration range 30–120 mg/L (Figures 7 and 8). Inhibitions of copper adsorption on the $Fe(OH)_3$ of the order of 80–90% are obtained from acetic acid. An inhibition for the copper adsorption of about 80% was obtained for EDTA at the lowest concentration 30 mg/L. A consistent decrease was observed in the inhibiting action on copper coprecipitation at the higher EDTA concentrations, 60 and 120 mg/L. The suggestion is offered that the copper/EDTA complex formed with increasing EDTA concentration may adsorb on the ferric hydroxide, diminishing the efficiency of this agent for providing selective precipitation.

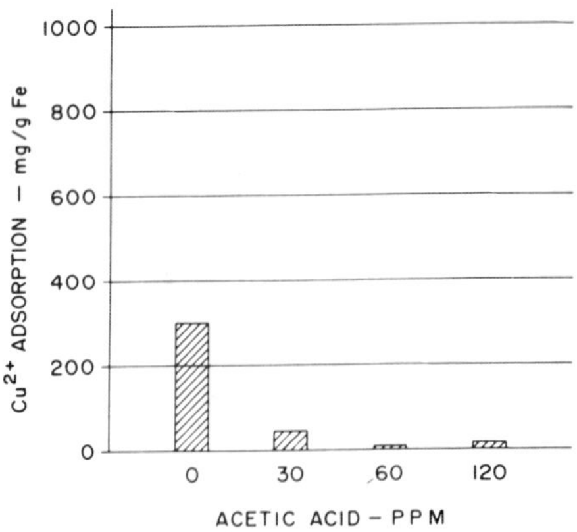

Figure 7. Cu^{2+} **Adsorption on** $Fe(OH)_3$ **in the presence of acetic acid at pH 4.0 Cu/Fe 0.44 atomic ratio.**

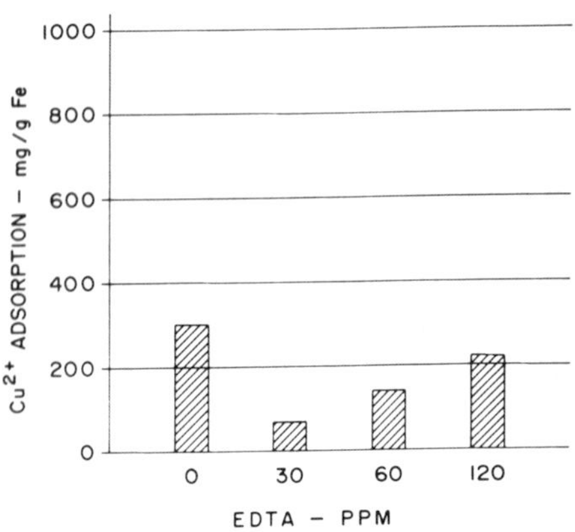

Figure 8. Cu^{2+} **Adsorption on** $Fe(OH)_3$ **in the presence of EDTA at pH 4.0 Cu/Fe 0.44 atomic ratio.**

Table I. Nickel Recovery from Industrial Wastes

Waste	Components	Solids Soluble in 20 wt.% H_2SO_4 at 20°C wt.%	Metals in Acid Solution PPM	Selective precipitation pH 4–6: Nickel Concentrated by Hydroxide Precipitation wt.%	Selective precipitation pH 4–6: Nickel Recovered by Oxalate Precipitation wt.%
ECM Filter Cake	Ni, Fe, Cr, Co, Nb, Ti, Silica	84	3350 Ni 1170 Fe	65[a] 80[b] 90[c]	99.0[c]
Nickel Catalyst	Ni Alumina Oil	26	4530 Ni 242 Al	49[a] 71[b]	99.9[c]

[a]No surfactant.
[b]200 mg/L EDTA.
[c]1000 mg/L EDTA.
[d]Data from Reference 8.

Selective Precipitation of Industrial Wastes

Results are presented here for two acid solubilized industrial wastes. One waste was an electrochemical milling waste collected on a drum filter with a siliceous filter aid from a milling brine. A second waste was a spent nickel/aluminum hydrogenation catalyst with residual oil (Table I).

The ECM filter cake was solubilized at room temperature with 20 wt.% H_2SO_4 leading to 16 wt% insolubles and 3550 mg/L nickel and 1170 mg/L iron in the acid solution. In addition to iron several parts per million of Cr(III), Ti, Co, and Nb were present. Adjustment of the pH with NaOH to the range 4–6 led to nickel concentrations after filtration separation of insolubles in the residual solutions of 2710 mg/L in the presence of 200 mg/L EDTA and 2840 mg/L in the presence of 1000 mg/L EDTA. These nickel concentrations correspond to 80 and 90%, respectively, of the nickel initially present in the acid solubilized waste. Alternatives for final recovery of the nickel as high grade chemicals with good marketability consist of concentration and separation as nickel sulfate or precipitation as nickel oxalate by addition of oxalic acid. The latter alternative would provide the most efficient separation from other residual metals such as Cr(III), Ti, Co, or Nb. Oxalate precipitation with prior preparation by solvent extraction of contaminants such as Cu, Fe, and Zn is capable of providing nickel recoveries of 99%.[8]

The acid solubilization of the nickel catalyst at room temperature with 20 wt.% H_2SO_4 provides a nickel concentration of 4530 mg/L in the acid solution. Separation from the alumina catalyst support occurring during acid solubilization of the nickel leads to 242 mg/L Al in the acid solution. A hydrocarbon extraction (toluene or xylene) can be used to separate the bulk of the oil residues prior to acid solubilization. Adjustment of the pH with NaOH to pH 5 in the presence of 200 mg/L EDTA precipitates the bulk of the remaining aluminum in solution leaving 2853 mg/L nickel in solution, corresponding to 71% of the initial nickel. As with the ECM waste this nickel can be recovered by concentration and separation as nickel sulfate. In this case negligible other metal contaminants will be present. In this system an efficient separation of 99.9% of the nickel can be obtained by oxalate precipitation without prior processing other than the acid solubilization.

Hydroxide precipitation in the absence of the EDTA surfactant provides appreciably lower nickel concentrations of 65 and 49%, respectively, for the ECM and nickel catalyst wastes. This indicates that there is a decided advantage in using a surfactant such as EDTA to inhibit nickel coprecipitation when using hydroxide precipitation to separate contaminants such as iron or aluminum in these waste systems.

CONCLUSIONS

Hydroxide precipitation, notably in the presence of low concentrations of EDTA, provides a technically feasible, and potentially economical, route for concentrating metals like nickel for recovery from acid solutions in the presence of contaminants like iron or aluminum.

REFERENCES

1. Brooks, C. S., "Adsorption/Co-Precipitation of Metal Cations and Surfactants on Ferric Hydroxide", *Proceedings of Second Engineering Foundation Conference on Fundamentals of Adsorption*, May 4–9, 1986, Santa Barbara, CA, Ed. A. I. Liapis Engineering Foundation, New York, NY (1987).
2. Dutrizac, J. E., *Proc. 3rd Int. Symp. Hydrometallurgy: Research, Development and Plant Practice*, K. Osseo-Asare and J. D. Miller, Eds., Met. Soc. AIME, Atlanta, GA (1983).
3. Saarinen, H. U. A., *U.S. Patent 4,042,474* (1977).
4. Chatman, T. E., S. D. Huang and D. J. Wilson, *Sep. Sci. Technol.*, 12, 461 (1977).
5. McIntyre, Gregory, J. J. Rodriguez, E. L. Thackston and D. J. Wilson, *J. Water Pollut. Control Fed.*, 55, 1144 (1983).
6. Okuda, T., I. Sugano and T. Tsuji, *Filtration & Sepn.*, 12, 472–478 (1975).
7. Howell, S. G., "Incineration and Treatment of Hazardous Waste," *Proc. 11th Annual Research Symposium*, Cincinnati, EPA/600/9-85/028 (1985).
8. Brooks, C. S., "Nickel Metal Recovery from Metal Finishing Industry Waste", Proceedings of 42nd Industrial Waste Conference, May 12–14, 1987, Lewis Publishers, Chelsea, MI (1988).

81 DESIGN, CONSTRUCTION AND OPERATION OF INDUSTRIAL WASTEWATER TREATMENT PLANT

Arul Ayyaswami, Project Engineer
Randall K. Drazba, Principal Engineer
Floyd Browne Asscciates, Inc.
Marion, Ohio 43302

INTRODUCTION

The United States Environmental Protection Agency (USEPA) effluent guideline limitations for the metal finishing point source categories limit the allowable discharge of heavy metals to navigable waters and to publicly owned treatment works (POTW). Communities have often placed additional restrictions in an effort to protect wastewater treatment plant (WWTP) performance and to extend usable life of sludge land application sites.

The following is a case history of design, construction and operation of metal finishing wastewater treatment facility for DTR Industries, a Japanese manufacturing firm, for their manufacturing plant located in Bluffton, Ohio. This company uses a zinc phosphating process to treat the metal portion of the anti-vibration rubber parts in preparation for painting. The spent rinse wastewater from this process is regulated by the metal finishing pretreatment standards for new sources, Federal Register 40 CFR 433.

Industrial wastewater enters the Industrial Wastewater Treatment Plant (IWTP) from the zinc phosphate treatment system. The phosphate system prepares metal parts for painting by immersing the parts in a series of tanks which are essentially used for degreasing, cleaning and treatment with chemicals. Rinse tanks which are included with the phosphate system provides the source of wastewater requiring pretreatment before discharge to the POTW.

Floyd Browne Associates, Inc. was contracted by DTR Industries to design, construct and place into operation an IWTP to pretreat this waste. Floyd Browne Associates, Inc. a wholly owned subsidiary of BMA Technologies served as a general contractor.

DTR Industries placed special emphasis on space constraints, and designed effluent quality to meet the pretreatment limits as established by the Village of Bluffton.

CONCEPTUAL DESIGN

DTR Industries provided FBA with chemical characteristics of wastewater from a similar plant which was built in Japan. From the review of the chemical characteristics, it was seen that zinc has exceeded the pretreatment limit established by Village of Bluffton's POTW. A treatment plant has to be designed to meet the limits without any conventional jar tests.

Initially FBA proposed a batch operating systems. The plant will receive an average flow of 10,000 gallons per day (gpd) from DTR's manufacturing plant. The pretreatment facility will consist of two painted steel batch tanks with mixers and pH controller. Decant from the batch tank will flow to the pH control tank with mixer and pH controller to adjust final pH before discharge to the Bluffton Sanitary Sewer System. Sludge from the batch tanks will be pumped by air operated diaphragm pumps to filter press. The filerate from the filter press will be pumped back to batch tanks. Dewatered sludge from the filter press will be discharged into 55 gallon drums for final disposal.

DTR Industries preferred a continuous flow system instead of original batch treatment system that was proposed. The design was modified to provide a continuous flow treatment system. The modified treatment process consists of flow equalization, chemical precipitation, sedimentation and filtration. The filtered effluent is stored in effluent storage tank and discharged to Bluffton's POTW. The solids are thickened using a sludge decant tank and dewatered using a filter press. The dewatered sludge cake is disposed at sanitary landfill. The block flow diagram of the process is shown on Figure 1. The basic design data for the treatment plant is as follows:

45th Purdue Industrial Waste Conference Proceedings, © 1991 Lewis Publishers, Inc., Chelsea, Michigan 48118. Printed in U.S.A.

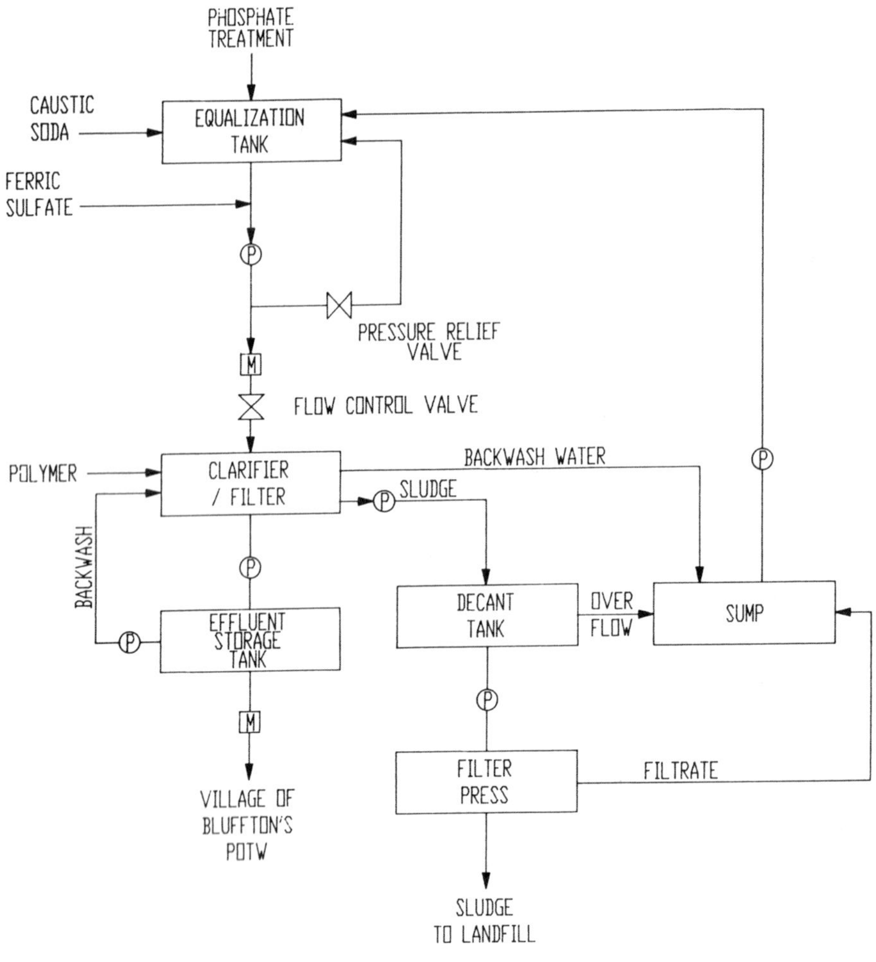

Figure 1. Industrial wastewater pretreatment plant block flow diagram.

BASIC DESIGN DATA

Treatment Required

These limits are based upon Village of Bluffton POTW pretreatment requirements.

Parameter	Concentration (mg/L)
COD	450
BOD	250
Cadmium	0.04
Copper	0.45
Cyanide – total	0.084
Iron (Soluble)	50
Lead	0.9
Nickel	2.5
Oils and grease	50
Silver	0.042
Suspended solids	300

Total chromium	0.28
Zinc	0.8
pH (Standard Units)	6.5–9.5

Treatment Units

Equalization Tank

Number and size	1@8 ft × 8 ft × 10 ft SWD
Capacity	4,800 gal
Mixing	Mechanical mixing

Clarifier/Filter

Number and size	1@7.5 ft × 2 ft × 7 ft SWD
Average Flow	12 gpm
Maximum Flow	16 gpm

Effluent Storage Tank

Number and size	1 @ 5.5 ft × 4.5 ft × 11 ft SWD
Capacity	2,000 gal
Mixing	Mechanical mixing
Detention Time at Peak Flow	2 hrs

Decant Tank

Number and size	1 @ 4 ft × 4 ft × 4 ft SWD (60° hoppered bottom)
Storage Capacity @ 2% Solids	890 gals 150 lbs

Filter Press

Number	1
Sludge Loading	160 lb/day
Feed Solids Concentration	3–5%
Cake Solids Concentration	30%
Cake Bulk Density	85 lb/cu ft
Size of Filter Press	3 cu ft
Number of Chambers	10

Influent Feed Pumps

Number and Type	2-Inline Process Pumps
Capacity	20 gpm at 28 ft TDH
Material of Construction	316 stainless steel

Sludge Transfer Pump

Number and Type	1 - Air-operated diaphragm pump
Capacity	20 gpm at 10 TDH

Clarifier/Filter Pump

Number and Type	2 - Motor mounted centrifugal pumps
Capacity	16 gpm at 32 ft TDH

Sandfilter Backwash Pump

Number and type	1 - Inline Process Pump
Capacity	75 gpm at 24 ft TDH

Sump

Number and size	1 @ 8 ft long × 5 ft wide × 2.5 SWD
Capacity	750 gal

Sump Pump

Number and Type	2 - Self-priming pump
Capacity	20 gpm at 20 ft TDH

Chemical Feed System

Caustic Feed System	
Number	1
Feed Concentration	50% solution
Caustic Source	55 gal drum
Feed Pump	LMI metering pump
Feed Range	0–7 gph
Ferric Sulfate Feed System	
Number	1
Feed Concentration	30%
Ferris Sulfate Source	55 gal drum
Feed Pump	LMI metering pump
Feed Range	0–7 gph
Polymer Feed System	
Number	1
Polymer Source	5 gal pail
Feed Pump	Stranco Polyblend
Feed Range	0–1 gph of neat polymer
Drum Pumps	
Number	2
Capacity	30 gpm @ 30 ft TDH

SYSTEM DESIGN

As indicated earlier, major problems during design is the plant layout. The client indicated that the plant should be constructed within 336 square feet and a minimum head room requirement as well. This changed the geometry of tanks from circular to rectangular tanks to make room for all the available space. Also the horizontal centrifugal pumps were changed to inline pumps for easy access, operation and maintenance. Figure 2 presents the layout of the plant.

Equalization Tank

Equalization tank is provided to minimize flow and concentration variations in the wastewater. This is accomplished by controlling the liquid volume and mixing the basin contents. Mixing is provided by the high efficiency turbine mixer. The mixer is operated manually by an ON/OFF selector switch.

Influent Pumps

Two influent pumps are provided by pumping wastewater from the equalization tank to the clarifier. The pumps are provided with HAND/OFF/AUTO selector switch. In AUTO mode the pumps are controlled by the levels in the equalization tank. The flow to the clarifier is designed to operate at a constant feed rate. A motor operated diaphragm valve is provided to maintain a selected flow rate. A pressure relief valve is provided between the pumps and motor operated valve to bypass excess flow back to equalization tank.

pH Control System

Two pH systems are provided. One is for caustic addition and second system is provided for future acid addition for plant effluent. The caustic addition can be operated by two optional modes. The system can be programmed to be controlled as a cascade type or as a feed back type control. When programmed for cascade control the clarifier pH control set point controls the equalization tank

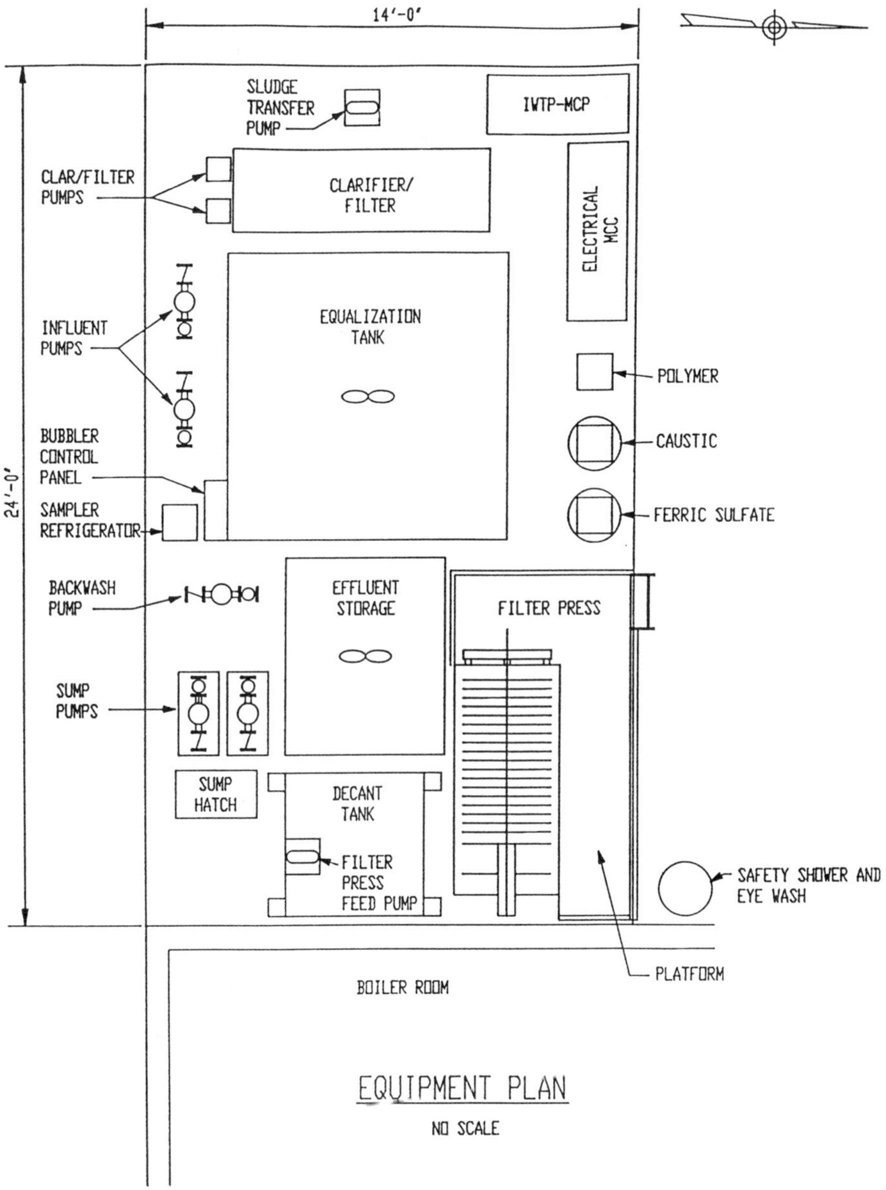

Figure 2. Industrial wastewater treatment plant block flow.

controller which ultimately controls the caustic addition. When Ferric Sulfate is added between two pH probes cascade control is useful in maintaining the desired pH control in the clarifier. This is suited for fluctuating influent pH.

The feedback control utilizes equalization tank controller and a pH probe to control the pH in the equalization tank. This type of control is best suited for small changes. The final pH controller is designed for future acid addition to lower the effluent pH.

The pH adjusted influent waste is pumped from equalization tank to the plate clarifier where chemicals are added. Ferric Sulfate is added to the suction of influent pumps and the polymer is added at the clarifier.

Clarifier/Filter

As the wastewater flows upwards to the clarification zone, the flocculated solids settle on the reclined plates and then drop to the bottom of the clarifier clear water flows over a weir at the top to the top of sand filter.

The sludge settles to the bottom of the clarifier (sludge hopper) and periodically drawn off using sludge pumps.

The clarifier water overflow onto a sandfilter. Th filter is designed to remove small concentrations of suspended solids. The filter is designed as a down flow type. The filter media consist of six inches of 6–10 mesh gravel at the bottom. Ten inches of 0–5 mm sand and top layer of 20 inches of anthracite coal.

The filtered water is drawn off the bottom of the filter by one of two clarifier/filter pumps. The float control valve controls the pump discharge to match the incoming flow when the filter level is low, the float level trips a proximity switch which stops the clarifier/filter pump. Periodically, the filter will require backwashing due to solids accumulation in the filter media causing the water level to rise in the filter. The filter can be backwashed automatically or manually by a push-button on the clarifier panel. A high level float switch located below the tank overflow initiates a backwash automatically.

The influent pump is interlocked with the backwash operation. The influent pump is shut off automatically until the backwash cycle has been completed.

Effluent Storage Tank

The clarifier/filter pumps discharge water to the effluent storage tank. This tank is designed to store water for backwashing the filter. A mixer is installed in the tank which mixes the filtered effluent and also the mixer can be used for future addition of acid. There are three float switches in the tank which are designed to provide high level alarm, start and stop levels for backwash pumps. The tank discharge flows by gravity to a discharge point in the boiler room and ultimately to the village's sewer.

A solenoid valve from the discharge effluent line is provided to collect an effluent sample.

Solids Handling

The sludge withdrawn from the clarifier is pumped to sludge decant tank for thickening.

Filter Press

The thickened sludge is pumped by a air driven diaphragm pump to the filter press for dewatering. A plate and frame filter press with a capacity of three cubic feet is used for dewatering. The press is capable of producing a sludge cake of approximately 30%.

The filter press has ten polypropylene filter plates which are supported in vertical position. During a press cycle the plates are pressed and sealed together between a fixed end and movable arm under pressure. As the sludge is pumped, it builds pressure and when it is pushed against the filter cloth, water is forced out of the sludge. The filtrate flows by gravity to the sump and is returned to the equalization basin for treatment.

The filter press feed pumps are air-operated pumps which receive air from plant air system. The filter press feed pumps are controlled at filter press control panel. The plates in the filter press are shifted manually to discharge sludge into a drum. A plastic scraper is provided to scrape sludge off the filter cloth as required. The sludge cake is collected in 55-gal drums and then sealed and transported by contract hauler for landfill.

Construction

FBA served as a prime contractor and to provide the related engineering services required for construction of the industrial wastewater treatment.

FBA performed the following services:

Table I. Jar Test Results

Parameter	Test No. 1	Test No. 2	Test No. 3	Test No. 4
Polymer I (mg/L)	5			
Polymer II (mg/L)		5	5	
Polymer III (mg/L)				50
Ferric Chloride (mg/L)			500	
Zinc (μg/L)	1600	1300	780	1110
Influent Zinc = 15,900 μg/L				

1. Scheduled the direct purchase of the following major equipment items.
   A. Process tanks
   B. Clarifier/filter
   C. Filter press
   D. Pumps
   E. Mixers
   F. Chemical Feed System
   G. Valves
   H. Sampler
   I. Electrical item (motor control panel and motor control center)
   J. pH system and flow meter
2. Subcontracted to prepare a structure foundation slab to support the IWTP equipment, new sump, sump hatch, trench drain and below slab piping.
3. Subcontracted to install the direct purchase equipment.
4. Subcontracted to furnish and install the remaining equipment, piping, platform, etc. required for IWTP.
5. The demolition of existing floor slab, design of sump and new floor slab, sump hatch, trench drain, floor drains and below slab was completed by others.

Listed below are the problems encountered and changes that were made in the field.

Concrete Floor Paint

Ameron 100 was the concrete paint that was specified. Because of the long curing time of 30 days a urethane paint manufactured by Coronada, Inc. was substituted. This essentially reduced the curing time by 28 days.

Polymer Feed

A dry polymer feed system replaced with a Stranco polymer feed system to reduce the required manpower for polymer solution preparation.

Clarifier/Filter

The clarifier/filter unit was purchased as a standard unit to reduce the delivery time and cost of the equipment. Modifications were made in the field to incorporate an additional clarifier/filter effluent pump.

Schedule

The direct purchase of equipment was initiated in February 1989. The delivery of equipment took about 8–10 weeks. The slab modification at the site was started in February 1989. The construction of the plant was started in May 1989 and completed by August 1989. The plant was started up in September 1989.

Startup

During startup of the plant, the wastewater was collected and several jar tests were conducted with caustic, polymer and ferric chloride. From the visual examination of the wastewater, very low suspended solids were noticed in the influent. The results of jar tests are shown in Table I.

From the review of jar test data it was seen that the removal of zinc was 99% with the addition of ferric chloride and polymer. (Test No. 3)

The same removal of zinc cannot be duplicated in the field because of the shearing of the floc in the clarifier/filter. Modifications were made in the field by changing the location of polymer application within the flocculation zone. Good removals of zinc were accomplished after the modification to meet the pretreatment requirements.

CONCLUSIONS

The above system is a compact, safe, easy to operate means of treating metal finishing wastewaters containing zinc. Effluent zinc concentrations are well below the limitations established by the Village of Bluffton's pretreatment requirements. The treatment plant was constructed within the limited space and head room available at the manufacturing facility and theoretical knowledge of the chemical treatability of the wastewater.

82 MANAGING SOLID RESIDUES FROM A TREATMENT PROCESS FOR PRIORITY METALS

Douglas T. Merrill, Managing Engineer
Brown and Caldwell Consultants
Pleasant Hill, California 94523

Mary E. McLearn, Senior Project Manager
Electric Power Research Institute
Palo Alto, California 94303

Iron adsorption/coprecipitation is an economical process that, under the appropriate conditions, can remove dissolved and particulate priority metals to the extremely low concentrations now being required by regulatory agencies.[1,2,3,4] Treatment involves adding an iron salt such as ferric chloride to the water, unless the water contains sufficient dissolved iron already. An amorphous iron oxyhydroxide precipitate then forms, as shown in Equation 1.

$$FeCl_3 + 3H_2O \longrightarrow Fe(OH)_3 \downarrow + 3H^+ + 3Cl^- \qquad (1)$$

The metals are trapped within (coprecipitated) and adsorbed onto the precipitate, which then settles out, leaving a purified effluent. Removals are pH-dependent, with anionic species, (for example, arsenic and selenium) removed best at pH 6 or below and cationic species (for example, zinc, cadmium, and nickel) best removed at pH 7 and above.

The technology's metal-removal effectiveness is due in part to iron oxyhydroxide's large adsorptive surface. The literature reports specific surface values of approximately 200 to 800 m^2/gr.[1] However, such large-surface solids are a challenge to concentrate, because they release water grudgingly. In our studies, underflow solids concentrations from sedimentation tanks were 1 to 2% at best. We have used gravity thickening to increase solids concentrations to 5 to 8%, but concentrations of 1 to 3% were more typical. Solids concentrations for cakes produced by pressure filtration varied widely. We have produced cakes with concentrations as high as 40% solids and as low as 6% solids.

This paper summarizes what we have learned about management of iron oxyhydroxide sludges from 12 years of bench- and pilot-scale testing. The paper covers:

- Methods for predicting solids production
- Solids thickening techniques
- Solids dewatering and drying procedures
- Results of hazardous waste toxicity evaluations
- Effects of sludge leachate on clay landfill liners and the effects of the liners on leachate composition
- Reuse of iron oxyhydroxide solids

ESTIMATING SOLIDS PRODUCTION

In this paper, we define solids production as the quantity of solids (free water excluded) generated by waste treatment operations. Total solids production includes native wastewater solids, solids generated by chemical precipitation, bound water* associated with hydroxide sludges, and sludge conditioners. The solids produced are discharged in the waste sludge and in the process effluent. If solids separators are performing adequately, solids production is approximately the weight of solids (free water excluded) that the sludge management system must process and dispose of.

Equation 2 is a simplified formula for estimating approximate solids production:

$$P_p = [S + H\ (1 + A)]\ [1 + C] \qquad (2)$$

45th Purdue Industrial Waste Conference Proceedings,

*Bound water is defined here as water that is not released from the solids at the temperature of the standard moisture analysis (105° C) but is released upon heating at 550° C. Bound water is counted as a solid in the standard moisture analysis.

where: P_p = Solids production, milligrams per liter of wastewater treated
S = Total suspended solids in the process influent, mg/L
H = Metal hydroxide production, mg/L
A = Bound water, as a fraction of H, mg/L
C = Sludge conditioners, as a fraction of all other solids, mg/L

H is defined as follows:

$$H = \sum (M_t - M_c) F \qquad (3)$$

where: M_t = Concentration (dissolved) of individual metals in the process influent *and* chemical reagents, based on milligrams per liter of wastewater processed.
M_c = Solubility of individual metals, mg/L, as shown in Figure 1
F = Conversion factor (see Table I)

If M_t is less than M_c for any metal, that metal will not precipitate, and thus will not contribute to solids production. Note that Figure 1 is not valid if strong complexing agents (EDTA, for example) are present in the wastewater.

As a calculation example, assume that a wastewater containing 20 mg/L of total suspended solids, 2 mg/L of aluminum, and 1 mg/L of iron is treated with 15 mg/L of iron. Bound water is 30% of the hydroxide solids production and the sludge container (lime) is 25% of all other solids. The reaction pH is 9.0. Then:

1. S = 20 mg/L
2. Components of H are:
   a. $Al(OH)_3$ = 0 (neglect, because M_t is less than M_c)
   b. $Fe(OH)_3$ = $(1 + 15 - 10^{-5})$ 1.91 = 31 mg/L
3. A = 0.30
4. C = 0.25
5. P_p = [20 + 31 (1 + 0.30)] [1 + 0.25] = 75 mg/L

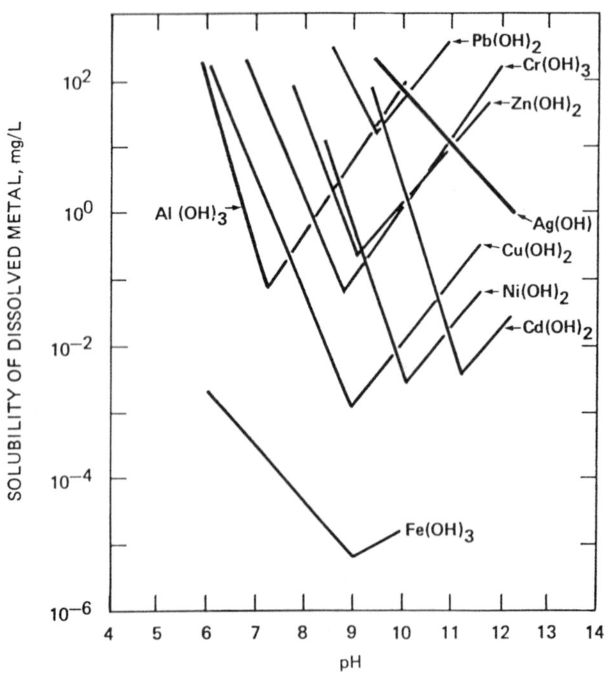

Figure 1. Solubility of metal hydroxides as a function of pH. Source: Reference 15.

Table I. Metal Hydroxide Conversion Factors for Solids Production Calculations

Metal	Metal Hydroxide Assumed Present	F (Converison Factor)[a]
Zinc	$Zn(OH)_2$	1.52
Copper	$Cu(OH)_2$	1.54
Nickel	$Ni(OH)_2$	1.58
Lead	$Pb(OH)_2$	1.16
Iron	$Fe(OH)_3$	1.91
Aluminum	$Al(OH)_3$	2.89

$$^{a}F = \frac{\text{Molecular weight metal hydroxide}}{\text{Molecular weight metal}}$$

Calcium carbonate and gypsum may precipitate if a relatively hard, alkaline water is treated at pH values exceeding approximately 7.5, particularly if lime is used for pH adjustment. Also, calcium phosphates and iron phosphates may precipitate in phosphate-bearing wastewaters. Equation 2 does not consider any of these compounds. Their production may be estimated with SEQUIL, a computerized water chemistry code[5] developed by the Electric Power Research Institute (EPRI).

SOLIDS THICKENING

Several devices can be used for thickening, including gravity thickeners, dissolved-air flotation thickeners, centrifugal thickeners, and belt thickeners. Results from gravity thickening tests are reported here. The objective of our tests was to determine thickened sludge concentrations that could be obtained at given solids loadings in continuous gravity thickeners. We used a batch thickener and solids flux theory described by Wilhelm and Naide[6] to make these determinations.

A continuously loaded gravity thickener from which solids are removed at a fixed rate has a maximum flux (lbs solids/day/ft^2) at which solids can settle to the thickener bottom. This rate is called the limiting solids flux, G_L. If the thickener is loaded at a rate greater than G_L, the thickening process will fail. Failure occurs because solids that cannot reach the thickener bottom accumulate in the sludge blanket, causing it to expand and eventually overflow from the top of the thickener. Ideally, the thickener would operate at G_L. In practice, we recommend operation somewhat below G_L to provide a safety factor against variability in thickener feed quality.

Several factors affect solids underflow concentrations.

1. *Applied Solids Flux*. The curves on Figure 2 were developed from batch thickening tests with iron sludges.[2,3] They predict results that would occur in flux-limited continuous gravity thickeners with sludges generated at two test sites. Underflow concentrations decrease as G_L increases, other factors being equal.

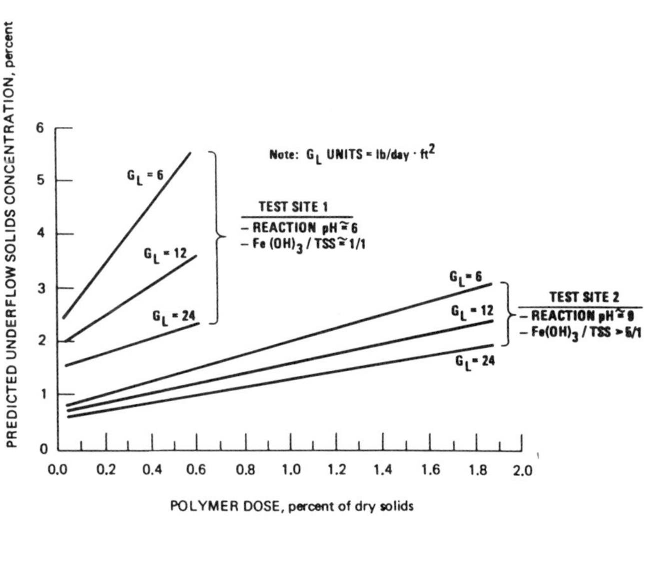

Figure 2. Effect of solids loading rate flux (G_L) and polymer dose on predicted solids underflow concentration for sludges from two different sites.

2. *Reaction pH.* Knocke et al.[7] showed that conditions governing formation of the chemical solids dramatically affect their thickening properties. Specifically, solids precipitated at relatively low pH values (for example, 6), thicken better than solids precipitated at higher values (for example, 9). They attributed the changes in sludge behavior to changes in the size and relative water content of the floc. Enmeshment coagulation, favored at high pH, tends to yield larger floc with significantly more incorporated water than floc formed at low pH, where adsorption is the coagulation mechanism. The sludges tested at Site 1 were mainly formed at pH 6. The sludges tested at Site 2 were precipitated at pH 9. Figure 2 shows that Site 1 sludges thickened better than Site 2 sludges. However, this improvement may have been partly caused by the lower ratio of lighter iron solids to denser native solids at Site 1.
3. *Iron Solids/Native Solids Ratio.* Knocke et al.[7] also demonstrated that thickening performance deteriorated when the concentration of iron solids increased relative to the water's initial turbidity. High iron/native solids ratios also encourage enmeshment coagulation, hence greater water retention. At Site 1, the ratio of iron solids ($Fe(OH_3)$) to native wastewater solids was typically 1 to 1; at Site 2, the ratio typically exceeded 5 to 1.
4. *Polymer Conditioning.* Figure 2 shows that the thickened sludge solids concentration increases as the polymer dose increases. Polymer conditioning produces a sludge with a visibly more open structure than unconditioned sludge. Hence water released from the conditioned sludge is displaced upward through the conditioned sludge matrix with less uplift on the downward-moving sludge particles. The result is better thickening.

Reference 4 provides a design example for gravity thickening.

DEWATERING

Mechanical devices such as centrifuges, belt filters, vacuum filters, and filter presses can be used to dewater sludges from water and wastewater treatment plants. Our tests focused on filter presses. They produce the driest product of any mechanical device, which is particularly important for difficult-to-dewater sludges like iron oxyhyroxide.

We successfully simulated sludge filtration with a model derived by Brown.[8] References 2, 3, and 4 provide design examples. The model showed that filtration performance was most greatly affected by filtering time, feed solids concentration, filtration pressure and sludge conditioning (as characterized by the sludge filtration constant, K).

Figure 3 illustrates the effects of these variables. We developed the curves from Brown's equations, assuming different values for filtering time, feed sludge solids concentration, filtering pressure, and filtration constant. The base case sludge has a feed solids concentration of 2% ($X = 0.02$) and is conditioned to have a filtration constant of 0.02 $lb.in^2/hr.ft^4$ ($K = 0.02$); the filter pressure is 100 psig ($\triangle P_m = 100$). The other curves were developed with all parameters held at base conditions, except for the parameter noted on each curve.

Figure 3 shows that dilute sludges take longer to process. With other factors being equal, filtering time to achieve a desired cake solids concentration is inversely proportional to feed solids concentration. For example, the base case sludge (2% solids; $X = 0.02$) must be filtered for about 360 minutes to achieve a cake solids concentration of 25%. In contrast, a 4% sludge ($X = 0.04$) must be filtered 165 minutes (or 46% as long) to obtain the same cake solids concentration. We note that Site 2 sludges took much longer to dewater than Site 1 sludges and attribute the reduced performance to the fact that Site 2 feed sludges were much more dilute.

Figure 3 also shows that filtration time required to achieve equal cake solids concentrations is inversely proportional to pressure, other factors being equal. We advise caution, however, in predicting results for operations at one pressure from data obtained at a different pressure. At high pressure, pores in some cakes can become compressed and constricted, offering greater resistance to filtrate flow. The effect is a reduction in the apparent filtration constant. Cakes demonstrating this characteristic are known as compressible cakes. Iron oxyhydroxide cakes are believed to be moderately compressible.

The filtration constant K characterizes solids in terms of filterability. Large values of K suggest filterable solids; smaller values of K connote less filterable solids. Other factors being equal, a low filtration constant lengthens the time required to get a given solids concentration in the cake. This situation is demonstrated on Figure 3, where the base case sludge ($K = 0.02$ $lb.in^2/hr.ft^4$) requires about twice as much filtering time to achieve a practical cake solids concentration (20 to 40%) as does

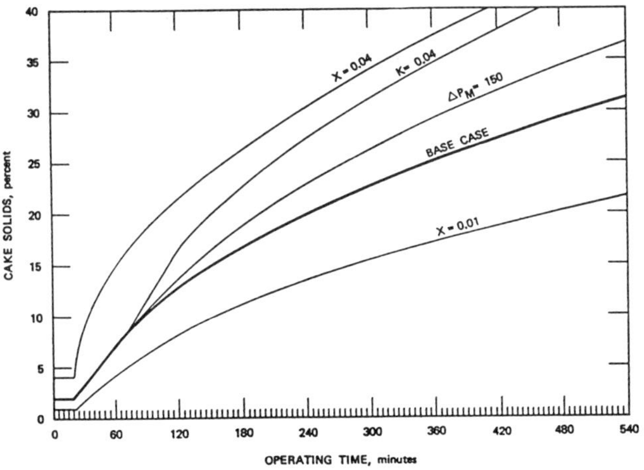

NOTE:

1. PARAMETERS ARE HELD AT BASE CONDITIONS EXCEPT AS NOTED ON THE CURVE (SEE TEXT FOR DEFINITION OF PARAMETERS)

2. BASE CASE CONDITIONS ARE:

   $K = 0.02$ lb. in^2/hr. ft^4

   $\Delta P_m = 100$ psi

   $X = 0.02$

Figure 3. Effect of filtration variables on cake solids concentration.

a sludge with $K = 0.04$ lb.in^2/hr.ft^4. Sludge filterability is usually controlled by chemical conditioning.

Figures 4 and 5 show the results of polymer conditioning and lime/calcium carbonate conditioning on the filtration constant, as measured during Site 1 pilot studies. The filtration constant increases as the conditioner dose increases.

RADIANT DRYING

A recently developed radiant batch dryer may be useful for dewatering of filter-pressed iron hydroxide sludges.[9] The unit consists of an enclosure containing plowshare-like mixing blades and

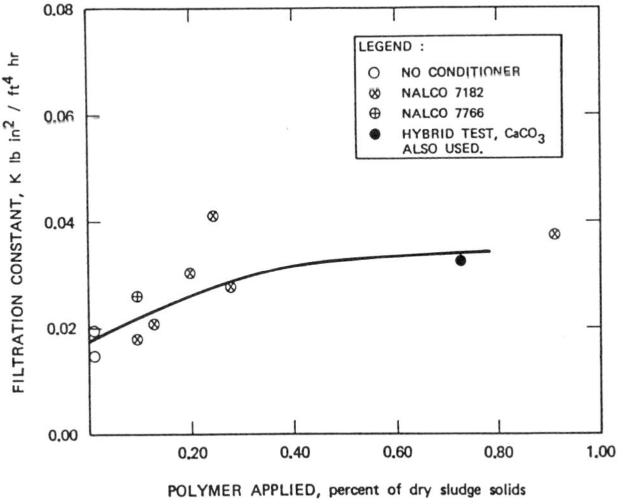

Figure 4. Effect of polymer in conditioning sludge for dewatering.

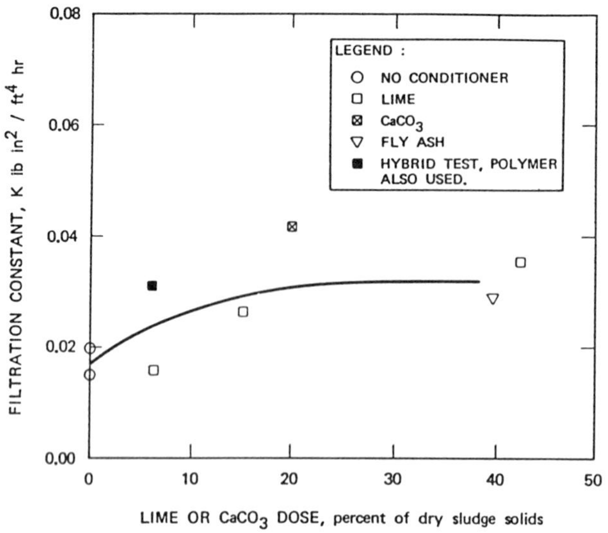

Figure 5. Effect of lime and calcium carbonate in conditioning sludge for dewatering.

infrared heaters. The heaters are mounted in the top of the enclosure and do not touch the sludge. The filter cake is loaded into the chamber and the enclosure is sealed. The heaters direct radiant energy to the sludge. The mixer blades break up the cake and continually expose new wetted surface to the radiant energy. The unit is shut down and unloaded once moisture has been reduced to the desired degree.

The vendor claims high heat transfer efficiency. The heat source can be tuned to emit infrared energy in the spectral band best absorbed by the surface moisture, and the dryer has no heat transfer surfaces to foul. Furthermore, dusting problems common to convective dryers do not occur. Drying time is primarily a function of the amount of water to be removed and the characteristics of the solid. Drying times vary from 4 to 5 hours for sludges containing more than 20% solids up to 8 to 10 hours for sludges with 8 to 10% solids.

HAZARDOUS WASTE CHARACTERIZATIONS

Metals removed from wastewater are concentrated in the sludge, which increases the possibility of the sludge being categorized as toxic. At Site 1, we tested a sludge generated by treating ash pond effluent and found it to be not toxic according to criteria of the Extraction Procedure (EP) toxicity test.[10] At Site 2, we tested four different sludges with four different procedures. The sludges were generated by the treatment of ash basin effluent, coal pile drainage, air-heater wash, and fly ash landfill leachate. All four sludges passed the EP Toxicity and Toxic Characteristics Leaching Procedure (TCLP)[11] tests. However, sludge generated by the treatment of coal pile drainage failed California's Waste Extraction Test[12] because cadmium, lead, and nickel concentrations in the sludge leachate exceeded the soluble threshold limiting concentrations of 1, 5 and 20 mg/L, respectively. This sludge would be designated as a Class I (hazardous) waste in California. Interestingly, this same sludge passed the Texas Distilled Water Leachate Test,[13] thereby qualifying as a Class III (inert) solid waste in Texas.

INTERACTIONS BETWEEN SLUDGE LEACHATE AND CLAY LANDFILL LINERS

Battelle Pacific Northwest Laboratories conducted limited testing to investigate the interactions between the iron sludge leachates and clay liners in sludge-only landfills.[3] The study provided information for understanding: 1) how sludge leachates affect the permeability of the clay liners; and 2) how the liners affect leachate composition as the leachate migrates through the liners.

Two clays were tested. The first (Lucky Mc clay) was an inorganic clay of high plasticity, native to the gas hills area of Wyoming. The second (Montour clay) was a sandy clay of low-to-medium plasticity taken from Site 2.

The Lucky Mc clay, which originates from a poorly weathered site, contains much higher levels of soluble components (calcium, magnesium, sodium, and sulfate) than the Montour clay, which comes from a highly weathered site. The Lucky Mc clay also has a higher calcium carbonate content, a higher soil paste pH, and a higher plastic index than the Montour clay. The higher plastic index means the Lucky Mc clay can withstand higher stresses than the Montour clay.

The sludge selected for testing was generated by treatment of air-heater wash. It was rich in iron (18%), calcium (14%), aluminum (9%), and sulfur (4%). Its chromium, manganese, and zinc levels exceeded 1,000 mg/kg. Its copper and nickel levels exceeded 650 mg/kg, and its soil paste pH was 12.5.

During the tests, researchers percolated synthetic rainwater through columns containing sludge only, clay one, and sludge overlying a layer of clay. Sludge to clay ratios in the clay-plus-sludge columns were 15/2 to simulate ratios that might occur in sludge landfills. Percolation rates and column effluent compositions were monitored with time. Comparison of the percolation rates of the clay-only and clay-plus-sludge columns indicated how the sludge leachate affected liner permeability. Comparison of the effluent compositions from the sludge-only, clay-only, and clay-plus-sludge columns showed how leachate composition was modified during leachate migration.

The average permeabilities of the Montour and Lucky Mc clays were slightly below 1×10^{-7} cm/sec, which satisfies USEPA permeability requirements for landfill liners. The permeabilities of both clay-plus-sludge columns were 1 to 2 orders of magnitude lower, indicating the liners were being sealed as the result of leachate/clay interactions.

Chromium appeared to break through both Montour and Lucky Mc clays, particularly the Lucky Mc clay. We do not know whether the chromium was trivalent or hexavalent chromium because the test did not include chromium speciation. No other priority metal that originated in the sludge migrated through either liner in significant quantities. Significant amounts of sodium and sulfate and some chloride migrated through the Montour clay. Iron was leached from the Montour clay.

Analyses of liners before and after the experiments showed few noticeable mineralogical changes. The amount of chromium in both liners increased, and the sodium content of the Montour clay increased. Results showed no evidence of particle migration from the sludge into either liner, suggesting that the sludge was physically stable.

When the column containing the Montour clay-sludge sample was opened, the clay was found to be very plastic (of slurry-like consistency). This result suggests that the Montour clay may not be suitable liner for a landfill containing this sludge. No such problem occurred with the Lucky Mc clay.

SLUDGE REUSE – AN ALTERNATIVE TO SLUDGE DISPOSAL

Sludge disposal can be expensive, particularly if the sludge is classified as a hazardous waste. Researchers at the University of Washington (UW) have recently investigated a sludge reuse concept with potential to reduce the amount of iron sludge for ultimate disposal, as well as the amount of reagent iron salts purchased.[14]

In conventional processing, iron solids are used only once. They adsorb metals from the water, then are separated from the water, thickened, dewatered, and disposed of. The reuse concept relies on the fact that adsorption reactions are reversible and pH-dependent. In the conceptual diagram shown on Figure 6, cationic metals adsorbed in the pH range 9 to 10, are subsequently stripped from the iron solids in an acid regeneration unit operated at pH less than 5. The pH in the regeneration unit is low enough to absorb the metals of interest, but not so low as to dissolve the iron solids.* The stripped iron solids are then separated from the liquid and recycled to the adsorption tank to treat more wastewater. The separated liquid, rich in metals, is returned to the regeneration unit. The desorbed metals accumulate within the recycled regenerant. Operators control the concentration of the desorbed metals by bleeding a small portion of the regenerant to waste.

The system on Figure 6 is designed for removal of cationic metals. A basic regeneration system could probably remove anionic metals.

The UW researchers simulated the treatment scheme on Figure 6 in bench-scale experiments using two synthetic wastewaters and a real waste from a metal plating shop. They simulated 50 operating cycles. The primary test objective was to determine whether repeated regeneration and reuse of the iron solids would affect their ability to absorb metals from the wastewater.

The metals of interest were zinc, copper, nickel, cadmium, lead, chromium (III), and chromium (VI). The wastewater metal levels were in the 2 to 6 mg/L range. The iron solids concentration was

*Iron solubility is low, less than 1 mg/L over the pH range 2.5 to 13.

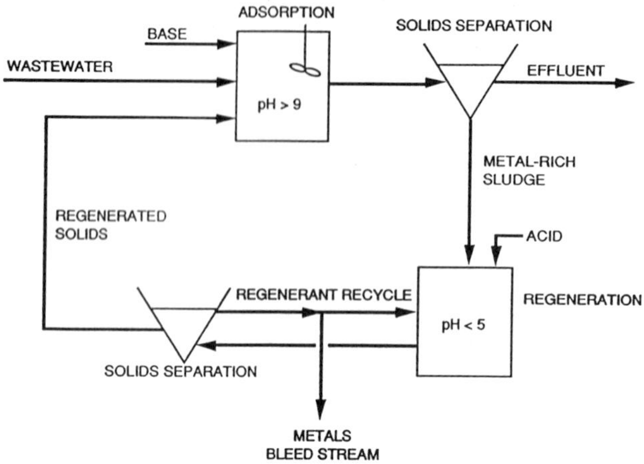

Figure 6. Conceptual diagram for reuse of iron solids—cationic metal removal system.

approximately 10^{-2} M (1,068 mg/L $Fe(OH)_3$). The metal residuals (except for Cr(VI), which was not expected to be removed) were low and constant over the entire 50 cycles. The average metal removals (except for Cr(VI)) were about 98%.

Solids regeneration was tested at two different pH values, 3.5 and 4.5. Cadmium, zinc, and nickel were essentially all released from the solids at both pH values. Copper was nearly all released at pH 3.5 and about 80% released at pH 4.5. Even though very little chromium (III) was released during regeneration, the recycled solids continued to remove chromium (III) from the wastewater.

The regeneration process does not eliminate the metals; it simply concentrates them. If they can be recycled in solution form for beneficial reuse (for example, back to the manufacturing process) so much the better. The fact that they are concentrated usually enhances their reuse value.

If the metals cannot be recycled, they must be removed from solution in a form suitable for ultimate disposal. Again, their greater concentration improves treatment economics, because the cost per pound of metal removed drops sharply as metal concentrations increase. A fairly simple system like hydroxide precipitation is effective in obtaining high metal removals from concentrated solutions.

SUMMARY

Sludge produced from the iron adsorption/coprecipitation process are often thin and difficult to dewater.

The amount of sludge produced (sludge production) includes native wastewater solids, hydroxides formed by the reactions of water with metals in the waste stream and reagents, bound water, sludge conditioners, and could, under certain conditions, contain calcium carbonate, gypsum, calcium phosphate, or iron phosphate.

Gravity thickening tests have produced iron sludges with solids concentrations of 5 to 8%, but solids concentrations of 1 to 3% are more typical. Factors that appear to affect underflow sludge solids concentrations include solids loading rate, polymer conditioning, pH at which the solids were generated, and ratio of iron solids formed to native solids.

Filter press tests produced iron sludges with solids concentrations as high as 40% and as low as 6%. Factors that appear to affect cake solids concentrations are filtration time, feed sludge solids concentration, filter pressure, and sludge conditioning. Use of polymers and lime as conditioners improved sludge concentrations to some extent. Sludges formed under relatively high pH conditions (for example, 9) and sludges containing high iron-to-native solids ratios have proved difficult to dewater. We attribute this behavior to the fact that such sludges do not thicken well, and the thickened sludges are thus too dilute to dewater well.

Batch radiant drying appears to have potential to further reduce the water content of iron hydroxide filter cakes.

All sludges tested for toxicity passed the EP Toxicity and TCLP tests. One sludge generated by treatment of coal pile drainage failed to pass California's Waste Extraction Test.

Laboratory tests conducted on a limited number of sludges and liners suggest that disposing of iron sludges in clay-lined landfills does not adversely affect liner permeability. In fact, leachates from the sludges appear to seal the liners. Except for chromium, no priority metal from the sludge migrated through the liners.

Iron solids reuse has the potential to reduce sludge disposal and chemical costs. The process concentrates the metals but does not prepare them for ultimate disposal, unless the concentrated metal stream can be recycled to the manufacturing process for reuse. In bench-scale simulations, University of Washington researchers found that recycled solids continued to remove high percentages of cations (zinc, copper, nickel, cadmium, and chromium (III)) from synthetic and real wastewaters over 50 cycles of regeneration and reuse. They regenerated the iron solids by desorbing the adsorbed metals at pH values less than 5. The metals accumulated in the liquid regenerant.

REFERENCES

1. *Adsorption/Coprecipitation of Trace Elements From Water With Iron Oxyhydroxide*, Palo Alto, Calif.: Electric Power Research Institute (September 1980).
2. *Field Evaluation of Arsenic and Selenium Removal by Iron Coprecipitation*, Palo Alto, Calif.: Electric Power Research Institute (May 1987).
3. *Trace Metal Removal by Iron/Coprecipitation: Field Evaluation*, Palo Alto, Calif.: Electric Power Research Institute (June 1989).
4. *Process Design Manual: Trace Element Removal by Iron Adsorption/Coprecipitation*, Palo Alto, Calif.: Electric Power Research Institute (In press).
5. *SEQUIL: An Inorganic Aqueous Chemical Equilibrium Code for Personal Computers*, Volume 1, User's Manual/Workbook, Version 1.0. Palo Alto, Calif.: Electric Power Research Institute) (April 1989).
6. J. H. Wilhelm and Y. Naide, "Sizing and Operating Continuous Thickeners," A paper presented at the AIME Meeting in New Orleans, Louisiana (February 18–22, 1979).
7. W. R. Knocke, J. R. Hamon, and B. E. Dulin, "Effects of Coagulation on Sludge Thickening and Dewatering," *Journal of American Water Works Association* (June 1987).
8. T. R. Brown, "Designing Batch Pressure Filters," *Chemical Engineering* (July 1982).
9. G. Picciotti and D. Lucas, "Thermal Reduction of Dewatered Hydroxide Sludge Using Direct Radiant Heat," *Plating and Surface Finishing* (April 1985).
10. *Code of Federal Regulations*, 40 CFR, Part 261, U.S. Government Printing Office, Washington, D.C. (1987).
11. *Federal Register*, Volume 51, p. 40636, Office of the Federal Register, Washington, D.C. (November 7, 1986).
12. *California Administrative Code*, Title 22, Section 66693–66723 (January 1985).
13. *Waste Evaluation/Classification, Technical Guide Number 1*, Texas Water Commission (revised May 1982).
14. M. Edwards and M. M. Benjamin, "Regeneration and Reuse of Iron Hydroxide Absorbents in Treatment of Metal-Bearing Wastes", *Journal Water Pollution Control Federation* (April 1989).
15. U.S. Environmental Protection Agency, *Treatability Manual: Treatability Data*, EPA-600/2-82-001C (1981).

83 DEVELOPMENT OF A TREATMENT SCHEME FOR SPENT CLEANING SOLUTIONS USING STATISTICAL PROCESS CONTROL CONCEPTS

Kenneth E. Kaszubowski, Manager
Solid Waste and Wastewater Compliance

James A. Bilgo, Senior Environmental Chemist
Kohler Co.
Kohler, WI 53044

INTRODUCTION

Kohler Co. is a world-wide manufacturer of commercial and household plumbing products, small internal combustion engines, generator sets, furniture, and other household goods. The manufacturing facility located in Kohler, Wisconsin, produces vitreous china, enameled cast-iron fixtures, brass plumbing fittings, and internal combustion engines.

The Brass Faucets and Fittings Division of the Kohler, Wisconsin, facility manufactures brass plumbing fittings, which are finished by a variety of metal finishing operations within the Brass Building. The wastewaters generated by the metal finishing operations present significant challenges with respect to pollution prevention, optimization of waste treatment methods, and compliance with permitted effluent discharge limits. The manufacturing processes generate wastewaters which are highly variable in composition, volume, and flow rate. Many of the wastewater streams generated contain complexing agents which prevent the precipitation of soluble metal ions from a solution. In some cases, these streams are shipped off-site for treatment and disposal, at significant expense. The remainder of the streams are treated by conventional metal hydroxide precipitation in the Brass Waste Treatment (BWT) facility to meet categorical metal finishing standards. A simplified schematic diagram of the BWT facility is shown in Figure 1.

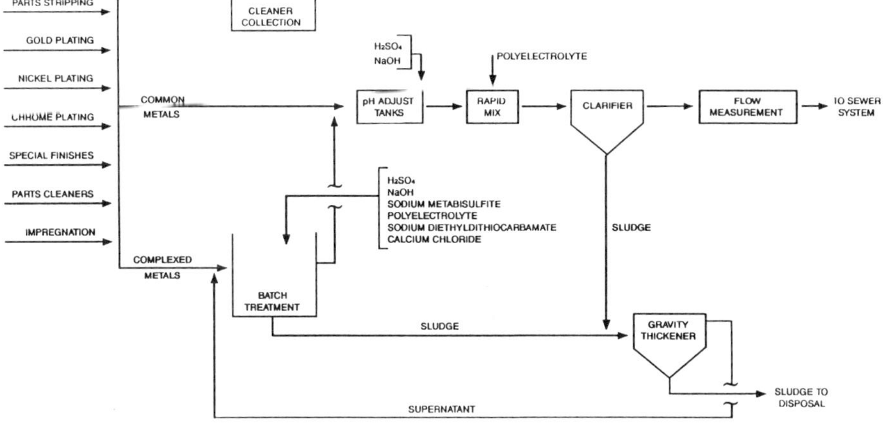

Figure 1. Brass waste treatment schematic diagram.

45th Purdue Industrial Waste Conference Proceedings, © 1991 Lewis Publishers, Inc., Chelsea, Michigan 48118. Printed in U.S.A.

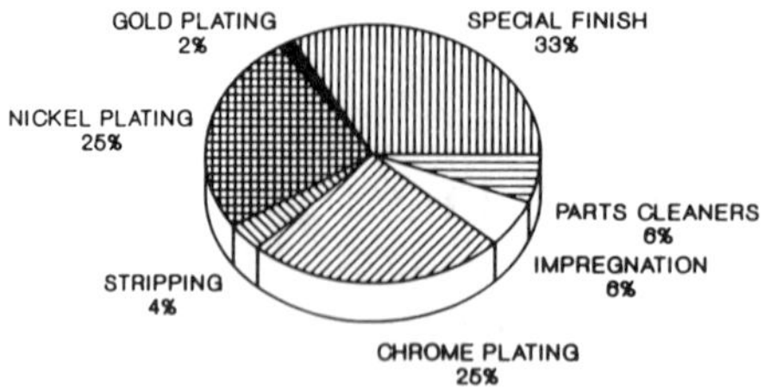

Figure 2. Brass waste treatment, wastewater sources (percent by volume).

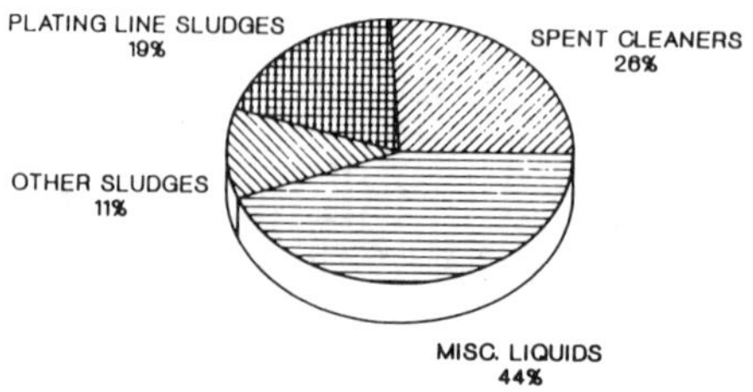

Figure 3. Brass waste treatment sources of disposal costs.

The Brass Division requested the assistance of Kohler's Corporate Environmental Engineering and Safety Department to help them reduce their operating costs, specifically, disposal costs associated with the BWT operation.

The goals of this paper are:

1. to present Kohler Co.'s approach to meet the objective of reduced operating cost for the BWT facility, and
2. to present the development of a strategy utilizing statistical methods for managing a specific waste stream generated in the Brass Division.

The paper is organized as follows:

- Preliminary Operations Audit
- Sources of Cleaner Waste and Management Alternatives
- Experimental Design and Methodology
- Experimental Results
- Discussion of Results
- Full-Scale Implementation

PRELIMINARY OPERATIONS AUDIT

Quantitative Review of Waste Sources and Disposal Costs

A comprehensive operations audit was conducted in order to identify the specific manufacturing operations generating wastewater and contributing to the substantial disposal costs being incurred by the BWT facility. A summary of the relative percentages of the major wastewater sources is presented in Figure 2, on a percent of total volume basis.

The waste disposal costs incurred by the BWT operation include transportation and disposal costs for the treatment process residues (sludges) and untreatable liquid wastes sources. The operations audit resulted in identification of the specific major contributors to waste disposal costs. A summary of these contributing sources of cost is presented in Figure 3, as a percentage of total annual disposal cost.

Conducting an audit of the wastewater generation sources and identifying the primary contributors to waste disposal costs was essential in developing a focused approach to cost reduction. The audit revealed the most likely targets for both disposal cost reduction and waste reduction, as presented in Figure 3. These target areas are described below.

Description of Waste Categories

1. *Miscellaneous Liquids* — The miscellaneous liquids consist of seven individual liquid wastes which are untreatable with technologies available in the BWT operation. These wastes represented the largest component of total annual disposal costs. Detailed information regarding the unit costs associated with each of the seven sources is not available.
2. *Cleaners* — This waste source is the focus of this paper. A detailed description of this waste source is provided in the next section. These wastes represented the largest individual source of total annual disposal costs.
3. *Plating Sludges* — These are metal hydroxide sludges generated by wastewater streams

from the plating operations. The disposal costs associated with these sludges could be reduced by further dewatering, or eliminated by metals recovery.

4. *Other Sludges* – These are metal hydroxide sludges generated from mixed metal waste streams. Disposal costs could be reduced by further dewatering.

Identification of Tasks and Project Plan

A schematic diagram of the project approach is presented in Figure 4. Each of the individual cost components identified in the operations audit provided significant opportunities for both waste reduction and operating cost savings. The project approach was structured to concurrently address all the sources of disposal cost in BWT. However, the efforts centered on the spent cleaner waste source. Since reduction in the disposal costs of other waste sources, required capital investment, the spent cleaners waste was selected for initial focus since it was a segregated stream, represented the individual greatest contributor to disposal costs, and had the potential to be eliminated or reduced without further capital investment.

SOURCES OF CLEANER WASTE AND MANAGEMENT ALTERNATIVES

Description of Spent Cleaner Waste Sources

Brass castings and machined brass parts are finished in the Brass Division by applying corrosion resistant and decorative finishes. The parts are prepared for application of protective and decorative finishes in process baths containing alkaline-based cleaning agents. Periodically, these process baths are replenished to make up for losses from dragout, or changed over in order to maintain effective cleaning capability and product quality. Process bath changeover generates a wastewater stream referred to as spent cleaners. Rinse baths follow the process cleaning bath to remove remaining cleaning material from the cleaned part. The rinsewater represents a second source of spent cleaners.

The spent cleaners are physically segregated from other process wastewater streams generated by production operations, and are conveyed to a segregated collection tank located in the BWT facility.

Management Alternatives for Spent Cleaners

A project team, consisting of members from the Brass Division, Corporate Environmental Engineering, and the Corporate Chemical and Metallurgical Laboratory identified several strategies for reducing the cost associated with disposal of the spent cleaners:

1. *Remove or replace the cleaners.*
   Removal or replacement of the cleaners currently being used could impact the quality of the parts produced in the plating operation. Removal or replacement of these processes would require significant resource allocations for research and development in order to implement a process modification. This strategy was dismissed due to the duration of the research and development effort needed, the low probability of success, and the development cost.
2. *Treat spent cleaners as single waste stream.*
   Treatment of the spent cleaners as a sole source waste stream had been pursued. The most promising process involved chemical treatment and precipitation of soluble metals as hydroxides, settling, sludge withdrawal, pH adjustment, and final effluent polishing by microfiltration. This strategy was pilot tested in August, 1986, and full-scale equipment was installed in April, 1987. Operation of the full-scale system was unsuccessful and ineffective in reliably attaining compliance with 40 CFR 433 discharge limits. Additionally, the BWT operators were reluctant to undertake operation and management of new treatment technologies. This strategy was not pursued a second time.
3. *Treating spent cleaners with other complexed metals waste streams*:
   The spent cleaners are a complexed metals wastewater source. A third strategy was identified involving treatment of the spent cleaners along with other complexed metals wastewaters in the existing batch treatment system. Treatment of all complexed metal waste sources in a single batch treatment mode is a practice commonly employed by most operations regulated under 40 CFR 433. This strategy is consistent with the recommended treatment technology as described in the Development Document for the Metal Finishing Point Source Category. The project team hypothesized that mixing the two complexed metal waste stream sources would reduce the impact of the complexing agents associated with the spent cleaners. An effective treatment might be attained by

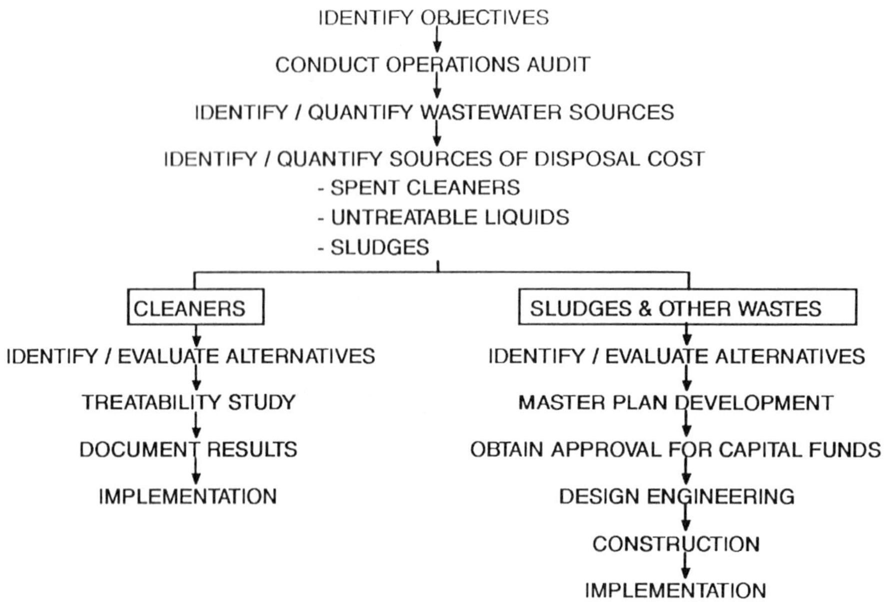

Figure 4. Project approach.

reducing the impact of the spent cleaner complexing agents as opposed to the unsuccessful efforts to treat the spent cleaners as a single waste stream.

The principal advantages for this strategy were:

- Minimal, if any, capital investment required;
- A treatment scheme consistent with current BWT technology;
- Minimal time required to train BWT operators.

The pursuit of this strategy also had the following disadvantages:

- The process might be viewed by regulations as dilution.
- A new treatment scheme might result in permit violations for an existing system which had a successful compliance track record.

The decision was made to pursue the third strategy and demonstrate that the strategy would consistently meet the permit limits for the facility. This was accomplished using sound experimental design and statistical methods. Additionally, the experimental design was structured to demonstrate that the strategy allowed improved metals precipitation from the spent cleaners as opposed to simple dilution.

EXPERIMENTAL DESIGN AND METHODOLOGY

Objectives of Experiment

A designed experiment was conducted to prove that spent cleaners could be treated with other complexed metal wastewaters generated in the Brass Division. There were three objectives of this experimental design:

Objective 1: Statistically demonstrate that a mix of cleaners and complexed metals can be treated to consistently meet permit requirements.

Objective 2: Identify the optimum ratio between cleaners and complexed metals waste streams.

Objective 3: Demonstrate that the treatment removes metals from the spent cleaners, rather than simply diluting them.

To satisfy Objective 1, an experiment was designed to quantify the after-treatment, day-to-day variability in the combined spent cleaners and complexed metal waste streams. This would be done by controlling and reducing the sources of variability that were not of interest. Then, the experiment would be conducted with time, duplicating the treatment process presently used in BWT (Figure 5). By controlling the sources that were not of interest, the assumption could be made that all the variability seen in the results was due to the spent cleaner and complexed metals waste streams changing with time. Two sources of variability that were identified up front and controlled were:

a) *Treatment from Day-to-Day.* This variable was controlled by keeping the treatment process identical from day-to-day. To do this a written procedure was followed that identified the concentrations of each chemical to add and standardized the mixing and settling times.
b) *Analysis of Samples.* Because of the large number of samples, analyses would be run over a month-long period. Therefore, not only would variability be introduced by the procedure and instrument, but some variability would be introduced with time, depending on how the procedure and calibration of the instrument was performed. This variability was reduced, or controlled, by following procedures approved in Wisconsin Administrative Code Chapter NR 219 and by running appropriate quality control checks.

BATCH TREATMENT FOR COMPLEXED METALS WASTE STREAMS

Chemicals Used

Sodium Hydroxide
Sulfuric Acid
Sodium Metabisulfite
Betz 1120 Anionic Polymer
Sodium Diethyldithiocarbamate (DX 2000)

Treatment Steps

1. Lower pH to below 1.0 via addition of sulfuric acid
2. Add bisulfite until the hexavalent chrome test is negative (use diphenlcarbazide test)
3. Raise pH to 9.8 to 10.2 range via addition of sodium hydroxide
4. Add Dx 2000 until flox formation occurs (typically 0.25 gallons per 3,500 gallon batch).
5. Reduce mixer speed to promote flocculation mix for approximately 15 minutes
6. Add Betz 1120 polymer and mix for three minutes
7. Shut off mixer, allow sludge to settle
8. Pump sludge to gravity thickener, check supernatant for compliance with discharge control limits
9. Pump supernatant to pH adjustment

Figure 5. Batch treatment for complexed metals waste streams.

To determine the variability contributed by the spent cleaners and complexed metals, the experiment could not be biased. The collected samples of each waste stream needed to be truly representative of the entire range of variation seen. Failure to do this could potentially lead to an unjustified optimistic projection of treatment capabilities. Therefore, new composite samples of each waste source would be collected each day the treatment was to be performed. By repeating the experiment over an extended period of time, the variability in the spent cleaners and complexed metal wastes would be identified. This would allow for the evaluation of the treatment performance under a range of normal production conditions.

The experiment was terminated after 18 data sets because the objectives of the designed experiment were met at that point.

To satisfy Objective 2, the experiment was designed to mix the spent cleaners with the complexed metals at different percentages. The percentages of cleaners to mix were established before starting the experiment. Two factors were considered when establishing the spent cleaner mix percentages.

1. The spent cleaners needed to be mixed in at a high enough percentage to allow for treatment of all cleaner waste generated in-house (approximately 10% by volume if compared to complexed metal waste).
2. Previous experience had shown that excessive cleaner addition would result in failure to meet discharge limits.

Therefore, an unbalanced experiment was conducted that concentrated on adding the spent cleaner at the lower percentages.

To satisfy Objective 3, a mass balance was performed to demonstrate that the metals were being removed from the spent cleaners and not just diluted. The mass balance identified the amount of metals contributed by the spent cleaners, the amount contributed by the complexed metals waste, and the amount removed during treatment. In addition, a second experiment was run to eliminate the effect of dilution and show that metals were being removed from the spent cleaners to levels below regulatory limits.

Description of Tests

Jar tests were conducted for a period of 18 sampling days. A daily composite sample of both spent cleaners and complexed metal wastes were collected by BWT personnel. Samples were stored in separate plastic bottles until treatment studies were performed.

Before treatment, the spent cleaner and complexed metal wastes were mixed at the percentages defined in the experimental design. Treatment of the samples was conducted over a 16-hour period to allow both the first-shift BWT operator and second-shift BWT operator to participate in the experiment. This was important in order to incorporate into the experiment the variance that would be seen in full scale operation due to multiple treatment operators. Samples of untreated spent cleaner waste and complexed metal wastes were submitted with each treated sample set, to determine background levels for the mass balance. Details regarding the experimental design are shown in Figure 6.

Jar Test Procedure

The following is a description of the batch chemical treatment procedure used in the jar test experiment:

1. A 500 mL sample of sample mixture was prepared with the designed proportion of spent cleaner and complexed metal wastes (i.e., 500 mL of 3% cleaner, 97% complexed metal waste) and placed in a 1000 mL beaker. The beaker was then connected to one of the four positions on a gang stirrer.
2. The waste mixtures were slowly stirred and conc. sulfuric acid was added to reduce the pH below 2. The stirring rate remained constant throughout the study. The amount of acid added and final pH reached were recorded.
3. Sodium meta-bisulfite was added to reduce the hexavalent chrome to the trivalent form. The addition was based on the quantity of hexavalent chrome present as determined by the diphenylcarbazide detection kit method.
4. 50% sodium hydroxide was added until the pH was approximately 10. The amount of caustic added and final pH were recorded.
5. 10 mL of a 10% solution of DX 2000 were added to each sample followed by the addition of 5 mL of 0.2% calcium chloride.
6. 6 mL of Betz 1120 polymer (0.2% stock solution) was added to each sample.

TREATMENT PROCESS	% SPENT CLEANERS ADDED	# OF DAYS SAMPLE COLLECTED
BATCH TREATMENT	0% CLEANER ADDED	18
OF COMPLEXED	3% CLEANER ADDED	18
METALS WASTE	5% CLEANER ADDED	18
STREAMS	7% CLEANER ADDED	18
	10% CLEANER ADDED	18
	15% CLEANER ADDED	18
	20% CLEANER ADDED	18
UNTREATED SPENT CLEANERS		18
UNTREATED COMPLEXED METALS WASTES		18

METAL ANALYSIS PERFORMED ON EACH SAMPLE
Cr,Cu,Pb,Ni,Zn

Figure 6. Experimental design for treating spent cleaners along with complexed metal wastes.

7. The samples were mixed for 15 minutes. The gang stirrer was turned off and the sludge generated allowed to settle for 20 minutes. The volume of sludge generated was recorded.
8. The supernatant was decanted into a 250 mL bottle and preserved for metals analysis.

All chemical additions, pH levels, and other observations made during the chemical treatment experiment were recorded on standard data forms developed at the beginning of the experiment. Chemical addition dosages were based on the full-scale treatment process operation.

Chemical Analysis

Each sample of supernatant was digested and analyzed for chromium, copper, lead, nickel, and zinc by the Kohler C & M Laboratory (Wisconsin Registered Laboratory #460032870). The metals were analyzed following the applicable methods in *Standard Methods For the Examination of Water and Wastewater*, 16th Edition. The methods are listed below:

Method 302D Preliminary Treatment of Samples, Nitric Acid Digestion
Method 310A Cadmium AA Spectrometric Method
Method 312A Chromium AA Method for Total Chromium
Method 313A Copper AA Spectrometric Method
Method 316A Lead AA Spectrometric Method
Method 321A Nickel AA Spectrometric Method
Method 328A Zinc AA Spectrometric Method

Additional Experiment on the Cleaners

To satisfy Objective 3, an additional experiment was designed and conducted to provide direct evidence that metals were being removed from the cleaners and not diluted.

Spent cleaner samples that had been treated in the previous experiment were mixed with water instead of the complexed metal waste streams. All the metals present in the resulting mixture were all contributed by the spent cleaner waste, and the effect of the complexed metal waste stream was eliminated.

The spent cleaners were mixed with water at 7% cleaner to 93% water. The mixtures were then treated as described previously and metals analysis run on the supernatant. Some samples of spent cleaner had initial concentrations below the regulatory limits prior to treatment and were not included in the analysis of experimental results.

EXPERIMENTAL RESULTS

The analytical results for each metal are summarized in Table I. Table I presents a statistical summary of the treated concentrations of each metal at the varying mixtures. Figures 7 through 11

Table I. Statistical Summary of Chromium Results (All Values in mg/L)

Cleaner Addition	Mean	Max.	Min.	Std. Dev.
0%	0.20	0.43	0.05	0.10
3%	0.22	0.45	0.06	0.12
5%	0.22	0.53	0.08	0.12
7%	0.39	1.81	0.06	0.52
10%	1.05	7.20	0.06	1.73
15%	3.79	32.9	0.06	8.00
20%	3.87	31.8	0.09	7.78

Statistical Summary of Copper Results
(All Values in mg/L)

Cleaner Addition	Mean	Max.	Min.	Std. Dev.
0%	0.12	0.20	0.09	0.03
3%	0.13	0.31	0.08	0.06
5%	0.17	0.72	0.08	0.15
7%	0.30	1.88	0.009	0.50
10%	0.74	4.60	0.09	1.41
15%	2.17	23.00	0.10	5.50
20%	2.66	22.30	0.11	5.31

Statistical Summary of Lead Results
(All Values in mg/L)

Cleaner Addition	Mean	Max.	Min.	Std. Dev.
0%	0.37	0.68	0.25	0.10
3%	0.37	0.57	0.25	0.08
5%	0.37	0.47	0.25	0.07
7%	0.36	0.51	0.25	0.08
10%	0.45	1.60	0.25	0.31
15%	0.52	2.72	0.25	0.56
20%	0.74	3.26	0.25	0.87

Statistical Summary of Nickel Results
(All Values in mg/L)

Cleaner Addition	Mean	Max.	Min.	Std. Dev.
0%	0.36	0.50	0.26	0.07
3%	0.36	0.55	0.27	0.08
5%	0.38	0.53	0.29	0.08
7%	0.45	1.09	0.29	0.21
10%	0.82	3.60	0.29	0.86
15%	1.28	7.50	0.28	1.90
20%	1.91	7.80	0.30	2.33

Statistical Summary of Zinc Results
(All Values in mg/L)

Cleaner Addition	Mean	Max.	Min.	Std. Dev.
0%	0.08	0.12	0.05	0.03
3%	0.08	0.14	0.05	0.03
5%	0.09	0.20	0.05	0.04
7%	0.14	0.70	0.05	0.15
10%	0.31	2.00	0.05	0.59
15%	0.90	9.20	0.05	2.24
20%	1.02	9.20	0.06	2.16

show the after-treatment concentrations found in the supernatant for chromium, copper, lead, nickel, and zinc, at each percentage of spent cleaner. The figures illustrate the mean concentration and the range of results, and provide a comparison to the regulatory daily maximum limit. The test results presented in Figure 7 through 11 demonstrate that a mixed waste stream of spent cleaners and complexed metals were treated to consistently meet permit requirements and indicate that the optimum ratio between cleaners and complexed metals waste stream is 7:93.

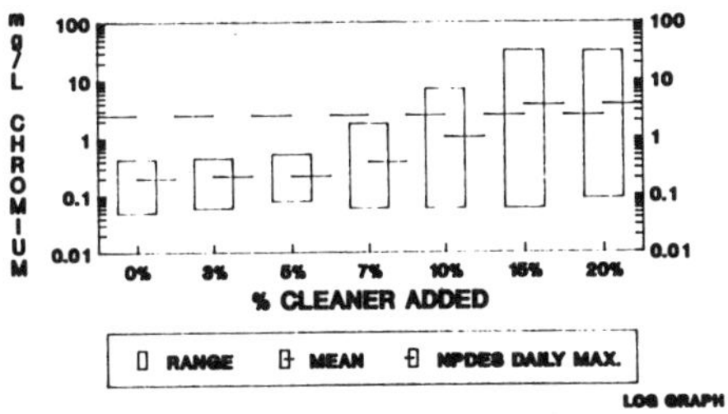

Figure 7. Chromium analytical results.

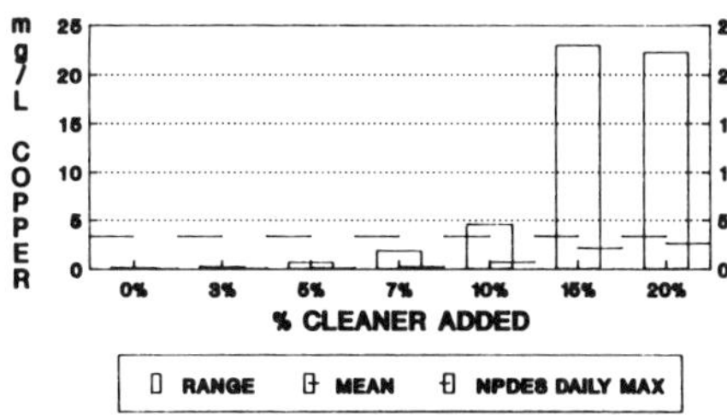

Figure 8. Copper anlytical results.

Figure 12 summarizes test results for the addition of 7% cleaner to water for each metal. The mean, maximum value, and 99% confidence value (three standard deviations from the mean) are compared to both regulatory limits; the daily maximum, and monthly average.

The results of experiments to evaluate the impact of dilution are presented in Figures 13, 14, and 15. Figure 13 illustrates a mass balance of the lead results in tests with complexed metals waste. Lead was the only metal found to be present at significant levels in the spent cleaners. In Figure 13, the mass of lead contributed by the spent cleaners are compared to the mass contributed by the complexed metals waste and to the actual mass of lead removed.

Figure 14 and Figure 15 illustrate the copper and lead results when the cleaners were mixed at 7% with water. The figures show the initial concentrations of each metal present, the concentration present after treatment, and the daily maximum regulatory limit adjusted to take into account the dilution. This adjusted maximum regulatory limit was calculated by multiplying the Daily Maximum limit by a dilution factor. In this case, the dilution factor was 0.07.

DISCUSSION OF RESULTS

The three objectives of the experimental design were met. Figures 7 through 11 identify the optimum ratio (7%) to mix the spent cleaners and complexed metals waste streams, which was listed as Objective 2.

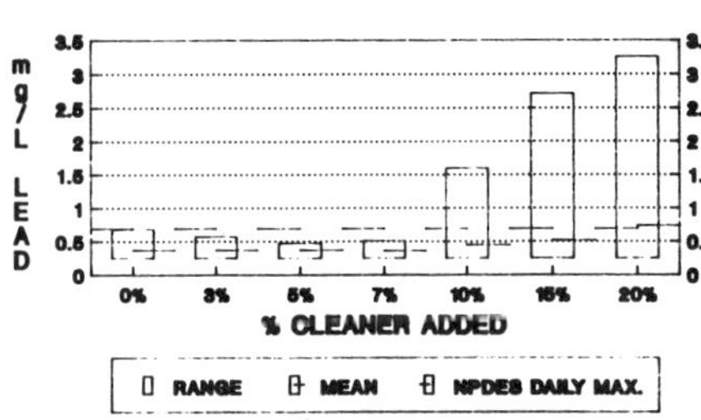

Figure 9. Lead analytical results.

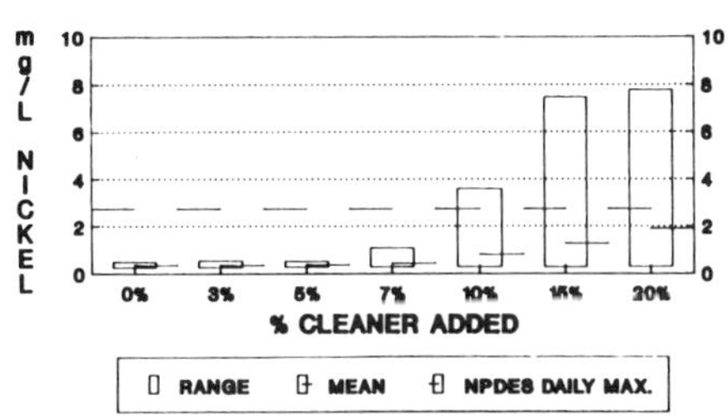

Figure 10. Nickel analytical results.

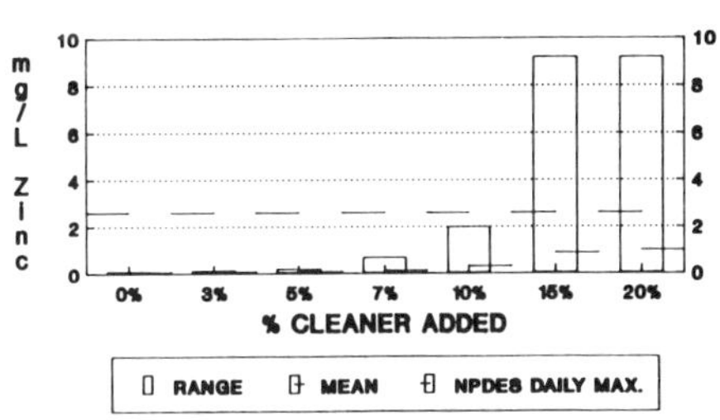

Figure 11. Zinc analytical results.

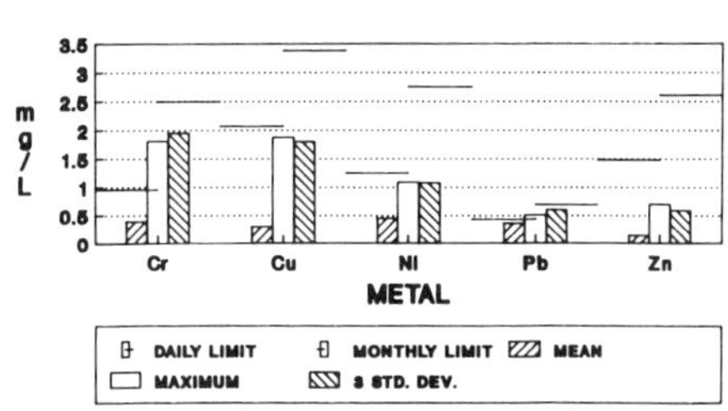

Figure 12. Statistical summary of results.

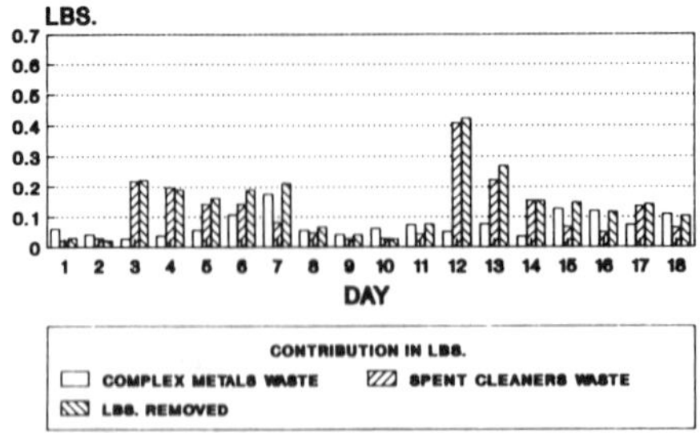

Figure 13. Lead removal in pounds.

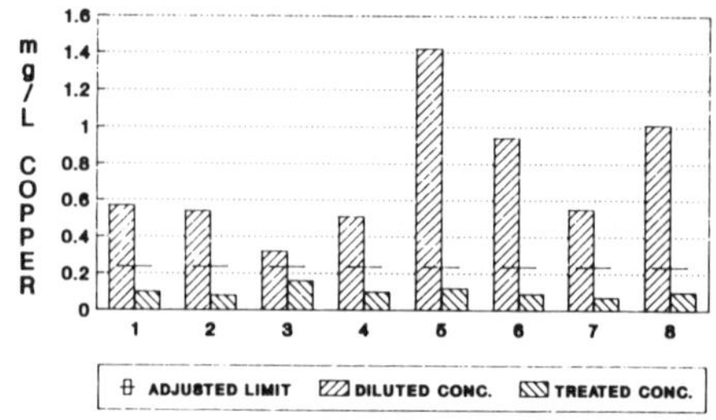

Figure 14. Treating cleaners mixed with tap water copper results.

The maximum concentration found in an individual test did not exceed the regulatory daily maximum limit until after the 7% cleaner addition level was exceeded. These figures clearly show that the daily maximum categorical standard can be consistently met at or below the 7% mixture.

To meet Objective 1, to statistically demonstrate that treatment could be attained to consistently meet permit requirements, standard deviation was calculated from the analytical data at the 7% mixture. The 99% confidence interval was then determined and compared to the regulatory limit. A statistical summary is presented in Table II and shown graphically in Figure 12.

From this analysis, a determination was made of the process capability in meeting the regulatory limit. Figure 16 illustrates a process that is capable and one that is not capable of meeting a limit. In statistical process control (SPC), a process is determined to be capable when it can meet a specification limit with a given level of confidence. If the tail of the normal distribution of the results extends beyond the limit, the process is considered not capable. In other words, there may be an exceedance of a regulatory limit and no identified assignable cause.

When the process capability analysis was performed for all metals at all mixtures, it was found that the treatment process was capable of meeting permit specifications for mixtures of 7% and less. The reason for the inability to demonstrate statistical capability at higher mixtures becomes apparent as the standard deviations are examined in Table I. In general, the standard deviations for each of the metals observed increased an order of magnitude as the mixtures increase from 3% up to 10%.

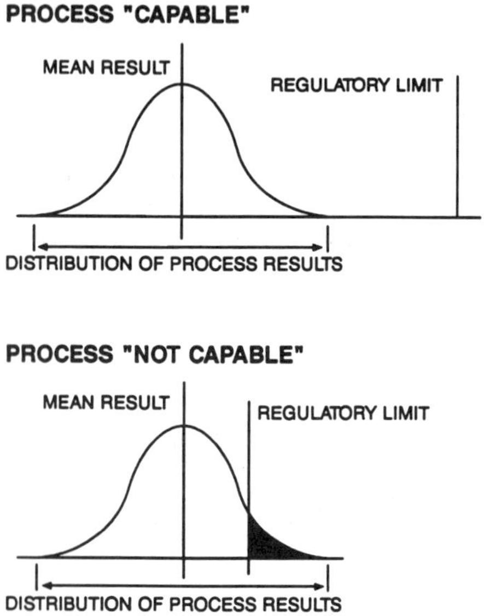

Figure 16. Interpreting process capability using SPC.

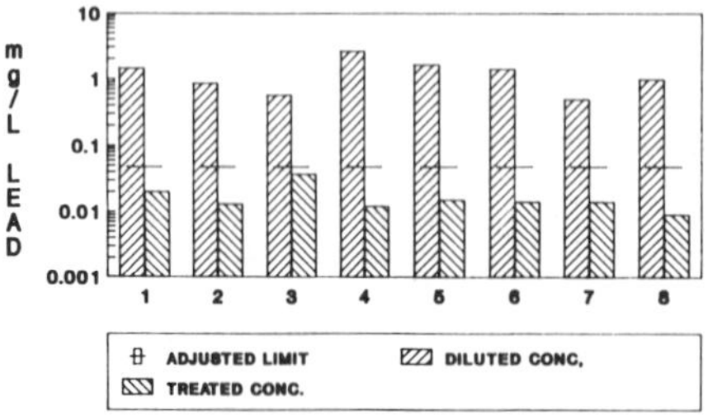

Figure 15. Treating cleaners mixed with tap water lead results.

Table II. Statistical Summary of Analytical Data for Each Metal at 7% Cleaner Addition. All Results in mg/L

Metal	Mean	Maximum	Standard Deviation	99% Confidence Level
Chromium	0.39	1.81	0.52	1.95
Copper	0.30	1.88	0.50	1.80
Lead	0.36	0.51	0.08	0.60
Nickel	0.45	1.09	0.21	1.08
Zinc	0.14	0.70	0.15	0.59

Although the means of the high concentration mixtures still met the permit requirements, the spread of the normal distribution curve would not allow consistent compliance with the permit requirements.

Figure 12 shows that for each metal, the daily maximum regulatory limit can be met at a 99% confidence level. Therefore, Objective 1, to statistically prove that a mixed waste stream can be treated to consistently meet the regulatory limit, was proven for mixtures at or below the 7% level.

Evidence for the removal of metals from the spent cleaners is given in Figure 13. In this figure, histogram bars are clustered in groups of three. The first bar in each cluster shows the mass of lead in the complexed metals wastewater, the second bar shows the mass contribution by the spent cleaners, and the third bar shows the total mass of lead removed. In the majority of cases, the mass of lead removed exceeded the total mass of lead in the complexed metals wastes. This is evidence that the lead in the cleaners is being removed from the spent cleaner waste. However, it doesn't prove that sufficient lead is removed from the cleaners to be able to say that the cleaners met the permit limits. To show this, the additional experiment previously described was conducted.

Figures 15 and 16 show the lead and copper results from the second experiment. Since the spent cleaner wastes were being diluted 7 parts to 93 parts, the categorical daily maximum discharge limits were lowered (limit multiplied by 0.07) to allow for the dilution. These adjusted limits were compared to the metals concentration found in the supernatant after treatment. Comparing the concentration of copper and lead found after treatment to the adjusted regulatory limit shows that the regulatory limit can be met, even though the cleaners were being diluted. This fulfilled the third objective of the experimental design.

FULL-SCALE IMPLEMENTATION

Kohler Co. submitted the results from this study to the Wisconsin Department of Natural Resources and obtained approval to begin implementation of the spent cleaner treatment strategy in September, 1989. Since implementation, the BWT facility has

1. experienced complete compliance with the discharge permit limits,
2. reduced the disposal costs associated with spent cleaners by 75%

Currently, the volume of spent cleaners generated by the manufacturing operations is greater than the volume which may be treated at the 7% maximum volume with other wastes in batch treatment. The spent cleaners which are not treated in batch treatment are collected and hauled to a licensed disposal facility. Kohler Co. is continuing to identify additional alternatives to further reduce operating costs while consistently maintaining compliance with permit requirements.

The ability to implement this change in treatment without a single violation of Kohler Co.'s permit is attributed to a well planned, statistically designed, set of experiments with clear objectives and which considered many factors that could influence the success of this project. Key steps involved were the analysis of the sources of variance, the control of undesirable sources of variance, the incorporation of sources of variances that will be present when the process is implemented, and finally the capability analysis performed at the 99% confidence level.

ACKNOWLEDGMENTS

The authors would like to acknowledge the efforts of the Brass Division personnel; specifically, Mitch Myers, Fred Bingman, and Brett Mulholland; the Chemical and Metallurgical Laboratory personnel, and the technical reviewers of the paper.

REFERENCES

1. Box, G.E.P., Hunter, W. G., Hunter, J.S. *Statistics for Experimenters*, John Wiley and Sons, 1978.
2. Braverman, J. D., *Fundamentals of Statistical Quality Control*, Reston Publishing Co., Inc., Reston, Virginia, 1981.
3. *Development Document for Effluent Limitations Guidelines and Standards for the Metal Finishing Point Source Category*, United States Environmental Protection Agency, June, 1983.
4. *Standard Methods for the Examination of Water and Wastewater*, 16th Edition, American Public Health Association, Washington, D.C., 1985.
5. 40 CFR 433, 32462 (July 15, 1983).

84 1988 GUIDELINE COMPLIANCE FOR MAJOR GRAY IRON FOUNDRY

Lloyd W. Curry, Project Engineer
Crawford, Murphy & Tilly, Inc.
Springfield, Illinois 62702

Carey French, Environmental Engineer
Caterpillar Inc.
Mapleton, Illinois 61547-9799

INTRODUCTION

This paper describes the actions taken by Caterpillar's gray iron foundry, located at Mapleton, Illinois, to comply with stringent discharge regulations imposed by United States Environmental Protection Agency (USEPA) on the Metal Molding and Casting Industry. The Mapleton foundry's only categorically regulated wastewater stream is from wet dust collection scrubbers. The process envisioned by the USEPA was recirculation of water to the scrubbers, with a 3% blowdown to control dissolved solids build-up.

Two pilot plants were operated to study the technical feasibility of two possible treatment methods:

(a) A 5 gallon per minute (GPM) recirculation system with treatment of the blowdown to achieve National Pollution Discharge Elimination System (NPDES) discharge requirements.
(b) A once-through system with treatment to achieve NPDES discharge requirements.

The pilot plant studies indicated that the recirculation system was both difficult to operate and did not consistently produce an effluent that met discharge limitations. Conversely, the once-through system consistently met discharge requirements.

A full-scale once-through treatment system, based on the pilot plant study, has been constructed and is in start-up. This paper outlines the more significant aspects of the study and construction phases of this project.

REGULATORY BACKGROUND

On October 30, 1985, USEPA promulgated Rules and Regulations for the Metal Molding and Casting Industry, limiting the discharge of certain pollutants to navigable waterways and publicly-owned treatment plants. This action was in response to requirements of the Clean Water Act Amendments of 1977, and a settlement agreement to a suit brought by citizen groups.

For the Mapleton Foundry, the only categorically regulated wastewater stream is the discharge from the dust collection scrubber system. Currently, the once-through water usage is approximately 2 million gallons per day (MGD). For the Mapleton foundry, the regulation's "normalized" dust collector water flow is 3 GPM per 1000 standard cubic feet per minute (SCFM) of air flow.

Although the guidelines were developed for an assumed blowdown of 3% of a "normalized" water recirculation rate, the regulations limit the mass (weight) of pollutants that may be discharged, but do not limit the water discharge volume. The regulated pollutants are: phenol, copper, lead, zinc, suspended solids, oil and grease, and pH. The maximum daily and monthly average limits, as stated in the regulations, are given in Table I. Table I mass values are given in terms of pounds of pollutant per billion cubic feet of dust collector air flow. Table I concentration values are given in terms of milligrams per liter (mg/L), for a normalized water flow of 3 GPM per 1000 SCFM of air with a 3% blowdown.

45th Purdue Industrial Waste Conference Proceedings, © 1991 Lewis Publishers, Inc., Chelsea, Michigan 48118. Printed in U.S.A.

Table I. BPT and BAT Dust Scrubber Effluent Limitations[a]

Pollutant	Max. Day		Max. 30 Day Avg.	
	lbs.[b]	mg/L[c]	lbs.[b]	mg/L[c]
Suspended Solids	22.5	30.0	7.51	10.0
Oil and Grease	26.5	34.0	11.3	15.0
Phenol	0.646	0.86	0.225	0.30
Copper	0.218	0.29	0.120	0.16
Lead	0.398	0.53	0.195	0.26
Zinc	0.736	0.98	0.278	0.37

[a]Values from BPT (Best Practicable Technology) and BAT (Best Available Technology) Effluent Limitations, 40 CFR Part 464 for gray iron foundries.

[b]Lbs. per 1 billion cubic feet of scrubber air flow.

[c]Mg/L based on water flow of 3 GPM per 1000 SCFM scrubber air flow, with 3% blowdown.

DESCRIPTION OF MAPLETON PLANT

The Mapleton Plant is situated on a 400 acre site, located on the Illinois River, some 10 miles downstream of Peoria, Illinois. The original foundry was completed in 1968 and operated through 1986, when it was shut down. The current foundry was completed in 1978, and has approximately 900,000 square feet under roof and employs approximately 900 people.

The foundry produces gray iron engine castings; including blocks, heads, and cylinder liners. All melting is electric, using arc and induction furnaces.

All water used at the facility, is obtained from and returned to the Illinois River. Due to the rural location, the facility operates its own potable water treatment, sanitary wastewater treatment, and industrial wastewater treatment plants. A drawing of the original wastewater treatment system is provided as Figure 1. The potable water treatment plant has a capacity of 0.75 MGD, while the sanitary treatment plant has a capacity of 0.2 MGD.

The industrial wastewater treatment facility was designed for two separate wastewater streams from the foundry:

(a) Dust Collector Waste Water (DCWW)—From wet scrubbers installed to remove particulates and fumes from foundry molding areas.
(b) Acid-Alkaline-Oil (AAO)—Industrial wastewater system.

The DCWW system was designed for a maximum flow of 17.28 MGD (12,000 GPM), while the AAO system was designed for a maximum flow of 0.72 MGD (500 GPM). When the originally constructed foundry was shut down in 1986, the wastewater discharges were greatly reduced. Therefore capacities of the treatment processes exceed current demands. The effluent from the sanitary treatment plant and AAO clarifier effluent was mixed with the DCWW clarifier effluent, at the influent to the traveling bridge sand filters.

A single system was provided to dewater the combined sanitary, AAO, DCWW, and potable backwash sludges. Over 90% of the sludge comes from the DCWW system. Sludge is preconditioned with lime to aid dewatering in the vacuum filters.

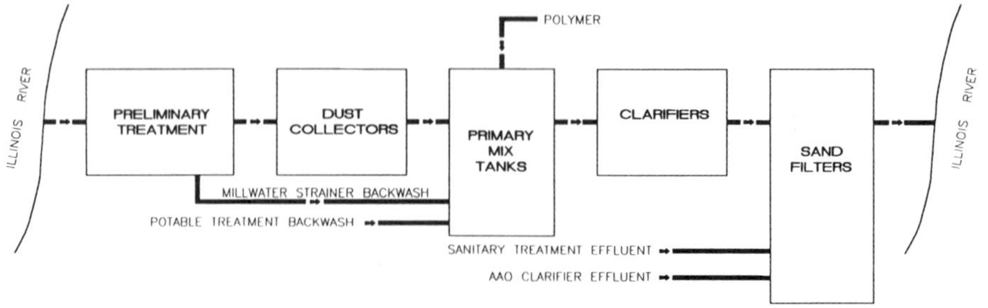

Figure 1. Existing DCWW, sanitary and AAO systems flow diagram.

In general, the original treatment facilities produced an effluent which complied with the facility's NPDES permit. However, phenol excursions were experienced occasionally. Some organic binders used in foundry processes contain phenolics. Phenol fumes are released and are absorbed by the dust collector scrubber water. A naturally occurring colony of pseudomonas bacteria has developed in the DCWW clarifiers, which reduces the phenol concentrations below the previous NPDES permit limit during warm weather. However, temperatures and plant vacations adversely effect the colony and its ability to reduce phenol levels.

SCOPE OF STUDY PROGRAM

Crawford, Murphy & Tilly, Inc. (CMT) was retained by the Mapleton Plant to study means of compliance with the 1988 Foundry Effluent Guidelines. The study consisted of five major phases:

Phase I – Sampling and analysis program of various waste streams.
Phase II – Pilot plant design.
Phase III – Pilot plant operation.
Phase IV – Feasibility design and estimate.
Phase V – Final report.

WASTEWATER CHARACTERIZATION

A sampling program was conducted for one week, with samples collected from 22 waste streams. Analysis was made for the following parameters: Biochemical Oxygen Demand (BOD), Suspended Solids (SS), Phenol, Copper, Lead, and Zinc. Table II summarizes the results of the sampling program.

ANALYSIS OF SAMPLE DATA

The sampling results for the DCWW system were considerably lower in metals than expected. The phenol levels generally agreed with earlier sampling programs conducted by plant staff.

Table III shows the BOD, suspended solids, phenol, and metals removal resulting from the clarification and sand filtration operations.

The above data indicates an exceptionally high removal efficiency in the DCWW clarifier for all pollutants. BOD and phenol are removed biologically. Suspended solids are removed by the use of polymer with sedimentation in the clarifiers. The removal of metals is not as easily explained. In order to better understand this removal phenomenon, low level detection analyses were made on both filtered and unfiltered samples. Table IV shows the results of these analyses.

The filtration was done according to *Standard Methods*, using a 1.5 micron retention filter, and essentially removed all non-soluble metals from the sample. The analysis performed on the filtrate, therefore, represented only dissolved pollutants.

Analysis of the data in Table IV indicates no soluble copper on zinc were added by the dust collector scrubbers, although an apparent 0.001 mg/L of soluble lead was added. As indicated in the sampling programs, high levels of phenol were found in the DCWW. While the DCWW clarifiers were quite efficient at reducing phenol, this efficiency can be expected to decrease during cold weather. As indicated by the filtered sample for the DCWW influent, the phenol content is essentially soluble.

We conclude from Table IV data that no soluble copper, lead, or zinc are added by the dust collection process. There is, however, significant addition of phenols.

ALTERNATE WATER TREATMENT CONCEPTS

Two treatment concepts were developed to achieve compliance with effluent guidelines.

Plan I – Recirculation

Plan I would consist of a closed water recirculation system supplying the dust collector scrubbers, with blowdown to control total dissolved solids. The blowdown stream would be treated to comply with NPDES discharge requirements.

Table II. Results of Initial Sampling Program (5-Day Average Values, mg/L)

	BOD[a]	SS[b]	Phenol	Coper	Lead	Zinc
Sanitary System:						
Sanitary Lift Station	284	516	1.4	0.14	0.03	0.64
Sanitary Influent	141.4	99.4	0.4	0.05	0.025	0.33
Sanitary Effluent	13.4	21	0.1	0.02	0.027	0.19
AAO System[c]:						
AAO Lift Station	2230	1164	0.9	0.25	0.115	1.08
AAO Mix Tank Influent	252	56.8	1.1	0.03	0.026	0.32
AAO Clarifier Effluent	250.6	20.6	0.1	0.02	0.033	0.1
DCWW System[d]:						
Dust Collector 1	20.6	1098	3.2	0.03	0.056	0.07
Dust Collector 2	17.4	2542	1.9	0.07	0.099	0.13
Dust Collector 3	65.6	1470	11.9	0.05	0.06	0.08
Dust Collector 4	33.4	1440	6.5	0.08	0.051	0.08
Dust Collector 5	81.4	2334	17.4	0.05	0.065	0.11
Dust Collector 6	12.4	7300	1.3	0.05	0.144	0.2
Dust Collector 7	37.2	9240	5.6	0.05	0.107	0.19
Dust Collector 8	98	572	0.5	0.02	0.045	0.05
Dust Collector 9	6.2	172	0.3	0.01	0.03	0.08
Slurry Tank	45.2	2300	7.9	0.07	0.083	0.14
Dust Collector 13	8	1866	0.1	0.04	0.069	0.13
Dust Collector 14	10.8	362.8	0.1	0.02	0.034	0.06
Dust Collector 15	10.5	4180	0.1	0.06	0.115	0.21
DCWW Clarifier Effluent	8.4	45.8	0.1	0.01	0.028	0.01
Combined Effluent:						
Sand Filter Effluent	15.6	3.6	0.1	0.02	0.028	0.02
River Water:						
Mill Water	3.2	12.6	0.1	0.01	0.027	0.02

[a]BOD = Biochemical Oxygen Demand.
[b]SS = Suspended Solids.
[c]AAO = Acid-Alkaline-Oil (industrial).
[d]DCWW = Dust Collector Waste Water.

Plan II – Once-Through

Plan II would consist of a once-through water system supplying the dust collector scrubbers. This system would isolate DCWW from the other wastewater streams. Additional treatment facilities would be added to achieve compliance with NPDES effluent requirements.

PILOT PLANT DEVELOPMENT

Two pilot treatment plants were developed to demonstrate the salient features of the two water system concepts.

Table III. DCWW Pollutant Removals by Treatment Methods

Pollutant	Untreated DCWW mg/L	Clarifier Effluent		Sand Filter Effluent[a]	
		mg/L	% Removal	mg/L	% Removal
BOD	45.2	8.4	81.4	15.6	–
Suspended Solids	2300	45.8	98.0	3.6	99.8
Phenol	7.9	<0.1	>98.7	<0.1	>98.7
Copper	0.07	0.01	85.7	0.02	71.4
Lead	0.083	0.028	66.3	0.028	66.3
Zinc	0.14	0.01	92.6	0.02	92.9

[a]These values were influenced by the pollutants contributed by the AAO and sanitary systems.

Table IV. Results of Analysis of Filtered and Unfiltered Samples mg/L

	Phenol	Copper	Lead	Zinc
DCWW Influent				
Unfiltered Sample	6.9	0.071	0.076	0.10
Filtered Sample	6.3	0.009[a]	0.003	0.001
DCWW Clarifier Effluent				
Unfiltered Sample	0.008	<0.010	0.085	0.083
Filtered Sample	0.003	0.008	0.003	0.002
AAO Clarifier Effluent				
Unfiltered Sample	<0.0025	0.030	0.017	0.19
Filtered Sample	<0.002	0.006	0.003	0.12
Sanitary System Effluent				
Unfiltered Sample	<0.0025	0.014	0.068	0.23
Filtered Sample	<0.002	0.016	<0.002	0.16
Mill Water				
Unfiltered Sample	<0.0025	<0.010	0.039	0.11
Filtered Sample	<0.002	0.008	<0.002	0.015

[a]Questionable result.

Plan I – Recirculation

The pilot plant schematic drawing, presented as Figure 2, was developed to model the water recirculation full scale system. Major features of the pilot plant included:

(a) A 1700 CFM dust collector, which was functionally identical to the full size collectors. Analysis of the results of the sampling program indicated the effluent from Dust Collector No. 4 most closely matched the combined dust collector system effluent.
(b) A parallel plate separator with an effective surface settling area of 100 square feet. While the DCWW clarifiers achieved almost complete removal of phenols during warm weather, the pilot plant parallel plate separator was not expected to realize any phenol reduction. Thus the recirculated flow to the dust collector would be higher in phenol than would occur for a full scale system, and should therefore, highlight any problems caused by high phenol levels.
(c) A 250 gallon reservoir and a 5 GPM recirculation pump.
(d) A 250 gallon cone bottom tank, for sludge phenol reduction.
(e) A pressure sand filter for the effluent. This effluent would represent the effluent from the full scale treatment facility.

Recirculation Pilot Plant Operation

The initial concept was to feed cationic polymer to aid sedimentation in the parallel plate separator. After about a week of operation, an ionic charge developed with the organic content (phenol, etc.) of the recirculated flow, which dispersed suspended solids instead of coagulating them. After observing the same phenomenon with anionic type polymer, the decision was made to operate without polymer usage. Lime was also tried as a coagulant and was effective at first, but the system soon stabilized and additional lime feed was ineffective.

Sludge was discharged from the solids storage portion of the parallel plate separator to a tank. The volume of the discharge was adjusted to 6% of the total volume of water recirculated to the pilot dust collector. The suspended solids in this flow ranged up to 74% and averaged 16.5%.

The sludge treatment tank was equipped with a mixer and draw-off connections at several levels. Following treatment to oxidize phenols, the solids in the tank were allowed to settle, the supernatant was drawn-off, and pumped to the sand filter. The effluent from the sand filter represented the effluent from the full scale system that would be discharged to the Illinois River.

The procedure for oxidation of phenols in the sludge treatment tank was time consuming and difficult. Potassium permanganate was used for treatment of the phenols. Previous tests on phenol-bearing wastewater by Caterpillar, suggested potassium permanganate/phenol treatment ratios of from 8:1 to 12:1. However, these ratios proved inadequate to oxidize the phenol contained in the sludge. Effective ratios were found to vary from 15:1 to 25:1. Even at these high ratios, the reduced phenol concentration goal of 0.1 mg/L was not consistently achieved.

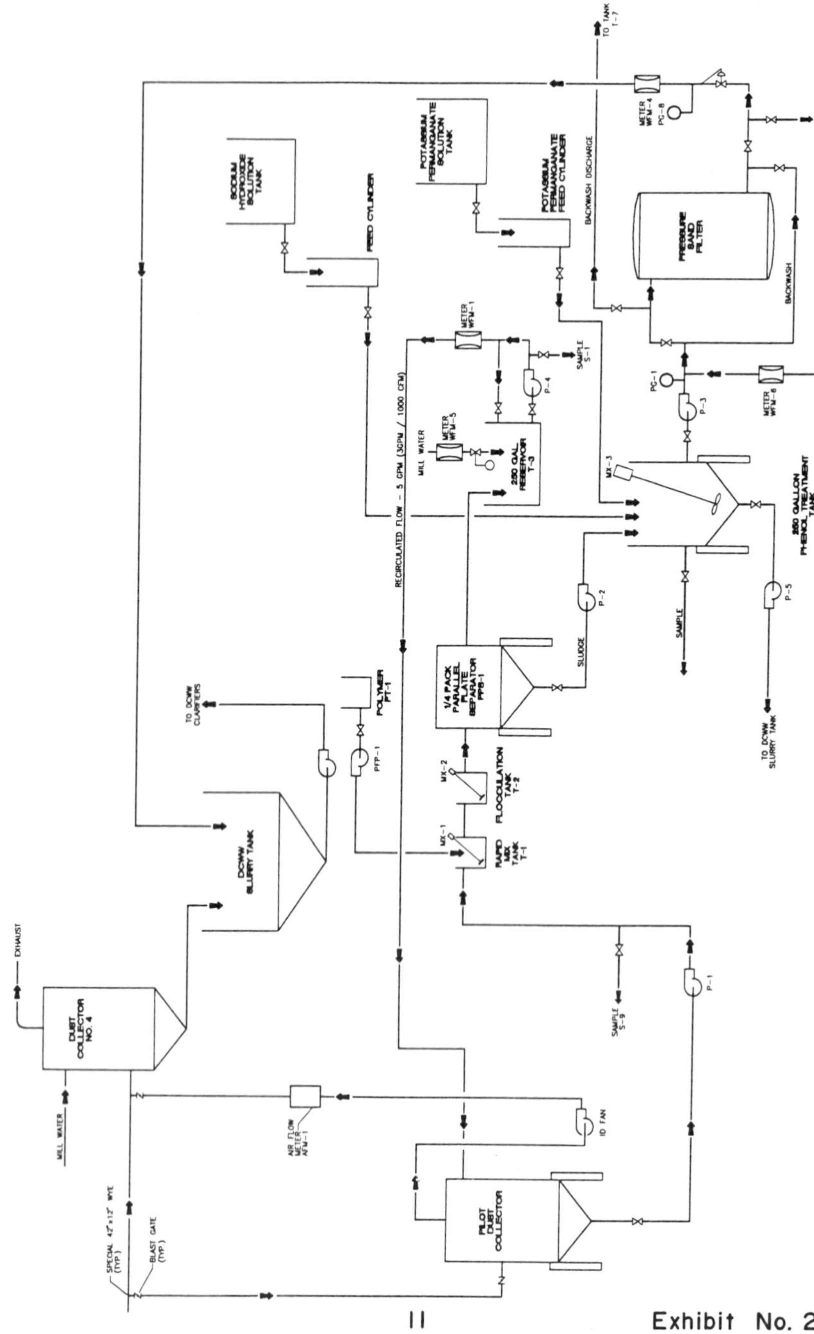

Figure 2. Pilot study for recirculated water use concept.

Plan II Once-Through

The pilot plant schematic drawing presented as Figure 3, was developed to model the once-through full scale system. Major features included:

(a) Parallel plate separator with effective surface area of 100 square feet.
(b) A pre-coat pressure filter, with 48 square feet of filter area that was similar in design to the proposed full size unit.
(c) Two granular activated carbon filters in the series to duplicate a full size system. In a full size system these filters would function as polishing units, but the pilot units were subjected to full phenol loading for the entire study.

Once-Through Pilot Plant Operation

A cationic polymer was used for this system to promote flocculation and sedimentation in the parallel plate separator. This polymer proved effective during the entire study. Following minor start-up problems, the diatomaceous earth filter and the granular activated carbon filters operated satisfactorily.

Flow Sampling and Analysis

Flow sampling was performed routinely at nine locations. Sample points S-1 through S-9 are indicated in Figures 2 and 3. Analyses were performed on both filtered and unfiltered samples.

RESULTS OF PILOT PLANT OPERATIONS

Figures 4 through 12 show graphs of the pilot plant sampling results. Figures 4 through 8 graph the pollutant levels for Plan I – Recirculation. Figures 9 through 12 graph the pollutant levels for Plan II – Once-Through. Figure 4 shows the total dissolved solids (TDS) and suspended solids (SS) concentrations, to illustrate the quality of the water returned to the pilot dust collector. Figures 5 through 8 show the required equivalent effluent pollutant levels; based on a water flow to the dust collectors of 3 GPM per 1000 CFM of air flow, and a blowdown of 3%, as presented in Table I.

Recirculation

As shown on Figure 5, the phenol content of the system effluent rose about halfway through the pilot program for the recirculation scheme, and exceeded the maximum monthly average limitation for 24 of the 27 day period. The reason for this increase in the phenol content is not fully understood, since the colormetric phenol test used in the jar test indicated a phenol content of 0.1 mg/L would be obtained. A possible explanation is that additional phenols continued to leach from the sludge following depletion of the potassium permanganate.

Once-Through

Table V presents equivalent effluent limitations for a once-through water usage. Note that units in Table V are micrograms per liter (μg/L). Figures 6 through 8 indicate full compliance with the equivalent effluent limitation for copper, lead, and zinc.

Note that the effluent from the once-through system consistently complied with the discharge limitations during the entire pilot plant operation.

ANALYSIS OF PILOT PLANT OPERATION

Operation of the pilot plants demonstrated that both treatment methods could meet USEPA's discharge requirements for the Metal Molding and Casting Industry.

However, the recirculation system would require the addition of a granular activated carbon filter to ensure that phenol levels consistently meet discharge limitations. Another significant consideration for the recirculation system would be the high phenol content of the water phase of the sludge, which could render the sludge hazardous under RCRA regulations. The determination of the phenol content of the sludge blowdown is difficult, and would require a high level of technical skill. The effectiveness of the phenol oxidation is subject to close operator attention at all times, and would be susceptible to upsets.

Operation of the once-through system was judged to be fairly routine, requiring a lower level of technical skill than the recirculation system. The major components of the once through system have

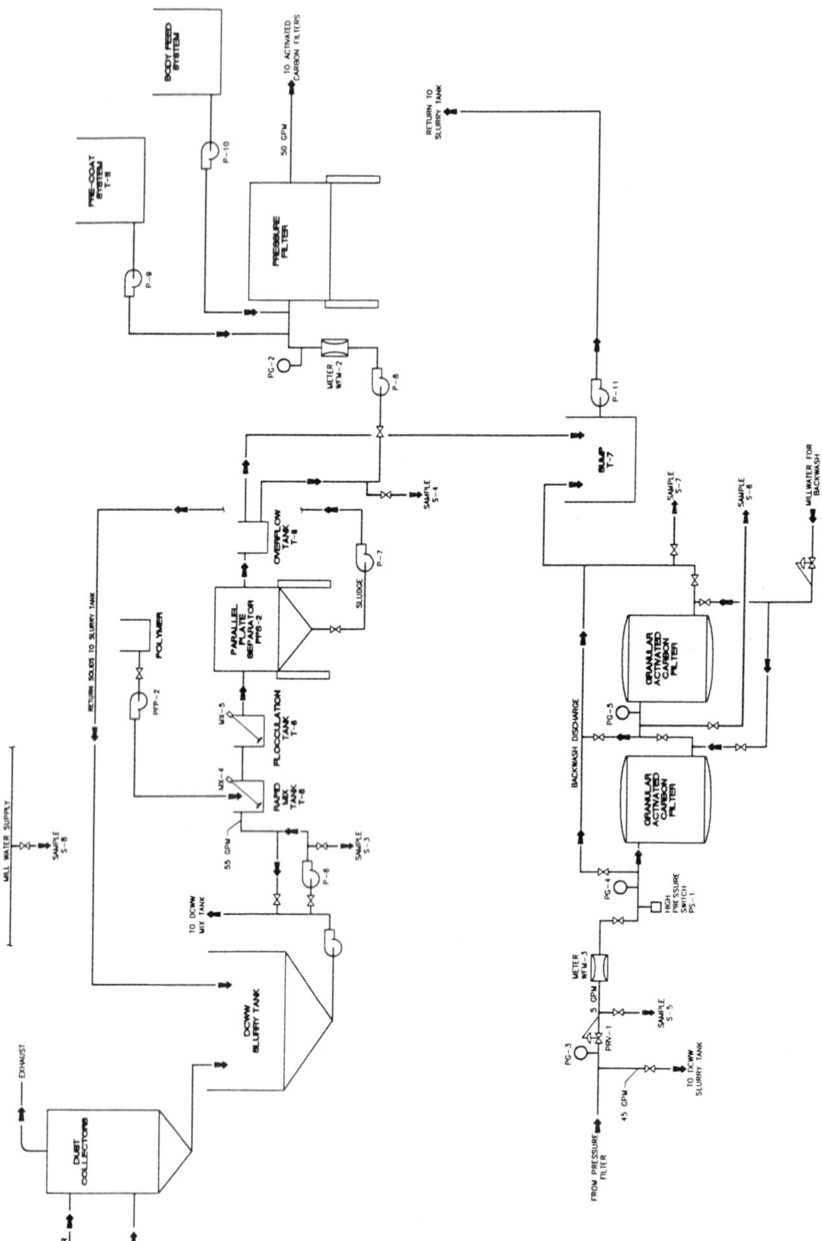

14 Exhibit No. 3

Figure 3. Pilot study for once-through water use concept.

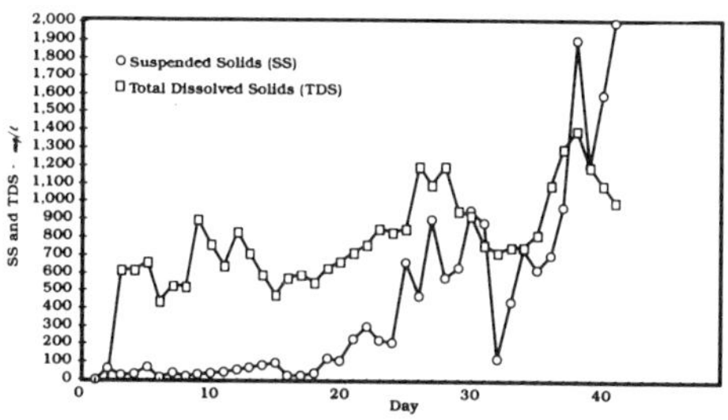

Figure 4. Recirculation system - SS and TDS.

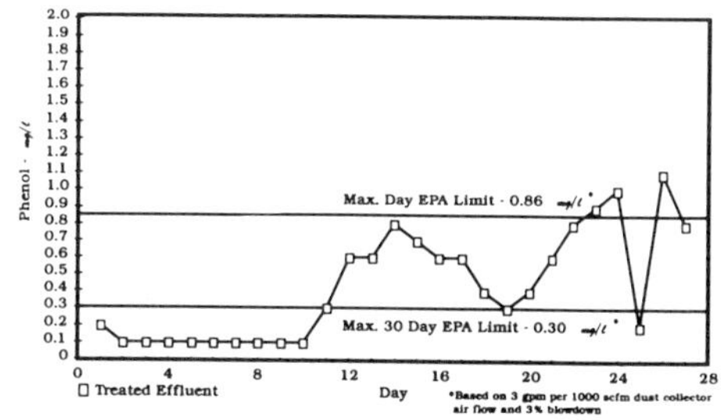

Figure 5. Recirculation system effluent-Phenol.

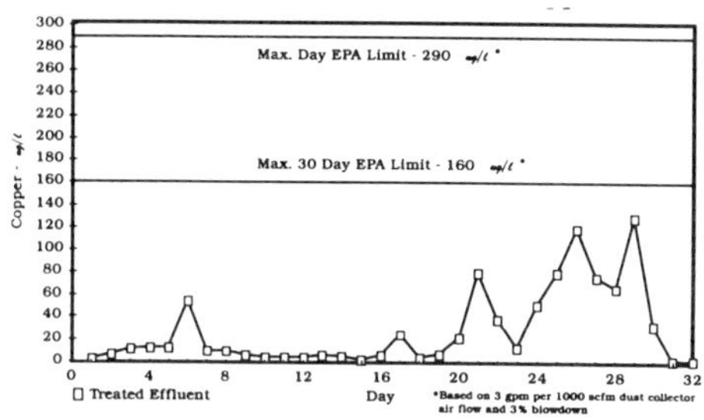

Figure 6. Recirculation system effluent-Copper.

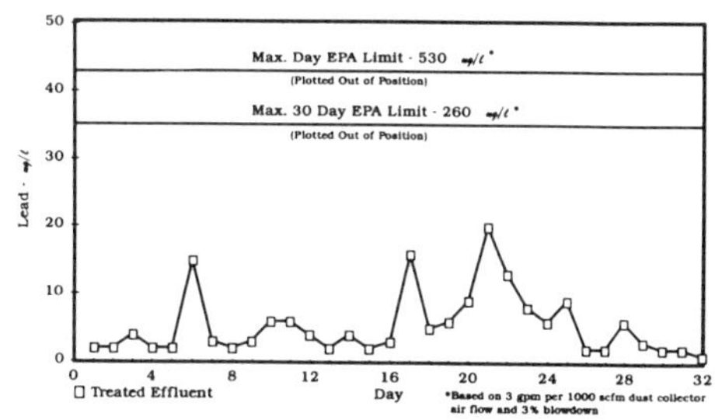

Figure 7. Recirculation system effluent-Lead.

been in industrial use for decades. Diatomaceous earth filters are used extensively for processing of beer and wine, and are used for other critical services. The use of activated carbon adsorption for organic pollutants in food grade products, as well as for pollution control, is well documented.

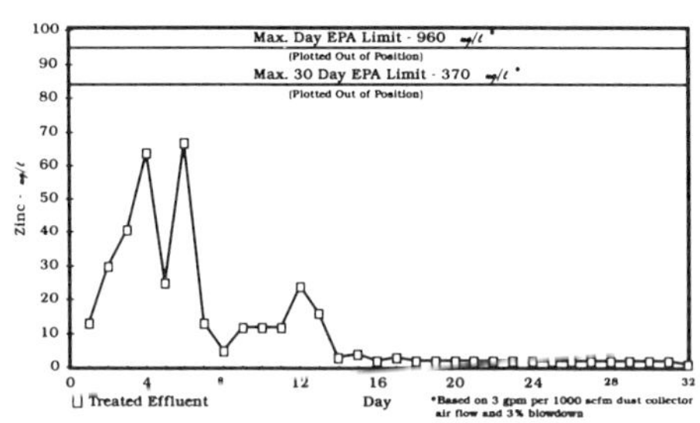

Figure 8. Recirculation system effluent-Zinc.

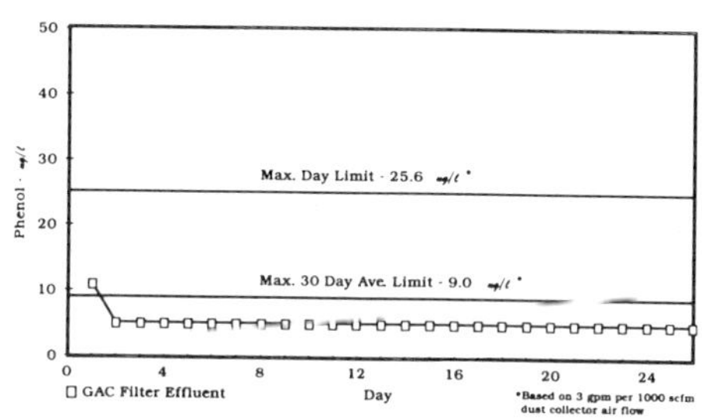

Figure 9. Once-through system-Phenol.

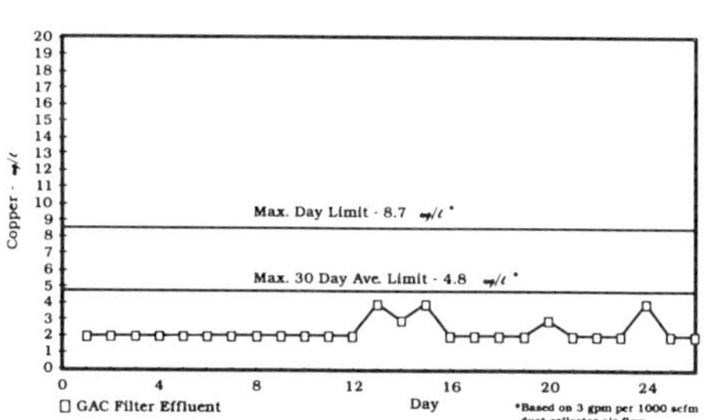

Figure 10. Once-through system-Copper.

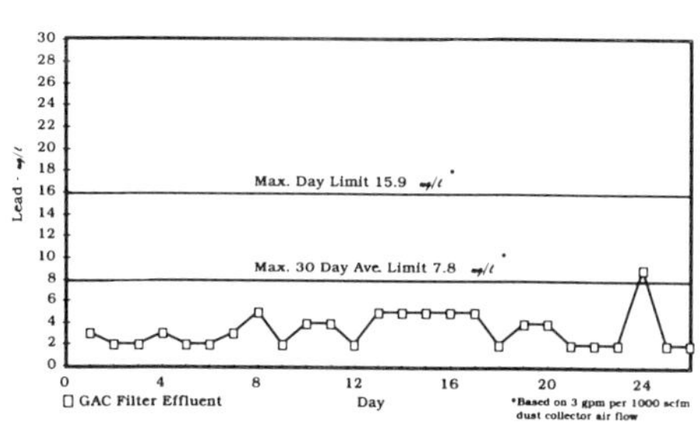

Figure 11. Once-through system-Lead.

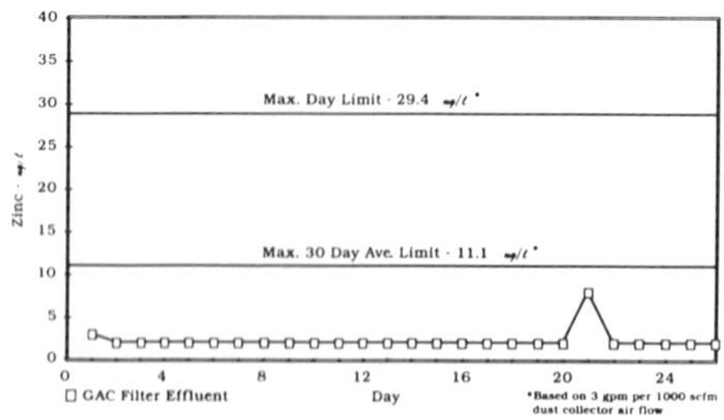

Figure 12. Once-through system-Zinc.

STUDY RECOMMENDATIONS

The study recommended the modification of the wastewater treatment plant to provide two separate discharges, each with its own treatment system and discharge limitations:

(a) DCWW System—Once-through treatment
(b) AAO/SAN System—Treatment of the combined discharge of the sanitary treatment system, AAO, backwash from the incoming river water strainers, and backwash from the portable water treatment plant.

Figure 13 shows a block drawing of the recommended plan. Major system components include:

DCWW System:

Three—1150 square foot vertical leaf pre-coat diatomaceous earth pressure filters.
Three—2—vessel granular activated carbon filter trains, each vessel filled with 20,000 lbs. of carbon.

AAO/SAN System:

One—2-vessel granular activated carbon filter train added to improve current system reliability.

FULL-SCALE PLANT CONSTRUCTION

Construction of the full-scale system began during the annual two week summer vacation period in 1988. Work included certain system tie-ins that could only be made during a shut-down period. The contract for the second phase construction work was awarded in January 1989, and was essentially completed a year later.

Major system components were purchased by Caterpillar to expedite project completion. These items included the diatomaceous earth filters, granular activated carbon filters, pumps, major piping components, and control system. The control system provides monitoring of the entire treatment system and control of the system using an industrial programmable logic controller (PLC), coupled with a personal computer (PC) for data logging and report generation. The PLC and PC are each provided with on-line redundant units and with real time supervisory monitoring for automatic switch-over in event of a malfunction. The PLC and PC also have uninteruptable power supplies.

Two 900 KW stand-by diesel generators are provided to furnish emergency electrical power for vital equipment operation.

Table V. Equivalent Concentration Limitations for Once-Through Water Usage[a]

	Max. Day µg/L	Max. Monthly Avg. µg/L
Suspended Solids	899.3	300.2
Oil and Grease	1059.2	451.6
Phenol	25.8	9.0
Copper	8.7	4.8
Lead	15.9	7.8
Zinc	29.4	11.1

[a]Based on "normalized" water flow of 3 GPM for 1000 SCFM Scrubber Air Flow.

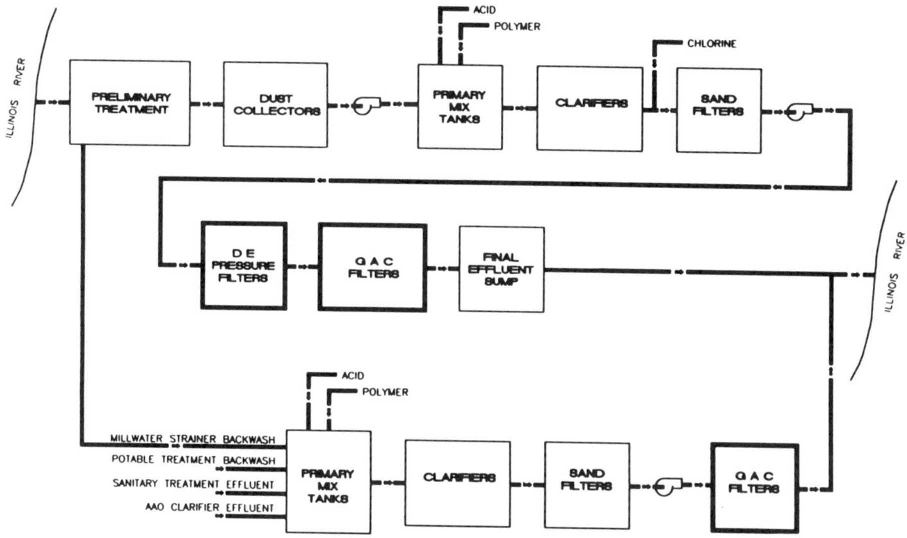

Figure 13. Revised wastewater treatment flow diagram.

The plant is now in start-up.

85 EVALUATING TOXICITY OF SPECIATED METALS FROM METAL DISCHARGING INDUSTRIES

Mujde Erten, Project Engineer
Malcolm Pirnie Inc.,
Newport News, Virginia 23606

Jerome L. Pitt, Environmental Engineer
U.S. EPA Region VII
Kansas City, Kansas 66101

Nord L. Gale, Chairman
Life Sciences Department
University of Missouri-Rolla
Rolla, Missouri 65401

Bobby G. Wixson, Dean
College of Sciences
Clemson University
Clemson, South Carolina 29634

Resit Unal, Assistant Professor
Old Dominion University
Norfolk, Virginia 23529-0456

INTRODUCTION

The aquatic toxicity of metals is a complicated phenomena involving interactions between the environment and the metal pollutants of concern. Predicting the toxic effect that metals have in natural waters requires evaluating the bioavailability of the metal pollutants. The term "bioavailability" reflects the premise that for some heavy metals organisms may be exposed to less than the total amount present in their habitat.[1]

Earlier studies to determine the toxicity of lead,[2,3] cadmium,[4-8] and zinc[9-12] to fish indicated that water quality characteristics such as pH and hardness affect bioavailability and toxicity. Further investigations[13-15] showed that evaluation of the toxic bioavailability of metals is dependent on the chemical complexation and speciation. The toxicity of heavy metals to aquatic biota is immensely influenced by the physicochemical factors of the specific environment. Many chemical characteristics of the water are important in defining both the chemical species of the metal present and physiological status of the organism being tested.[14]

The fractionation of metal species has been recognized as an essential step in the assessment of the potential biological uptake, and predicting the toxicity is dependent on understanding the limitations of laboratory toxicity tests and the effects of physicochemical variables at the field site. In recent years, increasing effort has been made to identify the ligands which dominate heavy metal speciation in natural waters, and to develop a speciation scheme for measuring the fraction of the trace metal concentration which is toxic to aquatic biota.[16] For a speciation scheme to be useful, however, it must reliably detect all the important toxic forms of the metals present. Also, information must be available about the relative toxicity of the various forms measured by the scheme.

The lack of precise and definitive information about the various chemical species of lead, cadmium and zinc which might be present or the relative toxicity of various compounds of these metals have frustrated the development of appropriate and adequate water quality standards. Because of the different species of metals occurring in the environment and the lack of definitive information about their relative toxicities, there is no ideal analytical measurement available for expressing aquatic life criteria for metals of lead, cadmium and zinc. A criterion or a standard for a metal that is based on only one set of environmental variables will be unnecessarily stringent for one type of environment

45th Purdue Industrial Waste Conference Proceedings, © 1991 Lewis Publishers, Inc., Chelsea, Michigan 48118. Printed in U.S.A.

and underprotective for other environments having different physicochemical properties.[17,18] Physicochemical environmental factors, such as pH, Eh, inorganic cations and anions, organic materials, clay minerals, and hydrous metal oxides, influence the speciation of heavy metals and, hence, their bioavailability and ultimately, their toxicity to the aquatic organisms. It is the lack of such data regarding influence of environmental factors on the toxicity of heavy metals that has hindered the development of criteria which are reflective of the different types of ecosystems.

No doubt, a better understanding of aqueous chemistry and improved methods of analysis are required to develop water quality standards which accurately relate toxicity to bioavailable species. Consideration of total metal concentration alone can be misleading. Uncomplexed metal ions (Cd^{2+}, Zn^{2+}, and Pb^{2+}) are apparently more readily assimilated by organisms than are complexed forms.[12,4,] It is therefore evident that the effects of chemical speciation such as; inorganic complexation, chelation, precipitation, and adsorption of metals should be taken into account to compensate for the toxic variability when developing guidelines for aquatic life criteria. Water quality criteria and effluent guidelines imposed through the NPDES permitting process should probably take into consideration site-specific environmental conditions when evaluating potential toxicity of vagrant metals.

There are several computer models now available that may offer considerable assistance in our understanding of complex chemical behavior and toxicity of heavy metals. MINTEQA2 (MINTEQ) was developed by EPA's Environmental Research Laboratory located in Athens, Georgia.[19] It is a thermodynamic equilibrium model that computes aqueous speciation, adsorption, and precipitation/dissolution of solids. For a defined set of environmental conditions, it will quickly predict the type and relative quantities of metal complexes possible. Previous attempts have indicated useful applications of MINTEQ to characterize streams in the New Lead Belt.[20] The purpose of the present study was to test the predictive capabilities of the chemical equilibrium model to identify the various species of lead, cadmium, and zinc metals that may exist under a given set of conditions and compare these results with the results of toxicity tests performed on fathead minnows (*Pimephales promelas*) and daphnia (*Daphnia magna*).

BACKGROUND

This study was initiated as a continuation of interests in environmental impacts from the lead mining industry in the New Lead Belt area of Missouri. The wastewaters discharged from mines, mills, and smelters which may contain residual heavy metals such as lead, cadmium, and zinc have posed potential environmental problems. Specifically, current efforts attempt to better define the toxic effects of heavy metals within the context of the hard, alkaline ground and surface waters.

Industrial effluents from mines, mills, and smelters within the New Lead Belt contain various amounts of lead, zinc and cadmium in a variety of site-specific chemical forms. Most of these metals in mine-mill effluents occur as particulate matter, or finely divided insoluble sulfides, or metal complexes bound to clays and other minerals. Groundwaters as well as surface receiving streams are characterized by high pH and hard alkaline conditions. Organic complexing agents are generally lower in abundance than inorganics, especially bicarbonates and carbonates. Moreover, it is likely that the inorganic complexes, especially carbonates, will predominate even in the presence of organics in such waters.[5] Abundant aquatic organisms within and beyond the wastewater treatment systems offer assurance that toxicity is low or unapparent under existing conditions.

MATERIALS AND METHODS

Toxicity tests were conducted with *Pimephales promelas* (fathead minnows) and *Daphnia magna* (daphnia) by using reagent grade salts of sulfide, carbonate, chloride, and sulfate of lead, cadmium and zinc. Monocultures of both test organisms were obtained from the U.S. Department of Interior National Fisheries research Laboratory, Columbia, Missouri. Cultures were propagated and maintained according to the procedures outlined by the U.S EPA.[21] Acute static toxicity tests were conducted in duplicate and test conditions were specified in accordance with EPA procedures.[21,22] Dechlorinated Rolla, Missouri, tap water was used for culturing and toxicity testing for fathead minnows. Extreme sensitivity of neonate daphnids to low concentrations of chlorine necessitated the routine use of non-chlorinated well water for culturing and toxicity tests.

Finely ground reagent grade lead salts were weighed out in appropriate amounts to create the desired metal content when added directly to the final volume of the test suspension. Tests involving zinc and cadmium salts required much smaller quantities, making separately weighed aliquots impractical. Stock suspensions were made using finely ground reagent grade salts of zinc and cadmium. Appropriate aliquots were diluted to the final volume for testing. Adjustments of pH were necessary only in suspensions of $PbCl_2$ and $PbSO_4$ by addition of $NaHCO_3$ which ultimately resulted in

Table I. Toxicity Test Results of Lead Salts on Fathead Minnows and Daphnia

Metal Salt	Conc. mg/L	Fathead Minnows Survival Range	Fathead Minnows pH Range	Daphnia % Survival Range	Daphnia pH Range
PbS	500	100	8.7	100	8.3–8.2
	1,000	100	8.6–8.7		
	2,000	90–100	8.4–8.8		
	3,000	55–100	8.3–8.8		
	4,000	50– 60	8.0–9.0	90	8.4–8.2
	5,000	40	8.2–8.8		
	6,000			80	8.4–8.2
	10,000			20	8.4–8.2
$PbCO_3$	500	100	8.7		
	1,000	100	8.7		
	1,500	100	8.6–8.7		
	2,000	100	8.6–8.7	90	8.3–8.2
	2,500	100	8.7		
	3,000	100	8.6–8.7		
	5,000	100	8.6–8.7		
	6,000				
	10,000	0	8.4–9.1	95	8.3–8.2
$PbCl_2$	500	100	8.2–9.0	60	8.0–8.4
	1,000	100	8.2–8.9	90	8.0–8.6
	1,500	100	8.1–8.6		
	2,000	95–100	7.8–8.5		
	2,500	90– 95	7.8–8.5		
	3,000	90– 95	7.6–8.3		
	3,500	90	7.9–8.9		
	4,000	0– 90	6.9–9.0		
	4,500	0	6.0–7.0		
	5,000	0	6.0–7.0	0	8.0–8.6
$PbSO_4$	500	100	8.1–8.6	100	7.9–7.7
	1,000	100	7.8–8.8	85	8.4–7.5
	1,500	100	8.0–8.8		
	2,000	90	7.8–8.7		
	2,500	90	7.7–8.6		
	3,000	55– 60	7.2–8.6		
	3,500	0	7.0–7.5		
	5,000			5	8.5–7.2

precipitation of $PbCO_3$. This also resulted in restoration of alkalinity removed by $PbCO_3$ precipitation. With the modest amounts of $ZnCl_2$ and $ZnSO_4$ required for toxic effects, final suspensions were aerated for 12 hours without further pH adjustment. Test solutions/suspensions were allowed to equilibrate for 24 hours. Test organisms were transferred to test vessels and controls using a wide bore glass pipette. pH measurements were made on all vessels at the end of the test period. Unless otherwise noted, concentrations of the various metal salts are reported as mg/L Pb, Zn, and Cd calculated from respective molecular weights.

RESULTS AND DISCUSSION

Toxicity Tests and MINTEQ Results

Results of toxicity tests for metal salts of lead, cadmium and zinc performed on fathead minnows and daphnia are displayed in Tables I, II, and III. Variations in acute toxicity to these two organisms were sometimes observed. In addition, variations appeared among the metal salts themselves. Tables IV, V and VI illustrate the predominant dissolved lead, cadmium and zinc species as predicted by MINTEQ at pH 8.9. Figures 1, 2 and 3 compare the predominant dissolved metal components predicted by MINTEQ at different pH levels of 7.7, 8.3 and 8.9. Even though absolute concentrations tested of lead, cadmium and zinc varied greatly, observed toxic effects showed similar trends depend-

Table II. Toxicity Test Results of Cadmium Salts on Fathead Minnows and Daphnia

Metal Salt	Conc. mg/L	Fathead Minnows Survival Range	Fathead Minnows pH Range	Daphnia % Survival Range	Daphnia pH Range
CdS	0.01			85	8.3
	0.05	95–100	8.9–9.2	61	8.0 –8.5
	0.10	95–100	8.7–8.8	63	8.2 –8.5
	0.20	60– 95	8.7–8.9	10	8.2 –8.5
	0.30	60–100	8.9–9.3	5	8.2 –8.5
	0.40			70	8.4
	0.50	0	8.7–8.9	15	8.2 –8.4
	0.80	0	8.7–8.9		
	1.00	0	8.7–8.9	5	8.3
	1.50	0	8.7–8.9		
	5.00			0	8.2 –8.4
	10.00			5	8.2 –8.4
$CdCO_3$	1	100	8.9		
	5	100	8.8–9.0		
	10	90	8.8–9.0		
	12	95–100	8.8–9.0		
	15	90–100	8.8–9.0		
	20	100	8.8–9.0		
	40	100	8.8–8.9		
	60				
	100	100	8.8–8.9	100	8.4 –8.25
	200			100	8.4 –8.25
	300	95	8.8–8.9		
	500	45	8.8–8.9		
	600			100	8.3
	1000			100	8.3
	2000			95	8.25–8.35
	4000			95	8.35–8.4
$CdCl_2$	0.1	45	8.7–8.8		
	0.3	0	8.7–8.8		
	0.5			25–100	8.35–8.45
	0.6	0	8.7–8.8		
	0.8	0	8.7–8.8		
	1.0	0	8.7–8.8	10–100	8.25–8.45
	5.0			0–100	8.35–8.45
	10.0	0	8.6–8.8	0– 80	8.25–8.45
	15.0			0–100	8.35–8.40
	30.0			20	8.35
	50.0			0– 35	7.85–8.30
	100.0			20– 45	7.50–8.95
	200.0			0	7.0 –8.0
$CdSO_4$	0.01	100	8.7–8.8		
	0.03	100	8.7–8.9		
	0.05	95	8.7–8.9		
	0.08	65	8.7–8.9		
	0.1	60	8.7–8.9		
	0.2	50	8.7–8.8		
	1.0			85	8.2
	3.0			55	8.15–8.2
	5.0			50	8.1 –8.2
	8.0			20	8.05–8.2
	10.0			30–100	8.0 –8.3
	15.0			60	8.0 –8.2
	20.0			10	7.95–8.15
	30.0			65	8.25–8.3
	50.0			0	7.65–8.1
	80.0			100	8.2
	100.0			50	8.2

Table III. Toxicity Test Results of Zinc Salts on Fathead Minnows and Daphnia

		Fathead Minnows		Daphnia	
Metal Salt	Conc. mg/L	Survival Range	pH Range	% Survival Range	pH Range
ZnS	10.0	100	8.7–8.9		
	20.0	100	8.7–9.0		
	50.0	100	8.7–9.0		
	80.0	100	8.7–9.0		
	100.0	100	8.7–9.0		
	120.0	100	8.7–9.0		
	400.0			100	8.3 –8.4
	500.0	100	8.8		
	600.0			100	8.3 –8.4
	1000.0	90	8.8	90	8.35–8.4
	3000.0	5	8.8		
	5000.0			100	8.3 –8.4
	10000.0			100	8.3 –8.35
$ZnCO_3$	10.0	100	8.7–9.0		
	40.0	100	8.7–9.1		
	60.0	100	8.7–9.1		
	80.0	100	8.7–9.1		
	100.0	100	8.7–9.0		
	200.0	95	8.7		
	300.0	85	8.7		
	500.0	100	8.7		
	1000.0	100	8.7	75	8.3 –8.35
	3000.0	100	8.7		
	6000.0			70	8.35–8.4
	10000.0			75	8.35–8.4
$ZnCL_2$	1.0			75	8.2 –8.3
	5.0			65	8.2 –8.3
	8.0			80	8.2 –8.3
	10.0	85	8.7–8.9	50	8.2 –8.3
	15.0			68	8.05–8.3
	17.0			65	8.1 –8.3
	20.0	85	8.7–8.9	88	8.1 –8.3
	30.0			55	8.1 –8.15
	40.0	90	8.6–8.8		
	50.0			25	8.0 –8.15
	60.0	50	8.2–8.8		
	80.0	0	7.6–8.5		
	100.0	0	7.5–8.3	32	7.85–8.15
$ZnSO_4$	10.0	0	8.3–8.7		
	20.0	50	7.9–8.8	25	8.3 –8.35
	40.0	0	8.0–8.7	45	8.25–8.3
	60.0	0	7.7–8.6	65	8.2 –8.3
	80.0	0	7.6–7.9	15	8.2
	100.0	0	7.5–7.6	25	7.7 –8.15
	200.0			40	6.85–8.1
	400.0			0	6.6 –6.7

ing on specific anions present. Comparison of the predicted MINTEQ results with the corresponding acute toxicity are described below.

Sulfides: Metal sulfides in general displayed a low acute toxic effect due to their insoluble nature. However, acute toxicity was evident with concentrations of cadmium sulfide of 0.2 mg/L Cd and higher, with 0% survival at concentrations tested between 0.5 and 1.5 mg/L Cd for fathead minnows (Table II). Similar acute toxicity was shown with daphnia with 0% survival at 5 mg/L Cd, and 5% survival at 10 mg/L Cd. Lead sulfide, as shown in Table I, displayed variable acute effects for fathead minnows at 4,000 mg/L Pb, with 5% and 50% survival for two independent tests. Meanwhile, zinc

Table IV. MINTEQ Model Results for Dissolved Lead Species, Chloride, and Sulfate Ions at pH 8.9

LEAD SALTS CONC. mg/L		LEAD SPECIES (MOLAL)							Chloride Ions (Molal)	Sulfate Ions (Molal)
		PbCl +	Pb(CO3)2-2	PbOH +	PbCO3 AQ	Pb(SO4)2-2	PbHCO3 +	Pb2+		
PbS	500	1.2870E-17	5.6200E-10	5.2840E-14	7.6200E-11	7.0500E-19	1.0400E-13	5.7080E-15	1.1000E-04	4.0000E-04
	1000	1.2870E-17	5.6200E-10	5.2840E-14	7.6200E-11	7.0500E-19	1.0400E-13	5.7080E-15	1.1000E-04	4.0000E-04
	1500	1.2870E-17	5.6200E-10	5.2840E-14	7.6200E-11	7.0500E-19	1.0400E-13	5.7080E-15	1.1000E-04	4.0000E-04
	2000	1.2870E-17	5.6200E-10	5.2840E-14	7.6200E-11	7.0500E-19	1.0400E-13	5.7080E-15	1.1000E-04	4.0000E-04
	2500	1.2870E-17	5.6200E-10	5.2840E-14	7.6200E-11	7.0500E-19	1.0400E-13	5.7080E-15	1.1000E-04	4.0000E-04
	3000	1.2870E-17	5.6200E-10	5.2840E-14	7.6200E-11	7.0500E-19	1.0400E-13	5.7080E-15	1.1000E-04	4.0000E-04
PbCO3	500	2.1080E-13	9.6450E-06	8.8810E-10	1.2790E-06	1.1570E-14	1.7480E-09	9.4670E-11	1.1000E-04	4.0000E-04
	1000	2.1080E-13	9.6440E-06	8.8810E-10	1.2790E-06	1.1570E-14	1.7480E-09	9.4680E-11	1.1000E-04	4.0000E-04
	1500	2.1080E-13	9.6430E-06	8.8810E-10	1.2790E-06	1.1570E-14	1.7480E-09	9.4680E-11	1.1000E-04	4.0000E-04
	2000	2.1080E-13	9.6420E-06	8.8810E-10	1.2790E-06	1.1580E-14	1.7480E-09	9.4690E-11	1.1000E-04	4.0000E-04
	2500	2.1090E-13	9.6420E-06	8.8810E-10	1.2790E-06	1.1580E-14	1.7480E-09	9.4690E-11	1.1000E-04	4.0000E-04
	3000	2.1090E-13	9.6410E-06	8.8810E-10	1.2790E-06	1.1580E-14	1.7480E-09	9.4690E-11	1.1000E-04	4.0000E-04
PbCl2	500	7.0970E-12	9.8980E-06	8.9390E-10	1.2780E-06	1.1130E-14	1.7590E-09	9.7180E-11	3.7000E-03	3.9000E-04
	1000	1.3970E-11	1.0140E-05	8.9930E-10	1.2770E-06	1.0740E-14	1.7700E-09	9.9540E-11	7.3000E-03	3.9000E-04
	1500	2.0850E-11	1.0360E-05	9.0430E-10	1.2760E-06	1.0380E-14	1.7800E-09	1.0180E-10	1.0900E-02	3.9000E-04
	2000	2.7700E-11	1.0580E-05	9.0910E-10	1.2750E-06	1.0060E-14	1.7890E-09	1.0400E-10	1.5000E-02	3.9000E-04
	2500	3.4570E-11	1.0640E-05	9.1030E-10	1.2750E-06	1.0090E-14	1.7920E-09	1.0460E-10	1.8000E-02	3.9000E-04
	3000	4.1390E-11	1.0990E-05	9.1790E-10	1.2740E-06	9.4690E-15	1.8070E-09	1.0810E-10	2.1700E-02	3.9000E-04
PbSO4	500	2.1060E-13	1.0040E-05	8.9710E-10	1.2780E-06	3.1410E-13	1.7660E-09	9.8560E-11	1.0100E-02	2.1300E-03
	1000	2.1030E-13	1.0400E-05	9.0510E-10	1.2760E-06	9.8350E-13	1.7810E-09	1.0210E-10	1.0100E-02	3.8400E-03
	1500	2.1010E-13	1.0740E-05	9.1240E-10	1.2750E-06	1.9730E-12	1.7960E-09	1.0550E-10	1.0100E-02	5.5400E-03
	2000	2.0990E-13	1.1050E-05	9.1900E-10	1.2730E-06	3.2480E-12	1.8090E-09	1.0860E-10	1.0100E-02	7.2000E-03
	2500	2.0970E-13	1.1220E-05	9.2250E-10	1.2730E-06	4.9240E-12	1.8160E-09	1.1030E-10	1.0100E-02	9.0000E-03
	3000	2.0940E-13	1.1630E-05	9.3090E-10	1.2710E-06	6.5340E-12	1.8320E-09	1.1440E-10	1.0100E-02	1.0500E-02

Table V. MINTEQ Model Results for Dissolved Cadmium Species, Chloride, and Sulfate Ions at pH 8.9

CADMIUM SALTS CONC. mg/L		CADMIUM SPECIES (MOLAL)								Chloride Ions (Molal)	Sulfate Ions (Molal)
		CdCl+	CdOH +	Cd(OH)2 AQ	CdSO4 AQ	CdHCO3 +	CdCO3 AQ	Cd(SO4)2-2	Cd2+		
CdS	0.1	2.9170E-16	2.1450E-15	7.5680E-17	1.6400E-15	1.5690E-13	1.0460E-11	7.1900E-18	5.4330E-14	1.1000E-04	4.0000E-04
	0.2	2.9170E-16	2.1450E-15	7.5680E-17	1.6400E-15	1.5690E-13	1.0460E-11	7.1900E-18	5.4330E-14	1.1000E-04	4.0000E-04
	0.5	2.9170E-16	2.1450E-15	7.5680E-17	1.6400E-15	1.5690E-13	1.0460E-11	7.1900E-18	5.4330E-14	1.1000E-04	4.0000E-04
	0.8	2.9170E-16	2.1450E-15	7.5680E-17	1.6400E-15	1.5690E-13	1.0460E-11	7.1900E-18	5.4330E-14	1.1000E-04	4.0000E-04
	1	2.9170E-16	2.1450E-15	7.5680E-17	1.6400E-15	1.5690E-13	1.0460E-11	7.1900E-18	5.4330E-14	1.1000E-04	4.0000E-04
	1.5	2.9170E-16	2.1450E-15	7.5680E-17	1.6400E-15	1.5690E-13	1.0460E-11	7.1900E-18	5.4330E-14	1.1000E-04	4.0000E-04
CdCO3	1	1.2410E-13	9.2990E-13	3.3330E-14	6.9430E-13	6.8010E-11	4.5280E-09	3.0440E-15	2.3230E-11	1.1000E-04	4.0000E-04
	5	1.2410E-13	9.2990E-13	3.3330E-14	6.9430E-13	6.8010E-11	4.5280E-09	3.0440E-15	2.3230E-11	1.1000E-04	4.0000E-04
	10	1.2410E-13	9.2990E-13	3.3330E-14	6.9430E-13	6.8010E-11	4.5280E-09	3.0440E-15	2.3230E-11	1.1000E-04	4.0000E-04
	12	1.2410E-13	9.2990E-13	3.3330E-14	6.9430E-13	6.8010E-11	4.5280E-09	3.0440E-15	2.3230E-11	1.1000E-04	4.0000E-04
	15	1.2410E-13	9.2990E-13	3.3330E-14	6.9430E-13	6.8010E-11	4.5280E-09	3.0440E-15	2.3230E-11	1.1000E-04	4.0000E-04
	20	1.2410E-13	9.2990E-13	3.3330E-14	6.9430E-13	6.8010E-11	4.5280E-09	3.0440E-15	2.3230E-11	1.1000E-04	4.0000E-04
CdCl2	0.05	1.2450E-13	9.2990E-13	3.3330E-14	6.9430E-13	6.8010E-11	4.5280E-09	3.0440E-15	2.3230E-11	1.1000E-04	4.0000E-04
	0.1	1.2470E-13	9.2990E-13	3.3330E-14	6.9430E-13	6.8010E-11	4.5280E-09	3.0440E-15	2.3230E-11	1.1000E-04	4.0000E-04
	0.2	1.2540E-13	9.2990E-13	3.3330E-14	6.9430E-13	6.8010E-11	4.5280E-09	3.0440E-15	2.3230E-11	1.1000E-04	4.0000E-04
	0.5	1.2720E-13	9.2990E-13	3.3330E-14	6.9430E-13	6.8010E-11	4.5280E-09	3.0440E-15	2.3230E-11	1.1000E-04	4.0000E-04
	1	1.3030E-13	9.2990E-13	3.3330E-14	6.9430E-13	6.8010E-11	4.5280E-09	3.0440E-15	2.3230E-11	1.2000E-04	4.0000E-04
	5	1.5490E-13	9.3000E-13	3.3330E-14	6.9420E-13	6.8010E-11	4.5280E-09	3.0440E-15	2.3240E-11	1.4000E-05	4.0000E-04
CdSO4	0.01	1.2410E-13	9.2990E-13	3.3330E-14	6.9430E-13	6.8010E-11	4.5280E-09	3.0440E-15	2.3230E-11	1.1000E-04	4.0000E-04
	0.03	1.2410E-13	9.2990E-13	3.3330E-14	6.9440E-13	6.8010E-11	4.5280E-09	3.0460E-15	2.3230E-11	1.1000E-04	4.0000E-04
	0.05	1.2410E-13	9.2990E-13	3.3330E-14	6.9460E-13	6.8010E-11	4.5280E-09	3.0470E-15	2.3230E-11	1.1000E-04	4.0000E-04
	0.08	1.2410E-13	9.2990E-13	3.3330E-14	6.9500E-13	6.8010E-11	4.5280E-09	3.0500E-15	2.3230E-11	1.1000E-04	4.0000E-04
	0.1	1.2410E-13	9.2990E-13	3.3330E-14	6.9510E-13	6.8010E-11	4.5280E-09	3.0520E-15	2.3230E-11	1.1000E-04	4.0000E-04
	0.3	1.2410E-13	9.2990E-13	3.3330E-14	6.9670E-13	6.8010E-11	4.5280E-09	3.0650E-15	2.3230E-11	1.1000E-04	4.0000E-04

Table VI. MINTEQ Model Results for Dissolved Zinc Species, Chloride, and Sulfate Ions at pH 8.9

ZINC SALTS CONC. mg/L		ZINC SPECIES (MOLAL)									Chloride Ions (Molal)	Sulfate Ions (Molal)
		ZnCl +	ZnOH +	Zn(OH)2 AQ	ZnSO4 AQ	Zn(SO4)2−2	ZnHCO3 +	ZnCO3 AQ	Zn(CO3)2−2	Zn2+		
ZnS	10	1.5850E-12	5.4520E-09	4.1190E-08	2.5630E-10	8.3300E-13	3.0240E-08	1.6050E-06	1.0090E-04	9.9380E-09	1.1000E-04	4.0000E-04
	20	1.5850E-12	5.4520E-09	4.1190E-08	2.5630E-10	8.3300E-13	3.0240E-08	1.6050E-06	1.0090E-04	9.9380E-09	1.1000E-04	4.0000E-04
	50	1.5850E-12	5.4520E-09	4.1190E-08	2.5630E-10	8.3300E-13	3.0240E-08	1.6050E-06	1.0090E-04	9.9380E-09	1.1000E-04	4.0000E-04
	80	1.5850E-12	5.4520E-09	4.1190E-08	2.5630E-10	8.3300E-13	3.0240E-08	1.6050E-06	1.0090E-04	9.9380E-09	1.1000E-04	4.0000E-04
	100	1.5850E-12	5.4520E-09	4.1190E-08	2.5630E-10	8.3300E-13	3.0240E-08	1.6050E-06	1.0090E-04	9.9380E-09	1.1000E-04	4.0000E-04
	120	1.5850E-12	5.4520E-09	4.1190E-08	2.5630E-10	8.3300E-13	3.0240E-08	1.6050E-06	1.0090E-04	9.9380E-09	1.1000E-04	4.0000E-04
ZnCO3	10	1.1850E-12	4.1510E-09	3.1860E-08	1.9070E-10	6.1990E-13	2.3030E-08	1.2200E-06	7.8420E-05	7.4580E-09	1.1000E-04	4.0000E-04
	40	4.7220E-12	1.6560E-08	1.2700E-07	7.5650E-10	2.4590E-12	9.1870E-08	4.8620E-06	3.1390E-04	2.9840E-08	1.1000E-04	4.0000E-04
	60	7.0630E-12	2.4790E-08	1.8990E-07	1.1280E-09	3.6660E-12	1.3750E-07	7.2720E-06	4.7080E-04	4.4740E-08	1.1000E-04	4.0000E-04
	80	9.3900E-12	3.2980E-08	2.5250E-07	1.4950E-09	4.8590E-12	1.8300E-07	9.6680E-06	6.2780E-04	5.9630E-08	1.1000E-04	4.0000E-04
	100	1.0570E-11	3.7150E-08	2.8420E-07	1.6810E-09	5.4620E-12	2.0610E-07	1.0880E-05	7.0790E-04	6.7220E-08	1.1000E-04	4.0000E-04
	120	1.0570E-11	3.7150E-08	2.8420E-07	1.6810E-09	5.4620E-12	2.0610E-07	1.0880E-05	7.0790E-04	6.7220E-08	1.1000E-04	4.0000E-04
ZnCL2	10	1.8170E-12	3.8200E-09	2.9320E-08	1.7550E-10	5.7040E-13	2.1190E-08	1.1230E-06	7.2190E-05	6.8650E-09	1.8000E-04	3.4000E-04
	20	5.0800E-12	7.6290E-09	5.8530E-08	3.4960E-10	1.1360E-12	4.2320E-08	2.2420E-06	1.4430E-04	1.3720E-08	2.6000E-04	3.4000E-04
	40	1.5920E-11	1.5230E-08	1.1670E-07	6.9470E-10	2.2580E-12	8.4470E-08	4.4690E-06	2.8880E-04	2.7450E-08	4.0000E-04	3.4000E-04
	60	3.2460E-11	2.2780E-08	1.7450E-07	1.0350E-09	3.3630E-12	1.2640E-07	6.6810E-06	4.3320E-04	4.1150E-08	5.5000E-04	3.4000E-04
	80	5.4650E-11	3.0310E-08	2.3190E-07	1.3710E-09	4.4540E-12	1.6810E-07	8.8790E-06	5.7770E-04	5.4850E-08	7.0000E-04	3.4000E-04
	100	8.1080E-11	3.7180E-08	2.8420E-07	1.6750E-09	5.4420E-12	2.0620E-07	1.0880E-05	7.1030E-04	6.7420E-08	8.4000E-04	3.4000E-04
ZnSO4	10	9.2020E-13	3.2240E-09	2.4740E-08	1.7090E-10	6.4130E-13	1.7880E-08	9.4750E-07	6.0940E-05	5.7950E-09	1.1000E-04	4.6000E-04
	20	1.8360E-12	6.4380E-09	4.9380E-08	3.8570E-10	1.6420E-12	3.5710E-08	1.8910E-06	1.2190E-04	1.1590E-08	1.1000E-04	5.2000E-04
	40	3.6560E-12	1.2830E-08	9.8300E-08	9.4510E-10	4.9740E-12	7.1190E-08	3.7640E-06	2.4380E-04	2.3170E-08	1.1000E-04	6.5000E-04
	60	5.4580E-12	1.9180E-08	1.4670E-07	1.6730E-09	1.0490E-11	1.0640E-07	5.6200E-06	3.6570E-04	3.4720E-08	1.1000E-04	7.2000E-03
	80	7.2460E-12	2.5490E-08	1.9480E-07	2.5660E-09	1.8660E-11	1.4140E-07	7.4600E-06	4.8760E-04	4.6270E-08	1.1000E-04	8.9000E-04
	100	9.0180E-12	3.1770E-08	2.4240E-07	3.6170E-09	2.9940E-11	1.7620E-07	9.2840E-06	6.0960E-04	5.7820E-08	1.1000E-04	1.0200E-03

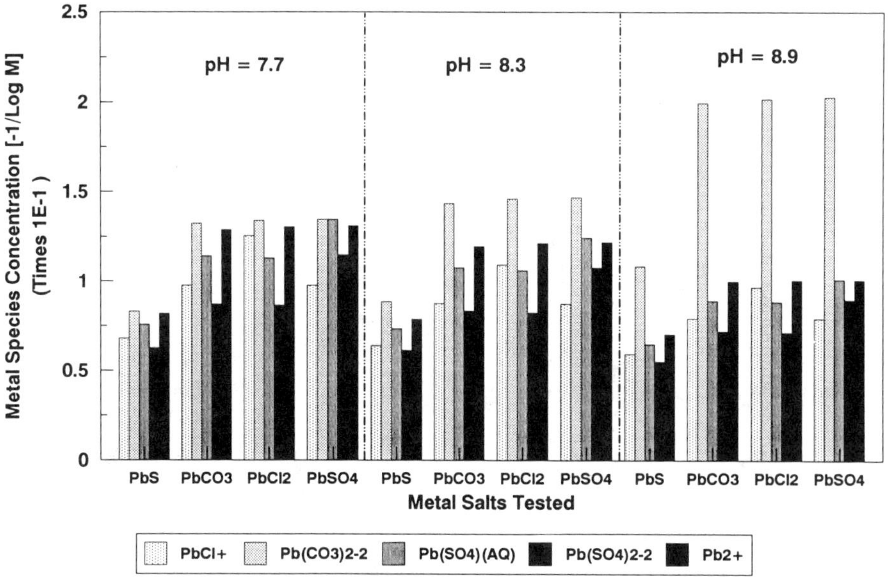

Figure 1. Predominant dissolved lead components of MINTEQ model results at concentration of 3000 mg/L and at pHs of 7.7, 8.3, and 8.9.

sulfide displayed no significant toxicity (Table III) for fathead minnows or daphnia as high as 500 mg/L Zn and 600 mg/L Zn, respectively. Independent tests of 3,000 mg/L Zn produced 5% survival for fathead minnows, yet a test of 10,000 mg/L Zn produced no acute toxicity for daphnia.

The variable acute toxicity occurring at very high concentrations is difficult to explain using the

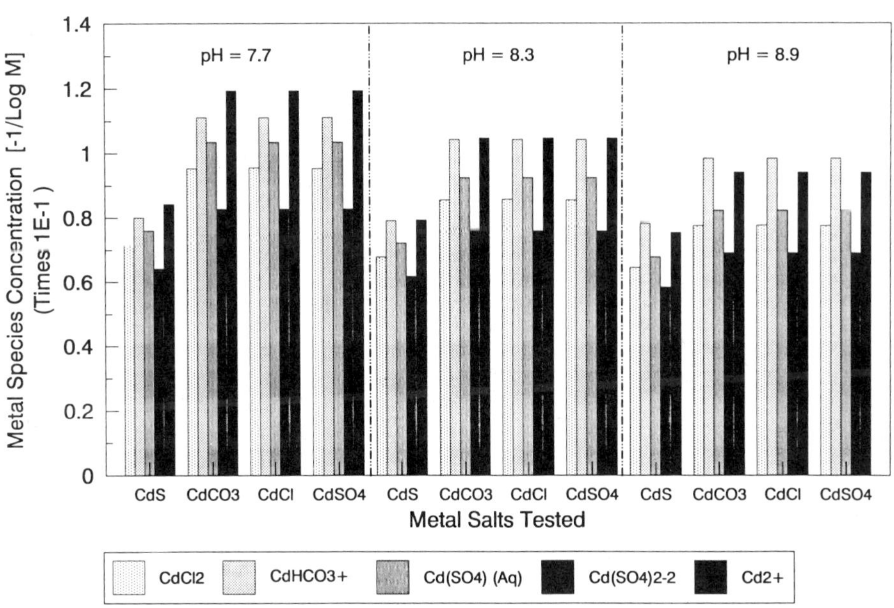

Figure 2. Predominant dissolved cadmium components of MINTEQ model results at concentration of 3000 mg/L and pHs of 7.7, 8.3, and 8.9.

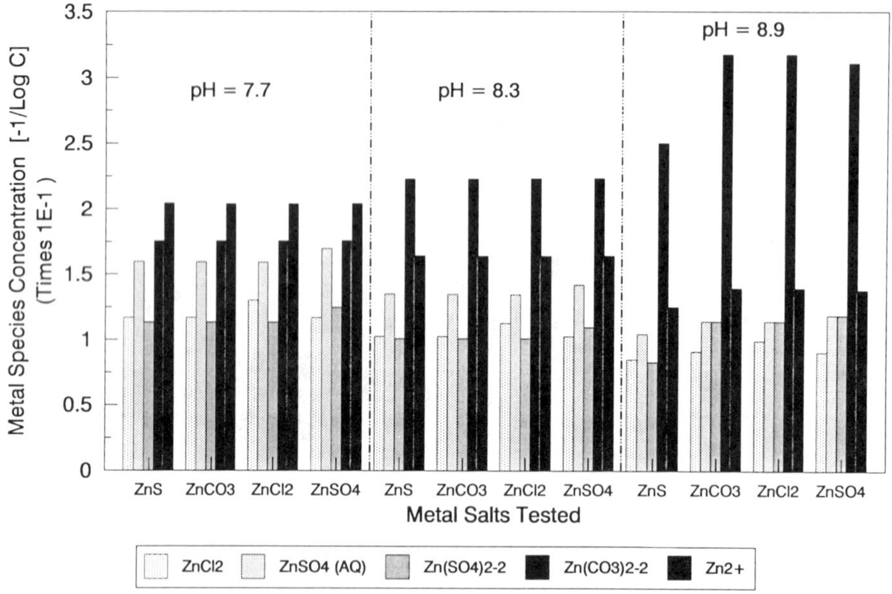

Figure 3. Predominant dissolved zinc components of MINTEQ model results at concentration of 3000 mg/L and at pHs of 7.7, 8.3, and 8.9.

MINTEQ predictions. Conditions placed on the MINTEQ input did not allow oxidation of sulfide ions and formation of other resulting complexes. Dissolved components that may be responsible for toxicity are S^{2-}, SO_4^{2-}or H_2S which may be present at higher concentrations upon oxidation of sulfide salts. A more likely source of toxicity at these high concentrations involved possible impurities present in the refined salts.

Carbonates: In all cases, test vessels containing Pb, Cd, and Zn carbonate salts all displayed little to no acute toxicity. MINTEQ data for carbonates was, therefore, depicted as a suggestive control for comparison to chloride and sulfate results since dissolution of metals initially bound to SO_4^{2-} or Cl^- eventually precipitated as carbonates under alkaline conditions. In the examination of lead carbonate at 5,000 mg/L Pb, cadmium carbonate at 100 mg/L Cd and zinc carbonate at 3,000 mg/L Zn, 100% survival resulted for fathead minnows. As shown in Figure 1, Concentrations of lead ion, Pb^{2+}, are corresponding low (less than 2.0E-10M) and similar for chloride, sulfate and carbonate salts, leading to the concept that the metal ion may not be independently responsible for acute toxicity. This same case appears in Figure 2 with cadmium and in Figure 3 with zinc, where Cd^{2+} and Zn^{2+} (metal ions) are of similar concentrations among carbonate, chloride and sulfate salts of these metals.

Chlorides and Sulfates: Lead, zinc and cadmium salts of chloride and sulfate displayed acute toxicity. Upon examination of lead salts, lead sulfate displayed acute toxicity more frequently and at lower concentrations as illustrated in Table I. At 3,500 mg/L $PbSO_4$-Pb and 5,000 mg/L $PbSO_4$-Pb, 0% survival and 5% survival were observed for fathead minnows and daphnia, respectively. On the other hand, a concentration of 4,000 mg/L $PbCl_2$-Pb displayed 90% survival in one test and 0% survival in an independent test of identical concentration with fathead minnows. In examining the MINTEQ predicted dissolved metal components (Table IV), 1.19E-19 M $PbSO_4$ (AQ) ($PbSO_4$ AQ) and 6.53E- 12 M $PbSO_4$2-2 ($Pb(SO_4)_2^{2-}$) were found with the addition 3,000 mg/L $PbSO_4$-Pb salt. However, predicted concentrations of $PbSO_4$ (AQ) and $Pb(SO_4)_2^{2-}$ at 3,000 mg/L Pb for lead carbonate and lead chloride salts never exceeded levels of 5.6E-12 M $PbSO_4$ (AQ) and 1.2E-14 M $Pb(SO_4)_2^{2-}$. This comparison is illustrated in Figure 1 at pH 7.7, 8.3 and 8.9. Figure 1 also displays decreasing concentrations of Pb^{2+} with increasing pH (7.7 to 8.9). A possible relationship may be noted between observed toxicity and presence of various sulfate bound dissolved metal complexes. However, wider ranges of pH were observed for test solutions of lead chloride and sulfate salts (Table I), implying toxicity or toxic variability may be attributed to altering the concentrations of dissolved metals consisting of lead ions and/or lead-sulfate bound complexes.

Variable toxicity occurred for daphnia throughout concentrations tested of 0.5–100 mg/L $CdCl_2$-Cd and 1.0–100 mg/L $CdSO_4$-Cd. A solution of 50 mg/L $CdSO_4$-Cd produced 100% survival for *Daphnia magna* whereas, an increased concentration of 100 mg/L $CdSO_4$-Cd resulted in 50% survival as shown in Table II. No survival was noted for fathead minnows at concentrations greater than 0.3 mg/L $CdCl_2$-Cd. 50% survival was found for 0.2 mg/L $CdSO_4$-Cd with fathead minnows, but no additional testing was performed. Table V indicates insignificant difference between dissolved cadmium levels in chloride and sulfate salts compared to that of carbonate, where dissolved Cd^{2+} never exceeded 2.4E-11 M.

No survival was noted for all concentrations tested for zinc sulfate (Table III) between 10 mg/L and 100 mg/L Zn for fathead minnows with the exception of one independent test of 20 mg/L Zn where 5% survival occurred. Corresponding MINTEQ results (Table VI) indicate increases in zinc- sulfate bound dissolved complexes, such as $Zn(SO_4)_2^{2-}$ and $ZnSO_4$ (AQ), but not in Zn2+ (zinc ion) concentration when compared to zinc carbonate salts.

Additional information displayed in Tables IV, V, and VI are concentrations of sulfate, chloride ions for metal salts at pH 8.9. Upon precipitation of metal carbonates from sulfate and chloride salts, the majority of SO_4^{2-} and Cl^- were predicted to remain in dissolved form which would increase with increasing concentrations of lead and zinc salts. Cadmium sulfate displayed no dramatic increase in SO_4^{2-}, however, initial concentrations tested were between 0.01–3.00 mg/L $CdSO_4$-Cd and were too low to note any difference.

CONCLUSIONS

Several factors are imperative to the varying toxicity results that can be explained by pH, hardness, alkalinity, equilibration and kinetics involved in chemical reactions. Other factors influencing variations in toxicity can not be solely explained by aqueous and sediment chemistry, but by specific biology or activities of the test organisms.

Daphnia magna, the invertebrate test organism, remained primarily in the water column and avoided intimate contact with existing sediments. On the other hand, the vertebrate test organism, fathead minnow, frequently interacted with sediment and pore water to a point that ingestion of sediments was occasionally observed. Both of these factors and others were suspected to influence toxicity, however, differences in pH stability between daphnia and fathead minnow tests appeared to have a more influential role (Table I, II and III). Where daphnia acute toxicity was observed at higher concentrations than that of fathead minnow acute toxicity, often the pH range was not near as broad.

Chloride and sulfate salts of lead and zinc showed variable acute toxicity in the higher concentrations tested for both fathead minnows and daphnia. Similar cadmium salts displayed toxic variability for daphnia. A major influence for toxic variability is indicated by fluctuating pH values. For example, as indicated in Table I, 4,000 mg/L $PbCl_2$-Pb solutions resulted in acute toxicity for fathead minnows of 0–90% survival. Corresponding pH levels ranged from 6.9 to 9.0. Metal ion and suspected toxic metal complex interactions increase with decreasing pH as shown in Figures 1, 2 and 3. Accordingly, acute toxicity was more prevalent when substantial pH changes occurred between the beginning and end of the toxicity tests. Overall, the mode of toxicity was different for each metal. Lead appeared to be several orders of magnitude less toxic to fathead minnows and daphnia than that of zinc or cadmium.

Use of the MINTEQ model to depict metal species' concentration is an effective tool and appears to be useful in the examination of the toxicity data. However, further work with the model's attributes is necessary for a better understanding of specific toxicity of metal species. In particular, future work should involve analysis of various redox potentials as MINTEQ inputs.

MINTEQ should have a future in the regulatory field for establishing water quality standards, waste load allocations, and deriving permit limits and laboratory procedures. In addition, the software would be helpful in establishing effectiveness of treatment methods to meet desirable metal species concentrations.

Future research in metal toxicity is highly recommended. The relationship of toxicity to chemical kinetics of carbonate precipitation and the synergistic effects of two or more metals are two important issues of concern. Likewise, future toxicity studies such as these will benefit greatly with an examination of metal species in dissolved and precipitated form utilizing a metal speciation model.

ACKNOWLEDGMENTS

This study has been supported by grant # C86043 A2 from the Missouri Department of Natural Resources. The cooperation and assistance of Doe Run Mining Company during the course of this study is also gratefully acknowledged.

REFERENCES

1. Schmitt, C. J., Finger, S. E., May, T. W., Kaiser, M. S., "Bioavailability of Lead and Cadmium from Mine Tailings to the Pocketbook Mussel (*Lampilis Ventricosa*)," *Proceedings of the Die-offs of Freshwater Mussels in the United States*, Richard J. Neves, editor. U.S. Fish and Wildlife Service and the Upper Mississippi River Conservation Committee, 115 (1987).
2. Davies, P. H., Goettl, J. P., Sinley, J. R., and Smith, N. F., "Acute and Chronic Toxicity of Lead to Rainbow Trout (*Salmo gairdneri*) in Hard and Soft Water," *Water Research*, 10, 199 (1976).
3. Pickering, Q. H., and Henderson, C., "The Acute Toxicity of Some Heavy Metals to Different Species of Warm Water Fishes. Part 2," *Int. J. Air Water Poll.*, 10, 453 (1966).
4. Sunda, W. G., Engel, D. W., and Thuette, R. M., "Effects of Chemical Speciation on Toxicity of Cadmium to Grass Shrimp (*Palaemantas pugio*): Importance of the Free Cadmium Ion," *Environmental Science and Technology*, 13, 409 (1978).
5. Borgman, U., "Metal Speciation and Toxicity of Free Metal Ions to Aquatic Biota," In: *Aquatic Toxicology, Advances in Environmental Science and Technology*, J. O. Nriagu (ed.), John Wiley and Sons, Volume 13, pp. 47–73 (1983).
6. Sherman, R. E., Gloss, S. P., and Lion, L. W., "A Comparison of Toxicity Tests Conducted in the Laboratory and in Experimental Ponds Using Cadmium and the Fathead Minnow (*Pimephales promelas*)," *Water Research*, 21, 317 (1987).
7. McCarty, L. S., and Houston, A. H., "Effects of Exposure to Sublethal Levels of Cadmium Upon Water Electrolyte Status in Goldfish (*Carassius auratus*)," *J. Fish Biology*, 9, 11 (1976).
8. Calamari, D., Alabastar, J. S., "An Approach to Theoretical Models in Evaluating the Effects of Toxicants in the Aquatic Environment," *Chemosphere*, 9, 533 (1980).
9. Mount, D. I., "The Effects of Total Hardness and pH on the Acute Toxicity of Zinc to Fish," *Int. J. Air Water Poll.*, 10, 49 (1966).
10. Sprague, J. B., "Lethal Concentrations of Copper and Zinc for Young Atlantic Salmon," *J. Fish. Res. Board Canada*, 21, 17 (1964).
11. Bradley, R. W., and Sprague, J. B., "The Influence of pH, Water Hardness, and Alkalinity on the Acute Lethality of Zinc to Rainbow Trout (*Salmo gairdneri*)," *Can. J. Fish. Aquat. Sci.*, 42, 731 (1985).
12. Sunda, W. G., and Guillard, R. L., "The Relationship Between Cupric Ion Activity and the Toxicity of Copper to Phytoplankton," *J. Mar. Res.*, 34, 511 (1976).
13. Vuceta, J., and Morgan, J. J., "Chemical Modeling of Trace Metals in Fresh Waters: Role of Complexation and Adsorption," *Environmental Science and Technology*, 12, 12 (1978).
14. Black, J. A., *et al.*, " The Significance of Physicochemical Variables in Aquatic Bioassays of Heavy Metals," *Bioassay Techniques and Environmental Chemistry*, G. E. Glass (ed), Ann Arbor Science, Ann Arbor, Michigan, pp. 259–275 (1973).
15. Stumm, W., and Morgan, J. J., "*Aquatic Chemistry*," Wiley-Interscience, 2nd Ed., New York (1981).
16. Florence, T. M., "Development of Physico-chemical Speciation Procedures to Investigate the Toxicity of Copper, Lead, Cadmium and Zinc Towards Aquatic Biota," *Analytica Chimica Acta*, 141, 73 (1982).
17. Babich, H., and Stotzky, G., "Physicochemical Factors That Affect Toxicity of Heavy Metals to Microbes in Aquatic Habitats," In: *Proceedings of the ASM Conference, Aquatic Microbial Ecology*, R. R. Colwell and J. Foster, Eds., University of Maryland Sea Grant Publication, College Park, Maryland, pp. 181–203 (1980).
18. Babich, H., and Stotzky, G., "Influence of Chemical Speciation of Heavy Metals," In: *Aquatic Toxicology, Advances in Environmental Science and Technology*, J. O. Nriagu (ed.), John Wiley and Sons, Volume 13, pp. 3–37 (1983).
19. Brown, D., and Allison, J., "MINTEQA1 Equilibrium Metal Speciation Model: A User's Manual," United States Environmental Protection Agency, Office of Research and Development, Athens, Georgia (1987).
20. Schlachter, J. D., "Modeling Chemical Speciation with the Introduction of Heavy Metals into Natural Streams in Missouri's New Lead Belt," M.S. Thesis, Southern Illinois University at Edwardsville, Edwardsville, Illinois (1989).
21. United States Environmental Protection Agency-EPA, "Methods For Measuring the Acute Toxicity of Effluents to Freshwater and Marine Organisms," Third Edition, EPA 600/4–85/013, Environmental Monitoring and Support Laboratory, Cincinnati, Ohio (March 1985).
22. Erten, Z. M., "Toxicity Evaluation and Biological Treatment of Lead Mine-Mill Effluents in Southeast Missouri," Ph.D. Dissertation, University of Missouri-Rolla, Rolla, Missouri (1988).

Section Nine
INDUSTRIAL WASTES – D. COKE, FOUNDRY, AND STEEL MILL WASTES

86 BIOLOGICAL TREATMENT OF WASTEWATER FROM BYPRODUCT COKING OPERATIONS: AN INNOVATIVE APPROACH

Paul M. Sutton, Principal
P. M. Sutton & Associates, Inc.
Bethel, Connecticut 06801

Igor J. Marvan, Director
Dearborn Chemical Co., Ltd.
Mississauga, Ontario L5A 3T5

Fraser Craig, Supervisor Environmental Control
Algoma Steel Corporation
Sault Ste. Marie, Ontario P6A 5P2

INTRODUCTION

The production of metallurgical coke for use in operation of iron-making blast furnaces is achieved through the pyrolysis of coal. Byproduct coking operations result in the generation of complex wastewaters containing high concentrations of conventional organic and inorganic contaminants, such as phenolics and ammonia, as well as trace organic contaminants.

The Algoma Steel Corporation operates a byproduct coke plant at its facility located in Sault Ste. Marie, Ontario, Canada on the bank of the St. Mary's river. The St. Mary's is an international waterway and as such discharges of certain compounds including phenolics to the river from industrial sources are limited in accordance with an international agreement developed between Canada and the United States. In order to meet the phenolic discharge limitations set forth by the agreement, Algoma is required to reduce the level of phenolics in the wastewater from its byproduct coking operations. In the past Algoma has reduced the level of phenolics by operation of a phenol extraction/recovery system. This dephenolization plant was recently upgraded and although costly to operate and maintain, generally performs adequately provided its capacity is not exceeded. Recently Algoma agreed to a control order issued by the Ministry of the Environment (MOE) of the Province of Ontario to reduce the level of phenolics discharged from the coke plant to equal or less than 22.7 kg/day (50 lb/day). It was determined by Algoma that coke production would have to be limited in order to meet this criteria or the level and consistency of phenol removal would have to be improved. As a result of the evaluation of alternative wastewater management options and treatment alternatives, a decision was made to replace the dephenolization plant with a biological treatment system. The aerobic fluidized bed (AFB) process configuration was selected for implementation.

The purpose of this paper is to briefly review the byproduct coking operations on-going at Algoma, to describe the design of the Algoma AFB system, to present the results from a biological treatability study conducted in support of the process design of the plant and to discuss the full scale AFB plant operating status.

ALGOMA BYPRODUCT COKING OPERATIONS

The Algoma byproduct coke plant consists of four cokeover batteries capable of producing approximately 3810 metric tons/day of coke. It is anticipated the coke plant wastewater flow associated with this production level will be approximately 47.3 m^3/h following certain byproduct plant modifications. The source of the major process wastewater stream from the coke plant is the excess flushing

45th Purdue Industrial Waste Conference Proceedings, © 1991 Lewis Publishers, Inc., Chelsea, Michigan 48118. Printed in U.S.A.

liquor or weak ammonia liquor produced during the initial cooling of the coke oven gases (Figure 1). This raw liquor stream is directed to a free leg ammonia still from the raw liquor storage tank. Ammonia is recovered through steam stripping at approximately neutral pH conditions. The ammonia still wastewater is collected in the south oil/water separating basin prior to being directed to an equalization tank (Figure 1). The south oil/water separating basin relies on gravity to achieve removal of any particulate material. After equalization the wastewater is directed to the biological treatment plant.

BIOLOGICAL TREATMENT PLANT DESIGN

Algoma selected the supported growth AFB process configuration, and specifically the Oxitron® system a commercial embodiment of the concept, for treatment of the wastewater from byproduct coking operations following partial ammonia removal through steam stripping. The Algoma fluidized bed system represents the first full scale application of the AFB technology for treatment of coke plant wastewater in North America and the first AFB plant to be installed in Canada.

In the Oxitron® system (Figure 2), wastewater or contaminated water passes upward through a rectangular or circular reactor containing a bed of sand or granular activated carbon media at a velocity sufficient to expand the bed, resulting in a fluidized state. Once fluidized, the media particles provide a vast surface area for biological growth, in part leading to the development of a biomass concentration approximately five to ten times greater than that normally maintained in conventional bioreactors. Design and operating information relative to the components of the system have been

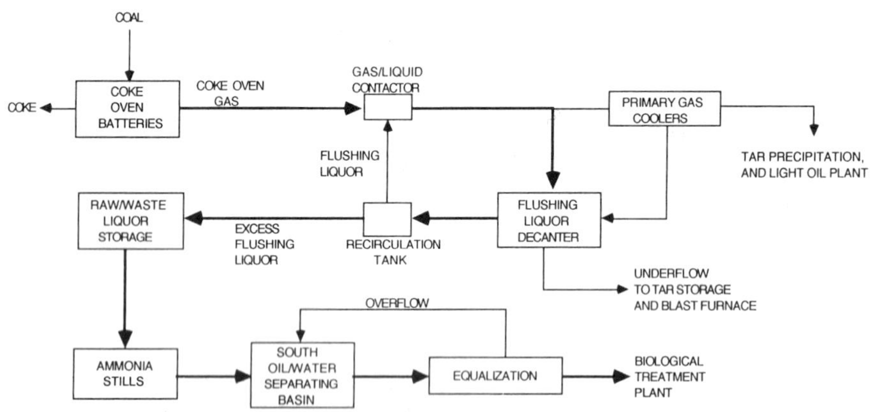

Figure 1. Simplified flowsheet representing algoma coke plant operation following start-up of AFB treatment system.

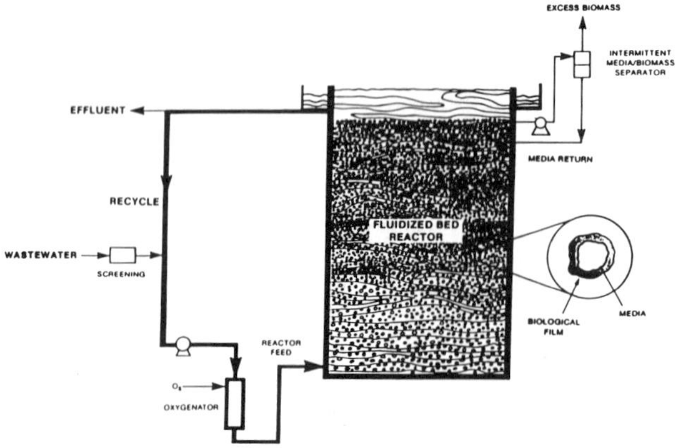

Figure 2. Oxitron® aerobic fluidized bed process schematic.[3]

presented elsewhere.[1,2] In the AFB reactors, sand-biomass separation is required to control the thickness of the biofilm that develops on the sand media, by control of the height of the fluidized bed. Bed height control is accomplished by monitoring the bed expansion optically, carrying out separation of the sand from the biomass if the maximum specified bed height is reached, and return of the sand to the reactor. Sand-biomass separation is normally carried out external to the reactor using a pump in combination with a gravity screen (Figure 1, Intermittent Media/Biomass Separator). The Algoma Oxitron® system relies on an external hydrocyclone operating in conjunction with a hydraulic disengagement tube located in the reactors to achieve phase separation of the sand and biomass.

Figure 3 represents a simplified flowsheet of the Algoma biological treatment system. Equalized wastewater from the byproduct coking operations is diluted prior to mixing with recycled effluent and entering the AFB reactors. The addition of dilution water is required for temperature control purposes and to reduce the concentration of ammonia in the feed. The AFB plant consists of 2, 9.14 m diameter reactors. The sizing of the reactors anticipated on average approximately 200 mg/L of phenolics in the diluted wastewater and a wastewater flow of 142.3 m^3/h. On this basis, the concentration of phenolics in the effluent must be equal to or less than 6.6 mg/L to meet the discharge criteria of 22.7 kg/day set by the Ontario MOE.

Sizing of the Algoma AFB reactors was essentially based on information developed in Holland at the Hoogovens steel mill. Hoogovens developed information for full scale process design of AFB reactors from extensive pilot plant studies.[4] At Hoogovens, two fluidized bed reactors designed to treat 6800 kg/day of COD contained in coke plant wastewater, are now operating.[4] The Algoma and Hoogovens AFB reactor design loading rates are comparable (Table I). Hoogovens interest in the use of the AFB process configuration was influenced by pilot studies undertaken by Dearborn Environmental at the Wastewater Technology Centre (WTC), Burlington, Ontario. At the WTC, a coupled two-stage carbonaceous oxidation-nitrification fluidized bed pilot plant (Figure 4) was operated treating coke plant wastewater from the DOFASCO steel mill located in Hamilton, Ontario.[5,6] Carbonaceous oxidation was achieved in the first stage reactor with nitrate versus oxygen serving as the terminal electron acceptor. Essentially this stage served as the phenolics removal step. It has been shown that many degradable organic carbon compounds can be oxidized through nitrate respiration at approximately the same rate as when oxygen is utilized.[7] This implies that after accounting for any fluidized bed scale-up factors, if the DOFASCO coke plant wastewater was treated in an AFB reactor, complete phenolics removal (effluent phenolics less than 1.0 mg/L) could be achieved at comparable COD and phenolics volumetric loading rates as determined from the WTC first stage results. Calculations indicate that on a volumetric loading rate basis, the Algoma and Hoogovens AFB reactors are designed conservatively (Table I) in terms of achieving complete phenolics removal, unless reactor scale-up is a consideration. It has been shown[8] that the only major scale-up factor effecting AFB performance is the level of active biomass concentration that can be achieved in pilot versus full-scale reactors. A fluidized bed volatile attached solids (VAS) concentration of 25 g/L was achieved in the first stage WTC pilot reactor. The Algoma and Hoogovens reactors have been designed based on achieving a VAS concentration of 12 g/L. Expressing the loading rates previously calculated as specific rates (Table I), the results still imply the Algoma and Hoogovens reactors have been designed conservatively.

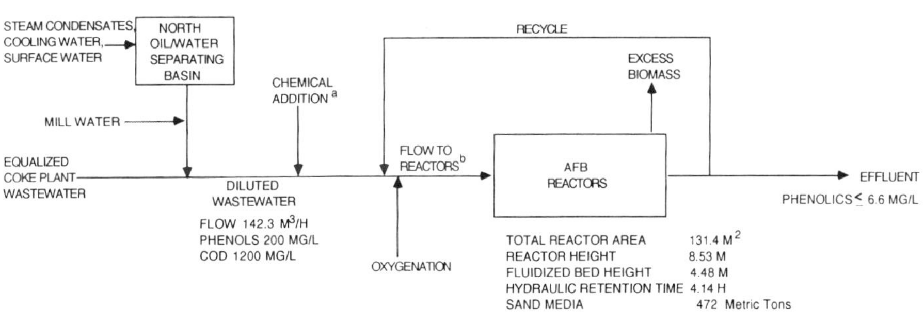

[a]Chemical addition in form of sodium hydroxide for pH control and phosphoric acid as a growth nutrient.

[b]Total flow to reactors is constant at 3532 m^3/h.

Figure 3. Process schematic representing Algoma biological treatment system.

Table I. Comparison of Design Loading Rates for AFB Treatment of Coke Plant Wastewater

Wastewater Source	Volumetric Rates, kg COD / m³ · day	kg Phenolics / m³ · day	Specific Rate, kg COD / kg VAS · day
Algoma-Full-Scale[a]	8.70	1.63	0.72
Hoogovens IJmuiden-Full-Scale[4]	7.03	1.45	0.59
DOFASCO-WTC Pilot Results[b5]	32.88	4.74	1.32

[a]Based on wastewater COD and phenolic values at 99% probability level (approximately 280 mg/L) after equalization and dilution (Figure 3, diluted wastewater).
[b]Based on COD and phenolic in wastewater to first stage reator and first stage reactor fluidized bed hydraulic retention time (empty bed volume basis) and volatile attached solids (VAS) concentration.

LABORATORY SCALE DESIGN VERIFICATION STUDIES

The Hoogovens pilot results and full-scale design information, provided a basis for design of the Algoma AFB reactors. Comparison of the Algoma and Hoogovens full-scale design loading rates to the WTC pilot results (Table I) implies, providing the aerobic treatability of the Algoma and DOFASCO coke plant wastewaters is comparable, the Algoma AFB reactors are sized adequately. The treatability of the Algoma and DOFASCO wastewaters was compared through laboratory testing over a two month period. A severe time constraint prevented completion of more lengthy studies. A description of the laboratory equipment and procedures follows together with a portion of the results derived during the study.

Laboratory Equipment, Operating and Analytical Procedures

In order to develop treatability information, two parallel, flow-through or fed-batch, suspended growth laboratory reactors were operated one treating representative wastewater from the Algoma coke plant (Algoma reactor) and one treating DOFASCO coke plant wastewater (DOFASCO reactor). The DOFASCO wastewater was representative of that previously used in the fluidized bed pilot plant studies conducted at the WTC under the direction of Dearborn Environmental. The parallel operating laboratory reactors were operated under two identical solids retention time (SRT) conditions in order to derive comparable information. The units were operated in the laboratory of the Dearborn Chemi-

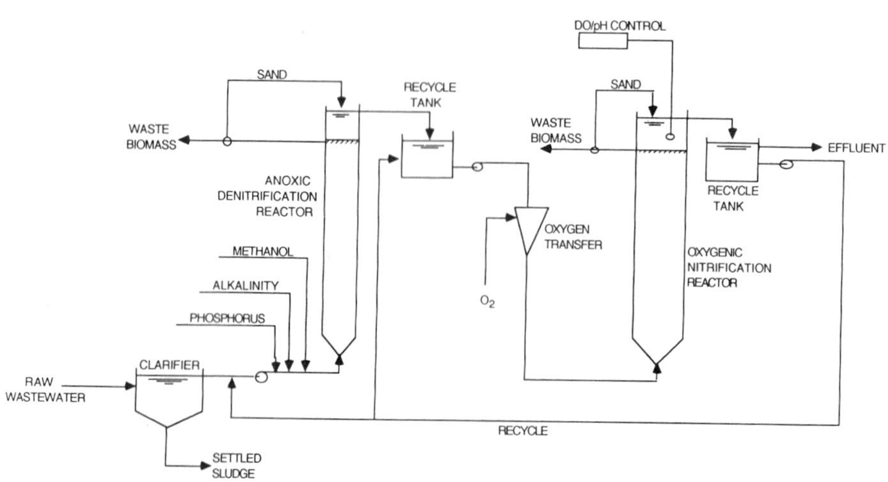

Figure 4. Schematic of TWC pilot scale two-stage AFB biological treatment system.[6]

cal Company in Mississauga, Ontario. Dearborn personnel carried out the day to day operation of the units, and completed all sampling and sample analysis. The treatability of study program was directed by P. M. Sutton & Associates. The laboratory reactors consisted of 7.5 L (maximum liquid volume) acrylic reactors equipped with provisions for air supply and temperature control (Figure 5). Although not indicated on the equipment schematic (Figure 5), fish aquarium heaters were used to adjust the reactor temperature as required. The dissolved oxygen concentration in the reactors was maintained above 2.0 mg/L. The temperature in the reactors was maintained at approximately 30°C, the level at which the full scale Algoma AFB reactors were anticipated to operate.

The experimental design developed for the treatability study called for operation of the two parallel units at two levels of SRT, 5 and 10 days. Since the units were designed to operate as flow-through reactors or chemostats, the reactor SRT was equilvalent to the hydraulic retention time (HRT). During operation at the 10 day SRT experimental design condition, the liquid volume in the reactors was maintained at a constant level by removing reactor contents continuously at a rate equivalent to the feed rates. For a 10 day SRT this equated to 700 mL/day at a constant reactor volume of 7 L. At the 5 day SRT experimental design condition, the reactors were operated as fed-batch units, by removing a certain volume of the reactor contents, normally once per day. The volume removed was equal to the feed added over the fed-batch period. During the period from Monday to Friday, the liquid volume in the reactors normally varied from 7.50 L to 6.35 L over a 24 h period, as feed was added at rates of 1.15 L/day. During Saturday and Sunday the fed batch period was increased and the liquid volume in the reactors and feed rates adjusted accordingly.

The Algoma reactor wastewater feed was obtained from the plant site at a point prior to the existing dephenolization unit. In order to simulate the feed characteristics anticipated once the AFB reactors are on-line, the NH_3-N concentration in the feed was reduced by elevating the pH to approximately 10.5 with NaOH, air stripping the material and adding dilution water as required. The DOFASCO reactor feed was obtained from the Hamilton mill site at the same sampling point from which the feed was taken during the fluidized bed studies completed at the WTC.

Phosphorus in the form of phosphoric acid was added to the Algoma and DOFASCO feed containers in sufficient quantity to ensure the the effluent contained a residual filtered phosphorus level. The pH in the reactors was maintained near neutral by adding HCl or NaOH as required.

During acclimation (non-equilibrium operation) to pseudo-steady state conditions (equilibrium SRT, constant feed rate but variable feed concentration), frequency of feed, reactor contents and effluent sampling was limited to one to two days per week depending on the sample location. Once pseudo-steady state (equilibrium operation) was achieved sampling frequency was increased to two or three times per week depending on the sample location. Analysis for various parameters was generally in accordance with standard procedures.[9] On occasion, the oxygen uptake rate of the contents of each reactor was determined by adding a sample to a 300 mL glass bottle fitted with an air tight oxygen sensor connected to an oxygen meter, and determining oxygen consumption as a function of time. At the conclusion of the treatability study, the mixed liquor suspended solids contents of the Algoma and DOFASCO reactors were allowed to settle, 5 L of the supernatant material was removed and an equivalent volume of feed material was added. The reactors were then operated under batch conditions for a period of 18 h, and samples were taken and analyzed in an attempt to derive further performance information.

Laboratory Study Results and Discussion

Start-up of the laboratory reactors was initiated under continuous feed flow conditions at a rate equivalent to approximately a 10 day HRT or SRT. The units were seeded with biomass obtained from the DOFASCO activated sludge system treating coke plant wastewater. After 25 days of operation at these conditions (non-equilibrium conditions), the concentration of mixed liquor volatile suspended solids (MLVSS) in the reactors appeared to have stabilized. The performance of the reactors was then assessed at an equilibrium SRT of approximately 10 days. Immediately after this period of equilibrium operation, conditions were altered and the reactors were operated to achieve an SRT of approximately 5 days. After 11 days of operation in order to achieve equilibrium SRT conditions, the performance of the reactors was assessed.

The phenol and the total (unfiltered) COD characteristics of the Algoma and DOFASCO wastewaters fed to the reactors during the entire study were determined by developing probability relationships for these analytical parameters.[10] The DOFASCO wastewater contained a higher level of phenolics but a lower level of COD relative to the Algoma wastewater (Figure 6). The concentration of phenolics in the effluent from both the Algoma and DOFASCO reactors was consistently below 1.0 mg/L throughout the entire study under both equilibrium and non-equilibrium conditions (Figure 7). The effluent COD from the Algoma reactor, was higher than the DOFASCO reactor value (Figure 7)

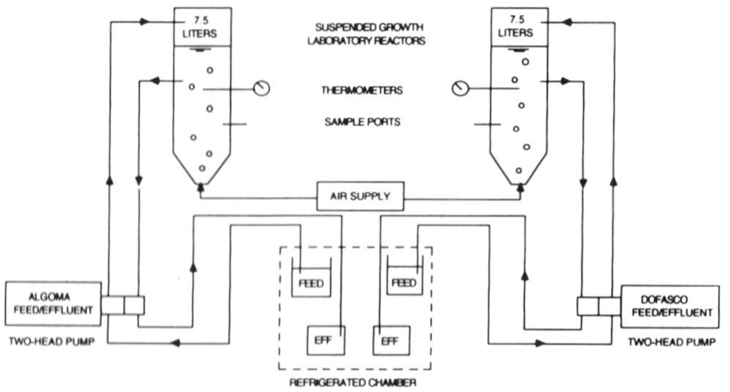

Figure 5. Schematic of laboratory scale suspended growth units in algoma study.

based on analysis of filtered samples, likely reflecting the higher COD concentration in the feed to the Algoma reactor. Effluent filtered samples were analyzed in order to separate out the reactor biomass material which will always be present in the effluent from flow-through or fed-batch chemostat reactors. During equilibrium operation at the experimental design point of 10 days, the actual SRT maintained in the reactors was 9.7 days (Table II). At the 5 day SRT design condition, the actual value was in the range from 4.8 to 6.0 days (Table II). The performance results derived under equilibrium conditions indicate the Algoma wastewater can be treated to an effluent phenolics level of less than 1.0 mg/L at a reactor SRT as low as 4.8 days (Table III). There are indications that the performance may be negatively affected by higher levels of NH_3-N in the feed. At the 9.7 day SRT condition, the effluent phenolics averaged 0.55 versus 0.10 mg/L at the 4.8 to 6.0 day SRT. At the 9.7 day SRT, the COD and phenolics in the feed were lower but the NH_3-N was higher than at 4.8 to 6.0 day SRT condition (Table III). The NH_3-N in the Algoma feed was higher than in the DOFASCO feed (Table III). This may explain the minor difference in performance of the two reactors relative to phenol oxidation.

The Algoma laboratory study results can be used to project the performance of the full-scale AFB reactors. The specific rate of phenol oxidation achieved during pseudo-steady state run 2 can be determined from the performance results (Table III) and the reactor MLVSS concentration results (Table II) observed during that period. Although this rate represents the maximum rate achieved

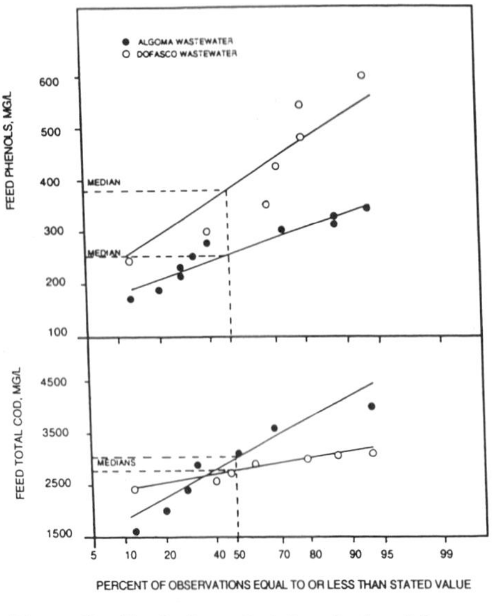

Figure 6. Feed characteristics during laboratory study under equilibrium and non-equilibrium operating conditions.

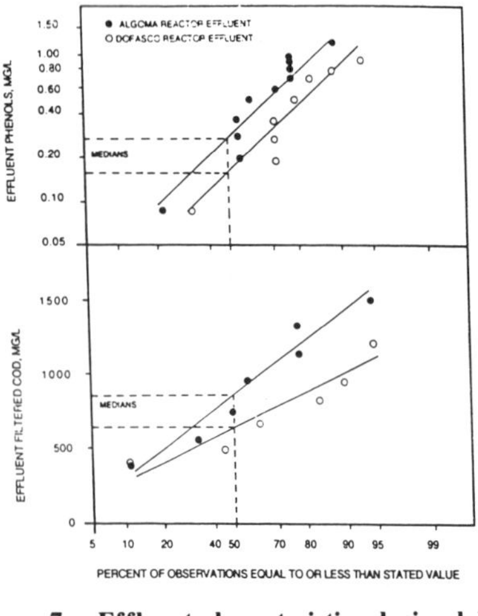

Figure 7. Effluent characteristics during laboratory study under equilibrium and non-equilibrium operating conditions.

Table II. Reactor Operating Results Under Equilibrium Conditions During Algoma Laboratory Study[a]

Pseudo-Steady State Run	Reactor	MLVSS, mg/L	SRT, days	pH Range	Mean Temperature (Range), °C
1	Algoma	280	9.7	7.3–7.8	31 (28–35)
	DOFASCO	321	9.7	7.2–7.7	32 (29–36)
2	Algoma	298	4.8–6.0[b]	7.1–7.9	30 (28–30)
	DOFASCO	359	4.8–6.0[b]	7.2–7.7	30 (29–30)

[a]all numbers are averages over entire pseudo-steady state run unless otherwise stated.
[b]Variation in SRT owing to difference in solids wasting procedure over weekends versus during week.

Table III. Performance Results Under Equilibrium Conditions During Algoma Laboratory Study[a]

Pseudo-Steady State Run	Reactor	Feed Values, mg/L			Effluent Values, mg/L		
		Total COD	Phenolics	NH_3-N	Filtered COD	Phenolics	NH_3-N
1	Algoma	2678	244	535	662	0.55	1011
	DOFASCO	2660	384	145	377	0.40	171
2	Algoma	3526	300	418	1090	0.10	407
	DOFASCO	2461	287	48	789	0.09	124

[a]All numbers are averages over entire pseudo-steady state run.

during the laboratory study based on the pseudo-steady state results, it is likely not the maximum possible rate, being limited by a reactor/effluent phenolics concentration of 0.1 mg/L. The rate does allow a projection of the minimum level of phenol oxidation achievable in the AFB reactors depending on the fluidized bed VAS concentration. Calculations indicate a minimum of 321 mg/L of the phenolics can be oxidized even if the AFB reactors only achieve 10 g/L of VAS (Table IV). This value is well above the 99 percentile phenol value (approximately 280 mg/L) anticipated in the wastewater feed to the Algoma AFB reactors.

FULL-SCALE PLANT OPERATING STATUS

Algoma Steel had originally planned on completing the full-scale AFB plant construction in January 1990 with start-up to follow shortly thereafter. Start-up is now planned for June 1990. Coke production is currently limited; therefore the regulatory criteria set by the Ontario MOE is being met.

Table IV. Minimum Projected Level of Phenol Oxidation Anticipated in the Algoma AFB Reactors

Pseudo-Steady State Run	Observed Specific Rate of Phenol Oxidation, kg Phenolics / kg MLVSS · day	VAS Concentration in AFB Reactors,[a] g/L	Projected Level of Phenolics Oxidized,[b] mg/L
		10	321
2	0.186	12	385
		14	449

[a]The activity of the VAS in the fluidized bed reactors is assumed equivalent to the activity of the MLVSS in the Algoma laboratory reactor on a equivalent dry weight basis.
[b]Based on Algoma FBR operating at fluidized bed HRT of 4.14 h. Number represents level of phenolics which potentially could be oxidized.

SUMMARY AND CONCLUSIONS

The Algoma Steel Corporation operates a byproduct coke plant at its facility located in Sault Ste. Marie, Ontario, Canada on the bank of the St. Mary's river. Faced with a limit on the discharge of phenolics, Algoma selected biological treatment for wastewater management and specifically the Oxitron® fluidized bed system for installation.

The Algoma fluidized bed reactors were designed based on process information developed by others. In order to ensure confidence in the adequacy of the Algoma Oxitron® system design, a treatability study was conducted in the laboratory of the Dearborn Chemical Company in Mississauga, Ontario under the direction of P. M. Sutton & Associates. The treatability study results provided assurance that the fluidized bed reactors will have adequate capacity to ensure achievement of an effluent phenol level of less than 22.7 kg/day (50 lb/day) at a diluted wastewater flow of 142 m^3/h. Start-up of the Oxitron® system is scheduled for June 1990.

REFERENCES

1. Sutton, P. M., Shieh, W. K., Kos, P. and Dunning, P. R., "Dorr-Oliver's Oxitron System Fluidized Bed Water and Wastewater Treatment Process". *Biological Fluidized Bed Treatment of Water and Wastewater*, P. F. Cooper and F. Atkinson (Eds.). Ellis Horwood, Chichester, England, pp 284–300 (1980).
2. *O & M of Trickling Filters, RBCs and Related Processes*. Manual of Practice OM-10, Water Pollution Control Federation, pp 100–102 (1988).
3. Hare, R. W., Sutton, P. M., Mishra, P. N. and Potochnik, K. F., "Utilization of Fluidized Bed Biological Treatment at General Motors Facilities: Pilot and Full Scale Results". *Proceedings of the Industrial Waste Symposia, 61st Annual Conference, Water Pollution Control Federation*, Dallas, TX (1988).
4. van Hoorn, A. I. "Treating Coke Oven Wastewater by Precoatfiltration and Fluidised Bed treatment", Presented at the Meeting of the American Institute of Metallurgical Engineers, Detroit, MI (1990).
5. Nutt, S. G. and Marvan I. J., "Biological Fluidized Bed Treatment of Coke Plant Wastewater and Blast Furnace Scrubber Blowdown". Environmental Protection Service Report, EPA 3/MM/1, Environment Canada (1984).
6. Melcer, H., Nutt, S. G., Marvan, I. J. and Sutton, P. M., "Combined Treatment of Coke Plant Wastewater and Blast Furnace Blowdown Water in a Coupled Biological Fluidized Bed System". *Journal Water Pollution Control Federation*, 56, 192 (1984).
7. McClintock, S. A., Sherrard, J. H. Novak, J. T. and Randall, C. W., "Nitrate Versus Oxygen Respiration in the Activated Sludge Process". *Journal Water Pollution Control Federation*, 60, 342 (1988).
8. Sutton, P. M. and Mishra, P. N., "Fluidized Bed Biological Wastewater Treatment: Effects of Scale-up on System Performance". *Water Science and Technology*, 22, 1/2 419 (1990).
9. *Standard Methods for the Examination of Water and Wastewater*, 17th Ed., Am. Public Health Assoc. Washington, D.C. (1989).
10. Chow, V. T., *Handbook of Applied Hydrology*, McGraw-Hill, New York, NY, pp. 8–13 (1964).

87 STEEL MILL SCALE DE-OILING: A NEW WASTE MINIMIZATION PROCESS

Natalie R. Blake, Senior Engineer

Kristine S. Siefert, Senior Research Chemist
Nalco Chemical Company
Naperville, Illinois 60566

INTRODUCTION

In 1988, our Waste Minimization Marketing identified a new market for our products in steel mill scale de-oiling. There appears to be no existing process in the U.S. for chemically cleaning mill scale, though solvent extraction and steam stripping have been proposed as alternate technologies. The use of high shear mechanical treatment, combined with non-volatile surface-active chemical treatment represents a new process for cleaning mill scale. A patent application has been submitted for this unique technology.

In high temperature steel-making operations, such as in the hot strip mill, the hot surface of steel sheets, bars, blooms, and slabs becomes oxidized, is cooled, and cleaned with high pressure water spray.[1] A layer of iron oxides is continually being formed on the surface of the steel during exposure to air at elevated temperature. This layer is referred to as "mill scale." The scale is composed primarily of oxides of iron, FeO, Fe_2O_3 and Fe_3O_4. Scale must be removed from the steel to insure satisfactory results in downstream processes involving deformation of the material, such as rolling or drawing.

As hot steel is processed, the layer of mill scale is broken away and replaced by a fresh layer of iron oxides. The formation of new scale takes place every time the shape or size of the hot steel is changed. As shown in Figure 1, the scale is removed from the steel with high pressure water, and the solids and water are collected in a sewer system. As the result of leaks from equipment hydraulics, direct use of oils in some milling operations, and accidental dumping of oils and lubricants, mill process waste water can contain a high level of dispersed oil which also contaminates the mill scale surfaces. Waste water flows into one of a number of pits where floating oil is removed by skimming, and solids can settle to the bottom. These mill scale solids, coated with oil, are periodically dredged or continuously scraped from the pits.

A Midwestern steel mill produces approximately 500 tons per day of mill scale from its normal steelmaking operations. Currently the mill uses only a small fraction (less than 10%) of the less oily scale in its sinter plant. The remaining, larger fraction of high oil content scale is not used because its use in sintering would result in unacceptable hydrocarbon emissions and fouling of the filter bags and fans in the baghouse. Iron ore is currently being purchased to replace unusable mill scale as an iron source. The steel industry as a whole is faced with increasing amounts of mill scale that cannot be used in sintering operations, resulting in both higher raw materials costs and higher disposal costs. This mill previously used a patented kiln process which vaporized the oil, but the resulting hydrocarbon emissions exceeded allowable air quality levels, and the kiln process was later discontinued.

EXPERIMENTAL

Process Requirements

It was difficult to define the Midwestern steel mill's requirements with regard to final oil content of the treated mill scale. Even mill personnel were not sure what level of oil would allow use of a larger fraction of the mill scale without causing problems in sinter plant operation. In addition, early problems were encountered with differences between our oil analysis methodology and the mill's methodology. A final oil content of 0.2–0.3 wt% was eventually chosen as the target, as determined by solvent extraction.

The de-oiling process should be capable of dealing with a range of oil concentrations, scale composition, and different particle size distributions, since mill scale is quite heterogeneous in composition.

45th Purdue Industrial Waste Conference Proceedings, © 1991 Lewis Publishers, Inc., Chelsea, Michigan 48118. Printed in U.S.A.

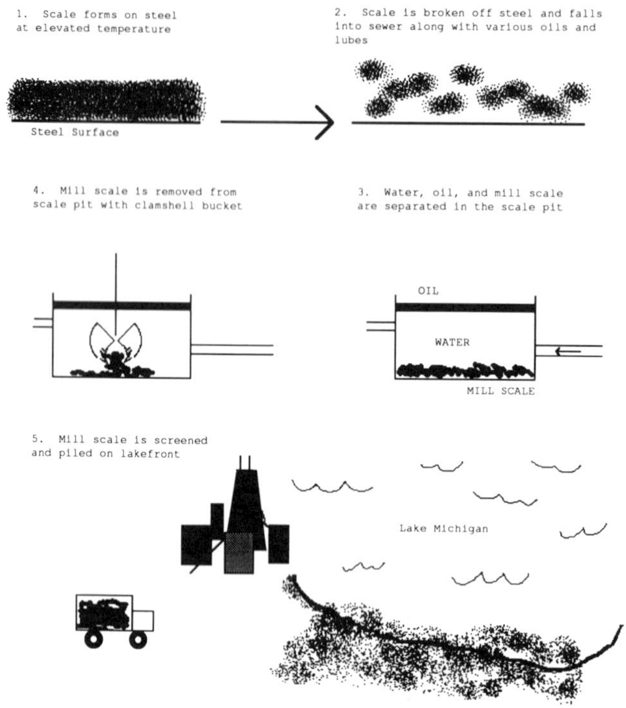

Figure 1. Mill scale problem at a Midwestern steel mill.

In an aqueous washing process, it was thought that these variables could have significant impact on the effectiveness of the chemical program.

The cleaning process will generate a volume of waste water that will need to be treated for removal of suspended solids and oil. It was necessary to conduct laboratory studies to select appropriate water clarification products for different de-oiling chemistries. At that time it was also necessary to determine the requirements for discharging the effluent into the mill's water recycle system.

Mill Scale

The oil on the mill scale is highly variable both in composition and in concentration. Information from the mill indicated that dozens of different oils, greases, hydraulic fluids, and so on, are used. Even though there are many forms of "oil" on the scale, they will all be referred to as oil for simplicity. However, characterization of the mill scale oil allows more efficient optimization of the treatment chemistry.

Mill scale from different sources within a mill can vary extensively in composition, particle size and oil content. The scale typically contains less than 0.1% metallic iron, but the percentages of FeO, Fe_2O_3, and $FeO{\cdot}Fe_2O_3$ ($=Fe_3O_4$) can vary greatly with steel processing conditions and mill scale age. The longer the scale is exposed to air, the higher the percentage of Fe_2O_3 (hematite, rust). FeO and Fe_3O_4 are magnetic while Fe_2O_3 is not.

Oil content on mill scale from different sources in this mill typically ranged from 0.1 to 4% by weight. Even from one source of scale, oil content was seen to vary. For example, oil contents from 0.2 to 0.9% have been measured to date from a particular location.

Within a sample of scale from one source, it was found that smaller particles contain more oil than do larger particles. For example, one scale sample was sieved into four size fractions, and the oil content of each fraction was determined. The fraction retained on a Tyler 5 mesh screen with a 4,000 micron diameter opening contained 0.112% oil, the material passing this screen but retained on a screen with a 1190 micron opening contained 0.315% oil. The fraction which passed a 1190 micron screen but was retained on a 590 micron screen contained a 0.667% oil, and the material passing the 590 micron screen, retained on a 297 micron diameter opening screen, contained 0.783% oil.

Analytical Techniques

Oil content of mill scale was determined by solvent extraction of the solids, followed by spectrophotometric determination of the extract's absorbance. Comparison of absorbance with a calibration curve for the mill scale oil allows the calculation of oil concentration in the extract and, ultimately, on the solids.

Laboratory Mixing Studies

Mixing studies were performed using a torque mixer and a baffled jar. The torque mixer was driven with a 1/8 hp motor. Shaft speed in rpm can be set and the resulting torque read directly from a meter. The torque can then be expressed in a variety of units which allows comprehensible transfer of engineering data between laboratories. Experiments were performed to determine the effects on oil removal of shear, mixing time, chemical dosage, and other variables.

The first step of the process was to make a water slurry of mill scale containing 30–80% solids by weight. This mill scale slurry was then subjected to mixing for 5 to 120 minutes. The mixing energy generally ranged from medium to very high shear, although low shear was also investigated. The shear conditions considered are easily attainable in the commercial equipment planned for use. Chemical dosage requirements under various mixing regimes (G, time) were also investigated. As testing progressed, chemical dosage of the original program was optimized, and then the product itself was reformulated and ameliorated to minimize overall costs.

Pilot Test Studies

Extensive laboratory testing showed the process to be feasible, and thus a pilot test was conducted to evaluate process scale-up parameters and economics for commercial operation. A pilot size version of the unit expected to be used in commercial operation was used for the pilot scale experiments. This equipment easily imparts energy density in the range of the commercial unit. A rough estimate for the G factor was calculated based on estimated viscosity values for the mill scale slurry.[2] The range calculated was 90–3360 sec^{-1}. Although the commercial unit will be operated continuously, the pilot unit was operated in a batch mode, for ease of operation.

All pilot scale testing was performed at a pre-set rpm, with variations in the amount of applied shear (G) achieved by changing mixing time and slurry solids. Most studies were performed at room temperature, although elevated temperatures were also investigated. At the end of mixing in the chamber, the temperature was slightly elevated, due to operation. The importance of these variables was later examined with a computer-based experimental design technique, as described below.

Experimental Design

An experimental design was developed using the RS/Discover® software program available from BBN Software[3] which helps to design and analyze industrial experiments. When running an experimental design, a goal for an experiment is selected, for example oil removed from mill scale, and both the controllable variables and the responses of interest can be defined. For the pilot test, quantitative controllable variables initially were mixing time, chemical dosage and slurry solids. A qualitative variable describing whether the scale was rinsed after treatment was also included.

The responses examined for oil content were final oil content (absolute difference), and percent oil removed. The model for percent oil removed did not give an accurate representation of the experimental data, and was not chosen for further examination. In a process which imparts shear, the physical degradation of the particles must be considered. Mill scale, made up of iron oxides, is a somewhat friable material. Thus determination of the weight % of scale passing a 100 mesh screen (149 micron diameter opening) was made for untreated and treated scale samples, and thus fines generation was included in the design as a response.

A D-optimal design with 3 variables was run. With a D-optimal design, the best set of runs is selected from all possible candidates, based on the design model and factor settings. This approach minimizes the number of experiments that must be run to obtain a statistically significant model. In our case with 3 variables, the D-optimal number of necessary runs is 12, as opposed to the 27 required for a full factorial design for 3 variables. Replicates should always be run to check reproducibility and improve model fit.

For the pilot test, RS/Explore®[4] was used to determine the importance of each variable on absolute oil removal and fines generation, and to generate visual representations of statistical results. Whether the scale was rinsed or not was included as a qualitative variable in the modeling, and its importance

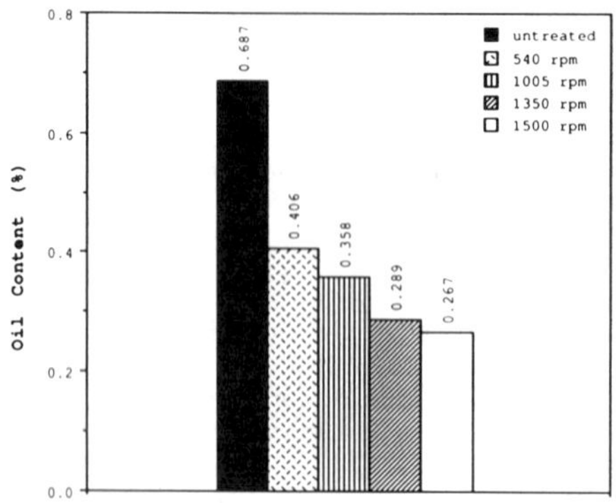

Figure 2. Oil content of raw and treated mill scale (constant mix time, chemical dosage).

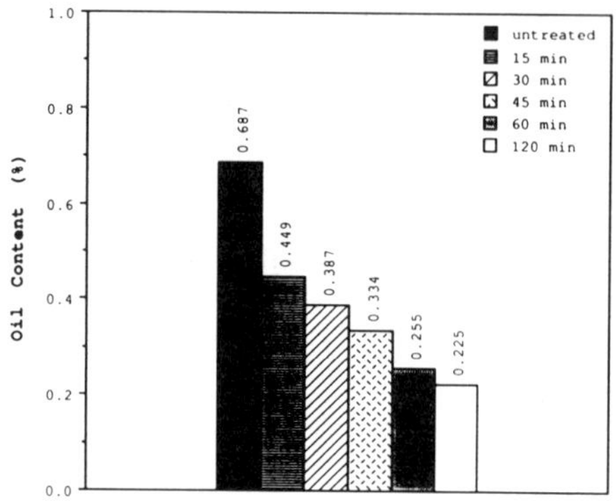

Figure 3. Oil content of mill scale at increasing mix time (fixed rpm, chemical dosage).

to oil removal was also determined. The results of the experimental design are presented graphically below in the results and discussion section.

RESULTS AND DISCUSSION

Oil Removal

Laboratory Studies

Shear, as measured by impeller speed or torque, was investigated to determine its effect on oil removal. At a constant mixing time and chemical dosage, increased shear resulted in greater oil removal. Oil contents of scale treated at different shear, expressed as impeller speeds, are shown in Figure 2.

Mixing time effects were also investigated. At a constant rpm and chemical dosage, increased mixing time (increased applied shear) resulted in greater oil removal. Oil contents of scale mixed for 15, 30, 45, 60, and 120 minutes are shown in Figure 3. In addition to greater oil removal at longer mixing times, the greatest quantity of fines was produced at the longest mixing time, a potentially negative effect.

The effect of initial oil content on oil removal with chemical treatment was examined. Increasing chemical dosage, all other parameters remaining fixed, increased oil removal. Incremental oil removal as a function of chemical dosage was observed to decrease above a certain product dosage. This indicates that if scale with a higher than average oil content is fed to the de-oiling process it will be treatable, but higher product usage will be required.

Trade-offs between chemical dosage and mechanical treatment were investigated. Results indicate that the same oil removal can be achieved at a lower mixing energy (shear) or shorter mixing time with proper chemical addition. This information can be used to optimize the process with respect to cost while minimizing fines production. In addition, recycle and re-use of a concentrated wash solution is possible. This option would allow the maximization of oil removal at high chemical dosage while controlling overall chemical costs. In addition, second generation chemical development has greatly reduced chemical dosages in laboratory scale testing.

Depending on the chemical treatment program used, there was sometimes a great advantage in rinsing the scale with fresh water after treatment. Up to 21% of the remaining oil may be removed through rinsing. This could reduce the oil content of a scale from unacceptable to within the requirements, reducing de-oiling costs. All other conditions being equal, increasing the number of times the scale was rinsed after treatment increased oil removal.

Some samples of scale with relatively low initial oil content were very hard to treat compared to other scales studied. This could indicate the existence of one of several possible situations: the scale may have very large internal pore volume which effectively traps the oils, even in the presence of a surface active agent. Another possibility is that the oil on this particular scale is a different type of oil that is more difficult to remove. Also, scale age may be important due to oil weathering.

Pilot Scale Studies

The extensive laboratory testing allowed excellent characterization of the process and process variables, and thus progression of the project to pilot scale testing. For each pilot test run, two oil analyses were performed on the treated scale. Part of the treated scale was rinsed in the laboratory with a weight of fresh water equal to the scale, and oil analysis of this rinsed scale was subsequently performed. Another fraction of the treated scale was analyzed without rinsing. A correlation was also established between the laboratory and pilot scale studies. Running the experiment for the same amount of time, at the same chemical dosage in both cases gave similar de-oiling results if the lab mixer was kept at 1350 rpm.

The pilot scale studies confirmed conclusions from laboratory testing such as the benefit of chemical addition in reducing the severity of the necessary mixing conditions. As chemical dosage was increased above a certain level, however, incremental oil removal per unit chemical dose was decreased, as was seen in the laboratory. Thus, there was a point of diminishing return for chemical dosage.

The importance of post-treatment rinsing was demonstrated in pilot tests. Up to 20% additional oil is removed by rinsing the scale *just once* after treatment compared to that removed without rinsing. If more rinses were performed, more oil is removed, although there was a decrease in the incremental oil removal per rinse. Even though some of this additional oil removal may be achieved through loss of a portion of the fines (smaller particles have a higher oil content), care was taken to ensure that very few fines were lost upon rinsing, and certainly not enough were lost to account for the additional oil removal. In addition, laboratory studies show that fines loss is not the major mechanism for additional oil removal when rinsing was used after washing treatment.

It can be postulated that the oil is loosened from the scale by the high shear mixing, and this oil readily enters the process water. Some of the oil, however, while being less tightly held compared to oil on the raw scale, is not completely removed from the vicinity of the scale surface, and may actually still be loosely bound to the surface. A fresh water rinse thus takes this loosely attached oil and displaces it from the surface, resulting in a scale with lower oil content.

The dependence of oil removal and fines generation on mix time was also verified. As mixing time was increased, all other variables fixed, more oil was removed from a given scale. This can be attributed to the mixing and extra shearing force to which the scale was subjected, providing greater exposure of scale surfaces to the wash water and subsequent release of oil. A greater quantity of fines was generated with increased mixing time, however. This reflects the extra beating the friable mill scale is subjected to as mixing time is lengthened.

Laboratory studies had indicated that process water could be reused with a fresh charge of mill scale and still show benefit of chemical addition, i.e., enhanced oil removal over mechanical-only treatment. The economic advantage of reuse or recycle of concentrated wash solution is reduction of product usage. Recycling experiments were investigated during the pilot tests and confirmed laboratory results.

Data generated in the pilot scale experiments were analyzed with RS/Discover®, and fit to a model. The model was then refined so that all remaining terms were "significant". Significance of a term indicates that the term has an effect on the results of the model. Factor effect (significance) plots and contour plots were generated for the model. The model fit the data with a correlation coefficient of 0.923, indicating very good fit.

The RS/Discover® software package transformed each factor so that it varied from -1 at its lowest value to +1 at the highest value. For example, if a factor F were to vary from 0 to 100, its transformed term, F′, would be expressed as

$$F' = \frac{(F-50)}{50} \tag{1}$$

Thus, when $F=0$, $F' = (0-50)/50 = -1$; $F = 100$ results in an F' value of $+1$, and the midpoint of the range, $F = 50$, gives an $F' = 0$. This scaling, an "orthogonal" representation, insures that the true significance of any particular term can be determined without interference from potential differences in the order of magnitude of different factor ranges.

Figure 4 shows the effects of the factors (mix time, dosage, slurry solids, and rinse) on the response (oil difference). The horizontal bar for each factor represents the possible range of effect that factor has if it is varied from its lowest to its highest value, with the other variables (factors) remaining constant. Only the linear terms from the model are used to generate the main effects plot. As observed experimentally, mix time (energy density) is the factor that has the most impact on oil removal. The

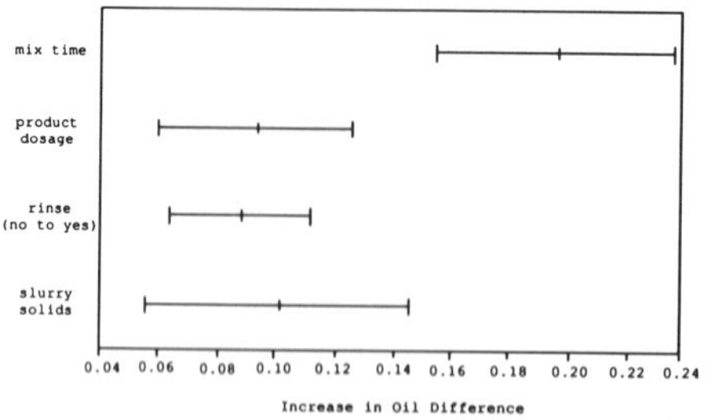

Figure 4. Main effects on the response "oil difference" for quantitative and qualitative variables.

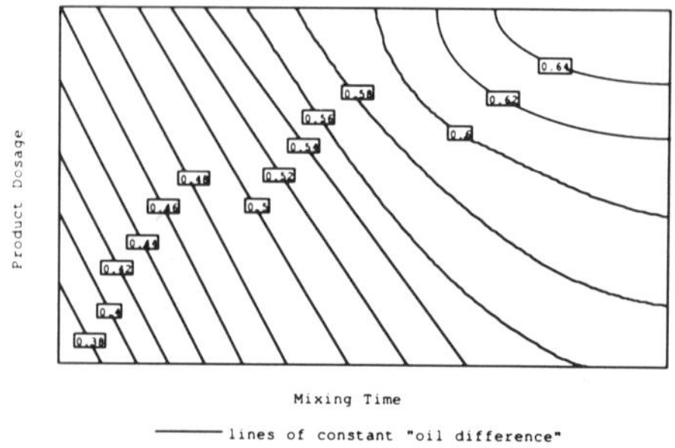

Figure 5. Contour pilot of constant "oil difference" as a function of mixing time and product dosage.

large slurry solids effects bar indicates that the effect caused by changing slurry solids from low to high is known with the least certainty.

Figure 5 is a contour plot of the refined model. The x-axis shows mixing time, the y-axis represents dosage. The solid lines represent lines of constant oil difference, the difference between initial (untreated) and final (treated) oil content. Again, mixing time is seen to be the most important factor of all those considered in the pilot test.

Wastewater Treatment

Wastewater treatment studies were initiated during the laboratory process characterization. Jar tests using various treatment wastewaters (o/w emulsions) were performed with various Nalco products. Initial work showed that increased product dosages in the de-oiling phase of the process result in increased chemical requirements for wastewater treatment. Depending on the chemical used during oil removal, the emulsion was sometimes more difficult to break.

During both the laboratory and the pilot phases of operation, process wastewater was successfully treated. Pilot scale wastewater containing approximately 1680 ppm of oil was treated to <50 ppm oil and <5 NTU turbidity. The oil removed from the wastewater was associated with settled solids, as no floating oil was observed. The weight of sludge produced was approximately 3% of the total scale weight treated.

MECHANISTIC CONSIDERATIONS

There are several possible mechanisms for oil removal. In one mechanism, it is assumed that oil is spread over the scale surface as a thin film, forming a solid-oil interface. The oil is in contact with the wash water, so this is a second interface. The driving force causing oil removal from the scale surface is an imbalance of interfacial tensions at the contact line of the three phases (solid, oil, water). This can be accomplished by adsorption of surface-active agents at the water/oil and solid/water interfaces so that the oil is "rolled up" on the scale surface. The high shear mixing environment promotes release of oil droplets from the surface and into the wash water. Emulsification of the oil into the water prevents readsorption of oil onto the scale surface.

A second mechanism postulates that a film of water is present between the solid and oil phases. In this case the surface-active agent molecules must diffuse between the oil and solid surfaces, causing a "significant disjoining pressure" in the film,[5] which promotes oil droplet separation into the wash water. Released oil droplets are then emulsified in the water by high energy mixing.

If this mechanism is correct, then mill scale stored on the land for long periods of time may be more difficult to treat than fresh scale, since water content appears to decrease over time. The loss of the water film between the scale and oil phases probably decreases the effectiveness of a chemical used on fresh scale. Thus, de-oiling of weathered mill scale may require different washing chemistry.

SUMMARY

The use of high shear mechanical processing in combination with non-volatile surface-active chemical treatment achieved oil removals of over 80% at room temperature in laboratory testing. Our high shear mill scale de-oiling process (patent pending) was also successfully demonstrated on a pilot scale producing mill scale with $\leq 0.2\%$ oil, which represents $>90\%$ removal. The low temperature process avoids the volatilization of oil which could be as much as 25% of the oil on mill scale at 60°C and could contribute to air emissions if this occurred.

Fresh water rinsing of the mill scale after high shear mixing was seen to increase oil removal by 20% relative to no rinsing. Rinsing, therefore, is a cost effective processing step for reducing chemical usage. Recycling process water showed some benefit in removing oil from additional charges of oily scale, a further means of overall program cost reduction. In addition, second generation chemical development has greatly reduced chemical dosages in laboratory scale testing.

A correlation was determined between the laboratory bench scale testing equipment and the pilot scale equipment. This allows the development of products whose performance on a commercial scale can be predicted with confidence. Process performance for mill scale from other sources can also be assessed, and commercial operation predicted.

Clarification of process wastewater from all pilot tests was successfully performed, resulting in supernatant with <50 ppm of oil and < 5 NTU turbidity. No floating oil was evident, as the oil was removed with the settled solids.

ACKNOWLEDGMENTS

The authors would like to thank Dave Picco, whose timely analytical work was crucial to the project.

REFERENCES

1. Anon., "Steel Industry Sludge is Being Reused," *Env. Sci. Technol.* 9(7), 624 (1975).
2. Perry's Handbook.
3. RS/Discover® User's Guide, BBN Software Products Corporation (1988).
4. RS/Explore® User's Guide, BBN Software Products Corporation (1988).
5. R. L. Kao, D. T. Wasan, A. D. Nikolov, D. A. Edwards, "Mechanisms of Oil Removal from a Solid Surface in the Presence of Anionic Micellar Solutions," *Colloids and Surfaces* 34, 389 (1988/89).

88 DEVELOPMENTAL STRATEGY FOR THE IMPLEMENTATION OF SOLID WASTE RECLAMATION: FOUNDRY INDUSTRY

Raymond W. Regan Sr., Associate Professor
Luke G. Contos, Research Assistant
Department of Civil Engineering and
Environmental Resources Research Institute
The Pennsylvania State University
University Park, Pennsylvania 16802

INTRODUCTION

The need for improved cost-effective technologies and management strategies for handling industrial solid wastes in an environmentally safe manner remains a challenge for U.S. industries and the engineering profession.[1] Over the past decade much emphasis has been put on efforts for hazardous and toxic waste (HTW) minimization, recovery and reuse.[2] Comparable needs for improved residual (non-hazardous) solid wastes (RSW) from industrial sources have received less attention.

PROGRAM

This paper describes an industry-wide approach by the Process Recovery Corporation, Inc. (PRC) for implementing many of the desired aspects of a modern solid wastes management strategy, including waste minimization and beneficial reuse of process by-products. In a way, the research approach was somewhat unorthodox in that significant efforts were made to include industry, government and university representatives in the on-going development of the research.

Also, the implementability of the technical findings as provided by the University researchers was suggested to be improved by the early and continuing input of foundrymen, regulators and government officials.

Therefore, the strategic development of the project presented includes at least three main components. Simply stated, these components describe the "what", "how" and "why" of the technical as well as the implementation of the PRC waste management facility.

BACKGROUND

The subject industries for this paper are relatively small, traditional type foundries operating in Southeastern Pennsylvania. Results presented were developed as part of a Ben Franklin partnership (BFP) between the Advanced Technology Center of Northern and Central Pennsylvania, the PRC and The Pennsylvania State University (PSU) (September 1, 1985 through August 31, 1989). PRC is headquartered in Reading, PA and represents about 33 foundries generally located in that area. A general goal of PRC was to establish a centralized facility for the collective management of RSW from its members providing options for the reclamation of their sand for reuse, finding alternative uses for other foundry wastes and managing ultimate residuals by landfilling.

Researchers from the PSU assisted the PRC in several aspects of the project. The descriptions in the following sections of the paper highlight the research conducted by the Environmental Resources Research Institute (ERRI). Overall, the research staff included expertise from environmental engineers (Civil Engineering Department and ERRI), industrial and management systems engineers (IMSE) and material scientists of the Material Research Laboratory (MRL).

45th Purdue Industrial Waste Conference Proceedings,

STATEMENT OF PROBLEM

The problem confronting the PRC foundries included the following elements:

- market pressure from foreign competitors limiting the selling price of domestic products;
- loss of production lines and management jobs associated with individual foundry closings;
- diminishing approved landfill space in the long term with availability limited by increased tipping fees in the short run;
- continuing pressure from state regulatory officials to comply with more restrictive environmental regulations associated with RSW; and
- being relatively small-scale operations, the costs for retrofitting and/or modifying basic pollution control processes were considered to be prohibitive.

On the other hand, the problem perspective of the Pennsylvania state Department of Environmental Resources (DER), included alternative concerns:

- the foundry industry statewide was a major source of industrial solid wastes;
- more comprehensive sampling of waste streams indicated the presence (albeit in reduced concentrations) of HTW heavy metals and organics;
- the increased use of "no-bake" chemical binders for core making in foundry operations to supplement the traditional green sand technology (labor-cost-saving) provided the possibility that the solid wastes could be more dangerous to the environment due to the organic contaminants present; and
- the historical use of the "back forty" for on-site disposal of production solids wastes should no longer be allowed.

Research and logistic aspects of the problem included the siting of a suitable parcel of land for the centralized PRC facility, equipment needs for the collection and transportation of solid wastes from individual foundries, the chemical characterizations of the solid wastes from a large number of sources, the investigation of the technical feasibility of commercially available sand reclamation equipment and the development of alternative uses or products from the remaining solid wastes.

OBJECTIVES

Addressing the multi-institutional aspects of the problem statement served as the main goal of the research effort to be presented. Emphasis of the paper is placed on the approach taken for the development of the research program, rather than the experimental and quantitative findings. Specific objectives included the following:

- use of a multi-institutional strategy for the implementation of foundry sand reclamation and by-product reuse program with the ultimate potential of receiving and reprocessing as much as 100,000 tons/year of an industrial solid wastes stream;
- as the industrial component of the strategy, to briefly summarize aspects of the research plan that served to enhance the technical feasibility of the design, and operation of the PRC facility;
- as the regulatory component of this strategy, to briefly summarize the roles of the DER personnel that in fact accelerated the site permit application approval process;
- as the component applicable to the public acceptability of the siting of the PRC facility, outcomes of significant meetings with local and state government officials are identified; and
- as the overall coordinator of the research plan, to describe the contributions of the various engineering and scientific groups within the University for the initiation of the PRC program to date.

APPROACH

The following developmental strategy represents the genesis of the research program and was presented at several foundry forums during the project period.[3] The elements of the plan include:

- Industry should take an affirmative position for resolving environmental issues and for implementing technical-solution approaches.
- Government regulations should encourage innovative solution approaches by "fast tracking" the permit application process, provided adequate safeguards are maintained for

protecting the environment, especially if the trial technology does not provide the desired results.

- Individual foundrymen should participate in industrial cooperatives, either regionally or statewide. Environmental control strategies may only prove to be "cost-effective" when combined total of the solid wastes from several foundries are managed at a common facility.
- During the initial technology evaluation and selection stage, foundrymen should continue to work closely with the regulatory staff to insure that all environmental issues are addressed satisfactorily.
- Traditional practices for handling wastes in a foundry (combining sand, dusts, slags, etc.) may have to be changed to maintain separate stockpiling areas at the foundry. Undesirable materials in the spent sand fraction, for example, may affect the quality of sand reclaimed, and seriously complicate the disposal problem if HTW are involved.
- For small amounts of potentially HTW, shipments could be coordinated so that a single licensed transporter could service all cooperating foundry members, at a lesser combined cost.
- For smaller and larger amounts of potentially HTW, on-site processing could be used to reduce costs and potential liability. The resultant RSW could be processed at the PRC centralized facility.
- On-going chemical testing of the various process streams at a given foundry may show how process changes can be incorporated into the present operation to reduce costs and to identify potentially valuable by-product materials.

RESULTS AND DISCUSSION

Multi-institutional Strategy

An introduction to the technical and economic needs of this industry for improved RSW disposal were crystalized at a meeting sponsored by the PSU and BFP.[4] At that meeting other factors identified as major challenges to the industry were (excerpted):

- individualization—mono sandfills under the direction and research of the industry it serves;
- sand reclamation—regional centers in conjunction and coordination with the mono sandfills;
- communication—maintain a dialogue with governmental representatives . . . to allow the industry to accept the challenge; and
- cooperation—challenges and employees and their business agents and union representatives to allow the foundry to incorporate the necessary technical changes.

Therefore, the needs of the research investigation were recognized as involving technical and institutional components from the outset.

The key institutional actors represented during the investigation was presented (Figure 1). In terms of cash flow, initially funding of the research was a cooperative effort among PRC (representing the member foundries), BFP and PSU. As the research progressed funding of specific tasks was provided to the private sector directly and indirectly, as indicated. The flow of technical information resulting from the investigation was generally complementary with the PRC serving as the focal point (Figure 2).

Industrial Component

The individual foundry members contributed technical and operating data to the PRC, who then coordinated the transferal of this information to the appropriate institutional element. The nature of information obtained by the PSU researchers by this cooperative effort were included in the summary of research tasks (Table I). Specific information included:

- the amounts of various categories of RSW produced, including used sand, slag, baghouse dust, sludges and used refractory linings;
- existing reports and technical information describing the chemical characterizations of RSW previously submitted to DER;
- surveys of the material handling of the various RSW fractions and the handling of new sand; and perhaps most significantly;

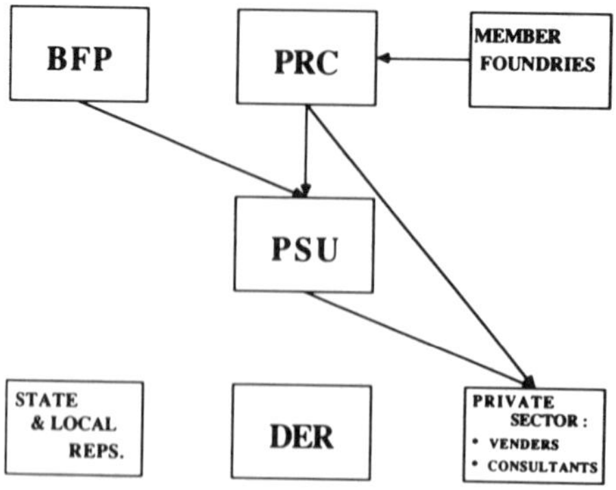

Figure 1. Institutional elements of the comprehensive strategy (Cash flow– – –>).

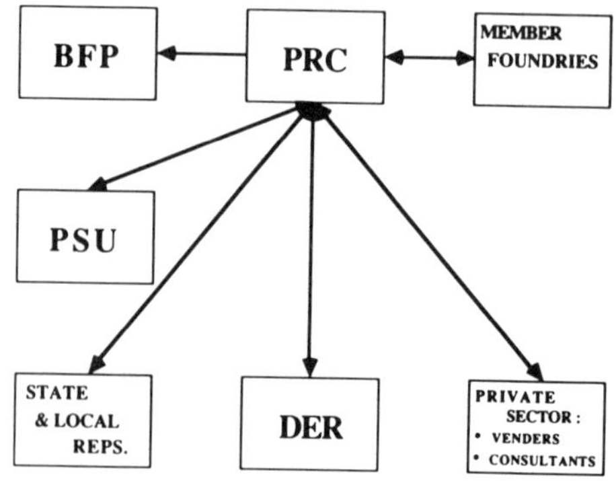

Figure 2. Institutional elements of the comprehensive strategy (Information transferred <– – –>).

- the in-foundry preparation, collection and shipment to PSU of representative samples of the desired RSW fractions, following guidelines prepared by the MRL.

Regulatory Component

The coordination of the PRC technical and economic objectives with the DER mission to "protect the waters and soils of the Commonwealth" was established as an immediate step in the development of the research. The initial meeting of the PRC and PSU representatives was held in the central headquarters of DER about one month into the project. Resulting from this meeting, significant direction was given to the project as represented by the following items:

- As a RSW classification under state authority, DER had a maximum flexibility in administrating the site permit; as a HTW, the US Environmental Protection Agency (EPA) would require their oversite and approval [with associated delays].
- If RSW were source separated at each foundry, spent sand would be classified by DER as "residual". This classification directly impacted about 70% of the foundry solid waste generated by the PRC members. The remaining fractions (slag, sludge and dust) would be evaluated by direct testing.
- The permit application would be considered a phased project, including sand reclamation, stock piling and landfill areas.
- Personnel from the regional DER office, having the final authority to approve the facility permit, would be requested to facilitate the permit application approval process by making informal inspections of the site during the conduct of the engineering and hydrogeologic studies by the consultant.

Source separation of the non-sand solid wastes from the spent sand at an individual foundry along with the sand reclamation objective of the PRC program provided desirable technical features for establishing a strong link of cooperation with the DER that has been maintained throughout the project.

Public Acceptability

There probably has been few solid waste disposal sites ever developed that have not been concerned about the NIMBY syndrome in terms of the local citizen's reaction. Siting of the PRC facility was no exception. In order to introduce the PRC objectives an initial meeting was held with the local government (county) representatives and planning staff within the first few months of the project before any site was selected. As a result of that meeting the PRC goals were then and continue to be reported in the local press in a favorable light. When a site was finally selected and the site application submitted, following a 90 day public comment period, the DER approved the facility (Phase 1–landfill) permit within approximately seven months. The inclusion of local government representatives in the program from the very early stages and the positive news reports aided in the relatively rapid actions by the DER.

Table I. Tabular Summary of Research Activities by the University Researchers and Cooperating Institutional Groups

Project Year	Task	Title	Univ. Team	Group Coop
1	(A)	*Site Evaluation, Selection and Regulatory Issues*		
		(A.1) Site Selection and Regulatory Issues	ERRI	Local Reps, DER
		(A.2) Transportation Issues in Site Evaluation	IMSE	
		(A.3) Site Evaluation Methodology	IMSE	
	(B)	*Waste Stream Inventory and Chemical Assessment*		
		(B.1) Sand Profile/Classification	IMSE	PRC
		(B.2) Material Testing (Spent Sand)	MRL	PRC
		(B.3) Toxicity Assessment	ERRI	
	(C)	*Material Handling Procedures and Considerations*		
		(C.1) Review of On-Site Material Handling	IMSE	PRC
		(C.2) Material Handling Equipment Review	IMSE	Venders
	(D)	*Technology Assessment*		
		(D.1) Reclamation Technology	IMSE	Venders
		(D.2) Landfill Modelling	IMSE	
		(D.3) Identification of Treatment and Recovery Systems	ERRI	
		(D.4) On-Site Inspection Report	MRL	Vender
2	(A)	*Site Selection and Regulatory Issues*		
		(A.1) Evaluation of Candidate Sites and Permit Application Activities	ERRI	DER, State Reps Consultant
	(B)	*Waste Stream Inventory, and Chemical Assessment*		
		(B.1) Survey of Waste Quantities for Landfilling		PRC
		(B.2) Material Testing (Sludge, Dust, Slag)	MRL	PRC
		(B.3) Toxicity Assessment	ERRI	
	(D)	*Technology Assessment*		
		(D.1) Performance Evaluation Criteria Survey Results	ERRI	PRC
		(D.2) Preliminaty Experimental Technology Development for Slag and Baghouse Dusts (New Products)	MRL	PRC
		(D.3) Preliminary Experimental Technology Development for Sludges (Treatment)	ERRI	PRC
3	(A)	*Site Selection and Regulatory Issues*		
		(A.1) Materials Testing for Landfill Permit	MRL ERRI	DER PRC, Consultant
	(D)	*Technology Assessment*		
		(D.1) Compatibility of Chemical Binders on Reclaimed Sand	ERRI	PRC
		(D.2) Materials Testing for Sand Reclamation	MRL	PRC, Vender Consultant
		(D.3) Detoxificaiton of Sludges	ERRI	PRC
		(D.4) Stabilization/Utilization of Slags	MRL	PRC
		(D.5) Stabilization/Utilization of Sludges	MRL	PRC
		(D.6) Utilization of Residual Monofill Waste	MRL	PRC

University Contribution

A tabular matrix of the research contributions by the ERRI, IMSE and MRL investigators was provided (Table I). A sampling of the research previously reported by the ERRI was available elsewhere.[5-8] A brief discussion of these research findings was presented in the following section.

SELECTION OF TECHNICAL DEVELOPMENTS

Main topics for the research conducted by the ERRI included the following:

Toxicity Assessment

The PRC has established that no HTW would be received at the centralized facility. Therefore, the need for a screening procedure to assess the toxicity of incoming waste deliveries was indicated.

Research was conducted to determine if RSW have the potential for leaching toxic materials, and to evaluate the usefulness of alternative biological toxicity tests as screening tools for monitoring incoming wastes. Waste samples were obtained from the PRC foundries and composited, by the estimated weight contribution, into four samples representing ferrous plus non-ferrous sand (33 foundries), non-ferrous sand only (11 foundries), baghouse dust (29 foundries), and sludge wastes (8 foundries). Samples were leached by four extraction procedures and evaluated for toxicity. Chemical analyses were performed on the bulk samples and leachates to determine the inorganic constituents present. Toxicity tests included two biological methods, namely, Microtox and *Daphnia magna*, and one chemical-specific method, the Toxicity Characteristic Leaching Procedure (TCLP, metals only). Further evaluation was performed to predict the speciation of metals in the leachates based on principles of chemical equilibrium to determine the sources of toxicity as being either organic or inorganic through chelation of the metals with EDTA, and to evaluate which of the metals were present in the leachates in toxic amounts as indicated by the concentrations that would elicit a toxic dose response by the biological methods.

The outcome of this phase of the research indicated the complexity of toxicity assessments with results that indicated implementation remained in the future.

Because of the high sensitivity of Microtox and *Daphnia magna* to zinc toxicity (a non-regulated metal), it was not recommended that these biological toxicity tests be used to screen for metals in foundry wastes. However, Microtox response for organics could be used to screen the incoming sand for reclamation suitability. Additional studies were recommended before Microtox be utilized for this purpose.

Detailed results of the research were reported.[8]

IDENTIFICATION OF TREATMENT TECHNOLOGIES

Performance evaluation criteria (PEC) surveys for the non-sand residual waste produced as a result of foundry operations (baghouse dust, slag, and sludges) were sent to each member of the PRC generating that type of waste. Briefly, the surveys were designed to rank, in order of importance, the PEC upon which candidate treatment technologies would be compared. A total of 12 PEC were defined for comparative purposes (Table II).

The purposes of the survey were two-fold, namely:

- To involve the PRC members in the technology development process; and
- to use the PEC rankings as a guide for determining treatment technologies to be tested on a laboratory scale (concept evaluation studies) by the ERRI.

A review of the survey results indicated the following:

- The survey findings[6] indicated that the PRC members as a whole did not consider the PEC "technology mobility" a major criteria for the development of treatment technologies for sludge, slag or baghouse dusts.
- In general, the grouping of the PRC members, for each of the waste streams did offer some preliminary insights into the differences in individual concerns among the PRC members.
- Foundries generation sludge ($n = 9$) were grouped by the amount generated (<25 tons/mo. and >50 tons/mo.). The larger generators ranked safety as the primary PEC followed by cost and environmental impact. The smaller generators placed regulatory impact and cost as the major criteria, with safety further down the list.
- Slag producing foundries ($n = 17$) were grouped by size and major product lines (large and small ferrous, steel castings). Cost of processing remained near the top of the concerns for all three groups. Steel foundries placed a greater emphasis on safety, than the ferrous foundries.
- Grinding room dust from baghouses was reported by the largest number of PRC members ($n = 21$), compared to the other two residual waste streams. The four groupings of

Table II. Definition of Performance Evaluation Criteria for Non-Sand Foundry Solid Wastes

1) Effectiveness	- Removal and/or stabilization of contaminants from selected waste stream has been demonstrated by the process.
2) Contaminant Interference	- The process effectiveness has been demonstrated to be unaffected by the presence of various other types of contaminants in a single waste stream.
3) Versatility of Technology	- The technology demonstrates the ability to remove or stabilize a variety of organic and inorganic contaminants in single waste stream.
4) Time for Decontamination	- The decontamination process will effectively treat the waste stream in an acceptable period of time.
5) Residuals Generated	- If residuals are generated as a result of the treatment process they will be nonhazardous and will not significantly increase the volume of waste to be disposed.
6) Safety	- The process does not pose a threat to workers or local residents during or after treatment.
7) Operability	- The process can be properly operated or implemented with minimal skill and/or technical training.
8) Cost of Processing	- Operating and maintenance costs must be competitive when compared to legitimate alternatives.
9) Technical Feasibility	- The commercially available process can be immediately utilized for the treatment of foundry wastes without substantial alterations to its design.
10) Mobility of Technology	- The technology is fully mobile or easily transportable from foundry to foundry to facilitate on-site treatment of selected solid wastes.
11) Environmental	- The technology will have no immediate adverse environmental impacts such as transforming a solid waste solution to an air or water pollution problem.
12) Regulatory Impact	- The technology will be acceptable to the state regulatory agency without a lengthy litigation process or extensive permitting requirements.

foundries used to establish the rankings were not as clearly differentiated as for the other waste streams.

- Based on the rankings for sludge, and a review of the technical literature the candidate technologies, the ERRI researchers selected the following sludge treatment technologies for "concept evaluation studies":
  a) heavy metal containment using various inorganic additives[7]; and
  b) dewatering characteristics using the Buckner funnel and filter leaf apparatus procedures.[9]

CLOSURE

Over the four-year period that this research investigation was active, many different people representing the PRC, BFP, DER, local and state government, consultants, venders and the PSU contributed to the development of the project goals. At times the overall tasks required a significant degree of risk-taking on the part of the professionals involved. The culmination of the immediate efforts may be viewed as the groundbreaking ceremony for Phase I (August 31, 1990). The reality of the situation, however, is that significant challenges remain before completion of the sand reclamation aspect can be claimed.

This effort does indicate the effectiveness of an industry taking an affirmative role in the development of modern waste management systems. Hopefully, the approach will serve as a more general model for other industrial groups to emulate.

ACKNOWLEDGMENT

Financial support for this research program was derived from the PRC, BFP and PSU.

REFERENCES

1. Regan, R. W. and C. E. Draper, "Sludge Disposal: Examining the Segregated Landfill Concept", *Journal of Plating and Surface Finishing*, 58-63 (1987).
2. Drabkin, M., "The Waste Minimization Assessment: A Useful Tool for the Reduction of Industrial Hazardous Wastes", *Journal APCA* 38, 12, 1530-1541 (1988).
3. Regan, R. W., "Foundry Waste Management: Doing Our Jobs", invited speaker, *Amer. Foundry Society—Philadelphia Chapter*, Plymouth Meeting, PA (1986).
4. The Pennsylvania State University and the Ben Franklin Partnership Program, Foundry Solid Waste Disposal Conference, University Park, PA 16802, proc. ed. J. W. Davis and K. Sprinkle, 95 pp (1984).
5. Regan, R. W. and M. C. Fontaine, "Pennsylvania Foundries: Management of Non-Sand Solid Wastes", *Proc. 19th Mid-Atlantic Industr. Wastes Conf.*, Lewisburg, PA, 257-275 (1987).
6. Fontaine, M. C., J. S. O'Hearn and R. W. Regan, "Systematic Approach for Selecting Treatment Technologies Best Suited for Residual Solid Wastes Produced by Pennsylvania Foundries", *Proc. 20th Mid-Atlantic Industr. Wastes Conf.*, Washington, DC, 343-350 (1988).
7. Contos, L. G., R. W. Regan and J. S. O'Hearn, "Detoxification of Foundry Air Pollution Control Sludges", *Proc. 21st Mid-Atlantic Industr. Wastes Conf.*, Harrisburg, PA, 30-41 (1989).
8. Westervelt, W. W., "A Toxicity Assessment of Foundry Residual Wastes", MS Thesis, Penn State University, Univ. Park, PA (1988).
9. O'Hearn, J. S., "A Study of Various Treatment Technologies for Non Sand Solid Waste," MS Thesis, Penn State University, Univ. Park, PA (1990).

Section Nine
INDUSTRIAL WASTES – E. ELECTRONIC INDUSTRY WASTES

89 WASTE ELIMINATION AT MEMC ELECTRONIC MATERIALS, INC.

Paul S. Dickens, Senior Environmental Engineer
MEMC Electronic Materials, Inc.
Spartanburg, South Carolina 29304

The 1980s changed forever the management of manufacturing industries. Companies that did not improve product quality lost market share. Companies that did not streamline product development lost market position. Companies that did not reduce product inventory and increase product yield were not able to compete on price.

The 1990s hold similar dramatic change. A growing requirement for competitive survival of manufacturing industries is the proactive management of toxic chemicals and hazardous waste. It is no longer sufficient to merely comply with environmental laws. Companies that eliminate chemical waste and associated management cost obtain a market advantage. These companies are no longer hostage to changing environmental laws and public perception that their manufacturing plants are "dirty". These companies can devote capital to new products and improved quality rather than end-of-pipe pollution controls. The successful companies in the 1990s will be those that win the race to implement "clean" manufacturing technology.

This paper describes successful efforts by MEMC Electronic Materials, Inc. to eliminate chemical emissions and hazardous waste by changing manufacturing technology.

WASTE ELIMINATION AND CHEMICAL USE

The *Resource Conservation and Recovery Act of 1976*[1] (RCRA) established a hierarchy of waste management. Source reduction is the preferred waste management practice followed in order of preference by on-site waste recycling, off-site waste recycling, waste treatment, and land disposal. The 1984 *Hazardous and Solid Waste Amendments*[2] (HSWA) established new requirements that eliminate land disposal as an option for most chemical waste. Regulations under HSWA substantially increase the cost of managing hazardous waste.

Title III of the 1986 *Superfund Amendments and Reauthorization Act* (SARA) requires public reporting of chemical use, chemical emissions, and hazardous waste generation by manufacturing industries.[3] The Community Right-To-Know provisions of SARA, more than any other environmental regulation, are driving industry to eliminate toxic chemical use. The electronics industry is a large user of chlorofluorocarbons (CFCs). The 1987 *Montreal Protocol on Substances That Deplete the Ozone Layer*[4] requires the phase out of CFC manufacture and use. The proposed *Declaration of London*[5] will end worldwide production of CFCs by the year 2000. Meanwhile, the cost of these chemicals has greatly increased.

There are significant regulatory and public relation incentives for industry to change manufacturing technology to eliminate chemical emissions and hazardous waste. Such changes can yield an economic benefit by eliminating process chemical and waste management costs.[6] However, little technical information is available on waste elimination techniques.

MEMC ELECTRONIC MATERIALS, INC.

MEMC Electronic Materials, Inc. (MEMC) manufactures polished and epitaxial silicon wafers. Silicon wafers are the substrate, or base, on which microelectronic circuits (microchips) are built. MEMC is a worldwide producer of silicon with manufacturing plants in the United States, Europe,

45th Purdue Industrial Waste Conference Proceedings, © 1991 Lewis Publishers, Inc., Chelsea, Michigan 48118. Printed in U.S.A.

and Asia. MEMC's customers are the manufacturers of logic and memory microchips. Logic and memory microchips are used in everything from computers and consumer electronics to automobiles and airplanes.

Global competition in the sale of silicon wafers is fierce. Quality and on-time delivery are critical. A single defect in a shipment of wafers can lose market share forever. A late shipment can lose a customer for good. Product lifetime is short, on the order of 1 to 2 years. Customer specifications constantly change becoming more and more stringent. To meet market demand and remain profitable, MEMC must continuously improve manufacturing technology to support new products and to increase product quality.

PROACTIVE WASTE MANAGEMENT

Silicon wafer manufacturing technology is a combination of chemical and mechanical steps. Acids, bases, surfactants, abrasives, and organic solvents are required. Chemical emissions and hazardous waste are byproducts of silicon manufacturing. These emissions and waste require special handling and treatment.

The market conditions in which MEMC operates are a significant barrier to manufacturing changes that eliminate emissions and hazardous waste. Product specifications and quality requirements are stringent. Any change in manufacturing technology must be proven *before* it is implemented. The change must be reviewed and qualified by MEMC's customers. MEMC cannot ship off-specification product resulting from an experimental process change. MEMC cannot afford manufacturing downtime and lost yield resulting from a process change. This creates a conflict between short-term and long-term needs.

MEMC cannot afford to spend limited resources on hazardous waste management and end-of-pipe pollution controls. As legal requirements for environmental protection become more stringent, the drain of these unproductive expenditures will increase. To free long-term resources for product quality and yield improvement, manufacturing operations that generate chemical waste must be changed.

To resolve these short and long-term needs, MEMC evolved to a proactive waste management strategy. Its focus is on changes in manufacturing technology that reduce or eliminate chemical emissions and the generation of hazardous waste. The result of these changes over time is a large reduction in unproductive expenditures for waste management and pollution control.

The MEMC management strategy has several components.

Company Goals

MEMC has established company goals for reduction of chemical emissions and hazardous waste. These are:

- Reduce hazardous air emissions by 80% by year end 1994.
- Eliminate use of ozone depleting chemicals by year end 1995.
- Reduce generation of hazardous wastes by 50% by year end 1996.

The base year for these goals is 1988. Where possible, the goals are to be achieved through investment in manufacturing process change rather than end-of-pipe pollution controls.

Targeted Effort

MEMC establishes task groups at the corporate and plant level to target waste reduction efforts. These tasks groups include environmental, safety, manufacturing, engineering, and research personnel. The task groups identify manufacturing steps with the greatest chemical emissions and/or generation of hazardous waste. They then identify the best opportunities for change to reduce or eliminate chemical waste. This integration of expertise is essential to obtain creative ideas for waste elimination. It is a departure from traditional industry management of regulatory compliance and pollution control.

End-of-pipe pollution controls (and the staff to operate them) evolved at most industrial plants as environmental regulations were established during the 1970s and 1980s. The "end-of-pipe" approach was the most expedient way to comply with changing laws. Every industry established an environmental control department to manage their regulatory compliance and pollution control equipment Unfortunately, these departments tend to communicate with manufacturing personnel only ir response to problems or when needed to obtain environmental permits. They typically have little or nc contact with research personnel. End-of-pipe pollution controls also discourage "ownership" of waste management by manufacturing and research groups. The fact that "someone else handles it" is a barrier to their understanding of waste management problems and opportunities.

The waste reduction task groups successfully overcome communication barriers between environmental, manufacturing, and research personnel. Each task group is assigned a specific charter and a deadline in which to achieve results. The charter and deadline insure that waste elimination ideas are evaluated on a timely basis and that definitive recommendations are made.

Manufacturing and Research Ownership

The waste reduction task groups have a second function. Manufacturing and research personnel take ownership of waste reduction goals. There is tremendous creative energy within manufacturing and research groups. Many of their ideas to eliminate chemical emissions and hazardous waste include process changes that improve product quality and yield. The opportunity for improved process and cost performance is a strong motivator for implementing waste reduction. Once shown these opportunities, manufacturing and research groups can achieve outstanding results.

Resource Commitment

Commitment of resources is a barrier to all waste reduction efforts. Many worthwhile projects compete for limited capital funds. Day-to-day problems with product quality and yield consume much of the attention of manufacturing and research personnel. Any manufacturing process change must be *thoroughly* tested. The test results must be formally documented and approved; manufacturing procedures must be revised; and manufacturing personnel must be trained. The process change must be qualified by MEMC's customers. These process change procedures take considerable time.

The problem of resource commitment can only be resolved by management decision. At MEMC, key managers are held accountable for obtaining chemical emission and hazardous waste reduction goals. Each waste reduction task group reports to mid and upper-level manufacturing managers. These managers are responsible for completing the task group charter within assigned deadlines.

The waste reduction task groups are a great help in establishing priorities for resource commitment. Many of the ideas generated by task groups are so worthwhile that the recommended projects sell themselves.

RESULTS

The remainder of this paper describes recent manufacturing technology changes implemented by MEMC. Individually, the changes are small. All but one required no expenditure of capital funds. However, the changes eliminated chemical use in several manufacturing steps. The changes significantly reduced chemical emissions and the generation of hazardous waste.

Chromic Acid

MEMC grows ingots of single crystal silicon from polycrystalline silicon in specially designed electric furnaces. Silicon wafers are then cut from these ingots, or rods as they are called by manufacturing personnel. The silicon rods must be evaluated to identify any defects in crystal structure. This evaluation is a chemical process using a "structural etchant." The structural etchant attacks defects in the single crystal structure leaving pits in the crystal surfaces that are visible upon inspection. Small scale defects in the silicon rod cannot be identified without structural etching. Prior to 1988, the only available structural etchants for silicon were based on chromic acid. Chromic acid is a suspected human carcinogen.[7]

A single crystal silicon rod is illustrated in Figure 1. The seed and base ends of the rods are cropped off. Evaluation slugs are then cut from the ends of the rod that remains. The base end of the silicon rod often contains a boundary between single and polycrystalline structure. Formerly, crystal evaluation at MEMC involved two structural etching steps. The base end of each silicon rod was etched to determine the boundary, if present, where single crystal structure was lost. The "Rod Etch" evaluation was a screening step to reduce errors in rod cropping. Slugs were then cut from the rod and etched to confirm crystal structure in the portion of the rod that was retained. The "Slug Etch" evaluation detects any errors in cropping silicon rods to remove the polycrystalline base.

It was recognized that experienced technicians could visually determine a polycrystalline boundary without Rod Etching. Since the Rod Etch process involved chromic acid and a lifting hazard (silicon rods are heavy), the process was eliminated. Crystal structure is now confirmed by etching slugs cut

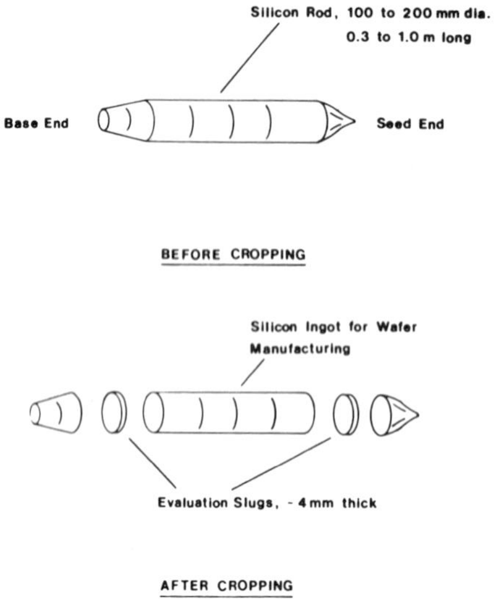

Figure 1. Single crystal silicon rod.

from the base end of each silicon rod. The Rod Etch process was eliminated in December, 1988. The elimination of Rod Etch also eliminated the potential for contaminating the silicon rods with chrome.

The chromium-based structural etchants for silicon were originally developed in the 1950s and work extremely well.[8,9] Unfortunately, these etchants yield hexavalent chromium waste which requires special and increasingly expensive treatment and disposal. In 1987, MEMC began research to identify alternative structural etchants that do not contain chromic acid. None were found, so MEMC developed its own.[10] "MEMC Etch" is a copper-based structural etchant for silicon. In November 1988, MEMC Etch was substituted for chromium-based etchants for the majority of silicon products produced at the MEMC Spartanburg, SC Plant.

The results of MEMC's chromium use reduction are outlined in Tables I and II. Elimination of the Rod Etch process and the substitution of copper-based etchants for chromium-based etchants eliminated 96% of chromic acid use at the MEMC Spartanburg, SC manufacturing plant. Elimination of the Rod Etch process eliminated 80% of chromic acid use at the MEMC St. Peters, MO plant. The volume of acid treatment sludge at the Spartanburg Plant was reduced by half. The chromium content

Table I. Chrome Use Reduction at MEMC Spartanburg, SC Plant[a]

Item	1988	1989
Chromic acid use, lbs as CrO_3	5,490	210
Acid treatment sludge, lbs[b]	60,590	30,850
Manufacturing volume[c]	1.00	1.10
Chrome use eliminated, lbs as CrO_3[d]		5,830
Acid treatment sludge eliminated, lbs[b,d]		35,800
Annual cost eliminated:[d]		
Process chemicals		$27,900
Personnel protective equipment[e]		16,800
Waste treatment chemicals		4,800
Sludge disposal		10,800
Total		$60,300

[a]Elimination of Rod Etch process. Substitution of copper-based etchants for chromium-based etchants in Crystal Evaluation Laboratory.
[b]Dewatered sludge resulting from chemical reduction, neutralization and precipitation of chrome and copper-based etchants. The dewatered acid treatment sludge is a D007 hazardous waste.
[c]Manufacturing capacity is proprietary. 1988 production assigned value of 1.00.
[d]Escalated for 1989 manufacturing volume.
[e]Acid gowns, gloves, face shields, and other expendable supplies required for Rod Etch process.

Table II. Chrome Use Reduction at MEMC St. Peters, MO Plant[a]

Item	1988	1989
Chromic acid use, lbs as CrO_3	8,220	1,890
Waste treatment sludge, lbs[b]	6,580,000	5,620,000
Manufacturing volume[c]	1.00	1.13
Chrome use eliminated, lbs as CrO_3[d]		7,400
Waste treatment sludge eliminated, lbs[b,d]		1,815,000
Annual cost eliminated:[d]		
Process chemicals		$23,700
Personnel protective equipment[e]		17,000
Waste treatment chemicals		5,900
Total		$46,600

[a]Elimination of Rod Etch process.
[b]Dewatered sludge containing treated acid waste and settled high solids wastewater. Treated acid waste from chemical reduction, neutralization, and precipitation of chrome-based etchants, copper-based etchants, and mixed acid etchants. The dewatered waste treatment sludge is a *non-hazardous* waste.
[c]Manufacturing capacity is proprietary. 1988 production assigned value of 1.00.
[d]Escalated for 1989 manufacturing volume.
[e]Acid gowns, gloves, face shields, and other expendable supplies required for Rod Etch process.

of the Spartanburg acid treatment sludge was reduced tenfold from 47,000 parts-per-million (ppm) as Cr^{+3} by weight to 3600 ppm as Cr^{+3} by weight. More than $106,000 in annual process chemical and waste management costs were eliminated.

Mixed Acid

In addition to evaluating silicon rods for crystal structure, the rods are examined for electrical and chemical properties. Slugs cut from the rods for evaluation must be polished to remove damage to the slug caused by the rod cropping saw. Formerly, slug polishing was done chemically with a mixture of nitric, acetic and hydrofluoric acids. The polishing etch required 10 to 15 minutes in a heated bath of "Mixed Acid Etchant."

MEMC manufacturing personnel were not satisfied with the quality of chemically polished slugs. The thickness and taper of the chemically polished slugs would vary causing variation in electrical and chemical property measurements. Extra slugs (representing lost yield) had to be cut, chemically polished, and tested to overcome sporadic bad evaluation results. The heated acid bath for chemical polishing was a large source of air emissions requiring special pollution control equipment.

A mechanical slug polishing process was developed in response to these concerns. The mechanical process eliminates variation in slug thickness and taper. Slugs are no longer chemically polished. Instead, mechanically polished slugs are etched for 5 minutes in a cold mixed acid bath to clean the slug surface prior to electrical and chemical property tests. Because of the greatly improved quality of mechanically polished slugs, the number of extra slugs required to obtain good test results is reduced by 50 to 60%. The reduced number of slugs required for testing is a substantial improvement in product cost and yield.

The combined effect of fewer slugs being etched for a shorter time at a lower bath temperature is summarized in Table III. Mechanical slug polishing was fully implemented in October 1989. It eliminates more than 70,000 lbs/year of Mixed Acid use at the MEMC Spartanburg, SC Plant. Similar results were obtained at the MEMC St. Peters, MO Plant.

Acid Air Emissions

The elimination of Rod Etch and the development of mechanical slug polishing had additional benefits in acid air emission reduction. These results are summarized in Table IV. These air emission reductions are the result of chemical use elimination and process operation change. No new end-of-pipe pollution controls were employed.

Table III. Mixed Acid (Nitric, Acetic, Hydrofluoric) Use Reduction at MEMC[a,b] Spartanburg, SC Plant

Item	Value
Mixed Acid use eliminated:	
lbs/year	71,500
percent of process use[c]	31.9
percent of total plant use	3.4
Waste acid eliminated:	
lbs/year	69,800
Annual cost eliminated:	
Process chemicals	$37,600
Waste treatment chemicals	1,100
Total	$38,700

[a]Substitution of mechanical slug polishing for chemical slug polishing in Crystal Evaluation Laboratory.
[b]Data based on 1989 Spartanburg Plant manufacturing volume.
[c]Crystal Evaluation Laboratory.

Freon 113

Freon 113 (Trichlorotrifluoroethane) is an ozone depleting chemical. It is a common solvent used in the electronics industry for degreasing and particle removal. At the MEMC Spartanburg Plant, Freon 113 was used to clean containers, called Tote Pans. These Tote Pans were used for temporary storage of polished silicon wafers between final cleaning and packaging steps. Freon 113 was also used to clean plastic cassettes. The plastic cassettes are a protective device designed to carry silicon wafers between manufacturing steps. Tote Pan Cleaning and Cassette Cleaning were large users of Freon 113. These uses were targeted for elimination.

MEMC manufacturing personnel knew that storage of polished wafers between cleaning and packaging was a bottleneck to smooth product flow. The opportunity to eliminate a large Freon 113 use provided extra incentive to switch to a "just-in-time" product flow in the Spartanburg Plant Clean Room. The temporary storage of wafers in Tote Pans was eliminated. The need for Tote Pan cleaning was eliminated. The improved product flow in the Clean Room has increased both product quality and yield.

Solvent cleaning of cassettes was required to remove a build up of chemical residue from wafer lapping and etching operations. If not removed, the residue caused edge bumps and scratches on the silicon wafers. These defects were a significant product yield loss. MEMC manufacturing personnel conducted tests and found that residue build up could be eliminated by changing the flow of cassettes through the Spartanburg Plant. Instead of using the same cassettes to carry wafers through lapping and etching operations, different cassettes are now used in each process. The cassettes are cleaned

Table IV. Acid Emission Reduction at MEMC Spartanburg, SC Plant[a]

Item	Value
Acid air emissions eliminated:	
– Oxides of Nitrogen[b]	
lbs NOx/hr	1.53
percent of total plant process emission	12.5
– Acetic Acid[b]	
lbs CH_3COOH/hr	1.00
percent of total plant process emission	5.1
– Hydrogen Fluoride[b,c]	
lbs HF/hr	0.175
percent of total plant process emission	32.5

[a]Data based on 1989 Spartanburg Plant manufacturing volume.
[b]Substitution of mechanical slug polishing for chemical slug polishing in Crystal Evaluation Laboratory.
[c]Elimination of Rod Etch process. Substitution of copper-based etchants for chrome-based etchants in Crystal Evaluation Laboratory.

Table V. Freon 113 (Trichlorotrifluoroethane) Use Reduction at MEMC[a,b] Spartanburg, SC Plant

Item	Value
Freon 113 use eliminated:	
lbs/year	88,300
percent of total plant use	40.6
Freon 113 air emission eliminated:	
lbs/year	85,300
percent of total plant emission	42.0
Waste Freon 113 eliminated:[c]	
lbs/year	3,800
percent of total plant waste	23.6
Annual cost eliminated:	
Process chemicals	$121,000
Freon tax[d]	97,100
Waste solvent disposal[e]	400
Total	$218,500

[a]Elimination of Tote Pan Cassette Degreasing.
[b]Data based on 1989 Spartanburg Plant manufacturing volume.
[c]F001 hazardous waste from operations using Freon 113. Weight includes water and contaminates picked up in degreasing operations.
[d]Federal excise tax on ozone depleting chemicals effective 1/1/90. For Freon 113, tax is $1.10/lb.
[e]Waste solvent is recycled for credit. Disposal cost represents transportation to off-site rcycling facility.

with soap and water. The change in cassette flow allowed elimination of cassette cleaning with Freon 113.

The results of Freon 113 elimination at the MEMC Spartanburg, SC Plant are outlined in Table V. Blends of Freon 113 with other solvents are used in silicon wafer cleaning and packaging steps. In November 1989, MEMC maintenance personnel modified solvent degreasing machines that use Freon 113 blends to reduce the volume of solvent lost to waste when solvent stills are cleaned. These modifications involved installation of valves to isolate the boiling sump from the recovery still in each vapor degreaser. Previously, the entire volume of solvent in the degreaser (100 to 200 gallons) was wasted every 90 days when the solvent stills and filters were cleaned. Now, maintenance personnel isolate the boiling sump from the still. They operate the still to recover as much solvent as possible before cleaning the still and filters. This change is expected to eliminate an additional 6,100 lbs/year of Freon 113 containing solvent waste at the MEMC Spartanburg, SC Plant.

SUMMARY

MEMC Electronic Materials, Inc. has learned that waste elimination cannot be achieved without a fundamental examination of each manufacturing step. This requires a large commitment of resources normally devoted to production. However, the results can be dramatic. In 1989, MEMC identified and eliminated several manufacturing steps that were unnecessary. Mechanical methods were substituted for chemicals in another operation. Alternative chemicals were substituted in still another. Maintenance procedures on solvent vapor degreasers were improved. The following were *eliminated* by change in manufacturing technology:

- 96% of the chromic acid use and chromium waste generation at the MEMC Spartanburg, SC Plant.
- 80% of chromic acid use and chromium waste generation at the MEMC St. Peters, MO Plant.
- 3% of mixed acid use and waste acid generation at MEMC's Spartanburg, SC and St. Peters, MO Plants.
- 32% of hydrogen fluoride air emission at the MEMC Spartanburg, SC Plant.
- 40% of Freon 113 use and air emission at the MEMC Spartanburg, SC Plant.

These changes were implemented during a period of increasing manufacturing output with no adverse effect on product quality or on-time delivery. The changes eliminated more than $360,000/year in process chemical and waste management costs.

MEMC's proactive waste management strategy is a model for other industries considering chemical emission and hazardous waste reduction. The proactive approach encourages ownership of waste reduction goals by manufacturing and research personnel. It encourages productive investment in manufacturing technology change rather than end-of-pipe pollution controls.

CREDITS

The results reported in this paper were made possible by the hard work of manufacturing and research personnel at MEMC's Spartanburg, SC and St. Peters, MO plants. Without their support, MEMC's proactive waste management strategy would not be a success.

REFERENCES

1. U.S. Congress, "Resource Conservation and Recovery Act of 1976." *Public Law 94-580* (1976).
2. U.S. Congress, "Hazardous and Solid Waste Amendments." *Public Law 98-616* (1984).
3. U.S. Congress, "Emergency Planning and Community Right-to-Know Act; Title III of Superfund Amendments and Reauthorization Act of 1976." *Public Law 99-499* (1986).
4. U.S. EPA, "EPA Stratospheric Ozone Protection Regulations." *Federal Register*, 52, 47488 (December 14, 1987).
5. Bureau of National Affairs, "UNEP Director General Predicts CFC Production to End by Year 2000." *Chemical Regulation Reporter*, 13, 1609 (March 23, 1990).
6. U.S. EPA, *Waste Minimization, Environmental Quality with Economic Benefits* (EPA/530-SW-87-026). Office of Solid Waste and Emergency Response, Washington, D.C. (October, 1987).
7. M. Sittig, *Handbook of Toxic and Hazardous Chemicals and Carcinogens*. 2nd Edition, Noyes Publications: Park Ridge, New Jersey, p. 243–248 (1985).
8. E. Sirtl and A. Alder, *Z. Metallkunde*. 52, 529 (1961).
9. M. W. Jenkins, "A New Preferential Etch for Defects in Silicon Crystals". *J. Electrochem. Soc.* 124, 757 (1977).
10. T. C. Chandler, "MEMC Etch – A Chromium Trioxide-Free Etchant for Delineating Dislocations and Slip in Silicon." *J. Electrochem. Soc.* 137, 944 (1980).

90 NON-CONVENTIONAL METAL PRECIPITATION AT A PRINTED CIRCUIT BOARD MANUFACTURER: A CASE STUDY

Charles Ficker, Senior Chemist
William S. Carter, Product Manager
DuBois Chemicals
Cincinnati, Ohio 45202

Jim Martin, Plant Chemist
Circuit Center, Inc.
Dayton, Ohio 45440

INTRODUCTION

Circuit Center, Inc. is a manufacturer of printed circuit boards located in Dayton, Ohio. The production of circuit boards is accomplished through a series of wet chemical processes, and a number of graphic art procedures. Details of production techniques will vary from one facility to another; however, the basic processes of board etching, electroless and electrolytic copper plating, and water rinsing are consistent.

The great majority of wastewater at Circuit Center, as at all board shops, is generated in the rinsing operations. Here water picks up organic material, suspended solids and metal ions, including: copper, tin, lead, nickel and other species. Another significant source of copper at Circuit Center is from the use of a proprietary ammonia etchant. This is used to dissolve away unnecessary copper from the circuit board laminate. This ammonia as well as the electroless plating process results in copper ions which are complexed and difficult to precipitate.

To protect human health and the environment, the toxic metal ions and other contaminants of concern must be removed prior to discharge. As a circuit board manufacturer, Circuit Center falls under the Federal Environmental Protection Agency (EPA) guidelines for metal finishers. Local ordinances for wastewater discharge must be no less stringent than the Federal EPA Guidelines. Also per federal law, any sludge from an electroplating rinse treatment is a categorical hazardous waste and subject to the regulations of the Resource Conservation and Recovery Act (RCRA) and its amendments.

When Circuit Center built a new facility in 1983, new EPA regulations were imminent, becoming effective in 1984 (See Table I). The management of Circuit Center made the conscious decision to minimize environmental impact. Among other actions, the shop was made solvent free, using only aqueous cleaning, and an engineering company was selected to install a state-of-the-art waste treatment system. However, when Circuit Center began operations in 1984, it became apparent there would be difficulties with compliance. The federal limit of 4.5 mg/L copper and 0.6 mg/L lead were attainable. However, the local control authority, Montgomery County Sewer District, planned limits

Table I. Printed Circuit Board Pretreatment Regulations 1984

Contaminant	1 Day Max	4 Day Average
Cyanide, Total	1.9 mg/L	1.0 mg/L
Copper	4.5 mg/L	2.7 mg/L
Nickel	4.1 mg/L	2.7 mg/L
Chromium, Total	7.0 mg/L	4.0 mg/L
Zinc	4.2 mg/L	2.6 mg/L
Lead	0.6 mg/L	0.4 mg/L
Cadmium	1.2 mg/L	0.7 mg/L
Total Metals	10.5 mg/L	6.8 mg/L

45th Purdue Industrial Waste Conference Proceedings, © 1991 Lewis Publishers, Inc., Chelsea, Michigan 48118. Printed in U.S.A.

Table II. Previous Waste Treatment at Circuit Center, Inc.

Treatment	Feed Point
Sulfuric Acid to pH 2.0	pH Adjustment Tank
Ferrous Sulfate	pH Adjustment Tank
Aluminum Sulfate	pH Adjustment Tank
Caustic Soda	2nd Reaction Tank
Anionic Polymer	Flocculator

of 1.45 mg/L copper and 0.36 mg/L lead for 1986, and effluent data showed these impossible to meet consistently. Circuit Center collaborated extensively with Montgomery County to determine the extent of the problems and what would be the most effective solution.

SITUATION ANALYSIS

It became obvious that the conceptual engineering and system hardware provided to Circuit Center was inappropriate for the treatment of waste from the production of printed circuit boards at a quick turn prototype shop. The equalization provided by receiving pits was terribly undersized. In order to prevent flooding, the 25 gpm design system had to operate at times at 50 to 60 gpm. This increased hydraulic loading resulted in insufficient reaction times and sludge carryover in the clarifier.

In the shop, a dry film operation generated waste streams containing 1.0 to 1.5 mg/L copper. These streams could be discharged without treatment under the federal regulations, but would have to be treated before discharge under the impending local limits. Since the treatment system capacity was already overloaded, the addition of 70% more water would prove impossible to treat. While Montgomery County was patient and extremely helpful, they eventually had to put Circuit Center on a compliance schedule to ensure discharge compliance.

Circuit Center was at a critical stage for survival. The company was committed to environmental awareness, but Circuit Center was competing in a national and international marketplace where differences in discharge regulations could result in economic disadvantages. Solutions had to be found which would allow Circuit Center to meet Montgomery County standards and still compete in the market.

ALTERNATIVES CONSIDERED

Several alternatives were considered to reduce the waste loading. For example, process changes can often dramatically reduce loading. In this case it was determined that there were no appropriate changes in the manufacture that could be made to lower solids loading to the system. Likewise, hydraulic loading could not be significantly reduced through water conservation, as usage was not considered excessive. Expansion of the wastewater treatment system itself was deemed impossible because the building which housed it was literally built around it. There simply was no room for expansion. Therefore, if the nature of the incoming wastewater could not be improved significantly, and the system which treated it could not be expanded, the only avenue of relief was to change the chemical treatment in the system.

At that time, Circuit Center treated its wastewater by adding sulfuric acid to ensure acidification of the water; adding ferrous sulfate to coprecipitate the complex copper; neutralizing with sodium hydroxide; adding alum to condition the floc; and adding an anionic polymer (See Table II and Figure 1). In practice, Circuit Center was adding 100 pounds of inorganic to remove 2 pounds of metal loading per day.

With the major problem being sludge carryover from the clarifier, Circuit Center focused attention on finding a chemical treatment without the need for so much inorganic. Technologies such as membrane separation or ion exchange did not prove practical for Circuit Center. The relatively high organic content of the wastewater would quickly foul most membranes or resins. These technologies would also have required Circuit Center to make a major capital investment replacing one relatively new waste treatment system with another.

DuBois Chemicals, Inc., a worldwide supplier of multiple lines of specialty chemicals, has been very active in the printed circuit board industry for many years. During the summer of 1986, a local DuBois representative was working with Circuit Center on a project involving aqueous circuit board cleaners. When the subject of waste treatment was raised, DuBois accepted the invitation to survey the system and conduct jar tests as necessary to make recommendations toward improving the chemical treatment program.

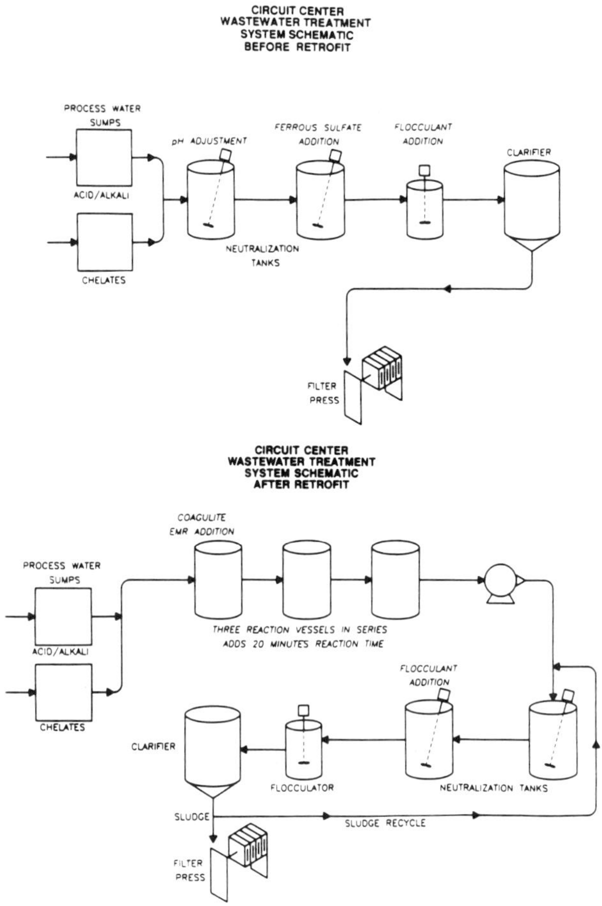

Figure 1. Circuit Center wastewater treatment system schematic before and after retrofit.

PROGRAM OBJECTIVES

After assessing the situation, a common set of goals was established between the three concerned parties: Circuit Center; DuBois Chemicals, and Montgomery County Sewer District. It was recognized that each had its own slightly different priorities in judging the success of any new treatment program.

Montgomery County Sewer District had perhaps the most straightforward definition of success. Montgomery County needed consistent compliance from Circuit Center in the long term, and a minimum of disruption to effluent quality during testing and startup of any new treatment.

The generator, Circuit Center, placed highest priority on continued compliance with existing regulations, and on being positioned to meet future, more stringent regulations. Any new treatment program would have to provide this. Operating in a competitive market, however, it was also important to Circuit Center that this new treatment program not raise the overall cost of waste treatment. Preferably it should result in a net reduction in the cost of waste treatment. This could take the form of reduced sludge volume for disposal, lower chemical consumptions, lower cost chemicals, or a simplified treatment requiring fewer man-hours of attention. Any new chemical treatment program would therefore have to be put in place with a minimum of system retrofits and no capital expenditures.

Given its role of chemical program supplier, DuBois Chemicals attached a high priority to the establishment of a long-term customer/supplier relationship. This required identifying the needs of the customer and fulfilling them. DuBois goals then really became an integration of all the other parties goals. DuBois thus had the challenge of providing a program to produce quality effluent on a consistent basis within the economic realities set forth by Circuit Center.

NEW TREATMENT TRIAL

Following the situation analysis of the existing treatment, several alternate treatments were considered and some were jar tested. Most were variations on the metal coprecipitation theme, using ferric iron, calcium, or a combination. While some were effective at metal removal, and all produced less sludge than the ferrous treatment, solids loading was too critical to allow for the addition of any extra sludge from inorganics.

At this same time in 1986, DuBois was field testing a new organic precipitating agent formulated with dithiocarbamate. This product had two distinct advantages—first, it effectively precipitated a wide range of heavy metals even in the presence of strong chelants. Secondly, the material was organic and did not contribute solids to the sludge. For these reasons this product, COAGULITE EMR, was chosen for a system trial at Circuit Center, following successful bench tests.

In addition to the aforementioned long-term goals for a new treatment, this trial had several specific conditions. First, the trial would run in the continuous flow system for a minimum of two weeks. It was agreed there would be no physical adjustments to the system before or during the trial. There would be a small, deliberate overdose of COAGULITE EMR to protect against excursions from discharge compliance. Following the trial period data would be reviewed and the decision made whether to continue the new program.

The system trial lasted from September 15, 1986 until October 30, 1986. Given the relatively short length of the test, some assumptions had to be made in judging its success. First, it was known that the COAGULITE EMR had been overfed, but the extent could only be estimated from experience regarding copper levels in the untreated water. The sludge handling and pressing was done without interruption. Therefore, the ferrous sulfate treated sludge and the carbamate treated sludge were mixed in the press hopper. The amount of each was estimated and from this information it was estimated the new treatment produced 33% less sludge than the old treatment. This estimation was hampered by the similarity in appearance of the carbamate sludge and the ferrous sludge.

During the course of the trial, metal levels were consistently low, well below discharge limits. The sludge settled well in the clarifier, even allowing a few inches of clear water at the surface. All of this was accomplished with no physical changes to the system itself. Although the short-term objectives of the trial were met, the longer-term goal of lowering the overall cost of treatment was not evident from the data studied. (See Table III.)

Over the course of the next several months, other chemical suppliers did bench testing and treatability testing at Circuit Center. The results of these screenings and the results of the DuBois trial were scrutinized closely by Circuit Center. Two things were apparent: first, something had to be done to improve the treatment program and second, to date, the DuBois program showed the best results.

GETTING STARTED

Consequently, in late 1987, DuBois was chosen as the supplier of a new chemical treatment program for Circuit Center. As with any new venture there was a period of learning and getting adjusted to the new program. During this time, the operators had to get a "feel" for the new carbamate reaction. They were accustomed to the ferrous treatment, and had the confidence to make adjustments based on qualitative observations of floc size, settling, etc. This could still be the case, but a new set of signals had to be learned. In the original treatment, pH control and an excess of ferrous were critical to metal removal. The new program made use of a complexing reaction between the carbamate molecule and the metal ion. Each metal requires a given amount of product as shown in Table IV. With this new treatment, incoming metal level quantification was necessary when previously it was not.

Because this treatment was relatively new to DuBois as well, there was much learned about practical application of the new product by both parties during this startup period. The first lesson learned was the importance of reaction time to the carbamate reaction. Without sufficient reaction time and energy, the carbamate-metal complex would form but the reaction would not go to completion and form the precipitate. In this condition it is almost impossible to remove through conventional sedimentation. This was not an issue at first, as only gross treatment efficiencies were desired. As the program was fine-tuned, however, it became obvious that to achieve the most efficient use of the product, additional reaction time would be necessary. To achieve this, Circuit Center decided to make one modification to their system. Three 650 gallon tanks were added ahead of the first reaction tank,

Table III. Cost Comparisons

Previous Treatment	
Daily Chemical Consumption	
10 gallons sulfuric acid	
50 pounds ferrous sulfate	
25 pounds alum	
20–25 pounds caustic beads	
1–2 pints emulsion polymer	
Monthly treatment costs	
Approximate chemical cost	800.00
Sludge disposal (1.75 tons/mo.)	437.00
Sludge hauling	192.00
	1429.00
System Trial Estimates	
Daily chemical consumption	
2.85 gallons coagulite EMR	
0.4 pints floculite 402	
Monthly treatment costs	
Approximate chemical costs	1205.00
Sludge disposal	287.50
Sludge hauling	127.05
	1619.55
Coagulite EMR Program After One Year	
Daily chemical consumption	
1.0 gallons coagulite EMR	
0.4 pints floculite 402	
Monthly treatment costs	
Monthly chemical costs	660.00
Sludge disposal	36.42
Sludge hauling	16.00
	712.42

and the COAGULITE EMR was fed to the first of these tanks. These tanks added about twenty minutes reaction time to the system. This modification is shown in Figure 1.

It was also learned that sludge recycle is very advantageous to the carbamate program for two reasons. During periods of low solids loading to the system, the floc had poor settling characteristics without the addition of solids from certain coagulants. The recycled sludge provides the solids necessary for building floc particles without adding additional solids to the system. Also within the sludge is a small residue of unreacted carbamate. Recycling this material allows for the maximum use of all the produce added to the system.

Table IV. Stoichiometric Relationships of Various Metals to Coagulite EMR

Metal – (1 mg/L)	Approximate Coagulite EMR Dose (mg/L)
Lead (Pb)	5
Cadmium (Cd)	8
Zinc (Zn)	13
Copper (Cu)	14
Nickel (Ni)	14
Mercury (Hg)	5
Silver (Ag)	5
Manganese (Mn)	15
Chromium +6 to Chromium +3 (Cr)	8

However, the most important lesson was that a successful chemical treatment program alone is not enough. Everyone involved must be comfortable with the program and committed to its success. Without the commitment from management at Circuit Center there would have been no investment of the time and money necessary to explore new treatments and make the chosen one work.

If Montgomery County had not been cooperative and given Circuit Center the freedom to work with their system and explore alternatives, no new program could have been started. All too often there exists a tense and confrontational relationship between the control authority and the regulated industry. This type of relationship breeds reluctance on the part of the regulator to allow disruptions in treatment while new treatments are evaluated. The industry is reluctant to try new treatments for fear of fines if the treatment is unsuccessful, or a tightening of regulations if successful.

For this type of program to be a success, the supplier must work within the economic and logistical constraints of the client. They must work closely with the control authority so that all parties are making decisions from an informed position. They must exercise the patience to work within a timetable which is agreeable to all parties.

FIRST YEAR REVIEW

After the new treatment program had been in place for about a year, some conclusions could be drawn regarding the success of the program in meeting the stated goals. These can best be described by discussing the goals one at a time.

1. Consistent Discharge Compliance—This was the priority objective for everyone. Discharge metal levels have consistently been at or below discharge limits. Figure 2 shows data from 1985 through March 1990.
2. No system changes or capital expenditures—This goal was met with the exception of the addition of the three reaction tanks at the head end of the system. It is estimated that Circuit Center spent approximately $2600 to purchase and install these tanks.
3. Sludge Characteristics—Sludge pressing has been as good or better with the new treatment as with the ferrous sludge. The same cycle times and filter cloths have been used. The ferrous sludge averaged 32% solids, while the carbamate sludge has averaged 46% to 51% solids. Sludge previously contained 0.5% copper by weight; the carbamate sludge has averaged 15%.
4. Lower overall costs of treatment—Circuit Center has realized reduction in the overall cost of waste treatment. This has been accomplished primarily through the reduction of sludge generated which requires hauling and disposal. On the ferrous treatment, Circuit Center averaged 12 to 15 metric tons per year of sludge. On the COAGULITE EMR program, they have averaged 1 metric ton per year. This represents a 93% reduction of sludge. A cost comparison is shown in Table III.

FUTURE

This paper has reported on a successful joint effort to improve effluent treatment at a circuit board manufacturer. Of course, the story is not over, and a word about current efforts and future plans is in order.

The development of a state-of-the-art program, to provide Circuit Center with a means to meet compliance standards was an excellent beginning. However, in the search for continuing improvement and in order to reduce variability, DuBois has evaluated a monitoring and control system which may provide the convenience, simplicity, and labor savings sought by today's customers.

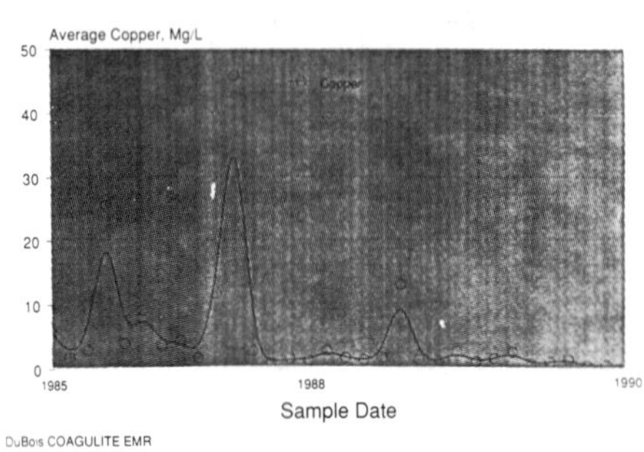

Figure 2. Circuit Center effluent copper values.

Through the use of strategically placed Oxidation Reduction Potential (ORP) instrumentation, and through the proper applications technology of the treatment chemicals, compliance has been maintained automatically. In the evaluations performed to data, DuBois has seen outstanding results on both a cost and performance basis.

A correlation is established between the millivolt reading of the waste stream and the approximate level of product in the water. Assuming proper sampling locations, chemical feedpoints, pH conditions, and knowledge of other contributing factors in the system, an effective monitor and control system may be implemented.

As long as these factors are known, the auto-control can be implemented to fit the need. The sample location(s) are then monitored for the ORP range which correlates with consistent metals levels, and the feed pump (either variable speed or staged pumps) can be set to feed product as needed and to shut down at the proper time.

The final stage in "closing the loop" is the installation of some form of data logging and reporting system. Through simple telephone touch tone programming techniques, this type of system can accept and store scores to hundreds of events prior to down-loading by modem to a computer for printout or for graphing purposes.

Many options are available to fit the needs of this area, but they are beyond the scope of this discussion. It is understood that the use of Oxidation Reduction Potential measurements is certainly nothing new to the waste treatment area. It is also understood that its absolute accuracy is not assumed, and the potential impact of extraneous and known contaminants and treatment chemicals in many systems will either preclude its use, or at least require a full awareness of potential impact. This is so that proper adjustments can be made. In short, this "closing of the loop" may not be for everyone.

However, in the search for convenience, consistency, and compliance, many facilities may find that with limited funds, equipment, and personnel, a method of automatically monitoring and controlling the "heart" of the metals removal program, (including complexed and chelated streams), may be justified.

The goal of all waste treatment facilities is to produce an effluent water which is at the lowest practical level of cost, metals level, and excess treatment chemical level in the water. This fine balance may be possible with this type of program.

CONCLUSION

As the benefits outlined earlier, (sludge volume reduction, discharge compliance consistency, etc.), are weighed against the other alternatives, the future considerations should include carbamate technology with automated control for complexed or chelated metals.

Section Ten
KEYNOTE ADDRESS

91 REMEMBRANCES OF THE PURDUE INDUSTRIAL WASTE CONFERENCES

James E. Etzel, Professor Emeritus
School of Civil Engineering
Purdue University
West Lafayette, Indiana 47907

I usually introduce the keynote speaker, so I am not quite sure how to do this. I thought I should talk about some of the things in the field that have occurred in the past which many times we do not give recognition to, and try to prognosticate some things that will happen in the future. I told Dr. Wukasch that there was a slight error in the program because this is the 44th, not the 45th Conference. The first conference was entitled the Purdue Industrial Waste Utilization Conference. By the time the second conference came around we had dropped the word "utilization" from the title. I also did a little study on the people who were on the program at the first conference, and to the best of my knowledge there is only one who remains alive, and that is Roy F. Weston. Roy at that time was the sanitary engineer for a refining company. When you read a list of who was on the first program, you find that there were many famous people. Wilhelm Rudolphs was one of the speakers; F. W. Mohlmann (many of you may remember something about a sludge volume index being taught to you by your professor); Treblor, who at that time was reporting all about dairy wastes; Sandborn, who at that time was the head of National Canners Association; and a guy by the name of George Erganian, who was a graduate student and showed slides at the conference. For those of you who do not remember George, he is one of the more recently retired senior partners of HNTB. But I think that the topic at the conference which was to me most interesting was when Treblor talked about the fact that if you're going to solve all the dairy wastes problems, you have to do it by having them stop losing the products in the plant. He wasn't talking about how you treated the waste from a milk processing plant. The one thing that dated his article was the statement that, "the worst thing they do in a dairy is the way they design milk cans. If we could only design a milk can that would drain better, we would be able to eliminate a lot of the waste problems." As you may know, we no longer put milk in cans to transport it. It was interesting, however, that his whole emphasis was on recovery and stopping losses in the plant. Then Mohlmann, who was the head of the Metropolitan Sanitary District of Greater Chicago at that time, talked about the fact that the Sanitary District had run more than 300 industrial waste surveys (each of two-week duration) with the sole purpose of going back into the industries and helping them cut their losses, rather than talking about how they should treat their wastewaters. The fact he presented that was most interesting was that they showed a processor of corn how to bottle up the industry to the point to where they were saving some 1,000 bushels a week of corn that had previously gone down the sewer. Now remember, this is in 1944, and we are talking about this kind of internal control. Rudolph gave a paper on the recovery of yeast and how you would be able to stop pollution problems if you would look at by-product recovery.

When I thought about keynote speakers, one of those I remember most, because he happened to be a personal friend, was Paul McCracken. Paul was a professor of economics at the University of Michigan, but he was famous for being the chairman of the Counsel of Economic Advisors to the President of the United States. He gave us some words of wisdom when he said that "Twenty-five percent of the budget of this country is discretionary; 75% is committed by law." When you stop and think about what he was saying, it is clear that if we want to solve all these pollution problems and other matters, we don't have the money to do it. If we want to do it, we either have to increase the budget or we have to unenact some of the laws that commit us to spending 75% of our budget. When I think of what Paul said, and then reflect on things that you read and things you say yourself, it is most interesting that no country ever has a pollution problem until they have discretionary income. When you are worried about feeding your people; when you are worried about the fact that your country is

45th Purdue Industrial Waste Conference Proceedings, © 1991 Lewis Publishers, Inc., Chelsea, Michigan 48118. Printed in U.S.A.

so underdeveloped that you can't keep the people from dying of starvation, no one is going to worry about how you solve the problems of water and air pollution. They are worried about how they are going to stay alive. When you reflect on this and on what Dr. McCracken said, it becomes obvious that our budget, our taxes, and our solution to pollution problems, which we have legislated without worrying about costs, are on a collision course and we have to figure out what we are going to do. We have enacted a lot of legislation that commits us to doing these various things, whether we have a recession or whatever the state of the economy is. I am sure most of you, were you to have an annual report from a series of companies in which you own stock, each of which said there would be no dividend this year because they had to clean up their pollution problems, might turn around and call your broker and say, "Sell that dog; it ain't making me any money."

If we go back, then, and reflect on what happened at the second conference, we have a rather interesting tie-up of names. The second conference was held in 1946, because in 1945 there were so many bans on travel they could not hold the conference. The luncheon was chaired by a man with the name of Anton Hulman, Jr. Those of you who know something about the history of Indiana and auto racing realize that this is Tony Hulman of the Indianapolis Speedway. And the reason he was chairman, of course, was because at that time the state sanitary engineer in Indiana was a man by the name of Joe Quinn, who was Tony's assistant. For years there has been a tie-in between Purdue's Environmental Engineering program, the State Board of Health, and the Speedway. Don Bloodgood served as the chief parking attendant at the raceway for almost all the years that he was on the staff at Purdue. Joe Quinn was the one who suggested that this conference start. If you read the preface to the first proceedings of the conference, Joe told Don Bloodgood that he thought he should have a conference on industrial wastes. One of the things it says in the first program, which is startling, is that registration was free. I am sure you all wish that were true now, but unfortunately, that is not the case. While presiding at the luncheon of the second conference, Tony introduced a gentlemen who at that time was the state health commissioner of Indiana by the name of Dr. Leroy Bernie. Many of you may remember Bernie because he became Surgeon General of the United States shortly after he had been luncheon speaker. They did an analysis of the attendance at the conference to determine who was from where. In order to get some idea of how we do today, I went back and analyzed the attendance using the same categories that were used in 1946, for the 1988 conference attendance. Believe it or not, the percentage attendance of various groups was almost identical. Ten percent of the people were from the food industry; 10% from equipment vendors; 9% from state regulatory agencies; 9% from sewage treatment plant operators (that percentage was down somewhat); and 8% metal products people. One of the categories that went up, for which we are very proud, is university students, which has gone up from 7% to 11% of the attendees. We have always been honored that many of our contemporaries think enough of the conference and the caliber of the papers to bring their students along, at either great personal expense, or at least the sacrifice of many hours of sleep, in order to get here.

I unearthed many facts as I was getting ready to give this talk that were interesting trivia. At the 5th industrial waste conference, there was a guy who registered as a graduate student from the University of Illinois. His name was Bob Baumann. Then there was Jack Borchardt, who at that time was an assistant professor at the University of Michigan. I can't believe that Borchardt was ever an assistant professor. People who attended the second conference that you may remember are Dick Houck, who was a senior fellow at Mellon Institute, and W. D. Hatfield, the man that the Hatfield award was named for. One of the key speakers was Colonel J. J. Gilbert, who later became famous when he was a salesman at Link Belt Company, but at that time he was military. If we look a little further, we find that at the 5th conference the COD test was introduced by Hoover and Porges. This was the first time it was ever made public. Another fact that I found fascinating was that the imminent W. W. Eckenfelder, Jr. gave his first paper on the "Application of ORP to Biological Waste Treatment." He and John Hood co-authored the paper, which was presented at the 6th conference. The person who holds the record of having more papers at the conference for more consecutive years than he would like to admit, gave his first paper at the 15th conference when he was an assistant professor at the University of Illinois, and that, of course, is Tony Gaudy. He has had more papers for more years than anyone else who has ever attended.

If we look at what we are discussing at these conferences, and where we have been and where we are going, it is rather interesting how many people coined phrases for what is pollution – my friend Jack Garrett of Monsanto had the most interesting, yet most useless phrase. Jack said, "Environmental pollution is anything that interferes with man's fruitful inhabitation of the planet" – sounds great until you examine it. You then realize that it doesn't say a damn thing, but it did sound good. I have always kidded Jack about that particular statement anyway.

We've come full circle, if you look at it. The first conference was Waste Utilization. We tried to solve the problems internally. In the '50s and '60s we were talking about how you treat wastewater.

Who had the better mousetrap? Were we going to use activated sludge in this form or that form? Was there something like endogenous respiration? I remember hearing Eckenfelder, Kountz, Buswell, and McKinney arguing with each other at some of the very earliest conferences I attended as a graduate student. They would run up and write things on the board, while one or the other one was giving a paper on whether or not you could run an activated sludge plant without having any need to dispose of solids. I won't tell you who advocated what because I'm sure all of you know the answers to the question now. At that time they didn't know the answers and they were vehement as to who was right, and the minor question of "what about the suspended solids that are in the overflow from the clarifier," of course, became a very interesting topic of discussion. But we had many, many vehement arguments over these things for a couple of years. It used to get heated enough so they didn't even talk to each other for maybe half a day after these sessions when people were presenting papers. And then they would get together and go out and drink beer. I remember one evening when some of the Purdue graduate students pulled a few pranks that sometimes used to be embarrassing to the staff. I was a very young assistant professor, and at that time if they were going to have any drinking affairs, they would always go across the river to Lafayette. One evening, quite late, a very august gentlemen who had been over heisting a few was coming back to the Union building and he called a taxi. Four of our graduate students got into the taxi with him. When the man got to the Union building, the cab driver wanted to charge each of them two bucks. Well, the four graduate students got out and ran away and left this guy holding the sack. The guy called me at home that night and complained about the incident. I told Clair Sawyer that I was sorry, but I laughed as well to think that some graduate students were able to get the best of him.

If we look at the fact that in the '50s and '60s we were treating waste, we are now back to the same point. People last year, this year, and the past few years, have been talking about how you solve the problem at the source. How do you eliminate the problem? Because that's what the EPA says we have to do; that's what the guidelines tell us to do; industries have to get religion. So we've come full circle. I always love to hear stories about how occasionally people win arguments with the EPA. Probably my favorite story in this regard was when I got a phone call from a man at Travinol Laboratories. The phone call went like this, "We have a problem with our plant in Kingstree, South Carolina. We can't meet the effluent standard and EPA is threatening to close the plant down. Would you please come and take a look at it?" And I said, "Gee, I'm very busy. It will probably be 3 or 4 weeks before I'd even find time." He said, "Oh, that's okay, don't worry. Nothin's gonna happen". I thought, "Ha ha ha, famous last words. The EPA won't let you get away with that." Well, about three and a half weeks later I walked into the plant, sat down at the plant manager's office and said, "How have you managed to keep the EPA from closing you down?" He said, "Oh, it's very simple." And with this he handed me a manila folder. When I opened it there were two letters in it. The gist of the letters was very simple. They were the only plant on the east coast at that time that was producing intravenous glucose solutions. Another producer had been closed down for a contamination problem. He had written to the head of the AMA in New York and Baltimore and had them write letters to him. The letters said as follows: "If this plant were to be closed down, there would be significant deaths in every hospital on the east coast within 48 hours." He said, "I called the EPA and said "Here are a copy of the letters. I am going to take an ad on the front page of the *Atlanta Constitution Journal* and publish these if you close me down. And the EPA said, 'Well, we don't think we will. We will give you a chance. Just don't take too long. Get the problem cleared up.' "

Many of you would love to have been in a position like that. Unfortunately, not too many of us have life-threatening products such as those, although I am sure all of you figure your product is life-threatening. At least *your* life. And then we go to the next step in the process. This one occurred just a few weeks ago and I couldn't believe my ears. I shall tell you this one because it is the kind of thing that, if someone had told you this even as short a time as five years ago, I don't think you would've believed it. I walked into a meeting with the New York State Department of Environmental Conservation. They were trying to enforce a new permit requirement on a General Motors plant. The NYDEC was asking that this plant accept a new limit in their NPDES permit for PCBs. Nine point six parts per trillion. I said to myself, "Why not a hundred ppt? Do you really think you can measure 9.6 ppt?" And their attitude was, "Well, this is what the biologists tell us. We as engineers know we can't enforce it, but you've got to provide us evidence to show why you can't meet it or measure it." It turns out that the level will be set lower than 9.6 ppt. The reason is very simple. Two of the other people who were involved with this limit decided to contest it and went to court and had open hearings. And when they had open hearings, many of the environmental groups, and some of the other groups who were interested in the problem, came to the public hearings. These people entered into the record the fact that if you could demonstrate problems at 9.6 ppt (they wanted zero), it is now possible that it may be necessary to set the limit *below* 9.6 ppt. The interesting anomaly is that the commissioner in New York will probably not be willing to set limits in the permit. He will let it go to court and let the courts decide what limit to set, because of the public pressure that was created when they decided to have public hearings, rather than to set a limit by arbitration. Stop and think about what we're saying and how far we've come in this area. I never even heard of "parts per trillion." Parts per billion was something that you heard years ago about phenols in water. But "parts per trillion?" I can't believe we can even measure that low. But when you look at it, what have we done? We've said we're going to set guidelines and the industries decided how they would treat their wastes to meet them. Prior to the time we set guidelines, everybody came to this conference (in the '50s and '60s), to learn what form of activated sludge they should use. What's the best way to aerate the activated sludge? What's the best way to run the system? What particular modification is best? All of my university colleagues were playing the "me too" games. You know, "If you can do something with activated sludge, so can I." Nobody in his right mind ever thought that there would be any other way to treat waste (after once you had minimized the problem in the plant) than to use activated sludge. Then along came the effluent regulations and suddenly, if you had to cut down the flow as these guidelines suggested and the waste strength went high, people began to say, "Hey, I'm having trouble transferring enough oxygen, the waste is too concentrated. We probably ought to look at anaerobic forms of treatment." Now, when you stop and think about it, just a few years ago all of our learned colleagues started to think anaerobic treatment was great. In fact, I want to cite some interesting points. We now know that loadings of BOD of 140 pounds per day per 1,000 cubic feet on an anaerobic system is considered reasonable; that 80%-90% reduction is achievable; and that you can run continuously and don't have to use batch operation. We know you get less sludge than with an aerobic process. I think all of you understand that our colleagues told us this just a few years ago in some of the new papers. Unfortunately, Art Buswell told us this in the 5th Industrial Waste Conference. In fact, he wrote it in an article in 1923 before he said it at the conference. So we really didn't discover a new process. We didn't *discover* that activated sludge was no longer any good and high strength waste demanded anaerobic treatment. It was well known. Our friend Art Buswell said it all in 1923–1947.

One of my most memorable trips to a Water Pollution Control Federation meeting may years ago was when I was in Atlantic City attending a meeting with a new man on our staff by the name of Ed Kirsch. Ed retired a few years ago. Many of you may remember his name. We were standing in the lobby of the hotel one evening, deciding where we wanted to go for dinner, and Art Buswell came walking through. He was looking for a place to eat, so all of us went to dinner together and, while we were having dinner, Art was telling tales about when he was doing research on anaerobic systems at the Illinois Water Survey. He related how difficult it was sometimes to measure pH because they didn't have meters, they had to do everything with color comparators. I'll never forget some of the things that he said, especially when you read the things which he has been able to put into the literature that were so far ahead of his time.

And then we come to the last part of the treatment system and we say, "How about this business of sludge disposal?" What are we going to do with all of the residue that we have left? And I went back and looked at some statements, and what I found was so unbelievable that I thought I'd better read them to you because you would never believe it. In 1938, there was a man who was with the City of New York by the name of E. J. Cleary. Many of you may remember Ed Cleary as the person who headed up ORSANCO for many years. One of his early jobs was with the City of New York. And it seems that in 1938 they had discovered that dumps weren't good, but landfills were the panacea. The Corps of Engineers and the City of San Francisco had found that they could use a sanitary landfill to get rid of solid wastes. And Cleary, in an article in the *Engineering News Record* in 1938, talks about the fact that the City of New York was going to use sanitary landfills because they were low in cost and they were adaptable to many locations where refuse disposal was expensive and troublesome. The interesting part about it was that he concluded that low marshy areas were the ideal place because nobody wanted them; they were great from the standpoint that you didn't have any heavy traffic in the area; and besides, if there was any pollution in the stuff that drained out, it was diluted in the water around the marshes quickly and thus wasn't a problem. When you look at what Cleary said, and you think that they shut down 11 incinerators in favor of landfills, where they buried over six million cubic yards of refuse in the first year, you wonder where the superfund sight is that Mr. Cleary created. And then you say, "Well, they didn't know any better at that time." You could read the comments pro and con on the subject that appeared in the *Engineering News Record* in September 1939. Cleary's article was, of course, dated September 1, 1938. And here is a man who was saying, "Why did we shut down 11 incinerators in order to dump this stuff in swamps? "And they hired a laboratory to take some samples of the leachate draining out of these swamps and the article says, "We found that in a couple of the samples, there weren't any coliform organisms. "Now you may wonder why there weren't. Well, if you look at it, there were 56 parts per million of copper. Maybe that's why there weren't any coliform organisms. Even before this, when people were just originally starting to think about what should be done on landfilling solid wastes, an ENR article of December

15th, 1938, was published by Harrison P. Eddy, Jr. of Metcalf & Eddy. He was raising the question whether or not we should be using these landfills for materials from industry. And he said, "I question whether it is right to say that the essential success in the operation of this kind of a fill is such that the seepage should flow into great bodies of water to be diluted."

When we look at what we're saying about the solids disposal problem, we are saying that there were many people who had visions even in the early '40s or late '30s that maybe this wasn't the right thing to do, but we did it anyway. I also remember an article by Clarence Clauson. For those of you who don't remember Clarence, he was the State Sanitary Engineer in Illinois. Clarence said, "Why is it man's given right to dig a hole in the ground called a 'well' and pump water out and expect to drink it without having to treat it. Yet, that same person would never think of trying to say, 'I want to drink the water out of a stream without having to treat it before I use it?" We apply totally different standards for one source than for the other.

If we look at the future, we can say that one of the things industry is starting to do is to include an environmental engineer on the group that plans new products, so they don't choose the wrong materials and thus create problems that cannot be solved. Ten years from now we won't have many of the things we use today. We are going to have to do away with some of the products because we can't produce them and stay within the environmental guidelines. We're going to have to be able to do this because we're not going to find places to bury this waste. Nobody wants it in their back yard, and unless we start trying to decide that we have to legislate landfill sites or materials, we're just not going to find places to put the wastes. One of the things that I hope we learn is that, when we start to look at building new industries, we group them so they are compatible with each other, so that the waste from one helps to treat the waste from the other. It is not sufficient to look only at the economics of where an industry should be built. We should look at the economics of what groupings of industry should be made when we start to develop an industrial area.

And finally, I would like to make some other comments. One of the men who used to attend this conference, who was probably the orneriest I have ever dealt with, was a guy by the name of M. B. Ettinger. Maury Ettinger was one of the earliest of the researchers in the industrial waste field with the Public Health Service. And I made the mistake one time of inviting him to be the luncheon speaker at the waste conference. Some of you may know that Maury has written a book of essays which is hilarious. Something about the one-humped camel which is the camel designed by a committee, and how to discuss sewage with women, and the other choice tidbits. There's one that isn't in the book. I have one of maybe two or three copies that exist. What happened was, when he got here and tried to register they told him that he was going to have to pay for a luncheon ticket. Of course, Ettinger did. And then he proceeded to castigate me about the incident, in spite of the fact that at registration headquarters, they had a free registration and a free luncheon ticket, he would not let us refund his money. But when he went back home, he wrote a satirical essay on "There is Nothing Like a Free Lunch," which of course, was a satire about me and about how we ran the conference. I keep that in my desk drawer. I get it out and read it every once in awhile when I think I really know what I am doing, and I realize what a nasty guy he was. The other thing that I remember most, and I will leave you with this parting comment, was that I was one of the first graduate students at Purdue who was ever allowed to give a paper at the Industrial Waste Conference. I had done my Master's research on coagulation of oil wastes with ferric sulfate, and it involved coagulate recovery. I presented the paper at this conference and I was just as proud of this work as I could be. When I finished, an elderly gentleman in the audience got up and said, "Would you tell me please, Mr. Etzel, what is the difference between what you are doing and what I have been doing for the past 12 years at the Chevrolet Division of General Motors, except that you use ferric sulfate and I use alum." The guy's name was Clark Hathaway. He had never published any articles. How he expected me to know what he was doing, I had no idea, but you can imagine the feeling that I had. If the earth could have opened up and swallowed me at that point, I wished it would do so, because how could I answer the man? He had me cold. I was proud of the work and all of a sudden, I went from being proud to being probably the smallest bump on the log that could ever exist. Whenever I sit in and listen to a graduate student give a paper, or when they make a presentation or take an exam, I always remember Clark and his question to me, and I try not to do that to other people. He and I eventually became friends, in spite of that fateful day. Well, these are the ramblings and the things I think about when I think about the waste conference. I thank you for putting up with me.

INDEX